# Table of Atomic Weights

| Element | Symbol | Atomic Number | Atomic Weight | Element | Symbol | Atomic Number | Atomic Weight |
|---|---|---|---|---|---|---|---|
| Actinium | Ac | 89 | (227)[a] | Manganese | Mn | 25 | 54.9380 |
| Aluminum | Al | 13 | 26.98154 | Mendelevium | Md | 101 | (258) |
| Americium | Am | 95 | (243) | Mercury | Hg | 80 | 200.59 |
| Antimony | Sb | 51 | 121.75 | Molybdenum | Mo | 42 | 95.94 |
| Argon | Ar | 18 | 39.948 | Neodymium | Nd | 60 | 144.24 |
| Arsenic | As | 33 | 74.9216 | Neon | Ne | 10 | 20.179 |
| Astatine | At | 85 | (210) | Neptunium | Np | 93 | 237.0482 |
| Barium | Ba | 56 | 137.34 | Nickel | Ni | 28 | 58.71 |
| Berkelium | Bk | 97 | (247) | Niobium | Nb | 41 | 92.9064 |
| Beryllium | Be | 4 | 9.01218 | Nitrogen | N | 7 | 14.0067 |
| Bismuth | Bi | 83 | 208.9804 | Nobelium | No | 102 | (255) |
| Boron | B | 5 | 10.81 | Osmium | Os | 76 | 190.2 |
| Bromine | Br | 35 | 79.904 | Oxygen | O | 8 | 15.9994 |
| Cadmium | Cd | 48 | 112.40 | Palladium | Pd | 46 | 106.4 |
| Calcium | Ca | 20 | 40.08 | Phosphorus | P | 15 | 30.97376 |
| Californium | Cf | 98 | (251) | Platinum | Pt | 78 | 195.09 |
| Carbon | C | 6 | 12.01115 | Plutonium | Pu | 94 | (244) |
| Cerium | Ce | 58 | 140.12 | Polonium | Po | 84 | (210) |
| Cesium | Cs | 55 | 132.9054 | Potassium | K | 19 | 39.098 |
| Chlorine | Cl | 17 | 35.453 | Praseodymium | Pr | 59 | 140.9077 |
| Chromium | Cr | 24 | 51.996 | Promethium | Pm | 61 | (147) |
| Cobalt | Co | 27 | 58.9332 | Protactinium | Pa | 91 | 231.0359 |
| Copper | Cu | 29 | 63.546 | Radium | Ra | 88 | 226.0254 |
| Curium | Cm | 96 | (247) | Radon | Rn | 86 | (222) |
| Dysprosium | Dy | 66 | 162.50 | Rhenium | Re | 75 | 186.2 |
| Einsteinium | Es | 99 | (254) | Rhodium | Rh | 45 | 102.9055 |
| Erbium | Er | 68 | 167.26 | Rubidium | Rb | 37 | 85.4678 |
| Europium | Eu | 63 | 151.96 | Ruthenium | Ru | 44 | 101.07 |
| Fermium | Fm | 100 | (257) | Samarium | Sm | 62 | 150.4 |
| Fluorime | F | 9 | 18.99840 | Scandium | Sc | 21 | 44.9559 |
| Francium | Fr | 87 | (223) | Selenium | Se | 34 | 78.96 |
| Gadolinium | Gd | 64 | 157.25 | Silicon | Si | 14 | 28.086 |
| Gallium | Ga | 31 | 69.72 | Silver | Ag | 47 | 107.868 |
| Germanium | Ge | 32 | 72.59 | Sodium | Na | 11 | 22.98977 |
| Gold | Au | 79 | 196.9665 | Strontium | Sr | 38 | 87.62 |
| Hafnium | Hf | 72 | 178.49 | Sulfur | S | 16 | 32.06 |
| Hahnium | Ha | 105 | (260)[b] | Tantalum | Ta | 73 | 180.9479 |
| Helium | He | 2 | 4.00260 | Technetium | Tc | 43 | 98.9062 |
| Holmium | Ho | 67 | 164.9304 | Tellurium | Te | 52 | 127.60 |
| Hydrogen | H | 1 | 1.00797 | Terbium | Tb | 65 | 158.9254 |
| Indium | In | 49 | 114.82 | Thallium | Tl | 81 | 204.37 |
| Iodine | I | 53 | 126.9045 | Thorium | Th | 90 | 232.0381 |
| Iridium | Ir | 77 | 192.22 | Thulium | Tm | 69 | 168.9342 |
| Iron | Fe | 26 | 55.847 | Tin | Sn | 50 | 118.69 |
| Krypton | Kr | 36 | 83.80 | Titanium | Ti | 22 | 47.90 |
| Kurchatovium | Ku | 104 | (260)[c] | Tungsten | W | 74 | 183.85 |
| Lanthanum | La | 57 | 138.9055 | Uranium | U | 92 | 238.029 |
| Lawrencium | Lr | 103 | (256) | Vanadium | V | 23 | 50.9414 |
| Lead | Pb | 82 | 207.19 | Xenon | Xe | 54 | 131.30 |
| Lithium | Li | 3 | 6.941 | Ytterbium | Yb | 70 | 173.04 |
| Lutetium | Lu | 71 | 174.97 | Yttrium | Y | 39 | 88.9059 |
| Magnesium | Mg | 12 | 24.305 | Zinc | Zn | 30 | 65.38 |
| | | | | Zirconium | Zr | 40 | 91.22 |

[a] Value in parentheses is the mass number of the most stable or best-known isotope.
[b] Suggested by American workers but not yet accepted internationally.
[c] Suggested by Russian workers. American workers have suggested the name Rutherfordium.

# GENERAL CHEMISTRY PRINCIPLES AND STRUCTURE

SECOND EDITION

JAMES E. BRADY
St. John's University

GERARD E. HUMISTON
Harcum Junior College

JOHN WILEY & SONS
New York • Chichester • Brisbane • Toronto

This book was printed and bound by Quinn and Boden.
It was set in Trump Medieval by Progressive Typographers.
The drawings were designed and executed by John Balbalis
with the assistance of the Wiley Illustration Department.
Peter Klein supervised production.

Text design by Suzanne G. Bennett.

Cover design by Edward A. Butler.



**Library of Congress Cataloging in Publication Data**

Brady, James E 1938–
General chemistry.

Includes index.
1. Chemistry. I. Humiston, Gerard E., 1939– joint author. II. Title.

QD31.2.B7 1978 540 77-11045
ISBN 0-471-01910-0

Printed in the United States of America

10 9

# PREFACE

We were both pleased and encouraged by the enthusiastic reception that greeted the first edition of our book. Therefore, our goal in revising this text has been to strengthen and refine those features that attracted instructors to the book in the first place, while correcting weaknesses or deficiencies pointed out by our students and some reviewers.

In preparing this revision we have examined the first edition carefully, line by line, with an eye toward improving the overall readability of the text and the clarity of presentations. We have continued to choose simple language to develop concepts and have provided a liberal assortment of worked out example problems to assist the student. As before, no prior knowledge of chemistry is assumed, and new terms have been carefully defined before using them in subsequent discussions. As in the first edition, new terms are set in boldface type and are indexed; important definitions are italicized. In an effort to heighten student interest and to provide an increased awareness of the role of chemistry in our earthly existence, we have added many more examples of chemicals that enter our everyday lives as products of both nature and technology. These appear in discussions of concepts, in example problems, and in review questions and problems.

Although the theme and basic features of the book remain the same, several key changes have been made. At the suggestion of many users of the first edition, the end-of-chapter exercises have been considerably expanded. They have been divided into Review Questions and Review Problems. Among the problems a range of difficulty has been provided, and the most difficult problems are marked with an asterisk. At the end of each chapter there is an index of subject areas covered by the exercises to aid the instructor in assigning homework and to assist students in planning their study.

Pictorial illustrations continue to be used generously throughout the text. We have retained the use of stereoscopic illustrations to enrich discussions of the three-dimensional aspects of chemistry since, in the experience of many adopters of the first edition, they were a useful pedagogical device. At the suggestion of a number of users of the first edition, we have replaced line drawings of orbitals with stereophotographs of models, so that both the orientation of the orbitals and their space-filling character are depicted.

While refinements have been made throughout the book, several chapters were singled out for extensive revision. In the stoichiometry chapter the entire introduction to the mole concept has been drastically changed and, we feel, pedagogically improved. We have found the present approach very effective in the classroom. The material on ionic equilibrium has been entirely reorganized and divided into two chapters. The first deals with acid-base equilibria, the second with solubility and complex ion equilibria. The general approach to equilibrium problems has also been modified to make it easier for students to follow. The chapter on organic chemistry has also been completely rewritten to provide an overview of types of organic compounds with examples of where they are encountered in everyday situations. In this chapter we attempt to give the student a feel for the breadth of this important and fascinating area of chemistry.

Another new feature of the book, appearing on the viewer pocket inside the cover of the book, is an index to important reference tables. This handy reference guide permits the quick location of useful tables distributed throughout the various chapters in the text.

A difficult decision to reach in preparing a textbook of this kind is how many SI units to embrace. We have chosen to retain the atmosphere and torr as the units of pressure because of ease of measurement in the laboratory, although the relationship of these units to their SI counterparts is pointed out. With energy units, we have employed a dual approach. Tables include energies in both joules (or kJ) and calories (or kcal). Numerical problems are worked out sometimes in joules and other times in calories. We have selected this approach because energies appearing in all but the most recent literature have generally been expressed in calories. Therefore, we feel that students must develop an ability to handle both joules and calories.

As in the past, we have assumed a mathematical background sufficient to handle only simple algebra; calculus is avoided entirely. A review of some mathematical concepts, including the use of logarithms, is found in Appendix C. In developing concepts we have tried to limit the use of mathematics to that needed to impress on the student the importance of quantitative concepts and why these are necessary in the evolution of scientific thought.

In this edition the overall sequence of topics remains unchanged. Concepts have been developed in a logical order, beginning with quantitative relationships involving atomic weights, formulas, and chemical equations in Chapters 1 and 2; this order permits an early introduction of quantitative experiments in the laboratory. These are followed by a discussion of atomic structure and the periodic table. A historical approach is taken here to give some perspective to current notions about atomic structure.

The treatment of chemical bonding once again is divided into two chapters. Chapter 4 deals with elementary concepts of covalent and ionic bonding, sufficient we feel, to carry students through approximately two-thirds of the book. Modern theories of bonding are presented in Chapter 17, just prior to the need to use them in discussions of descriptive chemistry. This division provides students with the opportunity to gain some maturity in chemistry before the more sophisticated bonding concepts are presented, and obviates the need to reteach this material in the second half of the course when it is finally needed. Users of the book who have not agreed with this division of bonding concepts, however, have found no difficulty in teaching the material in Chapter 17 immediately following Chapter 4.

Our treatment of solutions is also divided between two chapters. Chapter 5 focuses on solutions (particularly aqueous solutions) as a medium for carrying out chemical reactions. Chapter 9, which follows a discussion of liquids (Chapter 8), deals with the physical properties of solutions as they are affected by the interactions between solute and solvent.

Chapter 5 introduces many important concepts that are developed in greater detail in later chapters (for example, chemical equilibrium and acid-base reactions). The stoichiometry of solutions, the concepts of ionic reactions, and acid-base and redox titrations are also discussed in Chapter 5. This chapter, at a relatively early stage, prepares students for a variety of quantitative and qualitative laboratory experiments that deal with reactions in solution. It also reflects our approach to descriptive chemistry. There is a certain body of factual descriptive chemistry that students "must know" because they need it in other courses. There are other aspects of descriptive chemistry that students should "know about." We have attempted to compile much of this "must know" chemistry in Chapter 5.

Students who have had a high-school chemistry course may be familiar with a good deal of the material in Chapter 5, and the instructor may assign portions of it for review. We think, however, that every student who has had a course in general chemistry should know this material thoroughly.

Thermodynamics (Chapter 10) and kinetics (Chapter 11) are included sequentially to relate the importance of these two factors in determining the outcome of a chemical reaction. The interplay between thermodynamics and kinetics is discussed later in connection with descriptive chemistry.

In Chapter 12, on equilibrium, the equilibrium law is discussed first as an experimental phenomenon, and then it is analyzed in terms of kinetics and thermodynamics. This general chapter on equilibrium concentrates on gaseous and heterogeneous systems and includes a thorough discussion of Le Chatelier's principle. After a chapter on acids and bases, the discussion of equilibrium is concluded with the two-chapter treatment of ionic equilibrium mentioned earlier.

Electrochemistry is considered in Chapter 16, which includes practical applications to electroplating, energy production, and the electrochemical measurement of concentrations.

As in the first edition, the intention of the descriptive chemistry chapters (Chapters 18 to 20) is to display trends and similarities in the structure and reactivity of the elements and their compounds. These chapters serve to illustrate chemical relationships; they are not intended to be memorized by the student. Their dominant theme is structure, and the stereoscopic illustrations serve well to illustrate a variety of three-dimensional shapes encountered here. In revising these chapters we have made more frequent reference to familiar chemicals and their practical applications.

Chapter 21 on organic chemistry is followed by a separate chapter on biochemistry. Here we show how complex biomolecules are composed of relatively simple building blocks and how their structures and biological functions are accounted for.

The final chapter is on nuclear chemistry. It includes, in addition to the usual topics, illustrations of how chemists can take advantage of nuclear phenomena to aid them in their understanding of chemical processes.

For completeness, more information has been included here than can usually be presented in a two-semester course. What, then, can be cut away? This decision must be made by the instructor. Since it is often the descriptive chemistry that is pruned, we have made each section, as nearly as possible, a self-contained unit. Thus the instructor can stress the areas that he or she feels are important.

The order of chapters reflects our own bias about the sequence of topics in a general chemistry course. We realize, however, that there are other pedagogically sound orders of presentation. Therefore, in our development of what we have found to be an effective topic sequence, we have also attempted to make units sufficiently independent so that their order of presentation can be easily modified. For example, if an instructor prefers not to divide the discussions on bonding between two semesters, the topics in Chapter 17 easily can be presented after Chapter 4. Similarly, a great deal of the material in Chapter 5 can, if the instructor wishes, be included in Chapter 9.

Supplements available to accompany this textbook include a student Study Guide, which is keyed section by section to the text and contains solutions to approximately two-thirds of the even-numbered numerical problems from the text, and a Laboratory Manual for General Chemistry, which contains 51 experiments, both quantitative and qualitative. For the instructor there is an Instructor's Manual listing chapter objectives with answers to questions and worked-out solutions to numerical problems not found in the Study Guide. Transparency masters of important illustrations and typical problem solutions from the text are also available.

Finally, we wish to thank the reviewers who have contributed to this revision. Professors Charles Barr and Michael Imhoff of Austin College; Jo Beran, Texas A & I University; Philip Fuchs, Purdue University; Floyd Kelly, Casper College; Philip Lamprey, Lowell Technological Institute; Michael Peterson, North Seattle Community College; Jack Powell, Iowa State University; Don Roach, Miami-Dade Community College; Ted Sottery, University of Maine, Portland-Gorham; and Michael Wartell, Metropolitan State College have all helped us to develop a framework for the revision through their suggestions and criticisms of the first edition. We also thank Professors John Alexander, University of Cincinnati; I. C. Hisatsune, Pennsylvania State University; Delwin Johnson, St. Louis Community College at Forest Park; Joseph Long, Broome Community College; Ruth Sime, Sacramento City College; and Kenneth Watkins, Colorado State University all of whom provided many detailed suggestions for even further refinement of the final draft of this revision. We are particularly grateful to our colleagues and students for their suggestions, especially Drs. Ernest Birnbaum, Eugene Holleran, Eugene Kupchik, William Pasfield, John Skarulis, and Siao Sun, who served as sounding boards for our ideas and helped us to assess student reaction to the various modes of presentation in the text. Our continued appreciation is extended to Drs. Don Cromer and Carroll Johnson, who provided us with copies of some of the stereo illustrations. Special praise goes to the staff at Wiley, particularly our editor and friend, Gary Carlson, whose guidance, sound decisions, and sense of humor have made our task enjoyable. And, most important, we must thank our wives and children, who continue to be our inspiration.

James E. Brady

Gerard E. Humiston

# TO THE STUDENT

This textbook contains a substantial number of stereo (three-dimensional) illustrations that are intended to help you visualize some of the 3-D aspects of chemistry. Each stereo illustration, like the one below, consists of a pair of drawings that, at first glance, appear to be identical; actually they are slightly different. When viewed in such a way that the left eye focuses on the left drawing and the right eye focuses on the right drawing, your mind brings them together and creates a three-dimensional image.

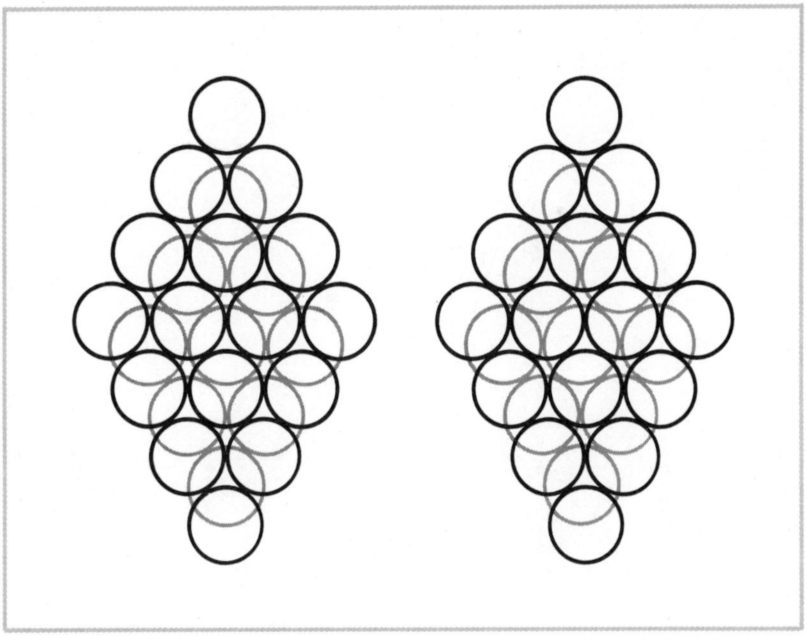

A viewer is included inside the back cover of the book to help you obtain a 3-D illusion. To get accustomed to using the viewer, assemble it according to the directions printed on the viewer and locate the bottom edge along the solid line under the drawing above. The viewer should be placed so that the folded support panel is placed between the two drawings. Now look through the lenses of the viewer, keeping both eyes open. Start with your eyes a few inches above the viewer. At first you may find that it takes a moment for the stereo image to fuse. You may have to move the viewer slightly if a double image persists. The drawing should appear to be two layers of tangent circles, one above the other.

In your study of chemistry you will encounter many new terms whose meaning you will need to know to understand the discussions that follow. These are set in **boldface** type the first time that they appear in the text. Each of these is also included in the index for later reference. Important definitions have been set in *italics* to call your attention to them.

## SUPPLEMENTARY MATERIALS

A **study guide** including worked-out problems, important terms, answers to problems, review material and detailed solutions to approximately one third of the numerical problems in this book is available for student use.

Brady and Sottery, Study Guide and Selected Problem Solutions to Accompany General Chemistry:
Principles and Structure. ISBN: 0-471-03498-3.

# CONTENTS

# 1 INTRODUCTION

Never before in history have people found themselves so able to influence their physical environment, for good or bad, as today. This has come about as a result of scientific discoveries. This book deals with a branch of physical science called chemistry, which concerns itself with the composition of substances, the ways in which their properties are related to their composition, and the interaction of these substances with one another to produce new materials.

The degree to which chemistry has changed civilization is evident everywhere. A good part of the clothing we wear, the automobiles we drive, and other products we encounter daily are composed of materials that simply did not exist at the turn of the century. In recent years the realization that a living organism is a complex chemical "factory" has generated a strong interest in the study of biochemistry and has brought great advances in our knowledge of the nature of life. Medicines created in the laboratory have made us healthier and, through the cure of disease, have prolonged our lives. It has been only recently, however, that we have also become painfully aware of a host of problems arising from this growth of technology. It is the solution of such problems that poses much of the challenge for chemistry in the future.

In this first chapter we consider how science operates, see the materials and concepts with which chemists work and about which they are concerned, and see how the concept of the atom became firmly established. We shall also introduce you to some of the jargon used by chemists. It is important to become familiar with chemical terminology (which will undoubtedly require some memorization), because many of the difficulties that students encounter in the study of chemistry can be traced to their inability to "speak the language."

## 1.1 THE SCIENTIFIC METHOD

Many of the most important advances in science, such as the discoveries of radioactivity by Henri Becquerel and penicillin by Alexander Fleming, have come about by accident. These discoveries were really only partly accidental, however, because the people involved had learned to think "scientifically" and were aware that they had observed something new and exciting.

Progress in chemistry, as well as in other sciences such as biology, physics, and psychology, is accomplished by applying a procedure called the **scientific method.** It can be divided into a series of steps that are followed, often unconsciously, in answering scientific questions. The first step can be called **observation.** The experiments that you, or any other scientist, perform in the laboratory are designed to observe nature under controlled conditions, and the bits of information that you gather are called **data.** For example, you might observe that when hydrogen gas and oxygen gas are heated together, a violent explosion results and water vapor is produced. This type of observation, which is devoid of numerical information, is said to be **qualitative.** A different chemist might make some measurements and find that, under the same conditions of temperature and pressure, one cubic foot of hydrogen gas will completely consume only one-half cubic foot of oxygen gas to produce one cubic foot of water vapor. This is a quantitative observation because it results in numerical data. We will see that quantitative measurements are generally more useful to a scientist than are qualitative observations because the former provide more information.

After a large amount of data has been collected, it is desirable to find a way to summarize the information in a concise way. Statements that accomplish this goal are called **laws** and, in a sense, simply serve as a convenient means of storage for vast quantities of experimental facts. They also provide a means of predicting the results of some as yet untried experiment. For instance, after a series of measurements regarding the relative quantities of hydrogen and oxygen that will react with one another, a chemist would conclude that when these two substances interact at the same temperature and pressure to form water, one volume of oxygen gas consumes two volumes of hydrogen gas. This simple statement is a law dealing with the reaction of hydrogen with oxygen. If we had five cubic feet of oxygen gas, we would predict that the optimum production of water would require 10 cubic feet of hydrogen.

A law may be expressed in a simple verbal statement, such as the law we just discussed regarding the reaction of hydrogen with oxygen. However, it is often more useful to have a law stated in the form of an equation. For instance, it is observed that the force of attraction between oppositely charged particles decreases as their distance of separation increases. This is more accurately stated by Coulomb's equation, or law,

$$F = \frac{q_1 q_2}{r^2}$$

in which $F$ is the force of attraction between two oppositely charged particles, $q_1$ and $q_2$ are the charges on the particles, and $r$ is their distance of separation. Laws quite commonly are expressed in equation form.

As we have noted, a law simply correlates large quantities of information. Laws in themselves do not explain why nature behaves as it does. Scientists, being human (despite what you may have heard to the contrary), are not satisfied with simple statements of fact and seek to explain their observations. Thus the second step in the scientific method is to propose tentative explanations, or **hypotheses,** that may be tested by experiment. If they are not disproven by repeated experiment, they develop into **theories.** Theories themselves always serve as guides to new experiments and are constantly being tested. When a theory is proven incorrect by experiment, it must either be discarded in favor of a new one or, as is often the case, modified so that all of the experimental observations may be accounted for. Science develops, then, through a constant interplay between theory and experiment.

It should be remembered that theories can seldom be *proven* to be correct. Usually the best we can do is to fail to find an experiment that disproves the theory. A scientist must always be careful not to confuse theory and experimental fact. Too many times in the past has incorrect theory been accepted as fact, and the progress of science been slowed down because of it.

## 1.2 MEASUREMENT

No science can proceed very far without resorting to quantitative observations. This means that scientists, and more specifically chemists, must make measurements. The process of measurement usually involves the reading of numbers from some device; because of this, there is nearly always some limitation on the number of meaningful digits that may be obtained in an experimentally determined quantity. For example, let us consider the measurement of the length of a piece of wood with two different rulers, as shown in Figure 1.1.

Using the ruler in Figure 1.1*a* we might read the length of the piece of wood as 3.2 inches (abbreviated 3.2 in.). To arrive at this number we are forced to estimate the second digit; that is, we must decide whether the length lies closer either to 3.2 or to 3.3 in. Because we are making an estimate, some uncertainty exists in the second digit (the 2), and the third digit, for all practical purposes, is completely unknown. Therefore, for measurements made with the ruler in Figure 1.1*a*, we are not justified in reporting numbers containing more than two figures.

Numbers (or digits) that arise as the result of measurement are called **significant figures.** When a number is written to represent the result of a measurement, it is always assumed that, unless stated otherwise, only the rightmost digit is uncertain. Thus, the measurement illustrated in Figure 1.1*a* yields a number with two significant figures.

In Figure 1.1*b* it is possible to estimate the length of the piece of wood to be 3.24 in. because of the greater number of calibration marks on the ruler. In this case there is no uncertainty about the 2, and the third digit is estimated to be equal to 4. The number 3.24 contains three significant figures.

The importance of significant figures is that they indicate to us the reliability of our measurements. In the determination of the length of the piece of wood described above, we saw two different values, obtained with two different measuring devices; our intuition tells us that we can place more confidence in the value with the greater number of significant figures. Laws and theories are derived from measured quantities, and our confidence in them is directly related to the quality of the data on which they are based.

In discussing measured quantities the words precision and accuracy are

Figure 1.1
*Measuring the length of a piece of wood with two different rulers. (a) Length = 3.2 in. (b) Length = 3.24 in.*

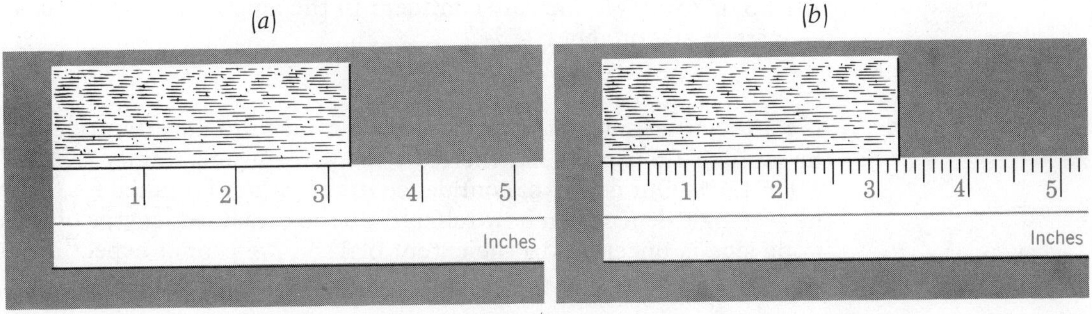

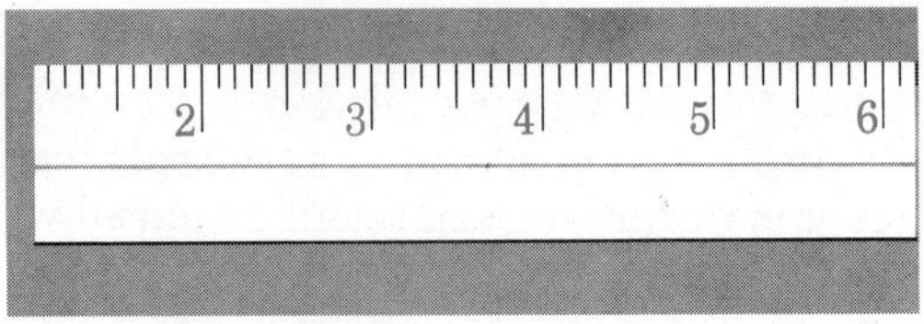


Figure 1.2
*An improperly calibrated ruler. All measurements made with this ruler will be in error by 1 in.*

often encountered. The term **precision** refers to how closely two measurements of the same quantity come to each other. In general, the more significant figures that there are in a measured quantity, the greater is the precision of measurement. The value 3.24 in. suggests that if the measurement is repeated, the result will lie within a few hundredths of an inch of 3.24 in. On the other hand, a reported value of 3.2 in. implies that if the measurement is made again, the result may differ from 3.2 in. by as much as a few tenths of an inch. The reported value of 3.24 in. thus implies a greater precision of measurement than 3.2 in.

The term **accuracy** refers to how close an experimental observation lies to the true value. Generally, a more precise measurement will also be a more accurate measurement. In our example above, the value 3.24 in. has greater precision than 3.2 in. and probably also lies closer to the true length (whatever that may be).

There are instances where a number may be precise but not particularly accurate. The ruler in Figure 1.2, for instance, is improperly calibrated. Failure to notice this error in calibration (hence, failure to adjust appropriately the measurements made with this ruler) would lead to results that would all be wrong by 1 in., even though three significant figures can be read from the scale.

In almost all cases, measured numbers are used to calculate other quantities. In these situations we must exercise care to assure ourselves of reporting the proper number of significant figures in the computed result. This is particularly important when an electronic calculator is used in performing arithmetic, since these usually give answers with eight or ten digits. Generally most of these digits are meaningless in the sense that most of them should not be counted as significant figures. To see how problems might arise, suppose that we wished to calculate the area of a rectangle whose sides have been measured to be 6 and 7.0 in. long. We know that the area is the product of these two numbers; that is, 6 in. $\times$ 7.0 in. = 42 in.$^2$ The question is, how many significant figures are justified in the answer? To determine this, let us consider the number of significant figures in each of the measured lengths.

The number 6 has one significant figure, which implies that there is some uncertainty in this digit (i.e., the length of the side could actually be either 5 or 7 in.). We thus are confident in the length of this side to, at best, one part in six, or about 17%.

$$\tfrac{1}{6} \times 100 = 17\%$$

As for the other side of the rectangle, the minimum uncertainty implied by the number 7.0 is $\pm 0.1$ in.; therefore the percent uncertainty is $(0.1/7.0) \times 100 = 1.4\%$. Our degree of confidence in the value computed for the area of the rectangle depends on how reliable are the measured lengths of the sides. If one side is uncertain to the extent of 17%, we cannot expect the uncertainty in the area to be less than 17%, which means that the number 42 in.$^2$, which implies an uncertainty of only $(1/42) \times 100 = 2.4\%$, must be rounded off to 40 in.$^2$ (which implies an uncertainty of 25%). To avoid this rather tedious analysis every time we perform a computation, we can use the general rule that *for multiplication or division, the product or quotient should not possess any more significant figures than the least precisely known factor in the calculation.*

For addition and subtraction the procedure used to determine the number of significant figures in an answer is slightly different. Here the number we write as the result of a calculation is determined by the figure with the fewest number of decimal places. Thus, in the sum

$$\begin{array}{r} 4.371 \\ \underline{302.5\phantom{00}} \\ 306.871 \end{array}$$

we must round off the answer to 306.9. The reason for this is that the two digits that follow the 5 in the number 302.5 are completely unknown; that is, they could conceivably have any value from 0 to 9. As a result, the last two digits in the answer 306.871 must also be completely uncertain. If we follow our rules for writing numbers such that *only* those digits with real significance are included, we are not justified in writing these last two digits and must round off the answer to 306.9. For addition and subtraction, then, the rule is that *the absolute uncertainty in a sum or difference cannot be smaller than the largest absolute uncertainty in any of the terms in the calculation.* In the example above, the absolute uncertainty implied by the number 4.371 is $\pm 0.001$, whereas the uncertainty in 302.5 is implied to be $\pm 0.1$. According to our rule, the sum of these two cannot have an uncertainty less than $\pm 0.1$; therefore the sum 306.871 must be rounded to 306.9 in order to suggest an uncertainty of $\pm 0.1$.

In some calculations we employ numbers that come from definitions (such as 3 feet = 1 yard) or that are the result of a direct count (such as the number of people in a room). These numbers are called **exact numbers** and contain no uncertainty (e.g., there are exactly 3 feet in 1 yard, or the number of people in a room must be a whole number). When such quantities are used in computation, they may be considered to possess an infinite number of significant figures. Thus, the conversion of a measured length of 4.27 yards into feet would be accomplished as

$$4.27 \text{ \cancel{yd}} \times \left(\frac{3 \text{ ft}}{1 \text{ \cancel{yd}}}\right) = 12.8 \text{ ft}$$

where the number of significant figures in the product is determined by the number of significant figures in the measured length. Note that in performing this calculation we have cancelled the units, yards. We will see more of this **factor-label method** of setting up arithmetic in future problems. The factor-label method is explained in detail in Appendix C.1 at the back of the book.

EXAMPLE 1.1 Perform the following computations and report the results to the proper number of significant figures:

(a) $3.142 \div 8.05$ (b) $29.3 + 213.87$ (c) $144.3 + (2.54 \times 8.3)$

SOLUTION (a) Using a calculator,

$$\frac{3.142}{8.05} = 0.390310559$$

The number 3.142 contains four significant figures; the number 8.05 contains three significant figures. The result must be rounded off to 0.390 because for division the answer should not contain more significant figures than are found in the factor with the fewest number of significant figures.

(b) In performing this computation, note that the 7 must be added to a question mark.

$$\begin{array}{r} 29.3? \\ +213.87 \\ \hline 243.1? \end{array}$$

The only thing that can be said about the question mark in the answer is that it must be at least 7, and therefore we should round off the answer to

243.2

In this kind of situation many students make the error of writing a zero after the 3 in 29.3 (that is, 29.30). The digit following the 3 is unknown. If it were known to be a zero, a zero would have been written there!

(c) In mixed computations such as this, we perform multiplications and divisions before additions and subtractions.

$$2.54 \times 8.3 = 21.082 \qquad \text{(using a calculator)}$$

Since 8.3 contains only two significant figures, the product must be rounded to 21. Now the appropriate addition is performed.

$$\begin{array}{r} 144.3 \\ +\ 21.? \\ \hline 165.? \end{array} = 165$$

## 1.3 UNITS OF MEASUREMENT

Units form an integral part of any measurement. For instance, to say that the distance between two points is "three" is meaningless unless a specific unit or units (inches, feet, miles, etc.) is associated with the number. Chemists have traditionally used the **metric system** of units in all their measurements. In 1964 the National Bureau of Standards adopted a slightly modified

Table 1.1
*The seven basic SI units*

| Physical Quantity | Name of Unit | Symbol |
|---|---|---|
| Mass | Kilogram | kg |
| Length | Meter | m |
| Time | Second | s, sec |
| Electric current | Ampere | A |
| Temperature | Kelvin | K |
| Luminous intensity | Candela | cd |
| Quantity of substance | Mole | mol |

Table 1.2
*Prefixes that modify the size of metric system units*

| Prefix | Multiplication Factor | Examples | Abbreviation |
|---|---|---|---|
| Kilo- | 1000 ($10^3$) | 1 kilometer = 1000 meter ($10^3$ m) | km |
| | | 1 kilogram = 1000 gram ($10^3$ g) | kg |
| Deci | 1/10 ($10^{-1}$) | 1 decimeter = 0.1 meter ($10^{-1}$ m) | dm |
| Centi- | 1/100 ($10^{-2}$) | 1 centimeter = 0.01 meter ($10^{-2}$ m) | cm |
| Milli- | 1/1,000 ($10^{-3}$) | 1 millimeter = 0.001 meter ($10^{-3}$ m) | mm |
| | | 1 milliliter = 0.001 liter ($10^{-3}$ l) | ml |
| | | 1 milligram = 0.001 gram ($10^{-3}$ g) | mg |
| Micro- | 1/1,000,000 ($10^{-6}$) | 1 micrometer = 1 micron | $\mu$m |
| | | 1 micron = 0.000,001 meter ($10^{-6}$ m) | $\mu$ |
| | | 1 microliter = 0.000,001 liter ($10^{-6}$ l) | $\mu$l |
| Nano- | 1/1,000,000,000 ($10^{-9}$) | 1 nanometer = 0.000,000,001 meter ($10^{-9}$ m) | nm |
| | | 1 nanogram = 0.000,000,001 gram ($10^{-9}$ g) | ng |

version of the metric system, which had been officially recommended in 1960 by an international body, the General Conference of Weights and Measures. This revised set of units is known as the **International System of Units** (abbreviated SI) and has as its foundation seven basic units. These are shown in Table 1.1. Other units of measurement are derived from the basic units by suitable combination. For example, the SI unit for volume is the cubic meter, $m^3$; the unit of force is the newton (N).

$$1 \text{ N} = 1 \text{ kg m/sec}^2$$

When you hold an object weighing approximately 3.6 oz. (e.g., a large lemon), gravity causes it to exert a force (its weight) equal to 1 N.

The SI units are slowly being accepted; however, the older metric system is slow to fade away and is still used by many practicing scientists. Furthermore, the existence of the older units in the scientific literature demands that we be aware of both the old and the new. Some of the SI units are used in this book, but in a number of cases the more familiar metric units are retained. In some places we will present both the old and the new units.

In chemistry it is necessary to measure, on a routine basis, mass, length, and volume. The units that chemists ordinarily use to express these quantities are based on the **gram** (abbreviated g), the **meter** (**m**), and the **liter** (**l**), respectively. These and other metric system units are modified in a decimal fashion by use of an appropriate prefix (see Table 1.2). Thus, for example, 1 meter (m) = 100 centimeters (cm) = 1000 millimeters (mm). A relationship

Table 1.3
*Comparison of the English and metric systems*

| | |
|---|---|
| Length | 1 meter = 39.37 inches (in.)<br>2.540 centimeters = 1 inch |
| Mass | 1 kilogram = 2.204 pounds (lb)<br>453.6 grams = 1 pound |
| Volume | 1 liter = 1.057 quarts (qt)<br>29.57 milliliters = 1 fluid ounce (oz)<br>28.32 liters = 1 cubic foot ($ft^3$) |

also exists between the units of length and of volume since the liter is currently defined as exactly 1000 $cm^3$, and therefore, 1 ml = 1 $cm^3$. Table 1.3 gives the approximate English equivalents of some of these units. You will probably find it to your advantage to develop a feel for the magnitude of these quantities.

EXAMPLE 1.2 Calculate the number of meters in 0.200 miles.

SOLUTION Unit conversions such as this are most easily handled using the factor-label method. First let us list the relationships that we know connecting the units.

1 mile = 5280 ft
1 ft = 12 in.
1 m= 39.37 in. (from Table 1.3)

Now we use these relationships to construct conversion factors that enable us to cancel unwanted units. Our problem involves converting the units "miles" into "meters".

$$0.200 \text{ miles} = ?\ \text{m}$$

$$0.200\ \cancel{\text{miles}} \times \left(\frac{5280\ \cancel{\text{ft}}}{1\ \cancel{\text{mile}}}\right) \times \left(\frac{12\ \cancel{\text{in.}}}{1\ \cancel{\text{ft}}}\right) \times \left(\frac{1\ \text{m}}{39.37\ \cancel{\text{in.}}}\right) = 322\ \text{m}$$

Note that the answer has been rounded to give three significant figures.

When we express measurements in some of these units, either very large or very small numbers are often encountered. To avoid having to write down a large number of zeros, it is convenient to express these quantities as the product of a number lying between 1 and 10 multiplied by 10 raised to some power. This type of representation is called **exponential notation** or **scientific notation.**[1] For example, using this notation we could write 1 kg = $1 \times 10^3$ g rather than 1 kg = 1000 g. If we write numbers in this fashion, the real usefulness of the metric system is apparent because converting from one unit to another merely involves changing the exponent on the 10. For example,

$$1 \text{ km} = 1 \times 10^3 \text{ m} = 1 \times 10^5 \text{ cm} = 1 \times 10^6 \text{ mm}$$

You might compare this to the conversions among miles, yards, feet, and inches in the English system.

There are occasions when the presence of zeros leads to difficulties in determining the number of significant figures in a number. The use of exponential notation allows us to eliminate any confusion that might arise. For example, suppose that we had measured the length of an object with a ruler and had found it to be 1.2 m long. This number possesses two significant figures and implies an uncertainty of about one part in 12 (0.1/1.2 = 1/12). We could also express this length as 1200 mm and, because it still represents the same measurement, it must still possess only two significant figures, the 1

[1] A complete discussion of exponential notation and how it can be applied to performing numerical calculations will be found in Appendix C.

and the 2. The two zeros in 1200 are used only to locate the decimal point. Unfortunately, someone unfamiliar with our experiment might think that all four digits are significant, implying an uncertainty of only one part in 1200, which is not at all what we intended to convey.

By writing numbers using exponential notation we can eliminate this ambiguity. Thus our 1200 mm could be written as $1.2 \times 10^3$ mm; here the first portion of the number expresses only two significant figures. Had we wanted to specify four significant figures, we could have done so by writing the number as $1.200 \times 10^3$ mm.

To summarize, then, the only time that zeros are considered as significant figures is when they are not present for the sole purpose of locating the decimal point. Thus, the quantity 0.0072 has only two significant figures while 0.007020 has four, since these are written as $7.2 \times 10^{-3}$ and $7.020 \times 10^{-3}$, respectively. In the latter example, the zero between the 7 and the 2, as well as the rightmost zero, are significant because they are not needed to position the decimal point.

## 1.4 MATTER

In their investigations, chemists study the properties and transformations of **matter.** Matter is anything that occupies space and possesses **mass.** In setting down this definition a careful distinction is made between the terms mass and weight, even though we generally use them as if they were interchangeable. The mass of a body is a measure of its resistance to a change in velocity. A Ping-Pong ball moving at 20 miles/hr, for example, is easily deflected by a soft breeze, while a cement truck is not. Quite clearly, the mass of the cement truck is considerably greater than that of the Ping-Pong ball. The term **weight** refers to the force with which an object is attracted to the earth. Force and mass are related to each other by Newton's equation (Newton's law),

$$F = ma$$

where $F$ = force, $m$ = mass, and $a$ = acceleration. In order to accelerate a body, a force must be applied to it. When an object is dropped, it accelerates because of the gravitational attraction of the earth. An object resting on the earth exerts a force (its weight, $W$) that is equal to its mass, $m$, multiplied by the acceleration due to gravity, $g$; that is,

$$W = mg$$

The weight of an object thus depends on $g$, a quantity that varies slightly from place to place on the earth's surface. Because the weight of an object depends on where the measurement is made, it is considered more desirable to express the amount of matter present in the object in terms of its mass.

The determination of mass (a process, oddly enough, called **weighing**) is actually performed by comparing the weights of two objects, one of known mass, the other of unknown mass. The apparatus used for this is called a **balance** (Figure 1.3). The object to be weighed is placed on the left pan of the balance and objects of known mass are added to the other until the two pans are balanced. At this point the contents of both pans weigh the same and, since they both experience the same gravitational acceleration, both pans contain equal masses. In chemistry, as mentioned previously, we generally measure mass in grams.

A modern single-pan analytical balance is shown in Figure 1.4. To use this balance the object to be weighed is placed on the pan and the weights

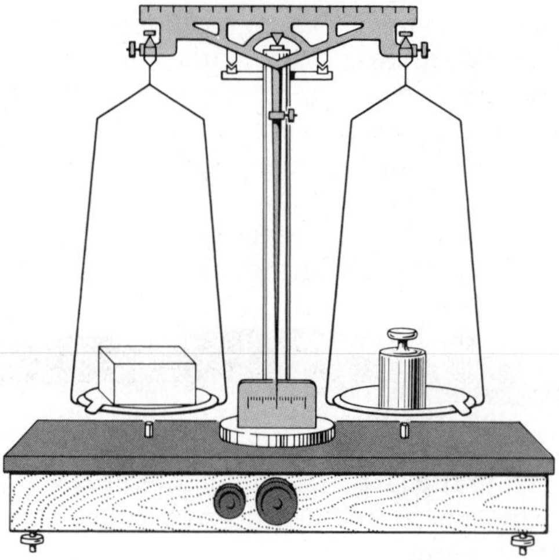

Figure 1.3
*A balance. Known masses are added to the right pan until the pointer is centered. The contents of each pan then have the same weight and therefore also possess the same mass.*

are adjusted internally by turning the knobs on the face until balance is achieved. Although this modern apparatus looks quite different from its older brother, they both operate on exactly the same principle, that is, a balance beam adjusted to have the same weight on either side of the knife-edge pivot.

## 1.5 PROPERTIES OF MATTER

The properties that we use to describe matter may be classified into two broad categories: **extensive properties,** which depend on the size of a sample of matter; and **intensive properties,** which are independent of sample size. Of the two, intensive properties are the more useful, because a substance[2] will exhibit the same intensive property regardless of how much of it we examine.

Examples of extensive properties are mass and volume, for as the quantity of a substance increases, its mass and volume also increase. Some examples of intensive properties are melting point, boiling point, and **density** (which is defined as the ratio of an object's mass to its volume). Water, for instance, has a density of 1 g/ml. This means that if we had 1 g of water, it would occupy a volume of 1 ml. If instead we had 20 g of water, we would find that it occupied a volume of 20 ml, but the ratio of the water's mass to volume, 20 g/20 ml, is still the same as 1 g/1 ml. Therefore we see that density is a property that does not depend on sample size.

Notice that the intensive property, density, is computed as the ratio of two extensive properties, mass and volume. Later in our discussion of chemistry we will encounter quite a few intensive properties defined in a similar fashion.

In speaking of the properties of substances, we also distinguish between physical properties and chemical properties. A **physical property** can be specified without reference to any other substance. All of the examples cited earlier in this section are physical properties. A **chemical property,** on the other hand, states some interaction between chemical substances. We say, for instance, that sodium is very reactive toward water. Reactivity is a

[2] The word *substance* is used frequently by the chemist and is taken to mean the material of which an object is composed. For example, an ice cube is composed of the substance water. Complicated objects, of course, can be composed of many substances.

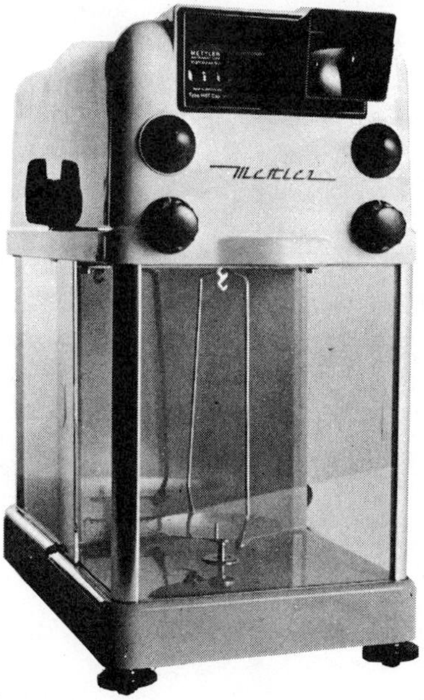

Figure 1.4
*A modern analytical balance.*

chemical property that refers to the tendency of a substance to undergo a particular chemical reaction. However, to say simply that a substance is very reactive, without specifying "with what" or under what conditions, is not particularly helpful. Sodium, for example, is very reactive with water but quite unreactive toward the gas, helium.

## 1.6 ELEMENTS, COMPOUNDS, AND MIXTURES

The three words that form the title to this section lie very close to the heart of chemistry, because it is with elements, compounds, and mixtures that laboratory chemists work. They must therefore understand what elements, compounds, and mixtures are and be able to distinguish among them before they can hope to make use of them.

**Elements** are the simplest forms of matter that can exist under conditions that we encounter in a chemical laboratory; thus they are the simplest forms of matter with which the chemist deals directly. Elements serve as the building blocks for all of the more complex substances with which we work, from common table salt to extremely complex proteins. All are composed of a limited set of elements. At present there are 105 known elements, but only a much smaller number will be of real interest to us.

Elements combine to form compounds. A **compound** is characterized by having its constituent elements always present in the same proportions. For example, water is composed of two elements, hydrogen and oxygen. All samples of pure water contain these two elements in the proportion of one part by weight hydrogen to eight parts by weight oxygen (for example, 1.0 g of hydrogen to 8.0 g of oxygen). Also, when hydrogen is reacted with oxygen to produce water, the relative amounts of hydrogen and oxygen that combine are always the same. Thus, whenever 1.0 g of hydrogen reacts, it is always observed that 8.0 g of oxygen are consumed.

**Mixtures** differ from elements and compounds in that they may be of variable composition. A solution of sodium chloride (table salt) in water is a mixture of two substances, and we know that by dissolving varying quantities of salt in water we can obtain solutions with a wide range of composi-

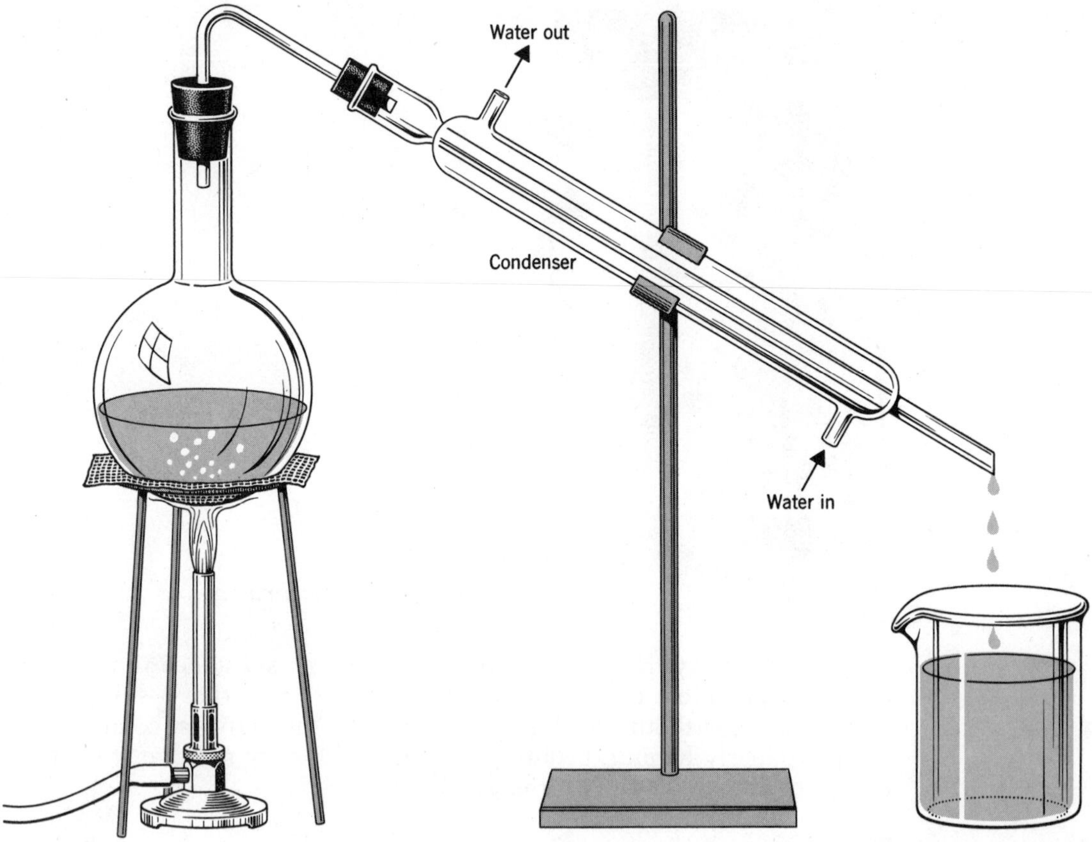


Figure 1.5
*Simple distillation apparatus. Sodium chloride solution is boiled in the flask. The steam is converted to the liquid in the condenser.*

tions. Most materials found in nature, or prepared in the laboratory, are not pure but, instead, are mixtures. One of the greatest problems faced by the chemist is the separation of mixtures into their components. This can usually be accomplished by some physical process (as opposed to a chemical reaction). Our solution of sodium chloride, for instance, if left to evaporate will leave the salt behind as a solid. Had we wished to recover the water as well, we could have boiled the solution in an apparatus similar to the one in Figure 1.5 and collected the water after it had condensed from the stream. This process is called **distillation.** It is one method, incidentally, that is used to desalinate sea water.

Another method of separating mixtures, called **chromatography,** makes use of the different tendencies that substances have for being adsorbed onto the surface of some stationary support. For example, in **thin-layer chromatography** (Figure 1.6), a sample of a mixture is spotted onto a material such as silica gel which thinly coats a glass plate. A solvent is then allowed to creep up the coating from a reservoir. As the solvent flows past the spot, the different components of the mixture move through the silica gel at different rates, the more strongly adsorbed components moving more slowly. The result is a separation of the original spot into a set of spots each containing (hopefully) one component. This technique is widely used today by chemists who synthesize new compounds.

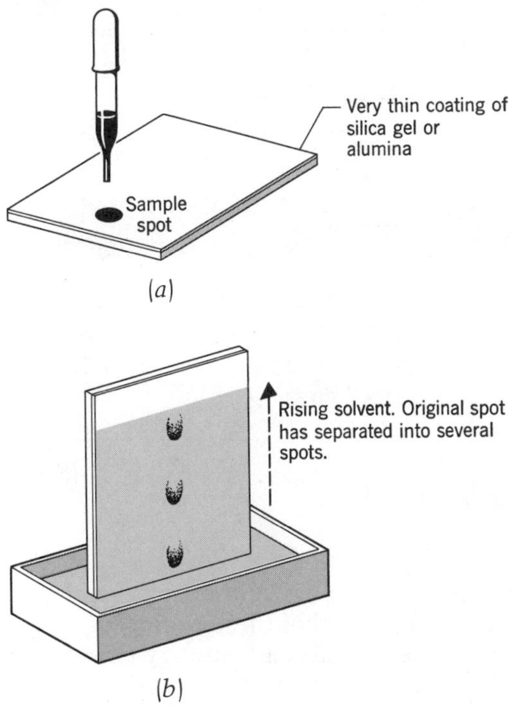


Figure 1.6
*Thin-layer chromatography (TLC). (a) A solution containing a mixture is spotted near one end of a glass plate coated with silica gel or alumina. (b) The plate is placed with the end nearest the spot in a tray of solvent that rises through the coating by capillary action. The mixture is separated, with substances least strongly adsorbed by the coating moving farthest.*

Mixtures can be described as being either **homogeneous** or **heterogeneous.** *A homogeneous mixture is called a* **solution** *and has uniform properties throughout.* If we were to sample any portion of a sodium chloride solution, we would find that it has the same properties (e.g., composition) as any other portion of the solution; we say that it consists of a single **phase.** Thus, we define a phase as any part of a system that has uniform properties and composition.

A heterogeneous mixture, such as oil and water, is not uniform (Figure 1.7). If we were to sample one portion of the mixture, it would have the properties of water, while some other part of the mixture would have the properties of oil. This mixture consists of two phases, the oil and the water. If we shook the mixture so that the oil was dispersed throughout the water as small droplets (as in a salad dressing), all of the oil droplets taken together would still constitute only a single phase, since the oil in one droplet has the same properties as the oil in another. If we added an ice cube to this "brew,"

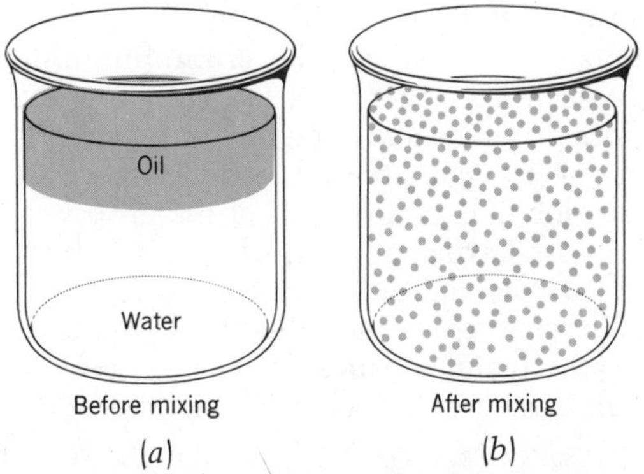


Figure 1.7
*Oil and water—a heterogeneous mixture. (a) Before mixing. (b) After mixing.*

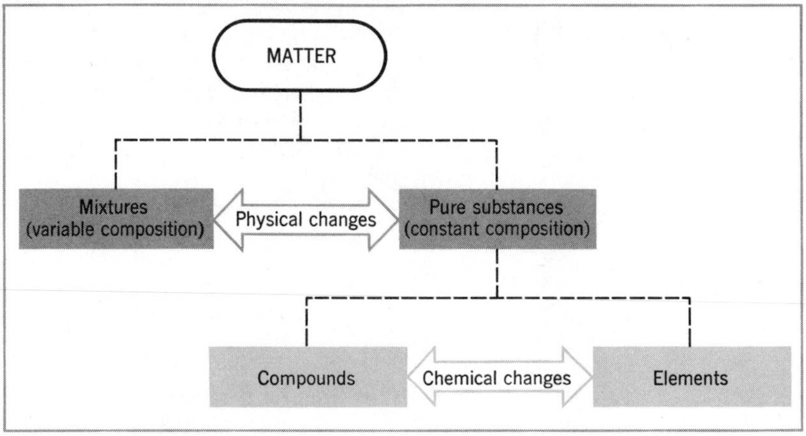


Figure 1.8
*Classification of matter.*

we would then have three phases: the ice (a solid), the water (a liquid), and the oil (another liquid). In all of these examples, we can detect the presence of two or more phases because a boundary exists between them.

A useful feature of pure compounds is that they undergo phase changes (e.g., solid to liquid or liquid to gas) at constant temperature. Ice, for instance, melts at a temperature of 32°F, a temperature that remains constant while the water undergoes the change from solid to liquid. When mixtures exhibit phase changes, they generally do so over a range of temperatures. This phenomenon provides us with one experimental test to determine when we have obtained a pure compound. The relationship among elements, compounds, and mixtures is summarized in Figure 1.8.

## 1.7 THE LAWS OF CONSERVATION OF MASS AND DEFINITE PROPORTIONS

It is not surprising that the early history of chemistry was marked by incorrect theories about what occurred during chemical reactions. It had long been observed, for example, that when the combustion of wood took place, the resulting ash was very light and fluffy. Metals, too, when heated and caused to react with air, changed their appearance. The resulting product was less dense than the original metal and thus appeared lighter. These observations led to the conclusion that "something," which the German chemists Becher and Stahl called **phlogiston**, was lost by substances when they burned. Even when it was pointed out that metals gained weight when they were heated in air, the theory was salvaged by concluding that phlogiston was simply lighter than air!

The reluctance to abandon the crumbling phlogiston theory demonstrates a general human phenomenon. New theories are difficult to come by, and old ones become so thoroughly entrenched that it is often tempting to try to shore up a sagging theory rather than to dream up a new one that will do a better job of explaining all of the observed facts.

It was Antoine Lavoisier, a French chemist, who finally laid the phlogiston theory to rest and set chemistry on the proper course again. He demonstrated by his experiments that the combustion process actually occurred by the reaction of substances with oxygen. He also showed, *through careful measurements,* that if a reaction is carried out in a closed container, so that none of the products of reaction escape, the total mass present after the reaction has occurred is the same as before the reaction began. These observations form the basis of the **law of conservation of mass,** which states that *mass is neither created nor destroyed in a chemical reaction.*[3]

The work of Lavoisier clearly demonstrated the importance of careful measurement. After his book, *Traité Eléméntaire de Chemie,* appeared in 1789, many chemists were inspired to investigate the quantitative aspects of chemical reactions. These investigations led to another important chemical law, the law of definite proportions.

The **law of definite proportions** (also called the **law of definite composition**) states that, *in a pure chemical substance, the elements are always present in definite proportions by mass.* In the substance water, for instance, the ratio of the mass of hydrogen to the mass of oxygen is always 1/8, regardless of the source of the water. Thus, if 9.0 g of water were decomposed, 1.0 g of hydrogen and 8.0 g of oxygen would be obtained, while if 18.0 g of water were broken down, 2.0 g of hydrogen and 16.0 g of oxygen would be produced. Furthermore, if 2.0 g of hydrogen are mixed with 8.0 g of oxygen and the mixture ignited, 9.0 g of water are formed and 1.0 g of hydrogen remains unreacted. Again, in the product the hydrogen-to-oxygen mass ratio is 1/8. Thus, varying the amounts of hydrogen and oxygen present during reaction does not alter the composition of the water produced.

## 1.8 THE ATOMIC THEORY OF DALTON

The real father of modern chemistry could well be considered to be the Englishman, John Dalton, who proposed his atomic theory of matter around 1803. The concept of the atom (from the Greek *atomos,* meaning indivisible) did not originate with Dalton. The Greek philosophers Leucippus and Democritus suggested, as early as 400 to 500 B.C., that matter cannot be forever divided into smaller and smaller parts and that ultimately particles would be encountered that would be indivisible. These early proposals, however, were not based on the results of experiments and were little more than exercises in thought. Dalton's theory was different because it was based on the laws of conservation of mass and definite proportions, laws that had been derived from many direct observations.

The theory Dalton proposed can be expressed by the following postulates:

1. Matter is composed of indivisible particles called atoms.
2. All atoms of a given element have the same properties (e.g., size, shape, and mass), which differ from the properties of all other elements.
3. Chemical reaction consists merely of a reshuffling of atoms from one set of combinations to another.

[3] Einstein has shown that there is a relationship between mass and energy, $E = mc^2$, where $c$ is the speed of light. Energy changes that take place during chemical reactions therefore are also accompanied by mass changes; however, the changes in mass are far too small to be detected experimentally. For example, the energy change associated with the reaction of 2 g of hydrogen with 16 g of oxygen happens to be equivalent to a mass change of approximately $10^{-9}$ g. Sensitive analytical balances can only detect mass differences of $10^{-6}$ to $10^{-7}$ g. Consequently, as far as the average chemist is concerned, there are no observable mass increases or decreases accompanying chemical reactions.

The test of any theory, of course, is how well it explains already existing fact and whether it can predict as yet undiscovered laws. Dalton's theory proved successful on both counts.

First, it accounts for the law of conservation of mass. If a chemical reaction does nothing more than redistribute atoms and no atoms are lost from the system, it follows that the total mass must remain constant when the reaction occurs.

Second, it explains the law of definite proportions. To see this, we imagine a substance formed from two elements, say, $A$ and $B$, in which each molecule of the substance is composed of one atom of $A$ and one atom of $B$. We shall define a **molecule** as *a group of atoms bound tightly enough together that they behave as, and can be recognized as, a single particle* (just as a car is composed of many parts held together tightly enough so that we identify them as *a car*). Let us also suppose that the mass of an atom of $A$ is twice the mass of an atom of $B$. Then, in one molecule of this substance, twice as much mass is contributed by $A$ as by $B$ and the ratio of the mass of $A$ to the mass of $B$ in this molecule is 2/1. If we take a large collection of these molecules, we will always have equal numbers of $A$ and $B$ atoms; therefore, regardless of the size of the sample, we still always have a mass ratio ($A$ to $B$) of 2/1. Also, if we were to react $A$ and $B$ together to form this compound, each atom of $A$ would combine with only one atom of $B$. If we were to mix 100 atoms of $A$ with 110 atoms of $B$, after the reaction was over, we would be left with 10 atoms of $B$ unreacted.

Third, Dalton's theory predicted the **law of multiple proportions.** This law states that, *when two different compounds are formed from the same two elements, the masses of one element, which react with a fixed mass of the other, are in a ratio of small whole numbers.* Actually, this sounds a lot more complicated than it really is. Let's consider the two compounds formed by carbon and oxygen. In one of them (carbon monoxide) we find 1.33 g of oxygen combined with 1.00 g of carbon, while in the second (carbon dioxide) there are 2.66 g of oxygen combined with 1.00 g of carbon. If we examine the ratio of the masses of oxygen (1.33 g/2.66 g) that combines with a fixed mass of carbon (1.00 g), we observe a ratio of small whole numbers,

$$\frac{1.33}{2.66} = \frac{1.33/1.33}{2.66/1.33} = \frac{1}{2}$$

This is consistent with the atomic theory if we consider that carbon monoxide contains *one* atom of carbon and *one* of oxygen whereas carbon dioxide contains *one* atom of carbon and *two* atoms of oxygen. Because carbon dioxide has twice as many oxygen atoms bound to a carbon atom as does carbon monoxide, the weight of oxygen in a molecule of carbon dioxide *must* be twice the weight of oxygen in a molecule of carbon monoxide.

EXAMPLE 1.3 Nitrogen forms several different compounds with oxygen. In one (called laughing gas), it is observed that 2.62 g of nitrogen are combined with 1.50 g of oxygen. In another (a major air pollutant), 0.655 g of nitrogen is combined with 1.50 g of oxygen. Show that these data demonstrate the law of multiple proportions.

SOLUTION In both cases we are dealing with a weight of nitrogen that combines with 1.50 g of oxygen. If these data do fit the law of multiple proportions, the ratio of the masses of nitrogen in the two compounds should be a ratio of small whole numbers. Let us take the ratio

$$\frac{2.62}{0.655}$$

Dividing the numerator and denominator by 0.655, we get

$$\frac{4.00}{1.00}$$

which is indeed a ratio of small whole numbers.

EXAMPLE 1.4 Sulfur forms two compounds with fluorine. In one of them it is observed that 0.447 g of sulfur is combined with 1.06 g of fluorine and in the other, 0.438 g of sulfur is combined with 1.56 g of fluorine. Show that these data illustrate the law of multiple proportions.

SOLUTION To work out this example we must compare the weights of one element (e.g., sulfur) that combine with the *same weight* of the other in the two compounds. For the first compound there is an "equivalence" between the weights of sulfur and fluorine that we can express as

$$1.06 \text{ g fluorine} \sim 0.447 \text{ g sulfur}$$

where the symbol ~ is taken to mean "equivalent to." The equivalence here stems from the experimental fact that these weights of fluorine and sulfur are chemically combined with each other. In this case we are speaking of a chemical equivalence.

For the second compound we can write

$$1.56 \text{ g fluorine} \sim 0.438 \text{ g sulfur}$$

For the purposes of calculation the equivalence implied by the ~ behaves the same as an equality, so that this sort of equivalence relationship can be used to construct conversion factors useful in mathematical computations. For the second compound, let us calculate the weight of sulfur that would have been combined with 1.06 g of fluorine.

$$1.06 \cancel{\text{ g fluorine}} \times \left(\frac{0.438 \text{ g sulfur}}{1.56 \cancel{\text{ g fluorine}}}\right) \sim 0.298 \text{ g sulfur}$$

We now know the weights of sulfur that combine with the *same weight* of fluorine (1.06 g) in these two compounds. The next step is to look at the ratio of these weights, that is,

$$\frac{0.447}{0.298} = \frac{0.447/0.298}{0.298/0.298} = \frac{1.50}{1.00}$$

which is the same as 3/2, a ratio of small whole numbers.

The use of chemical equivalencies in computations will be found extensively in later chapters. You should make a special effort now to begin to understand the concept of "chemical equivalence." You can do this by carefully working all of the problems on Dalton's theory at the end of the chapter.

## 1.9 ATOMIC WEIGHTS

Dalton's atomic theory proved so successful at explaining the laws of chemistry that it was accepted almost immediately. Since the key to the success of the theory had been the concept that each element had a characteristic atomic mass, chemists at once set out to measure them. It was at this point that a serious problem arose.

Because of their small size, there certainly was no way to determine the masses of individual atoms. All that scientists could hope to do was to arrive at a set of relative atomic weights. (Note that the terms atomic mass and atomic weight are used interchangeably.) We might determine, for instance, that in one compound formed between carbon and oxygen, 3.0 g of carbon were combined with 4.0 g of oxygen. In other words, oxygen contributed $\frac{4}{3}$ (or $1\frac{1}{3}$) times as much mass toward the compound as did the carbon. If this substance is made up of molecules, each containing one atom of carbon and one atom of oxygen, then it follows that each oxygen atom must weigh $1\frac{1}{3}$ times as much as one carbon atom; thus it appears that we have established the relative masses of these two elements.

The arguments just presented rest on a very critical assumption, that is, that the compound we were discussing is composed of one atom of carbon and one atom of oxygen. If this assumption about the number of carbon and oxygen atoms in a molecule is false, we have arrived at the wrong relative weights. Because of this kind of difficulty a self-consistent table of atomic weights was not developed until some 60 years after Dalton had introduced his theory.

A complete table of atomic weights appears on the inside front cover of the book. This is a table of relative atomic weights with the masses of atoms expressed in units called **atomic mass units** (amu). The size of this unit is chosen in a rather arbitrary way. To see this, let us return to our discussion of the compound between carbon and oxygen.

Let us assume, for the moment, that molecules of the substance are in fact composed of one atom of carbon and one of oxygen; then, as mentioned, one oxygen atom weighs 1.33 times as much as one carbon atom. If we were to define the atomic mass unit as equal to the mass of one carbon atom, we would then assign carbon an atomic weight of 1.00 amu and oxygen an atomic weight of 1.33 amu. Since we might find it more convenient to have these numbers as whole numbers (or as near to whole numbers as possible), we might define the amu as equal to one-third of the mass of a carbon atom. If we did this, we would assign carbon an atomic weight of 3.00 amu. Oxygen, which is $\frac{4}{3}$ times heavier, would have an atomic weight equal to 4.00 amu. Thus the size of the atomic mass unit, and hence the values that appear in a table of atomic weights, are really quite arbitrary.

As originally defined by chemists, the amu was taken to be $\frac{1}{16}$ of the mass of naturally occurring oxygen. In this way the masses of nearly all of the elements come out to be approximately whole numbers. Oxygen was chosen as a standard because it formed compounds with nearly all of the elements. However, as we will discuss in Chapter 3, not all atoms of an element have precisely the same mass, and naturally occurring oxygen was later found to be composed of a mixture of atoms of slightly different masses. Since the relative proportion of these different atoms (called **isotopes**) could conceivably change over a period of time, the entire atomic weight table could also change because the size of the amu was based on this mixture. To avoid this problem, the atomic mass unit is currently defined as $\frac{1}{12}$ of the mass of one particular isotope of carbon.

Since atoms are so very tiny, the actual mass of the amu is extremely small. One atom of the reference isotope of carbon (called carbon-12) has a mass of $1.992 \times 10^{-23}$ g. The mass of the amu is one-twelfth of this, or $1.660 \times 10^{-24}$ g. You might write this out in decimal form to get a feel for how really small atoms are.

## 1.10 SYMBOLS, FORMULAS, AND EQUATIONS

In a certain sense, learning chemistry is like learning a language such as Greek (in fact, some students have even been heard to say, "Chemistry is Greek to me!") We might compare the chemical symbols for the elements with an alphabet, the chemical formulas that we construct from the symbols with words, and the chemical equations that we write with sentences. In learning any new language we must start at the beginning with the alphabet.

At the present time a total of 105 different elements are known. Each element is identified by its name and can also be represented by its **chemical symbol.** Usually, the symbol bears a resemblance to the English name for the element. For instance, carbon = C, chlorine = Cl, nitrogen = N, and zinc = Zn. Some elements, however, have symbols that do not seem to correspond at all with their names. In nearly all of these cases, the elements have been known since the early history of chemistry when Latin was used as the universal language among scientists. For this reason the symbols are derived from their Latin names, for example, potassium (L. *kalium*) = K, sodium (*natrium*) = Na, silver (*argentum*) = Ag, mercury (*hydrargyrum*) = Hg, and copper (*cuprum*) = Cu. Regardless of the origin of the symbol, the first letter is always capitalized. A complete list of the elements with their chemical symbols appears on the inside front cover of the book.

A chemical compound is represented symbolically by its **chemical formula.** Thus, water is represented by $H_2O$, carbon dioxide by $CO_2$, methane (natural gas) by $CH_4$, and aspirin by $C_9H_8O_4$. Chemical formulas show the composition of substances. The subscripts in a formula denote the relative number of atoms of each element that appear in the compound; when no subscript is written, a subscript of 1 is implied. The formula $H_2O$, then, describes a substance containing two hydrogen atoms for every one oxygen atom. Similarly, the compound $CH_4$ contains one atom of carbon for every four atoms of hydrogen. Some compounds are more complex, and their formulas are written containing parentheses. For example $(NH_4)_2SO_4$ denotes the presence of two $NH_4$ units, for a total of two nitrogen atoms and eight hydrogen atoms, plus one sulfur atom and four oxygen atoms. As we shall see later, there are good reasons for writing this formula as $(NH_4)_2SO_4$ rather than $N_2H_8SO_4$, although both represent the same number of atoms.

There are certain substances that form crystals containing water molecules when their aqueous solutions are evaporated. Copper sulfate, for instance, forms beautiful blue crystals that contain five water molecules for every one copper sulfate ($CuSO_4$). This substance is an example of a **hydrate.** Its formula is written $CuSO_4 \cdot 5H_2O$. If the blue crystals are heated, water can be driven off to leave pure $CuSO_4$, which appears white.

**Chemical equations** are written to show the chemical changes that occur during chemical reactions. For example, the equation

$$Zn + S \longrightarrow ZnS$$

describes a reaction in which zinc (Zn) reacts with sulfur (S) to produce zinc sulfide (ZnS).[4] The substances on the left side of the arrow are known as the **reactants,** and the substances on the right of the arrow are called the **products.** (In the example above there is only one product). The arrow is read as "react to yield" or simply "yield." This equation is read as, "zinc plus sulfur react to yield zinc sulfide" or "zinc plus sulfur yield zinc sulfide." Sometimes it is also desirable (or necessary) to indicate whether the reactants and products in a chemical reaction are solids, liquids, gases, or are dissolved in a solvent such as water. This is accomplished by placing the letters s = solid, l = liquid, g = gas, aq = aqueous (water) solution in parentheses following the formulas of the substances in the equation. For instance, the equation

$$CaCO_3\ (s) + H_2O\ (l) + CO_2\ (g) \longrightarrow Ca(HCO_3)_2\ (aq)$$

describes a reaction between solid calcium carbonate (limestone), liquid water, and gaseous carbon dioxide to give an aqueous solution of calcium bicarbonate. This is the reaction responsible for the dissolving of limestone by ground water containing carbon dioxide. It is one of the causes of "hard" water and the formation of limestone caves.

Many of the equations that we write will contain **coefficients** preceding the chemical formulas. For example,

$$2H_2 + O_2 \longrightarrow 2H_2O$$

This equation is interpreted to mean that two hydrogen molecules plus one oxygen molecule (a coefficient of 1 is assumed when none is written) react to yield two molecules of $H_2O$. Such an equation is said to be **balanced** because it contains the same number of atoms of each element on both sides of the arrow. The techniques that we employ to help us balance equations will be discussed at a later time.

## 1.11 ENERGY

When a chemical change occurs, it almost always is accompanied by either an absorption or release of energy. In fact, these energy changes are an important factor in the driving force of a reaction, that is, the tendency or drive that the reaction has to occur. Energy changes tell us a great deal about the nature of the substances that react, and for this reason they are important to us.

Energy is usually defined as the capacity to do work. When an object possesses energy, it can affect other objects by doing work on them. A moving car possesses energy because it can do work on another car by moving it some distance in a collision. Because energy can be transferred from one object to another as work, the units of energy and work are the same.

There are two kinds of energy, kinetic energy and potential energy. **Kinetic energy** (K.E.) is associated with motion and is equal to one-half of an object's mass, *m*, multiplied by its speed, *s*, squared.

$$\text{K.E.} = \tfrac{1}{2}ms^2$$

Thus we see that the amount of work a moving body can do depends on both its mass and its velocity. For example, a truck moving at 20 miles/hr can do more "work" on the rear end of a car than a bicycle moving at the same

[4] The naming of chemical compounds is discussed in Chapter 4 and in Appendix B. For now, we use these names simply as labels.

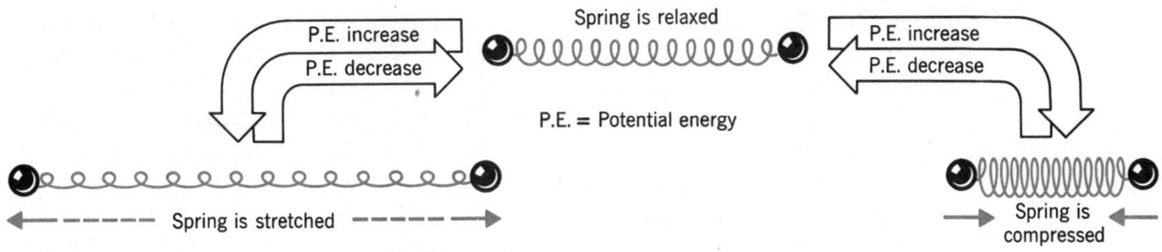


Figure 1.9
*Potential energy exists between objects that either attract or repel each other. When the spring is either stretched or compressed, the P.E. of the two balls increases.*

speed. We also know that a truck moving at 80 miles/hr can do more "work" on a car than one traveling at only 5 miles/hr.

**Potential energy** (P.E.) represents energy a body possesses because of the attractive or repulsive forces it experiences with other objects. If there are no attractive or repulsive forces, the body does not possess potential energy.

To see how potential energy is affected by these forces, consider two balls connected by a coiled spring as shown in Figure 1.9. When the balls are pulled apart, the spring is stretched and the energy spent in separating the balls is stored in the stretched spring. We say, therefore, that the potential energy of the two balls has increased. This energy is released and converted to kinetic energy when the balls are allowed to move toward one another. In this case there was an attractive force (the stretched spring) between the two objects.

Using this same apparatus, energy can also be stored by compressing the spring, thereby producing a repulsive force between the two balls. This stored energy is released as kinetic energy as the balls are permitted to move away from one another.

The kinds of changes in potential energy that accompany changes in the relative position of objects that either attract or repel one another are very important to remember, since they will help us analyze many physical and chemical changes in the chapters ahead.

As we have indicated, energy contained in chemical substances can be released through chemical reaction. Wood, for example, can react with oxygen, present in the air, in the process known as combustion. As the products of the reaction are formed, rather sizable quantities of energy are released. This chemical energy is initially present in the wood and oxygen and is released as the reaction proceeds. A process that results in the release of energy, such as the combustion of wood, is said to be **exothermic.** Processes that absorb energy, on the other hand, are termed **endothermic.**

The amount of energy released or absorbed in a chemical reaction depends on the quantity of materials that react. The burning of a match, for example, releases only a very small amount of energy while a large bonfire produces much more. Energy, therefore, is an extensive quantity.

The total amount of energy that a body possesses is equal to the sum of its kinetic energy and its potential energy. In an isolated system, such as our universe, the total energy is constant and *energy is neither created nor destroyed but, instead, can only be transformed from one kind of energy to another.* This statement is called the **law of conservation of energy** and is another example of the many conservation laws that appear to govern our physical world.

Energy is transmitted from one body to another in a variety of ways, for example, as light, sound, electricity, and/or heat. These various forms of energy can be converted from one to another and therefore are ultimately equivalent. The SI unit for energy is the **joule** (J). The joule is defined as 1 kg $m^2/sec^2$ in terms of the basic SI units. It represents the kinetic energy possessed by an object with a mass of 2.0 kg traveling at 1m/sec (in English units, an object weighing 4.4 lb traveling at a speed of 197 ft/min, or about 2.2 miles/hr).[5] A smaller unit, the **erg,** is equal to $1 \times 10^{-7}$ J. In referring to energies involved in chemical reactions it is useful to use the term **kilojoule** (kJ). One kilojoule is equal to one thousand joules (1 kJ = 1000 J).

All forms of energy ultimately end up as heat, and when chemists measure energy it is usually in the form of heat. For example, we might measure the heat liberated during the combustion of a gas such as methane. The measurement of heat energy involves the concept of **temperature.** It is important to remember that heat is not the same as temperature. Heat *is* energy, while temperature is a measure of the intensity of heat or hotness. Another useful definition of temperature is that it is an intensive quantity that defines the *direction* and *rate* of heat flow. We know that heat always flows from a warm object to a cool one. We also know that the rate of heat transfer depends on the difference in temperature between two objects. If we want to warm something slowly, we place it in contact with an object just slightly warmer; if we wish to heat it quickly, we place it in contact with a very hot object.

The measurement of temperature is usually accomplished with a thermometer consisting of a narrow capillary tube connected to a thin-walled reservoir filled with some liquid (usually mercury). As the temperature is raised, the liquid expands and its volume increases which causes the fluid to rise in the capillary. The height of fluid in the capillary then is proportional to temperature.

A number of temperature scales have been devised. The **Fahrenheit** scale, in common use in the United States, is defined by the freezing point and boiling point of water.

At its freezing point (which is also its melting point), a pure substance such as water can exist as both solid (ice) and liquid in contact with each other *at the same temperature.* If heat is added to this mixture, some solid melts and more liquid is formed, but the temperature remains constant while the solid is melting. If heat is removed from the mixture, liquid freezes to produce more solid, once again without a temperature change. At its boiling point a pure liquid can exist in contact with its vapor at the same temperature. In this case, removal of heat causes more liquid to evaporate, all at a constant temperature. The constancy of temperature during these phase changes makes them ideal temperatures to serve as calibration points on a temperature scale.

On the Fahrenheit scale the freezing point of water is assigned a temperature of 32°F and the boiling point a temperature of 212°F. The difference between these two reference points, 180 Fahrenheit degrees, thus defines the size of the degree unit. In the sciences the temperature scale that is employed is the **Celsius** scale (also called the **centigrade** scale), which defines 0°C as the freezing point of water and 100°C as the boiling point of water. Therefore, we see that 100 Celsius degrees are equal to 180 Fahrenheit degrees, so that the Celsius degree is nearly twice as large as that on the Fahr-

[5] Remember, K.E. = $\frac{1}{2}ms^2$. In this case K.E. = $\frac{1}{2}(2\text{ kg})(1\text{ m/sec})^2$, or K.E. = 1 kg $m^2/sec^2$ = 1 J.

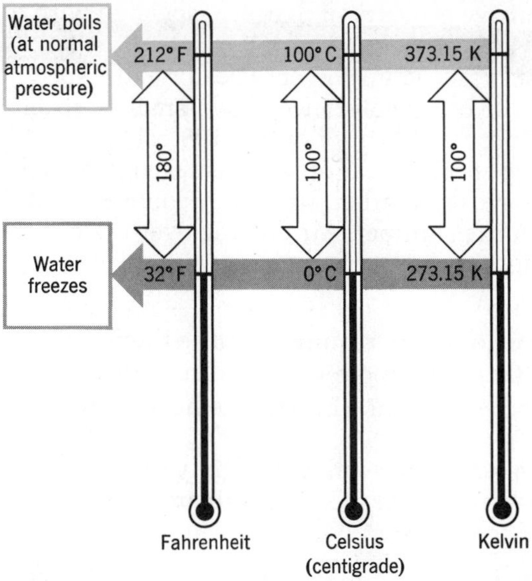


Figure 1.10
*Comparison of temperature scales.*

enheit scale. Temperatures in °C are related to temperatures in °F by the equation

$$°C = \tfrac{5}{9}(°F - 32)$$

The SI temperature unit is the kelvin (K). The size of the degree on the kelvin scale is the same as that on the Celsius scale; the difference between them is the location of the zero point. On the kelvin scale water freezes at 273.15 K (note that the degree symbol, °, is not used in the SI unit) so that the relationship between the Celsius and kelvin scales is

$$°C = K - 273.15$$

Figure 1.10 illustrates the differences between these three temperature scales.

Using the concept of temperature, the unit of heat energy, the **calorie** (abbreviated **cal**) is defined as the amount of energy required to raise the temperature of 1 g of water at 15°C by 1°C.[6] The **kilocalorie** (**kcal**), like the kilojoule, is a convenient size unit for dealing with energy changes in chemical reactions. The Calorie (which should be written with a capital C) used in nutrition is really the same as the kilocalorie. Thus when we read that a serving of mashed potatoes contains 230 Cal, we are being told that 230 kcal of energy are liberated when the body metabolizes this food.

Until recently, nearly all of the chemical scientific literature has used the calorie (or kilocalorie) in reporting energy changes. With the introduction of SI units, the joule (or kilojoule) is now preferred. Unfortunately, at this time in the sciences we must be able to handle both sets of units. The conversion between calories and joules, or kilocalories and kilojoules, is given by

$$1 \text{ cal} = 4.1840 \text{ J}$$
$$1 \text{ kcal} = 4.1840 \text{ kJ}$$

[6] It is necessary to specify the temperature of the water because the amount of heat required to raise the temperature of 1 g of water 1°C varies slightly with the temperature of the water.

An intensive property of matter associated with energy is **specific heat,** the amount of heat required to raise the temperature of 1 g of a substance by 1°C. Because of the definition of the calorie, the specific heat of water is 1.00 cal/g °C. Most other substances have smaller specific heats. Iron, for example, has a specific heat of only 0.108 cal/g °C.

The large specific heat of water is responsible for the moderating effect the oceans have on weather, since they cool very slowly in winter and warm up slowly in the summer. Air moving over the oceans in winter never gets very cold, and in the summer the air never gets very hot.

EXAMPLE 1.5 How many calories are required to raise the temperature of a 3-in. iron nail weighing 7.05 g from room temperature (25°C) to 100°C? How many joules does this correspond to? The specific heat of iron is 0.108 cal/g °C.

SOLUTION To solve this problem we must multiply the specific heat by mass (g) and by the temperature change (°C) to eliminate these units and obtain calories as the units of the answer.

$$\text{specific heat} \times \text{mass} \times \text{temperature change} = \text{heat energy}$$
$$\left(\frac{\text{cal}}{\cancel{\text{g}}\,\cancel{°\text{C}}}\right) \times (\cancel{\text{g}}) \times (\cancel{°\text{C}}) = \text{cal}$$

Using our data (the temperature change is 75°C),

$$\left(\frac{0.108\ \text{cal}}{\cancel{\text{g}}\,\cancel{°\text{C}}}\right) \times (7.05\ \cancel{\text{g}}) \times (75\cancel{°\text{C}}) = 57\ \text{cal}$$

Notice that the answer has been rounded to two significant figures. Do you know why? If not, review the significant figure rules for multiplication.

To answer the final part of this question we must convert calories to joules.

$$57\ \cancel{\text{cal}} \times \left(\frac{4.184\ \text{J}}{1\ \cancel{\text{cal}}}\right) = 240\ \text{J}$$

(Again, the answer is rounded to two significant figures.)

## INDEX TO QUESTIONS AND PROBLEMS (problem numbers in italics)

Scientific method 1, 2
Measurement 3
Scientific notation *20, 21, 22*
Significant figures *18, 19, 23, 24, 25, 26, 27*
Units and conversions 4, *28, 29, 30, 31, 32*
Properties of matter 5
Elements, compounds, mixtures 6, 7, 8
Definite proportions *33, 39, 40*
Atomic theory 9, 10
Atomic weights *35, 43*
Multiple proportions *34, 41, 42*
Symbols, formulas, equations 11, 12, 13
Energy 14, 15, 16, *36, 37, 47, 48, 49, 50, 51, 53*
Temperature 17, *38, 44, 45, 46, 52*

## REVIEW QUESTIONS

**1.1** What is the difference between a theory and a law?

**1.2** Why is it important for scientists to be concerned about reporting the correct number of significant figures in their measurements?

**1.3** What is the difference between accuracy and precision?

**1.4** The SI units for mass and volume are the kilogram and cubic meter, respectively. Why do we generally use the units gram and milliliter in the laboratory instead of the SI units?

**1.5** What is the difference between an extensive and an intensive property? Can you think of any examples that were not mentioned in the text?

**1.6** There are many examples of homogeneous and heterogeneous mixtures in the world around us. How would you classify sea water, air (unpolluted), smog, smoke, homogenized milk, black coffee, a penny, a ham sandwich?

**1.7** Identify the phases that exist in a copper pan containing two iron nails, a quart of water, and four glass marbles.

**1.8** Distinguish among the terms element, compound, and mixture.

**1.9** What is the most important atomic property in Dalton's atomic theory?

**1.10** Distinguish between the terms atom and molecule.

**1.11** Write the chemical symbols for the elements iron, sodium, potassium, antimony, tin, lead, mercury, gold, silver, tungsten, and copper.

**1.12** How many atoms of each kind are indicated in each of the following formulas: $K_2S$, $Na_2CO_3$, $(NH_4)_3PO_4$, $K_4Fe(CN)_6$, $Na_3Ag(S_2O_3)_2$?

**1.13** Which of the following equations are not balanced?
(a) $ZnCl_2 + NaOH \rightarrow Zn(OH)_2 + NaCl$
(b) $CuCO_3 + 2HCl \rightarrow CuCl_2 + CO_2 + H_2O$
(c) $Fe_2O_3 + 2CO \rightarrow 2Fe + 2CO_2$
(d) $NH_4NO_3 \rightarrow N_2O + 2H_2O$

**1.14** How does potential energy differ from kinetic energy?

**1.15** What is meant by endothermic and exothermic?

**1.16** If the potential energy of an object decreases as it is moved away from another object, what kind of force (attractive or repulsive) must exist between the two?

**1.17** What is the difference between heat and temperature?

## REVIEW PROBLEMS (More difficult problems are marked by an asterisk.)

**1.18** How many significant figures are there in the following numbers: 1.0370, 0.000417, 0.00309, 100.1, 9.0010?

**1.19** Perform the following computations, rounding answers to the proper number of significant figures.
(a) $2.41 \times 3.2$
(b) $4.025 \times 18.2$
(c) $81.8 \div 104.2$
(d) $3.476 + 0.002$
(e) $81.4 - 0.002$

**1.20** Express each of the following in scientific notation.
(a) 1,250
(b) 13,000,000
(c) 60,230,000,000,000,000,000,000
(d) 214,570
(e) 31.47

**1.21** Express each of the following in scientific notation.
(a) 0.00040
(b) 0.0000000003
(c) 0.002146
(d) 0.0000328
(e) 0.00000000000091

**1.22** Write the following numbers in decimal notation.
(a) $3 \times 10^{10}$
(b) $2.54 \times 10^{-5}$
(c) $122 \times 10^{-2}$
(d) $3.4 \times 10^{-7}$
(e) $0.0325 \times 10^{6}$

**1.23** Perform the following computations, expressing the answers in scientific notation rounded to the proper number of significant figures.
(a) $(3.42 \times 10^{8}) \times (2.14 \times 10^{6})$
(b) $(1.025 \times 10^{6}) \times (14.8 \times 10^{-3})$
(c) $(143.7) \times (84.7 \times 10^{16})$
(d) $(5274) \times (0.33 \times 10^{-7})$
(e) $(8.42 \times 10^{-7}) \times (3.211 \times 10^{-19})$

**1.24** Perform the following computations, expressing the answers in scientific notation rounded to the proper number of significant figures.
(a) $(12.45 \times 10^{6}) \div (2.24 \times 10^{3})$
(b) $822 \div 0.028$
(c) $(635.4 \times 10^{-5}) \div (42.7 \times 10^{-14})$
(d) $(31.3 \times 10^{-12}) \div (8.3 \times 10^{-6})$
(e) $(0.74 \times 10^{-9}) \div (825.3 \times 10^{18})$

**1.25** Perform the following computations, expressing the answers in scientific notation rounded to the proper number of significant figures.

(a) $(2.047 \times 10^8) + (14.33 \times 10^8)$
(b) $(12.4 \times 10^8) + (92.3 \times 10^7)$
(c) $(42.003 \times 10^5) - (3.25 \times 10^3)$
(d) $118.45 - (0.033 \times 10^3)$
(e) $1.00 + (3.75 \times 10^{-8})$

**1.26** Perform the following computations, rounding the answers to the correct number of significant figures.

(a) $(341.7 - 22) + (0.00224 \times 814{,}005)$
(b) $(82.7 \times 143) + (274 - 0.00653)$
(c) $(3.53 \div 0.084) - (14.8 \times 0.046)$
(d) $(324 \times 0.0033) + (214.2 \times 0.0225)$
(e) $(4.15 + 82.3) \times (0.024 + 3.000)$

**1.27** Perform the following computations, expressing the answers in scientific notation rounded to the proper number of significant figures.

(a) $(8.3 \times 10^{-6}) \times (4.13 \times 10^{-7}) \div (5.411 \times 10^{-12})$
(b) $[(3.125 \times 10^{-6}) + (5.127 \times 10^{-5})] \times (6.72 \times 10^8)$
(c) $[14.39 + (2.43 \times 10^1)] \div 1275$
(d) $[(1.583 \times 10^{-4}) - (0.00255)] \times [(142.3) + (0.257 \times 10^2)]$
(e) $(0.0000425) \div [0.0008137 + (2.65 \times 10^{-5})]$

**1.28** Perform the following conversions.

(a) 1.40 m to cm
(b) 2800 mm to m
(c) 185 ml to liters
(d) 18 g to kg
(e) 10 sq yards (10 $yd^2$) to $m^2$
(f) 100 miles to inches
(g) 20 ft/sec to mi/hr
(h) 1 $mi^3$ to $m^3$
(i) 40 mi/hr to cm/sec
(j) 25 liters to $dm^3$

**1.29** After shopping for a sports car, a certain wealthy freshman decided that she would choose between a new two-passenger, six-cylinder Smokebelcher, which gives 21 miles/gal, and an old 1974, eight-cylinder, 10-passenger Pferdburper (with automatic transmission), which the owner guaranteed would deliver 10 km/liter. On the basis of gas mileage, which car will our friend find more economical to operate?

**1.30** A student has just returned from Germany with a car that he purchased while on vacation. The speedometer is calibrated in kilometers per hour (km/hr). As he drives away from the pier, he notices a sign that posts the speed limit at 35 mph. What is the maximum speed that he can reach, in km/hr, without having to worry about receiving a speeding ticket?

**1.31** After receiving a speeding ticket, the student in the preceding question is informed by the police officer that the courthouse is located "4.3 miles straight down the road. You can't miss it!" The odometer in the student's car measures kilometers (km). How far must he travel, in km, before arriving at the courthouse?

***1.32** In the country of Ferdovia the Ferds thrive on potatoes. The average Ferd earns 142 thrubs (the local currency of Ferdovia) per week and spends approximately $\frac{1}{14}$ of this yearly income on potatoes. If potatoes cost 2 thrubs per pound, how many pounds of potatoes does the average Ferd consume each year?

**1.33** Cyclopropane, a very effective anesthetic, contains the elements carbon and hydrogen combined in a ratio of 1.0 g of hydrogen to 6.0 g of carbon. If a given sample of cyclopropane was found to contain 20 g of hydrogen, how many grams of carbon would it contain?

**1.34** Copper forms two oxides. In one of them there are 1.26 g of oxygen combined with 10 g of copper. In the other there are 2.52 g of oxygen combined with 10 g of copper. Show that these data illustrate the law of multiple proportions.

**1.35** If the atomic mass unit were defined such that a single fluorine atom weighed 1 amu, what would be the atomic weights of carbon and hydrogen?

**1.36** A truck having a mass of 4500 kg is moving at a speed of 1.79 m/sec. Calculate its kinetic energy in joules. How many calories does this correspond to?

***1.37** Calculate the kinetic energy, in joules, of a 140-lb athlete running at a speed of 15 miles/hr.

**1.38** Gallium metal has the largest liquid range of any element. It melts at 30°C and boils at 1983°C. What are its melting and boiling points in degrees Fahrenheit?

***1.39** Two samples of Freon (a coolant used in refrigerators and air conditioners) were analyzed. In one sample 1.00 g of carbon was found to be combined with 6.33 g of fluorine and 11.67 g of chlorine. In the second sample 2.00 g of carbon were found to be combined with 12.66 g of fluorine and 23.34 g of chlorine. What are the ratios of the weights of carbon to fluorine, carbon to chlorine, and fluorine to chlorine in the two samples? Show that these data support the law of definite composition.

***1.40** Three samples of a solid substance composed of elements X and Y were prepared. The first was found to contain 4.31 g X and 7.69 g Y; the second was composed of 35.9% X and 64.1% Y; it was observed that 0.718 g X reacted with Y to form 2.00 g of the third sample. Show how these data demonstrate the law of definite composition.

***1.41** In Example 1.3, the first compound (2.62 g of N, 1.50 g O) has the formula $N_2O$ (the substance is nitrous oxide). Suggest a possible formula for the second compound.

***1.42** Two compounds are formed between phosphorus and oxygen. 1.50 g of one compound was found to contain 0.845 g of phosphorus while a 2.50-g sample of the other contained 1.09 g of phosphorus. Show that these data are consistent with the law of multiple proportions.

***1.43** In a certain compound, 6.92 g of X were found combined with 0.584 g of carbon. If the atomic weight of carbon = 12.0 amu and if four atoms of X are combined with one atom of carbon, calculate the atomic weight of X.

**1.44** Tungsten, used as filaments in electric light bulbs, has a melting point of 6152°F. What is its melting point expressed in degrees Celsius?

**1.45** Solid carbon dioxide (Dry Ice) has a temperature of −78°C. What is its temperature on the Fahrenheit scale? What is its temperature on the kelvin scale?

**1.46** Normal human body temperature is 98.6°F. In Europe, and in many hospitals in the United States, the centigrade temperature scale is used almost universally so that clinical thermometers are calibrated in °C. If you had one of these thermometers, what temperature would you expect, in °C, for a normal healthy individual? If your temperature registered 39°C, what would your temperature be in °F?

**1.47** How many calories are required to raise the temperature of 500 g of water by 24°C?

**1.48** How much will the temperature of 150 g of water change if it loses 35 cal?

***1.49** An automobile weighing 2.4 tons is traveling at 35 miles/hr. What is its kinetic energy in joules? If all this energy were converted to heat, how much water could have its temperature raised by 10°C?

**1.50** A car, whose mass is 1500 kg, skids to a halt from a speed of 60 m/sec. How many calories of heat are generated by friction between the car and the pavement?

***1.51** A packing crate full of machinery weighs 1400 kg. It is moved a distance of 40 m by a man exerting a force of 510 N. How many joules of work has been accomplished?

***1.52** Naphthalene (used in moth balls) has a melting point of 80°C and a boiling point of 218°C. Suppose that this substance were used to define a new temperature scale on which the melting point of naphthalene was 0°N and the boiling point of naphthalene was 100°N. What would be the freezing point and boiling point of water in °N? What general equation could we use to relate temperatures in °C and °N?

***1.53** A copper penny weighing 3.14 g is heated to 100°C in boiling water. It is quickly dried and placed into a Styrofoam cup containing 10.00 g of water at 25.0°C. Copper has a specific heat of 0.0920 cal/g °C. What will be the final temperature of the penny and water? (Assume that a negligible amount of heat is absorbed by the cup.) Hint: The law of conservation of energy requires that the number of calories lost by the metal be equal to the number of calories gained by the water.

# 2 STOICHIOMETRY: CHEMICAL ARITHMETIC

In Chapter 1 you were introduced to many of the important basic concepts in chemistry: the ideas of atoms and atomic weight, elements and compounds, and several major chemical laws. Chemists, physicists, and biologists have developed these ideas so that today they routinely think of chemical, physical, and biological processes taking place between atoms and molecules on a microscopic, atomic scale. However, we cannot see individual atoms or molecules, and in the laboratory it is necessary to work with huge numbers of these small particles. Chemists have devised ways of carrying out chemical reactions so that observations on a large (macroscopic) scale can be readily translated into the language of the atomic world. That this can be done with relative ease is really quite fascinating.

In order to study chemical compounds in the laboratory, it is necessary for a chemist (or even a chemistry student) to have a knowledge of the quantitative relationships that exist among the substances that enter into chemical reactions. **Stoichiometry** (derived from the Greek *stoicheion* = element and *metron* = measure) is the term used to refer to all of the quantitative aspects of chemical composition and reaction. We will now see how chemical formulas are determined and how chemical equations prove useful for predicting the proper amounts of reactants that must be mixed together to get a complete reaction.

## 2.1 THE MOLE

We know that atoms react to form molecules in simple whole-number ratios. Hydrogen and oxygen, for instance, combine in a 2-to-1 ratio to form water; carbon and oxygen combine in a 1-to-1 ratio to form carbon monoxide, CO. In any real-life laboratory situation we must increase the size of these quantities so that we can see them and weigh them, because individual atoms are too tiny to work with.

One way of enlarging the reaction is to work with dozens, instead of individual atoms.

$$\begin{array}{ccccc} \text{1 atom C} & + & \text{1 atom O} & \longrightarrow & \text{1 molecule CO} \\ \text{1 dozen C} & + & \text{1 dozen O} & \longrightarrow & \text{1 dozen CO} \\ \textbf{(12 atoms C)} & & \textbf{(12 atoms O)} & & \textbf{(12 molecules CO)} \end{array}$$

A dozen atoms or molecules is still too small to work with, however, so we must find a still larger unit. The "chemist's dozen" is called the **mole.** It is composed of $6.023 \times 10^{23}$ objects (we shall talk more about the origin of this number, called **Avogadro's number,** later).

$$1 \text{ dozen} = 12 \text{ objects}$$
$$1 \text{ mole} = 6.023 \times 10^{23} \text{ objects}$$

The same reasoning that we can use with the dozen applies equally to the mole. The mole is simply a larger "box."

$$\underset{\substack{\textbf{(6.023} \times \textbf{10}^{23} \\ \textbf{atoms C)}}}{1 \text{ mole C}} + \underset{\substack{\textbf{(6.023} \times \textbf{10}^{23} \\ \textbf{atoms O)}}}{1 \text{ mole O}} \longrightarrow \underset{\substack{\textbf{(6.023} \times \textbf{10}^{23} \\ \textbf{molecules CO)}}}{1 \text{ mole CO}}$$

We see that when we take 1 mole of carbon and 1 mole of oxygen we have equal numbers of carbon and oxygen atoms and can construct enough CO molecules to fill our mole "box" exactly.

A very important thing to notice here is that the same whole-number ratios that apply to the individual atoms and molecules also apply exactly to the numbers of *moles* of atoms and molecules. Everything is simply increased by the same factor. For example, to form carbon tetrachloride, $CCl_4$, we know that

$$1 \text{ atom C} + 4 \text{ atoms Cl} \longrightarrow 1 \text{ molecule } CCl_4$$

We can immediately enlarge this to moles.

$$1 \text{ mole C} + 4 \text{ moles Cl} \longrightarrow 1 \text{ mole } CCl_4$$

To repeat this very important idea, the *ratio* by which moles of substances react is the same as the *ratio* by which their atoms and molecules react.

EXAMPLE 2.1 What mole ratio of carbon to chlorine must be chosen to prepare the substance $C_2Cl_6$?

SOLUTION The atom ratio in $C_2Cl_6$ is

$$\frac{2 \text{ atoms C}}{6 \text{ atoms Cl}} = \frac{1}{3}$$

To prepare $C_2Cl_6$ the carbon to chlorine *atom* ratio must be maintained at 1:3. *The mole ratio must also be 1:3.* Carbon and chlorine must be combined in a ratio of 1 mole of C to 3 moles of Cl.

EXAMPLE 2.2 How many moles of carbon are in 2.65 moles of $C_2Cl_6$?

SOLUTION The formula tells us that 1 molecule of $C_2Cl_6$ contains 2 atoms of C. This can be immediately translated to moles; 1 mole of $C_2Cl_6$ contains 2 moles of C. If you have trouble understanding this, answer this question: How many dozen C atoms are in 1 dozen $C_2Cl_6$ molecules? Then substitute "mole" for "dozen."

For the purpose of computation we can express the mole relationship as an equivalence,

$$1 \text{ mole } C_2Cl_6 \sim 2 \text{ moles C}$$

which can be used to construct a conversion factor to set up the arithmetic.

$$2.65 \cancel{\text{moles } C_2Cl_6} \times \left(\frac{2 \text{ moles C}}{1 \cancel{\text{mole } C_2Cl_6}}\right) \sim 5.30 \text{ moles C}$$

Thus, 2.65 moles $C_2Cl_6$ contains 5.30 moles C.

A mole of atoms or molecules is large enough to work with experimentally. The mole, however, is a chemical unit, because moles react in the same ratios as atoms and molecules. We still must have a way to translate this to laboratory units, something that can be measured directly in the laboratory.

Earlier it was stated that a mole consists of $6.023 \times 10^{23}$ objects. This rather odd number is chosen because this number of atoms of any element has a weight in grams that is precisely numerically equal to its atomic weight.[1] For example, the atomic weights of C and O are 12.01115 and 15.9994, respectively. Therefore,

$$1 \text{ mole C} = 12.01115 \text{ g C}$$
$$1 \text{ mole O} = 15.9994 \text{ g O}$$

The balance, then, serves as our tool for measuring moles. If we take 12.01115 g C and 15.9994 g O, we know that we have taken 1 mole of C and 1 mole of O which can produce exactly 1 mole of CO. With this as background, let us look at some sample problems dealing with the mole.

**EXAMPLE 2.3** How many moles of Si are in 30.5 grams of Si?

**SOLUTION** Our problem is one of converting the units of grams of Si to moles of Si, that is, 30.5 g Si = (?) moles Si.

We know from the table of atomic weights that

$$1 \text{ mole Si} = 28.1 \text{ g Si}$$

To convert from g Si to moles we must multiply 30.5 g Si by a factor that contains the units g Si in the denominator, that is,

$$\frac{1 \text{ mole Si}}{28.1 \text{ g Si}} = 1$$

When we set up this problem,

$$30.5 \cancel{\text{g Si}} \times \left(\frac{1 \text{ mole Si}}{28.1 \cancel{\text{g Si}}}\right) = 1.09 \text{ moles Si}$$

Thus 30.5 g Si = 1.09 moles Si.

**EXAMPLE 2.4** How many grams of Cu are there in 2.55 moles of Cu?

**SOLUTION** From the table of atomic weights,

$$1 \text{ mole Cu} = 63.5 \text{ g Cu}$$

[1] Historically the mole was first defined as the amount of an element having a mass in grams numerically equal to its atomic weight. Later it was discovered experimentally that there are $6.023 \times 10^{23}$ atoms in 1 mole of an element. We will see how Avogadro's number can be measured in later chapters.

Since there are equal numbers of atoms when the weights of elements are taken in proportion to their atomic weights, we can also speak of a *pound-mole* or a *ton-mole*. Thus, 12 lb of C (1 lb-mole C) contain the same number of atoms as 16 lb of O (1 lb-mole O) because the weights of C and O are taken in the same ratio as their atomic weights. Similarly, 12 tons of C (1 ton-mole C) contain the same number of atoms as 16 tons of O (1 ton-mole O). These are useful ideas when dealing with very large quantities of reactants in industrial processes.

Our conversion factor must have the units, moles Cu in the denominator, that is,

$$\frac{63.5 \text{ g Cu}}{1 \text{ mole Cu}} = 1$$

Setting up the problem, we have

$$2.55 \cancel{\text{moles Cu}} \times \left(\frac{63.5 \text{ g Cu}}{1 \cancel{\text{mole Cu}}}\right) = 162 \text{ g Cu}$$

EXAMPLE 2.5 How many moles of Ca are required to react with 2.50 moles of Cl to produce the compound $CaCl_2$?

SOLUTION Our problem: 2.50 moles Cl ~ (?) moles Ca.

We know from the formula that 1 atom of Ca combines with 2 atoms of Cl. Thus,

$$1 \text{ atom Ca} \sim 2 \text{ atoms Cl}$$

We also know that 1 mole Ca and 1 mole Cl contain the same number of atoms; therefore, to maintain a ratio of 1 atom of Ca to 2 atoms of Cl, we state that 1 mole of Ca combines with 2 moles of Cl.

$$1 \text{ mole Ca} \sim 2 \text{ moles Cl}$$

We obtain our answer as follows:

$$2.50 \cancel{\text{moles Cl}} \times \left(\frac{1 \text{ mole Ca}}{2 \cancel{\text{moles Cl}}}\right) \sim 1.25 \text{ moles Ca}$$

EXAMPLE 2.6 How many grams of Ca must react with 41.5 g Cl to produce $CaCl_2$?

SOLUTION Our problem: 41.5 g Cl ~ (?) g Ca.
What quantities do we need and what do we know?
We know that

$$1 \text{ mole Ca} \sim 2 \text{ moles Cl} \qquad \text{(Why?)}$$

Also,

$$1 \text{ mole Cl} = 35.5 \text{ g Cl} \qquad \text{(atomic weight table)}$$
$$1 \text{ mole Ca} = 40.1 \text{ g Ca} \qquad \text{(atomic weight table)}$$

Here is the solution to the problem:

$$41.5 \cancel{\text{g Cl}} \times \left(\frac{1 \text{ mole Cl}}{35.5 \cancel{\text{g Cl}}}\right) = 1.17 \text{ moles Cl}$$

$$1.17 \cancel{\text{moles Cl}} \times \left(\frac{1 \text{ mole Ca}}{2 \cancel{\text{moles Cl}}}\right) \sim 0.585 \text{ mole Ca}$$

$$0.585 \cancel{\text{mole Ca}} \times \left(\frac{40.1 \text{ g Ca}}{1 \cancel{\text{mole Ca}}}\right) = 23.5 \text{ g Ca}$$

And here are all of these steps written together.

$$41.5 \cancel{\text{g Cl}} \times \left(\frac{1 \cancel{\text{mole Cl}}}{35.5 \cancel{\text{g Cl}}}\right) \times \left(\frac{1 \cancel{\text{mole Ca}}}{2 \cancel{\text{moles Cl}}}\right) \times \left(\frac{40.1 \text{ g Ca}}{1 \cancel{\text{mole Ca}}}\right) \sim 23.5 \text{ g Ca}$$

In this last example we have strung together a series of conversion factors. When we set up the problem, we link these together by considering what units must be eliminated by cancellation. Thus, the first factor had to have g Cl in the denominator; the second factor had to have moles Cl in the denominator; the third was required to have moles Ca in the denominator. At this point we stop since our units are now those of the answer, and we have only to perform the arithmetic to obtain the correct result.

EXAMPLE 2.7 What is the mass of 1 atom of calcium?

SOLUTION Avogadro's number provides us with a means of relating the submicroscopic world of atoms and molecules with the large world of the laboratory. Our problem can be stated as

$$1 \text{ atom Ca} \sim (?) \text{ g Ca}$$

We have the relationships,

$$1 \text{ mole Ca} = 40.1 \text{ g Ca}$$
$$1 \text{ mole Ca} = 6.02 \times 10^{23} \text{ atoms Ca}$$

The solution to the problem can then be set up as

$$1 \cancel{\text{atom Ca}} \times \left(\frac{1 \cancel{\text{mole Ca}}}{6.02 \times 10^{23} \cancel{\text{atoms Ca}}}\right) \times \left(\frac{40.1 \text{ g Ca}}{1 \cancel{\text{mole Ca}}}\right) \sim 6.66 \times 10^{-23} \text{ g Ca}$$

## 2.2 MOLECULAR WEIGHTS AND FORMULA WEIGHTS

As with elements, the balance can also be used to measure moles of compounds. The simplest way of obtaining the weight of 1 mole of a substance is merely to add up the atomic weights of all of the elements present in the compound. If the substance is composed of molecules (for example, $CO_2$, $H_2O$, or $NH_3$), the sum of the atomic weights is called the **molecular weight.** Thus the molecular weight of $CO_2$ is obtained as

| | | |
|---|---|---|
| C | $1 \times 12.0$ amu = | 12.0 amu |
| 2O | $2 \times 16.0$ amu = | 32.0 amu |
| $CO_2$ | total | 44.0 amu |

Similarly, the molecular weight of $H_2O$ = 18.0 amu and that of $NH_3$ = 17.0 amu. The weight of 1 mole of a substance is obtained simply by writing its molecular weight followed by the units, grams. Thus,

$$1 \text{ mole } H_2O = 18.0 \text{ g}$$
$$1 \text{ mole } NH_3 = 17.0 \text{ g}$$

In later chapters we shall encounter many compounds that do not contain separate, distinct molecules. Often, when certain atoms react, they gain or lose negatively charged particles called **electrons.** Sodium and chlorine happen to react in this way. When sodium chloride, NaCl, is formed from the elements, each Na atom loses one electron and each Cl atom gains one. Since Na and Cl are electrically neutral to start, these atoms acquire a charge when NaCl is formed. These are written as $Na^+$ (positive because Na has lost a negatively charged electron) and $Cl^-$ (negative because Cl has gained an electron). *Atoms or groups of atoms that have acquired an electrical charge are called* **ions.** Since solid NaCl is composed of $Na^+$ and $Cl^-$ ions, this compound is said to be *ionic.*

This entire topic is explored further in Chapters 3 and 4. For now, it is only necessary for you to know that compounds that are ionic do not contain molecules. Their formulas simply state the ratio of the different atoms in the substance. In NaCl the atoms are in a one-to-one ratio. In the ionic compound $CaCl_2$ the ratio of Ca to Cl atoms is 1 to 2 (relax—at this point you were not expected to know that $CaCl_2$ is ionic). Rather than refer to molecules of NaCl or $CaCl_2$ we use the term **formula unit** to specify the pair of ions ($Na^+$ and $Cl^-$) in NaCl, or the set of three ions in $CaCl_2$.

For ionic compounds the sum of the atomic weights of the elements present in an ionic solid is known as the **formula weight.** For NaCl this is $22.99 + 35.45 = 58.44$. One mole of NaCl would contain 58.44 g NaCl ($6.023 \times 10^{23}$ formula units of NaCl). It is also true that $6.023 \times 10^{23}$ formula units (1 mole) of NaCl contains $6.023 \times 10^{23}$ $Na^+$ ions and $6.023 \times 10^{23}$ $Cl^-$ ions. Use of the term formula weight, of course, is not restricted to ionic compounds. It can also be applied to molecular substances, in which case the terms formula weight and molecular weight mean the same thing.

## 2.3 PERCENTAGE COMPOSITION

A very simple and also often very useful computation is the calculation of the percentage composition of a compound, that is, the percentage of the total mass contributed by each element. The procedure to determine the percent composition is illustrated by Example 2.8.

EXAMPLE 2.8 What is the percentage composition of chloroform, $CHCl_3$?

SOLUTION The total mass of 1 mole of $CHCl_3$ is obtained from the molecular weight,

$$(12.01 + 1.008 + 3 \times 35.45)\text{ amu} = 119.37\text{ amu}$$

or,

$$(12.01\text{ g} + 1.008\text{ g} + 3 \times 35.45\text{ g}) = 119.37\text{ g}$$

$$\%\text{ C} = \frac{\text{weight carbon}}{\text{weight CHCl}_3} \times 100$$

$$\%\text{ C} = \frac{12.01\text{ g C}}{119.37\text{ g CHCl}_3} \times 100 = 10.06\%\text{ C}$$

$$\%\text{ H} = \frac{1.008\text{ g H}}{119.37\text{ g CHCl}_3} \times 100 = 0.844\%\text{ H}$$

$$\%\text{ Cl} = \frac{106.35\text{ g Cl}}{119.37\text{ g CHCl}_3} \times 100 = 89.09\%\text{ Cl}$$

$$\text{total }\% = 100.00$$

A similar calculation can be used to determine the quantity of an element in a given sample of a compound. This is illustrated in Example 2.9.

EXAMPLE 2.9 Calculate the weight of iron (Fe) in a 10.0-g sample of rust, $Fe_2O_3$.

SOLUTION To solve this problem we can first calculate the fraction of $Fe_2O_3$ that is Fe. From the formula we see that one formula unit of $Fe_2O_3$ contains 2 atoms of Fe. Therefore we know that 1 mole of $Fe_2O_3$ contains 2 moles of Fe. Also,

$$1\text{ mole Fe} = 55.85\text{ g Fe}$$
$$1\text{ mole Fe}_2\text{O}_3 = 159.7\text{ g Fe}_2\text{O}_3$$

(Remember, we can always calculate the weight of 1 mole of a substance if we know its chemical formula!) Therefore 1 mole $Fe_2O_3$ (159.7 g $Fe_2O_3$) contains 2 moles Fe (111.7 g Fe). The fraction of $Fe_2O_3$ that is Fe is thus

$$\frac{111.7 \text{ g Fe}}{159.7 \text{ g } Fe_2O_3}$$

Finally, the weight of Fe in the 10-g sample $Fe_2O_3$ is equal to the sample weight multiplied by the fraction of the sample that is Fe.

$$10.0 \text{ g } Fe_2O_3 \times \left(\frac{111.7 \text{ g Fe}}{159.7 \text{ g } Fe_2O_3}\right) \sim 6.99\text{g Fe}$$

## 2.4 CHEMICAL FORMULAS

A formula conveys to us certain kinds of information, including elemental composition, relative numbers of each kind of atom present, the actual numbers of each kind of atom in a molecule of the substance, or the structure of the compound. We can classify formulas according to the amount of information they provide.

A formula that simply gives the relative number of atoms of each element present is called a **simplest formula.** It is also called an **empirical formula** because it is invariably obtained as the result of some experimental analysis. The formulas NaCl, $H_2O$, and $CH_2$ are empirical formulas.

A formula that states the actual number of each kind of atom found in a molecule is called a **molecular formula.** $H_2O$ is a molecular formula (as well as an empirical formula) since a molecule of water contains 2 atoms of H and 1 atom of O. The formula $C_2H_4$ is a molecular formula for a substance (ethylene) containing 2 atoms of carbon and 4 atoms of hydrogen. Note that the simplest formula of this compound is $CH_2$ because the carbon-to-hydrogen ratio is 1:2. A substance whose empirical formula is $CH_2$ could have a molecular formula $CH_2$, $C_2H_4$, $C_3H_6$, and so on. Molecular formulas for ionic substances do not exist, of course, because such compounds do not contain molecules.

A third type of formula is a **structural formula,** for example,

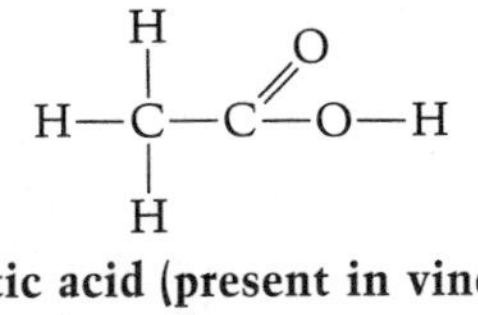

**acetic acid (present in vinegar)**

In a structural formula the dashes between the different atomic symbols represent the "chemical bonds" that hold the atoms together in the molecule. We will see more of them in Chapter 4. A structural formula gives us information about the way in which the atoms in a molecule are linked together, and provides information that allows us to write the molecular and empirical formulas. Thus, for acetic acid shown above we can also write its molecular formula ($C_2H_4O_2$) and its empirical formula ($CH_2O$).

The most desirable kind of formula to have, of course, is the structural formula since it also contains all of the information provided by the other two types. However, in chemistry, as in the rest of life, we never get something for nothing. The more information a formula conveys, the more difficult it is to arrive at experimentally. We shall see how empirical and molecular formulas are derived; however, most of the procedures involved for the determination of structural formulas are beyond the scope of this book.

## 2.5 EMPIRICAL FORMULAS

Since the simplest formula gives the relative numbers of atoms present in a compound, it must also give the relative number of moles of each element. Here are some examples that show how we might obtain this information.

EXAMPLE 2.10 A sample of a brown-colored gas that is a major air pollutant is found to contain 2.34 g of N and 5.34 g of O. What is the simplest formula of the compound?

SOLUTION We proceed by calculating the number of moles of each element present. We know that

$$1 \text{ mole N} = 14.0 \text{ g N} \qquad \text{(Why?)}$$
$$1 \text{ mole O} = 16.0 \text{ g O} \qquad \text{(Why?)}$$

Therefore,

$$2.34 \, \cancel{\text{g N}} \times \left(\frac{1 \text{ mole N}}{14.0 \, \cancel{\text{g N}}}\right) = 0.167 \text{ mole N}$$

$$5.34 \, \cancel{\text{g O}} \times \left(\frac{1 \text{ mole O}}{16.0 \, \cancel{\text{g O}}}\right) = 0.334 \text{ mole O}$$

We might write our formula $N_{0.167}O_{0.334}$. It does indeed tell us the relative number of moles of N and O; however, since the formula should have meaning on a molecular level, we prefer the subscripts to be integers. If we divide each subscript by the smallest one, we obtain

$$N_{\frac{0.167}{0.167}}O_{\frac{0.334}{0.167}} = NO_2$$

EXAMPLE 2.11 What is the empirical formula of a compound composed of 43.7% P and 56.3% O by weight?

SOLUTION It is quite common to have a chemical analysis in the form of percent composition by weight. The simplest way to proceed is to imagine having a 100-g sample of the compound. From the analysis, such a sample would contain 43.7 g P and 56.3 g O (notice that the percents become grams of compound). We now convert these weights to moles and proceed as before.

$$43.7 \, \cancel{\text{g P}} \times \left(\frac{1 \text{ mole P}}{31.0 \, \cancel{\text{g P}}}\right) = 1.41 \text{ moles P}$$

$$56.3 \, \cancel{\text{g O}} \times \left(\frac{1 \text{ mole O}}{16.0 \, \cancel{\text{g O}}}\right) = 3.52 \text{ moles O}$$

Our formula is

$$P_{1.41}O_{3.52} = P_{\frac{1.41}{1.41}}O_{\frac{3.52}{1.41}} = PO_{2.50}$$

Whole numbers may be obtained by doubling each of these values. Thus the empirical formula is $P_2O_5$.

EXAMPLE 2.12 0.1000 g of ethyl alcohol (grain alcohol), known to contain only carbon, hydrogen, and oxygen, was completely reacted with oxygen to produce the products $CO_2$ and $H_2O$. These products were trapped separately and weighed. 0.1910 g of $CO_2$ and 0.1172 g of $H_2O$ were found. What is the empirical formula of the compound?

SOLUTION To many students this problem at first appears impossible; however, let us consider what we know and what we can calculate.

Following the same procedure used in Example 2.9, we can compute the weight of carbon and the weight of hydrogen in the $CO_2$ and $H_2O$, respectively. Since the only source of C and H was the original compound, the difference between the weight of the compound taken (0.1000 g) and the total weight of carbon and hydrogen must be the weight of oxygen in the original 0.1000 g. We must obtain the weight of oxygen in this way, rather than from the total weight of oxygen in the $H_2O$ and $CO_2$, since only a portion of the oxygen in the products came from the original compound. Once we know the weights of carbon, hydrogen, and oxygen, we can calculate how many moles of each are in the 0.1000 g and, hence, the empirical formula of the unknown compound. Thus, having planned our course of action, we now proceed with the computations.

The formula weights of $CO_2$ and $H_2O$ are 44.0 and 18.0, respectively. The fraction of the mass of $CO_2$ that is carbon is

$$\frac{12.0 \text{ g C}}{44.0 \text{ g CO}_2}$$

Likewise, the fraction of $H_2O$ that is hydrogen is equal to

$$\frac{2.01 \text{ g H}}{18.0 \text{ g H}_2\text{O}}$$

The weight of carbon in the original compound is equal to the weight of $CO_2$ multiplied by the fraction of the weight that is due to carbon.

$$0.1910 \cancel{\text{ g CO}_2} \times \left(\frac{12.0 \text{ g C}}{44.0 \cancel{\text{ g CO}_2}}\right) \sim 0.0521 \text{ g C}$$

Similarly, the weight of hydrogen in the original sample is

$$0.1172 \cancel{\text{ g H}_2\text{O}} \times \left(\frac{2.01 \text{ g H}}{18.0 \cancel{\text{ g H}_2\text{O}}}\right) \sim 0.0131 \text{ g H}$$

The total weight contributed by the carbon and hydrogen is

$$0.0521 \text{ g C} + 0.0131 \text{ g H} = 0.0652 \text{ g}$$

The weight of oxygen = 0.1000 g − 0.0652 g = 0.0348 g O. Next, we calculate the number of moles of C, H, and O.

$$0.0521 \cancel{\text{ g C}} \times \left(\frac{1 \text{ mole C}}{12.0 \cancel{\text{ g C}}}\right) = 4.34 \times 10^{-3} \text{ mole C}$$

Similar calculations for hydrogen and oxygen give $1.31 \times 10^{-2}$ mole H and $2.17 \times 10^{-3}$ mole O. The empirical formula, then, is

$$C_{0.00434}H_{0.0131}O_{0.00217} = C_{\frac{0.00434}{0.00217}}H_{\frac{0.0131}{0.00217}}O_{\frac{0.00217}{0.00217}}$$

or

$$C_2H_6O$$

Table 2.1
*Molecular weights as multiples of the empirical formula weight*

| Formula | Molecular Weight |
|---|---|
| $CH_2$ | $14.0 = 1 \times 14.0$ |
| $C_2H_4$ | $28.0 = 2 \times 14.0$ |
| $C_3H_6$ | $42.0 = 3 \times 14.0$ |
| $C_4H_8$ | $56.0 = 4 \times 14.0$ |
| $C_nH_{2n}$ | $n \times 14.0$ |

## 2.6 MOLECULAR FORMULAS

Not only does the molecular formula provide the information contained in the empirical formula, but it also tells us how many atoms of each element are present in a molecule of a substance. Remember that an empirical formula of $CH_2$ is found for any molecule that possesses twice as many hydrogen atoms as carbon atoms. To distinguish among all of the possible choices, we require the molecular weight of the compound. This is because the molecular weight is an integral multiple of the empirical formula weight (see Table 2.1). To find the number of times that the empirical formula is repeated in the molecular formula, we simply divide the experimentally determined molecular weight[2] by the empirical formula weight.

EXAMPLE 2.13 A colorless liquid used in rocket engines, whose empirical formula is $NO_2$, has a molecular weight of 92.0. What is its molecular formula?

SOLUTION The formula weight of $NO_2$ is 46.0.

The number of times the empirical formula, $NO_2$, occurs in the compound is

$$\frac{92.0}{46.0} = 2$$

The molecular formula is then $(NO_2)_2 = N_2O_4$ (dinitrogen tetroxide).

$N_2O_4$ is the preferred answer because $(NO_2)_2$ implies a knowledge of the structure of the molecule (i.e., that two $NO_2$ units are somehow joined together).

## 2.7 BALANCING CHEMICAL EQUATIONS

Recall that a chemical equation is a shorthand description of the changes that occur during a chemical reaction. One of the most useful properties of a chemical equation is that it allows us to determine the quantitative relationships that exist between reactants and products. To be helpful in this way, however, the equation must be **balanced;** that is, it must obey the law of conservation of mass by having the same number of atoms of each kind on both sides of the arrow.[3]

In order to minimize errors, writing a balanced chemical equation should always be considered a two-step process.

[2] We will see how molecular weights may be determined in Chapters 6 and 9.

[3] We shall see that chemical equations are sometimes written to show the reaction between ions. These equations must demonstrate the law of conservation of electrical charge by having the total charge on both sides of the equation the same.

1. We first write an unbalanced equation with correct formulas for all of the reactants and products.
2. We balance the equation by adjusting the coefficients that precede the formulas. *During this step we are not permitted to change the subscripts in any of the formulas!* To do so would change the nature of the substances.

There is never any excuse for having an improperly balanced equation, since it is always possible, by counting atoms on each side of the equation, to determine whether the equation is, in fact, balanced.

Most simple chemical equations can be easily balanced by inspection. This involves examining the equation and adjusting the coefficients until equal numbers of each element are present among both the reactants and products. For example, consider the reaction of sodium carbonate with hydrochloric acid (HCl) to produce sodium chloride, carbon dioxide, and water.

To obtain a properly balanced chemical equation we proceed as follows:

1. We write the unbalanced equation,

$$Na_2CO_3 + HCl \longrightarrow NaCl + H_2O + CO_2$$

2. Coefficients are introduced to balance the equation. In this step, since there are two Na atoms on the left, we begin by placing a 2 in front of the formula NaCl. This gives us

$$Na_2CO_3 + HCl \longrightarrow 2NaCl + H_2O + CO_2$$

Since there are two Cl atoms on the right but only one on the left, we place a 2 before the HCl to get

$$Na_2CO_3 + 2HCl \longrightarrow 2NaCl + H_2O + CO_2$$

A fast inspection reveals that this equation is now balanced.

It was stated above that a balanced equation obeys the law of conservation of mass. For the equation that we have just balanced, and in fact for any equation, there are an infinite number of sets of coefficients that fulfill this requirement. Thus, the equations

$$\tfrac{1}{2}Na_2CO_3 + HCl \longrightarrow NaCl + \tfrac{1}{2}CO_2 + \tfrac{1}{2}H_2O$$
$$5Na_2CO_3 + 10HCl \longrightarrow 10NaCl + 5CO_2 + 5H_2O$$

are also balanced. The usual practice, however, is to use the smallest possible set of whole-number coefficients, although there are occasions, as we shall see, where other choices are advantageous.

EXAMPLE 2.14 Balance the following equation for the combustion of octane, $C_8H_{18}$, a component of gasoline:

$$C_8H_{18} + O_2 \longrightarrow CO_2 + H_2O$$

SOLUTION Inspection of the equation quickly suggests that we must adjust the coefficients preceding $CO_2$ and $H_2O$ to balance C and H. Carbon can be balanced by placing an 8 before the $CO_2$; hydrogen can be balanced by placing a 9 before the $H_2O$ ($9H_2O$ contains 18 H atoms because each $H_2O$ contains 2 H atoms). This gives

$$C_8H_{18} + O_2 \longrightarrow 8CO_2 + 9H_2O$$

Now we can work on the oxygen. On the right there are 25 O atoms $(2 \times 8 + 9 = 25)$. On the left the O atoms come in pairs. This means that we must have $12\frac{1}{2}$ pairs ($O_2$ molecules) to have 25 O atoms on the left. This gives us

$$C_8H_{18} + 12\tfrac{1}{2}O_2 \longrightarrow 8CO_2 + 9H_2O$$

Finally, we can eliminate the fractional coefficient by doubling each of the coefficients.

$$2C_8H_{18} + 25O_2 \longrightarrow 16CO_2 + 18H_2O$$

## 2.8 CALCULATIONS BASED ON CHEMICAL EQUATIONS

A chemical equation can be interpreted in several ways. For example, on a molecular level the balanced equation for the combustion of ethylene,

$$C_2H_4 + 3O_2 \longrightarrow 2CO_2 + 2H_2O$$

can be read as

$$1 \text{ molecule } C_2H_4 + 3 \text{ molecules } O_2 \longrightarrow 2 \text{ molecules } CO_2 + 2 \text{ molecules } H_2O$$

Alternatively, in terms of laboratory-sized quantities, we can express the reaction as

$$1 \text{ mole } C_2H_4 + 3 \text{ moles } O_2 \longrightarrow 2 \text{ moles } CO_2 + 2 \text{ moles } H_2O$$

In this case we have merely enlarged everything by a factor equal to Avogadro's number. The key point here is that *the coefficients in a chemical equation provide the* **ratio** *in which moles of one substance react with moles of another.* In this equation, the number of moles of $O_2$ consumed is always equal to three times the number of moles of $C_2H_4$ that react. For example, if 5 moles of $C_2H_4$ were available, 15 moles of $O_2$ would be needed for complete reaction. We could also use the coefficients and say that if 5 moles of $C_2H_4$ were reacted, 10 moles of $CO_2$ and 10 moles of $H_2O$ would be formed. The following are some sample problems involving calculations based on chemical equations.

EXAMPLE 2.15 Ethyl alcohol, $C_2H_5OH$, is produced in large quantities by fermentation of sugars such as glucose.

$$C_6H_{12}O_6 \xrightarrow{\text{yeast}} \underset{\textbf{ethyl alcohol}}{2C_2H_5OH} + 2CO_2$$

How many moles of $C_2H_5OH$ are produced from 1.40 moles of glucose?

SOLUTION The question is:

$$1.40 \text{ moles } C_6H_{12}O_6 \sim (?) \text{ moles } C_2H_5OH$$

From the equation we can say that 2 moles of $C_2H_5OH$ are formed for every 1 mole of $C_6H_{12}O_6$ that is consumed. This can be stated as an equivalence,

$$2 \text{ moles } C_2H_5OH \sim 1 \text{ mole } C_6H_{12}O_6$$

We can now use this relationship to construct a conversion factor that can be applied to the solution of the problem.

$$1.40 \cancel{\text{moles } C_6H_{12}O_6} \times \left(\frac{2 \text{ moles } C_2H_5OH}{1 \cancel{\text{mole } C_6H_{12}O_6}}\right) \sim 2.80 \text{ moles } C_2H_5OH$$

**EXAMPLE 2.16** Freshly exposed aluminum surfaces react with oxygen to form a tough oxide coating that protects the metal from further corrosion. The reaction is

$$4Al + 3O_2 \longrightarrow 2Al_2O_3$$

How many grams of $O_2$ are required to react with 0.300 mole of Al?

**SOLUTION** Our problem:

$$0.300 \text{ mole Al} \sim (?) \text{ g } O_2$$

According to the balanced equation,

$$4 \text{ moles Al} \sim 3 \text{ moles } O_2$$

Therefore, the number of moles of $O_2$ required is

$$0.300 \cancel{\text{mole Al}} \times \left(\frac{3 \text{ moles } O_2}{4 \cancel{\text{moles Al}}}\right) \sim 0.225 \text{ mole } O_2$$

To convert moles of $O_2$ to grams of $O_2$ we use the relationship

$$1 \text{ mole } O_2 = 32.0 \text{ g } O_2$$

$$0.225 \cancel{\text{mole } O_2} \times \left(\frac{32.0 \text{ g } O_2}{1 \cancel{\text{mole } O_2}}\right) = 7.20 \text{ g } O_2$$

We could have combined these two steps.

$$0.300 \cancel{\text{mole Al}} \times \left(\frac{3 \cancel{\text{moles } O_2}}{4 \cancel{\text{moles Al}}}\right) \times \left(\frac{32.0 \text{ g } O_2}{1 \cancel{\text{mole } O_2}}\right) \sim 7.20 \text{ g } O_2$$

**EXAMPLE 2.17** From the chemical equation in the last example, calculate the number of grams of $Al_2O_3$ that could be produced if 12.5 g of $O_2$ completely react with aluminum.

**SOLUTION** Our problem:

$$12.5 \text{ g } O_2 \sim (?) \text{ g } Al_2O_3$$

The solution to this problem requires that we establish the chemical equivalence between $O_2$ and $Al_2O_3$. This kind of equivalence is *always* found from the coefficients in the balanced equation and is expressed in our chemical units, moles. From the equation.

$$3 \text{ moles } O_2 \sim 2 \text{ moles } Al_2O_3$$

We now need the relationships permitting us to translate between laboratory units, grams, and chemical units, moles. These are

$$1 \text{ mole } O_2 = 32.0 \text{ g } O_2$$
$$1 \text{ mole } Al_2O_3 = 102 \text{ g } Al_2O_3$$

These three relationships can now be applied as conversion factors. We arrange them so that the proper units cancel.

$$12.5 \cancel{\text{g } O_2} \times \left(\frac{1 \cancel{\text{mole } O_2}}{32.0 \cancel{\text{g } O_2}}\right) \times \left(\frac{2 \cancel{\text{moles } Al_2O_3}}{3 \cancel{\text{moles } O_2}}\right) \times \left(\frac{102 \text{ g } Al_2O_3}{1 \cancel{\text{mole } Al_2O_3}}\right) \sim 26.6 \text{ g } Al_2O_3$$

## 2.9 LIMITING-REACTANT CALCULATIONS

If arbitrary quantities of reactants are chosen when a chemical reaction is carried out, it is quite likely that one of the reactants will be completely consumed before the others are used up. For example, if 4 moles of $H_2$ and 1 mole of $O_2$ are mixed and allowed to react following the equation

$$2H_2 + O_2 \longrightarrow 2H_2O$$

only 2 moles of $H_2$ will react, completely consuming the 1 mole of $O_2$. After the $O_2$ is gone, no further reaction can occur and no further product can be formed. The amount of product is therefore limited by the reactant that disappears first. This reactant is called the **limiting reactant.** The following two examples illustrate how we can determine which is the limiting reactant and then calculate how much product is formed.

EXAMPLE 2.18 Zinc and sulfur react according to the equation

$$Zn + S \longrightarrow ZnS$$

How many grams of ZnS can be formed when 12.0 g of Zn are reacted with 6.50 g of S? Which is the limiting reactant? How much of which element will remain unreacted?

SOLUTION The equation gives us the mole relationship,

$$1 \text{ mole Zn} \sim 1 \text{ mole S}$$

To determine the limiting reactant we first convert our reactant quantities to moles.

$$12.0 \not{\text{g Zn}} \times \left(\frac{1 \text{ mole Zn}}{65.4 \not{\text{g Zn}}}\right) = 0.183 \text{ mole Zn}$$

$$6.50 \not{\text{g S}} \times \left(\frac{1 \text{ mole S}}{32.1 \not{\text{g S}}}\right) = 0.202 \text{ mole S}$$

Because these elements combine in a 1-to-1 mole ratio, 0.183 mole Zn requires 0.183 mole S. We see that there is more S than is required and that all of the Zn is able to react. *Zn is therefore the limiting reactant.*

The amount of product formed depends only on the amount of limiting reactant. After the Zn has been used up, no more ZnS can form. From the equation,

$$1 \text{ mole Zn} \sim 1 \text{ mole ZnS}$$

Therefore, 0.183 mole Zn will form 0.183 mole ZnS. The weight of product is

$$0.183 \not{\text{mole ZnS}} \times \left(\frac{97.5 \text{ g ZnS}}{1 \not{\text{mole ZnS}}}\right) = 17.8 \text{ g ZnS}$$

To calculate the weight of unreacted sulfur we first subtract the number of moles of sulfur that reacted from the number of moles of sulfur initially available.

$$\text{moles S unreacted} = (0.202 - 0.183) = 0.019 \text{ mole S}$$

Now we convert to grams.

$$0.019 \not{\text{mole S}} \times \left(\frac{32.1 \text{ g S}}{1 \not{\text{mole S}}}\right) = 0.61 \text{ g S}$$

EXAMPLE 2.19 Ethylene, $C_2H_4$, burns in air to form $CO_2$ and $H_2O$ according to the equation

$$C_2H_4 + 3O_2 \longrightarrow 2CO_2 + 2H_2O$$

How many grams of $CO_2$ will be formed when a mixture containing 1.93 g $C_2H_4$ and 5.92 g $O_2$ is ignited?

SOLUTION The relationship between $C_2H_4$ and $O_2$, as specified by the chemical equation, is in the units, moles.

$$1 \text{ mole } C_2H_4 \sim 3 \text{ moles } O_2$$

To solve our problem we first convert our given quantities to moles.

$$1.93 \text{ g } C_2H_4 \times \left(\frac{1 \text{ mole } C_2H_4}{28.0 \text{ g } C_2H_4}\right) = 0.0689 \text{ mole } C_2H_4 \text{ available}$$

$$5.92 \text{ g } O_2 \times \left(\frac{1 \text{ mole } O_2}{32.0 \text{ g } O_2}\right) = 0.185 \text{ mole } O_2 \text{ available}$$

Let's see now whether there is sufficient $O_2$ to react with all of the $C_2H_4$ (alternatively, we could see if there is enough $C_2H_4$ to react with all the $O_2$).

$$0.0689 \text{ mole } C_2H_4 \times \left(\frac{3 \text{ moles } O_2}{1 \text{ mole } C_2H_4}\right)$$

$$\sim 0.207 \text{ mole } O_2 \text{ needed to consume all the } C_2H_4$$

The calculation tells us that 0.207 mole of $O_2$ is required, but only 0.185 mole of $O_2$ is available. Therefore, $O_2$ is the limiting reactant.

We now use the limiting reactant to calculate the amount of product formed.

$$0.185 \text{ mole } O_2 \times \left(\frac{2 \text{ moles } CO_2}{3 \text{ moles } O_2}\right) \times \left(\frac{44.0 \text{ g } CO_2}{1 \text{ mole } CO_2}\right) \sim 5.43 \text{ g } CO_2$$

There is an interesting postscript to the last problem. Although not stated explicitly, our assumption was that even without sufficient oxygen to consume all of the $C_2H_4$, any $C_2H_4$ that did react was converted completely to $CO_2$ and $H_2O$. If this were the case, one aspect of automotive air pollution would be removed. What actually occurs when the hydrocarbon (in this case $C_2H_4$) is present in excess is that some of it is converted to CO. In the internal-combustion engine, gasoline (which is composed of a mixture of hydrocarbons) is burned in a limited supply of oxygen and the incomplete combustion therefore produces a mixture of CO, $CO_2$, and $H_2O$.

## 2.10 THEORETICAL YIELD AND PERCENTAGE YIELD

Sometimes a given set of reactants are able to produce more than one set of products, depending on reaction conditions. In the last paragraph, for instance, it was pointed out that the combustion of hydrocarbons in a limited supply of oxygen produces a mixture of products. Usually the formation of side products (products other than those being sought) is undesirable, and two quantities that chemists are concerned with under these circumstances are the theoretical yield and the actual percentage yield.

The **theoretical yield** *of a given product is the maximum yield that could be obtained if the reactants gave only that product.* In Example 2.19 we calculated the theoretical yield of $CO_2$, assuming that all of the $C_2H_4$ that burned was converted entirely to $CO_2$ and $H_2O$.

The percentage yield is a measure of the efficiency of the reaction. It is defined as

$$\text{percentage yield} = \frac{\text{actual yield}}{\text{theoretical yield}} \times 100$$

For example, suppose that in the reaction in Example 2.19 we had obtained only 3.48 g $CO_2$, with the remainder of the carbon as either CO or elemental carbon. The percentage yield of $CO_2$ would then be

$$\text{percent yield } CO_2 = \frac{3.48 \text{ g } CO_2}{5.43 \text{ g } CO_2} \times 100$$

$$\text{percent yield } CO_2 = 64.1\%$$

## INDEX TO QUESTIONS AND PROBLEMS (problem numbers in italics)

The mole 1, *11, 14, 15, 16, 17, 18, 19, 20, 25*
Mole relationships *21, 22, 23, 24, 26, 27*
Avogadro's number *28, 29, 30, 31*
Formula weights 2, *12, 13*
Percent composition *32, 33, 34, 35*
Empirical formulas 3, 4, *37, 38, 39, 40, 42, 44*
Molecular formulas *36, 41, 43*
Balancing equations 6, 7, 8
Calculations from equations 5, *45, 46, 47, 48, 49, 50, 51, 52, 53, 59*
Ton mole problems *62, 63*
Limiting reactant 9, *54, 56, 57, 58*
Theoretical yield, percent yield 10, *55, 60, 61*

## REVIEW QUESTIONS

**2.1** In what sense are the mole, the dozen, and the gross related to each other?

**2.2** Why is the term formula weight preferred for some substances, rather than molecular weight?

**2.3** Stated in words, what is the difference between structural formulas, molecular formulas, and empirical formulas?

**2.4** What is the empirical formula of each of the following?
$(NH_4)_2S_2O_8$ $Fe_2O_3$ $Al_2Cl_6$ $C_6H_6$
$C_3H_8O_3$ $C_6H_{12}O_6$ $Hg_2SO_4$

**2.5** What important chemical law is obeyed by a balanced chemical equation?

**2.6** Balance the following equations by inspection.
(a) $ZnS + HCl \rightarrow ZnCl_2 + H_2S$
(b) $HCl + Cr \rightarrow CrCl_3 + H_2$
(c) $Al + Fe_3O_4 \rightarrow Al_2O_3 + Fe$
(d) $H_2 + Br_2 \rightarrow HBr$
(e) $Na_2S_2O_3 + I_2 \rightarrow NaI + Na_2S_4O_6$

**2.7** Balance the following equations.
(a) $FeCl_3 + Na_2CO_3 \rightarrow Fe_2(CO_3)_3 + NaCl$
(b) $NH_4Cl + Ba(OH)_2 \rightarrow BaCl_2 + NH_3 + H_2O$
(c) $Ca(OH)_2 + H_3PO_4 \rightarrow Ca_3(PO_4)_2 + H_2O$
(d) $Fe_2(CO_3)_3 + H_2SO_4 \rightarrow Fe_2(SO_4)_3 + H_2O + CO_2$
(e) $Na_2O + (NH_4)_2SO_4 \rightarrow Na_2SO_4 + H_2O + NH_3$

**2.8** Balance the following equations.
(a) $C_4H_{10} + O_2 \rightarrow CO_2 + H_2O$
(b) $C_7H_6O_2 + O_2 \rightarrow CO_2 + H_2O$
(c) $P_4O_{10} + H_2O \rightarrow H_3PO_4$
(d) $FeS_2 + O_2 \rightarrow Fe_2O_3 + SO_2$
(e) $NH_3 + O_2 \rightarrow NO + H_2O$

**2.9** What is meant by the term, limiting reactant? In words, describe how the limiting reactant is identified.

**2.10** What is the theoretical yield in a given reaction? What is the meaning of percent yield?

## REVIEW PROBLEMS

**2.11** Give the weight of 1 mole of each of the following:
(a) Mg (b) C (c) Fe (d) Cl (e) Sr

**2.12** Give the formula weight of the following:
(a) $MgO$ (d) $S_2Cl_2$
(b) $CaCl_2$ (e) $Na_3PO_4$
(c) $PCl_5$

**2.13** Give the formula weight of the following:
(a) $SiO_2$ (quartz)
(b) $Mg(OH)_2$ (milk of magnesia)
(c) $MgSO_4{\cdot}7H_2O$ (epsom salts)
(d) $CaMg_3Si_4O_{12}$ (asbestos)
(e) $C_6H_8O_6$ (vitamin C)
(f) $C_{12}H_{22}O_{11}$ (sucrose—cane sugar)

**2.14** What is the weight of 1.35 moles of caffeine, $C_8H_{10}N_4O_2$?

**2.15** What is the weight of 2.33 moles of penicillin, $C_{16}H_{18}O_4N_2S$?

**2.16** What is the weight of 6.30 moles of lead sulfate, $PbSO_4$?

**2.17** What is the weight of 0.144 moles of $TiO_2$, a pigment used in white paint?

**2.18** How many moles are in 242 g of sodium bicarbonate, $NaHCO_3$ (baking soda)?

**2.19** How many moles are in $1.40 \times 10^3$ g of butane, $C_4H_{10}$?

**2.20** How many moles are in 85.3 g of sulfuric acid, $H_2SO_4$?

**2.21** How many moles of S are in 1.00 mole of $As_2S_3$?

**2.22** How many moles of O are in 1.50 moles of $Cr_2O_3$?

**2.23** How many moles of $CO_2$ could be liberated from 1.00 mole of limestone, $CaCO_3$?

**2.24** How many moles of $BaSO_4$ could be made from 1.25 moles of $Al_2(SO_4)_3$?

**2.25** How many moles of potassium are in 120 g of KCl?

**2.26** How many moles of S are in 632 g of iron pyrite, $FeS_2$?

**2.27** When coal containing iron pyrite, $FeS_2$, is burned, sulfur dioxide, $SO_2$, is produced. How many moles of $FeS_2$ would have to react to produce 1 kg of $SO_2$?

**2.28** What would be the weight of 1 molecule of sucrose ($C_{12}H_{22}O_{11}$)?

**2.29** How many atoms of carbon are in $4.00 \times 10^{-8}$ g of propane, $C_3H_8$?

**2.30** Carbon atoms have a diameter of approximately $1.5 \times 10^{-8}$ cm. If carbon atoms were laid in a row 3 cm long, what would be the total mass of carbon?

**2.31** Calculate the mass of Cu required to react with $5.00 \times 10^{20}$ molecules of $S_8$ to form $Cu_2S$.

**2.32** Calculate the percent composition of each of the following:
(a) $FeCl_3$ (d) $(NH_4)_2HPO_4$
(b) $Na_3PO_4$ (e) $Hg_2Cl_2$
(c) $KHSO_4$

**2.33** Calculate the percentage composition of each of the following: (a) (benzene) $C_6H_6$; (b) (ethyl alcohol) $C_2H_5OH$; (c) (potassium dichromate) $K_2Cr_2O_7$; (d) (xenon tetrafluoride) $XeF_4$; (e) (calcium carbonate) $CaCO_3$.

**2.34** Calculate the weight of nitrogen in 30 g of the amino acid, glycine, $CH_2NH_2COOH$.

**2.35** Calculate the weight of hydrogen in 12.0 g of $NH_3$.

***2.36** Polystyrene has a structure similar to that below.

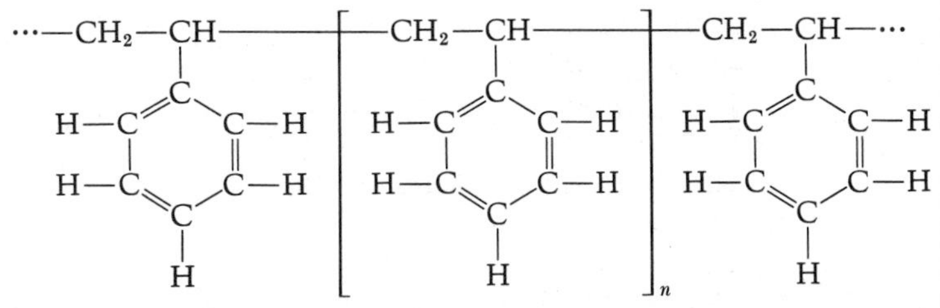


Here the basic styrene unit (within brackets) is repeated a large number, $n$, times to give long chainlike molecules. A particular sample of this plastic was found to have an average molecular weight of 1 million. What is the average number of styrene units in a chain?

**2.37** A 1.31 g sample of sulfur was reacted with an excess of chlorine to produce 4.22 g of a product that contains only sulfur and chlorine. What is the empirical formula of the compound?

**2.38** A substance was found to be composed of 60.8% sodium, 28.5% boron, and 10.5% hydrogen. What is the empirical formula of the compound?

**2.39** Vanillin is composed of carbon, 63.2%; hydrogen, 5.26%; and oxygen, 31.6%. What is the empirical formula of vanillin?

**2.40** A 0.537 g sample of an organic compound containing only carbon, hydrogen, and oxygen was burned in air to produce 1.030 g of $CO_2$ and 0.632 g $H_2O$. What is the empirical formula of the compound?

**2.41** The following are empirical formulas and molecular weights for five compounds. What are their molecular formulas?

(a) $NaS_2O_3$, MW = 270.4
(b) $C_3H_2Cl$, MW = 147.0
(c) $C_2HCl$, MW = 181.4
(d) $Na_2SiO_3$, MW = 732.6
? (e) $NaPO_3$, MW = 305.9

***2.42** A 1.35 g sample of a substance containing carbon, hydrogen, nitrogen, and oxygen was burned to produce 0.810 g $H_2O$ and 1.32 g $CO_2$. In a separate reaction, 0.735 g of the substance yielded 0.284 g of $NH_3$. Determine the empirical formula of the substance.

***2.43** A 0.5000 g sample of citric acid, containing only C, H, and O, was burned. It produced 0.6871 g of $CO_2$ and 0.1874 g of $H_2O$. The molecular weight of the compound is 192. What is the molecular formula for citric acid?

***2.44** An organic compound was synthesized and a sample of it was analyzed and found to contain C, H, N, O, and Cl. It was observed that when a 0.150 g sample of the compound was burned, it produced 0.138 g of $CO_2$ and 0.0566 g of $H_2O$. All of the nitrogen in a different 0.200 g sample of the compound was converted to $NH_3$, which was found to weigh 0.0238 g. Finally, the chlorine in a 0.125 g sample of the compound was converted to $Cl^-$ and precipitated as AgCl. The AgCl, when dried, was found to weigh 0.251 g.

(a) Calculate the weight percent of each element in the compound.
(b) Determine the empirical formula for the compound.

**2.45** The insecticide DDT (which ecologists now recognize as a serious environmental pollutant) is manufactured in a reaction between chlorobenzene and chloral,

$$\underset{\textbf{chlorobenzene}}{2C_6H_5Cl} + \underset{\textbf{chloral}}{C_2HCl_3O} \longrightarrow \underset{\textbf{DDT}}{C_{14}H_9Cl_5} + H_2O$$

How many kilograms of DDT can be produced from 1000 kg of chlorobenzene?

**2.46** Aspirin (which many students take after working on chemistry problems) is prepared by the reaction of salicylic acid ($C_7H_6O_3$) with acetic anhydride ($C_4H_6O_3$) according to the reaction

$$C_7H_6O_3 + C_4H_6O_3 \longrightarrow \underset{\textbf{(aspirin)}}{C_9H_8O_4} + C_2H_4O_2$$

How many grams of salicylic acid must be used to prepare two 5 grain aspirin tablets (1 g = 15.4 grains)?

**2.47** Acetylene, which is used as a fuel in welding torches, is produced in a reaction between calcium carbide and water,

$$\underset{\textbf{calcium carbide}}{CaC_2} + 2H_2O \longrightarrow Ca(OH)_2 + \underset{\textbf{acetylene}}{C_2H_2(g)}$$

(a) How many moles of $C_2H_2$ would be produced from 2.50 moles of $CaC_2$?
(b) How many grams of $C_2H_2$ would be formed from 0.500 mole of $CaC_2$?
(c) How many moles of water would be consumed when 3.20 moles of $C_2H_2$ are formed?
(d) How many grams of $Ca(OH)_2$ are produced when 28.0 g of $C_2H_2$ are formed?

**2.48** Consider the following balanced equation:

$$6ClO_2 + 3H_2O \longrightarrow 5HClO_3 + HCl$$

(a) How many moles of $HClO_3$ are produced from 14.3 g $ClO_2$?
(b) How many grams of $H_2O$ are needed to produce 5.74 g HCl?
(c) How many grams of $HClO_3$ are produced when 4.25 g of $ClO_2$ are added to 0.853 g of $H_2O$?

**2.49** Dimethylhydrazine, $(CH_3)_2NNH_2$, has been used as a fuel in the Apollo lunar descent module, with liquid $N_2O_4$ as the oxidizer. The products of the reaction between these two in the rocket engine are $H_2O$, $CO_2$, and $N_2$.

(a) Write a balanced chemical equation for the reaction.

(b) Calculate the mass of $N_2O_4$ required to burn 50 kg of dimethylhydrazine.

**2.50** The fermentation of sugar to produce ethyl alcohol follows the equation

$$C_6H_{12}O_6 \xrightarrow{\text{yeasts}} 2C_2H_5OH + 2CO_2$$

What is the maximum weight of alcohol that can be obtained from 500 g of sugar?

**2.51** Hydrazine, $N_2H_4$, and hydrogen peroxide, $H_2O_2$, have been used as rocket propellants. They react according to the equation

$$7H_2O_2 + N_2H_4 \longrightarrow 2HNO_3 + 8H_2O$$

(a) How many moles of $HNO_3$ are formed from 0.0250 moles of $N_2H_4$?
(b) How many moles of $H_2O_2$ are required if 1.35 moles of $H_2O$ are to be produced?
(c) How many moles of $H_2O$ are formed if 1.87 moles of $HNO_3$ are produced?
(d) How many moles of $H_2O_2$ are required to react with 22.0 g of $N_2H_4$?
(e) How many grams of $H_2O_2$ are needed to produce 45.8 g of $HNO_3$?

**2.52** Phosgene, $COCl_2$, was once used as a war gas. It is poisonous because when it is inhaled, it reacts with water in the lungs to produce hydrochloric acid, HCl, which causes severe lung damage, leading ultimately to death. The chemical reaction is

$$COCl_2 + H_2O \longrightarrow CO_2 + 2HCl$$

(a) How many moles of HCl are produced by the reaction of 0.430 mole of $COCl_2$?
(b) How many grams of HCl are produced when 11.0 g of $CO_2$ are formed?
(c) How many moles of HCl will be formed if 0.200 mole of $COCl_2$ are mixed with 0.400 mole of $H_2O$?

**2.53** White phosphorus, composed of $P_4$ molecules, is used in military incendiary devices because it ignites spontaneously when exposed to air. The product of reaction with oxygen is $P_4O_{10}$.
(a) Write a balanced chemical equation for the reaction of $P_4$ with $O_2$.
(b) How many moles of $P_4O_{10}$ can be produced using 0.500 mole $O_2$?
(c) How many grams of $P_4$ are needed to produce 50.0 g of $P_4O_{10}$?
(d) How many grams of $P_4$ will react with 25.0 g $O_2$?

**2.54** Under appropriate conditions acetylene, $C_2H_2$, and HCl react to form vinyl chloride $C_2H_3Cl$. This substance is used to manufacture polyvinyl chloride (PVC) plastics and has recently been shown to be carcinogenic (cancer producing). The equation for the reaction is

$$C_2H_2 + HCl \longrightarrow C_2H_3Cl$$

In a given instance, 35.0 g of $C_2H_2$ are mixed with 51.0 g of HCl.
(a) Which is the limiting reactant?
(b) How many grams of $C_2H_3Cl$ are formed?
(c) How many grams of the reactant in excess remain after the reaction is completed?

**2.55** In a typical experiment a student reacted benzene, $C_6H_6$, with bromine, $Br_2$, in an attempt to prepare bromobenzene, $C_6H_5Br$. This reaction also produced, as a by-product, dibromobenzene, $C_6H_4Br_2$. On the basis of the equation,

$$C_6H_6 + Br_2 \longrightarrow C_6H_5Br + HBr$$

(a) What is the maximum amount of $C_6H_5Br$ that the student could have hoped to obtain from 15.0 g of benzene?
(b) In this experiment the student obtained 2.50 g of $C_6H_4Br_2$. How much $C_6H_6$ was not converted to $C_6H_5Br$?
(c) What was the student's actual yield of $C_6H_5Br$?
(d) Calculate the percentage yield for this reaction.

***2.56** Freon-12, a gas used as a refrigerant, is prepared by the reaction

$$3CCl_4 + 2SbF_3 \longrightarrow \underset{\textbf{Freon-12}}{3CCl_2F_2} + 2SbCl_3$$

If 150 g of $CCl_4$ are mixed with 100 g of $SbF_3$,
(a) How many grams of $CCl_2F_2$ can be formed?
(b) How many grams of which reactant will remain after reaction has ceased?

***2.57** Acetylene, $C_2H_2$, can be reacted with two molecules of $Br_2$ to form $C_2H_2Br_4$ by the series of reactions

$$C_2H_2 + Br_2 \longrightarrow C_2H_2Br_2$$
$$C_2H_2Br_2 + Br_2 \longrightarrow C_2H_2Br_4$$

If 5.00 g of $C_2H_2$ are mixed with 40.0 g of $Br_2$, what weights of $C_2H_2Br_2$ and $C_2H_2Br_4$ will be formed? Assume that all of the $C_2H_2$ has reacted.

*2.58 Silver tarnishes in the presence of hydrogen sulfide (rotten egg odor) because of the reaction

$$\underset{\text{hydrogen sulfide}}{4Ag + 2H_2S} + O_2 \longrightarrow \underset{\text{(silver sulfide, black)}}{2Ag_2S} + 2H_2O$$

How much $Ag_2S$ could be obtained from a mixture of 0.950 g Ag, 0.140 g $H_2S$, and 0.0800 g $O_2$?

*2.59 White lead, a pigment used in lead-based paints, is manufactured by the reactions

$$2Pb + 2HC_2H_3O_2 + O_2 \longrightarrow 2Pb(OH)C_2H_3O_2$$

$$6Pb(OH)C_2H_3O_2 + 2CO_2 \longrightarrow \underset{\text{white lead}}{Pb_3(OH)_2(CO_3)_2} + 2H_2O + 3Pb(C_2H_3O_2)_2$$

(a) Starting with 20.0 g of Pb, how many grams of white lead can be prepared?
(b) How many grams of $CO_2$ will be required if 14.0 g of $O_2$ are consumed in the first reaction?

Assume that all of the Pb in the first reaction is completely converted to the products of the second reaction.

*2.60 A chemist wishes to synthesize a certain compound that has a molecular weight of 100. The synthesis requires six consecutive steps, each giving a 50% yield. If he begins with 30 g of starting material having a molecular weight of 80, how many grams of final product will he obtain? How many grams of starting material will he require to produce 10.0 g of final product?

*2.61 In a reaction between methane, $CH_4$, and chlorine, $Cl_2$, four products can be formed: $CH_3Cl$, $CH_2Cl_2$, $CHCl_3$, and $CCl_4$. In a particular instance 20.8 g of $CH_4$ were reacted and gave 5.0 g $CH_3Cl$, 25.5 g $CH_2Cl_2$, and 59.0 g $CHCl_3$.
(a) How many grams of $CCl_4$ were formed?
(b) On the basis of available $CH_4$, what is the theoretical yield of $CCl_4$?
(c) What is the percent yield of $CCl_4$?
(d) How many grams of $Cl_2$ reacted with the $CH_4$?

2.62 Phosphate rock, $Ca_3(PO_4)_3$, is treated with sulfuric acid to produce phosphate fertilizer,

$$Ca_3(PO_4)_2 + 2H_2SO_4 + 4H_2O \longrightarrow \underset{\text{phosphate fertilizer}}{Ca(H_2PO_4)_2 + 2CaSO_4 \cdot 2H_2O}$$

How many tons of sulfuric acid are required to react with 25.0 tons of phosphate rock?

2.63 How many pounds of ammonia can be produced from 650 lb of $H_2$ by the reaction

$$N_2 + 3H_2 \rightarrow 2NH_3$$

# 3 ATOMIC STRUCTURE AND THE PERIODIC TABLE

The atomic theory as proposed by Dalton represented a major breakthrough in the development of chemistry. All of the computations you learned to perform in the last chapter, in fact, are based on his idea that atoms of each element have a characteristic atomic mass. Although Dalton's theory accounted for the mass relationships observed in chemical reactions, it was not able to explain why substances react the way they do. It could be determined, for example, that one oxygen atom was able to react with a maximum of only two atoms of hydrogen, but no one understood why. Furthermore, as additional evidence came to light, it became increasingly clear that the simple picture of an indivisible atom was just not sufficient to account for all of the facts, and it was by a rather fascinating piecing together of bits of information that our current picture of the atom developed. In this chapter we will see how our present knowledge of atomic structure was developed, and how an understanding of atomic structure can help to explain and correlate many chemical and physical properties of the elements.

## 3.1 THE ELECTRICAL NATURE OF MATTER

In 1834, Michael Faraday reported the results of experiments in which he showed that chemical change could be caused by the passage of electricity through water solutions of chemical compounds. These experiments demonstrated that matter was electrical in nature and led G. J. Stoney, 40 years later, to propose the existence of particles of electricity that he called **electrons.**

Toward the end of the nineteenth century, physicists began to investigate the flow of electric current in **gas discharge tubes.** These tubes were made of glass with a piece of metal (called an electrode) sealed into each end (Figure 3.1). It was found that when a high voltage was applied across the electrodes and air was partially removed from the tube, a flow of electricity was observed and the residual air within the tube began to glow (a neon sign is a common example of a gas discharge tube in which the gas is the element neon). If virtually all of the gas was removed, the glow was no longer observed but the electrical discharge continued. A screen, coated with a zinc sulfide phosphor (similar to that found in a TV picture tube), placed between the electrodes was found to glow on the side facing the negative electrode (the **cathode**). This demonstrated that the discharge originates at the cathode and flows toward the positive electrode (the **anode**). These "rays" were thus called **cathode rays.**

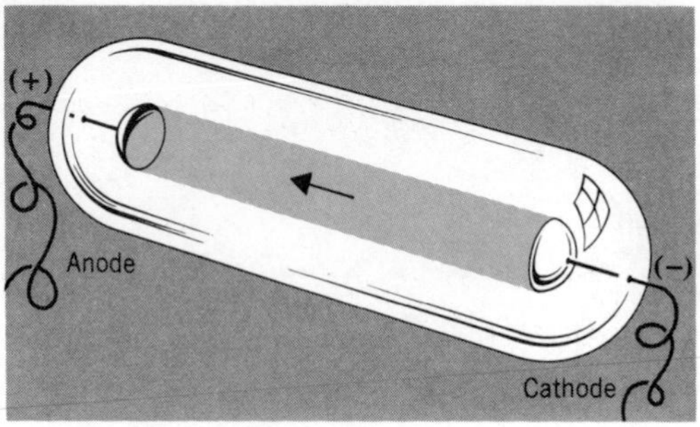


Figure 3.1
*The gas discharge tube.*

Further investigations showed that cathode rays:

1. Normally travel in straight lines
2. Cast shadows
3. Can turn a pinwheel placed in their path, suggesting that cathode rays are composed of particles
4. Heat a metal foil placed between the electrodes
5. Can be bent by an electric or magnetic field in such a direction as to indicate that the particles are electrically charged and that the charge is negative (Figure 3.2)
6. Are always the same regardless of the nature of the material composing the electrodes or the kind of residual gas within the tube, suggesting that cathode rays are a fundamental part of all matter.

These observations showed that cathode rays are composed of highly energetic, negatively charged fundamental particles of matter. They were, in fact, the electrons described by Stoney.

J. J. Thomson, in 1897, used a cathode-ray tube, quite similar to present-day TV picture tubes, to measure the ratio of the charge to the mass of an electron. This device is shown schematically in Figure 3.3. Electrons generated at the cathode are accelerated toward the anode, which has a hole in it. Some electrons pass through the hole and continue on, striking the face of the tube, which is coated with a phosphor, at *B*. If oppositely charged plates are placed above and below the tube, the beam is deflected toward the

Figure 3.2
(*a*) *Cathode rays are deflected by an electric field and* (*b*) *by a magnetic field.*

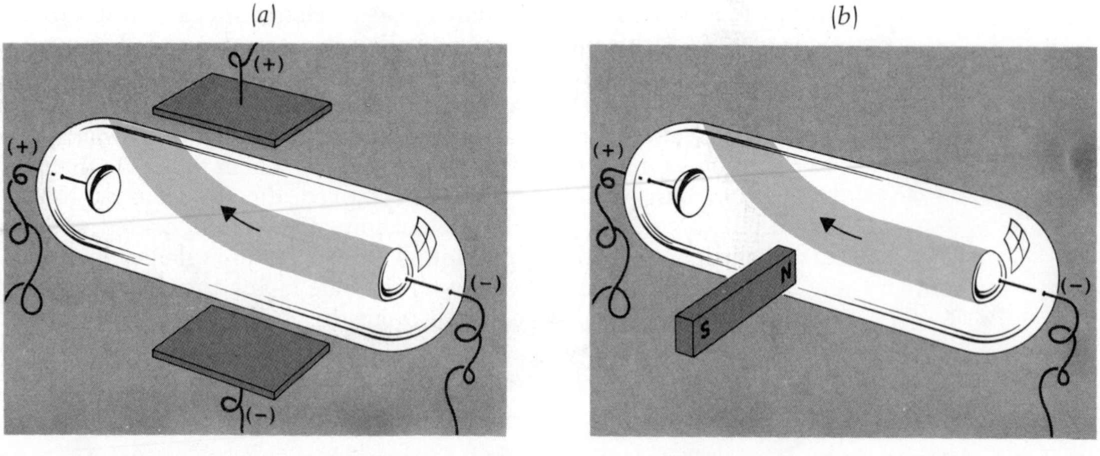

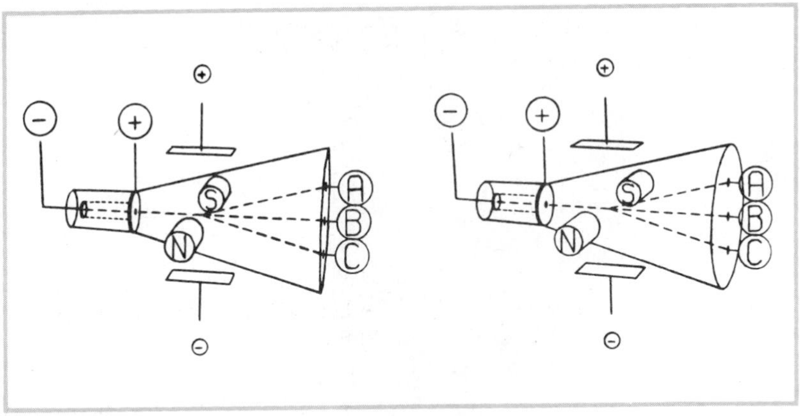


Figure 3.3
*A cathode-ray tube used to measure the charge-to-mass ratio of the electron.*

positive plate and strikes the face at *A*. The amount of deflection that the particle undergoes will be directly proportional to its charge, since a highly negatively charged particle will be attracted to the positive plate more than one with a small charge. The amount of deflection will also be inversely proportional to the mass of the particle because a heavy particle will be less affected by the electrostatic attraction, as it passes between the plates, than a particle of smaller mass. Thus the deflection that is observed depends on the ratio of charge to mass ($e/m$) of the particle.

If a magnetic field is generated at right angles to the electric field, as shown in Figure 3.3, the electrons are deflected in a direction exactly opposite to that caused by the charged plates. In the absences of the electric field, the electron beam is bent by the magnetic field so that it collides with the surface of the tube at *C*.

In practice, Thomson applied a magnetic field of known strength across the tube and noted the deflection of the electron beam. Charge was then applied to the plates until the beam was brought back to its original point of impact, *B*. From the magnitude of the electric and magnetic fields, Thomson calculated the charge-to-mass ratio, $e/m$, for the electron to be $-1.76 \times 10^8$ coulombs/gram. The **coulomb** is a unit of electric charge. It is equal to the amount of charge that moves past a given point in a wire when an electric current of 1 ampere (1 A) flows for 1 sec. The results of Thomson's experiment indicated that the electron had either a very large electric charge or that its mass was very small.

## 3.2 THE CHARGE ON THE ELECTRON

The charge on the electron was determined by means of a rather clever experiment performed in 1908 by R. A. Millikan at the University of Chicago. In his apparatus, illustrated in Figure 3.4, a fine mist of oil droplets was sprayed above a pair of parallel metal plates. As droplets settled through a hole in the upper plate, the air between the plates was briefly irradiated with X-rays. Electrons, knocked from gas atoms by the X-rays, were picked up by the oil droplets, thereby giving them a negative charge. Millikan found that by placing an electric charge on the plates (upper plate positive, lower plate negative), the downward motion of negatively charged drops could be stopped. A knowledge of the mass of a drop (measured by observing its rate of fall in the absence of the electric field) and the amount of charge on the

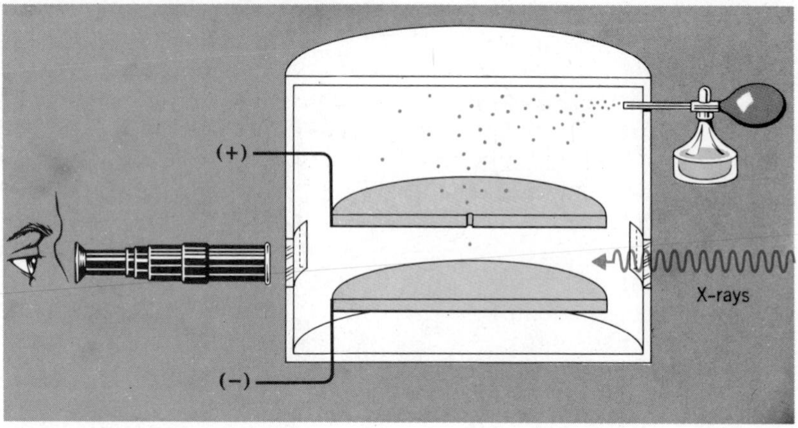


Figure 3.4
*Millikan's oil drop experiment.*

plates required to keep that drop suspended permitted Millikan to calculate the amount of charge on the drop.

After he performed this experiment many times, Millikan observed that the charge on the oil drops was always a multiple of $-1.60 \times 10^{-19}$ coulombs. He reasoned that since the oil drops could pick up one, two, three, or more electrons, the total charge on any drop must be a multiple of the charge on a single electron. This suggested that the charge on the electron is $-1.60 \times 10^{-19}$ coulombs. Once the charge on the electron had been measured, its mass, $9.11 \times 10^{-28}$ g, was obtained from the already known charge-to-mass ratio.

## 3.3 POSITIVE PARTICLES, THE MASS SPECTROMETER

Because negative particles (electrons) are present in a gas discharge, and because ordinary matter is observed to be electrically neutral, it seems reasonable to suspect the existence of positively charged particles in a gas discharge as well. When a perforated cathode was used in a discharge tube, streamers of light were observed coming from the holes at the rear of the negative electrode (Figure 3.5). These were called "canal rays."

During an electric discharge, electrons, emitted from the cathode, collide with neutral gas atoms, knocking electrons off them. The atoms, by losing electrons, become positively charged particles (positive **ions**[1]) and are

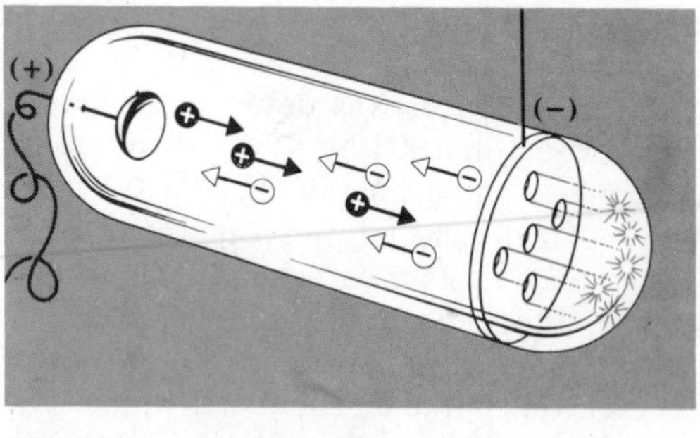


Figure 3.5
*Canal rays. Positive ions passing through holes in the cathode appear as canal rays at the rear of the electrode. Bursts of light are seen when they collide with a phosphor on the end of the tube.*

[1] Charged particles that result from the addition or removal of electrons from atoms or molecules are called **ions.** We shall come across many examples of ions during future discussions.

attracted toward the cathode. Most of them collide with the cathode; however, some travel through the perforations and emerge at the rear where they are observed as the "canal rays." If the rear wall of the discharge tube is coated with a phosphor, flashes of light can also be seen when these positive particles, which have passed through the cathode, hit the wall.

An instrument designed to determine the charge-to-mass ratio of positive ions is called a **mass spectrometer** (Figure 3.6). Gaseous material is introduced at *A* and ionized by an electric discharge across electrodes *B* and *C*. The positive ions thus produced are accelerated through the wire grid, *E*, and through the slits *F* and *G* that form a narrow beam which is fed between the poles of a powerful magnet. The magnetic field acts to deflect the particles into a circular path, with the degree of curvature determined by the charge-to-mass ratio of the ions. For ions with the same charge the extent to which their paths are bent depends on their masses, with lighter particles being deflected more than heavier ones. For ions with the same mass, the degree to which their paths are curved is directly proportional to their charge. By adjusting the strength of the magnetic field, ions with any desired $e/m$ ratio may be focused on the detector at *H*. Ions with higher $e/m$ ratios are deflected more (for example, to *P*) while those of lower $e/m$ ratios are deflected less (for example, to *Q*).

The measurement of $e/m$ for positive particles reveals the following information.

1. Positive ions always have $e/m$ ratios that are *much smaller* than that of the electron. This means either that they are much more massive than the electron (that is, $m$ is very large) or that they carry very small positive charges (that is, $e$ is small). Since they are formed from neutral atoms by the loss of electrons, the charge that they carry is some integral (that is, whole-number) multiple of the charge on the electron. This means, in effect, that in order to have a much smaller $e/m$ than the electron their masses must be much greater.

Figure 3.6
*The mass spectrometer.*

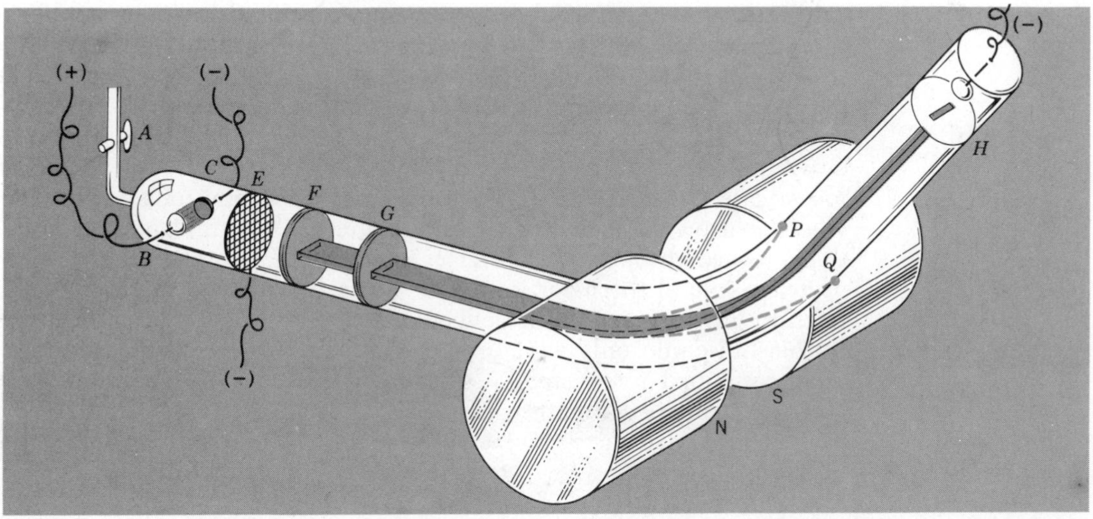

2. The $e/m$ ratio is dependent on the nature of the gas introduced into the mass spectrometer, which shows that not all positive ions have the same $e/m$.

When hydrogen, the lightest of all gases, is placed in the mass spectrometer, the $e/m$ ratio of the hydrogen ion is found to be $+9.63 \times 10^4$ coulombs/g. This is the largest $e/m$ observed for any positive ion, and it is thus assumed that the hydrogen ion represents a fundamental particle of positive charge, the **proton.** A neutral hydrogen atom, therefore, is composed of 1 electron and 1 proton. If we compare the charge-to-mass ratios of the proton and electron, we find the proton to be 1836 times as heavy as the electron. Thus nearly all of the atom's mass is associated, somehow, with its positive charge.

Because ions are formed by the addition or loss of electrons, each of which either adds or removes $1.60 \times 10^{-19}$ coulombs of charge, it is convenient to express the charges on particles in units of this size. For example, the electron would possess 1 unit of negative charge. On this scale, this constitutes a charge of $-1e$. Two units of positive charge would be represented as $+2e$. We commonly indicate the charge on an ion formed from an atom by writing the number of units of positive or negative charge as a superscript on the right side of the chemical symbol. Thus the ion $He^{2+}$ is an ion formed from the helium atom by the loss of two electrons. The ion, $O^{2-}$, is an ion formed from oxygen by the addition of two electrons.

## 3.4 RADIOACTIVITY

Further evidence that the atom is not indivisible is the spontaneous emission of particles and radiation by certain unstable elements. This phenomenon, called **radioactivity,** was discovered by Henri Becquerel in 1896. Radioactive substances emit three important types of radiation.

1. Alpha radiation, composed of $He^{2+}$ ions called alpha particles ($\alpha$-particles).
2. Beta radiation, consisting of electrons, in this instance, called beta particles ($\beta$-particles).
3. Gamma radiation ($\gamma$-rays), highly energetic, very penetrating light waves similar to X-rays.

## 3.5 THE NUCLEAR ATOM

One of the most significant developments in our understanding of the structure of the atom was provided by Ernest Rutherford in 1911. Prior to this time it was thought that the atom had a nearly uniform density throughout, with the electrons being buried in a glob of positive charge, much like raisins in a pudding. With this picture of a rather mushy atom in mind, one of Rutherford's students made an astonishing discovery when he set out to investigate the scattering of $\alpha$-particles by thin metal foils. He found that when a narrow beam of $\alpha$-particles was directed at a thin gold foil, not all of them passed through the metal essentially undisturbed as had been expected. Some of the $\alpha$-particles were scattered at large angles and a small fraction, in fact, were even deflected back toward the source (Figure 3.7). Rutherford could only explain such observations by concluding that the atom contained a very small, extremely dense positive **nucleus** containing all of the protons and nearly all of the atom's mass. Since the nucleus contains the positive charge in the atom, it follows that the electrons must be distributed somewhere in the remaining volume of the atom.

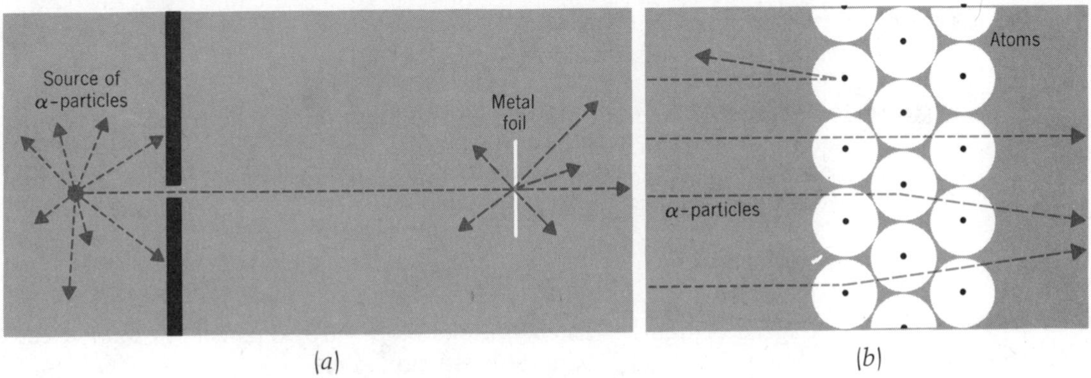


Figure 3.7
*(a) Scattering of α-particles by a metal foil. (b) Deflection of α-particles due to repulsions between positively charged α-particles and positively charged nuclei.*

## 3.6 ELECTROMAGNETIC RADIATION

X-rays, visible light, infrared and ultraviolet radiation, and also radio waves are all forms of **electromagnetic radiation** that travel as waves through space at a constant speed $c$ (called the speed of light, $3.00 \times 10^{10}$ cm/sec). These waves are characterized by their **wavelength,** $\lambda$ (Figure 3.8), which is the distance between consecutive peaks (or troughs) in the wave, and by their **frequency,** $\nu$, which is the number of peaks that pass by a given point per second. Wavelength and frequency are related to each other by the equation

$$\lambda \cdot \nu = c \tag{3.1}$$

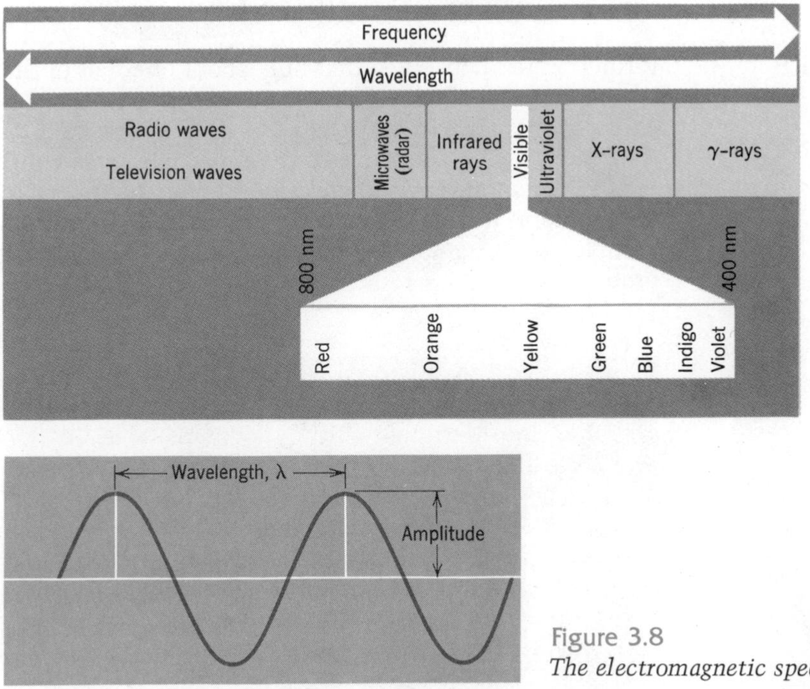


Figure 3.8
*The electromagnetic spectrum.*

EXAMPLE 3.1 X-rays emitted by copper have a wavelength of $1.54 \times 10^{-8}$ cm. What is the frequency of this radiation?

SOLUTION The SI unit of frequency is called the **hertz** (Hz), where 1 Hz = 1 cycle per second. This is written as 1 Hz = 1/sec, or 1 Hz = 1 $sec^{-1}$. Note that this treatment of units is the same as expressing $\frac{1}{10}$ as $10^{-1}$. To solve this problem we can first solve Equation 3.1 for frequency,

$$\nu = \frac{c}{\lambda}$$

The speed of light, $c$, has a value of $3.00 \times 10^{10}$ cm/sec, or $3.00 \times 10^{10}$ cm $sec^{-1}$; $\lambda = 1.54 \times 10^{-8}$ cm. Substituting gives

$$\nu = \frac{3.00 \times 10^{10}\ \text{cm sec}^{-1}}{1.54 \times 10^{-8}\ \text{cm}}$$

$$\nu = 1.95 \times 10^{18}\ \text{sec}^{-1}$$

or

$$\nu = 1.95 \times 10^{18}\ \text{Hz}$$

## 3.7 X-RAYS AND ATOMIC NUMBER

Roentgen, in 1895, discovered that when high energy electrons in a discharge tube collide with the anode, penetrating radiation, which he called X-rays, are produced (Figure 3.9).

Henry Moseley, in 1914, found that he could assign a number, called the **atomic number,** to an element based on the frequency of the X-rays that the element emits when it is used as the anode of an X-ray tube. Experiments by Rutherford and his students, in which they were able to measure the charge on the nucleus, enabled Moseley to conclude that this atomic number represented the number of protons in the nucleus.

## 3.8 THE NEUTRON

Rutherford had observed that only about one-half of the nuclear mass could be accounted for by protons. He therefore suggested that particles of zero charge and of mass nearly the same as that of the proton are also present in the nucleus. The existence of these particles was confirmed in 1932 by an English scientist, J. Chadwick, who bombarded beryllium with $\alpha$-particles and found that highly energetic, uncharged particles were emitted. These particles, called neutrons, have a mass only slightly larger than that of the proton.

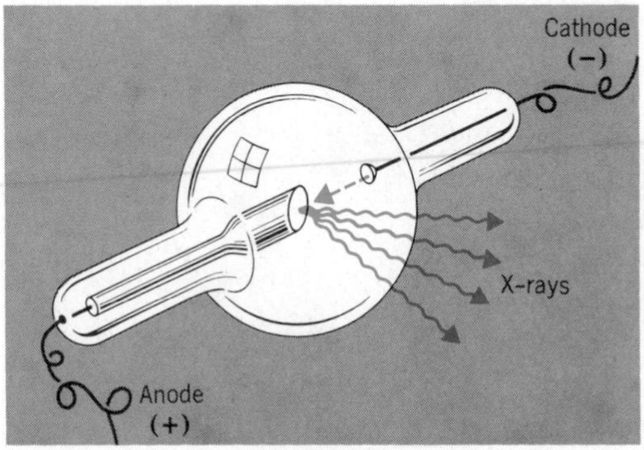


Figure 3.9
*Production of X-rays. High-voltage electric discharge across the electrodes causes the anode to emit X-rays.*

Table 3.1
*Some properties of subatomic particles*

| | Mass | | Charge | |
|---|---|---|---|---|
| | Grams | Atomic Mass Units (amu) | Coulombs | Electronic Charge Units |
| Proton | $1.67 \times 10^{-24}$ | 1.007274 | $+1.602 \times 10^{-19}$ | +1 |
| Neutron | $1.67 \times 10^{-24}$ | 1.008665 | 0 | 0 |
| Electron | $9.11 \times 10^{-28}$ | 0.000549 | $-1.602 \times 10^{-19}$ | −1 |

In summary, then, the atom is composed of a dense nucleus, containing protons and neutrons that provide nearly all of the atom's mass, surrounded in some fashion by electrons distributed throughout the remaining volume of the atom. The nucleus is extremely small, having a diameter of approximately $10^{-13}$ cm, compared to the atom itself whose diameter is of the order of $10^{-8}$ cm, or 1 Å (the **angstrom,** Å, is a unit of length that is convenient for expressing atomic dimensions; 1 Å = $1 \times 10^{-8}$ cm). The properties of the three major particles found in the atom are summarized in Table 3.1.

## 3.9 ISOTOPES

In Chapter 1 we noted that, contrary to Dalton's original hypothesis, not all atoms of the same element have identical masses. We referred to these different kinds of atoms as **isotopes.** The existence of isotopes is a common phenomenon, with most of the elements occurring naturally as mixtures of isotopes.

As we shall see, the properties of an element are almost entirely determined by the number and distribution of its electrons. *Therefore, it is the atomic number (or number of protons) that serves, indirectly, to distinguish an atom of one element from an atom of another, since the number of electrons must equal the number of protons in an electrically neutral atom.* Any mass differences that exist between atoms of the same element, then, must arise from different numbers of neutrons.

A particular isotope of an element is identified by specifying its atomic number, $Z$, and its **mass number,** $A$, which is the *sum* of the number of protons and neutrons in the atom. The number of neutrons present can be obtained from the difference, $A - Z$. We indicate the atom symbolically by writing the mass number as a superscript and the atomic number as a subscript. Both precede the atomic symbol,

$$^{A}_{Z}\mathrm{X}$$

For example, the carbon atom (atomic number = 6) containing six neutrons would have a symbol of $^{12}_{6}\mathrm{C}$. It is this isotope of carbon, incidentally, that serves as the basis of the current scale of atomic weights; that is, the mass of one atom of $^{12}_{6}\mathrm{C}$ is defined as exactly 12 atomic mass units (amu).

The fact that atoms of a given element can differ in the number of neutrons they contain has been put to practical use. One example is in archeological dating. Radioactive isotopes, such as $^{14}\mathrm{C}$, undergo nuclear transformations that cause them to be transformed into other elements. The rate of this decay is known, and from the relative abundance of $^{14}\mathrm{C}$ and $^{12}\mathrm{C}$ (a nonradioactive isotope of carbon) in both living and fossil material the age of the fossil can be estimated. The details of radioactive decay will be discussed in greater detail in Chapter 23.

Another application of radioactive isotopes is in chemotherapy. Treatment of thyroid cancer, for instance, is accomplished by administering carefully controlled doses of radioactive $^{131}I$, which tends to concentrate in the thyroid gland where the radiation produced by the $^{131}I$ causes destruction of cancerous cells. In this case use is made of the body's natural tendency to concentrate iodine (either radioactive or nonradioactive) in the thyroid gland.

As we noted above, nearly all elements as found in nature occur as mixtures of isotopes. For example, the element copper is found to contain the naturally occurring isotopes $^{63}_{29}Cu$ and $^{65}_{29}Cu$ whose masses have been accurately determined to be 62.9298 and 64.9278 amu, respectively. Their relative abundances are 69.09% and 30.91%. The observed atomic weight of copper, 63.55, is obtained as an average that is weighted according to the mass contributed by each isotope, as shown in Example 3.2.

**EXAMPLE 3.2** Using the data supplied in the paragraph above, calculate the average atomic mass of copper.

**SOLUTION** The amount of mass that each isotope contributes toward the average atomic mass is equal to the product of its mass multiplied by its fractional abundance.

| | | |
|---|---|---|
| $^{63}Cu$ | 62.9298 amu × 0.6909 = | 43.48 amu |
| $^{65}Cu$ | 64.9278 amu × 0.3091 = | 20.07 amu |
| | total | 63.55 amu |

In conclusion, observe the distinction between the mass number of an isotope and its actual mass. The mass number is simply the total count of protons plus neutrons and is not quite equal to the mass of the atom.

## 3.10 THE PERIODIC LAW AND THE PERIODIC TABLE

Scientists, even as early as 1800, had accumulated a significant amount of information concerning the physical and chemical properties of the known elements. This knowledge, however, existed for the most part as isolated and unrelated facts that needed to be correlated in some fashion before their total significance could be grasped. Early attempts at classification of the elements met with only limited success. It wasn't until 1869 that the forerunner of our modern periodic table was devised. This resulted from the efforts of two chemists, a Russian, Dmitri Mendeleev, and a German, Julius Lothar Meyer, each of whom worked independently of the other and produced similar tables about the same time. Mendeleev presented the results of his work to the Russian Chemical Society in the early part of 1869, while Meyer's table did not appear until December of that same year.

Mendeleev's table, in the revised form that appeared in 1871 (Figure 3.10), is very similar to the one we use today and is essentially the "short form" of the current table except for the absence of the noble gases, helium, neon, argon, krypton, zenon, and radon, which had not yet been discovered.

In constructing the table the elements were first arranged in order of increasing atomic weights. The elements were then placed in the table on the basis of similarities in properties among elements in a column, or **group.** For example, tellurium and iodine, whose atomic weights in 1869 were thought to be 128 and 127 amu, respectively, were placed into the table in

| | Group I | Group II | Group III | Group IV | Group V | Group VI | Group VII | Group VIII |
|---|---|---|---|---|---|---|---|---|
| 1 | H 1 | | | | | | | |
| 2 | Li 7 | Be 9.4 | B 11 | C 12 | N 14 | O 16 | F 19 | |
| 3 | Na 23 | Mg 24 | Al 27.3 | Si 28 | P 31 | S 32 | Cl 35.5 | |
| 4 | K 39 | Ca 40 | — 44 | Tc 48 | V 51 | Cr 52 | Mn 55 | Fe 56, Co 59<br>Ni 59, Cu 63 |
| 5 | (Cu 63) | Zn 65 | — 68 | — 72 | As 75 | Se 78 | Br 80 | |
| 6 | Rb 85 | Sr 87 | ?Yt 88 | Zr 90 | Nb 94 | Mo 96 | — 100 | Ru 104, Rh 104<br>Pd 105, Ag 100 |
| 7 | (Ag 108) | Cd 112 | In 113 | Sn 118 | Sb 122 | Te 125 | I 127 | |
| 8 | Cs 133 | Ba 137 | ?Di 138 | ?Ce 140 | — | — | — | — —<br>— — |
| 9 | — | — | — | — | — | — | — | |
| 10 | — | — | ?Er 178 | ?La 180 | Ta 182 | W 184 | — | Os 195, Ir 517<br>Pt 198, Au 199 |
| 11 | (Au 199) | Hg 200 | Tl 204 | Pb 207 | Bi 208 | — | | |
| 12 | — | — | — | Th 231 | — | U 240 | — | — —<br>— — |

Figure 3.10
*Mendeleev's Periodic Table (1871). The numbers appearing with the symbols are atomic weights.*

reverse order (according to atomic weight) because their properties dictated that tellurium be placed in Group VI and iodine be placed in Group VII. Because of this insistence upon placing elements of similar properties in the same group, Mendeleev was forced to leave empty spaces in the table which, presumably, would eventually be occupied by new elements after they had been discovered.

An advantage of Mendeleev's table was that it was possible to predict the properties of the missing elements because elements in any particular column had to have similar properties. For example, germanium, which lies below silicon and above tin in Group IV, had not been discovered when Mendeleev constructed his table. Therefore, a blank space appears at this spot in the chart. On the basis of its position in the table, Mendeleev predicted that the properties of germanium, which he called "eka-silicon," should lie intermediate between those of silicon and tin. Table 3.2 shows how closely he predicted the properties that were found for germanium when it was discovered in 1886.

The necessity of reversing the order of atomic weights, which occurred when placing tellurium and iodine in the periodic table, was repeated after the noble gases were discovered. It was found that the atomic weight of argon (39.9 amu) was greater than that of potassium (39.1 amu); however, on the basis of physical and chemical properties, potassium clearly belonged in Group I (following argon) while argon had to be included in a separate group with the other noble gases. These reversals present no problems at all and the proper arrangement is obtained when the elements are placed in order of

Table 3.2
*Predicted properties of eka-silicon and observed properties of germanium*

| Property | Silicon | Tin | Predicted for Eka-silicon (Es) | Currently Accepted for Germanium |
|---|---|---|---|---|
| Atomic weight (amu) | 28 | 118 | 72 | 72.59 |
| Density (g/ml) | 2.33 | 7.28 | 5.5 | 5.3 |
| Melting point (°C) | 1410 | 232 | High | 947 |
| Physical form at room temperature | Gray nonmetal | White metal | Gray metal | Gray metal |
| Reaction with acids and alkalies | Acid—no reaction Alkalies—slow reaction | Slow attack by acids; no reaction with alkalies | Very slow attack by acids; no reaction with alkalies | No reaction with acids; will react only with concentrated alkalies |
| Number of chemical bonds usually formed | 4 | 4 | 4 | 4 |
| Formula of chloride | $SiCl_4$ | $SnCl_4$ | $EsCl_4$ | $GeCl_4$ |
| Density of chloride (g/ml) | 1.50 | 2.23 | 1.9 | 1.88 |
| Boiling point of chloride (°C) | 57.6 | 114 | 100 | 84 |

increasing atomic number (see Section 3.7) rather than atomic weight. This leads us to the modern statement of the **periodic law:** *When the elements are arranged in order of increasing atomic number, there occurs a periodic repetition of physical and chemical properties.* Thus it is the charge on the nucleus, which we associate with the atomic number, and the number of electrons in the neutral atom that are important in determining the sequence in which the elements occur and that are responsible for their properties.

The periodic table in use today (sometimes called the "long" form of the periodic table) is shown in Figure 3.11. We see that, like Mendeleev's table, it is constructed of a number of vertical columns, called **groups,** each containing a *family of elements.* These groups are identified by a Roman numeral and a letter, either A or B. Groups IA through VIIA and Group O are referred to collectively as the **representative elements,** while Groups IB through VIIB and Group VIII (actually composed of the three short columns in the center of the table) constitute the **transition elements.** Similarities between properties of the A- and B-group elements exist, although the similarities are often very weak.

Finally, we see that there are two long rows of elements lying just below the main part of the table. These elements, called the **inner transition elements,** actually belong in the body of the table but are placed where they are simply to conserve space. The first of these rows, elements 58 through 71, fit into the chart following lanthanum and is collectively called the **lanthanides** or the **rare earths.** The second row, elements 90 through 103, belongs between actinium ($Z = 89$) and kurchatovium ($Z = 104$). The elements in this latter series are termed the **actinides.**

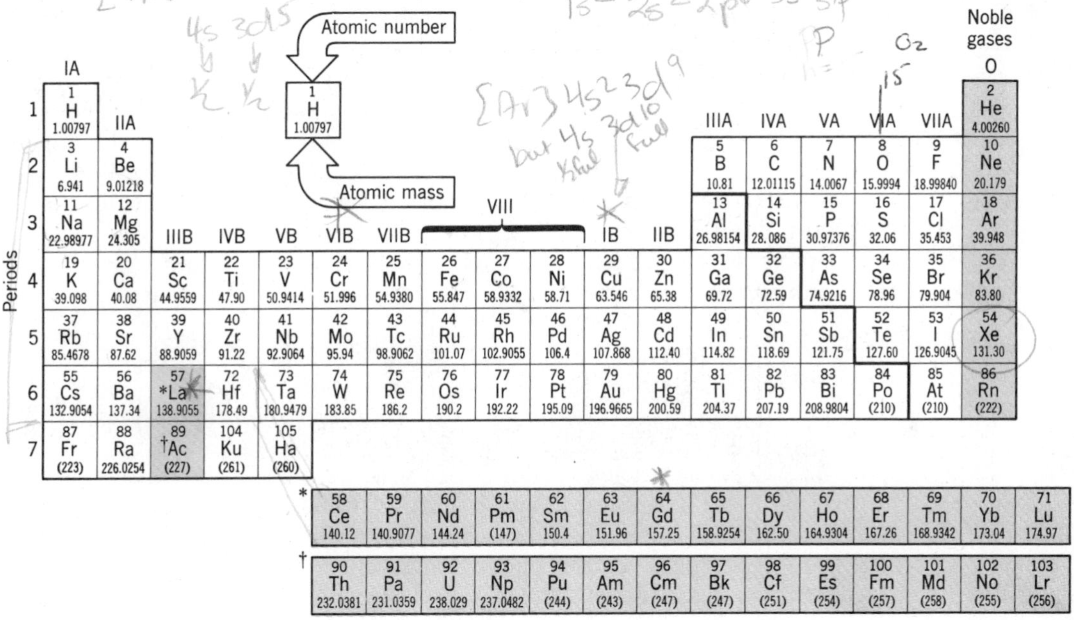


Figure 3.11
*The modern periodic table of the elements.*

The horizontal rows in the periodic table are called **periods** and are designated by means of Arabic numerals. The elements hydrogen and helium are members of the first period; lithium through neon are known as second-period elements; and so on.

Certain families of elements are characterized by names as well as by their group number. For example, the Group IA elements are frequently spoken of as the **alkali metals.** Similarly, the Group IIA elements are called the **alkaline earth metals;** the Group VIIA elements, the **halogens;** and the Group O elements, the **noble gases** (these are also sometimes called the *inert gases* because of their extremely limited ability to react chemically).

The elements can also be broadly classified as **metals, nonmetals,** or **metalloids.** You are probably familiar with most of the physical properties that serve to identify metals: high electrical conductivity, luster, generally high melting points, **ductility** (ability to be drawn into wires), and **malleability** (ability to be hammered into thin sheets). Nonmetals, on the other hand, are uniformly very poor conductors of electricity, do not possess that characteristic luster found with metals and, as solids, are brittle. Metalloids have properties that lie intermediate between those of metals and nonmetals. For example, they are useful as semiconductors of electricity.

In the periodic table, elements on the left are metals and those on the right are nonmetals. The heavy jagged line drawn from boron to astatine approximately represents the boundary between metallic and nonmetallic behavior, with elements lying immediately adjacent to the line generally having metalloid properties. Hydrogen, at the top of Group IA, is the only element that really fails to fit into this division, since it displays only nonmetallic behavior under normal conditions. It is interesting, however, that at extremely high pressures it has been found that hydrogen shows metallic properties similar to the other members of its group.

The periodic table is probably the most useful aid that chemists have at their disposal. We will see how it can be used to correlate much of the theoretical and factual information of chemistry that you should take with you when you leave this course. From the standpoint of the development of theory concerning the structure of the atom, the periodic table represents a compilation of experimental data that must be explained. A successful theory must somehow account for the way the table is structured. For example, it must explain why there are only two elements in period 1, why there are eight each in periods 2 and 3, and so on. It must also explain why elements in any given group exhibit similar properties at all.

## 3.11 ATOMIC SPECTRA

When atoms combine during chemical reactions, it is the electrons surrounding the nucleus that interact, because only the outer parts of the atoms come in contact with each other. Therefore, the chemical properties of the elements are determined by the way in which the electrons are arranged. The nucleus serves mainly to determine the number of electrons that must be present to give a neutral atom. The key that has permitted the deduction of the electronic structure of the elements is atomic spectra.

If sunlight, or the light from an incandescent lamp, is formed into a narrow beam by a slit and is passed through a prism onto a screen, a rainbow of colors is observed (Figure 3.12). This spectrum is composed of visible light of all wavelengths and is called a **continuous spectrum.** However, if the source of light is a discharge tube containing a gas such as hydrogen, the observed spectrum consists of a number of *lines* projected on the screen. These lines are the image of the slit and the spectrum is called a **line spectrum** (Figure 3.13). Obviously the visible light emitted by hydrogen does not contain radiation of all wavelengths, as sunlight does, but rather only a relatively few wavelengths. Similar, yet distinctive line spectra are produced by all of the elements when they are caused to emit light. The wavelengths of the lines are characteristic of a particular element, and can be used to identify new elements. Atomic spectra can also be used to identify the compositions of mixtures. For example, as you have probably seen and heard on TV crime dramas, a sample of paint taken from the clothing of a hit-and-run victim can be analyzed for the elements it contains, as well as the relative amounts

Figure 3.12
*Production of a continuous spectrum.*

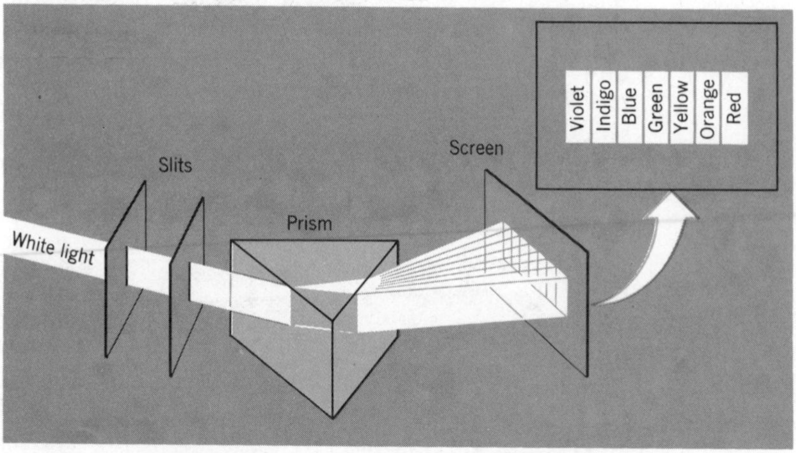

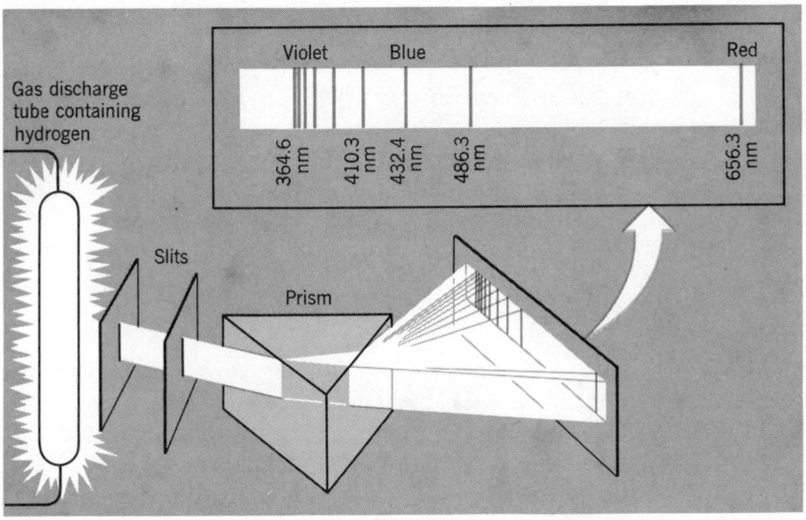


Figure 3.13
*The line spectrum of hydrogen.*

of each; this is done by using atomic spectra. The results can be compared with a similar analysis of a paint scraping taken from a suspect's car. If they match, there is fairly strong evidence that the suspect is indeed the hit-and-run driver.

The hydrogen spectrum in Figure 3.13 shows only the lines that appear in the visible region of the spectrum. Light is also emitted by hydrogen in the infrared and ultraviolet regions. The occurrence of line spectra baffled physicists for many years. In 1885 Balmer found that there was a relatively simple equation that could be used to calculate the wavelengths of all of the lines in the visible spectrum of hydrogen:

$$\frac{1}{\lambda} = 109{,}678\ \text{cm}^{-1}\left(\frac{1}{2^2} - \frac{1}{n^2}\right) \tag{3.2}$$

where $\lambda$ is the wavelength and $n$ is an integer that can have values of 3, 4, 5, 6, . . . , $\infty$. By choosing a particular value of $n$, the wavelength of a line in the spectrum can be calculated. Thus, when $n = 3$,

$$\frac{1}{\lambda} = 109{,}678\ \text{cm}^{-1}\left(\frac{1}{4} - \frac{1}{9}\right)$$

$$\frac{1}{\lambda} = 15233\ \text{cm}^{-1}$$

or

$$\lambda = 6.564 \times 10^{-5}\ \text{cm} = 656.4\ \text{nm}$$

Similarly, when $n = 4$, 5, and 6, we compute $\lambda$ to be 486.3, 432.4, and 410.3 nm, respectively. These values, as we can see from Figure 3.13, are equal to the wavelengths of the lines in the visible portion of the hydrogen spectrum. If higher values of $n$ are substituted into Equation 3.2, wavelengths of lines in the ultraviolet are obtained. All of the lines related by Equation 3.2 constitute what is called the **Balmer series.** Other series of lines are also

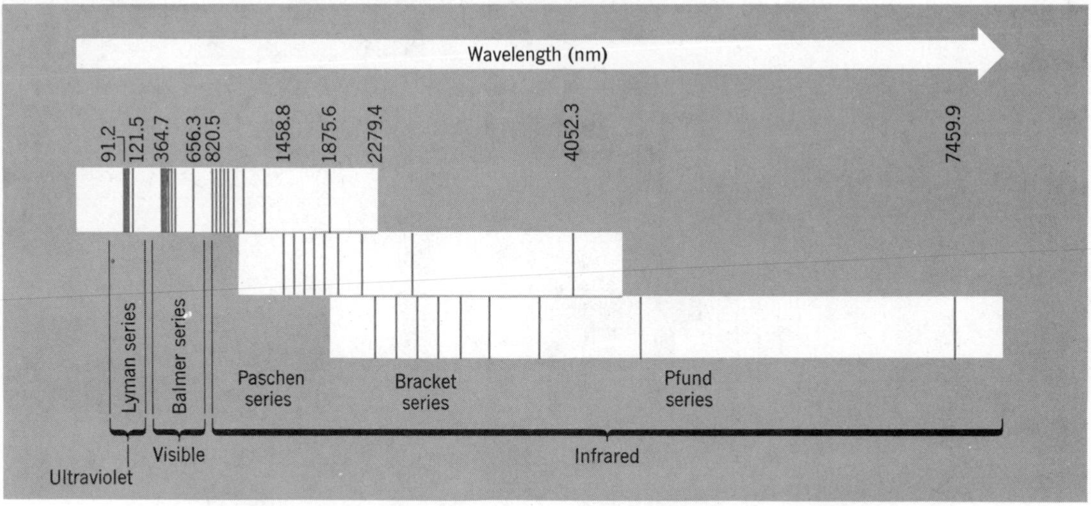


Figure 3.14
*Series of lines in the hydrogen spectrum.*

observed for hydrogen and can be fitted to the general equation (called the Rydberg equation)

$$\frac{1}{\lambda} = 109{,}678\ \text{cm}^{-1}\left(\frac{1}{n_1^2} - \frac{1}{n_2^2}\right) \tag{3.3}$$

where $n_1$ and $n_2$ are integers that may assume values of 1, 2, 3, . . . , ∞ with the requirement that $n_2$ is always greater than $n_1$. Thus, when $n_1 = 1$, the values of $n_2$ can be 2, 3, 4, . . . , ∞ and the lines in the **Lyman series** are obtained. When $n_1 = 2$ and $n_2 = 3, 4, 5, \ldots, \infty$, we get the Balmer series. These and other series are summarized in Figure 3.14 and Table 3.3.

Table 3.3
*Series of lines in the hydrogen spectrum*

| Series | $n_1$ | $n_2$ |
|---|---|---|
| Lyman | 1 | 2, 3, 4, . . . , ∞ |
| Balmer | 2 | 3, 4, 5, . . . , ∞ |
| Paschen | 3 | 4, 5, 6, . . . , ∞ |
| Brackett | 4 | 5, 6, 7, . . . , ∞ |
| Pfund | 5 | 6, 7, 8, . . . , ∞ |

EXAMPLE 3.3 Calculate the wavelength of the third line in the Brackett series for hydrogen.

SOLUTION The Rydberg equation gives the reciprocal of the wavelength

$$\frac{1}{\lambda} = 109{,}678\ \text{cm}^{-1}\left(\frac{1}{n_1^2} - \frac{1}{n_2^2}\right)$$

For the Brackett series (Table 3.3), $n_1 = 4$. The third line in the series would correspond to $n_2 = 7$. Substituting gives

$$\frac{1}{\lambda} = 109{,}678\ \text{cm}^{-1}\left(\frac{1}{4^2} - \frac{1}{7^2}\right)$$

$$\frac{1}{\lambda} = 109{,}678\ \text{cm}^{-1}(0.0420918)$$

$$\frac{1}{\lambda} = 4616.55\ \text{cm}^{-1}$$

Taking the reciprocal gives

$$\lambda = 2.16612 \times 10^{-4}\ \text{cm}$$

Expressed in nanometers (1 nm = $10^{-9}$ m)

$$\lambda = 2166.12\ \text{nm}$$

## 3.12 THE BOHR THEORY OF THE HYDROGEN ATOM

Early attempts to account for the existence of line spectra on the basis of the motion of electrons in the atom met with complete failure. If an electron is moving about a nucleus, it must be following a curved path; otherwise it would simply leave the atom. A particle following a curved path undergoes an acceleration[2] and, according to the accepted laws of physics at that time, a charged particle (such as the electron) undergoing an acceleration should continuously lose energy by radiating electromagnetic radiation. This means that the electron should gradually spiral in toward the nucleus, resulting in the collapse of the atom. Since atoms do not collapse, physicists were faced with a problem that challenged their most fundamental theories; they called it a "catastrophe."

The way out of this problem found its origin in the work of Max Planck (1900) and Albert Einstein (1905). They had demonstrated that, in addition to possessing wave properties, light also has particle properties. Thus, there are instances where light behaves as if it were composed of tiny packets, or **quanta,** of energy (later called **photons**). The energy, $E$, of the photon emitted or absorbed by a substance is proportional to the frequency of the light, $\nu$. These two quantities are related by the equation

$$E_{\text{photon}} = h\nu \tag{3.4}$$

[2] For example, consider what happens to an object, such as a coin, when it is placed on the edge of a rapidly spinning phonograph record. Anyone who has attempted this has found that the object is thrown outward, away from the center of the record (Figure 3.15). The coin obviously experiences a force (called centrifugal force) that causes it to fly off the edge. Since the coin possesses mass, and since there is the relationship that *force = mass × acceleration,* the coin must also be experiencing an acceleration as it moves in its *circular* path at the edge of the record.

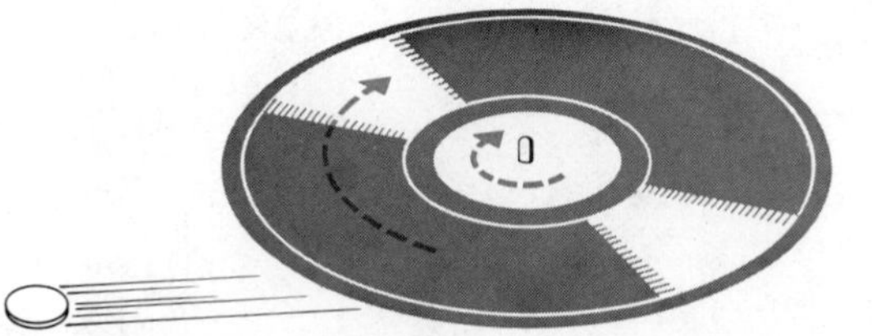

Figure 3.15
*A coin is thrown outward from the center of a spinning record. The coin experiences an acceleration directed away from the center of the record.*

where $h$, the proportionality constant, called **Planck's constant,** has a value of $6.63 \times 10^{-27}$ erg sec.

In 1913 Niels Bohr developed a theory incorporating the ideas of Planck and Einstein that successfully accounted for the spectrum of hydrogen. Unfortunately, the theory failed for atoms more complicated than hydrogen and it has since been replaced by a more successful one. Bohr's theory is instructive to look at briefly, though, because it illustrates how theories about the submicroscopic world of atoms develop and how they are tested.

Bohr's approach to the structure of the atom was simply to postulate that because atoms do not collapse and because light is emitted by an atom only at certain frequencies (which means that only certain specific energy changes occur), the electron in an atom can possess only certain, restricted quantities of energy. This is often phrased in a somewhat more esoteric way by saying that the energy of the electron is **quantized.** This means that the electron can have only certain discrete amounts of energy and none in between. We express this by saying that the electron is restricted to specific **energy levels** in the atom.

Bohr treated the electron in the hydrogen atom as if it traveled about the nucleus in circular orbits of fixed, or quantized energy; he derived an equation for the energy of the electron that had the form

$$E = -A\frac{1}{n^2} \tag{3.5}$$

in which the constant $A$ could be evaluated from a knowledge of the mass and charge of the electron and Planck's constant. The value of $A$ is $2.18 \times 10^{-11}$ erg. The quantity $n$ is an integer, called a **quantum number,** that can have values of 1, 2, 3, and so on, up to infinity. The energy of an electron in a particular orbit depends on the value of $n$ (Figure 3.16), the lowest energy level being obtained when $n = 1$, since this yields the largest value for the fraction $1/n^2$ and thus the most negative (and therefore lowest) $E$. The idea of a negative energy seems rather odd at first glance. Actually, the minus sign occurs because of a rather arbitrary choice of the zero point

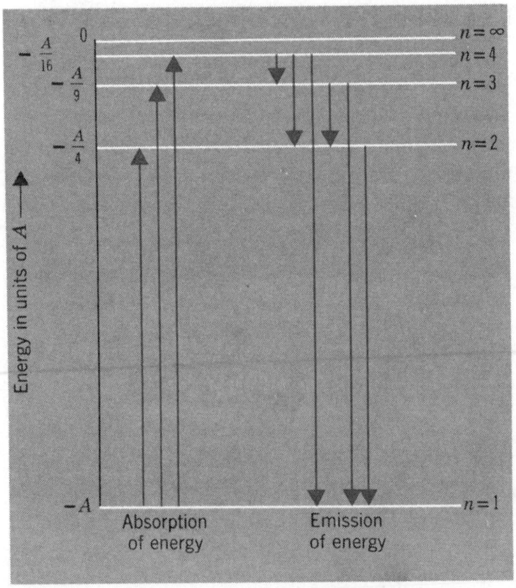


Figure 3.16
*Energies of Bohr orbits.*

on the energy scale. Since we can only measure differences in energy, the choice of where the zero point is placed is really unimportant.

With his theory Bohr had created a model of how the electron behaves in the atom. His theory, just like any other theory, must be able to be checked experimentally. Of course, there is no way actually to observe the electron. Therefore indirect evidence must be used to check the validity of the model. To do this Bohr mathematically derived an equation for the energy of the electron. According to Bohr, when energy is absorbed by an atom, for example in an electric discharge, the electron is raised in energy from one level to another, and when the electron returns to a lower energy level, a photon is emitted whose energy is equal to the difference between the two levels (see Figure 3.16). If we take $n_2$ to be the quantum number of the upper level and $n_1$ to be that of the lower (so that $n_2 > n_1$), the difference in energy, $\Delta E$, between the two is

$$\Delta E = E_{n_2} - E_{n_1} \tag{3.6}$$

$$\Delta E = \left(-A\frac{1}{{n_2}^2}\right) - \left(-A\frac{1}{{n_1}^2}\right)$$

which can be written as

$$\Delta E = A\left(\frac{1}{{n_1}^2} - \frac{1}{{n_2}^2}\right) \tag{3.7}$$

If this energy difference appears as a photon, it would have a frequency, $\nu$, that could be calculated from Equation 3.4,

$$\Delta E = h\nu$$

which, by incorporating Equation 3.1, can be expressed as

$$\Delta E = h\frac{c}{\lambda} = hc\frac{1}{\lambda}$$

Substituting this into Equation 3.7, we get

$$hc\frac{1}{\lambda} = A\left(\frac{1}{{n_1}^2} - \frac{1}{{n_2}^2}\right) \tag{3.8}$$

which, upon rearrangement, yields

$$\frac{1}{\lambda} = \frac{A}{hc}\left(\frac{1}{{n_1}^2} - \frac{1}{{n_2}^2}\right) \tag{3.9}$$

The quantity $A/hc$ has a value of 109,730 cm$^{-1}$, so our final equation is

$$\frac{1}{\lambda} = 109{,}730\ \text{cm}^{-1}\left(\frac{1}{{n_1}^2} - \frac{1}{{n_2}^2}\right) \tag{3.10}$$

Comparing Equations 3.3 and 3.10, we see that they are virtually identical. The Rydberg equation (3.3) is obtained from experimental observation while Equation 3.10 is derived from theory. This match between theory and experiment would suggest that Bohr was on the right track. Unfortunately, his approach was not at all successful with atoms more complex than hydrogen; however, his introduction of the notion of quantum numbers and quantized energy levels played a significant role in the development of our understanding of atomic structure.

EXAMPLE 3.4 Calculate the energy required to remove an electron from the lowest energy level of the hydrogen atom to produce the $H^+$ ion.

SOLUTION The lowest energy level has $n = 1$. The electron becomes free of the atom if it is raised to the level with $n = \infty$. The energy required to raise the electron from $n = 1$ to $n = \infty$ is given by Equation 3.6,

$$\Delta E = E_\infty - E_1$$

or, as shown in Equation 3.7,

$$E = A\left(\frac{1}{1^2} - \frac{1}{\infty^2}\right)$$

Substituting the value of $A$, $2.18 \times 10^{-11}$ erg, gives

$$\Delta E = 2.18 \times 10^{-11} \text{ erg}\left(\frac{1}{1^2} - \frac{1}{\infty^2}\right)$$

The value of $1/\infty^2$ is zero and $1^2 = 1$. Therefore,

$$\Delta E = 2.18 \times 10^{-11} \text{ erg}$$

EXAMPLE 3.5 Calculate the energy liberated when an electron drops from the fifth to the second energy level in hydrogen.

SOLUTION We have $n_2 = 5$, $n_1 = 2$.

$$\Delta E = 2.18 \times 10^{-11} \text{ erg}\left(\frac{1}{2^2} - \frac{1}{5^2}\right)$$

$$\Delta E = 2.18 \times 10^{-11} \text{ erg}\left(\frac{1}{4} - \frac{1}{25}\right)$$

$$\Delta E = 4.58 \times 10^{-12} \text{ erg}$$

The energy liberated is $4.58 \times 10^{-12}$ erg.

## 3.13 WAVE MECHANICS

The currently accepted theory explaining the behavior of subatomic particles is called **wave mechanics,** which has its roots in a hypothesis put forward by Louis de Broglie in 1924. De Broglie suggested that if light can behave in some instances as if it were composed of particles, perhaps particles, at times, exhibit properties that we normally associate with waves.

De Broglie's argument proceeded as follows. Einstein had shown that the energy equivalent, $E$, of a particle of mass, $m$, is equal to

$$E = mc^2 \qquad (3.11)$$

where $c$ is the speed of light. A photon whose energy is $E$ could thus be said to have an effective mass equal to $m$. Max Planck had shown that the energy of a photon is given by Equation 3.4,

$$E = h\nu = \frac{hc}{\lambda}$$

Equating these two gives

$$\frac{hc}{\lambda} = mc^2$$

When we solve for λ, the wavelength, we obtain

$$\lambda = \frac{h}{mc}$$

If this equation also applies to particles, such as the electron, the equation can be written as

$$\lambda = \frac{h}{ms} \tag{3.12}$$

where we have replaced $c$, the speed of light, by $s$, the speed of the particle.

Experimental evidence for this dual wave-particle nature of matter exists. Some particles, including electrons, protons, and neutrons, have been observed to exhibit diffraction,[3] a property that can only be explained by wave motion. The reason that the wave nature of matter was not discovered earlier is that objects large enough to see, either with the naked eye or with the aid of a microscope, possess so much mass that their wavelengths are much too short to be observed (see Example 3.6).

EXAMPLE 3.6 What is the wavelength of a grain of sand that weighs 0.000010 g and is moving at a speed of 1.0 cm/sec (approximately 0.02 miles/hr)?

[3] *Diffraction.* If light is allowed to pass through a pinhole whose diameter is about the same as the wavelength of the light, the hole behaves as if it were a tiny light source, scattering light in all directions. This phenomenon is called **diffraction.** If two such pinholes are placed along side of one another, each hole behaves as a separate source of light, sending out light waves in all directions. When a screen is placed such that this light falls on it, we observe a pattern, called a **diffraction pattern,** that consists of light and dark areas as shown in Figure 3.17. In the bright areas light waves that arrive from each hole are **in phase;** that is, the peaks and troughs of the two waves are lined up so that the amplitudes of the waves add together to produce a resultant wave of greater intensity. This is illustrated in Figure 3.18*a on page 70.* In the darkened areas, waves that arrive from the two pinholes are **out of phase** with each other, which means that the peaks of one wave coincide with the troughs of the other. When this happens, the amplitudes of the waves cancel each other so that zero intensity and hence darkness, is observed (Figure 3.18*b*). The factors that determine where the waves are in phase and where they are out of phase in a diffraction pattern will be considered in Chapter 7.

Figure 3.17
*Production of a diffraction pattern.*

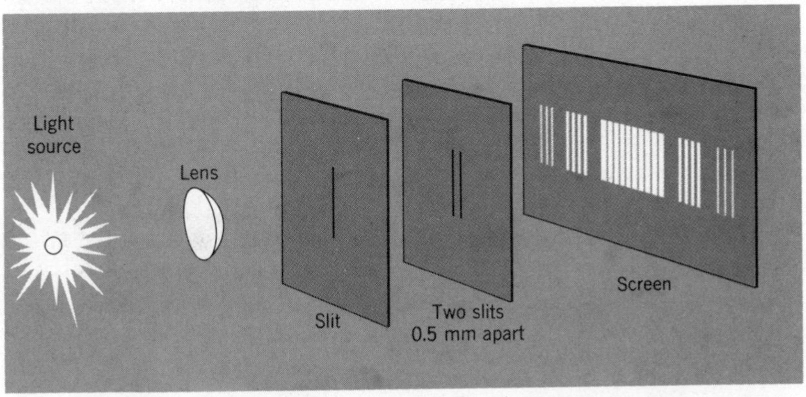

SOLUTION From Equation 3.12,

$$\lambda = \frac{h}{ms}$$

Planck's constant has a value of

$$h = 6.63 \times 10^{-27} \text{ erg sec}$$

Since 1 erg = 1 g cm²/sec², we can write $h$ as

$$h = 6.63 \times 10^{-27}(\text{g cm}^2/\text{sec}^2) \text{ sec} = 6.63 \times 10^{-27} \text{ g cm}^2/\text{sec}$$

We are given that

$$m = 1.0 \times 10^{-5} \text{ g}$$
$$s = 1.0 \text{ cm/sec}$$

Substituting these quantities into our equation, we have

$$\lambda = \frac{6.63 \times 10^{-27} \text{ g cm}^2/\text{sec}}{(1.0 \times 10^{-5}\text{g})(1.0 \text{ cm/sec})} = 6.63 \times 10^{-22} \text{ cm}$$

This wavelength is far too small to be detected by any device existing at this time. Larger objects have even larger masses and therefore still smaller wavelengths.

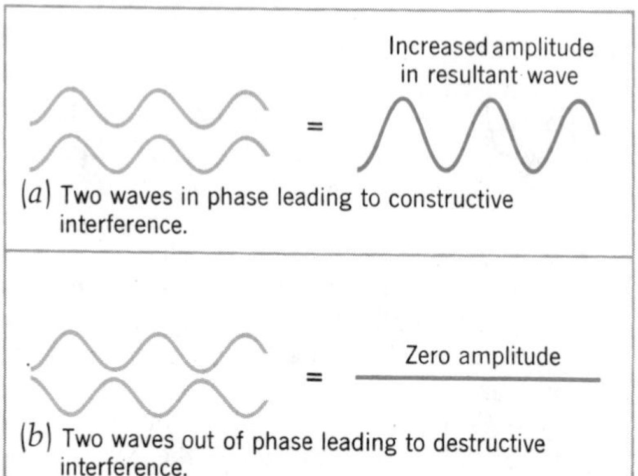


(*a*) Two waves in phase leading to constructive interference.

(*b*) Two waves out of phase leading to destructive interference.

Figure 3.18
*Constructive and destructive interference.*

We saw in Section 3.12 that one of the significant results of the Bohr theory of the atom was the introduction of integer quantum numbers. If we consider the electron to be moving as a wave about the nucleus, we find that the appearance of integers occurs in a very natural way. For example, let us imagine a wave moving about the nucleus along the circumference of a circle. Unless the wavelength of the wave is properly chosen, the wave will be out of phase with itself after it has completed one revolution (Figure 3.19*a*). If this were to occur, cancellation of the wave would result and the wave would simply disappear. Therefore, stable waves will result only if they stay in phase with themselves (Figure 3.19*b*). This requirement means that the wavelength must be repeated an integral number (whole number) of times along the circumference of the circle. Stated another way, the length of the circumference must be equal to an integral number of wavelengths; that is,

$$2\pi r = n\lambda$$

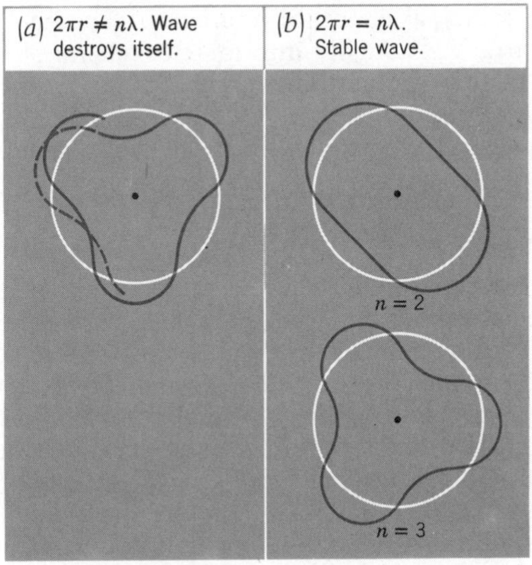


Figure 3.19
*Destructive and constructive interference produced by waves traveling along the circumference of the circle. (a) $2\pi r \neq n\lambda$. Wave destroys itself. (b) $2\pi r = n\lambda$. Stable wave.*

where $2\pi r$ is the length of the circumference, $\lambda$ is the wavelength, and $n$ is an integer having values of $n = 1, 2, 3, \ldots, \infty$. Waves that satisfy this condition for $n = 2$ and 3 are illustrated in Figure 3.19*b*.

Because the electron has wavelike properties, it is not surprising that its motion in the atom can be described by means of a wave equation. Physicists had already developed equations to treat wave motion in general and Erwin Schrödinger, in 1926, employed the ideas of de Broglie to arrive at a wave equation that could be applied to the hydrogen atom. The branch of physics that deals with the solution of wave equations is called either **wave mechanics** or **quantum mechanics.**

The solution of the wave equation is very complicated and is certainly far beyond the scope of this book. We can, however, look at some of the results that come from the theory. To begin with, the wave equation can really be solved only for a very few species. Fortunately, one of them is the hydrogen atom, and the results obtained for hydrogen, as it turns out, can be extended quite successfully to the other elements in the periodic table.

In the Bohr theory we saw that the electron was restricted to certain energy levels and was thought to move in certain prescribed circular orbits about the nucleus. We were able to identify these energy levels by means of an integer that we called a quantum number. In the course of solving the wave equation, a series of **wave functions** emerge (usually designated by the Greek letter psi, $\psi$), each of which is characterized by a set of three integer quantum numbers. *Each wave function corresponds to a certain energy and describes a region about the nucleus* (called an **orbital,** to distinguish it from Bohr's orbits) *where an electron having that energy may be found.*

According to wave mechanics, the energy levels in the atom are composed of one or more orbitals and in atoms that contain more than one electron, the distribution of the electrons about the nucleus is determined by the number and kind of energy levels that are occupied. Therefore, in order to investigate the way the electrons are arranged in space, we must first examine the energy levels in the atom. This is best accomplished through a discussion of the quantum numbers mentioned above.

1. The **principal quantum number,** *n*. The energy levels in the atom are arranged roughly into main levels, or **shells,** as determined by the principal quantum number, *n*. As in the Bohr theory, *n* may have values of 1, 2, 3, . . . , and so on up to infinity. Letters are also frequently associated with these shells as shown below.

| Principal quantum number | 1 | 2 | 3 | 4 | . . . |
|---|---|---|---|---|---|
| Letter designation | *K* | *L* | *M* | *N* | . . . |

For example, we would refer to the shell with $n = 1$ as the *K* shell.

2. The **azimuthal quantum number,** *l*. Wave mechanics predicts that each main shell is composed of one or more **subshells,** or sublevels, each of which is specified by a secondary quantum number, *l*, that is called the azimuthal quantum number. For any given shell this quantum number may have values of 0, 1, 2, and so on, up to a maximum of $n - 1$. Thus when $n = 1$, the largest (and only) value of *l* that is allowed is $l = 0$; therefore the *K* shell consists of only one subshell. When $n = 2$, two values of *l* occur, $l = 0$ and $l = 1$; hence the *L* shell is made up of two subshells. The values of *l* that occur for each value of *n* are summarized below.

| *n* | *l* |
|---|---|
| 1 | 0 |
| 2 | 0, 1 |
| 3 | 0, 1, 2 |
| 4 | 0, 1, 2, 3 |
| . | . |
| . | . |
| . | . |
| *n* | 0, 1, 2, . . . , $n - 1$ |

We see that *the number of subshells in any given shell is simply equal to its value of n.*

For the purposes of discussing the distribution of electrons in the atom, it is common practice to associate letters with the various values of *l*:

| *l* | 0 | 1 | 2 | 3 | 4 | 5 | 6 | . . . |
|---|---|---|---|---|---|---|---|---|
| Subshell designation | *s* | *p* | *d* | *f* | *g* | *h* | *i* | . . . |

The first four letters find their origin in the atomic spectra of the alkali metals (lithium through cesium). In these spectra four series of lines were observed and were termed the "sharp," "principal," "diffuse," and "fundamental" series, hence the letters *s*, *p*, *d*, and *f*. For $l = 4, 5, 6$, and so on, we just continue on with the alphabet; for our purpose, however, we will be interested only in *s*, *p*, *d*, and *f* subshells, since these are the only ones that are populated by electrons in atoms in their **ground state** (state of lowest energy).

To describe a subshell within a given shell, we write the value of *n* for the shell followed by the letter designation of the subshell. For example, the *s* subshell of the second shell ($n = 2, l = 0$) would be called the 2*s* subshell. In like manner, the *p* subshell of the second shell ($n = 2$, $l = 1$) would be the 2*p* subshell.

3. The **magnetic quantum number,** *m*. Each subshell is composed of one or more orbitals, where an orbital within a particular subshell is distinguished by its value of *m*, the magnetic quantum number. This quantum number, which derives its name from the fact that it can be used to explain the appearance of additional lines in atomic spectra produced when atoms emit light while in a magnetic field, has integer values that range between $-l$ and $+l$. When $l = 0$, only one value of *m* is permitted, $m = 0$; therefore an *s* subshell consists of only one orbital (we call it an *s* orbital). A *p* subshell ($l = 1$) contains three orbitals corresponding to *m* equal to $-1$, 0, and $+1$. In a similar fashion, we find that a *d* subshell ($l = 2$) is composed of five orbitals and an *f* subshell ($l = 3$), seven. This is summarized in Table 3.4.

   If we wished, we could identify a particular orbital in the atom by its set of three quantum numbers. Thus, an orbital for which $n = 3$, $l = 1$, and $m = 0$ corresponds to one of the three orbitals that belong to the *p* subshell of the third shell.

The energies of these shells, subshells, and orbitals in atoms having more than one electron are perhaps best illustrated by means of Figure 3.20 (page 74). There are several points about this diagram that are worth noting. First, we see that the energy of the shells increases with increasing value of the principal quantum number, *n*. Thus the *K* shell, with $n = 1$, lies lowest in energy; above that there is the *L* shell with $n = 2$ (composed of the 2*s* and 2*p* subshells); higher still we find the *M* shell ($n = 3$), and so on.

Also note that as *n* becomes larger, the spacing between successive shells becomes less, as illustrated on the right side of Figure 3.20. Because of this narrowing energy separation, we begin to observe overlap among the subshells of the third and higher shells. The 4*s* subshell, for example, lies lower in energy than does the 3*d* subshell. This overlap is even more pro-



Table 3.4
*Summary of quantum numbers*

| Principal Quantum Number, *n* (Shell) | Azimuthal Quantum Number, *l* (Subshell) | Subshell Designation | Magnetic quantum Number, *m* (Orbital) | Number of Orbitals in Subshell |
|---|---|---|---|---|
| 1 | 0 | 1*s* | 0 | 1 |
| 2 | 0 | 2*s* | 0 | 1 |
| | 1 | 2*p* | −1 0 +1 | 3 |
| 3 | 0 | 3*s* | 0 | 1 |
| | 1 | 3*p* | −1 0 +1 | 3 |
| | 2 | 3*d* | −2 −1 0 +1 +2 | 5 |
| 4 | 0 | 4*s* | 0 | 1 |
| | 1 | 4*p* | −1 0 +1 | 3 |
| | 2 | 4*d* | −2 −1 0 +1 +2 | 5 |
| | 3 | 4*f* | −3 −2 −1 0 +1 +2 +3 | 7 |

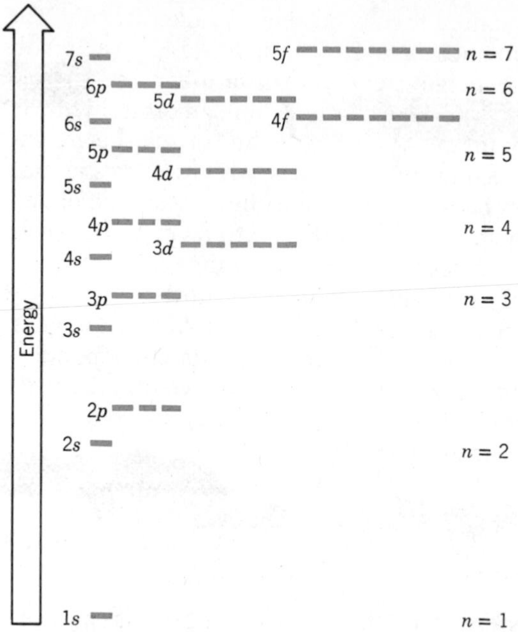


Figure 3.20
*Electronic energy-level diagram for atomic orbitals.*

nounced in higher shells, where the 5*s* subshell lies below the 4*d*, the 6*s* and 4*f* below the 5*d*, and the 7*s* and 5*f* below the 6*d*.

Also in Figure 3.20 we have indicated each orbital by means of a dash. Each *s* subshell is shown as a single dash to stress that it is composed of only one orbital. Likewise, *p* subshells are shown as three dashes, *d* subshells as five dashes, and *f* subshells as seven dashes. Observe that each orbital of a given subshell is shown to have the same energy. This applies for isolated atoms but not always for atoms in chemical compounds.

The sequence of energy levels described by Figure 3.20 turns out to be of critical importance in determining the arrangement of electrons in the atom. Before discussing this, however, we must look at yet another quantum number.

## 3.14 ELECTRON SPIN AND THE PAULI EXCLUSION PRINCIPLE

In addition to the three quantum numbers, $n$, $l$, and $m$, which come directly from the solution of the wave equation, there is yet another number, the **spin quantum number,** $s$. This quantum number arises because the electron behaves as if it were spinning (in much the same manner as the earth spins about its axis). The circular motion of electric charge that results causes the electron to act as a tiny electromagnet, just as passing an electric current through a wire wrapped about a nail causes the nail to become magnetic (see Figure 3.21). Since the electron can only spin in either of two directions, $s$ may have only two values. These turn out to be $+\frac{1}{2}$ and $-\frac{1}{2}$.

We find, then, that each electron in an atom can be assigned a set of values for its four quantum numbers, $n$, $l$, $m$, and $s$, which determine the orbital in which the electron will be found and the direction in which the electron will be spinning. There is a restriction, however, on the values that we may assign to these quantum numbers. This is expressed as the **Pauli exclusion principle,** which states that *no two electrons in any one atom may have all four quantum numbers the same.* This means that if we choose a particular set of values for $n$, $l$, and $m$ corresponding to a particular orbital

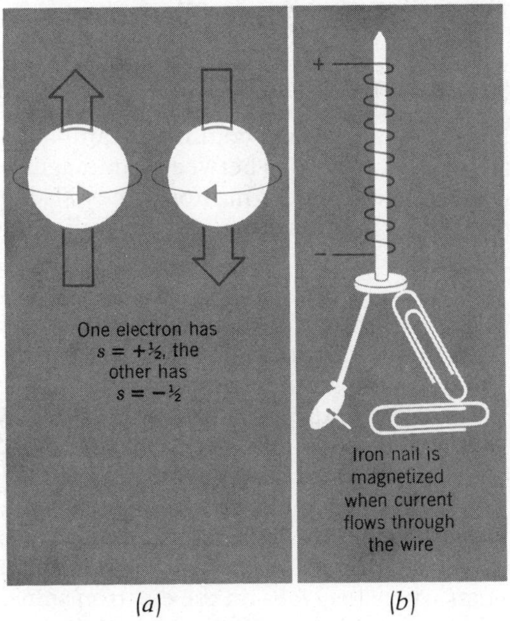

One electron has $s = +\frac{1}{2}$, the other has $s = -\frac{1}{2}$
Iron nail is magnetized when current flows through the wire
(a)
(b)


Figure 3.21
*The spin of the electron. (a) The electron behaves as if it is spinning about an axis through its center. Since there are two directions of spin, there are two values of the spin quantum number. (b) Spinning charge produces a magnetic field just as circulation of charge through a wire wrapped about a nail causes the nail to become magnetic.*

(for example, $n = 1, l = 0, m = 0$; the 1*s* orbital), we are able to have only two electrons with different values of the spin quantum number, *s* (that is, either $s = +\frac{1}{2}$ or $s = -\frac{1}{2}$). In effect, this limits the number of electrons in any given orbital to two, and it also requires that the spins of these two electrons be in opposite directions.

Because the Pauli exclusion principle leads to a restriction of a maximum of two electrons in any orbital, the maximum number of electrons that can be accommodated in *s*, *p*, *d*, and *f* subshells can be summarized as follows.

| Subshell | Number of Orbitals | Maximum Number of Electrons |
|---|---|---|
| *s* | 1 | 2 |
| *p* | 3 | 6 |
| *d* | 5 | 10 |
| *f* | 7 | 14 |

The maximum number of electrons permitted in any shell is equal to $2n^2$. Thus for example, the *K* shell ($n = 1$) can hold up to two electrons and the *L* shell ($n = 2$) a maximum of eight.

The spin of the electron is also responsible for most of the magnetic properties that we find associated with atoms and molecules. Materials that are **diamagnetic** experience no attraction for another magnet.[4] In these substances there are the same number of electrons of each spin so that their magnetic effects cancel. **Paramagnetic** substances, on the other hand, are weakly attracted to a magnetic field. In these materials there are more elec-

[4] They are, in fact, repelled slightly by a magnetic field. This is a result of the motion of the electrons in the atom and is not associated with the electron's spin.

trons of one spin than the other (as will always be true when an atom or molecule has an odd number of electrons) and total cancellation does not occur. The extra electrons of one spin cause the atom or molecule, as a whole, to behave as if it were itself a tiny magnet. **Ferromagnetic** substances, of which iron is the most common example, owe their very strong magnetic behavior to interactions between paramagnetic atoms in the solid state. Ferromagnetism is about 1 million times stronger than usually observed for paramagnetism. This phenomenon is discussed in more detail in Chapter 20.

## 3.15 THE ELECTRON CONFIGURATIONS OF THE ELEMENTS

As we have intimated earlier, the way that the electrons are arranged in an atom (i.e., its **electron configuration**) is determined by the order in which the subshells occur on the scale of increasing energy. This is so because, in an atom in its ground state, the electrons will be found in the lowest energy levels available. In hydrogen, for instance, the single electron will be located in the 1*s* subshell because it is this level that has the lowest energy. To indicate that the 1*s* subshell is populated by one electron, we use a superscript (in this case, 1), on the subshell designation. Thus we would denote the electron configuration of hydrogen as $1s^1$. As we proceed in this discussion, it will also be necessary to keep tabs on the electron spins. One method that is often employed is to symbolize an electron with its spin in one direction by an arrow pointing up, ↑, and an electron with opposite spin as an arrow pointing down, ↓. To indicate the distribution of electrons among the orbitals of the atom, we then place the arrows over bars that symbolize orbitals. Hydrogen, for example, is represented as

$$\text{H}\quad \underset{1s}{\underline{\uparrow}}$$

This kind of representation of the electron configuration is usually called an **orbital diagram.**

To obtain the electron configuration of the other elements in the periodic table, let us imagine that we are able to proceed from one atom to the next by adding a proton (plus necessary neutrons) to the nucleus followed by an electron, which we place into the lowest available energy level. As you follow this discussion, refer both to the periodic table (Figure 3.11) and the energy-level diagram in Figure 3.20. A complete table of the electron configurations of the elements is contained in Table 3.5.

Hydrogen is the simplest element, consisting of just a single proton and one electron. The next element, atomic number, 2, is helium. Here there are two electrons to consider and, because the 1*s* orbital can accommodate both of them, the electronic structure of helium is $1s^2$ and its orbital diagram is

$$\text{He}\quad \underset{1s}{\underline{\uparrow\downarrow}}$$

Note that in placing the electrons in the same orbital we have indicated that their spins are in opposite directions as required by the Pauli exclusion principle. We refer to this by saying that their spins are *paired,* or simply that the electrons are paired.

The next two elements following He are Li and Be, which have three and four electrons, respectively. In each of these, the first two electrons will enter the 1*s* subshell and, since no more than two electrons can occupy an *s*

Table 3.5

*The electron configurations of the elements*

| Atomic Number | | | | Atomic Number | | | | Atomic Number | | | |
|---|---|---|---|---|---|---|---|---|---|---|---|
| 1 | H | $1s^1$ | | 36 | Kr | [Ar] | $4s^23d^{10}4p^6$ | 71 | Lu | [Xe] | $6s^24f^{14}5d^1$ |
| 2 | He | $1s^2$ | | 37 | Rb | [Kr] | $5s^1$ | 72 | Hf | [Xe] | $6s^24f^{14}5d^2$ |
| 3 | Li | [He] | $2s^1$ | 38 | Sr | [Kr] | $5s^2$ | 73 | Ta | [Xe] | $6s^24f^{14}5d^3$ |
| 4 | Be | [He] | $2s^2$ | 39 | Y | [Kr] | $5s^24d^1$ | 74 | W | [Xe] | $6s^24f^{14}5d^4$ |
| 5 | B | [He] | $2s^22p^1$ | 40 | Zr | [Kr] | $5s^24d^2$ | 75 | Re | [Xe] | $6s^24f^{14}5d^5$ |
| 6 | C | [He] | $2s^22p^2$ | 41 | Nb | [Kr] | $5s^14d^4$ | 76 | Os | [Xe] | $6s^24f^{14}5d^6$ |
| 7 | N | [He] | $2s^22p^3$ | 42 | Mo | [Kr] | $5s^14d^5$ | 77 | Ir | [Xe] | $6s^24f^{14}5d^7$ |
| 8 | O | [He] | $2s^22p^4$ | 43 | Tc | [Kr] | $5s^24d^5$ | 78 | Pt | [Xe] | $6s^14f^{14}5d^9$ |
| 9 | F | [He] | $2s^22p^5$ | 44 | Ru | [Kr] | $5s^14d^7$ | 79 | Au | [Xe] | $6s^14f^{14}5d^{10}$ |
| 10 | Ne | [He] | $2s^22p^6$ | 45 | Rh | [Kr] | $5s^14d^8$ | 80 | Hg | [Xe] | $6s^24f^{14}5d^{10}$ |
| 11 | Na | [Ne] | $3s^1$ | 46 | Pd | [Kr] | $4d^{10}$ | 81 | Tl | [Xe] | $6s^24f^{14}5d^{10}6p^1$ |
| 12 | Mg | [Ne] | $3s^2$ | 47 | Ag | [Kr] | $5s^14d^{10}$ | 82 | Pb | [Xe] | $6s^24f^{14}5d^{10}6p^2$ |
| 13 | Al | [Ne] | $3s^23p^1$ | 48 | Cd | [Kr] | $5s^24d^{10}$ | 83 | Bi | [Xe] | $6s^24f^{14}5d^{10}6p^3$ |
| 14 | Si | [Ne] | $3s^23p^2$ | 49 | In | [Kr] | $5s^24d^{10}5p^1$ | 84 | Po | [Xe] | $6s^24f^{14}5d^{10}6p^4$ |
| 15 | P | [Ne] | $3s^23p^3$ | 50 | Sn | [Kr] | $5s^24d^{10}5p^2$ | 85 | At | [Xe] | $6s^24f^{14}5d^{10}6p^5$ |
| 16 | S | [Ne] | $3s^23p^4$ | 51 | Sb | [Kr] | $5s^24d^{10}5p^3$ | 86 | Rn | [Xe] | $6s^24f^{14}5d^{10}6p^6$ |
| 17 | Cl | [Ne] | $3s^23p^5$ | 52 | Te | [Kr] | $5s^24d^{10}5p^4$ | 87 | Fr | [Rn] | $7s^1$ |
| 18 | Ar | [Ne] | $3s^23p^6$ | 53 | I | [Kr] | $5s^24d^{10}5p^5$ | 88 | Ra | [Rn] | $7s^2$ |
| 19 | K | [Ar] | $4s^1$ | 54 | Xe | [Kr] | $5s^24d^{10}5p^6$ | 89 | Ac | [Rn] | $7s^26d^1$ |
| 20 | Ca | [Ar] | $4s^2$ | 55 | Cs | [Xe] | $6s^1$ | 90 | Th | [Rn] | $7s^26d^2$ |
| 21 | Sc | [Ar] | $4s^23d^1$ | 56 | Ba | [Xe] | $6s^2$ | 91 | Pa | [Rn] | $7s^25f^26d^1$ |
| 22 | Ti | [Ar] | $4s^23d^2$ | 57 | La | [Xe] | $6s^25d^1$ | 92 | U | [Rn] | $7s^25f^36d^1$ |
| 23 | V | [Ar] | $4s^23d^3$ | 58 | Ce | [Xe] | $6s^24f^15d^1$ | 93 | Np | [Rn] | $7s^25f^46d^1$ |
| 24 | Cr | [Ar] | $4s^13d^5$ | 59 | Pr | [Xe] | $6s^24f^3$ | 94 | Pu | [Rn] | $7s^25f^6$ |
| 25 | Mn | [Ar] | $4s^23d^5$ | 60 | Nd | [Xe] | $6s^24f^4$ | 95 | Am | [Rn] | $7s^25f^7$ |
| 26 | Fe | [Ar] | $4s^23d^6$ | 61 | Pm | [Xe] | $6s^24f^5$ | 96 | Cm | [Rn] | $7s^25f^76d^1$ |
| 27 | Co | [Ar] | $4s^23d^7$ | 62 | Sm | [Xe] | $6s^24f^6$ | 97 | Bk | [Rn] | $7s^25f^9$ |
| 28 | Ni | [Ar] | $4s^23d^8$ | 63 | Eu | [Xe] | $6s^24f^7$ | 98 | Cf | [Rn] | $7s^25f^{10}$ |
| 29 | Cu | [Ar] | $4s^13d^{10}$ | 64 | Gd | [Xe] | $6s^24f^75d^1$ | 99 | Es | [Rn] | $7s^25f^{11}$ |
| 30 | Zn | [Ar] | $4s^23d^{10}$ | 65 | Tb | [Xe] | $6s^24f^9$ | 100 | Fm | [Rn] | $7s^25f^{12}$ |
| 31 | Ga | [Ar] | $4s^23d^{10}4p^1$ | 66 | Dy | [Xe] | $6s^24f^{10}$ | 101 | Md | [Rn] | $7s^25f^{13}$ |
| 32 | Ge | [Ar] | $4s^23d^{10}4p^2$ | 67 | Ho | [Xe] | $6s^24f^{11}$ | 102 | No | [Rn] | $7s^25f^{14}$ |
| 33 | As | [Ar] | $4s^23d^{10}4p^3$ | 68 | Er | [Xe] | $6s^24f^{12}$ | 103 | Lw | [Rn] | $7s^25f^{14}6d^1$ |
| 34 | Se | [Ar] | $4s^23d^{10}4p^4$ | 69 | Tm | [Xe] | $6s^24f^{13}$ | | | | |
| 35 | Br | [Ar] | $4s^23d^{10}4p^5$ | 70 | Yb | [Xe] | $6s^24f^{14}$ | | | | |

subshell, the remaining electron(s) must occupy the $2s$ subshell. The electron configurations of Li and Be, then, are Li, $1s^22s^1$ and Be, $1s^22s^2$. We could also show this as

Li $\underline{\uparrow\downarrow}$ $\underline{\uparrow}$

Be $\underline{\uparrow\downarrow}$ $\underline{\uparrow\downarrow}$
$\quad\;\; 1s \quad 2s$

Since both Li and Be have a completed $1s$ subshell, which corresponds to the electron configuration of He, they can also be written as

Li [He] $\underline{\uparrow}$

Be [He] $\underline{\uparrow\downarrow}$
$\qquad\qquad 2s$

Here we focus our attention on the electronic structure of the outermost shell which, in chemical reactions, is responsible for chemical changes. In this example, the inner filled $1s$ subshell is called the helium **core.** We shall frequently find it useful to consider only those electrons that occur outside a core of electrons corresponding to one of the noble gases.

At beryllium, which has four electrons, the $2s$ subshell is completed. The fifth electron of boron ($Z = 5$), then, must enter the next lowest available subshell, which is the $2p$, to give boron the configuration $1s^2 2s^2 2p^1$. Likewise, the fifth and sixth electrons of carbon must enter the $2p$ subshell; thus, we represent carbon as $1s^2 2s^2 2p^2$. However, if we examine the distribution of the electrons over the various orbitals, we must face a choice: the electrons could be arranged in the following three ways;[5]

C [He] ⇅ ⇅ __ __

or

C [He] ⇅ ↑ ↓ __

or

C [He] ⇅ (2s) ↑ ↑ __ (2p)

The last two electrons can be paired in the same orbital, paired in different orbitals, or arranged so that their spins are in the same direction (unpaired).

As it turns out, experiments show that the last diagram is correct. **Hund's rule,** summarizes this experimental evidence: *electrons, entering a subshell containing more than one orbital, will be spread out over the available orbitals with their spins in the same direction.* For nitrogen ($Z = 7$), then, the electron configuration would be written as $1s^2 2s^2 2p^3$, and it would have the orbital diagram

N [He] ⇅ (2s) ↑ ↑ ↑ (2p)

Finally, the elements oxygen, fluorine, and neon ($Z$ = 8, 9, and 10, respectively) lead to the completion of the $2p$ subshell.

O [He] ⇅ ⇅ ↑ ↑

F [He] ⇅ ⇅ ⇅ ↑

Ne [He] ⇅ (2s) ⇅ ⇅ ⇅ (2p)

After the $2p$ subshell is filled at Ne, the next lowest available energy level is the $3s$. This becomes populated with Na and Mg ($Z$ = 11 and 12). After this the $3p$ subshell is gradually filled by the next six electrons as we complete the atoms Al through Ar ($Z$ = 13 through 18). Then, since the $4s$ subshell lies at lower energy than the $3d$, it is occupied next by the nineteenth and twentieth electrons of K and Ca ($Z$ = 19 and 20).

[5] These are the only three possibilities that we have to consider because in an isolated atom, each of the $p$ orbitals is equivalent in energy. Thus the arrangements,

⇅ __ __ (2p)    __ ⇅ __ (2p)    __ __ ⇅ (2p)

are indistinguishable from one another experimentally.

Examination of Figure 3.20 reveals that after the 4*s* subshell is completed, additional electrons begin to populate the 3*d* subshell. Scandium, therefore, will have the electron configuration $1s^22s^22p^63s^23p^64s^23d^1$ or

$$\text{Sc}\quad [\text{Ar}]\quad \underset{4s}{\underline{\uparrow\downarrow}}\quad \underset{3d}{\underline{\uparrow}\ \underline{\quad}\ \underline{\quad}\ \underline{\quad}\ \underline{\quad}}$$

As we proceed through Ti and V ($Z = 22$ and 23), two more electrons are added to the 3*d* subshell; however, when we get to Cr ($A = 24$), we find the structure

$$\text{Cr}\quad [\text{Ar}]\quad \underset{4s}{\underline{\uparrow}}\quad \underset{3d}{\underline{\uparrow}\ \underline{\uparrow}\ \underline{\uparrow}\ \underline{\uparrow}\ \underline{\uparrow}}$$

instead of

$$\text{Cr}\quad [\text{Ar}]\quad \underset{4s}{\underline{\uparrow\downarrow}}\quad \underset{3d}{\underline{\uparrow}\ \underline{\uparrow}\ \underline{\uparrow}\ \underline{\uparrow}\ \underline{\quad}}$$

This unexpected result occurs because a half-filled or completely filled subshell possesses an extra, added stability. The origin of this extra stability is very complex; so we cannot discuss it here. Nevertheless, the phenomenon is quite important and should be kept in mind. We see it again in period 4, for example, when we get to copper. On the basis of our energy-level diagram in Figure 3.20, we would predict copper to have the electron configuration

$$\text{Cu}\quad [\text{Ar}]\quad \underset{4s}{\underline{\uparrow\downarrow}}\quad \underset{3d}{\underline{\uparrow\downarrow}\ \underline{\uparrow\downarrow}\ \underline{\uparrow\downarrow}\ \underline{\uparrow\downarrow}\ \underline{\uparrow}}$$

The actual structure is given by

$$\text{Cu}\quad [\text{Ar}]\quad \underset{4s}{\underline{\uparrow}}\quad \underset{3d}{\underline{\uparrow\downarrow}\ \underline{\uparrow\downarrow}\ \underline{\uparrow\downarrow}\ \underline{\uparrow\downarrow}\ \underline{\uparrow\downarrow}}$$

By promoting an electron from the 4*s* to the 3*d* subshell, one filled and one half-filled subshell is produced, instead of the filled 4*s* and the neither filled nor half-filled 3*d* subshell that we initially would predict. Because the electron configurations of Cr and Cu are not predictable by our rules, they must be remembered as exceptions.

After the 3*d* subshell is completed at atomic number 30 (zinc), the 4*p* subshell is filled as we proceed from Ga to Kr ($Z = 31$ to 36). This is followed by the completion of the 5*s* subshell from Rb to Sr ($Z = 37, 38$); the 4*d* subshell as we progress across the second row of transition elements ($Z = 39$ to 48); the 5*p* from In to Xe ($Z = 49$ to 54); and the 6*s* with Cs and Ba ($Z = 55, 56$).

Based on the energy-level sequence in Figure 3.20, we would expect that, after the 6*s* subshell had been filled, we would begin to populate the 4*f* subshell next. Actually, at La ($Z = 57$) the last electron enters the 5*d* subshell instead. The 4*f* subshell is filled afterward, with only one other irregularity at Gd, where the 5*d* subshell is again the recipient of an electron so as to provide Gd with a half-filled subshell. As we go to higher and higher shells, these irregularities become more frequent because the spacing between subshells becomes smaller and smaller. As we proceed from atom to atom, the energy of the various subshells shift about somewhat as the nuclear charge increases. The result is that it is difficult to predict accurately the electron configuration of elements of very high atomic number. Nevertheless, we can account for the occurrence of the lanthanide elements by the filling of the 4*f* subshell (an *f* subshell can accommodate 14 electrons and the lanthanide series consists of 14 elements). Likewise, we can account for the actinide elements as the result of the filling of the 5*f* subshell.

## 3.16 THE PERIODIC TABLE AND ELECTRON CONFIGURATIONS

In the last section we saw that the results of wave mechanics could be used to predict the electron configurations of the elements. These electron configurations are based on theory and, to be considered useful and valid, they must somehow manifest themselves in obvious ways. One of the strongest supports for the assignment of electron configurations is the periodic table itself. Recall that, in constructing the current periodic table, elements were arranged under each other in groups because of their similar chemical properties. For example, all of the elements in Group IA are metals that, when they react, form ions with a charge of +1. If we examine the electron configurations of these elements, we see that the outer shell (shell of highest *n*) for each has only one electron in an *s* subshell. Similarly, all of the elements in Group IIA have an outer-shell electron configuration that we might generalize as $ns^2$. In fact, by examining any group within the periodic table, we see that all of the elements in the group possess essentially identical outer-shell electronic structures. It is not really surprising that similar electronic structures lead to similar chemical and physical properties.

Because the properties of the elements depend on their electron configurations, it is important that you develop the ability to write them down. There are a variety of ways to remember the sequence in which the various levels are filled; however, the best aid is the periodic table itself. As we have just seen, the order of filling the energy levels can be used to account for the structure of the periodic table. We can also work in the other direction and use the periodic table to deduce electronic structure.

If we look back over the procedure for determining the electronic structure of the elements, we find that for any element in Groups IA and IIA the final electron was added to an *s* subshell, and that the principal quantum number of that subshell was the same as the period number. Sodium, for instance, a period 3 element, has its outer electron in the 3*s* subshell. For elements in Groups IIIA to O the last electron is added to a *p* subshell whose value of *n* is also the same as the period number. In the case of the transition elements, the final electron that we add is placed into a *d* subshell with *n* equal to one less than the period number. For example, with iron (a fourth-period element), the last electron enters a 3*d* subshell. Finally, the electronic structure of an inner transition element (i.e., one from the lanthanide or actinide series) is completed by an electron in an *f* subshell whose principal quantum number is *two* less than the period number.

In arriving at the electron configuration of an element, let us again imagine that we start at hydrogen and proceed through the periodic table, in order of increasing atomic number, until we arrive at the element in which we are interested. We soon realize that an *s* subshell is filled when we cross Groups IA and IIA; a *p* subshell when we go through Groups IIIA to O; a *d* subshell as we pass across a row of transition elements; and an *f* subshell when we pass through a row of inner transition elements. We also realize that for the *s* and *p* subshells their value of *n* is equal to the period number; for the *d* subshells, *n* is equal to the period number minus one; and for the *f* subshells *n* is equal to the period number minus two.

We can see how this knowledge allows us to derive electron configurations by choosing lead ($Z = 82$) as an example. Figure 3.22 shows how we determine which subshells become filled. These are obtained from the

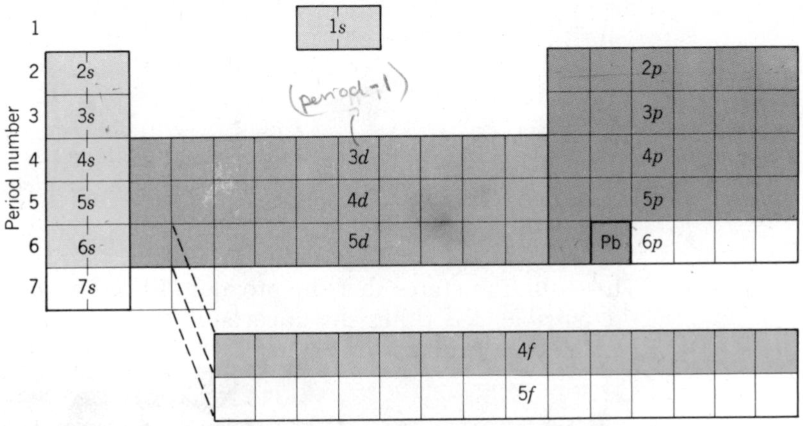


Figure 3.22
*The use of the periodic table for predicting electron configurations.*

regions in the table occupied by all of the elements up to and including the atom, lead. This means that in building up the lead atom we cross through periods 1 to 5 and part of 6, and that upon proceeding from left to right across one period after another, we fill, in order, the subshells 1*s*, 2*s*, 2*p*, 3*s*, 3*p*, 4*s*, 3*d*, 4*p*, 5*s*, 4*d*, 5*p*, 6*s*, 4*f*, 5*d*, and finally end by placing two electrons into the 6*p* subshell. Taking into account the maximum population of each subshell, then, we obtain as the electron configuration of lead,

$$1s^22s^22p^63s^23p^64s^23d^{10}4p^65s^24d^{10}5p^66s^24f^{14}5d^{10}6p^2$$

Sometimes it is preferred to write all subshells of a given shell together. Thus for lead we would have

$$1s^22s^22p^63s^23p^63d^{10}4s^24p^64d^{10}4f^{14}5s^25p^65d^{10}6s^26p^2$$

In Chapter 4 we will see that often we are only interested in the electron population of the outer shell of the atom. In this case we do not have to work through the entire electron configuration. Instead we can locate the element in the periodic table and immediately find the information we need. For instance, suppose that we wished to know the outer-shell electron configuration for antimony, Sb. We find this element in Group VA and in period 5. The outer shell of Sb is the fifth shell ($n = 5$). Crossing period 5 as far as Sb, we would place two electrons in the 5*s* subshell, ten in the 4*d*, and finally three in the 5*p*. If we are interested only in the subshells with highest *n*, we would conclude that the outer-shell configuration of antimony is

$$\text{Sb} \qquad 5s^25p^3$$

## 3.17 THE SPATIAL DISTRIBUTION OF ELECTRONS

In wave mechanics our view of where the electron may be found about the nucleus is far different from the idea of circular orbits imagined by Bohr. This is a consequence of the **uncertainty principle** of Heisenberg, which states that if we attempt to measure, at the same time, both the position and momentum of a particle, our measurements will be subject to errors that are related to one another by the equation

$$\Delta x \cdot \Delta(mv) \geq \frac{h}{4\pi} \tag{3.13}$$

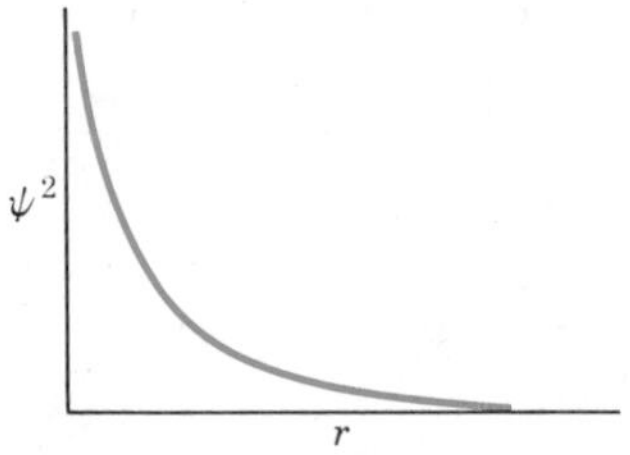


Figure 3.23
*Probability of finding the electron as a function of distance from the nucleus for the 1s orbital of hydrogen.*

This equation states that the product of the uncertainty in the position of the particle, $\Delta x$, times the uncertainty in its momentum, $\Delta(mv)$, must be greater than or equal to Planck's constant divided by $4\pi$. This statement actually says that we are limited in our ability to know simultaneously where the electron is and where it is going. It leads us, instead, to refer to the probability of finding the electron in some small element of volume at various places around the nucleus. More specifically, it is the square of the wave function, $\psi^2$, which is taken to specify this probability.

On the basis of this concept, let us look at the **probability distribution** (the way the probability varies throughout the volume of an atom) for the single electron in the 1*s* orbital of the hydrogen atom. A graph of $\psi^2$ as a function of distance from the nucleus, *r*, is shown in Figure 3.23. It can be seen that those regions in which the probability of observing the electron is greatest lies close to the nucleus and that, as we might expect, the probability decreases as we move away from the nucleus, gradually approaching zero as *r* approaches infinity. In other words, we might expect to find the electron in the hydrogen atom almost anywhere, but most of the time it stays fairly close to the nucleus, effectively surrounding it in a cloud of electronic charge. The electron will spend most of its time in those regions where the electron probability is high; thus the concentration of charge, which we might call **electron density,** will be large. In other areas the charge is thinly spread and the electron density is small.

There are several ways to indicate the distribution of charge in an orbital. One way is to plot $\psi^2$ as we have already done. Another, in two dimensions, is to illustrate the charge cloud as shown in Figure 3.24, where the darker shaded areas represent regions of high electron density. In three dimensions (Figure 3.25) we see this as a scatter of dots in which regions of high electron density are illustrated by the occurrence of a large number of points.

Higher-energy *s* orbitals differ in some respects from the 1*s* orbital. Figure 3.26 compares the way the electron density varies for 1*s*, 2*s*, and 3*s* orbitals. Notice that with the 2*s* orbital, as we move out from the nucleus, the electron density drops to zero, then increases again before gradually de-

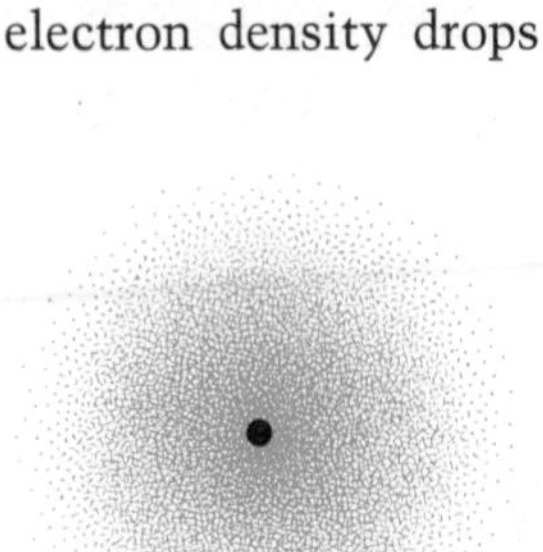

Figure 3.24
*A representation of the charge cloud in hydrogen.*

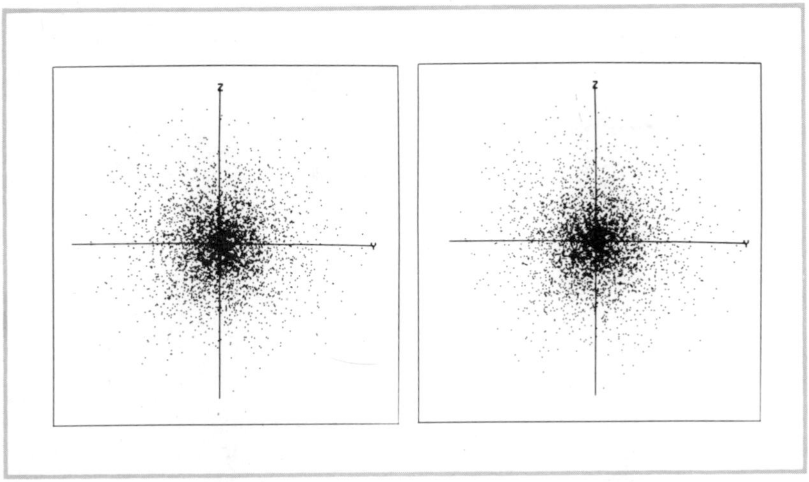


Figure 3.25
*Three-dimensional representation of charge cloud. From D. T. Cromer,* Journal of Chemical Education. *Vol. 45, p. 626, October 1968, used by permission.*

creasing once more. The region where the electron density drops to zero is called a **node.** The 2*s* orbital contains one node; the 3*s* orbital (Figure 3.26*c*) contains two nodes.

Despite these differences, all *s* orbitals have an important property in common. If we draw a surface on which the probability of observing the electron is constant, the surface has the shape of a sphere. *All s orbitals have a spherical "shape."* The main difference is that as we go to higher values of *n*, the sphere within which we find most (for example, 90%) of the electron density becomes larger. In other words, the size of the charge cloud gets larger with increasing principal quantum number, not only for *s* orbitals, but also for *p*, *d*, and *f* orbitals. This means that electrons in orbitals of

Figure 3.26
*Electron density distribution in the 1s, 2s, and 3s orbitals of an atom.*

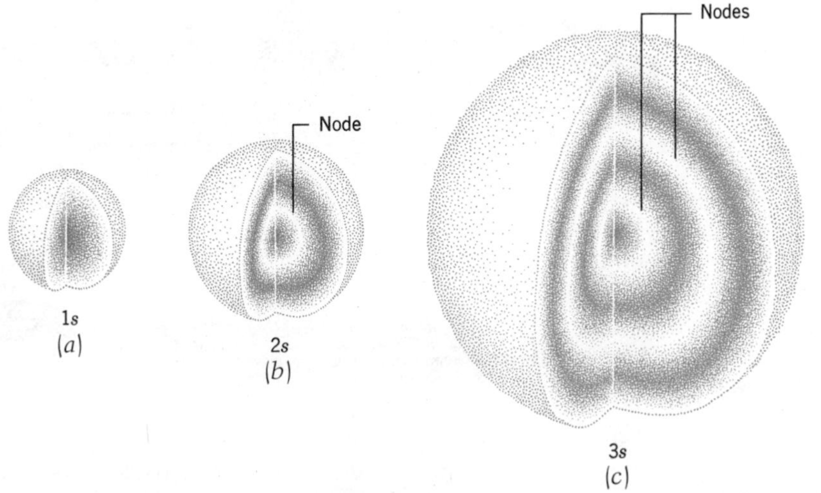

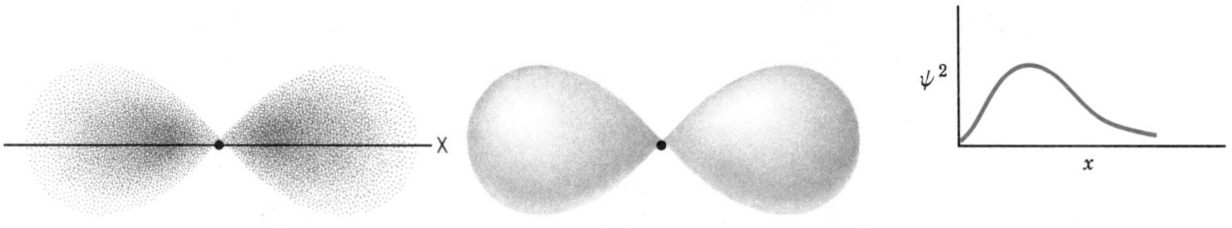


(a)

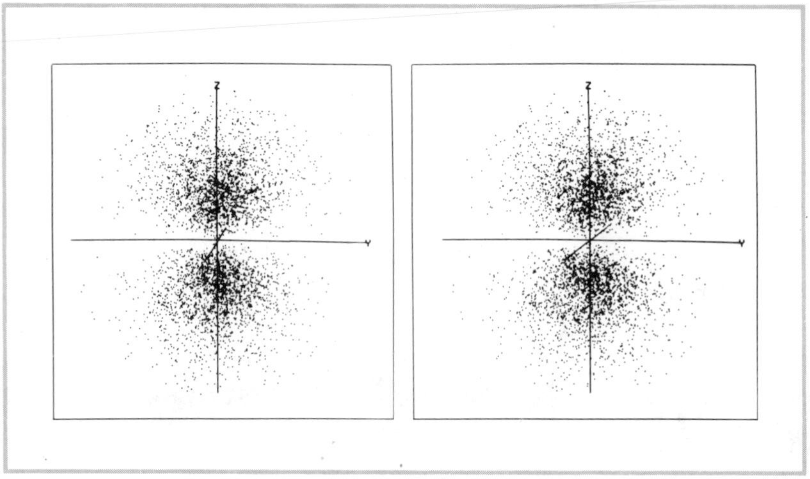


(b)

Figure 3.27

*The shape of a 2p orbital. (a) Two-dimensional representations of a 2p orbital. (b) A three-dimensional representation. From D. T. Cromer,* Journal of Chemical Education, *Vol. 45, p. 626, October 1968, used by permission.*

higher $n$ will be at a greater average distance from the nucleus and that the atom gets larger as its higher energy subshells become populated.

The "shape" of the electron cloud characteristic of a $2p$ orbital is illustrated in Figure 3.27. We see that for a $2p$ orbital the electron density is not distributed symmetrically about the nucleus as in an $s$ orbital, but rather is concentrated in particular regions along a straight line passing through the nucleus. Electron density occurs on both sides of the nucleus so that an electron in a $2p$ orbital spends part of its time on each side of the atom.

Higher-energy $p$ orbitals contain nodes just as do the higher-energy $s$ orbitals. A cross section of a $3p$ orbital is illustrated in Figure 3.28. Despite these differences there is a concentration of electron density along certain specific directions in all $p$ orbitals. As a result all $p$ orbitals have definite directional properties which, as we shall see, allow us to understand why molecules have the shapes they do. Because we are interested only in the directional properties of the orbitals, we will represent all $s$ orbitals as spheres and all $p$ orbitals as a pair of dumbbell-shaped lobes pointing in opposite directions from the nucleus.

A $p$ subshell is composed of three $p$ orbitals, each one having the same "shape." They differ from one another only in the directions in which their electron density is concentrated. These directions lie at right angles to one

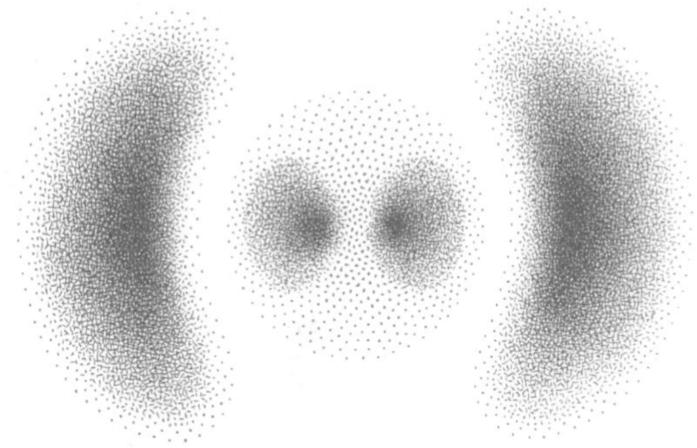

Figure 3.28
*Cross section of a 3p orbital.*

Figure 3.29
*Stereo photographs showing the directional properties of p orbitals. (a) A single p orbital (a $p_y$ orbital). (b) Three p orbitals ($p_x$, $p_y$, $p_z$) on a single set of axes. Atomic-Molecular Orbital Models by Science Related Materials, Inc., Janesville, Wisconsin.*

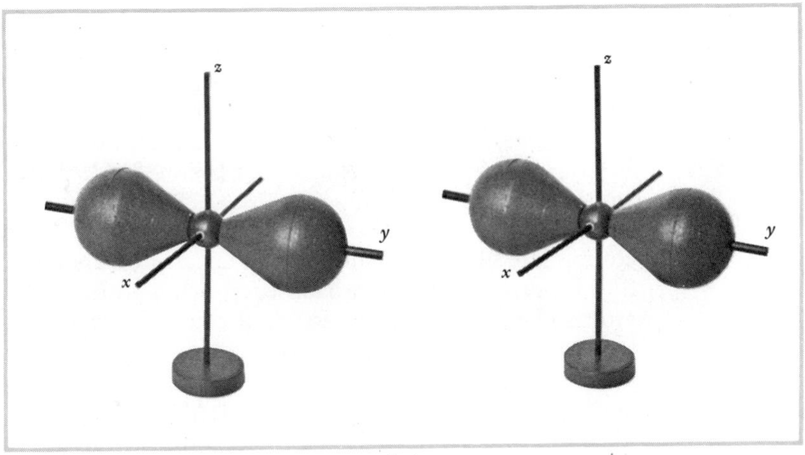


(a)

(b)

another as shown in Figure 3.29. Because the *p* orbitals can be drawn on a set of *xyz* axes, we identify the orbitals by the notation $p_x$, $p_y$, $p_z$.

We have seen that *s* and *p* orbitals differ in the "shapes" of their electron clouds. Orbitals in *d* and *f* subshells also have characteristic "shapes"; however, they are considerably more complicated than *p* orbitals. The *d* orbitals play an important role in the chemistry of the transition elements, and we shall discuss them in some detail when we get to Chapter 20. The shapes of *f* orbitals are very complex and, since they are only required to discuss the chemistry of the inner transition elements, we will not attempt to describe them further.

## 3.18 THE VARIATION OF PROPERTIES WITH ATOMIC STRUCTURE

Many of the properties of the elements vary in a more or less regular fashion as we proceed from left to right within a period or from top to bottom within a group in the periodic table. Most of these variations can be accounted for directly in terms of variations in the electronic structures of the elements. Let us look briefly at three of these properties: atomic size, ionization energy, and electron affinity.

**ATOMIC SIZE.** We have seen that the electron density in an atom does not end abruptly at some particular distance from the nucleus but, instead, trails off gradually, approaching zero at very large distances from the center of the atom. Because of this, it is difficult to define precisely what is meant by the size of the atom. Since atoms never occur all by themselves in chemical systems, but are always in the neighborhood of other atoms, the radius of an atom could be taken to be half the distance between neighboring atoms when the element is present in its most dense form (i.e., most highly compacted form, which is usually the solid). Even this definition, however, is complicated because when atoms enter into a bond, as they do in molecules like $H_2$ or $Cl_2$, they approach each other more closely than do nonbonded atoms (e.g., the noble gases when they are frozen). Also, the atomic radius that we measure for atoms of a pure element will not necessarily be the same in compounds. For example, carbon atoms in diamond (pure carbon) are separated by a distance of 1.54 Å[6] and we would thus assign carbon a radius of 0.77 Å. In the ethane molecule, $C_2H_6$, the carbon–carbon distance is also 1.54 Å; however, in ethylene, $C_2H_4$, and acetylene, $C_2H_2$, we find carbon–carbon distances equal to 1.37 and 1.20 Å, respectively. These lead to atomic radii for carbon of 0.69 and 0.60 Å, both considerably smaller than 0.77 Å.

Despite this difficulty of definition, we can compare atomic radii of the elements if they are measured under circumstances that lead to essentially similar kinds of bonds between their atoms. In Figure 3.30 we illustrate the variation of atomic radius with atomic number. We see that as we proceed down within a group, the size of atoms generally increases, and that as we proceed from left to right across a period, a gradual decrease in size is observed.

In order to interpret these trends within the periodic table in terms of electronic structure, we must look at the factors that determine the size of the outer shells of atoms, that is, the average distance at which electrons in the outer shell occur. As we have discussed earlier (Section 3.17), one of these is the principal quantum number of the outer shell (you will recall

[6] We shall see how interatomic distances are obtained in Chapter 7.

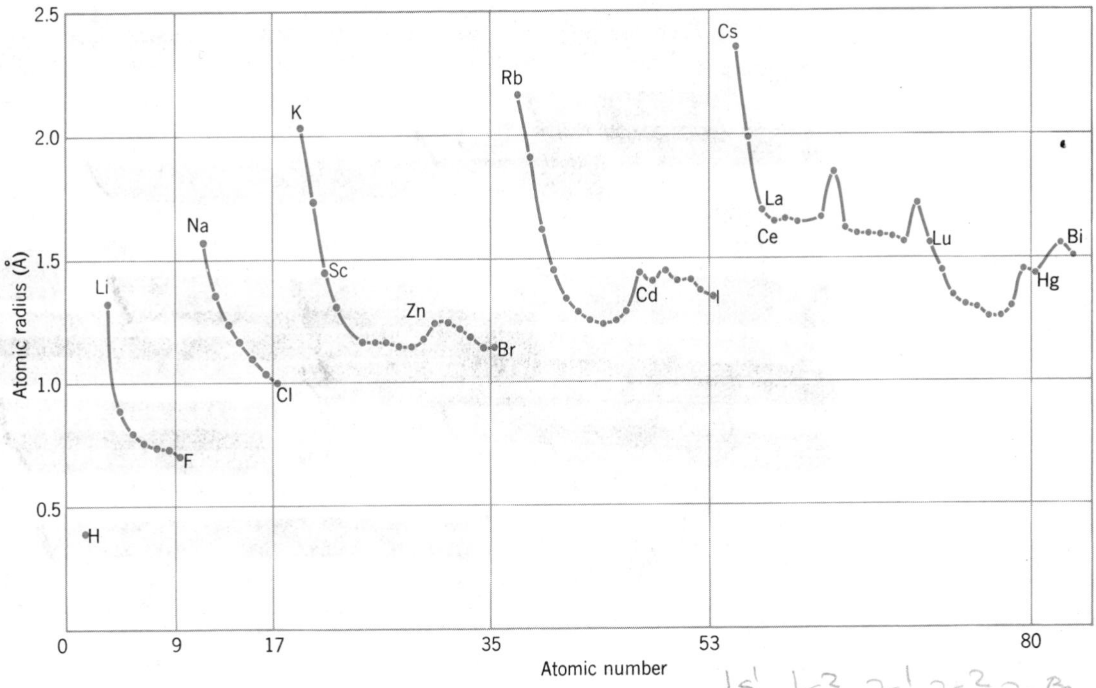


Figure 3.30
*A graph of atomic radius versus atomic number.*

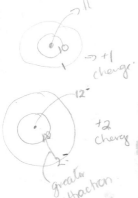

that the electron occurs at increasingly larger distances from the nucleus with increasing value of $n$).

The size of the outer shell also depends on the **effective nuclear charge** that an electron in that shell experiences. Electrons in inner shells tend to lie between the nucleus and those in the outer shell and thus shield the outer shell from the nuclear charge. In a sodium atom, for example, the 10 electrons of the neon core shield the outer $3s$ electron from the positive charge of the 11 protons in the nucleus. Therefore, the outer $3s$ electron feels an effective charge of only about $+1$. Electrons within the same shell also provide some shielding for one another; however, their ability to do so is not very great.

As we proceed from one atom to the next down within a group, each successive element has its outer electron in a shell with a larger value of $n$. The effective nuclear charge experienced by the outer electron(s) remains nearly the same so that the net effect is an increase in size with increase in atomic number within a group. For instance, among the alkali metals, Li through Cs, the principal quantum number of the outer electron increases from 2 for Li to 6 for Cs. The single outer electron in each of these elements, however, experiences a nearly constant effective nuclear charge of $+1$. Therefore the increase in size that occurs from Li to Cs is a result of the electron being in a shell with progressively higher $n$.

For the representative elements, as we move from left to right across a period, we add electrons to the same shell and simultaneously increase the nuclear charge. Since the outer-shell electrons do not shield each other from the nucleus very well, the effective nuclear charge experienced by any one electron in the outer shell increases. This increase in effective nuclear charge leads to a greater attraction for the outer-shell electrons. As a result, they are pulled in closer to the nucleus. Hence a decrease in the size of the atom occurs.

The variation in size as we pass through a row of transition or inner transition elements is much less than among the representative elements. This is so because electrons are being added to an inner shell as the nuclear charge gets larger. In the first row of the transition elements, for instance, the outer electrons occur in a 4*s* subshell, but each successive electron is added to the inner 3*d* subshell as we proceed across the table. The inner-shell electrons are nearly completely effective at shielding the outer shell from the nuclear charge so that the outer 4*s* electrons experience only a very gradual increase in effective nuclear charge across this region of the periodic table. Hence small changes in size occur.

The gradual decrease in size that occurs upon the filling of the 4*f* subshell in the lanthanides, which is termed the **lanthanide contraction,** has some very marked effects on the chemistry of the transition elements that follow the lanthanides in the sixth period. For example, because of the lanthanide contraction, the size of Hf is the same as the size of Zr. Since their outer-shell electron configurations are virtually identical, the chemistry of these two elements is very similar and it is extremely difficult to separate them from one another.

As we shall discuss in the next chapter, when many of the elements react to produce compounds, they do so by the formation of ions. We find that positive ions are smaller than the neutral atoms from which they are formed, while negative ions are larger than neutral atoms (Table 3.6). The decrease in size that accompanies the creation of a positive ion is often a result of the removal of all of the electrons from the outer shell of the atom so that a noble gas electron configuration is attained. For example, the sodium

Table 3.6
*Atomic and ionic radii (in angstroms)*

| Positive Ions | | Atomic Radius | Ionic Radius | Charge |
|---|---|---|---|---|
| Group IA | Li | 1.35 | 0.60 | (+1) |
| | Na | 1.54 | 0.95 | (+1) |
| | K | 1.96 | 1.33 | (+1) |
| | Rb | 2.11 | 1.48 | (+1) |
| | Cs | 2.25 | 1.69 | (+1) |
| Group IIA | Be | 0.90 | 0.31 | (+2) |
| | Mg | 1.30 | 0.65 | (+2) |
| | Ca | 1.74 | 0.99 | (+2) |
| | Sr | 1.92 | 1.13 | (+2) |
| | Ba | 1.98 | 1.35 | (+2) |
| Group IIIA | Al | 1.43 | 0.50 | (+3) |
| | Ga | 1.22 | 0.62 | (+3) |
| | In | 1.62 | 0.81 | (+3) |

| Negative Ions | | Atomic Radius | Ionic Radius | Charge |
|---|---|---|---|---|
| Group VIIA | F | 0.64 | 1.36 | (−1) |
| | Cl | 0.99 | 1.81 | (−1) |
| | Br | 1.14 | 1.95 | (−1) |
| | I | 1.33 | 2.16 | (−1) |
| Group VIA | O | 0.66 | 1.40 | (−2) |
| | S | 1.04 | 1.84 | (−2) |
| | Se | 1.17 | 1.98 | (−2) |
| | Te | 1.37 | 2.21 | (−2) |
| Group VA | N | 0.70 | 1.71 | (−3) |
| | P | 1.10 | 2.12 | (−3) |

Elements That Form More Than One Ion

| | | | | | |
|---|---|---|---|---|---|
| Fe | 1.26 | $Fe^{2+}$ | 0.76 | $Fe^{3+}$ | 0.64 |
| Co | 1.25 | $Co^{2+}$ | 0.78 | $Co^{3+}$ | 0.63 |
| Cu | 1.28 | $Cu^{+}$ | 0.96 | $Cu^{2+}$ | 0.69 |

atom loses its single 3*s* electron to produce an $Na^+$ ion whose electronic structure consists of the neon core. The outer shell at this point has its principal quantum number equal to 2 and, thus, the outer-shell electrons in the $Na^+$ ion are at a smaller average distance from the nucleus than the 3*s* electron in the Na atom.

When negative ions are produced from neutral atoms, electrons are added to the outer shell without any change in the nuclear charge. An additional electron will provide some degree of shielding for other electrons originally present; therefore the effective nuclear charge felt by any one electron in the outer shell will decrease. At the same time, the presence of an additional electron in the outer shell will increase interelectron repulsions (that is, repulsions between electrons). Both of these factors tend to cause the outer shell to expand in size, causing the negative ion to be larger than the neutral atom.

IONIZATION ENERGY. *The* **ionization energy** *is defined as the energy required to remove an electron from an isolated gaseous atom in its ground state.* We can represent this process by an equation such as

$$Na\ (g) \longrightarrow Na^+\ (g) + e^-$$

This is an endothermic process because the electron is attracted to the positive nucleus; therefore, energy must be supplied to remove it. Since all atoms other than hydrogen possess more than one electron, they also have more than one ionization energy. The amount of energy required to remove the first electron from the neutral atom is termed the first ionization energy and that required to remove the second is called the second ionization energy. As we might expect, successive ionization energies increase in magnitude because the species from which the electron is removed becomes progressively more positively charged. For example, the first ionization energy involves the removal of an electron from a neutral atom, while the second ionization energy involves the removal of an electron from an ion whose charge is +1.

Table 3.7 contains successive ionization energies, measured in kilojoules per mole (kJ/mole) for the first 20 elements in the periodic table. Entries in the table represent energies needed to remove electrons from 1 mole of atoms. An examination of the data in this table points out the great stability associated with an electron core having a noble gas electron configuration. We see, for example, that for a Group IA element the first ionization energy is relatively low and that the second ionization energy is very much greater. For the Group IIA elements a large increase in ionization energy occurs after two electrons have been removed, while for Group IIIA elements the break occurs after the third electron has been lost. In fact, we see that in general a very large jump in ionization energy always occurs after an atom has lost a number of electrons that is numerically equal to its group number. Since a Group IA element contains one electron outside a noble gas electron configuration, a Group IIA element, two, and so on, these large increases in ionization energy must reflect the extreme difficulty that is encountered in trying to break into the noble gas structure that lies below the outer shell.

The variation of the first ionization energy across periods and down groups, illustrated in Figure 3.31, quite closely parallels the trends in atomic size. This really should not be too surprising, since the energy required to remove an electron from an atom completely should depend in part on how far

Table 3.7
*Ionization energies of the first 20 elements (kJ/mole)*

| | First | Second | Third | Fourth | Fifth | Sixth | Seventh | Eighth |
|---|---|---|---|---|---|---|---|---|
| H | 1,312 | | | | | | | |
| He | 2,371 | 5,247 | | | | | | |
| Li | 520 | 7,297 | 11,810 | | | | | |
| Be | 900 | 1,757 | 14,840 | 21,000 | | | | |
| B | 800 | 2,430 | 3,659 | 25,020 | 32,810 | | | |
| C | 1,086 | 2,352 | 4,619 | 6,221 | 37,800 | 47,300 | | |
| N | 1,402 | 2,857 | 4,577 | 7,473 | 9,443 | 53,250 | 64,340 | |
| O | 1,314 | 3,391 | 5,301 | 7,468 | 10,980 | 13,320 | 71,300 | 84,050 |
| F | 1,681 | 3,375 | 6,045 | 8,418 | 11,020 | 15,160 | 17,860 | 92,000 |
| Ne | 2,080 | 3,963 | 6,276 | 9,376 | 12,190 | 15,230 | — | — |
| Na | 495.8 | 4,565 | 6,912 | 9,540 | 13,360 | 16,610 | 20,110 | 25,490 |
| Mg | 737.6 | 1,450 | 7,732 | 10,550 | 13,620 | 18,000 | 21,700 | 25,660 |
| Al | 577.4 | 1,816 | 2,744 | 11,580 | 15,030 | 18,370 | 23,290 | 27,460 |
| Si | 786.2 | 1,577 | 3,229 | 4,356 | 16,080 | 19,790 | 23,780 | 29,250 |
| P | 1,012 | 1,896 | 2,910 | 4,954 | 6,272 | 21,270 | 25,410 | 29,840 |
| S | 999.6 | 2,260 | 3,380 | 4,565 | 6,996 | 8,490 | 28,080 | 31,720 |
| Cl | 1,255 | 2,297 | 3,850 | 5,146 | 6,544 | 9,330 | 11,020 | 33,600 |
| Ar | 1,520 | 2,665 | 3,947 | 5,770 | 7,240 | 8,810 | 11,970 | 13,840 |
| K | 418.8 | 3,069 | 4,600 | 5,879 | 7,971 | 9,619 | 11,380 | 14,950 |
| Ca | 589.5 | 1,146 | 4,941 | 6,485 | 8,142 | 10,520 | 12,350 | 13,830 |

away it is from the nucleus. In addition, the same factors that are responsible for causing an outer shell to contract in size as we proceed across a period will also lead to the electron being held more tightly. Thus, as we proceed down within a group (e.g., the alkali metals), the increase in size that occurs is accompanied by a decrease in ionization energy. As we move across a period, from left to right, the increased effective nuclear charge experienced by the outer-shell electrons causes the shell to shrink in size and also makes it more difficult to remove an electron.

If we examine more closely the trend in ionization energy across a period, we note some irregularities. In period 2, for example, we expect a uniform increase in ionization energy as we go from Li to Ne. We observe, however, that the ionization energy of beryllium is higher than that for boron. In addition, the energy required to remove an electron from nitrogen is greater than for oxygen. These reversals can also be explained by the electronic structures of the elements.

In the case of beryllium, the first electron that is removed (recall that we are comparing first ionization energies) lies in the filled 2*s* subshell, while the electron that is removed first from boron lies in the singly occupied 2*p*

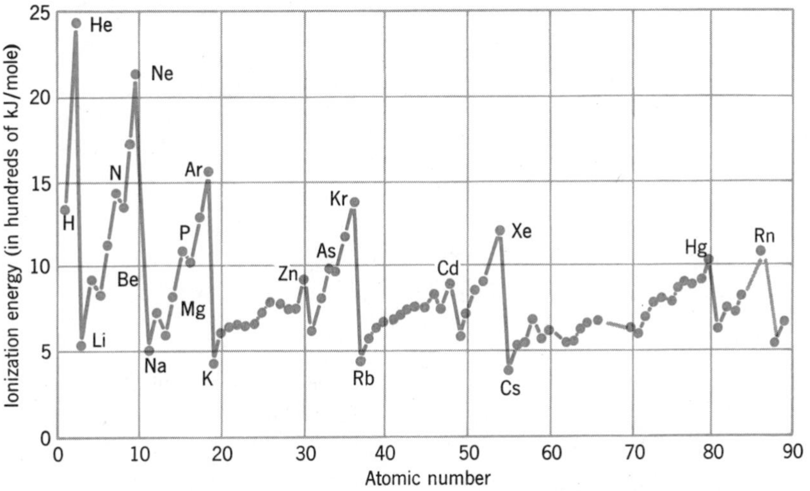


Figure 3.31
*The variation of first ionization energy with atomic number.*

subshell. The $2p$ subshell is higher in energy than the $2s$; thus the $2p$ electron of boron is more easily removed than a $2s$ electron of beryllium.

When we get to nitrogen, we find that we have a half-filled $2p$ subshell (electronic structure of nitrogen is $1s^22s^22p^3$), while in oxygen the $2p$ subshell is occupied by four electrons. The fourth electron in this $2p$ subshell must enter an orbital already occupied by an electron. It therefore experiences considerable repulsion. As a result this electron is more easily removed than one of the electrons in a singly occupied orbital in the nitrogen atom. Note that the same inverted order of values for the ionization energy also occurs in periods 3 and 4, where the ionization energy of phosphorus is greater than for sulfur and that for arsenic is greater than for selenium.

**ELECTRON AFFINITY.** *The* **electron affinity** *is the energy released or absorbed when an electron is added to a neutral gaseous atom in its ground state.* Such a process occurs, for example, when a chlorine atom picks up an electron to become a negative ion

$$Cl\ (g) + e^- \longrightarrow Cl^-\ (g)$$

The electron affinity (like the ionization energy) applies to isolated atoms and usually represents an exothermic process. This is so because we are placing the electron into an environment where it experiences the attraction of the nucleus. We can see how the addition of an electron to an atom would release energy by considering the reverse process, pulling the electron away from the attractive force of the nucleus. If removing the electron requires work (i.e., is endothermic), the opposite process would release energy.

There are instances where more than one electron is added to the outer shell of the atom. For example, oxygen reacts to form the ion, $O^{2-}$, in which an oxygen atom picks up two electrons. The second electron that is added to give the ion a charge of $-2$ must be forced onto an already negative ion. This requires work; therefore we find that the second electron affinity of an atom is an endothermic quantity. Table 3.8 contains electron affinities for some of the representative elements. The table is not complete because electron

Table 3.8

*Electron affinities[a] for the representative elements (kJ/mole)*

| IA | IIA | IIIA | IVA | VA | VIA | VIIA |
|---|---|---|---|---|---|---|
| H −73 | | | | | | |
| Li −59 | Be ≈+100 | B −30 | C −121 | N ≈+9 | O −140 | F −334 |
| Na ≈−50 | Mg ≈+30 | Al −50 | Si −140 | P −75 | S −200 | Cl −349 |
| K ≈−40 | Ca — | Ga −17 | Ge −120 | As −58 | Se −160 | Br −325 |
| Rb ≈−20 | Sr — | In — | Sn — | Sb — | Te — | I −296 |
| Cs — | Ba — | Tl — | Pb — | Bi — | Po — | At — |

[a] Negative values mean that the process $M + e^- \rightarrow M^-$, is exothermic.

affinities are difficult to measure and, for many elements, have not been determined.

As with the ionization energy, the variations in electron affinity generally parallel the variations in atomic size. This is because we are considering the placement of an electron into the outer shell of the atom. The closer the electron can get to the nucleus, the greater will be the effect of the nuclear charge. Atoms that are very small and that have outer shells that experience a high effective nuclear charge (e.g., elements in the upper right of the periodic table) therefore have very large electron affinities. On the other hand, atoms that are large and whose outer shells feel the effect of a small effective nuclear charge (such as the elements in Groups IA and IIA) are expected to have small electron affinities.

If we examine the data in Table 3.8, we find that fluorine, which is smaller than chlorine, also has a smaller electron affinity. Apparently the greater interelectron repulsions experienced by an entering electron in the small outer shell of fluorine more than compensate for the fact that the added electron in fluorine lies closer to the nucleus than it does in chlorine.

Finally, carbon has a rather substantial exothermic electron affinity, while that of nitrogen is actually endothermic. With carbon the entering electron can occupy a vacant 2*p* orbital and therefore experiences only minimal interelectron repulsion. With nitrogen, however, an additional electron must be placed into an orbital that is already occupied by an electron. The greater interelectron repulsion that results causes the electron affinity to be an endothermic quantity for nitrogen.

## INDEX TO QUESTIONS AND PROBLEMS (problem numbers in italics)

Nuclear particles 8, 10, 11, 12, 13, 14, *67*, *68*

Isotopes 15, 16, 17, *56*, *57*, *58*, *70*

Periodic Table 18, 19, 20, 21, 22, 23, 24, 25, 26, 27

Atomic spectra 28, *61*, *62*, *65*, *73*

Bohr Theory 29, *63*

Wave properties of matter 30, 31, 32, *66*, *71*, *72*

$E = mc^2$ *69*

Shells, subshell and orbitals; quantum numbers 33, 34, 35, 41, 44

Electron configurations 36, 37, 38, 39, 40

Orbital shapes 42, 43, 45, 46, 47

Periodic trends in
- size: 48, 49, 50
- ionization energy: 51, 52
- electron affinity: 53, 54



## REVIEW QUESTIONS

**3.1** What properties are observed for cathode rays?

**3.2** What is the definition of the coulomb?

**3.3** In Millikan's experiment, the charge on the oil droplets was always found to be a multiple of $-1.60 \times 10^{-19}$ coulomb. Suppose that this experiment were repeated and the following values were obtained:

$-3.20 \times 10^{-19}$ coulomb
$-5.60 \times 10^{-19}$ coulomb
$-6.40 \times 10^{-19}$ coulomb
$-2.40 \times 10^{-19}$ coulomb
$-7.20 \times 10^{-19}$ coulomb

On the basis of these data, what would be the charge on the electron?

**3.4** What are canal rays? What are they composed of?

**3.5** Why was the hydrogen ion assumed to be a fundamental particle (a particle that cannot be broken into something simpler)?

**3.6** Describe the three important kinds of radiation that are observed to be emitted by radioactive substances.

**3.7** How are particles of different mass separated in the mass spectrometer?

**3.8** Why did Rutherford conclude that the positive charge must be concentrated in a very dense nucleus within the atom?

**3.9** What are the meanings of the terms, wavelength and frequency? For electromagnetic radiation, how are they related?

**3.10** What are the properties of the electron, the proton, and the neutron?

**3.11** What are the numbers of protons, neutrons, and electrons in each of the following: ${}^{132}_{55}Cs$, ${}^{115}_{48}Cd^{2+}$, ${}^{194}_{81}Tl$, ${}^{105}_{47}Ag^{+}$, ${}^{78}_{34}Se^{2-}$?

**3.12** What are the number of protons, neutrons, and electrons in each of the following: ${}^{131}_{56}Ba$, ${}^{109}_{48}Cd^{2+}$, ${}^{36}_{17}Cl^{-}$, ${}^{63}_{28}Ni$, ${}^{170}_{69}Tm$?

**3.13** Write the appropriate symbol for each of the following isotopes: (a) $Z = 26, A = 55$; (b) $Z = 37, A = 86$; (c) $Z = 81, A = 204$; (d) $Z = 81, A = 170$; (e) $Z = 70, A = 169$.

**3.14** How many neutrons are in each of the atoms in Question 3.13?

**3.15** What isotope serves as the current standard for the atomic mass scale?

**3.16** Can you explain why the atomic masses of some elements (such as Cl or Cu, for instance) are so far from whole numbers?

**3.17** What is the difference between mass number and atomic mass?

**3.18** State the modern version of the periodic law.

**3.19** Why were there blanks left in Mendeleev's periodic table?

**3.20** What was the basis on which Mendeleev constructed his periodic table?

**3.21** Which of the following are representative elements: Mg, Ti, Fe, Se, Ni, Br?

**3.22** Which of the following are transition elements: Sr, Ru, As, W, Ag, Al?

**3.23** Which elements constitute the inner transition elements?

**3.24** Which elements constitute the halogens? The noble gases? The alkali metals? The alkaline earth metals?

**3.25** Which of the following are metals: Ta, Nd, Se, F, Cs?

**3.26** What properties are observed for nonmetals?

**3.27** What significance does the periodic table have with regard to the development of theory about the electronic structure of the atom?

**3.28** What is a line spectrum? How does it differ from a continuous spectrum? From the point of view of atomic structure, what is the significance of the occurrence of line spectra?

**3.29** Describe Bohr's model of the atom. Why was his theory discarded? What initial evidence was there that Bohr's theory might be correct?

**3.30** What is a diffraction pattern? How is it produced?

**3.31** Why don't we observe the wave properties of large objects such as baseballs or airplanes?

**3.32** What direct evidence is there for the wave properties of the electron?

**3.33** How many electrons can be accommodated in each of the following types of subshells: *s, p, d, f, g, h*? What is the lowest value of *n* for a shell that has an *h* subshell? What are the allowed values of *m* for an *h* subshell?

**3.34** Give the values of *n, l, m*, and *s* for each electron in a filled *L* shell.

**3.35** How many electrons can be placed into the *M* shell of an atom?

**3.36** Draw orbital diagrams for each element in the first row of transition elements ($Z = 21$ through 30). Indicate which of these are paramagnetic and which are diamagnetic.

**3.37** Use the periodic table as a guide in writing the complete electron configurations of these elements: P, Ni, As, Ba, Rh, Ho, Sn.

**3.38** Use the periodic table to arrive at the electronic structure of the outer shells of the atoms of Si, Se, Sr, Cl, O, S, As, and Ga.

**3.39** Write the complete electron configuration for Rb, Sn, Br, Cr, and Cu.

**3.40** Give the outer-shell electron configuration for K, Al, F, S, Tl, and Bi.

**3.41** What is the major difference between a Bohr orbit and an orbital? In what way is the Heisenberg uncertainty principle involved in this comparison?

**3.42** On a single set of Cartesian coordinate axes, sketch the "shapes" of the three *p* orbitals. Label them $p_x$, $p_y$, and $p_z$.

**3.43** How do the 1*s* and 2*s* orbitals differ? How are they alike?

**3.44** What is the Pauli exclusion principle? What is Hund's rule?

**3.45** How does the shape of an *s* orbital differ from that of a *p* orbital?

**3.46** What is a node?

**3.47** On the basis of interelectron repulsion and the spatial arrangement of the *p* orbitals, the observed orbital diagram for nitrogen seems quite reasonable. Why?

**3.48** Explain the variation in ionic size observed for the series, $N^{3-}$, $O^{2-}$, and $F^-$ (Table 3.6) in terms of the effective nuclear charge and interelectron repulsions experienced by the outer-shell electrons.

**3.49** What is the lanthanide contraction? How might this be used to explain why the elements in the sixth period following the lanthanides have higher ionization energies than the elements directly above them in the fifth period (e.g., the ionization energy of Pt = 870 kJ/mole, while that of Pd = 805 kJ/mole)?

**3.50** In Table 3.6 we find some elements that form more than one positive ion. In each case the ion with the greater positive charge is smaller. Why is this so?

**3.51** How can we explain the variation of ionization energy across a period in the periodic table?

**3.52** Draw a graph, on a set of axes like that below, of the ionization energy versus the number of electrons removed from the atom for each of the elements Li, C, O, S, and Ne.

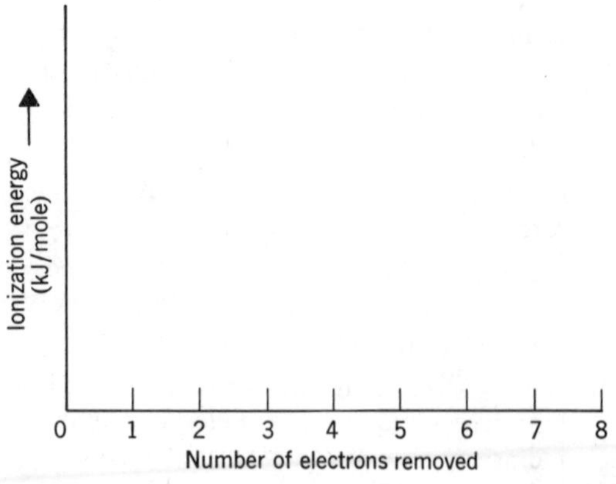


**3.53** Why is the second electron affinity for an element always an endothermic quantity?

**3.54** The Group VIIA elements have electron affinities that are considerably larger than those of the Group VIA elements. What does this suggest about the stability of the noble gas electron configuration?

**3.55** Describe the contributions made by each of the following scientists toward the development of atomic theory: Faraday, Thomson, Rutherford, Millikan, Moseley.

## REVIEW PROBLEMS

**3.56** The element Eu occurs naturally as a mixture of 47.82% $^{151}_{63}Eu$, whose mass is 150.9 amu, and 52.18% $^{153}_{63}Eu$, whose mass is 152.9 amu. Calculate the average atomic mass of Eu.

**3.57** Naturally occurring boron consists of two isotopes, $^{10}B$ with a mass of 10.01294 and $^{11}B$ with a mass of 11.00931. The abundance of $^{10}B$ is 19.6% and that of $^{11}B$ is 80.4%. Calculate the average atomic mass of B.

**3.58** Naturally occurring lead is composed of four isotopes. Their abundances and masses are given below. Calculate the average atomic mass of lead.

| Isotope | Mass (amu) | Abundance (%) |
|---|---|---|
| $^{204}Pb$ | 203.973 | 1.48 |
| $^{206}Pb$ | 205.9745 | 23.6 |
| $^{207}Pb$ | 206.9759 | 22.6 |
| $^{208}Pb$ | 207.9766 | 52.3 |

**3.59** Verify the value for the mass of the electron using the experimentally determined charge-to-mass ratio of $-1.76 \times 10^8$ coulombs/g and the measured charge, $-1.60 \times 10^{-19}$ coulomb.

***3.60** The charge-to-mass ratio ($e/m$) of the proton (a hydrogen nucleus) is $9.65 \times 10^4$ coulombs/g. The proton has a charge equal to $1.60 \times 10^{-19}$ coulomb. Calculate the value of Avogadro's number.

**3.61** Use the Rydberg equation (Equation 3.3) to calculate the wavelengths of the first two lines in the Pfund series of the hydrogen spectrum.

**3.62** Use the Rydberg equation to calculate the wavelength of the spectral line in hydrogen that would result when an electron drops from the fourth Bohr orbit to the second, and from the sixth Bohr orbit to the third.

**3.63** How much energy must be supplied to raise an electron from the first Bohr orbit to the third?

**3.64** Calculate the wavelength of light whose frequency is $8.0 \times 10^{15}$ Hz. Calculate the frequency of light whose wavelength is 200.0 nm.

**3.65** Use the Rydberg equation (Equation 3.3) to calculate the ionization energy of hydrogen and compare your answer to the results of Example 3.4.

**3.66** Calculate the energy of a photon having a frequency of $3 \times 10^{15}$ Hz. If a photon has an energy of $2 \times 10^{-13}$ erg, what is its wavelength?

***3.67** The proton (the nucleus of a hydrogen atom) has a mass of $1.67 \times 10^{-24}$ g and a diameter of $10^{-13}$ cm. Calculate the density of the nucleus assuming it to be spherical in shape.

***3.68** The earth has a mass of $6.59 \times 10^{21}$ tons and a diameter of approximately 8000 miles. What would be the diameter of the earth (in miles) if it had the same mass but was composed entirely of nuclear material? (Use the density of nuclear material calculated in Question 3.67).

***3.69** How many tons of water could be heated from 0°C to 100°C by converting 1.0 g of matter entirely into energy. Recall that it takes 1 cal of energy to raise the temperature of 1.0 g of water by 1°C. (See also Section 1.11.)

***3.70** Naturally occurring chlorine is composed of $^{35}Cl$ with atomic mass of 34.96885 and $^{37}Cl$ with atomic mass of 36.96590. The average atomic mass of Cl is 35.453. What are the percentages of each isotope in naturally occurring chlorine?

***3.71** Calculate the kinetic energy of an electron with a wavelength of 0.10 nm.

***3.72** How long would it take a 2.0 g bullet to travel the length of a 10-cm gun barrel if it had a wavelength of 0.10 nm?

***3.73** If all of the energy required to remove the electrons from 1 mole of H atoms was used instead to heat water, how many grams of water could have their temperature increased by 25°C?

# 4 CHEMICAL BONDING: GENERAL CONCEPTS

In Chapter 3 we spent considerable time discussing electronic structures of atoms and their relationship to some properties of atoms. A property possessed by almost all atoms is their ability to combine with other atoms to produce more complex species. The forces of attraction that hold atoms together in their combined states are called chemical bonds.

The theories and language used to describe chemical bonds have evolved from very simple theories to more elaborate ones based on wave mechanics. In this chapter we examine some of the simpler ideas about chemical bonding that should serve to carry you through the discussions in the first part of this book. In Chapter 16 we delve deeper into the more modern, sophisticated theories about bonding.

## 4.1 LEWIS SYMBOLS

When atoms interact to form a bond, only the outer portions of the atoms come in contact; consequently, only their outer electron configurations are usually important. To keep tabs on the outer-shell (also called **valence-shell**) electrons, we use a special type of notation called **Lewis symbols,** named after the American chemist, G. N. Lewis (1875–1946). To construct the Lewis symbol for an element, we write down its atomic symbol surrounded by a number of dots (or X's or circles, etc.), each of which represents one electron in the atom's valence shell.

For example, the element hydrogen, which has one electron in its valence shell, is given the Lewis symbol, H·. Any atom, in fact, with one electron in its outer shell has a similar Lewis symbol. This includes any element in Group IA of the periodic table, so that each of the elements Li, Na, K, Rb, Cs, and Fr has a Lewis symbol that we might generalize as X· (where X = Li, Na, etc.). Generalized Lewis symbols for the representative elements are given in Table 4.1.

We shall see that Lewis symbols are useful in discussing bonds between atoms. The formulas we draw with them are called either *Lewis structures* or *electron-dot formulas.*

Table 4.1
*Lewis symbols for A-Group elements*

| Group | IA | IIA | IIIA | IVA | VA | VIA | VIIA |
|---|---|---|---|---|---|---|---|
| Symbol | $X\cdot$ | $\cdot X\cdot$ | $\cdot\dot{X}\cdot$ | $\cdot\dot{\underset{\cdot}{X}}\cdot$ | $\dot{\underset{\cdot}{X}}:$ | $\cdot\ddot{\underset{\cdot}{X}}:$ | $\cdot\ddot{\underset{\cdot\cdot}{X}}:$ |

## 4.2 THE IONIC BOND

Chemical bonds can be divided into two general categories: **ionic** (or **electrovalent**) **bonds** and **covalent bonds.**

An ionic bond occurs when one or more electrons are transferred from the valence shell of one atom to the valence shell of another. The atom that loses electrons becomes a positive ion (**cation**) while the atom that acquires electrons becomes negatively charged (an **anion**). The ionic bond results from the attraction between the oppositely charged ions.

An example of the formation of an ionic substance is the reaction between atoms of lithium and fluorine. The electronic structures of these are

$$\text{Li} \quad 1s^2 2s^1$$

and

$$\text{F} \quad 1s^2 2s^2 2p^5$$

When they react the lithium atom loses the electron from its 2*s* subshell to become $Li^+$, thereby assuming an electron configuration that is the same as the noble gas He.

$$\text{Li}(1s^2 2s^1) \longrightarrow \text{Li}^+(1s^2) + e^-$$

The electron lost by Li is picked up by the fluorine atom, which thereby acquires an electron configuration identical to that of the noble gas Ne.

$$\text{F}(1s^2 2s^2 2p^5) + e^- \longrightarrow \text{F}^-(1s^2 2s^2 2p^6)$$

Once formed, the $Li^+$ and $F^-$ ions attract one another because of their opposite charges. It is this attraction between the ions that constitutes the ionic bond. The $Li^+$ and $F^-$ ions that are produced are attracted to each other and pack themselves together to form the ionic solid, LiF. It is important to remember that an ionic solid such as this does not contain discrete molecules, but instead contains ions packed so that the attractive forces between ions of opposite charge are maximized while repulsive forces between ions of the same charge are minimized. In LiF, for example, each cation ($Li^+$) is surrounded by, and attracted equally to six anions ($F^-$), as shown in Figure 4.1.

Figure 4.1
*The structure of LiF. Small circles = $Li^+$; large circles = $F^-$.*

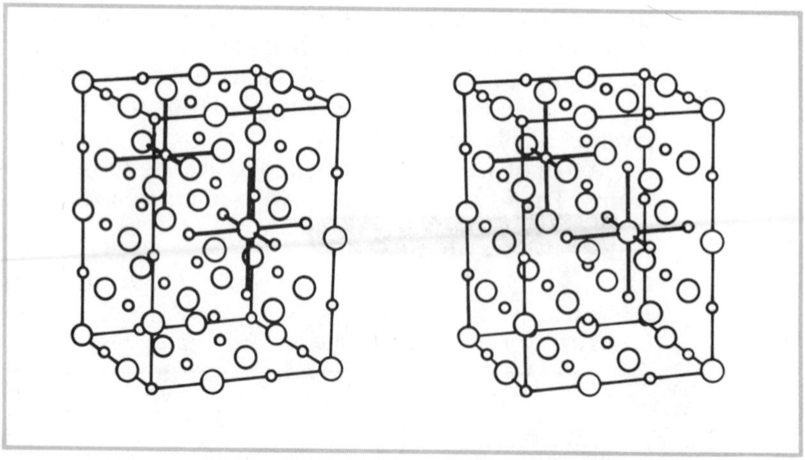

In a similar fashion, each anion is attracted equally to the six cations surrounding it.

The Lewis symbols that were introduced in the last section can be used to illustrate the transfer of electrons that occurs during the formation of an ionic compound. For instance, we can show the reaction of Li and F as

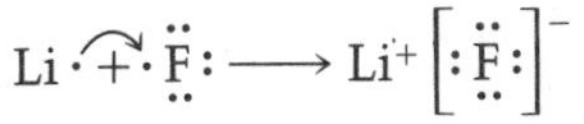

The brackets appearing around the fluorine on the right are intended to show that all eight electrons are the exclusive property of the fluoride ion, $F^-$.

When Li and F react, electrons are lost or gained until a noble gas electron configuration is reached. Except for He, this corresponds to $ns^2np^6$ (a total of eight electrons in the outer shell). As you will recall from the last chapter, the electronic structures of the noble gases possess a great deal of stability. The tendency for atoms to achieve this electronic arrangement forms the basis of the so-called **octet rule,** which simply states that *atoms tend to gain or lose electrons until there are eight electrons in their valence shell.*

The octet rule applies particularly to the representative elements, although there are some exceptions. The octet rule does not hold for the transition elements, however. In general, when a positive ion is formed from an atom, the electrons are always lost first from the shell with the largest $n$. The transition elements therefore lose their outer $s$ electrons before any of the underlying $d$ electrons are removed. The $Zn^{2+}$ ion, for instance, is formed when a zinc atom loses its outer $4s$ electrons. The outer-shell electron configuration of the $Zn^{2+}$ ion is therefore $3s^23p^63d^{10}$. The $ns^2np^6nd^{10}$ outer-shell configuration is often called the **pseudonoble gas configuration.** Another outer-shell electron distribution that is relatively stable is $ns^2np^6nd^5$, found in ions such as $Fe^{3+}$ and $Mn^{2+}$.

Some ions occur with none of the electronic structures just discussed. Since it is very difficult to form a highly charged ion (ions with a charge greater than +3 are rare), electron loss sometimes ceases before a noble gas electron configuration is achieved. Some examples are $Ti^{2+}$ ([Ar]$3d^2$), $V^{2+}$ ([Ar]$3d^3$), and $Cr^{2+}$ ([Ar]$3d^4$). Ions of this type are common among the transition elements. A list of some common cations and anions, including the type of electron configuration they possess, is given on Table 4.2 (page 100).

The ratio in which two elements react to form an ionic substance is usually determined by the numbers of electrons that must be lost or gained by the respective reactant atoms in order to attain a stable electron configuration. For instance, in the reaction between calcium (Group IIA) and chlorine (Group VIIA), each calcium atom must lose two electrons to achieve a noble gas structure, while each chlorine atom needs to acquire only one electron to obtain an octet. As we can see,

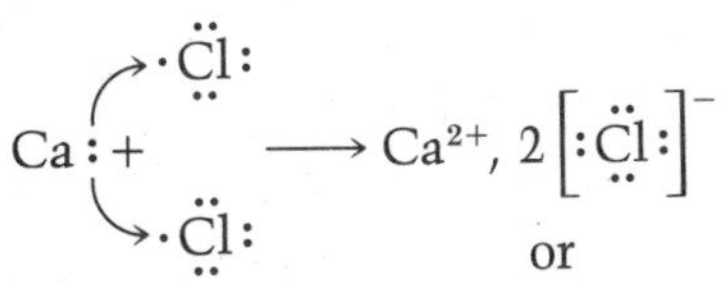

or

$CaCl_2$

Table 4.2
*Some common cations and anions*

(a) Ions with noble gas electron configurations

| +1 | +2 | +3 | −3 | −2 | −1 |
|---|---|---|---|---|---|
| $Li^+$ | | $Al^{3+}$ | $N^{3-}$ | $O^{2-}$ | $F^-$ |
| $Na^+$ | $Mg^{2+}$ | | $P^{3-}$ | $S^{2-}$ | $Cl^-$ |
| $K^+$ | $Ca^{2+}$ | | | $Se^{2-}$ | $Br^-$ |
| $Rb^+$ | $Sr^{2+}$ | | | $Te^{2-}$ | $I^-$ |
| $Cs^+$ | $Ba^{2+}$ | | | | |

(b) Ions with pseudonoble gas electron configurations

| +1 | +2 | +3 |
|---|---|---|
| $Cu^+$ | $Zn^{2+}$ | $Ga^{3+}$ |
| $Ag^+$ | $Cd^{2+}$ | $In^{3+}$ |
| $Au^+$ | $Hg^{2+}$ | $Tl^{3+}$ |

(c) Ions with other electron configurations

| | |
|---|---|
| $Cr^{2+}$ | $[Ne]3s^23p^63d^4$ |
| $Cr^{3+}$ | $[Ne]3s^23p^63d^3$ |
| $Mn^{2+}$ | $[Ne]3s^23p^63d^5$ |
| $Fe^{2+}$ | $[Ne]3s^23p^63d^6$ |
| $Fe^{3+}$ | $[Ne]3s^23p^63d^5$ |
| $Co^{2+}$ | $[Ne]3s^23p^63d^7$ |
| $Ni^{2+}$ | $[Ne]3s^23p^63d^8$ |
| $Cu^{2+}$ | $[Ne]3s^23p^63d^9$ |

the result is that two chlorine atoms must react with each calcium atom to produce one $Ca^{2+}$ and two $Cl^-$ ions. The neutral compound has the formula $CaCl_2$. Similar reasoning leads us to expect a compound between Li and O to have the formula $Li_2O$.

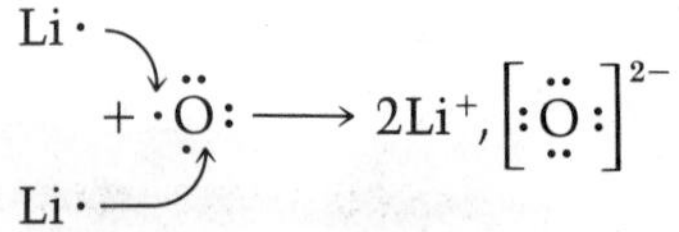


A slightly more complex situation occurs with Al and O. Aluminum, in Group IIIA, loses three electrons to achieve a noble gas structure and produces the ion $Al^{3+}$. Oxygen, on the other hand, forms the ion $O^{2-}$. To produce a neutral compound two $Al^{3+}$ ions must be combined with three $O^{2-}$ ions; hence aluminum oxide has the formula $Al_2O_3$.

EXAMPLE 4.1 What is the formula of an ionic compound formed between Mg and P?

SOLUTION Magnesium is in Group IIA and therefore will lose two electrons to achieve a noble gas configuration. The ion formed is $Mg^{2+}$.

Phosphorus is found in Group VA and must acquire three electrons to attain a noble gas structure. This produces the ion $P^{3-}$.

The formula of the compound can be obtained in a very simple way by exchanging superscripts for subscripts.

$Mg_3P_2$

The total positive charge is +6; the total negative charge is −6. If you use this method to obtain the formula, remember that ionic compounds are always represented by empirical formulas. Be sure that the subscripts are the simplest set of whole numbers. For example, using this method with $Mg^{2+}$ and $O^{2-}$ gives $Mg_2O_2$ initially. This must be reduced to MgO.

There are many substances that contain ions composed of more than one atom (i.e., **polyatomic ions**). The formulas of these compounds are determined by the relative numbers of cations and anions that must be present in order to achieve a neutral solid. In Table 4.3 on page 102 some of the common polyatomic ions are listed. The atoms within a polyatomic ion are held to each other by covalent bonds, which we will discuss later. Some of these ions are highly colored and impart their characteristic colors to compounds (and aqueous solutions) containing them. Let us now look at some examples of how the formulas of ionic compounds containing this type of ion are obtained.

EXAMPLE 4.2 What is the formula for the ionic substance containing the ions $Na^+$ and $CO_3^{2-}$?

SOLUTION In order for the compound to be neutral the number of positive charges must equal the number of negative charges. This requires two $Na^+$ ions per $CO_3^{2-}$ ion. The compound (called sodium carbonate) therefore has the formula $Na_2CO_3$.

EXAMPLE 4.3 An ionic compound contains the ions $Ca^{2+}$ and $PO_4^{3-}$. What is its formula?

SOLUTION The total number of positive or negative charges represented in the formula must be divisible by both 2 and 3. The smallest number that meets this requirement is $2 \times 3 = 6$. Thus, there must be six positive and six negative charges in the formula. This is achieved by taking three $Ca^{2+}$ ions and two $PO_4^{3-}$ ions. The formula of this compound (calcium phosphate), then, is $Ca_3(PO_4)_2$.

We get this same answer using the method shown in Example 4.1:

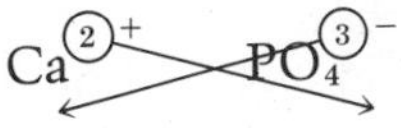


$Ca_3(PO_4)_2$

Table 4.3
*Some common polyatomic ions*

(a) Cations

| | |
|---|---|
| $NH_4^+$ | Ammonium |
| $H_3O^+$ | Hydronium |

(b) Anions (alternate names in parentheses)

| | |
|---|---|
| $CO_3^{2-}$ | Carbonate |
| $HCO_3^-$ | Hydrogen carbonate (bicarbonate) |
| $C_2O_4^{2-}$ | Oxalate |
| $NO_3^-$ | Nitrate |
| $NO_2^-$ | Nitrite |
| $OH^-$ | Hydroxide |
| $SO_4^{2-}$ | Sulfate |
| $HSO_4^-$ | Hydrogen sulfate (bisulfate) |
| $SO_3^{2-}$ | Sulfite |
| $HSO_3^-$ | Hydrogen sulfite (bisulfite) |
| $ClO_4^-$ | Perchlorate |
| $ClO_3^-$ | Chlorate |
| $ClO_2^-$ | Chlorite |
| $ClO^-$ ($OCl^-$) | Hypochlorite |
| $PO_4^{3-}$ | Orthophosphate (phosphate) |
| $HPO_4^{2-}$ | Hydrogen orthophosphate (hydrogen phosphate) |
| $H_2PO_4^-$ | Dihydrogen orthophosphate (dihydrogen phosphate) |
| $CrO_4^{2-}$ | Chromate |
| $Cr_2O_7^{2-}$ | Dichromate |
| $MnO_4^-$ | Permanganate |
| $C_2H_3O_2^-$ | Acetate |

## 4.3 FACTORS INFLUENCING THE FORMATION OF IONIC COMPOUNDS

The driving force in the formation of an ionic bond is a lowering of the energy of the particles that come together to form the compound. In general, we associate a lowering of the energy of a system with an increase in its stability. For example, a ruler standing on end will fall on its side. In the process, its potential energy has decreased and it has achieved a more stable position. As a rule, any system will seek its most stable configuration spontaneously, that is, without outside help. This applies to chemical reactions as well as to rulers.

The actual reaction between lithium and fluorine is not as simple as we have pictured it above. Lithium does not exist as simple atoms, but as a solid; fluorine occurs as a gas composed of $F_2$ molecules. Nevertheless, an important feature of the reaction is the transfer of electrons from lithium to fluorine. Let us look briefly at the energy changes in this reaction to see why it occurs.

The actual reaction is

$$Li\ (s) + \tfrac{1}{2}F_2(g) \longrightarrow LiF\ (s)$$

A coefficient of $\frac{1}{2}$ is used here because we will consider the reaction taking place between laboratory-sized quantities, that is,

$$1 \text{ mole } Li\ (s) + \tfrac{1}{2} \text{ mole } F_2\ (g) \longrightarrow 1 \text{ mole } LiF\ (s)$$

To analyze this reaction we can break it down into a number of simple steps, the net result of which corresponds to the overall reaction. For example, we can examine the following steps:

1. Vaporization of 1 mole of Li (*s*) to lithium atoms in a vapor where they are so far apart that they are effectively isolated atoms. The energy required for this process has been measured to be 37.1 kcal. (In this example we shall use kilocalories rather than kilojoules. Most tabulated data in the scientific literature have energies listed in kilocalories.)
2. Decomposition of $\frac{1}{2}$ mole of $F_2$ (*g*) to give 1 mole of fluorine atoms. The energy required for this process is 18.9 kcal.
3. Removal of the outer electrons from the Li atoms. This is the first ionization energy of Li and has a value of 124.3 kcal. This produces 1 mole of $Li^+$ (*g*). This step is also endothermic.
4. Addition of an electron to each F atom. This is the electron affinity of F and is equal to 79.5 kcal. This exothermic step produces 1 mole of $F^-$ (*g*).
5. Bringing together the mole of $Li^+$ and $F^-$ to give 1 mole of LiF (*s*). Energy is also released in this step[1] and is called the **lattice energy.** It is equal to 242.8 kcal.

Figure 4.2 illustrates these steps schematically. This is called a **Born-Haber cycle.** Just as the overall reaction is the net result of steps 1 through 5, the overall energy change is the net result of the energy changes in steps 1 through 5. Steps 1, 2, and 3 are endothermic—energy must be put into the system to vaporize Li (*s*), decompose $F_2$ (*g*), and remove electrons from Li (*g*). The total input is 180.3 kcal. The last two steps (4 and 5) are exothermic—energy is released from the system when electrons are added to F (*g*) and when $Li^+$ and $F^-$ come together to form LiF (*s*). The total energy released is 322.3 kcal.

The net energy change is the difference between the energy put into the system and the energy released. Here we see that 142.0 kcal more are re-

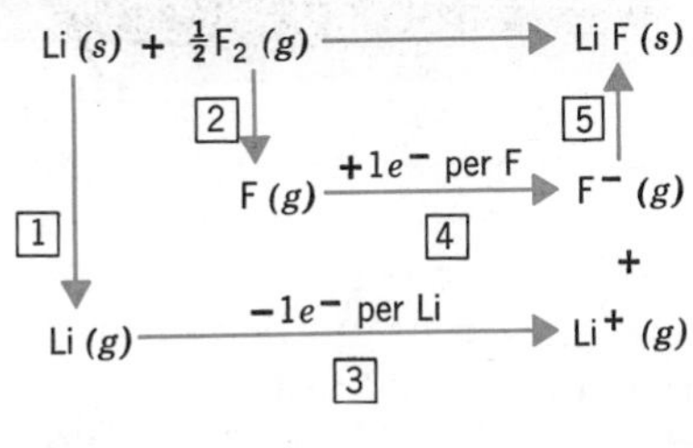


Figure 4.2
*Born-Haber cycle for LiF. Numbers correspond to processes described in the text.*

[1] Since energy is required to separate positive and negative ions, energy must obviously be released when they are brought together. The reaction $Li^+ + F^- \rightarrow LiF$ is, therefore, exothermic.

leased than must be added. Therefore the net reaction is exothermic. Because it is exothermic, the system's energy is decreasing and therefore the LiF (*s*) is more stable than Li (*s*) + $F_2$ (*g*). *Thus the formation of the ionic solid, with the release of a large amount of energy, leads to the stability of an ionic compound.*

We can also understand why electron transfer ceases once we have formed $Li^+$ and $F^-$. The octet rule referred to in the last section results because once a metal atom such as lithium has emptied its valence shell the removal of additional electrons from the noble gas core requires a very large input of energy, more than can be recovered through the lattice energy. Similarly, addition of electrons to a nonmetal that has achieved a noble gas configuration requires that they be added to the *s* orbital of the next higher shell. This also requires an input of a very large quantity of energy. As a result, electron gain or loss ceases once a noble gas configuration has been reached.

At this point we can ask: What conditions most favor the formation of an ionic substance? From our analysis we see that the most stable compounds will result when elements of low ionization energy combine with elements of high electron affinity, or when the lattice energy of the resulting compound is very large, or both. Under these conditions more energy is given off by the exothermic processes than is absorbed by the endothermic ones, with the net result that the total energy contained within the reacting species decreases.

Since metals generally have rather low ionization energies and electron affinities, they tend to lose electrons to form cations; nonmetals, on the other hand, have large ionization energies and electron affinities, so they usually acquire electrons to produce anions. For this reason, most compounds formed between metals and nonmetals are ionic, particularly the substances formed when an element from Group IA or IIA reacts with an element in the upper right corner of the periodic table (excluding Group O).

## 4.4 THE COVALENT BOND

In many instances the formation of an ionic substance is not energetically favorable. For example, the creation of a cation may require too large an energy input (ionization energy) to be recovered by the energy released when the anion is formed and the ionic solid is produced (electron affinity and lattice energy). In these situations a **covalent bond** is formed.

A covalent bond results from the *sharing* of a *pair* of electrons between atoms. The binding force results from the attraction between these shared electrons and the positive nuclei of the atoms entering into the bond. In this sense, the electrons serve as a sort of glue cementing the atoms together. As with the ionic bond, the stability of the covalent bond results from a lowering of the energy of the pair of atoms that are joined. Analysis of the energy changes in the formation of a covalent bond is complicated. At least in part, the energy released results from the bringing together of electrons of one atom and the nucleus of the others to which they are also attracted. In addition to the attractive forces between the electrons and both nuclei, there are repulsive forces between the electrons on the two atoms as well as repulsive forces between the positive nuclei. The distance that separates the atoms in the bond after it is formed is controlled by a balancing of these attractive and repulsive forces.

The simplest covalent bond that we can examine exists between H atoms in the molecule $H_2$. Each H atom completes its valence shell by

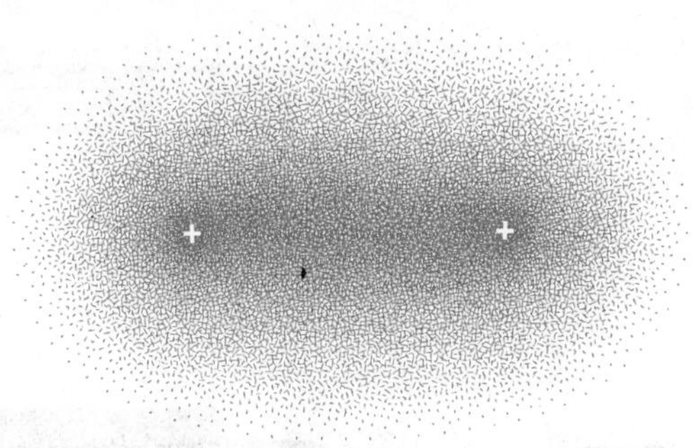

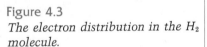

Figure 4.3
*The electron distribution in the $H_2$ molecule.*

acquiring a share of an electron from another atom. We can indicate the formation of $H_2$, using Lewis symbols, as

$$\mathrm{H\cdot + H\cdot \longrightarrow H{:}H}$$

in which the pair of electrons in the bond is shown as a pair of dots between the two H atoms. Often a dash is used instead of the pair of dots, so that the $H_2$ molecule may be represented as H—H. The actual distribution of electronic charge in this molecule is illustrated in Figure 4.3. Notice that the electron density is concentrated between the positive nuclei.

As with the ionic bond, the number of covalent bonds that an atom will form can frequently be predicted by counting the number of electrons required to achieve a stable electron configuration (usually that of a noble gas). For example, the carbon atom has four electrons in its valence shell; to attain a noble gas configuration it usually acquires, through sharing, four additional electrons. The carbon atom, therefore, is capable of forming four bonds with H atoms to form the molecule $CH_4$ (methane).

$$\cdot\dot{\underset{\cdot}{\mathrm{C}}}\cdot + 4\mathrm{H}\times \longrightarrow \begin{array}{c}\mathrm{H}\\ \times\cdot\\ \mathrm{H}\overset{\cdot}{\times}\mathrm{C}\overset{\times}{\cdot}\mathrm{H}\\ \cdot\times\\ \mathrm{H}\end{array}$$

Nitrogen, which has five valence electrons, has to gain only three electrons through sharing to complete an octet; therefore, nitrogen forms three covalent bonds with hydrogen to form the ammonia molecule, $NH_3$. In a similar fashion it is easy to see why the formula for water is $H_2O$ and that for hydrogen fluoride is HF.

$$\begin{array}{c}\mathrm{H}\overset{\times}{\cdot}\ddot{\mathrm{N}}\overset{\times}{\cdot}\mathrm{H}\\ \times\cdot\\ \mathrm{H}\end{array}\qquad \begin{array}{c}\mathrm{H}\overset{\times}{\cdot}\ddot{\mathrm{O}}:\\ \times\cdot\\ \mathrm{H}\end{array}\qquad \mathrm{H}\overset{\times}{\cdot}\ddot{\underset{\cdot\cdot}{\mathrm{F}}}:$$

Unfortunately, it is not always possible to predict the formula of a covalent molecule on the basis of these simple rules. There are many examples of covalent compounds that fail to obey the octet rule. For instance, the molecule $BeCl_2$ is formed by the pairing of the two Be valence electrons with electrons on two chlorine atoms:[2]

$$:\ddot{\underset{\cdot\cdot}{\mathrm{Cl}}}\cdot + \times\mathrm{Be}\times + \cdot\ddot{\underset{\cdot\cdot}{\mathrm{Cl}}}: \longrightarrow :\ddot{\underset{\cdot\cdot}{\mathrm{Cl}}}\overset{\cdot}{\times}\mathrm{Be}\overset{\times}{\cdot}\ddot{\underset{\cdot\cdot}{\mathrm{Cl}}}:$$

[2] Even though $BeCl_2$ is formed from elements in Groups IIA and VIIA, it is covalent rather than ionic.

In this molecule the Be atom has only four electrons in its valence shell. Since this is less than the usual octet, $BeCl_2$ is said to be **electron-deficient.** Another example of an electron-deficient molecule is $BCl_3$:

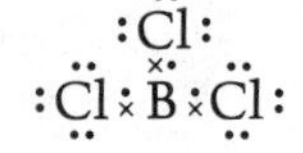

Besides electron-deficient compounds, there are many more examples of molecules in which the central atom has more than eight electrons in its valence shell. Two typical examples are $PCl_5$ and $SF_6$. To form covalent bonds between the central atom (generally the one written first in the formula) and each of the surrounding atoms, more than four pairs of electrons (an octet) are needed. In $PCl_5$, for example, there are five covalent bonds; in $SF_6$ there are six. The central atom in each of these molecules uses all of its valence electrons to form covalent bonds.

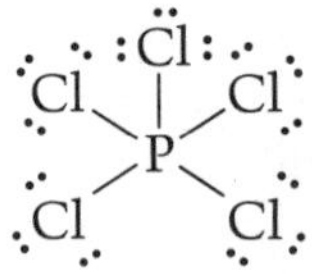

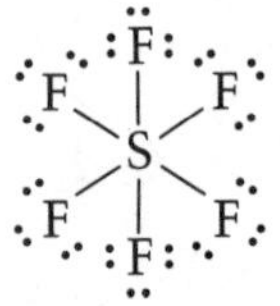

In these compounds, both phosphorus and sulfur have exceeded the number of electrons required for a noble gas electron configuration. This can occur with these elements because, in each case, the valence shell can accommodate more than eight electrons (both P and S are in the third period and the third shell can contain up to 18 electrons because of the availability of the relatively low-energy $3d$ subshell). Elements in the second period (Li to Ne) never form compounds with more than eight electrons in their valence shell because the second shell cannot accommodate more than an octet.

Let us look now at an example of how we arrive at the dot structure of a molecule or ion. Before we can proceed, however, we must first know the relative positions of the atoms that are bound together. For example, in $CO_2$ we must know that there are two O atoms bound to the C atom and that the molecule does not have a structure such as O—O—C. In many instances the arrangement of atoms can be inferred from the formula, since it is common practice to write the central atom of a molecule first in the formula, followed by the atoms that surround the central atom. This is so with $CO_2$, for example. It is also true for species such as $NH_3$, $NO_2$, $NO_3^-$, $SO_3$, $CO_3^{2-}$, and $SO_4^{2-}$. However, it is not true for $H_2O$ and $H_2S$ (in which H atoms are bound to O and S, respectively). Nor is it true for molecules such as HClO (in which the O is the central atom) or ions like $SCN^-$ (in which C is central). The structure of the molecule is therefore not always obvious. If you must guess, the most symmetrical arrangement of atoms has the greatest chance of being correct. Once we know the arrangement of atoms in the molecule, however, we can then go about distributing the valence electrons. This is usually done so that after the electrons are distributed each atom in the molecule is surrounded by an octet.

**EXAMPLE 4.4** Draw the Lewis structure for the $ClO_3^-$ ion, which has the structure

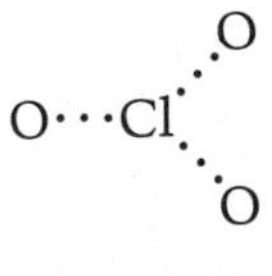

**SOLUTION** This ion contains three atoms from Group VIA (each contributing 6 valence electrons) and one atom from Group VIIA (which contributes 7 valence electrons). In addition, there is still another electron needed to account for the negative charge.

| | | |
|---|---|---|
| 3 O | $3 \times 6 =$ | $18e^-$ |
| 1 Cl | $1 \times 7 =$ | $7e^-$ |
| −1 charge | | $1e^-$ |
| | total | $26e^-$ |

There are 26 electrons to distribute. First we place a pair in each bond, thereby accounting for 6 electrons.

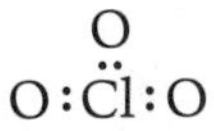

Now we place the remaining 20 electrons in pairs, keeping in mind the octet rule. This gives

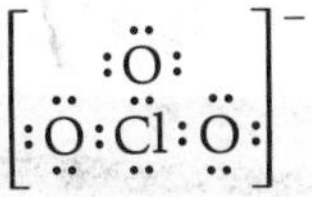

Writing Lewis structures for molecules containing only single bonds (i.e., one electron pair shared between two atoms) is usually relatively simple and straightforward. There are, however, a very large number of molecules in which atoms complete their octet by sharing more than one pair of electrons in a bond. For example, $CO_2$ has a Lewis structure that we can write as

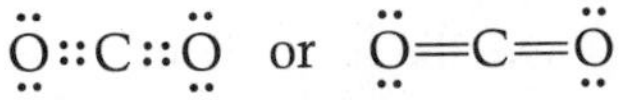

Let us take a look at how to draw electron-dot formulas of this type.

For $CO_2$ we proceed as follows. Oxygen is in Group VIA of the periodic table and each oxygen atom, therefore, supplies 6 valence electrons to the molecule; carbon, in Group IVA, provides 4. The total number of valence electrons that we must account for, then, is 16. For the oxygen atoms to be bound to the carbon, at least one pair of electrons must be shared between the carbon and each oxygen. Placing these into the molecule gives us

O:C:O

This accounts for 4 electrons, leaving 12 more to play with. These 12 electrons must be distributed so that each atom is surrounded by 8 electrons (an octet). One possible arrangement is

$$:\underset{\cdot\cdot}{\ddot{O}}:C:\underset{\cdot\cdot}{\ddot{O}}:$$

In this structure each oxygen atom is surrounded by an octet of electrons; however, the carbon atom has only 4 electrons around it. We can give carbon an octet by allowing it to share an additional pair of electrons from each oxygen atom. The oxygen atom still retains a share of this pair of electrons, but now the carbon atom shares it too. Here we have eight electrons about each atom and a satisfactory Lewis structure. Thus

When two pairs of electrons are shared between two atoms, we call the bond a **double bond. Triple bonds** are also possible. In the $N_2$ molecule, for example, each N atom completes its valence shell by sharing three electrons with another nitrogen atom.

$$:N:::N: \quad \text{or} \quad :N{\equiv}N:$$

In conclusion, then, the steps to be followed in drawing Lewis structures can be summarized as follows:

1. Calculate the total number of *valence* electrons available from all of the atoms in the molecule or ion.
2. Bond all of the appropriate atoms together using one pair of electrons per bond.
3. Distribute the remaining electrons in pairs so that each atom has an octet.
4. If you run short of electrons, make multiple bonds.

EXAMPLE 4.5 Draw the electron-dot formula for the poisonous gas, hydrogen cyanide, HCN (C is the central atom).

SOLUTION There are 10 valence electrons to distribute (1 from H, 4 from C, 5 from N). First we place a pair in each bond.

$$H:C:N$$

This accounts for 4 electrons. As we add the remaining $6e^-$, we must keep in mind that the valence shell of H can hold only $2e^-$. No more electrons can be placed around H because it already has two. One way of adding the other $6e^-$ is

$$H:\ddot{C}:\ddot{N}:$$

but C and N do not have octets. This can be corrected as follows.

The HCN molecule contains a triple bond.

## 4.5 RESONANCE

It often happens that a single satisfactory electron-dot formula for a molecule or ion cannot be drawn. For example, two electron-dot structures that obey the octet rule can be drawn for sulfur dioxide (a substance that is a major air pollutant). These are shown below as structures **1** and **2.**

(1) (2)

These structures have their nuclei in identical positions but differ in the arrangement of electrons. In both, one oxygen atom is bound through a single bond to the sulfur while the other oxygen atom is connected via a double bond.

All experimental evidence suggests that the two sulfur–oxygen bonds are identical; therefore neither of the electron-dot structures that we have drawn for $SO_2$ is satisfactory. In fact, it is impossible to draw a single electron-dot formula for $SO_2$ that obeys the octet rule and that is, at the same time, consistent with all of the experimental facts.

We circumvent this problem by the concept of **resonance.** We say that the actual electronic structure of $SO_2$ does not correspond to either **1** or **2** but, instead, to a structure somewhere in between that has properties of both. This true structure is known as a **resonance hybrid** *of the contributing structures* **1** *and* **2.**

It is really quite unfortunate that the term resonance was used to describe this phenomenon, because the impression is often received that the structure of $SO_2$ fluctuates between **1** and **2**. This is definitely not the case. The structure of $SO_2$ is never **1** or **2** but a structure in between that we cannot draw satisfactorily using Lewis symbols. The problem is somewhat like trying to describe the beast you would obtain if you were able to cross a cat and a dog. When you try to picture this hypothetical offspring, you visualize it having characteristics of both parents. However, you do not think of it as being a cat one instant and a dog the next.

Some species cannot be adequately explained with only two resonance structures. For example, even though $SO_2$ can be represented by two structures, the $SO_3$ molecule requires three.

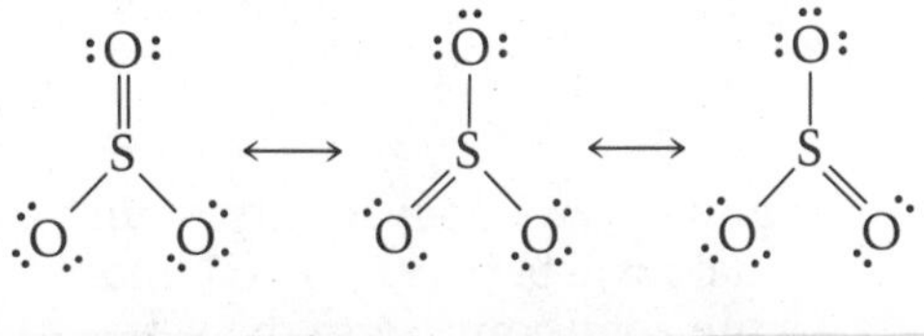

A double-headed arrow is used here to indicate resonance.

The electronic structure of some ions must also be represented by resonance. For example, the nitrate ion, $NO_3^-$, and the carbonate ion, $CO_3^{2-}$, have the same number of valence electrons as $SO_3$ and therefore have similar resonance structures.

Resonance is certainly not restricted to inorganic compounds. In proteins, for example, amino acids are linked together in long chains by "peptide bonds."

$$2H-\overset{\overset{\displaystyle H}{|}}{\underset{\underset{\displaystyle H}{|}}{N}}-\overset{\overset{\displaystyle H}{|}}{C}-\overset{:O:}{\overset{\|}{C}}-\overset{..}{O}-H \longrightarrow H-\overset{\overset{\displaystyle H}{|}}{N}-\overset{\overset{\displaystyle H}{|}}{C}-\overset{:O:}{\overset{\|}{C}}-\overset{..}{N}-\overset{\overset{\displaystyle H}{|}}{C}-\overset{:O:}{\overset{\|}{C}}-\overset{..}{O}-H + H_2O$$

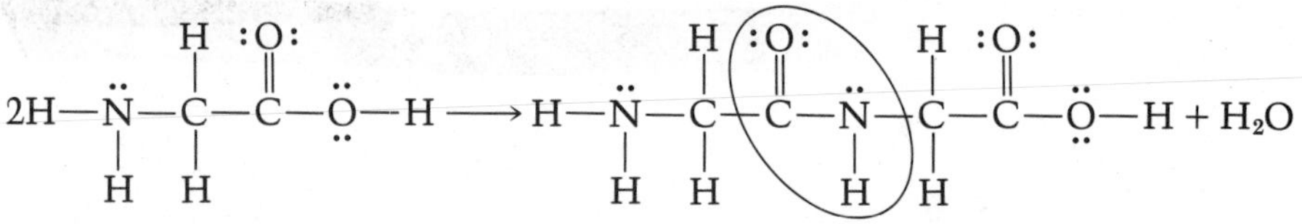

**glycine**
**(an amino acid)**

**peptide bond**
**(a peptide)**

There is evidence that the C—N bond in the peptide linkage actually lies somewhere between a single bond and a double bond. To explain this it is suggested that the peptide bond is a resonance hybrid of structures such as

$$-\overset{\overset{\displaystyle ..}{}}{\underset{\underset{\displaystyle H}{|}}{N}}-\overset{\overset{\displaystyle \overset{..}{O}:}{\|}}{C}- \quad \text{and} \quad -\underset{\underset{\displaystyle H}{|}}{N}=\overset{\overset{\displaystyle :\overset{..}{O}:}{|}}{C}-$$

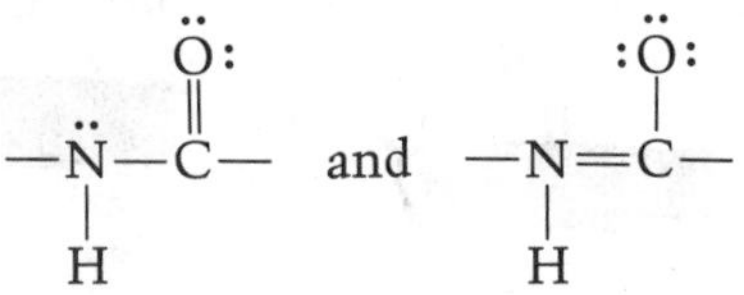

$$(1) \qquad\qquad\qquad (2)$$

Structure **2** is obtained by rearranging the electrons in structure **1** in this way:

$$-\overset{}{\underset{\underset{\displaystyle H}{|}}{N}}\overset{\overset{\displaystyle \overset{..}{O}:}{\|}}{C}-$$

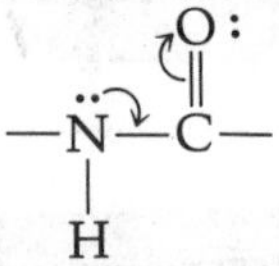

**4.6**

**COORDINATE**
**COVALENT**
**BONDS**

When a nitrogen atom combines with three hydrogen atoms to form the molecule $NH_3$, the N atom has completed its octet. We might expect, therefore, that the maximum number of covalent bonds that we would observe an N atom to form is three. There are instances, however, where N may have more than three covalent bonds. In the ammonium ion, $NH_4^+$, which is formed in the reaction

$$H:\overset{\overset{\displaystyle H}{..}}{\underset{\underset{\displaystyle H}{..}}{N}}: + H^+ \longrightarrow \left[ H:\overset{\overset{\displaystyle H}{..}}{\underset{\underset{\displaystyle H}{..}}{N}}:H \right]^+$$

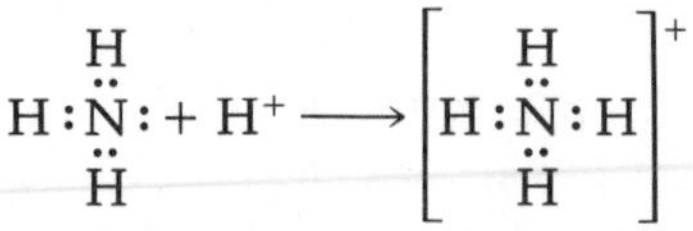

the nitrogen is covalently bound to four hydrogen atoms. When the additional bond between the $H^+$ and the N atom is created, both of the electrons in the bond come from the nitrogen. *This type of bond, where a pair of electrons from one atom is shared by two atoms, is called either a* **coordinate covalent bond**, *or a* **dative bond**. It is important that you remember that the coordinate covalent bond is really no different, once formed, than any other covalent bond and that our distinction is primarily aimed at keeping track of electrons; that is, it is "bookkeeping."

When Lewis structures are written using dashes to represent electron pairs, the coordinate covalent bond is sometimes indicated by means of an arrow pointing away from the atom supplying the electron pair. For example, the product of the reaction of boron trichloride, $BCl_3$, and ammonia, $NH_3$, is a substance known as an **addition compound** (because it is formed by the simple addition of two molecules).

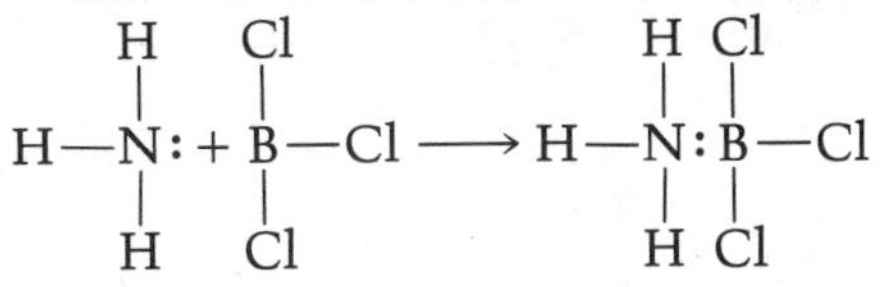

To show that the electron pair shared between the B and N originates on the nitrogen, the Lewis structure of this addition compound can be written

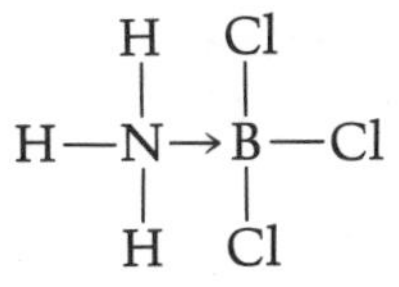

Using this type of notation we are tempted to write the structure of the $NH_4^+$ ion as

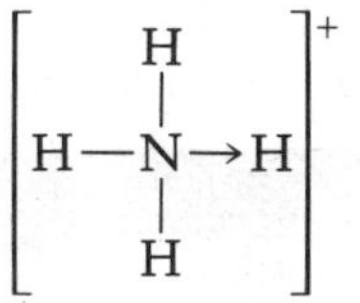

This gives the impression that one of the N—H bonds is different from the other three. It has been shown experimentally, however, that all four N—H bonds are identical. Therefore, to avoid conveying false impressions, the $NH_4^+$ ion is written simply as

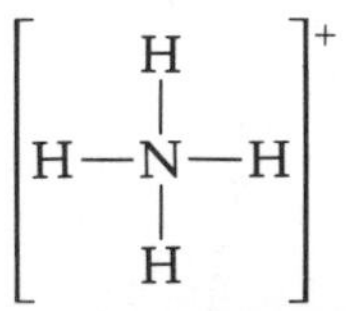

## 4.7 BOND ORDER AND SOME BOND PROPERTIES

The term **bond order** refers to *the number of covalent bonds that exist between a pair of atoms.* For example, the carbon–carbon bond order in acetylene, $C_2H_2$, is 3; in ethylene, $C_2H_4$, it is 2; and in ethane, $C_2H_6$, it is 1. Fractional bond orders are also possible, as in the case of $SO_2$. Each SO bond in the two resonance structures we draw for $SO_2$ is shown as a single bond in one structure and a double bond in the other. As we might expect, the bond order in sulfur dioxide is intermediate between 1 and 2.

The concept of bond order arises as a result of our description of the bonding in covalent molecules. Since we cannot view electrons directly, we have no firsthand way of knowing whether or not the theory is sound. As a result, we must examine properties that are directly related to bond order in an effort to test the validity of our theories. One of these properties is the **bond length,** *the distance between the nuclei of two bonded atoms.* As the bond order between a pair of atoms increases, additional electron density is placed between the two nuclei, causing them to be pulled together. Consequently, we expect bond length to decrease with increasing bond order, and we see that this is true in Table 4.4.

A second property related to bond order is the **bond energy,** *the energy required to break the bond to produce neutral fragments.* For a diatomic molecule such as $H_2$ this represents the process

$$H:H \ (g) \longrightarrow H \cdot (g) + H \cdot (g)$$

while in a molecule such as $C_2H_6$ the carbon–carbon bond energy represents the energy needed to cause the reaction

$$\begin{matrix} H & H \\ | & | \\ H-C-C-H \\ | & | \\ H & H \ (g) \end{matrix} \longrightarrow \begin{matrix} H \\ | \\ H-C \cdot \\ | \\ H \ (g) \end{matrix} + \begin{matrix} H \\ | \\ \cdot C-H \\ | \\ H \ (g) \end{matrix}$$

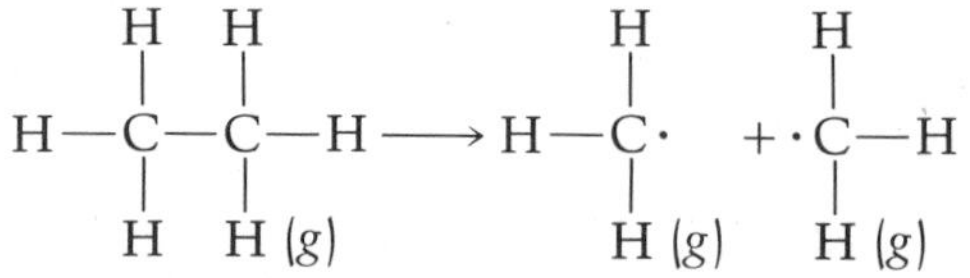

In general the bond energy increases with increasing bond order. As the electron density between the nuclei is increased, the nuclei are held together more tightly; therefore, more energy (work) must be supplied to pull the nuclei apart. The variation of this bond property with bond order is also shown in Table 4.4.

The last property that we shall examine that is related to the bond order is the **vibrational frequency** of the atoms joined by the bond. The atoms within a molecule are not stationary; they are in constant motion. This motion can be resolved into two basic types: vibration in which a pair of atoms move toward and away from each other along a line joining their centers, much as two balls connected by a spring (Figure 4.4*a*); and bending in which the angle between the three atoms alternately increases and decreases (Figure 4.4*b*). For simplicity we shall restrict our discussion to vibrational motion.

Table 4.4
*Variation of bond properties with bond order*

| Bond | Bond Order | Average Bond Length (Å) | Average Bond Energy (kcal/mole) | (kJ/mole) | Average Vibrational Frequency (Hz) |
|------|-----------|------------------------|--------------------------------|-----------|-----------------------------------|
| C—C | 1 | 1.54 | 88 | 370 | $3.0 \times 10^{13}$ |
| C=C | 2 | 1.37 | 167 | 699 | $4.9 \times 10^{13}$ |
| C≡C | 3 | 1.20 | 230 | 960 | $6.6 \times 10^{13}$ |
| C—O | 1 | 1.43 | 83 | 350 | $3.2 \times 10^{13}$ |
| C=O | 2 | 1.23 | 180 | 750 | $5.2 \times 10^{13}$ |
| C—N | 1 | 1.47 | 71 | 300 | $3.7 \times 10^{13}$ |
| C≡N | 3 | 1.16 | 175 | 730 | $6.8 \times 10^{13}$ |

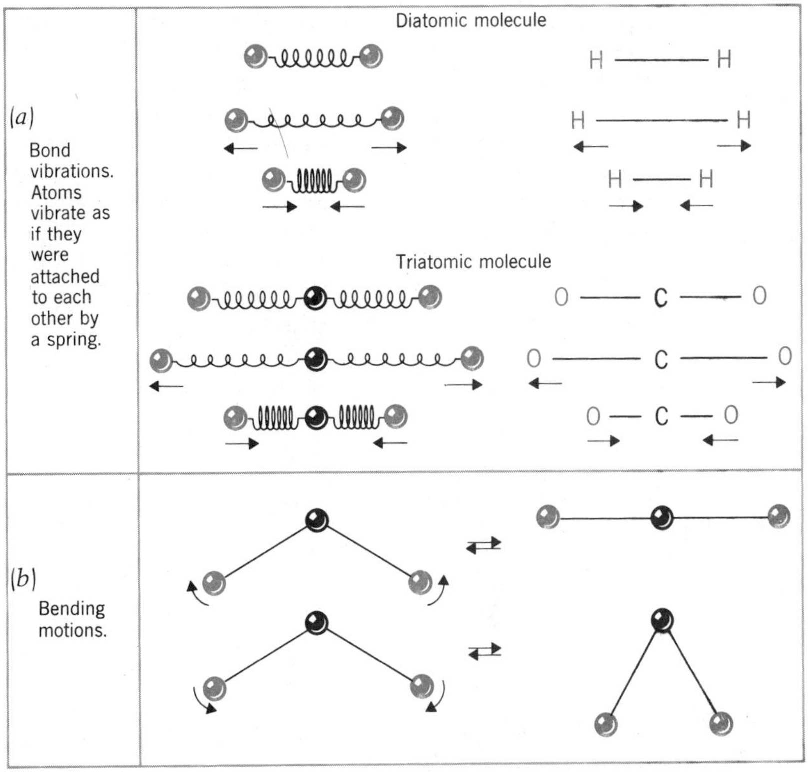


Figure 4.4
*The motion of atoms within molecules.*

There are two factors that affect the frequency of vibration (i.e., the number of vibrations per second). One of these is the masses of the atoms bonded together and the other is the bond order. *For a given pair of atoms, as the bond order increases, so does the vibrational frequency.* This is because increasing the bond order increases the attractive forces holding the nuclei together, in effect, stiffening the "spring" between the two atoms.

The measurement of the vibrational frequencies of bonds is really quite simple today. It happens that these vibrational frequencies are about the same as the frequency of infrared radiation and when infrared light is shined on a substance, radiation having the same frequencies as the vibrational frequencies of the bonds is absorbed. By observing which frequencies are selectively removed from the infrared spectrum, we can deduce these vibrational frequencies. The data recorded in the right column of Table 4.4 were obtained in this way.

In complex molecules there are many different vibrational modes available to the atoms, and many different frequencies are absorbed from the infrared "rainbow." The infrared absorption spectra of any but the most simple molecules are therefore quite complicated. Nevertheless, an experienced chemist often finds such absorption spectra extremely valuable as an aid in deducing molecular structure. In addition, each molecule, because of its unique structure, gives rise to its own characteristic absorption spectrum, which can be used to identify the compound and thus serves as a sort of "fingerprint." Examples of infrared absorption spectra of some drugs are shown in Figure 4.5.

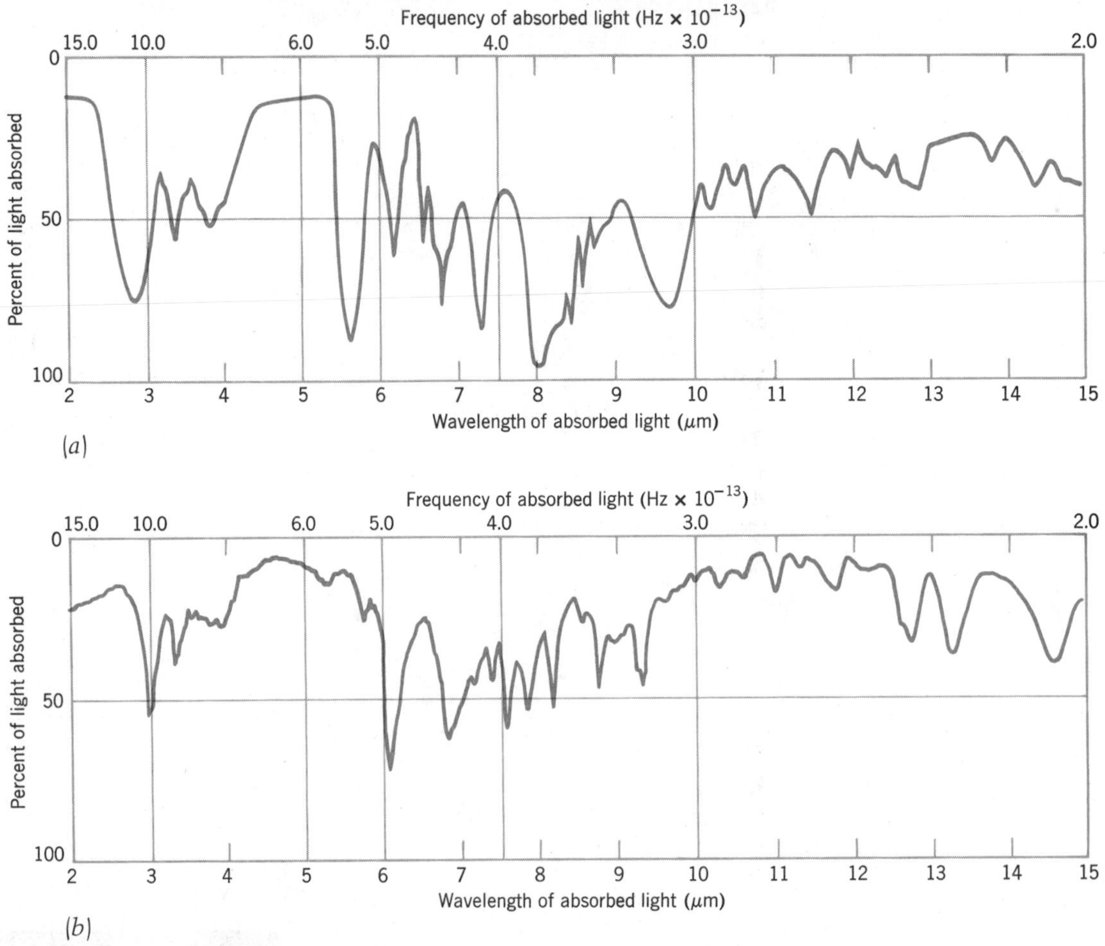


Figure 4.5
*Infrared absorption spectra of two drugs. (a) Heroin. (b) LSD.*

## 4.8 POLAR MOLECULES AND ELECTRONEGATIVITY

Generally, we expect atoms of different elements to have different abilities to attract electrons when they enter into a chemical bond. Consequently, we introduce the term, **electronegativity,** which is defined as *the attraction an atom has for electrons in a chemical bond.* It is important not to confuse this term with electron affinity, which is an energy term and refers to an isolated atom.

When two identical atoms combine, as for instance in $H_2$, both atoms have the same electronegativity. Since each atom is equally capable of attracting the electron pair in the bond, the pair will be shared equally and will spend, on the average, 50% of its time in the vicinity of each nucleus. Each H atom has around it, then, two electrons 50% of the time that when averaged out is the same as one electron all of the time. The "averaged" one electron will completely neutralize the positive charge on each nucleus, and each atom in $H_2$ carries a net charge of zero.

If the electronegativities of the two atoms in a bond are different, as is the case with HCl, the electron pair will spend more of its time around the more electronegative element (in this instance, Cl). For HCl, this means that the Cl atom acquires a slight negative charge and the H atom, a slight positive charge. We indicate this as

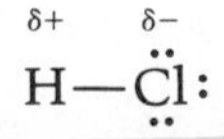

where $\delta+$ and $\delta-$ are meant to indicate partial positive and negative charges, respectively.

Equal positive and negative charges separated by a distance constitute a **dipole.** The HCl molecule, with its centers of positive and negative charge, is, therefore, a dipole and is said to be **polar.** It follows that any diatomic molecule formed between two atoms of different electronegativity will be polar.

A dipole is defined quantitatively by its **dipole moment,** the product of the charge on either end of the dipole times the distance between the charges. A very polar molecule is one with a large dipole moment, while a nonpolar molecule will have no dipole moment at all.

When three or more atoms are bonded together, it is possible to have a nonpolar molecule even though there are polar bonds present. Carbon dioxide is an often-used example. The $CO_2$ molecule is linear and may be represented as

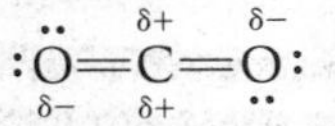

The overall dipole moment of a molecule arises as a sum of the individual bond dipoles within the molecule, which add together like vectors. In $CO_2$ these bond dipoles are oriented in opposite directions, and they exactly cancel each other.

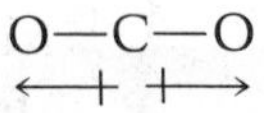

(An arrow with a plus sign on one end is used to represent the bond dipole.)

In the water molecule, which happens to have a bent or angular shape, the two bond dipoles do not cancel each other entirely, but rather are partially additive. As a result, the $H_2O$ molecule does have a net dipole moment (the heavy arrow below) and is polar.

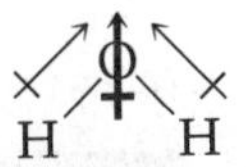

In general, a molecule $AX_n$ (where $n$ $X$'s are bound to $A$) will have polar bonds if $A$ and $X$ differ in electronegativity. In order for such a molecule not to be polar, the individual bond dipoles must completely cancel. Some structures that satisfy this condition are listed in Table 4.5. We will go into detail about predicting molecular structures in Chapter 17. For now, however, you can use a simple rule that a molecule will possess one of these symmetrical shapes when all of the valence-shell electron pairs of the central atom are involved in chemical bonds to atoms of the same element. In the dot structure of $CO_2$, for example, all of the valence electrons of carbon are used in bonds to oxygen. The molecule is linear and therefore nonpolar.

Table 4.5
*Symmetric molecular shapes*

| Formula | Example | Shape |
|---|---|---|
| $AX_2$ | $CO_2$ | Linear |
| $AX_3$ | $BCl_3$ | Planar triangle |
| $AX_4$ | $CH_4$ | Tetrahedral[a] |
| $AX_5$ | $PCl_5$ | Trigonal bipyramidal[a] |
| $AX_6$ | $SF_6$ | Octahedral[a] |

[a] See Appendix A.

With $H_2O$ and $NH_3$, however, the central atom in each case contains at least one pair of electrons not being used in a chemical bond. Neither of these has a shape described in Table 4.5, and both are therefore polar molecules.

We would like to have some quantitative measure of electronegativity in order to be in a position to make predictions concerning the polarity of bonds. One approach toward this, taken by R. S. Mulliken in 1934, uses the average of the ionization energy and electron affinity. A very electronegative element has a very high ionization energy, so that it is difficult to remove its electrons, and a very high electron affinity, so that a very stable species results when electrons are added. On the other hand, an element of low electronegativity will have a low ionization energy and low electron affinity so that it loses electrons readily and has little tendency to pick them up. Unfortunately, it is very difficult to measure the electron affinity of an element. Therefore, this method of assigning electronegativities is not universally applicable.

The most widely used scale of electronegativities was developed by Linus Pauling (a winner of two Nobel prizes and an advocate of using vitamin C to ward off the common cold). He observed that when atoms of different electronegativities are combined, their bonds are stronger than expected. Presumably two factors contribute to the bond strength. One of them is the covalent bonding between the atoms. The other is an additional binding produced by an attraction between the oppositely charged ends of the bond dipole. The extra bond strength, then, was attributed to this additional binding and Pauling used this concept to develop his table of electronegativities (Table 4.6). In constructing the table the most electronegative element, fluorine, was arbitrarily assigned an electronegativity of 4.0. The electronegativities of the other elements were obtained by comparison with the fluorine standard.

Electronegativity trends within the periodic table are worth noting. We see that the most electronegative elements are located in the upper right portion of the table; the least electronegative are found in the lower left. This is consistent with trends in ionization energy (IE) and electron affinity (EA), where we find the highest IE and EA in the upper right region of the periodic table and the lowest IE and EA in the lower left.

We have described electronegativity as an atom's ability to attract electrons in a bond. When the difference between the electronegativities of two combining atoms is very large, the electron pair will spend virtually 100% of its time about the more electronegative element. This is the same as saying that an electron is transferred from the atom of low electronegativity to that of high electronegativity. The result, of course, is an ionic bond. It is

Table 4.6
*The complete electronegativity scale*[a]

| | | | | | | | | | | | | | | | | |
|---|---|---|---|---|---|---|---|---|---|---|---|---|---|---|---|---|
| Li 1.0 | Be 1.5 | | | | | | H 2.1 | | | | | B 2.0 | C 2.5 | N 3.0 | O 3.5 | F 4.0 |
| Na 0.9 | Mg 1.2 | | | | | | | | | | | Al 1.5 | Si 1.8 | P 2.1 | S 2.5 | Cl 3.0 |
| K 0.8 | Ca 1.0 | Sc 1.3 | Ti 1.5 | V 1.6 | Cr 1.6 | Mn 1.5 | Fe 1.8 | Co 1.8 | Ni 1.8 | Cu 1.9 | Zn 1.6 | Ga 1.6 | Ge 1.8 | As 2.0 | Se 2.4 | Br 2.8 |
| Rb 0.8 | Sr 1.0 | Y 1.2 | Zr 1.4 | Nb 1.6 | Mo 1.8 | Tc 1.9 | Ru 2.2 | Rh 2.2 | Pd 2.2 | Ag 1.9 | Cd 1.7 | In 1.7 | Sn 1.8 | Sb 1.9 | Te 2.1 | I 2.5 |
| Cs 0.7 | Ba 0.9 | La–Lu 1.1–1.2 | Hf 1.3 | Ta 1.5 | W 1.7 | Re 1.9 | Os 2.2 | Ir 2.2 | Pt 2.2 | Au 2.4 | Hg 1.9 | Tl 1.8 | Pb 1.8 | Bi 1.9 | Po 2.0 | At 2.2 |
| Fr 0.7 | Ra 0.9 | Ac 1.1 | Th 1.3 | Pa 1.5 | U 1.7 | Np–No 1.3 | | | | | | | | | | |

[a] Reprinted from Linus Pauling, *The Nature of the Chemical Bond.* Copyright 1939 and 1940 by Cornell University. Third edition © 1960 by Cornell University. Used by permission of Cornell University Press. British Commonwealth rights by permission of Oxford University Press.

clear that the degree of ionic character in a bond can vary between zero (for example, $H_2$) to essentially 100% depending on the electronegativity difference between the bonded atoms. There is no sharp dividing line between ionic and covalent bonding. When the electronegativity difference is 1.7, the bond is about 50% ionic.

## 4.9 OXIDATION AND REDUCTION, OXIDATION NUMBERS

In the formation of the ionic bond between Li and F, we saw that an electron was transferred from Li to F to produce $Li^+$ and $F^-$. With HCl, we saw that a polar covalent bond was formed in which an electron was only partially transferred from the H to the Cl atom. Very many chemical reactions are of this type; that is, they involve some transfer of electronic charge from one atom to another. Because this is such a common and important process, we define terms that apply specifically to these changes. These are

**Oxidation**—a loss of electrons
**Reduction**—a gain of electrons

Thus in the formation of LiF, Li undergoes oxidation by losing an electron and F undergoes reduction by acquiring an electron. In a similar fashion, when the HCl molecule is formed by the reaction of hydrogen and chlorine, the H atom, by losing some electronic charge to the Cl atom, is oxidized while the Cl atom becomes reduced.

Lithium, in its reaction with fluorine, is said to be a **reducing agent** because it supplies the electron that the fluorine requires in order to be reduced—that is, it is the agent that has allowed reduction to occur. The fluorine, on the other hand, by accepting the electron from Li, permits oxidation to take place and thus is said to be an **oxidizing agent.** In a similar manner, we would consider hydrogen the reducing agent and chlorine the oxidizing agent when these two elements react to produce HCl. In general, *oxidizing agents acquire electrons and become reduced, while reducing agents lose electrons and become oxidized.*

Oxidation and reduction are always discussed together because, in any reaction, whenever one substance loses electrons, another substance picks them up. We know that this is true because we never observe electrons as a product of a chemical reaction, nor are electrons ever consumed when a chemical change occurs. Thus, oxidation is always accompanied by reduction. As a result, the simple abbreviated term, **redox**, is often used in discussing such reactions.

Chemists have devised a bookkeeping system using what are called **oxidation numbers** to keep track of electrons during chemical reactions. An oxidation number can be defined as *the charge that an atom would have if both of the electrons in each bond were assigned to the more electronegative element.* The term **oxidation state** is also used, interchangeably, with the term oxidation number.

In the substance LiF, because an electron has, in fact, been transferred to the F atom, the oxidation number assigned to $Li^+$ is $+1$. The oxidation number of fluorine in the $F^-$ ion is $-1$.

In the HCl molecule, because Cl is more electronegative than hydrogen, we assign an oxidation number of $+1$ to H and $-1$ to Cl, as if the electron pair were in the sole possession of the Cl atom.

In a nonpolar molecule such as $H_2$, where both atoms are the same and therefore have the same electronegativity, it is senseless to assign the electron pair to either atom since no electron transfer has occurred. In this case each H atom is assigned an oxidation number of zero.

To aid us in assigning oxidation numbers to the various atoms in a compound the following set of rules has been developed.

1. The oxidation number of any element in its elemental form is zero, regardless of the complexity of the molecule in which it occurs. Thus the atoms in Ne, $F_2$, $P_4$, and $S_8$ all have oxidation numbers of zero.
2. The oxidation number of any simple ion (one atom) is equal to the charge on the ion. The ions $Na^+$, $Al^{3+}$, and $S^{2-}$ have oxidation numbers of $+1$, $+3$, and $-2$, respectively.
3. The sum of all of the oxidation numbers of all of the atoms in a neutral compound is zero. For a complex ion (more than one atom), the algebraic sum of the oxidation numbers must be equal to the ion's charge.

In addition to these basic rules, the rules below also prove useful.

4. Fluorine, in compounds, always has an oxidation number of $-1$.
5. The elements in Group IA (except hydrogen) always have an oxidation number of $+1$ in compounds.
6. The elements in Group IIA always have an oxidation number of $+2$ in compounds.
7. A Group VIIA element has an oxidation number of $-1$ in **binary compounds** with metals (compounds that contain only two different elements). For example, Cl has an oxidation number of $-1$ in $FeCl_2$, $CrCl_3$, and NaCl.
8. Oxygen usually has an oxidation number of $-2$, with three exceptions:
   (a) In binary compounds with fluorine, where it must have a positive oxidation number.

(b) In peroxides (which contain an O—O bond, for example, $O_2^{-2}$ and $H_2O_2$), where it has an oxidation number of $-1$.
(c) In the superoxide ion, $O_2^-$, in which the oxidation number is $-\frac{1}{2}$.

9. Hydrogen has an oxidation number of $+1$ except in binary compounds with metals, where it has an oxidation number of $-1$.

Let us look at some examples of how these rules are applied.

**EXAMPLE 4.6** What are the oxidation numbers of all of the atoms in $KNO_3$ (potassium nitrate)?

**SOLUTION** We know that the sum of the oxidation numbers of all of the atoms must be equal to zero (the charge on $KNO_3$—rule 3).

| | | |
|---|---|---|
| K | $1 \times (+1) = +1$ | (rule 5) |
| N | $1 \times (x) = x$ | |
| O | $3 \times (-2) = -6$ | (rule 8) |
| *sum of oxidation numbers* | $= 0$ | (rule 3) |

$x$ must equal $+5$ in order for the sum to be zero.

**EXAMPLE 4.7** What is the oxidation number of sulfur in $Na_2S_4O_6$ (sodium tetrathionate)?

**SOLUTION** Again, the sum of the oxidation numbers must be zero.

| | | |
|---|---|---|
| Na | $2 \times (+1) =$ | $+2$ |
| S | $4 \times (x) =$ | $4x$ |
| O | $6 \times (-2) =$ | $-12$ |
| | sum $=$ | $0$ |

$$4x = +10 \quad \text{or} \quad x = \frac{+10}{4}$$

Note that the oxidation number of an atom need not be an integer.

**EXAMPLE 4.8** What is the oxidation number of Cr in the $Cr_2O_7^{2-}$ ion?

**SOLUTION** This time the sum of the oxidation numbers must equal $-2$ (rule 3).

| | | |
|---|---|---|
| Cr | $2 \times (x) =$ | $2x$ |
| O | $7 \times (-2) =$ | $-14$ |
| | sum $=$ | $-2$ |

Therefore,

$$2x = +12$$
$$x = +6$$

It is very important to keep in mind that oxidation numbers were developed simply for bookkeeping. Except for simple ions like $Na^+$ or $F^-$, the charges are fictitious. For example, the Cl in HCl does not carry a $-1$ charge as its oxidation number would imply; it carries only a partial negative charge (the actual charge is only about $-0.17$).

## 4.10 THE NAMING OF CHEMICAL COMPOUNDS

Up to now we have been using chemical names simply as labels. When chemists speak to each other, they find it necessary to convey information about chemical substances by giving them names. Ideally, these should transmit a maximum amount of information about the composition and structure of the compounds being discussed. Toward this end a systematic procedure for naming chemical compounds has been developed and is presented, in a somewhat abbreviated form, in Appendix B. Some substances are so familiar to chemists that their common (or *trivial*) names are always used, for example, water ($H_2O$) and ammonia ($NH_3$). Trivial names are often used for extremely complex compounds where the names derived on a systematic basis are very long, complex, and cumbersome.

## 4.11 OTHER BINDING FORCES

The ionic and covalent bond represent very strong interactions. In molecular substances it is the covalent bonds that determine chemical reactivity and thereby control the chemical properties of these substances. In addition to these bonds within molecules there are other, weaker attractive forces that exist between neutral atoms and molecules. These weaker intermolecular forces are responsible for controlling the physical properties of molecular compounds. We shall now take a brief look at the origin of these forces, and in later chapters we shall examine in greater detail how they may be used to explain a variety of physical properties.

**DIPOLE INTERACTIONS.** When two polar molecules approach one another, they tend to line up so that the positive end of one dipole is directed toward the negative end of the other. When this occurs, there is an electrostatic attraction between the two dipoles. This is a much weaker attraction than between oppositely charged ions for several reasons: first, there are only partial charges on the ends of the dipoles; second, because atoms and molecules are in constant motion,[3] collisions prevent the dipoles from becoming perfectly aligned; and third, there is a repulsive force between the end of the dipoles that carry like charges (see Figure 4.6). Dipole interactions are only about 1% as strong as covalent and ionic bonds. They are responsible for controlling such properties as the boiling points and melting points of polar substances, which tend to be higher than those of nonpolar substances of similar molecular weights.

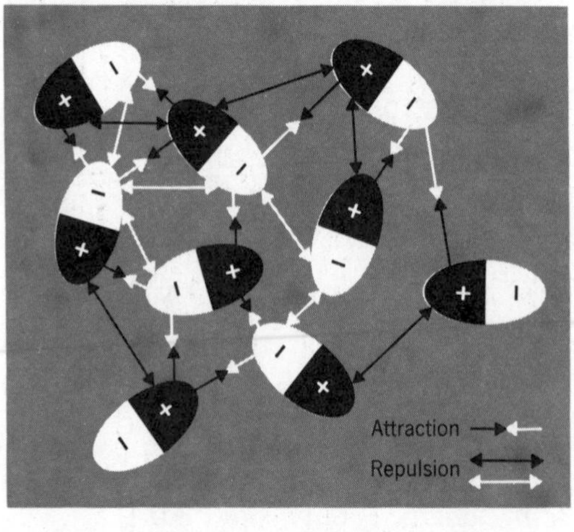


Figure 4.6
*Electrostatic interactions between dipoles.*

[3] This is discussed more completely in Chapter 6.

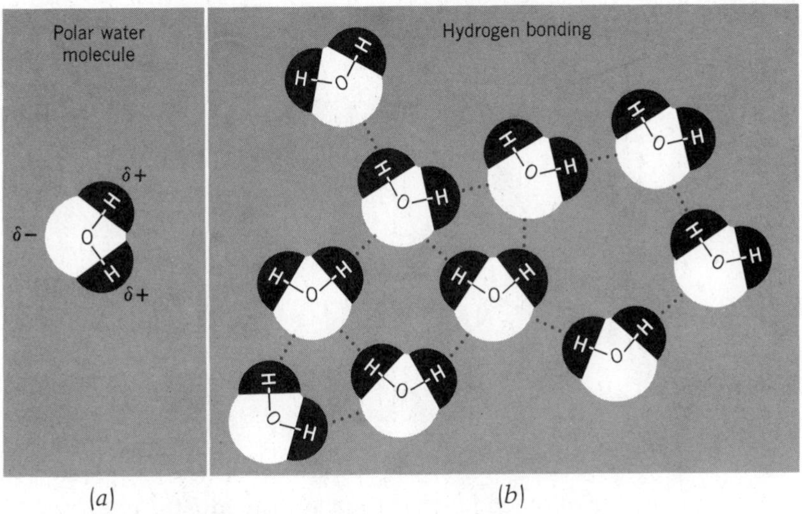


Figure 4.7
*Hydrogen bonding in a substance such as* $H_2O$. *(a) Polar water molecule. (b) Hydrogen bonding.*

**HYDROGEN BONDING.** A particularly strong dipole–dipole attraction occurs when hydrogen is covalently bound to a very small electronegative element such as fluorine, oxygen, or nitrogen. In these instances very polar molecules result in which the extremely small hydrogen atom carries a substantial positive charge. Because the positive end of this dipole can approach close to the negative end of a neighboring dipole, the force of attraction between the two is quite large (Figure 4.7). This special kind of dipole interaction is called a hydrogen bond and is about 5 to 10% as strong as an ordinary covalent bond.

Figure 4.8
*The crystal structure of ice. Large circles are oxygen. Small circles are hydrogen. Only half of the small circles are occupied by H atoms (see text). From J. M. Williams,* Journal of Chemical Education, *Vol. 52, p. 210, April 1975, used by permission.*

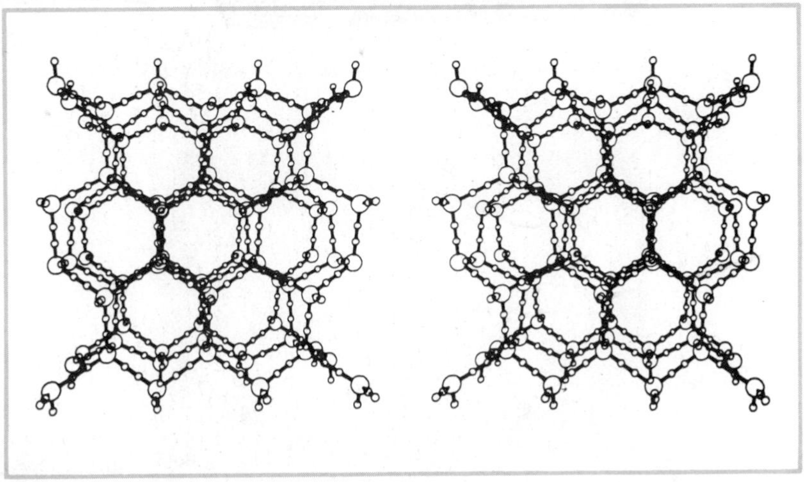

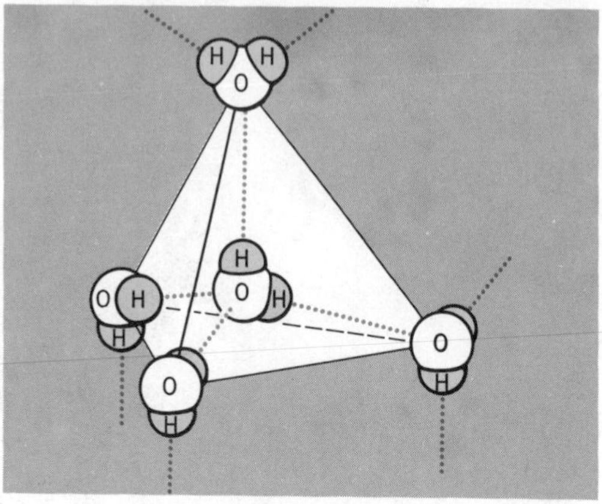


Figure 4.9
*Hydrogen bonding (dotted lines) between water molecules in ice.*

Hydrogen bonds are one of the most important types of weak attractive forces. For instance, hydrogen bonds are responsible for water being a liquid at room temperature, rather than a gas. It is also responsible for controlling the orientation of water molecules in ice, which gives rise to a very "open" kind of crystal structure. The structure of ice is depicted in three dimensions in Figure 4.8. In this illustration each oxygen atom of the water molecules is surrounded by four others at the corners of a tetrahedron. Hydrogen atoms (small circles) are located along the lines joining the oxygen atoms. In this illustration there are twice as many small circles as there actually are H atoms in the structure because the arrangement of H atoms is random and their locations cannot be specified exactly. In Figure 4.9 we have a close-up of the arrangement of H atoms around an O atom. Two H atoms are close to the O atom (they are covalently bound to it) while the other two are further away (they are held by hydrogen bonding to the central O atom, but are covalently bound to the other O atom). This open structure causes ice to be less dense than liquid water. That is why icebergs float (much to the distress of the captain of the *Titanic*).

**LONDON FORCES.** Even uncombined atoms and nonpolar molecules experience weak attractions. These interactions, called London forces (after the physicist, Fritz London), are considered to arise from the random motion of the electrons in an atom or molecule.

When electrons move about an atom, there is a chance that at some instant more electrons will be on one side of the nucleus than on the other and

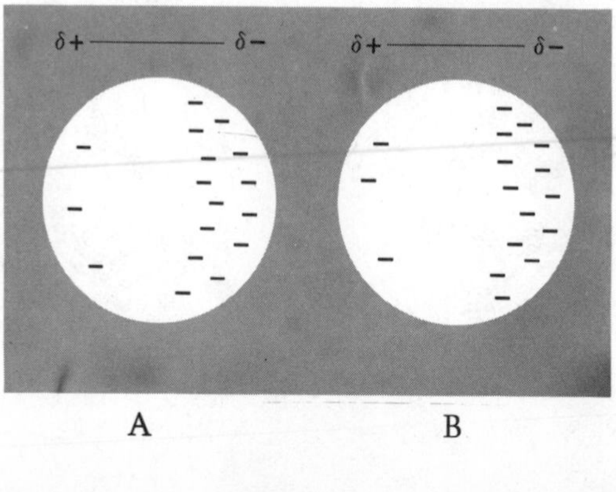


Figure 4.10
*London forces. The instantaneous dipole in atom* A *induces a dipole in atom* B.

for this brief moment a dipole exists because of the imbalance in charge (Figure 4.10). The positive end of the instantaneous dipole will attract electrons in a neighboring atom, thereby inducing another instantaneous dipole. These dipoles attract one another briefly before they disappear. Although attractions between the instantaneous dipoles may be fairly strong, their duration is very short, and the average attraction over a period of time is generally very small.

The overall strengths of London forces depend on a number of factors, including molecular shape and size. This will be discussed further in Chapters 6 and 8.

## INDEX TO QUESTIONS AND PROBLEMS (problem numbers in italics)



## REVIEW QUESTIONS

**4.1** Write Lewis symbols for Se, Br, Al, K, Ba, Ge, and P.

**4.2** What is the purpose of using Lewis symbols?

**4.3** Draw Lewis structures for the ionic compounds, BaO, $Na_2O$, KF, $MgBr_2$.

**4.4** Why is KF (*s*) more stable than K (*s*) and $F_2$ (*g*)?

**4.5** Indicate how the electron configuration changes for each atom when the following ionic compounds are formed from the elements: $K_2O$, $Mg_3N_2$, $Na_2S$, $BaBr_2$.

**4.6** Give the electron configuration of each of the following ions: $Ba^{2+}$, $Se^{2-}$, $Al^{3+}$, $Li^+$, $Br^-$, $Fe^{2+}$, $Cu^+$, $Ni^{2+}$.

**4.7** What is the octet rule? Show that it is not obeyed in $SF_4$.

**4.8** Why do elements from period 2 never exceed an octet in their valence shells?

**4.9** What is a pseudonoble gas electron configuration? List some ions that possess this kind of electronic structure.

**4.10** Write formulas for compounds composed of the following pairs of ions: (a) $Na^+$, $CO_3^{2-}$; (b) $Ca^{2+}$, $ClO_3^-$; (c) $Sr^{2+}$, $S^{2-}$; (d) $Cr^{3+}$, $Cl^-$; (e) $Ti^{4+}$, $ClO_4^-$.

**4.11** Write formulas for compounds composed of the following pairs of ions:
(a) $Fe^{3+}$, $HPO_4^{2-}$
(b) $K^+$, $N^{3-}$
(c) $Ni^{2+}$, $NO_3^-$
(d) $Cu^{2+}$, $C_2H_3O_2^-$
(e) $Ba^{2+}$, $SO_3^{2-}$

**4.12** Construct a Born-Haber cycle for the formation of KBr (*s*) from K (*s*) and $Br_2$ (*l*). Indicate which steps are endothermic and which are exothermic.

**4.13** What is the name of the energy term associated with the reaction, $Na^+$ (*g*) + $Cl^-$ (*g*) → NaCl (*s*)? Is the process indicated in this chemical equation endothermic or exothermic?

**4.14** Use Lewis symbols to diagram the reaction for the formation of the covalently bonded molecules $NH_3$, $H_2O$, and HF.

**4.15** Draw Lewis structures for the molecules $PCl_3$, $SiH_4$, $BCl_3$, $H_2S$, $C_3H_8$, and CO.

**4.16** Draw Lewis structures for the molecules $Cl_2$, $SO_2$, $OF_2$, $SnH_4$, $C_2H_4$, and $SCl_2$.

**4.17** Draw Lewis structures for the ions $Cl^-$, $S^{2-}$, $ClO^-$, $ClO_4^-$, $SO_3^{2-}$, and $PO_4^{3-}$.

**4.18** Draw Lewis structures for the ions $NO_3^-$, $NO^+$, $NO_2^-$, and $CO_3^{2-}$.

**4.19** Draw Lewis structures for $SeF_6$, $SeF_4$, $ICl_3$, $AsCl_5$, $ICl_2^-$, $ICl_4^-$, and $XeF_4$.

**4.20** Which of the following compounds do not obey the octet rule: $ClF_3$, $OF_2$, $SF_4$, $SO_2$, $IF_7$, $NO_2$, $BCl_3$?

**4.21** Draw the resonance structures of $SO_3$, $NO_3^-$ and $CO_3^{2-}$; $SO_2$ and $NO_2^-$.

**4.22** Draw the resonance structures of $HNO_3$, $SeO_2$, $SeO_3$, and $N_2O_4$.

Structures: 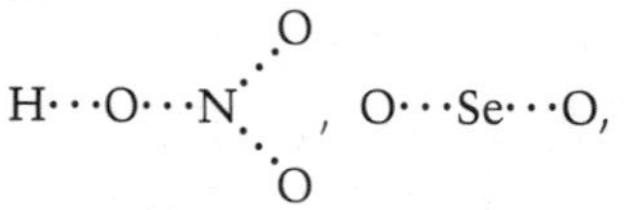

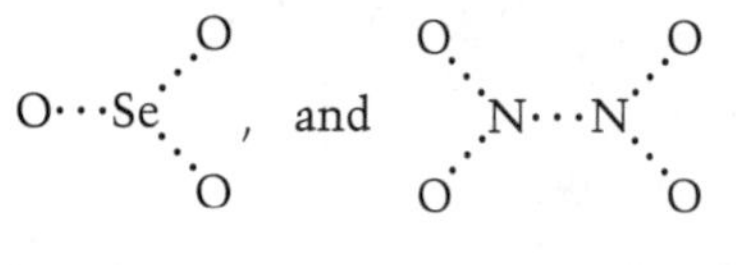

**4.23** Draw resonance structures for $C_2O_4^{2-}$, $CH_3COO^-$, and $N_3^-$.

Structures: 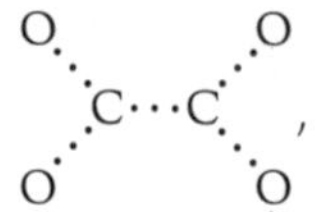

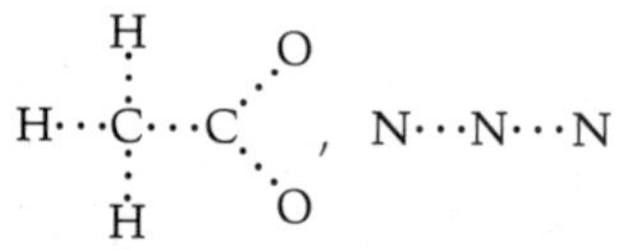

**4.24** What is a resonance hybrid? Why is the concept of resonance used?

**4.25** What is a coordinate covalent bond? How does it differ from other covalent bonds?

**4.26** Which of the following have Lewis structures that would be considered to contain coordinate covalent bonds: $CCl_4$, $HNO_3$ (see Question 4.22), $NO_3^-$, $BH_4^-$, $OF_2$?

**4.27** In Section 4.5 we noted that there was evidence that the C—N bond in the peptide linkage possessed some double-bond character. What kind of experimental evidence would be expected to confirm this?

**4.28** The C—C bond length in a series of compounds was found to be as follows: compound 1, 1.54 Å; compound 2, 1.37 Å; compound 3, 1.46 Å; and compound 4, 1.40 Å. Arrange these in order of increasing C—C bond order. How would you expect the C—C bond energies to vary?

**4.29** What would you expect the value of the bond order in $SO_3$ to be?

**4.30** For the molecules $SO_2$ and $SO_3$, compare the SO bond energies, bond lengths, and S—O vibrational frequencies.

**4.31** How would you expect the N—O bond distance in $NO_2^-$ to compare with that in $NO_3^-$?

**4.32** Compare the C—O bond properties in the following:

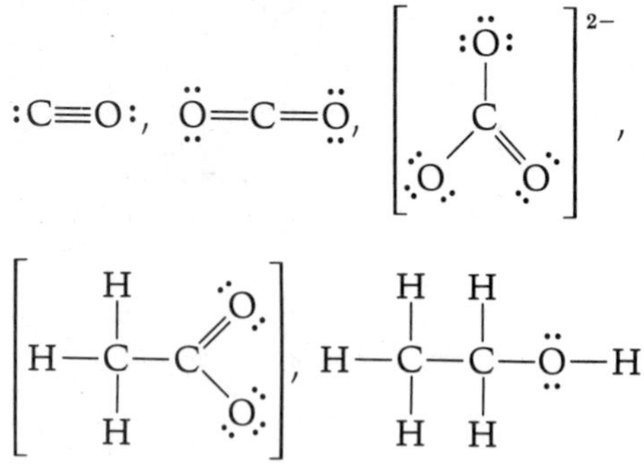

**4.33** What kind of information is provided by an infrared absorption spectrum of a molecule?

**4.34** Define polar, dipole, and dipole moment.

**4.35** What is the difference between electronegativity and electron affinity?

**4.36** What trends in electronegativity occur in the periodic table? What correlation, if any, exists between ionization energy and electronegativity?

**4.37** Which of the following contain bonds that are predominantly ionic: $AlCl_3$, $MgO$, $Al_2O_3$, $NF_3$, $CsF$, $FeCl_2$, $SO_2$, $Ca_3P_2$, $Mg_2Si$?

**4.38** Which of the following have bonds that are predominantly covalent: $NH_3$, $MnF_2$, $BCl_3$, $MgCl_2$, $BeI_2$, $NaH$?

**4.39** Arrange the following compounds in order of increasing ionic character of their bonds: $SO_2$, $H_2S$, $SF_2$, $OF_2$, $ClF_3$, $H_2Se$, $F_2$.

**4.40** Define oxidation, reduction, oxidation state, oxidizing agent, and reducing agent.

**4.41** Assign oxidation numbers to each atom in $KClO_2$, $BaMnO_4$, $Fe_3O_4$, $O_2F_2$, $IF_5$, $HOCl$, $CaSO_4$, $Cr_2(SO_4)_3$, $Na_2S_2O_8$ (contains one O—O bond), $O_3$, and $Hg_2Cl_2$.

**4.42** Many biological processes involve oxidation and reduction. For example, ethyl alcohol (grain alcohol) is metabolized in a series of oxidation steps that involve the following carbon-containing compounds.

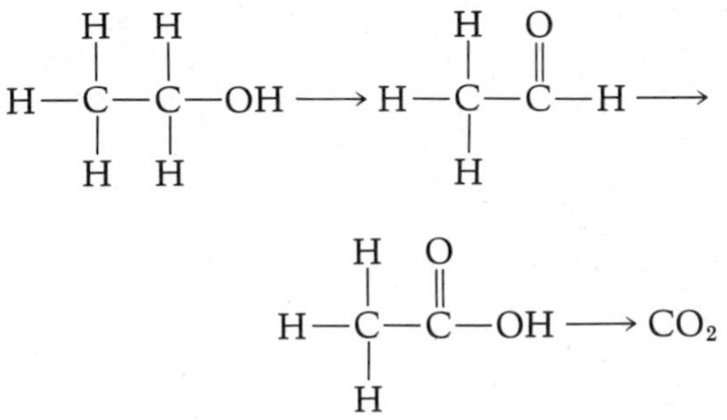

Follow the oxidation number of carbon as your body oxidizes the alcohol.

**4.43** Assign oxidation numbers to each atom in $H_2SO_4$, $CBr_4$, $OF_2$, $H_2O_2$ (hydrogen peroxide), $CrCl_3$, $Mn_2O_7$, $KMnO_4$, $H_2C_2O_4$, $KClO_3$, and $LiNO_3$.

**4.44** Identify the following changes as either oxidation or reduction:
(a) $MnO_2$ to $MnO_4^-$
(b) $BiO_3^-$ to $Bi^{3+}$
(c) $SO_2$ to $SO_3$
(d) $OCl^-$ to $ClO_3^-$
(e) $N_2O_4$ to $N_2O$

**4.45** Name the following:
$NaBr$
$CaO$
$FeCl_3$
$CuCO_3$
$CBr_4$
$P_4O_6$
$AsCl_5$
$Mn(HCO_3)_2$
$NaMnO_4$
$O_2F_2$
(Refer to Appendix B)

**4.46** Write chemical formulas for the following compounds.
Aluminum nitrate
Iron(II) sulfate
Ammonium dihydrogen phosphate
Iodine pentafluoride
Phosphorus(III) chloride
Dinitrogen tetroxide
Potassium permanganate
Magnesium hydroxide
Hydrogen selenide
Sodium hydride

**4.47** Name the following:
$Cr_2O_3$
$Mg(H_2PO_4)_2$
$Cu(NO_3)_2$
$CaSO_4$
$Ba(OH)_2$
$AlPO_4$
$Mg_3N_2$
$PbC_2O_4$
$(NH_4)_2CO_3$
$K_2Cr_2O_7$

**4.48** Write formulas for the following:
Titanium(IV) oxide
Silicon tetrachloride
Calcium selenide
Potassium nitrate
Aluminum sulfate
Nickel(II) bicarbonate
Sodium bisulfate
Ammonium dichromate
Calcium acetate
Strontium hydroxide

**4.49** What are dipole–dipole attractions, hydrogen bonding, and London forces?

**4.50.** How does hydrogen bonding affect the properties of water?

**4.51** Indicate the type of intermolecular attractive force found in the following substances: $HCl$, $Ar$, $CH_4$ (a tetrahedral molecule), $HF$, $NO$, $CO_2$ (a linear molecule), $H_2S$ (a nonlinear molecule), $H_2O$.

## REVIEW PROBLEMS

**4.52** Use a Born-Haber cycle to show that the reaction

$$K\ (s) + \tfrac{1}{2}Cl_2\ (g) \longrightarrow KCl\ (s)$$

is exothermic. The following energies are known: For $K\ (s) \rightarrow K\ (g)$, 21.5 kcal; for $\tfrac{1}{2}Cl_2\ (g) \rightarrow Cl\ (g)$, 28.5 kcal; for $K\ (g) \rightarrow K^+\ (g)$, 100.1 kcal; for $Cl\ (g) \rightarrow Cl^-\ (g)$, −83.3 kcal; for $K^+\ (g) + Cl^-\ (g) \rightarrow KCl\ (s)$, −168.3 kcal.

**4.53** Below are calculated and experimental bond energies for the hydrogen halides. Use these data to show that the electronegativities of the halogens decrease from F to I.

| | Calculated (kcal/mole) | Experimental (kcal/mole) |
|---|---|---|
| HF | 70.5 | 135 |
| HCl | 80.5 | 103 |
| HBr | 75 | 87 |
| HI | 70 | 71 |

**4.54** Below are the ionization energies (IE) and electron affinities (EA) of the halogens. Show that the electronegativities of the halogens decrease from F to I.

| | IE( kcal/mole) | EA (kcal/mole) |
|---|---|---|
| F | 402 | −79.5 |
| Cl | 300 | −83.3 |
| Br | 273 | −77.6 |
| I | 241 | −70.9 |

*4.55 Given the following data, calculate the lattice energy of $CaCl_2$ in kilojoules per mole. Energy needed to vaporize 1 mole of Ca (*s*) = 46.0 kcal; first ionization energy of Ca = 140.9 kcal/mole; second ionization energy of Ca = 273.8 kcal/mole; electron affinity of Cl = 83.6 kcal/mole; bond energy of $Cl_2$ = 57.0 kcal/mole of Cl—Cl bonds; energy released by the reaction, $Ca\ (s) + Cl_2\ (g) \rightarrow CaCl_2\ (s)$, 190.0 kcal/mole of $CaCl_2$ (*s*) formed.

*4.56 Given the following data, calculate the electron affinity of Br. The energy released by the reaction, $Na\ (s) + \frac{1}{2}Br_2\ (l) \rightarrow NaBr\ (s)$, is 86.0 kcal. The energy needed to vaporize 1 mole of Na (*s*) is 26.0 kcal. The energy needed to vaporize 1 mole of $Br_2$ (*l*) is 7.3 kcal. The ionization energy of Na (*g*) is 118.5 kcal/mole. The bond energy of $Br_2$ is 46.0 kcal/mole of Br—Br bonds. The lattice energy of NaBr is 175.5 kcal/mole.

*4.57 Hydrogen is more electronegative than any of the Group IA elements. Based on this statement and the data below, show that the electronegativities of the alkali metals decrease from Li to Rb.

| | Calculated Bond Energy (kcal/mole) | Experimental Bond Energy (kcal/mole) |
|---|---|---|
| Li—H | 65.1 | 56.9 |
| Na—H | 61.1 | 48 |
| K—H | 58.4 | 44 |
| Rb—H | 57.8 | 39 |

# 5 CHEMICAL REACTIONS IN AQUEOUS SOLUTION

As we might expect, in order for a chemical reaction to occur between two substances, the ions or molecules comprising the reactants must come into contact with one another. For this reason the speed at which a reaction takes place depends on how freely the reacting species are able to intermingle. For instance, if crystals of NaCl and $AgNO_3$ are mixed together, no noticeable chemical changes are observed. However, if the NaCl and $AgNO_3$ are first dissolved in water, and then their solutions mixed, a white solid, AgCl, is produced. Here, the formation of silver chloride requires that silver ions and chloride ions meet. When the two solids are mixed, this does not occur, except at the surfaces where the crystals touch one another. Because of the homogeneous nature of solutions, however, dissolved substances are intimately mixed at the molecular or ionic level, and chemical changes can occur rapidly. For this reason, chemists routinely use solutions to carry out chemical reactions.

Water is one of the most abundant chemicals in nature and serves as a good solvent for many substances, both ionic and molecular. The chemist's preoccupation with reactions in aqueous systems stems from the general availability of water as a solvent and, particularly in recent times, the recognition of the importance of water as a medium in which biochemical reactions take place. In this chapter we will discuss various types of chemical reactions that occur in aqueous solution and learn how the quantitative principles developed in Chapter 2 can be applied to such reactions.

## 5.1 SOLUTION TERMINOLOGY

There are certain terms that apply to all kinds of solutions and that should be understood before we proceed further. The words **solvent** and **solute** are two of these. The general practice is to refer to the substance present in greatest proportion in a solution as the solvent, with all of the other substances in the solution considered solutes. In solutions that contain water, however, the solvent is nearly always considered to be the water, even when it is present in relatively small amounts. For example, a mixture of 96% $H_2SO_4$ and 4% $H_2O$ by weight is called "concentrated sulfuric acid," which implies that a large quantity of sulfuric acid is *dissolved* in a small amount of water; that is, $H_2O$ is taken to be the solvent and $H_2SO_4$ the solute.

It is often necessary to express the proportions of solute and solvent in a solution. This is done by specifying the concentration of the solute in the mixture. Concentration can be stated quantitatively in a variety of ways, as we shall see. The terms **concentrated** and **dilute** are used when we wish to speak, in qualitative terms, of the relative proportions of solvent and solute. In a concentrated solution there is a relatively large amount of solute present in the solvent; a dilute solution, on the other hand, possesses only a small quantity of solute. These two terms have meaning only in relationship to one another; they do not imply any specific quantities of solute in solvent. For example, concentrated sulfuric acid contains, as we said above, 96% $H_2SO_4$ and 4% $H_2O$. By comparison, a solution containing 20% $H_2SO_4$ would be dilute. This latter solution would be considered concentrated in comparison to a 5% $H_2SO_4$ solution.

In most cases there is a limit to the amount of solute that will dissolve in a fixed quantity of solvent at any particular temperature. For example, if we add sodium chloride to 100 ml of water at 0°C, only 35.7 g of the salt will dissolve, regardless of the total amount that we place into the water. A solution, such as this, which contains as much dissolved solute as it can hold while in contact with excess solute, is said to be **saturated.** The **solubility** of the solute is taken to be the amount required to achieve a saturated solution. Thus the solubility of NaCl in water at 0°C is 35.7 g/100 ml; at 100°C its solubility is 39.1 g/100 ml. Solutions that contain less solute than required for saturation are said to be **unsaturated.**

The terms saturated and unsaturated are in no way directly related to the terms concentrated and dilute. A saturated solution of silver chloride, for example, contains only 0.000089 g AgCl per 100 ml of water and is certainly considered dilute. On the other hand, it would take 500 g of lithium chlorate, $LiClO_3$, to form a saturated solution in 100 ml of water. A solution containing 400 g of $LiClO_3$ in 100 ml of water is unsaturated but, nevertheless, is quite concentrated. Thus a saturated solution can be dilute and an unsaturated solution, concentrated.

Finally, there are some substances, such as sodium acetate, which frequently form **supersaturated** solutions—solutions that contain more solute than ordinarily required for saturation. Sodium acetate is soluble to the extent of 119 g/100 ml of water at 0°C, and becomes more soluble at higher temperatures. If an unsaturated hot solution containing more than 119 g of sodium acetate per 100 ml is cooled slowly to 0°C, the excess solute remains dissolved and the solution becomes supersaturated. Such solutions are unstable; if a small crystal of the solute is added, additional solute crystallizes on this "seed" crystal until the concentration drops to the point of saturation (see Figure 5.1).

Figure 5.1
*Supersaturation. (a) Supersaturated solution. (b) Introduction of a seed crystal. (c) Excess solute crystallizes on the seed.*

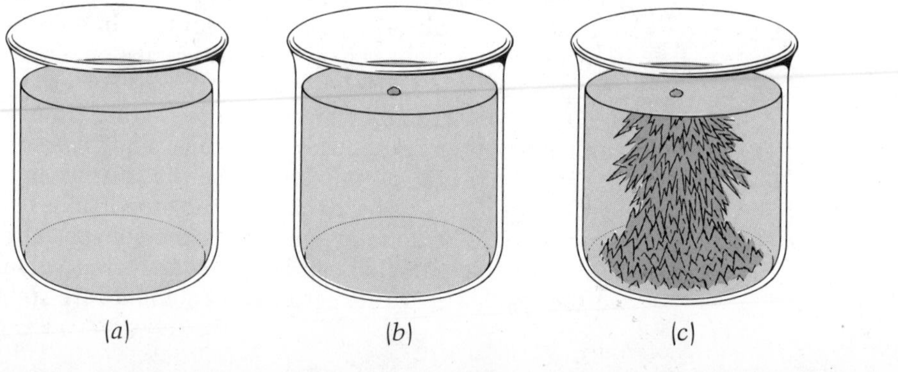

## 5.2 ELECTROLYTES

As mentioned in the introduction to this chapter, water is generally a good solvent for ionic compounds. In the solid state these substances are composed of positive and negative ions held together in a rigid framework by electrostatic forces. When they dissolve in water they break apart, or **dissociate,** to produce ions that are more or less free to roam about in the solution. The presence of ions imparts to the water the ability to conduct electricity, which we can demonstrate using an apparatus such as that shown in Figure 5.2. When electrical contact is made across the two electrodes, an electric current can flow and the light bulb will light. If we immerse these electrodes in pure water, no conductivity is observed (i.e., the bulb will not light) because water is a very poor conductor of electricity. However, if a typical ionic solid such as NaCl is added to the water, the bulb will begin to burn brightly as soon as the NaCl begins to dissolve. Substances like NaCl, which dissociate in solution to produce ions, yield solutions that conduct electricity and are called **electrolytes.**

The dissociation of NaCl that occurs when the solid is dissolved can be represented by the equation

$$NaCl\,(s) \longrightarrow Na^+\,(aq) + Cl^-\,(aq)$$

where the symbols *s* and *aq* in parentheses denote that the solid (*s*) produces ions in aqueous (*aq*) solution. As the solid dissolves, the ions become surrounded by water molecules and are said to be **hydrated.** We shall examine the solution process in more detail in Chapter 9.

The production of ions in solution is not limited to ionic compounds. There are many covalent substances that, when they dissolve in water, react with the solvent to produce ions and therefore yield conducting solutions. Hydrogen chloride is a typical example. When HCl gas is dissolved in water, the following reaction takes place:

$$HCl + H_2O \longrightarrow H_3O^+\,(aq) + Cl^-\,(aq)$$

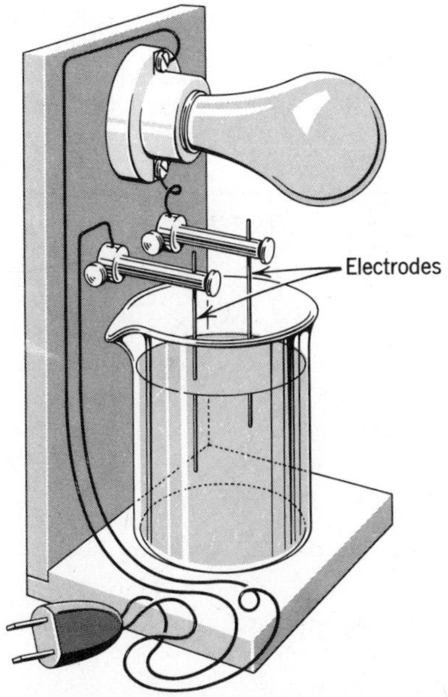


Figure 5.2
*Conductivity apparatus.*

This reaction occurs by the transfer of a proton from the HCl molecule to the water molecule to produce a **hydronium ion,** $H_3O^+$, and a chloride ion.

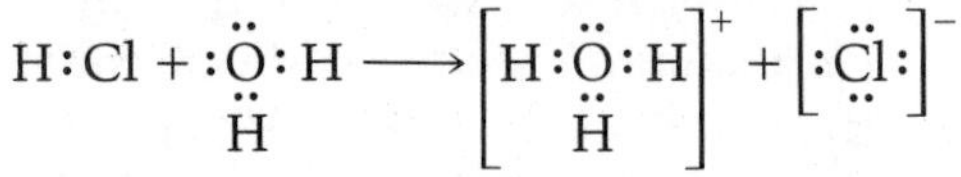

Thus, even though hydrogen chloride exists by itself as discrete molecules of HCl (liquid HCl does not conduct electricity), when it dissolves in water it produces ions and becomes an electrolyte.

The hydronium ion might be thought of as a proton ($H^+$) that has associated itself with a water molecule (we can think of the hydronium ion as a *hydrated proton*), and we will see shortly that when the hydronium ion reacts it loses a proton to regenerate the molecule of $H_2O$. In chemical reactions, therefore, the $H_2O$ of the hydronium ion merely serves as a carrier for the $H^+$ ion and for that reason the $H_3O^+$ ion is very often written as $H^+$. Thus, the dissociation that occurs when HCl is dissolved in water is often written as simply

$$HCl \longrightarrow H^+ + Cl^-$$

Even though we write $H^+$, always remember that there is at least one, and probably several more, $H_2O$ molecules associated with the proton in aqueous solution.[1]

The two examples of electrolytes that we have discussed above, NaCl and HCl, are essentially completely dissociated in aqueous solution; that is, 1 mole of NaCl gives 1 mole of $Na^+$ and 1 mole of $Cl^-$. Substances such as NaCl and HCl that, for all practical purposes, are completely dissociated in aqueous solution, are said to be **strong electrolytes.** There are many compounds, such as acetic acid ($HC_2H_3O_2$, found in vinegar), for example, that

Table 5.1
*Some weak electrolytes*

| Substance | Dissociation Reaction | Percent Dissociation of Solution Containing 1.00 Mole of Solute per Liter of Solution |
|---|---|---|
| Water | $H_2O + H_2O \longrightarrow H_3O^+ + OH^-$ | $1.8 \times 10^{-7}$ (55.5 moles of $H_2O$ per liter) |
| Acetic acid | $HC_2H_3O_2 + H_2O \longrightarrow H_3O^+ + C_2H_3O_2^-$ | 0.42 |
| Ammonia | $NH_3 + H_2O \longrightarrow NH_4^+ + OH^-$ | 0.42 |
| Hydrogen cyanide | $HCN + H_2O \longrightarrow H_3O^+ + CN^-$ | $2.0 \times 10^{-3}$ |
| Mercury(II) chloride | $HgCl_2 \longrightarrow HgCl^+ + Cl^-$ | 1 |
| Cadmium sulfate | $CdSO_4 \longrightarrow Cd^{2+} + SO_4^{2-}$ | 7 |

[1] There is, in fact, evidence that suggests that the $H^+$ ion may exist as $H_9O_4^+$, that is, $H_3O(H_2O)_3^+$, in aqueous solution.

dissociate to only a limited extent in water. Only a small fraction of all of the acetic acid molecules present in solution actually react with the solvent,

$$HC_2H_3O_2 + H_2O \longrightarrow H_3O^+ + C_2H_3O_2^-$$

Such substances are called **weak electrolytes** because their solutions contain relatively few ions. Additional examples are given in Table 5.1. In the table we see that water itself is a very weak electrolyte by virtue of the reaction,

$$H_2O + H_2O \longrightarrow H_3O^+ + OH^-$$

This slight dissociation of water plays a very important role in many chemical reactions in which water is the solvent. Special attention will be given to this topic in Chapter 13.

Finally, there are many molecular compounds that do not dissociate into ions at all when they are dissolved in water. Sugar and ethyl alcohol are two common examples. These are called **nonelectrolytes.** Since solutions of nonelectrolytes contain no ions, they do not conduct an electric current.

## 5.3 CHEMICAL EQUILIBRIUM

The reason for the limited degree of dissociation of weak electrolytes is worth discussing at this point because it illustrates one of the most important concepts in chemistry, one to which we shall devote three chapters at a later time (Chapters 12, 14, and 15).

In a solution of acetic acid, molecules of $HC_2H_3O_2$ are constantly colliding with molecules of water and, in each encounter, there is a certain probability that a proton will be transferred from an $HC_2H_3O_2$ molecule to a water molecule to yield $H_3O^+$ and $C_2H_3O_2^-$ ions. There are also encounters, in this solution, between acetate ions and hydronium ions. When these ions meet, there is a high probability that an $H_3O^+$ ion will lose a proton to a $C_2H_3O_2^-$ ion to reform $HC_2H_3O_2$ and $H_2O$ molecules. Thus, in this solution we have two reactions occurring simultaneously:

$$HC_2H_3O_2 + H_2O \longrightarrow H_3O^+ + C_2H_3O_2^-$$

and

$$H_3O^+ + C_2H_3O_2^- \longrightarrow HC_2H_3O_2 + H_2O$$

When the rate at which the ions are disappearing to reform acetic acid and water molecules becomes equal to the rate at which the ions are formed, the concentrations of all of the species in the solution will not change with time. Such a state of affairs is called **equilibrium.** It is said to be a *dynamic equilibrium* in that things are continually happening in the solution; two reactions are taking place, ions reacting to yield molecules and molecules reacting to produce ions.

To indicate chemical equilibrium in a reacting system, we use a double arrow in the chemical equation. Thus, the equilibrium that we have been discussing is expressed as

$$HC_2H_3O_2 + H_2O \rightleftharpoons H_3O^+ + C_2H_3O_2^-$$

The use of this notation implies that the forward reaction (as read from left to right) is occurring at the same rate as the reverse reaction (the reaction from right to left). In a solution of acetic acid these rates become equal when only a small fraction of all of the acetic acid exists as ions. We say that the extent of dissociation is small and that the **position of equilibrium** lies in the direction of the molecular form of the substance. In other words, almost everything is present in undissociated form.

For strong electrolytes the reaction of the ions to produce molecules has very little tendency to occur. The position of equilibrium, therefore, lies almost completely toward the ions, and the strong electrolyte is essentially fully dissociated in the solution. When we write an equation to represent what takes place when a strong electrolyte is dissolved in water, we omit the arrow for the reverse reaction because, for all practical purposes, the reverse reaction does not occur. For the strong electrolyte NaCl, then, we would write simply

$$NaCl\ (s) \longrightarrow Na^+\ (aq) + Cl^-\ (aq)$$

The concept of a dynamic equilibrium is very important. All processes, both chemical and physical, tend to move toward a state of equilibrium. We will use this concept at numerous times in later chapters to analyze physical changes as well as chemical reactions.

## 5.4 IONIC REACTIONS

Most of the chemical reactions encountered in the laboratory portion of an introductory chemistry course involve reactions between ions in solution. In fact, any chemist who uses water as a solvent eventually encounters such reactions. What we wish to consider now is what occurs when two solutions, each containing an electrolyte, are mixed.

We shall begin by considering the reaction that takes place when a solution containing 1 mole of sodium chloride is added to a solution of 1 mole of silver nitrate. When these two are combined, 1 mole of the white solid, silver chloride, is formed (we refer to a solid that is formed in a solution as the result of chemical reaction as a **precipitate**), and the solution that remains contains 1 mole of sodium nitrate. If we wished, we could separate the AgCl from the solution by filtering the mixture. When the clear solution is evaporated, sodium nitrate remains behind.

The chemical equation that represents the change that has occurred during this reaction is

$$NaCl\ (aq) + AgNO_3\ (aq) \longrightarrow AgCl\ (s) + NaNO_3\ (aq)$$

This kind of reaction, in which cations and anions have changed partners, is known as **metathesis, or double replacement** ($Cl^-$ has replaced $NO_3^-$ and $Na^+$ has replaced $Ag^+$). The equation that we have written to describe the reaction is called a **molecular equation**, because all of the reactants and products are written as if they were molecules. A more accurate representation of the reaction as it actually occurs in solution is given by the ionic equation. It is known that a solution of NaCl does not contain molecules but, instead, consists of $Na^+$ and $Cl^-$ ions dispersed throughout the solvent. Similarly, a silver nitrate solution contains $Ag^+$ and $NO_3^-$ ions. When these two solutions are mixed, solid AgCl is formed by a combination of the $Ag^+$ and $Cl^-$ ions. The solution of sodium nitrate that remains contains $Na^+$ and $NO_3^-$ ions and we can write the **ionic equation** as

$$Na^+\ (aq) + Cl^-\ (aq) + Ag^+\ (aq) + NO_3^-\ (aq) \longrightarrow AgCl\ (s) + Na^+\ (aq) + NO_3^-\ (aq)$$

In this equation we have shown all of the soluble ionic substances as being dissociated in solution.

If we examine the ionic equation that we have just written, we see that $Na^+$ and $NO_3^-$ do not actually undergo any change during the course of the reaction. The same $Na^+$ and $NO_3^-$ ions are present after the chemical reaction as before and they have, in a sense, just "gone along for the ride." For this reason, ions that do not change during reaction are frequently called **spectator ions.** Since they do not take part in the reaction, we may eliminate them from the equation to arrive at the **net ionic equation**, that is, the equation for the net change that has taken place:

$$Ag^+ (aq) + Cl^- (aq) \longrightarrow AgCl (s)$$

This net ionic equation is useful in more than one way. First, it focuses our attention on the species that participate in the important changes that are occurring in the solution. Second, it tells us that any substance that produces $Ag^+$ ions in solution will react with any other substance that gives $Cl^-$ ions in solution to yield a precipitate of AgCl. Thus, after mixing together solutions of silver fluoride (AgF is soluble in water even though AgCl is not[2]) and potassium chloride, we would predict that a precipitate of AgCl would be formed. This is, in fact, precisely what takes place. The molecular, ionic, and net ionic equations for this reaction are, respectively,

$$AgF (aq) + KCl (aq) \longrightarrow AgCl (s) + KF (aq) \qquad \text{(molecular)}$$

$$Ag^+ (aq) + F^- (aq) + K^+ (aq) + Cl^- (aq) \longrightarrow AgCl (s) + K^+ (aq) + F^- (aq) \qquad \text{(ionic)}$$

$$Ag^+ (aq) + Cl^- (aq) \longrightarrow AgCl (s) \qquad \text{(net ionic)}$$

Each of these kinds of equations is useful in its own way; none is the "best" way of representing the reaction. The form that we use in any particular instance depends on what aspect of the reaction we wish to focus attention.

In the two metathesis reactions that we have considered, the driving force was provided by the formation of an insoluble precipitate of AgCl. If silver chloride were soluble in water, no chemical reaction of this type would occur. For instance, if solutions of KCl and $NaNO_3$ are combined, we might be tempted to write the equation,

$$KCl + NaNO_3 \longrightarrow KNO_3 + NaCl$$

Both $KNO_3$ and NaCl are soluble, and in almost all instances ionic compounds **(salts)** undergo virtually complete dissociation when they dissolve. If we write this equation in ionic form, we have

$$K^+ + Cl^- + Na^+ + NO_3^- \longrightarrow K^+ + NO_3^- + Na^+ + Cl^-$$

When we compare the left and right sides of this equation, we see that they are identical except for the sequence in which we have written the ions. If we cross out all spectator ions, there is nothing left; that is, there is no net chemical change when solutions of KCl and $NaNO_3$ are mixed.

We have seen now that when a precipitate is formed on mixing two solutions of electrolytes, a net chemical change takes place. In principle, by

[2] The solubility of AgCl is very low, 0.000089 g/100 ml, and AgCl can be considered, for most purposes, to be insoluble; that is, the amount of AgCl in solution can generally be considered to be negligible. For comparison, the solubility of AgF in water is approximately 185 g/100 ml at room temperature.

knowing the solubilities of all of the compounds that can be formed between pairs of cations and anions, we could *predict*, based on the formation of a precipitate, when chemical reactions would occur. This is made complicated because there is no sharp distinction between soluble and insoluble compounds. Substances such as lithium chlorate or sodium chloride, both of which were mentioned in Section 5.1, are certainly considered soluble. Silver chloride, on the other hand, would undoubtedly be termed insoluble. Some salts, such as $PbCl_2$ and $AgC_2H_3O_2$, are of intermediate solubility and are referred to as *partially soluble*, or *slightly soluble* compounds. Whether or not a precipitate of a particular salt will form upon mixing solutions of reactants depends on whether the concentration of the ions that comprise the salt exceeds that required to achieve a saturated solution of the salt. If dilute solutions of potential reactants are mixed, this might not occur and no precipitate would be formed in the reaction mixture. Generally, a compound is considered to be insoluble when the combination of even very dilute solutions of its constituent ions leads to the formation of a precipitate.

This entire discussion has been on a qualitative level. A quantitative treatment of the solubilities of ionic solids is reserved until Chapter 15. For now, we will use the following solubility rules as a rough guide in predicting the course of metathesis reactions.

*Solubility Rules*

1. All salts of the alkali metals are *soluble*.
2. All ammonium salts are *soluble*.
3. All salts containing the anions, $NO_3^-$, $ClO_3^-$, $ClO_4^-$, and $C_2H_3O_2^-$ are *soluble* ($AgC_2H_3O_2$ and $KClO_4$, however, are slightly soluble).
4. All chlorides, bromides, and iodides are *soluble except* those of $Ag^+$, $Pb^{2+}$, and $Hg_2^{2+}$ (note that mercury in the +1 oxidation state exists as the ion $Hg_2^{2+}$). $PbCl_2$, you may recall, is slightly soluble.
5. All sulfates are *soluble except* those of $Pb^{2+}$, $Sr^{2+}$, and $Ba^{2+}$. The sulfates of $Ca^{2+}$ and $Ag^+$ are slightly soluble.
6. All metal oxides *except* those of the alkali metals and $Ca^{2+}$, $Sr^{2+}$, and $Ba^{2+}$ are *insoluble*. Metal oxides, when they dissolve, react with the solvent to form hydroxides; for example,

$$CaO + H_2O \longrightarrow Ca^{2+} + 2OH^-$$

7. All hydroxides are *insoluble except* those of the alkali metals, $Ba^{2+}$ and $Sr^{2+}$. $Ca(OH)_2$ is slightly soluble.
8. All carbonates, phosphates, sulfides, and sulfites are *insoluble except* those of $NH_4^+$ and the alkali metals.

know it!

EXAMPLE 5.1 Would a chemical reaction occur if solutions of $FeCl_3$ and KOH were mixed? If so, give the net ionic equation.

SOLUTION A metathesis reaction between these two salts would be represented by the equation

$$FeCl_3 + 3KOH \longrightarrow 3KCl + Fe(OH)_3$$

On the basis of our solubility rules, KCl is soluble but $Fe(OH)_3$ is not. Therefore, we would expect a precipitate of $Fe(OH)_3$ to form in the mixture and the KCl would remain dissociated in solution. The ionic equation for the reaction, then, is

$$Fe^{3+} + 3Cl^- + 3K^+ + 3OH^- \longrightarrow Fe(OH)_3 + 3K^+ + 3Cl^-$$

Cancelling spectator ions ($K^+$ and $Cl^-$), we have the net ionic equation

$$Fe^{3+} + 3OH^- \longrightarrow Fe(OH)_3$$

EXAMPLE 5.2 Would a chemical reaction be expected to occur if solutions containing $NH_4NO_3$ and $Pb(C_2H_3O_2)_2$ were mixed?

SOLUTION To answer the question we must attempt to write a net ionic equation. First we start with the molecular equation, exchanging $NO_3^-$ for $C_2H_3O_2^-$ and vice versa when we write the products.

$$2NH_4NO_3 + Pb(C_2H_3O_2)_2 \longrightarrow 2NH_4C_2H_3O_2 + Pb(NO_3)_2$$

Rules 2 and 4 tell us that all of the reactants *and* products are soluble. The ionic equation can then be written

$$2NH_4^+ + 2NO_3^- + Pb^{2+} + 2C_2H_3O_2^- \longrightarrow 2NH_4^+ + 2C_2H_3O_2^- + Pb^{2+} + 2NO_3^-$$

If we cancel ions that are the same on both sides of the equation, everything disappears. The answer to the question, then, is that there is no net chemical reaction.

For ionic reactions that do not involve oxidation-reduction there are two other factors, in addition to the formation of a precipitate, that can lead to a net chemical change. One of these is *the formation of a weak electrolyte;* the other is *the formation of a gaseous product.*

When HCl is added to a solution of $NaC_2H_3O_2$, the following reaction takes place:

$$H^+ + C_2H_3O_2^- \longrightarrow HC_2H_3O_2$$

Here the product of the reaction is the slightly dissociated weak electrolyte, acetic acid. $HC_2H_3O_2$ has very little tendency to dissociate; and when $H^+$ and $C_2H_3O_2^-$ ions are mixed, the molecular compound is formed. The ionic equation for this reaction is therefore

$$H^+ + Cl^- + Na^+ + C_2H_3O_2^- \longrightarrow Na^+ + Cl^- + HC_2H_3O_2$$

A net change has occurred because the particles present in the reaction mixture after the reaction is over are not the same as before it began. Thus the formation of a slightly dissociated product from ionic reactants leads to a net chemical change.

Frequently, the molecular species formed in a reaction can escape as a gas, either directly or by decomposing to produce a gaseous product. For example, when HCl is added to $Na_2S$, the reaction that occurs is

$$2HCl + Na_2S \longrightarrow H_2S\ (g) + 2NaCl$$

for which the net ionic equation is

$$2H^+ + S^{2-} \longrightarrow H_2S\ (g)$$

Hydrogen sulfide gas has a limited solubility in water and when that solubility is exceeded, the $H_2S$ bubbles out of the solution.

Table 5.2
*Gases partially soluble in water*

| Gas | Typical Reaction in Which It Is Produced |
|---|---|
| $CO_2$ | $Na_2CO_3 + 2HCl \longrightarrow H_2CO_3 + 2NaCl$<br>$H_2CO_3 \longrightarrow H_2O + CO_2\ (g)$<br>*Net equation:* $CO_3^{2-} + 2H^+ \longrightarrow CO_2\ (g) + H_2O$ |
| $SO_2$ | $Na_2SO_3 + 2HCl \longrightarrow H_2SO_3 + 2NaCl$<br>$H_2SO_3 \longrightarrow H_2O + SO_2\ (g)$<br>*Net equation:* $SO_3^{2-} + 2H^+ \longrightarrow H_2O + SO_2\ (g)$ |
| $NH_3$ | $NH_4Cl + NaOH \longrightarrow NH_3\ (g) + H_2O + NaCl$<br>*Net equation:* $NH_4^+ + OH^- \longrightarrow NH_3\ (g) + H_2O$ |
| $H_2S$ | $Na_2S + 2HCl \longrightarrow H_2S\ (g) + 2NaCl$<br>*Net equation:* $S^{2-} + 2H^+ \longrightarrow H_2S\ (g)$ |
| $NO$<br>$NO_2$ | $NaNO_2 + HCl \longrightarrow HNO_2 + NaCl$<br>$2HNO_2 \longrightarrow H_2O + NO_2\ (g) + NO\ (g)$<br>*Net equation:* $2NO_2^- + 2H^+ \longrightarrow H_2O + NO_2\ (g) + NO\ (g)$ |

Another example is the reaction between HCl and $Na_2CO_3$,

$$2HCl + Na_2CO_3 \longrightarrow H_2CO_3 + 2NaCl$$

for which the net ionic equation is

$$2H^+ + CO_3^{2-} \longrightarrow H_2CO_3$$

Carbonic acid, $H_2CO_3$, is not stable and decomposes readily to produce $CO_2$ and $H_2O$,

$$H_2CO_3 \longrightarrow H_2O + CO_2\ (g)$$

Therefore, we can write the net ionic equation for the overall reaction between hydrogen ion and carbonate ion as

$$2H^+ + CO_3^{2-} \longrightarrow H_2O + CO_2\ (g)$$

A list of some other substances that are gases having a limited solubility in water, or that decompose to produce gaseous products, is given in Table 5.2.

EXAMPLE 5.3 Write a net ionic equation for the reaction between $NH_4NO_3$ and $Ba(OH)_2$.

SOLUTION We begin as usual by writing a double replacement reaction.

$$2NH_4NO_3 + Ba(OH)_2 \longrightarrow 2NH_4OH + Ba(NO_3)_2$$

The molecular species $NH_4OH$ (ammonium hydroxide) does not really exist; it is fictitious. It actually is $NH_3 + H_2O$ (note that there are one N, five H, and one O in both $NH_4OH$ and $NH_3 + H_2O$). We should therefore rewrite the equation as

$$2NH_4NO_3 + Ba(OH)_2 \longrightarrow 2NH_3 + 2H_2O + Ba(NO_3)_2$$

The ionic equation is obtained by applying the solubility rules, and also noting that $NH_3$ is a gas (Table 5.2).

$$2NH_4^+ + 2NO_3^- + Ba^{2+} + 2OH^- \longrightarrow 2NH_3 + 2H_2O + Ba^{2+} + 2NO_3^-$$

Finally, cancelling spectator ions gives

$$2NH_4^+ + 2OH^- \longrightarrow 2NH_3 + 2H_2O$$

which becomes

$$NH_4^+ + OH^- \longrightarrow NH_3 + H_2O$$

when the coefficients are reduced to the simplest set of whole numbers.



In this last section we have developed methods of predicting the outcome of a large number of ionic reactions. To apply these methods you must learn the solubility rules and the contents of Tables 5.1 and 5.2. It would be well for you to learn these now, because we shall use these methods again later in this chapter.

## 5.5 ACIDS AND BASES IN AQUEOUS SOLUTION

The concept of acids and bases is, without a doubt, one of the most important and useful in all of chemistry. In fact, nearly all chemical reactions can be broadly classified as either reactions between acids and bases, or as reactions involving oxidation and reduction. Because the acid-base concept is so important, an entire chapter (Chapter 13) is devoted to a detailed discussion of acid-base behavior. For now, we will limit ourselves to a simple definition of acids and bases, adequate for the treatment of reactions in aqueous solution.

We shall define an **acid** to be *any substance that increases the concentration of hydronium ion in solution.* Thus, HCl is an acid because when it is dissolved in water, it reacts with the solvent to produce $H_3O^+$.

$$HCl + H_2O \longrightarrow H_3O^+ + Cl^-$$

Aqueous solutions of hydrogen chloride are called *hydrochloric acid.* Because it is a strong electrolyte, HCl is called a **strong acid.**

Carbon dioxide is also an acid because its aqueous solutions contain more $H_3O^+$ than does pure water. In the last section we described the decomposition of $H_2CO_3$ that occurs when it is produced in large quantities as the result of a chemical reaction. The reverse reaction occurs to a limited extent when $CO_2$ is dissolved in water.

$$CO_2 + H_2O \longrightarrow H_2CO_3$$

Carbonic acid is able to dissociate slightly (actually by further reaction with water).

$$H_2CO_3 + H_2O \rightleftharpoons H_3O^+ + HCO_3^-$$

Carbonic acid is a weak electrolyte and is also called a **weak acid.**

We see that acids do not necessarily have to contain hydrogen; HCl does but $CO_2$ does not. The only requirement is that solutions of these substances contain more $H_3O^+$ than is found in pure $H_2O$. Table 5.3 contains a list of common acids and the reactions that they undergo to produce acidic solutions. You might note that nonmetal oxides, when dissolved in water, produce acids that dissociate by further reaction with the solvent. This pro-

Table 5.3
*Some acids and bases*

*Acids* (Those followed by an asterisk are weak electrolytes and exist primarily as undissociated molecules in aqueous solution.)

| | | | | |
|---|---|---|---|---|
| *Monoprotic acids:* $HX \longrightarrow H^+ + X^-$ | $HF$<br>$HCl$<br>$HBr$<br>$HI$<br>$HOCl$<br>$HClO_2$ | Hydrofluoric acid (*)<br>Hydrochloric acid<br>Hydrobromic acid<br>Hydriodic acid<br>Hypochlorous acid (*)<br>Chlorous acid (*) | $HClO_3$<br>$HClO_4$<br>$HIO_4$<br>$HNO_3$<br>$HNO_2$<br>$HC_2H_3O_2$ | Chloric acid<br>Perchloric acid<br>Periodic acid<br>Nitric acid<br>Nitrous acid (*)<br>Acetic acid (*) |
| *Diprotic acids:*<br>$H_2X \longrightarrow H^+ + HX^{2-}$<br>$HX^- \longrightarrow H^+ + X^{2-}$ | $H_2SO_4$<br>$H_2SO_3$<br>$H_2CO_3$<br>$H_2C_2O_4$ | Sulfuric acid[a]<br>Sulfurous acid (*)<br>Carbonic acid (*)<br>Oxalic acid (*) | $H_2S$<br>$H_3PO_3$ | Hydrosulfuric acid (*)<br>Phosphorous acid (only two hydrogens can be removed as protons) (*) |
| *Triprotic acids:*<br>$H_3X \longrightarrow H^+ + H_2X^-$<br>$H_2X^- \longrightarrow H^+ + HX^{2-}$<br>$HX^{2-} \longrightarrow H^+ + X^{3-}$ | $H_3PO_4$ | Orthophosphoric acid (*) | | |
| *Typical acidic oxides (nonmetal oxides)* | $SO_2$<br>$SO_3$<br>$N_2O_3$<br>$N_2O_5$<br>$P_4O_6$<br>$P_4O_{10}$ | $SO_2 + H_2O \longrightarrow H_2SO_3$<br>$SO_3 + H_2O \longrightarrow H_2SO_4$<br>$N_2O_3 + H_2O \longrightarrow 2HNO_2$<br>$N_2O_5 + H_2O \longrightarrow 2HNO_3$<br>$P_4O_6 + 6H_2O \longrightarrow 4H_3PO_3$<br>$P_4O_{10} + 6H_2O \longrightarrow 4H_3PO_4$ | | |

*Bases*

| | | | |
|---|---|---|---|
| *Molecular bases* | $NH_3$<br>$N_2H_4$<br>$NH_2OH$ | Ammonia[b] (*)<br>Hydrazine (*)<br>Hydroxylamine (*) | $(NH_3 + H_2O \rightleftharpoons NH_4^+ + OH^-)$<br>$(N_2H_4 + H_2O \rightleftharpoons N_2H_5^+ + OH^-)$<br>$(NH_2OH + H_2O \rightleftharpoons NH_3OH^+ + OH^-)$ |
| *Ionic bases* | Metal hydroxides (see solubility rules—Section 5.4) | | $M(OH)_n \longrightarrow M^{n+} + nOH^-$ |
| *Typical basic oxides (metal oxides)* | $Na_2O$, $K_2O$ } | $M_2O + H_2O \longrightarrow 2MOH$ | |
| | $CaO$, $SrO$, $BaO$ } | $MO + H_2O \longrightarrow M(OH)_2$ | |

[a] The second dissociation of $H_2SO_4$ is that of a weak acid.
[b] Aqueous solutions of ammonia are sometimes referred to as ammonium hydroxide. Commercially prepared solutions of $NH_3$, for example, are labeled in this way.

vides us with one chemical distinction between metals and nonmetals (we will shortly see that soluble metal oxides react with water to yield bases).

A **base** will be defined as *any substance that increases the concentration of hydroxide ion in aqueous solutions.* Sodium hydroxide, an example of an ionic compound containing metal ions and hydroxide ions, is a base because when it dissolves in water it dissociates to yield $Na^+$ and $OH^-$ ions.

$$NaOH \longrightarrow Na^+ + OH^-$$

The dissociation of NaOH is complete, and therefore NaOH is said to be a **strong base.** Ammonia, $NH_3$, is also a base because it reacts with water to produce hydroxide ions.

$$NH_3 + H_2O \rightleftharpoons NH_4^+ + OH^-$$

In this case, a proton is transferred from a water molecule to the ammonia molecule.

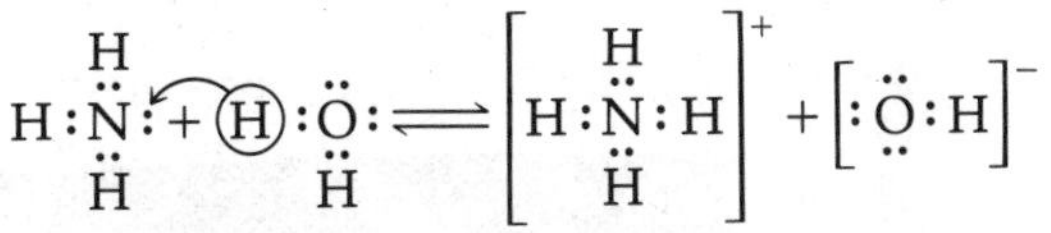

This reaction is indicated as an equilibrium because only a small fraction of the $NH_3$ in solution reacts to give $NH_4^+$ and $OH^-$ at any given instant. Ammonia is therefore called a **weak base** because its solutions contain only relatively small amounts of $OH^-$.

Metal oxides, when they are dissolved in water, produce basic solutions. We saw, for example, that when CaO is treated with water, it undergoes reaction to yield calcium hydroxide.

$$CaO + H_2O \longrightarrow Ca(OH)_2$$

When $Ca(OH)_2$ dissolves, it dissociates fully to $Ca^{2+}$ and $OH^-$.

$$Ca(OH)_2 \longrightarrow Ca^{2+} + 2OH^-$$

With $NH_3$ and CaO, we see that a base does not necessarily have to contain hydroxide ions in order to produce a basic solution. Furthermore, the presence of the OH unit in a compound does not guarantee that the substance will behave as a base. For example, the compounds, sodium hydroxide, sulfuric acid, and methyl alcohol can be represented by the following structures and formulas:

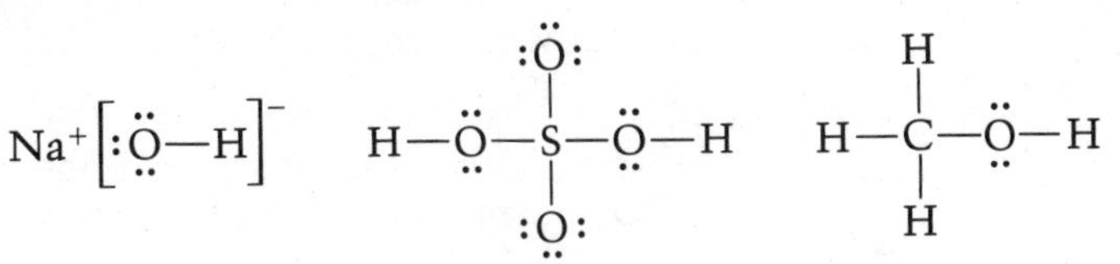

**NaOH** **$SO_2(OH)_2$** **$CH_3OH$**

Each of these contains an OH group; however, only the first compound is basic. Sulfuric acid dissociates by the breakage of O—H bonds and the transfer of protons to water molecules, while methyl alcohol does not dissociate at all in water. Therefore, we must exercise care in drawing conclusions about the acid-base behavior of compounds based on their chemical formulas or structures.

Acids and bases, in general, exhibit certain properties that lead us to characterize them as acids or bases. One of these properties is that they react with one another in a process called **neutralization.** In aqueous solution the neutralization reaction takes the form of the net ionic equation,

$$H_3O^+ + OH^- \longrightarrow 2H_2O$$

As mentioned previously, the hydronium ion can be considered a hydrated proton; it is therefore common practice to leave out the water molecule that is attached to the proton and simply to write the equation as

$$H^+ + OH^- \longrightarrow H_2O$$

A typical reaction between an acid and a base occurs between hydrochloric acid and sodium hydroxide.

$$HCl + NaOH \longrightarrow NaCl + H_2O$$

We see that the products of the neutralization reaction are a salt and water. The hydrochloric acid used in this reaction is said to be a **monoprotic acid** because 1 mole of HCl produces only 1 mole of $H^+$ ions in solution. There are also many acids that are capable of furnishing more than 1 mole of $H^+$ per mole of acid; as a group, these are called **polyprotic acids.** Examples are $H_2SO_4$ and $H_3PO_4$. These substances dissociate in a series of steps, each providing one proton. Sulfuric acid, for example, dissociates as follows:

$$H_2SO_4 \longrightarrow H^+ + HSO_4^-$$
$$HSO_4^- \longrightarrow H^+ + SO_4^{2-}$$

By controlling the quantity of base available for the neutralization of a diprotic acid, such as $H_2SO_4$, it is possible to form either salts that are the product of complete neutralization or salts in which only one of the available protons of the acid has been neutralized. For example, when 2 moles of NaOH are reacted with 1 mole of $H_2SO_4$, 2 moles of $H_2O$ and 1 mole of $Na_2SO_4$ are produced.

$$2NaOH + H_2SO_4 \longrightarrow 2H_2O + Na_2SO_4$$

However, if only 1 mole of NaOH is available to react with 1 mole of $H_2SO_4$, the products are 1 mole of water and 1 mole of $NaHSO_4$ (called sodium bisulfate or sodium hydrogen sulfate).

$$NaOH + H_2SO_4 \longrightarrow H_2O + NaHSO_4$$

Similar reactions can occur with a triprotic acid such as $H_3PO_4$, and salts such as the following can be obtained.

| | |
|---|---|
| $Na_3PO_4$ | trisodium phosphate (sodium phosphate) |
| $Na_2HPO_4$ | disodium hydrogen phosphate |
| $NaH_2PO_4$ | sodium dihydrogen phosphate |

Salts that are the product of the partial neutralization of a polyprotic acid are called **acid salts** because they can furnish protons for the neutralization of additional base. Thus $NaHSO_4$ can react with NaOH in a neutralization reaction.

$$NaHSO_4 + NaOH \longrightarrow Na_2SO_4 + H_2O$$

EXAMPLE 5.4 Write the chemical equation for the reaction of 1 mole of $H_3PO_4$ with 2 moles of NaOH.

SOLUTION

$$H_3PO_4 + 2NaOH \longrightarrow Na_2HPO_4 + 2H_2O$$

## 5.6 THE PREPARATION OF INORGANIC SALTS BY METATHESIS REACTIONS

In the past two sections we have described typical metathesis reactions that occur in aqueous solutions. These reactions can often be of use to us in preparing inorganic salts. For example, suppose that we were faced with the problem of preparing relatively pure NaBr using other inorganic compounds as starting materials. How would we proceed? We cannot achieve success if we mix together solutions of NaCl and KBr because, although the mixture would contain the constituents of NaBr, all of the ions, $Na^+$, $K^+$, $Cl^-$, and $Br^-$, are in solution together, and it would be nearly impossible to isolate pure NaBr from the mixture. *What we are looking for are reactions in which one product is easily separated from the other, a condition that is met if one of the products is either a precipitate, a gas, or the solvent, water.*

**PRECIPITATION REACTIONS.** Precipitation reactions are particularly useful when the desired product is insoluble in water. For instance, if we wish to prepare silver chloride, we need only mix a solution containing a soluble silver salt, such as $AgNO_3$, with one containing a soluble chloride, such as KCl. The reaction

$$Ag^+ + Cl^- \longrightarrow AgCl\ (s)$$

gives us our product as a solid that can be separated from the reaction mixture by filtration while the other product, $KNO_3$, remains in solution. In this preparation it is not necessary to worry too much about stoichiometry because, if we want to prepare 1 mole of AgCl, we can add 1 mole of $AgNO_3$ to a solution containing at least 1 mole of KCl. Any excess KCl beyond that required to react completely with the $AgNO_3$ remains dissolved and does not contaminate the AgCl precipitate.

When the desired product remains in solution, greater attention must be paid to the stoichiometry of the reaction. For example, to prepare NaBr (which is water-soluble) by a precipitation reaction, we might mix solutions of $K_2SO_4$ and $BaBr_2$, since $BaSO_4$ is insoluble. The net reaction is

$$Ba^{2+} + SO_4^{\ 2-} \longrightarrow BaSO_4\ (s)$$

and the molecular equation is

$$Na_2SO_4 + BaBr_2 \longrightarrow BaSO_4 + 2NaBr$$

To prepare pure NaBr it is necessary to react equal numbers of moles of $Na_2SO_4$ and $BaBr_2$. The solid $BaSO_4$ could then be filtered from the mixture and the volume of the resulting solution could then be reduced by evaporation to the point where nearly all of the desired product crystallizes.

There are some instances where use can be made of relative degrees of insolubility in synthetic procedures. For instance, the silver halides decrease in solubility in this order: AgCl, AgBr, and AgI. If a solution containing KBr is stirred with solid AgCl, the solid is converted to the less soluble AgBr, and $Cl^-$ replaces $Br^-$ in solution. The net effect is that AgCl is converted to AgBr.

EXAMPLE 5.5 How could lead sulfate, $PbSO_4$, be prepared by a precipitation reaction?

SOLUTION *Answering questions of this kind requires a thorough knowledge of the solubility rules.* $PbSO_4$ is insoluble. To prepare it we need a soluble lead salt and a soluble sulfate. Further, the other product in the metathesis must be soluble. There are undoubtedly a number of alternatives. One of them is

$$Na_2SO_4 + Pb(NO_3)_2 \longrightarrow PbSO_4\,(s) + 2NaNO_3$$

$Pb(NO_3)_2$ was chosen because all nitrates are soluble. This ensures a soluble lead salt as reactant *and* a soluble second product. After choosing $Pb(NO_3)_2$ we can pick any soluble sulfate as the other reactant.

**NEUTRALIZATION REACTIONS.** Another useful method of preparing salts is by the neutralization of an acid with a base. For example, we could prepare NaBr by the reaction of NaOH with HBr,

$$NaOH + HBr \longrightarrow H_2O + NaBr$$

If we are careful to react precisely 1 mole of NaOH with 1 mole of HBr, the only product in the aqueous solution is NaBr.

Metal oxides can be used in place of hydroxides in similar reactions. For example, the reaction

$$ZnO + 2HClO_4 \longrightarrow Zn(ClO_4)_2 + H_2O$$

could be used to prepare zinc perchlorate.

EXAMPLE 5.6 How could we prepare $Ca(C_2H_3O_2)_2$ by a neutralization reaction?

SOLUTION Since metal oxides react with acids, we could accomplish our goal by reacting CaO with acetic acid, $HC_2H_3O_2$

$$CaO + 2HC_2H_3O_2 \longrightarrow Ca(C_2H_3O_2)_2 + H_2O$$

**REACTIONS IN WHICH ONE PRODUCT IS A GAS.** A very convenient way to prepare inorganic salts is provided by the reaction between an acid and a metal carbonate. Recall that the carbonate ion reacts with hydrogen ion to produce carbon dioxide and water,

$$CO_3^{2-} + 2H^+ \longrightarrow H_2O + CO_2\,(g)$$

This reaction proceeds readily regardless of whether the source of carbonate ion is a soluble or insoluble salt.

The fact that most metal carbonates are insoluble leads to a very simple method for the preparation of pure salts. The synthesis of copper(II) bromide, for example, can be accomplished by the addition of HBr to a suspension of the insoluble $CuCO_3$ until nearly all of the starting material reacts according to the equation

$$CuCO_3\,(s) + 2HBr\,(aq) \longrightarrow CuBr_2\,(aq) + H_2O + CO_2\,(g)$$

When the mixture is filtered to remove the undissolved $CuCO_3$, the solution that remains contains $CuBr_2$, which is virtually free of contaminants. The product can then be recovered by evaporation of the water.

The preparation of salts from soluble carbonates is also feasible; once again, however, attention must be paid to the stoichiometry of the reaction to obtain pure products. For example, NaBr can be made from $Na_2CO_3$ and HBr.

$$Na_2CO_3\,(aq) + 2HBr\,(aq) \longrightarrow 2NaBr\,(aq) + H_2O + CO_2\,(g)$$

The quantity of *pure* NaBr that can be crystallized from the reaction mixture depends on how closely a 1:2 mole ratio is maintained between the reactants $Na_2CO_3$ and HBr.

---

**EXAMPLE 5.7** How could we prepare $Ca(ClO_4)_2$ by a reaction in which one product is a gas?

**SOLUTION** We can react $CaCO_3$ with $HClO_4$.

$$CaCO_3\ (s) + 2HClO_4\ (aq) \longrightarrow Ca(ClO_4)_2\ (aq) + CO_2\ (g) + H_2O$$

To obtain a pure product we use an excess of $CaCO_3$, which is filtered from the mixture after reaction has ceased.

---

**EXAMPLE 5.8** How could we prepare $Cu(ClO_4)_2$ from $CuSO_4$?

**SOLUTION** Often we can combine a series of reactions to give a final desired product. In this case, we know that we can prepare $Cu(ClO_4)_2$ from $Cu(OH)_2$ and $HClO_4$. Also, $Cu(OH)_2$ is insoluble and can be prepared from $CuSO_4$ by a precipitation reaction. One set of reactions, then, is

$$CuSO_4\ (aq) + 2NaOH\ (aq) \longrightarrow Cu(OH)_2\ (s) + Na_2SO_4\ (aq)$$
$$Cu(OH)_2\ (s) + 2HClO_4\ (aq) \longrightarrow Cu(ClO_4)_2\ (aq) + 2H_2O$$

In practice, we would add the NaOH to the $CuSO_4$ and filter the insoluble $Cu(OH)_2$ (*s*). This solid would then be reacted with less than a stoichiometric amount of $HClO_4$ so that a little excess $Cu(OH)_2$ remains unreacted. This would then be filtered to give a solution containing nearly pure $Cu(ClO_4)_2$.

---

## 5.7 OXIDATION-REDUCTION REACTIONS

Oxidation-reduction processes form a very important class of chemical reactions. They take place between many inorganic and organic compounds and, as implied earlier, they are extremely important in biochemical systems where they provide the mechanism for energy transfer in living organisms.

As a rule, the stoichiometry of redox reactions tends to be more complicated than for reactions that do not involve electron transfer. As a result, chemical equations for oxidation-reduction reactions are often complex and are difficult to balance by inspection. Fortunately, there are methods that can be applied to aid in balancing these equations.

In Section 4.9 it was noted that during redox reactions electrons are never produced as a product, nor are they necessary as a reactant. *When a chemical reaction involves oxidation-reduction, the total number of electrons lost in the oxidation process must equal the total number gained during reduction.* We can use this fact to help us balance equations for this type of reaction. Here is one procedure that can be used, called the **oxidation-number-change** method.

1. Assign oxidation numbers to each atom in the equation (we shall write them directly below the symbol for the element to avoid ever confusing oxidation numbers and charges). For example,

| | | | | | | | | | | |
|---|---|---|---|---|---|---|---|---|---|---|
| HCl | + | $K_2Cr_2O_7$ | $\longrightarrow$ | KCl | + | $Cl_2$ | + | $CrCl_3$ | + | $H_2O$ |
| +1 −1 | | +1 +6 −2 | | +1 −1 | | 0 | | +3 −1 | | +1 −2 |

gain n electrons
get less positive

2. Note which atoms change oxidation number and calculate the number of electrons transferred, per atom, during oxidation and reduction.
Each Cr atom gains three electrons during reduction

$$\underset{+6}{Cr} \xrightarrow{+3e^-} \underset{+3}{Cr}$$

Each $Cl^-$ that changes oxidation number loses 1 electron during oxidation.

$$\underset{-1}{Cl^-} \xrightarrow{-1e^-} \underset{0}{Cl}$$

3. When more than one atom of an element that changes oxidation number is present in a formula, we next calculate the number of electrons transferred per formula unit.
$K_2Cr_2O_7$ contains two Cr atoms, each of which gains 3 electrons for a total of $6e^-$/formula unit.

$$K_2Cr_2O_7 \longrightarrow 2CrCl_3$$

$$+6e^-/K_2Cr_2O_7$$

The coefficient 2 must be placed in front of the $CrCl_3$ to balance the chromium, which is the species being reduced.
When one $Cl_2$ molecule is formed, two electrons are lost by two $Cl^-$ ions. Therefore the number of electrons transferred per formula unit of $Cl_2$ produced = $2e^-$/formula unit. Note that a 2 must be placed before the HCl to balance chlorine.

$$2HCl \longrightarrow Cl_2$$

$$2e^-/Cl_2 \text{ formed}$$

4. Our next step is to make the number of electrons gained equal to the number lost by realizing that *three* $Cl_2$ formula units must be formed for every *one* $K_2Cr_2O_7$ formula unit reduced. These numbers become coefficients in our equation (Remember also that a 2 must be placed before the $CrCl_3$ to balance the chromium.) We then write

$$HCl + K_2Cr_2O_7 \longrightarrow KCl + 3Cl_2 + 2CrCl_3 + H_2O$$

(We have not placed the coefficient 6 in front of the HCl because some Cl has not undergone change and appears in several places on the right, thereby making the total number of Cl on the right greater than six.)
5. In step 4 we have established some of the coefficients. Once these have been obtained, the remainder of the equation is balanced by inspection. In this example, the sequence in which coefficients are introduced could be (a) a 2 is placed in front of KCl to balance the potassium; (b) since there are 14 Cl atoms on the right, place 14 in front of the HCl; (c) 14HCl on the left requires $7H_2O$ on the right to balance hydrogen (this also balances oxygen at the same time). The final equation is

$$14HCl + K_2Cr_2O_7 \longrightarrow 2KCl + 3Cl_2 + 2CrCl_3 + 7H_2O$$

The procedure outlined above will work on any oxidation-reduction equation, although simple equations (for example, $H_2 + Cl_2 \rightarrow 2HCl$) are more easily balanced by inspection.

## 5.8 BALANCING REDOX EQUATIONS: THE ION-ELECTRON METHOD

In addition to the oxidation-number-change method discussed above, there is still another method that is particularly well suited for balancing net ionic equations for oxidation-reduction reactions in solution. The procedure is called the **ion-electron method** and involves breaking the overall reaction into two **half-reactions,** one for the oxidation step and one for reduction. Each half-reaction is balanced materially (that is, in terms of atoms) and then electrically, by adding electrons to the side of the half-reaction deficient in negative charge. Finally, the balanced half-reactions are added together so that the electrons cancel from both sides of the final equation.

As a simple example, let us consider the reaction of $Sn^{2+}$ with $Hg^{2+}$ in the presence of chloride ion to produce $Hg_2Cl_2$ and $Sn^{4+}$ as products. The unbalanced equation is

$$Sn^{2+} + Hg^{2+} + Cl^- \longrightarrow Hg_2Cl_2 + Sn^{4+}$$

In this reaction, tin is undergoing a change in oxidation number. The unbalanced half-reaction corresponding to the change experienced by Sn is

$$Sn^{2+} \longrightarrow Sn^{4+}$$

To balance the half-reaction according to charge (it obviously is already balanced according to atoms), we must add two electrons to the right side of the equation so that the net charge on both sides is the same, +2.

$$Sn^{2+} \longrightarrow Sn^{4+} + 2e^- \qquad (5.1)$$

This is the balanced half-reaction for the oxidation (loss of electrons) of $Sn^{2+}$ to $Sn^{4+}$.

The next step is to write a half-reaction for the reduction of mercury(II). We begin by writing the formulas of the reactants and products that contain mercury.

$$Hg^{2+} \longrightarrow Hg_2Cl_2$$

We balance the equation materially by placing two chloride ions on the left and by writing the coefficient 2 in front of the $Hg^{2+}$,

$$2Hg^{2+} + 2Cl^- \longrightarrow Hg_2Cl_2$$

The next step is to balance the charge. On the left, the net charge is +2 $(2(+2) + 2(-1) = +2)$; on the right it is zero. We must therefore add two electrons, each with a charge of $-1$, to the left to give

$$2Hg^{2+} + 2Cl^- + 2e^- \longrightarrow Hg_2Cl_2 \qquad (5.2)$$

This is the balanced half-reaction for the reduction (gain of electrons) of $Hg^{2+}$.

To obtain the final balanced equation we add Equations 5.1 and 5.2 together so that there are the same number of electrons gained as lost.

$$\begin{array}{r} Sn^{2+} \longrightarrow Sn^{4+} + 2e^- \\ 2Hg^{2+} + 2Cl^- + 2e^- \longrightarrow Hg_2Cl_2 \\ \hline Sn^{2+} + 2Hg^{2+} + 2Cl^- + 2e^- \longrightarrow Sn^{4+} + Hg_2Cl_2 + 2e^- \end{array}$$

Since two electrons appear on both sides of the equation, we can cancel them out so that the final balanced equation is

$$Sn^{2+} + 2Hg^{2+} + 2Cl^- \longrightarrow Sn^{4+} + Hg_2Cl_2$$

In many oxidation-reduction reactions that take place in aqueous solution, water plays an active role. Any aqueous solution contains the species $H_2O$, $H^+$, and $OH^-$. In acidic solution the predominant species are $H_2O$ and $H^+$; in basic solution they are $H_2O$ and $OH^-$. When balancing half-reactions that occur in solution, we can use these species to achieve material balance. For example, let us consider the reaction between dichromate ion, $Cr_2O_7^{2-}$, and hydrogen sulfide in acidic solution to produce chromium(III) ion and elemental sulfur.

$$Cr_2O_7^{2-} + H_2S \longrightarrow Cr^{3+} + S$$

To balance the equation we write individual half-reactions for the changes that occur for chromium and sulfur. For the sulfur we write

$$H_2S \longrightarrow S$$

The half-reaction is balanced materially by adding two $H^+$ ions to the right.

$$H_2S \longrightarrow S + 2H^+$$

Electrical balance is achieved by adding two electrons to the right so that the net charge on both sides is zero.

$$H_2S \longrightarrow S + 2H^+ + 2e^- \qquad (5.3)$$

For chromium we begin by writing

$$Cr_2O_7^{2-} \longrightarrow Cr^{3+}$$

The chromium atoms are balanced by placing the coefficient 2 before the $Cr^{3+}$ on the right. The oxygens that appear on the left are balanced by adding $H_2O$ (seven of them) to the right and the hydrogen imbalance that results is removed by placing $14H^+$ on the left.

$$14H^+ + Cr_2O_7^{2-} \longrightarrow 2Cr^{3+} + 7H_2O$$

To balance the half-reaction electrically we must add six electrons to the left side to give

$$6e^- + 14H^+ + Cr_2O_7^{2-} \longrightarrow 2Cr^{3+} + 7H_2O \qquad (5.4)$$

The next step is to add these half-reactions (Equations 5.3 and 5.4) so that all of the electrons in the final equation cancel. This is accomplished by multiplying Equation 5.3 through by 3 before adding.

$$
\begin{array}{r}
3(H_2S \longrightarrow S + 2H^+ + 2e^-) \\
6e^- + 14H^+ + Cr_2O_7^{2-} \longrightarrow 2Cr^{3+} + 7H_2O \\
\hline
3H_2S + \underset{8}{\cancel{14}}H^+ + Cr_2O_7^{2-} + \cancel{6e^-} \longrightarrow 3S + \cancel{6H^+} + 2Cr^{3+} + 7H_2O + \cancel{6e^-}
\end{array}
$$

The six electrons cancel as do six of the $H^+$, so that the final balanced equation reads

$$3H_2S + 8H^+ + Cr_2O_7^{2-} \longrightarrow 3S + 2Cr^{3+} + 7H_2O$$

By way of summary, when balancing half-reactions in acid solution:

(a) *To balance a hydrogen atom we add a hydrogen ion, $H^+$, to the other side of the equation.*
(b) *To balance an oxygen atom we add a water molecule to the side deficient in oxygen and then two $H^+$ ions to the opposite side to remove the hydrogen imbalance.*

Applying the ion-electron method to reactions that occur in basic solution is perhaps a bit more difficult. The reason for this is that the species that can be used to balance the half-reactions, $H_2O$ and $OH^-$, both contain oxygen. We use the following rules.

(a) *To balance a hydrogen atom we add* **one** *$H_2O$ molecule to the side of the half-reaction deficient in hydrogen, and to the other side we add* **one** *hydroxide ion.*
(b) *To balance one oxygen atom we add* **two** *hydroxide ions to the side deficient in oxygen and* **one** *$H_2O$ molecule to the other side.* For example, consider the oxidation of Pb to PbO in basic media.

$$Pb \longrightarrow PbO$$

The left side of the equation needs an oxygen atom; so we add two $OH^-$.

$$2OH^- + Pb \longrightarrow PbO$$

Now we add one $H_2O$ to the right.

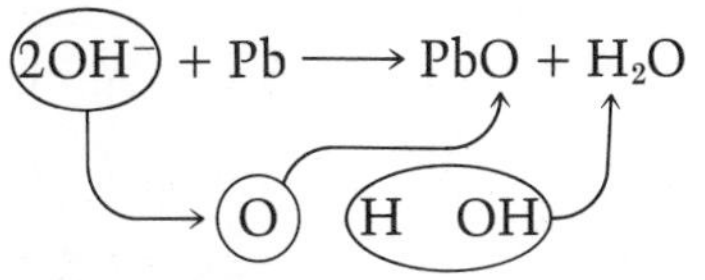


As indicated above, $2OH^-$ ions become one $H_2O$ molecule and one oxygen atom in some other formula.

As an example of how these rules are applied, let us balance the equation for the oxidation of methyl alcohol (wood alcohol), $CH_4O$, with permanganate ion, $MnO_4^-$, in basic solution, to yield $CO_2$ and $MnO_2$ as products.

$$CH_4O + MnO_4^- \longrightarrow MnO_2 + CO_2$$

As before, we break the equation into two half-reactions.

$$CH_4O \longrightarrow CO_2$$
$$MnO_4^- \longrightarrow MnO_2$$

The first equation is deficient in hydrogen on the right and deficient in oxygen on the left. To balance the hydrogen we place four $H_2O$ on the right and four $OH^-$ on the left.

$$4OH^- + CH_4O \longrightarrow CO_2 + 4H_2O$$

Now, to balance the oxygen we place two more $OH^-$ on the left and another $H_2O$ on the right.

$$6OH^- + CH_4O \longrightarrow CO_2 + 5H_2O$$

The half-reaction is now balanced materially. *Note that we balanced hy-*

*drogen and oxygen in separate operations.* Finally, it is necessary to balance the half-reaction electrically by placing six electrons on the right.

$$6OH^- + CH_4O \longrightarrow CO_2 + 5H_2O + 6e^- \quad (5.5)$$

Next we turn our attention to the reduction of the $MnO_4^-$ to $MnO_2$. This half-reaction is deficient in oxygen on the right, and balance is achieved by supplying four $OH^-$ to the right and two $H_2O$ to the left.

$$2H_2O + MnO_4^- \longrightarrow MnO_2 + 4OH^-$$

Electrical balance is provided by adding three electrons to the left.

$$3e^- + 2H_2O + MnO_4^- \longrightarrow MnO_2 + 4OH^- \quad (5.6)$$

In order to obtain the final equation we must add the two balanced half-reactions together so that all of the electrons will cancel. This requires that Equation 5.6 be multiplied by 2 before they are added.

$$\begin{array}{r} 6OH^- + CH_4O \longrightarrow CO_2 + 5H_2O + 6e^- \\ 2(3e^- + 2H_2O + MnO_4^- \longrightarrow MnO_2 + 4OH^-) \\ \hline 6OH^- + CH_4O + 4H_2O + 2MnO_4^- \longrightarrow CO_2 + 5H_2O + 2MnO_2 + 8OH^- \end{array}$$

We finally cancel six $OH^-$ and four $H_2O$ from both sides to obtain

$$CH_4O + 2MnO_4^- \longrightarrow CO_2 + 2MnO_2 + H_2O + 2OH^-$$

An alternative method of balancing reactions for basic solution is first to balance the equation as if it took place in acid media and then convert to basic solution as follows. If we balance the last equation for acid solution, we have

$$CH_4O + 2MnO_4^- + 2H^+ \longrightarrow CO_2 + 2MnO_2 + 3H_2O$$

To convert this to basic solution we add to *each side* the same number of $OH^-$ as there are $H^+$ in the equation. The $H^+$ and $OH^-$ that appear together on the same side then are assumed to produce $H_2O$. In this example there are $2H^+$ on the left, so we add $2OH^-$ *to each side,*

$$CH_4O + 2MnO_4^- + \underbrace{2H^+ + 2OH^-}_{\text{forms } 2H_2O} \longrightarrow CO_2 + 2MnO_2 + 3H_2O + 2OH^-$$

which gives

$$CH_4O + 2MnO_4^- + 2H_2O \longrightarrow CO_2 + 2MnO_2 + 3H_2O + 2OH^-$$

Finally, we can cancel $2H_2O$ from each side,

$$CH_4O + 2MnO_4^- \longrightarrow CO_2 + 2MnO_2 + H_2O + 2OH^-$$

This method gives the same result we got before. *If you use this method you must remember that the $OH^-$ is added to* **both** *sides of the equation.*

## 5.9 QUANTITATIVE ASPECTS OF REACTIONS IN SOLUTION

In Section 5.6 we saw that the preparation of pure compounds often depends on the mixing together of reactants in the proper ratio, as determined by the stoichiometry of the reaction. For example, if we wish to prepare $NaHSO_4$ from solutions containing NaOH and $H_2SO_4$, the two must be combined in such a way as to maintain a 1:1 mole ratio between the base and the acid. This means that we must have a means at our disposal of expressing the concentrations of reactants so that appropriate quantities of the various solutions may be mixed.

There are many ways of expressing the concentration of a solute in solution, and each has its advantages for specific applications. One way to express concentration, for instance, is percent composition by weight. Concentrated sulfuric acid, we have said, is composed of 96% $H_2SO_4$ and 4% $H_2O$. This tells us the composition of the solution in parts per hundred by weight. A somewhat similar unit that is frequently used to express very small concentrations (e.g., the concentration of impurities or pollutants in air or water) is **parts per million, ppm,** by volume or by weight. For instance, a typical carbon monoxide level in heavy smog is approximately 40 ppm by volume, whereas a typical nitrogen oxide concentration is about 0.2 ppm by volume.

EXAMPLE 5.9 Atmospheric $SO_2$, produced by the combustion of high-sulfur fuels, is an important air pollution problem. The amount of $SO_2$ in the air may be determined by bubbling the air through an acidic solution of $KMnO_4$, which oxidizes the $SO_2$ to $SO_4^{2-}$. In a typical analysis, 500 liters of air having a density of 1.20 g/liter are passed through the $KMnO_4$ solution and $1.50 \times 10^{-5}$ mole of $KMnO_4$ are reduced to $Mn^{2+}$. What is the concentration of $SO_2$ in parts per million (by weight)?

SOLUTION We first must have a balanced equation. The ion-electron method gives us

$$2MnO_4^- + 5SO_2 + 2H_2O \longrightarrow 2Mn^{2+} + 5SO_4^{2-} + 4H^+$$

If $1.50 \times 10^{-5}$ mole of $MnO_4^-$ is reduced, this requires (based on the stoichiometry of the reaction)

$$1.5 \times 10^{-5}\ \cancel{\text{mole MnO}_4^-} \times \left(\frac{5\ \text{moles SO}_2}{2\ \cancel{\text{moles MnO}_4^-}}\right) \sim 3.75 \times 10^{-5}\ \text{mole SO}_2$$

The weight of $SO_2$ in the air, then, is

$$3.75 \times 10^{-5}\ \cancel{\text{mole SO}_2} \times \left(\frac{64.1\ \text{g SO}_2}{1\ \cancel{\text{mole SO}_2}}\right) \sim 2.40 \times 10^{-3}\ \text{g SO}_2$$

The 500-liter sample of air contained $2.40 \times 10^{-3}$ g $SO_2$.
The concentration in parts per million is obtained as

$$\frac{\text{weight of SO}_2}{\text{total weight of air sample}} \times 10^6$$

The weight of the air sample can be obtained from its density

$$500\ \cancel{\text{liters}} \times \left(\frac{1.20\ \text{g}}{\cancel{\text{liter}}}\right) \sim 600\ \text{g air}$$

Therefore, the concentration of the $SO_2$ pollutant is

$$\frac{2.40 \times 10^{-3}\ \text{g SO}_2}{600\ \text{g air}} \times 10^6 = 4.00\ \text{ppm}$$

Expressing concentration in percent (as in the $H_2SO_4$ solution described earlier) or in parts per million (as in the solution of gases in Example 5.9) conveys information about the relative proportions of solute and solvent. However, for many applications to stoichiometry these concentration units are inconvenient and therefore not very useful. This leads us, then, to define concentration units that more closely suit problems in solution stoichiometry.

**MOLARITY. Molarity** is defined as *the number of moles of solute per liter of solution*[3]; that is, it represents a *ratio* in which the quantity of solute, in moles, is divided by the total volume of the mixture, in liters. A solution prepared by dissolving 58.44 g of NaCl (1 mole) in sufficient water to give a total volume of 1.00 liter, therefore, is a 1 *molar* solution of sodium chloride, written 1.00 *M* NaCl. If we dissolve 0.500 mole of NaCl in enough water to make 0.500 liter of solution, the concentration is 0.500 mole/0.500 liter which, of course, is the same as 1.00 mole/liter, or 1.00 *M*.

EXAMPLE 5.10 How many grams of NaOH are required to prepare 350 ml of 1.25 *M* NaOH solution?

SOLUTION We must first determine how many moles of NaOH are in 350 ml of 1.25 *M* NaOH. This can be accomplished by multiplying molarity by volume (in liters).

$$\left(\frac{1.25 \text{ moles NaOH}}{1.00 \text{ liter}}\right) \times 0.350 \text{ liter} \sim 0.437 \text{ mole NaOH}$$

Note that molarity functions here as a conversion factor that allows us to convert *liters of solution* to *moles of NaOH.* (It should be obvious that this same conversion factor, turned upside down, can also be used to change *moles* to *liters.*)

The weight of NaOH can then be obtained from the formula weight.

$$0.437 \text{ mole NaOH} \times \left(\frac{40.0 \text{ g NaOH}}{1 \text{ mole NaOH}}\right) \sim 17.5 \text{ g NaOH}$$

EXAMPLE 5.11 What is the molarity of a solution prepared by adding water to 10.0 g of KCl until 200 ml of solution is achieved?

SOLUTION The final solution contains 10.0 g/200 ml. Since our definition of molarity is the number of moles of solute per liter of solution, it is necessary to convert the 10.0 g of KCl into moles and the 200 ml of solution into liters.

We convert grams of KCl into moles of KCl by our usual method.

$$10.0 \text{ g KCl} \times \left(\frac{1 \text{ mole KCl}}{74.6 \text{ g KCl}}\right) \sim 0.134 \text{ mole KCl}$$

The number of liters of solution is 0.200 liter, so that the molarity is

$$\frac{0.134 \text{ mole}}{0.200 \text{ liter}} = 0.670\ M$$

EXAMPLE 5.12 What is the molar concentration of concentrated sulfuric acid that contains 96% by weight $H_2SO_4$ and has a density of 1.84 g/ml?

SOLUTION We must determine the number of moles of $H_2SO_4$ in 1 liter of the concentrated acid. We know, from the density, that

$$1.00 \text{ ml} \sim 1.84 \text{ g solution}$$

[3] Formality, another concentration unit, is defined in essentially the same way as the number of gram formula weights of solute per liter of solution. Since 1 gram formula weight contains Avogadro's number of formula units and 1 mole of solute is taken to mean the same thing, molarity and formality are identical.

or

$$1000 \text{ ml} = 1.00 \text{ liter} \sim 1840 \text{ g solution}$$

Only 96.0% of this weight is contributed by $H_2SO_4$. Thus

$$1840 \text{ g solution} \times \left(\frac{96.0 \text{ g } H_2SO_4}{100 \text{ g solution}}\right) \sim 1770 \text{ g } H_2SO_4$$

The formula weight of $H_2SO_4 = 98.08$; therefore,

$$1770 \text{ g } H_2SO_4 \times \left(\frac{1 \text{ mole } H_2SO_4}{98.08 \text{ g } H_2SO_4}\right) \sim 18.0 \text{ moles } H_2SO_4$$

The concentration, then, is

$$\frac{18.0 \text{ moles } H_2SO_4}{1.00 \text{ liter}} = 18.0\ M\ H_2SO_4$$

Molarity is a useful concentration unit because it allows us to meter out quantities of solute merely by dispensing volumes of solution. For instance, if we wished to use 0.300 mole of NaCl in some reaction and we had at our disposal a solution containing 1.00 mole of NaCl per liter (i.e., a 1.00 *M* NaCl solution), we could measure out 0.300 liter of this solution and know that it contained the desired 0.300 mole of NaCl. It is this aspect of molarity that is useful in problems dealing with the stoichiometry of solutions.

EXAMPLE 5.13 How many milliliters of 2.00 *M* NaCl solution are required to react with exactly 5.37 g of $AgNO_3$ to form AgCl?

SOLUTION The first step in any problem involving the stoichiometry of a chemical reaction should be to write a chemical equation (balanced, of course).

$$NaCl + AgNO_3 \longrightarrow AgCl + NaNO_3$$

We see that 1 mole of NaCl reacts with 1 mole of $AgNO_3$. Therefore, 5.37 g $AgNO_3$ requires

$$5.37 \text{ g } AgNO_3 \times \left(\frac{1 \text{ mole } AgNO_3}{170 \text{ g } AgNO_3}\right) \times \left(\frac{1 \text{ mole NaCl}}{1 \text{ mole } AgNO_3}\right) \sim 3.16 \times 10^{-2} \text{ mole NaCl}$$

Molarity has the units, mole/liter; therefore the concentration of the NaCl solution can be written as

$$2.00\ M \text{ NaCl} = \frac{2.00 \text{ moles NaCl}}{1.00 \text{ liter solution}}$$

This can be used as a conversion factor to find the volume of solution that contains $3.16 \times 10^{-2}$ mole NaCl.

$$3.16 \times 10^{-2} \text{ mole NaCl} \times \left(\frac{1.00 \text{ liter solution}}{2.00 \text{ mole NaCl}}\right) \sim 1.58 \times 10^{-2} \text{ liter solution}$$

Expressed in milliliters, the volume is 15.8 ml.

**EXAMPLE 5.14** How many grams of $Mg(OH)_2$ (found in milk of magnesia) are required to neutralize 50.0 ml of 0.0950 *M* HCl (stomach acid)?

**SOLUTION** Again we begin with a balanced equation.

$$Mg(OH)_2 + 2HCl \longrightarrow MgCl_2 + 2H_2O$$

The equation tells us that

$$1 \text{ mole } Mg(OH)_2 \sim 2 \text{ moles HCl}$$

In 50.0 ml of 0.0950 *M* HCl there is

$$50.0 \cancel{\text{ml}} \times \left(\frac{1 \cancel{\text{liter}}}{1000 \cancel{\text{ml}}}\right) \times \left(\frac{0.0950 \text{ mole HCl}}{1 \cancel{\text{liter}}}\right) \sim 0.00475 \text{ mole HCl}$$

Using the information from the equation,

$$0.00475 \cancel{\text{mole HCl}} \times \left(\frac{1 \text{ mole } Mg(OH)_2}{2 \cancel{\text{moles HCl}}}\right) \sim 0.00238 \text{ mole } Mg(OH)_2$$

Finally we calculate the weight of $Mg(OH)_2$.

$$0.00238 \cancel{\text{mole } Mg(OH)_2} \times \left(\frac{58.3 \text{ g } Mg(OH)_2}{1 \cancel{\text{mole } Mg(OH)_2}}\right) \sim 0.139 \text{ g } Mg(OH)_2$$

## 5.10 EQUIVALENT WEIGHTS AND NORMALITY

Certain kinds of calculations concerning stoichiometry may be performed without a balanced chemical equation. To accomplish this chemists have invented the concept of an **equivalent,** *defined in such a way that 1 equivalent of some substance, A, will react precisely with 1 equivalent of another substance, B.*

$$1 \text{ equivalent A} \sim 1 \text{ equivalent B}$$

*For oxidation-reduction an equivalent is defined as the quantity of substance that either gains or loses 1 mole of electrons.* Thus 1 equivalent of a reducing agent loses 1 mole of electrons; 1 equivalent of an oxidizing agent gains 1 mole of electrons. Since in redox reactions the number of electrons gained equals the number lost, 1 equivalent of oxidizing agent reacts precisely with 1 equivalent of reducing agent.

*In acid-base reactions an equivalent of an acid is defined as the quantity of acid that supplies 1 mole of $H^+$. An equivalent of base supplies 1 mole of $OH^-$.* In neutralization reactions 1 mole of $H^+$ reacts with 1 mole of $OH^-$. Therefore, 1 equivalent of acid reacts with exactly 1 equivalent of base.

In applying these concepts we employ the **equivalent weight,** that is, the weight of an equivalent. This is shown in Example 5.15.

**EXAMPLE 5.15** $FeSO_4$ reacts with $KMnO_4$ in $H_2SO_4$ to produce $Fe_2(SO_4)_3$ and $MnSO_4$. How many grams of $FeSO_4$ react with 3.71 g $KMnO_4$?

**SOLUTION** In this reaction,

1 mole of $FeSO_4$ loses 1 mole of electrons; $\quad \underset{+2}{Fe} \xrightarrow{-1e^-} \underset{+3}{Fe}$

1 mole of $KMnO_4$ gains 5 moles of electrons; $\quad \underset{+7}{Mn} \xrightarrow{+5e^-} \underset{+2}{Mn}$

The equivalent weight of $FeSO_4$ in this reaction would be the same as the weight of 1 mole because each mole $FeSO_4$ loses 1 mole of electron.

$$1 \text{ equiv } FeSO_4 = \frac{1 \text{ mole } FeSO_4}{1} = 152 \text{ g } FeSO_4$$

The equivalent weight of $KMnO_4$ in this reaction is one-fifth of the weight of 1 mole, since each mole of $KMnO_4$ gains 5 moles of electrons:

$$1 \text{ equiv } KMnO_4 = \tfrac{1}{5} \text{ mole } KMnO_4 = \tfrac{1}{5}(158.0 \text{ g } KMnO_4)$$
$$1 \text{ equiv } KMnO_4 = 31.6 \text{ g } KMnO_4$$

Since 1 equiv of $FeSO_4$ must react with precisely 1 equiv of $KMnO_4$,

$$152 \text{ g } FeSO_4 \sim 31.6 \text{ g } KMnO_4$$

The solution to the problem can then be obtained as

$$3.71 \text{ g } KMnO_4 \times \left(\frac{152 \text{ g } FeSO_4}{31.6 \text{ g } KMnO_4}\right) \sim 17.8 \text{ g } FeSO_4$$

From Example 5.15 we see that the equivalent weight is simply the weight of a mole divided by the number of electrons, $n$, gained or lost by the substance.

$$\text{equivalent weight} = \frac{\text{mole weight}}{n}$$

For acid-base reactions this same relationship applies, except that $n$ is the number of $H^+$ provided by a molecule of the acid or the number of $OH^-$ provided by one formula unit of the base. One mole of HCl can supply only 1 mole of $H^+$; therefore 1 equivalent of HCl is the same as 1 mole of HCl. For $H_2SO_4$, or any other polyprotic acid, the number of equivalents per mole is equal to the number of hydrogen ions each molecule of the acid furnishes when it undergoes neutralization by a base. One mole of $H_2SO_4$ consists of 2 equivalents if it is completely neutralized. It consists of 1 equivalent if it is only neutralized to $HSO_4^-$, since in this case 1 mole of $H_2SO_4$ has furnished only 1 mole of $H^+$. For complete neutralization the equivalent weight of $H_2SO_4$ would be half the weight of 1 mole; $n$ is equal to 2.

$$\text{equiv wt } H_2SO_4 = \frac{98.0 \text{ g}}{2} = 49.0 \text{ g } H_2SO_4$$

For bases the number of equivalents per mole is the same as the number of $OH^-$ provided by one formula unit of the base. One mole of $Ba(OH)_2$ would constitute 2 equivalents. The equivalent weight of $Ba(OH)_2$ is half the weight of 1 mole.

Finally, note that the equivalent weight of a substance can vary from one reaction to another. Potassium permanganate, for example, has an equivalent weight equal to $\frac{1}{5}$ mole when it reacts, as above, to produce manganese in the +2 oxidation state. In some reactions, $KMnO_4$ produces $MnO_2$ as a product. In this case 1 mole of $KMnO_4$ gains only 3 moles of electrons, and its equivalent weight is $\frac{1}{3}$ the weight of 1 mole.

**NORMALITY.** Another concentration unit that can be used in the treatment of problems concerning stoichiometry is **normality**, defined as *the number of equivalents of solute per liter of solution.* Again, we have a ratio—the quantity of solute, expressed in equivalents, divided by the total volume of the solution. Above, the equivalent for oxidizing and reducing agents was defined as the quantity that either gains or loses 1 mole of electrons. We saw that when the permanganate ion in $KMnO_4$ is reduced to $Mn^{2+}$, five electrons are gained by each formula unit of $KMnO_4$. One mole of $KMnO_4$, therefore, consumes 5 moles of electrons and the quantity of $KMnO_4$ that acquires only 1 mole of electrons (that is, an equivalent of $KMnO_4$) is $\frac{1}{5}$ mole of the salt.

$$1 \text{ equiv } KMnO_4 = \tfrac{1}{5} \text{ mole } KMnO_4$$

Thus a 1.00 normal (1.00 *N*) solution of $KMnO_4$ contains 1.00 equivalent of $KMnO_4$ per liter, or 0.200 mole of $KMnO_4$ per liter. A 1.00 *N* $KMnO_4$ solution is also a 0.200 *M* $KMnO_4$ solution. We see that there is a very simple relationship between normality and molarity—the normality is always an integral multiple of the molarity,

$$N = n \times M$$

where $n$ is an integer. For substances undergoing oxidation or reduction the integer, $n$, corresponds to the number of electrons transferred per formula unit. When $KMnO_4$ is reduced to $Mn^{2+}$, $n = 5$. If the product of the reduction is $MnO_2$, only three electrons are acquired by each $KMnO_4$ formula unit, and $n$ would equal 3. Similarly, when oxalic acid, $H_2C_2O_4$, is oxidized to produce $CO_2$, two electrons are lost by each $H_2C_2O_4$ formula unit and the normality of a solution of $H_2C_2O_4$ is twice its molarity.

For solutions of acids and bases, $n$ is once again the number of $H^+$ provided by a formula unit of acid or the number of $OH^-$ made available by a formula unit of base. For complete neutralization, a 1.00 *M* $H_3PO_4$ solution could also be labeled 3.00 *N* $H_3PO_4$; a 1.00 *M* $Ba(OH)_2$ solution is also 2.00 *N* $Ba(OH)_2$.

Normality is useful because the equivalent is defined so that 1 equivalent of an oxidizing agent will react with *exactly* 1 equivalent of a reducing agent (1 mole of electrons gained and 1 mole of electrons lost). Similarly, 1 equivalent of acid completely neutralizes precisely 1 equivalent of base, since 1 mole of $H^+$ reacts with 1 mole of $OH^-$.

$$H^+ + OH^- \longrightarrow H_2O$$

This means that mixing equal volumes of solutions having the same normality will lead to complete reaction between their solutes; 1 liter of 1 *N* acid will completely neutralize 1 liter of 1 *N* base because 1 equivalent of acid reacts with 1 equivalent of base. In the same manner, 1 liter of 1 *N* oxidizing agent will completely oxidize the contents of 1 liter of 1 *N* reducing agent.

EXAMPLE 5.16 How many milliliters of 0.150 *N* $K_2Cr_2O_7$ are required to react with 75.0 ml of 0.400 *N* $H_2C_2O_4$?

SOLUTION In 75.0 ml of 0.400 *N* $H_2C_2O_4$ there are

$$0.075 \text{ liter} \times \left(\frac{0.400 \text{ equiv } H_2C_2O_4}{1.00 \text{ liter}}\right) \sim 3.00 \times 10^{-2} \text{ equiv } H_2C_2O_4$$

This will require $3.00 \times 10^{-2}$ equivalents of $K_2Cr_2O_7$ for complete oxidation. The volume of potassium dichromate that is required is

$$3.00 \times 10^{-2} \text{ equiv} \times \left(\frac{1.00 \text{ liter}}{0.150 \text{ equiv}}\right) \sim 2.00 \times 10^{-1} \text{ liter} \quad \text{or} \quad 200 \text{ ml}$$

In Example 5.16 we used the idea that equal numbers of equivalents of oxidizing and reducing agent must react. Since the product of volume ($V$) × normality ($N$) gives equivalents, when any two substances in solution undergo complete reaction (let us call them A and B),

$$V_A N_A = V_B N_B$$

which means,

$$(\text{number of equivalents of A}) = (\text{number of equivalents of B}).$$

## 5.11 CHEMICAL ANALYSIS

Whether we synthesize chemicals in the laboratory or isolate them from natural sources, we must know their composition before we can proceed very far in understanding their properties. When a substance appears as the product of a reaction, it does not stand up and declare its composition—it must be analyzed. One very important facet of chemistry is the investigation of new and better ways to achieve chemical analyses, particularly of compounds present in mixtures, where difficulties in separating the components present a major obstacle to success.

Chemical reactions in solution can frequently be used to advantage in performing chemical analyses. Example 5.17 illustrates how a precipitation reaction might be put to use.

EXAMPLE 5.17 A mixture of NaCl and $BaCl_2$ weighing 0.200 g was dissolved in water, and sulfuric acid was added until no further precipitate was formed. The solid $BaSO_4$ was filtered, dried, and found to weigh 0.0643 g. What percent of the mixture was $BaCl_2$?

SOLUTION In this example the barium is precipitated as $BaSO_4$.

$$BaCl_2 + H_2SO_4 \longrightarrow 2HCl + BaSO_4$$

From the weight of $BaSO_4$ we can compute the weight of $BaCl_2$ in the mixture and hence the fraction that was $BaCl_2$.

$$0.0643\ \cancel{\text{g } BaSO_4} \times \left(\frac{1 \text{ mole } BaSO_4}{233.4\ \cancel{\text{g } BaSO_4}}\right) \sim 2.75 \times 10^{-4} \text{ mole } BaSO_4$$

According to the chemical equation, 1 mole of $BaSO_4$ is produced from 1 mole of $BaCl_2$; therefore, the original mixture must have contained $2.75 \times 10^{-4}$ mole of $BaCl_2$. The weight of $BaCl_2$ in the mixture was

$$2.75 \times 10^{-4}\ \cancel{\text{mole } BaCl_2} \times \left(\frac{208.4 \text{ g } BaCl_2}{1\ \cancel{\text{mole } BaCl_2}}\right) \sim 5.73 \times 10^{-2} \text{ g } BaCl_2$$

$$\text{weight of } BaCl_2 = 0.0573 \text{ g}$$

$$\text{percent } BaCl_2 = \frac{\text{weight } BaCl_2}{\text{total weight of mixture}} \times 100$$

$$= \frac{0.0573}{0.200} \times 100 = 28.7\%$$

Often, the quantitative relationships introduced in our discussion of concentration can be employed directly in what are termed **volumetric analyses.** For example, the quantity of base in some sample of an unknown can be determined by a procedure called a **titration,** in which a solution of an acid is dispensed into the solution of the base from a **buret** (Figure 5.3) until

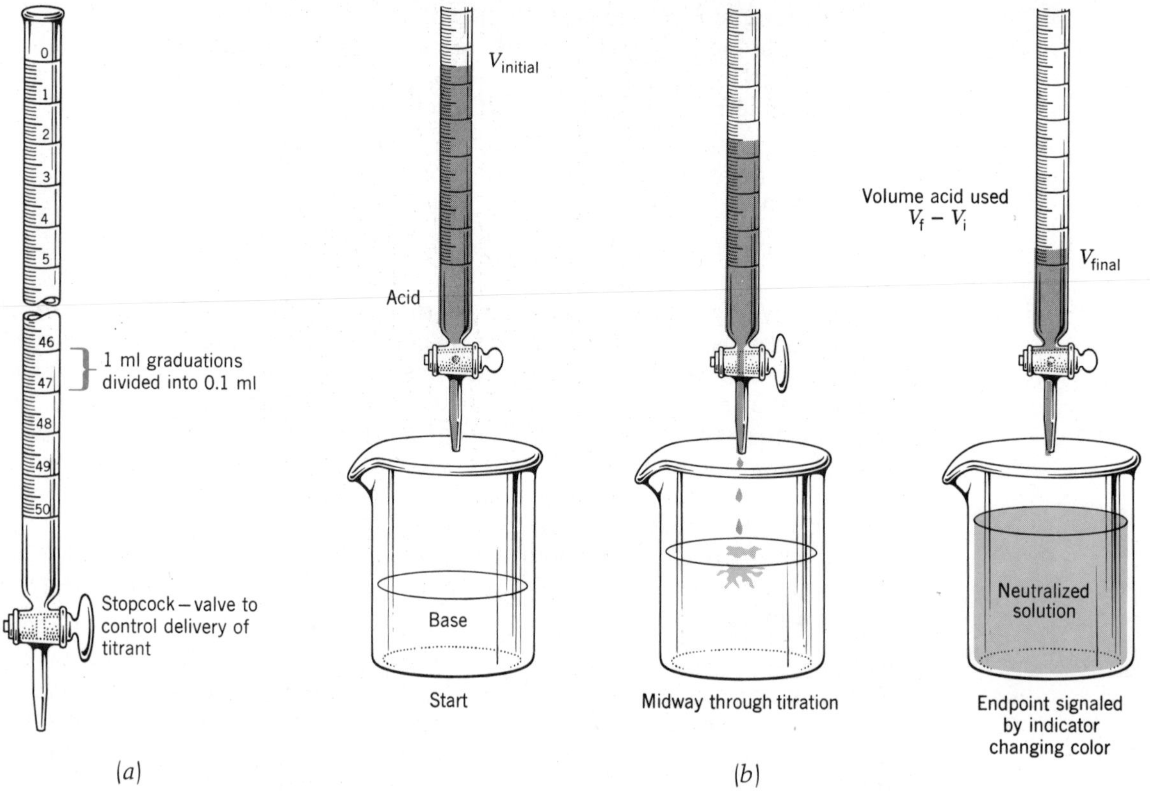


Figure 5.3
*(a) A buret. (b) Titration of a base with an acid.*

reaction is complete, as signaled by some form of **indicator.** There are a series of organic dyes that have one color in acid solution and a different color in basic solution that can be used for this purpose. Litmus is a common example—it is red in acid solution and blue in the presence of base. In a typical titration, the base is dissolved and the indicator added. Acid is delivered gradually from the buret until the **end point** (or **equivalence point**) is marked by a color change signifying that all of the base has been neutralized. At the equivalence point an equal number of equivalents of acid and base have been combined. From the total volume of acid added and its concentration, the number of equivalents of acid and, hence, the number of equivalents of base can be deduced.

EXAMPLE 5.18 A sample of an analgesic drug was analyzed for aspirin, a monoprotic acid, $HC_9H_7O_4$, by titration with base. A 0.500-g sample of the drug required 21.5 ml of 0.100 *N* NaOH. What percentage of the drug was aspirin?

SOLUTION From the volume of base and its concentration we can calculate the number of equivalents of base consumed.

$$0.0215\ \cancel{\text{liter}} \times \left(\frac{0.100\ \text{equiv NaOH}}{1\ \cancel{\text{liter}}}\right) \sim 0.00215\ \text{equiv NaOH}$$

From the concept of equivalents we can write

$$0.00215 \text{ equiv NaOH} \sim 0.00215 \text{ equiv aspirin}$$

Since aspirin is a monoprotic acid

$$0.00215 \text{ equiv aspirin} = 0.00215 \text{ mole aspirin}$$

The weight of aspirin in the sample is

$$0.00215 \text{ mole aspirin} \times \left(\frac{180.2 \text{ g aspirin}}{1 \text{ mole aspirin}}\right) \sim 0.387 \text{ g aspirin}$$

The percent aspirin can be found as

$$\text{percent aspirin} = \frac{\text{weight aspirin}}{\text{weight sample}} \times 100$$

Substituting,

$$\text{percent aspirin} = \frac{0.387 \text{ g}}{0.500 \text{ g}} \times 100 = 77.4\%$$

Similar titrations are frequently performed in which the reaction involves oxidation-reduction. $KMnO_4$ is a useful **titrant** (the solution delivered from the buret) to use because it has a deep purple color whereas the product of the reduction of the $MnO_4^-$ ion in acid solution is the almost colorless $Mn^{2+}$ ion. The end point in the titration is signaled by the appearance of a pale purple color in the reaction mixture due to excess $MnO_4^-$ ion.

EXAMPLE 5.19 A 1.000-gram sample of an iron ore containing $Fe_2O_3$ was dissolved in acid and all of the iron converted to $Fe^{2+}$. The solution was titrated with 90.4 ml of 0.100 *N* $KMnO_4$ to give $Mn^{2+}$ and $Fe^{3+}$ as products. What percentage of the ore is $Fe_2O_3$?

SOLUTION The number of equivalents of $KMnO_4$ reduced was

$$9.04 \times 10^{-2} \text{ liter} \times \left(\frac{0.100 \text{ equiv KMnO}_4}{\text{liter}}\right) \sim 9.04 \times 10^{-3} \text{ equiv KMnO}_4$$

This amount of $KMnO_4$ has oxidized $9.04 \times 10^{-3}$ equiv $Fe^{2+}$ to $Fe^{3+}$. Since 1 mole of $Fe^{2+}$ loses 1 mole of electrons,

$$1 \text{ mole Fe}^{2+} = 1 \text{ equiv Fe}^{2+}$$

Therefore, $9.04 \times 10^{-3}$ equiv $Fe^{2+} = 9.04 \times 10^{-3}$ mole $Fe^{2+}$. Each mole of $Fe_2O_3$ contains 2 moles of Fe.

$$1 \text{ mole Fe}_2\text{O}_3 \sim 2 \text{ moles Fe}$$

The number of moles of $Fe_2O_3$ in the sample was

$$9.04 \times 10^{-3} \text{ mole Fe} \times \left(\frac{1 \text{ mole Fe}_2\text{O}_3}{2 \text{ moles Fe}}\right) \sim 4.52 \times 10^{-3} \text{ mole Fe}_2\text{O}_3$$

The number of grams of $Fe_2O_3$ was

$$4.52 \times 10^{-3} \text{ mole Fe}_2\text{O}_3 \times \left(\frac{160 \text{ g Fe}_2\text{O}_3}{1 \text{ mole Fe}_2\text{O}_3}\right) \sim 0.723 \text{ g Fe}_2\text{O}_3$$

Since the sample weighed 1.000 g,

$$\text{percent } Fe_2O_3 = \frac{0.723}{1.000} \times 100 = 72.3\%$$

In the course of performing routine laboratory operations as well as volumetric analyses, it is not uncommon to have to dilute solutions. Many of the common laboratory reagents are bottled in concentrated form (Table 5.4) and usually must be diluted for use. The process of dilution actually involves simply spreading a given quantity of solute throughout a larger volume of solution; when a concentrated solution is diluted by the addition of solvent, the number of moles of solute in the mixture remains constant. This means that the solution's molar concentration multiplied by its volume must remain constant as solvent is added,[4] because

$$M \cdot V = \frac{\text{moles}}{\text{liter}} \times \text{liter} = \text{moles}$$

In other words, the product of the initial molarity and volume ($M_iV_i$) must equal the product of the final molarity and volume ($M_fV_f$).

$$M_iV_i = M_fV_f$$

For dilution problems in which the concentration unit is normality, the similar equation, $N_iV_i = N_fV_f$, applies.

EXAMPLE 5.20 How much water must be added to 25.0 ml of 0.500 *M* KOH solution to produce a solution whose concentration is 0.350 *M*?

SOLUTION Our equation is

$$M_iV_i = M_fV_f$$

$$M_i = 0.500\ M \qquad M_f = 0.350\ M$$

$$V_i = 25.0\ \text{ml} \qquad V_f = ?$$

Table 5.4
*Concentrated laboratory reagents*

| Reagent | Density (g/ml) | Weight Percent | Molarity |
|---|---|---|---|
| Sulfuric acid | 1.84 | 96 | 18 |
| Hydrochloric acid | 1.18 | 36.5 | 12.0 |
| Phosphoric acid | 1.7 | 85 | 15 |
| Nitric acid | 1.42 | 70.7 | 16.0 |
| Acetic acid | 1.05 | 100 | 17.5 |
| Aqueous ammonia | 0.90 | 28 | 14.8 |

[4] **Caution!** When diluting concentrated reagents, the proper procedure is to *add the reagent to the solvent. Never* add water to a concentrated acid.

Solving for $V_f$ and substituting, we get

$$V_f = \frac{(0.500\ M)(25.0\ \text{ml})}{0.350\ M}$$

$$V_f = 35.7\ \text{ml}$$

Since the initial volume was 25.0 ml, we must *add* 10.7 ml. (We are assuming that volumes are additive. When working with dilute solutions this assumption is generally quite valid.)

EXAMPLE 5.21 How many milliliters of concentrated $H_2SO_4$ (18.0 $M$) are required to prepare 750 ml of 3.0 $M$ $H_2SO_4$ solution?

SOLUTION Again,

$$M_iV_i = M_fV_f$$

$$M_i = 18.0\ M \qquad M_f = 3.0\ M$$

$$V_i = ? \qquad V_f = 750\ \text{ml}$$

Solving for $V_i$ gives

$$V_i = \frac{M_fV_f}{M_i}$$

$$V_i = \frac{(3.0\ M)(750\ \text{ml})}{18.0\ M}$$

$$V_i = 125\ \text{ml}$$

To prepare the solution, 125 ml of the concentrated $H_2SO_4$ is diluted to a total final volume of 750 ml.

## INDEX TO QUESTIONS AND PROBLEMS (problem numbers in italics)



## REVIEW QUESTIONS

**5.1** Why are solutions usually employed for carrying out chemical reactions?

**5.2** Give the meaning of the terms: solvent, solute, concentrated, dilute, saturated, supersaturated, unsaturated.

**5.3** What is the meaning of the term, solubility?

**5.4** How can one distinguish between a strong and a weak electrolyte? What are present in solutions of electrolytes that are absent in solutions of nonelectrolytes?

**5.5** Write chemical equations for the dissociation of the following electrolytes: KCl, $(NH_4)_2SO_4$, $Na_3PO_4$, NaOH, HCl.

**5.6** Why is the hydronium ion often abbreviated $H^+$?

**5.7** What is a dynamic equilibrium? Write an equation to represent the dissociation of the weak electrolyte, $CdSO_4$.

**5.8** Write an equation to represent the equilibrium dissociation of water.

**5.9** What can be said about the position of equilibrium in (a) the dissociation of water; (b) the dissociation of HCl?

**5.10** What is the term used to describe a solid that is formed in a solution during a chemical reaction?

**5.11** What is a metathesis reaction? What other name is often applied to this kind of reaction?

**5.12** Write chemical equations for the equilibria that exist when the following weak electrolytes undergo dissociation: HCN, $H_2S$, $NH_3$, $HgCl_2$.

**5.13** Write ionic and net ionic equations for the following:

(a) $CuCl_2\ (aq) + Pb(NO_3)_2\ (aq) \rightarrow Cu(NO_3)_2\ (aq) + PbCl_2\ (s)$
(b) $FeSO_4\ (aq) + 2NaOH\ (aq) \rightarrow Fe(OH)_2\ (s) + Na_2SO_4\ (aq)$
(c) $ZnSO_4\ (aq) + BaCl_2\ (aq) \rightarrow ZnCl_2\ (aq) + BaSO_4\ (s)$
(d) $2AgNO_3\ (aq) + K_2SO_4\ (aq) \rightarrow Ag_2SO_4\ (s) + 2KNO_3\ (aq)$
(e) $(NH_4)_2CO_3\ (aq) + CaCl_2\ (aq) \rightarrow 2NH_4Cl\ (aq) + CaCO_3\ (s)$

**5.14** Write ionic and net ionic equations for the following:

(a) $Al(OH)_3\ (s) + 3HCl\ (aq) \rightarrow AlCl_3\ (aq) + 3H_2O$
(b) $CuCO_3\ (s) + H_2SO_4\ (aq) \rightarrow CuSO_4\ (aq) + H_2O + CO_2\ (g)$
(c) $Cr_2(CO_3)_3\ (s) + 6HNO_3\ (aq) \rightarrow 2Cr(NO_3)_3\ (aq) + 3H_2O + 3CO_2(g)$
(d) $3Cu\ (s) + 8HNO_3\ (aq) \rightarrow 3Cu(NO_3)_2\ (aq) + 2NO\ (g) + 4H_2O$
(e) $MnO_2\ (s) + 4HCl\ (aq) \rightarrow MnCl_2\ (aq) + Cl_2\ (g) + 2H_2O$

**5.15** Without looking at the solubility rules, indicate whether the following are soluble or insoluble: KCl, $(NH_4)_2SO_4$, $AgNO_3$, $PbSO_4$, $Mn(OH)_2$, $FePO_4$, $CaCO_3$, $Zn(ClO_4)_2$, $Ba(C_2H_3O_2)_2$, NiO.

**5.16** Write net ionic equations for each of the following?

(a) $NaBr + AgNO_3 \rightarrow AgBr + NaNO_3$
(b) $CoCO_3 + 2HNO_3 \rightarrow Co(NO_3)_2 + CO_2 + H_2O$
(c) $NaC_2H_3O_2 + HNO_3 \rightarrow NaNO_3 + HC_2H_3O_2$
(d) $Pb(NO_3)_2 + (NH_4)_2SO_4 \rightarrow PbSO_4 + 2NH_4NO_3$
(e) $H_2S + Cu(NO_3)_2 \rightarrow 2HNO_3 + CuS$
(f) $NaOH + NH_4Cl \rightarrow NH_3 + H_2O + NaCl$

**5.17** Write molecular, ionic, and net ionic equations for the reaction (if any) between:

(a) $Na_2SO_4$ and $BaCl_2$
(b) $Ca(NO_3)_2$ and $(NH_4)_2CO_3$
(c) $NaC_2H_3O_2$ and $HNO_3$
(d) NaOH and $CuCl_2$
(e) $(NH_4)_2CO_3$ and $HNO_3$

**5.18** Write molecular, ionic, and net ionic equations for the reaction (if any) between:

(a) $H_2SO_4$ and $BaSO_4$
(b) $NH_4Br$ and $MnSO_4$
(c) $K_2S$ and $HC_2H_3O_2$
(d) $MgSO_4$ and $Li(OH)_2$
(e) $AgC_2H_3O_2$ and KCl

**5.19** Write molecular, ionic, and net ionic equations for any reaction that would occur between:

(a) AgBr and KI
(b) $BaCl_2$, $SO_2$, and $H_2O$
(c) $Na_2C_2O_4$ and HCl
(d) $K_2SO_3$ and HCl
(e) $BaCO_3$ and $H_2SO_4$

**5.20** Define the terms *acid* and *base* as applied to aqueous solutions.

**5.21** Write typical reactions for the dissociation of typical monoprotic, diprotic, and triprotic acids.

**5.22** Write chemical equations for the following neutralization reactions:

(a) HCl + KOH
(b) $H_2SO_4$ + NaOH
(c) $H_3AsO_4 + Ba(OH)_2$

**5.23** What is the net ionic equation characteristic of neutralization reactions in aqueous solution? Use Lewis symbols to diagram this reaction.

**5.24** What is an acid salt? Give some examples.

**5.25** A newly discovered element is found to react with oxygen to form an oxide that, when dissolved in water, causes litmus to turn blue. Would the element be classed as a metal or a nonmetal? Why?

**5.26** Describe how you would prepare the following by a metathesis reaction in which one product is a precipitate:
(a) $NH_4C_2H_3O_2$, (b) $Fe_3(PO_4)_2$, (c) $CuCO_3$, (d) $Na_2SO_4$, (e) $PbSO_4$.

**5.27** Describe how you would prepare the following by a metathesis reaction in which one product is a precipitate: (a) $Mg(OH)_2$, (b) $BaCl_2$, (c) $Fe(C_2H_3O_2)_2$, (d) $Ni(ClO_4)_2$, (e) $BaSO_3$.

**5.28** Describe how you would prepare the following by a metathesis reaction in which one product is a gas:
(a) $CaCl_2$, (b) $Mn(ClO_4)_2$, (c) $BaSO_4$, (d) $NaNO_3$, (e) $NH_4C_2H_3O_2$.

**5.29** Describe how you would prepare the following by a neutralization reaction:
(a) $Ca(NO_3)_2$, (b) $Na_2C_2O_4$, (c) $Fe(HSO_4)_2$, (d) $Al(ClO_4)_3$, (e) $NiBr_2$.

**5.30** How could you prepare
(a) $CuCl_2$ from $Cu(NO_3)_2$
(b) $BaCl_2$ from $BaBr_2$
(c) $NaClO_4$ from $Na_2SO_4$
(d) $Mg(C_2H_3O_2)_2$ from $MgCl_2$
(e) $Na_2CO_3$ from $Na_2SO_3$

**5.31** Balance the following equations by the oxidation-number-change method.
(a) $HNO_3 + Zn \rightarrow$
$Zn(NO_3)_2 + H_2O + NH_4NO_3$
(b) $K + KNO_3 \rightarrow N_2 + K_2O$
(c) $C_3H_7OH + Na_2Cr_2O_7 + H_2SO_4 \rightarrow$
$Cr_2(SO_4)_3 + Na_2SO_4 + H_2O + HC_3H_5O_2$
(d) $H_2S + HNO_3 \rightarrow S + NO + H_2O$
(e) $Fe(OH)_2 + O_2 + H_2O \rightarrow Fe(OH)_3$

**5.32** Balance the following equations by the oxidation-number-change method.
(a) $Cu + HNO_3 \rightarrow$
$Cu(NO_3)_2 + NO + H_2O$
(b) $MnO_2 + HBr \rightarrow Br_2 + MnBr_2 + H_2O$
(c) $(CH_3)_2CHOH + Cr_2O_3 + H_2SO_4 \rightarrow$
$(CH_3)_2CO + Cr_2(SO_4)_3 + H_2O$
(d) $PbO_2 + Sb + NaOH \rightarrow$
$PbO + NaSbO_2 + H_2O$
(e) $NO_2 + H_2O \rightarrow HNO_3 + NO$ ($NO_2$ is both oxidized *and* reduced)

**5.33** Identify the oxidizing agent in each reaction in Question 5.31.

**5.34** Balance the following equations by the ion-electron method. All reactions are in acid solution.
(a) $Cu + NO_3^- \rightarrow Cu^{2+} + NO$
(b) $Zn + NO_3^- \rightarrow Zn^{2+} + NH_4^+$
(c) $Cr + H^+ \rightarrow Cr^{3+} + H_2$
(d) $Cr_2O_7^{2-} + H_3AsO_3 \rightarrow Cr^{3+} + H_3AsO_4$
(e) $I^- + SO_4^{2-} \rightarrow I_2 + H_2S$
(f) $Ag^+ + AsH_3 \rightarrow H_3AsO_4 + Ag$
(g) $Cu + Cl^- + As_4O_6 \rightarrow CuCl + As$
(h) $MnO_2 + Br^- \rightarrow Br_2 + Mn^{2+}$
(i) $S_2O_3^{2-} + I_2 \rightarrow I^- + S_4O_6^{2-}$
(j) $IO_3^- + SO_3^{2-} \rightarrow I^- + SO_4^{2-}$

**5.35** Balance the following equations by the ion-electron method. All reactions are in acid solution.
(a) $Cr_2O_7^{2-} + CH_3CH_2OH \rightarrow$
$Cr^{3+} + CH_3CHO$
(b) $NO_2 \rightarrow NO_3^- + NO$
(c) $Mn^{2+} + BiO_3^- \rightarrow MnO_4^- + Bi^{3+}$
(d) $HSO_3NH_2 + NO_3^- \rightarrow N_2O + SO_4^{2-}$
(e) $PH_3 + I_2 \rightarrow H_3PO_2 + I^-$
(f) $MnO_4^- + S_2O_3^{2-} \rightarrow S_4O_6^{2-} + Mn^{2+}$
(g) $ClO_3^- + Cl^- \rightarrow Cl_2 + ClO_2$
(h) $As_2O_3 + NO_3^- \rightarrow H_3AsO_4 + N_2O_3$
(i) $P + Cu^{2+} \rightarrow Cu + H_2PO_4^-$
(j) $P + Cu^{2+} \rightarrow Cu_3P + H_3PO_3$

**5.36** Balance the following equations by the ion-electron method. All reactions are in basic solution.
(a) $Zn \rightarrow Zn(OH)_4^{2-} + H_2$
(b) $CrO_2^- + HO_2^- \rightarrow CrO_4^{2-}$
(c) $Br_2 \rightarrow Br^- + BrO_3^-$
(d) $ClO_2 \rightarrow ClO_2^- + ClO_3^-$
(e) $Zn + NO_3^- \rightarrow Zn(OH)_4^{2-} + NH_3$
(f) $Cu(NH_3)_4^{2+} + S_2O_4^{2-} \rightarrow$
$SO_3^{2-} + Cu + NH_3$
(g) $Fe(OH)_2 + O_2 \rightarrow Fe(OH)_3$
(h) $MnO_4^- + C_2O_4^{2-} \rightarrow MnO_2 + CO_2$
(i) $P_4 + OH^- \rightarrow PH_3 + H_2PO_2^-$
(j) $Al + OH^- \rightarrow AlO_2^- + H_2$

**5.37** Balance the following equations by the ion-electron method:
(a) $I_2O_4 \rightarrow IO_3^- + I^-$ (basic solution)
(b) $IPO_4 \rightarrow I_2 + IO_3^- + H_2PO_4^-$ (acid solution)
(c) $S + NO_3^- \rightarrow SO_4^{2-} + NO_2$ (acid solution)
(d) $MnO_4^- + H_2S \rightarrow Mn^{2+} + S$ (acid solution)
(e) $Se \rightarrow Se^{2-} + SeO_3^{2-}$ (basic solution)
(f) $ICl \rightarrow IO_3^- + I_2 + Cl^-$ (acid solution)
(g) $FNO_3 \rightarrow O_2 + F^- + NO_3^-$ (basic solution)
(h) $Ag^+ + AsH_3 \rightarrow H_3AsO_4 + Ag$ (acid solution)
(i) $Mn^{2+} + PbO_2 \rightarrow MnO_4^- + Pb^{2+}$ (acid solution)

**5.38** Describe in detail how you would prepare a 1.50 *M* KCl solution.

**5.39** Define molarity, normality, equivalence point.

**5.40** Why is $KMnO_4$ a convenient titrant in redox reactions?

## REVIEW PROBLEMS

**5.41** What is the molarity of each of the following?
(a) 1.50 moles of NaCl in 2.00 liters of solution
(b) 0.248 mole of KCN in 250 ml of solution
(c) 0.750 mole of $H_2SO_4$ in 1.35 liters of solution
(d) 85.5 g of $HNO_3$ in 1.00 liter of solution
(e) 44.5 g of $NH_4C_2H_3O_2$ in 600 ml of solution

**5.42** How many moles of solute are in: (a) 250 ml of 0.10 *M* KCl; (b) 1.65 liters of 1.40 *M* $HClO_4$; (c) 0.025 liter of 0.010 *M* $HC_2H_3O_2$.

**5.43** How many grams of $Na_2CO_3$ are required to prepare 300 ml of a 0.150 *M* solution?

**5.44** How many grams of $Ba(OH)_2$ are required to prepare 250 ml of a solution that has a hydroxide ion concentration of 0.300 *M*?

**5.45** Pure nitric acid has a density of 1.513 g/ml. What is its molar concentration?

**5.46** A solution of $MgSO_4$ contains 22.0% $MgSO_4$ by weight and contains 273.8 g of the salt per liter. What is the density of the solution? What is the molarity of the solution?

**5.47** How many milliliters of 0.300 *M* NaOH are required to react with 500 ml of 0.170 *M* $H_3PO_4$ to yield (a) $Na_3PO_4$, (b) $Na_2HPO_4$, (c) $NaH_2PO_4$?

**5.48** What weight of AgCl will be formed if 25.0 ml of 0.050 *M* HCl is added to 100 ml of 0.050 *M* $AgNO_3$?

**5.49** How many milliliters of 0.100 *M* $BaCl_2$ are required to react with 25.0 ml of 0.200 *M* $H_2SO_4$?

**5.50** How many milliliters of 0.1000 *M* $BaCl_2$ are required to react completely with 25.0 ml of 0.200 *M* $Fe_2(SO_4)_3$?

**5.51** A 0.244-g sample of benzoic acid (a monoprotic acid) requires 20.0 ml of 0.100 *M* NaOH for complete neutralization. Calculate the molecular weight of the acid.

**5.52** Caproic acid, which is found in certain excretions of goats, has an empirical formula of $C_3H_6O$. A 0.100-g sample of the acid required 17.2 ml of 0.0500 *M* NaOH for complete reaction. Assuming that the acid is monoprotic, calculate (a) its molecular weight; (b) its molecular formula.

**5.53** If 380 ml of 0.273 *M* $Ba(OH)_2$ is added to 500 ml of 0.520 *M* HCl, will the mixture be acidic or basic? Calculate the concentration of $H^+$ (or $OH^-$ if the solution is basic) in the final mixture. Assume that volumes are additive.

**5.54** 50.0 ml of 0.240 *M* $BaCl_2$ were added to 45.0 ml of 0.180 *M* $Fe_2(SO_4)_3$.
(a) What weight of $BaSO_4$ was formed?
(b) What are the concentrations of the remaining ions in the final solution?

**5.55** 20.0 ml of 0.200 *M* $AgNO_3$ were added to 30.0 ml of 0.200 *M* NaCl.
(a) What chemical reaction occurs?
(b) How many moles of precipitate are formed?
(c) How much does the precipitate weigh?
(d) What are the concentrations of each of the remaining ions in the final solution?

**5.56** A 0.249-g sample of a compound containing titanium and chlorine was dissolved in water and treated with silver nitrate solution. The silver chloride that formed was found to weigh 0.694 g after being filtered, washed, and dried. What is the empirical formula of the original compound?

**5.57** Mercury is an extremely toxic substance that deactivates enzyme molecules that promote biochemical reactions. A 25.0-g sample of tuna fish taken from a large shipment was analyzed for this substance and found to contain $2.1 \times 10^{-5}$ mole of Hg. By law, foods having a mercury content above 0.50 ppm cannot be sold (they cannot even be given away!). Determine whether this shipment of tuna must be confiscated.

*5.58 40.0 ml of 0.270 *M* $Ba(OH)_2$ are added to 25.0 ml of 0.330 *M* $Al_2(SO_4)_3$.
(a) Write the chemical equation for the reaction that occurs.
(b) What total weight of precipitate is formed?
(c) What is the concentration of each of the ions remaining in solution?

*5.59 A 1.850-g sample of a mixture of $CuCl_2$ and $CuBr_2$ was dissolved in water and mixed thoroughly with a 1.800-g portion of AgCl. After the reaction the solid, which now consisted of a mixture of AgCl and AgBr, was filtered, washed, and dried. Its mass was found to be 2.052 g. What percent of the original mixture was $CuBr_2$?

5.60 What is the equivalent weight of $MnSO_4$ when it is oxidized to produce (a) $Mn_2O_3$, (b) $MnO_2$, (c) $K_2MnO_4$, (d) $KMnO_4$?

5.61 How many grams of $NaBiO_3$ are required to react with 0.500 g of $Mn(NO_3)_2$ to produce $NaMnO_4$ and $Bi(NO_3)_3$?

5.62 What is the equivalent weight of
(a) $H_3PO_4$ when neutralized to $HPO_4^{2-}$
(b) $HClO_4$
(c) $NaIO_3$ when reduced to $I^-$
(d) $NaIO_3$ when reduced to $I_2$
(e) $Al(OH)_3$

5.63 What is the normality of each of the following solutions?
(a) 22.0 g of $Sr(OH)_2$ in 800 ml of solution
(b) 500 ml of 0.25 *M* $H_2SO_4$, for complete neutralization
(c) 0.150 *M* $H_3PO_4$, when neutralized to $HPO_4^{2-}$
(d) 41.7 g of $K_2Cr_2O_7$ in 600 ml of solution when used in a reaction where one product is $Cr^{3+}$
(e) 25.0 g of $Na_2O$ dissolved in sufficient water to give 1.50 liters of solution
(f) 0.135 equivalents of $H_2SO_4$ in 400 ml of solution

5.64 A volume of 129 ml of 0.850 *N* $Ba(OH)_2$ was required to completely neutralize a 4.93-g sample of an acid. What is the equivalent weight of the acid?

*5.65 In acid solution, 45.0 ml of $KMnO_4$ solution is required to react with 50.0 ml of 0.250 *N* $H_2C_2O_4$ to give $Mn^{2+}$ and $CO_2$ as products. How many milliliters of this same $KMnO_4$ solution is required to oxidize 25.0 ml of 0.250 *N* $K_2C_2O_4$ to yield, in basic solution, $MnO_2$ and $CO_2$ as products?

*5.66 A 10.0-ml portion of a solution of HCl was diluted to exactly 50.0 ml. If 5.0 ml of this solution required 41.0 ml of 0.255 *N* NaOH for complete neutralization, what was the concentration of the original HCl solution before dilution?

*5.67 Ascorbic acid (vitamin C) is a diprotic acid having the formula, $H_2C_6H_6O_6$. A sample of a vitamin supplement was analyzed by titrating a 0.1000-g sample dissolved in water with 0.0200 *M* NaOH. A volume of 15.2 ml of the base was required to completely neutralize the ascorbic acid. What was the percent ascorbic acid in the sample?

*5.68 A mixture of the monoprotic acids, lactic acid, $HC_3H_5O_3$ (found in sour milk) and caproic acid, $HC_6H_{11}O_2$, (found in excretions from the goat) was titrated with 0.0500 *M* NaOH. A 0.1000-g sample of the mixture required 20.4 ml of the base. What is the weight of each acid in the sample?

*5.69 A mixture of $MgCO_3$ and $CaCO_3$ (a dolomitic limestone used in agriculture) was heated to produce MgO and CaO. A 2.000-g sample of this oxide mixture was reacted with 100 ml of 1.00 *M* HCl. The excess HCl required 19.6 ml of 1.00 *M* NaOH for complete neutralization. What were the percentages of $CaCO_3$ and $MgCO_3$ in the original limestone sample?

*5.70 A sample of rock containing limestone ($CaCO_3$) was heated, converting the $CaCO_3$ to CaO. This was treated with $H_2O$ to give $Ca(OH)_2$, which was then titrated with HCl. In one analysis, a 0.2000-g sample, taken *after* converting the $CaCO_3$ to CaO as described above, required 30.3 ml of 0.1000 *M* HCl for complete neutralization. What was the weight percent $CaCO_3$ in the original rock?

*5.71 A 0.1000-g sample of a mixture of $FeSO_4$ and $Fe_2(SO_4)_3$ was dissolved in water and titrated with 0.00400 *M* $KMnO_4$. The titration required 15.8 ml of the $KMnO_4$ solution. What percent (by weight) of the mixture was $FeSO_4$?

5.72 What volume of 18.0 *M* $H_2SO_4$ must be added to 100 ml of $H_2O$ to give a solution of 5.0 *M* $H_2SO_4$?

*5.73 How many milliliters of 1.00 *M* HCl must be added to 50.0 ml of 0.500 *M* HCl to give a solution whose concentration is 0.600 *M*?

**5.74** What volume of concentrated $NH_3$ must be used to prepare 250 ml of 0.500 *M* $NH_3$?

**5.75** What volume of concentrated $H_2SO_4$ must be used to produce 400 ml of 3.0 *M* $H_2SO_4$ solution?

**5.76** To what volume must 100 ml of 0.500 *M* $H_2SO_4$ be diluted to give 0.200 *M* $H_2SO_4$?

**5.77** How many milliliters of 0.500 *N* $K_2Cr_2O_7$ must be used to completely oxidize the contents of 120 ml of 0.850 *N* $H_2C_2O_4$?

**5.78** How much water must be added to 85.0 ml of 1.00 *N* $H_3PO_4$ to produce 0.650 *N* $H_3PO_4$?

# 6 GASES

Matter is capable of existing in three different physical forms or **states:** solid, liquid, and gas. In the next three chapters we will examine the physical and chemical characteristics of these states and the transformations that occur among them. This chapter deals with the gaseous state, in which the intermolecular forces of attraction (the attraction one molecule experiences toward another) are sufficiently small to allow rapid, independent movement of the molecules.

The physical behavior of a gas is, to a first approximation, independent of its chemical composition and is determined, instead, by the variables volume, pressure, temperature, and the number of moles of the substance. Since these variables are of paramount importance, we will begin our discussion by taking a close look at them as they apply to the gaseous state.

## 6.1 VOLUME AND PRESSURE

When a gas is introduced into a container the molecules move freely within the walls, occupying its entire volume. The volume of a gas, then, is given simply by specifying the volume of the vessel in which it is held. Since gases mix freely with one another, if several gases were present each would occupy the same volume, that is, the entire volume of the container.

Pressure is defined as force per unit area; thus it is an intensive quantity formed as a ratio of two extensive quantities, force and area. For example, if a 100-lb force is exerted on a piston (Figure 6.1) whose total area is 100 in.$^2$, the pressure acting on each square inch is only 100 lb/100 in.$^2$ or 1 psi. If the same force is applied to a piston whose total area is 1.0 in.$^2$, the pressure exerted by the piston is 100 lb/1 in.$^2$, or 100 psi. In these examples we have considered the unit area to be 1 in.$^2$ The dependence of pressure on both force and the area over which it is spread has been experienced firsthand by anyone who has ever stepped on a nail. A 110-lb person stepping on even a dull nail having a point area of 0.01 in.$^2$ will experience a pressure of *11,000 psi!* This is more than enough to cause the nail to puncture the skin.

If the pressure generated by a piston is applied to a fluid (a gas or liquid), as illustrated in Figure 6.1, it is transmitted uniformly in all directions so that all of the walls of the container experience the same pressure. If the piston is supported by the fluid, then the fluid also exerts an equal pressure on the piston as well as the other walls of the container.

The ability of trapped gases to exert a pressure is demonstrated when an automobile tire is inflated. In most cases the four tires that support the car are inflated to a pressure of approximately 28 lb/in.$^2$ (psi). The reason a tire

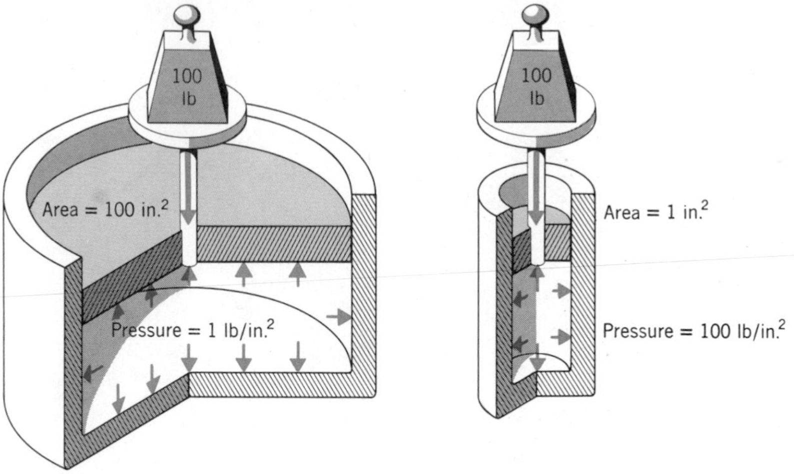


Figure 6.1
*Pressure. The fluid in the cylinder exerts the same pressure on all of the walls of the container.*

becomes flat when it "springs a leak" is because gases flow from a region of high pressure to a region of lower pressure; in our example, this flow is from inside the tire to the atmosphere.

The atmosphere of the earth is a mixture of gases that exert a pressure that, logically enough, we call the atmospheric pressure. We measure this pressure using a device called a **barometer.** A barometer, Figure 6.2, can be constructed by filling a glass tube with mercury and inverting it (without spilling any) into a mercury reservoir so that the open end is submerged. In Figure 6.2 we see that the mercury in the tube does not completely pour out when it is inverted; rather it maintains a particular height ($h$) above the reservoir. The height of this column is found to be independent of the diameter and length of the glass tube, as long as a space appears over the

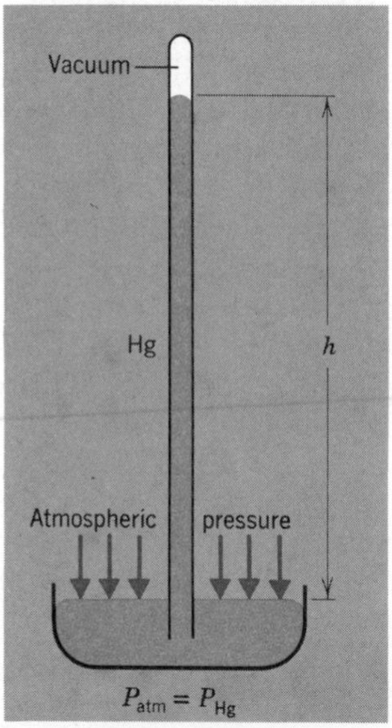


Figure 6.2
*A barometer. The atmospheric pressure supports a column of mercury of height* h.

mercury in the column. This space, for all practical purposes, is a vacuum ($P \approx 0$).

To determine atmospheric pressure we compare the various pressures acting along a reference level, which we choose as the surface of the reservoir. At this level outside the inverted tube, the pressure is caused by the downward force of the gases in the atmosphere ($P_{atm}$). Inside the tube the pressure at the reference level is caused by the downward pull of gravity on the mercury in the column ($P_{Hg}$). When these two opposing pressures are exactly equal ($P_{Hg} = P_{atm}$), the mercury in the column remains stationary. Atmospheric pressure, then, is directly related to the length ($h$) of the column of mercury in a barometer and can, therefore, be expressed in units of centimeters (cm Hg) or, more commonly, millimeters of mercury (mm Hg). A standard unit of pressure, the **standard atmosphere (atm)**, is defined as the pressure that will support a column of mercury 760 mm in length measured at 0°C and at sea level; that is, 1 atm = 760 mm Hg.[1] In English units this corresponds to a pressure of 14.7 lb/in.$^2$

Another unit of pressure is the **torr**, named after Evangelista Torricelli, the inventor of the barometer. One torr is equal to the pressure exerted by a column of mercury 1 mm high; that is, 1 torr = 1 mm Hg.

The SI unit of pressure is the *pascal* (Pa), defined as 1 newton per square meter (1 N/m$^2$). The standard atmosphere equals 101,325 Pa. In the chemistry laboratory we find torr and atm to be more convenient units with which to work than the pascal. For simplicity, therefore, we shall not be using this SI unit in our computations.

A barometer similar to that described above could also be constructed using water as a liquid. The length of the column of water would be considerably greater than that of a column of mercury, because the atmospheric pressure would be supporting a less dense liquid ($d_{water}$ = 1.00 g/ml, $d_{Hg}$ = 13.6 g/ml).

EXAMPLE 6.1 If water were the liquid in a barometer, what would be the length ($h$) of the water column at 1 atm of pressure?

SOLUTION The density of Hg is 13.6 times that of $H_2O$. This means that in order to have the same mass of mercury and water, the volume of water must be 13.6 times as great as the volume of mercury. Since we are comparing columns of the same diameter, the water column would have to be 13.6 times as long as the Hg column to contain 13.6 times as much volume. Thus,

$$1 \text{ mm Hg} \sim 13.6 \text{ mm } H_2O$$

Then

$$1 \text{ atm} \sim 760 \text{ \sout{mm Hg}} \times \left(\frac{13.6 \text{ mm } H_2O}{1 \text{ \sout{mm Hg}}}\right) \sim 1.03 \times 10^4 \text{ mm } H_2O$$

In general, as the density of the liquid being supported in a column by some external pressure increases, the length of the column of liquid decreases.

[1] The length of the column of mercury, which is supported by atmospheric pressure, varies with both the density of the mercury and the pull of gravity on the mercury in the column. Since density varies with temperature and the pull of gravity varies with altitude, then, in the definition of the standard atmosphere, it is necessary to specify a reference temperature (0°C) as well as a reference altitude (sea level).

Often it is desirable to know the pressure of a gas present in a closed system (for example, the pressure of gases produced during a chemical reaction). The instrument used for these pressure measurements is called a **manometer.** An open-end manometer (Figure 6.3) is simply a U-shaped tube containing some liquid, such as mercury. One arm of the tube is connected to a system whose pressure is to be measured while the other arm remains open to the atmosphere. When the pressure of the gas inside the system ($P_{gas}$) is equal to $P_{atm}$, the level of the liquids in both arms will be the same as shown in Figure 6.3*a*. If the pressure of the gas is greater than $P_{atm}$, the mercury in the left arm will be forced downward, causing the mercury in the right arm to rise (Figure 6.3*b*). We obtain the pressure of the gas in this system by comparing the pressures exerted in both arms at some reference level, $h_0$ (in a manometer this reference level is chosen to be the height of the shortest column). The pressure exerted on the left column when $P_{gas} > P_{atm}$ is simply $P_{gas}$, while at the same level in the right arm the pressure is $P_{atm}$ plus the pressure exerted by the column of mercury that rises above the reference level, $P_{Hg}$. When the levels are stationary,

$$P_{gas} = P_{atm} + P_{Hg}$$

The atmospheric pressure ($P_{atm}$) is found with a barometer, and $P_{Hg}$ is simply the difference in the heights of the two mercury columns ($\Delta h$). Similarly, when $P_{gas} < P_{atm}$, shown in Figure 6.3*c*, the pressure in the left arm at the reference level is $P_{gas} + P_{Hg}$, while in the right column the pressure is $P_{atm}$. In this case, when the columns are stationary,

$$P_{gas} + P_{Hg} = P_{atm}$$

so that

$$P_{gas} = P_{atm} - P_{Hg}$$

Therefore, when $P_{gas} < P_{atm}$ the pressure of the gas in the system is found by subtracting the difference in the heights of the columns from atmospheric pressure.

A closed-end manometer is generally used for the measurement of low pressures (usually much smaller than atmospheric pressure). This manom-

Figure 6.3
*Open-end manometer.*

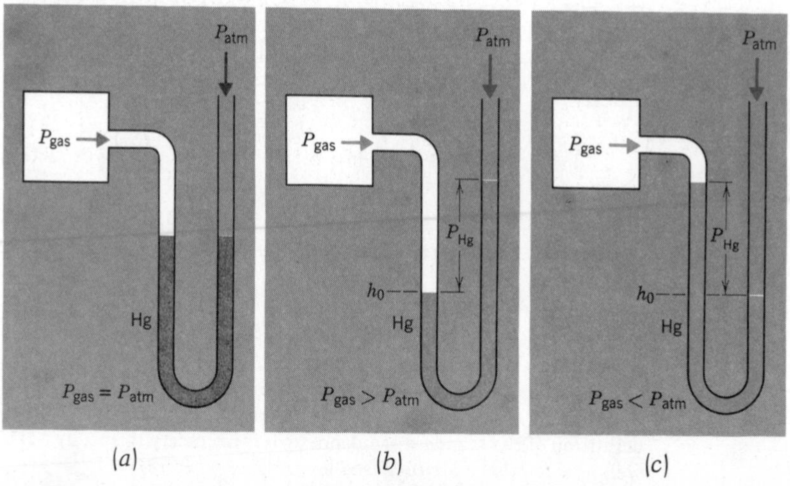

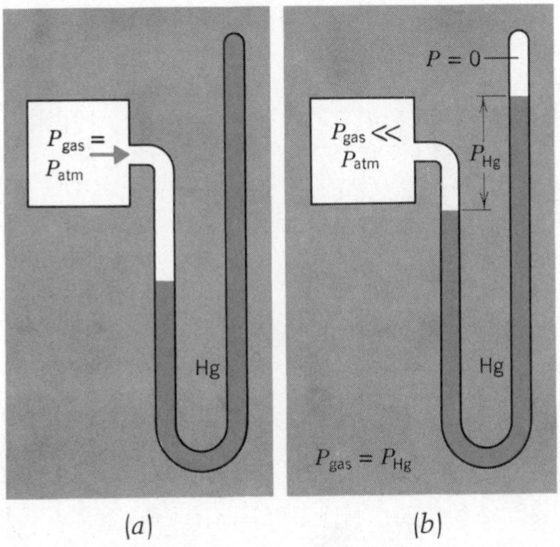


Figure 6.4
*A closed-end manometer.*

eter consists of a U-shaped tube with one arm closed and the other connected to the system as shown in Figure 6.4. When the pressure of the gas in the system is equal to $P_{atm}$, the right arm is completely filled while the left arm is only partially filled. If the pressure of the gas in the system is reduced, the level in the left arm will increase, which will cause the level in the right arm to decrease, as shown in Figure 6.4*b*. At the reference level the pressure exerted on the left arm is $P_{gas}$, while on the right arm the pressure is $P_{Hg}$ (the space above the mercury is a vacuum). When the columns are stationary, $P_{gas} = P_{Hg}$, and the pressure exerted by the gas in the system is simply found as the difference in the heights of liquid in the two arms of the manometer.

## 6.2 BOYLE'S LAW

By using an apparatus similar to that in Figure 6.5, Robert Boyle found that at constant temperature the volume of a fixed quantity of trapped gas decreases as the pressure on the gas is increased. Anyone who has ever used a bicycle pump is aware of this inverse relationship between the pressure and volume of a gas. As the piston of the pump is forced downward, the gas is compressed into a smaller volume while its pressure is raised (Figure 6.6). If we allow the compressed gas to escape, we can, for instance, use it to inflate a tire.

Boyle repeated his experiments many times with several different gases and found his initial observation to be a universal property of all gases. The results of his experiments can be formulated into **Boyle's law**, which states that *at a constant temperature, the volume occupied by a fixed quantity of gas is inversely proportional to the applied pressure.* This can be expressed mathematically as

$$V \propto \frac{1}{P} \tag{6.1}$$

The proportionality can be made into an equality by the introduction of a proportionality constant. Thus

$$V = \text{constant} \cdot \frac{1}{P}$$

or

$$PV = \text{constant} \tag{6.2}$$

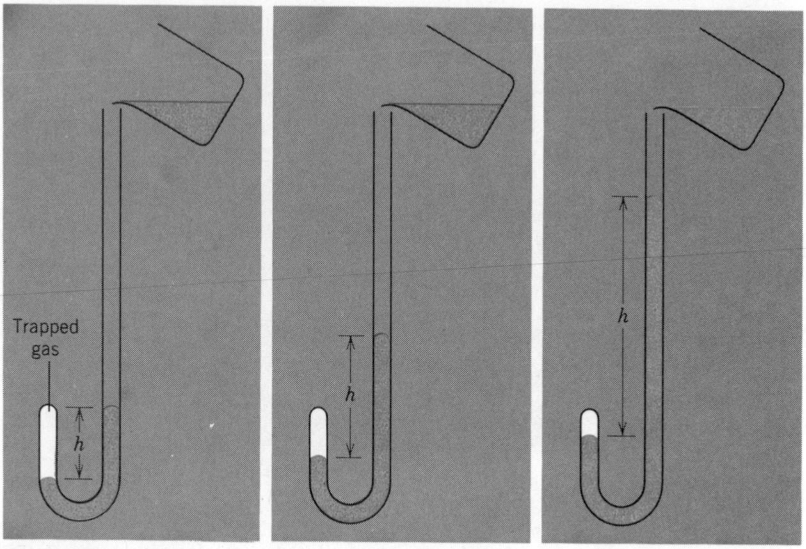


Figure 6.5
*Boyle's law apparatus. Pressure on the trapped gas is increased by adding mercury to U-tube. The volume of trapped gas decreases as the pressure is raised.*

Equation 6.2 states that for a given quantity of a gas at constant temperature, the product of its pressure and volume is a constant. If the pressure increases, the volume must decrease to keep the product of $P \times V$ constant.

The inverse relationship between $P$ and $V$ predicted from Boyle's law is shown graphically by the solid line in Figure 6.7. Real gases, however, do not follow this predicted behavior exactly, as shown by the dashed line. At very high pressures the measured volume is always somewhat larger than that calculated from Boyle's law. At low pressures, however, a real gas generally does follow Boyle's law quite closely, and we call its behavior ideal. A hypothetical gas that would follow Boyle's law under all conditions is called an **ideal gas.** Deviations from Boyle's law that occur with real gases therefore

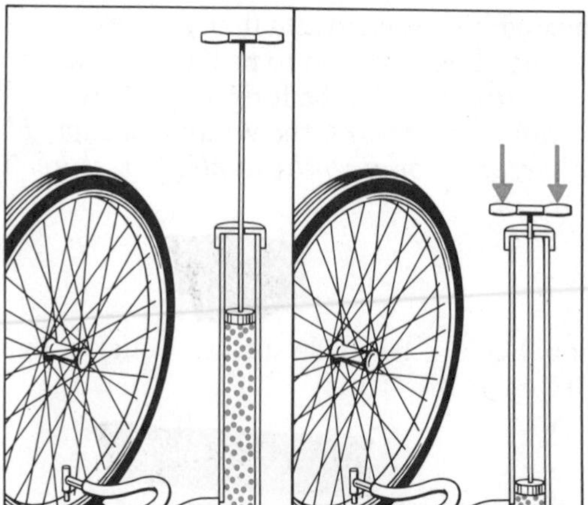

Figure 6.6
*A bicycle pump, an example of Boyle's law.*

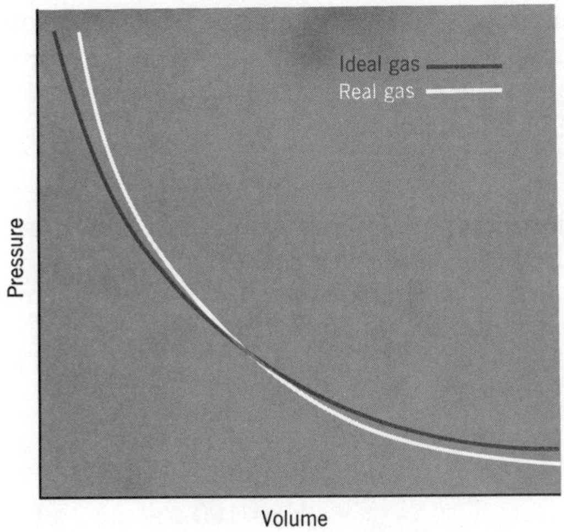


Figure 6.7
*Boyle's law. The colored line illustrates how volume is affected by pressure for an ideal gas. The behavior of a real gas is shown by the white line.*

represent *nonideal behavior.* The degree of nonideality differs for different gases.

Under conditions usually encountered in the laboratory, and to the degree of precision of most of our calculations, gases generally behave ideally; that is, they obey Boyle's law. Equation 6.1, therefore, is useful in calculating the effect of a pressure change on the volume of a gas at constant temperature.

EXAMPLE 6.2 If 100 ml of a gas, originally at 760 torr, is compressed to a pressure of 800 torr, at a constant temperature, what would be its final volume?

SOLUTION 1 In problems of this type, which deal with two sets of conditions, it is helpful to set up a table containing all of the data.

| | Initial ($i$) | Final ($f$) |
|---|---|---|
| Pressure ($P$) | 760 torr | 800 torr |
| Volume ($V$) | 100 ml | ? |

According to Boyle's law, the $PV$ products in both initial and final states are equal to the same constant; therefore they must be equal to each other.

$$P_iV_i = P_fV_f$$

Solving for the unknown final volume,

$$V_f = V_i\left(\frac{P_i}{P_f}\right) \qquad (6.3)$$

We now substitute values from our table.

$$V_f = 100\text{ ml} \times \left(\frac{760\text{ torr}}{800\text{ torr}}\right)$$

and

$$V_f = 95.0\text{ ml}$$

SOLUTION 2

The same result can be obtained following the reasoning of Boyle's law. If the pressure is increased on a quantity of gas at constant temperature, its volume must decrease. We have just seen (Equation 6.3) that $V_f$ is related to $V_i$ by a ratio of pressures,

$$V_f = V_i \times (\text{ratio of pressures})$$

We know that in this problem the final volume must be smaller than the initial volume because the pressure on the gas is increasing as we go from the initial to the final state. This requires that the initial volume be multiplied by a factor smaller than one; that is,

$$V_f = V_i \times \left(\frac{760 \text{ torr}}{800 \text{ torr}}\right)$$

Since

$$V_i = 100 \text{ ml}$$

then

$$V_f = 100 \text{ ml} \times \left(\frac{760 \text{ torr}}{800 \text{ torr}}\right)$$

$$V_f = 95.0 \text{ ml}$$

## 6.3 CHARLES' LAW

Jacques Charles investigated the effect of temperature changes on the volume of a given quantity of gas held at constant pressure. He found that, if a gas is heated so that the pressure remains constant, the gas will expand. If we plot data gathered in such experiments, we obtain a graph similar to that shown in Figure 6.8, where the volume is plotted against temperature in degrees Celsius. Each line represents the results of a series of measurements performed on a different quantity of gas. The straight lines obtained show that there is a direct proportionality between the volume of a gas and its temperature. All real gases ultimately condense if they are cooled to a sufficiently low temperature, and the solid portions of the lines in Figure 6.8 represent the temperature region above the condensation (liquefaction) point, where the substance is in the gaseous state. When these straight lines are extended back (extrapolated), they all intersect at the same point corre-

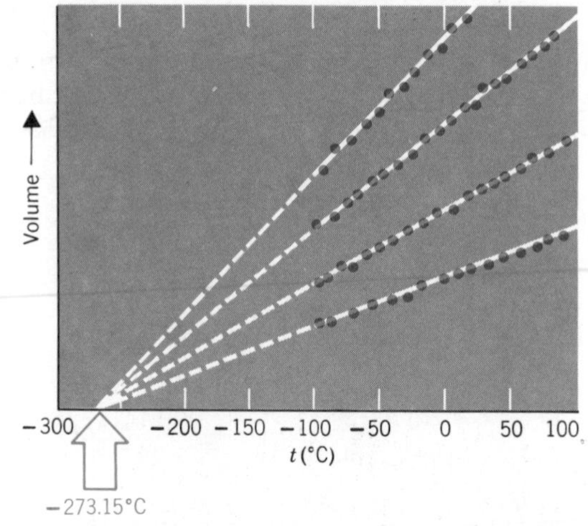


Figure 6.8
*Charles law plot of* V *versus* t(°C).

sponding to zero volume and at a temperature of −273.15°C. It is also found that, regardless of the gas, similar extrapolations yield the same temperature. This point represents a temperature at which all gases, if they did not condense, would have a volume of zero, and below which they would have a negative volume. Since negative volumes are impossible, this temperature must represent the lowest temperature possible and is called **absolute zero.**

Absolute zero represents the zero point on the kelvin temperature scale, described in Chapter 1. The kelvin scale is also called the **absolute temperature scale.** For general purposes we will use only three significant figures, so that 0 K = −273°C. To obtain the kelvin temperature we add 273 to the Celsius temperature,

$$T(\text{K}) = T(°\text{C}) + 273$$

The absolute temperature scale is *always* used when the temperature enters numerically into a computation involving pressures and volumes of gases.

The relationship that exists between the volume of a gas and its absolute temperature, represented in Figure 6.8, is summarized by Charles' law, which states that, *at constant pressure, the volume of a given quantity of a gas varies directly with its absolute temperature.* Writing Charles' law mathematically, we have

$$V \propto T \tag{6.4}$$

Making the proportionality an equality and rearranging, we obtain

$$\frac{V}{T} = \text{constant} \tag{6.5}$$

If Charles' law were strictly obeyed, gases would not condense when they are cooled. Condensation is, therefore, considered to be nonideal behavior, and all real gases behave more and more nonideally as their condensation temperatures are approached. This means that *gases behave in an ideal fashion only at relatively high temperatures and low pressures.* An example of the application of Charles' law is shown below.

EXAMPLE 6.3 A sample of a gas occupies 250 ml at 27°C. What volume will it occupy at 35°C if there is no change in pressure?

SOLUTION Once again we should set up our table of data.

| | Initial $(i)$ | Final $(f)$ |
|---|---|---|
| Volume $(V)$ | 250 ml | ? |
| Temperature $(T)$ | 27 + 273 = 300 K | 35 + 273 = 308 K |

Note that we have converted to absolute temperatures.

Equation 6.5 implies that

$$\frac{V_i}{T_i} = \frac{V_f}{T_f}$$

If we solve for $V_f$, we find that the initial and final volumes are related as

$$V_f = V_i \times (\text{ratio of absolute temperatures})$$

Without actually solving the equation, we can obtain the correct temperature ratio through reasoning. Gases expand when heated. Since the temperature is increasing from 300 K to 308 K, the final volume must be larger than the initial volume. The temperature ratio must be a fraction having a value greater than one. This requires the larger temperature in the numerator. Therefore,

$$V_f = 250 \text{ ml} \times \left(\frac{308 \text{ K}}{300 \text{ K}}\right)$$

$$V_f = 257 \text{ ml}$$

When working with gases it is useful to define a reference set of conditions of temperature and pressure. These conditions, known as **standard temperature and pressure,** or simply **STP,** are 0°C (273 K) and 1 atm (760 torr).

The equations corresponding to Boyle's law and Charles' law can be incorporated into one single equation that is useful for many computations. This is

$$\frac{P_iV_i}{T_i} = \frac{P_fV_f}{T_f} \tag{6.6}$$

Notice that if $T_i = T_f$, the temperature may be dropped and Equation 6.6 reduces to a statement of Boyle's law (that is, $P_iV_i = P_fV_f$ at constant temperature). Similarly, if $P_i = P_f$, the equation reduces to Charles' law. The following example illustrates how this equation can be applied.

EXAMPLE 6.4 What would be the volume of a gas at STP if it was found to occupy a volume of 255 ml at 25°C and 650 torr?

SOLUTION In this problem both temperature and pressure are changing. To compute the final volume we must combine Boyle's law and Charles' law. Equation 6.6 does this,

$$\frac{P_iV_i}{T_i} = \frac{P_fV_f}{T_f}$$

If we solve this equation for $V_f$, we find

$$V_f = V_i \times \text{(pressure ratio)} \times \text{(temperature ratio)}$$

Once again we can use reasoning to set up these ratios. First let us tabulate the data.

| | Initial ($i$) | Final ($f$) |
|---|---|---|
| $V$ | 255 ml | ? |
| $P$ | 650 torr | 760 torr } STP |
| $T$ | 298 K | 273 K } STP |

There is a pressure increase, which should tend to cause the volume of the gas to decrease. The pressure ratio should therefore be smaller than one, which requires the larger pressure in the denominator.

$$V_f = 255 \text{ ml} \times \left(\frac{650 \text{ torr}}{760 \text{ torr}}\right) \times (\text{temperature ratio})$$

The temperature is decreasing which, according to Charles' law, should cause a further decrease in the volume. The temperature ratio must also be smaller than one and the larger temperature must be in the denominator.

$$V_f = 255 \text{ ml} \times \left(\frac{650 \text{ torr}}{760 \text{ torr}}\right) \times \left(\frac{273 \text{ K}}{298 \text{ K}}\right)$$

$$V_f = 200 \text{ ml}$$

The volume at STP is 200 ml.

## 6.4 DALTON'S LAW OF PARTIAL PRESSURES

When two or more gases, which do not react chemically, are placed in the same container, the pressure exerted by each gas in the mixture is the same as it would be if it were the only gas in the container. The pressure exerted by each gas in a mixture is called its **partial pressure** and, as observed by John Dalton, the total pressure exerted by a mixture of gases is equal to the sum of the partial pressures of each gas in the mixture. This statement, known as **Dalton's law of partial pressures,** can be expressed as

$$P_T = p_a + p_b + p_c + \cdots$$

where $P_T$ is the total pressure of the mixture (which could be measured with a manometer) and $p_a$, $p_b$, and $p_c$ are the partial pressures of gases $a$, $b$, and $c$, respectively. For example, if nitrogen, oxygen, and carbon dioxide were placed in the same vessel, the total pressure of the mixture would be

$$P_T = p_{N_2} + p_{O_2} + p_{CO_2}$$

Thus, if the partial pressure of nitrogen were 200 torr, that of oxygen 250 torr, and that of carbon dioxide 300 torr, the total pressure of the mixture would be

$$P_T = 200 \text{ torr} + 250 \text{ torr} + 300 \text{ torr}$$
$$P_T = 750 \text{ torr}$$

Dalton's law can be useful in determining the pressure resulting from the mixing of two gases that were originally in separate containers, as shown by the following example.

EXAMPLE 6.5 If 200 ml of $N_2$ at 25°C and a pressure of 250 torr are mixed with 350 ml of $O_2$ at 25°C and a pressure of 300 torr, so that the resulting volume is 300 ml, what would be the final pressure of the mixture at 25°C?

SOLUTION From Dalton's law we know that we can treat each gas in the mixture as if it were the only gas present. Therefore, we can calculate *independently* the new pressures of $N_2$ and $O_2$ when they are placed in the 300-ml container. Since there is no temperature change, we have simply a Boyle's law calculation for each gas.

Following the methods used in Examples 6.2 through 6.4, we first set up our tables of data.

For $N_2$

| | $(i)$ | $(f)$ |
|---|---|---|
| $p$ | 250 torr | ? |
| $V$ | 200 ml | 300 ml |

For $O_2$

| | $(i)$ | $(f)$ |
|---|---|---|
| $p$ | 300 torr | ? |
| $V$ | 350 ml | 300 ml |

For each calculation we can write

$$p_f = p_i \times \text{(ratio of volumes)}$$

Since the volume of the $N_2$ is increasing, its pressure must decrease; $p_f$ must be less than $p_i$. This requires a volume ratio smaller than one, which means that the larger volume must be in the denominator. Thus

$$p_{N_2} = 250 \text{ torr} \times \left(\frac{200 \text{ ml}}{300 \text{ ml}}\right)$$

$$p_{N_2} = 167 \text{ torr}$$

For $O_2$ the volume is decreasing; $p_f$ must be greater than $p_i$. This requires a volume ratio larger than one.

$$p_{O_2} = 300 \text{ torr} \times \left(\frac{350 \text{ ml}}{300 \text{ ml}}\right)$$

$$p_{O_2} = 350 \text{ torr}$$

The total pressure of the mixture is the sum of the partial pressures.

$$P_T = p_{N_2} + p_{O_2} = 167 \text{ torr} + 350 \text{ torr}$$
$$P_T = 517 \text{ torr}$$

Gases prepared in the laboratory are quite often collected by the displacement of water, as shown in Figure 6.9. A gas collected in this manner becomes "contaminated" with water molecules that evaporate into the gas. These water molecules also exert a pressure called the **vapor pressure.** For reasons that we will discuss in Chapter 8, the vapor pressure of water depends *only* on the temperature of the liquid water (Table 6.1). The vapor pressure of the water contributes to the total pressure of the "wet" gas, and we can write

$$P_T = p_{\text{gas}} + p_{H_2O}$$

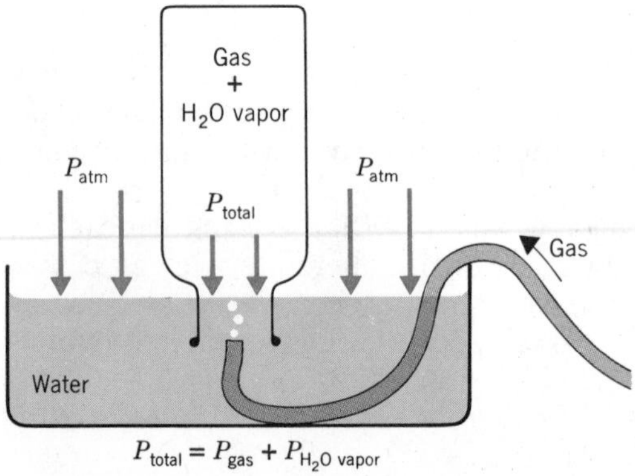


Figure 6.9
*Collection of a gas by displacement of water.*

Table 6.1
*Vapor pressure of water as a function of temperature*

| Temp. (°C) | Press. (torr) | Temp. (°C) | Press. (torr) | Temp. (°C) | Press. (torr) |
|---|---|---|---|---|---|
| 0 | 4.6 | 18 | 15.5 | 40 | 55.3 |
| 1 | 4.9 | 19 | 16.5 | 45 | 71.9 |
| 2 | 5.3 | 20 | 17.5 | 50 | 92.5 |
| 3 | 5.7 | 21 | 18.7 | 55 | 118.0 |
| 4 | 6.1 | 22 | 19.8 | 60 | 149.4 |
| 5 | 6.5 | 23 | 21.1 | 65 | 187.5 |
| 6 | 7.0 | 24 | 22.4 | 70 | 233.7 |
| 7 | 7.5 | 25 | 23.8 | 75 | 289.1 |
| 8 | 8.0 | 26 | 25.2 | 80 | 355.1 |
| 9 | 8.6 | 27 | 26.7 | 85 | 433.6 |
| 10 | 9.2 | 28 | 28.3 | 90 | 525.8 |
| 11 | 9.8 | 29 | 30.0 | 95 | 634.1 |
| 12 | 10.5 | 30 | 31.8 | 96 | 657.6 |
| 13 | 11.2 | 31 | 33.7 | 97 | 682.1 |
| 14 | 12.0 | 32 | 35.7 | 98 | 707.3 |
| 15 | 12.8 | 33 | 37.7 | 99 | 733.2 |
| 16 | 13.6 | 34 | 39.9 | 100 | 760.0 |
| 17 | 14.5 | 35 | 42.2 | 101 | 787.6 |

If the level of the water is the same inside the collection flask as outside, as shown in Figure 6.9, then the pressure inside must also be the same as outside, namely, atmospheric pressure. The atmospheric pressure can be determined with a barometer, and the vapor pressure of water can be obtained from Table 6.1 if the temperature of the liquid is known. The partial pressure of the pure gas is therefore

$$p_{gas} = P_T - p_{H_2O}$$

EXAMPLE 6.6 A student generates oxygen gas in the laboratory and collects it in a manner similar to that shown in Figure 6.9. He collects the gas at 25°C until the levels of the water inside and outside the flask are equal. If the volume of the gas is 245 ml and the atmospheric pressure is 758 torr:

(a) What is the partial pressure of $O_2$ gas in the "wet" gas mixture at 25°C?
(b) What would be the volume of dry oxygen at STP?

SOLUTION (a) For gases collected in this manner we know that

$$P_T = p_{gas} + p_{H_2O}$$

Substituting $p_{O_2}$ for $p_{gas}$ and rearranging, we have

$$p_{O_2} = P_T - p_{H_2O}$$

According to Table 6.1 the partial pressure of water at 25°C is 23.8 torr. The atmospheric pressure was given as 758 torr. Therefore, the partial pressure of $O_2$ is

$$p_{O_2} = 758 \text{ torr} - 23.8 \text{ torr}$$
$$p_{O_2} = 734 \text{ torr} \quad \text{(rounded to the correct number of significant figures)}$$

This is the pressure exerted by the oxygen alone.

(b) This part of the problem is a combined Boyle's law–Charles' law calculation. First we set up our table of data.

| | (*i*) | (*f*) |
|---|---|---|
| *V* | 245 ml | ? |
| *P* | 734 torr | 760 torr } STP |
| *T* | 298 K | 273 K } STP |

We now recall that

$$V_f = V_i \times \text{(ratio of pressures)} \times \text{(ratio of temperatures)}$$

The pressure change should tend to decrease the volume; therefore, the pressure ratio should be less than one. The temperature decrease should also decrease the volume; therefore, the temperature ratio should be less than one. Thus

$$V_f = 245 \text{ ml} \times \left(\frac{734 \text{ torr}}{760 \text{ torr}}\right) \times \left(\frac{273 \text{ K}}{298 \text{ K}}\right)$$

$$V_f = 217 \text{ ml at STP}$$

## 6.5 LAWS OF GAY-LUSSAC

When a gas is heated in a container of constant volume, its pressure increases. For example, the pressure in an automobile tire increases when the tire gets warm after being driven. In experiments similar to those performed by Charles, Joseph Gay-Lussac observed that if the volume of a given quantity of gas is held constant (e.g., in any closed vessel), the pressure of the gas varies directly with the absolute temperature; that is,

$$P \propto T$$

or

$$\frac{P}{T} = \text{constant} \tag{6.7}$$

This is a mathematical statement of the **law of Gay-Lussac.**

EXAMPLE 6.7 What would be the pressure of a gas, originally at 760 torr, if the temperature is lowered from 35°C to 25°C at a constant volume?

SOLUTION We should now be able to write that

$$P_f = P_i \times \text{(ratio of temperatures)}$$

Tabulating our data,

| | (i) | (f) |
|---|---|---|
| P | 760 torr | ? |
| T | 308 K | 298 K |

Since the temperature is decreasing, the pressure will decrease. This requires that the temperature ratio be smaller than one.

$$P_f = 760 \text{ torr} \times \left(\frac{298 \text{ K}}{308 \text{ K}}\right)$$

$$P_f = 735 \text{ torr}$$

Gay-Lussac also studied the volume changes that occurred when two or more gases were allowed to react with one another to form gaseous products at a constant temperature and pressure. He observed that the volumes of the gases that reacted and of those that were produced were related to one another in a simple fashion. For example, when hydrogen and oxygen are placed in a vessel and allowed to react with each other to form gaseous water, two volumes of hydrogen always react with one volume of oxygen to form two volumes of gaseous water. This can be expressed in the form of an equation as

$$\text{2 volumes hydrogen} + \text{1 volume oxygen} \longrightarrow \text{2 volumes gaseous water}$$

Similarly, when one volume of hydrogen reacts with one volume of chlorine, two volumes of hydrogen chloride gas are produced; that is,

$$\text{1 volume hydrogen} + \text{1 volume chlorine} \longrightarrow \text{2 volumes hydrogen chloride}$$

It could further be shown that the reaction between hydrogen and nitrogen to form ammonia is

$$\text{1 volume nitrogen} + \text{3 volumes hydrogen} \longrightarrow \text{2 volumes ammonia}$$

Such observations form the basis of **Gay-Lussac's law of combining volumes,** which states that *when reactions take place in the gaseous state, under conditions of constant temperature and pressure, the volumes of reactants and products can be expressed as ratios of small whole numbers.*

The significance of Gay-Lussac's observation was later recognized by Amadeo Avogadro. He proposed what is now known as **Avogadro's principle:** *Under conditions of constant temperature and pressure, equal volumes of gas contain equal numbers of molecules.* Since equal numbers of molecules mean equal numbers of moles (recall from Chapter 2 that there are $6.023 \times 10^{23}$ molecules in 1 mole of *any* substance), the number of moles of any gas is related directly to its volume:

$$V \propto n \qquad (6.8)$$

where $n$ is the number of moles of gas. On this basis Gay-Lussac's law is easily understood because the volumes of gaseous reactants and products occur

Table 6.2
*Molar volumes of several real gases at STP*

| Substance | Molar Volume (liters) |
|---|---|
| Oxygen, $O_2$ | 22.397 |
| Nitrogen, $N_2$ | 22.402 |
| Hydrogen, $H_2$ | 22.433 |
| Helium, He | 22.434 |
| Argon, Ar | 22.397 |
| Carbon dioxide, $CO_2$ | 22.260 |
| Ammonia, $NH_3$ | 22.079 |

in the same ratios as the coefficients in the balanced equation.[2] For example,

$$2H_2\ (g) + O_2\ (g) \longrightarrow 2H_2O\ (g)$$
$$2 \text{ volumes} + 1 \text{ volume} \longrightarrow 2 \text{ volumes}$$

From Avogadro's principle we expect 1 mole of any gas to occupy the same volume at a given temperature and pressure. It has been found experimentally that, at STP, 1 mole of gas occupies, on the average, 22.4 liters. We shall assume this to be the **molar volume** of an ideal gas at STP. For real gases the molar volume actually fluctuates about this average, as shown in Table 6.2.

EXAMPLE 6.8 What volume of $O_2$, at STP, is required for the complete combustion of 4.50 liters of $C_2H_6$, at STP?

SOLUTION As with any problem in stoichiometry, we should first write a balanced equation for the reaction. This is

$$2C_2H_6 + 7O_2 \longrightarrow 4CO_2 + 6H_2O$$

To solve the problem we can compute the number of moles of $C_2H_6$ using the molar volume at STP.

$$4.50 \text{ } \cancel{\text{liters } C_2H_6} \times \left(\frac{1 \text{ mole } C_2H_6}{22.4 \text{ } \cancel{\text{liters } C_2H_6}}\right) \sim 0.201 \text{ mole } C_2H_6$$

Next we calculate the number of moles of $O_2$ required, using the coefficients in the equation,

$$0.201 \text{ } \cancel{\text{mole } C_2H_6} \times \left(\frac{7 \text{ moles } O_2}{2 \text{ } \cancel{\text{moles } C_2H_6}}\right) \sim 0.704 \text{ mole } O_2$$

Finally we can calculate the volume of $O_2$, again using the molar volume of a gas at STP.

[2] In fact, these observations were used to show that gases such as $H_2$, $O_2$, and $Cl_2$ must be *at least* diatomic. The only way *two* volumes of hydrogen chloride could be formed from *one* volume of hydrogen and *one* volume of chlorine is if each molecule of hydrogen and chlorine contained two atoms of H and two atoms of Cl, respectively.

$$0.704 \text{ mole } O_2 \times \left(\frac{22.4 \text{ liters } O_2}{1 \text{ mole } O_2}\right) \sim 15.8 \text{ liters } O_2$$

The volume of $O_2$ required is 15.8 liters.

In problems involving gaseous reactants or products at the *same temperature and pressure,* we can take a shortcut to the answer. If we set up the calculation above with all of the conversion factors strung together, we have

$$4.50 \text{ liters } C_2H_6 \times \left(\frac{1 \text{ mole } C_2H_6}{22.4 \text{ liters } C_2H_6}\right) \times \left(\frac{7 \text{ moles } O_2}{2 \text{ moles } C_2H_6}\right) \times \left(\frac{22.4 \text{ liters } O_2}{1 \text{ mole } O_2}\right) \sim 15.8 \text{ liters } O_2$$

The number 22.4 appears in both numerator and denominator and therefore cancels. The volumes of reactants (or products) are simply related by the coefficients in the equation. Thus in this problem we could state that

$$2 \text{ liters } C_2H_6 \sim 7 \text{ liters } O_2$$

This is a consequence of Gay-Lussac's law. Realizing this, the solution to the problem could have been obtained as

$$4.50 \text{ liters } C_2H_6 \times \left(\frac{7 \text{ liters } O_2}{2 \text{ liters } C_2H_6}\right) \sim 15.8 \text{ liters } O_2$$

EXAMPLE 6.9 The drain cleaner, Drano, contains small bits of aluminum, which react with NaOH (the main ingredient in this product) to produce bubbles of hydrogen. These bubbles presumably are designed to stir the mixture and hasten its action. How many milliliters of $H_2$, measured at STP, will be released when 0.150 g of Al are dissolved? The chemical equation is

$$2Al + 2OH^- + 2H_2O \longrightarrow 3H_2 + 2AlO_2^-$$

SOLUTION First we calculate the number of moles of Al that react.

$$0.150 \text{ g Al} \times \left(\frac{1 \text{ mole Al}}{27.0 \text{ g Al}}\right) \sim 0.00556 \text{ mole Al}$$

Next we calculate the number of moles of $H_2$ produced.

$$0.00556 \text{ mole Al} \times \left(\frac{3 \text{ moles } H_2}{2 \text{ moles Al}}\right) \sim 0.00834 \text{ mole } H_2$$

Since 1 mole $H_2$ ~ 22.4 liters $H_2$ at STP,

$$0.00834 \text{ mole } H_2 \times \left(\frac{22.4 \text{ liters } H_2}{1 \text{ mole } H_2}\right) \sim 0.187 \text{ liter } H_2$$

Expressed in milliliters the answer is 187 ml $H_2$.

## 6.6 THE IDEAL GAS LAW

We have thus far discussed three volume relationships that an ideal gas obeys. These are

| | |
|---|---|
| Boyle's law | $V \propto \frac{1}{P}$ |
| Charles' law | $V \propto T$ |
| Avogadro's law | $V \propto n$ |

We can combine these to obtain

$$V \propto n \left(\frac{1}{P}\right) (T)$$

or

$$V \propto \frac{nT}{P} \qquad (6.9)$$

Equation 6.9 reduces to an expression of Boyle's law during a process where the volume changes as a result of a pressure change only. Since $n$ and $T$ remain constant, volume is proportional only to pressure; that is,

$$V \propto \frac{1}{P} \qquad \text{at constant } n \text{ and } T$$

Similarly, we can see that Equation 6.9 becomes Charles' law for a volume change at constant $n$ and $P$.

$$V \propto T \qquad \text{at constant } n \text{ and } P$$

Avogadro's principle is seen as

$$V \propto n \qquad \text{at constant } T \text{ and } P$$

The proportionality in Equation 6.9 can be made into an equality by the introduction of a proportionality constant, $R$, called the **universal gas constant.** Equation 6.9 then becomes

$$V = \frac{nRT}{P}$$

or

$$PV = nRT \qquad (6.10)$$

Equation 6.10 is obeyed exactly only by the hypothetical ideal gas and is a mathematical statement of the **ideal gas law.** It is also called the **equation of state for an ideal gas** because it relates those variables ($P$, $V$, $n$, $T$) that specify the physical properties of the gas. If any three of these are given, the fourth variable can have only one value as determined by Equation 6.10.

When a real gas comes very close to obeying the ideal gas law its behavior is said to be ideal. Fortunately, for most real gases under conditions of temperature and pressure encountered in the laboratory, the ideal gas law can be used quite accurately to describe their behavior. However, if extremely accurate computations are desired, Equation 6.10 cannot be used.

In order to use the ideal gas law we must have a value for the gas constant, $R$. This can be computed by inserting appropriate values for $P$, $V$, $n$, and $T$ into Equation 6.10 and solving for $R$. For 1 mole of an ideal gas at STP, $P = 1$ atm, $V = 22.4$ liters, $n = 1$ mole, and $T = 273$ K. Solving for $R$,

$$R = \frac{PV}{nT}$$

$$R = \frac{(1 \text{ atm})(22.4 \text{ liters})}{(1 \text{ mole})(273 \text{ K})}$$

$$R = 0.0821 \frac{\text{liter atm}}{\text{mole K}}$$

or

$$R = 0.0821 \text{ liter atm mole}^{-1} \text{ K}^{-1}$$

The constant $R$ may have other numerical values depending on the units used to express pressure and volume. Example 6.10 illustrates how we can convert $R$ from one set of units to another. The most useful values of $R$, with their corresponding units, are included on the inside back cover of this book.

EXAMPLE 6.10 What is the value of $R$ when pressure is expressed in torr and volume is expressed in milliliters?

SOLUTION This is simply a unit-conversion problem making use of the relationships

$$1 \text{ liter} = 1000 \text{ ml}$$
$$1 \text{ atm} = 760 \text{ torr}$$

$$R = \left(\frac{0.0821 \cancel{\text{liter}}\ \cancel{\text{atm}}}{\text{mole K}}\right) \times \left(\frac{1000 \text{ ml}}{1 \cancel{\text{liter}}}\right) \times \left(\frac{760 \text{ torr}}{1 \cancel{\text{atm}}}\right)$$
$$= \frac{6.24 \times 10^4 \text{ ml torr}}{\text{mole K}}$$

The choice of the value of $R$ to be used in a given computation is governed by the units of $P$ and $V$. Most people find it best, however, to learn one value of $R$ and to convert $P$ and $V$ to units that can be used with that value of $R$.

The following are some examples of the application of the ideal gas law.

EXAMPLE 6.11 What volume will 25.0 g of $O_2$ occupy at 20°C and a pressure of 0.880 atm?

SOLUTION From the ideal gas law,

$$V = \frac{nRT}{P}$$

We will use $R = 0.0821$ liter atm mole$^{-1}$ K$^{-1}$. Tabulating our data,

| | |
|---|---|
| $P$ | 0.880 atm |
| $V$ | ? |
| $n$ | $25.0 \text{ g of } O_2 \times \dfrac{1 \text{ mole of } O_2}{32.0 \text{ g of } O_2} = 0.781 \text{ mole}$ |
| $T$ | $20 + 273 = 293$ K |

Substituting,

$$V = \frac{(0.781 \cancel{\text{mole}}) \times (0.0821 \text{ liter } \cancel{\text{atm}}\ \cancel{\text{mole}^{-1}}\ \cancel{\text{K}^{-1}}) \times (293 \cancel{\text{K}})}{(0.880 \cancel{\text{atm}})}$$

$$V = 21.3 \text{ liters}$$

EXAMPLE 6.12 A student collected natural gas from a laboratory gas jet at 25°C in a 250-ml flask until the pressure of the gas was 550 torr. She then determined that the gas sample weighed 0.118 g at a temperature of 25°C. From these data, calculate the molecular weight of the gas.

SOLUTION We can determine the number of moles of gas present using the ideal gas law,

$$n = \frac{PV}{RT}$$

Again we shall use $R = 0.0821$ liter atm mole$^{-1}$ K$^{-1}$. Our data are

| | |
|---|---|
| $P$ | $550 \text{ torr} \times \frac{1 \text{ atm}}{760 \text{ torr}} = 0.724 \text{ atm}$ |
| $V$ | $250 \text{ ml} \times \frac{1 \text{ liter}}{1000 \text{ ml}} = 0.250 \text{ liter}$ |
| $n$ | ? |
| $T$ | $25 + 273 = 298$ K |

Substituting,

$$n = \frac{(0.724 \cancel{\text{atm}}) \times (0.250 \cancel{\text{liter}})}{(0.0821 \cancel{\text{liter}}\ \cancel{\text{atm}}\ \text{mole}^{-1} \cancel{\text{K}^{-1}}) \times (298 \cancel{\text{K}})}$$

$$n = \frac{0.00740}{\text{mole}^{-1}} = 0.00740 \text{ mole}$$

To calculate the molecular weight we must determine the weight of 1 mole of the substance. We now know that

$$0.118 \text{ g} \sim 0.00740 \text{ mole}$$

The weight of 1 mole, therefore, is

$$1 \cancel{\text{mole}} \times \left(\frac{0.118 \text{ g}}{0.00740 \cancel{\text{mole}}}\right) = 15.9 \text{ g}$$

Thus the molecular weight is 15.9. Natural gas is really methane, $CH_4$, having a molecular weight of 16.0.

EXAMPLE 6.13 A student determined the density of a gas to be 1.340 g/liter at 25°C and 760 torr. In a separate experiment he determined that the gas was composed of 79.8% carbon and 20.2% hydrogen.

(a) What is the empirical formula of the compound?
(b) What is its molecular weight?
(c) What is the molecular formula of the compound?

SOLUTION (a) Following the procedure outlined in Section 2.5, we find the empirical formula of the carbon–hydrogen compound.

$$79.8 \cancel{\text{g C}} \times \left(\frac{1 \text{ mole C}}{12.0 \cancel{\text{g C}}}\right) = 6.65 \text{ moles C}$$

$$20.2 \cancel{\text{g H}} \times \left(\frac{\text{mole H}}{1.01 \cancel{\text{g H}}}\right) = 20.0 \text{ moles H}$$

The empirical formula is $C_{\frac{6.65}{6.65}}H_{\frac{20.0}{6.65}}$ or $CH_3$ which would give an empirical formula weight of $CH_3$ = 15.0.

(b) The density gives the weight of 1 liter of the gas.

$$1 \text{ liter} \sim 1.340 \text{ g}$$

To calculate molecular weight we need a relationship between mass and moles. Following the procedure in Example 6.11,

| | |
|---|---|
| $P$ | 760 torr = 1 atm |
| $V$ | 1 liter |
| $n$ | ? |
| $T$ | 25 + 273 = 298 K |

$$n = \frac{PV}{RT} = \frac{(1 \cancel{\text{atm}}) \times (1 \cancel{\text{liter}})}{(0.0821 \cancel{\text{liter}}\ \cancel{\text{atm}}\ \text{mole}^{-1} \cancel{\text{K}^{-1}}) \times (298 \cancel{\text{K}})}$$

$$n = 0.0409 \text{ mole}$$

Thus,

$$0.0409 \text{ mole} \sim 1.340 \text{ g}$$

The weight of 1 mole is

$$1 \cancel{\text{mole}} \times \left(\frac{1.340 \text{ g}}{0.0409 \cancel{\text{mole}}}\right) = 32.8 \text{ g}$$

The molecular weight is 32.8.

(c) We see that the molecular weight is approximately twice the empirical formula weight, which means that the molecular formula must be

$$(CH_3)_2 \quad \text{or} \quad C_2H_6$$

This is a substance called ethane.

## 6.7 GRAHAM'S LAW OF EFFUSION

The ability of a gas to mix spontaneously with and spread throughout another gas, a process known as diffusion, is demonstrated every time we drive near a skunk that didn't quite make it across the road. It isn't long before the occupants of the car are in need of "a little fresh air." Effusion, on the other hand, is the process by which a gas, under pressure, escapes from one chamber of a vessel to another by passing through a very small opening, or orifice. This process is demonstrated in Figure 6.10.

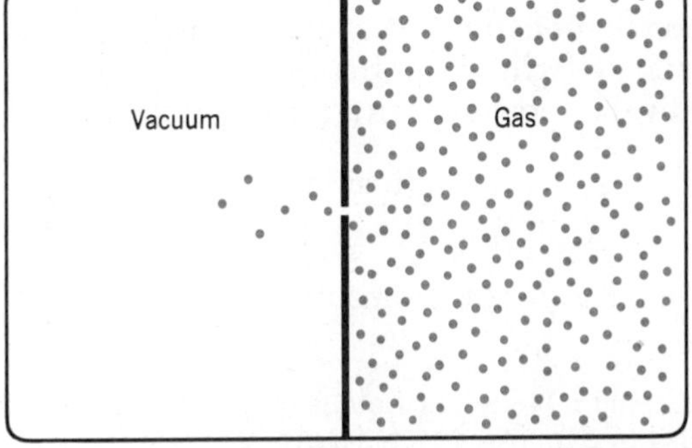


Figure 6.10
*Effusion of a gas into a vacuum.*

Thomas Graham demonstrated that when the rates of effusion of several gases are compared, the less dense gases (lighter gases) always effuse faster than the more dense ones. When the rates are compared under identical conditions of temperature and pressure, the best agreement between the rate of effusion and density is obtained when the rate is expressed as being inversely proportional to the square root of the density of the gas. This statement, known as **Graham's law**, can be expressed mathematically as

$$\text{rate of effusion} \propto \sqrt{\frac{1}{d}}$$

The rate of effusion of two gases (labeled simply $A$ and $B$) can be compared by dividing the rate of one by the other; that is,

$$\frac{\text{rate of effusion } (A)}{\text{rate of effusion } (B)} = \sqrt{\frac{d_B}{d_A}}$$

With the aid of the ideal gas equation, we can show that the density of a gas is directly proportional to its molecular weight. We know that the number of moles of a substance is obtained by dividing the weight, in grams, of that substance by its molecular weight; that is,

$$\text{number of moles } (n) = \frac{\text{weight in grams } (g)}{\text{molecular weight } (M)}$$

or simply

$$n = \frac{g}{M} \tag{6.11}$$

Making this substitution into the ideal gas equation, we obtain

$$PV = \left(\frac{g}{M}\right) RT \tag{6.12}$$

which can be rearranged to solve for $M$.

$$M = \left(\frac{g}{V}\right) \frac{RT}{P} \tag{6.13}$$

Density ($d$) is equal to $g/V$ (mass/volume) and, therefore, the above equation can be written as[3]

$$M = d\,\frac{RT}{P} \tag{6.14}$$

This equation demonstrates that at any particular temperature and pressure the molecular weight of a gas is directly proportional to its density. Graham's law can now be written to include this relationship, as

$$\frac{\text{rate of effusion } (A)}{\text{rate of effusion } (B)} = \sqrt{\frac{d_B}{d_A}} = \sqrt{\frac{M_B}{M_A}} \tag{6.15}$$

where $M_A$ and $M_B$ are the molecular weights of gases $A$ and $B$, respectively.

[3] Equations 6.12 and 6.14 are other forms of the ideal gas law that can be used in solving problems similar to Examples 6.12 and 6.13.

EXAMPLE 6.14 Which gas will effuse faster, ammonia or carbon dioxide? What are their relative rates of effusion?

SOLUTION The molecular weight of $CO_2$ is 44 and that of $NH_3$ is 17. Therefore $NH_3$ will effuse faster. We can calculate how much faster with Equation 6.15.

$$\frac{\text{rate of effusion } (NH_3)}{\text{rate of effusion } (CO_2)} = \sqrt{\frac{M_{CO_2}}{M_{NH_3}}} = \sqrt{\frac{44}{17}} = 1.6$$

Therefore the rate of effusion of $NH_3$ is 1.6 times faster than the rate of $CO_2$.

## 6.8 THE KINETIC MOLECULAR THEORY

Thus far in our discussion of the gaseous state we have seen that several laws governing gaseous behavior were formulated to account for experimental observations. We now need a theory to explain how gases function in obeying these laws. The kinetic molecular theory, which treats gases as being composed of randomly moving molecules, adequately explains these laws.

The observations of Robert Brown in 1827 laid the groundwork for the idea that the molecules comprising a gas are in continuous random motion. Brown observed that when microscopic particles are suspended in water, they undergo continuous erratic motion. He first observed this motion by following the path of grains of pollen that were suspended in water. This random motion of a microscopic particle when suspended in a liquid or gas is called **Brownian motion.** This motion, which becomes even more violent with increasing temperature, is believed to be caused by the suspended particle being constantly battered about by the molecules of the medium (liquid or gas). Brownian motion, therefore, is strong support for the idea that gases are composed of molecules that are constantly and randomly moving.

**POSTULATES OF THE KINETIC MOLECULAR THEORY.** The following are the basic postulates of the kinetic molecular theory. Since the real test of any theory manifests itself in how well it agrees with experimental observations, we shall look at the foundation on which each postulate is based and then see how the theory accounts for the observed gas laws.

**POSTULATE 1.** A gas is composed of tiny particles that have a negligibly small volume and that are separated from each other by relatively large distances. Gases therefore are mostly empty space.

**BASIS.** One of the fundamental characteristics of a gas is its very high compressibility (particularly when compared to liquids and solids). According to this postulate, it is this empty space that becomes smaller when gases are compressed.

**POSTULATE 2.** There are no intermolecular forces of attraction (the attraction of one molecule toward another) between the molecules of a gas. The molecules in a gas, therefore, travel in straight lines independently of each other.

**BASIS.** This postulate finds its support in the fact that even in those situations where the intermolecular forces of attraction are considered to be a maximum, for example, a highly compressed gas, the container holding them is completely filled. Since all gases spontaneously fill a container, there must be no appreciable intermolecular forces of attraction between the molecules.

**POSTULATE 3.** The molecules of a gas undergo constant motion, and their normally straight-line paths are interrupted only by collision with each other or with the walls of the container. All collisions occur with no net loss in kinetic energy of the molecules, and are said to be **elastic collisions.**

**BASIS.** Since it has been postulated that there are no intermolecular forces present to alter the paths of molecules, they should travel in straight lines. Because there are so many molecules present that are traveling extremely fast (the average speed of an $O_2$ molecule can be calculated to be over 1000 miles/hr at 0°C), many collisions between the molecules occur each second. A single gas molecule can undergo several million collisions each second, and each collision could cause the molecule to change its direction. If randomly moving molecules were constantly bombarding a microscopic particle, they could cause it to move randomly. Brownian motion, then, is explained by considering that the suspended particle is being hit by several gas molecules at the same time, as shown in Figure 6.11. If the particle is struck equally on all sides (Figure 6.11*a*), it will not move. However, there is a great probability that, because of its small size, the particle will receive an unequal number of impacts on opposite sides (Figure 6.11*b*), thereby causing the particle to move. The next instant its direction could similarly be changed, and the process continues with the particle moving in a random fashion.

The fact that molecular motion is continuous means that the intermolecular collisions are *elastic*. When an elastic collision occurs, there is no net loss in the kinetic energies of the two molecules. This means that the total kinetic energy of the two colliding molecules before the collision must equal their total kinetic energy after the collision. If, as a result of a collision, one molecule loses energy, the other molecule must gain it in order to keep the total kinetic energy constant. In 1 mole of gas at STP approximately $10^{30}$ collisions occur each second, and each could involve an exchange of kinetic energy. However, *overall* there is no net change in the total kinetic energy of the gas. If this were not the case and each collision resulted in a decrease in

Figure 6.11
*Brownian motion.*

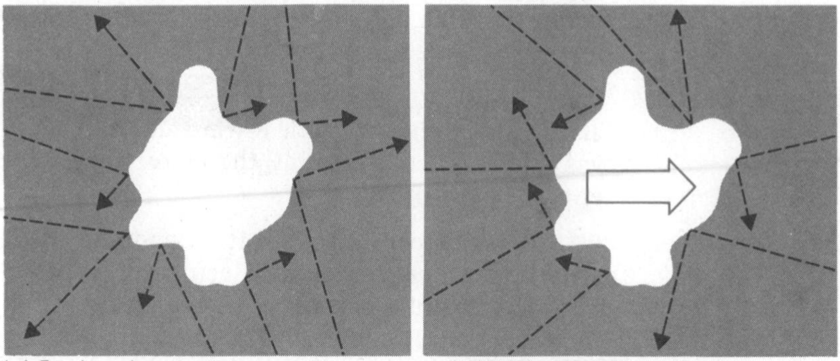

(*a*) Equal number of collisions on all sides – no displacement of particle.

(*b*) Unequal number of collisions on opposite sides – particle moves as shown by large arrow.

kinetic energy of the two molecules (inelastic collision), then each molecule would slow down upon each collision, because K.E. = $\frac{1}{2}ms^2$. For example, an inelastic collision occurs when a golf ball is thrown against a sidewalk. Initially the ball bounces very high in the air. But with each successive collision with the cement it loses energy and before long comes to rest. In a gas, inelastic collisions mean that the molecules would slow down and settle to the bottom of the container. Since this never occurs, the intermolecular collisions must be elastic.

**POSTULATE 4.** The pressure of a gas is the result of collisions between the gas molecules and the walls of the container.

**BASIS.** We have defined pressure as a force per unit area. When the molecules in a gas collide with the walls of a container, they each exert a force on that wall. The total force acting on a unit area is related directly to both the force of each collision and the frequency of the collisions with the wall. According to this postulate, an increase in pressure can be achieved by increasing either the force or frequency of molecule–wall collisions.

An increase in the number of collisions per second can be brought about by packing the molecules closer together so that there are more molecules in the vicinity of any given area of a wall. A decrease in the volume of a container does just this by forcing the molecules closer together; therefore, a decrease in volume is accompanied by an increase in pressure—Boyle's law.

**POSTULATE 5.** In a gas individual molecules possess different kinetic energies; however, the average kinetic energy of all the molecules collectively is directly proportional to the absolute temperature of the gas; that is, K.E. $\propto T$. The average kinetic energy of *any* gas is the same at the same temperature.

**BASIS.** This postulate leads directly to Graham's law. If two gases, which have different molecular weights, are at the same temperature, then, according to Postulate 5, their average kinetic energies are the same. This means that

$$\overline{\text{K.E.}_A} = \overline{\text{K.E.}_B}$$

or

$$\tfrac{1}{2}m_A\overline{s_A^2} = \tfrac{1}{2}m_B\overline{s_B^2} \tag{6.16}$$

where $\overline{s^2}$ is called the **mean square speed** of the molecules, and is the average of the speeds-squared of all the molecules; that is,

$$\overline{s^2} = \frac{s_1^2 + s_2^2 + s_3^2 + \cdots}{n_T}$$

where $s_1$, $s_2$, $s_3$, etc., represent the speeds of molecules 1, 2, 3, etc. and $n_T$ is the total number of molecules present. Equation 6.16 can be rearranged to give

$$\frac{\overline{s_A^2}}{\overline{s_B^2}} = \frac{m_B}{m_A}$$

Taking the square root of both sides, we have that

$$\frac{\bar{s}_A}{\bar{s}_B} = \sqrt{\frac{m_B}{m_A}} \qquad (6.17)$$

where $\bar{s}$ is called the root-mean-square speed. We have seen earlier that, for a given number of moles of a gas, the weight of the gas present is directly related to the molecular weight by Equation 6.11, which means that

$$M \propto m$$

Therefore, we can substitute $M_A$ and $M_B$ for $m_A$ and $m_B$ into Equation 6.17 and arrive at

$$\frac{\bar{s}_A}{\bar{s}_B} = \sqrt{\frac{M_B}{M_A}}$$

The rate at which gases effuse should be directly proportional to the velocity of their molecules, with faster molecules effusing at a higher rate. Thus we are led to conclude that

$$\frac{\text{rate of effusion }(A)}{\text{rate of effusion }(B)} = \frac{\bar{s}_A}{\bar{s}_B} = \sqrt{\frac{M_B}{M_A}}$$

or simply

$$\frac{\text{rate of effusion }(A)}{\text{rate of effusion }(B)} = \sqrt{\frac{M_B}{M_A}}$$

which is Graham's law.

We have seen that the kinetic molecular theory accounts for Boyle's law and Graham's law. The other gas laws also can be interpreted in terms of this theory.

**LAW OF GAY-LUSSAC ($P \propto T$ AT CONSTANT $V$).** According to Postulate 5, the average kinetic energy of a gas increases with increasing temperature. This means that the molecules of the gas are, on the average, moving faster at a higher temperature and hence collide with the walls of a container with greater force and with greater frequency. If the volume of the gas is held constant, therefore, the pressure must increase when the temperature of the gas is raised.

**CHARLES' LAW ($V \propto T$ AT CONSTANT $P$).** In order for the pressure to remain constant when the temperature of a gas is increased, the volume must increase as well in order to reduce the number of molecules colliding with any given area of the wall. In other words, if we allow the gas to expand as the temperature is raised, there are fewer collisions per square centimeter, and even though each occurs with greater force, the pressure (the total force per unit area) remains constant.

**DALTON'S LAW OF PARTIAL PRESSURES ($P_T = p_A + p_B + \cdots + p_N$).** This gas law follows as a direct consequence of Postulate 2. If there are no intermolecular forces of attraction, each gas molecule behaves independently of all of the others in the container. The pressure exerted by a particular gas in a mixture, therefore, depends only on the number of molecules of that gas present in the container. Thus, each gas exerts a partial pressure that is independent of the other gases and the total pressure of the mixture is the sum of the pressures exerted by each gas.

It also follows that the partial pressure of a particular gas in a mixture depends on its **mole fraction** (usually given the symbol, $X$); that is, the fraction of the total number of moles of gas that are contributed by the gas in question. For example, in a mixture of 1 mole of $O_2$ and 3 moles of $N_2$ at a total pressure of 2 atm, the mole fraction of $O_2$ is $\frac{1}{4} = 0.25$ and the partial pressure of $O_2$ is (0.25)(2 atm) = 0.50 atm. Similarly, for $N_2$ the mole fraction is $\frac{3}{4} = 0.75$ and its partial pressure is (0.75)(2 atm) = 1.50 atm. In general, we can write

$$p_A = X_A P_T \tag{6.18}$$

where $p_A$ is the partial pressure of $A$, $P_T$ is the total pressure, and $X_A$ is the mole fraction of gas $A$, given as

$$X_A = \frac{\text{number of moles of } A}{\text{total number of moles of gas in the mixture}}$$

**AVOGADRO'S PRINCIPLE ($V \propto n$ AT CONSTANT $P$ AND $T$).** If a gas is contained in a given volume at a particular temperature and pressure, there are a certain number of collisions per second with each square centimeter of the walls. If more gas is introduced into the container, the number of collisions per second per square centimeter must increase. The only way for the pressure to stay the same is if there is a volume increase accompanying the addition of gas.

## 6.9 DISTRIBUTION OF MOLECULAR SPEEDS

One of the most important and useful concepts to evolve from the kinetic theory is the idea that in a gas (or, in fact, in a liquid or a solid) there is a distribution of kinetic energies, and hence molecular speeds, that is dependent on temperature. Figure 6.12 on page 192 is a graphical display of this distribution (called a Maxwell-Boltzmann distribution) for a gas at three different temperatures, in which the fraction of the total number of molecules possessing a particular kinetic energy is plotted against kinetic energy.

At zero K.E., which corresponds to molecules standing still, this fraction is essentially zero since very few, if any, molecules are standing still at any instant. The fraction having a particular K.E. increases as we move to higher energies (higher speeds) and eventually becomes a maximum. At still higher kinetic energies the fraction decreases and gradually approaches zero again at kinetic energies corresponding to very fast-moving molecules. The curve does not go all the way to zero, however, because there is virtually no upper limit to molecular speeds, other than the speed of light.

The maximum on this curve represents the kinetic energy possessed by the largest fraction of molecules in the gas. This kinetic energy would be found most frequently (that is, with the greatest probability) if we examine molecules at random; hence it is called the *most probable kinetic energy.* The average kinetic energy occurs at a higher value than the most probable kinetic energy because the curve is not symmetrical. Just as a few "curve breakers" in a chemistry class tend to raise the class average on exams, the high-velocity molecules shift the average K.E. above the most probable K.E.

When the temperature of a gas is raised, the curve changes so that the average K.E. increases, as shown in Figure 6.12. More molecules have high speeds and fewer have low ones and, on the average, the molecules of the gas move faster. Thus when heat is added to a substance to raise its tempera-

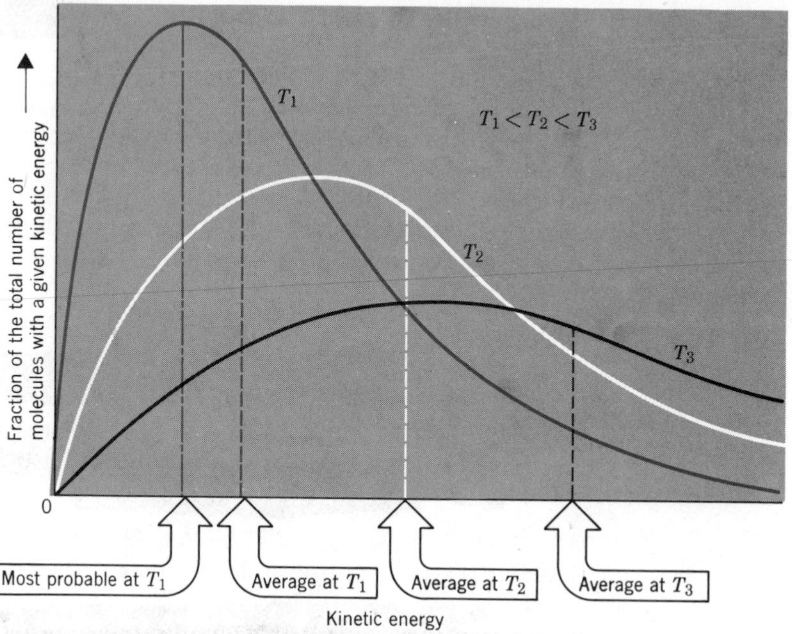


Figure 6.12
*Distribution of kinetic energies.*

ture, the energy goes into increasing the average kinetic energy and increases the motion of the particles.

The relationship between kinetic energy and temperature also leads quite naturally to the concept of an absolute zero. As kinetic energy is removed from a substance, its molecules move more and more slowly. If all of the molecules were to cease moving, their average kinetic energy would be zero and, since negative kinetic energies are impossible (a molecule cannot be going slower than when it is standing still), the temperature of the substance would also be at its lowest point. It is this temperature that we refer to as absolute zero—the temperature when all molecular motion has ceased.[4] It should be understood, however, that electronic motion would still continue at absolute zero. Even though the molecules would be motionless, the electrons would still be "whizzing" about their respective nuclei.

## 6.10 REAL GASES

A number of times it has been stated that real gases fail to obey the gas laws under all conditions of temperature and pressure. In Figure 6.7, for example, it was shown that at high pressures real gases occupy a larger volume than would be calculated from Boyle's law. Another example of nonideal behavior is observed when a substance, initially in the gas phase, is gradually cooled to temperatures approaching absolute zero. If substances were ideal

[4] In fact, even at absolute zero there is still a residual molecular motion that is required by the Heisenberg uncertainty principle (Chapter 3). This principle states that we cannot simultaneously know exactly the position and momentum of a particle. We shall see in the next chapter that the average positions of particles in the solid state can be determined, and if the particles were motionless we would also know that their momentum was zero, thereby violating the uncertainty principle.

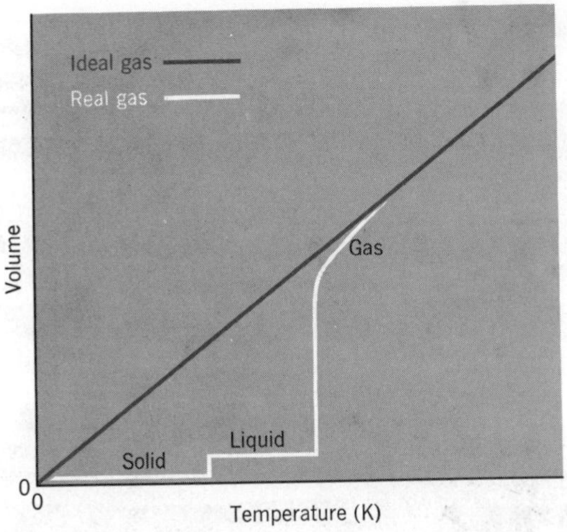


Figure 6.13
*Variation of volume with temperature for real and ideal gases.*

gases, their volumes would follow Charles' law. Instead, their behavior is similar to that shown in Figure 6.13. As a gas is cooled, its volume begins to fall below the Charles' law value. Then, suddenly the substance condenses to a liquid having a much smaller volume. At still lower temperatures it freezes to a solid. This nonideal behavior reveals much about the nature of the particles comprising the gas. To see this we must examine the postulates of the kinetic theory more closely.

According to Postulate 1, the volume of the gas molecules is negligibly small compared to the volume of the container in which they are held. This is not completely valid. Let us imagine, for simplicity, that gas molecules could be stopped in a container and allowed to settle to the bottom. We would see, contrary to Postulate 1, that part of the volume of the container is occupied by the gas molecules. The remaining free space is somewhat less than the volume of the container. If, in this hypothetical case, another gas molecule were introduced, it could move in the free space, but not the entire volume of the container. This same situation exists when the molecules are moving. The volume within which the molecules may not move is called the **excluded volume.** Thus the real volume (that is, the volume of the container) is actually slightly larger than the ideal volume (the volume the gas would occupy if the molecules themselves occupied no space). According to J. D. van der Waals, the real volume is

$$V_{\text{real}} = V_{\text{ideal}} + nb$$

where $b$ is the correction due to the excluded volume per mole and $n$, as usual, is the number of moles of gas. Solving for the ideal gas volume we have

$$V_{\text{ideal}} = V_{\text{real}} - nb \tag{6.19}$$

Contrary to Postulate 2, there are attractive forces between molecules in a gas. These must exist because as the gas is cooled the molecules begin to cling together and the gas becomes a liquid. Another manifestation of the attractive forces between gas molecules is the cooling that occurs when a compressed gas is allowed to expand freely into a vacuum. As the gas expands, the average distance of separation between the molecules increases. Since there are forces of attraction between them, moving the molecules apart requires work (energy). The source of this energy is the kinetic energy of the gas—in the process of expansion kinetic energy is converted to potential energy. This removal of kinetic energy leads, of course, to a decrease in the average kinetic energy of the gas and, since the average kinetic energy is directly related to temperature, the gas becomes cooler.

Van der Waals included a correction to the pressure of a gas that takes into account the attractive forces. A molecule in a gas that is just about to make a collision with the wall feels the attraction of all the molecules surrounding it. Since there are no molecules in front of it (or relatively very few), the greatest concentration of these forces is in a direction away from the wall (Figure 6.14). When the collision does take place, it is less energetic than it would be if there were no attractive forces present. The overall effect of these forces is a lowering of the pressure. The extent to which the pressure will be reduced is directly proportional to (1) the number of impacts with the wall, which in turn is directly proportional to the concentration of the molecules ($n/V$), and (2) the decrease in the force of each impact, which is also proportional to the concentration of the molecules. The decrease in the pressure, therefore, is directly proportional to the square of the concentration, or $n^2/V^2$. The ideal pressure, that is, the pressure the gas could exert in the absence of intermolecular attractive forces, is higher than the actual pressure by an amount that is directly proportional to $n^2/V^2$ or

$$P_{\text{ideal}} = P_{\text{real}} + \frac{n^2a}{V^2} \tag{6.20}$$

where $a$ is a proportionality constant that depends on the strength of the intermolecular attractions.

Substituting these corrected pressures and volumes (from Equations 6.19 and 6.20, respectively) into the ideal gas equation gives us

$$\left(P + \frac{n^2a}{V^2}\right)(V - nb) = nRT \tag{6.21}$$

which is the **van der Waals equation of state for a real gas.** This equation is much more complex than the ideal equation, but it does work well for many gases over fairly wide ranges of temperature and pressure.

The values of the constants $a$ and $b$ depend on the nature of the gas because the molecular volumes and the molecular attractions vary from gas to

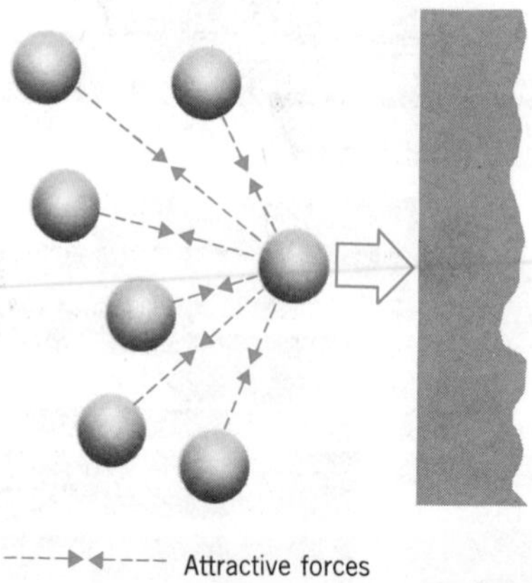


Figure 6.14
*Effect of attractive forces in real gases on pressure.*

Table 6.3
*van der Waals constants for real gases*

| | $a$ (liters$^2$ atm/mole$^2$) | $b$ (liters/mole) |
|---|---|---|
| He | 0.034 | 0.0237 |
| $O_2$ | 1.36 | 0.0318 |
| $NH_3$ | 4.17 | 0.0371 |
| $H_2O$ | 5.46 | 0.0305 |
| $CH_4$ | 2.25 | 0.0428 |
| $C_2H_6$ | 5.489 | 0.06380 |
| $CH_3OH$ | 9.523 | 0.06702 |
| $C_2H_5OH$ | 12.02 | 0.08407 |

gas. Some typical values of $a$ and $b$ are found in Table 6.3. We see that molecules containing many atoms, such as $C_2H_5OH$, have large values of $b$. This is not surprising, since such molecules would be expected to be larger than molecules containing only a few atoms.

The variation among the values of $a$ is also consistent with the trends expected for the intermolecular attractive forces. These kinds of forces were discussed in Chapter 4. Here we see that small nonpolar molecules, such as $O_2$ and monatomic He, have very small values of $a$. This is because there are only very weak London forces between the particles in the gas. Polar molecules, such as $NH_3$ and $H_2O$, have considerably larger values of $a$ because of the greater strengths of the intermolecular dipole–dipole attractions.

Comparing molecules such as $CH_4$ and $C_2H_6$, or $CH_3OH$ and $C_2H_5OH$, we find the larger molecule in each pair to have a larger value of $a$. This reflects an increase in the strengths of the London forces with an increase in the number of atoms in the molecule. These forces increase because as the number of atoms increase the chances for instantaneous dipole-induced dipole attractions to occur increase.

From the discussion above, we see that the values of $a$ and $b$ for a real gas enable us to expand our knowledge about the molecules of which it is composed. The van der Waals constants for a gas are obtained by making careful measurements of $P$, $V$, and $T$ and then choosing values of $a$ and $b$ that make the van der Waals equation produce the best match with the experimental data. In this sense, $a$ and $b$ are experimentally determined quantities that enable us to check our theories on molecular size and attractions.

## INDEX TO QUESTIONS AND PROBLEMS (problem numbers in italics)

## REVIEW QUESTIONS

6.1 Define pressure. Explain why the height of mercury in a barometer is independent of the cross-sectional area of the tube.

6.2 An open-end manometer containing mercury was connected to a vessel containing a gas at a pressure of 740 torr. The atmospheric pressure was 765 torr. Sketch a diagram showing the relative heights of the mercury in each arm of the manometer.

6.3 State Boyle's law in words. Is the law obeyed exactly by all gases? What is a gas that obeys Boyle's law called?

6.4 State Charles' law in words. In terms of Charles' law, why is $-273°C$ the lowest possible temperature?

6.5 What is STP?

6.6 What is Dalton's law of partial pressures? Can partial pressures actually be measured directly?

6.7 What is Gay-Lussac's law of combining volumes? What is Avogadro's principle?

6.8 Suppose that the ideal gas law were $PV^2 = nR/T^2$. What would be the units of $R$ if $P$ is in atm and $V$ in liters?

6.9 What determines the choice of the value of $R$ that you should use in an ideal gas law calculation?

6.10 What is Graham's law of effusion?

6.11 Without referring to the text, state in your own words the five basic postulates of the kinetic molecular theory.

6.12 If a warm object is placed in contact with a cool object, heat transfer occurs until they both come to the same temperature. How can the kinetic molecular theory account for this heat transfer and these temperature changes that occur?

6.13 Sketch a graph showing the distribution of kinetic energies for a gas at two different temperatures. For each temperature indicate the most probable K.E. and the average K.E. Why is the curve not symmetrical?

6.14 What is meant by nonideal behavior of a gas? Under what conditions is this behavior most evident? What errors are present in the postulates of the kinetic theory that are exposed by the nonideal character of real gases?

6.15 Explain, on the basis of the kinetic theory, why most gases cool upon expansion into a vacuum.

6.16 What physical significance do the constants $a$ and $b$ have in the van der Waals equation of state for real gases?

6.17 The $CH_3OH$ molecule contains fewer atoms than $C_2H_6$, yet it has a larger value of $a$. What do you think might explain this?

6.18 Carbon dioxide and ammonia have molar volumes significantly smaller than the other gases in Table 6.2. What is a plausible explanation for this? Does this also explain why $NH_3$ has a much smaller molar volume than $CO_2$?

## REVIEW PROBLEMS

6.19 An open-end manometer was connected to a flask containing a gas at an unknown pressure. The mercury in the arm open to the atmosphere was 65 mm higher than in the closed end. The atmospheric pressure was 733 torr. What was the pressure of the gas in the flask?

6.20 A manometer connecting two flasks (labeled A and B) contains an oil having a density of 0.847 g/ml. The oil in the arm connected to flask A is 74 cm higher than the oil in the arm connected to flask B. The gas in flask A has a pressure of 836 torr. What is the pressure of the gas in flask B?

6.21 A man digs a well and finds water 35 ft below the ground. If the average atmospheric pressure at the well is 1 atm, will the man be able to draw water from the well using a pump, mounted at ground level, that works by suction? Explain your answer.

6.22 A gas has a volume of 350 ml at 740 torr. What will its volume be at 900 torr if the temperature remains constant?

6.23 A sample of $SO_2$ occupies 1.45 liters at 2.75 atm. Assuming no temperature change, what volume will this gas occupy at 800 torr?

**6.24** A gas is compressed at constant temperature from a volume of 540 ml to 320 ml. If the initial pressure was 475 torr, what is the final pressure?

**6.25** A gas exerts a pressure of 20.0 lb/in.$^2$ in a container having a volume of 35.0 ft$^3$. What will its pressure be if the gas is transferred to a 40.0-ft$^3$ container at the same temperature?

**6.26** At 25°C and 1 atm a gas occupies a volume of 1.50 liters. What volume will it occupy at 100°C and 1 atm?

**6.27** A balloon has a volume of 2.0 liters indoors at a temperature of 25°C (77°F). If it is taken outdoors on a very cold winter day when the temperature is −28.9°C (−20°F), what will its volume be? Assume constant air pressure within the balloon.

**6.28** What would be the final volume of a 2.00-liter sample of a gas that is heated from 26 to 100°C at constant pressure?

**6.29** A sample of $O_2$ occupies 285 ml at 25.0°C. At what temperature will it occupy 350 ml if the pressure remains constant?

**6.30** At what temperature will a gas sample occupy 0.850 liter at 1 atm pressure if it occupies 400 ml at 32°C and 1 atm?

**6.31** If a gas, originally in a 50-ml container at a pressure of 645 torr, is transferred to another container whose volume is 65 ml, what would be its new pressure if
(a) There were no temperature change?
(b) The temperature of the first container was 25°C and that of the second was 35°C?

**6.32** A 300-ml sample of a gas exerts a pressure of 450 torr at 27°C. What pressure would it exert in a 200-ml container at 20°C?

**6.33** A 2.00-liter sample of a gas originally at 25°C, and a pressure of 700 torr, is allowed to expand to a volume of 5.00 liters. If the final pressure of the gas is 585 torr, what is its final temperature?

**6.34** The density of $CO_2$ is 1.96 g/liter at 0°C and 1 atm. Determine its density at 650 torr and 25°C.

**6.35** A gas exerts a pressure of 350 torr at 20°C. What pressure will it exert if its temperature is raised to 40°C without a change in volume?

**6.36** An automobile tire is inflated to a pressure of 29 lb/in.$^2$ at 65°F. After a trip the temperature of the tire has risen to 130°F. What will be the pressure in the tire, assuming that no air has leaked out and that the volume of the tire hasn't changed?

**6.37** A 50.0-ml sample of gas exerts a pressure of 450 torr at 35°C. What is its volume at STP?

**6.38** A 1.00-liter mixture of gases is produced from 1.00 liter of $N_2$ at 200 torr, 1.00 liter of $O_2$ at 500 torr, and 1.00 liter of Ar at 150 torr. What is the pressure of the mixture?

**6.39** A 1.00-liter flask is filled by placing in it the contents of a 2.00-liter flask of $N_2$ at 300 torr and a 2.00-liter flask of $H_2$ at 80 torr. What is the pressure of the mixture in the 1.00-liter flask?

**6.40** What would be the total pressure of a mixture prepared by adding 20.0 ml of $N_2$ at 0°C and 740 torr plus 30.0 ml of $O_2$ at 0°C and 640 torr to a 50.0-ml container at 0°C?

**6.41** A mixture of $N_2$ and $O_2$ has a volume of 100 ml at a temperature of 50°C and a pressure of 800 torr. It was prepared by adding 50 ml of $O_2$ at 60°C and 400 torr with $X$ ml of $N_2$ at 40°C and 400 torr. What is the value of $X$?

**6.42** A gas is collected by the displacement of water until the total pressure inside a 100-ml flask is 700 torr at 25°C. Calculate the volume of dry gas at STP.

**6.43** A mixture of $N_2$ and $O_2$ in a 200-ml vessel exerts a pressure of 720 torr at 35°C. If there are 0.0020 mole of $N_2$ present:
(a) What is the mole fraction of $N_2$? (See Equation 6.18.)
(b) What is the partial pressure of $N_2$?
(c) What is the partial pressure of $O_2$?
(d) How many moles of $O_2$ are present?

**6.44** How many milliliters of $CO_2$ at 30°C and 700 torr must be added to a 500-ml container of $N_2$ at 20°C and 800 torr to give a mixture having a pressure of 900 torr at 20°C?

**6.45** Calculate the volume occupied, *at STP*, by
(a) 0.200 mole $O_2$
(b) 12.4 g $Cl_2$
(c) A mixture of 0.100 mole $N_2$ and 0.050 mole $O_2$

**6.46** Calculate the weight of 245 ml of $SO_2$ at STP.

6.47 What is the density of butane, $C_4H_{10}$, at STP?

6.48 The density of a gas was found to be 1.96 g/liter at STP. What is its molecular weight?

6.49 What is the value of the gas constant in the units Pa $m^3$/mole K?

6.50 In the laboratory a student filled a 250-ml container with an unknown gas until a pressure of 760 torr was obtained. He then found that the sample of gas weighed 0.164 g. Calculate the molecular weight of the gas if the temperature in the laboratory was 25°C.

6.51 Calculate the pressure, in torr and in atmospheres, that would be exerted by 25 kg of steam ($H_2O$) in a 1000-liter boiler at 200°C assuming ideal gas behavior.

6.52 The density of a gas was found to be 1.81 g/liter at 30°C and 760 torr. What is its molecular weight?

6.53 Calculate the volume occupied by 0.234 g of $NH_3$ at 30°C and a pressure of 0.847 atm.

6.54 A chemist observed a gas being evolved in a chemical reaction and collected some of it for analysis. It was found to contain 80.0% carbon and 20.0% hydrogen. It was also observed that 500 ml of the gas at 760 torr and 0°C weighed 0.6695 g.
(a) What is the empirical formula of the gaseous compound?
(b) What is its molecular weight?
(c) What is its molecular formula?

6.55 In the reaction, $N_2(g) + 3H_2(g) \rightarrow 2NH_3(g)$, how many milliliters of $N_2$, measured at STP, are required to produce 400 ml of $NH_3$, measured at STP? How many milliliters of $H_2$ at STP are required?

6.56 In the reaction,

$$2NO\ (g) + 2H_2\ (g) \rightarrow 2H_2O\ (g) + N_2\ (g)$$

how many milliliters of $N_2$, measured at STP, would be produced from (a) 0.00140 mole NO, (b) $1.3 \times 10^{-3}$ g $H_2$?

6.57 Oxygen gas, generated in the reaction, $KClO_3 \rightarrow KCl + O_2$ (unbalanced), was collected over water at 30°C in a 150-ml vessel until the total pressure was 600 torr.
(a) How many grams of dry $O_2$ were produced?
(b) How many grams of $KClO_3$ were consumed in the reaction?

6.58 Nitric acid is produced by dissolving $NO_2$ in water according to the equation,

$$3NO_2\ (g) + H_2O\ (l) \rightarrow 2HNO_3\ (l) + NO\ (g)$$

How many milliliters of $NO_2$ at 25°C and 770 torr are required to produce 10.0 g of $HNO_3$?

6.59 Compare the rates of effusion of He and Ne. Which gas effuses faster, and how much faster?

6.60 If, at a particular temperature, the average speed of $CH_4$ molecules is 1000 miles/hr, what would be the average speed of $CO_2$ molecules at the same temperature?

6.61 The rate of effusion of an unknown gas was determined to be 2.92 times faster than that of $NH_3$. What is the approximate molecular weight of the unknown gas?

6.62 Use the van der Waals equation to calculate the pressure exerted by 1.000 mole of He at 0°C in a volume of 22.400 liters. Compare this to the pressure an ideal gas would exert under these same conditions.

6.63 Use the van der Waals equation to calculate the pressure of 1.000 mole of $C_2H_6$ at 0°C in a volume of 22.400 liters. Compare this to the pressure of an ideal gas under these same conditions.

*6.64 Mercury has a density of 13.6 g/ml. Calculate the value of the standard atmosphere in the units, lb/in.$^2$.

6.65 Calculate the volume occupied by 0.0244 g of $O_2$ if it were collected over water at 23°C and at a total pressure of 740 torr.

*6.66 Three gases were added to the same 10-liter container to give a total pressure of 800 torr at 30°C. If the mixture contained 8.0 g of $CO_2$, 6.0 g of $O_2$, and an unknown amount of $N_2$, calculate the following.
(a) The total number of moles of gas in the container
(b) The mole fraction of each gas
(c) The partial pressure of each gas
(d) The weight of $N_2$ in the container

*6.67 A gas, at a total pressure of 800 torr and a volume of 500 ml over water at 35°C, is compressed to a volume of 250 ml, also over water at 35°C. Calculate the final pressure of the wet gas.

*6.68 During a rainstorm in July in New York City the humidity was found to be 100%. The atmospheric pressure was 740 torr and the temperature was 31°C. Dry air has an average molecular weight of 28.8. Calculate the weight of water in 1.00 liter of the air during the storm.

*6.69 280 ml of gas are collected over water at 20°C. The water level inside the collection bottle is 28.4 mm higher than the water level outside. The atmospheric pressure is 763 torr. What would be the volume of dry gas at STP?

*6.70 The product $PV$ has the dimensions of energy. Given the data, 1 J = 1 N m, 1 atm = 101,325 Pa, and 1 Pa = 1 N/m$^2$, calculate the number of joules equal to 1 liter atm. What is the value of the gas constant, $R$, in $J\ mole^{-1}\ K^{-1}$ and $cal\ mole^{-1}\ K^{-1}$?

*6.71 Calculate the maximum volume of $CO_2$, at 750 torr and 28°C, that could be produced by reacting 500 ml of CO, at 760 torr and 15°C, with 500 ml of $O_2$ at 770 torr and 0°C.

*6.72 Ozone, $O_3$, is an important species in the chain of reactions that lead to the production of smog. In an ozone analysis, $2.0 \times 10^4$ liters of air at STP were drawn through a solution of NaI where the $O_3$ undergoes the reaction,

$$O_3 + 2I^- + H_2O \longrightarrow O_2 + I_2 + 2OH^-$$

The $I_2$ formed was titrated with 0.0100 $M$ $Na_2S_2O_3$ with which it reacts.

$$I_2 + 2S_2O_3^{2-} \longrightarrow 2I^- + S_4O_6^{2-}$$

In the analysis, 0.042 ml of the $Na_2S_2O_3$ solution was required to completely react with all of the $I_2$.

(a) Calculate the number of moles of $I_2$ that were reacted with the $S_2O_3^{2-}$ solution.
(b) How many moles of $I_2$ were produced in the first reaction?
(c) How many moles of $O_3$ were contained in the 20,000 liters of air?
(d) What volume would the $O_3$ occupy at STP?
(e) What is the concentration of $O_3$, in parts per million by volume, in the air sample?

*6.73 An important reaction in the production of nitrogen fertilizers is the oxidation of ammonia,

$$4NH_3\ (g) + 5O_2\ (g) \xrightarrow{500°C} 4NO\ (g) + 6H_2O\ (g)$$

How many liters of $O_2$, measured at 25°C and 0.895 atm, must be used to produce 100 liters of NO at 500°C and 750 torr?

*6.74 A student collected 35.0 ml of $O_2$ over water at 25°C and a total pressure of 745 torr from the decomposition of a 0.2500-g sample known to contain a mixture of KCl and $KClO_3$. The reaction that produced the oxygen was

$$2KClO_3 \longrightarrow 2KCl + 3O_2$$

(a) How many moles of $O_2$ were collected?
(b) How many grams of $KClO_3$ were decomposed?
(c) What percentage of the sample was $KClO_3$?

*6.75 Calculate the molar volume of $O_2$ at STP from the van der Waals equation and compare it to the value in Table 6.2. (Hint: The volume can be obtained by successive approximations if you solve for the $V$ in the term, $V - nb$.)

# 7 SOLIDS

As you know, solids differ from gases in many ways: (1) They retain their shape and volume when transferred from one container to another, (2) they are virtually incompressible, and (3) they exhibit extremely slow rates of diffusion. What features do solids possess that cause them to be so different from gases?

We saw in the last chapter that the particles comprising a gas move about freely and are essentially independent of one another. Just the opposite is true in a solid, where the attractive forces between the atoms, molecules, or ions are relatively strong. Here the particles are held in a rigid structural array, wherein they exhibit only vibrational motion.

There are two types of solids, **amorphous** and **crystalline.** If the temperature at which the solid is formed is approached slowly so as to allow the array of particles to become well ordered, a crystalline solid results. If, on the other hand, the temperature is lowered very rapidly, there is a chance that the particles will be "frozen" in a chaotic state. In this case the particles are arranged in a random fashion and the resulting solid is said to be amorphous (that is, without form). Glass, rubber, and most plastics are examples of amorphous substances.

## 7.1 CRYSTALLINE SOLIDS

The surface of a well-formed crystalline solid reveals, upon examination, flat planes that intersect at angles characteristic of the particular substance under investigation (Figure 7.1). These flat planes are called **faces,** and the characteristic angles are called **interfacial angles.** These faces and characteristic interfacial angles are present no matter how the crystal was formed and are completely independent of the size of the crystal. Furthermore, if a crystal is cleaved, or even crushed into a powder, each resulting particle will possess identical interfacial angles.

This may be contrasted with what happens when glass, a typical amorphous solid, is broken. The surfaces of the broken pieces are generally not flat and intersect at random angles.

## 7.2 X-RAY DIFFRACTION

It is clear, even from a visual observation, that there is much external order to a crystal. How can the internal order of a crystal, which is implied by this external regularity, be confirmed?

A German physicist, Max von Laue, pointed out in 1912 that a crystal could serve as a three-dimensional diffraction grating if the wavelength of

Figure 7.1
*Photograph of some typical crystals. On the left, potassium alum, $KAl(SO_4)_2 \cdot 12H_2O$. On the right, calcite, $CaCO_3$.*

the incident radiation were of the same order of magnitude as the distance between particles in the solid. This condition is fulfilled by X-rays, which have wavelengths of approximately 1 Å (0.1 nm).

When a crystal is bathed in X-rays, each atom of the crystal within the path of an X-ray absorbs some of its energy and then reemits it in all directions. Thus each atom is a source of secondary wavelets, and the X-rays are said to be scattered by the atoms. These secondary wavelets from the different sources interfere with each other, either by reinforcing or by cancelling each other. In certain directions the waves emanating from nearly all of the atoms in any orderly array are in phase (that is, the peaks and troughs of the waves coincide as shown in Figure 7.2*a*) and intense beams of X-rays are observed in these directions. In all other directions the waves from various atoms are out of phase (Figure 7.2*b*) and cancel each other; thus no intensity is detected.

Two English scientists, William Bragg and his son Lawrence, treated the diffraction of X-rays as if the process were reflection. In Bragg's treatment the X-rays that penetrate a crystal are thought of as being reflected by successive layers of particles within the substance (Figure 7.3). We can see from this diagram that beams reflected from deeper layers must travel further to reach the detector. For there to be any intensity at the detector these waves have to be in phase, which must mean that the extra distance traveled by the more penetrating beam has to be some integral multiple of the wavelength of the X-rays.

Bragg showed that in order to observe any intensity in the emerging X-rays, a relatively simple relationship had to be fulfilled. This relationship, known as the **Bragg equation,** is

$$2d \sin \theta = n\lambda \tag{7.1}$$

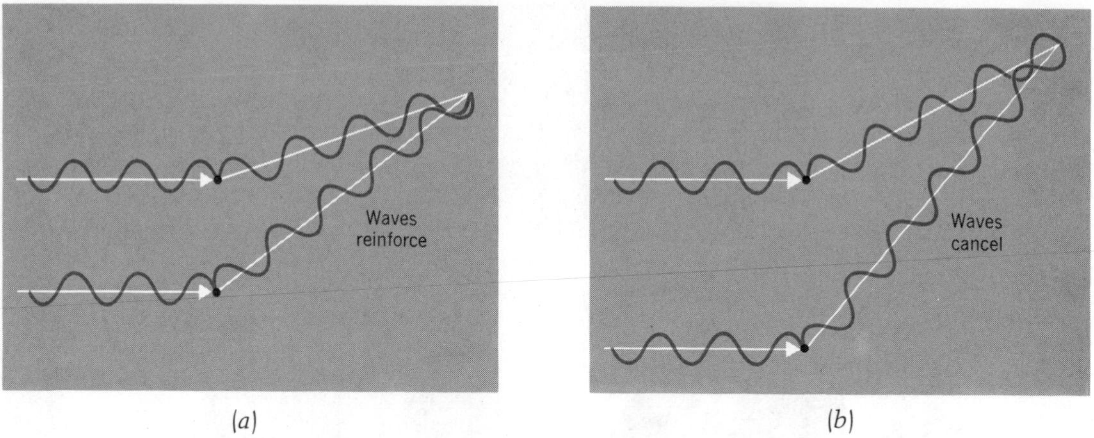


Figure 7.2
*Diffraction. Scattered waves reinforce each other only at certain angles. (a) In phase. (b) Out of phase.*

where $d$ is the spacing between the successive layers that are reflecting the X-rays, $\theta$ is the angle at which the X-rays enter and leave the particular set of layers, $\lambda$ is the wavelength of the X-rays, and $n$ is an integer (that is, $n = 1$, or 2, or 3, etc.). The Bragg equation serves as the basis for the study of crystalline structure by X-ray diffraction.

In practice, X-rays of known wavelength are directed at a crystal and the angles at which they are reflected are recorded, for example, on a piece of photographic film (Figure 7.4). By measuring the angles at which the X-rays are reflected it is a simple matter to calculate the distances between planes of atoms within a crystal, as illustrated in Example 7.1. If, in addition, the intensities of the reflected X-rays are measured, a crystallographer may be able to deduce, through a rather complex procedure, the actual positions of atoms within the solid. In this way the molecular structures of many sub-

Figure 7.3
*Braggs law. Bragg showed that the waves from different planes of atoms are in phase only when* $2d \sin\theta = n\lambda$.

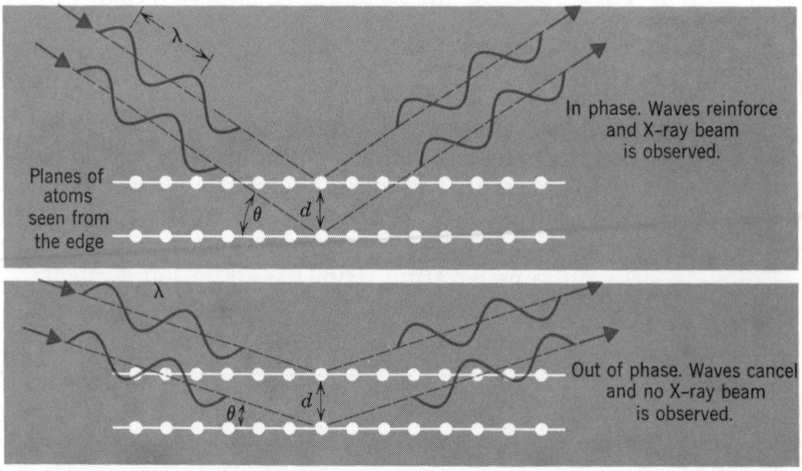

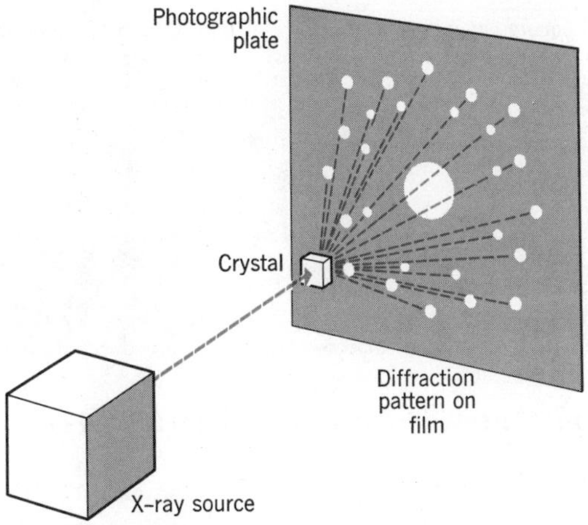


Figure 7.4
*Production of an X-ray diffraction pattern.*

stances have been found. In recent years, X-ray diffraction has become a powerful tool in biochemistry by which the structures of even very complex molecules have been investigated. James Watson and Francis Crick, for example, shared the Nobel Prize with Maurice Wilkins in 1962 for their use of Rosalind Franklin's X-ray data in the deduction of the double-helix structure of DNA.

EXAMPLE 7.1 X-rays of wavelength 1.54 Å strike a crystal and are observed to be reflected at an angle of 22.5°. Assuming that $n = 1$, calculate the spacing between the planes of atoms that are responsible for this reflection.

SOLUTION We wish to calculate $d$. Solving Equation 7.1 for $d$, we have

$$d = \frac{n\lambda}{2 \sin \theta}$$

From the data, $n = 1$, $\lambda = 1.54$ Å, $\theta = 22.5°$. Substituting,

$$d = \frac{(1)(1.54 \text{ Å})}{2 \sin(22.5)}$$

$$d = \frac{1.54 \text{ Å}}{2(0.383)}$$

$$d = 2.01 \text{ Å}$$

## 7.3 LATTICES

There are only a very limited number of ways in which atoms, molecules, or ions may be packed together in a crystalline substance. We can characterize these packing arrangements by what are called **space lattices.**

In the strictest sense, a lattice is merely a regular or repetitive pattern of points that may be either one-, two-, or three-dimensional. A two-dimensional lattice is illustrated in Figure 7.5*a*. What we see there is called a square lattice since it consists of a set of points arranged so that each lattice point lies at the corner of a square.

If we were wallpaper designers we could create, with this one simple square lattice, an unlimited number of wallpaper patterns. For example, each lattice point could be assigned some geometrical design such as a dia-

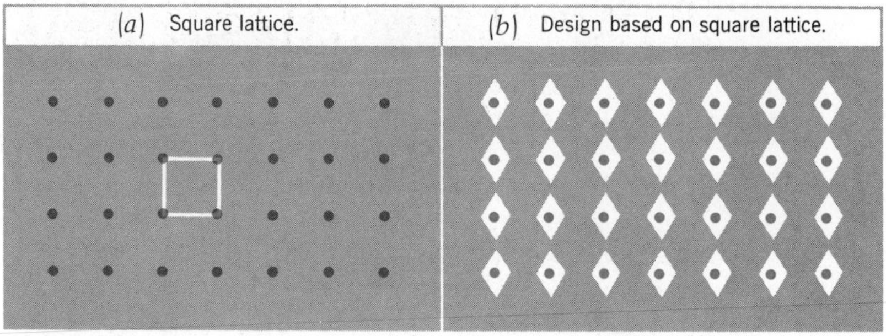


Figure 7.5
*Two-dimensional lattice. (a) Square lattice. (b) Design based on square lattice.*

mond or a flower (Figure 7.5*b*). By varying the size of the squares defined by the lattice points and by changing the motif about each point, an infinite number of designs could be formed.

We can think of a lattice as extending to infinity in all directions. However, in order to completely describe the lattice, or any pattern derived from it, it is not necessary to specify explicitly the positions of each and every point. Instead, we only have to describe a portion of the lattice, called the **unit cell,** which can be used to generate the entire lattice. In the lattice of Figure 7.5*a* the unit cell corresponds to a square drawn by connecting four points. By moving this unit cell repeatedly to the left and right and up and down by a distance equal to the length of the unit cell, we can create the entire square lattice.

The extension of the lattice concept to three dimensions is quite straightforward. In Figure 7.6 we see an example of a simple cubic space lattice in which the unit cell is drawn with heavy lines. By associating a particular chemical environment with each lattice point on the three-dimensional lattice we can arrive at a chemical structure, and by varying the chemical environment about each point we can create an infinite number of chemical structures all based on the same lattice. It is easy to see, then, how a very small number of lattices could be sufficient to describe the crystal structures of all known chemical substances.

In addition to simple, or **primitive lattices,** which have lattice points located only at the corners of the unit cell, it is also possible to have **body-**

Figure 7.6
*A simple cubic space lattice.*

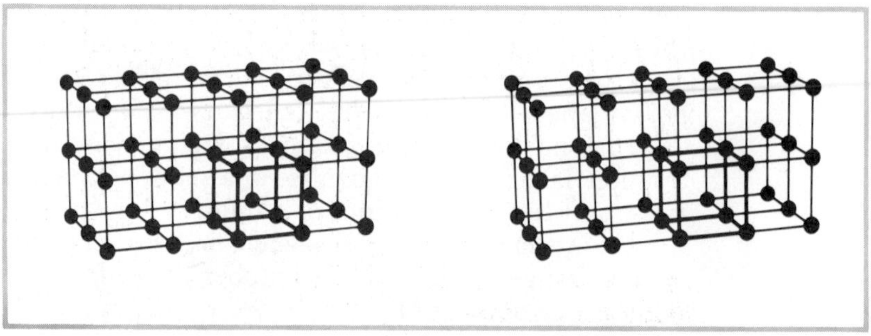

**centered, face-centered,** and **end-centered lattices** (Figure 7.7). In a body-centered lattice there are points at each corner of the unit cell plus an additional point in the very center. A face-centered lattice has an additional lattice point in the center of each of the six faces of the unit cell, and an end-centered lattice has additional lattice points in only two opposite faces. It was shown by A. Bravais in 1848 that the total number of possible space lattices is 14. They belong to seven basic crystal systems: cubic, tetragonal, orthorhombic, monoclinic, triclinic, rhombohedral, and hexagonal.

The unit cell for each lattice may be described by specifying the quantities $a$, $b$, and $c$, which correspond to the lengths of the edges of the cell, and the angles $\alpha$, $\beta$, and $\gamma$, which are the angles at which these edges intersect one another (Figure 7.8). The properties of the seven crystal systems are given in Table 7.1.

In the various unit cells that we have just seen, there are three kinds of lattice points: points located at the corners, points in the face centers, and points that lie entirely within the unit cell. In a crystal, atoms located at the corner and face center of a unit cell are shared by other cells, and only a portion of such an atom actually lies within a given unit cell. In certain instances we also find a fourth type of lattice point, where an atom or ion lies along an edge of a unit cell. We might wish to know how many chemical units (atoms, molecules, or groups of ions) are found in one unit cell of a crystalline substance. This is related to the density of the substance and is controlled by the attractive and repulsive forces that the particles experience in the solid.

A point that lies at the corner of a unit cell is shared among eight unit cells and, therefore, only one-eighth of each such point lies within the given unit cell (see Figure 7.6).

A point along an edge is shared by four unit cells, and only one-fourth of it lies within any one cell.

A face-centered point (which we have in both face-centered and end-centered lattices) lies partly in two cells, and only one-half of it is present in a given unit cell.

A body-centered point lies entirely within the unit cell, and contributes one complete point to the cell.

| Type of Lattice Point | Contribution to One Unit Cell |
|---|---|
| Corner | $\frac{1}{8}$ |
| Edge | $\frac{1}{4}$ |
| Face center | $\frac{1}{2}$ |
| Body center | 1 |

A simple (primitive) lattice has points only at the eight corners of the unit cell (Figure 7.7*a*). The number of lattice points that lie within the cell is

$$8 \text{ corners} \times \frac{1}{8}\left(\frac{\text{point}}{\text{corner}}\right) = 1 \text{ point}$$

A face-centered unit cell has points at the eight corners and, in addition, points centered in the six faces (Figure 7.7*c*). It therefore contains a total of four lattice points.

(a)

(b)

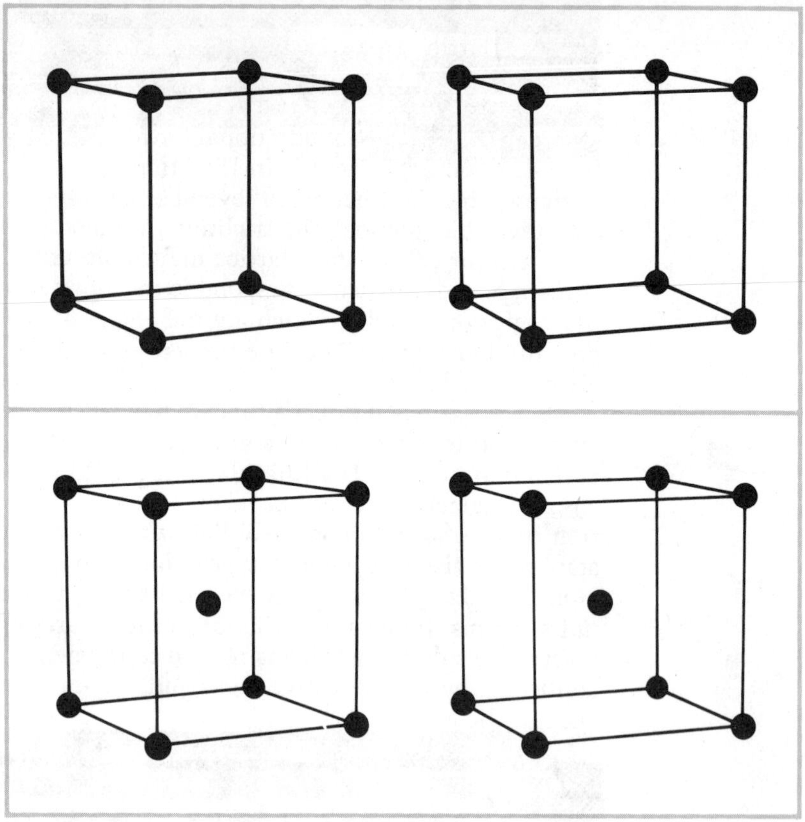

Figure 7.7
*The four types of unit cells. (a) Simple cubic. (b) Body-centered cubic. (c) Face-centered cubic. (d) End-centered orthorhombic.*

$$8\ \cancel{\text{corners}} \times \frac{1}{8}\left(\frac{\text{point}}{\cancel{\text{corner}}}\right) = 1 \text{ point}$$

$$6\ \cancel{\text{faces}} \times \frac{1}{2}\left(\frac{\text{point}}{\cancel{\text{face}}}\right) = 3 \text{ points}$$

$$\text{total} = 4 \text{ points}$$

It is quite easy to verify that the body-centered and end-centered unit cells each contain two lattice points.

Table 7.1
*Properties of the seven crystal systems*

| System | Edge Lengths | Angles |
|---|---|---|
| Cubic | $a = b = c$ | $\alpha = \beta = \gamma = 90°$ |
| Tetragonal | $a = b \neq c$ | $\alpha = \beta = \gamma = 90°$ |
| Orthorhombic | $a \neq b \neq c$ | $\alpha = \beta = \gamma = 90°$ |
| Monoclinic | $a \neq b \neq c$ | $\alpha = \beta = 90° \neq \gamma$ |
| Triclinic | $a \neq b \neq c$ | $\alpha \neq \beta \neq \gamma$ |
| Rhombohedral | $a = b = c$ | $\alpha = \beta = \gamma \neq 90°$ |
| Hexagonal | $a = b \neq c$ | $\alpha = \beta = 90°;\ \gamma = 120°$ |

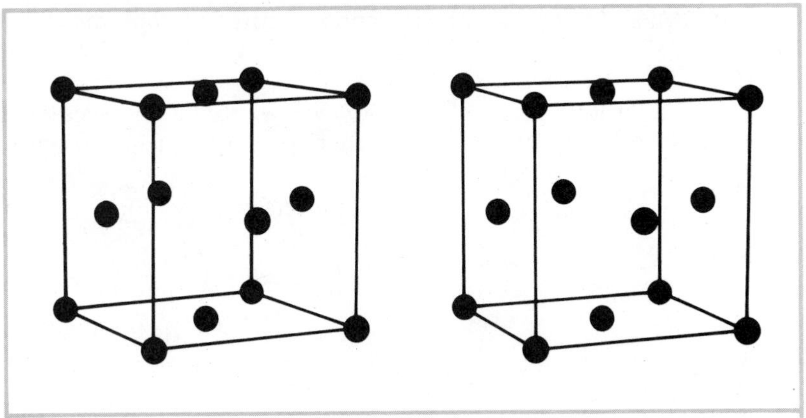
(c)

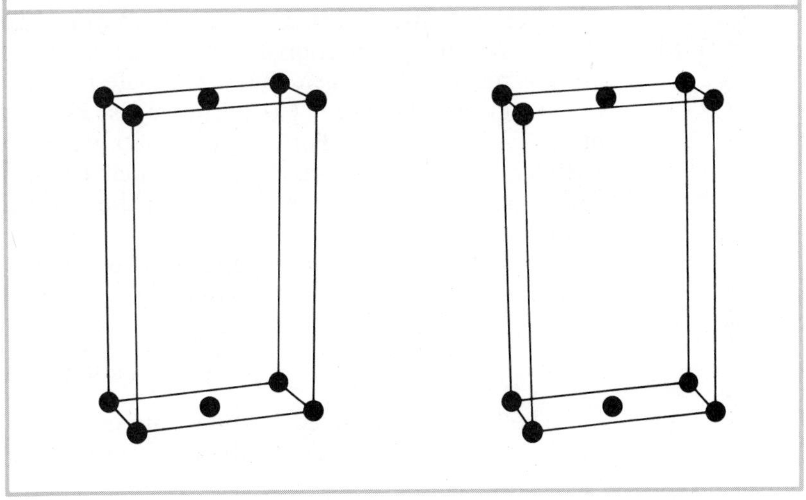
(d)

Figure 7.8
*The unit cell. Edges intersect at characteristic angles $\alpha$, $\beta$, and $\gamma$.*

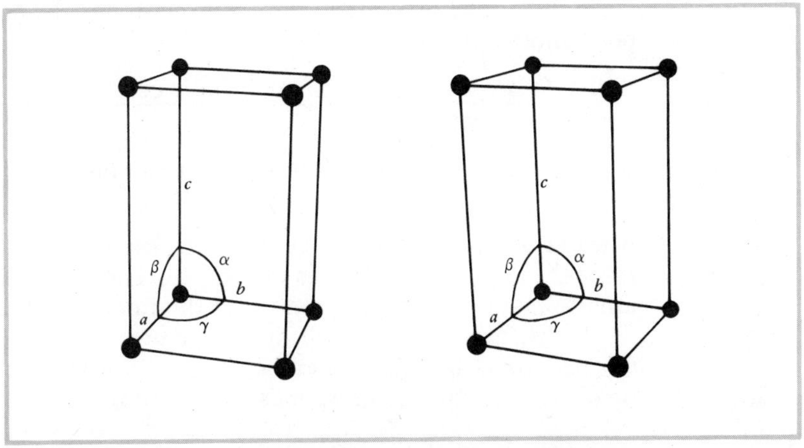

EXAMPLE 7.2 Show that a body-centered unit cell contains two lattice points.

SOLUTION Referring to Figure 7.7, we see that a body-centered unit cell has points located at the eight corners plus a point in the center. This gives

$$8 \text{ corners} \times \frac{1}{8}\left(\frac{\text{point}}{\text{corner}}\right) = 1 \text{ point}$$

$$1 \text{ center} \times \left(\frac{1 \text{ point}}{\text{center}}\right) = 1 \text{ point}$$

$$\text{total} = 2 \text{ points}$$

## 7.4 AVOGADRO'S NUMBER

There are a number of methods of computing Avogadro's number. One of them makes use of the charge-to-mass ratio of the proton, its charge, and the atomic mass of hydrogen (Review Problem 3.60, Page 95). Knowledge of the crystalline structure of a substance also allows us to compute the value of Avogadro's number. For example, metallic copper is found to crystallize with a face-centered cubic lattice in which a copper atom is located at each lattice point. The length of the edge of the unit cell is found to be 3.615 Å (from X-ray diffraction).

Copper has an atomic mass of 63.54 and a density of 8.936 g/cm³. The volume occupied by 1 mole of copper is

$$1 \text{ mole Cu} \times \left(\frac{63.54 \text{ g Cu}}{1 \text{ mole Cu}}\right) \times \left(\frac{1 \text{ cm}^3}{8.936 \text{ g Cu}}\right) \sim 7.111 \text{ cm}^3$$

The volume occupied by one unit cell is found by multiplying the height times the width times the depth of the cube.

$$\text{volume of unit cell} = (3.615 \times 10^{-8} \text{ cm})^3$$
$$1 \text{ unit cell} = 4.724 \times 10^{-23} \text{ cm}^3$$

The number of unit cells in 1 mole of copper is

$$1 \text{ mole Cu} \times \left(\frac{7.111 \text{ cm}^3}{1 \text{ mole Cu}}\right) \times \left(\frac{1 \text{ unit cell}}{4.724 \times 10^{-23} \text{ cm}^3}\right) \sim 1.505 \times 10^{23} \text{ unit cells}$$

Since there are four atoms of copper in the face-centered unit cell and since there are $1.505 \times 10^{23}$ unit cells per mole, the number of atoms of Cu per 1 mole of Cu is

$$\left(\frac{4 \text{ atoms}}{\text{unit cell}}\right) \times \left(\frac{1.505 \times 10^{23} \text{ unit cells}}{1 \text{ mole}}\right) \sim \frac{6.02 \times 10^{23} \text{ atoms}}{1 \text{ mole}}$$

The value of Avogadro's number, therefore, is $6.02 \times 10^{23}$.

## 7.5 ATOMIC AND IONIC RADII

If we know the crystal structure of a substance, we are also able to determine the radii of the particles in the solid. For example, we can calculate the atomic radius of the Cu atom using the information in the last section.

In the face-centered cell of Cu, Figure 7.9, the length $AC$ = 3.615 Å. Copper atoms are in contact along the line joining points $A$ and $B$ (face diagonal). This distance corresponds to four times the radius of a copper atom.

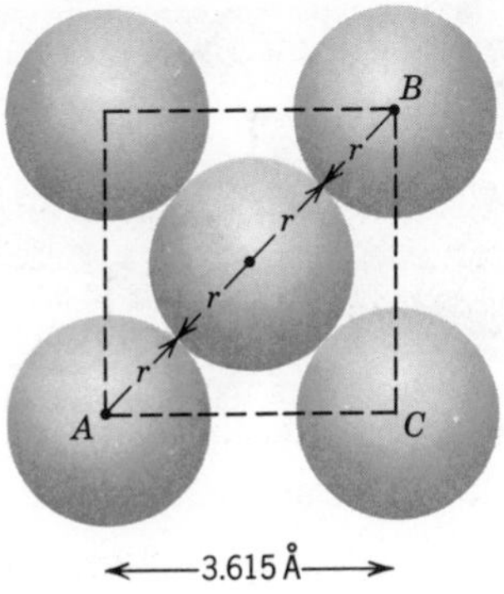


Figure 7.9
*Copper atoms in the face of a unit cell.*

From geometry we know that

$$\overline{AB} = \sqrt{2}(AC) = \sqrt{2}(3.615 \text{ Å})$$
$$\overline{AB} = 5.12 \text{ Å}$$

Therefore,

$$4r = 5.12 \text{ Å}$$
$$r = 1.28 \text{ Å}$$

The radius of a copper atom in metallic copper is therefore 1.28 Å. In a similar fashion we can also use the results of X-ray diffraction to determine the radii of ions in ionic crystals.

## 7.6 THE FACE-CENTERED CUBIC LATTICE

In Section 7.3 it was pointed out that a single lattice could be used to describe a large number of crystal structures. The face-centered cubic lattice serves as a convenient example of this. We know that copper crystallizes with a face-centered structure, and, in fact, so do many other metals, including such familiar ones as aluminum, lead, silver, and gold.

Sodium chloride (ordinary table salt) is characteristic of many ionic solids in which the ratio of cations to anions is 1:1. The structure of NaCl, termed the **rock salt structure,** is shown in Figure 7.10. The solid is constructed of positive and negative ions arranged so that each positive ion is

Figure 7.10
*Sodium chloride structure.*

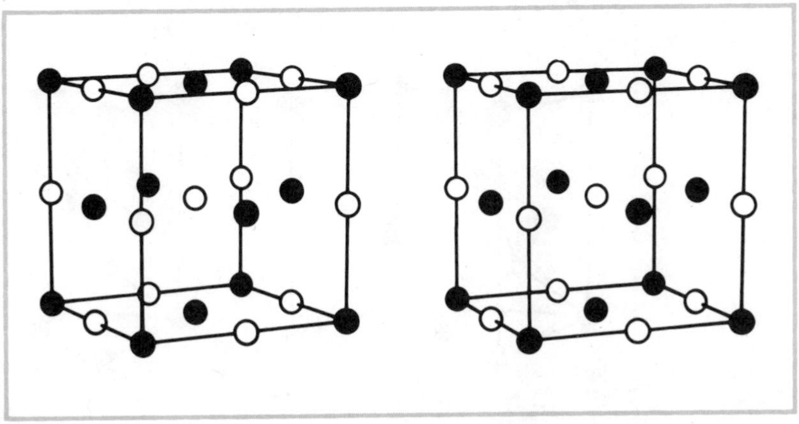

surrounded by six negative ions at the vertices of an octahedron; similarly, each negative ion is surrounded octahedrally by positive ions. If the solid spheres in Figure 7.10 represent $Na^+$ ions located at the lattice points in a face-centered lattice, then there are four $Na^+$ ions per unit cell. Chloride ions are located along each of the 12 edges and in the center of the unit cell. Only one-fourth of each edge ion is within the cube and therefore these contribute a total of three chloride ions to the unit cell. The chloride ion in the center adds one more, so that there are a total of four $Cl^-$ ions. Therefore, the ratio of $Na^+$ to $Cl^-$ is 4:4, which, of course, is the same as 1:1. Other substances possessing the rock salt structure include:

1. All alkali halides except CsCl (below 450°C), CsBr, and CsI
2. AgF, AgCl, AgBr
3. Oxides and sulfides of the alkaline earth metals
4. NiO
5. Some alloys

In the **zinc blende structure** (Figure 7.11), which is observed for ZnS, ZnO, CuCl, CuBr, and BeO, we also have a 1:1 ratio of cations to anions. In this case, we have one type of ion, for example an anion, located at the lattice points in a face-centered cubic unit cell. Each of the cations, shown as open circles within the unit cell, are surrounded tetrahedrally by four anions. Each anion is also surrounded tetrahedrally by four cations.

The **fluorite structure** (Figure 7.12), possessed by $CaF_2$, is similar to the zinc blende structure. Here we have cations at lattice points in the face-centered cube and eight anions, surrounded tetrahedrally by cations, located entirely within the cube. The cations at the lattice points contribute a total of four positive ions to the cube. Since all eight anions are within the unit cell, they contribute eight negative ions to the cube. The ratio of positive to negative ions in the fluorite structure is therefore 1:2. Other solids exhibiting the fluorite structure are $BaF_2$, $SrCl_2$, and $ThO_2$. Another structure, called the antifluorite structure, is precisely the same as the fluorite structure except that the positions of positive and negative ions are reversed to give a cation-to-anion ratio of 2:1. Solids displaying this type of structure include the oxides, sulfides, selenides, and tellurides of sodium and potassium (for example, $M_2S$, $M_2Se$ and $M_2Te$, where $M = Na^+$ or $K^+$).

Figure 7.11
*Zinc blende structure.*

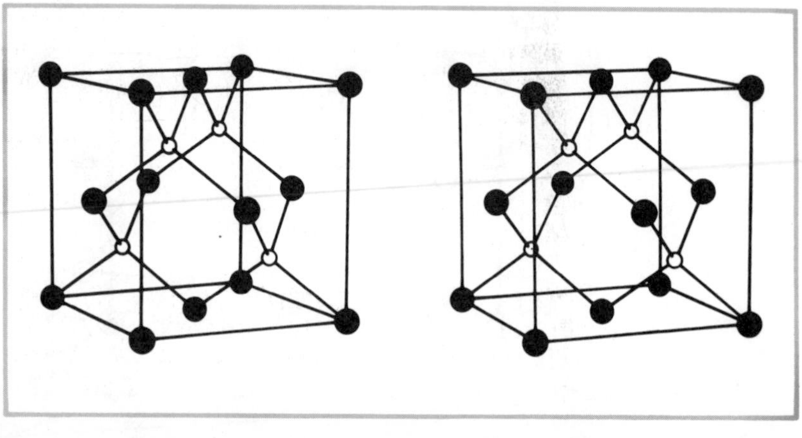

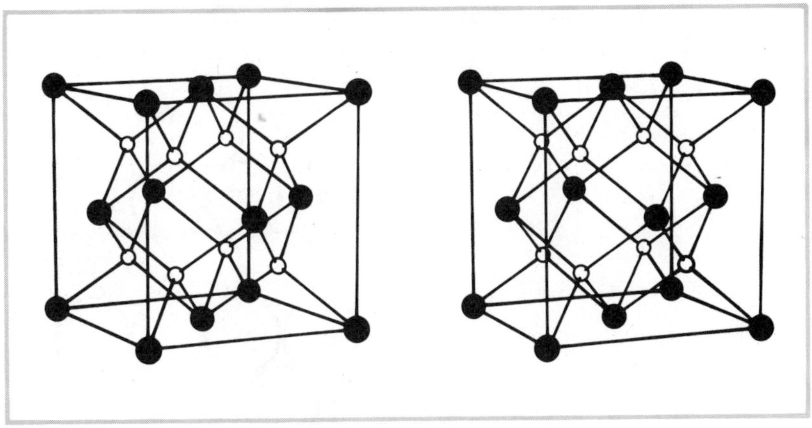

Figure 7.12
*Fluorite structure.*

## 7.7 CLOSEST-PACKED STRUCTURES

When all atoms in a substance are the same, its crystal structure is often determined by the most efficient packing of spheres, that is, the arrangement of atoms that gives rise to the smallest amount of unoccupied space. There are two ways in which spheres can be packed most efficiently; the resulting arrangements are called closest-packed structures. One of these is called **cubic closest-packed** (*ccp*) and the other, **hexagonal closest-packed** (*hcp*). We can consider them to be constructed in the following way.

A first layer of spheres is packed together as tightly as possible (Figure 7.13*a*). A second layer of spheres is placed on the first with each sphere in the second layer resting in a depression, or hole, between spheres in the first layer (Figure 7.13*b*). There are now two ways of arranging the spheres in the third layer. One is to place them directly above those in the first layer (Figure 7.13*c*). If we arrange them in this manner, we arrive at the *hcp* structure. We can continue this process with the fourth layer directly above the second, the fifth above the third, and so on, and we can think of this as an alternating *ABAB. . .* pattern.

The second way of arranging the third layer of spheres is shown in Figure 7.13*d* and leads to the *ccp* structure. In this case the third layer lies above holes in the first layer; if we continue building the structure by laying the fourth layer directly above the first, the fifth above the second, and so on, we arrive at an *ABCABC. . .* pattern. The lattice that corresponds to this arrangement is face-centered cubic, although it is not particularly evident from Figure 7.13*d*. Actually, in this case we are looking down at a corner of the face-centered cube along a *body diagonal* (the line joining one corner of the cube, the center, and an opposite corner).

In both the *ccp* and the *hcp* structures each atom is in contact with 12 other "nearest neighbors." We can compare this with a body-centered cubic structure where each atom would be in contact with its eight nearest neighbors and with a simple cubic structure where each atom would have only six nearest neighbors.

There is a relatively large number of substances that crystallize with these closest-packed structures (for example, most metals). It is also possible to interpret a variety of other structures as being derived from the closest-packed arrangements. For example, the *ccp* structure, we have said, has a face-centered cubic lattice. We have also discussed several ionic

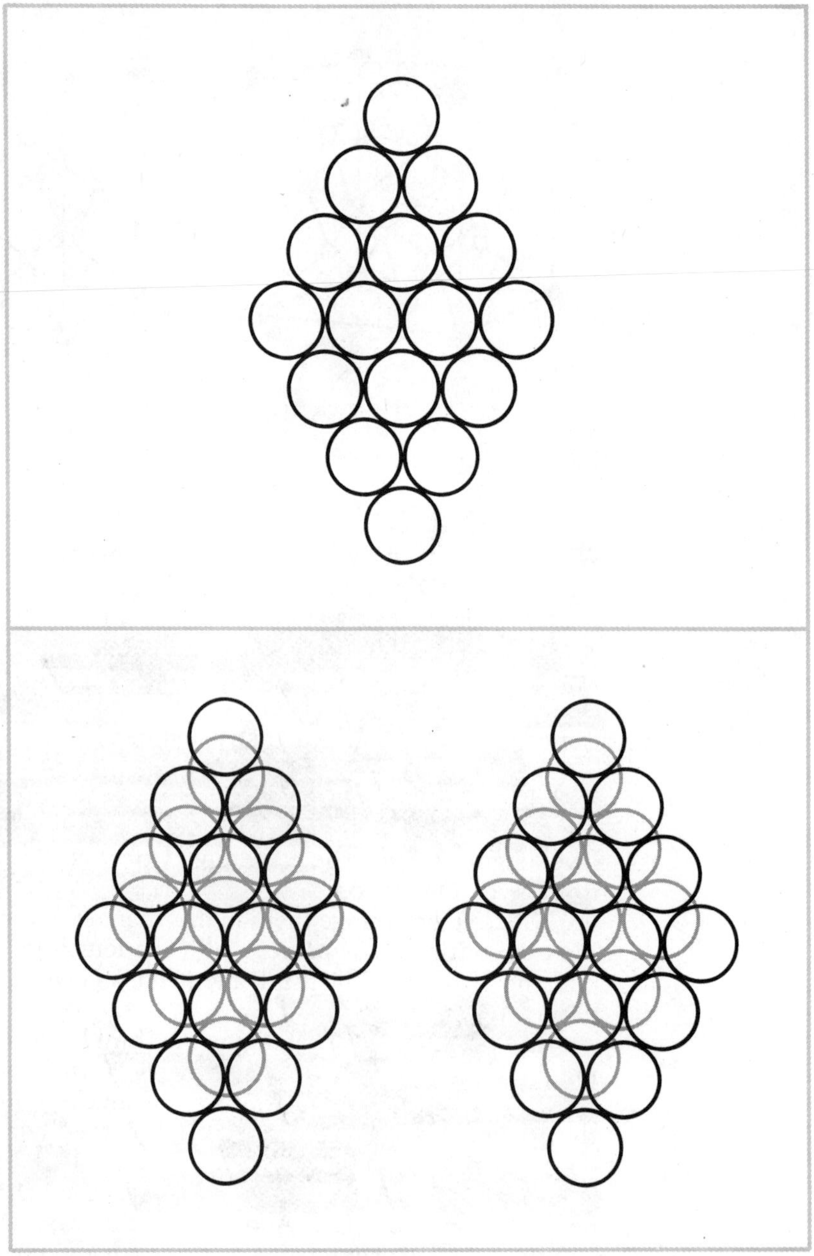

(a)

(b)

Figure 7.13

*Closest packing of spheres. (a) First layer of tangent spheres. (b) Second layer of spheres resting in depressions in first layer. (c) Third layer of spheres resting in depressions in second layer over spheres in first layer—*hcp *structure. (d) Third layer of spheres resting in depressions in second layer over unused depressions in first layer—*ccp *structure.*

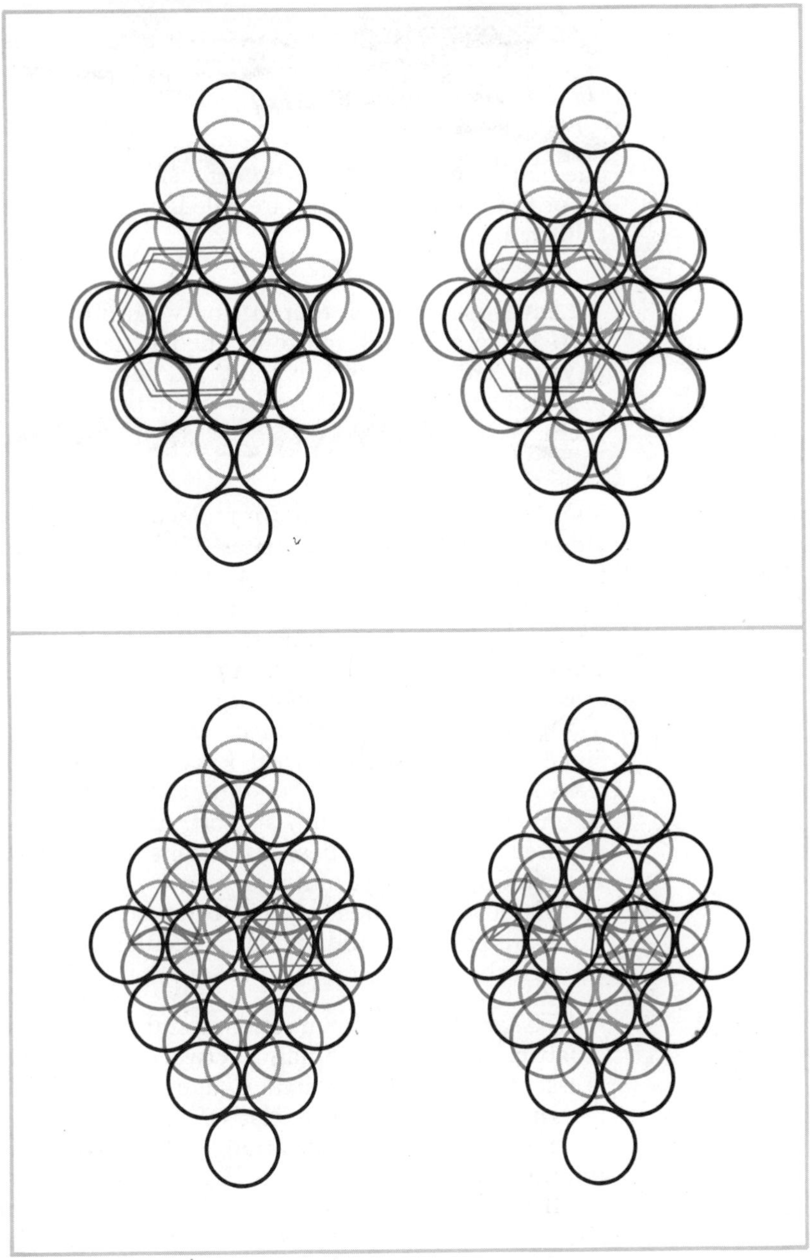

(c)

(d)

crystals in terms of this same lattice. These substances can be considered to be derived from the *ccp* structure as follows.

In the *ccp* structure there are two types of vacant sites: tetrahedral and octahedral. A **tetrahedral site** is an empty space surrounded by four spheres and an **octahedral site** is an empty space surrounded by six spheres. These are shown in Figure 7.13*d*. There are twice as many tetrahedral sites as there are octahedral sites. These vacant "holes" may be used to accommodate other atoms and, when filled in this way, give rise to a variety of different structures.

Sodium chloride, for example, may be considered as a *ccp* arrangement of $Cl^-$ ions with $Na^+$ ions located in octahedral sites (recall that in NaCl each cation is surrounded octahedrally by anions).

The zinc blende structure may be looked upon as a *ccp* arrangement of anions with cations in tetrahedral sites (Figure 7.11). In this structure only one-half of the available tetrahedral sites are filled. In the fluorite structure, all of the tetrahedral sites are filled, and a 1:2 ratio of cation to anions is achieved.

## 7.8 TYPES OF CRYSTALS

We have just seen that there are only a limited number of ways of arranging particles in a crystalline solid. The particular arrangements, as well as the physical properties of the solid, are determined by the types of particles present at the lattice points and the nature of the attractive forces between them. We can divide crystals into types: molecular, ionic, covalent, and metallic.

**MOLECULAR CRYSTALS.** In molecular crystals either molecules or individual atoms occupy lattice sites. The attractive forces between them are of the type described in Section 4.11 and are much weaker than the covalent bonds that exist within individual molecules. London forces are present in crystals of nonpolar substances such as Ar, $O_2$, naphthalene (moth crystals), and $CO_2$ (Dry Ice). In crystals of polar molecules such as $SO_2$, the dominant forces are a result of dipole–dipole attractions; and in solids such as ice ($H_2O$), $NH_3$, and HF, the molecules are held in place by hydrogen bonding. Since these are relatively weak forces (compared to covalent or ionic attractions), molecular crystals tend to have small lattice energies and are easily deformed; we say that they are soft. Also, relatively little thermal energy is required to overcome these attractions, and molecular solids generally tend to have low melting points.

Molecular crystals are poor conductors of electricity because the electrons are bound to individual molecules and are not free to move freely through the solid.

**IONIC CRYSTALS.** In an ionic crystal such as NaCl there are ions located at lattice sites and the binding between them is mainly electrostatic (which is essentially nondirectional). As a result, the kind of lattice that is formed is determined mostly by the relative sizes of the ions and their charges. When the crystal forms, the ions arrange themselves to maximize attractions and minimize repulsions.

Because electrostatic forces are strong, ionic crystals have large lattice energies. They are hard and are characterized by high melting points. They are also very brittle. When struck with a blow they tend to shatter, because as planes of ions slip by one another they pass from a condition of mutual attraction to one of mutual repulsion. This is illustrated in Figure 7.14.

In the solid state ionic compounds are poor conductors of electricity because the ions are held rigidly in place. When melted, however, the ions are free to move about and ionic substances become good conductors.

**COVALENT CRYSTALS.** In a covalent crystal there is a network of covalent bonds between the atoms that extends throughout the entire solid. An example of such a substance is diamond (Figure 7.15), where each atom is covalently bonded to four nearest neighbors. Other common examples are

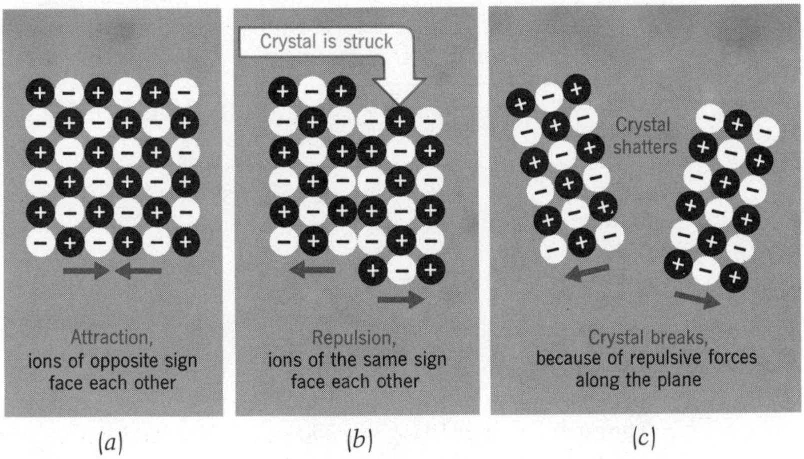


Figure 7.14
*An ionic crystal breaks when struck. (a) Attraction between ions opposite each other. (b) When struck, part of the crystal slips past the rest. Ions of the same sign face each other. (c) The repulsive forces push the crystal apart.*

carborundum (silicon carbide, SiC) and quartz ($SiO_2$, shown in Figure 19.16). The highly directional nature of the covalent bonds usually prevents these substances from assuming one of the closest-packed structures, and we usually observe somewhat open structures like that in diamond.

Because of the interlocking framework of covalent bonds, covalent crystals have very high melting points and are usually extremely hard. Diamond, of course, is the hardest substance known and is used in grinding and cutting tools. Silicon carbide is like diamond, except that half of the carbon atoms in the structure have been replaced by silicon atoms. It too is very hard and is used as an abrasive in sandpaper, as well as in other grinding and cutting applications.

Covalent crystals are poor conductors of electricity because the electrons in the solid are localized in the covalent bonds and are not free to move through the crystal.

Figure 7.15
*Diamond.*

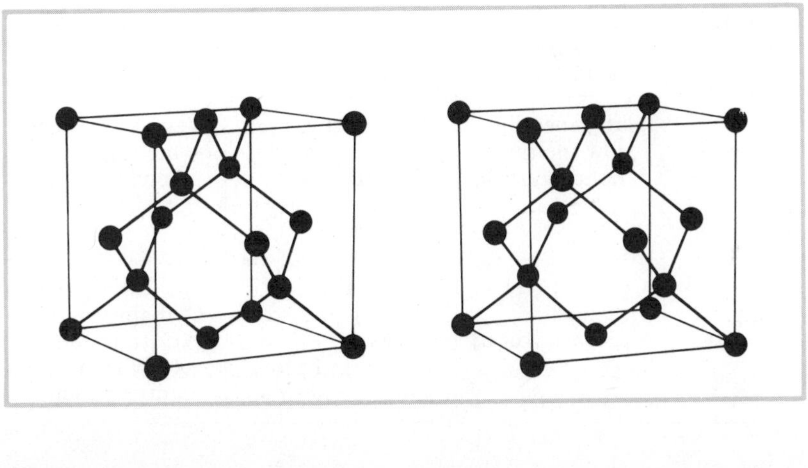

Table 7.2
*Properties of some typical metals*

| Metal | Melting Point (°C) | Relative Hardness (room temperature) |
|---|---|---|
| Mercury | −38.4 | (Liquid) |
| Sodium | 97.8 | 0.07 |
| Lead | 327 | 4.2 |
| Magnesium | 650 | 30 |
| Aluminum | 660 | 16 |
| Nickel | 1453 | 90 |
| Iron | 1536 | 77 |
| Platinum | 1769 | 64 |
| Molybdenum | 2610 | 150 |
| Tungsten | 3410 | 350 |

**METALLIC CRYSTALS.** The simplest picture of a metallic crystal has positive ions (nuclei plus core electrons) situated at lattice points, with the valence electrons belonging to the crystal as a whole instead of to any single atom. The solid is held together by the electrostatic attraction between the lattice of positive ions and this sort of "sea of electrons." These electrons may move freely; hence, we find metals to be good conductors of electricity. Since the melting points and hardness of metals vary over wide ranges (see Table 7.2), there must, at least in some cases, also be some degree of covalent bonding between atoms in the solid.

Table 7.3 summarizes the important properties of these different kinds of solids.

Table 7.3
*Types of solids*

| | Molecular | Ionic | Covalent | Metallic |
|---|---|---|---|---|
| Chemical units at lattice sites | Molecules or atoms | Positive and negative ions | Atoms | Positive ions |
| Forces holding the solid together | London forces, dipole–dipole, hydrogen bonds | Electrostatic attraction between + and − ions | Covalent bonds | Electrostatic attraction between + ions and electron "sea" |
| Some properties | Soft, generally low melting, nonconductors | Hard, brittle, high melting, nonconductors (but conduct when melted) | Very hard, high melting, nonconductors | Hard to soft, low to high melting, high luster, good conductors |
| Some examples | $CO_2$ (Dry Ice)<br>$H_2O$ (ice)<br>$C_{12}H_{22}O_{11}$ (sugar)<br>$I_2$ | NaCl (salt)<br>$CaCO_3$ (limestone; chalk)<br>$MgSO_4$ (in Epsom salt | SiC (carborundum)<br>C (diamond)<br>WC (tungsten carbide, used in cutting tools) | Na, Fe, Cu, Hg |

## 7.9 BAND THEORY OF SOLIDS

We have remarked that certain types of crystals (molecular, ionic, and covalent) are poor conductors of electricity, while metallic crystals are very good conductors. There are some substances, such as germanium and silicon, that have conductivities between these two extremes. To explain these properties the **band theory** of solids was developed.

In a solid an **energy band** is composed of a very large number of closely spaced energy levels, formed by combining atomic orbitals (of similar energy) from each of the atoms within the substance. For example, in sodium the 1*s* atomic orbitals, one from each atom, combine to form a single 1*s* band that extends in three dimensions throughout the entire solid. The same thing occurs with the 2*s*, 2*p*, etc., orbitals, so that we also have 2*s*, 2*p*, etc., bands within the lattice.

Sodium atoms have filled 1*s*, 2*s*, and 2*p* orbitals; therefore the corresponding bands in the solid are also filled. The 3*s* orbital of sodium, however, is only half-filled, which leads to a half-filled 3*s* band. Following the same logic, it is clear that the 3*p* and higher energy bands are completely empty.

When a voltage is applied across sodium, electrons in filled bands may not move through the solid because orbitals in the same band on neighboring atoms are already filled and thus cannot acquire an additional electron. In the 3*s* band, which is half-filled, an electron may hop from atom to atom with ease. In Figure 7.16 the localized nature of electrons in filled bands is illustrated. We can also see that the 3*s* band extends continuously through the solid, which accounts for the high conductivity of Na.

We refer to the band containing the outer-shell (valence-shell) electrons

Figure 7.16
*Bands in metallic sodium.*

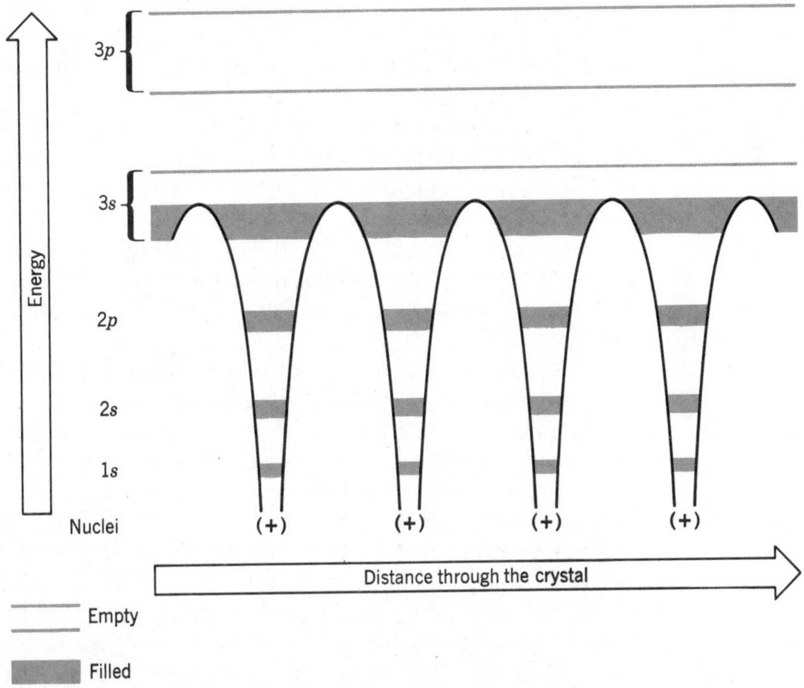

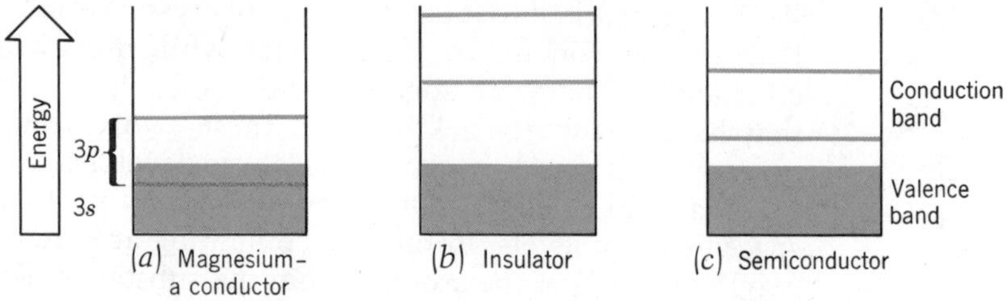


Figure 7.17
*Energies of the valence band and conduction band in three kinds of solids. (a) Magnesium—a conductor. (b) Insulator. (c) Semiconductor.*

as the **valence band.** Any band that is either vacant or partially filled and is uninterrupted throughout the lattice is called a **conduction band.**

In metallic sodium the valence band and conduction band are the same. In magnesium the 3*s* valence band is filled and therefore cannot be used to transport electrons. However, the vacant 3*p* conduction band actually overlaps the valence band and can easily be populated by electrons when a voltage is applied (Figure 7.17*a*). This causes Mg to be a conductor.

In an insulator (for example, glass, diamond, or rubber) the energy separation between the filled valence band and the conduction band is very large (Figure 7.17*b*). This large separation prevents electrons from populating the conduction band under an applied voltage.

A semiconductor, such as Si and Ge, has a relatively small gap between the valence and conduction bands (Figure 7.17*c*). Thermal energy promotes some electrons to the conduction band where they are then able to move through the solid. This is why the conductivity of these substances increases with temperature.

## 7.10 DEFECTS IN CRYSTALS

We have described a crystal as an ordered array of particles. In all real crystals this order is not perfect, and any deviation from perfection is termed a **lattice defect.**

If only a few atoms are causing this disorder, a **point defect** results. There exist in crystals two main types of point defects. The first, called a **Frenkel defect,** is caused by cations that are not present in their normal lattice sites but are found between layers in *interstitial* positions, as shown in Figure 7.18. This type of defect can occur only in solids where the difference in size between the cation and anion provides interstitial sites large enough to accommodate the cation. Frenkel defects are found in such substances as AgCl and AgBr.

The second major type of point defect is the **Schottky defect.** This results when an equal number of anionic and cationic sites are left vacant in a crystal (an equal number so as to maintain electrical neutrality). This type of defect is common in the alkali halides (e.g., NaCl and KBr).

Another variety of defect occurs in **nonstoichiometric compounds.** The formula for iron(II) sulfide is best represented as $Fe_{(1-x)}S$, where $x$ is the fraction of unoccupied cationic sites. In order to maintain an electrically neutral solid, some of the iron is present as $Fe^{3+}$, which compensates for the loss of $Fe^{2+}$. This type of defect is often found in solids where the cation may have more than one oxidation state (other examples are FeO, $Cu_2O$, and $Cu_2S$).

Other nonstoichiometric compounds can be prepared from substances

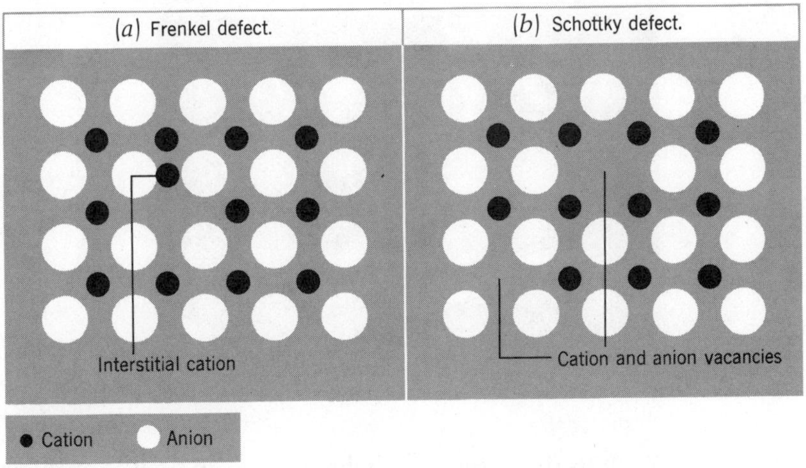


Figure 7.18
*Defects in crystals.* (a) *Frenkel defect.* (b) *Schottky defect.*

that are ordinarily stoichiometric. For example, NaCl can be reacted with Na vapor to form $Na_{(1+x)}Cl$. In this case the extra sodium is present as $Na^+$ ions in cationic sites, and the extra anionic sites are occupied by the electrons liberated from the sodium. These trapped electrons in anionic sites are called F-centers.

Perhaps the most important type of defect results when an impurity, often amounting to merely several parts per million, is introduced into a crystal. Here we find that a certain number of atoms in a solid are replaced by atoms of a different kind. For example, Ga or As atoms may be introduced into very pure germanium to yield substances with enhanced semiconductor properties.

Germanium has the diamond structure (Figure 7.15) with four covalent bonds to each Ge atom. When a Ga atom ($4s^24p^1$) replaces a Ge atom ($4s^24p^2$), one of the Ga—Ge bonds is deficient by one electron. Under an applied voltage an electron from a neighboring atom may move to fill this deficiency and thus leave a positive "hole" behind. When this "hole" is filled, another is created elsewhere and electrical conduction results from the migration of this positive "hole" through the crystal. Because of the positive nature of the charge carrier, the substance is said to be a **p-type semiconductor.**

Arsenic ($4s^24p^3$) has one more electron in its valence shell than germanium. When arsenic is added to germanium as an impurity (we say that the Ge is "doped" with As), the extra electrons supplied are able to move through the solid when a voltage is applied. Because the conduction is due to negative electrons, the crystal is called an **n-type semiconductor.**

Since their discovery, *n*- and *p*-type semiconductors have served as the nucleus for the explosive growth of solid-state electronics. The transistor, used in so many electronic gadgets that we now take for granted, is made of *n*- and *p*-type semiconductors. Your pocket calculator, for instance, contains thousands of transistors. Perhaps of even more current interest, these materials may also be used to harness the sun's energy in solar batteries.

A silicon solar battery is composed of a silicon wafer doped with arsenic (giving an *n*-type semiconductor) over which is placed a thin layer of silicon doped with boron (a *p*-type semiconductor). This is illustrated in Figure

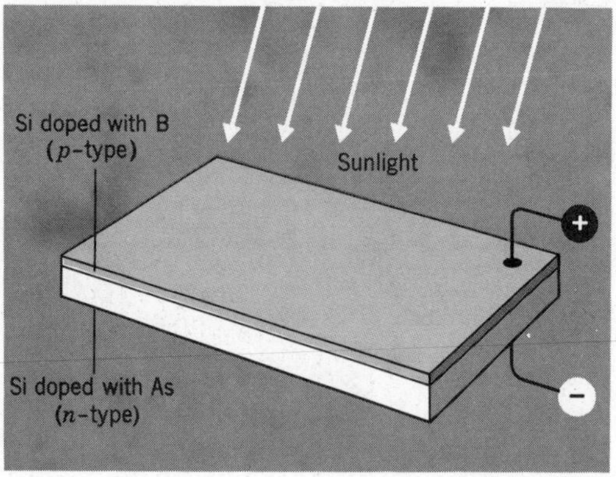


Figure 7.19
*A solar battery.*

7.19. In the absence of light there is an equilibrium between electrons and holes at the interface between the two layers (called a *p-n junction*). Some electrons from the *n*-type layer diffuse into the holes in the *p*-layer and are trapped. This leaves positive holes behind in the *n*-layer. Equilibrium is achieved when the positive holes in the *n*-layer prevent further movement of electrons to the *p*-layer. When light is allowed to fall on the surface of the cell, the equilibrium is upset. Energy is absorbed which permits electrons that were trapped in the *p*-layer to return to the *n*-layer. As these released electrons move across the *p-n* junction, other electrons leave the *n*-layer through the wire, pass through the electrical circuit, and enter the *p*-layer. Thus an electrical current flows when light falls on the cell and the external circuit is completed.

## INDEX TO QUESTIONS AND PROBLEMS (problem numbers in italics)

Crystalline features 1, 2

Bragg equation 4, *31, 32, 33, 41*

Lattices 3, 5, 7

Cubic lattices 6, 8, 9, 10, 11, 12, *42, 44, 45*

Closest packed structures 13, 14, 15

Atomic and ionic radii *34, 36, 37, 38, 39, 40, 43 46, 47, 48, 49*

Avogadro's number *35*

Types of solids 16, 17, 18, 19, 20, 21, 22, 23

Band theory 24, 25, 30

Defects in solids 26, 27, 28, 29

## REVIEW QUESTIONS

**7.1** What external features do crystals exhibit?

**7.2** What is an amorphous solid? How does it differ from a crystalline solid?

**7.3** What is a lattice? What is a unit cell? Why can one kind of lattice be used to describe many different chemical structures?

**7.4** What is the Bragg equation? What do the symbols in the equation stand for?

**7.5** How many different space lattices are there? What quantities are used to describe a particular lattice?

**7.6** Sketch and name the three types of cubic lattices.

**7.7** Show that there are two atoms per unit cell in a substance that crystallizes in (a) a body-centered lattice and (b) an end-centered lattice.

**7.8** Describe the rock salt structure. What kind of lattice does it belong to? How many formula units are there per unit cell? Could a salt like $K_2S$ crystallize in the rock salt structure? Explain your answer.

**7.9** Describe the zinc blende structure. What is the cation-to-anion ratio in this structure? What geometric arrangement of cations is there about each anion?

**7.10** Describe the fluorite structure. Why can't NaCl crystallize in this kind of structure? How is the antifluorite structure related to the fluorite structure?

**7.11** How many formula units are there in the unit cell of $CaF_2$ (fluorite structure)?

**7.12** In the zinc blende structure the cation and anion sites are interchangeable; that is, the open circles in Figure 7.11 could be either cations or anions. This is not the case for the fluorite structure as typified by $CaF_2$. Why?

**7.13** How do the *ccp* and *hcp* structures differ?

**7.14** In the closest-packed structures, what is a tetrahedral site? What is an octahedral site?

**7.15** Use the rock salt and fluorite structures to prove that there are twice as many tetrahedral sites as octahedral sites in the face-centered cubic lattice.

**7.16** Identify the kinds of chemical units associated with each of the following kinds of crystals: molecular, metallic, covalent, ionic. Describe the properties of each of them. What kinds of attractive forces exist between the chemical units in these crystals?

**7.17** Indicate which type of crystal (ionic, covalent, etc.) each of the following would form upon solidification: (a) $O_2$ (b) $H_2S$ (c) Pt (d) KCl (e) Ge (f) $Al_2(SO_4)_3$ (g) Ne

**7.18** Indicate which type of crystal (ionic, covalent, etc.) each of the following would form upon solidification: (a) $Br_2$ (b) LiF (c) MgO (d) Cr (e) $SiO_2$ (f) $PH_3$ (g) NaOH.

**7.19** $SnCl_4$ is a colorless liquid having a boiling point of 114°C and a melting point of −33°C. $SnCl_2$, on the other hand, is a white solid that melts at 246°C. What type of solid (ionic, covalent, etc.) is most likely formed when $SnCl_4$ solidifies?

**7.20** Elemental boron is extremely hard (nearly as hard as diamond) and has a melting point of 2300°C. It is a poor conductor of electricity at room temperature. What kind of solid would you expect for boron based on these properties?

**7.21** Parafin (wax) is generally low-melting, soft, and is a nonconductor in both the solid and liquid states. What kind of solid is expected for parafin?

**7.22** $OsO_4$ has a melting point of 39.5°C and is a nonconductor of electricity in the molten state. It boils at 130°C. What kind of solid is expected for $OsO_4$?

**7.23** $CaCO_3$ (calcite, Figure 7.1) is hard and brittle. It decomposes, before it melts, at a temperature of about 900°C. What kind of solid is likely for calcite?

**7.24** On the basis of the band theory of solids, how do conductors, semiconductors, and nonconductors differ?

**7.25** Construct a diagram to illustrate the band structure of potassium.

**7.26** What are (a) Frenkel defects, (b) Schottky defects?

**7.27** What is an *n*-type semiconductor? What is a *p*-type semiconductor?

**7.28** (a) Give two elements that would make Si a *p*-type semiconductor. (b) Give two elements that would make Si an *n*-type semiconductor.

**7.29** Describe what happens when light falls on a silicon solar cell.

**7.30** Why does the electrical conductivity of a semiconductor increase with increasing temperature?

## REVIEW PROBLEMS

**7.31** Calculate the angles at which X-rays of wavelength 2.29 Å will be observed to be reflected from crystal planes spaced (a) 10 Å apart, (b) 2 Å apart. Assume that $n = 1$.

**7.32** Calculate the interplanar spacings that correspond to reflections at $\theta = 20°$, 27.4°, and 35.8° by X-rays of wavelength 1.41 Å. Assume that $n = 1$.

**7.33** From the following list of angles, determine the angles at which X-rays of wavelength 1.41 Å, diffracted from planes of atoms 2.0 Å apart, are in phase: $\theta = 17.3°$, 20.5°, 44.4°, and 55.3°. Assume that $n = 1$.

**7.34** Chromium, used to protect and beautify other metals, crystallizes in a body-centered cubic structure in which the Cr atoms are in contact along the body diagonal of the unit cell. The edge of the unit cell is 2.884 Å. Calculate the atomic radius of a Cr atom.

**7.35** Chromium crystallizes with a body-centered lattice. Its density is 7.19 g/ml and the unit cell edge is 2.884 Å. Use these data to compute Avogadro's number. How does it compare with the value calculated on page 208?

**7.36** Gold crystallizes with a face-centered cubic lattice. The length of the unit cell edge is 4.0786 Å. What is the atomic radius of a gold atom?

**7.37** Aluminum crystallizes in a cubic closest-packed (face-centered cubic) structure. If the Al atom has an atomic radius of 1.43 Å, what is the length of the unit cell edge in Al?

**7.38** CsCl forms a simple cubic lattice in which there are $Cs^+$ ions at the corners of the unit cell and a $Cl^-$ ion in the center of the cell. The cation/anion contact occurs along the body diagonal of the unit cell. The length of the unit cell edge is 4.123 Å. The $Cl^-$ ion has a radius of 1.81 Å. What is the radius of the $Cs^+$ ion?

**7.39** RbCl had the rock salt structure shown in Figure 7.10. The unit-cell edge length is 6.58 Å. Cations and anions are in contact along the edges. The ionic radius of the chloride ion is 1.81 Å. Calculate the ionic radius of the $Rb^+$ ion.

**7.40** Silver has an atomic radius of 1.44 Å. What would be the density of Ag if it were to crystallize in the following structures: (a) simple cubic, (b) body-centered cubic, (c) face-centered cubic? The actual density of Ag is 10.6 g/ml. Which of these corresponds to the correct structure for Ag?

***7.41** Refer to the diagram at the right to derive the Bragg equation. Remember that the extra distance traveled by the more penetrating beam must be an integral multiple of the wavelength in order to have constructive interference.

***7.42** Calculate the amount of vacant (unoccupied) space in a primitive cubic, a body-centered cubic, and a face-centered cubic packing of identical spheres of diameter 1.0 Å.

***7.43** LiBr has the rock salt structure in which $Br^-$ ions, centered at lattice points, are in contact. Calculate the ionic radii of $Br^-$ and $Li^+$ if the unit-cell edge is 5.50 Å. Why is the accepted value for the ionic radius of $Li^+$ (0.60 Å) smaller than the value that you just computed?

***7.44** CsCl crystallizes with a cubic unit cell of edge length 4.123 Å. The density of CsCl is 3.99 g/cm³. Show that the unit cell cannot be face-centered or body-centered.

***7.45** Metallic sodium crystallizes with a body-centered lattice. The element has a density of 0.97 g/ml. What is the length of the edge of the unit cell in Na?

***7.46** $CaF_2$ (fluorite) has a unit-cell edge of 5.4626 Å. The fluoride ion has a radius of 1.33 Å. What is the radius of $Ca^{2+}$?

***7.47** NaCl (which has the rock salt structure) has a density of 2.165 g/ml. The ionic radius of $Cl^-$ is 1.81 Å. What is the ionic radius of $Na^+$?

***7.48** Solid germanium has the diamond structure (Figure 7.15) with a unit-cell edge of 5.6576 Å. Calculate the Ge—Ge bond length in this solid.

***7.49** The C—C bond length in diamond is 1.54 Å. Calculate: (a) the length of the unit cell edge; (b) the density of diamond.

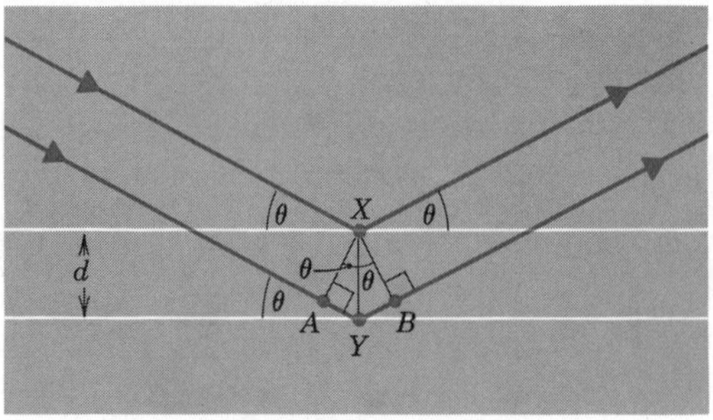


$\angle XAY = \angle XBY = 90°$

*Reflection of X-rays from layers of crystal planes.*

# 8 LIQUIDS AND CHANGES OF STATE

As heat energy is added to a solid there is an increase in the kinetic energy of the molecules or ions that occupy the lattice sites. The particles move about more and more violently, until finally the attractive forces are no longer able to hold them in the lattice and the solid melts, forming a liquid. A gas can also be condensed to a liquid by sufficiently lowering its temperature or, under appropriate conditions, sufficiently increasing its pressure. Lowering the temperature of a gas brings about a decrease in the kinetic energy of the molecules. This causes the molecules to slow down and, at the condensation temperature, the intermolecular attractive forces are able to allow groups of molecules to cling together. Increasing the pressure on a gas causes the molecules to move closer together; if the attractive forces become strong enough, condensation occurs.

In this chapter we will examine the properties of liquids in terms of the nature of the particles that make up the liquid. We will also look at factors that influence the transition between the three states—solid, liquid and gas.

## 8.1 GENERAL PROPERTIES OF LIQUIDS

A liquid is composed of molecules that are constantly and randomly moving about, each undergoing many billions of collisions per second. However, strong attractive forces of the dipole–dipole, hydrogen bond, or London type prevent them from moving as freely and as far apart as in a gas. On the other hand, the molecules of a liquid are not as close together or as structured as they are in a solid.[1] For these reasons liquids exhibit characteristics that place them somewhere between the completely chaotic gaseous state and the well-ordered solid state.

**VOLUME AND SHAPE.** In a liquid the attractive forces are strong enough to restrict the molecules to move about within a definite volume, but they are not strong enough to cause the molecules to maintain a definite position within the liquid. In fact the molecules, within the limits of the liquid's volume, are free to move over and around one another, thus allowing liquids to flow. Liquids, therefore, maintain a definite volume but, because of their ability to flow, their shape depends on the contour of the container holding them.

**COMPRESSION AND EXPANSION.** In a liquid the attractive forces hold the molecules close together, and increasing the pressure has little effect on

[1] One exception to this is ice, which is less dense than water (Section 4.11).

the volume because there is little free space into which the molecules may be crowded. Liquids are therefore virtually incompressible. Similarly, changes in temperature cause only small volume changes (compared to a gas). The increasing molecular motion that accompanies rising temperature tends to increase the intermolecular distances, but this is opposed by the strong attractive forces.

**DIFFUSION.** When two liquids mix, the molecules of one liquid diffuse throughout the molecules of the other liquid at a rate much slower than is observed when two gases are mixed. We can observe the diffusion of two liquids by dropping a small quantity of ink into some water. As demonstrated in Figures 8.1*a* and 8.1*b*, when the ink drop strikes the water we see it as a concentrated "dot," which slowly spreads throughout the water. Diffusion takes place because the molecules in both liquids are able to move throughout the container. However, because the molecules in both liquids are so close together, each molecule undergoes billions of collisions before traveling very far. The average distance between collisions, called the *mean free path,* is much shorter in liquids than in gases, where the molecules are relatively far apart. Because of the constant interruptions in their molecular paths, liquids diffuse much more slowly than gases.

**SURFACE TENSION.** Each molecule in a liquid moves about, always under the influence of its neighboring molecules. A molecule near the middle of a quantity of liquid feels its attracting neighbors nearly equally in all directions (Figure 8.2*a*). A molecule at the surface of the liquid, however, is not completely surrounded and, as a result, feels attractions only by those molecules below and beside it (Figure 8.2*b*). The molecules along the surface thus feel an attraction in a direction toward the interior of the liquid, which causes the surface molecules to be drawn in. The most stable situation arises when the number of molecules experiencing these unequal attractive forces at the surface is a minimum, a condition that is fulfilled when the surface area of the liquid is as small as possible. The tendency of liquids to minimize their surface area explains why water, for example, beads when splashed on a clean polished solid surface and why the shape of raindrops is spherical. In the laboratory you use this phenomenon when you "fire-polish" glass tubing. As the glass softens, sharp angular edges become

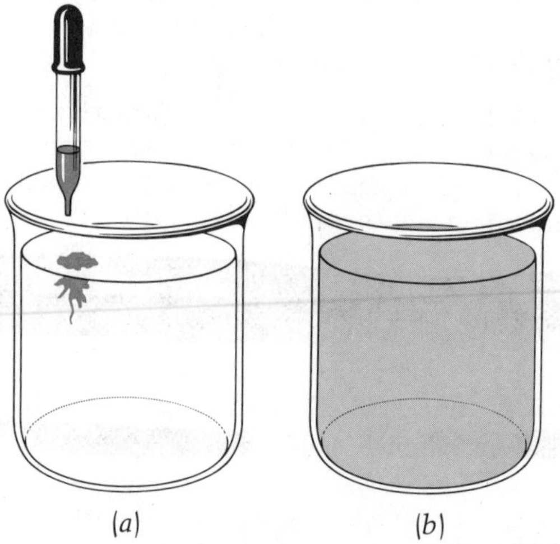


Figure 8.1
*Diffusion in liquids. (a) Ink drop placed into water. (b) Ink has spread throughout the liquid.*

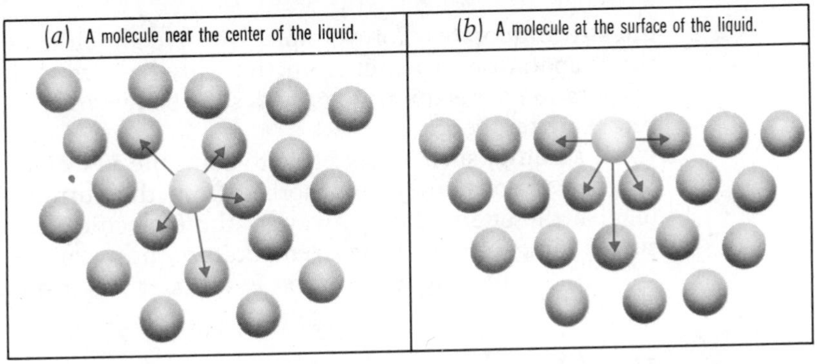


Figure 8.2
*Intermolecular attractive forces in liquids. (a) A molecule near the center of the liquid. (b) A molecule at the surface of the liquid.*

rounded because the attractive forces within the glass tend to reduce the surface area. The amount of work needed to expand the surface of a liquid is dependent on the strength of the inward forces and is called the liquid's **surface tension.** Surface tension is also a function of the temperature of the liquid. Since increasing the temperature (which increases the kinetic energy of the individual molecules) decreases the effectiveness of the intermolecular attractive forces, surface tension decreases as the temperature is raised.

**EVAPORATION.** In a liquid the molecules are constantly undergoing elastic collisions, giving rise to a distribution in individual molecular velocities and, of course, kinetic energies. Following reasoning similar to that described in Section 6.9, it follows that in a liquid, even at room temperature, a small percentage of the molecules are moving with relatively high kinetic energies. If some of these faster-moving molecules possess enough kinetic energy to overcome the attractive forces operating within the liquid and can escape through the surface into the gaseous state, the liquid evaporates. Figure 8.3 represents a typical distribution of the kinetic energies of the molecules in a liquid; the shaded area corresponds to the fraction of the total number of molecules that possess sufficient kinetic energy to evaporate. Just as removing the smart students from a chemistry class will lower the class average on exams, the loss of the higher-energy fraction because of evaporation leads to a lowering of the average kinetic energy of the remaining molecules. Since the temperature is directly proportional to the average kinetic energy, this results in a decrease in the temperature of a liq-

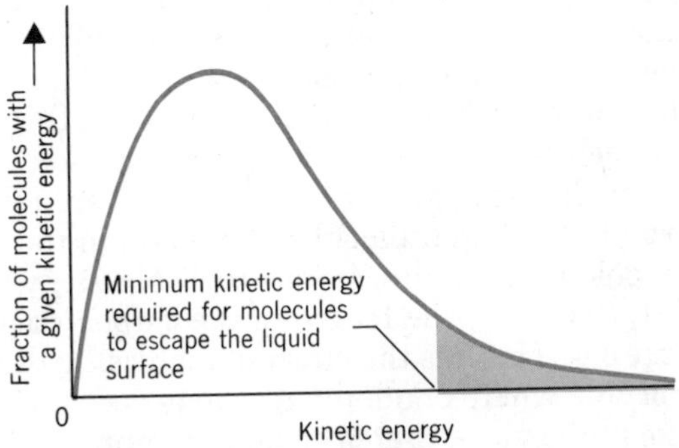


Figure 8.3
*Kinetic energy distribution in a liquid.*

uid as it evaporates. For example, we have all felt cool after a bath, because the evaporation of water from the body has drawn heat from us. In fact, evaporation of perspiration provides the body with a mechanism for controlling body temperature.

If a liquid, such as water, is to continue to evaporate from a container, heat must constantly be absorbed from the surroundings in order to replenish the energy taken away by the molecules leaving the liquid. If the surroundings are at a high temperature, heat can be supplied faster than when they are cool. Thus water evaporates faster on dry hot days than on dry cold days.

## 8.2 HEAT OF VAPORIZATION

The **molar heat of vaporization,** which we shall refer to as $\Delta H_{\text{vaporization}}$ (or simply $\Delta H_{\text{vap}}$), *represents the amount of energy that must be supplied to 1 mole of liquid to convert it into 1 mole of vapor at the same temperature.* The Greek letter, Δ, is usually used to symbolize a change, in this case, a change in the heat content (the total amount of heat energy) of a substance as it undergoes a change from liquid to vapor. This change in heat content is equal to the energy contained in the substance in its final state (vapor) minus the energy that the substance possessed in its initial state (liquid). Thus.

$$\Delta H_{\text{vaporization}} = H_{\text{vapor}} - H_{\text{liquid}}$$

In actual practice, neither $H_{\text{vapor}}$ nor $H_{\text{liquid}}$ can be measured; however, their difference ($\Delta H_{\text{vap}}$) can be.

EXAMPLE 8.1 The heat of vaporization of water is 9.71 kcal/mole. How much heat energy is required to convert 1.0 liter of water to steam?

SOLUTION Water has a density of 1 g/ml. Therefore, 1.0 liter of water weighs 1000 g. The problem, as is generally the case, is one of unit conversion.

$$1000 \text{ g } H_2O \sim (?) \text{ kcal}$$

This can be set up as

$$1000 \text{ g } H_2O \times \left(\frac{1 \text{ mole } H_2O}{18.0 \text{ g } H_2O}\right) \times \left(\frac{9.71 \text{ kcal}}{1 \text{ mole } H_2O}\right) \sim 540 \text{ kcal}$$

540 kcal of heat are required.

The magnitude of $\Delta H_{\text{vap}}$ provides a good measure of the strengths of the attractive forces operative in a liquid. In Table 8.1 we find values of $\Delta H_{\text{vap}}$ for several substances. If we look at the series of hydrocarbons, $CH_4$ through $C_{10}H_{22}$, we observe a steady increase in $\Delta H_{\text{vap}}$ with an increase in molecular weight. These compounds are nonpolar; therefore the only attractive forces that exist between their molecules are London forces. In Chapter 6 it was pointed out that the strengths of London forces are related, at least in part, to the number of atoms in molecules that contain the same elements. If we take a closer look at the hydrocarbons in Table 8.1, we find that as we proceed from $CH_4$ to $C_{10}H_{22}$, the length of the carbon chain increases, as illustrated in Figure 8.4. This has the effect of increasing the number of locations along the molecule where London forces may occur with other molecules. A long chainlike molecule is therefore held in more places than a short mole-

Table 8.1
*Heats of vaporization and boiling points*

| Compound | $\Delta H_{vap}$ kcal/mole (kJ/mole) | Boiling Point (°C) |
|---|---|---|
| $CH_4$ | 2.20 (9.20) | −161 |
| $C_2H_6$ | 3.3 (14) | −89 |
| $C_3H_8$ | 4.32 (18.1) | −30 |
| $C_4H_{10}$ | 5.32 (22.3) | 0 |
| $C_6H_{14}$ | 6.83 (28.6) | 68 |
| $C_8H_{18}$ | 8.10 (33.9) | 125 |
| $C_{10}H_{22}$ | 8.56 (35.8) | 160 |
| $F_2$ | 1.56 (6.52) | −188 |
| $Cl_2$ | 4.88 (20.4) | −34.6 |
| $Br_2$ | 7.34 (30.7) | 59 |
| HF | 7.21 (30.2) | 17 |
| HCl | 3.60 (15:1) | −84 |
| HBr | 3.90 (16.3) | −70 |
| HI | 4.34 (18.2) | −37 |
| $H_2O$ | 9.71 (40.6) | 100 |
| $H_2S$ | 4.49 (18.8) | −61 |
| $NH_3$ | 5.63 (23.6) | −33 |
| $PH_3$ | 3.49 (14.6) | −88 |
| $SiH_4$ | 2.95 (12.3) | −112 |

cule, and more energy must be supplied to remove such long-chain molecules from the liquid. The result is that as the chain length increases, $\Delta H_{vap}$ increases.

Another factor that influences the strengths of London forces is molecular size. If we examine molecules of the same general formula, such as the halogens ($F_2$, $Cl_2$, $Br_2$), we find that large molecules have a greater $\Delta H_{vap}$ than small molecules. As we proceed from $F_2$ to $Br_2$, the atoms that make up the molecules become larger; hence the molecules also become larger. As the

Figure 8.4
*Attractive forces increase with increasing chain length. There are more points along the molecule that can be attracted to other molecules nearby.*

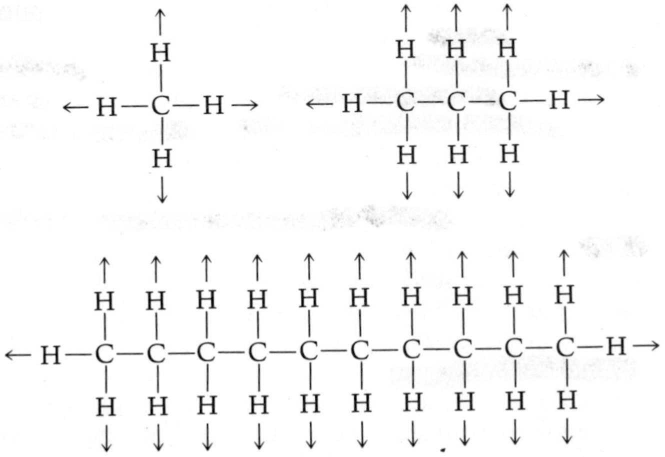

size increases, the outer electrons are further from the nuclei and are not held as tightly. Because of this the electron cloud of a large molecule is more easily distorted and it is easier to create the instantaneous dipoles that are responsible for the London forces. The ease of distortion of the electron cloud is referred to as **polarizability.** The result of this is that the London forces are stronger between molecules composed of large, easily polarized atoms such as Br than between molecules composed of small atoms such as F. Hence $\Delta H_{vap}$ increases from $F_2$ to $Br_2$.

When we look at the hydrogen halides, HF through HI, however, we find that the expected variation of $\Delta H_{vap}$ with molecular size is reversed between HF and HCl. In fact, HF has a considerably higher heat of vaporization than any of the other HX compounds. This anomalous behavior is attributed to the presence of hydrogen bonding. As discussed in Chapter 4, this bonding is a particularly strong dipole–dipole interaction that can occur when hydrogen is bound covalently to a small, very electronegative element. We see the same inverted order of $\Delta H_{vap}$ for $H_2O$ and $H_2S$ and for $NH_3$ and $PH_3$, where again, hydrogen bonding is significant for $H_2O$ and $NH_3$ but not for $H_2S$ and $PH_3$. Oxygen, fluorine, and nitrogen are all very small and are the most electronegative elements in the periodic table, while the elements below them are much larger and much less electronegative. Thus we predict that hydrogen bonding is important only for $H_2O$, HF, and $NH_3$. "Normal" behavior is reached in Group IVA hydrids, where $\Delta H_{vap}$ for $CH_4$ is less than $\Delta H_{vap}$ for $SiH_4$. Here neither $CH_4$ nor $SiH_4$ have any tendency to hydrogen bond because they are nonpolar.

## 8.3 VAPOR PRESSURE

If a liquid evaporates in an open container, eventually all the liquid will disappear because the molecules that have escaped from the liquid into the vapor phase diffuse readily into the atmosphere. If the same quantity of liquid at the same temperature is placed in a closed container, what will happen? In this case the volume of the liquid will initially decrease and then eventually become constant. If we monitored the pressure of the gas above the liquid, we would find that it initially increases and then it too levels off at a constant value. These observations can be explained in the following way. The molecules with higher kinetic energies begin to leave the liquid, evaporating into the vapor phase, where they become trapped. The loss of molecules from the liquid must, of course, be accompanied by a volume decrease. In time the space above the liquid becomes occupied with more and more gaseous molecules, and the pressure of the vapor increases. With the increasing number of chaotically moving gas molecules the number of collisions with the walls in this restricted volume also increases. One of these walls is the surface of the liquid itself, which will trap any bombarding molecules having low kinetic energies. Thus condensation as well as evaporation (vaporization) take place at the surface of the liquid. Eventually the number of molecules in the vapor becomes large enough so that the rate at which the gas condenses exactly equals the rate at which the liquid evaporates, and no further change in either the volume of the liquid or the pressure exerted by its vapor is observed. To emphasize this point again, vaporization and condensation are still taking place, but with no change in the liquid volume or the vapor pressure. At this point the liquid is said to be in **dynamic equilibrium** with its vapor. The pressure exerted by the quantity of vapor above the liquid, when equilibrium is established, is called the **equilibrium vapor pressure** of the liquid. The vapor pressure of a liquid, quite naturally, depends on the ease with which its molecules can leave the liquid and enter the vapor state. In liquids where the intermolecular attractive forces are strong, the vapor pressure will be low, and in liquids where the attractive forces are weak, the vapor pressure will be high. Since increasing the temperature of a liquid increases the number of molecules possessing sufficient energy to overcome the attractive forces, the vapor pressure must increase with increasing temperature. Thus, whenever the vapor pressure is given, the temperature at which it was measured must also be specified.

One way that we can determine the vapor pressure of a liquid is by the use of a barometer, as shown in Figure 8.5. The height of the mercury in the barometer, before any liquid is added, is measured accurately. A liquid whose vapor pressure is to be determined is carefully added to the barometer by means of an eye dropper, and allowed to rise to the top of the mercury in the column, as shown in Figures 8.5*b*, 8.5*c*, and 8.5*d* (most liquids are less dense than mercury and will therefore float on the mercury surface). The space above the mercury column in Figure 8.5*a* is, for all practical purposes a vacuum[2] and exerts no downward force. The space above the mercury in Figures 8.5*b*, 8.5*c*, and 8.5*d* is filled with a small amount of liquid and its vapor. As the liquid begins to evaporate, the pressure of the trapped vapor causes the level of the mercury in the column to decrease; when the liquid and vapor are in equilibrium, the height of the mercury column becomes stationary. The total pressure exerted at the reference level outside each barometer will be the atmospheric pressure, $P_{atm}$. The total pressure exerted within the barometer is $P_{Hg}$, the pressure due to the pull of gravity on the mercury in the column, plus $P_{vapor}$, the pressure exerted by the vapor in

**Figure 8.5**
*Measurement of vapor pressure. When a small amount of liquid is introduced above the mercury in a barometer, the vapor pressure of the liquid forces the mercury down. (a) No liquid above the mercury. (b) $H_2O$. (c) Ethyl alcohol, $C_2H_5OH$. (d) Diethyl ether, $(C_2H_5)_2O$.*

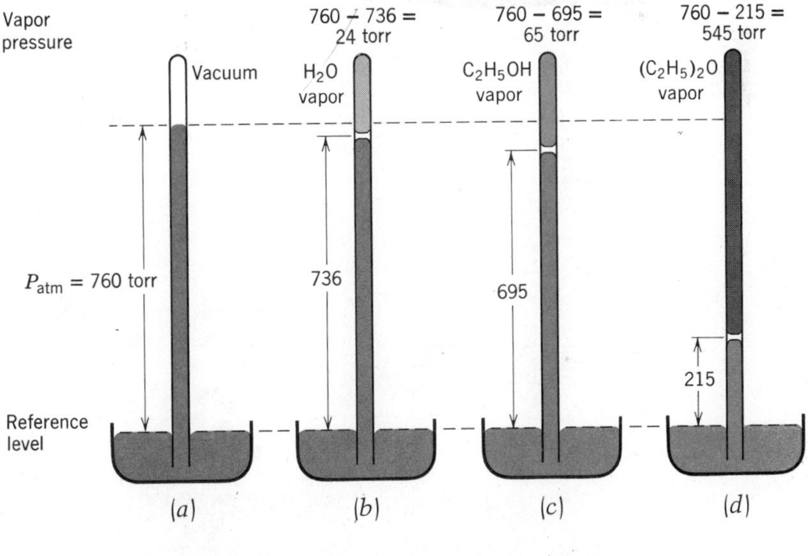


[2] Mercury itself does have a finite vapor pressure (about $10^{-3}$ torr at room temperature) and, therefore, should never be left in an open container because of its high toxicity.

equilibrium with its liquid. The additional pressure exerted by the weight of the small amount of liquid on top of the column is negligibly small. Therefore, at equilibrium in each barometer

$$P_{atm} = P_{Hg} + P_{vapor}$$

In Figure 8.5*a*, $P_{vapor} = 0$; therefore, $P_{atm} = P_{Hg} = 760$ torr. In Figure 8.5*b* 8.5*c*, and 8.5*d*, $P_{Hg} = 736$ mm, 695 mm, and 215 mm, respectively. Therefore, at 25°C the vapor pressure of water is 24 torr, that of ethyl alcohol is 65 torr, and that of diethyl ether is 545 torr. Water has the lowest vapor pressure of the three liquids in our example; therefore, it must have the strongest intermolecular attractive forces. Diethyl ether, on the other hand, has the highest vapor pressure of the three liquids, which means that relatively weak attractive forces exist in it.

We have seen that the vapor pressure of a liquid is dependent on the nature of the liquid and its temperature. What happens to the vapor pressure when the volume or pressure of the vapor is changed? In Figure 8.6 we see an illustration of an apparatus that can be used to demonstrate the effect of volume and pressure changes on a liquid–vapor equilibrium. At a constant temperature equilibrium is established, as shown by Figure 8.6*a*. If, at this same constant temperature we allow the vapor to expand by rapidly raising the piston (Figure 8.6*b*), then the system is no longer in equilibrium. As the piston is withdrawn, creating the larger volume, the number of molecule–wall collisions decreases, causing a decrease in the pressure exerted by the vapor. The rate of condensation at the surface, which depends on the number of collisions between the vapor molecules and the surface of the liquid, must also decrease. The rate of evaporation, however, remains essentially the same. This means that, at a constant temperature, an increase in the volume of a vapor in equilibrium with its liquid causes more molecules to leave the liquid state than return to it. As the process continues, more and more molecules enter the vapor phase, causing an increase in the pressure exerted by the vapor and a corresponding increase in the rate of condensation. After a while enough molecules will be present in the vapor phase so

Figure 8.6
*Effect of volume changes on vapor pressure. (a) Equilibrium between liquid and vapor. (b) No equilibrium. Rate of evaporation is greater than rate of condensation. (c) No equilibrium. Rate of condensation is greater than rate of evaporation.*

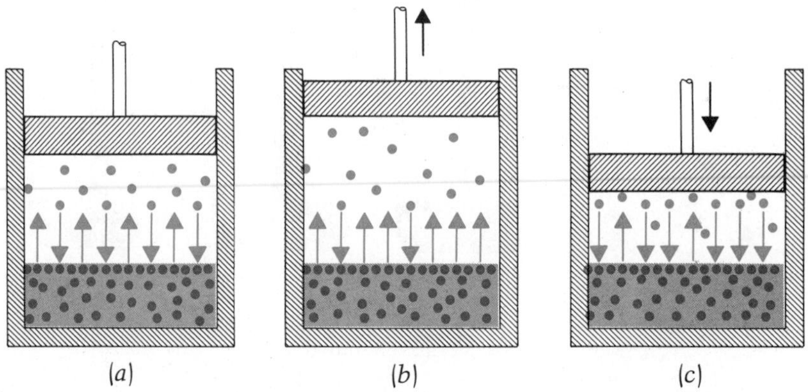

that the rate of condensation will again exactly equal the rate of evaporation, and equilibrium will be reestablished. At this newly established equilibrium the larger volume of gas is now occupied by more molecules. The vapor pressure will be the same as before the volume change occurred, but the volume of the liquid will be slightly smaller.

Decreasing the volume of the vapor by lowering the piston (Figure 8.6*c*) will also disturb the equilibrium. Increasing the pressure of the vapor will cause an increase in the number of molecule–wall collisions. This, in turn, will lead to an increase in the rate of condensation but will have very little effect on the rate of evaporation. The rate at which the molecules leave the vapor phase, then, will be greater than the rate at which molecules leave the liquid phase. This imbalance in rates causes the pressure exerted by the vapor to decrease and the volume of the liquid to increase. Eventually, the rate of condensation will decrease to a point where it exactly equals the rate of evaporation, reestablishing equilibrium. At this new equilibrium the smaller vapor volume, caused by the movement of the piston, will be occupied by fewer gaseous molecules. The vapor pressure will have returned to its initial value, and the volume of the liquid will have increased slightly.

The net result of this discussion is that the *vapor pressure of a liquid is independent of the volume of the container, provided that there is some liquid present so that an equilibrium can be established.*

**LE CHATELIER'S PRINCIPLE.** The dynamic equilibrium between a liquid and its vapor can be represented by the equation,

$$\text{liquid} \rightleftharpoons \text{vapor} \tag{8.1}$$

Here the double half-arrows mean that the rates of evaporation and condensation are equal. If we in any way disturb this system so that it is no longer at equilibrium (we say that a "stress" is applied to the system), a change occurs that will, if possible, bring the system back to equilibrium. In our example above, an increase in the volume of the vapor caused the system to no longer be at equilibrium. We saw that more liquid evaporated until equilibrium was reestablished. In Equation 8.1, this corresponds to the process as read from left to right, that is, liquid → vapor, and results in a new "position of equilibrium" in which there is less liquid and more vapor. In this sense the position of equilibrium has shifted to the right when we applied the stress. The action taken by any system at equilibrium when a stress is applied can be described by the **principle of Le Chatelier,**[3] which states that *when a system in a state of dynamic equilibrium is acted upon by some outside stress, the system will, if possible, shift to a new position of equilibrium in order to minimize the effect of the stress.*

For example, let us apply Le Chatelier's principle to describe what effect pressure changes have on a liquid–vapor equilibrium. When the stress applied is a pressure decrease caused by an increase in the volume of the container, the system attempts to undergo a change that will return the pressure to its initial value. In this example, the pressure can be increased if more molecules enter the vapor phase, that is, if some additional liquid evaporates. After equilibrium has been reestablished, there will be less liquid and more vapor present in the container, and we say that the position of equilibrium represented by Equation 8.1 has shifted to the right. However, if

[3] Henri Le Chatelier, a professor in Paris, proposed his important "law of reaction" in 1888. He died in 1936.

the volume is increased sufficiently, all of the liquid will evaporate and equilibrium will not be reestablished. This will occur, for example, if the piston in Figure 8.6 is removed entirely so that the liquid is open to the atmosphere.

In a similar fashion, we predict that raising the pressure leads to a decrease in the quantity of vapor and, of course, a corresponding increase in the amount of liquid. Thus, we might conclude in general that an increase in pressure on a system at equilibrium favors the production of the more dense phase, while a decrease in pressure favors the formation of the less dense phase.

The effect of temperature changes on an equilibrium can also be described using Le Chatelier's principle. Increasing the temperature of a system at equilibrium favors the absorption of energy (an endothermic change). In a liquid–vapor equilibrium system this means that a temperature increase causes more liquid to evaporate, because this process absorbs heat. Decreasing the temperature (removing heat), on the other hand, favors the release of energy, an exothermic change. As the temperature is decreased in a liquid–vapor equilibrium, more molecules condense into the liquid phase, releasing heat and thereby minimizing the effect of the applied stress.

In summary, Le Chatelier's principle predicts that a temperature increase will shift the position of equilibrium in the direction of the endothermic process. Similarly, a decrease in temperature will favor the exothermic change.

**VAPOR PRESSURE CURVES FOR LIQUIDS.** We can determine the vapor pressure of liquids as a function of temperature by using the same apparatus

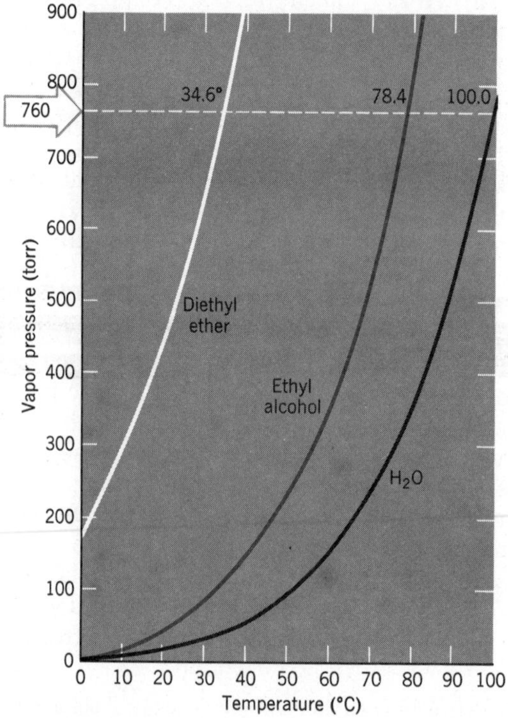


Figure 8.7
*Vapor pressure curves.*

described in Figure 8.5 and varying the surrounding temperature. Data accumulated in such experiments performed on water, ethyl alcohol, and diethyl ether are illustrated graphically in Figure 8.7. We see from the shapes of the curves in the figure that at lower temperatures vapor pressure changes relatively slowly with increasing temperature, while at higher temperatures the changes are more rapid. Points along the curve in Figure 8.7 represent combinations of pressures and temperatures that must be satisfied in order for the liquid to be in equilibrium with its vapor. These curves terminate at a temperature, called the **critical temperature** ($T_c$), above which molecular motion is so violent that the substance can exist only as a gas. In other words, *the critical temperature is that temperature above which a substance can no longer exist as a liquid, regardless of the applied pressure.* The pressure that must be applied to a substance at its critical temperature in order to achieve a liquid–vapor equilibrium is called its **critical pressure.** In Table 8.2 a few substances are listed with their corresponding critical temperatures and critical pressures.

Vapor pressure (along with its variation with temperature) and the heat of vaporization are both controlled by the intermolecular attractive forces within a liquid. It should not be surprising, therefore, to find that there is a quantitative relationship among vapor pressure, temperature, and $\Delta H_{vap}$.

If the logarithm of the vapor pressure ($\log p$) is plotted versus the reciprocal of the absolute temperature ($1/T$), a straight line is obtained, at least over relatively short temperature ranges, as shown in Figure 8.8.[4] In general, any straight line can be described by an equation,

$$y = b + mx$$

where $m$ is the slope of the line and $b$ is the intercept of the line with the vertical axis. In the present case we can write

$$\log p = b + m\left(\frac{1}{T}\right)$$

It can be shown that the slope of the line is related to the heat of vaporization.

$$m = \frac{-\Delta H_{vap}}{2.303R}$$

Table 8.2
*Some critical temperatures and pressures*

| Compound | $T_c$ (°C) | $P_c$ (atm) |
|---|---|---|
| Methane ($CH_4$) | −82.1 | 45.8 |
| Ethane ($C_2H_6$) | 32.2 | 48.2 |
| Benzene ($C_6H_6$) | 288.9 | 48.6 |
| Ammonia | 132.5 | 112.5 |
| Carbon dioxide | 31 | 72.9 |
| Water | 374.1 | 217.7 |
| Helium | −267.8 | 2.3 |

[4] You can review logarithms in Appendix C.

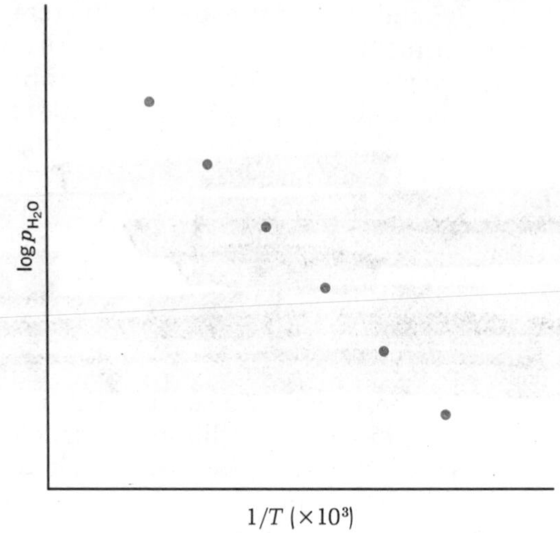


Figure 8.8
*Plot of log(vapor pressure of $H_2O$) versus 1/T in the temperature range from 25°C to 50°C.*

When the value of $R$ used in the calculation is 8.31 J/mole K, the units of $\Delta H_{vap}$ will be J/mole; if $R$ = 1.99 cal/mole K, then $\Delta H_{vap}$ will have units of cal/mole.

A convenient form of the relationship between $\Delta H_{vap}$ and vapor pressure is given by the *Clausius-Clapeyron equation* (named after a German physicist, R. Clausius, and a French engineer, B. P. E. Clapeyron).

$$\log\left(\frac{p_1}{p_2}\right) = \frac{-\Delta H_{vap}}{2.303R}\left(\frac{1}{T_1} - \frac{1}{T_2}\right) \tag{8.2}$$

In this equation $p_1$ is the vapor pressure when the absolute temperature of the liquid is $T_1$; $p_2$ is the vapor pressure at temperature $T_2$.

Equation 8.2 can be used to calculate $\Delta H_{vap}$ if the vapor pressure is known at two different temperatures. It can also be used to calculate the vapor pressure at a specific temperature provided that $\Delta H_{vap}$ and the vapor pressure at some other temperature are known.

EXAMPLE 8.2 At 25°C the vapor pressure of carbon tetrachloride is 115 torr; at 40°C the vapor pressure is 216 torr. Calculate $\Delta H_{vap}$ for $CCl_4$ in kJ/mole.

SOLUTION To obtain $\Delta H_{vap}$ we must substitute $p$'s and $T$'s into Equation 8.2 and solve for $\Delta H_{vap}$. First let's organize the data.

$$p_1 = 115 \text{ torr} \qquad T_1 = 25 + 273 = 298 \text{ K}$$
$$p_2 = 216 \text{ torr} \qquad T_2 = 40 + 273 = 313 \text{ K}$$

Substituting into Equation 8.2 using $R$ = 8.31 J mole$^{-1}$ K$^{-1}$ gives

$$\log\left(\frac{115 \text{ torr}}{216 \text{ torr}}\right) = \frac{-\Delta H_{vap}}{(2.303)(8.31 \text{ J mole}^{-1} \text{ K}^{-1})}\left(\frac{1}{298 \text{ K}} - \frac{1}{313 \text{ K}}\right)$$

$$\log(0.532) = \frac{-\Delta H_{vap}}{19.1 \text{ J mole}^{-1} \text{ K}^{-1}}(1.61 \times 10^{-4} \text{ K}^{-1})$$

$$-0.274 = -\Delta H_{vap}(8.43 \times 10^{-6} \text{ J}^{-1} \text{ mole})$$

$$\Delta H_{vap} = 3.25 \times 10^{4} \text{ J mole}^{-1}$$

Since

$$1 \text{ kJ} = 10^3 \text{ J}$$
$$\Delta H_{vap} = 32.5 \text{ kJ/mole}$$

## 8.4 BOILING POINT

That temperature at which the vapor pressure of a liquid is equal to the atmospheric pressure is known as the **boiling point** of the liquid. At this temperature the vapor pressure is high enough to cause vaporization to occur at various points throughout the interior of the liquid. Thus boiling is accompanied by the formation of bubbles, which form simultaneously at many spots in the liquid.[5]

When a bubble is formed within the liquid, the liquid that originally occupied this space is pushed aside and the level of the liquid in the container is forced to rise against the downward pressure exerted by the atmosphere. In other words, it is the pressure exerted by the vapor inside the bubble that pushes the surface of the liquid up against the atmospheric pressure. This can occur only when the vapor pressure of the liquid becomes equal to the prevailing atmospheric pressure. If it were less, the atmospheric pressure would cause the bubble to collapse.

As long as bubbles are forming within the liquid, that is, as long as the liquid is boiling, the vapor pressure of the liquid is equal to the atmospheric pressure. Since the vapor pressure remains constant, the temperature of the boiling liquid also stays the same. An increase in the rate at which heat is supplied to the boiling liquid simply causes bubbles to form more rapidly. The liquid boils away more quickly, but the temperature does not increase.

It is obvious, from the discussion above, that the boiling point of a liquid depends on the prevailing atmospheric pressure. The boiling point of a liquid at 1 atm (760 torr) is referred to as its **standard** or **normal boiling point.** For water, the normal boiling point is 100°C. At higher pressures its boiling point is greater; at lower pressures (for example, on a mountain top) its boiling point is less. Boiling points given in reference tables are always normal boiling points, unless otherwise stated.

The constant temperature maintained by a boiling liquid is utilized when we employ water for cooking foods. Once water boils, its temperature remains at 100°C, which is ideal for cooking foods evenly and at a rapid rate. The pressure cooker also takes advantage of the fact that the boiling point changes with pressure. These cookers are time savers because they allow foods to be prepared at a much faster rate than they could be in an open pot. The lid on a pressure cooker forms a tight seal on the pot and is equipped with a pressure-relief valve to prevent the pot from exploding. The heat supplied by the stove causes more and more liquid water to evaporate; as a result, the pressure inside the kettle increases until steam begins to exit from the relief valve. Since the pressure inside the cooker at this point is higher than 760 torr, the water boils at a higher temperature and foods cook faster.

The temperature at which a liquid boils is another example of a property that gives a good estimation of the strength of the attractive forces operating within the liquid. Liquids whose attractive forces are relatively high have

[5] When heat is first applied to a liquid, many small bubbles begin to form. These small bubbles are due to the expulsion of dissolved gases and do not mean that the liquid has begun to boil. We refer here to the larger bubbles formed during boiling.

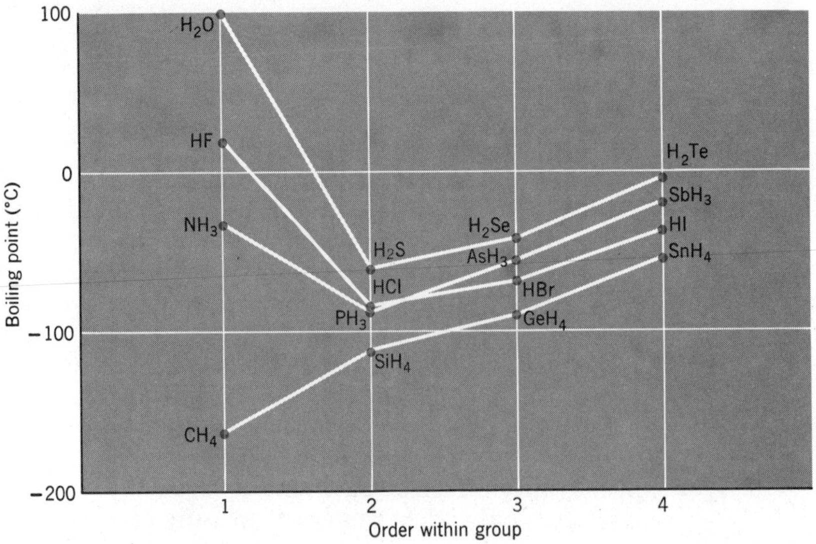


Figure 8.9
*Boiling points of hydrogen compounds of Group IVA, VA, VIA, and VIIA elements.*

correspondingly high boiling points, while liquids with weak attractive forces boil at a relatively low temperature. In Table 8.1 (page 227), for example, we see that the trends in boiling points follow trends in $\Delta H_{vap}$. The dependence of boiling point on intermolecular attractive forces is also demonstrated by Figure 8.9, in which the boiling points of some hydrogen compounds of the elements in Groups IVA, VA, VIA, and VIIA are compared. Let us look at the compounds of Group IVA first because they form a nearly ideal pattern. We see from the figure that as the atomic weights of the elements in Group IVA increase, so do the boiling points of their hydrogen compounds. Using reasoning similar to that developed in Section 8.2 in the discussion of the molar heat of vaporization, we know that as the size of the molecules increase from $CH_4$ to $SnH_4$, the London forces also increase. We expect, therefore, that the boiling points of this series of compounds increase, as they actually do, in the direction of increasing molecular weights.

Except for the first members of the hydrogen compounds of Groups VA, VIA, and VIIA, the same trend is also observed—that is, increasing boiling point with increasing atomic weight of the element in the group. The first member of each of these groups, however, has a relatively high boiling point, much higher than expected from molecular weights alone. They, therefore, must possess attractive forces in addition to those of the London type. The position of each of the first members in relation to the rest of the group can be attributed to the presence of hydrogen bonding. We saw in Section 8.2 that the strongest hydrogen bonds form when hydrogen is bonded to a very small electronegative element. Therefore, the contribution of hydrogen bonding is expected to be the strongest for the first member of Groups VA, VIA, and VIIA, and becomes relatively unimportant for the remaining members. Methane, $CH_4$, which is nonpolar and cannot hydrogen bond (the electronegativity of carbon is too low and there are no lone electron pairs on the carbon to which hydrogen bonds can be formed) follows the normal, nearly straight-line pattern, within its group.

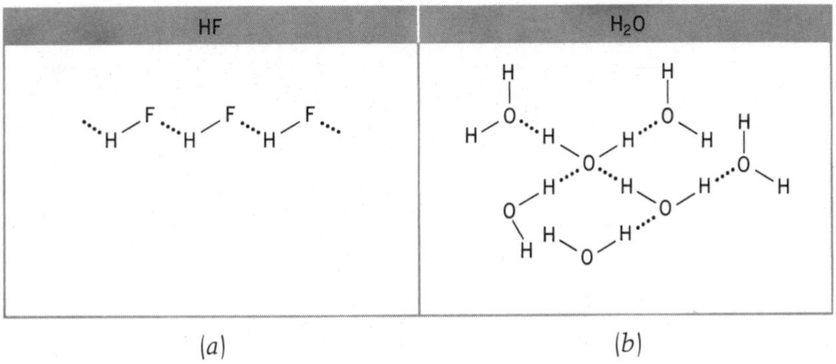


Figure 8.10
*Hydrogen bonding in HF and $H_2O$. (a) Each HF molecule has only one hydrogen atom that can hydrogen bond to something else. (b) Each $H_2O$ molecule has two hydrogen atoms that can hydrogen bond to other $H_2O$ molecules.*

The fact that water has a higher boiling point than HF, even though fluorine is more electronegative than oxygen, seems to be because each water molecule is capable of forming four hydrogen bonds with other $H_2O$ molecules while an HF molecule forms only two hydrogen bonds with two other HF molecules (Figure 8.10). The strength of the four hydrogen bonds in water exceeds that of the two hydrogen bonds in HF even though an HF hydrogen bond is stronger than a single hydrogen bond between $H_2O$ molecules. The hydrogen bonding in $NH_3$ is much weaker than in $H_2O$ or HF because of the considerably lower electronegativity of nitrogen. Thus, even though $NH_3$ could conceivably form three or four hydrogen bonds, their total strength is so small that $NH_3$ has a lower boiling point than either HF or $H_2O$.

## 8.5 FREEZING POINT

Liquids become solids by the removal of heat. As the temperature is lowered, more molecules in the liquid slow down. If the temperature is sufficiently lowered so that the attractive forces cause the slower-moving molecules to become rigidly held in position, the liquid begins to freeze. As pointed out in Chapter 7, if the molecules are "frozen" in a well-ordered lattice, a crystalline solid is formed, while if the molecules are frozen in a random fashion, an amorphous solid is produced. As a solid forms, the average kinetic energy of the molecules remaining in the liquid increases. This is because the molecules with low kinetic energies have been lost to the solid. Therefore, heat must continually be removed if freezing is to continue.

In a solid, as well as in a liquid, we have a distribution of kinetic energies. When a solid is in contact with its liquid, high-energy molecules at its surface can break away from the solid and enter the liquid state. At some particular temperature, called the **freezing point** (or **melting point**), the rate at which molecules leave the solid to enter the liquid is the same as the rate at which molecules are leaving the liquid state to become part of the solid. Thus, at the melting point of a solid or the freezing point of a liquid, equilibrium exists between the liquid and solid.

*The total amount of heat that must be removed in order to freeze 1 mole of a liquid is called its* **molar heat of crystallization.** The **molar heat of**

Table 8.3
*Heats of fusion and vaporization*

| Substance | $\Delta H_{fus}$ kcal/mole (kJ/mole) | $\Delta H_{vap}$ kcal/mole (kJ/mole) |
|---|---|---|
| Water | 1.43 (5.98) | 9.71 (40.6) |
| Benzene | 2.37 (9.92) | 7.35 (30.7) |
| Chloroform | 2.97 (12.4) | 7.62 (31.9) |
| Diethyl ether | 1.64 (6.86) | 6.21 (26.0) |
| Ethanol | 1.82 (7.61) | 9.22 (38.6) |

**fusion,** $\Delta H_{fus}$, on the other hand, *is equal in magnitude but opposite in sign to the molar heat of crystallization and is defined as the amount of heat that must be supplied to melt 1 mole of a solid.* Since heat must be added to melt a solid, the attractive forces must be slightly higher in a solid than in a liquid. This is not surprising because it is expected that greater attractive forces are necessary to hold the particles rigidly in place in a solid than are required to keep them within the liquid where they are free to roam about. The magnitude of the molar heat of fusion provides us with a measure of the differences between the intermolecular attractive forces in the solid and the liquid; that is,

$$\Delta H_{fus} = H_{liquid} - H_{solid}$$

As before, we cannot actually measure $H_{liquid}$ or $H_{solid}$ but, instead, only their difference, $\Delta H_{fus}$. $\Delta H_{fus}$ is always much smaller than the molar heat of vaporization, as shown in Table 8.3. The reason for this is as follows. When a solid melts there are relatively small changes in the distances between the molecules. As a result, only small energy changes are involved. When a liquid is converted to a gas, however, the intermolecular distances increase tremendously and large energy changes occur. This means that the amount of energy ($\Delta H_{fus}$) required to cause the molecules of a solid to overcome their attractive forces and form a liquid is small compared to the energy ($\Delta H_{vap}$) required for liquid molecules to move apart, forming a gas.

EXAMPLE 8.3 Calculate the energy, in kilojoules, necessary to melt 1.00 g of ice.

SOLUTION As before, we have a unit conversion requiring $\Delta H_{fus}$. The problem reduces to

$$1.00 \text{ g } H_2O \sim (?) \text{ kJ}$$

From Table 8.3, $\Delta H_{fus} = 5.98$ kJ/mole for $H_2O$. Therefore,

$$1.00 \, \cancel{\text{g } H_2O} \times \left(\frac{1 \, \cancel{\text{mole } H_2O}}{18.0 \, \cancel{\text{g } H_2O}}\right) \times \left(\frac{5.98 \text{ kJ}}{1 \, \cancel{\text{mole } H_2O}}\right) \sim 0.332 \text{ kJ}$$

The energy needed is 0.332 kJ.

## 8.6 HEATING AND COOLING CURVES: CHANGES OF STATE

The data obtained by following the temperature of a solid as the heat supplied to it causes it to melt, forming a liquid, and continuing until the liquid boils into a gas, gives rise to a heating curve. A graph produced from such an experiment is represented in Figure 8.11, where we have plotted temperature versus time for 1 mole of a substance as it is heated. If the rate of heating is constant during the entire experiment, then any length of time (distance along the abscissa) also represents an amount of heat that has been added. As heat is added to the solid, the average kinetic energy of the molecules in the solid increases; that is, they vibrate more vigorously about their lattice positions, and an increase in the temperature of the solid is observed. This is represented as line $A$ in Figure 8.11. When enough heat has been added to cause the molecules to break away from their lattice positions, the solid begins to melt and we have arrived at point $B$ on the curve. The temperature corresponding to point $B$ is the melting point, or freezing point, $T_f$.

As the solid melts along line $BC$ in Figure 8.11, the temperature of both the solid and liquid remains constant. This means that as the solid melts, the average kinetic energy of this system does not change; therefore, the energy added between points $B$ and $C$ must raise the potential energy of the substance. Where does this potential energy appear?

Recall that potential energy depends on position and, in this case, the position of the molecules of the substance relative to one another. In general, as a substance melts, there is an increase in the average distance of separation between the particles as they pass from the solid to the liquid state.[6] Energy must be absorbed for the particles to move apart against the attractive forces that exist between them. In other words, the energy that is absorbed during the melting process increases the potential energy of the particles with respect to one another as they become separated. The total amount of energy that must be absorbed to melt the 1 mole of material, its molar heat of fusion, is represented as the interval from points $B$ to $C$.

After the entire sample has melted (at point $C$), additional heat goes to increasing the kinetic energy of the liquid molecules. Since the average kinetic energy is increasing, we see a corresponding increase in the temperature from points $C$ to $D$. At point $D$ the liquid begins to boil. As the liquid boils, the temperature of both the liquid and its gas remains constant (line $DE$), which means that the heat supplied during the boiling process does not further increase the kinetic energy of the molecules. Therefore, when point $D$ is reached, the energy supplied goes to increasing the potential energy of

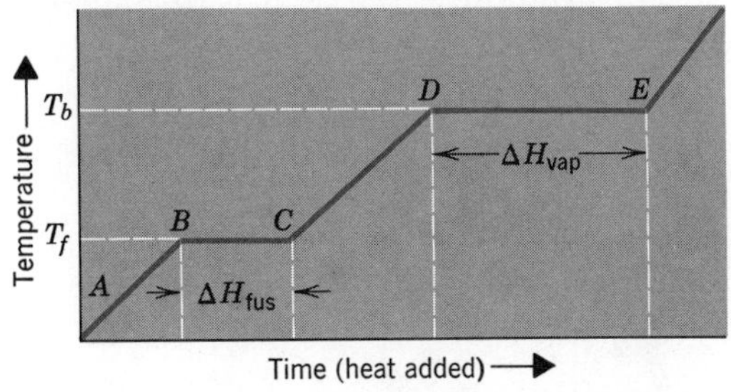


Figure 8.11
*A typical heating curve for 1 mole of a substance.*

[6] A most important exception to this, of course, is ice, in which the solid is less dense (more expanded) than the liquid. Here energy supplied to the solid disrupts some of the hydrogen bonding that exists in the solid ice and the "open" structure of ice collapses to give a more dense liquid.

the molecules as they pass from the liquid to the gas. The total amount of energy absorbed to boil the 1 mole of material (its molar heat of vaporization) is represented as the interval between points *D* and *E*, and the boiling point ($T_b$) is the temperature corresponding to line *DE* on the graph. After the entire sample is in the gaseous state, the absorption of additional heat causes an increase in the average kinetic energy of the gas and an increase in temperature.

A cooling curve, shown in Figure 8.12, can be obtained by following the temperature of a gas as it is cooled. The decrease in the temperature of the gas (line *A*) is caused by a decrease in the average kinetic energy of the gaseous molecules as heat is removed. At point *B* the kinetic energy has been lowered to the extent that the attractive forces can cause condensation to occur. During condensation (line *BC*) the temperature remains constant, which means once again that the average kinetic energy of the molecules in both the gas and the liquid must also stay the same. The removal of heat between points *B* and *C*, therefore, goes to decreasing the potential energy of the molecules as the gas condenses into a liquid. The total amount of heat that must be removed for the 1 mole of gas to condense completely is equal in magnitude (but opposite in sign) to its molar heat of vaporization and is given by the length of the line *BC*. The temperature at which condensation occurs is the same as the boiling point ($T_b$) of the liquid.

When point *C* is reached, the material is entirely in the liquid state. Any further cooling will bring about a decrease in the average kinetic energy of the molecules and a corresponding decrease in its temperature. At point *D* enough heat has been removed to cause the liquid to begin to crystallize. While crystallization is taking place (line *DE*), there is no further decrease in the temperature, which means that the average kinetic energy of the molecules, both liquid and solid, must remain the same. The removal of heat during the crystallization process must, therefore, go to decreasing the potential energy of the molecules as the solid forms from the liquid. The total amount of heat that must be removed in order to crystallize the 1 mole of material is numerically equal (but opposite in sign) to the molar heat of fusion. The temperature corresponding to line *DE* is the freezing point of the liquid and, of course, is the same as the melting point of the solid. Once point *E* is reached, any further removal of heat causes a decrease in the average kinetic energy of the molecules in the solid, and the temperature decreases.

Some liquids do not follow a smooth transition into the solid state, but instead give rise to a cooling curve such as that shown in Figure 8.13. As the

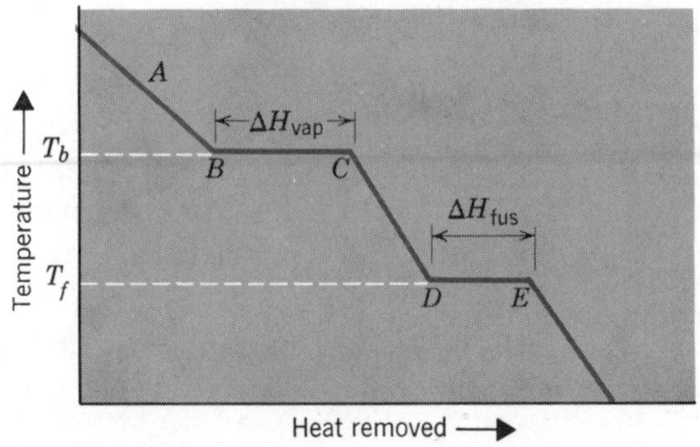


Figure 8.12
*A typical cooling curve for 1 mole of a substance.*

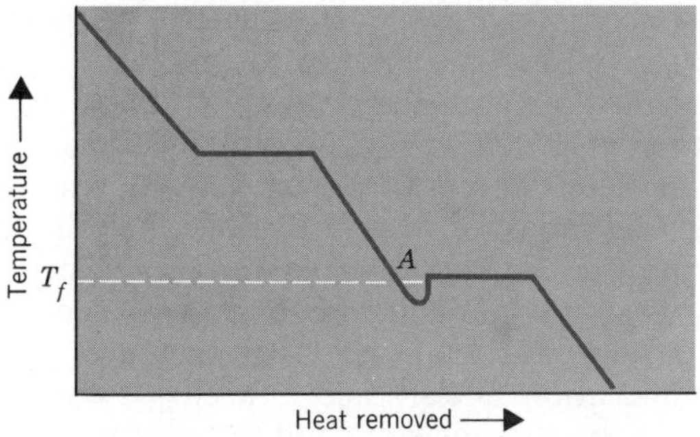


Figure 8.13
*Supercooling. As the liquid is cooled, its temperature drops below the freezing point. After a short time, freezing begins and the temperature rises to the freezing point.*

temperature of the liquid drops, it eventually reaches point $A$, the expected freezing point of the substance. The molecules, however, may not be oriented properly to fit into the crystalline lattice and random motion continues as heat is further withdrawn from the liquid. Consequently, the temperature of the liquid drops below its expected freezing point and the liquid is said to be **supercooled.** Once a small number of molecules have achieved the correct pattern, a tiny crystal is formed that serves as a seed on which additional molecules may rapidly accumulate. Potential energy is suddenly released as this crystal quickly grows, and the energy that is evolved increases the average kinetic energy of the molecules in the liquid and solid. As a result, the temperature of the system rises again until it returns to the freezing point, after which the substance behaves normally. Further removal of heat eventually leads to complete conversion of the liquid to a solid.

Some substances, such as glass, rubber, and many plastics, never do achieve a crystalline state when their liquids solidify upon cooling. These compounds consist of long chainlike molecules that intertwine in the liquid. As they are cooled their molecules move so slowly that they never do find the proper orientation to form a crystalline solid, and an amorphous solid results instead. Thus, these amorphous solids are actually supercooled liquids and in fact continue to flow, although very slowly to be sure, even at room temperature. For example, very old glass shows greater crystallinity, when examined by X-ray diffraction, than does freshly formed glass, showing that molecules are slowly finding their way into a crystalline lattice. A supercooled liquid familiar to many children is Silly Putty. In many ways it behaves as a solid, particularly when forced to flow rapidly (for example, it breaks when pulled apart suddenly). In other ways it flows like a liquid. Supercooled liquids such as glass, Silly Putty, and plastics in general, do not have sharp, well-defined melting points but instead gradually soften when heated.

## 8.7 VAPOR PRESSURE OF SOLIDS

Like liquids, solids too undergo evaporation and therefore exhibit a vapor pressure. The molecules in a crystalline solid are vibrating about their lattice positions and are continually undergoing collisions with their nearest neighbors, giving rise to a distribution of kinetic energies. A small fraction of the molecules at the surface of a solid possess large enough kinetic energies for them to overcome the attractive forces within the solid and break away from the surface, entering the gaseous phase above. The process whereby molecules go directly from the solid into the gaseous state is known as **sublimation.** Anyone who has seen Dry Ice (solid carbon dioxide) disappear knows that no liquid puddles are left behind because the $CO_2$ evaporates directly from the solid to the gas.

Freeze-dried coffee is manufactured by first freezing a batch of brewed coffee and then removing the ice component by vacuum. The vacuum creates an atmosphere of diminished pressure in which ice readily sublimes, thereby removing the water vapor rapidly. Once the water is removed, the dried solid ("brewed") coffee that remains is ready to be placed into a cup, along with some hot water and perhaps a little cream and sugar (or anisette or brandy) for your consumption.

When sublimation takes place in a closed container, more and more molecules enter the gaseous state and the pressure exerted by the vapor increases. The slower-moving gaseous molecules that are colliding with the surface of the solid become trapped and return to the solid state. In time the leaving rate will exactly equal the returning rate, and a dynamic equilibrium will be established. The pressure exerted by a vapor in equilibrium with its solid is known as the **equilibrium vapor pressure of the solid.** Just as in liquids, the vapor pressure of a solid is dependent on the ease with which its molecules enter the gaseous state. For example, the attractive forces are stronger in ionic solids than in molecular solids and, as expected, we find that the vapor pressures of ionic solids are generally very much lower than those of molecular solids.

## 8.8 PHASE DIAGRAMS

The vapor pressure of a solid, like that of a liquid, is a function of its temperature. Increasing the temperature on a solid–vapor equilibrium, according to Le Chatelier's principle, leads to a shift in the position of the equilibrium that will occur with the absorption of heat. The production of vapor from the solid is an endothermic process; therefore as the temperature rises, more of the solid will evaporate and more of the vapor will be produced until equilibrium is once again attained. Hence, the equilibrium vapor pressure of a solid increases with increasing temperature until eventually a temperature is reached at which the solid melts. Further increases in temperature, beyond this point, will then give rise to a liquid–vapor equilibrium curve that terminates at the critical temperature of the substance. If, using water as an example, we plotted the vapor pressure versus temperature for the solid–vapor equilibrium and for the liquid–vapor equilibrium on the same graph, we would produce Figure 8.14. Each point along the "solid" curve

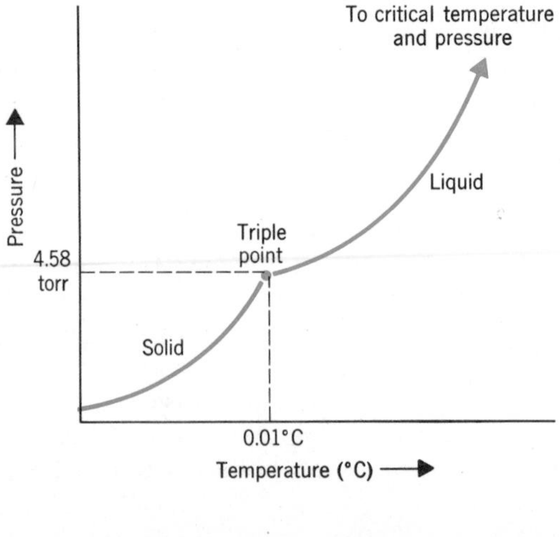


Figure 8.14
*Solid and liquid vapor pressure curves for water.*

represents the specific combinations of temperatures and pressures that must be achieved in order for the solid to be in equilibrium with its vapor. Likewise, every point along the "liquid" curve represents combinations of temperatures and pressures required for the liquid to be in equilibrium with its vapor. The point of intersection of these two curves, the **triple point**, corresponds to a unique temperature and pressure where all *three* states of matter (solid, liquid, and gas) coexist in equilibrium with each other. The triple point occurs at a temperature and pressure that depend on the nature of the substance in question. For example, the triple point of water occurs at a temperature of 0.01°C and a pressure of 4.58 torr, while the triple-point temperature of carbon dioxide is −57°C and the triple-point pressure is 5.2 atm.

There is still another equilibrium that can be represented on the same graph. This line corresponds to the combinations of temperatures and pressures that must be maintained in order to achieve a solid–liquid equilibrium. At a pressure of 1 atm, the melting point of water is 0°C; therefore, the solid–liquid equilibrium line passes through both the triple point and the normal melting point as shown in Figure 8.15. The resulting drawing is called a **phase diagram** because it allows us to pinpoint temperatures and pressures at which the various phases exist, as well as those conditions under which equilibrium can occur. For instance, at a pressure of 1 atm, water exists as a solid at all temperatures below 0°C, and in fact the region bounded by the solid–liquid[7] and solid–vapor equilibrium lines corresponds to all of the temperatures and pressures at which water exists as a solid. Similarly, in the region bounded by the solid–liquid and liquid–vapor equilibrium lines, the substance can exist only as a liquid, while to the right of both the solid–vapor and liquid–vapor lines the substance must be a gas.

In Table 8.4 are some randomly chosen temperatures and pressures and the physical states of water that we can predict from its phase diagram. You might verify these predictions to illustrate how to use a phase diagram.

To gain a further insight into the meaning of a phase diagram, let us

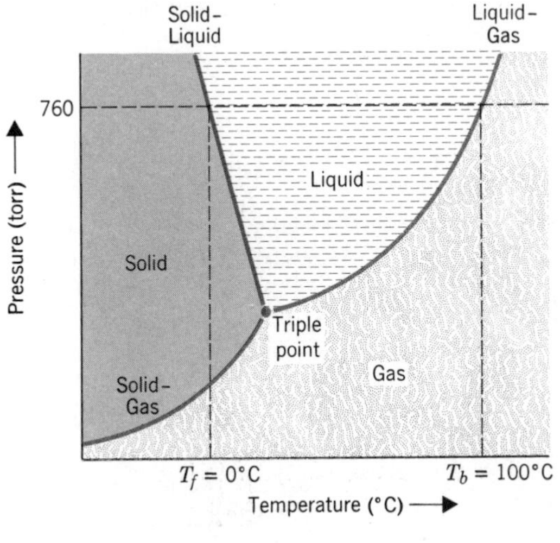


Figure 8.15
*Phase diagram for water (somewhat distorted).*

[7] The slope of this line, as drawn in Figure 8.15, is exaggerated for our discussion. The actual slope is much less to the left (33 atm is required to lower the melting point of ice by only 1°C).

Table 8.4
*Physical state of water at random temperatures and pressures*

| Temperature (°C) | Pressure (atm) | State |
|---|---|---|
| 25 | 1.0 | Liquid |
| 0 | 2.0 | Liquid |
| 0 | 0.5 | Solid |
| 100 | 0.5 | Gas |

follow the changes that take place as we move along a line of constant pressure, say 1 atm, by varying the temperature. In Figure 8.16, point *A* lies in the region of the diagram where a sample of the substance would exist entirely as a solid, as shown in Figure 8.17*a*. When the temperature rises to point *B* in Figure 8.16, the solid begins to melt and an equilibrium between the solid and liquid can occur (Figure 8.17*b*). At a still higher temperature, point *C*, all of the solid will have been converted to a liquid (Figure 8.17*c*); and, when the liquid–vapor line is encountered at point *D* in Figure 8.16, vapor may at last begin to form and an equilibrium can exist (Figure 8.17*d*). Finally at a sufficiently high temperature, such as point *E*, all of the water will exist in the vapor state (Figure 8.17*e*).

We could also proceed with a similar analysis in which the temperature is held constant and the pressure is permitted to change. For example, at point *F* in Figure 8.16, the water would exist entirely as a gas (Figure 8.18*a*). At a higher pressure, point *G* in Figure 8.16, a solid–vapor equilibrium would exist (Figure 8.18*b*), and above that pressure, at point *H*, all of the water would be converted to a solid (Figure 8.18*c*). As the pressure is increased further, we encounter the solid–liquid line at point *B* in Figure 8.16, where we again have an equilibrium as represented by Figure 8.18*d*. At still higher pressures, the water will melt so that at point *I* all of the water is present in the liquid state (Figure 8.18*e*).

In the phase diagram for water we see that the solid–liquid equilibrium

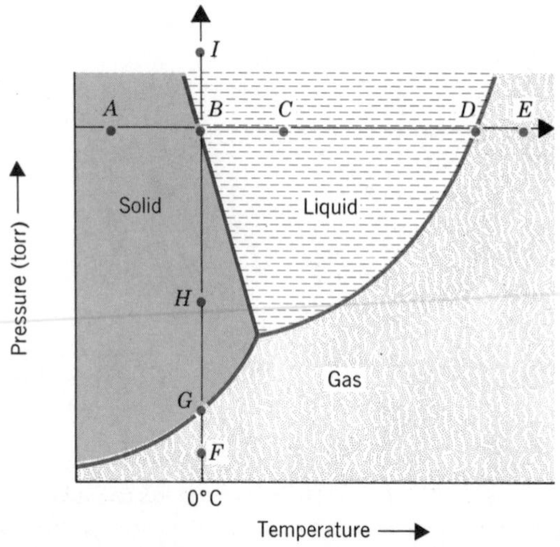


Figure 8.16
*Phase diagram for water (not drawn to scale).*

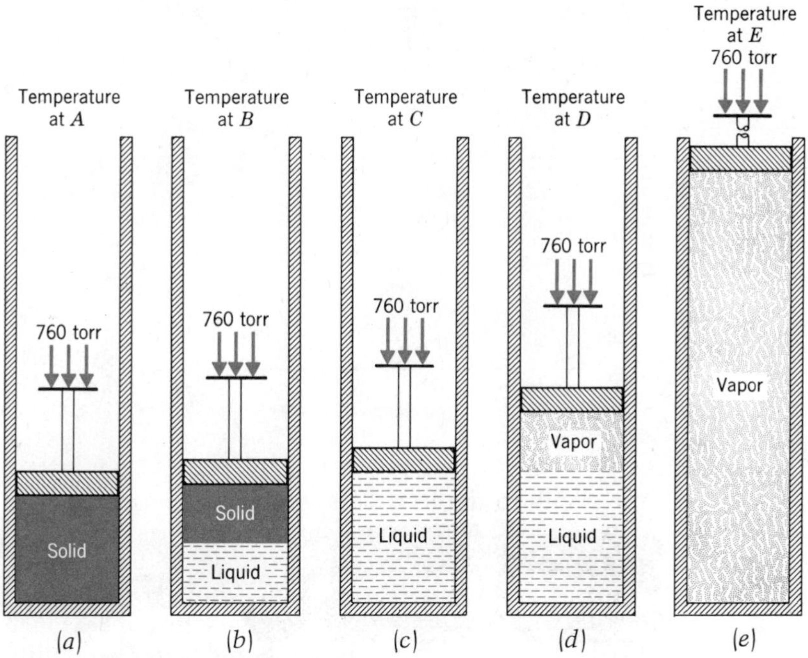


Figure 8.17
*Raising the temperature at a constant pressure of 760 torr. Temperatures correspond to points* A *to* E *in Figure 8.16.*

Figure 8.18
*Raising the pressure at a constant temperature of 0°C. Pressures correspond to points on Figure 8.16.*

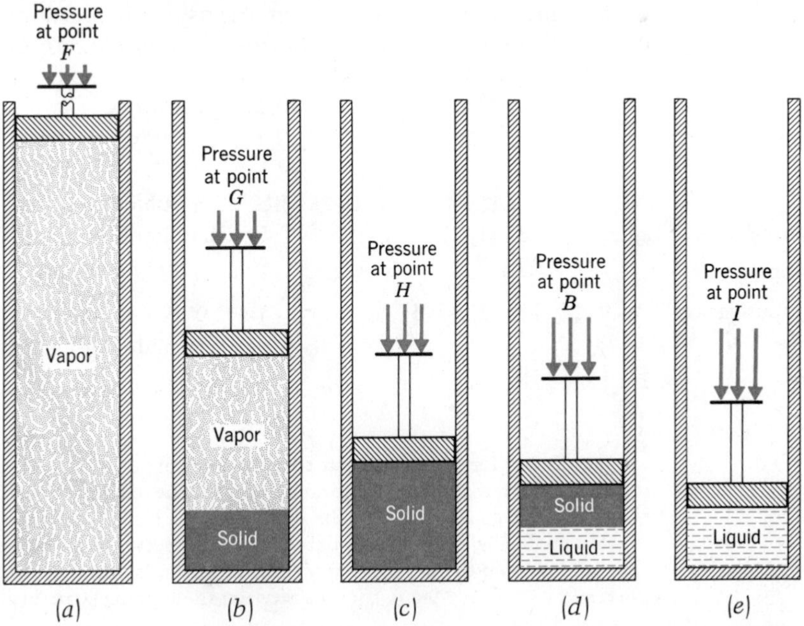

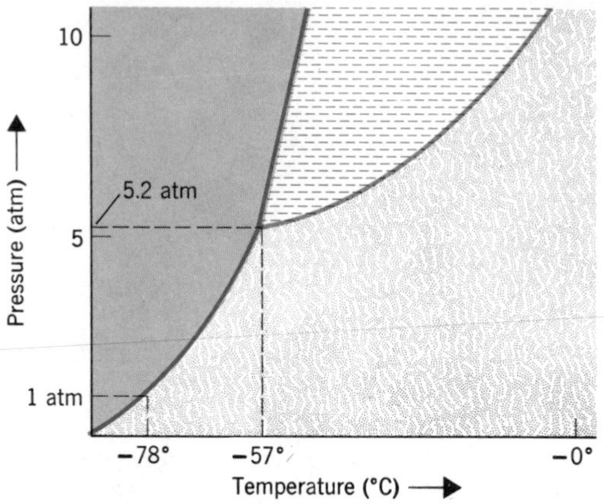


Figure 8.19
*Phase diagram for carbon dioxide.*

line slants to the left. This is a direct consequence of the fact that liquid water at 0°C has a higher density than does the solid. Le Chatelier's principle requires that an increase in pressure on a system at equilibrium will lead to the production of the more dense phase; that is, a rise in pressure favors the packing together of molecules—quite a reasonable expectation. This means that if we have solid and liquid water at equilibrium and we increase the pressure while holding the temperature at 0°C, we should produce the higher-density, liquid phase.[8] On the phase diagram a rise in pressure at constant temperature amounts to moving upward along a vertical line. We can move from the solid–liquid equilibrium line upward into a region of all liquid only if the solid–liquid line leans to the left.

Water is quite an unusual substance. For nearly all other compounds the solid phase is more dense than the liquid and for these substances the solid–liquid line slants to the right, as is shown in the phase diagram for $CO_2$ that appears in Figure 8.19. An interesting feature of this phase diagram is that the entire liquid range lies above a pressure of 1 atm; therefore, it is impossible to form liquid $CO_2$ at atmospheric pressure. Instead, as the gas is cooled, the solid–vapor equilibrium line is encountered at −78°C and the vapor is converted directly to the solid. This also explains why Dry Ice sublimes rather than melts at ordinary pressures.

## INDEX TO QUESTIONS AND PROBLEMS (problem numbers in italics)

General properties of liquids 1, 2, 3, 21, 22

Evaporation 4, 5, 6

Heat of vaporization 7, 8, 9, 10, 11, 12, 16, 17, *41, 42, 44, 46, 49, 50, 51, 52*

Vapor pressure 13, 14, 15, 16, 18, 19

Le Chatelier's principle 20, 40

Boiling 23, 24, 25, 26

Heat of fusion 27, *43, 45, 47, 48*

Heating and cooling curves 28, 33, 35, 36, 38

Phase diagrams 29, 30, 31, 32, 34, 37, 39

[8] In fact, it had long been thought that this melting of water that occurs at high pressure is responsible for our ability to skate on ice. It was believed that the high pressure produced by the skater's weight concentrated on the sharp edge of a blade caused the ice just beneath the blade to melt, producing a thin film of liquid water that serves as a lubricant and allows the skate to slide smoothly on the ice. Current feeling, however, is that this film of water is most probably the result of melting due to friction between the moving skate blade and the ice.

## REVIEW QUESTIONS

**8.1** State how the following physical properties differ for the three states of matter—solid, liquid, and gas: (a) density, (b) rate of diffusion, (c) compressibility, (d) ability to flow.

**8.2** Explain why the rate of diffusion in a liquid is less than in a gas.

**8.3** Sketch the kinetic energy distribution for a liquid at two temperatures. Indicate the minimum K.E. required for molecules to escape the liquid. On the basis of this diagram, explain why liquids evaporate faster at higher temperatures.

**8.4** Why does evaporation lead to a lowering of temperature?

**8.5** Clothes dry more rapidly on a dry day than on a humid day. They also dry more rapidly if there is a breeze blowing than if the air is still. Why?

**8.6** A glass of cola, or other beverage, with ice in it often becomes wet on the outside. Explain why this happens.

**8.7** If you were asked to compare the strengths of the attractive forces operative in liquid A with those in liquid B, what types of data would you collect?

**8.8** Suppose that two substances, X and Y, have heats of vaporization equal to 9.0 and 6.5 kcal/mole, respectively. Which compound would you expect to have the higher boiling point? Which compound would be less likely to exhibit hydrogen bonding?

**8.9** Among the hydrocarbons, $CH_4$ through $C_{10}H_{22}$, both $\Delta H_{vap}$ and boiling points increase with increasing molecular weight, even though the molecules are composed of the same kinds of atoms. How can this be explained?

**8.10** How would you expect the heats of vaporization to vary among the compounds, $PH_3$, $AsH_3$, $SbH_3$?

**8.11** How would you expect the heats of vaporization to vary among the compounds, $H_2S$, $H_2Se$, $H_2Te$?

**8.12** What is a reasonable explanation for the fact that $\Delta H_{vap}$ is larger for $H_2O$ than for HF?

**8.13** What is meant by the term, vapor pressure?

**8.14** At 0°F there is less water in 1 liter of air at 100% humidity than at 75°F when the humidity is 100%. Why is this so?

**8.15** When warm moist air is forced to rise over mountains, clouds form and rain frequently falls. On the basis of the concepts in this chapter, explain why this happens.

**8.16** What is the source of energy in a thunderstorm?

**8.17** Steam at 100°C produces a much more severe burn than an equivalent amount of water at 100°C. Why?

**8.18** Why is the vapor pressure a measure of the strengths of intermolecular attractive forces?

**8.19** Explain why decreasing the volume of a container does not alter the vapor pressure of a liquid.

**8.20** Using Le Chatelier's principle, predict the effect of a change in temperature and pressure on the equilibria:
(a) solid + heat $\rightleftharpoons$ liquid
(b) liquid + heat $\rightleftharpoons$ vapor

**8.21** Define the terms: critical temperature, critical pressure.

**8.22** Define: boiling point, normal boiling point.

**8.23** From Figure 8.7, estimate the boiling point of water at a pressure of 500 torr.

**8.24** At the top of Mount Everest in the Himalayas, which is 29,000 ft above sea level, the atmospheric pressure is approximately 270 torr. At what temperature would water boil at that altitude?

**8.25** Explain why compounds with strong intermolecular attractive forces have higher boiling points than compounds with weak intermolecular attractive forces.

**8.26** What evidence is there for hydrogen bonding in $H_2O$, HF, and $NH_3$?

**8.27** Explain why, for any given substance, $\Delta H_{fus}$ is smaller than $\Delta H_{vap}$.

**8.28** Aluminum has a melting point of 660°C and a boiling point of 1800°C. Its $\Delta H_{fus}$ = 2.55 kcal/mole and $\Delta H_{vap}$ = 53.8 kcal/mole. Construct a heating curve for aluminum.

**8.29** Iodine, $I_2$, sublimes without melting when heated in an open container at atmospheric pressure. What can be said about the triple point of $I_2$?

**8.30** Can you think of any common household products that sublime?

**8.31** At a pressure of 760 torr a new compound was found to melt at 25°C and boil at 95°C. The triple point of the substance was determined to occur at a pressure of 150 torr and at a temperature of 20°C. Sketch the phase diagram for this substance. Label, on your drawing, the solid, liquid, and vapor regions as well as the solid–liquid, liquid–vapor, and solid–vapor equillibrium lines.

**8.32** On the basis of the phase diagram in Question 8.31, describe the changes that you would observe if, at a constant temperature of 22°C, the pressure on a sample of the compound is gradually increased from 10 to 1000 torr. What would be observed if the same process were to occur at a constant temperature of 10°C?

**8.33** Sketch the heating curve that you would expect to find when 1 mole of the compound described in Question 8.31 is heated at a constant rate under a constant pressure of 1.00 atm. On your drawing, indicate the melting point and boiling point of the substance. Also, label the intervals that correspond to $\Delta H_{fus}$ and $\Delta H_{vap}$.

**8.34** What can we conclude about the relative densities of the liquid and solid phases of the compound in Question 8.31?

**8.35** When a supercooled liquid begins to freeze, its temperature rises. Why doesn't the temperature ever rise above the melting point of the substance?

**8.36** Why doesn't glass have a sharp melting point?

**8.37** Is it possible to have only *liquid* water in a container at 32°F (0°C)?

**8.38** Explain, on a molecular level, why the temperature remains constant as heat is added to vaporize a liquid at its boiling point.

**8.39** With the aid of the phase diagram in Figure 8.19, predict the physical state of carbon dioxide under the following conditions of temperature and pressure:

| Temperature (°C) | Pressure (atm) |
|---|---|
| −80 | 1.0 |
| −60 | 1.0 |
| −56 | 10.0 |
| −56 | 2.0 |
| −65 | 5.0 |
| −40 | 10.0 |

**8.40** Use Le Chatelier's principle to predict how variations in pressure will affect the melting point of: (a) water, (b) carbon dioxide.

## REVIEW PROBLEMS

**8.41** Trouton's rule states that the ratio of the heat of vaporization to the boiling point (in degrees kelvin) is approximately a constant. Verify this for the hydrocarbons $CH_4$ through $C_{10}H_{22}$ in Table 8.1. What conclusions can you draw concerning the relationship between $\Delta H_{vap}$ and boiling point?

**8.42** Calculate the heat necessary to convert 55.0 g of ethanol (ethyl alcohol, $C_2H_5OH$) from liquid to vapor. $\Delta H_{vap} = 9.22$ kcal/mole.

**8.43** How many kilojoules of energy are necessary to melt 35.0 g of benzene ($C_6H_6$)? $\Delta H_{fus} = 2.37$ kcal/mole.

**8.44** A 14.5-g sample of liquid mercury required 4.29 kJ to completely convert it to vapor at the same temperature. What is $\Delta H_{vap}$ of Hg in kJ/mole? What is it in kcal/mole?

***8.45** A 150-lb (68.2-kg) skater skids to a halt from a speed of 10 miles/hr. Assuming that all of his energy appears as frictional heat transferred to ice at 0°C, how many grams of ice will be melted?

***8.46** A student (with very slow reflexes) holds his hand in a stream of steam at 100°C until exactly 1.00 g of water has condensed. If this water then cools to 40°C, how many calories have been absorbed by the student's hand?

*8.47 A cube of solid benzene ($C_6H_6$) at its melting point weighing 10.0 g is introduced into 50.0 g of $H_2O$ at 30°C. Given that $\Delta H_{fus}$ for $C_6H_6$ is 2.37 kcal/mole, to what temperature will the water have cooled by the time all of the benzene has melted?

*8.48 A 50.0-g ice cube at 0°C is added to 10.0 g of steam at 100°C. What will be the final temperature of the 60.0 g of water?

8.49 The vapor pressure of liquid nitrogen is 21.8 torr at −216°C and 47.0 torr at −213°C. Calculate the heat of vaporization of $N_2$ in kJ/mole. Calculate the normal boiling point of $N_2$.

8.50 Diethyl ether, often used as an anesthetic, has a vapor pressure of 185 torr at 0°C. At 10°C its vapor pressure is 0.384 atm. What is $\Delta H_{vap}$ for diethyl ether in kcal/mole?

8.51 The heat of vaporization of toluene (used to make trinitrotolune, TNT) is 38.0 kJ/mole. Its vapor pressure is 20.0 torr at 18.4°C. What will its vapor pressure be at 25°C? What must the temperature of toluene be to have a vapor pressure of 40 torr?

*8.52 At 25°C, liquid A has a vapor pressure of 100 torr while liquid B has a vapor pressure of 200 torr. The heat of vaporization of liquid A is 8.50 kcal/mole and that of liquid B is 4.00 kcal/mole. At what temperature will both A and B have the same vapor pressure?

# 9 PROPERTIES OF SOLUTIONS

In the previous three chapters we discussed the properties associated with the three states of matter. For the most part, however, these discussions applied to pure substances, and it is only rarely that we work with pure materials. Usually our chemicals occur in mixtures, and very often these are solutions. The presence of a solute in a solution has very marked effects on the properties of the substance in which it is dissolved, and many times these effects can provide us with useful information about the way substances interact with one another.

In this chapter we will take a close look at the solution process to explore the changes that occur when one substance dissolves in another. We will also focus our attention on the way in which the solute affects the physical properties of the solution. Many of these phenomena have very useful applications, such as the determination of molecular weights. They can also be applied to many practical problems, such as refining crude oil and desalination of sea water.

## 9.1 TYPES OF SOLUTIONS

The most common type of solution that we come across in the laboratory consists of a solute dissolved in a liquid; for this reason most of our attention will be directed toward solutions of this type. Liquid solutions can be prepared by dissolving a solid in a liquid (for example, NaCl in water), a liquid in a liquid (for example, ethylene glycol in water—antifreeze solution), or a gas in a liquid (for example, any carbonated beverage contains dissolved carbon dioxide).

In addition to liquid solutions it is possible to have solutions of gases, such as the atmosphere that surrounds the earth, and solid solutions, formed when a substance is dissolved in a solid. The properties of gaseous solutions were discussed in Section 6.4 under the heading "Dalton's Law of Partial Pressures," and nothing more need be said about them here. Solid solutions, of which many **alloys** (mixtures of metals) are examples, are of two types. **Substitutional solid solutions** exist in which atoms, molecules, or ions of one substance take the place of particles of another substance in a crystalline lattice, as shown in Figure 9.1*a*. Zinc sulfide and cadmium sulfide form such mixtures in which cadmium ions randomly replace zinc ions in the ZnS lattice. Another example is provided by brass, which is a substitutional solid solution of copper and zinc.

**Interstitial solid solutions** constitute the other type and are formed by

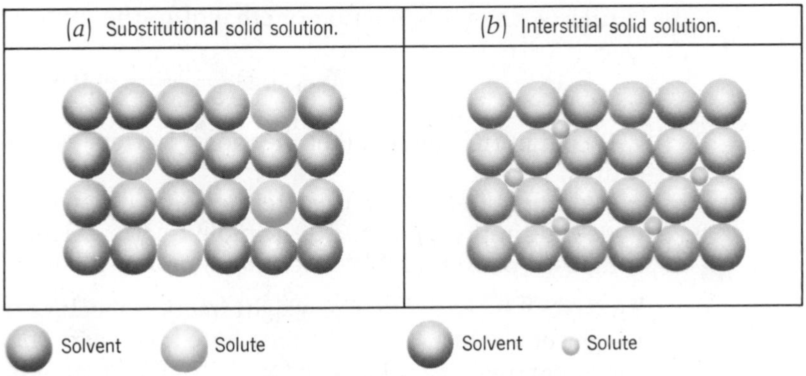


Figure 9.1
*Solid solutions. (a) Substitutional solid solution. (b) Interstitial solid solution.*

placing atoms of one kind into voids, or interstices, that exist between atoms in the host lattice, as illustrated by Figure 9.1*b*. These voids occur, for example, as the tetrahedral or octahedral holes in the closest-packed structures described in Chapter 7. Tungsten carbide, WC, an extremely hard substance that has found many uses in cutting tools designed for machining steels, is an example of an interstitial solid solution in which the tungsten atoms are arranged in a cubic closest-packed pattern with carbon atoms in octahedral holes.

## 9.2 CONCENTRATION UNITS

The physical properties of solutions are determined by the relative proportions of the various components of which they are composed. We have already seen that there are a variety of ways of expressing the concentration of one substance in another. For example, in Chapter 5 we discussed the concentration units, molarity and normality, and we saw that they are useful for the purpose of dealing with the stoichiometry between reactants in solution. Molarity and normality, defined again below for the sake of completeness, were created to satisfy a need. In a similar fashion, it has been found that certain other concentration units are most convenient for interpreting the physical properties of solutions. A point to remember about all concentration units is that they are *ratios*. The way to remember them is to learn the units associated with numerator and denominator.

**MOLE FRACTION.** The mole fraction unit of concentration appeared in our discussion of Dalton's law of partial pressures in Section 6.8. It is defined as the number of moles of a particular component of the solution divided by the total number of moles of all of the substances present in the mixture,

$$X_A = \frac{n_A}{n_A + n_B + n_C + \cdots}$$

For example, a solution composed of 2 moles of water and 3 moles of ethanol ($C_2H_5OH$) has a mole fraction of water given by

$$X_{H_2O} = \frac{2 \text{ moles } H_2O}{2 \text{ moles } H_2O + 3 \text{ moles } C_2H_5OH} = \frac{2 \text{ moles}}{5 \text{ moles}}$$

$$X_{H_2O} = 0.40$$

Similarly, the mole fraction of ethanol in the mixture is

$$X_{C_2H_5OH} = \frac{3 \text{ moles}}{5 \text{ moles}} = 0.60$$

We see that the sum of all of the mole fractions is equal to one, as of course it must be.

Another frequently used term is **mole percent** (abbreviated mol %), which is simply equal to 100 × mole fraction. Thus, the mixture above is composed of 40 mol % water and 60 mol % ethanol.

**WEIGHT FRACTION.** The weight fraction specifies the fraction of the total weight of a solution that is contributed by a particular component. A mixture composed of 25.0 g of water and 75.0 g of ethanol has a weight fraction of water, $w_{H_2O}$, given by

$$w_{H_2O} = \frac{25.0 \text{ g } H_2O}{25.0 \text{ g } H_2O + 75.0 \text{ g } C_2H_5OH} = \frac{25.0 \text{ g}}{100.0 \text{ g}}$$

$$w_{H_2O} = 0.250$$

In a similar fashion, we find that the weight fraction of ethanol in the mixture is 0.750.

**Weight percent,** which is equal to the weight fraction multiplied by 100, is more frequently used than weight fraction. The solution above, then, is described as composed of 25.0% water and 75.0% ethanol, by weight.

**MOLARITY.** Molarity, you should recall, is the ratio of the number of moles of solute to the *total volume* of the solution. It is expressed in moles per liter. If 0.500 mole of HCl is dissolved in 250 ml of solution, the molarity is

$$\frac{0.500 \text{ mole}}{0.250 \text{ liter}} = 2.00 \text{ moles/liter} = 2.00\ M$$

**NORMALITY.** Normality is expressed as the ratio of the number of equivalents of solute to the volume of solution, in liters.

**MOLALITY.** Molality is defined as the number of moles of solute per 1000 g (1.00 kg) of the solvent; that is, it is a ratio of *moles of solute to mass of solvent.* A 1.00 molal (written 1.00 *m*) solution would therefore contain 1.00 mole of solute for every 1.00 kg of the solvent. It is very important not to confuse molality with molarity—they are quite different. To see this difference let us consider how to prepare typical 1.00 *M* and 1.00 *m* solutions using, for example, sucrose ($C_{12}H_{22}O_{11}$) as the solute and water as the solvent.

To prepare the 1.00 *M* solution we place exactly 1.00 mole of sucrose (342 g) into a flask (called a **volumetric flask**) that is calibrated to contain precisely 1.00 liter when filled to a line around its neck (see Figure 9.2). Water is added while the mixture is stirred to dissolve the solute, until the flask is filled to the mark. At this point we have exactly 1.00 mole of solute in a total volume of 1.00 liter of solution and the concentration is 1.00 mole/liter, or 1.00 molar (1.00 *M*).

To prepare the 1.00 *m* solution, we place 1.00 mole of sucrose into a flask or beaker and *add to it* 1000 g of water. Since the density of water is nearly 1 g/ml (at room temperature its density is 0.9982 g/ml), we are adding very nearly 1 liter of water to the 342 g (1 mole) of solute; however, the total final volume of this 1.00 *m* solution is somewhat larger than 1 liter (it

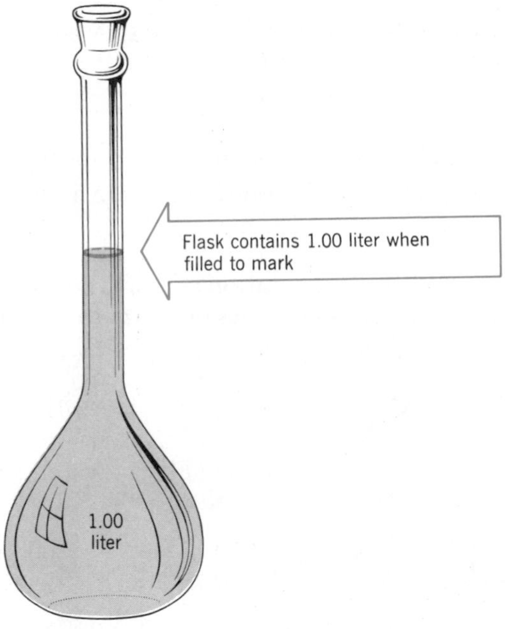


Figure 9.2
*A volumetric flask.*

is actually 1110 ml), because part of the volume of the final solution is taken up by molecules of the sucrose. The molarity of this 1 *m* solution is 1.00 mole/1.110 liter = 0.901 *M*. Because the mole of solute is contained in a larger volume in this 1 *m* solution, a 1-ml portion will contain a smaller amount of solute than a 1-ml portion of the 1 *M* solution of sucrose.

The difference between molarity and molality becomes even more apparent if we choose a solvent whose density is far from 1 g/ml. For instance, as illustrated in Figure 9.3, a 1 *m* solution of a solute in carbon tetrachloride[1] contains 1 mole of solute in a total volume of 1 liter; however, because $CCl_4$ has a density of 1.59 g/ml (considerably greater than water), a 1 *m* solution would contain the 1 mole of solute in a volume of only about 630 ml. In other words, it takes much less than 1 liter of $CCl_4$ to weigh 1000 g.

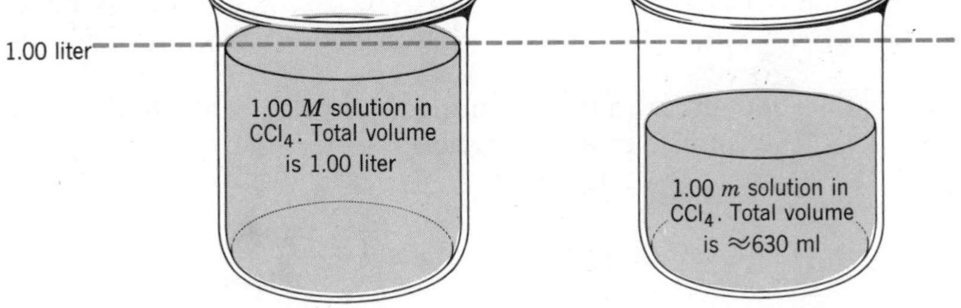


Figure 9.3
*Molar and molal solutions in* $CCl_4$.

[1] Carbon tetrachloride is a very toxic solvent that should always be handled with care. It is absorbed through the skin and is a cumulative poison.

The three concentration units, mole fraction, weight fraction, and molality, are closely related and can easily be converted from one to another as shown in Example 9.1.

EXAMPLE 9.1 An aqueous solution is composed of 20.0% by weight magnesium sulfate. What is (a) the molality of the solution; (b) the mole fraction of $MgSO_4$ and $H_2O$?

SOLUTION (a) We know that 100 g of the solution contains 20.0 g of $MgSO_4$ and 80.0 g of water. Since we wish a ratio of moles of solute to mass of solvent, let us first convert 20.0 g of $MgSO_4$ to moles.

$$20.0\ \text{g MgSO}_4 \times \left(\frac{1\ \text{mole MgSO}_4}{120.4\ \text{g MgSO}_4}\right) \sim 0.166\ \text{mole MgSO}_4$$

Thus, there is 0.166 mole of $MgSO_4$ in 80 g of water. All we must do now is calculate how many moles of the solute are in 1.00 kg of water.

$$\left(\frac{0.166\ \text{mole MgSO}_4}{80.0\ \text{g H}_2\text{O}}\right) \times \left(\frac{1000\ \text{g H}_2\text{O}}{1.00\ \text{kg H}_2\text{O}}\right) \sim \frac{2.08\ \text{moles MgSO}_4}{1.00\ \text{kg H}_2\text{O}}$$

Hence the concentration of this solution is 2.08 *m*.

(b) to calculate mole fraction we must know the number of moles of each component in the solution. We have already found that 100 g of the solution contains 0.166 mole of $MgSO_4$. The number of moles of water in the solution is

$$80.0\ \text{g H}_2\text{O} \times \left(\frac{1\ \text{mole H}_2\text{O}}{18.0\ \text{g H}_2\text{O}}\right) \sim 4.44\ \text{moles H}_2\text{O}$$

The mole fraction of $MgSO_4$ is found by dividing the number of moles of $MgSO_4$ by the total number of moles of both the solute and solvent.

$$X_{\text{MgSO}_4} = \frac{0.166\ \text{mole}}{4.44\ \text{moles} + 0.166\ \text{mole}} = \frac{0.166\ \text{mole}}{4.61\ \text{moles}}$$

$$X_{\text{MgSO}_4} = 0.0360$$

It follows that the mole fraction of water must be

$$X_{\text{H}_2\text{O}} = 1.000 - 0.0360 = 0.9640$$

EXAMPLE 9.2 In a solution of benzene ($C_6H_6$) and chloroform ($CHCl_3$), the mole fraction of $C_6H_6$ is 0.450. What is the weight percent of $C_6H_6$ in this mixture?

SOLUTION Let us begin by imagining that we had enough of this mixture so that we had exactly 1 mole of particles. In this 1 mole of mixture we must have 0.450 mole of $C_6H_6$.

$$1.00\ \text{mole mixture} \times \left(\frac{0.450\ \text{mole C}_6\text{H}_6}{1\ \text{mole mixture}}\right) \sim 0.450\ \text{mole C}_6\text{H}_6$$

The amount of $CHCl_3$ must be the difference between the total number of moles and the number of moles that are $C_6H_6$.

$$\begin{aligned}\text{moles CHCl}_3 &= \text{total moles} - \text{moles C}_6\text{H}_6\\ &= 1.00\ \text{mole} - 0.450\ \text{mole}\\ &= 0.550\ \text{mole}\end{aligned}$$

To compute weight percent $C_6H_6$, we need the weight of $C_6H_6$ and the total weight of the mixture.

$$0.450 \text{ mole } C_6H_6 \times \left(\frac{78.1 \text{ g } C_6H_6}{1 \text{ mole } C_6H_6}\right) \sim 35.1 \text{ g } C_6H_6$$

$$0.550 \text{ mole } CHCl_3 \times \left(\frac{119 \text{ g } CHCl_3}{1 \text{ mole } CHCl_3}\right) \sim 65.5 \text{ g } CHCl_3$$

The total weight of the mixture is

$$35.1 \text{ g} + 65.5 \text{ g} = 100.6 \text{ g}$$

The percent of $C_6H_6$ can be found as

$$\% C_6H_6 = \frac{\text{weight } C_6H_6}{\text{weight of mixture}} \times 100$$

$$\% C_6H_6 = \frac{35.1 \text{ g}}{100.6 \text{ g}} \times 100 = 34.9\%$$

To perform conversions among mole fraction, weight percent, and molality, the only data required are molecular weights. In order to convert any of these concentration units to molarity, the density of the solution is necessary.

EXAMPLE 9.3 The painful sting of ant bites is caused by formic acid injected under the skin by the ant. Calculate the weight percent of formic acid ($HCO_2H$) in a solution that is 1.099 *M* $HCO_2H$. The density of the solution is 1.0115 g/ml.

SOLUTION To compute weight percent we need the weight of $HCO_2H$ and the weight of the solution.

If we take 1.000 liter of solution, its weight is

$$1.000 \text{ liter} \times \left(\frac{1000 \text{ ml}}{1 \text{ liter}}\right) \times \left(\frac{1.0115 \text{ g}}{1 \text{ ml}}\right) \sim 1011.5 \text{ g solution}$$

From the molarity we know that this solution contains 1.099 moles of $HCO_2H$. The weight of solute is

$$1.099 \text{ moles } HCO_2H \times \left(\frac{46.03 \text{ g } HCO_2H}{1 \text{ mole } HCO_2H}\right) \sim 50.59 \text{ g } HCO_2H$$

The percent solute in the solution, then, is

$$\% HCO_2H = \frac{50.59 \text{ g}}{1011.5 \text{ g}} \times 100$$

$$\% HCO_2H = 5.00\%$$

## 9.3 THE SOLUTION PROCESS

Experience has taught us that substances differ widely in their solubilities in various solvents. For instance, we all know that oil and water "don't mix," and that to remove an oil stain from clothing a solvent such as naptha must be used. It is also generally known that sodium chloride (table salt) will dissolve in water but not in gasoline. What accounts for these differences in behavior? The answer lies in a close examination of the solution process.

When one substance dissolves in another, particles of the solute (either molecules or ions, depending on the nature of the solute) must be distributed throughout the solvent and, in a sense, the solute particles in the solution occupy positions that are normally taken by solvent molecules. In a liquid (we restrict this discussion to liquid solutions), molecules are packed together very closely and interact strongly with their neighbors. The ease with which a solute particle may replace a solvent molecule depends on the relative forces of attraction of solvent molecules for each other, solute particles for each other, and the strength of the solute–solvent interactions. For example, in a solution formed between benzene ($C_6H_6$) and carbon tetrachloride ($CCl_4$), both species are nonpolar and, therefore, experience only relatively weak London forces. As it happens, the strengths of the attractive forces between pairs of benzene molecules and between pairs of carbon tetrachloride molecules are of nearly the same magnitude as between molecules of benzene and carbon tetrachloride. For this reason, molecules of benzene can replace $CCl_4$ molecules in solution with ease; as a consequence, these two substances are completely **miscible** (soluble in all proportions).

What happens when we attempt to dissolve water in $CCl_4$? Water is a very polar substance that interacts with other water molecules through the formation of hydrogen bonds. By comparison, the strength of the attractive forces between water and the nonpolar $CCl_4$ molecules is much weaker. If we attempt to disperse water molecules throughout $CCl_4$, we find that when the water molecules encounter one another, they tend to stick together simply because they attract each other much more strongly than they do molecules of the solvent. This "clumping together" continues until the two substances have formed two distinct phases, one consisting of water with a very small amount of $CCl_4$ in it and the other, $CCl_4$ containing a small quantity of $H_2O$.

When two polar substances, such as ethanol and water, are mixed, we again have a situation in which the solute-solute forces of interaction are of comparable strength to the solvent-solvent attractive forces, and where the solute and solvent molecules interact strongly with each other. Once again a condition exists where the solute particles can readily replace those of the solvent and, hence, water and ethanol are miscible.

Between the two extremes of complete miscibility (benzene-carbon tetrachloride; water-ethanol) and virtually total immiscibility ($CCl_4$-$H_2O$), we have many substances that are only partially soluble in one another. For instance, in Table 9.1 the solubilities of a series of different alcohols are listed in moles of solute (alcohol) per 100 g of water. As we proceed to higher molecular weights, the polar OH group represents an ever smaller portion of the molecules; as a result, alcohol molecules become less like water as they become larger. Paralleling the increasing size of the nonpolar hydrocarbon portions of these alcohols, we observe a corresponding decrease in their solubilities in water.

When a solid dissolves in a liquid, somewhat different factors must be considered. In a solid, the molecules or ions are arranged in a very regular pattern and the attractive forces are at a maximum. In order for the solute particles to enter into a solution, the solute–solvent forces of attraction must be sufficient to overcome the attractive forces that hold the solid together. In molecular crystals these attractive forces are relatively weak, being of dipole–dipole or London type, and are rather easily overcome. Substances whose crystals are held together by London forces will, therefore,

Table 9.1
*Solubilities of some alcohols in water*

| Substance | Formula | Solubility (moles of solute/ 100 g $H_2O$) |
|---|---|---|
| Methanol | $CH_3OH$ | $\infty$ |
| Ethanol | $C_2H_5OH$ | $\infty$ |
| Propanol | $C_3H_7OH$ | $\infty$ |
| Butanol | $C_4H_9OH$ | 0.12 |
| Pentanol | $C_5H_{11}OH$ | 0.031 |
| Hexanol | $C_6H_{13}OH$ | 0.0059 |
| Heptanol | $C_7H_{15}OH$ | 0.0015 |

dissolve to appreciable extents in nonpolar solvents. They are not, however, soluble to any great degree in polar solvents for the same reasons that nonpolar liquids are not soluble in polar solvents—that is, the polar solvent molecules attract each other too strongly to be replaced by the molecules to which they are only weakly attracted. For example, solid iodine, which is composed of nonpolar $I_2$ molecules, is appreciably soluble in $CCl_4$ (giving rise to a beautiful violet solution) but only very slightly soluble in water (where it produces a pale yellow-brown solution).

By similar reasoning we expect that very polar solutes and ionic solids are not soluble in nonpolar solvents. The weak solute–solvent interactions, compared to the attractions between solute particles, are not sufficient to tear apart the lattice. Ionic solids, in particular, are held together by the very strong electrostatic forces between ions; therefore a very polar solvent, such as water, is required to rip apart an ionic lattice. Hence NaCl is soluble in water but not in gasoline, which is a mixture of nonpolar hydrocarbons (compounds containing only C and H).

When an ionic substance dissolves in water, the ions that are adjacent to one another in the solid become separated and are surrounded by water molecules. In Chapter 5 we represented this dissociation of the solute by an equation such as

$$NaCl\ (s) \longrightarrow Na^+\ (aq) + Cl^-\ (aq)$$

In Figure 9.4 we take a closer look at what takes place during this process. In the immediate vicinity of a positive ion, the surrounding water molecules are oriented so that the negative ends of their dipoles point in the direction of the positive charge, while surrounding a negative ion the water molecules have their positive ends directed at the ion. An ion enclosed within this "cage" of water molecules is said to be **hydrated** and, in general, when a solute particle becomes surrounded by molecules of a solvent we say that it is **solvated;** hydration is a special case of the more general phenomenon of solvation.

The layer of oriented water molecules that surrounds an ion (actually this orientation may extend through several layers) helps to neutralize the ion's charge and serves to keep ions of opposite charge from attracting each other strongly over large distances within the solution. Nonpolar solvents do not dissolve ionic compounds because they can neither tear an ionic lat-

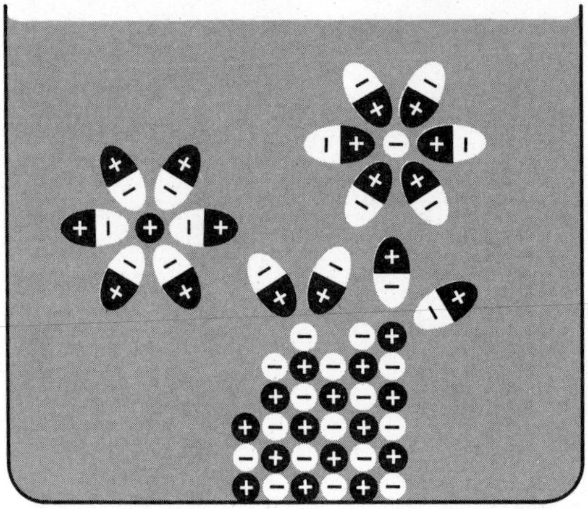

Figure 9.4
*Hydration of ions in solution.*

tice apart nor offer any shielding for the ions. In a nonpolar solvent ions quickly congregate and separate from the solution as the solid.

In summary, substances that exhibit *similar* intermolecular attractive forces tend to be soluble in one another. This observation is often stated very simply as "like dissolves like." Nonpolar substances are soluble in nonpolar solvents, while polar or ionic compounds dissolve in polar solvents.

## 9.4 HEATS OF SOLUTION

The solution process nearly always occurs with either an absorption or a release of energy. For example, if potassium iodide is dissolved in water the mixture becomes cool, indicating that for potassium iodide the solution process is endothermic. On the other hand, when lithium chloride is added to water the mixture becomes warm, signifying that, in this case, the solution process evolves heat and is therefore exothermic. *The quantity of energy that is absorbed or released when a substance enters solution is called the* **heat of solution** and is given the symbol, $\Delta H_{soln}$. As in our definitions of the heat of vaporization and heat of fusion, $\Delta H_{soln}$ represents a difference between the energy possessed by the solution after it has been formed and the energy that the components of the solution possessed before they were mixed; that is,

$$\Delta H_{soln} = H_{soln} - H_{components}$$

Neither $H_{soln}$ nor $H_{components}$ can actually be measured; however, their difference, $\Delta H_{soln}$, can be. When energy is evolved during the solution process, the resulting solution possesses less energy than did the components from which it was prepared and the difference represented by $\Delta H_{soln}$ is a negative number. Conversely, an endothermic solution process would have a positive $\Delta H_{soln}$. Heats of solution for some typical ionic solids in water are shown in Table 9.2.

The magnitude of the heat of solution provides us with information about the relative forces of attraction between the various particles that make up a solution. To analyze the various factors that contribute to this absorption or evolution of energy, let us imagine that we could create the solution in a stepwise fashion.

Table 9.2
*Heats of solution*

| Substance | Heat of Solution[a] (kcal/mole of solute) |
|---|---|
| KCl | 4.12 |
| KBr | 4.75 |
| KI | 4.86 |
| LiCl | −8.85 |
| LiI | −14.1 |
| $LiNO_3$ | −0.3 |
| $AlCl_3$ | −76.8 |
| $Al_2(SO_4)_3 \cdot 6H_2O$ | −56 |
| $NH_4Cl$ | −3.9 |
| $NH_4NO_3$ | 6.2 |

[a] At "infinite" dilution. The heat of solution depends, to an extent, on the concentration of the solution produced. A negative sign signifies an exothermic process.

**SOLUTIONS OF LIQUIDS IN LIQUIDS.** When one liquid dissolves in another, we can imagine that the molecules of the solvent are caused to move apart so as to allow room for the solute molecules. Similarly, for the solute to enter solution, its molecules must also become separated so that they can take their places in the mixture. Since there are attractive forces between molecules in both the solvent and solute, the process of separating their molecules requires an input of energy; that is, work must be done on both the solute and solvent to separate their molecules from one another. Finally, as the solute and solvent, in their "expanded" states, are brought together, energy is released because of the attractions that exist between the solute and solvent molecules.[2] This sequence of steps we have just described is illustrated in Figure 9.5 on page 260.

In some substances, such as benzene and carbon tetrachloride, the intermolecular attractive forces are of very nearly the same magnitude; therefore, these compounds form solutions with virtually no evolution or absorption of heat. Solutions in which the solute–solute, solute–solvent, and solvent–solvent interactions are all the same are called **ideal solutions.** The energy changes that occur along the series of steps that we have devised to arrive at the solution are shown graphically in Figure 9.6. We see that for an ideal solution the energy released in the final step is the same as that absorbed in the first two; thus the net change is zero.

When the solute and solvent molecules are strongly attracted to each other, more energy can be released in the final step, when the solution is formed, than is initially required to separate the molecules of the solute and the molecules of the solvent (Figure 9.7). Under these circumstances the overall solution process results in the evolution of heat and is therefore ex-

[2] Recall from Chapter 1 that when particles that attract one another are pulled apart, potential energy (work) must be supplied. When these particles are brought together again, the same amount of energy is released.

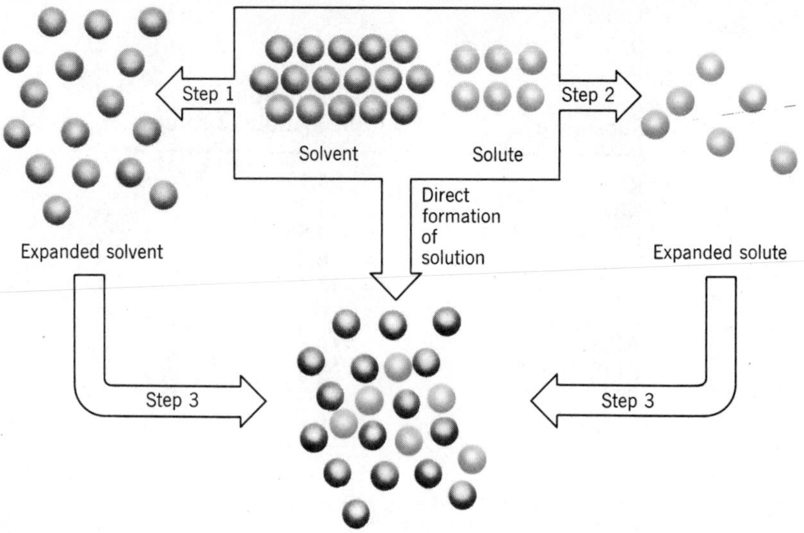


Figure 9.5
*Formation of a liquid–liquid solution.*

othermic. This occurs when acetone, an important solvent (nail polish remover), and water are mixed.

When the solute–solvent attractive forces are weaker than those between pure solute or pure solvent, the formation of a solution requires a net input of energy. This is because more energy is absorbed in separating the molecules in the first two steps than is recovered when the solute and solvent are brought together in the third (Figure 9.8). In this case the solution becomes cool as it is formed, signifying that an endothermic change has occurred. This happens when ethyl alcohol and water are mixed.

Figure 9.6
*Energy changes to produce an ideal solution.*

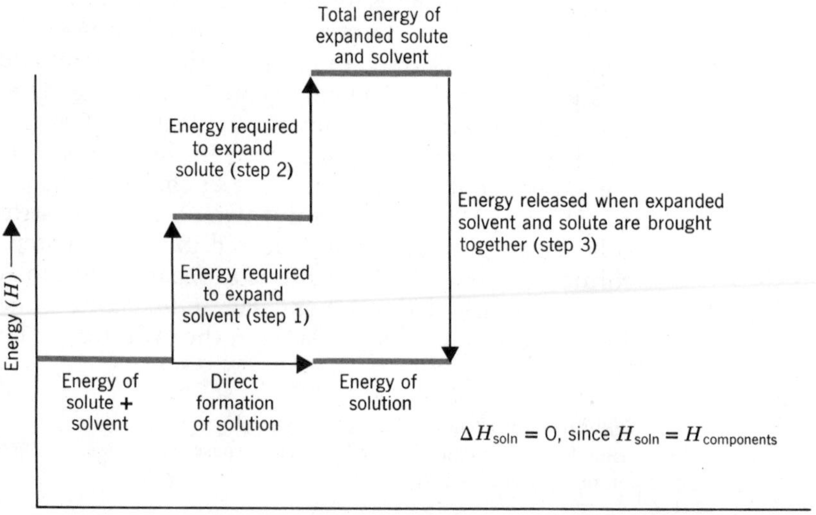

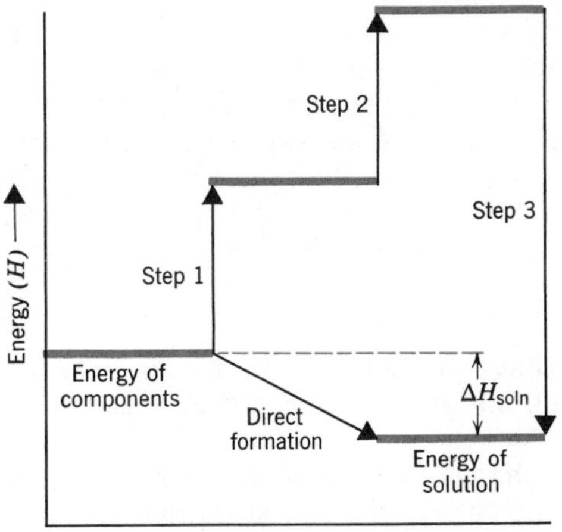


Figure 9.7
*Exothermic solution process.*

**SOLUTIONS OF SOLIDS IN LIQUIDS.** We can approach the energetics of the solution of a solid in a liquid in much the same way as we have for a liquid in a liquid. We recognize that the act of dissolving a compound such as KI in water involves removing the $K^+$ and $I^-$ ions from the solid and placing them in an environment where they are surrounded by water molecules. Suppose, now, that we could separate these two steps. The first step involves separating the ions so that they are infinitely far apart, and the second amounts to taking these now isolated ions and placing them into water where they become surrounded by molecules of the solvent. To accomplish the first step, it is necessary to break apart the lattice by pulling the ions away from one another, so that the process

$$\mathrm{KI}\ (s) \longrightarrow \mathrm{K}^+\ (g) + \mathrm{I}^-\ (g)$$

requires an input of energy; that is, it is endothermic. Recall that the amount of energy required to tear apart the solid to obtain isolated particles is called the lattice energy. Ionic substances, because of the very strong electrostatic attractions between oppositely charged ions, have quite large lattice energies. Molecular solids, however, have small lattice energies because of their relatively weak intermolecular attractive forces.

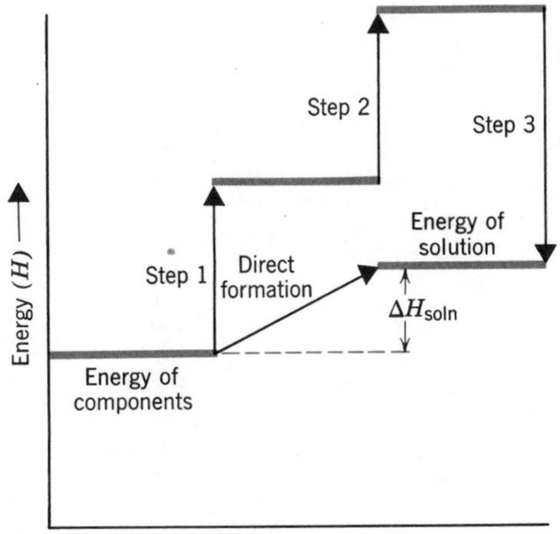


Figure 9.8
*Endothermic solution process.*

In the second step of our solution process we imagine that we place the $K^+$ and $I^-$ ions into water where they become hydrated. As described in the last section, an ion in water is surrounded by a "cage" of oriented water molecules; therefore a hydrated ion experiences a net attraction for the solvent dipoles. Because of this attraction, a quantity of energy, known as the **hydration energy,** is released when an ion is placed into water.[3] We indicate the change that occurs in this hydration step as

$$K^+ (g) + I^- (g) + xH_2O \longrightarrow K^+ (aq) + I^- (aq)$$

which tells us that some number, *x*, of water molecules becomes "bound" to the potassium and iodide ions as they become hydrated in aqueous solution.

The overall change that takes place when KI dissolves in water can be represented as the sum of the two steps that we have just considered.

$$\begin{array}{r} KI (s) \longrightarrow K^+ (g) + I^- (g) \\ K^+ (g) + I^- (g) + xH_2O \longrightarrow K^+ (aq) + I^- (aq) \\ \hline KI (s) + K^+ (g) + I^- (g) + xH_2O \longrightarrow K^+ (g) + I^- (g) + K^+ (aq) + I^- (aq) \end{array}$$

or, after cancelling species that appear on both sides,

$$KI (s) + xH_2O \longrightarrow K^+ (aq) + I^- (aq)$$

The heat of solution, $\Delta H_{soln}$, corresponds to the net energy change that occurs and is equal to the difference between the amount of energy supplied during the first step and the amount of energy evolved during the second (Figure 9.9). If the lattice energy is greater than the hydration energy, a net input of energy is required when the substance dissolves and the process is endothermic. This occurs with KI. Conversely, when the hydration energy exceeds the lattice energy, a situation that occurs with LiCl, more energy is released when the ions become hydrated than is required to break up the ionic lattice, and an exothermic change is observed when the solid dissolves.

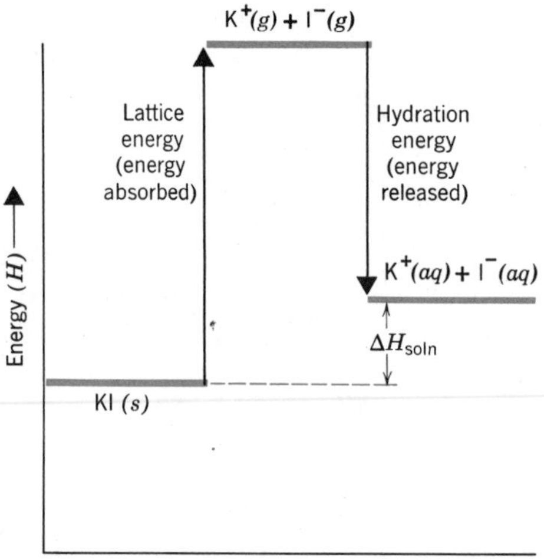


Figure 9.9
*Energy changes that occur when KI dissolves in $H_2O$.*

[3] In general, for a solvent other than water, this energy is called the solvation energy.

Table 9.3
*Lattice energy, hydration energy, and heats of solution for some alkali halides*

| Compound | Lattice Energy (kcal/mole) | Hydration Energy (kcal/mole) | Calculated $\Delta H_{soln}$ (kcal/mole) | Measured $\Delta H_{soln}$ (kcal/mole) |
|---|---|---|---|---|
| LiCl | +199 | −211 | −12 | −8.85 |
| LiBr | +188 | −204 | −16 | −11.7 |
| NaCl | +183 | −184 | −1 | +0.93 |
| NaBr | +174 | −177 | −3 | −0.144 |
| KCl | +165 | −164 | +1 | +4.12 |
| KBr | +159 | −157 | +2 | +4.75 |
| KI | +151 | −148 | +3 | +4.86 |

The relationship between the lattice energy and hydration energy can be seen in Table 9.3. Note that the agreement between calculated and experimentally determined heats of solution is far from perfect. This is a result of inaccuracies in the theoretical models used in computing these energy quantities. Nevertheless, we still can see that when theory predicts a large exothermic change the experimental quantity corresponds to a large release of energy. Similarly, when the theory predicts a trend in $\Delta H_{soln}$, as with KCl, KBr, and KI, the measured values follow the same trend. The agreement is therefore sufficient to support the arguments made above.

The preceding explanations allow us to understand the energy changes that take place during the solution process. Unfortunately, however, it is very difficult to predict ahead of time, in any particular case, whether the formation of a solution will be exothermic or endothermic, because the same factors leading to a high hydration energy also tend to produce a high lattice energy. The extent to which an ion is attracted to a solvent dipole increases as the ion becomes smaller, because a smaller ion can get closer to a solvent molecule than can a larger one. The interaction of the solvent with an ion also increases as the charge on the ion becomes greater. However, the degree to which ions attract each other in the solid also grows with decreasing size and increasing charge, so that as the hydration energy becomes larger so does the lattice energy. Thus we have two factors that are affected in the same way by changes in size and charge, and it is virtually impossible to predict in advance which effect will predominate.

## 9.5 SOLUBILITY AND TEMPERATURE

In Chapter 5 we defined the solubility of a substance as the amount of solute required to produce a saturated solution in some particular quantity of solvent. At a given temperature a saturated solution in contact with undissolved solute represents another example of a state of dynamic equilibrium. As illustrated in Figure 9.10, particles of the solute are constantly passing into the solution and, at the same time, the solute particles already in the solution are continually colliding with and sticking to the undissolved solute. Although we show this equilibrium here for a solid dissolved in a liquid, the same concept applies to any type of solution (except gases—all gases are completely miscible).

Since a saturated solution in contact with excess solute constitutes a state of dynamic equilibrium, when the system is disturbed the effect of the disturbance can be predicted on the basis of Le Chatelier's principle. A

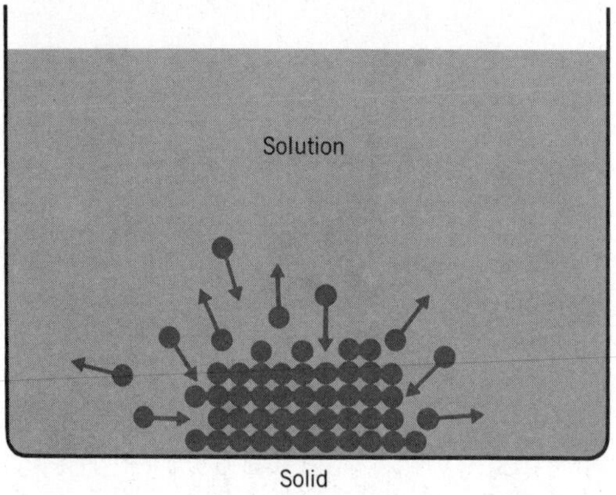


Figure 9.10
*Solubility as an equilibrium state. Solute dissolves at the same rate as it crystallizes.*

change in temperature corresponds to such a disturbance; in Chapter 8 we saw that a rise in temperature favors a shift in the position of equilibrium in a direction that will absorb heat. Therefore, if the dissolving of additional solute into an already saturated solution will absorb energy, the solubility of that substance increases as the temperature is raised. Conversely, if placing additional solute into the saturated solution is an exothermic process, the solute will become less soluble as the temperature is increased.

In general, the solubility of most solid and liquid substances in a liquid solvent increases with increasing temperature. For gases in liquids, the opposite behavior has been observed. The solution process for a gas in a liquid is nearly always exothermic, because the solute particles are already separated from each other and the dominant heat effect arises from the solvation that occurs when the gas dissolves. Le Chatelier's principle predicts that a rise in temperature will favor an endothermic change which, for a gas, takes place when it leaves solution. Therefore, we expect gases to become less soluble as the temperature of the liquid in which they are dissolved becomes higher. For example, in bringing water to a boil, tiny bubbles appear on the surface of the pot before boiling begins. These bubbles contain air that is driven out of solution as the water becomes hot. We also use this general solubility behavior of gases when we store opened bottles of carbonated beverages in the refrigerator. These liquids retain their dissolved $CO_2$ longer when they are kept cold because $CO_2$ is more soluble in them at low temperatures. Analysis of the quantity of dissolved gases in streams, lakes, and rivers reveals still another example of this phenomenon. The concentration of dissolved oxygen, which is imperative to marine life, decreases in the summer months, compared to when similar analyses are performed during the winter months—all other conditions being equal, of course.

## 9.6 FRACTIONAL CRYSTALLIZATION

Figure 9.11 illustrates graphically the way in which solubility changes with temperature for a variety of typical solids in water. From these solubility curves it is evident that the variation of solubility with temperature is quite different for different substances. For some substance such as $KNO_3$, the solubility changes very rapidly, while for others the change is much less. These differences in solubility behavior provide the basis for a useful laboratory technique, called **fractional crystallization,** which is frequently used for the separation of impurities from the products of a chemical reaction.

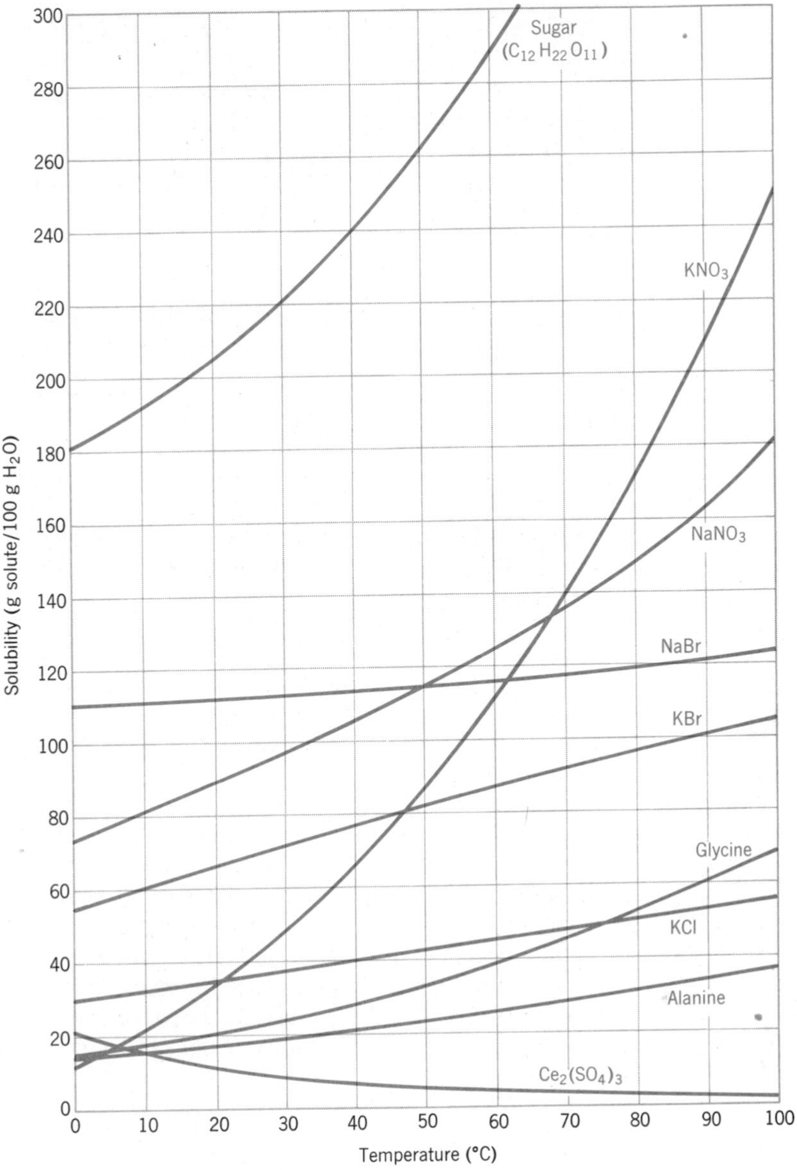


Figure 9.11
*Solubility curves for typical solids in $H_2O$.*

In this technique the impure product is first dissolved in a small amount of hot solvent (generally one in which the desired product is less soluble than the impurities). As the hot solution is allowed to cool, the pure product separates from the mixture, leaving the impurities behind. Finally, the crystals of the product are filtered from the cool solution and dried. The quantity of pure product that can be recovered in this fashion depends on the concentration of the impurities and their solubility relative to that of the desired material.

To illustrate how this technique might be applied, let us suppose that we had a mixture of two amino acids[4] and that the composition of this mixture was 80% glycine (Gly) and 20% alanine (Ala). How much pure Gly could we expect to be able to recover from 100 g of this mixture?

The first step in the procedure is to dissolve the solids in hot water. After the solution has been cooled we want all of the Ala to remain in solution and as much solid Gly as possible to have been formed. According to Figure 9.11, the solubilities of alanine and glycine at 0°C and 100°C are as follows:

| | Solubility in g/100 g of $H_2O$ at: | |
|---|---|---|
| Amino Acid | 0°C | 100°C |
| Alanine | 13 | 37 |
| Glycine | 14 | 67 |

Our mixture contains 20 g of Ala, so that the minimum amount of water required to keep the Ala dissolved at 0°C is

$$20 \text{ g Ala} \times \left(\frac{100 \text{ g } H_2O}{13 \text{ g Ala}}\right) \sim 154 \text{ g } H_2O$$

We must next check to see if this is enough water to also dissolve, when hot, the 80 g of Gly that is present in the mixture. At 100°C the solubility of Gly is 67 g/100 g of $H_2O$; therefore, 154 g of water at 100°C is capable of dissolving

$$\left(\frac{67 \text{ g Gly}}{100 \text{ g } H_2O}\right) \times 154 \text{ g } H_2O \sim 103 \text{ g Gly}$$

Thus we see that 154 g of hot water is *more* than sufficient to dissolve not only the alanine but all of the glycine as well.

The amount of pure Gly that will separate from the solution after it has been cooled to 0°C is the difference between the amount of Gly in the original mixture, 80 g, and the amount that will remain in the solution at 0°C. The solubility of Gly at 0°C is 14 g/100 g $H_2O$ and therefore in 154 g of water at 0°C we have

$$\left(\frac{14 \text{ g Gly}}{100 \text{ g } H_2O}\right) \times 154 \text{ g } H_2O \sim 22 \text{ g Gly}$$

The quantity of solid Gly that is precipitated, then, is 80 g − 22 g = 58 g. This represents the maximum amount of pure glycine that we can recover from our mixture. If we use more than 154 g of $H_2O$, we will obtain less than 58 g of Gly when the solution is cooled. If we use *less than* 154 g of $H_2O$, more solid will be obtained but it will be contaminated by some alanine, which will also precipitate out.

In actual practice, computations such as this are rarely carried out. This is because generally the necessary solubility information simply is not available and the composition of the mixture isn't known. The purification of substances by fractional crystallization is something of an art, which can be learned only through repeated practice.

[4] Amino acids are constituents of biologically important compounds called proteins, and are discussed in detail in Chapter 22.

## 9.7 THE EFFECT OF PRESSURE ON SOLUBILITY

In general, pressure has very little effect on the solubility of liquids or solids in liquid solvents. The solubility of gases, however, always increases with increasing pressure. Carbonated beverages, for example, are bottled under pressure to ensure a high concentration of $CO_2$; once the bottle has been opened, the beverage quickly loses its carbonation unless it is recapped.

Let us imagine that a liquid is saturated with a gaseous solute and that this solution is in contact with the gas at some particular pressure. Once again we have a dynamic equilibrium where molecules of the solute are leaving the solution and entering the vapor phase at the same rate at which molecules from the gas are entering the solution, as shown in Figure 9.12*a*. As we might expect, the rate at which molecules go into solution depends on the number of collisions per second that the gas experiences with the surface of the liquid and, similarly, the rate at which the solute molecules leave the solution depends on their concentration.[5] If we suddenly increase the pressure of the gas, we pack the molecules closer together and the number of collisions per second that the gas molecules make with the surface of the liquid gets larger. When this occurs, the rate at which molecules of the solute (gas) enter the solution also gets larger without a corresponding increase in the rate at which they leave (Figure 9.12*b*). As a result, the concentration of solute molecules in solution rises until the rate at which they are leaving the solution once again equals the rate at which they enter; at this point we have reestablished equilibrium (Figure 9.12*c*).

The solubility behavior of gases with respect to pressure also can be explained easily in terms of Le Chatelier's principle. We might represent the equilibrium by the following equation:

$$\text{solute } (g) + \text{solvent } (l) \rightleftharpoons \text{solution } (l)$$

According to Le Chatelier's principle, an increase in pressure on this system at equilibrium favors a shift in the position of equilibrium that leads to a decrease in pressure. If the reaction were to proceed to the right, so that more of the gaseous solute dissolves, the quantity of solute in the gas phase would

Figure 9.12
*Effect of pressure on the solubility of a gas. (a) Solution equilibrium. (b) System not at equilibrium. (c) Equilibrium again.*

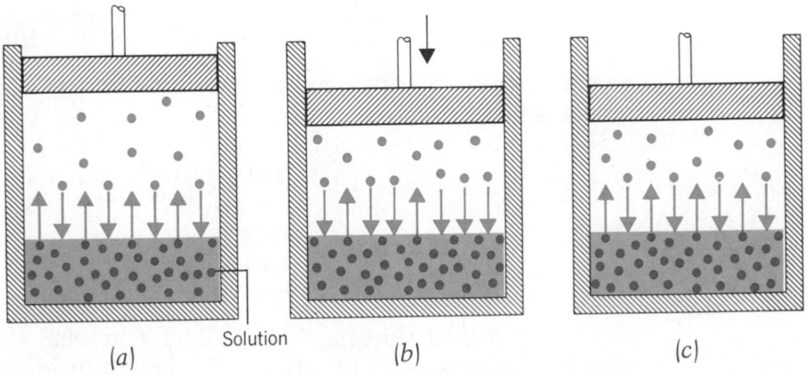


[5] At a given temperature and concentration, the rate at which gaseous solute molecules leave a solution also depends on the nature of the gas. Some gases are very soluble, and at a given concentration their rate of escape from the liquid is much less than that for a gas of low solubility.

decrease. As you recall from Chapter 6, a decrease in the number of moles of gas leads to a drop in pressure. Thus, an externally imposed increase in pressure favors an increase in the solubility of a gas because it is this process that tends to bring the pressure down in the direction of its initial value.

Quantitatively, the influence of pressure on the solubility of a gas is given by **Henry's law,** which states that the concentration of the gaseous solute in the solution, $C_g$, is directly proportional to the partial pressure of the gas above the solution; that is,

$$C_g = k_g p_g \tag{9.1}$$

where the proportionality constant, $k_g$, is called the **Henry's law constant.** This relationship allows us to compute the solubility of a gas at some particular pressure, provided that we know its solubility at some other pressure, as demonstrated in Example 9.4. Actually, Henry's law is accurate only for relatively low concentrations and pressures and for gases that do not react significantly with the solvent.

EXAMPLE 9.4 At 25°C, oxygen gas collected over water at a *total* pressure of 1.00 atm is soluble to the extent of 0.0393 g/liter. What would its solubility be if its partial pressure over water were 800 torr?

SOLUTION The data given us permits the calculation of the Henry's law constant if we know the partial pressure of oxygen above the solution. The total pressure is the sum of the partial pressures of the $H_2O$ vapor and the oxygen,

$$P_{\text{total}} = p_{H_2O} + p_{O_2}$$

From Table 6.1 we find the vapor pressure of water to be 23.8 torr at 25°C; therefore, the partial pressure of oxygen is

$$p_{O_2} = P_{\text{total}} - p_{H_2O}$$
$$p_{O_2} = 760 \text{ torr} - 24 \text{ torr} = 736 \text{ torr}$$

The Henry's law constant is obtained as the ratio

$$k_{O_2} = \frac{C_{O_2}}{p_{O_2}}$$

$$k_{O_2} = \frac{0.0393 \text{ g/liter}}{736 \text{ torr}} = 5.34 \times 10^{-5} \frac{\text{g}}{\text{liter torr}}$$

Now we can use Henry's law to determine that at a partial pressure of 800 torr the solubility of oxygen is

$$C_{O_2} = \left(5.34 \times 10^{-5} \frac{\text{g}}{\text{liter } \cancel{\text{torr}}}\right)(800 \ \cancel{\text{torr}})$$

$$C_{O_2} = 0.0427 \frac{\text{g}}{\text{liter}}$$

## 9.8 VAPOR PRESSURES OF SOLUTIONS

We have discussed, thus far, some of the factors that affect the solubilities of solutes in various types of solvents. When a solution is formed, its physical properties are no longer the same as the solvent or solute but, instead, depend on the concentrations of the components that make up the mixture. One property that we might examine, which is rather easily measured, is the vapor pressure of the solution.

For a solution in which a nondissociating, nonvolatile solute is dissolved in a solvent (that is, the solute itself has very little tendency to dissociate or to escape from the solution and enter the gas phase), the vapor pressure is due only to the vapor of the *solvent* above the solution. This vapor pressure is given by **Raoult's law,** which states that *the vapor pressure of the solution at a particular temperature is equal to the mole fraction of the solvent in the liquid phase multiplied by the vapor pressure of the pure solvent at the same temperature;* that is,

$$P_{\text{solution}} = X_{\text{solvent}} P^0_{\text{solvent}} \tag{9.2}$$

Thus, for example, a solution that contains 95 mol % water and 5 mol % of a nonvolatile solute such as sugar will have a vapor pressure only 95% as great as would the pure solvent. Stated qualitatively, the vapor pressure of the solution is lowered by the addition of a nonvolatile solute.

Let us see why Raoult's law holds. Figure 9.13*a* illustrates the condition in which the pure solvent is in equilibrium with its vapor. This vapor, as described in Section 8.3, exerts a pressure (which we call the vapor pressure) that is ultimately determined by the fraction of the total number of molecules at the surface that have enough kinetic energy to escape the liquid and enter the gas phase. If we now look at a solution containing a nonvolatile solute (Figure 9.13*b*), we find that a portion of the solvent molecules at the surface has been replaced by molecules of the solute. Since the entire system, solvent plus solute, is at a single temperature, all of the molecules in the solution belong to a single distribution of kinetic energies. In both the solution and the pure solvent, the same fraction of surface molecules has more than that minimum kinetic energy that solvent molecules need in order to break away from the liquid, but in the solution only a *portion* of that fraction is actually composed of molecules of the solvent. The others are solute molecules. The result is that there are fewer molecules at the surface of the solution capable of leaving than at the surface of the pure solvent. Consequently, the rate of evaporation of solvent molecules from a solution is less than from the pure solvent.

The magnitude of the equilibrium vapor pressure is determined by the rate of evaporation from the surface of the liquid. If the rate is high, a large concentration of molecules must be present in the vapor at equilibrium so that the rate of return to the liquid can also be high. Conversely, if the rate

Figure 9.13
*Molecular view of Raoult's law. (a) Pure solvent (b) Solution.*

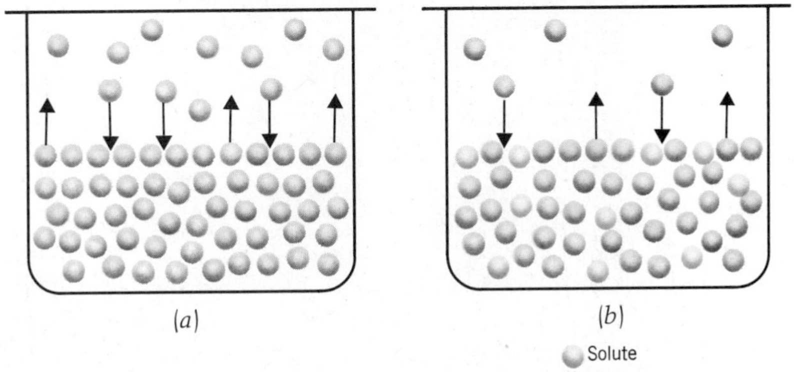

of evaporation is low, so must be the concentration in the vapor phase. Since the rate of evaporation from the solution is less than from the pure solvent, the equilibrium concentration of molecules in the vapor is smaller over the solution than over the solvent alone; therefore, the equilibrium vapor pressure is lower for the solution than for the pure solvent.

Because only the solvent can evaporate, the fraction of molecules at the surface of the solution that can escape from the liquid depends on that *fraction* of all of the molecules at the surface that are solvent molecules—that is, the ratio of the number of moles of solvent particles to the total number of moles of particles that comprise the surface. This ratio, of course, is the mole fraction of the solvent. If the solution were composed of 95 mol % solvent (we expect to find only 95% of the molecules at the surface to belong to the solvent), then the rate of evaporation from the solution is expected to be only 95% of that for the solvent alone. The equilibrium vapor pressure should therefore be reduced to 95% of that for the pure solvent, which is the same result as we obtain by the application of Raoult's law.

The lowering of the vapor pressure of the solvent by the addition of a solute occurs at all temperatures. The result is that the vapor pressure curve of the solvent in a solution falls below that of the pure solvent, as shown in Figure 9.14. At any particular temperature there is a linear relationship, predicted by Raoult's law, between vapor pressure and mole fraction. This is shown in Figure 9.15.

In many solutions, such as benzene and carbon tetrachloride, for example, both solute and solvent have appreciable tendencies to undergo evaporation. In this case, the vapor will contain both solute and solvent molecules, and the vapor pressure of the solution will be the sum of the partial pressures exerted by each component. If we follow the same line of reasoning as above, we conclude that the partial pressure of any component above such a mixture is also given by Raoult's law. Thus the partial pressure of component $A$, $p_A$, is given by

$$p_A = X_A P_A{}^0 \tag{9.3}$$

where $P_A{}^0$ is the vapor pressure of pure $A$ and $X_A$ is its mole fraction in the

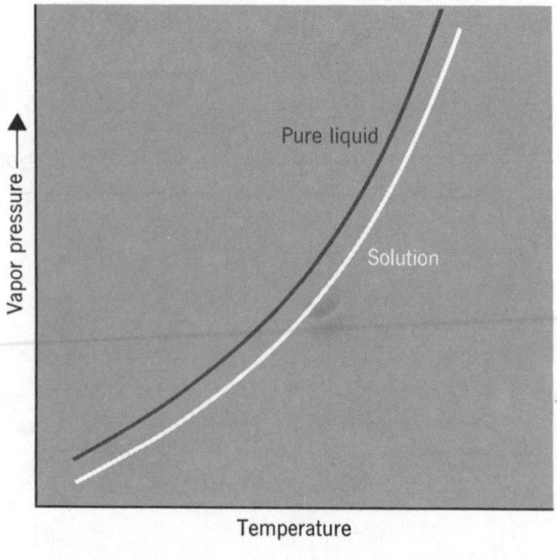


Figure 9.14
*The lowering of the vapor prssure of a liquid by the addition of a solute.*

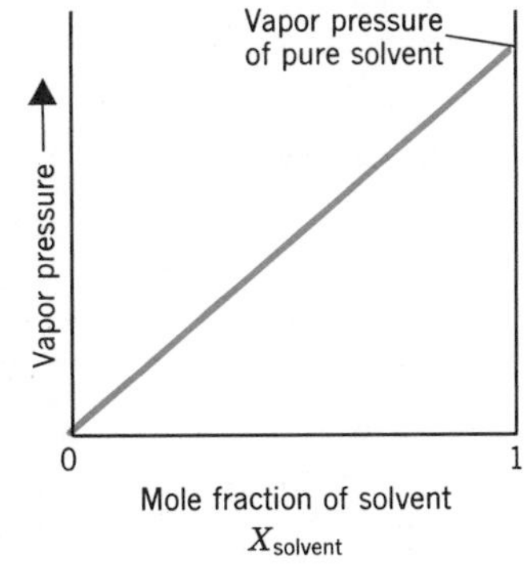


Figure 9.15
*Vapor pressure of solvent as a function of concentration.*

solution. In the same manner, the partial pressure of a second component, $p_B$, is given as

$$p_B = X_B P_B^0 \tag{9.4}$$

Finally, the total vapor pressure of a binary mixture of $A$ and $B$, according to Dalton's law, is

$$P_T = p_A + p_B \tag{9.5}$$

Substituting Equations 9.3 and 9.4 into Equation 9.5 gives

$$P_T = X_A P_A^0 + X_B P_B^0$$

Figure 9.16*a* is a plot of the partial pressures of $A$ and $B$, and the total vapor pressure as a function of solution composition for such a two-component mixture.

Actually, very few mixtures really obey Raoult's law very closely over wide ranges of composition. Benzene and carbon tetrachloride, a pair of substances that do form such mixtures, are said to yield **ideal solutions.** Mixtures that deviate from Raoult's law are called **nonideal.** When the vapor pressure of a mixture is greater than that predicted, it is said to exhibit a **pos-**

Figure 9.16
*Vapor pressure of a two-component system (a) Ideal solution. (b) Positive deviations from Raoult's law. (c) Negative deviations from Raoult's law.*

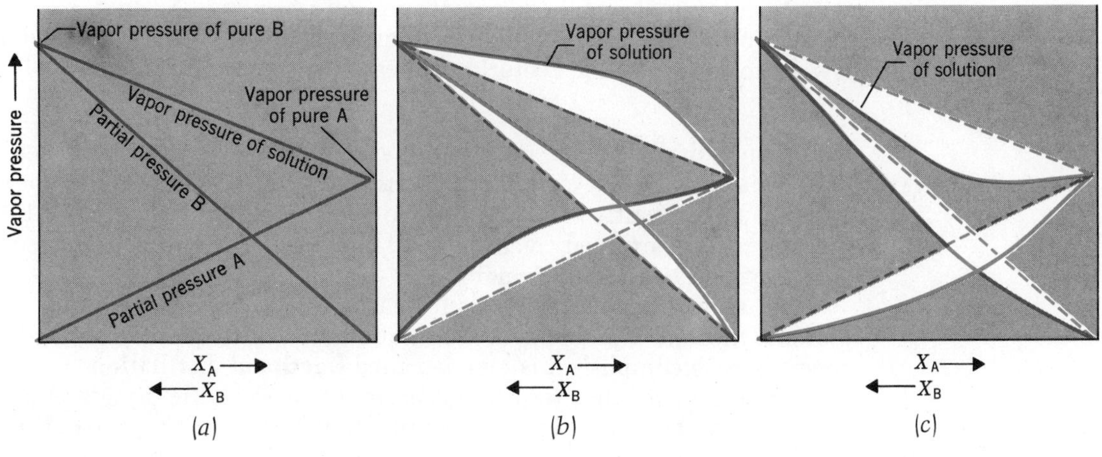

Table 9.4
*Summary of solution properties*

| Relative Attractive Forces | $\Delta H_{soln}$ | Temperature Change when Solution Is Formed | Deviations from Raoult's Law | Example |
|---|---|---|---|---|
| A–A, B–B = A–B | Zero | None | None (ideal solution) | Benzene–chloroform |
| A–A, B–B < A–B | Negative (exothermic) | Increase | Negative | Acetone–water |
| A–A, B–B > A–B | Positive (endothermic) | Decrease | Positive | Ethyl alcohol–water |

**itive deviation** from Raoult's law (Figure 9.16*b*); conversely, when a solution gives a lower vapor than we would expect from Raoult's law, it is said to show a **negative deviation** (Figure 9.16*c*).

The origin of nonideal behavior lies in the relative strengths of the interactions between molecules of the solute and solvent. When the attractive forces between the solute and solvent molecules are weaker than those between solute molecules or between solvent molecules, neither the solute nor solvent particles are held as tightly in the solution as they are in the pure substances. The escaping tendency of each is therefore greater in the solution than in the solute or solvent alone. As a result, the partial pressures of both of them over the solution are greater than predicted by Raoult's law. Consequently, the total effect is that the solution exhibits a larger vapor pressure than expected.

Just the opposite effect is produced when the solute–solvent interactions are stronger than the solute–solute or solvent–solvent interactions. Each substance, in the presence of the other, is held more tightly than in the pure materials, and their partial pressures over a solution are therefore less than Raoult's law would predict. The result is that such a solution exhibits a negative deviation from ideality.

Since, in a solution that shows positive deviations from ideal behavior, the forces of attraction between solute and solvent are weaker than those between both solute molecules and solvent molecules, the formation of these solutions occurs with the absorption of energy (Section 9.4). Conversely, of course, mixtures that exhibit negative deviations from Raoult's law are formed with the evolution of heat. This is summarized in Table 9.4.

## 9.9 FRACTIONAL DISTILLATION

In a simple distillation process, one that could be used to separate sodium chloride and water, for example, a volatile solvent is vaporized from a solution and subsequently condensed to provide a pure liquid (see Figure 1.4). If the process is continued, eventually all of the solvent will be removed and only the solid solute will remain.

The separation of mixtures of volatile liquids into their components presents more of a problem. A technique that can frequently be used successfully to accomplish this task is called **fractional distillation.**

Let us suppose that we had a mixture of two volatile liquids, *A* and *B*, that form an ideal solution. This mixture will boil when the sum of the par-

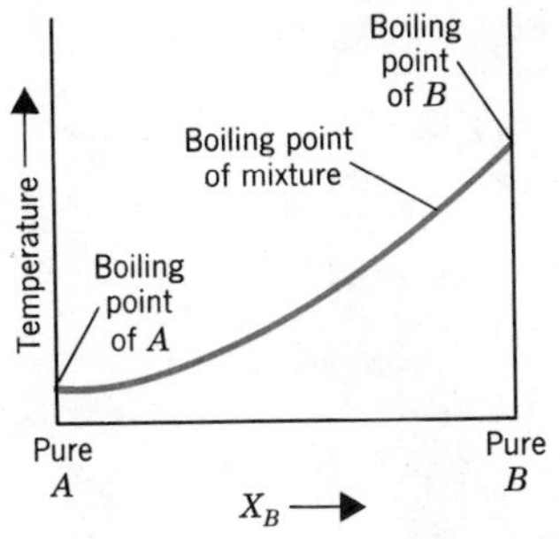


Figure 9.17
*Boiling-point curve for a mixture of* A *and* B.

tial pressures of $A$ and $B$ equals the prevailing atmospheric pressure; that is, when

$$P_{atm} = p_A + p_B$$

The boiling points of various mixtures of $A$ and $B$ will increase gradually from that of the more volatile component (let us say, $A$) to that of the less volatile one, $B$, as shown in Figure 9.17.

Suppose, now, that when 1 mole of $A$ is mixed with 2 moles of $B$, the resulting mixture boils (at 1 atm) at a temperature at which the vapor pressure of pure $A$ is 1140 torr and that of pure $B$ is 570 torr. Under these conditions the partial pressure of $A$ is

$$p_A = X_A P_A{}^0$$
$$p_A = \left(\frac{1 \text{ mole } A}{1 \text{ mole } A + 2 \text{ moles } B}\right) 1140 \text{ torr}$$
$$p_A = \left(\frac{1 \text{ mole}}{3 \text{ moles}}\right) (1140 \text{ torr})$$
$$p_A = (0.333)(1140 \text{ torr}) = 380 \text{ torr}$$

Similarly, the partial pressure of $B$ would be

$$p_B = \left(\frac{2 \text{ moles}}{3 \text{ moles}}\right) (570 \text{ torr})$$
$$p_B = 380 \text{ torr}$$

The sum of $p_A$ and $p_B$ is 760 torr as, of course, it must be if the solution is to boil.

What can we say about the composition of the vapor? In Section 6.8, under our discussion of Dalton's law of partial pressures, it was stated that the partial pressure of a gas in a mixture is equal to its mole fraction multiplied by the total pressure *exerted by the gas.*

$$p_A = X_A P_T$$

In the vapor over our solution, the partial pressure of each gas is 380 torr and the total pressure is 760 torr. This means that the mole fraction of both $A$ and $B$ *in the vapor* must be 0.500. In the liquid the mole fraction of $A$ was only 0.333. Thus the vapor contains a greater amount of the more volatile component ($A$) than does the solution. In fact, any time we boil a mixture of these two substances, the vapor will be richer than the solution in the more volatile compound. On our boiling-point diagram we can indicate the com-

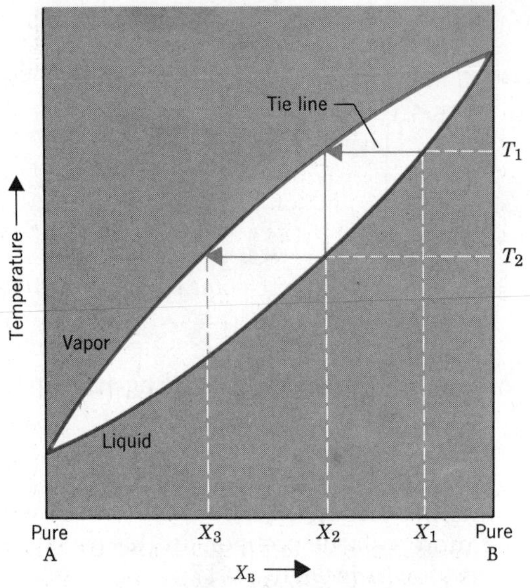


Figure 9.18
*Boiling-point diagram for a two-component mixture.*

position of the vapor by the upper curve drawn in Figure 9.18. Here, points corresponding to the composition of liquid and vapor in equilibrium can be obtained by drawing a horizontal line, called a **tie line,** between the curve for the liquid and that for the vapor. When the composition of the mixture is $X_1$, it boils at a temperature $T_1$ to provide a vapor that has a composition $X_2$. If this vapor is condensed and then reheated, it will boil at a temperature $T_2$ and give a vapor whose composition is $X_3$. Repetition of this process will produce fractions ever richer in $A$. This procedure is called fractional distillation. This technique is useful not only in the laboratory, where it is employed for purifying the products of chemical reactions, but also industrially. For instance, the petroleum industry uses fractional distillation to separate crude oil into its various components, which include gasoline, kerosene, oils, and paraffin.

There are some solutions that exhibit very large deviations from ideality; as a result they cannot be totally separated into their components even by fractional distillation. Ethyl alcohol (grain alcohol) and water form such a mixture. Solutions of these two substances have such large positive deviations from Raoult's law that there is a maximum in the vapor pressure curve and hence a minimum in the boiling-point diagram as shown in Figure 9.19. A solution with such a minimum boiling point is called a **minimum-boiling azeotrope.** Fractional distillation of solutions lying on either side of this azeotropic composition is capable of separating them into, at best, one pure component plus a solution having the minimum boiling point. As any "moonshiner" will agree, ethyl alcohol–water mixtures (obtained by fermentation of sugars, for example) are rich in water. Fractional distillation is able to concentrate the alcohol to, at best, the azeotropic composition of approximately 95% by volume of ethyl alcohol.[6] Once this composition has been achieved, the liquid and vapor have the same composition, and no additional fractionation takes place.

[6] This is too strong to consume without dilution. A 95% solution of ethyl alcohol is 190 proof. Good aged whiskey that is 86 proof is only 43% alcohol, by volume.

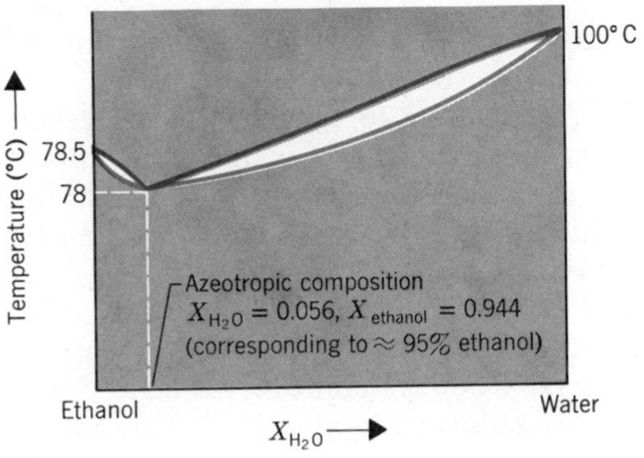


Figure 9.19
*Boiling-point diagram for water—ethanol mixtures (not drawn to scale).*

There are also solutions that show large negative deviations from ideality and therefore have a minimum in their vapor pressure curves. This leads to a maximum on the boiling point diagram and hence to a **maximum-boiling azeotrope.** Hydrochloric acid, for instance, forms a maximum-boiling azeotrope having the approximate composition, 20% HCl and 80% $H_2O$ by weight, with a boiling point of 109°C.

## 9.10 COLLIGATIVE PROPERTIES OF SOLUTIONS

Properties that depend on the number of particles of solute in a solution, instead of on their specific chemical nature, are called **colligative properties.**[7] Vapor pressure is one of these. We have seen that, according to Raoult's law, the addition of a nonvolatile solute to a substance causes its vapor pressure to be lowered. In our explanation of how this happened, nothing was said about the specific nature of the solute other than that it was incapable of escaping from the solution and that it was undissociated.

What effect does this vapor pressure lowering have on the phase diagram of a solvent such as water? In Figure 9.20, we see again that the vapor pressure of the solution lies below that of the pure solvent at every temperature, as indicated by the dashed line. Because of this the solution must attain a higher temperature in order to have the same vapor pressure as water alone.

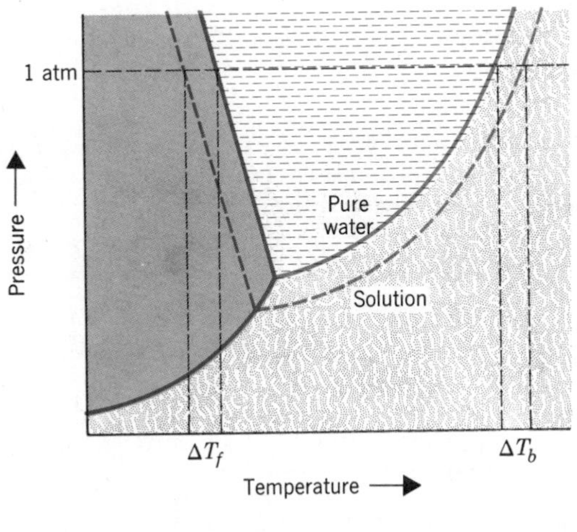


Figure 9.20
*Effect of a nonvolatile solute on the phase diagram of $H_2O$.*

[7] Derived from the latin, *colligare,* to collect. These properties are determined by the number of particles in the entire "collection," not by what the particles are composed of.

Therefore the normal boiling point of the solution is higher, by an amount $\Delta T_b$, than that of water by itself.

We also find, in Figure 9.20, that the vapor pressure curve of the solution intercepts the solid–vapor line of water at a temperature below the triple point of the solvent.[8] The solid–liquid equilibrium line of the solution must pass through this new triple point and consequently lies to the left of the solid–liquid line for pure water. As a result, the freezing point of the solution, at 1 atm, lies below that of the solvent by $\Delta T_f$.

In summary, then, a nonvolatile solute increases the liquid range of a solution and results in a boiling-point elevation and freezing-point depression. In effect the solute reduces the escaping tendency of the solvent molecules both in the direction of the vapor and the solid. A common example of this is the effect that automotive antifreeze has on the liquid range of an antifreeze solution in the radiator of an automobile. The solute (usually ethylene glycol) lowers the freezing point so that the radiator does not freeze in cold weather and it also raises the boiling point so that the radiator does not "boil over" as readily in very hot weather.

For dilute solutions it has been found that the extent to which the boiling point is raised, and the freezing point is lowered, depends on the *molality* of the solute in the solution,

$$\Delta T_b = K_b m \tag{9.6}$$

and

$$\Delta T_f = K_f m \tag{9.7}$$

where $K_b$ and $K_f$ are referred to as the molal boiling-point elevation constant and the molal freezing-point depression constant, respectively. The magnitudes of $K_b$ and $K_f$ are characteristic of each solvent. Table 9.5 contains a list of some typical solvents and their values of $K_b$ and $K_f$.

If we know the concentration of the solution, in moles of solute per kilogram of solvent, the relationships expressed in Equations 9.6 and 9.7 permit us to calculate the extent to which the boiling point and freezing point are changed. For example, a solution containing 1 mole of sugar in 1000 g of water will have its freezing point lowered by 1.86°C and its boiling point raised by 0.51°C. The solution will therefore freeze at −1.86°C and boil at 100.51°C when the atmospheric pressure is 1 atm. Example 9.5 provides another sample calculation.

[8] There is almost never a solid–vapor curve for the solution, because when it freezes the solid that is formed nearly always is composed of the pure solvent. Solute particles can only rarely be accommodated in the solid solvent lattice.

Table 9.5

*$K_b$ and $K_f$ for some solvents*

| Solvent | Boiling Point (°C) | $K_b$ (°C/m) | Melting Point (°C) | $K_f$ (°C/m) |
|---|---|---|---|---|
| Water | 100.0 | 0.51 | 0.0 | 1.86 |
| Benzene | 80.1 | 2.53 | 5.5 | 5.12 |
| Camphor | — | — | 179 | 39.7 |
| Acetic acid | 118.2 | 2.93 | 17 | 3.90 |

EXAMPLE 9.5 What would be the freezing point and boiling point of a solution containing 6.50 g of ethylene glycol ($C_2H_6O_2$), commonly used as an automotive antifreeze, in 200 g of water?

SOLUTION To determine $\Delta T_f$ and $\Delta T_b$ we must know the molality of the solution—the ratio of moles of solute to kilograms of solvent. The number of moles of $C_2H_6O_2$ is

$$6.50\ \text{g } C_2H_6O_2 \times \left(\frac{1\ \text{mole } C_2H_6O_2}{62.0\ \text{g } C_2H_6O_2}\right) \sim 0.105\ \text{mole } C_2H_6O_2$$

The number of kilograms of solvent is

$$200\ \text{g } H_2O \times \left(\frac{1\ \text{kg}}{1000\ \text{g}}\right) \sim 0.200\ \text{kg } H_2O$$

The molality is therefore

$$\frac{0.105\ \text{mole } C_2H_6O_2}{0.200\ \text{kg } H_2O} = 0.525\ m\ C_2H_6O_2$$

For $H_2O$, $K_f = 1.86°\text{C}/\text{m}$ and $K_b = 0.51°\text{C}/\text{m}$. Hence the changes in the freezing point and boiling point are

$$\Delta T_f = \left(1.86\ \frac{°\text{C}}{m}\right) \times (0.525\ m) = 0.98°\text{C}$$

$$\Delta T_b = \left(0.51\ \frac{°\text{C}}{m}\right) \times (0.525\ m) = 0.27°\text{C}$$

The freezing and boiling points of the solution are then $-0.98°\text{C}$ and 100.27°C, respectively. We see that solutions considerably more concentrated than this (approximately 3%) are necessary to protect an automobile's cooling system in frigid weather.

If a knowledge of the molal concentration permits us to determine the extent to which the boiling point and freezing point differ from those of the pure solvent, then it should also be possible to calculate the molal concentration of a solution from $\Delta T_b$ and $\Delta T_f$. This aspect of these colligative properties proves particularly useful because with it we can measure molecular weights experimentally. This, combined with analytical data on percent composition, enables us to find molecular formulas. Example 9.6 illustrates how to apply this concept.

EXAMPLE 9.6 A 5.50-g sample of a compound, whose empirical formula is $C_3H_3O$, dissolved in 250 g of benzene, gives a solution whose freezing point is 1.02°C below that of pure benzene. Determine (a) the molecular weight and (b) the molecular formula of this compound.

SOLUTION (a) From Table 9.5, $K_f$ for benzene is 5.12°C/m. If we solve Equation 9.7 for the concentration, we obtain

$$m = \frac{\Delta T_f}{K_f}$$

Upon substituting the values for the freezing-point depression and $K_f$, we have

$$m = \frac{1.02°C}{5.12°C/m}$$

$$\text{molality} = 0.199\ m = \frac{0.199 \text{ mole solute}}{1.00 \text{ kg benzene}}$$

From the data given, we have 5.50 g of solute per 250 g of benzene, from which we can obtain the number of grams of solute per kilogram of solvent.

$$\left(\frac{5.50 \text{ g solute}}{250 \text{ }\cancel{\text{g benzene}}}\right) \times \left(\frac{1000 \text{ }\cancel{\text{g benzene}}}{1.00 \text{ kg benzene}}\right) = \frac{22.0 \text{ g solute}}{1.00 \text{ kg benzene}}$$

We now have two expressions giving the quantity of solute per 1.00 kg of solvent. They must be equivalent because the solution cannot have two different concentrations at the same time. Therefore,

$$0.199 \text{ mole solute} = 22.0 \text{ g solute}$$

Finally, the weight of 1 mole (which is numerically equal to the molecular weight) is

$$1 \text{ }\cancel{\text{mole solute}} \times \left(\frac{22.0 \text{ g solute}}{0.199 \text{ }\cancel{\text{mole solute}}}\right) = 111 \text{ g solute}$$

(b) Now that we know the molecular weight, we can determine the molecular formula as in Chapter 2. The molecular formula must contain the empirical formula repeated an integral number of times. The molecular weight is therefore an integral multiple of the empirical formula weight, which for $C_3H_3O$ is 55.0 amu. Since the molecular weight that we have found is twice this value, the molecular formula must be $(C_3H_3O)_2$ or $C_6H_6O_2$.

In practice, molecular weights cannot be determined by this method as accurately as we have implied. However, an error even as large as 10% (that is, measured molecular weights ranging, in this case, from about 100 to 120 amu) certainly still permits us to choose among the possibilities, $C_3H_3O$, $C_6H_6O_2$, and $C_9H_9O_3$, with their corresponding molecular weights, 55.0, 110, and 165 amu. For this reason, this technique has proven to be a valuable tool for the chemist.

**SOLUTIONS OF ELECTROLYTES.** For simplicity, we have limited our discussion thus far to solutions that do not contain an electrolyte. The reason for this is that the freezing-point depression and boiling-point elevation depend on the number of particles present in the solution. One mole of a nonelectrolyte, such as sugar, when placed in water yields 1 mole of particles, and a solution labeled "1 *m* sucrose" would, therefore, have a freezing point 1.86°C lower than pure water. However, a solution containing 1 mole of an electrolyte such as NaCl contains 2 moles of particles—1 mole of $Na^+$ ions and 1 mole of $Cl^-$ ions. As a result, a solution labeled "1 *m* NaCl" actually contains 2 moles of particles per 1000 g of water and theoretically should have a freezing-point depression of 2 × 1.86°C = 3.72°C. In a similar fashion, a 1 *m* solution of $CaCl_2$, which contains 3 moles of ions per 1000 g of water, would have a freezing-point depression three times as great as a 1 *m* solution of sucrose. In fact, neither the prediction for NaCl nor for $CaCl_2$ is entirely accurate (see section 9.12). In each case the observed depression is slightly less than expected.

For a weak electrolyte, such as acetic acid, we expect the freezing-point depression and boiling-point elevation to be intermediate between that of a nonelectrolyte and that of a strong electrolyte. Acetic acid, which we have said in Chapter 5 undergoes reaction with water to establish the equilibrium

$$HC_2H_3O_2 + H_2O \rightleftharpoons H_3O^+ + C_2H_3O_2^-$$

is only partially dissociated. Thus, at equilibrium, 1 mole of this solute exists as more than 1 mole of particles but less than 2. We can use the boiling-point elevation and freezing-point depression to determine the extent to which such a weak electrolyte is dissociated.

EXAMPLE 9.7 Very careful measurement reveals that a 1.00 $m$ solution of HF has a freezing point of $-1.91°C$. What percent of the HF is dissociated into $H^+$ and $F^-$ ions in this solution?

SOLUTION The dissociation of hydrogen fluoride can be represented by

$$HF \rightleftharpoons H^+ + F^-$$

If HF were a nonelectrolyte, a 1.00 $m$ solution would have a freezing point of $-1.86°C$; if it were a strong electrolyte, the solution should freeze at $-3.72°C$. Because the measured freezing point lies between these two extremes, only a part of the 1 mole of HF has dissociated. We want to know how much. Since this is presently unknown, let us give it a name—that is, let us call the amount of HF that has undergone dissociation in 1.00 kg of water by the name, $x$.

$$\text{number of moles of HF dissociated} = x$$

The number of moles of HF remaining, then, must be the difference between the amount of HF that we put into the solution and the amount dissociated; that is, $1.00 - x$.

$$\text{number of moles of HF remaining at equilibrium} = 1.00 - x$$

We see from the chemical equation above that for every mole of HF that dissociates, we produce 1 mole of $H^+$ and 1 mole of $F^-$. When $x$ moles of HF dissociate, we must therefore form $x$ moles of $H^+$ and $x$ moles of $F^-$, so that at equilibrium we also have

$$\text{number of moles of } H^+ = x$$
$$\text{number of moles of } F^- = x$$

The *total* number of moles of particles in 1 kg of solvent is the sum of the moles contributed by the HF, $H^+$, and $F^-$. This total is

$$\begin{aligned}\text{total number of moles of particles} &= (1.00 - x) + x + x \\ &= 1.00 + x\end{aligned}$$

Now, from the freezing-point depression of this solution, 1.91°C, we can calculate the molal concentration of *particles*.

$$m = \frac{\Delta T_f}{K_f} = \frac{1.91°C}{1.86°C/m} = 1.03\ m$$

This number *also* represents the total number of moles of particles per 1.00 kg of water. Therefore,

$$1.03 \text{ moles of particles} = (1.00 + x) \text{ moles of particles}$$

and thus

$$x = 0.03 \text{ mole}$$

The fraction of HF dissociated is equal to the number of moles that have broken apart (0.03 mole) divided by the total number of moles of HF placed in the solution (1.00 mole).

$$\text{fraction dissociated} = \frac{0.03}{1.00} = 0.03$$

The percentage of the HF dissociated, then, is 3%.

Solutions of electrolytes have larger values of $\Delta T_f$ and $\Delta T_b$ than we might initially have expected because of dissociation. There are also instances where the freezing-point depression and boiling-point elevation are smaller than we would at first predict. This occurs when **association** (the opposite of dissociation) takes place between solute particles in the solution. For instance, a solution of 1 mole (122 g) of benzoic acid in 1.00 kg of benzene produces a freezing-point depression only slightly more than half the expected depression, implying that there are only about half as many particles in the solution as we anticipated. Since this approximately $\frac{1}{2}$ mole of particles weighs 122 g, the apparent molecular weight is about 240. Therefore, when association takes place, the measured molecular weights are actually higher than we would predict.

In this particular example, the anomalous behavior of benzoic acid in benzene is attributed to hydrogen bonding between benzoic acid molecules to form a **dimer** (a particle created from two identical simpler units).

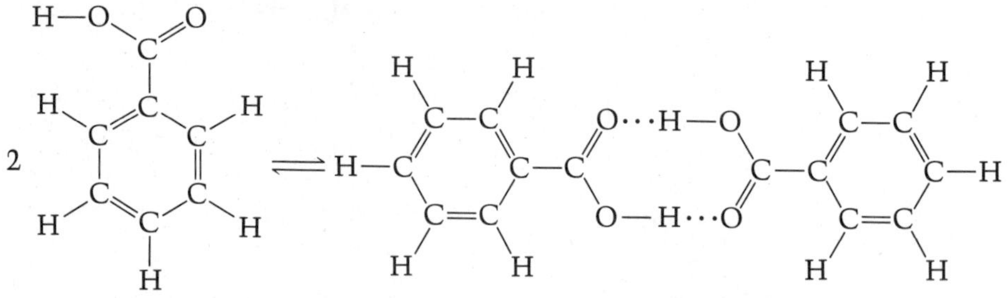

**benzoic acid** **benzoic acid dimer—hydrogen bonds shown as dotted lines**

## 9.11 OSMOTIC PRESSURE

**Osmosis** is a process whereby a solvent passes from a dilute solution into a more concentrated one by moving through a thin film that selectively permits the passage of the solvent, but restricts the passage of the solute. Such films are called **semipermeable membranes,** typical examples of which include certain types of parchment paper, some gelatinlike inorganic substances, and the cell walls of living organisms.

In the process of osmosis there is a drive toward equalization of concentrations between the two solutions in contact with one another across the

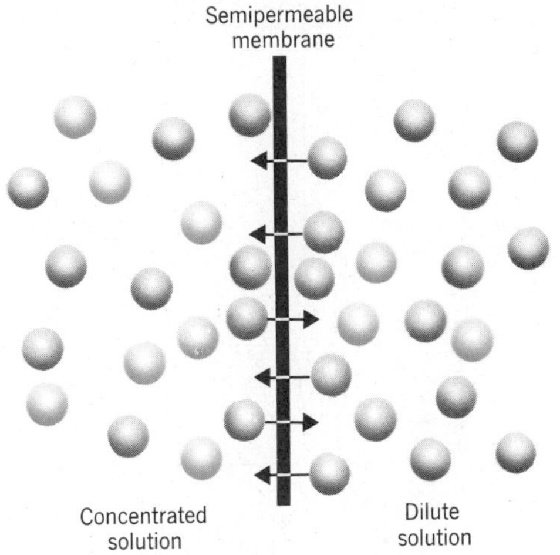


Figure 9.21
*Osmosis.*

membrane. The rate of passage of solvent molecules through the membrane into the more concentrated solution is greater than their rate of passage in the opposite direction, persumably because, at the surface of the membrane, the solvent concentration is greatest in the more dilute solution (Figure 9.21). We observe a similar effect if two solutions, with unequal concentrations of a nonvolatile solute, are placed in a sealed enclosure, as shown in Figure 9.22. The rate of evaporation from the dilute solution is greater than that from the concentrated solution, but the rate of return to each is the same (both solutions are in contact with the same gas phase). As a result, neither solution is in equilibrium with the vapor. In the dilute solution molecules are evaporating faster than they are condensing, while in the concentrated solution the reverse occurs. Consequently, there is a gradual net transfer of solvent from the dilute solution into the more concentrated one until they both achieve the same concentration.

If we perform an osmosis experiment using the apparatus in Figure 9.23, in which we have a solution in compartment $A$ and pure water in $B$, the passage of solvent from $B$ to $A$ will slowly increase the volume of $A$ and decrease the volume of $B$. As this occurs, the height of the liquid in the capillary of compartment $A$ will rise while the height of the liquid in the other capillary will drop, and there will be a pressure difference between the two

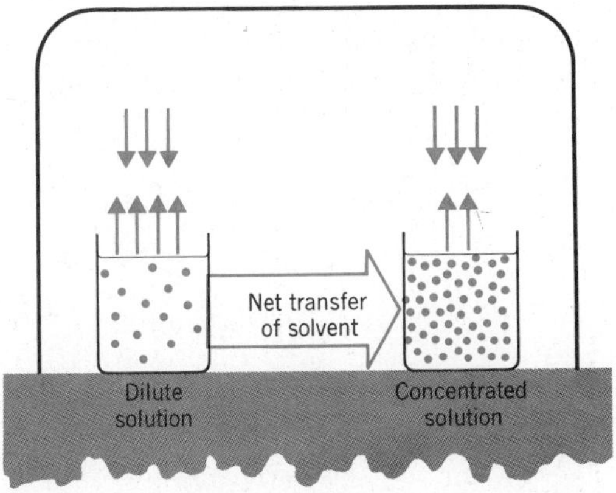


Figure 9.22
*Unequal vapor pressures lead to a net transfer of solvent.*

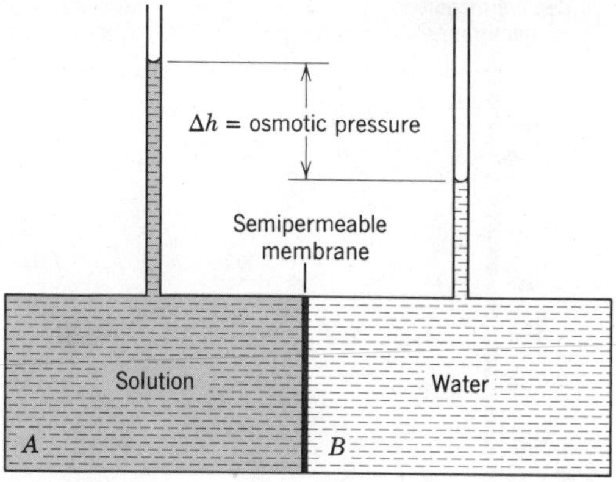


Figure 9.23
*Measurement of osmotic pressure.*

solutions that depends on the difference in these heights, $\Delta h$. Now, the ease with which a solvent molecule can be transferred from $B$ to $A$, and from $A$ to $B$, also depends on this pressure difference. As the pressure on side $A$ increases, it becomes increasingly more difficult to squeeze another solvent molecule into this solution. It also becomes easier to "pop out" a solvent molecule and push it into the pure water in $B$. Hence as $\Delta h$ increases, the rate of transfer of solvent from $B$ to $A$ decreases while that from $A$ to $B$ gets larger, until finally a value of $\Delta h$ is achieved at which both rates are the same and equilibrium is established. The pressure difference between the two compartments at equilibrium is called the **osmotic pressure** of the *solution* and is symbolized by the Greek letter, $\Pi$.

The magnitude of the osmotic pressure developed in a solution is directly proportional to the *molar* concentration of the solute.

$$\Pi \propto M$$

or

$$\Pi = kM \tag{9.8}$$

The proportionality constant, $k$, is a function of temperature and, for very dilute solutions, can be quite closely approximated by $RT$, where $R$ is the universal gas constant and $T$ is the absolute temperature. Thus Equation 9.8 becomes

$$\Pi = MRT \tag{9.9}$$

Another form of this equation can be obtained by recognizing that molarity is a ratio of moles to volume; that is, $M = n/V$. Equation 9.9 can therefore be written

$$\Pi = \frac{n}{V}\mathrm{RT}$$

or

$$\Pi V = nRT \tag{9.10}$$

It is interesting to note the similarity between Equation 9.10 (called the van't Hoff equation) and the ideal gas law,

$$PV = nRT$$

The magnitude of the osmotic pressure, even in very dilute solutions, is quite large. For instance, with a concentration of 0.010 mole of solute particles per liter (0.010 $M$) at room temperature (298 K), the osmotic pressure would be

$$\Pi = MRT$$

$$\Pi = \left(\frac{0.010\ \cancel{\text{mole}}}{1\ \cancel{\text{liter}}}\right) \times \left(\frac{0.0821\ \cancel{\text{liter}}\ \text{atm}}{\cancel{\text{mole}}\ \cancel{\text{K}}}\right) 298\ \cancel{\text{K}}$$

$$\Pi = 0.24\ \text{atm}$$

This pressure is sufficient to support a column of water 8.1 ft high!

Because the osmotic pressure that can be developed between solutions of only slightly different concentrations is so great, it is very important that fluids added to the body intravenously not alter significantly the osmotic pressure of the blood. If the blood fluids become too dilute, the osmotic pressure that develops within the blood cells can cause them to rupture. On the other hand, if the fluids are too concentrated, water will diffuse out of the cells and they will no longer function properly. For this reason care is taken to use solutions with the same osmotic pressure as the solution within the cells. Solutions that have the same osmotic pressure are called **isotonic** solutions.

The large differences in pressure developed between solutions of very similar concentrations provide us with a method of measuring the very large molecular weights of polymers (both of synthetic and biological origin). Freezing-point lowering and boiling-point elevation just won't work in these cases. For instance, a solution containing even as much as 150 g of a solute whose molecular weight is 30,000 in 1000 g of water produces a solution with a concentration of only 0.005 $m$. The freezing-point depression of this solution is approximately 0.009 degrees. Moreover, in most cases it is not even possible to dissolve this much solute, so that in practice the freezing-point changes are virtually undetectable.

A solution containing only 15 g of this solute per 1000 g of water at 25°C ($5.0 \times 10^{-4}\ m$), however, would nevertheless have a measurable osmotic pressure.[9]

$$\Pi = \left(\frac{5.0 \times 10^{-4}\ \text{mole}}{\text{liter}}\right) \times \left(\frac{0.0821\ \text{liter atm}}{\text{mole K}}\right) \times 298\ \text{K}$$

$$\Pi = 1.2 \times 10^{-2}\ \text{atm}$$

This is 9.1 mm Hg or 9.1 torr. If the solution is assumed to have a density of 1.0 g/ml, this pressure will support a column of the liquid 12.3 cm high (about 4.8 in.). A height such as 12.3 cm is very easily measured with accuracy which makes osmotic pressure measurements a useful tool for the determination of molecular weights of large molecules.

The osmosis process can be stopped and even reversed by the application of pressure equal to or greater than the osmotic pressure of the solution. This reversal of osmosis is used to desalinate sea water. An apparatus for this process is shown in Figure 9.24. If pressure were not applied to this system, osmosis would occur from left to right; that is, molecules would be transferred from fresh water to salt water (Figure 9.24$a$). However, with the

[9] In this calculation we assume that $5.0 \times 10^{-4}\ m \approx 5.0 \times 10^{-4}\ M$. In dilute aqueous solutions this introduces only a very small error.

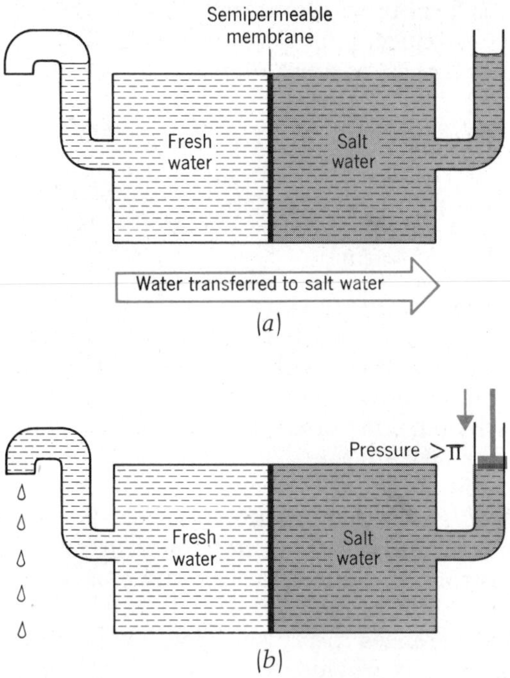


Figure 9.24
*Desalination by reverse osmosis. (a) Osmosis with no pressure applied to saline solution. (b) Reverse osmosis when pressure >Π is applied to saline solution.*

application of pressure exceeding Π, we drive the osmosis in the reverse direction (Figure 9.24*b*), forcing water molecules out of the saline solution, thereby leaving the impurities behind. One type of membrane that is strong enough to withstand these pressures is a film of cellulose acetate placed over a suitable support. Cellulose acetate is permeable to water but impermeable to the ions and impurities in the saline solution. Desalination plants have been constructed thus far that can produce as much as 45,000 gallons of fresh water daily.

## 9.12 INTERIONIC ATTRACTIONS

In Section 9.10 it was mentioned that the freezing-point depression produced by a solution of an electrolyte such as NaCl is not quite twice that expected for a nonelectrolyte. The ions in solution apparently are not totally independent particles. Very careful measurements made by the Dutch scientist, van't Hoff, on the colligative properties of solutions of electrolytes revealed that as these solutions become more and more concentrated, the ions in a solution become less independent. As a result, their effectiveness at altering the properties of the solution (boiling point, freezing point, osmotic pressure) diminishes as the concentration of the solute gets larger. Ionic compounds, then, behave as if they are less fully dissociated in concentrated solutions than when they are dilute.

Quantitatively, the degree to which an electrolyte behaves as if it were dissociated can be expressed by the van't Hoff factor, *i*. This quantity may be defined as the ratio of the observed freezing-point depression produced by a solution to the freezing point that the solution would exhibit if the solute were a nonelectrolyte.

$$i = \frac{(\Delta T_f) \text{ measured}}{(\Delta T_f) \text{ calculated as nonelectrolyte}}$$

Table 9.6
*Values of the van't Hoff factor at various concentrations*

| Salt | Concentration (moles/kg of $H_2O$) | | | *i* Factor if Completely Dissociated |
|---|---|---|---|---|
| | 0.1 | 0.01 | 0.001 | |
| NaCl | 1.87 | 1.94 | 1.97 | 2.00 |
| KCl | 1.85 | 1.94 | 1.98 | 2.00 |
| $K_2SO_4$ | 2.32 | 2.70 | 2.84 | 3.00 |
| $MgSO_4$ | 1.21 | 1.53 | 1.82 | 2.00 |

For NaCl, the *i* factor approaches a value of 2.00 at very high dilutions (essentially infinite dilution). As Table 9.6 reveals, however, the *i* factor diminishes at higher concentrations. For nonelectrolytes, of course, the *i* factor has a value of 1.00.

## INDEX TO QUESTIONS AND PROBLEMS (problem numbers in italics)



## REVIEW QUESTIONS

**9.1** Describe substitutional and interstitial solid solutions.

**9.2** State the definitions of the following concentration units: mole fraction, mole percent, weight fraction, weight percent, molarity, molality.

**9.3** As applied to solubility, what is the significance, on a molecular level, of the phrase "like dissolves like"?

**9.4** Frequently a liquid that is soluble in water can be made less soluble (and can therefore be made to separate as a distinct phase) by addition of salt to the solution. How can this be explained?

**9.5** Small amounts of water in the fuel tank of an automobile can cause severe difficulties in engine performance. The problem can be overcome by adding "dry gas" to the fuel. The "dry gas" consists mostly of methyl alcohol, $CH_3OH$, which allows the water to dissolve in the gasoline. Can you explain how the dry gas accomplishes this?

**9.6** What is meant when an ion is said to be hydrated?

**9.7** Compare the definitions of an ideal gas and an ideal solution.

**9.8** Discuss the relationship between lattice energy and hydration energy in determining the magnitude and sign of the heat of solution.

**9.9** On the basis of size and charge, choose the ion in each of the following pairs with the largest hydration energy:
(a) $Na^+$ or $K^+$ (d) $Fe^{2+}$ or $Fe^{3+}$
(b) $F^-$ or $Cl^-$ (e) $S^{2-}$ or $Cl^-$
(c) $K^+$ or $Ca^{2+}$

**9.10** The solution process for KI in water is endothermic ($\Delta H_{soln}$ is positive). Would you expect KI to become more or less soluble as the temperature is increased? Explain.

**9.11** Why is $\Delta H_{soln}$ for gases nearly always negative?

**9.12** Describe, qualitatively, the procedure called fractional crystallization.

**9.13** On the basis of the information provided in Figure 9.11, predict which solid will separate first from a solution containing equal weights of $KNO_3$ and KBr when the solution is gradually evaporated at a temperature of 70°C. What will occur if the solution is gradually evaporated at 20°C?

**9.14** Why do gases become more soluble in liquids as the pressure is increased?

**9.15** Why do pressure changes have very little effect on the solubility of a solid in a liquid?

**9.16** Explain, on a molecular level, why the vapor pressure of the solvent is expected to be directly proportional to its mole fraction in the solution (Raoult's law).

**9.17** What are meant by positive and negative deviations from Raoult's law?

**9.18** How is the sign of $\Delta H_{soln}$ related to positive and negative deviations from Raoult's law?

**9.19** Describe, qualitatively, the procedure called fractional distillation.

**9.20** Referring to Figure 9.18, approximately how many times must boiling, followed by condensation of the resulting vapor, be repeated in order to obtain a portion of liquid having a mole fraction of $A$ of at least 0.80 if the original mole fraction of $A$ was 0.20?

**9.21** Benzene has a boiling point of 80.1°C; carbon tetrachloride boils at 76.8°C. Sketch a boiling-point diagram for benzene–carbon tetrachloride mixtures. Assume ideal solution behavior.

**9.22** Water and butyl alcohol form an azeotrope that boils at 92.4°C at 760 torr. At this same pressure butyl alcohol boils at 117.8°C. The composition of the azeotrope is 28.4 mol % butyl alcohol, 71.6 mol % water. Sketch the boiling-point diagram for butyl alcohol–water mixtures. Do these substances show positive or negative deviations from ideality?

**9.23** We found that the addition of a nonvolatile solute to a solvent reduces the escaping tendency of the solvent from the solution, and in Section 9.10 we saw that this leads to a boiling-point elevation. On a molecular level, account for the fact that the presence of a solute also reduces the tendency of the solvent to escape from the liquid into the solid. Explain why a lower temperature must be achieved to establish equilibrium between the solid solvent and the solution than between the pure solid and liquid solvent.

**9.24** Based on the data in Table 9.6, which 1:1 electrolyte (one positive ion to one negative ion) appears to be *least* fully dissociated in concentrated solutions? How does this agree (or disagree) with what might be predicted based on the charges on the ions involved?

**9.25** On the basis of what you have learned in this chapter, how would you interpret an $i$ factor having a value less than 1.00?

## REVIEW PROBLEMS

**9.26** Calculate the mole fraction, weight fraction, weight percent, and molality of glycerin in a solution prepared by dissolving 45.0 g of glycerin, $C_3H_5(OH)_3$, in 100.0 g of $H_2O$.

**9.27** A mixture is prepared from 45.0 g of benzene ($C_6H_6$) and 80.0 g of toluene ($C_7H_8$). Calculate (a) the weight percent of each component, (b) the mole fraction of each component, (c) the molality of the solution if toluene is taken to be the solvent.

**9.28** A solution containing 121.8 g of $Zn(NO_3)_2$ per liter has a density of 1.107 g/ml. Calculate (a) the weight percent of $Zn(NO_3)_2$

in the solution, (b) the molality of the solution, (c) the mole fraction of $Zn(NO_3)_2$, (d) the molarity of the solution.

**9.29** What are the mole fraction, molality, and weight precent of a solution prepared by dissolving 0.30 mole of $CuCl_2$ in 40.0 moles of $H_2O$?

**9.30** An antifreeze solution is prepared from 222.6 g of ethylene glycol, $C_2H_4(OH)_2$, and 200 g of water. Its density is 1.072 g/ml. Calculate the molality and molarity of the solution.

**9.31** A 4.03 *M* solution of ethylene glycol, $C_2H_4(OH)_2$, has a density of 1.045 g/ml. Calculate the weight percent $C_2H_4(OH)_2$, mole fraction of $C_2H_4(OH)_2$ and the molality of the solution.

**9.32** A solution of iso-propyl alcohol (rubbing alcohol), $C_3H_7OH$, in water has a mole fraction of alcohol equal to 0.250. What is the weight percent alcohol and the molality of the alcohol in the solution?

**9.33** The solubility of baking soda, $NaHCO_3$, in water at 20°C is 9.6 g/100 g of $H_2O$. What is the mole fraction of $NaHCO_3$ in a saturated solution? What is the molality of the solution?

**9.34** A saturated solution of NaCl at 30°C has a molality of 6.25 *m*. What is the mole fraction and weight fraction of NaCl in the solution?

**9.35** A solution of sodium carbonate was prepared containing 14.0% $Na_2CO_3$ by weight. What is the mole fraction and molality of $Na_2CO_3$ in this solution?

**9.36** Use the data in Table 9.2 to calculate the amount of heat liberated by dissolving 10.0 g of $AlCl_3$ in 1.00 liter water.

**9.37** What is the maximum amount of pure $KNO_3$ that can be obtained by fractional crystallization of a mixture containing 65 g of $KNO_3$ and 25 g of KBr?

**9.38** How many grams of $NaNO_3$ will precipitate if a saturated solution of $NaNO_3$ in 200 g of $H_2O$ at 70°C is cooled to 25°C?

**9.39** How many grams of water at 80°C are required to dissolve 35.0 g of NaBr?

**9.40** The partial pressure of ethane over a saturated solution containing $6.56 \times 10^{-2}$ g of ethane is 751 torr. What is its partial pressure when the saturated solution contains $5.00 \times 10^{-2}$ g of ethane?

**9.41** The Henry's law constant for a gas dissolved in water was found to be $6.50 \times 10^{-5}$ g/liter torr at 25°C. In an experiment the gas was collected over water and its concentration was found to be 0.0478 g/liter. What was the *total* pressure of gas above the solution?

**9.42** The vapor pressure of benzene ($C_6H_6$) at 25°C is 93.4 torr. What will be the vapor pressure, at 25°C, of a solution prepared by dissolving 56.4 g of the nonvolatile solute, $C_{20}H_{42}$, in 1000 g of benzene?

**9.43** The vapor pressure of pure methyl alcohol at 30°C is 160 torr. What mole fraction of glycerol (a nonvolatile nondissociating solute) would be required to lower the vapor pressure to 130 torr?

**9.44** Heptane ($C_7H_{16}$) has a vapor pressure of 791 torr at 100°C. At this same temperature, octane ($C_8H_{18}$) has a vapor pressure of 352 torr. What will be the vapor pressure of a mixture of 25.0 g of heptane and 35.0 g of octane? Assume ideal solution behavior.

**9.45** What will be the freezing point and boiling point of an aqueous solution containing 55.0 g of glycerol, $C_3H_5(OH)_3$, dissolved in 250 g of water? Glycerol is a nonvolatile, undissociated solute.

**9.46** What is the molecular weight and molecular formula of a nondissociating compound whose empirical formula is $C_4H_2N$ if 3.84 g of the compound in 500 g of benzene gives a freezing-point depression of 0.307°C?

**9.47** A solution containing 16.9 g of a nondissociating substance in 250 g of water has a freezing point of −0.744°C. The substance is composed of 57.2% C, 4.77% H, and 38.1% O. What is the molecular formula of the compound?

**9.48** How many grams of glucose, $C_6H_{12}O_6$ (a nondissociating solute), are required to lower the temperature of 150 g of $H_2O$ by 0.750°C? What will be the boiling point of this solution?

**9.49** An aqueous solution freezes at −2.47°C. What is its boiling point?

**9.50** Calculate the freezing point of a 0.100 *m* aqueous solution of a weak electrolyte HX that is 7.5% dissociated.

**9.51** Calculate the osmotic pressure, in torr, of an aqueous solution containing 5.0 g of sucrose, $C_{12}H_{22}O_{11}$, per liter at 25°C.

**9.52** A solution of 0.40 g of a polypeptide in 1.00 liter of an aqueous solution has an osmotic pressure at 27°C of 3.74 torr. What is the approximate molecular weight of this polymer?

**9.53** What would be the osmotic pressure of a 0.010 *M* aqueous solution of the electrolyte, NaCl, at 25°C? (Assume 100% dissociation of NaCl in water).

**9.54** Calculate the *i* factor for the weak electrolyte, HF, in Example 9.7. What conclusions would you draw about the *i* factors for weak electrolytes?

**9.55** Below is a list of the most abundant ions in sea water.

| Ion | Molality |
|---|---|
| Chloride | 0.566 |
| Sodium | 0.486 |
| Magnesium | 0.055 |
| Sulfate | 0.029 |
| Calcium | 0.011 |
| Potassium | 0.011 |
| Bicarbonate | 0.002 |

Calculate the weight, in grams, of each component contained in 3.78 liters (1.00 gallon) of sea water having a density of 1.024 g/ml. What is the total weight of ions in this sample?

***9.56** Suppose that you wish to prepare a solution containing 10% $Na_2CO_3$ by weight. The bottle of chemical that you have lists the contents as $Na_2CO_3{\cdot}10\,H_2O$. How many grams of the hydrate would be needed to prepare 50.0 g of the 10% $Na_2CO_3$ solution?

***9.57** Air contains approximately 20% $O_2$ by volume. The Henry's law constant for $O_2$ at 25°C is $5.34 \times 10^{-5}$ g/liter torr. Calculate the weight of $O_2$ per liter of water in a stream that has a temperature of 25°C if the atmospheric pressure is 760 torr. (Assume equilibrium with the atmosphere.)

***9.58** At 25°C the vapor pressures of benzene ($C_6H_6$) and toluene ($C_7H_8$) are 93.4 and 26.9 torr, respectively. At what applied pressure will a solution prepared from 60 g of benzene and 40 g of toluene boil at 25°C?

***9.59** The vapor pressure of a mixture containing 400 g of carbon tetrachloride and 43.3 g of an unknown substance is 137 torr at 30°C. The vapor pressure of pure carbon tetrachloride at 30°C is 143 torr, while that of the pure unknown is 85 torr. What is the approximate molecular weight of the unknown?

***9.60** A solution containing 8.3 g of a nonvolatile nondissociating substance dissolved in 1 mole of chloroform, $CHCl_3$, has a vapor pressure of 511 torr. The vapor pressure of pure $CHCl_3$ at the same temperature is 526 torr. Calculate (a) the mole fraction of the solute, (b) the number of moles of solute, (c) the molecular weight of the solute.

***9.61** The cooling system of an automobile usually contains a solution of antifreeze prepared by mixing equal volumes of ethylene glycol, $C_2H_4(OH)_2$, and water. The density of ethylene glycol is 1.113 g/ml. Calculate the freezing point of this mixture. On the label of the antifreeze container it is said that this mixture will protect your engine to a temperature of −34°F. How does your computed freezing point compare with this?

***9.62** What is the percent dissociation of a weak electrolyte HX in water if a 0.250 *m* solution of it has a freezing point of −0.500°C?

# 10 CHEMICAL THERMODYNAMICS

In the study of chemistry it's natural to question why certain chemical reactions take place and why others do not. Certainly, it would be nice if we could predict what will occur when several chemicals are mixed. This chapter and the next discuss the two factors that ultimately determine whether one is able to observe a particular chemical reaction, either in the laboratory or elsewhere. A study of thermodynamics reveals whether or not a given process can occur spontaneously (that is, without outside help) and what will be the position of equilibrium after reaction has ceased. Chemical kinetics, the subject of Chapter 11, is concerned with the speeds at which chemical changes take place. Both of these factors, spontaneity and speed, must be in our favor if we hope to observe the formation of products of a chemical change. For example, thermodynamics predicts that at room temperature hydrogen gas and oxygen gas should react to produce water. However, a mixture of $H_2$ and $O_2$ is stable virtually indefinitely (provided that no one strikes a match). This is so because, at room temperature, hydrogen and oxygen react at such an extremely slow rate that even though their reaction to produce water is spontaneous, it takes nearly forever for the reaction to proceed to completion.

Thermodynamics is basically concerned with the energy changes that accompany chemical and physical processes. Historically, it evolved without a detailed knowledge of the structure of matter; in fact, this is one of its strongest points. In this chapter we will take a rather informal approach to the subject in an effort to avoid mathematical formalism, and we will develop many of the concepts of thermodynamics by considering changes that take place on a molecular level.

## 10.1 SOME COMMONLY USED TERMS

Before we proceed, let us establish the meaning of some frequently used terms. A word that has been used rather loosely in previous sections is **system.** By system we mean *that particular portion of the universe on which we wish to focus our attention.* Everything else we call the **surroundings.** For example, if we wished to consider the changes taking place in a solution of sodium chloride and silver nitrate, our system is the solution, while the beaker and everything else around the solution is considered the surroundings.

If a change occurs so that heat cannot be transferred across the interface, or boundary, between the system and its surroundings, we speak of it as an **adiabatic** process. An example is a reaction carried out in an insulated container, such as a Thermos bottle. Explosive reactions are also examples of adiabatic processes. Such reactions occur so rapidly that the heat energy produced cannot be readily dissipated. The heat build-up that occurs raises the products to very high temperatures, and these products fly apart rapidly, pushing walls, ceilings, and so on (the surroundings) before them.

When thermal contact is maintained between system and surroundings, heat can flow between them and it is frequently possible to keep the system at a constant temperature while a change takes place. In this case the process is said to be **isothermal.** The human body possesses an elaborate temperature-control system that maintains a constant body temperature. Biochemical reactions within us are therefore essentially isothermal.

To discuss the changes that occur in a system it is necessary to define its properties very precisely before and after the change occurs. We do this by specifying the **state** of the system, that is, some particular set of conditions of pressure, temperature, number of moles of each component, and their physical form (for example, gas, liquid, solid, or crystalline form). When these variables are specified, all of the properties of the system are fixed. Thus a knowledge of these quantities permits us to define unambiguously the properties of our system. For instance, if we have two samples of pure liquid water, each consisting of 1 mole and each at the same temperature and pressure, we know that all of the properties of each sample will be identical (volume, density, surface tension, vapor pressure, etc.).

The quantities $P$, $T$ (and $V$) are called **state functions.** This is because (1) they serve to determine the state of any given system, and (2) in a particular state their values do not depend on the prior history of the sample. Furthermore, upon going from one state to another the changes in these quantities do not depend on how the sample is treated. For example, the volume of 1 mole of water at 25°C and 1 atm does not depend on what its temperature or pressure might have been at some time in the past. For the same reason, if the temperature of this sample is changed to 35°C, it does not matter if the sample were first cooled to 0°C and then warmed to 35°C, or whether the temperature were increased directly from 25 to 35°C. In the final state the temperature is the same regardless of the path taken between the initial and final conditions, and the change in temperature, $\Delta T$, is therefore dependent *only* on the temperatures of those initial and final states.

There are some instances where the interrelationships between the state functions can be expressed in equation form, to give an **equation of state.** The equation of state for an ideal gas, $PV = nRT$, is an example. We have also seen the van der Waals equation of state, which can be applied with reasonable success to real gases.

Another quantity that we shall use is called the **heat capacity**—*the amount of heat energy required to raise the temperature of a given quantity of a substance one degree Celsius.* The **specific heat** represents the heat capacity per gram; that is, it is *the amount of heat necessary to raise the temperature of 1 g of a substance by 1.0°C.* The specific heat of water is 1.00 cal/g °C. We also speak of the **molar heat capacity:** *the heat necessary to raise the temperature of 1 mole of a substance 1 degree.*

## 10.2 THE FIRST LAW OF THERMODYNAMICS

In thermodynamics we study the energy changes that occur when systems pass from one state to another. Repeated observations by many scientists over many years have led to the conclusion that, in any process, energy is neither created nor destroyed. Another way of saying this is that energy is conserved. The **first law of thermodynamics** merely puts this meaning into the form of a simple equation,

$$\Delta E = q - w \tag{10.1}$$

Here, $E$ represents the **internal energy** of the system—the total of all of the energies possessed by the system as a consequence of the kinetic energy of its atoms, ions, or molecules, plus the potential energy that arises from the binding forces between the particles that make up the system. $\Delta E$ is the difference between the energy contained in a system in some final state and the energy it possessed in an initial state. It corresponds to the change in the internal energy of a system that occurs when the system goes from an initial to a final state.

$$\Delta E = E_{\text{final}} - E_{\text{initial}}$$

Note that we have used the same convention here as in our previous discussions of energy changes (heats of vaporization, heats of solution). Here, too, we cannot actually determine $E$, but instead only $\Delta E$.

The quantity $q$ in Equation 10.1 represents the amount of heat that is *added* to the system as it passes from the initial to the final state, and $w$ denotes the work done *by* the system on its surroundings. Thus Equation 10.1 simply states that the change in the internal energy is equal to the difference between the energy *supplied to* the system as heat and the energy *removed from* the system as work performed on the surroundings.[1]

Since the first law deals with the transfer of quantities of energy, it is necessary to establish sign conventions to avoid confusion in our bookkeeping. Heat *added to* a system and work *done by* a system are considered positive quantities. Thus, if a certain change is accompanied by the absorption of 50 cal of heat and the expenditure of 30 cal of work, $q = +50$ cal and $w = +30$ cal. The change in internal energy of the system is

$$\Delta E_{\text{system}} = (+50\text{ cal}) - (+30\text{ cal})$$

or

$$\Delta E_{\text{system}} = +20\text{ cal}$$

Thus the system has undergone a net increase in energy amounting to +20 cal. How about the surroundings?

When the system gains 50 cal, the surroundings lose 50 cal; therefore $q = -50$ cal for the surroundings. When the system performs work, it does so on the surroundings. We say that the surroundings have done negative work, and $w = -30$ cal for the surroundings. The change in the internal energy of the surroundings is thus

$$\Delta E_{\text{surroundings}} = (-50\text{ cal}) - (-30\text{ cal})$$
$$\Delta E_{\text{surroundings}} = -20\text{ cal}$$

[1] Energy, you recall, is the capacity to do work. When the system performs work, its capacity to do additional work diminishes, which means that its energy has diminished. Energy, equal to the work performed, has been lost by the system and, in the process, gained by the surroundings.

The change in internal energy of the system is thus equal, but opposite in sign, to $\Delta E$ for the surroundings, so that the net change for the universe (system plus surroundings) is zero. This is what we mean by the law of conservation of energy,

In summary,

$q$ positive ($q > 0$); heat is added to the system
$q$ negative ($q < 0$); heat is evolved by (removed from) the system
$w$ positive ($w > 0$); the system performs work—energy is removed
$w$ negative ($w < 0$); work is done on the system—energy is added

The internal energy happens to be a state function, and the magnitude of $\Delta E$ therefore depends only on the initial and final states of the system and not on the path taken between them. This is very much the same as the change in your bank balance that occurs between the beginning and the end of a month. During any given month the change in the balance is brought about as the combined results of some number of deposits and withdrawals. If the total number of dollars provided by the deposits exceeds those removed by the withdrawals, your balance increases. However, the net change in your balance at the end of the month depends only on the initial and final amounts of money in the bank, not on the individual transactions during the month. There is an infinite number of combinations of deposits and withdrawals that could lead to the same change in your balance. The same sort of relationship exists among $\Delta E$, $q$, and $w$. The sign and magnitude of $\Delta E$ is controlled only by the values of $E$ in the initial and final state. For any given change, $\Delta E$, there are many different paths that can be followed with their own characteristic values of $q$ and $w$. However, for the same initial and final states, the difference between $q$ and $w$ is always the same. Let us consider, now, some concrete examples to illustrate the meaning of "path" and to show how $q$ and $w$ can differ for various paths even though $\Delta E$ remains the same.

We shall first look at the change that takes place when 1.0 liter of a gas at an initial pressure of 10 atm is permitted to push back the piston (assumed to be frictionless) in the cylinder shown in Figure 10.1, isothermally, against a uniform opposing pressure of 1.00 atm. Clearly, this is a spontaneous process, and the expansion continues until the internal pressure of the gas is the same as the external pressure exerted on the piston, 1.00 atm. From Boyle's law, we find that the final volume is 10 liters.

According to our definition of an ideal gas, there are no attractive forces present between the gas particles. Consequently, as these particles move apart during the expansion, there is no change in their potential energy. Since the temperature is held constant during the expansion, the average kinetic energy of the gas also stays the same. Thus, during the expansion there is no change in either kinetic or potential energy; therefore, for an isothermal expansion (or compression) of an *ideal gas*, $\Delta E = 0$. This means that

$$q - w = 0$$

or

$$q = w$$

In other words, when the expansion takes place, any heat absorbed by the system is returned to the surroundings by way of the system performing work on the surroundings.

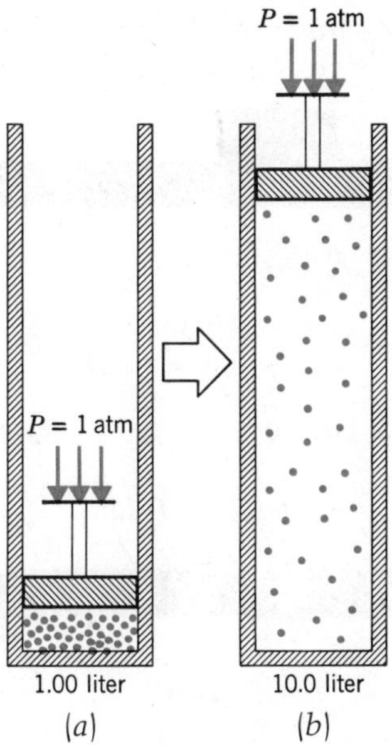


Figure 10.1
*Expansion of an ideal gas against a constant opposing pressure of one atmosphere. (a) Initial state: $P_{gas} = 10.0$ atm, $V_{gas} = 1.00$ liter. (b) Final state: $P_{gas} = 1.00$ atm, $V_{gas} = 10.0$ liter.*

Before we can proceed further we must ask ourselves, how does the system do work? We know that work is accomplished by moving an opposing force through some distance:

$$\text{work} = \text{force} \times \text{distance}$$

Pressure is defined as force per area. In Figure 10.2 the external pressure on the piston corresponds to a certain total force, $F$, spread over the area of the piston, $A$.

$$P = \frac{F}{A}$$

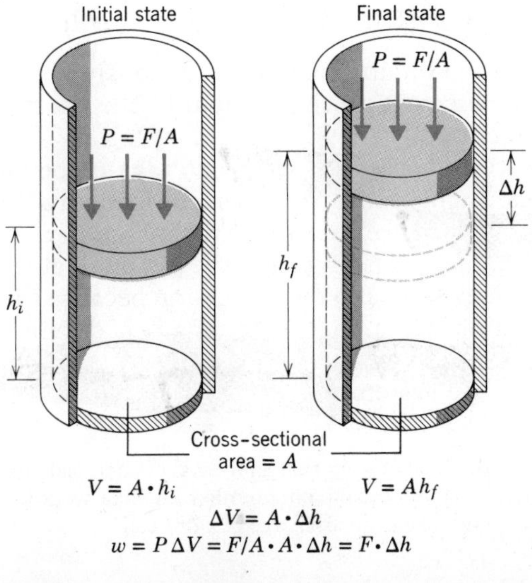


Figure 10.2
*Pressure-volume work.*

The volume of the gas in the cylinder is equal to its cross-sectional area, $A$, multiplied by the height of the column of gas, $h$.

$$V = Ah$$

When the gas expands, $A$ remains the same but $h$ changes. The volume change is therefore

$$\Delta V = V_f - V_i$$
$$\Delta V = Ah_f - Ah_i$$
$$\Delta V = A(h_f - h_i) = A(\Delta h)$$

The product of pressure times volume change, then, is

$$P\ \Delta V = \frac{F}{A}\ A(\Delta h) = F\ \Delta h$$

We see that $P\ \Delta V$ is equivalent to force ($F$) times distance ($\Delta h$) and, therefore, equals work.

$$w = P\ \Delta V \qquad (10.2)$$

This expansion work[2] is performed by a system (any system—it doesn't have to be an ideal gas) when it expands against an external pressure imposed by the surroundings. Conversely, when the system contracts under the influence of an external pressure, work is performed on the system. If the volume change is measured in liters and the pressure in atmospheres, $P\ \Delta V$ has the units, liter atmosphere (liter atm). We could, if we wish, convert this to the more familiar energy units, calories or joules. These relationships are

$$1 \text{ liter atm} = 24.2 \text{ cal}$$
$$1 \text{ liter atm} = 101.3 \text{ J}$$

In our example, the ideal gas expands from its initial volume of 1.0 liter to a final volume of 10 liters, hence $\Delta V = 9$ liters. The external pressure is constant at 1.00 atm; therefore, the gas does work on the surroundings amounting to

$$w = P\ \Delta V$$
$$w = (1.00 \text{ atm}) \times (9 \text{ liters}) = 9 \text{ liter atm}$$

Since $\Delta E = 0$, the system must simultaneously absorb heat from the surroundings in an amount precisely equal to the energy it expends by doing work. Consequently, $q = +9$ liter atm.

Let's now consider a second path to take us between the same initial and final states. Suppose that the apparatus in Figure 10.1 is modified so that a perfect vacuum ($P = 0$) exists above the piston in the cylinder. Once again the gas will expand by pushing back the piston (Figure 10.3). However, this time there is no resistance to the expansion because there is a zero opposing pressure (assuming, again, that the piston is frictionless). Since no pressure opposes the expansion, $P = 0$ and the $P\ \Delta V$ product is equal to zero. No work is performed by the system. Since $\Delta E = 0$ (constant $T$),

[2] $P\ \Delta V$ work is only one kind of work that a physical or chemical change can produce. It is also possible (depending on conditions) to obtain other kinds of work from a changing system, for example, electrical work from the discharge of a dry cell.

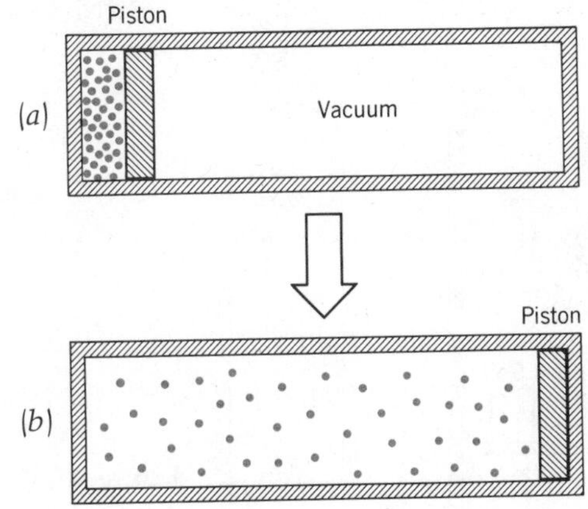


Figure 10.3
*Expansion of a gas against an opposing pressure equal to zero. (a) Before expansion. P = 10 atm; V = 1 liter. (b) After expansion, final state. P = 1 atm; V = 10 liter.*

$$q = w = 0$$

Thus no heat is exchanged with the surroundings either.

We have now considered two processes whereby an ideal gas is taken from an initial state of $P = 10$ atm, $V = 1$ liter to a final state of $P = 1$ atm and $V = 10$ liters. In both cases $\Delta E$ was the same, namely zero. However, $q$ and $w$ were not identical along the two paths; following the first, $q = +9$ liter atm and $w = +9$ liter atm, while along the second path, $q = 0$ and $w = 0$. Thus, although $E$ is a state function, the magnitudes of $q$ and $w$ depend on how the process is carried out.

## 10.3 REVERSIBLE AND IRREVERSIBLE PROCESSES

In the examples just presented we saw that the work obtained from the expansion of an ideal gas depends on the external pressure resisting the expansion. When $P = 0$, the work done was also zero, while, when $P = 1$ atm, work equal to 9 liter atm was performed as the gas expanded from 1 to 10 liters. What would be the *maximum* amount of work that we could obtain when a gas expands from an initial state of $P = 10$ atm and $V = 1$ liter to a final state in which $P = 1$ atm and $V = 10$ liters?

Let's suppose that we carried out the process in two steps: first with the opposing pressure equal to 5 atm and then, in the second step, with an opposing pressure equal to 1 atm (Figure 10.4). During the first step the gas expands until its pressure drops from 10 atm to that of the opposing pressure, 5 atm. At this point the gas would occupy a volume of 2 liters (why?) and $\Delta V$ would equal 1 liter. The work performed, let's call it $w_1$, is

$$w_1 = P\,\Delta V = 5 \text{ atm } (1 \text{ liter})$$
$$w_1 = 5 \text{ liter atm}$$

In the second step the gas expands from a volume of 2 liters to the final volume, 10 liters ($\Delta V = 8$ liters), against an opposing pressure of 1 atm. The work performed here is

$$w_2 = P\,\Delta V = 1 \text{ atm } (8 \text{ liters})$$
$$w_2 = 8 \text{ liter atm}$$

The total work performed by the gas is the sum of that performed in each step along the way.

$$w_{\text{total}} = w_1 + w_2$$
$$w_{\text{total}} = (5 + 8) \text{ liter atm} = 13 \text{ liter atm}$$

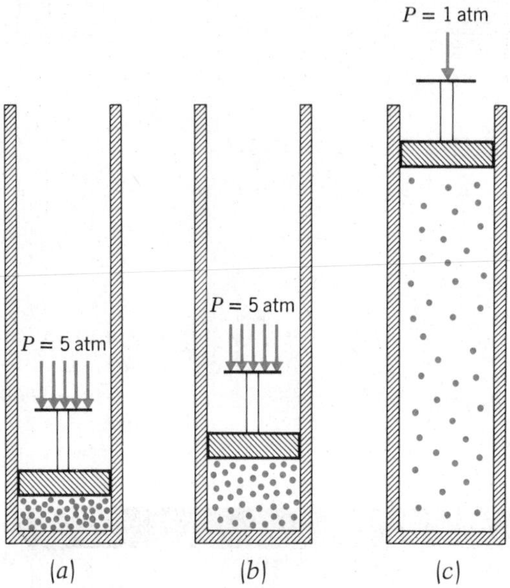


Figure 10.4
*Two-step expansion. (a) Initial state. External pressure equals 5 atm. $V = 1.00$ liter; $P = 10.0$ atm. (b) After step 1. $V = 2.00$ liter; $P = 5.00$ atm. (c) After step 2. $V = 10.0$ liter; $P_{gas} = 1.00$ atm.*

Note that we have obtained more work in this two-step expansion than when the pressure is kept constant at 1 atm throughout the entire change. We might conclude (correctly) that even more work could be obtained by carrying out the expansion in still more steps, in which the opposing pressure is kept as high as possible. We might also conclude that the greatest amount of work could be extracted if an infinite number of steps were employed in which the external pressure is always just barely below that exerted by the gas. This process could be approximated if we used a piston-cylinder apparatus such as that shown in Figure 10.5. As the water slowly evaporates, the external pressure gradually decreases and the gas gradually expands. In this case we have to use calculus to derive the expression for the total work. The result for 1 mole of a gas is

$$w_{\text{maximum}} = 2.30\ RT \log \frac{V_f}{V_i}$$

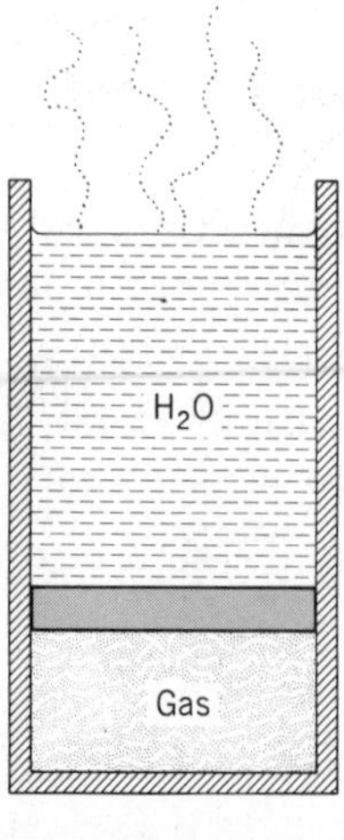


Figure 10.5
*Reversible expansion. As $H_2O$ gradually evaporates, pressure decreases and the gas expands. The process will be reversed if water molecules begin to condense into the liquid instead of evaporate.*

where $V_i$ and $V_f$ are the initial and final volumes and $R$ is the universal gas constant (0.0821 liter atm/mole K). In this example the maximum work is (at 0°C)

$$w_{max} = 2.30\ (0.0821\ \text{liter atm/mole}\ \cancel{K})(273\ \cancel{K})\ \log\left(\frac{10\ \text{liters}}{1\ \text{liter}}\right)$$

$$w_{max} = 51.5\ \text{liter atm/mole}$$

The number of moles of gas is (10 liters/22.4 liter mole$^{-1}$) = 0.446 mole. Therefore,

$$w_{max} = \left(51.5\ \frac{\text{liter atm}}{\cancel{\text{mole}}}\right)(0.446\ \cancel{\text{mole}}) = 23.0\ \text{liter atm}$$

The expansion of a gas in the manner just described, where the opposing pressure is virtually equal to the pressure exerted by the gas, is one example of a reversible process. It is reversible because any slight increase in the external pressure will reverse the process and cause compression to occur. Any change that is resisted by an opposing "force" essentially equal to the driving force of the process constitutes a reversible change.

The key point in this section is that *the maximum work derived from any change will be obtained only if the process is carried out in a reversible manner.* This requires, however, an infinite number of steps, and therefore a truly reversible process takes forever to occur. All real, spontaneous changes are therefore not reversible, and the work that can be derived from an irreversible change is always less than the theoretical maximum.

## 10.4 HEATS OF REACTION: THERMOCHEMISTRY

When chemical reactions occur, they do so with either the absorption or evolution of energy. These changes reflect the differences between the potential energies associated with the bonds in the reactants and the products. For instance, when two hydrogen atoms come together to form an $H_2$ molecule, energy is evolved because the total potential energy of the nuclei and electrons in the $H_2$ molecule is lower than the total potential energy of these particles in two isolated H atoms. This same energy, when added to an $H_2$ molecule, can break it apart; this energy thus represents the bond energy discussed in Chapter 4. We see, therefore, that the measurement of the amount of energy evolved or absorbed when a chemical reaction occurs has the potential of providing us with very fundamental information concerning the stability of molecules and the strengths of chemical bonds.

If we carry out a chemical reaction in a closed container of fixed volume, the system undergoing reaction cannot perform pressure-volume work on the surroundings because $\Delta V = 0$; hence $P\ \Delta V = 0$. Any heat absorbed or evolved under these circumstances (let's call it $q_v$) is precisely equal to the change in the internal energy of the system,

$$\Delta E = q_v \tag{10.3}$$

Stated another way, $\Delta E$ is equal to the heat absorbed or evolved by the system under conditions of constant volume (i.e., the *heat of reaction at constant volume*). When the reaction is endothermic, both $q$ and $\Delta E$ are positive. For an exothermic process $\Delta E$ is negative.

Experimentally, $\Delta E$ can be measured by using a device called a bomb calorimeter (Figure 10.6). The apparatus consists of a strong steel "bomb" into which the reactants (for example, $H_2$ and $O_2$) are placed. The bomb is then immersed in an insulated bath containing a precisely known quantity

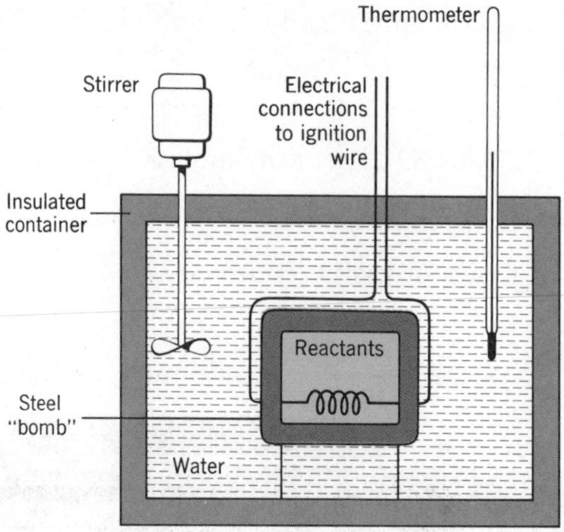


Figure 10.6
*Bomb calorimeter.*

of water. There the reaction is set off by a small heater wire within the bomb, and heat is evolved. The entire system is permitted to come to thermal equilibrium, at which point the calorimeter (bomb and water) will be at a higher temperature than before reaction. By carefully measuring the temperature of the water before and after reaction, and by knowing the heat capacity of the calorimeter (including the bomb and the water), the quantity of heat evolved by the chemical reaction can be computed.

EXAMPLE 10.1 Hydrogen (0.100 g) and oxygen (0.800 g) are compressed into a 1.00-liter bomb, which is then placed into the water in a calorimeter. Before the reaction is set off, the temperature of the water is 25.000°C; after reaction the temperature rises to 25.155°C. The heat capacity of the calorimeter (bomb, water, etc.) is 21,700 cal/°C. What is $\Delta E$, in kilocalories, for this reaction?

SOLUTION The change in temperature that occurs is 0.155°C. From the heat capacity we can find the number of calories evolved.

$$\text{Heat evolved, } q = \left(21{,}700\ \frac{\text{cal}}{°\text{C}}\right)(0.155°\text{C})$$

Therefore

$$q_v = -3360 \text{ cal}$$

Hence,

$$\Delta E = -3.36 \text{ kcal}$$

In Example 10.1, the magnitude of $\Delta E$ depends on the quantity of $H_2$ and $O_2$ reacted; that is, $\Delta E$ is an extensive quantity. We can convert this to an intensive property, one that is characteristic of the reaction between any amounts of $H_2$ and $O_2$, by calculating the heat evolved per mole of product formed. In Example 10.1 we produced 0.0500 mole of $H_2O$. Therefore, we say that $\Delta E = -3.36$ kcal/0.0500 mole of $H_2O$ or $\Delta E = -67.2$ kcal/mole.

In the past it was common practice always to express heats of reaction in calories or kilocalories. Since the acceptance of the SI system, joules and kilojoules are preferred. At the present time science is in the midst of a transition between these units, and therefore to provide familiarity with both we will work out some examples in joules, others in calories. Regardless of the units, the methods employed in solving the problems are the same. By now you should be able to convert readily between joules and calories.

The heat evolved at constant volume permits us to compute $\Delta E$. However, most changes that are of practical interest to us take place in open containers at essentially constant atmospheric pressure. Under these conditions rather sizable volume changes can occur. For example, when 2 moles of gaseous $H_2$ react with 1 mole of gaseous $O_2$ to produce 2 moles of liquid water, at a constant pressure of 1 atm, the volume changes from about 67 liters to 0.036 liter. Imagine that this change takes place in a cylinder with a piston exerting a constant pressure of 1 atm; as the reaction proceeds to completion, the surroundings perform work on the system, the magnitude of which is the $P\,\Delta V$ product of very nearly 67 liter atm.

In order to avoid the necessity of considering $PV$ work when heats of reaction are measured at constant pressure, we define a new thermodynamic function called the **heat content,** or **enthalpy** (from the German, *enthalen,* to contain). This is denoted by the symbol $H$, as

$$H = E + PV \tag{10.4}$$

For a change at constant pressure,

$$\Delta H = \Delta E + P\,\Delta V \tag{10.5}$$

If only $PV$ work is involved in the change, we know that

$$\Delta E = q - P\,\Delta V$$

Substituting this into Equation 10.5, we have

$$\Delta H = (q - P\,\Delta V) + P\,\Delta V$$
$$\Delta H = q_p \tag{10.6}$$

Thus we see that $\Delta H$ is the heat, $q_p$, absorbed or evolved at constant pressure.

The enthalpy, like the internal energy, is a state function and thus the magnitude of $\Delta H$ depends only on the heat contents of the initial and final states. Thus we can write

$$\Delta H = H_{\text{final}} - H_{\text{initial}}$$

Here we use the same symbolism as in our earlier discussions of $\Delta H_{\text{vap}}$, $\Delta H_{\text{fus}}$, and so on. Those quantities, in fact, correspond to enthalpy changes associated with vaporization, fusion, and so on.

In many instances the differences between $\Delta H$ and $\Delta E$ are small, particularly for chemical reactions. When a reaction occurs in which all of the reactants and products are liquids or solids, only very small changes in volume take place. As a result, $P\,\Delta V$ is very small and $\Delta H$ has very nearly the same magnitude as $\Delta E$. When chemical reactions occur in which gases are either consumed or produced, much larger volume changes occur, and the $P\,\Delta V$ product is also much greater. Even in these cases, however, $\Delta E$ is usually so large compared to the $P\,\Delta V$ term that $\Delta E$ and $\Delta H$ are still nearly the same.

EXAMPLE 10.2 When 2.0 moles of $H_2$ and 1.0 mole of $O_2$, at 100°C and 1 atm, react to produce 2.0 moles of gaseous water at 100°C and 1 atm, a total of 484.5 kJ are evolved. What are (a) $\Delta H$, and (b) $\Delta E$ for the production of a single mole of $H_2O\ (g)$?

SOLUTION (a) Since the reaction

$$2H_2\ (g) + O_2\ (g) \longrightarrow 2H_2O\ (g)$$

is occurring at constant pressure,

$$q = \Delta H = \frac{-484.5\ \text{kJ}}{2\ \text{moles}\ H_2O}$$

The minus sign, remember, signifies that the reaction is exothermic. For the production of 1 mole of water,

$$\Delta H = -242.3\ \text{kJ/mole}$$

(b) If we assume ideal behavior of the gaseous reactants and products, we have for the reactants (the initial state) at a given $P$ and $T$,

$$PV_i = n_iRT$$

where $n_i$ corresponds to the number of moles of gaseous reactants. In a similar fashion, for the final state.

$$PV_f = n_fRT$$

The pressure-volume work in the process is given by

$$PV_f - PV_i = P(V_f - V_i) = P\ \Delta V$$

This is equal to

$$P\ \Delta V = n_fRT - n_iRT$$
$$P\ \Delta V = (n_f - n_i)RT = (\Delta n)RT$$

The quantity $\Delta n$ = (number of moles of gaseous products) − (number of moles of gaseous reactants). In this example

$$\Delta n = 2.0\ \text{moles} - 3.0\ \text{moles} = -1.0\ \text{mole}$$

Therefore, using $R = 8.31$ J/mole K,

$$P\ \Delta V = (-1.0\ \text{mole})(8.31\ \text{J/mole K})(373\ \text{K})$$
$$P\ \Delta V = -3100\ \text{J} = -3.10\ \text{kJ}$$

For each *1* mole of $H_2O$ produced, $P\ \Delta V = -3.10$ kJ. Hence, solving Equation 10.5 for the change in internal energy,

$$\Delta E = \Delta H - P\ \Delta V$$
$$\Delta E = -242.3\ \text{kJ/mole} - (-3.10\ \text{kJ/mole})$$
$$\Delta E = -239.2\ \text{kJ/mole}$$

In this last example we see that the $PV$ work that is involved in a chemical reaction in which gases, which we assume to be ideal, are either consumed or produced, can be calculated by the simple expression,

$$\text{pressure-volume work} = (\Delta n)RT \qquad (10.7)$$

Remember that $\Delta n$ *is the change in the number of moles of gas on going from reactants to products.*

Equation 10.7 does not apply to reactions where only liquids or solids are involved. For such reactions the volume changes are extremely small and the $P\ \Delta V$ work is usually negligible compared to other energy changes that take place. For such reactions $\Delta E$ and $\Delta H$ are therefore essentially identical.

## 10.5 HESS' LAW OF HEAT SUMMATION

Since enthalpy is a state function, the magnitude of $\Delta H$ for a chemical reaction does not depend on the path taken by the reactants as they proceed to form the products. Let's consider, for example, the conversion of 1 mole of liquid water at 100°C and 1 atm to 1 mole of vapor at 100°C and 1 atm. This process absorbs 9.7 kcal of heat for each mole of $H_2O$ vaporized and, hence, $\Delta H = +9.7$ kcal. We can represent this "reaction" as

$$H_2O\ (l) \longrightarrow H_2O\ (g) \qquad \Delta H = +9.7 \text{ kcal}$$

An equation written in this manner, in which the energy change is also shown, is called a **thermochemical equation** and is nearly always interpreted on a mole basis. Here, for instance, we see that 1 mole of $H_2O$ $(l)$ is converted to 1 mole of $H_2O$ $(g)$ by the absorption of 9.7 kcal.

The value of $\Delta H$ for this process will always be +9.7 kcal, provided that we refer to the same pair of initial and final states. We could even go so far as first to decompose the 1 mole of liquid into gaseous hydrogen and oxygen and then to recombine the elements to produce $H_2O$ $(g)$ at 100°C and 1 atm. The net change in enthalpy would still be the same, +9.7 kcal. Consequently, it is possible to look at some overall change as the net result of a sequence of chemical reactions. The net value of $\Delta H$ for the overall process is merely the sum of all of the enthalpy changes that take place along the way. These last statements constitute **Hess' law of heat summation.**

Thermochemical equations serve as a useful tool for applying Hess' law. For example, the thermochemical equations that correspond to the indirect path just described for the vaporization of water are[3]

$$\begin{aligned} H_2O\ (l) &\longrightarrow H_2\ (g) + \tfrac{1}{2}O_2\ (g) \qquad \Delta H = +67.6 \text{ kcal} \\ H_2\ (g) + \tfrac{1}{2}O_2\ (g) &\longrightarrow H_2O\ (g) \qquad\qquad\quad \Delta H = -57.9 \text{ kcal} \end{aligned}$$

These equations tell us that 67.6 kcal are required to decompose 1 mole of $H_2O$ $(l)$ into its elements and that 57.9 kcal are evolved when they recombine to produce 1 mole of $H_2O$ $(g)$. The sum of the two equations, after cancelling quantities that appear on both sides of the arrow, gives us the equation for the vaporization of 1 mole of water,

$$H_2O\ (l) + \cancel{H_2\ (g)} + \cancel{\tfrac{1}{2}O_2\ (g)} \longrightarrow H_2O\ (g) + \cancel{H_2\ (g)} + \cancel{\tfrac{1}{2}O_2\ (g)}$$

or

$$H_2O\ (l) \longrightarrow H_2O\ (g)$$

[3] Note that fractional coefficients are permitted in thermochemical equations. This is because a coefficient such as ½ is taken to mean ½ mole. In ordinary equations fractional coefficients are avoided because they are meaningless on a molecular level. One cannot have half an atom or molecule and still retain the chemical identity of the species.

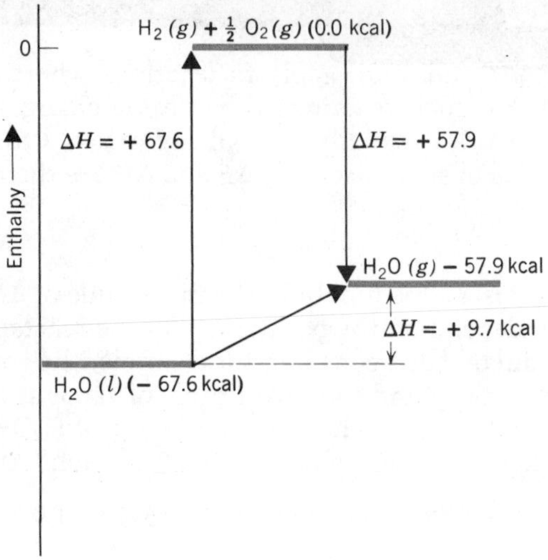


Figure 10.7
*Enthalpy diagram for the reaction $H_2O$ (l) → $H_2O$ (g).*

We also find that the heat of the overall reaction is equal to the algebraic sum of the heats of reaction for the two steps.

$$\Delta H = +67.6 \text{ kcal} + (-57.9 \text{ kcal})$$
$$\Delta H = +9.7 \text{ kcal}$$

Thus, *when we add thermochemical equations to obtain some net change, we also add their corresponding heats of reaction.*

To illustrate the nature of these thermochemical changes, we can also demonstrate them graphically (Figure 10.7). This type of figure is frequently called an **enthalpy diagram.** Notice that we have chosen as our zero on the energy scale the enthalpy of the free elements. This choice is entirely arbitrary because we are interested only in determining differences in $H$. In fact, we have no way at all of knowing absolute enthalpies; we can only measure $\Delta H$. We used diagrams similar to Figure 10.7 in the last chapter in our discussion of heats of solution, so the concept of Hess' law should already be familiar.

**HEATS OF FORMATION.** A particulary useful type of thermochemical equation corresponds to the formation of a substance from its elements. The enthalpy changes associated with these reactions are called **heats of formation** and are denoted as $\Delta H_f$. For example, thermochemical equations for the formation of liquid and gaseous water at 100°C and 1 atm are, respectively,

$$H_2\,(g) + \tfrac{1}{2}O_2\,(g) \longrightarrow H_2O\,(l) \qquad \Delta H_f = -67.6 \text{ kcal/mole}$$
$$H_2\,(g) + \tfrac{1}{2}O_2\,(g) \longrightarrow H_2O\,(g) \qquad \Delta H_f = -57.9 \text{ kcal/mole}$$

How can we use these equations to obtain the heat of vaporization of water? Clearly, we must reverse the first equation and then add it to the second. When we reverse this equation, we must also change the sign of $\Delta H$. If the formation of $H_2O$ ($l$) is exothermic, as indicated by a negative $\Delta H_f$, the reverse process must be endothermic.

$$\text{(Exothermic)} \quad H_2\,(g) + \tfrac{1}{2}O_2\,(g) \longrightarrow H_2O\,(l) \qquad \Delta H = \Delta H_f = -67.6 \text{ kcal}$$
$$\text{(Endothermic)} \quad H_2O\,(l) \longrightarrow H_2\,(g) + \tfrac{1}{2}O_2\,(g) \qquad \Delta H = -\Delta H_f = +67.6 \text{ kcal}$$

When this last equation is added to that for the formation of $H_2O$ (*g*), we obtain

$$H_2O\ (l) \longrightarrow H_2O\ (g)$$

and the heat of reaction is

$$\Delta H = \Delta H_f[H_2O\ (g)] - \Delta H_f[H_2O\ (l)]$$
$$\Delta H = -57.9 \text{ kcal} - (-67.6 \text{ kcal}) = +9.7 \text{ kcal}$$

Notice that the heat of reaction for the overall change is equal to the heat of formation of the product *minus* the heat of formation of the reactant. In general, we can write that for any overall reaction

$$\Delta H_{\text{reaction}} = (\text{sum of } \Delta H_f \text{ of products}) - (\text{sum of } \Delta H_f \text{ of reactants}) \qquad (10.8)$$

## 10.6 STANDARD STATES

The magnitude of $\Delta H_f$ depends on the conditions of temperature, pressure, and the physical state (gas, liquid, solid, crystalline form) of the reactants and products. For instance, at 100°C and 1 atm, the heat of formation of liquid water is −67.6 kcal/mole, while at 25°C and 1 atm, $\Delta H_f$ for $H_2O$ (*l*) is −68.3 kcal/mole. To avoid the necessity of always having to specify the conditions for which $\Delta H_f$ is recorded, and to permit comparisons between $\Delta H_f$ for various compounds, a standard set of conditions is chosen, usually 25°C and a pressure of 1 atm.[4] Under these conditions a substance is said to be in its **standard state.** Heats of formation of substances in their standard states are indicated as $\Delta H_f^0$. For example, the standard heat of formation of liquid water, $\Delta H_f^0[H_2O\ (l)] = -68.3$ kcal/mole, and represents the heat liberated when $H_2$ and $O_2$, each in their natural form at 25°C and 1 atm, react to produce $H_2O$ (*l*) at 25°C and 1 atm.

Table 10.1 contains standard heats of formation for a variety of different substances at 25°C. Such a table is very useful because it permits us to calculate, using Equation 10.8, the standard heats of reaction, $\Delta H^0$, for a very large number of different chemical changes. In performing these calculations, we arbitrarily take the $\Delta H_f^0$ for an element in its natural, most stable form at 25°C and 1 atm to be equal to zero. Thus, in our computations, the zero point on the energy scale is again chosen to be that of the free elements. As mentioned before, because we speak only of *changes* in energy, the actual location of this zero point is unimportant. The following examples illustrate how the principles developed in the preceding two sections can be applied.

EXAMPLE 10.3 Determine $\Delta H^0$ for the reaction,

$$2Na_2O_2\ (s) + 2H_2O\ (l) \longrightarrow 4NaOH\ (s) + O_2\ (g)$$

SOLUTION Equation 10.8 implies that

$$\Delta H^0 = (\text{sum } \Delta H_f^0 \text{ products}) - (\text{sum } \Delta H_f^0 \text{ reactants})$$

This means that we must add up all of the heat evolved during the formation of the products from their elements and then subtract the heat evolved by the formation of the reactants from their elements.

[4] Note that the choice of the standard temperature here differs from the standard temperature of 0°C used in calculations involving gases in Chapter 6.

Table 10.1
*Standard heats of formation of some substances at 25°C and 1 atm*

| Substance | $\Delta H_f^0$ kcal/mole (kJ/mole) | Substance | $\Delta H_f^0$ kcal/mole (kJ/mole) |
|---|---|---|---|
| $Al_2O_3$ (*s*) | −400.5 (−1676) | HCl (*g*) | −22.1 (−92.5) |
| $Br_2$ (*l*) | 0.00 (0.00) | HBr (*g*) | −8.7 (−36) |
| $Br_2$ (*g*) | +7.39 (+30.9) | HI (*g*) | +6.3 (+26) |
| C (*s*, diamond) | +0.45 (+1.88) | KCl (*s*) | −104.2 (−436.0) |
| CO (*g*) | −26.4 (−110) | LiCl (*s*) | −97.7 (−408.8) |
| $CO_2$ (*g*) | −94.1 (−394) | $MgCl_2$ (*s*) | −153.4 (−641.8) |
| $CH_4$ (*g*) | −17.9 (−74.9) | $MgCl_2 \cdot 2H_2O$ (*s*) | −306.0 (−1280) |
| $C_2H_6$ (*g*) | −20.2 (−84.5) | $Mg(OH)_2$ (*s*) | −221.0 (−924.7) |
| $C_2H_4$ (*g*) | +12.4 (+51.9) | $NH_3$ (*g*) | −11.0 (−46.0) |
| $C_2H_2$ (*g*) | +54.2 (+227) | $N_2O$ (*g*) | +19.5 (+81.5) |
| $C_3H_8$ (*g*) | −24.8 (−104) | NO (*g*) | +21.6 (+90.4) |
| $C_6H_6$ (*l*) | +11.7 (+49.0) | $NO_2$ (*g*) | +8.1 (+34) |
| $CH_3OH$(*l*) | −57.0 (−238) | NaF (*s*) | −136.5 (−571) |
| HCOOH (*g*) | −86.7 (−363) | NaCl (*s*) | −98.6 (−413) |
| $CS_2$ (*l*) | +21.4 (+89.5) | NaBr (*s*) | −86.0 (−360) |
| $CS_2$ (*g*) | +28.0 (+117) | NaI (*s*) | −68.8 (−288) |
| $CCl_4$ (*l*) | −32.1 (−134) | $Na_2O_2$ (*s*) | −120.6 (−504.6) |
| $C_2H_5OH$ (*l*) | −66.4 (−278) | NaOH (*s*) | −102.0 (−426.8) |
| $CH_3CHO$ (*g*) | −39.8 (−167) | $O_3$ (*g*) | +34.1 (+143) |
| $CH_3COOH$ (*l*) | −116.4 (−487.0) | $PbO_2$ (*s*) | −66.3 (−277) |
| CaO (*s*) | −151.9 (−635.5) | $PbSO_4$ (*s*) | −219.9 (−920.1) |
| $Ca(OH)_2$ (*s*) | −235.8 (−986.6) | $SO_2$ (*g*) | −70.9 (−297) |
| $CaSO_4$ (*s*) | −342.4 (−1433) | $SO_3$ (*g*) | −94.6 (−396) |
| CuO (*s*) | −37.1 (−155) | $H_2SO_4$ (*l*) | −194.5 (−813.8) |
| $Fe_2O_3$ (*s*) | −196.5 (−822.2) | $SiO_2$ (*s*) | −217.7 (−910.9) |
| $H_2O$ (*l*) | −68.3 (−286) | $SiH_4$ (*g*) | +8.2 (+34) |
| $H_2O$ (*g*) | −57.8 (−242) | ZnO (*s*) | −83.2 (−348) |
| HF (*g*) | −64.8 (−271) | Zn(OH) (*s*) | −153.5 (−642.2) |

For the products, the total enthalpy of formation is

$$4 \text{ moles } \cancel{NaOH} \times \left(\frac{-102.0 \text{ kcal}}{1 \text{ mole } \cancel{NaOH}}\right) = -408.0 \text{ kcal}$$

$$1 \text{ mole } \cancel{O_2} \times \left(\frac{0.0 \text{ kcal}}{1 \text{ mole } \cancel{O_2}}\right) = 0.0 \text{ kcal}$$

$$\text{total of } \Delta H_f^0 \text{ products} = -408.0 \text{ kcal}$$

For the reactants, we have

$$2 \text{ moles } \cancel{Na_2O_2} \times \left(\frac{-120.6 \text{ kcal}}{1 \text{ mole } \cancel{Na_2O_2}}\right) = -241.2 \text{ kcal}$$

$$2 \text{ moles } \cancel{H_2O}\,(l) \times \left(\frac{-68.3 \text{ kcal}}{1 \text{ mole } \cancel{H_2O}\,(l)}\right) = -136.6 \text{ kcal}$$

$$\text{total of } \Delta H_f^0 \text{ reactants} = -377.8 \text{ kcal}$$

We have said that

$$\Delta H^0 = (\text{sum } \Delta H_f^0 \text{ products}) - (\text{sum } \Delta H_f^0 \text{ reactants})$$

Therefore

$$\Delta H^0 = -408.0 \text{ kcal} - (-377.8 \text{ kcal})$$

or

$$\Delta H^0 = -30.2 \text{ kcal}$$

Notice that in computing $\Delta H^0$ for the overall reaction we have multiplied each $\Delta H_f^0$ by the appropriate coefficient from the equation. This gives the total heat of reaction for the numbers of moles specified by the chemical equation.

EXAMPLE 10.4 On the basis of the previous example, how many kilocalories of heat are evolved when 25.0 g of $Na_2O_2$ are treated with water to produce NaOH and $O_2$?

SOLUTION In the preceding example we found that 30.2 kcal are evolved when 2 moles of $Na_2O_2$ are reacted with water. Thus we write for this reaction,

$$2 \text{ moles } Na_2O_2 \sim 30.2 \text{ kcal}$$

or

$$1 \text{ mole } Na_2O_2 \sim 15.1 \text{ kcal}$$

Since the formula weight of $Na_2O_2$ is 78.0,

$$25.0 \cancel{\text{ g } Na_2O_2} \times \left(\frac{1 \cancel{\text{ mole } Na_2O_2}}{78.0 \cancel{\text{ g } Na_2O_2}}\right) \times \left(\frac{15.1 \text{ kcal}}{1 \cancel{\text{ mole } Na_2O_2}}\right) \sim 4.84 \text{ kcal}$$

Hence the consumption of 25.0 g of $Na_2O_2$ releases 4.84 kcal.

It is frequently impossible to measure directly the heat of formation of a compound. For example, we cannot get hydrogen, oxygen, and graphite (the most stable crystalline form of carbon) to react directly together to produce ethyl alcohol, $C_2H_5OH$. This is also true with many, if not most, other compounds. In order to determine $\Delta H_f^0$ for these substances, then, an indirect method must be applied. One technique, which can be applied to most organic materials, is to burn the substance in a calorimeter to produce products whose heats of formation are known, as shown in Example 10.5.

EXAMPLE 10.5 The combustion of 1 mole of benzene, $C_6H_6$ (*l*), to produce $CO_2$ (*g*) and $H_2O$ (*l*) liberates 3271 kJ when the products are returned to 25°C and 1 atm. What is the standard heat of formation of $C_6H_6$ (*l*) expressed in kilojoules per mole?

SOLUTION The equation for the combustion of 1 mole of $C_6H_6$ is

$$C_6H_6\,(l) + 7\tfrac{1}{2}O_2\,(g) \longrightarrow 6CO_2\,(g) + 3H_2O\,(l)$$

The standard heat of reaction, $\Delta H^0 = 3271$ kJ. From Equation 10.8, we know that

$$\Delta H^0 = 6\ \Delta H_f^0[CO_2\,(g)] + 3\ \Delta H_f^0[H_2O\,(l)] - \Delta H_f^0[C_6H_6\,(l)]$$

Solving for the heat of formation of benzene,

$$\Delta H_f^0[C_6H_6\,(l)] = 6\ \Delta H_f^0[CO_2\,(g)] + 3\ \Delta H_f^0[H_2O\,(l)] - \Delta H^0$$

From Table 10.1 we can obtain the heats of formation of $CO_2$ and $H_2O$. Therefore,

$$\Delta H_f^0[C_6H_6\ (l)] = 6(-394)\ kJ + 3(-286)\ kJ - (-3271\ kJ)$$
$$\Delta H_f^0[C_6H_6\ (l)] = +49\ kJ$$

Since 1 mole of $C_6H_6$ is involved,

$$\Delta H_f^0 = +49\ kJ/mole$$

## 10.7 BOND ENERGIES

We stated earlier that it should be possible to relate heats of reaction to changes in the potential energy associated with chemical bonds. Strictly speaking, we should use $\Delta E$ for this purpose; however, since $P\ \Delta V$ contributions to $\Delta H$ are relatively small for chemical reactions, we can use $\Delta H$ in place of $\Delta E$ and still expect to obtain quite reasonable results. Consequently, we shall use the terms bond energy and bond enthalpy interchangeably.

In Chapter 4 the bond energy was defined as the energy required to break a bond to produce neutral fragments. For a complex molecule, the energy needed to reduce the gaseous molecule to neutral gaseous atoms, called the **atomization energy,** is the sum of all of the bond energies in the molecule. Simple diatomic molecules, such as $H_2$, $O_2$, $Cl_2$, or HCl, possess only one bond; therefore the atomization energy is the same as the bond energy. For these simple cases the atomization energy can be obtained by studying the spectra produced when these molecules absorb or emit light. For more complex molecules, however, we employ an indirect method that makes use of measured heats of formation.

As an example, let's consider the molecule $CH_4$. If we use the same approach that was followed in Example 10.5, the standard heat of formation of $CH_4\ (g)$ can be determined experimentally to be $-17.9$ kcal/mole. This corresponds to the enthalpy change, $\Delta H_f^0$, for the reaction,

$$C\ (s,\ graphite) + 2H_2\ (g) \longrightarrow CH_4\ (g)$$

We can envision an alternative path to take us from the free elements to the compound, methane, that follows the succession of reactions,

$$(1)\ C\ (s,\ graphite) \longrightarrow C\ (g) \qquad \Delta H_1$$
$$(2)\ 2H_2\ (g) \longrightarrow 4H\ (g) \qquad \Delta H_2$$
$$(3)\ C\ (g) + 4H\ (g) \longrightarrow CH_4\ (g) \qquad \Delta H_3$$

The sum of these three will give us our desired overall reaction.

Steps 1 and 2 each involve the heat of formation of gaseous atoms from an element in its standard state. For hydrogen the heat of formation of each mole of gaseous hydrogen atoms is half the atomization energy of $H_2\ (g)$. For carbon it amounts to the sublimation energy of graphite. Table 10.2 contains some heats of formation of gaseous atoms from typical elements in their standard states.

Applying Hess' law, we know that $\Delta H$ for the overall reaction [that is, $\Delta H_f^0$ for $CH_4\ (g)$] can be obtained by adding up the enthalpy changes for each step.

$$\Delta H_f^0 = \Delta H_1 + \Delta H_2 + \Delta H_3 \qquad (10.9)$$

Table 10.2

*Heats of formation of gaseous atoms from the elements in their standard states*

| Atom | $\Delta H_f$ kcal/mole of atoms (kJ/mole of atoms) |
|---|---|
| H | 52.1 (218) |
| Li | 38.4 (161) |
| Be | 78.2 (327) |
| B | 132.6 (555) |
| C | 170.9 (715) |
| N | 113.0 (473) |
| O | 59.6 (249) |
| F | 18.9 (79.1) |
| Na | 25.8 (108) |
| Si | 108.4 (454) |
| Cl | 28.9 (121) |
| Br | 26.7 (112) |
| I | 25.5 (107) |

From the data in Table 10.2, we find that $\Delta H_1 = +170.9$ kcal, the heat of formation of gaseous carbon atoms. Similarly, $\Delta H_2 = 4(+52.1 \text{ kcal})$, that is, four times the heat of formation of 1 mole of H $(g)$. The quantity $\Delta H_3$ is the negative of the atomization energy of $CH_4$ $(g)$.

$$\Delta H_3 = -\Delta H_{\text{atom}}[CH_4\ (g)]$$

Solving for the atomization energy of methane in Equation 10.9 gives

$$\Delta H_{\text{atom}}[CH_4\ (g)] = \Delta H_1 + \Delta H_2 - \Delta H_f^0$$

Substituting numerical values, we have

$$\Delta H_{\text{atom}}[CH_4\ (g)] = (+170.9 + 208.4 + 17.9)\ \text{kcal}$$
$$\Delta H_{\text{atom}}[CH_4\ (g)] = +397.2\ \text{kcal}$$

This quantity is the total amount of energy that must be absorbed to break all 4 moles of C—H bonds in 1 mole of $CH_4$. Division by 4, then, provides us with an average bond energy of 99.3 kcal/mole of C—H bonds. This value, along with some other bond energies, appears in Table 10.3.

A very important fact is that the average bond energies found in Table 10.3 can be used, in many cases, to compute heats of formation with a fair degree of accuracy, as illustrated below in Example 10.6. It is very significant that a bond between two atoms has very nearly the same strength in one molecule as it does in another. This implies, for example, that nearly all C—H bonds are pretty much alike, whether in a small molecule such as $CH_4$ or in a large complex molecule such as $C_{42}H_{86}$. The same applies to many other bonds as well. This phenomenon has greatly simplified the development of the modern theories about chemical bonding that we shall discuss in Chapter 17.

Table 10.3
*Average bond energies*

| Bond | Bond Energy kcal/mole (kJ/mole) |
|---|---|
| H—C | 99.3 (415) |
| H—O | 110.6 (463) |
| H—N | 93.4 (391) |
| H—F | 134.6 (563) |
| H—Cl | 103.2 (432) |
| H—Br | 87.5 (366) |
| H—I | 71.4 (299) |
| C—O | 85.0 (356) |
| C=O | 173 (724) |
| C—N | 69.7 (292) |
| C=N | 148 (619) |
| C≡N | 210 (879) |
| C—C | 83.1 (348) |
| C=C | 145 (607) |
| C≡C | 199 (833) |

EXAMPLE 10.6 Use the data in Tables 10.2 and 10.3 to compute the heat of formation of liquid ethyl alcohol in kilocalories per mole. This compound has a heat of vaporization, $\Delta H^0_{\text{vap}} = 9.4$ kcal/mole and the structural formula,

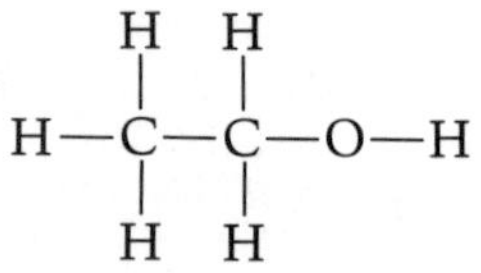

SOLUTION We wish to determine $\Delta H^0$ for the reaction,

$$2\text{C}\ (s,\ \text{graphite}) + 3\text{H}_2\ (g) + \tfrac{1}{2}\text{O}_2\ (g) \longrightarrow \text{C}_2\text{H}_5\text{OH}\ (l)$$

To compute $\Delta H_f$, we follow an alternative path from reactants to products as illustrated in Figure 10.8. Using the data in Table 10.2, we can compute the energy required to convert the reactants to gaseous atoms, that is, $\Delta H_A$ in Figure 10.8.

$$2\text{C}\ (s,\ \text{graphite}) \longrightarrow 2\text{C}\ (g) \qquad \Delta H_1{}^0 = 2\ \Delta H_f{}^0[\text{C}\ (g)]$$
$$= 2(+170.9\ \text{kcal}) = +341.8\ \text{kcal}$$
$$3\text{H}_2\ (g) \longrightarrow 6\text{H}\ (g) \qquad \Delta H_2{}^0 = 6\ \Delta H_f{}^0[\text{H}\ (g)]$$
$$= 6(+52.1\ \text{kcal}) = +312.6\ \text{kcal}$$
$$\tfrac{1}{2}\text{O}_2\ (g) \longrightarrow \text{O}\ (g) \qquad \Delta H_3{}^0 = \Delta H_f{}^0[\text{O}\ (g)] = +59.6\ \text{kcal}$$

The total energy needed to give gaseous atoms, $\Delta H_A{}^0 = \Delta H_1{}^0 + \Delta H_2{}^0 + \Delta H_3{}^0 = +714.0$ kcal.

Next we can compute the energy liberated when these atoms combine to form 1 mole of gaseous $C_2H_5OH$. This is the negative of the atomization energy of $C_2H_5OH$, which involves five C—H bonds, one C—C bond, one C—O bond, and one O—H bond. The energies are obtained from Table 10.3.

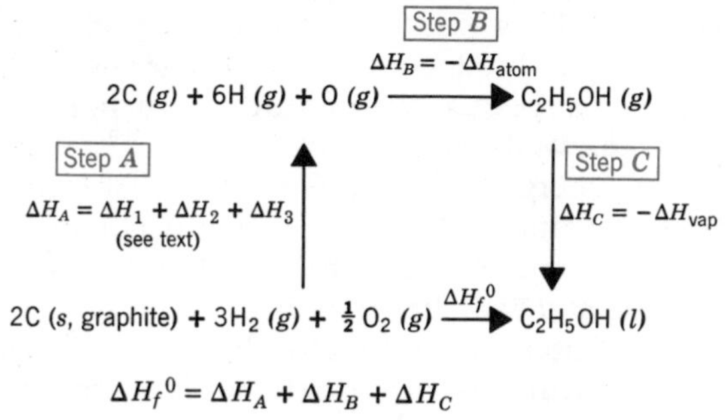


Figure 10.8
*An alternative path for the formation of $C_2H_5OH$ (l).*

| | |
|---|---|
| 5 (C—H) | 5(99.3 kcal) |
| 1 (C—C) | 83.1 kcal |
| 1 (C—O) | 85.0 kcal |
| 1 (O—H) | 110.6 kcal |
| $\Delta H^0_{atom}$ [$C_2H_5OH$ (g)] = | 775.2 kcal |

Therefore, for step $B$, $\Delta H_B{}^0 = -775.2$ kcal

Finally, energy is liberated when $C_2H_5OH$ (g) is condensed to a liquid. The $\Delta H^0$ for this process is the negative of $\Delta H^0_{vap}$. Therefore $\Delta H_C{}^0 = -9.4$ kcal.

Now we can compute $\Delta H_f{}^0$ for $C_2H_5OH$ (l) by adding the $\Delta H^0$ values of each step in the alternative path. This is justified by Hess' law, which says that the energy change is the same regardless of the path we follow from reactants to products.

$$\begin{aligned}\Delta H_f{}^0 &= \Delta H_A{}^0 + \Delta H_B{}^0 + \Delta H_C{}^0 \\ &= +714.0 \text{ kcal} + (-775.2 \text{ kcal}) + (-9.4 \text{ kcal}) \\ &= -70.6 \text{ kcal}\end{aligned}$$

Since the computation was performed for 1 mole, we can write

$$\Delta H_f{}^0 = -70.6 \text{ kcal/mole}$$

Comparing this to the value reported in Table 10.1,

$$\Delta H_f{}^0 = -66.4 \text{ kcal/mole}$$

we see that the agreement is not really too bad, considering that we have assumed that any particular bond has the same energy in *all* compounds.

## 10.8* SPONTANEITY OF CHEMICAL REACTIONS

At the beginning of this chapter we indicated that thermodynamics is able to tell us when a reaction will proceed without outside help. To see how this can be accomplished, let's begin by finding out what factors are involved in determining whether a particular process, either physical or chemical, will occur spontaneously. One spontaneous process that we have all observed is a ball rolling down a hill. When it finally comes to rest at the bottom, its potential energy has decreased and it is in a more stable, lower energy state than before. We might conclude, therefore, that a process leading to a decrease in the energy of a system (here, a ball) should tend to be spontaneous. Indeed, many processes that are spontaneous do occur with the evolution of energy. For example, a mixture of hydrogen and oxygen, when ignited, reacts very rapidly to produce water. This chemical change is accompanied by the release of a large quantity of heat, so much, in fact, that the hot water vapor produced expands explosively.

Evolution of energy, however, is not the only criterion to be considered. There are many examples of processes that occur with the absorption of energy and yet are spontaneous. In the last chapter, for example, we discussed heats of solution and saw that in many instances when a salt (for example, KI) dissolves in water, energy is absorbed. The formation of the solution, although endothermic, is nevertheless spontaneous. What, then, is the driving force for this process that is capable of outweighing the endothermic energy effect that occurs?

When a solid such as KI dissolves in water, the particles of the solute leave the well-ordered crystalline state and gradually diffuse throughout the liquid to produce a solution. In this final state the particles of the solute are in a more random condition than they were before they dissolved, as shown in Figure 10.9. Similarly, the solvent is in a more random state in the solution because the solvent molecules are, in a sense, dispersed throughout those of the solute as well.

In any process there is a natural tendency or drive toward increased randomness because a highly random distribution of particles represents a condition of higher statistical probability than an ordered one. To see this, suppose that we had 1 mole of a gas in the left compartment of the apparatus in Figure 10.10, and a vacuum in the other compartment. If the stopcock between them is opened, we know intuitively that the gas will rush into the right compartment, and that after a time each side will contain equal numbers of molecules, provided that the volume of each is the same. Why does this change occur spontaneously?

To understand this, imagine that there are only two molecules in the compartment on the left. When the stopcock is opened, both molecules are free to wander into the other bulb and, at this point, there are four different particle distributions possible for the system (Figure 10.10*b*). Since both particles reside in the left compartment in only one of these four distributions, the probability of finding both on the left is one out of four, or 1/4. Similarly, since there are two ways of having an even distribution of particles between the two compartments, the probability of an even distribution is two out of four, or 1/2. From these probabilities we would conclude that there will be an even distribution of particles 50% of the time and that there will be two particles on the left only 25% of the time.

Figure 10.9
*Increase in disorder occurs when solution is formed.*

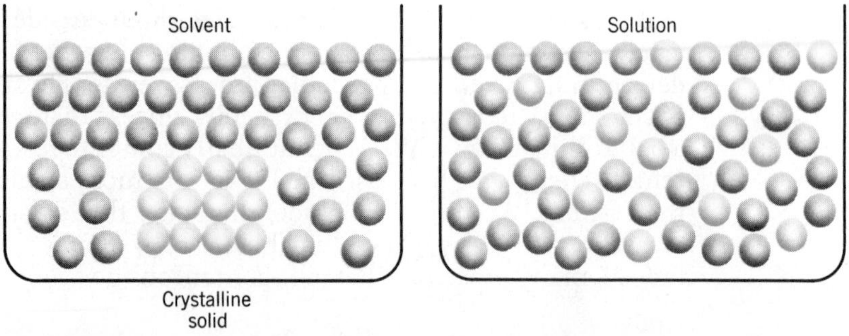

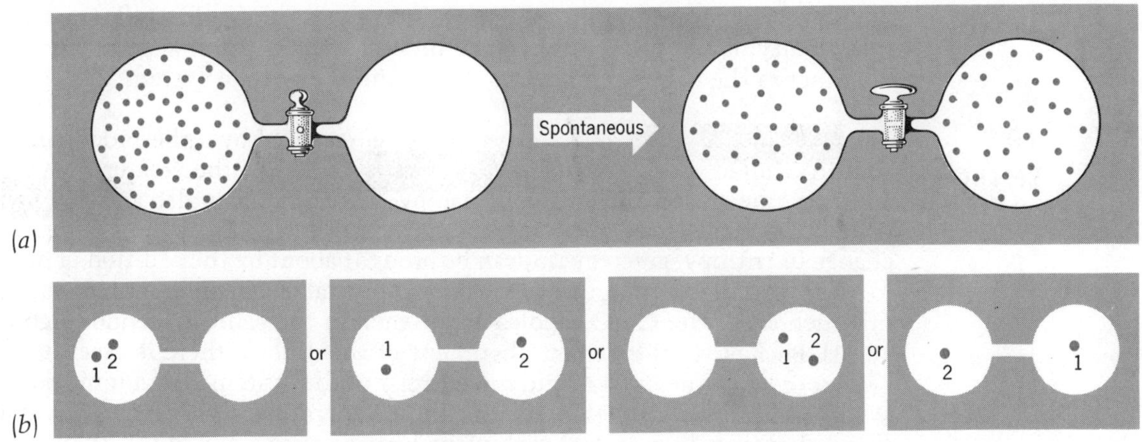


Figure 10.10
*(a) A gas spontaneously fills a vacuum. (b) Two molecules with four possibilities.*

Even in the two-particle case we feel intuitively that when the stopcock is opened the gas will expand in such a way as to occupy the entire volume of both compartments. In light of the discussion above, this simply means that the system proceeds from a state of low statistical probability (both particles on one side) to a state of higher probability.

When there are more particles in our system, the difference between these two types of molecular distributions is even more dramatic. With six particles, for instance, the probability of finding all of them in one compartment is only 1/64; whereas, an even three-to-three distribution has a probability of 20/64.

In general, the probability of finding all of the molecules in a gas in one compartment of this apparatus is $1/2^n$, where $n$ is the number of particles. For 1 mole of gas the probability that all of the gas will be found in one bulb (that is, the probability that the gas will *not* spontaneously expand) is

$$p = \frac{1}{2^{(6.02\times10^{23})}}$$

This can be shown to be the same as

$$p = \frac{1}{10^{(1.81\times10^{23})}} = 10^{(-1.81\times10^{23})}$$

It is nearly impossible to appreciate how small this probability is. If we were to write this as a decimal fraction, it would be

$$p = \underbrace{0.0000.......00001}_{1.81\times10^{23}\text{ zeros}}$$

and if each zero were 0.10 inches in diameter, we would need a piece of paper 2,800,000,000,000,000,000 *miles* long just to write the number! This very small probability tells us that a state having all the molecules in one compartment cannot be maintained. The expansion of the gas that occurs when the stopcock is opened is therefore simply the result of the system proceeding from a state of very low probability to one of high probability. The final state is one of greater disorder, of randomness, because the particles of the gas are more widely distributed and have a greater degree of freedom of movement.

By way of summary, we see that there are two factors that influence the spontaneity of a physical or chemical process. One is the change in energy; the other is the change in randomness or disorder of the system.

## 10.9 ENTROPY

The degree of randomness of a system is represented by a thermodynamic quantity called the **entropy**, denoted by the symbol $S$; the greater the randomness, the greater the entropy. Entropy, like $E$ and $H$, is also a state function; $\Delta S$ depends only on the initial and final entropies of the system. A change in entropy, given by $\Delta S$, can be brought about by the addition of heat to a system. For example, consider a perfect crystal of carbon monoxide at 0 K in which all of the C—O dipoles are aligned in the same direction (Figure 10.11). Because of the perfect alignment of the dipoles there is essentially perfect order in the crystal, and the entropy of the system is at a minimum. When heat is added to this crystal, the temperature rises above 0 K and thermal motion (vibrations) within the lattice cause some of the dipoles to become oriented in the opposite direction (Figure 10.11*b*). As a result, there is less order (more disorder), and the entropy of the crystal has obviously increased. Logically, the more heat added to the system, the greater will be the extent of disorder afterwards; therefore we expect the entropy change, $\Delta S$, to be directly proportional to the quantity of heat, $q_{rev}$, added to the system,[5]

$$\Delta S \propto q_{rev}$$

The magnitude of $\Delta S$ is also *inversely* proportional to the temperature at which the heat is added. At low temperatures a given quantity of heat makes a large change in the relative degree of order. Near absolute zero the system goes from essentially perfect order to some extent of randomness—a very substantial change. If the same amount of heat is added to a system at high temperature, the system goes from an already highly random state to one just slightly more random. This constitutes only a very slight change in the *relative* degree of disorder and, hence, to only a small entropy change.

Figure 10.11
*(a) Perfect crystal of CO at 0 K. (b) Crystal of CO above 0 K.*

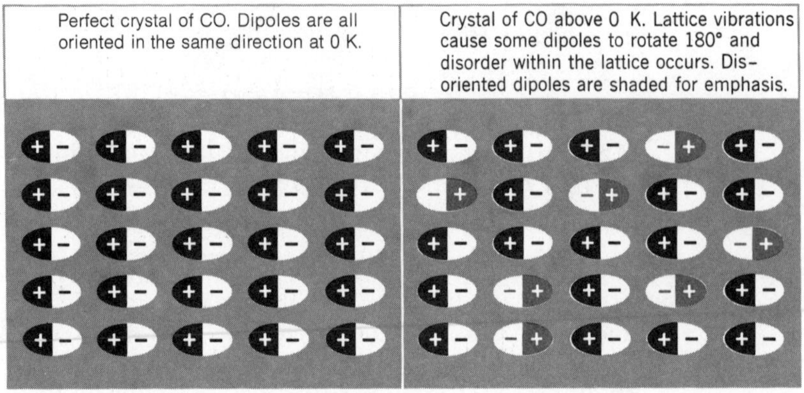

Perfect crystal of CO. Dipoles are all oriented in the same direction at 0 K.
Crystal of CO above 0 K. Lattice vibrations cause some dipoles to rotate 180° and disorder within the lattice occurs. Dis-oriented dipoles are shaded for emphasis.


[5] In Section 10.2 we saw that, in general, $q$ is not a state function. For a reversible change, however, $q_{rev}$ does depend only on the initial and final states because the path (a reversible one) is clearly defined. $\Delta S$ can be a state function only if the value of $q$ used to compute it is itself a quantity that depends only on the initial and final states, and therefore $q_{rev}$ must be used.

Thus, it can be shown that a change in entropy is finally given as

$$\Delta S = \frac{q_{rev}}{T} \tag{10.10}$$

where $T$ is the absolute temperature at which $q$ is transferred to the system. Note that entropy has units of *energy/temperature,* for example, calories per kelvin (cal/K) or joules per kelvin (J/K).

## 10.10 THE SECOND LAW OF THERMODYNAMICS

The second law of thermodynamics provides us with a way of comparing the effects of the two driving forces involved in a spontaneous process—changes in energy and changes in entropy. One statement of the second law is that *in any spontaneous process there is always an increase in the entropy of the universe* ($\Delta S_{total} > 0$). This increase takes into account entropy changes in both the system and its surroundings,

$$\Delta S_{total} = \Delta S_{system} + \Delta S_{surroundings}$$

The entropy change that occurs in the surroundings is brought about by the heat added to the surroundings divided by the temperature at which it is transferred. For a process at constant $P$ and $T$, the heat added to the system is $\Delta H_{system}$ and is equal to the *negative* of the heat added to the surroundings. In other words,

$$-q_{surroundings} = \Delta H_{system}$$

or

$$q_{surroundings} = -\Delta H_{system}$$

The entropy change for the surroundings is therefore

$$\Delta S_{surroundings} = \frac{-\Delta H_{system}}{T}$$

The total entropy change for the universe is thus

$$\Delta S_{total} = \Delta S_{system} - \frac{\Delta H_{system}}{T}$$

or

$$\Delta S_{total} = \frac{T\,\Delta S_{system} - \Delta H_{system}}{T}$$

This can be rearranged to give

$$T\,\Delta S_{total} = -(\Delta H_{system} - T\,\Delta S_{system})$$

Since $\Delta S_{total}$ must be a positive number for a spontaneous change, the product $T\,\Delta S_{total}$ must also be positive. This means that the quantity in parentheses on the right, $(\Delta H_{system} - T\,\Delta S_{system})$, must be negative so that $-(\Delta H_{system} - T\,\Delta S_{system})$ may be positive. Thus, in order for a spontaneous change to take place, the expression $(\Delta H_{system} - T\,\Delta S_{system})$ must be negative.

At this point it is convenient to introduce another thermodynamic state function, $G$, called the **Gibbs free energy.** This is defined as

$$G = H - TS$$

For a change at constant $T$ and $P$, we write

$$\Delta G = \Delta H - T\,\Delta S \tag{10.11}$$

From the argument presented in the preceding paragraph, we see that $\Delta G$ must be less than zero for a spontaneous process; that is, $\Delta G$ must have a negative value at constant $T$ and $P$.

The Gibbs free energy change, $\Delta G$, represents a composite of the two factors contributing to spontaneity, $\Delta H$ and $\Delta S$. For systems in which $\Delta H$ is negative (exothermic) and $\Delta S$ is positive (increased disorder accompanying the change), both factors favor spontaneity and the process will occur spontaneously at all temperatures. Conversely, if $\Delta H$ is positive (endothermic) and $\Delta S$ is negative (increase in order), $\Delta G$ will always be positive and the change cannot occur spontaneously at any temperature.

In situations where $\Delta H$ and $\Delta S$ are both positive, or both negative, Equation 10.11 shows that temperature plays the determining role in controlling whether or not a reaction will take place. In the first case ($\Delta H$, $\Delta S > 0$), $\Delta G$ will be negative only at high temperatures, where $T\,\Delta S$ is greater in magnitude than $\Delta H$; as a consequence, the reaction will be spontaneous only at elevated temperatures. On the other hand, when $\Delta H$ and $\Delta S$ are both negative, $\Delta G$ will be negative only at low temperatures. An example of this is the freezing of water. We know that heat must be removed from the liquid to produce ice; hence the process is exothermic with a negative $\Delta H$. Freezing is also accompanied by an ordering of the water molecules as they leave the random liquid state and become part of the crystal. As a result, $\Delta S$ is also negative. The sign of $\Delta G$ is determined both by $\Delta H$, which in this case is negative, and by $T\,\Delta S$, which is also negative. To compute $\Delta G$ we must subtract a negative $T\,\Delta S$ from a negative $\Delta H$. The result will be negative only at low temperature. Consequently, at 1 atm we observe $H_2O$ to freeze spontaneously only below 0°C. Above 0°C the magnitude of $T\,\Delta S$ is greater than $\Delta H$, and $\Delta G$ becomes positive. As a result, freezing is no longer spontaneous. Instead, the reverse process (melting) occurs.

## 10.11 FREE ENERGY AND USEFUL WORK

One of the most important applications of chemical reactions is in the production of energy in the form of useful work. This can, for example, take the form of combustion, in which the heat generated is used to create steam for the production of mechanical work, or perhaps electrical work drawn from a dry cell or storage battery. The quantity $G$ is called the *free energy* because $\Delta G$ represents the *maximum* amount of energy released in a process occurring at constant temperature and pressure that is free to perform useful work. We have already associated $\Delta G$ with the factors that lead to a drive for spontaneity. What we see now is that this driving force in a chemical change can be harnessed to perform work for us.

The actual amount of work obtained from any real spontaneous process is always less than the maximum predicted by $\Delta G$. This is because real processes are always irreversible, and we saw earlier that the maximum work can be extracted only from a truly reversible change. The free-energy change gives us a goal at which to aim. The closer a given process is to reversibility, the greater will be the amount of available work that can be used. However, even relatively efficient systems are able to harness only a small fraction of the available free energy. Living systems, for example, are able to convert only about 40% of the free energy available in the oxidation of glucose to other forms of stored chemical energy (for example, ATP).

## 10.12 FREE ENERGY AND EQUILIBRIUM

In the last paragraph we said that $\Delta G$ determines the maximum amount of energy that is available to perform useful work as a system passes from one state to another. As a reaction proceeds, its capacity to perform work, as measured by $G$, diminishes until finally, at equilibrium, the system is no longer able to supply additional work. This means that both reactants and products possess the same free energy and therefore $\Delta G = 0$. We see, then, that the value of $\Delta G$ for a particular change determines the approach toward equilbrium. When $\Delta G$ is negative, the reaction is spontaneous. When $\Delta G$ is zero, the system is in a state of dynamic equilibrium, and when $\Delta G$ is positive, the reaction is really spontaneous in the reverse direction.

At this point, it should be reemphasized that although $\Delta G$ may predict that a particular process is spontaneous, nothing is implied about how rapid the change will be.

## 10.13 STANDARD ENTROPIES AND FREE ENERGIES

The **third law of thermodynamics** states that the entropy of any pure crystalline substance at absolute zero is equal to zero. This makes sense because in a perfect crystal at absolute zero there is perfect order. Because of this it is possible, by summing $q_{rev}/T$ increments from 0 K to 298 K (25°C), to determine the absolute entropy of a substance in its standard state. Table 10.4 contains a number of such **standard entropies.**

From standard heats of formation and standard entropies we can also calculate **standard free energies of formation,** $\Delta G_f^0$. For example, consider the formation of $CO_2$ from the elements, all reactants and products in their standard states,

$$C\ (s,\ \text{graphite}) + O_2\ (g) \longrightarrow CO_2\ (g)$$

Table 10.1 gives us the standard enthalpy of formation, $\Delta H_f^0$, as −94.1 kcal/mole. From the data in Table 10.4 we can calculate $\Delta S_f^0$.

$$\Delta S_f^0 = S^0_{CO_2} - (S^0_C + S^0_{O_2})$$
$$\Delta S_f^0 = 51.1 - (1.4 + 49.0)\ \text{cal/mole K}$$
$$\Delta S_f^0 = +0.7\ \text{cal/mole K}$$

We can then obtain $\Delta G_f^0$ as

$$\Delta G_f^0 = \Delta H_f^0 - T\,\Delta S_f^0$$

At 25°C (298 K), then,

$$\Delta G_f^0 = -94.1\ \text{kcal/mole} - (298\ \text{K})(0.7\ \text{cal/mole K})$$
$$\Delta G_f^0 = -94.1\ \text{kcal/mole} - 200\ \text{cal/mole}$$

Converting entirely to kilocalories per mole gives

$$\Delta G_f^0 = (-94.1 - 0.2)\ \text{kcal/mole}$$
$$\Delta G_f^0 = -94.3\ \text{kcal/mole}$$

This and other standard free energies of formation are given in Table 10.5.

Earlier in this chapter we saw that $\Delta H^0$ for a reaction can be computed from standard heats of formation. The same rules also apply for the calculation of $\Delta G^0$ using standard free energies of formation; that is,

$$\Delta G^0 = (\text{sum of } \Delta G_f^0 \text{ of products}) - (\text{sum of } \Delta G_f^0 \text{ of reactants})$$

For example, consider the reaction

$$SiH_4\ (g) + 2O_2\ (g) \longrightarrow SiO_2\ (s) + 2H_2O\ (g)$$

Table 10.4
*Absolute entropies at 25°C and 1 atm*

| Substance | $S°$ cal/mole K (J/mole K) | Substance | $S°$ cal/mole K (J/mole K) |
|---|---|---|---|
| Al (*s*) | 6.77 (28.3) | Mg (*s*) | 7.77 (32.5) |
| $Al_2O_3$ (*s*) | 12.19 (51.0) | $Mg(OH)_2$ (*s*) | 15.09 (63.1) |
| $Br_2$ (*l*) | 36.38 (152.2) | $N_2$ (*g*) | 45.77 (191.5) |
| $Br_2$ (*g*) | 58.65 (245.4) | $NH_3$ (*g*) | 46.01 (192.5) |
| C (*s*, graphite) | 1.36 (5.69) | $N_2O$ (*g*) | 52.58 (220.0) |
| C (*s*, diamond) | 0.58 (2.4) | NO (*g*) | 50.34 (210.6) |
| CO (*g*) | 47.30 (197.9) | $NO_2$ (*g*) | 57.47 (240.5) |
| $CO_2$ (*g*) | 51.06 (213.6) | Na (*s*) | 12.2 (51.0) |
| $CH_4$ (*g*) | 44.50 (186.2) | NaF (*s*) | 12.3 (51.5) |
| $C_2H_6$ (*g*) | 54.9 (230) | NaCl (*s*) | 17.4 (72.8) |
| $C_2H_4$ (*g*) | 52.5 (220) | NaBr (*s*) | 20.0 (83.7) |
| $C_2H_2$ (*g*) | 48.0 (201) | NaI (*s*) | 21.8 (91.2) |
| $C_3H_8$ (*g*) | 64.51 (269.9) | $O_2$ (*g*) | 49.00 (205.0) |
| $CCl_4$ (*l*) | 51.25 (214.4) | Pb (*s*) | 15.5 (64.9) |
| $Cl_2$ (*g*) | 53.29 (223.0) | $PbO_2$ (*s*) | 16.4 (68.6) |
| $F_2$ (*g*) | 48.44 (202.7) | $PbSO_4$ (*s*) | 35.5 (149) |
| $H_2$ (*g*) | 31.21 (130.6) | S (*s*, rhombic) | 7.60 (31.8) |
| $H_2O$ (*l*) | 16.72 (70.0) | $SO_2$ (*g*) | 59.3 (248) |
| $H_2O$ (*g*) | 45.11 (188.7) | $SO_3$ (*g*) | 61.3 (256) |
| HF (*g*) | 41.47 (173.5) | $H_2SO_4$ (*l*) | 37.5 (157) |
| HCl (*g*) | 44.62 (186.7) | Si (*s*) | 4.5 (19) |
| HBr (*g*) | 47.44 (198.5) | $SiO_2$ (*s*) | 10.0 (41.8) |
| HI (*g*) | 49.3 (206) | Zn (*s*) | 10.0 (41.8) |
| $I_2$ (*s*) | 27.76 (116.1) | ZnO (*s*) | 10.4 (43.5) |

The standard free-energy change is given by

$$\Delta G^0 = (\Delta G_f^0[SiO_2\ (s)] + 2\ \Delta G_f^0[H_2O\ (g)]) - (\Delta G_f^0[SiH_4\ (g)] + 2\ \Delta G_f^0[O_2\ (g)])$$

Let's calculate $\Delta G^0$ in kilojoules per mole. As with enthalpy calculations, we take $\Delta G_f^0$ for any free element to be equal to zero. Therefore, using the data in Table 10.5, we have

$$\Delta G^0 = 1\ \cancel{\text{mole}} \times \left(\frac{-805\ \text{kJ}}{\cancel{\text{mole}}}\right) + 2\ \cancel{\text{moles}} \times \left(\frac{-228\ \text{kJ}}{\cancel{\text{mole}}}\right) - 1\ \cancel{\text{mole}} \times \left(\frac{-39\ \text{kJ}}{\cancel{\text{mole}}}\right)$$

$$\Delta G^0 = -1222\ \text{kJ}$$

The reason that this type of calculation is important is because *the value of $\Delta G^0$ for a reaction determines the position of equilibrium,* that is, the relative numbers of moles of reactants and products that will be present when the chemical system achieves equilibrium. We will deal with this quantitatively in Chapter 12, but for now let's look qualitatively at the relationship between the computed value of $\Delta G^0$ and the free energy of a chemical system as it passes from reactants to products, as illustrated in Figure 10.12. This graph allows us to make a very important distinction between $\Delta G^0$, the free-energy difference between the reactants and products *in their standard states,* and the *direction* of the free-energy change, which we will call $\Delta G$, at various intervals along the reaction path.

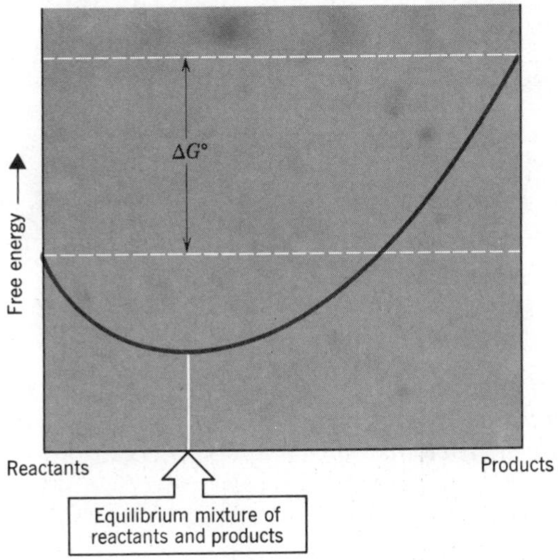


**Figure 10.12**
*The variation of free energy in a homogeneous chemical system as the reaction proceeds from pure reactants on the left to pure products on the right. The minimum in the curve represents the extent of reaction required for the system to achieve equilibrium.*

In Figure 10.12 we see that for a reaction in which reactants and products are in the same phase there is a minimum in the free-energy curve that lies below both $G^0_{\text{reactants}}$ and $G^0_{\text{products}}$. Since a system always seeks the minimum free energy, some reaction always occurs, whether we begin with pure reactants and proceed in the direction of the products or vice versa. In each case the chemical change is accompanied by a free energy decrease as $G$ heads toward the minimum; hence $\Delta G$ for *reaction toward the minimum* is negative. A spontaneous chemical change, therefore, always takes place regardless of whether we begin with pure reactants or products.

Another important aspect of Figure 10.12 is the minimum itself. Once this minimum free energy has been achieved by the system, the composition of the system can no longer change, because such a change involves going "uphill" on the free energy curve (that is, $\Delta G$ positive). We already know that this is not spontaneous. As a result, the minimum on the curve represents the system in a state of equilibrium.

A third important point to note is that the direction in which the reaction will proceed for a given composition (that is, a given number of moles of reactants and products) is not controlled by the sign of $\Delta G^0$ but, instead, depends on the slope of the free-energy curve at that point along the reaction path. For the reaction depicted by Figure 10.12, for instance, if we were to begin with pure reactants and proceed in the direction of the products, $\Delta G$ would be negative but $\Delta G^0 = G^0_{\text{prod}} - G^0_{\text{react}}$ would be positive. It is the sign of $\Delta G$, not $\Delta G^0$, that determines spontaneity.

Although the slope of the free-energy curve determines the direction in which the reaction will proceed at a given point along the reaction path, it is $\Delta G^0$ that controls the position of the minimum in the curve. This is shown in Figure 10.13. Notice that the position of the minimum in the free energy curve depends on the relative values of $G^0_{\text{react}}$ and $G^0_{\text{prod}}$, always lying closer to the one that is lowest in energy. Since these minima correspond to different system compositions, we are therefore observing different positions of equilibrium that depend on the sign and magnitude of $\Delta G^0$. Although the computed value of $\Delta G^0$ for a reaction is not an indicator of spontaneity, it does serve as a guide to the *feasibility* of a reaction. Consider, for example,

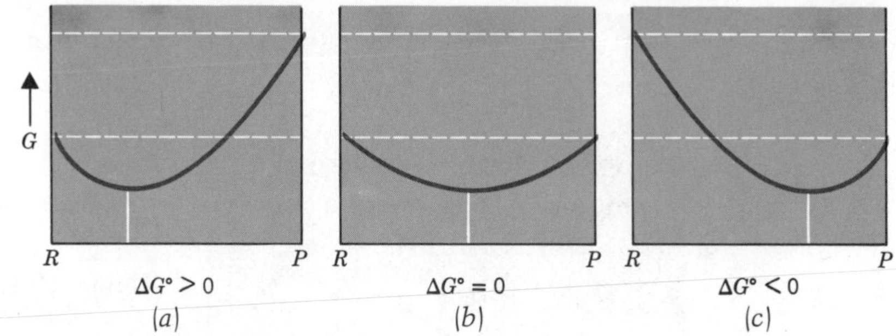


Figure 10.13

*The position of equilibrium changes as the value of $\Delta G^0$ changes. (a) Position of equilibrium in favor of reactants. (b) Position of equilibrium intermediate between reactants and products. (c) Position of equilibrium in favor of products.*

the compounds listed in Table 10.5. Some of these compounds possess positive standard free energies of formation. In these cases the reaction

$$\text{elements} \longrightarrow \text{compound}$$

will not proceed far to the right (i.e., not much product will be formed) because the position of equilibrium favors the reactants. Consequently, substances having positive free energies of formation are not usually prepared directly from the free elements because the equilibrium reaction mixture contains only a very small quantity of the desired product. From a practical standpoint, then, such a reaction is not very feasible. Since we can use values of $\Delta G_f^0$ to compute $\Delta G^0$ for any reaction, we have a guide to feasibility.

Table 10.5

*Standard free energies of formation at 25°C and 1 atm*

| Substance | $\Delta G_f^0$ kcal/mole (kJ/mole) | Substance | $\Delta G_f^0$ kcal/mole (kJ/mole) |
|---|---|---|---|
| $Al_2O_3$ (*s*) | −376.8 (−1577) | HBr (*g*) | −12.7 (−53.1) |
| $AgNO_3$ (*s*) | −7.7 (−32) | HI (*g*) | +0.31 (+1.30) |
| C (*s*, diamond) | +0.69 (+2.9) | $H_2O$ (*l*) | −56.7 (−237) |
| CO (*g*) | −32.8 (−137) | $H_2O$ (*g*) | −54.6 (−228) |
| $CO_2$ (*g*) | −94.3 (−395) | $MgCl_2$ (*s*) | −141.6 (−592.5) |
| $CH_4$ (*g*) | −12.1 (−50.6) | $Mg(OH)_2$ (*s*) | −199.3 (−833.9) |
| $C_2H_6$ (*g*) | −7.9 (−33) | $NH_3$ (*g*) | −4.0 (−17) |
| $C_2H_4$ (*g*) | +16.3 (+68.2) | $N_2O$ (*g*) | +24.8 (+104) |
| $C_2H_2$ (*g*) | +50.0 (+209) | NO (*g*) | +20.7 (+86.8) |
| $C_3H_8$ (*g*) | −5.6 (−23) | $NO_2$ (*g*) | +12.4 (+51.9) |
| $CCl_4$ (*l*) | −15.6 (−65.3) | $HNO_3$ (*l*) | −19.1 (−79.9) |
| $C_2H_5OH$ (*l*) | −41.8 (−175) | $PbO_2$ (*s*) | −52.3 (−219) |
| $CH_3COOH$ (*l*) | −93.8 (−392) | $PbSO_4$ (*s*) | −193.9 (−811.3) |
| CaO (*s*) | −144.4 (−604.2) | $SO_2$ (*g*) | −71.8 (−300) |
| $Ca(OH)_2$ (*s*) | −214.3 (−896.6) | $SO_3$ (*g*) | −88.5 (−370) |
| $CaSO_4$ (*s*) | −315.6 (−1320) | $H_2SO_4$ (*l*) | −164.9 (−689.9) |
| CuO (*s*) | −30.4 (−127) | $SiO_2$ (*s*) | −192.4 (−805) |
| $Fe_2O_3$ (*s*) | −177.1 (−741.0) | $SiH_4$ (*g*) | −9.4 (−39) |
| HF (*g*) | −64.7 (−271) | ZnO (*s*) | −76.1 (−318) |
| HCl (*g*) | −22.8 (−95.4) | | |

EXAMPLE 10.7 Under standard conditions, would the reaction,

$$2HI\ (g) + Cl_2\ (g) \longrightarrow 2HCl\ (g) + I_2\ (s)$$

be thermodynamically feasible?

SOLUTION To answer this question, we must compute $\Delta G^0$ for the reaction,

$$\Delta G^0 = 2\ \Delta G_f^0[HCl\ (g)] - 2\ \Delta G_f^0[HI\ (g)]$$

We can compute $\Delta G^0$ in either kilocalories or kilojoules. The choice isn't really important, because we are interested only in the sign of $\Delta G^0$. From Table 10.5,

$$\Delta G^0 = 2\ \cancel{\text{moles}} \times \left(\frac{-95.4\ \text{kJ}}{\cancel{\text{mole}}}\right) - 2\ \cancel{\text{moles}} \times \left(\frac{+1.30\ \text{kJ}}{\cancel{\text{mole}}}\right)$$

$$\Delta G^0 = -193.4\ \text{kJ}$$

From the sign of $\Delta G^0$, we conclude that the position of equilibrium in this system lies far to the right and that when HI and $Cl_2$ gases are mixed, we ought to observe a significant amount of $I_2$ being formed. This, in fact, is precisely what is observed when these two gases are mixed.

## 10.14 APPLICATIONS OF THE PRINCIPLES OF THERMODYNAMICS

The thermodynamic concepts of enthalpy, entropy, and spontaneity developed in this chapter have wide applicability to all areas of chemistry. We will have occasion to apply them to chemical equilibrium, electrochemistry (where we look at the relationship between electrical energy and chemical change), and to many of the chemical characteristics of the elements and their compounds.

The impact of thermodynamics can also be seen in the practical world around us. A classic example of this is in the production of synthetic diamonds. People had been fascinated by this problem ever since 1797, when it was found that diamond was simply a form of carbon. Over the years many experiments were devised in an attempt to convert graphite, the common form of carbon, into its much more valuable counterpart. However, as of 1938 no one had yet been able to accomplish this feat. At that time a careful thermodynamic analysis of the problem was performed, the results of which are summarized in Figure 10.14.

In this figure, $-\Delta G/T$ is plotted along the vertical axis and temperature, in degrees kelvin, is plotted along the horizontal axis. Since $\Delta G$ for the reaction,

$$C\ (s,\ \text{graphite}) \longrightarrow C\ (s,\ \text{diamond})$$

must be negative in order for the process to be spontaneous, diamond can be produced only at temperatures and pressures that lie *above* the zero on the $-\Delta G/T$ scale. For example, at 470 K the conversion of graphite into diamond can only take place at pressures greater than or equal to 20,000 atm. We can also conclude that at a constant pressure of 20,000 atm the reaction is not spontaneous above 470 K.

This analysis, then, served to define the limits of temperature and pressure that would permit the conversion to take place. That was not the end of the problem, however, because suitable materials had to be found that would allow the reaction to proceed at a measurable rate. In fact, it was not until 1955 that success was finally achieved.

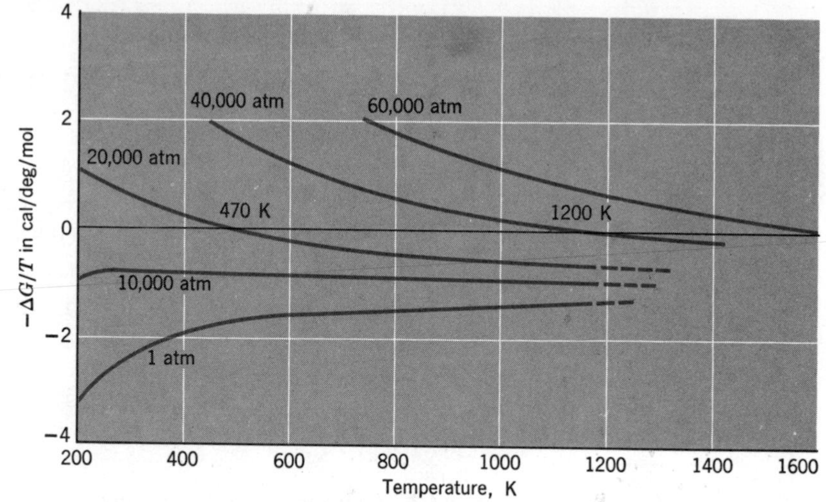


Figure 10.14
*Thermodynamics of graphite to diamond conversion. From* Chemical and Engineering News, *April 5, 1971, p. 51. Used by permission.*

The conversion of graphite to diamond represents but one example of how thermodynamics can serve as a guide to answering practical questions. The principles of thermodynamics have been applied over the years to such diverse problems as the design of steam engines and the development of fuel cells that are now used as a source of electric power in spacecraft. Dr. Frederick Rossini, a noted thermodynamicist, has pointed out that the balance between the simultaneous opposing drives toward security (low energy) and freedom (high entropy) that control chemical equilibrium also seem to determine equilibrium in a stable society, illustrating, perhaps, the truly wide scope of thermodynamics.

Another illustration of the impact of thermodynamics on our lives is the problem of thermal pollution that has become so important in our efforts to save what is left of our environment. It is a consequence of the second law of thermodynamics that any heat engine that is designed to convert heat into useful work must absorb heat from a high-temperature source. Some of this heat can be converted to work; however, some of it must also be deposited in a low-temperature reservoir. Therefore, any device that tries to convert heat into useful work also discharges some heat into the environment. An automobile engine burns gasoline (when it's available), converts part of the resultant heat energy into work that propels the car, while the remainder is discharged to the radiator. Large power stations burn fuels (either fossil fuels such as coal or oil, or nuclear fuels such as uranium) to generate electrical power. In the process they too deposit relatively large quantities of heat into the environment.

Cooling water from such power plants is heated to such an extent that it has been known to raise the temperature of surrounding rivers or lakes by as much as 25°F or more. These higher temperatures decrease the solubility of oxygen and kill some species of fish. It has also been found, however, that clams and oysters thrive in these warmer waters.

Several years ago a nuclear power station in New Jersey was forced to shut down for repairs during the winter. It is interesting that in this case, the halting of thermal pollution caused a large fish kill because many fish that normally live in warm water were suddenly "put out in the cold."

Perhaps closest to home, however, is the energetics of biochemical reactions. Solar energy is captured by green plants through photosynthesis in which $CO_2$ and $H_2O$ are combined to produce the glucose molecule, $C_6H_{12}O_6$. This reaction,

$$6CO_2 + 6H_2O \longrightarrow C_6H_{12}O_6 + 6\ O_2$$

is highly endothermic, with $\Delta G^0 = +686$ kcal/mole (2870 kJ/mole). At a later time plants and animals release the stored energy in glucose by oxidation, which regenerates $CO_2$ and $H_2O$. As mentioned earlier, only about 40% of the available free energy from this oxidation finds its way into useful work. The remainder is released as heat that goes toward increasing the entropy of the universe. Thus, through a series of reactions having $CO_2$ and $H_2O$ as both reactants and ultimate products, life on the earth is able to capture and make use of the energy released from thermonuclear reactions in the sun.

## INDEX TO QUESTIONS AND PROBLEMS (problem numbers in italics)

Terminology 1

First law 2, 3, 4, 5, 6, *23, 24*

Reversible and irreversible processes 8

Heats of reaction *42, 43, 44, 45, 46, 47*

Calorimetry *27, 28*

Enthalpy 9

$\Delta E$ and $\Delta H$ *25, 26, 29, 30, 49*

Hess' law 10, *31, 32, 33, 34, 35, 36, 37, 38, 39, 40, 41, 48*

Atomization energy and bond energies 11, 12, *50, 51*

Spontaneity 7, 13, 16

Entropy 14, 17, *52, 54, 56*

Free energy 18, 19, 20, 21, 22, *53, 55, 57, 58, 59*

Third law 15

## REVIEW QUESTIONS

**10.1** Give definitions for the following: system, surroundings, adiabatic process, isothermal process, state function, molar heat capacity, specific heat.

**10.2** State the first law of thermodynamics. State, on a molecular basis, why $E$ is a state function.

**10.3** Can you explain why it is not possible to measure or calculate the total amount of internal energy possessed by a system?

**10.4** Why is $\Delta E$ equal to zero for an isothermal expansion or compression of an ideal gas?

**10.5** When an ideal gas undergoes an isothermal expansion into an evacuated container, $\Delta E$, $q$, and $w$ are all equal to zero. Real gases, however, generally cool upon expansion into a vacuum. What conclusions can you draw about $\Delta E$, $q$ and $w$ for the *isothermal* expansion of a real gas into a vacuum?

**10.6** What are some different kinds of work that a system can perform on its surroundings?

**10.7** What is meant by the term, spontaneous change?

**10.8** What is a reversible process?

**10.9** Why are chemists usually concerned about changes in enthalpy rather than changes in internal energy?

**10.10** What is Hess' law of heat summation? What is meant by the term, standard state?

**10.11** What is meant by atomization energy? Write an equation representing the process for which $\Delta H$ is the atomization energy of $H_2O$. Indicate the physical state (gas, liquid, or solid) for each substance in the equation.

**10.12** Why don't values of $\Delta H_f^0$ computed from tabulated bond energies agree precisely with $\Delta H_f^0$ measured experimentally?

**10.13** In what way is statistical probability related to spontaneity? How is entropy related to statistical probability?

**10.14** In computing $\Delta S$ for a system, why must $q_{rev}$ be used?

**10.15** Explain why the entropy of a pure substance is zero at 0 K. Would the entropy of a mixture be zero at 0 K? Explain your answer.

**10.16** What two criteria must be met in order for a process to be spontaneous, regardless of the temperature?

**10.17** What is the sign of the entropy change for each of the following processes?
(a) A solute crystallizes from a solution.
(b) Water evaporates.
(c) A deck of playing cards is shuffled.
(d) A card player is dealt 13 spades.
(e) Solid AgCl precipitates from a solution of $AgNO_3$ and NaCl.
(f) $^{235}_{92}U$ is extracted from a mixture of $^{235}_{92}U$ and $^{238}_{92}U$.
(g) $Na_2CO_3$ $(aq)$ + HCl $(aq) \rightarrow$ 2NaCl $(aq)$ + $H_2O$ + $CO_2$ $(g)$

**10.18** Why is it possible for a chemical reaction to occur spontaneously even though $\Delta G^0$ for the reaction is positive?

**10.19** Describe the relationship between $\Delta G^0$ and the position of equilibrium in a chemical reaction.

**10.20** Referring to Figure 10.14, what is the minimum pressure necessary for the conversion of graphite to diamond at a temperature of 200 K? Is it theoretically possible to change graphite to diamond at 1 atm?

**10.21** Explain why $\Delta G = 0$ for a system that is in a state of equilibrium.

**10.22** $\Delta H$ and $\Delta S$ are nearly independent of temperature. Why is this not true for $\Delta G$?

## REVIEW PROBLEMS

**10.23** A gas, initially under a pressure of 15.0 atm and having a volume of 10.0 liters, is permitted to expand isothermally in two steps. In the first step the external pressure is held constant at 7.50 atm, and in the second step the external pressure is maintained at 1.00 atm. What are $q$ and $w$ for each step? What is the net change in the internal energy for both the system and its surroundings? Assume ideal gas behavior.

**10.24** What is the maximum work, in joules, that could be obtained if the expansion described in Problem 10.23 were carried out reversibly?

**10.25** At 25°C and a constant pressure of 1.00 atm, the reaction of $\frac{1}{2}$ mole of $OF_2$ with water vapor, according to the equation

$$OF_2\ (g) + H_2O\ (g) \longrightarrow O_2\ (g) + 2HF\ (g)$$

liberates 38.6 kcal. Calculate $\Delta H$ and $\Delta E$ per mole of $OF_2$.

**10.26** At 25°C and 1.00 atm, the reaction of 1.00 mole of CaO with water (shown below) evolves 15.6 kcal.

$$CaO\ (s) + H_2O\ (l) \longrightarrow Ca(OH)_2\ (s)$$

What are $\Delta H$ and $\Delta E$, per mole of CaO, for this process given that the densities of CaO $(s)$, $H_2O$ $(l)$, and $Ca(OH)_2$ $(s)$ at 25°C are 3.25 g/ml, 0.997 g/ml, and 2.24 g/ml, respectively? What does this tell you about the relative values of $\Delta H$ and $\Delta E$ when all substances are either liquids or solids?

***10.27** At 25°C, burning 0.20 mole of $H_2$ with 0.1 mole of $O_2$ to produce $H_2O$ $(l)$ in a bomb calorimeter raises the temperature of the apparatus 0.880°C. When 0.0100 mole of toluene, $C_7H_8$, is burned in this calorimeter, the temperature is raised by 0.615°C. The equation for the combustion reaction is

$$C_7H_8\ (l) + 9\ O_2\ (g) \longrightarrow 7CO_2\ (g) + 4H_2O\ (l)$$

Calculate $\Delta E$ for this reaction. Use $\Delta H_f^0$ for $H_2O$ $(l)$ found in Table 10.1 to compute $\Delta E_f^0$ for $H_2O$ $(l)$.

***10.28** Based on the results of Question 10.27, compute the standard heat of formation, $\Delta H_f^0$, of toluene. (Assume that the reaction was carried out at 25°C.)

**10.29** The reaction, $Ca\ (s) + O_2\ (g) + H_2\ (g) \rightarrow Ca(OH)_2\ (s)$, has $\Delta H^0 = -214.3$ kcal. What is $\Delta E$ for this reaction?

**10.30** The heat of vaporization $\Delta H_{vap}$ of $H_2O$ at 25°C is 10.5 kcal/mole. Calculate $q$, $w$, and $\Delta E$ for the process.

**10.31** Use Hess' law to calculate $\Delta H^0$ for each of the following reactions:

(a) $2Al\ (s) + Fe_2O_3\ (s) \rightarrow Al_2O_3\ (s) + 2Fe\ (s)$

(b) $SiH_4\ (g) + 2O_2\ (g) \rightarrow SiO_2\ (s) + 2H_2O\ (g)$

(c) $CaO\ (s) + SO_3\ (g) \rightarrow CaSO_4\ (s)$

(d) $CuO\ (s) + H_2\ (g) \rightarrow Cu\ (s) + H_2O\ (g)$

(e) $C_2H_4\ (g) + H_2\ (g) \rightarrow C_2H_6\ (g)$

**10.32** Calculate $\Delta H^0$ for each of the following reactions:

(a) $C_2H_2\ (g) + H_2\ (g) \rightarrow C_2H_4\ (g)$

(b) $SO_3\ (g) + H_2O\ (l) \rightarrow H_2SO_4\ (l)$

(c) $Mg(OH)_2\ (s) + 2HCl\ (g) \rightarrow MgCl_2{\cdot}2H_2O\ (s)$

(d) $CO_2\ (g) + H_2\ (g) \rightarrow CO\ (g) + H_2O\ (g)$

(e) $10N_2O\ (g) + C_3H_8\ (g) \rightarrow 10N_2\ (g) + 3CO_2\ (g) + 4H_2O\ (g)$

**10.33** Given the following thermochemical equations,

$$Fe_2O_3\ (s) + 3CO\ (g) \longrightarrow 2Fe\ (s) + 3CO_2\ (g) \qquad \Delta H = -28 \text{ kJ}$$

$$3Fe_2O_3\ (s) + CO\ (g) \longrightarrow 2Fe_3O_4\ (s) + CO_2\ (g) \qquad \Delta H = -59 \text{ kJ}$$

$$Fe_3O_4\ (s) + CO\ (g) \longrightarrow 3FeO\ (s) + CO_2\ (g) \qquad \Delta H = +38 \text{ kJ}$$

calculate $\Delta H$ for the reaction,

$$FeO\ (s) + CO\ (g) \longrightarrow Fe\ (s) + CO_2\ (g)$$

without referring to the data in Table 10.1.

**10.34** Use the results of Question 10.33 and the data in Table 10.1 to compute the standard heat of formation of FeO.

**10.35** Acetylene, a gas used in welding torches, is produced by the action of water on calcium carbide, $CaC_2$. Given the following thermochemical equations, calculate $\Delta H_f^0$ for acetylene.

$$CaO\ (s) + H_2O\ (l) \longrightarrow Ca(OH)_2\ (s) \qquad \Delta H° = -15.6 \text{ kcal}$$

$$CaO\ (s) + 3C\ (s) \longrightarrow CaC_2\ (s) + CO\ (g) \qquad \Delta H° = +110.5 \text{ kcal}$$

$$CaC_2\ (s) + 2H_2O\ (l) \longrightarrow Ca(OH)_2\ (s) + C_2H_2\ (g) \qquad \Delta H° = -30.0 \text{ kcal}$$

$$2C\ (s) + O_2\ (g) \longrightarrow 2CO\ (g) \qquad \Delta H° = -52.8 \text{ kcal}$$

$$2H_2O\ (l) \longrightarrow 2H_2\ (g) + O_2\ (g) \qquad \Delta H° = +136.6 \text{ kcal}$$

**10.36** Aerosol propellants are often chlorofluoromethanes (CFMs) such as Freon-11 ($CFCl_3$) and Freon-12 ($CF_2Cl_2$). It has been suggested that continued use of these may ultimately deplete the ozone shield in the stratosphere, with catastrophic results to the inhabitants of our planet. In the stratosphere CFMs absorb high-energy radiation and produce Cl atoms that have a catalytic effect on removing ozone.

$$O_3 + Cl \longrightarrow O_2 + ClO \qquad \Delta H^0 = -30 \text{ kcal}$$

$$ClO + O \longrightarrow Cl + O_2 \qquad \Delta H^0 = -64 \text{ kcal}$$

$$\text{net} \qquad O_3 + O \longrightarrow 2\ O_2$$

The O atoms are present due to dissociation of $O_2$ molecules by high-energy radiation. Calculate $\Delta H^0$ for the net reaction for the removal of the ozone.

**10.37** Plaster of Paris, $CaSO_4{\cdot}\frac{1}{2}H_2O$, is mixed with water with which it combines to produce gypsum, $CaSO_4{\cdot}2H_2O$. The reaction is exothermic, which explains why a plaster cast on a broken arm becomes warm as the cast hardens. Given that for $CaSO_4{\cdot}\frac{1}{2}H_2O$, $\Delta H_f^0 = -1573$ kJ/mole, and for $CaSO_4{\cdot}2H_2O$, $\Delta H_f^0 = -2020$ kJ/mole, calculate $\Delta H^0$ for the reaction,

$$CaSO_4{\cdot}\tfrac{1}{2}H_2O\ (s) + \tfrac{3}{2}H_2O\ (l) \longrightarrow CaSO_4{\cdot}2H_2O\ (s)$$

**10.38** Important reactions in the production of ozone in polluted air are

$$2NO\ (g) + O_2\ (g) \longrightarrow 2NO_2\ (g)$$

$$NO_2\ (g) \xrightarrow{h\nu} NO\ (g) + O\ (g)$$

$$O_2\ (g) + O\ (g) \longrightarrow O_3\ (g)$$

Calculate $\Delta H^0$ (in kilojoules) for each of these processes using the data in Table 10.1 and 10.2.

**10.39** Solid sodium bicarbonate (baking soda) is easily decomposed to produce $Na_2CO_3\ (s)$, $CO_2\ (s)$, and $H_2O\ (g)$. This property makes it useful in baking because the evolved $CO_2$ produces tiny bubbles in the dough, thereby causing it to "rise" during baking. Given that $\Delta H_f^0 = -226.5$

kcal/mole for $NaHCO_3$ (s) and $\Delta H_f^0 = -270.3$ kcal/mole for $Na_2CO_3$ (s), use the data in Table 10.1 to compute $\Delta H^0$ for the reaction.

$$2NaHCO_3\,(s) \longrightarrow Na_2CO_3\,(s) + CO_2\,(g) + H_2O\,(g)$$

**10.40** The body eliminates ethyl alcohol, $C_2H_5OH$, by oxidation to give water and the following series of carbon-containing products,

$$C_2H_5OH \xrightarrow{O_2} CH_3CHO \xrightarrow{O_2} CH_3COOH \xrightarrow{O_2} CO_2$$

Write balanced equations for each step in the oxidation and calculate its $\Delta H^0$. What is the overall $\Delta H^0$ for complete oxidation to $CO_2$ and $H_2O$?

**10.41** Use the data in Table 10.1 to determine how many calories are evolved in the combustion of 45.0 g of $C_2H_6$ (g) to produce $CO_2$ (g) and $H_2O$ (g) under a constant pressure of 1.00 atm.

**10.42** The average adult expends about 2000 kcal of energy per day for normal activity. If 1 g of carbohydrate provides 4 kcal of usable energy, how many grams of carbohydrates must be consumed to meet these caloric demands?

**10.43** The evaporation of perspiration is one mechanism whereby the body disposes of excess thermal energy and manages to maintain a constant temperature. How much energy is removed from the body by the evaporation of 10.0 g of $H_2O$?

***10.44** Calculate the energy liberated during the combustion of 1 gal (3.8 liters) of octane (gasoline). The density of octane is 0.703 g/ml. What weight of hydrogen would have to be burned [giving $H_2O$ (*l*)] to produce this same amount of heat? If this $H_2$ were compressed to a pressure of 2500 lb/in.$^2$ (170 atm) at 25°C, what volume would it occupy? What does this suggest about the feasibility of a hydrogen fuel economy for the automobile? For octane, $C_8H_{18}$ (*l*), $\Delta H_f^0 = -49.82$ kcal/mole.

***10.45** It is estimated that the body generates up to 1400 kcal of thermal energy per hour during heavy physical exercise. If the only way that this excess energy could be dissipated was through evaporation of water, how much water would have to evaporate per hour to keep the body temperature constant?

***10.46** How many grams of glucose, $C_6H_{12}O_6$, would have to be metabolized per hour (to give $CO_2$ and $H_2O$) in order to generate the excess thermal energy described in the preceding problem if it is assumed that 60% of the available energy appears as excess body heat? (The remaining 40% is used by the body to do mechanical work—moving of limbs, pumping of blood, etc.). For $C_6H_{12}O_6$, $\Delta H_{\text{combustion}} = -674$ kcal/mole.

***10.47** How many liters of natural gas ($CH_4$) at 25°C and 1 atm must be burned to provide sufficient energy to convert 250 ml of $H_2O$ at 20°C (approx. 8 fluid ounces) into steam at 100°C?

***10.48** The ionization energy of Na is 118.1 kcal/mole and the electron affinity of Cl is 83.3 kcal/mole. Use this information, along with the heats of formation of gaseous Na and Cl atoms as well as the heat of formation of NaCl, to calculate $\Delta H$ for the reaction,

$$NaCl\,(s) \longrightarrow Na^+\,(g) + Cl^-\,(g)$$

The answer corresponds to the lattice energy of sodium chloride. (*Hint:* It will help if you write thermochemical equations for each process described in the problem.)

***10.49** An important photochemical reaction in the production of smog is

$$NO_2\,(g) + h\nu \longrightarrow NO\,(g) + O\,(g)$$

If one quantum of energy is required to cause this reaction to occur, what must the wavelength of the light be? (Use the data in Tables 10.1 and 10.2, calculate $\Delta E$ for the process, then calculate $\nu$ from the Planck relationship $\Delta E = h\nu$.)

***10.50** Benzene is often written as a resonance hybrid of two equivalent structures,

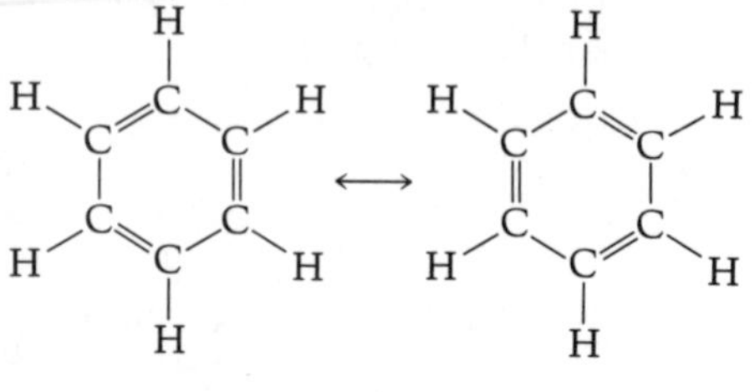

The $\Delta H_f^0$ for gaseous benzene has been determined from its heat of combustion to be +82.8 kJ/mole.

$$6C\ (s) + 3H_2\ (g) \longrightarrow C_6H_6\ (g) \qquad \Delta H_f^0 = +82.8\ \text{kJ/mole}$$

Use the data in Tables 10.2 and 10.3 to calculate $\Delta H_f^0$. How does your calculated value compare with the experimental value? The difference between the calculated and experimental values is called the resonance energy. What might you conclude about the stability of species that exist as a composite of two or more resonance structures?

**10.51** Use the average bond energies in Table 10.3 to compute the standard heat of formation of $C_3H_8$. Its structure is

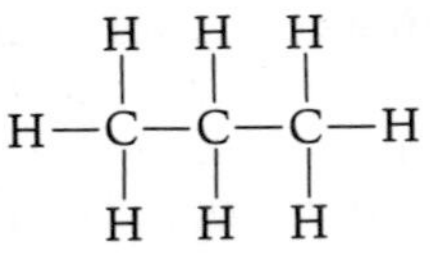

How well does your computed value compare with that reported in Table 10.1?

**10.52** The heat of fusion of water at 0°C is 1.44 kcal/mole; its heat of vaporization is 9.72 kcal/mole at 100°C. What are $\Delta S$ for the melting and boiling of 1 mole of water? Can you explain why $\Delta S_{\text{vap}}$ is greater than $\Delta S_{\text{melting}}$?

**10.53** From the data in Tables 10.1 and 10.4, calculate the boiling point of liquid bromine [i.e., the temperature at which $Br_2$ (*l*) and $Br_2$ (*g*) can coexist in equilibrium with each other].

**10.54** Which of the following reactions is accompanied by the greatest entropy change?
(a) $SO_2\ (g) + \frac{1}{2}O_2\ (g) \rightarrow SO_3\ (g)$
(b) $CO\ (g) + \frac{1}{2}O_2\ (g) \rightarrow CO_2\ (g)$

**10.55** Compute $\Delta G^0$ for each of the reactions in Problem 10.31.

**10.56** Use the results of Problems 10.31 and 10.55 to calculate $\Delta S^0$ for each of the reactions in Problem 10.31.

**10.57** The standard free energy of formation of glucose is $\Delta G_f^0 = -217.54$ kcal/mole. Calculate $\Delta G^0$ for the reaction,

$$C_6H_{12}O_6\ (s) + 6O_2\ (g) \longrightarrow 6CO_2\ (g) + 6H_2O\ (l)$$

**10.58** What is the maximum amount of useful work that could be obtained by the oxidation of propane, $C_3H_8$, according to the equation,

$$C_3H_8\ (g) + 5O_2\ (g) \longrightarrow 3CO_2\ (g) + 4H_2O\ (g)$$

Why is it that we always get less than this maximum amount of work in any real process that uses propane as a fuel?

**10.59** Which of the following reactions could *potentially* serve as a practical method for the preparation of $NO_2$? (*Note:* The equations are not balanced.)
(a) $N_2\ (g) + O_2\ (g) \rightarrow NO_2\ (g)$
(b) $HNO_3\ (l) + Ag\ (s) \rightarrow AgNO_3\ (s) + NO_2\ (g) + H_2O\ (l)$
(c) $NH_3\ (g) + O_2\ (g) \rightarrow NO_2\ (g) + NO\ (g) + H_2O\ (g)$
(d) $CuO\ (s) + NO\ (g) \rightarrow NO_2\ (g) + Cu\ (s)$
(e) $NO\ (g) + O_2\ (g) \rightarrow NO_2\ (g)$
(f) $H_2O\ (g) + N_2O\ (g) \rightarrow NH_3\ (g) + NO_2\ (g)$

# 11 CHEMICAL KINETICS

It does not take long to find a reaction that thermodynamics predicts should proceed nearly to completion but yet is not observed to occur. We know from the last chapter that hydrogen and oxygen can be kept in contact with one another almost forever without forming noticeable amounts of water, even though their reaction to produce water is accompanied by a free-energy decrease. This is an example of a chemical change where the speed of the reaction governs whether the formation of the products will or will not be observed.

Chemical kinetics, also referred to as chemical dynamics, concerns itself with the speed, or rates, of chemical reactions. In this area of chemistry we study the factors that control how rapidly chemical changes occur. These include the following:

1. The nature of the reactants and products
2. The concentration of reacting species (related to this is the pressure of gaseous reactants)
3. The effect of temperature
4. The influence of outside agents (catalysts)

By studying the factors that influence rates of reaction we can begin to make educated guesses about the *detailed sequence of steps* (called the **mechanism** of the reaction) that are followed along the path from reactants to products. This is important because it allows us to gain insight into some very fundamental aspects of why substances react the way they do.

## 11.1 REACTION RATES AND THEIR MEASUREMENT

Before we examine the factors that influence rates of reaction, let's be sure that we know what is meant by "rate." To determine the rate of a given chemical reaction, we must measure how fast the concentration of a reactant or product changes during the course of the investigation. In practice, the species whose concentration is easiest to follow is determined at various time intervals. The simplest example is a reaction where only one reactant undergoes a change to form a single product. An example of this type of reaction is the conversion of cyclopropane,

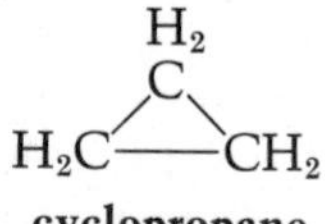

**cyclopropane**

into the molecule propylene:

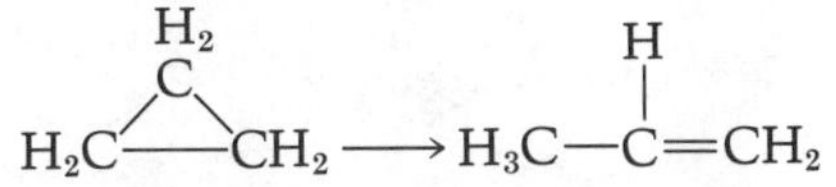

In general, the balanced equation for this type of reaction would be

$$A \longrightarrow B \qquad (11.1)$$

When the reaction is carried out, no product ($B$) is present initially and, as time goes on, the concentration of $B$ increases with a corresponding decrease in the concentration of $A$ (Figure 11.1).

In general, the rate (or speed) of any chemical reaction can be expressed as the ratio of the change in the concentration of a reactant (or product) to a change in time. This is exactly analogous to giving the speed of an automobile as the change in position (that is, the distance traveled) divided by its time of travel. Here the speed might be given in the units, miles per hour. With chemical reactions the rate is usually expressed in the units, moles per liter per second:

$$\text{speed of auto} = \frac{\text{distance}}{\text{time}} = \frac{\text{miles}}{\text{hour}}$$

$$\text{rate of chemical reaction} = \frac{\text{change in concentration}}{\text{time}}$$

$$= \frac{\text{moles/liter}}{\text{sec}}$$

An inspection of Figure 11.1 reveals that the rate of this chemical reaction changes with time. For instance, near the start of the reaction the concentration of $A$ is decreasing rapidly and the concentration of $B$ is rising rapidly. Much later during the reaction, however, only small changes in concentration occur with time, and the rate is therefore much less. In general, this type of behavior is observed with nearly every chemical reaction; as the reactants are consumed, the rate of reaction gradually decreases.

An accurate, quantitative estimate of the rate of reaction at any given moment during the reaction can be obtained from the slope of the tangent to the concentration–time curve at that particular instant. This is shown in

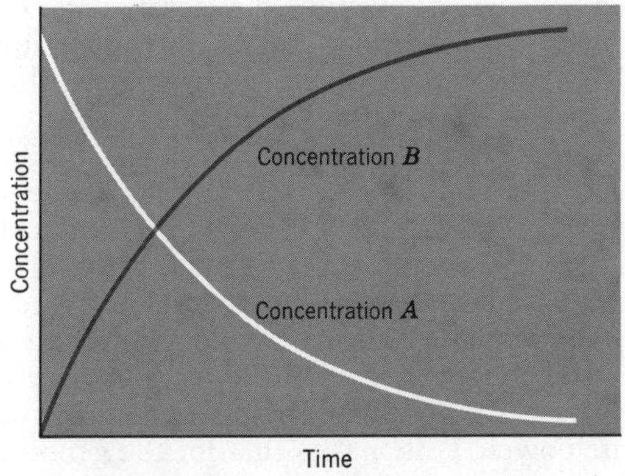


Figure 11.1
*Change in the concentration of reactants and products with time for the reaction* A → B.

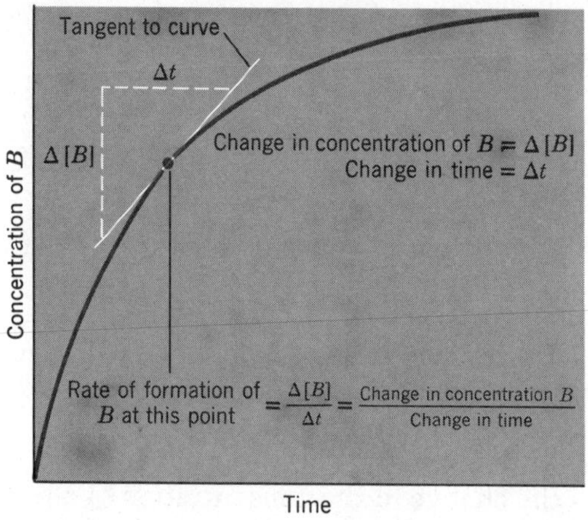


**Figure 11.2**
*Estimation of the rate of reaction based on the change in concentration of* B *with time.*

Figure 11.2. Square brackets, [ ], are used here to denote concentration in moles per liter. From the tangent to the curve we can write that

$$\text{rate} = \frac{\Delta[B]}{\Delta t} \tag{11.2}$$

We can also express the rate of the above reaction in terms of the concentration of the reactant $A$, because its concentration is also changing with time. The rate measured in terms of the concentration of $A$ would be

$$\text{rate} = \frac{-\Delta[A]}{\Delta t} \tag{11.3}$$

The minus sign indicates that the concentration of $A$ is decreasing with time. A minus sign is always used whenever reactants are employed to express the rate.

When measuring the rate of any chemical reaction, the concentration that is monitored and the technique that is used to measure the change depends on the nature of the reaction. For example, for gaseous reactions the pressure can be followed, provided that there is a change in the number of moles of gas as the reaction proceeds. On the other hand, if a colored reactant or product is involved, the intensity of color can be monitored during the reaction. Whatever method of analysis is employed, it must be fast, accurate, and in no way interfere with the normal course of the reaction being studied.

## 11.2 RATE LAWS

In this section we begin to examine the factors that control the rate of reaction. Not all reactions take place at the same rate. Ionic reactions like those described in Chapter 5 are virtually instantaneous; the speed is determined by how rapidly we can mix the chemicals. Other reactions, such as the digestion of food, take place more slowly. These different rates exist primarily because of chemical differences among the reacting substances.

For any given reaction, one of the most important controlling influences is the concentrations of the reactants. Generally, if we follow a chemical reaction over a period of time, we find that its rate gradually decreases as the reactants are consumed. From this we conclude that the rate is related, in some way, to the concentrations of the reacting species. In fact, the rate is nearly always directly proportional to the concentration of the reactants raised to some power. This means that for the general reaction

$$A \longrightarrow B$$

the rate can be written as

$$\text{rate} \propto [A]^x \tag{11.4}$$

where the exponent, $x$, is called the **order of the reaction.** When $x = 1$ we have a first order reaction. An example is the decomposition of cyclopropane mentioned earlier,

$$\text{rate} \propto [\text{cyclopropane}]^1$$

Second ($x = 2$), third ($x = 3$), and higher-order reactions are also possible, as are reactions in which $x$ is a fraction. There are also examples of zero order reactions, where $x = 0$. For a zero order reaction the rate is constant and does not depend on the concentration of the reactants. An example of this type of reaction is the decomposition of ammonia on a platinum or tungsten metal surface. The rate at which the ammonia decomposes is always the same, regardless of its concentration. Another example of a zero order process is the elimination of ethyl alcohol by the body. Regardless of how much alcohol is present in the bloodstream, its rate of expulsion from the body is constant. Thus the rate is independent of concentration.

A very important fact is that there is not necessarily any direct relationship between the coefficients in the chemical equation for a reaction and the order of the reaction. *The value of* $x$ *can only be determined from experiment.*

For the slightly more complex reaction,

$$A + B \longrightarrow \text{products}$$

the rate is usually dependent on both the concentration of $A$ and the concentration of $B$. Generally, decreasing the concentration of either $A$ or $B$ will decrease the reaction rate. The rate is therefore proportional to the concentrations of both $A$ and $B$, each raised to some power.

$$\text{rate} \propto [A]^x[B]^y \tag{11.5}$$

From this equation we say that the order of the reaction with respect to $A$ is $x$ and that the order with respect to $B$ is $y$, and that the overall order (i.e., the sum of the individual orders) is $x + y$. Once again, $x$ and $y$ can have whole-number, fractional, or zero values. When one of the exponents is zero, this simply means that the rate of reaction is independent of the concentration of that substance. For example, for the reaction

$$NO_2\ (g) + CO\ (g) \longrightarrow CO_2\ (g) + NO\ (g)$$

at temperatures below 225°C, the relationship between concentration and rate is

$$\text{rate} \propto [NO_2]^2$$

The rate is independent of the CO concentration but depends on the *square* of the $NO_2$ concentration. We say that the reaction is second order with respect to $NO_2$ and zero order with respect to CO. Notice that there is no relationship between the coefficients and exponent. As mentioned above, $x$ and $y$ can only be determined experimentally.

The proportionality represented by Equation 11.5 can be converted to an equality by introducing a proportionality constant, which we call the **rate constant.** The resulting equation, termed the **rate law** for the reaction, is

$$\text{rate} = k[A]^x[B]^y$$

For example, the rate law for the reaction between ICl and $H_2$,

$$2ICl\ (g) + H_2\ (g) \longrightarrow I_2\ (g) + 2HCl\ (g)$$

at 230°C has been found experimentally to be

$$\text{rate} = 0.163\ \frac{\text{liter}}{\text{mole sec}}\ [ICl][H_2]$$

This reaction, therefore, is *first order* with respect to both ICl and $H_2$ (hence *second order*, overall) and has as its rate constant, $k = 0.163$ liter/mole sec.

How can a rate law such as this be determined? One way is to perform a series of experiments in which the initial concentration of each reactant is systematically varied. Once again we can use as our example the simple reaction

$$A \longrightarrow B$$

The rate law for this reaction would take the form

$$\text{rate} = k[A]^x$$

If the reaction were first order, the value of $x$ would be 1, and the rate expression would then be

$$\text{rate} = k[A]$$

This means that the rate of the reaction varies directly with the concentration of $A$ raised to the first power. As a result, if we were to double the concentration of $A$ from one experiment to another, we would also find that the rate would increase by a factor of 2. We conclude, then, that *when the reaction rate is doubled by doubling the concentration of a reactant, the order with respect to that reactant is 1.*

Suppose, now, that the rate law were, instead,

$$\text{rate} = k[A]^2$$

In this instance, a twofold increase in the concentration would cause a *fourfold* increase in rate. To see this, let's imagine that the initial rate were measured with the concentration of $A$ equal to, say $a$ moles/liter. This rate would be given by

$$\text{rate} = k(a)^2$$

Now, if the reaction were repeated with $[A] = 2a$, the rate would be

$$\text{rate} = k(2a)^2$$

or

$$\text{rate} = 4ka^2$$

which is four times the previous rate. Thus, *if the rate is increased by a factor of four when the concentration of a reactant is doubled, the reaction is second order with respect to that component. In a similar fashion we predict that the rate of a third order reaction would undergo an eightfold increase when the concentration is doubled* ($2^3 = 8$).

The following examples illustrate how we can use these ideas to obtain the rate law for a reaction by varying the concentration of reactants.

EXAMPLE 11.1 Below are some data collected in a series of experiments on the reaction of nitric oxide with bromine:

$$2NO\ (g) + Br_2\ (g) \longrightarrow 2NOBr\ (g)$$

at 273°C.

| Experiment | Initial Concentration (moles/liter) | | Initial Rate (moles/liter sec) |
|---|---|---|---|
| | NO | $Br_2$ | |
| 1 | 0.1 | 0.1 | 12 |
| 2 | 0.1 | 0.2 | 24 |
| 3 | 0.1 | 0.3 | 36 |
| 4 | 0.2 | 0.1 | 48 |
| 5 | 0.3 | 0.1 | 108 |

Determine the rate law for the reaction and compute the value of the rate constant.

SOLUTION In experiments 1 to 3, the concentration of NO is constant and the concentration of $Br_2$ is varied. When the concentration of $Br_2$ is doubled (experiments 1 and 2), the rate is increased by a factor of 2; when it is tripled (experiments 1 and 3), the rate is increased by a factor of 3. We conclude, therefore, that the concentration of $Br_2$ appears to the first power in the rate law.

Comparing experiments 1 and 4, we see that by holding the $Br_2$ concentration constant, the rate increases by a factor of 4 when the NO concentration is multiplied by 2. Similarly, raising the concentration of NO by a factor of 3 causes a ninefold increase in rate (experiments 1 and 5). Thus the exponent of the NO concentration in the rate law is 2. Therefore,

$$\text{rate} = k[NO]^2[Br_2]$$

The rate constant can be evaluated using the data from any of these experiments. Working with experiment 1, we have

$$12\ \frac{\text{moles}}{\text{liter sec}} = k(0.10\ \text{mole/liter})^2(0.10\ \text{mole/liter})$$

$$12\ \frac{\text{moles}}{\text{liter sec}} = k(0.0010\ \text{mole}^3/\text{liter}^3)$$

Solving for $k$, we get

$$k = \frac{12\ \text{moles/liter sec}}{1.0 \times 10^{-3}\ \text{mole}^3/\text{liter}^3} = 1.2 \times 10^4\ \text{liter}^2/\text{mole}^2\ \text{sec}$$

You might wish to verify for yourself that the same rate constant is obtained from the other data.

EXAMPLE 11.2 The following data were collected for the reaction of *t*-butyl bromide, $(CH_3)_3CBr$, with hydroxide ion at 55°C.

$$(CH_3)_3CBr + OH^- \longrightarrow (CH_3)_3COH + Br^-$$

| Experiment | Initial Concentration $(CH_3)_3CBr$ | Initial Concentration $OH^-$ | Initial Rate moles $(CH_3)_3CBr$/liter sec |
|---|---|---|---|
| 1 | 0.10 | 0.10 | 0.0010 |
| 2 | 0.20 | 0.10 | 0.0020 |
| 3 | 0.30 | 0.10 | 0.0030 |
| 4 | 0.10 | 0.20 | 0.0010 |
| 5 | 0.10 | 0.30 | 0.0010 |

What is the rate law and rate constant for this reaction?

SOLUTION Let's first examine experiments 1, 2, and 3. In each of these the $OH^-$ concentration is the same. Doubling the $(CH_3)_3CBr$ concentration doubles the rate; tripling it triples the rate. The order with respect to $(CH_3)_3CBr$ must therefore be 1.

In experiments 1, 4, and 5, the $(CH_3)_3CBr$ concentration is the same. Changing the $OH^-$ concentration has no effect on the rate. This means that the reaction is zero order with respect to $OH^-$.

$$\text{rate} = k[(CH_3)_3CBr]^1[OH^-]^0$$

Since anything raised to the zero power is 1,

$$\text{rate} = k[(CH_3)_3CBr]^1 \cdot 1$$

The final rate law contains only the concentration of $(CH_3)_3CBr$, because this is the only concentration that affects the rate. To solve for the rate constant we can use the results of any of the experiments. Using experiment 1 and substituting the rate and concentration into the rate law gives

$$0.0010 \text{ mole/liter sec} = k(0.10 \text{ mole/liter})$$

$$k = \frac{0.0010 \text{ mole/liter sec}}{0.10 \text{ mole/liter}} = 0.010 \left(\frac{1}{\text{sec}}\right)$$

This is usually written as

$$k = 0.010 \text{ sec}^{-1}$$

In Example 11.1, the exponents in the rate law happen to be the same as the coefficients in the balanced equation. This is not true in Example 11.2. Please keep in mind that the only way that we can find the exponents in the rate law for a chemical reaction is by experimentally measuring the way that the concentrations of the reactants affect the rate. It is also important to remember that since temperature is another factor that influences the rate, a given value of $k$ applies only at *one* temperature (the temperature at which it was measured).

## 11.3 COLLISION THEORY

In order for a chemical reaction to occur, the reacting molecules must collide with each other. This idea forms the basis of the **collision theory** of chemical kinetics. Basically, this theory states that the rate of a reaction is proportional to the number of collisions occurring each second between the reacting molecules:

$$\text{rate} \propto \frac{\text{number of collisions}}{\text{second}} \tag{11.6}$$

As we will see shortly, this permits us to explain the dependence of reaction rate on the concentration of the reactants. In Section 11.5 we will also see that the number of collisions that are effective is dependent on the nature of the reactants and the temperature.

At this point, let's see how collision theory accounts for the concentration dependence observed in the last section. Suppose that we have a reaction that occurs by the collision of two molecules, such as

$$A + B \longrightarrow \text{products}$$

In this case we are assuming that we know precisely what occurs between $A$ and $B$; that is, we assume for the sake of this discussion that the products are formed in **bimolecular** (two-molecule) collisions between $A$ and $B$.

According to our theory, the rate of the reaction is proportional to the number of collisions each second between molecules of $A$ and $B$. If the concentration of $A$ is doubled, then the number of $A$–$B$ collisions would also be doubled because there would be twice as many $A$ molecules that can collide with $B$. Hence the rate is increased by a factor of 2. Similarly, if the concentration of $B$ were doubled, there would be a twofold increase in the number of $A$–$B$ collisions and the rate would increase by a factor of 2. From our previous discussion we conclude that the order with respect to each reactant is 1 and the rate law for this bimolecular collision process is

$$\text{rate} = k[A][B]$$

What would happen if we had a reaction of the type,

$$2A \longrightarrow \text{products}$$

where reaction occurs by the collision of two $A$ molecules? In this instance, if we double the concentration of $A$, we double the number of collisions that each *single* $A$ molecule makes with its neighbors, because we have doubled the number of neighbors. We have also, however, doubled the number of $A$ molecules that are colliding. The number of $A$–$A$ collisions has therefore "doubly doubled," that is, increased by a factor of 2 squared. Consequently, the rate law for this bimolecular reaction between identical molecules is

$$\text{rate} = k[A]^2$$

What we find, then, is that *if* we know what collision process is involved in the production of products, we can predict, on the basis of collision theory, what the rate law for that process will be. *The exponents in the rate law are equal to the coefficients in the balanced equation for that collision process.* We might ask, then, why is it necessary to determine the rate law for a reaction experimentally? Why can we not simply use the coefficients of the balanced overall equation to deduce the rate law?

## 11.4 REACTION MECHANISM

The overall balanced equation for a reaction represents the net chemical change that occurs as the reaction proceeds to completion. This does not mean, however, that all of the reactants must come together simultaneously to undergo a change that produces the products. In fact, the net change can actually represent the sum of a series of simple reactions. These simple reactions are referred to as **elementary processes.** The sequence of elementary processes that ultimately leads to the formation of the products is called the **reaction mechanism.**

For example, it appears that the reaction

$$2NO + 2H_2 \longrightarrow 2H_2O + N_2$$

proceeds by the three-step mechanism,

$$\begin{aligned} 2NO &\longrightarrow N_2O_2 \\ N_2O_2 + H_2 &\longrightarrow N_2O + H_2O \\ N_2O + H_2 &\longrightarrow N_2 + H_2O \end{aligned}$$

The sum of these steps in the sequence does give us the overall balanced equation. Such a mechanism is arrived at by bringing together both theory and experiment.

Suppose now that we wished to study the general reaction,

$$2A + B \longrightarrow C + D$$

We first determine the rate law, perhaps by varying the concentration of $A$ and $B$; let us say it turned out to be

$$\text{rate} = k[A]^2[B]$$

Next, we attempt to propose a mechanism that, by the application of the principles of collision theory, gives us a predicted rate law that is the same as the one found by experiment.

Since we are beginners at proposing mechanisms, we might be tempted to propose a one-step mechanism in which two molecules of $A$ and one of $B$ come together simultaneously, that is, a three-body or **termolecular** collision. This process,

$$2A + B \longrightarrow C + D$$

indeed leads to the rate law

$$\text{rate} = k[A]^2[B]$$

which is the same as that found from experiment. We must now ask ourselves, is this a realistic mechanism? A simultaneous three-body (termolecular) collision is a very unlikely event, and it has been generally found that reactions that must proceed by such a path are very slow. As a result, a third-order reaction such as this, if it is fairly rapid, is usually interpreted as taking place by way of a series of simple bimolecular processes. (Back to the drawing board!)

One possible sequence of reactions is

$$\begin{aligned} 2A &\longrightarrow A_2 \\ A_2 + B &\longrightarrow C + D \end{aligned}$$

Here we have two steps in which we propose that some relatively unstable intermediate, $A_2$, is first formed by the collision of two molecules of $A$. In a second step a reaction between $A_2$ and $B$ produces the products $C$ and $D$.

Again the sum of these elementary processes gives us our net overall change.

Both of these reactions are unlikely to occur at the same rate. Let's suppose that the first reaction was slow and that once the intermediate, $A_2$, is formed it rapidly reacts with $B$ in the second step to produce the products. If this were true, the rate at which the final products appear is actually determined by how fast $A_2$ is produced. This first step, then, serves as a "bottleneck" in the reaction path. We refer to this slowest step as the **rate determining step** in the reaction because it governs how rapidly the overall reaction takes place. Because the rate-determining step is an elementary process (in this instance a bimolecular collision between two $A$ molecules), we can predict, with the aid of collision theory, that the rate law should be

$$\text{rate} = k[A]^2$$

*If this is the rate law for the rate-determining step, it will also be the rate law for the overall reaction.* However, this rate law cannot be the correct one because it is not the same as the one determined from experiment. This does not necessarily mean that our mechanism is wrong. Let us see what we would expect to observe if the second step, instead of the first, were the slow step. In this instance the rate law is

$$\text{rate} = k[A_2][B]$$

However, this rate law contains the concentration of the proposed intermediate ($A_2$) and the experimental rate law contains only the concentration of reactants $A$ and $B$. How can we express the concentration of $A_2$ in terms of $A$ and $B$?

Once $A_2$ has been formed, it can react in either of two ways. Since we propose that $A_2$ is unstable (if it were stable we could isolate it and there would be no question at all about the path of the overall reaction) it can undergo decomposition to reform two molecules of $A$. The other possibility is that it undergoes a collision with $B$ that leads to the formation of the products, $C$ and $D$. Our mechanism therefore should include a reaction that allows $A_2$ to decompose; that is,

$$A_2 \longrightarrow 2A$$

Our total mechanism is now

$$\begin{aligned} 2A &\longrightarrow A_2 \\ A_2 &\longrightarrow 2A \\ A_2 + B &\longrightarrow C + D \end{aligned}$$

If the rate at which the intermediate is formed from reactant $A$ is equal to the rate at which $A$ is formed from intermediate $A_2$, then these two reactions represent a state of dynamic equilibrium. We could therefore write our first two equations as an equilibrium, which would take the form,

$$2A \rightleftharpoons A_2$$

We are now back to a two-step mechanism in which the first step is an equilibrium. Our mechanism is now

$$\begin{aligned} 2A &\rightleftharpoons A_2 && \text{fast} \\ A_2 + B &\longrightarrow C + D && \text{slow} \end{aligned}$$

Since, in an equilibrium situation, the rate of the forward reaction ($rate_f$) is equal to the rate of the reverse reaction ($rate_r$),

$$rate_f = k_f[A]^2 = rate_r = k_r[A_2]$$

or simply

$$k_f[A]^2 = k_r[A_2]$$

Solving this equation for $[A_2]$, we have

$$[A_2] = \frac{k_f[A]^2}{k_r}$$

Combination of these constants yields still another constant, let us say $k'$; therefore our rate expression is

$$rate = k'[A]^2[B]$$

which does agree with the rate law found from experiment. Our proposed mechanism, therefore, appears to be a good one. However, a mechanism is, in essence, a theory. It is a sequence of steps that we dream up to explain the chemistry and to provide a rate law that agrees with experiment. It frequently happens, though, that more than one mechanism can be written to satisfy both criteria, so that we can never be certain we have truly discovered the actual path of the reaction. We can only hope to gather further information that either supports (or proves wrong) our guess.

EXAMPLE 11.3 The decomposition of $NO_2Cl$ is believed to involve the two-step mechanism,

$$NO_2Cl \longrightarrow NO_2 + Cl$$
$$NO_2Cl + Cl \longrightarrow NO_2 + Cl_2$$

What would be the observed experimental rate law if the first step were slow and the second were fast?

SOLUTION If the first reaction is the slow step, it is also the rate-determining step. The rate law for the overall reaction should be the same as the rate law for the rate-determining step. Since only one molecule of $NO_2Cl$ is involved, the rate law for the first reaction, as well as for the overall reaction, is

$$rate = k[NO_2Cl]$$

## 11.5 EFFECTIVE COLLISIONS

If all collisions that take place in a reaction vessel were effective in producing chemical change, all chemical reactions, including biochemical ones, would be over almost instantaneously. Since living creatures have finite life spans, it is clear that some factor (or factors) must intervene to decrease reaction rates to a reasonable level. Consider, for example, the decomposition of hydrogen iodide,

$$2HI\ (g) \longrightarrow H_2\ (g) + I_2\ (g)$$

At a concentration of only $10^{-3}$ mole/liter of HI there are approximately $3.5 \times 10^{28}$ collisions per liter per second at 500°C. This is equivalent to $5.8 \times 10^4$ moles of collisions per liter per second; if each of these collisions were effective, we would expect a rate of reaction of $5.8 \times 10^4$ moles/liter sec. Actually the rate under these conditions is only about $1.2 \times 10^{-8}$ mole/liter sec; smaller by a factor of approximately $5 \times 10^{12}$ than we would observe if all collisions led to reaction! Clearly, not all encounters between HI molecules result in the production of $H_2$ and $I_2$. In fact, only a very small fraction of the total number of collisions are effective. If we let $Z$ be the total number of collisions that occur per second and $f$ be the fraction of the total number of collisions that are effective, the rate of a reaction, according to collision theory, is

$$\text{rate} = fZ \tag{11.7}$$

The fraction, $f$, is determined by the energies of the molecules that collide and, as we will see shortly, a certain minimum energy is required in order to cause a reaction to occur. In addition to this, in many instances the molecules must also collide with the proper orientation. The decomposition of the hypothetical $AB$ molecule, whose collisions result in the formation of $A_2$ and $B_2$, can serve as an example.

$$2AB \longrightarrow A_2 + B_2$$

In order to have the products $A_2$ and $B_2$ produced, the two atoms of $A$ and two atoms of $B$ must approach each other very closely so that $A—A$ and $B—B$ bonds can be formed. Suppose, now, that two $A—B$ molecules come together in a collision oriented as shown in Figure 11.3*a*. We certainly do not expect this collision to be effective in forming the products. However, a collision in which the $AB$ molecules are aligned as shown in Figure 11.3*b* can lead to the creation of $A—A$ and $B—B$ bonds and, hence, to a net chemical change. Thus the number of effective collisions and, therefore, the rate of the reaction, is further decreased by a factor, $p$, that is a measure of the importance of the molecular orientations during collision:

$$\text{rate} = pfZ \tag{11.8}$$

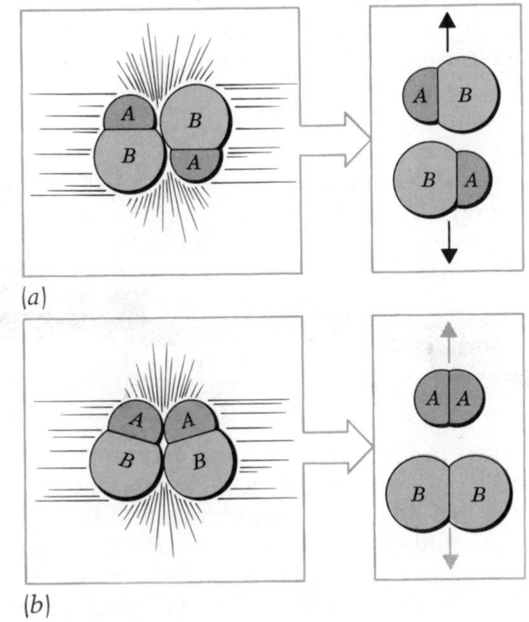


Figure 11.3
*Collisions between* A—B *molecules. (a) A collision that cannot produce a net chemical change. (b) A collision that can lead to net reaction.*

We have already seen that $Z$, the collision frequency, is proportional to the concentrations of the reacting molecules; therefore, in general,

$$Z = Z_0[A]^n[B]^m \ldots$$

where $Z_0$ is the collision frequency when all of the reactants are at unit concentration. Substituting this into Equation 11.8 gives us

$$\text{rate} = pfZ_0[A]^n[B]^m \ldots$$

or

$$\text{rate} = k[A]^n[B]^m \ldots$$

where $k = pfZ_0$. This then is the rate law derived from the principles of collision theory.

## 11.6 TRANSITION STATE THEORY

For a chemical change to occur during a collision between two molecules, some bonds must be broken and new ones formed. This requires that the electron clouds of the reacting molecules interpenetrate one another very substantially so that the needed electron reshuffling can take place. The collision theory discussed earlier focused attention primarily on the relationship between reaction rate and the numbers of collisions per second between reactant molecules. **Transition state theory** is concerned with the energy and geometry of the reactants when they collide to give the products. A collision between two molecules is quite unlike a collision between two billiard balls. The electron cloud of a molecule has no sharp boundary and, as molecules approach each other, they experience a gradual increase in their mutual repulsion that causes them to slow down, stop, and then fly apart again. Let us follow the kinetic and potential energies of a pair of molecules in a gas as they undergo such a collision.

As the molecules approach each other closely, they begin to slow down and their kinetic energy decreases, with a corresponding rise in their potential energy. If the molecules were not moving very fast when they entered into this collision, they would stop and reverse direction before any appreciable interpenetration of their electron clouds had taken place. As a result, slow-moving molecules simply bounce off one another without reacting. Very rapidly moving molecules, on the other hand, can penetrate each other enough so that the necessary bond breaking and forming can occur. These high-speed molecules have large kinetic energies that yield large increases in potential energy during collision. When the products fly apart, this potential energy decreases as the product molecules gain velocity (and hence kinetic energy). Thus, under these circumstances, only fast-moving molecules are able to react. In fact, there must be some minimum kinetic energy possessed jointly by the two molecules that is available to be transformed into potential energy. The minimum energy that must be available in a collision to cause reaction is called the **activation energy, $E_a$.**

The change in potential energy that takes place during the course of a reaction is shown in Figure 11.4. The horizontal axis is called the **reaction coordinate** and positions along this axis represent the extent to which the reaction has progressed toward completion. On the left of this potential energy diagram we find two molecules of $AB$. As they approach each other, their potential energy increases to a maximum. As we continue toward the right along the reaction coordinate, the potential energy of the system decreases as the products, $A_2$ and $B_2$, move apart. When the $A_2$ and $B_2$ mole-

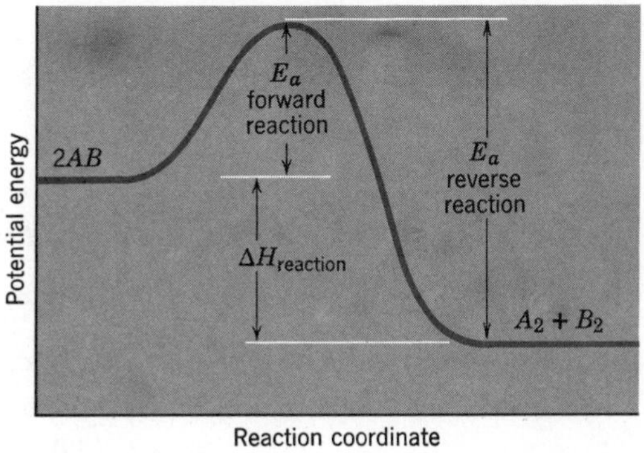


Figure 11.4
*Potential energy diagram for an exothermic reaction.*

cules are finally separated from one another, the total potential energy drops to essentially a constant value.

The activation energy for the decomposition of $AB$ corresponds to the difference between the energy of the reactants and the maximum on the potential energy curve. Slow-moving molecules of $AB$ do not possess sufficient energy to overcome this potential energy barrier, while fast-moving ones do.

In Figure 11.4 we have drawn the potential energy of the products to be lower than that of the reactants. The difference between them corresponds to the heat of reaction. In this case, because the products are at a lower energy than the reactants, the reaction is exothermic. The energy released appears as an increase in the kinetic energy of the products; therefore, the temperature of the system rises as the reaction progresses.

In the reaction mixture there are also collisions between $A_2$ and $B_2$ molecules. Such collisions, if energetic enough, can reform $AB$ molecules. In Figure 11.4 the activation energy for the reaction,

$$A_2 + B_2 \longrightarrow 2AB$$

is indicated as the difference in energy between the products and the top of the potential energy hill. Since the forward reaction is exothermic, the reverse reaction is endothermic.

Figure 11.5 depicts the energy changes for a reaction that is endothermic in the forward direction. In this case the products are at a higher potential energy than the reactants. The net absorption of energy that takes place as the products are formed occurs at the expense of kinetic energy. Consequently, there is a net overall decrease in the average kinetic energy as the reaction proceeds and the reaction mixture becomes cool.

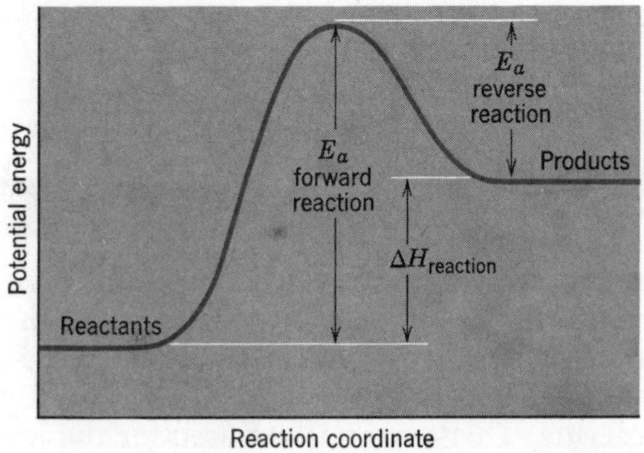


Figure 11.5
*Potential energy diagram for an endothermic reaction.*

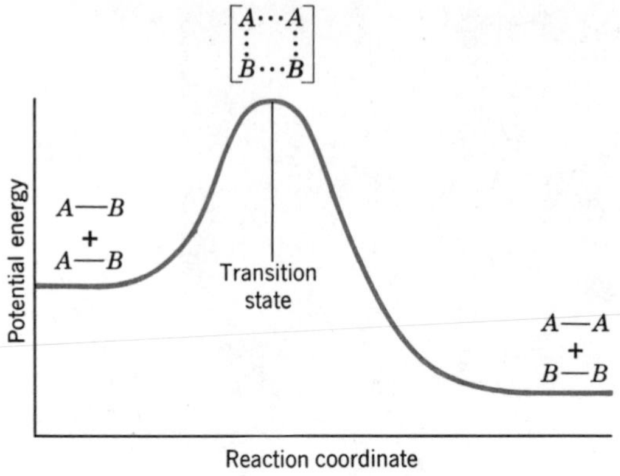

Figure 11.6
*Transition state theory and the potential energy diagram for a reaction.*

The species that exists at the top of the potential energy barrier (Figure 11.4) during an effective collision corresponds to neither the reactants nor the products but, instead, to some highly unstable combination of atoms that we speak of as the **activated complex.** This activated complex is said to exist in a **transition state** along the reaction coordinate (hence the name transition state theory).

Transition state theory views chemical kinetics in terms of the energy and geometry of the activated complex which, once it has formed, can come apart to yield the reactants again or go on to produce the products. For example, let us examine again the decomposition of hypothetical $AB$ molecules to produce $A_2$ and $B_2$. The change that takes place along the reaction coordinate can be represented as

$$\begin{matrix} A & A \\ | & | \\ B & B \end{matrix} \rightleftharpoons \begin{bmatrix} A\cdots A \\ \vdots \quad \vdots \\ B\cdots B \end{bmatrix} \longrightarrow \begin{matrix} A-A \\ + \\ B-B \end{matrix}$$

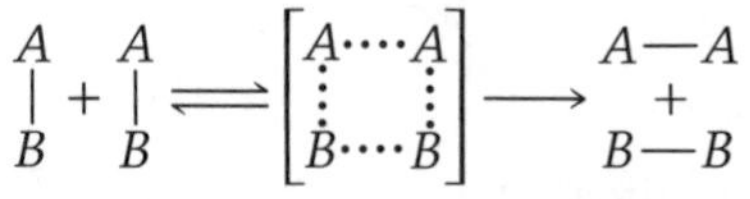

where we have used solid dashes to denote ordinary covalent bonds and dotted lines to symbolize the partially broken and partially formed bonds in the transition state (which is enclosed within brackets). Figure 11.6 illustrates this change as it occurs on our potential energy diagram for the reaction.

If the potential energy of the transition state is very high, then a great deal of energy must be available in a collision to form the activated complex. This results in a high activation energy and consequently a slow reaction. If it were possible somehow to produce an activated complex whose energy was closer to that of the reactants, the decreased activation energy would lead to a faster reaction rate.

## 11.7 EFFECT OF TEMPERATURE ON REACTION RATE

In nearly every instance an increase in temperature causes an increase in the rate of reaction and, as a very general rule of thumb, the rate is about doubled, for many reactions, by a ten-degree rise in temperature. How can this behavior be explained?

According to kinetic theory, in any system there is a distribution of kinetic energies. In the last section we interpreted the activation energy to be the minimum kinetic energy required for a collision to be effective. All molecules having kinetic energies higher than this minimum are, therefore, capable of reacting. This can be illustrated for the kinetic energy distribu-

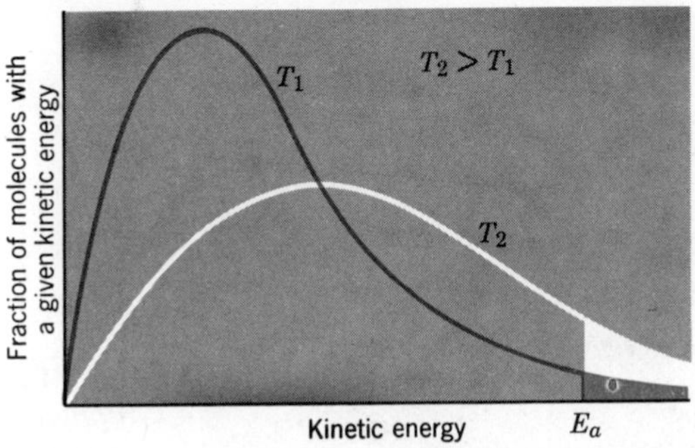


Figure 11.7
*The effect of temperature on the number of molecules having kinetic energies greater than* $E_a$.

tion in a system, as shown in Figure 11.7. The total fraction of all of the molecules having energies equal to or greater than $E_a$ corresponds to the shaded portion of the area under the curve. If we compare this area for two different temperatures, we see that the total fraction of molecules with sufficient kinetic energy to undergo effective collisions is greater at the higher temperature. As a result, the number of molecules that are capable of undergoing reaction increases with increasing temperature and, consequently, so does the reaction rate.

The magnitude of the rate constant, which is the rate of the reaction when all of the concentrations have a value of one, depends on the size of $E_a$ and also on the absolute temperature. For instance, $k$ is small when the activation energy is very large or when the temperature of the reaction mixture is low. Quantitatively, $k$ is related to $E_a$ and $T$ by the equation,

$$k = Ae^{-E_a/RT} \tag{11.9}$$

where $A$ is a proportionality constant, $R$ is the gas constant, and $e$ is the base of the natural logarithms (see Appendix C). This relationship is known as the Arrhenius equation after its discoverer, the Swedish chemist, Svante Arrhenius.[1]

To measure the activation energy, the rate constant must be measured at at least two different temperatures. Taking the natural logarithm of Equation 11.9 gives

$$\ln k = \ln A - \frac{E_a}{RT} \tag{11.10}$$

We can compare this equation to the equation for a straight line.

$$\begin{array}{ccccc} \ln k & = \ln A & - \frac{E_a}{R} & \left(\frac{1}{T}\right) \\ \updownarrow & \updownarrow & \updownarrow & \updownarrow \\ y & = b & + m & x \end{array}$$

Thus, a plot of ln $k$ versus $1/T$ gives a straight line whose slope is equal to $-E_a/R$ and whose intercept with the ordinate (the vertical axis) is ln $A$ (Figure 11.8).

We can also obtain $E_a$ from $k$ at two temperatures by direct computation. For any temperature, $T_1$, Equation 11.9 becomes

$$k_1 = Ae^{-E_a/RT_1}$$

[1] Arrhenius received the third Nobel Prize ever awarded in chemistry in 1903.

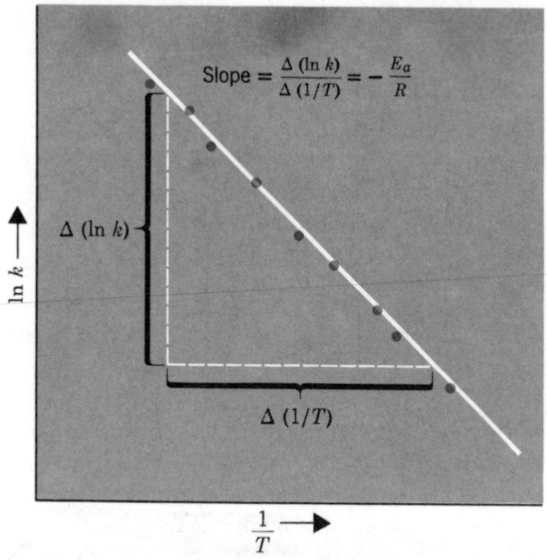


Figure 11.8
*Graphical determination of the activation energy, $E_a$. Points on the line represent the natural logarithms of experimentally measured rate constants at various temperatures. We determine the slope of the straight line that best fits the experimental data.*

and for any other temperature, $T_2$, we can write

$$k_2 = Ae^{-E_a/RT_2}$$

Dividing $k_1$ by $k_2$, we have

$$\frac{k_1}{k_2} = \frac{Ae^{-E_a/RT_1}}{Ae^{-E_a/RT_2}}$$

or

$$\frac{k_1}{k_2} = e^{(E_a/R)[(1/T_2)-(1/T_1)]} \tag{11.11}$$

Taking the natural logarithm of both sides, we get

$$\ln\left(\frac{k_1}{k_2}\right) = \frac{E_a}{R}\left(\frac{1}{T_2} - \frac{1}{T_1}\right) \tag{11.12}$$

Converting to common logarithms (base 10 logarithms) gives

$$\log\left(\frac{k_1}{k_2}\right) = \frac{E_a}{2.303R}\left(\frac{1}{T_2} - \frac{1}{T_1}\right) \tag{11.13}$$

Equation 11.13 can be used to compute $E_a$ if rate constants at two different temperatures are known. It can also be used to calculate the rate constant at some specific temperature if $E_a$ and $k$ at some other temperature are known.

EXAMPLE 11.4 At 300°C the rate constant for the reaction

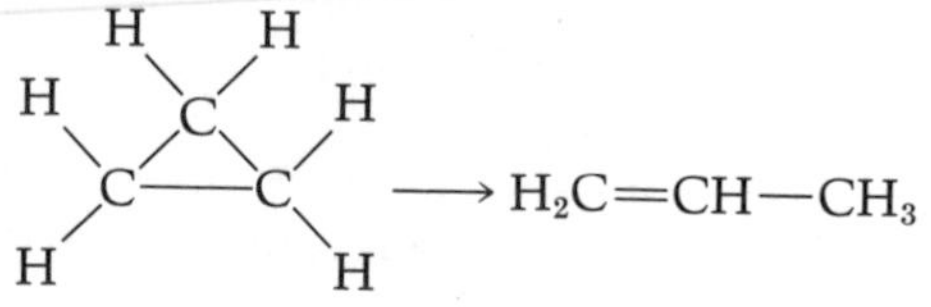


is $2.41 \times 10^{-10}$ sec$^{-1}$. At 400°C, $k$ equals $1.16 \times 10^{-6}$ sec$^{-1}$. What are the values of $E_a$ (in kilojoules per mole) and $A$ for this reaction?

SOLUTION We can obtain $E_a$ by substituting values of $k_1$, $k_2$, $T_1$ and $T_2$ into Equation 11.13 and then solving for $E_a$. To avoid confusion let's tabulate our data.

| | $k$ | $T$ |
|---|---|---|
| 1 | $2.41 \times 10^{-10}$ sec$^{-1}$ | $300 + 273 = 573$ K |
| 2 | $1.16 \times 10^{-6}$ sec$^{-1}$ | $400 + 273 = 673$ K |

Substituting into Equation 11.13,

$$\log\left(\frac{2.41 \times 10^{-10}\ \text{sec}^{-1}}{1.16 \times 10^{-6}\ \text{sec}^{-1}}\right)$$

$$= \frac{E_a}{2.303(8.314\ \text{J mole}^{-1}\ \text{K}^{-1})}\left(\frac{1}{673\ \text{K}} - \frac{1}{573\ \text{K}}\right)$$

$$\log(2.08 \times 10^{-4}) = \frac{E_a}{19.15\ \text{J mole}^{-1}\ \text{K}^{-1}}(0.00149\ \text{K}^{-1} - 0.00175\ \text{K}^{-1})$$

$$-3.68 = E_a(-1.36 \times 10^{-5}\ \text{J}^{-1}\ \text{mole})$$

$$E_a = \frac{-3.68}{-1.36 \times 10^{-5}\ \text{J}^{-1}\ \text{mole}}$$

$$= 2.71 \times 10^5\ \text{J/mole} = 271\ \text{kJ/mole}$$

We can now compute $A$ from the equation,

$$k = Ae^{-E_a/RT}$$

Taking the natural logarithm of both sides of the equation,

$$\ln k = \ln A - \frac{E_a}{RT}$$

or, in terms of common logarithms,

$$2.303 \log k = 2.303 \log A - \frac{E_a}{RT}$$

Solving for $\log A$,

$$\log A = \log k + \frac{E_a}{2.303RT}$$

Substituting the values for 300°C,

$$\log A = \log(2.41 \times 10^{-10}) + \frac{2.71 \times 10^5\ \text{J mole}^{-1}}{(2.303)(8.314\ \text{J mole}^{-1}\ \text{K}^{-1})(573\ \text{K})}$$

$$= -9.62 + 24.7$$

$$= 15.2$$

Taking the antilogarithm,

$$A = 1.6 \times 10^{15}\ \text{sec}^{-1}$$

Note that $A$ must have the same units as $k$.

## 11.8 CATALYSTS

A **catalyst** is a substance that increases the rate of a reaction without being consumed; after the reaction has ceased, it can be recovered from the reaction mixture chemically unchanged. The catalyst participates in the reaction by providing a lower energy alternative mechanism for the production of the products. In Figure 11.9 note that the energy curve of the catalyzed reaction is drawn along a different reaction coordinate to emphasize that a different mechanism is involved. In addition, the energy barrier for the catalyzed path is lower than for the uncatalyzed reaction. This smaller activation energy means that in the reaction mixture there is a greater total fraction of molecules possessing sufficient kinetic energy to react (Figure 11.10). Therefore in the presence of the catalyst there are an increased number of effective collisions. Of course, an increased number of effective collisions means a greater reaction rate.

Since a catalyst emerges chemically unchanged from a reaction, it does not appear either as a reactant or a product in the overall balanced chemical equation. Instead, its presence is indicated by writing its name or formula over the arrow. For example, in the preparation of oxygen by the thermal decomposition of $KClO_3$, manganese dioxide ($MnO_2$) is added to speed up the reaction and to allow the decomposition to proceed rapidly at a relatively low temperature. By contrast, in the absence of $MnO_2$ the reaction is slow and the $KClO_3$ must be heated to high temperatures to cause it to decompose. Analysis of the reaction mixture after the evolution of oxygen has ceased reveals that all of the $MnO_2$ added initially is still present, showing that $MnO_2$ has served as a catalyst. The equation for the catalyzed reaction is given as

$$2KClO_3 \xrightarrow{MnO_2} 2KCl + 3O_2$$

Figure 11.9
*Effect of a catalyst on the potential energy diagram. The catalyst changes the reaction mechanism by providing a different, low-energy mechanism for the formation of the products. ΔH is same for each path (it must be so because ΔH is a state function).*

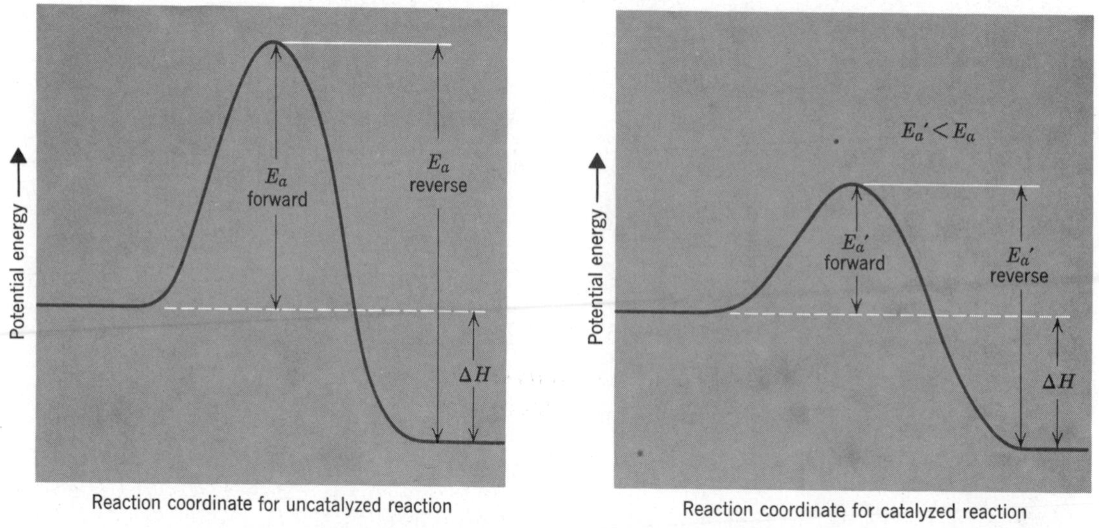

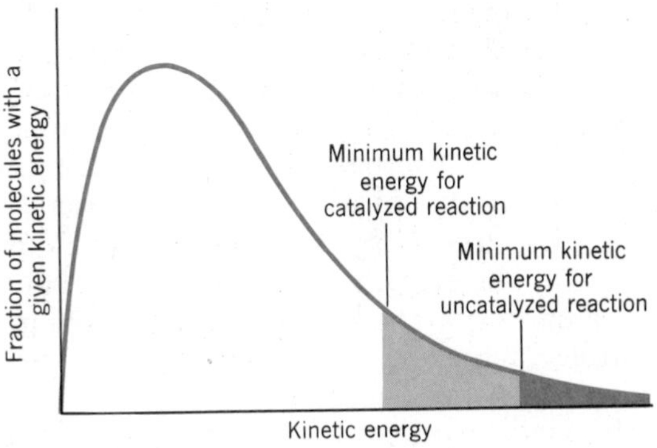


Figure 11.10
*More molecules possess the minimum kinetic energy for a reaction when the catalyst is present.*

Even though a catalyst does not change the overall stoichiometry of a reaction, it does participate chemically by being consumed at one stage in the mechanism and being produced again at a later stage. This regeneration of the catalyst permits the same catalyst to be used over and over again; therefore even a small amount of catalyst can have very profound effects on reaction rate. This phenomenon is particularly significant in biological systems, where practically every reaction is catalyzed by very small quantities of highly specific biochemical catalysts called enzymes.

Catalysts may be broadly classified into two categories: homogeneous and heterogeneous catalysts. A **homogeneous catalyst** is present in the same phase as the reactants and can serve to speed up the reaction by forming a reactive intermediate with one of the reactants. For example, the decomposition of *t*-butyl alcohol, $(CH_3)_3COH$, to produce water and isobutene, $(CH_3)_2C{=}CH_2$,

$$(CH_3)_3COH \longrightarrow (CH_3)_2C{=}CH_2 + H_2O$$

is catalyzed by the presence of small amounts of HBr. In the absence of HBr the activation energy for the reaction is 65.5 kcal/mole, and below 450°C the reaction takes place at a barely perceptible rate. In the presence of HBr an activation energy of only 30.4 kcal/mole is found, and it is possible that the catalyzed reaction proceeds by an attack of HBr on the alcohol,

$$(CH_3)_3COH + HBr \longrightarrow (CH_3)_3CBr + H_2O$$

followed by the rapid decomposition of the *t*-butyl bromide,

$$(CH_3)_3CBr \longrightarrow (CH_3)_2C{=}CH_2 + HBr$$

Thus HBr provides an alternative low-energy path for the reaction and the mechanism for the reaction when HBr is present is different than when it is absent.

A **heterogeneous catalyst** is not in the same phase as the reactants, but provides a favorable surface on which the reaction can take place. An example of a reaction whose rate is increased by the presence of a heterogeneous catalyst is the reaction between hydrogen and oxygen to produce water. In the introduction to this chapter we pointed out that this reaction proceeds at a very slow rate when the two gases are mixed at room temperature. However, it has been found that the reaction proceeds at an appreciable rate when metals such as nickel, copper, or silver are present.

Heterogeneous catalysts appear to function through a process whereby reactant molecules are adsorbed on a surface where the reaction then takes place. The high reactivity of hydrogen, in the presence of certain metals, for example, is thought to occur by the adsorption of $H_2$ molecules onto the catalytic surface. On the surface of the metal the bonds between hydrogen atoms are apparently stretched or broken, as shown in Figure 11.11, so that, the metal surface actually behaves as if it contained highly reactive hydrogen atoms.

Unless the reactant molecules can be adsorbed on the catalyst, no increase in reaction rate can occur. A substance whose presence during a reaction interferes with the adsorption process will therefore reduce the effectiveness of the catalysts and is thus called an **inhibitor.** These substances, by being strongly adsorbed on the catalytic surface, decrease the available space on which the reaction can occur. In some cases the catalyst eventually becomes useless and is said to be "poisoned." The destruction of catalytic activity by poisoning is very important in biological systems, as we shall see in Chapter 21.

There have been many commercial and industrial applications of heterogeneous catalysts. For example, small portable flameless heaters can be purchased (for heating camping tents and the like) in which the fuel and oxygen combine on a catalytic surface. The flameless combustion evolves the same amount of heat that would be generated if the fuel were burned directly. Since no flame is involved, however, the heaters are much safer to operate. There are drawbacks, however. Unfortunately, the catalytic oxidation of the fuel is not totally efficient, and small amounts of carbon monoxide are produced. As a result, catalytic heaters must be used with care.

Another very important application is in the control of auto exhaust emissions. Catalytic mufflers are in use that employ a mixed metal oxide bed over which the exhaust gases pass after they are mixed with additional air (Figure 11.12). The catalyst quite effectively promotes the oxidation of CO and hydrocarbons to harmless $CO_2$ and $H_2O$. Unfortunately, it does not reduce the emission of nitrogen oxides, another important pollutant. Catalytic mufflers suffer from the disadvantage that they are poisoned by lead. As a result, lead-free fuels must be employed in autos fitted with this type of antipollution device. Another disadvantage is that they also catalyze the oxidation of $SO_2$ to $SO_3$, which then reacts with water vapor to produce a mist

Figure 11.11

*Production of H atoms on a metal surface. $H_2$ molecule collides with the surface where it is adsorbed and dissociates to produce H atoms.*

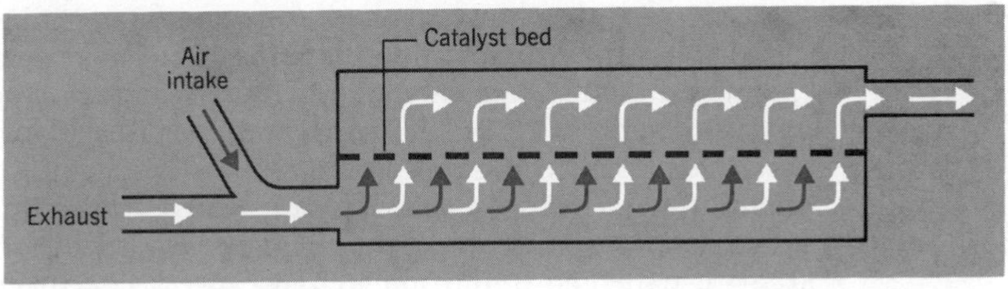


Figure 11.12
*The catalytic muffler.*

of $H_2SO_4$. Since $SO_2$ is produced from the combustion of high-sulfur fuels, this problem is a serious one and may ultimately lead to forced discontinuation of the use of catalytic mufflers.

## 11.9 CHAIN REACTIONS

The reactions that we have discussed up to now have been rather simple and straightforward, with uncomplicated rate laws. There are some reactions, usually with very complex kinetics, that take place by way of an extremely reactive intermediate, such as a free atom or a **free radical** (a neutral group of atoms or an ion containing one or more unpaired electrons). These species can be produced either thermally (at high temperature) or by the absorption of light of an appropriate wavelength; once created they can sometimes react with other molecules to produce the products *plus* yet another free atom or radical. This process, once initiated, can be repeated over and over, making the reaction *self-propagating.* The entire series of reactions following the production of the very reactive intermediate is called a **chain reaction.**

A chain reaction mechanism has been suggested to explain the rate equation observed for the reaction between hydrogen and bromine. The overall reaction is

$$H_2 + Br_2 \longrightarrow 2HBr$$

If this reaction proceeded simply by a bimolecular collision between $H_2$ and $Br_2$, the expected rate law would be

$$\text{rate} = k[H_2][Br_2]$$

However, the actual rate law turns out to be

$$\text{rate} = k\frac{[H_2][Br_2]^{1/2}}{1 + [HBr]/k'[Br_2]}$$

which is very complex indeed. A mechanism that has been proposed to account for this rate law is the chain reaction shown below. A dot is used to represent the unpaired electron on a highly reactive atom or free radical.

| | |
|---|---|
| 1. $Br_2 \rightarrow 2Br\cdot$ | initiation |
| 2. $Br\cdot + H_2 \rightarrow HBr + H\cdot$ | propagation |
| 3. $H\cdot + Br_2 \rightarrow HBr + Br\cdot$ | propagation |
| 4. $H\cdot + HBr \rightarrow H_2 + Br\cdot$ | inhibition |
| 5. $2Br\cdot \rightarrow Br_2$ | termination |

Reaction 1 is the thermal decomposition of diatomic bromine molecules to produce bromine atoms (the reactive intermediate). The overall reaction proceeds very slowly when the two reactants are mixed at room temperature. However, at high temperature reaction 1 takes place to an appreciable extent and rapidly sets off the remaining reactions. Step 1 is, therefore, the **initiation step.** In steps 2 and 3 the product HBr is formed as well as additional free atoms that serve to keep the reaction going. These steps then are **propagation steps** in the chain. Step 5, which leads only to the formation of a stable species, serves to end the chain and is known as the **termination step.** Step 4 is called an **inhibition step,** because its occurrence removes product and thus decreases the overall rate of production of HBr. It is included in the mechanism because the presence of HBr decreases the reaction rate (note the appearance of [HBr] in the denominator of the rate law).

In general, chain reactions are very rapid; in fact, many explosive reactions appear to occur by chain mechanisms. The production of a single reactive intermediate produces many product molecules before the chain is terminated. Consequently, the rate of production of the products is many times greater than the rate of the initiation step alone.

## INDEX TO QUESTIONS AND PROBLEMS (problem numbers in italics)

General 1, 2, 4, 5, 6

Reaction rates 3, *32, 33, 34, 35*

Rate laws 11, *36, 37*

Order of reaction 7, 8, 10

Rate constant 9, *44*

Collision theory 12, 13

Mechanisms 14, 15, 16, 17, 18

Effective collisions 19, 20

Transition state theory 23, 24

Effect of temperature on rate 21, 22, 25

Arrhenius equation *38, 39, 40, 41, 42, 43, 45*

Catalysis 26, 27, 28, 29

Chain reactions 30, 31

## REVIEW QUESTIONS

**11.1** What are the four factors that control the rates of chemical reactions?

**11.2** What are meant by reaction rate, order of reaction, rate law, and rate constant?

**11.3** For each of the following reactions, how would we express the reaction rate in terms of the disappearance of the reactants and the appearance of the products? Predict the role of the coefficients in the balanced overall equation in determining the relative rates of disappearance of reactants and formation of products.

(a) $2H_2 + O_2 \rightarrow 2H_2O$

(b) $2NOCl \rightarrow 2NO + Cl_2$

(c) $NO + O_3 \rightarrow NO_2 + O_2$

(d) $H_2O_2 + H_2 \rightarrow 2H_2O$

**11.4** What criteria must be met by methods used to study the rate of a reaction?

**11.5** Make a list of five reactions that occur in the world around you and compare their rates. Try to think of some that are fast and some that are slow.

**11.6** Why do we say that one of the factors that influences reaction rate is the nature of the reactants?

**11.7** The rate at which CO is removed from the earth's atmosphere by fungi in the soil is constant. What is the apparent order of this process?

**11.8** What is the order with respect to each reactant and the overall order of the reactions described by the following rate laws:

(a) Rate $= k_1[A][B]$

(b) Rate $= k_2[E]^2$

(c) Rate $= k_3[G]^2[H]^2$

**11.9** What would be the units of each of the rate constants in the preceding question if rate has the units moles liter$^{-1}$ second$^{-1}$?

**11.10** When the concentration of a reactant is doubled, by what factor would the rate of reaction be increased if the order with respect to that reactant were (a) 1 (b) 2 (c) 3 (d) 4 (e) $\frac{1}{2}$?

**11.11** Suppose that when the concentration of a reactant was doubled the rate of reaction decreased by a factor of 2. What would be the exponent on the concentration term for that reactant in the rate law?

**11.12** How does collision theory account for the dependence of rate on the concentration of the reactants?

**11.13** Why can't we use the ideas of collision theory to predict in general the rate laws of chemical reactions? What must we know in order to predict the rate law?

**11.14** What is meant by reaction mechanism?

**11.15** A mechanism for the reaction, $2NO + Br_2 \rightarrow 2NOBr$, has been suggested to be

Step 1 $\quad NO + Br_2 \longrightarrow NOBr_2$
Step 2 $\quad NOBr_2 + NO \longrightarrow 2NOBr$

(a) What would be the rate law for the reaction if the first step in this mechanism were slow and the second fast?
(b) What would be the rate law if the second step were slow, with the first reaction being a rapidly established dynamic equilibrium?
(c) Experimentally, the rate law has been found to be

$$\text{rate} = k[NO]^2[Br_2]$$

What can we conclude about the relative rates of steps 1 and 2?
(d) Why do we not prefer a simple, one-step mechanism,

$$NO + NO + Br_2 \longrightarrow 2NOBr$$

(e) Can we, on the basis of the experimental rate law, definitely exclude the mechanism in part (d)?

**11.16** The reaction, $NO_2\ (g) + CO\ (g) \rightarrow CO_2\ (g) + NO\ (g)$, appears to have the mechanism (at low temperature),

$$NO_2 + NO_2 \longrightarrow NO_3 + NO \qquad \text{slow}$$
$$NO_3 + CO \longrightarrow NO_2 + CO_2 \qquad \text{fast}$$

Explain why the reaction is zero-order with respect to CO.

**11.17** The reaction of methyl bromide, $CH_3Br$, with $OH^-$ appears to occur through a one-step mechanism involving collision of $CH_3Br$ with $OH^-$,

$$CH_3Br + OH^- \longrightarrow CH_3OH + Br^-$$

The rate law for the reaction is found to be

$$\text{rate} = k[CH_3Br][OH^-]$$

In Example 11.2 we found that the rate law of the reaction of $(CH_3)_3CBr$ with $OH^-$ has the rate law

$$\text{rate} = k[(CH_3)_3CBr]$$

Try to propose a mechanism that can account for the rate law for the reaction of $(CH_3)_3CBr$ with $OH^-$.

**11.18** Suppose that the following sequence of reactions were proposed for a reaction.

Step 1 $\quad 2A \longrightarrow A_2$
Step 2 $\quad A_2 + B \longrightarrow C + 2D$

(a) What would be the overall net chemical reaction?
(b) What would the rate law be if step 1 were slow and step 2 were fast?
(c) What would the rate law be if step 2 were slow and step 1 were fast?

**11.19** How do we know that not all collisions between reactant molecules lead to chemical change? What determines whether a particular collision will be effective?

**11.20** How do the orientations of molecules influence whether a collision between them can be effective at producing chemical change?

**11.21** Define the term, activation energy.

**11.22** Explain qualitatively, in terms of the kinetic theory, why an increase in temperature leads to an increase in reaction rate.

**11.23** Draw a potential energy diagram for an endothermic reaction. Indicate on the drawing (a) the potential energy of the reactants, (b) the potential energy of the products, (c) the energies of activation for the forward and reverse reaction, (d) the heat of reaction.

**11.24** What are meant by the terms, transition state and activated complex? Where on the potential energy diagram for a reaction will we find the transition state?

**11.25** Insects, which are cold-blooded animals whose changes in body temperature tend to follow changes in the temperature of their environment, become quite sluggish in cool weather. On the basis of chemical kinetics, explain this phenomenon.

**11.26** What is the difference between a homogeneous and a heterogeneous catalyst?

**11.27** What is a heterogeneous catalyst? How does it function? What is an inhibitor?

**11.28** How does a catalyst play a part in lowering the activation energy for a reaction?

**11.29** What effect does a catalyst have on
(a) The heat of reaction
(b) The potential energy of the reactants
(c) The transition state

**11.30** Why are chain reactions often so fast?

**11.31** The decomposition of acetaldehyde, $CH_3CHO$, follows the overall reaction

$$CH_3CHO \longrightarrow CH_4 + CO$$

with small amounts of $H_2$ and $C_2H_6$ also being produced. The reaction is thought to proceed by a chain reaction involving free radicals (note again that a free radical is indicated by using a dot to represent its unpaired electron). A proposed mechanism is
(1) $CH_3CHO \rightarrow CH_3\cdot + CHO\cdot$
(2) $2CH_3\cdot \rightarrow C_2H_6$
(3) $CHO\cdot \rightarrow H\cdot + CO$
(4) $H\cdot + CH_3CHO \rightarrow H_2 + CH_3CO\cdot$
(5) $CH_3\cdot + CH_3CHO \rightarrow CH_4 + CH_3CO\cdot$
(6) $CH_3CO\cdot \rightarrow CH_3\cdot + CO$

Identify (a) the initiation step, (b) the propagation step(s), and (c) the termination step(s).

## REVIEW PROBLEMS

**11.32** The following data were collected for the reaction,

$$2A \longrightarrow 4B + C$$

| Time (min) | Concentration of $A$ (moles/liter) | Concentration of $B$ (moles/liter) |
|---|---|---|
| 0 | 1.000 | 0.000 |
| 10 | 0.800 | 0.400 |
| 20 | 0.667 | 0.667 |
| 30 | 0.571 | 0.858 |
| 40 | 0.500 | 1.000 |
| 50 | 0.444 | 1.112 |

Make a graph of the concentrations of $A$ and $B$ versus time (concentrations along the vertical axis, time along the horizontal axis). Estimate the rate of disappearance of $A$ and the rate of formation of $B$ at $t = 25$ min and at $t = 40$ min. Compare the rates of disappearance of $A$ and formation of $B$. What would you expect the rate of formation of $C$ to be at $t = 25$ min and $t = 40$ min?

**11.33** The rate law for a reaction was found to be

$$\text{rate} = (2.35 \times 10^{-6}\ \text{liter}^2\ \text{mole}^{-2}\ \text{sec}^{-2})[A]^2[B]$$

What would the rate of reaction be if:
(a) The concentrations of $A$ and $B$ were 1 mole/liter?
(b) $[A] = 0.25\ M$, $[B] = 1.30\ M$?

**11.34** The rate constant for the reaction

$$2ICl + H_2 \longrightarrow I_2 + 2HCl$$

is $1.63 \times 10^{-1}$ liter/mole sec. The rate law is given by

$$\text{rate} = k[ICl][H_2]$$

What is the rate of the reaction for each of the sets of concentrations given below?

| ICl Concentration (moles/liter) | $H_2$ Concentration (moles/liter) |
|---|---|
| 0.25 | 0.25 |
| 0.25 | 0.50 |
| 0.50 | 0.50 |

**11.35** For the decomposition of dinitrogen pentoxide,

$$2N_2O_5 \longrightarrow 4NO_2 + O_2$$

the following data were collected:

| $N_2O_5$ Concentration (moles/liter) | Time (sec) |
|---|---|
| 5.00 | 0 |
| 3.52 | 500 |
| 2.48 | 1000 |
| 1.75 | 1500 |
| 1.23 | 2000 |
| 0.87 | 2500 |
| 0.61 | 3000 |

(a) Make a graph of the concentration of $N_2O_5$ versus time. Draw tangents to the curve at $t = 500$, 1000, and 1500 sec. Determine the rate at these different reaction times.

(b) Determine the value of the rate constant at 500, 1000, and 1500 sec, given the rate law, rate = $k[N_2O_5]$.

**11.36** At 27°C, the reaction, $2NOCl \rightarrow 2NO + Cl_2$, is observed to exhibit the following dependence of rate on concentration.

| Initial NOCl Concentration (moles/liter) | Initial Rate (moles/liter sec) |
|---|---|
| 0.30 | $3.60 \times 10^{-9}$ |
| 0.60 | $1.44 \times 10^{-8}$ |
| 0.90 | $3.24 \times 10^{-8}$ |

(a) What is the rate law for the reaction?
(b) What is the rate constant?
(c) By what factor would the rate increase if the initial concentration of NOCl were increased from 0.30 to 0.45 *M*?

**11.37** The reaction of NO with $Cl_2$ follows the equation,

$$2NO + Cl_2 \longrightarrow 2NOCl$$

The following data were collected:

| Initial NO Concentration (moles/liter) | Initial $Cl_2$ Concentration (moles/liter) | Initial Rate (moles/liter sec) |
|---|---|---|
| 0.10 | 0.10 | $2.53 \times 10^{-6}$ |
| 0.10 | 0.20 | $5.06 \times 10^{-6}$ |
| 0.20 | 0.10 | $10.1 \times 10^{-6}$ |
| 0.30 | 0.10 | $22.8 \times 10^{-6}$ |

(a) What is the rate law for the reaction?
(b) What is the value of the rate constant (be sure to give the proper units)?

**11.38** The rate constants for the reaction between ICl and $H_2$ (Question 11.34) at 230 and 240°C have been found to be 0.163 and 0.348 liter/mole sec, respectively. What are the values of $E_a$ (in kilocalories per mole) and $A$ for this reaction?

**11.39** The rate constant for the reaction

$$CH_3I\ (g) + HI\ (g) \longrightarrow CH_4\ (g) + I_2\ (g)$$

at 200°C is $1.32 \times 10^{-2}$ liter mole$^{-1}$ sec$^{-1}$. At 275°C the rate constant is 1.64 liter mole$^{-1}$ sec$^{-1}$. What is the activation energy (in kilojoules per mole) and the value of $A$?

**11.40** The activation energy for the decomposition of HI

$$2HI\ (g) \longrightarrow H_2\ (g) + I_2\ (g)$$

is 182 kJ/mole. The rate constant for the reaction at 700°C is $1.57 \times 10^{-3}$ liter mole$^{-1}$ sec$^{-1}$. What is the value of the rate constant at 600°C?

**11.41** The activation energy for the reaction, $HI + CH_3I \rightarrow CH_4 + I_2$, is 33.1 kcal/mole. At 200°C the rate constant has a value of $1.32 \times 10^{-2}$ liter/mole sec. What is the rate constant at 300°C?

**11.42** The decomposition of $C_2H_5Cl$ is a first-order reaction having $k = 3.2 \times 10^{-2}$ sec$^{-1}$ at 550°C and $k = 9.3 \times 10^{-2}$ sec$^{-1}$ at 575°C. What is the activation energy, in kilocalories per mole, for this reaction?

***11.43** A chemist was able to determine that the rate of a particular reaction at 100°C was four times faster than at 30°C. Calculate the approximate energy of activation for the reaction.

***11.44** For a first-order reaction, a graph of $\log[A]$ versus time (where $A$ is a reactant) gives a straight line having a slope equal to $-k/2.303$. On the other hand, if the reaction is second-order with respect to $A$, a straight line is obtained when $1/[A]$ is plotted against time. In this case the slope of the line is equal to $k$. From this information, determine whether the reaction in Question 11.32 is first-order or second-order. Calculate the rate constant for the reaction.

***11.45** The development of a photographic image on film is a process controlled by the kinetics of the reduction of silver halide by a developer. The time required for development at a particular temperature is inversely proportional to the rate constant for the process. Below are published data on development times for Kodak's Tri-X film using Kodak D-76 developer. From these data, estimate the activation energy for the development process in kilocalories per mole.

| Temperature (°C) | Time for Development (min) |
|---|---|
| 18 | 10 |
| 20 | 9 |
| 21 | 8 |
| 22 | 7 |
| 24 | 6 |

Estimate the development time at 15°C.

# 12 CHEMICAL EQUILIBRIUM

When chemical reactions occur spontaneously, they continue to proceed until a state of dynamic equilibrium is achieved. At equilibrium both the forward and reverse reactions are taking place at the same rate, and the concentrations of the reactants and products no longer change with time, as illustrated in Figure 12.1.

All chemical systems tend toward equilibrium. In this chapter we shall explore the quantitative relationships that can be used to describe the equilibrium state, and we shall see how the principles of kinetics and thermodynamics can be applied to a description of equilibrium.

## 12.1 THE LAW OF MASS ACTION

It was discovered experimentally some time ago that a very simple relationship governs the relative proportions of reactants and products in an equilibrium system. For the general reaction,

$$aA + bB \rightleftharpoons eE + fF$$

it is observed that, at constant temperature, the condition that is fulfilled at equilibrium is

$$\frac{[E]^e[F]^f}{[A]^a[B]^b} = K_c \tag{12.1}$$

where the quantities written within square brackets denote *equilibrium molar concentrations.* The quantity, $K_c$, is a constant, called the **equilibrium constant** and the entire relationship, discovered in 1866 by the Norwegian chemists Guldberg and Waage, is known as the **law of mass action.**

The fraction appearing to the left of the equal sign in Equation 12.1 is called the **mass action expression,** and it is constructed using the coefficients in the balanced chemical equation as exponents on the appropriate concentrations. For instance, if we consider the nitrogen fixation reaction used industrially in the production of ammonia and nitrogen fertilizers,

$$N_2\,(g) + 3H_2\,(g) \rightleftharpoons 2NH_3\,(g)$$

the mass action expression would be written as

$$\frac{[NH_3]^2}{[N_2][H_2]^3}$$

This fraction will, of course, always have some numerical value for a system containing these three gases. For example, if $N_2$ and $H_2$ are introduced into a

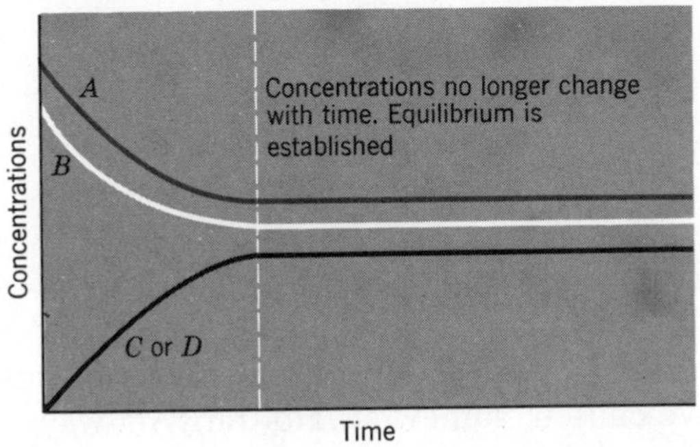


Figure 12.1
*The approach to equilibrium for the reaction* A + B → C + D.

container and are permitted to react, we would initially have no $NH_3$ and the value of the mass action expression would be zero. As $NH_3$ is produced the fraction grows larger until, when equilibrium is reached, the fraction becomes equal to a value that we call the equilibrium constant, $K_c$.

The important point about the law of mass action, however, is that, at a given temperature, any system containing $N_2$, $H_2$, and $NH_3$ in equilibrium will have its mass action expression equal to the same number. *There are no restrictions on the individual concentrations of any reactant or product.* The only requirement for equilibrium is that when these concentrations are substituted into the mass action expression, the fraction is equal to $K_c$. The data in Table 12.1 illustrate this point.

For reactions involving gases, the partial pressures of the reactants and products are proportional to their concentrations. The equilibrium constant expression for these reactions can therefore be written using partial pressures instead of concentrations. For example, the equilibrium condition for the reaction between $N_2$ $(g)$ and $H_2$ $(g)$ can be expressed as

$$\frac{p_{NH_3}^2}{p_{N_2}p_{H_2}^3} = K_P$$

We will use the symbol, $K_P$, to denote equilibrium constants derived from partial pressures and $K_c$ to indicate equilibrium constants having molar concentrations in the mass action expression. In general, $K_c$ and $K_P$ are not numerically equal. We will discuss this further in Section 12.5.

We have written the mass action expression with the concentrations (or partial pressures) of the products in the numerator and those of the reactants in the denominator. Since this fraction is equal to a constant at equilibrium,

Table 12.1
*Equilibrium concentrations (in moles per liter) at 500°C and the mass action expression for the reaction:* $3H_2$ (g) + $N_2$ (g) ⇌ $2NH_3$ (g)

| $[H_2]$ | $[N_2]$ | $[NH_3]$ | $\frac{[NH_3]^2}{[H_2]^3[N_2]} = K_c$ |
|---|---|---|---|
| 1.150 | 0.750 | $1.23 \times 10^{-2}$ | $5.98 \times 10^{-2}$ |
| 0.500 | 1.00 | $8.66 \times 10^{-2}$ | $6.00 \times 10^{-2}$ |
| 1.35 | 1.15 | $4.12 \times 10^{-1}$ | $6.00 \times 10^{-2}$ |
| 2.43 | 1.85 | 1.27 | $6.08 \times 10^{-2}$ |
| 1.47 | 0.750 | $3.76 \times 10^{-1}$ | $5.93 \times 10^{-2}$ |

its reciprocal must also be a constant. Thus,

$$\frac{[NH_3]^2}{[N_2][H_2]^3} = K_C \qquad \frac{p_{NH_3}^2}{p_{N_2}p_{H_2}^3} = K_P$$

and

$$\frac{[N_2][H_2]^3}{[NH_3]^2} = \frac{1}{K_c} = K_c' \qquad \frac{p_{N_2}p_{H_2}^3}{p_{NH_3}^2} = \frac{1}{K_P} = K_P'$$

Either form is a valid description of the equilibrium state. However, chemists have chosen, somewhat arbitrarily, always to write the equilibrium expression with the concentrations or partial pressures of the products appearing in the numerator. This allows us then to tabulate equilibrium constants without the necessity of always having to state explicitly the form of the mass action expression. It is only necessary to specify whether we are dealing with $K_c$ or $K_P$.

## 12.2 THE EQUILIBRIUM CONSTANT

The equilibrium constant is a quantity that must be calculated from experimental data. One method, involving the use of standard free energies of formation to determine a *thermodynamic equilibrium constant,* is outlined in Section 12.4. Another method involves the direct measurement of equilibrium concentrations that can then be inserted into the equilibrium expression to obtain a numerical value for $K$. We shall look at a sample calculation of this type in Section 12.7.

Simply examining the magnitude of $K$ provides us with information about the extent to which a reaction proceeds toward completion. For example, consider the simple reaction,

$$A \rightleftharpoons B$$

for which we could write

$$\frac{[B]}{[A]} = K_c$$

If $K_c = 10$, then

$$\frac{[B]}{[A]} = 10 = \frac{10}{1}$$

This tells us that at equilibrium the concentration of $B$ will be ten times larger than the concentration of $A$. On the other hand, if $K_c = 0.1$, then

$$\frac{[B]}{[A]} = 0.1 = \frac{1}{10}$$

In this case the concentration of A would be ten times larger than the concentration of $B$ at equilibrium. It is a general rule, then, that when $K$ is large the position of equilibrium lies far to the right. Conversely, when $K$ is small, only relatively small amounts of the products are present in the system at equilibrium.

Let's look at two examples of real chemical reactions: first, the reaction of hydrogen with chlorine,

$$H_2\ (g) + Cl_2\ (g) \rightleftharpoons 2HCl\ (g)$$

for which $K_c = 4.4 \times 10^{32}$ at 25°C. This very large value of $K$ tells us that at equilibrium the reaction will have proceeded far toward completion. If 1 mole each of $H_2$ and $Cl_2$ are combined, very little $H_2$ and $Cl_2$ will remain unreacted at equilibrium. Similarly, we conclude that the decomposition of water vapor at room temperature (25°C),

$$2H_2O\ (g) \rightleftharpoons 2H_2\ (g) + O_2\ (g)$$

which has $K_c = 1.1 \times 10^{-81}$, takes place to only a very small degree, because, in order to have such a very small value of $K_c$, the concentrations of the products must also be very small.

## 12.3 KINETICS AND EQUILIBRIUM

Guldberg and Waage, in 1866, realized that the rate of a chemical reaction depends on the concentrations of the reactants. In fact, they believed that the rate of a chemical reaction was proportional to the concentrations of the reactants raised to powers *equal* to their coefficients in the balanced equation. That is, they believed that for the reaction,

$$aA + bB \rightleftharpoons eE + fF$$

the rate of the forward reaction is

$$\text{rate}_f = k_f[A]^a[B]^b$$

while the rate of the reverse reaction is

$$\text{rate}_r = k_r[E]^e[F]^f$$

Once a dynamic equilibrium is established in this system, the rate of the forward and reverse reactions must, of course, be equal, so that we can write

$$k_f[A]^a[B]^b = k_r[E]^e[F]^f$$

This can be easily rearranged to give

$$\frac{k_f}{k_r} = \frac{[E]^e[F]^f}{[A]^a[B]^b}$$

Since the ratio of two constants, $k_f/k_r$, is also a constant, let's call it $K_c$, we obtain the equilibrium condition for the reaction,

$$K_c = \frac{[E]^e[F]^f}{[A]^a[B]^b}$$

which is the law of mass action.

We know now, of course, that one cannot predict the rate law for a chemical reaction based on the balanced overall equation. Does this affect our kinetic interpretation of the law of mass action?

The answer to this question is no. Let us suppose we had the reaction

$$2A + B \rightleftharpoons C + D$$

The equilibrium condition, we know, would be written as

$$\frac{[C][D]}{[A]^2[B]} = K_c$$

We also know that, *if* the reaction proceeds by a one-step mechanism, the rate of the forward reaction is

$$\text{rate}_f = k_f[A]^2[B]$$

while the rate of the reverse reaction is given as

$$\text{rate}_r = k_r[C][D]$$

Following the same argument as before, we find that

$$\frac{[C][D]}{[A]^2[B]} = \frac{k_f}{k_r} = K_c$$

Suppose, now, that we arbitrarily assume that this reaction proceeds by the following two-step mechanism, each step of which is reversible.

$$\text{Step 1} \qquad 2A \underset{k_{1r}}{\overset{k_{1f}}{\rightleftharpoons}} A_2$$

$$\text{Step 2} \qquad A_2 + B \underset{k_{2r}}{\overset{k_{2f}}{\rightleftharpoons}} C + D$$

At equilibrium the rates of the forward and reverse reactions in *each* step must be the same; therefore, for step 1 we have

$$K_1 = \frac{k_{1f}}{k_{1r}} = \frac{[A_2]}{[A]^2}$$

and similarly for step 2,

$$K_2 = \frac{k_{2f}}{k_{2r}} = \frac{[C][D]}{[A_2][B]}$$

If we now multiply $K_1$ and $K_2$ together, we can eliminate the concentration of the intermediate $A_2$ and obtain

$$K_1K_2 = K_c = \frac{[\cancel{A_2}]}{[A]^2} \times \frac{[C][D]}{[\cancel{A_2}][B]}$$

or

$$K_c = \frac{[C][D]}{[A]^2[B]}$$

Thus we see that regardless of the mechanism of the reaction, we always arrive at the same requirement for equilibrium.

## 12.4 THERMODYNAMICS AND CHEMICAL EQUILIBRIUM

In Section 10.13 we saw, qualitatively, that there is a relationship between $\Delta G°$ for a reaction and the position of equilibrium. In addition, the direction in which a reaction proceeds toward equilibrium is determined by where the system lies with respect to the free-energy minimum. The reaction proceeds spontaneously only in a direction that gives rise to a decrease in free energy, that is, when $\Delta G$ is negative.

All of this is summed up quantitatively by the equation (which we will not attempt to justify),

$$\Delta G = \Delta G^0 + 2.303RT \log Q \qquad (12.2)$$

The symbol, $Q$, represents the mass action expression for the reaction. For gases, $Q$ is written with partial pressures; for reactions in solution, molar concentrations are used.[1]

[1] Actually, to make Equation 12.2 fit exactly, "effective pressures" or "effective concentrations" must be used in $Q$. These are called **activities.** At low pressures in gaseous reactions, and in dilute solutions, the use of actual pressures and concentrations leads to only small errors.

Equation 12.2 **tells us** how $\Delta G$ varies with temperature and with the relative proportions of reactants and products. For example, for the reaction,

$$2NO_2\ (g) \rightleftharpoons N_2O_4\ (g)$$

Equation 12.2 would take the form

$$\Delta G = \Delta G^0 + 2.303RT \log\left(\frac{p_{N_2O_4}}{p^2_{NO_2}}\right) \quad (12.3)$$

At equilibrium the products and reactants have the same total free energy and $\Delta G = 0$ (Section 10.12). Equation 12.3 becomes

$$0 = \Delta G^0 + 2.303RT \log\left(\frac{p_{N_2O_4}}{p^2_{NO_2}}\right)$$

or

$$\Delta G^0 = -2.303RT \log\left(\frac{p_{N_2O_4}}{p^2_{NO_2}}\right)$$

At equilibrium for this reaction,

$$\frac{p_{N_2O_4}}{p^2_{NO_2}} = K_P$$

Therefore,

$$\Delta G^0 = -2.303RT \log K_P \quad (12.4)$$

Equation 12.4, derived here for this specific example, applies to all reactions involving gases. For reactions in solution,

$$\Delta G^0 = -2.303RT \log K_c \quad (12.5)$$

We now have a quantitative relationship between $\Delta G^0$ and the equilibrium constant. The $K$ computed using Equations 12.4 or 12.5 is sometimes called the **thermodynamic equilibrium constant.**

EXAMPLE 12.1 What is the thermodynamic equilibrium constant for the reaction,

$$2SO_2\ (g) + O_2\ (g) \rightleftharpoons 2SO_3\ (g)$$

at 25°C?

SOLUTION From the data in Table 10.5 we can obtain the standard free energies of formation of $SO_3$ and $SO_2$:

$$\Delta G_f^0(SO_3) = -88.5\ \frac{\text{kcal}}{\text{mole}}$$

$$\Delta G_f^0(SO_2) = -71.8\ \frac{\text{kcal}}{\text{mole}}$$

By definition, $\Delta G_f^0(O_2) = 0.0$ kcal/mole.

Using these data, we can compute $\Delta G^0$ for the reaction:

$$\Delta G^0 = 2\ \cancel{\text{moles}} \times \left(-88.5\ \frac{\text{kcal}}{\cancel{\text{mole}}}\right) - 2\ \cancel{\text{moles}} \times \left(-71.8\ \frac{\text{kcal}}{\cancel{\text{mole}}}\right)$$

$$= -33.4\ \text{kcal}$$

Solving Equation 12.4 for log $K_P$ (we are dealing with a gaseous reaction),

$$\log K_P = \frac{-\Delta G^0}{2.303RT}$$

We must express $\Delta G^0$ in calories ($\Delta G^0 = -33{,}400$ cal), $T$ in degrees kelvin (298 K), and use $R = 1.99$ cal/mole K). Substituting numerical values gives

$$\log K_P = \frac{-(-33{,}400)}{2.303(1.99)(298)}$$
$$= 24.46$$

Taking the antilogarithm gives

$$K_P = 2.9 \times 10^{24}$$

The magnitude of $K$ for this reaction tells us that the position of equilibrium in the system should lie far in the direction of $SO_3$ and that at room temperature $SO_2$ should react almost completely with oxygen to form $SO_3$. This reaction is extremely slow at room temperature, but with a catalyst it becomes an important step in the industrial preparation of $H_2SO_4$. A similar reaction takes place in the exhaust of an automobile equipped with a catalytic converter, but in this case the $H_2SO_4$ produced presents a health problem.

Thermodynamic data can also be used to calculate equilibrium constants at temperatures other than 25°C. This is shown in Example 12.2.

EXAMPLE 12.2 For the reaction, $2NO_2\ (g) \rightleftharpoons N_2O_4\ (g)$, $\Delta H^0_{298\,K} = -13.6$ kcal and $\Delta S^0_{298\,K} = -41.9$ cal/K. Calculate $K_p$ at 100°C.

SOLUTION To calculate $K_p$ we need to have the value of $\Delta G^0$ at 100°C. In Chapter 10 we saw that

$$\Delta G^0 = \Delta H^0 - T\,\Delta S^0$$

It happens that $\Delta H^0$ and $\Delta S^0$ vary only slightly with temperature, and for the purposes of our calculation we will assume that they are independent of temperature. Therefore, at 100°C (373 K),

$$\Delta G^0_{373} = -13{,}600 \text{ cal} - (373 \text{ K})(-41.9 \text{ cal/K})$$
$$= +2030 \text{ cal}$$

Solving Equation 12.4 for log $K_p$ gives us

$$\log K_p = \frac{-\Delta G^0}{2.303RT}$$

Once again we use $R = 1.99$ cal/mole K because $\Delta G^0$ is in calories. Substituting numerical values, using $T = 373$ K, gives

$$\log K_p = \frac{-2030}{2.303(1.99)(373)} = -1.19$$

Taking the antilogarithm gives

$$K_p = 6.5 \times 10^{-2}$$

The measurement of equilibrium constants also provides a very convenient method for obtaining thermodynamic data. This is illustrated in the next example

EXAMPLE 12.3 At 25°C it was found that $K_P = 7.13$ for the reaction

$$2NO_2\ (g) \rightleftharpoons N_2O_4\ (g)$$

What is $\Delta G^0$ for this reaction in kilojoules?

SOLUTION We can calculate $\Delta G^0$ by substituting appropriate values into Equation 12.4,

$$\Delta G^0 = -2.303RT \log K_P$$

Since we wish $\Delta G^0$ in kilojoules, we must use $R = 8.314$ J/mole K. As usual, $T$ is the absolute temperature ($T = 298$ in this example). Substituting numerical values,

$$\Delta G^0 = -2.303(8.314)(298) \log(7.13)$$
$$= -4870 \text{ J}$$

The value of $\Delta G^0$ is in joules because $R$ was in joules. To convert to kilojoules simply divide by 1000:

$$\Delta G^0 = -4.87 \text{ kJ}$$

## 12.5 THE RELATIONSHIP BETWEEN $K_P$ AND $K_C$

It was stated earlier that, in general, for reactions involving gases, $K_P$ and $K_c$ are not necessarily equal. For the general equation,

$$aA + bB \rightleftharpoons eE + fF$$

$$K_P = \frac{p_E{}^e p_F{}^f}{p_A{}^a p_B{}^b}$$

and

$$K_c = \frac{[E]^e[F]^f}{[A]^a[B]^b}$$

Concentration, you recall, has the units moles per liter, or $n/V$. Assuming ideal gas behavior, we can use the ideal gas law,

$$PV = nRT$$

to obtain the concentration of a gas, $X$, in a mixture as

$$[X] = \frac{n_X}{V} = \frac{p_X}{RT}$$

where $p_X$ is its partial pressure. From this it follows that

$$p_X = [X]RT$$

Substituting this relationship into the expression for $K_P$, we have

$$K_P = \frac{p_E{}^e p_F{}^f}{p_A{}^a p_B{}^b} = \frac{[E]^e(RT)^e[F]^f(RT)^f}{[A]^a(RT)^a[B]^b(RT)^b}$$

This can be rearranged to give

$$K_P = \frac{[E]^e[F]^f}{[A]^a[B]^b}(RT)^{(e+f)-(a+b)}$$

or

$$K_P = K_c(RT)^{\Delta n_g} \qquad (12.6)$$

where $\Delta n_g$ *is the change in the number of moles of* **gas** *upon going from reactants to products.* Thus, $K_P$ and $K_c$ are related in a very simple fashion for reactions between ideal gases, a relationship that also holds adequately for many real gases.

EXAMPLE 12.4 In Example 12.1 we determined the value of $K_P$ for the reaction of $SO_2$ with $O_2$ to produce $SO_3$. What is $K_c$ for this equilibrium at 25°C?

SOLUTION Solving Equation 12.6 for $K_c$, we obtain

$$K_c = \frac{K_P}{(RT)^{\Delta n_g}} = K_P(RT)^{-\Delta n_g}$$

For the reaction in Example 12.1, we find $\Delta n_g = -1$, because there are 3 moles of gaseous reactants and only 2 moles of gaseous products. Using the result of Example 12.1, we have

$$K_P = \frac{p_{SO_3}^2}{p_{SO_2}^2 p_{O_2}} = 2.9 \times 10^{24}\ \text{atm}^{-1}$$

From the equilibrium constant expression, we see that $K_P$ in this case has the units 1/atm, or atm$^{-1}$. As a result, we must use $R = 0.0821$ liter atm mole$^{-1}$ K$^{-1}$ to obtain the proper units for $K_c$ (why?). Thus

$$K_c = (2.9 \times 10^{24}\text{atm}^{-1})[(0.0821\ \text{liter atm mole}^{-1}\ \text{K}^{-1})(298\ \text{K})]^{-(-1)}$$

Therefore,

$$K_c = 7.1 \times 10^{25}\ \text{liter mole}^{-1}$$

## 12.6 HETEROGENEOUS EQUILIBRIA

Up to now our discussion has focused on homogeneous reactions in which all of the reactants and products are in the same phase. Heterogeneous reactions, of which there are many examples, also eventually arrive at a state of equilibrium. A typical reaction that we might consider is the decomposition of solid $NaHCO_3$ to produce solid $Na_2CO_3$, gaseous $CO_2$, and gaseous $H_2O$.[2]

$$2NaHCO_3\ (s) \rightleftharpoons Na_2CO_3\ (s) + CO_2\ (g) + H_2O\ (g)$$

Applying the law of mass action, we can write the equilibrium expression as

$$\frac{[Na_2CO_3\ (s)][CO_2\ (g)][H_2O\ (g)]}{[NaHCO_3\ (s)]^2} = K_c'$$

For reasons that will be apparent shortly, we have temporarily indicated the equilibrium constant as $K_c'$.

In this reaction we have an equilibrium between the gases, $CO_2$ and $H_2O$, and the two pure solid phases, $NaHCO_3$ and $Na_2CO_3$. We know that a

[2] $Na_2CO_3$ is produced commercially by this reaction. It is one of the most industrially important chemicals, ranking tenth in total production (about 15 billion pounds produced annually). It is used in the manufacture of glass and many other important products.

pure solid substance such as $NaHCO_3$ is characterized by a density that is the same for all samples of $NaHCO_3$, regardless of their size. In addition, this density is unaffected by the nature of the chemical reaction. This means that even during a chemical reaction the amount of $NaHCO_3$ in a given volume of the pure solid is always the same. As a result, the concentration of $NaHCO_3$ in pure solid $NaHCO_3$ is a constant. We cannot alter the number of moles per liter of $NaHCO_3$ in the pure solid, nor can we change the concentration of $Na_2CO_3$ in pure solid $Na_2CO_3$. Consequently, the concentrations of these two substances in the equilibrium expression take on constant values and can be incorporated into the equilibrium constant.

$$[CO_2\ (g)][H_2O\ (g)] = K_c' \frac{[NaHCO_3\ (s)]^2}{[Na_2CO_3\ (s)]}$$

or

$$[CO_2\ (g)][H_2O\ (g)] = K_c$$

Thus we find that for heterogeneous reactions, *the equilibrium constant expression does not include the concentrations of pure solids.* Similarly, in reactions in which a reactant or product occurs as a pure liquid phase, the concentration of that substance in the pure liquid is also constant. As a result, *the concentrations of pure liquid phases also do not appear in an equilibrium constant expression.* These simplifications apply *only* when we are dealing with *pure* condensed phases. When substances occur in liquid or solid solutions, their concentrations are variable and their concentration terms in the mass action expression therefore cannot be incorporated into $K$.

If we wish to work with $K_P$ rather than $K_c$, we again need take into account only the substances present in the gas phase. For the decomposition of $NaHCO_3$, therefore, we have

$$K_P = p_{CO_2(g)}p_{H_2O(g)}$$

As noted in the last section, if we know $K_c$, we can evaluate $K_P$ as

$$K_P = K_c(RT)^{\Delta n_g} \qquad (12.6)$$

where, for this reaction, $\Delta n_g = +2$.

EXAMPLE 12.5 What are the values of $K_P$ and $K_c$ for the "reaction"

$$H_2O\ (l) \rightleftharpoons H_2O\ (g)$$

at 25°C given that the vapor pressure of water at 25°C equals 23.8 torr?

SOLUTION Since liquid water is a pure liquid phase, we can write

$$K_P = p_{H_2O(g)}$$

and

$$K_c = [H_2O\ (g)]$$

(a) If we express the vapor pressure of water in atmospheres,

$$p_{H_2O} = 23.8\ \cancel{\text{torr}} \times \left(\frac{\text{atm}}{760\ \cancel{\text{torr}}}\right) = 0.0313\ \text{atm}$$

Therefore,

$$K_P = p_{H_2O} = 3.13 \times 10^{-2} \text{ atm}$$

Note that this equilibrium expression states that the partial pressure of water must be a constant when the liquid and vapor are in equilibrium.

(b) We can evaluate $K_c$ as

$$K_c = K_P(RT)^{-\Delta n_g}$$

For this "reaction," $\Delta n_g = 1$; therefore,

$$K_c = K_P(RT)^{-1} = \frac{K_P}{RT}$$

$$= \frac{3.13 \times 10^{-2} \text{ atm}}{(0.0821 \text{ liter atm/mole K})(298 \text{ K})}$$

or

$$K_c = 1.28 \times 10^{-3} \frac{\text{mole}}{\text{liter}}$$

## 12.7 LE CHATELIER'S PRINCIPLE AND CHEMICAL EQUILIBRIUM

The equilibrium expression, in the form of either $K_P$ or $K_c$, can be used to perform numerical computations of various kinds dealing with equilibrium systems. This is discussed in the next section. Often, however, it is desirable simply to be able to predict how some disturbance imposed on a system from outside will influence the position of equilibrium. For instance, we may wish to predict, in a qualitative way, the conditions that favor the greatest production of products. Should we run our reaction at high or low temperature? Should the pressure on the system be high or low? These are questions that we would like to answer quickly without having to perform tedious computations. We have already seen how Le Chatelier's principle can be applied to dynamic equilibria involving such phenomena as the vapor pressure of a liquid and solubility. Changes in the position of equilibrium in chemical systems can also be understood by applying the same concepts.

**CHANGES IN THE CONCENTRATION OF A REACTANT OR PRODUCT.** In a system such as

$$H_2\,(g) + I_2\,(g) \rightleftharpoons 2HI\,(g)$$

any change in the concentration of a reactant or product will cause the system no longer to be at equilibrium. As a result, a chemical reaction will occur that will return the system to equilibrium. From Le Chatelier's principle we know that if a system at equilibrium is disturbed, it will attempt to undergo some change to diminish the effect of the disturbance. For example, the addition of $H_2$ to an equilibrium mixture of $H_2$, $I_2$, and HI upsets the equilibrium, and the system responds by using up part of the additional $H_2$ by reaction with $I_2$ to produce more HI. When equilibrium has finally been reestablished, there will be a greater concentration of HI than before, and we say that for this reaction the position of equilibrium has been shifted to the right. This is illustrated in Figure 12.2. Notice that after equilibrium has been reestablished there continues to be more $H_2$ present than in the original reaction mixture. The system is never able to overcome completely the

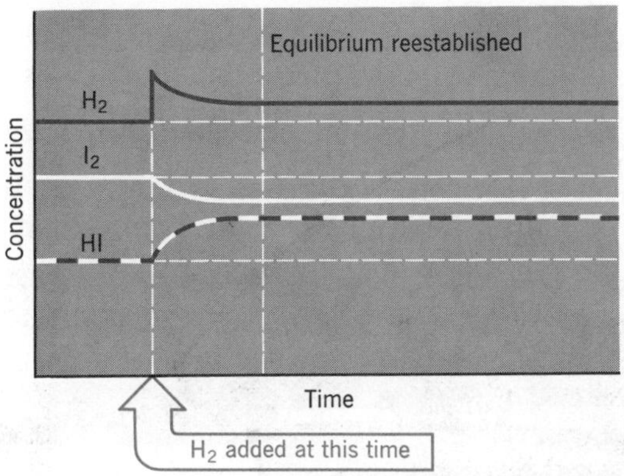


Figure 12.2
*Addition of $H_2$ to the equilibrium, $H_2 + I_2 \rightleftharpoons 2HI$, increases the amount of HI and decreases the amount of $I_2$.*

effect of a change in concentration. The final position of equilibrium differs from the original.

We can arrive at this same conclusion by considering the effect of added $H_2$ on the value of the mass action expression. For this reaction at equilibrium we have

$$\frac{[HI]^2}{[H_2][I_2]} = K_c$$

If $H_2$ were suddenly added to the system, the value of the denominator of the mass action expression would become larger, and the entire fraction would therefore become smaller than the equilibrium constant. The reaction that occurs to return the system to equilibrium must increase the value of the mass action expression until it once again is equal to $K_c$. In order for this to take place, the numerator must become larger and the denominator smaller. In other words, more HI will be formed at the expense of $H_2$ and $I_2$, and again we conclude that the addition of $H_2$ shifts the position of equilibrium in this reaction to the right.

By applying Le Chatelier's principle we can also predict the effect that removing a reactant or product will have on a system at equilibrium. For instance, if $H_2$ is somehow removed from the reaction vessel, the system will adjust by having some HI decompose in an effort to replenish the lost reactant. Consequently, we conclude that the position of equilibrium is shifted to the left when $H_2$ is removed.

Therefore we may conclude that to drive a reaction far toward completion we can either add a large excess of one of the reactants or remove the products as they are formed. Recall that it is the latter that serves as the driving force in ionic reactions (Chapter 5) when the product is either a precipitate, a gas, or a weak electrolyte. The creation of these products removes ions from solution and therefore forces the reaction to proceed toward completion.

**THE EFFECT OF TEMPERATURE ON EQUILIBRIUM.** Up to now we have been careful to imply that the equilibrium constant for a reaction has a fixed numerical value only as long as the temperature remains constant. This is because temperature, as well as the concentrations of reactants and products, affects the position of equilibrium. However, the temperature, unlike the concentrations of reactants and products, affects the value of the equilibrium constant itself.

The reaction between $H_2$ and $N_2$ to form $NH_3$ is exothermic, and the equation for the formation of ammonia is written as

$$3H_2\,(g) + N_2\,(g) \rightleftharpoons 2NH_3\,(g) + 92.0\text{ kJ}$$

where the heat of reaction is indicated as a product. If we have a system of these gases in equilibrium and wish to raise its temperature, we do so by adding heat to it from the surroundings. Le Chatelier's principle tells us that when we add this heat, the system will attempt to undergo a change that tends to use some of it up and, as before, we conclude that a rise in temperature should cause an endothermic change to take place. Since the production of $NH_3$ is exothermic, its decomposition is endothermic. Consequently, raising the temperature of this system will cause the position of equilibrium to shift to the left. An increase in temperature, then, drives the reaction to a new position of equilibrium in which there is more $N_2$ and $H_2$ and less $NH_3$. *In general, an increase in temperature causes the position of equilibrium of an exothermic reaction to be shifted to the left, while that of an endothermic reaction is shifted to the right.*

In Section 12.1, it was stated that at equilibrium the mass action expression is always equal to the same *K* provided that the temperature remains the same. A change in temperature, however, alters *K*. We have just seen that an increase in temperature leads to a decrease in the concentration of $NH_3$ and an increase in the concentration of both $N_2$ and $H_2$. This means that *at equilibrium* at the higher temperature the value of the mass action expression,

$$\frac{[NH_3]^2}{[H_2]^3[N_2]}$$

will have decreased. Thus we find that for this exothermic reaction, *K* decreases with rising temperature. By the same token, for a reaction that is endothermic in the forward direction, *K* increases with increasing temperature.

**EFFECT OF PRESSURE AND VOLUME CHANGES ON EQUILIBRIUM.** At constant temperature a change in the volume of a system also causes a change in pressure. We quite logically expect, therefore, that an increase in the external pressure on a system should favor any change that leads to a smaller volume (recall Boyle's law). We would not expect pressure changes to have any marked effect on the position of equilibrium in reactions where all of the reactants and products are either solids or liquids, because these phases are virtually incompressible. However, pressure changes can have a very dramatic effect on equilibria that involve reactions in which gases are consumed or produced.

Let us again choose as an example the reaction for the formation of $NH_3$. If we have this system at equilibrium and suddenly decrease the volume of the container, we know that the pressure will go up. By applying Le Chatelier's principle, we expect that a change in the system should occur that will reduce the pressure. How can this be brought about?

Let us suppose that we had initially placed into our reaction vessel 1 mole of nitrogen and 3 moles of hydrogen. Thus 4 moles of gas are initially in the container. If the reaction were to proceed entirely to completion, 2 moles of $NH_3$ would be produced; therefore as the reaction,

$$N_2\,(g) + 3H_2\,(g) \rightleftharpoons 2NH_3\,(g)$$

proceeds from left to right, the number of moles of gas in the container decreases. Since fewer moles of gas means lower pressure, it is clear that the production of additional $NH_3$ in a vessel of fixed volume leads to a reduction in the pressure. Consequently, when the volume of an equilibrium mixture of $H_2$, $N_2$, and $NH_3$ is reduced, and then held constant, more $NH_3$ will be formed and the position of equilibrium will shift to the right.[3] Note, however, that the value of $K$ remains unchanged.

In short, *increasing the pressure on a system at equilibrium will cause a shift in the position of equilibrium in the direction of the fewest number of moles of gaseous reactants or products.*

Finally, note that when there are the same number of moles of gaseous reactants and products, as in the reaction between $H_2$ and $I_2$,

$$H_2\,(g) + I_2\,(g) \rightleftharpoons 2HI\,(g)$$

pressure changes will not influence the quantities of the various substances present in the reaction mixture at equilibrium.

**ADDITION OF AN INERT GAS.** If an inert (nonreacting) gas is introduced into a reaction vessel containing other gases at equilibrium, it will cause an increase in the total pressure within the container. This kind of pressure increase, however, will not affect the position of equilibrium because it will not alter the partial pressures or the concentrations of any of the substances already present.

**EFFECT OF A CATALYST ON THE POSITION OF EQUILIBRIUM.** In Chapter 11 we saw that a catalyst affects a chemical reaction by lowering the activation energy barrier that must be overcome in order for the reaction to proceed. A catalyst affects the rate of a chemical change. It does not, however, affect the heat of reaction, and it is the heat of reaction, $\Delta H^0$, along with the entropy change, $\Delta S^0$, that determine $\Delta G^0$, which in turn fixes the position of equilibrium at any given temperature. A catalyst merely speeds the approach to the position of equilibrium that is determined by $\Delta G^0$.

## 12.8 EQUILIBRIUM CALCULATIONS

This section is intended to illustrate the type of computations that one might perform either to evaluate an equilibrium constant from measured concentrations or to use the equilibrium constant to calculate the concentrations of the reactants and products in a particular equilibrium mixture. First let's see how we might evaluate $K$ in a typical experiment.

EXAMPLE 12.6 The brown gas $NO_2$ and the colorless gas $N_2O_4$ exist in equilibrium as indicated by the equation,

$$2NO_2 \rightleftharpoons N_2O_4$$

In an experiment, 0.625 mole of $N_2O_4$ was introduced into a 5.0-liter vessel and permitted to come to equilibrium with $NO_2$. At equilibrium the concentration of $N_2O_4$ was 0.0750 $M$. What is $K_c$ for this reaction?

[3] From application of Le Chatelier's principle, we know that the production of ammonia from $H_2$ and $N_2$ is favored by high pressures and low temperatures. At low temperatures, however, the reaction is very slow; therefore in the industrial preparation of $NH_3$, pressures of $10^2$ to $10^3$ atm and temperatures from 400 to 550°C are employed. Even though there is less $NH_3$ produced at equilibrium at these high temperatures, the speed of reaction is boosted to the point where the production of $NH_3$ is economically worthwhile.

SOLUTION The equilibrium constant expression for this reaction is

$$\frac{[N_2O_4]}{[NO_2]^2} = K_c$$

In order to calculate $K_c$, we must know the equilibrium concentrations of $N_2O_4$ and $NO_2$. In working out equilibrium problems, we will generally find it useful to set up a table like that below in order to establish quantities that correspond to equilibrium concentrations. The entries in the table are obtained by reasoning from the data provided in the problem. Remember that when using $K_c$, molar concentrations (that is, moles/liter) must be used. In this example the concentration of $N_2O_4$ initially was 0.625 mole/5.0 liter = 0.125 $M$; the initial concentration of $NO_2$ was zero. From the equilibrium concentration of $N_2O_4$ given to us in the problem, we can conclude that in arriving at equilibrium 0.050 mole/liter of $N_2O_4$ was decomposed. Based on the stoichiometry of the equation, this means that 0.10 mole/liter of $NO_2$(2 × 0.050 mole/liter of $N_2O_4$) was formed.

| | Initial Concentrations | Change | Equilibrium Concentrations |
|---|---|---|---|
| $N_2O_4$ | 0.125 $M$ | −0.050 $M$ | 0.075 $M$ |
| $NO_2$ | 0.00 $M$ | +0.10 $M$ | 0.10 $M$ |

Substituting the equilibrium concentrations into the mass action expression, we have

$$\frac{(0.075\ M)}{(0.10\ M)^2} = K_c$$

and, finally,

$$K_c = 7.5 \text{ liters/mole}$$

Knowledge of the equilibrium constant for a reaction allows us to calculate the concentrations or partial pressures of the substances present in a reaction mixture at equilibrium. The ease with which these computations can be carried out depends on the complexity of the mass action expression, the concentrations of the various species in the reaction mixture, and the magnitude of the equilibrium constant. We shall look only at some of the more simple examples of problems of this type. The following sample problems, however, illustrate the type of reasoning that is employed in these computations, as well as some of the concepts that have been presented up to this point.

EXAMPLE 12.7 At 25°C, $K_P = 7.13$ atm$^{-1}$ for the reaction

$$2NO_2\ (g) \rightleftharpoons N_2O_4\ (g)$$

At equilibrium the partial pressure of $NO_2$ in a container is 0.15 atm. What is the partial pressure of $N_2O_4$ in the mixture?

SOLUTION The first step in the solution of any equilibrium problem is to write down the equilibrium expression. For $K_P$ we have

$$K_P = \frac{p_{N_2O_4}}{p^2_{NO_2}} = 7.13 \text{ atm}^{-1}$$

We are given the equilibrium partial pressure of $NO_2$ ($p_{NO_2} = 0.15$ atm). There is only one unknown quantity, $p_{N_2O_4}$. Substituting,

$$\frac{p_{N_2O_4}}{(0.15 \text{ atm})^2} = 7.13 \text{ atm}^{-1}$$

$$p_{N_2O_4} = 7.13 \text{ atm}^{-1}(0.15 \text{ atm})^2$$
$$= 0.16 \text{ atm}$$

The partial pressure of $N_2O_4$ at equilibrium is 0.16 atm.

EXAMPLE 12.8 At a temperature of 500°C, the equilibrium constant, $K_c$, for the nitrogen fixation reaction for the production of ammonia,

$$3H_2\,(g) + N_2\,(g) \rightleftharpoons 2NH_3\,(g)$$

has a value of $6.0 \times 10^{-2}$ liter²/mole². If, in a particular reaction vessel at this temperature, there are 0.250 mole/liter of $H_2$ and 0.0500 mole/liter of $NH_3$ present at equilibrium, what is the concentration of $N_2$?

SOLUTION Let's first write down the equilibrium constant expression. For this reaction we have

$$K_c = \frac{[NH_3]^2}{[H_2]^3[N_2]} = 6.0 \times 10^{-2}\,\frac{\text{liter}^2}{\text{mole}^2}$$

We wish to calculate the concentration of $N_2$. This can be accomplished if we know the values of the equilibrium concentrations of both $NH_3$ and $H_2$ and, in this problem, these are given to us.

$$[NH_3] = 0.0500\ M$$
$$[H_2] = 0.250\ M$$

Substituting these numerical values into the mass action expression gives us

$$\frac{(0.0500)^2}{(0.250)^3[N_2]} = 6.0 \times 10^{-2}$$

If we solve for $[N_2]$,

$$[N_2] = \frac{(0.0500)^2}{(0.250)^3(6.0 \times 10^{-2})}$$
$$= 2.7\ M$$

The equilibrium concentration of $N_2$ is thus 2.7 moles/liter.

EXAMPLE 12.9 At 440°C the equilibrium constant for the reaction,

$$H_2\,(g) + I_2\,(g) \rightleftharpoons 2HI\,(g)$$

is 49.5. If 0.200 mole of $H_2$ and 0.200 mole of $I_2$ are placed into a 10.0-liter vessel and permitted to react at this temperature, what will be the concentration of each substance at equilibrium?

SOLUTION Our equilibrium expression is

$$\frac{[HI]^2}{[H_2][I_2]} = 49.5$$

In this example we are given the *initial* concentrations of the reactants and products. These are

$$[H_2] = \frac{0.200 \text{ mole}}{10.0 \text{ liters}} = 0.0200\ M$$

$$[I_2] = \frac{0.200 \text{ mole}}{10.0 \text{ liters}} = 0.0200\ M$$

$$[HI] = 0.0\ M$$

Since no HI is present initially, we know that it will be formed from the violet-colored mixture of the $H_2$ and $I_2$ (the color being due to the $I_2$). Let's approach the problem, then, by allowing $x$ to be equal to the number of moles per liter of $H_2$ that react. From the stoichiometry of the reaction we realize that this $x$ moles/liter of $H_2$ will react with $x$ moles/liter of $I_2$ to produce $2x$ moles/liter of HI. Thus the concentrations of $H_2$ and $I_2$ decrease by $x$; the concentration of HI increases by $2x$. The equilibrium concentrations are obtained by applying these changes to the initial concentrations.

| | Initial Concentrations | Change | Equilibrium Concentrations |
|---|---|---|---|
| $H_2$ | 0.0200 $M$ | $-x$ | $(0.0200 - x)\ M$ |
| $I_2$ | 0.0200 $M$ | $-x$ | $(0.0200 - x)\ M$ |
| HI | 0.0 $M$ | $+2x$ | $0.0 + 2x = 2x\ M$ |

Substituting the equilibrium quantities into the mass action expression gives

$$\frac{(2x)^2}{(0.0200 - x)(0.0200 - x)} = 49.5$$

or

$$\frac{(2x)^2}{(0.0200 - x)^2} = 49.5$$

In this case, we can take the square root of both sides of the equation to obtain

$$\frac{2x}{0.0200 - x} = 7.04$$

Solving for $x$,

$$2x = 7.04(0.0200 - x) = 0.141 - 7.04x$$
$$2x + 7.04x = 0.141$$
$$9.04x = 0.141$$
$$x = 0.0156$$

Finally, the equilibrium concentrations are

$$[H_2] = 0.0200 - 0.0156 = 0.0044\ M$$
$$[I_2] = 0.0200 - 0.0156 = 0.0044\ M$$
$$[HI] = 2(0.0156) = 0.0312\ M$$

In this last problem we employed some relatively simple algebra to help us arrive at the solution. Let us look at another example of this type.

EXAMPLE 12.10 A 10-liter vessel is filled with 0.40 mole of HI at 440°C. What will be the concentration of $H_2$, $I_2$, and HI at equilibrium?

SOLUTION In this example we are concerned with the same equilibrium as in the previous problem. Initially we have no $H_2$ or $I_2$ and the reaction mixture is colorless. In order to have an equilibrium, some $H_2$ and $I_2$ must be formed from the decomposition of HI. Let's let $x$ be the number of moles per liter of HI that decomposes. Since 1 mole each of $H_2$ and $I_2$ are produced from every 2 moles of HI that break down, the HI concentration decreases by $x$ and the $H_2$ and $I_2$ concentrations each increase by $0.5x$. We can use our table now to write the equilibrium concentrations. The initial concentration of HI is 0.40 mole/10.0 liter = 0.040 $M$.

| | Initial Concentrations | Change | Equilibrium Concentrations |
|---|---|---|---|
| $H_2$ | 0.0 $M$ | $+0.5x$ $M$ | $0.0 + 0.5x = 0.5x$ $M$ |
| $I_2$ | 0.0 $M$ | $+0.5x$ $M$ | $0.0 + 0.5x = 0.5x$ $M$ |
| HI | 0.040 $M$ | $-x$ $M$ | $(0.040 - x)$ $M$ |

Substituting equilibrium quantities into the mass action expression gives us

$$\frac{(0.040 - x)^2}{(0.50x)(0.50x)} = 49.5$$

or

$$\frac{(0.040 - x)^2}{(0.50x)^2} = 49.5$$

Taking the square root of both sides of the equation, we have

$$\frac{0.040 - x}{0.50x} = 7.04$$

Solving for $x$, we get

$$0.040 - x = (0.50x)(7.04)$$
$$x = 0.00885$$

We now calculate the equilibrium concentrations to be

$$[H_2] = 0.50x = 0.0044\ M$$
$$[I_2] = 0.50x = 0.0044\ M$$
$$[HI] = 0.040 - x = 0.031\ M$$

Observe that we have obtained essentially the same answers in both Examples 12.9 and 12.10. If all of the $H_2$ and $I_2$ in Example 12.9 had completely reacted, it would have produced 0.40 mole of HI, the same amount of HI that we began with in Example 12.10. We find, therefore, that the same position of equilibrium can be approached from either direction.

In the last two examples the solution of the algebra was simple because we were able to take the square root of both sides of the equation. In cases where the equilibrium constant is either extremely large, or extremely small, it is frequently possible to make some approximations that greatly simplify the kind of calculations we have just seen.

EXAMPLE 12.11 The equilibrium constant for the decomposition of water at 500°C is $6.0 \times 10^{-28}$. If 2.0 moles of $H_2O$ are placed into a 5.0-liter container, what will be the equilibrium concentrations of $H_2$, $O_2$, and $H_2O$ (*g*) at 500°C?

SOLUTION The equation for the reaction is

$$2H_2O \rightleftharpoons 2H_2 + O_2$$

Therefore we can write

$$\frac{[H_2]^2[O_2]}{[H_2O]^2} = 6.0 \times 10^{-28}$$

The initial $H_2O$ concentration is 2.0 moles/5.0 liter $\doteq$ 0.40 *M*. If we let *x* equal the number of moles per liter of $H_2O$ that decomposes, we will get *x* moles of $H_2$ and 0.5*x* mole of $O_2$. Constructing our table,

| | Initial Concentrations | Change | Equilibrium Concentrations |
|---|---|---|---|
| $H_2O$ | 0.40 *M* | $-x$ *M* | $(0.40 - x)$ *M* |
| $H_2$ | 0.0 *M* | $+x$ *M* | $x$ *M* |
| $O_2$ | 0.0 *M* | $+0.5x$ *M* | $0.5x$ *M* |

Substituting equilibrium quantities into the mass action expression gives

$$\frac{(x)^2(0.50x)}{(0.40 - x)^2} = 6.0 \times 10^{-28}$$

Unless we can somehow simplify this equation, we have a real mess on our hands. Fortunately, in this case the problem can be made easy to solve.

Since *K* is very small, the reaction does not proceed very far toward completion. This means that very little $H_2$ and $O_2$ are formed. To simplify the algebra, then, we will make the assumption that *x* will be *much* smaller than 0.40 so that when *x* is subtracted from 0.40, the difference will still be very nearly 0.40. If we neglect *x* when we compute the $H_2O$ concentration, the equilibrium values become

$$[H_2O] = 0.40 - x \approx 0.40$$
$$[H_2] = x$$
$$[O_2] = 0.5x$$

Substituting these into the mass action expression gives

$$\frac{(x)^2(0.50x)}{(0.40)^2} = 6.0 \times 10^{-28}$$

from which we get

$$\frac{0.50x^3}{0.16} = 6.0 \times 10^{-28}$$

or

$$x^3 = \frac{0.16}{0.50}(6.0 \times 10^{-28}) = 1.9 \times 10^{-28}$$

At this point $x$ can be obtained by extracting the cube root. Many hand-held calculators can perform this operation by raising $1.9 \times 10^{-28}$ to the $\frac{1}{3}$ power.

$$\sqrt[3]{1.9 \times 10^{-28}} = (1.9 \times 10^{-28})^{1/3} = 5.7 \times 10^{-10}$$

If you use a slide rule, you must first make the exponent divisible by 3.

$$x^3 = 190 \times 10^{-30}$$
$$x = 5.7 \times 10^{-10}$$

We see that $x$ is, in fact, very much smaller than 0.40, thus justifying our initial assumption. The final equilibrium concentrations are

$$[H_2] = x = 5.7 \times 10^{-10}\ M$$
$$[O_2] = 0.50x = 2.8 \times 10^{-10}\ M$$
$$[H_2O] = 0.40 - (5.7 \times 10^{-10}) = 0.40\ M$$

In working out a problem of this sort, look for any assumption that will make the algebra easier to handle. If the assumption you make is invalid, you will discover this when you check the assumption after obtaining a value for $x$. Sometimes no assumption of the kind we made above will be valid and some other method of solving the equation for $x$ will have to be sought.

## INDEX TO QUESTIONS AND PROBLEMS (problem numbers in italics)

General 1, 5

Law of mass action 2, 3, 4, *22*

Significance of K 6, 7

Kinetics and equilibrium 8, 9

Thermodynamics and equilibrium 10, 11, 12, *24, 25, 26, 27, 28, 29, 30, 32*

$K_p$ and $K_c$ 13, 14, *23*

Heterogeneous equilibria 15, 16

Le Chatelier's principle 17, 18, 19, 20, 21

Equilibrium calculations *31, 33, 34, 35, 36, 37, 38, 39, 40, 41, 42, 43, 44*

## REVIEW QUESTIONS

**12.1** What is meant by a dynamic equilibrium?

**12.2** Write the mass action expression in terms of molar concentrations for each of the following reactions:

(a) $N_2\ (g) + O_2\ (g) \rightleftharpoons 2NO\ (g)$
(b) $2NO\ (g) + O_2\ (g) \rightleftharpoons 2NO_2\ (g)$
(c) $2H_2\ (g) + S_2\ (g) \rightleftharpoons 2H_2S\ (g)$
(d) $2N_2O_5\ (g) \rightleftharpoons 4NO_2\ (g) + O_2\ (g)$
(e) $P_4O_{10}\ (g) + 6PCl_5\ (g) \rightleftharpoons 10POCl_3\ (g)$

**12.3** Write equilibrium constant expressions for $K_P$ and $K_c$ for each of the following reactions:

(a) $CO\ (g) + 2H_2\ (g) \rightleftharpoons CH_3OH\ (g)$
(b) $CO\ (g) + H_2O\ (g) \rightleftharpoons CO_2\ (g) + H_2\ (g)$
(c) $PCl_3\ (g) + Cl_2\ (g) \rightleftharpoons PCl_5\ (g)$
(d) $2NO_2\ (g) + 2H_2\ (g) \rightleftharpoons N_2\ (g) + 2H_2O\ (g)$
(e) $2H_2S\ (g) + 3O_2\ (g) \rightleftharpoons 2H_2O\ (g) + 2SO_2\ (g)$

**12.4** Write equilibrium constant expressions for the reactions as written below.
(a) $H_2(g) + Cl_2(g) \rightleftharpoons 2HCl(g)$
(b) $\frac{1}{2}H_2(g) + \frac{1}{2}Cl_2(g) \rightleftharpoons HCl(g)$
How would the magnitude of the $K$ for reaction (a) compare with that for reaction (b)?

**12.5** Why do we always write the concentrations (or partial pressures) of the products in the numerator and those of the reactants in the denominator in the mass action expression?

**12.6** What general information can be gathered by observing the magnitude of the equilibrium constant?

**12.7** Arrange the following reactions in order of their increasing tendency to proceed toward completion:
(a) $4NH_3(g) + 3O_2(g) \rightleftharpoons 2N_2(g) + 6H_2O(g)$ $\quad K = 1 \times 10^{228}$
(b) $N_2(g) + O_2(g) \rightleftharpoons 2NO(g)$ $\quad K = 5 \times 10^{-31}$
(c) $2HF(g) \rightleftharpoons H_2(g) + F_2(g)$ $\quad K = 1 \times 10^{-13}$
(d) $2NOCl(g) \rightleftharpoons 2NO(g) + Cl_2(g)$ $\quad K = 4.7 \times 10^{-4}$

**12.8** What effect would altering the mechanism of a reaction have on the form of the equilibrium constant expression?

**12.9** The reaction, $2NO(g) + 2H_2(g) \rightleftharpoons N_2(g) + 2H_2O(g)$, is believed to occur by the mechanism,

$$2NO(g) \rightleftharpoons N_2O_2(g) \qquad \text{fast}$$
$$N_2O_2(g) + H_2(g) \longrightarrow N_2O(g) + H_2O(g) \qquad \text{slow}$$
$$N_2O(g) + H_2(g) \longrightarrow N_2(g) + H_2O(g) \qquad \text{fast}$$

On the basis of kinetics, derive the mass action law for the overall reaction.

**12.10** What value would $\Delta G^0$ have for a reaction if $K = 1$?

**12.11** For reactions between gases, what kind of equilibrium constant is calculated from $\Delta G^0$?

**12.12** Using Equations 10.11 (p. 314) and 12.4, show that a straight line should be obtained if $\log K_P$ is plotted against $1/T$ (that is, $\log K_P$ along the vertical axis, $1/T$ along the horizontal axis). What does the slope of this line give? What does the value of $\log K_P$ at $1/T = 0$ (the $y$ intercept of the line) give?

**12.13** For which of the reactions in Questions 12.2 and 12.3, would $K_P = K_c$?

**12.14** What would be the units for $K_P$ and $K_c$ for each of the reactions in Question 12.3?

**12.15** Why is it *not* necessary to include the concentration of pure liquid or solid phases in the equilibrium constant expression?

**12.16** Write equilibrium expressions for each of the following reactions:
(a) $CaCO_3(s) \rightleftharpoons CaO(s) + CO_2(g)$
(b) $Ni(s) + 4CO(g) \rightleftharpoons Ni(CO)_4(g)$
(c) $5CO(g) + I_2O_5(s) \rightleftharpoons I_2(g) + 5CO_2(g)$
(d) $Ca(HCO_3)_2(aq) \rightleftharpoons CaCO_3(s) + H_2O(l) + CO_2(g)$
(e) $AgCl(s) \rightleftharpoons Ag^+(aq) + Cl^-(aq)$

**12.17** Consider the equilibrium, $PCl_3(g) + Cl_2(g) \rightleftharpoons PCl_5(g)$. How would the following affect the position of equilibrium?
(a) Addition of $PCl_3$
(b) Removal of $Cl_2$
(c) Removal of $PCl_5$
(d) Decrease in the volume of the container
(e) Addition of He without a change in volume

**12.18** Which, if any, of the changes in Question 12.17 will change the value of the equilibrium constant for the reaction?

**12.19** Indicate how each of the following changes affects the concentration of $H_2$ in the system,
$$9.9 \text{ kcal} + H_2(g) + CO_2(g) \rightleftharpoons H_2O(g) + CO(g)$$
(a) Addition of $CO_2$
(b) Addition of $H_2O$
(c) Addition of a catalyst
(d) Increase in temperature
(e) Decrease in the volume of the container

**12.20** How will each of the changes in Question 12.19 affect the equilibrium constant?

**12.21** Sketch a graph to show how the concentrations of $H_2$, $N_2$, and $NH_3$ would change with time after $N_2$ had been added to a mixture of these gases initially at equilibrium.

## REVIEW PROBLEMS

**12.22** Show that the following data, obtained for the reaction,

$$PCl_5\ (g) \rightleftharpoons PCl_3\ (g) + Cl_2\ (g)$$

demonstrate the law of mass action. What is $K_c$ for this reaction?

| Experiment | $[PCl_5]$ | $[PCl_3]$ | $[Cl_2]$ |
|---|---|---|---|
| 1 | 0.0023 | 0.23 | 0.055 |
| 2 | 0.010 | 0.15 | 0.37 |
| 3 | 0.085 | 0.99 | 0.47 |
| 4 | 1.00 | 3.66 | 1.50 |

**12.23** At a certain temperature, the equilibrium constant ($K_c$) for the reaction,

$$2SO_2\ (g) + O_2\ (g) \rightleftharpoons 2SO_3\ (g)$$

is 35.5 liters/mole. Give the value of $K_P$ for the reaction.

**12.24** At 700 K, $\Delta G^0_{700\,K} = -3.22$ kcal for the reaction, $CO\ (g) + 2H_2\ (g) \rightleftharpoons CH_3OH\ (g)$. Calculate the value of $K_P$ for the reaction at 700 K.

**12.25** The equilibrium constant, $K_P$, for the reaction, $COCl_2\ (g) \rightleftharpoons CO\ (g) + Cl_2\ (g)$, has a value of $4.56 \times 10^{-2}$ atm at 395°C. What is the value of $\Delta G^0_{668\,K}$ (in kilojoules) for this reaction?

**12.26** At 527°C the reaction, $CO\ (g) + H_2O\ (g) \rightleftharpoons CO_2\ (g) + H_2\ (g)$, has $K_P = 5.10$. What is $\Delta G^0_{800\,K}$ for this reaction?

**12.27** Use the data in Tables 10.1 and 10.4 to compute $\Delta G^0_{773\,K}$ and $K_P$ at 500°C for the reaction,

$$2HCl\ (g) \rightleftharpoons H_2\ (g) + Cl_2\ (g)$$

Assume that $\Delta H^0$ and $\Delta S^0$ are independent of temperature.

**12.28** Use the data in Table 10.5 to calculate $K_P$ at 25°C for the reaction, $2HCl\ (g) + F_2\ (g) \rightleftharpoons 2HF\ (g) + Cl_2\ (g)$.

**12.29** Use the data in Tables 10.1 and 10.4 to compute the temperature at which $K_P = 1$ for the reaction,

$$C_2H_4\ (g) + H_2\ (g) \rightleftharpoons C_2H_6\ (g)$$

Assume that $\Delta H^0$ and $\Delta S^0$ are independent of temperature.

**12.30** A container at 700 K contains a mixture of CO, $H_2$, and $CH_3OH$ at the following pressures: $p_{CO} = 2 \times 10^{-3}$ atm, $p_{H_2} = 1 \times 10^{-2}$ atm, $p_{CH_3OH} = 3 \times 10^{-6}$ atm. For the reaction, $CO\ (g) + 2H_2\ (g) \rightleftharpoons CH_3OH\ (g)$ at 700 K, $\Delta G^0_{700\,K} = -3.22$ kcal. Use Equation 12.2 to determine whether the system is at equilibrium. If not, will the reaction proceed spontaneously from left to right?

**12.31** At 25°C, in a mixture of $N_2O_4$ and $NO_2$ in equilibrium at a total pressure of 0.844 atm, the partial pressure of $N_2O_4$ is 0.563 atm. Calculate for the reaction,

$$N_2O_4\ (g) \rightleftharpoons 2NO_2\ (g)$$

(a) $K_P$, (b) $K_c$, (c) $\Delta G^0_{298\,K}$ in kcal.

**12.32** The following thermodynamic data apply at 25°C:

| Substance | $\Delta G_f^0$ (kcal/mole) |
|---|---|
| $NiSO_4 \cdot 6H_2O\ (s)$ | −531.0 |
| $NiSO_4\ (s)$ | −184.9 |
| $H_2O\ (g)$ | −54.6 |

(a) What is $\Delta G^0$ for the reaction,

$$NiSO_4 \cdot 6H_2O\ (s) \rightleftharpoons NiSO_4\ (s) + 6H_2O\ (g)$$

(b) What is $K_P$ for this reaction?

(c) What is the equilibrium vapor pressure of $H_2O$ over solid $NiSO_4 \cdot 6H_2O$?

**12.33** At a certain temperature the following equilibrium concentrations were found for the reactants and products in the reaction,

$$2HI\ (g) \rightleftharpoons H_2\ (g) + I_2\ (g)$$

$[H_2] = 1.0 \times 10^{-3}\ M$
$[I_2] = 2.5 \times 10^{-2}\ M$
$[HI] = 2.2 \times 10^{-2}\ M$

What is the value of $K_c$ for this reaction?

**12.34** In a particular experiment the following partial pressures were determined for the reaction,

$$2NO\ (g) + Cl_2\ (g) \rightleftharpoons 2NOCl\ (g)$$

$p_{NO} = 0.65$ atm $\quad p_{Cl_2} = 0.18$ atm
$p_{NOCl} = 0.15$ atm

What is $K_P$ for this reaction at the temperature at which the experiment was performed? What is the value of $K_c$ at this same temperature?

**12.35** For the reaction, $PCl_5\ (g) \rightleftharpoons PCl_3\ (g) + Cl_2\ (g)$, $K_c = 33.3$ at 760°C. In a container at equilibrium there are $1.29 \times 10^{-3}$ mole/liter of $PCl_5$ and $1.87 \times 10^{-1}$ mole/liter of $Cl_2$. Calculate the equilibrium concentration of $PCl_3$ in the vessel.

**12.36** The reaction, $2CO_2 \rightleftharpoons 2CO + O_2$, has $K_c = 6.4 \times 10^{-7}$ at 2000°C. If there is $1 \times 10^{-3}$ mole of $CO_2$ placed into a 1.0-liter vessel at this temperature,

(a) What will be the equilibrium concentrations of CO and $O_2$?

(b) What fraction of the $CO_2$ will have decomposed?

**12.37** For the reaction, $H_2\ (g) + CO_2\ (g) \rightleftharpoons CO\ (g) + H_2O\ (g)$, $K_c = 0.771$ at 750°C. If 1.00 mole of $H_2$ and 1.00 mole of $CO_2$ are placed into a 5.00-liter container and permitted to react, what will be the equilibrium concentration of all species?

**12.38** At 100°C the equilibrium constant, $K_c$, for the reaction, $CO\ (g) + Cl_2\ (g) \rightleftharpoons COCl_2\ (g)$, has a value of $4.6 \times 10^9$ liters/mole. If 0.20 mole of $COCl_2$ are placed into a 10.0-liter flask at 100°C, what will be the concentration of all species at equilibrium?

**12.39** Sodium bicarbonate (baking soda) has many useful properties. Among them is the ability to serve as a fire extinguisher because of thermal decomposition to produce $CO_2$, which smothers the fire,

$$2NaHCO_3\ (s) \rightleftharpoons Na_2CO_3\ (s) + CO_2\ (g) + H_2O\ (g)$$

At 125°C the value of $K_P$ for this reaction is 0.25 atm². What are the partial pressures of $CO_2\ (g)$ and $H_2O\ (g)$ in this system at equilibrium? Can you explain why $NaHCO_3$ is used in baking?

***12.40** In a 10.0-liter mixture of $H_2$, $I_2$, and HI at equilibrium at 425°C there are 0.100 mole of $H_2$, 0.100 mole of $I_2$, and 0.740 mole of HI. If 0.50 mole of HI are now added to this system, what will be the concentration of $H_2$, $I_2$, and HI once equilibrium has been reestablished?

***12.41** In Question 12.38 it was stated that at 100°C the value of $K_c$ for the reaction, $CO\ (g) + Cl_2\ (g) \rightleftharpoons COCl_2\ (g)$, is $4.6 \times 10^9$. Suppose that 0.15 mole of CO and 0.30 mole of $Cl_2$ were placed into a 1.0-liter vessel and allowed to react. What would the concentration be of each of the gases in the system at equilibrium? (*Hint:* First assume 100% reaction; then work backwards toward equilibrium.)

***12.42** The production of NO by reaction of $N_2$ and $O_2$ in an automobile engine is an important source of nitrogen oxide pollution. At 1000°C the reaction, $N_2\ (g) + O_2\ (g) \rightleftharpoons 2NO\ (g)$, has $K_P = 4.8 \times 10^{-7}$. Suppose that the partial pressures of $N_2$ and $O_2$ in the cylinder of an engine after the gasoline vapor has been ignited are $p_{N_2} = 33.6$ atm and $p_{O_2} = 4.0$ atm. Assume that the temperature of the mixture is 1000°C. Calculate the partial pressure of NO in the mixture if the system has time to reach equilibrium.

***12.43** If it is assumed that the reactants and products in the preceding question are unable to react further when the exhaust gases are suddenly cooled as they exit the engine, calculate the partial pressure of the NO when the partial pressure of $N_2$ has dropped to 0.80 atm and the temperature has dropped to 150°C.

***12.44** At a certain temperature $K_c = 7.5$ liters/mole for the reaction,

$$2NO_2 \rightleftharpoons N_2O_4$$

If 2.0 moles of $NO_2$ are placed in a 2-liter container and permitted to react, what will be the concentrations of $NO_2$ and $N_2O_4$ at equilibrium?

# 13 ACIDS AND BASES

In Chapter 5 the concept of acids and bases was introduced, although at that time we restricted our discussion to reactions in aqueous solution. It was pointed out, however, that the acid-base concept is extremely useful because it permits the correlation of large amounts of what at first glance may appear to be widely different types of chemical reactions. This idea is so useful, in fact, that most chemical reactions can be broadly classified into two categories, acid-base reactions and oxidation-reduction reactions. Often there is an overlap between them, so that for some reactions it is sometimes convenient to view them as acid-base reactions, while at other times it seems best to describe them in terms of oxidation-reduction.

In this chapter we will see that whether a particular substance behaves as an acid or a base depends on the way in which acids and bases are defined. There are a variety of ways of approaching this problem, with some definitions being more restrictive than others. Nevertheless, several properties are characteristic of acids and bases in general. These include:

1. *Neutralization.* Acids and bases react with one another so as to cancel, or neutralize, their acidic and basic characters.
2. *Reaction with indicators.* Certain organic dyes, called indicators, give different colors depending on whether they are in an acidic or basic medium.
3. *Catalysis.* Many chemical reactions are catalyzed by the presence of acids or bases.

## 13.1 THE ARRHENIUS DEFINITION OF ACIDS AND BASES

In its modern version, the **Arrhenius concept** (sometimes referred to as the aqueous concept) of acids and bases defines *an acid as any substance that can increase the concentration of hydronium ion, $H_3O^+$, in aqueous solution.* A *base,* on the other hand, *is a substance that increases the hydroxide ion concentration in water.* Thus, it is this concept of acids and bases that was first presented in Section 5.5. Let's review briefly some of the ideas that were developed there. You will recall, for example, that HCl is an acid because it reacts with water according to the equation,

$$HCl + H_2O \longrightarrow H_3O^+ + Cl^-$$

Similarly, $CO_2$ is an acid because it reacts with water to form carbonic acid, $H_2CO_3$,

$$CO_2 + H_2O \rightleftharpoons H_2CO_3$$

which then undergoes further reaction to produce $H_3O^+$ and $HCO_3^-$.

$$H_2CO_3 + H_2O \rightleftharpoons H_3O^+ + HCO_3^-$$

In general, nonmetal oxides react with water to yield acidic solutions and are said to be **acid anhydrides** (Greek, *anydros,* waterless).

An example of an Arrhenius base is NaOH, an ionic compound containing $Na^+$ and $OH^-$ ions. In water it undergoes dissociation.

$$NaOH\ (s) \xrightarrow{H_2O} Na^+\ (aq) + OH^-\ (aq)$$

Other examples of bases include substances such as $NH_3$ and $N_2H_4$ that react with water to produce $OH^-$.

$$NH_3 + H_2O \rightleftharpoons NH_4^+ + OH^-$$
$$\underset{\textbf{(hydrazine)}}{N_2H_4} + H_2O \rightleftharpoons \underset{\textbf{(hydrazinium ion)}}{N_2H_5^+} + OH^-$$

You will also recall that metal oxides **(basic anhydrides)** undergo reaction with water to give the corresponding hydroxides,

$$Na_2O + H_2O \longrightarrow 2NaOH$$
$$BaO + H_2O \longrightarrow Ba(OH)_2$$

Finally, in aqueous solution the neutralization of an acid by a base takes the form of the ionic reaction,

$$H_3O^+ + OH^- \longrightarrow 2H_2O$$

## 13.2 BRØNSTED-LOWRY DEFINITION OF ACIDS AND BASES

The definition of acids and bases in terms of the hydronium ion and hydroxide ion in water is very restricted because it limits us to discussing acid-base phenomena in aqueous solutions only. A somewhat more general approach was that proposed independently in 1923 by the Danish chemist, J. N. Brønsted, and the British chemist, T. M. Lowry. They defined an *acid as a substance that is able to donate a proton* (i.e., a hydrogen ion, $H^+$) *to some other substance.* A *base,* then, is defined as *a substance that is able to accept a proton from an acid.* Stated more simply, an acid is a proton donor and a base is a proton acceptor.

A typical example of a Brønsted-Lowry acid-base reaction occurs when HCl is added to water.

$$HCl + H_2O \longrightarrow H_3O^+ + Cl^-$$

In this reaction HCl is functioning as an acid because it is donating a proton to the water molecule. Water, on the other hand, is behaving as a base by accepting a proton from the acid.

If we have a solution of concentrated HCl and heat it, we drive off HCl gas. In other words, we can reverse the above reaction so that $H_3O^+$ and $Cl^-$ react with each other to produce HCl and $H_2O$. This reverse reaction is also a Brønsted-Lowry reaction, with hydronium ion serving as an acid by giving up its proton, and with the chloride ion functioning as a base by accepting it. Thus, we might view our reaction as an equilibrium where we have two acids and two bases, one of each on either side of the arrow.

$$\underset{\textbf{acid}}{HCl} + \underset{\textbf{base}}{H_2O} \rightleftharpoons \underset{\textbf{acid}}{H_3O^+} + \underset{\textbf{base}}{Cl^-}$$

When the acid, HCl, reacts it yields the base, $Cl^-$. These two substances are related to one another by the loss or gain of a single proton and constitute a **conjugate acid-base pair.** We say that $Cl^-$ is the **conjugate base** of the acid, HCl, and similarly that HCl is the **conjugate acid** of the base, $Cl^-$. In this reaction we also find that $H_2O$ and $H_3O^+$ form a conjugate pair. Water is the conjugate base of $H_3O^+$, and $H_3O^+$ is the conjugate acid of $H_2O$.

Another example of a Brønsted-Lowry acid-base reaction occurs in aqueous solutions of ammonia.

$$NH_3 + H_2O \rightleftharpoons NH_4^+ + OH^-$$

In this case water serves as an acid by giving up a proton to a molecule of $NH_3$, which thereby acts as a base. In the reverse reaction, on the other hand, $NH_4^+$ is the acid and $OH^-$ is the base. Again we have two acid-base conjugate pairs: $NH_3$ and $NH_4^+$ plus $H_2O$ and $OH^-$.

In general, we can represent any Brønsted-Lowry acid-base reaction as

$$\text{acid } (X) + \text{base } (Y) \rightleftharpoons \text{base } (X) + \text{acid } (Y)$$

where acid $(X)$ and base $(X)$ represent one conjugate pair and acid $(Y)$ and base $(Y)$ the other. Notice that the members of a conjugate pair differ *only* by one proton. They are otherwise the same. Also, within a conjugate pair, the acid has one more hydrogen than the base.

In the two examples that we examined above, water, in one instance, functioned as a base, and in the other it behaved as an acid. Such a substance, which can serve in either capacity depending on conditions, is said to be **amphiprotic** or **amphoteric.** Water is not the only substance to behave in this fashion. For example, the autoionization of water, pure acetic acid, and liquid ammonia can be represented by the equations,

$$H_2O + H_2O \rightleftharpoons H_3O^+ + OH^-$$
$$HC_2H_3O_2 + HC_2H_3O_2 \rightleftharpoons H_2C_2H_3O_2^+ + C_2H_3O_2^-$$
$$NH_3\,(l) + NH_3\,(l) \rightleftharpoons NH_4^+ + NH_2^-$$
$$\text{(acid)} + \text{(base)} \rightleftharpoons \text{(acid)} + \text{(base)}$$

These *autoionization* reactions can also be illustrated using Lewis structures as follows:

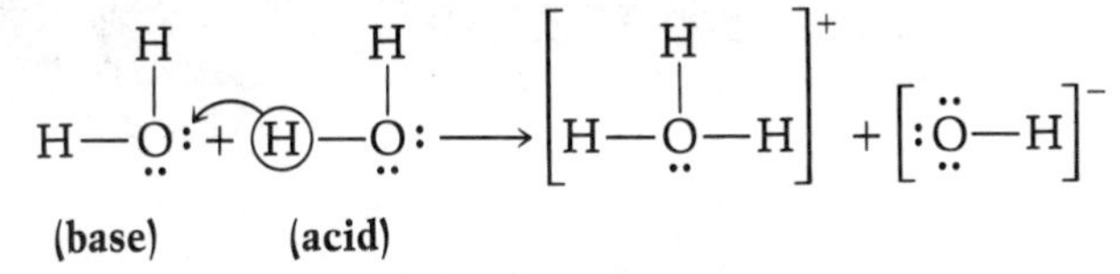


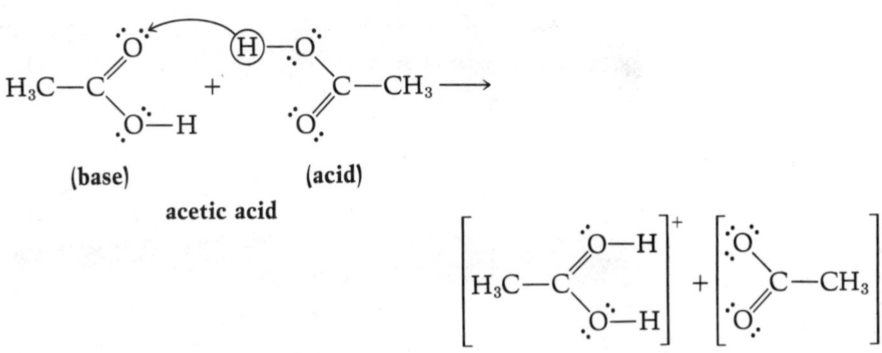

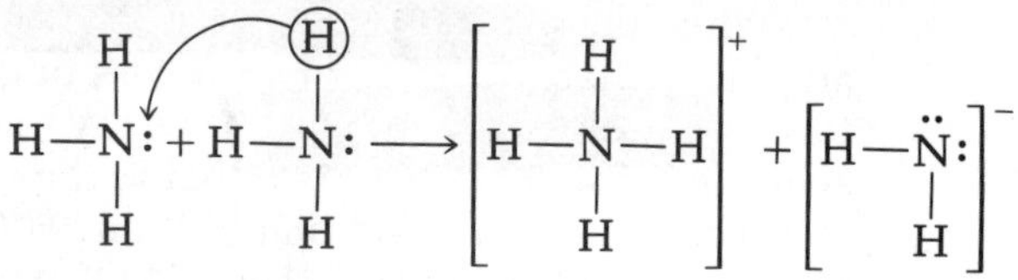

(base) (acid)

In each case the solvent (either water, acetic acid, of ammonia) is playing the role of both acid and base.

The Brønsted-Lowry concept is more general than the Arrhenius concept because it does not limit us to water solutions. In fact, we can find acid-base reactions that occur in the absence of any solvent whatsoever. For instance, when HCl and $NH_3$ gases are mixed, they react immediately to form the white ionic solid, $NH_4Cl$.

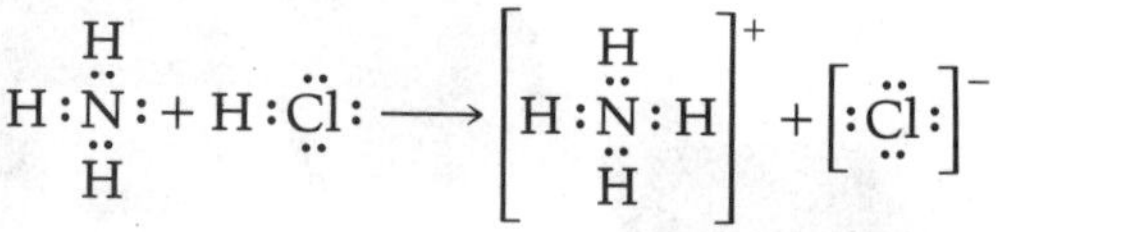

Since a proton is transferred from HCl to $NH_3$, this is clearly an acid-base reaction in the Brønsted-Lowry sense. However, because neither hydronium ion nor hydroxide ion ever enter the picture, the Arrhenius view of acids and bases ignores this reaction completely.

## 13.3 STRENGTHS OF ACIDS AND BASES

Any Brønsted-Lowry acid-base reaction can be viewed as two opposing or competing reactions between acids and bases and, in one sense, the two bases can be considered to be competing for a proton. When HCl reacts with water, for example, we find from conductivity measurements and freezing-point depression that essentially all of the HCl has reacted with water. This means that the position of equilibrium in the reaction,

$$HCl + H_2O \rightleftharpoons H_3O^+ + Cl^-$$

lies very far to the right. In turn, this tells us that $H_2O$ has a much stronger affinity for a proton than does a chloride ion because the water is able to capture essentially all of the available $H^+$. We express this relative ability to pick up a proton by saying that water is a stronger base than chloride ion.

We can also speak of the relative strengths of the two acids in this reaction, HCl and $H_3O^+$. Here we see that HCl is better able to donate its proton than is $H_3O^+$, because all of the HCl has lost its protons to give $H_3O^+$.

The position of equilibrium in an acid-base reaction tells us of the relative strengths of the acids and bases involved. Hydrogen chloride is a strong acid in water because the position of equilibrium in the ionization reaction lies far to the right. Hydrogen fluoride, on the other hand, is said to be a weak acid because in water it is only very slightly dissociated (actually about 3% for a 1 *M* solution at room temperature):

$$HF + H_2O \rightleftharpoons H_3O^+ + F^-$$

In general, when a Brønsted-Lowry reaction proceeds very far toward completion, both reactants are strong and both products are weak.

$$\text{strong acid (HA)} + \text{strong base (B)} \longrightarrow \text{weak base } (A^-) + \text{weak acid } (HB^+)$$

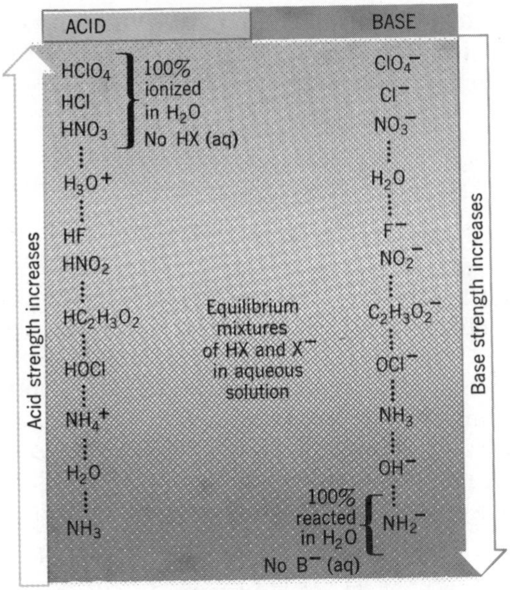


Figure 13.1
*Relative strengths of acid-base pairs.*

A useful generalization to remember is that a strong acid has a weak conjugate base. For example, the strong acid HCl has $Cl^-$ as its very weak conjugate base. Similarly, if a substance is a strong base, its conjugate acid is very weak. The amide ion, $NH_2^-$, is a very strong base while its conjugate acid, $NH_3$, is a very weak acid. The relationship between the relative strengths of members of an acid-base conjugate pair is illustrated in Figure 13.1.

If the relative position of an acid-base conjugate pair in Figure 13.1 is known, we can make some judgement as to the position of equilibrium in an acid-base reaction. For example, if a strong acid such as $HNO_3$ is added to a strong base such as liquid $NH_3$, the reaction will proceed very nearly to completion.

$$HNO_3 + NH_3 \longrightarrow NH_4^+ + NO_3^-$$
$$\text{(virtually 100\% complete)}$$

On the other hand, if a weak acid such as $HC_2H_3O_2$ is reacted with a weak base such as $Cl^-$ (from NaCl, for instance), essentially no reaction will be observed.

$$HC_2H_3O_2 + Cl^- \longrightarrow \text{no reaction}$$

Between these extremes, we can establish equilibria with varying amounts of both reactants and products. We saw earlier, for example, that in a 1 *M* solution about 3% of the HF is reacted to give $F^-$.

If we set about to compare the strengths of acids, such as HCl and HF, it is best to use the same reference base. For example, we conclude that HCl is a stronger acid than HF because HCl is better able to protonate the base, $H_2O$, than is HF. With a given base, however, it is not possible to compare the strengths of *all* acids. For example, HCl, $HNO_3$, and $HClO_4$ all appear to be 100% ionized in water. With the reference base water, then, each of these acids appears to be of equal strength. Their differences are removed, or leveled out, and we speak of this phenomenon as the **leveling effect.**

If acetic acid (the substance that gives vinegar its sour taste) is used as a solvent instead of water, we find that there is an appreciable difference between the extents to which the reactions,

$$HCl + HC_2H_3O_2 \rightleftharpoons H_2C_2H_3O_2^+ + Cl^-$$
$$HNO_3 + HC_2H_3O_2 \rightleftharpoons H_2C_2H_3O_2^+ + NO_3^-$$
$$HClO_4 + HC_2H_3O_2 \rightleftharpoons H_2C_2H_3O_2^+ + ClO_4^-$$

proceed toward completion. In this instance, acetic acid is a much weaker base than water and is not as easily protonated. Consequently, by using acetic acid as the solvent, it is possible to distinguish among the strengths of these three acids and, in fact, we find that their acidity increases in the order, $HNO_3 < HCl < HClO_4$. For these substances water is a **leveling solvent,** while acetic acid serves as a **differentiating solvent.**

The leveling effect is not restricted to acids alone. Strong bases such as the oxide ion, $O^{2-}$, amide ion, $NH_2^-$, and hydride ion, $H^-$, react completely with water to give hydroxide ion:

$$O^{2-} + H_2O \longrightarrow OH^- + OH^-$$
$$NH_2^- + H_2O \longrightarrow NH_3 + OH^-$$
$$H^- + H_2O \longrightarrow H_2 + OH^-$$

In these examples, water is such a strong acid that it is able to protonate all three bases completely. Therefore, with water as a solvent, it is impossible to differentiate among their base strengths.

In general, any basic solvent tends to exert a leveling effect on acids, while an acidic solvent tends to level the strengths of bases. In water for example, HF and HCl are clearly of different strengths as measured by their ability to protonate water molecules. In liquid ammonia (a basic solvent), however, the reaction

$$HX + NH_3 \longrightarrow NH_4^+ + X^-$$

proceeds essentially to completion for X = Cl and X = F. In ammonia, then, HF and HCl appear to be of equal strength.

A similar phenomenon is observed with ammonia in the solvents $H_2O$ and $HC_2H_3O_2$. Ammonia is only slightly protonated in water, while it becomes completely protonated when it is added to pure acetic acid. In the latter solvent, $NH_3$ behaves as a strong base while in water it is weak.

## 13.4 FACTORS INFLUENCING THE STRENGTHS OF ACIDS

By this time you have perhaps begun to wonder why some acids are strong and others are weak. How do we explain the trends in acid strength that we can observe experimentally?

Let's first consider the oxoacids such as $H_2SO_4$ and $H_2SO_3$. The structures of these two particular compounds can be represented as

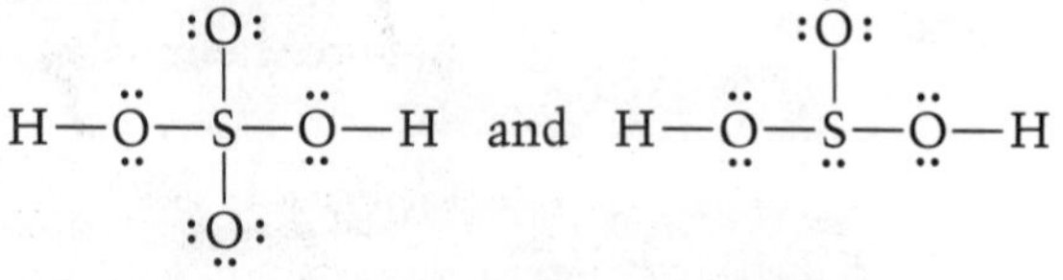

Each undergoes ionization in water by transferring its protons, in two successive steps, to water molecules. Thus, for $H_2SO_4$ we have in the first step,

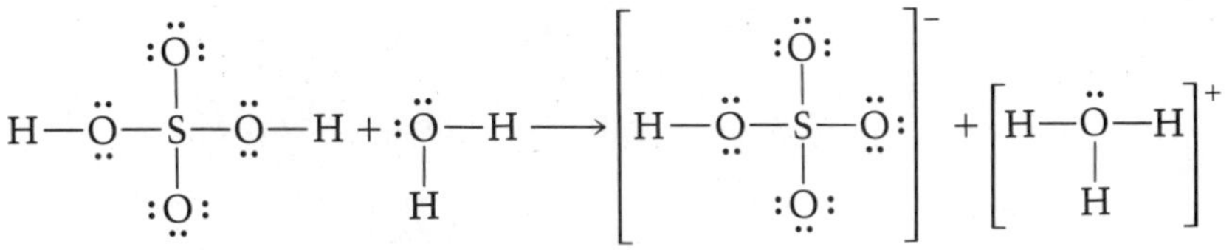

With $H_2SO_4$ this step proceeds essentially to completion, while for $H_2SO_3$, on the other hand, the first step,

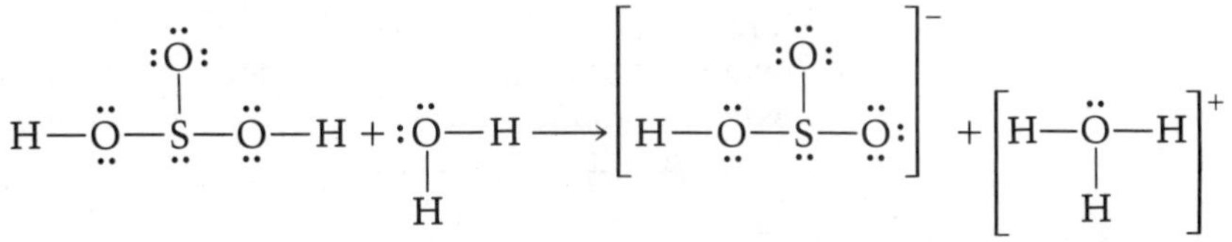

takes place only to a very limited degree (about 11% for a 1 *M* solution). In each of these substances the sulfur is bonded to two O—H groups. In $H_2SO_4$, however, the sulfur is also attached to two other lone oxygen atoms, while the sulfur in $H_2SO_3$ is bonded to only one such O atom. We know that oxygen is a very electronegative element, and we expect these S—O bonds to be polar, with the negative charge concentrated about the oxygen end of the dipole. In other words, the oxygen atoms, attached to the sulfur, draw electron density away so that the sulfur acquires a partial positive charge, the magnitude of which will be greater in $H_2SO_4$ than in $H_2SO_3$,

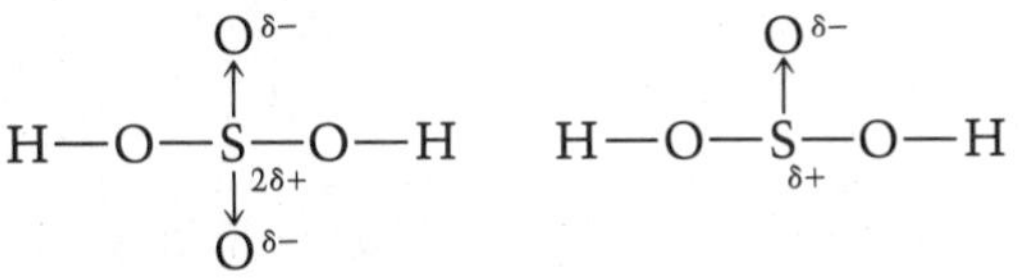

This positive charge on the sulfur tends to draw electron density from the S—OH bonds. Since the electronegative oxygen does not wish to lose electrons, the overall net effect is that some electronic charge will be withdrawn from the O—H bonds. This electron-withdrawing effect will be greater in $H_2SO_4$ than in $H_2SO_3$ because the sulfur in $H_2SO_4$ bears a greater positive charge and is therefore better able to draw electrons to itself. Consequently, in $H_2SO_4$ the hydrogen atoms carry a higher positive charge than those in $H_2SO_3$ and are thus more easily removed as $H^+$. As a result, $H_2SO_4$ is a stronger acid than $H_2SO_3$.

This phenomenon is illustrated even more graphically if we consider the oxoacids of chlorine:

| | | |
|---|---|---|
| hypochlorous acid | HOCl | 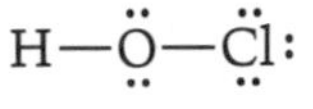 |
| chlorous acid | $HClO_2$ | 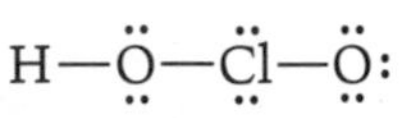 |
| chloric acid | $HClO_3$ | 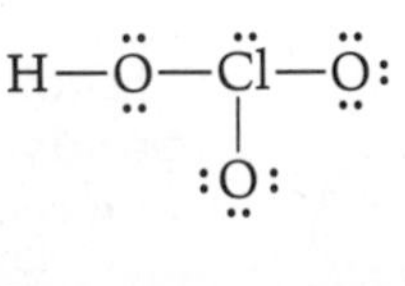 |

perchloric acid $HClO_4$ 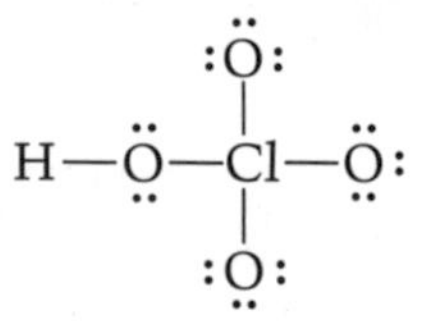

As the number of oxygen atoms surrounding the chlorine increases, so does the acidity. Thus HOCl is a relatively weak acid, $HClO_2$ is stronger, $HClO_3$ is fully dissociated in water, and $HClO_4$ is just about the strongest protonic acid there is.

This reasoning can also be extended to other compounds. For example, the substance ethanol (ethyl alcohol) is, for all practical purposes, undissociated in water. Its structure is

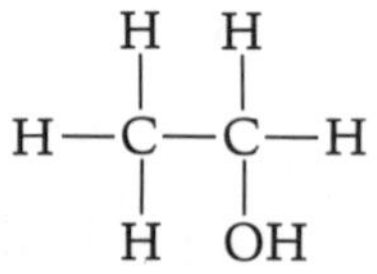

and there is virtually no tendency for the O—H bond to split off a proton. However, if the two hydrogen atoms that are attached to the carbon, which is also bound to the OH group, are replaced by oxygen to give

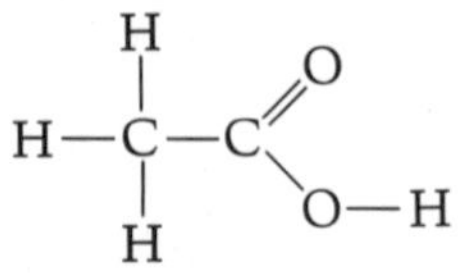

the acidity of the molecule is greatly enhanced (the compound is acetic acid). This new oxygen atom draws electron density from the carbon atom which, in turn, draws electronic charge from the C—OH bond. As before, the oxygen atom, not wishing to lose electron density, removes negative charge from the O—H bond and the hydrogen becomes more positively charged. As a result it can be more readily removed as $H^+$. This acidity can, in fact, be further increased by substituting a very electronegative element, such as Cl or F, for one or more of the hydrogen atoms of the $CH_3$ group. Thus chloroacetic acid is a stronger acid than ordinary acetic acid.

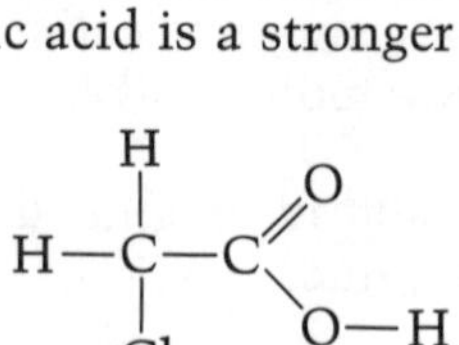

**chloroacetic acid**

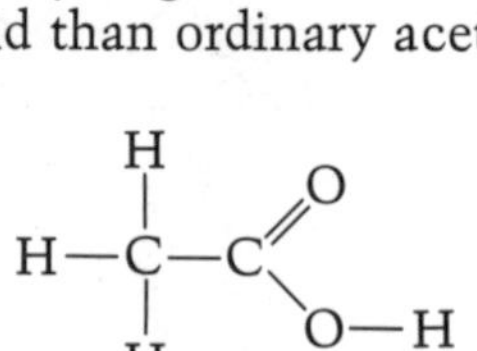

**acetic acid**

Following this same reasoning we can also understand why certain solutions of metal salts show acid properties. The acidity of these solutions is explained by the ability of the metal ion to polarize the water molecules that surround it in solution. For instance, in solutions of aluminum salts there is evidence that the $Al^{3+}$ ion is surrounded by six water molecules that are quite tightly bound to the metal ion, as shown in Figure 13.2. The high positive charge on the metal tends to draw electron density from the oxygen atoms of the water molecules and, as we have already seen, this leads to an

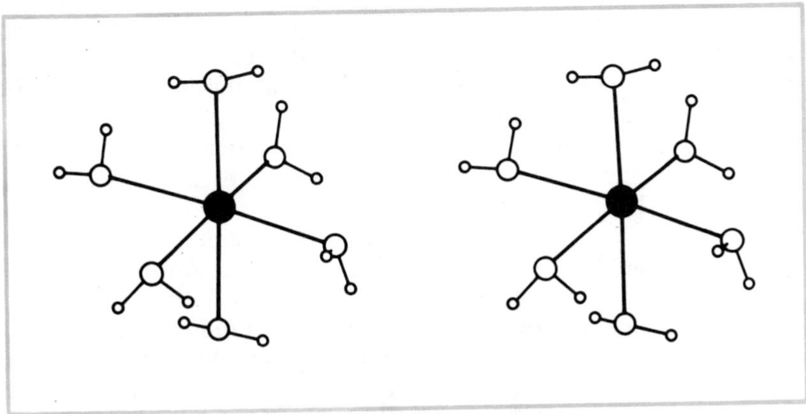

Figure 13.2
*The hydrated aluminum ion, $Al(H_2O)_6^{3+}$. The aluminum ion (large solid sphere) is surrounded octahedrally by six water molecules.*

increase in the positive charge on the hydrogen atoms. Consequently, when this hydrated aluminum ion collides with a water molecule, a proton can be rather easily transferred to $H_2O$ to give the hydronium ion,

$$Al(H_2O)_6^{3+} + H_2O \rightleftharpoons Al(H_2O)_5OH^{2+} + H_3O^+$$

This reaction is sometimes viewed as simply a reaction of the metal ion with a water molecule,

$$Al^{3+} + H_2O \rightleftharpoons AlOH^{2+} + H^+$$

In this last equation we have left out the five water molecules that have not changed, as well as the one that picked up the proton. Such a reaction is called **hydrolysis** (reaction with water). In general, *when a substance undergoes reaction with the solvent, the process is referred to as* **solvolysis.**

In summary, the acidity of an OH bond in M—O—H depends on the ability of M to draw electrons to itself. It is not surprising to find, then, that as the oxidation number of an element covalently bound to an OH group increases, so does the acidity of the compound. In $H_2SO_3$ the oxidation number of S is +4; in $H_2SO_4$ it is +6. As we have said, $H_2SO_4$ is a stronger acid than $H_2SO_3$. Similarly, in the sequence, HOCl, $HClO_2$, $HClO_3$, $HClO_4$, the acid strength parallels the increase in oxidation number of the chlorine. With metal ions, too, the acidity of their solutions depends on the charge on the metal ion. Aqueous solutions of $FeCl_2$ are only slightly acidic, while $FeCl_3$ solutions show extensive hydrolysis.

Thus far we have followed the trends in acidity for several series of acids in which the central atom remains the same as we vary those surrounding it. We can also examine the change in acid strength as the atom to which the OH groups are bound is varied. For instance, in the series, $H_2SO_4$, $H_2SeO_4$, $H_2TeO_4$, the acidity decreases as we go from sulfur to selenium to tellurium. In this case, as we proceed down within Group VIA, the atoms become larger and the positive charge induced by the oxygen atoms is spread out over a larger volume. As a result, the central atom becomes less effective in polarizing the O—H bonds and the acidity of the compounds decreases. It is thus a general observation that as we go down within a group in the periodic table, the strengths of the oxoacids decrease.

In addition to the oxoacids, we also must try to account for the relative strengths of the binary, or hydroacids in which one or more hydrogen atoms are bound directly to a nonmetal, for example, HF, HCl, and $H_2S$. When we consider binary acids derived from elements belonging to the same group, an increase in acidity with increasing atomic number is observed experimentally. Thus the hydrohalides increase in acidity in the order $HF < HCl < HBr < HI$.

At first glance this order of acidity seems opposite to what we might predict. We know, for instance, that fluorine is more electronegative than chlorine; therefore the HF bond is more polar than the HCl bond. Consequently, the H in HF is more positively charged than the H in HCl. Thus we are tempted to predict that HF should lose a proton more readily than HCl. This, however, is precisely the reverse of their acid strengths in water, where HF is weak and HCl is essentially 100% ionized.

The resolution of this dilemma can be understood by realizing that there are actually two opposing factors that contribute to the acidity of these compounds. One is, indeed, the ionic character of the H—X bond; the other is the H—X bond strength. As we proceed down within a group, the nonmetal becomes progressively larger and there is an accompanying rapid decrease in the strength of the H—X bond. This weakening of the H—X bond turns out to be more than sufficient to compensate for the decrease in the polarity of the bonds, and a net increase in acid strength is observed.

When we look at the acidity of the binary hydrogen compounds of elements in the same period, for example $NH_3$, $H_2O$, and HF, the dominant factor becomes the polarity of the H—X bond. As we go from left to right within the period there is little change in the size of the nonmetal and relatively little change in the H—X bond energy. There is, however, a very dramatic increase in the ionic character of the H—X bond, which is reflected in a rapid increase in acidity from ammonia to hydrogen fluoride.

## 13.5 LEWIS ACIDS AND BASES

The Brønsted-Lowry definition of acids and bases is more general than the Arrhenius definition because it removes the restriction of having to deal with reactions in aqueous solution. However, even the Brønsted-Lowry concept is restricted in scope, since it limits discussion of acid-base phenomena to proton-transfer reactions. There are many reactions that have all the earmarks of acid-base reactions but that do not fit the Brønsted-Lowry mold. The approach taken by G. N. Lewis further extends the acid-base concept to cover these cases.

In the Lewis definition of acids and bases, primary attention is focused on the base. *A* **base** *is defined as a substance that can donate a* **pair** *of electrons to the formation of a covalent bond. An* **acid,** *then, is a substance that can accept a pair of electrons to form the bond.*

A simple example of an acid-base neutralization is the reaction of a proton with hydroxide ion,

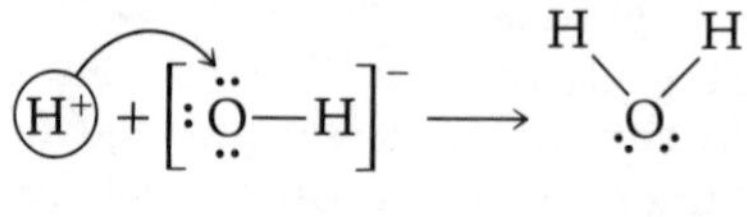


The hydroxide ion is the Lewis base because it furnishes the pair of electrons that become shared with the hydrogen. The hydrogen ion, on the other hand, is the Lewis acid because it accepts a share of the pair of electrons when the O—H bond is created.

Another example of neutralization is the reaction between $BF_3$ and ammonia,

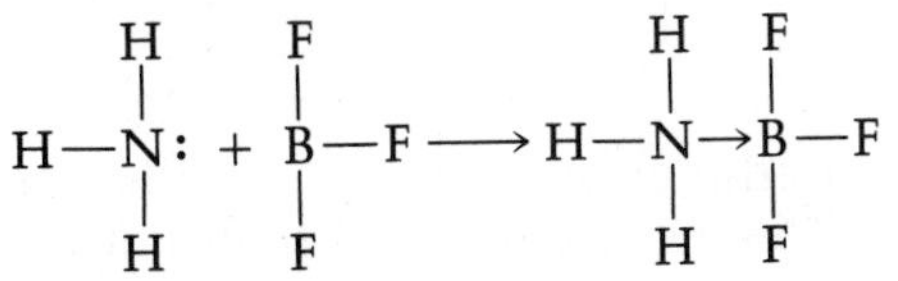

In this case the $NH_3$ functions as the base and $BF_3$ serves as the acid. Compounds containing elements with incomplete valence shells, such as $BF_3$ or $AlCl_3$, tend to be Lewis acids, while compounds or ions that have lone pairs of electrons can behave as Lewis bases. When the neutralization reaction occurs, a coordinate covalent bond is formed.

Still other examples of Lewis acid-base reactions are provided by the reaction of metal oxides with nonmetal oxides. Recall that metal oxides, in water, produce hydroxides. For instance,

$$Na_2O + H_2O \longrightarrow 2NaOH$$

Nonmetal oxides react to form acids as illustrated by the reaction,

$$SO_3 + H_2O \longrightarrow H_2SO_4$$

When these two solutions are mixed, neutralization occurs, with the production of the solvent plus a salt,

$$2NaOH + H_2SO_4 \longrightarrow 2H_2O + Na_2SO_4$$

The production of $Na_2SO_4$ from $Na_2O$ and $SO_3$ can take place directly without the introduction of any water whatsoever, as shown by the equation,

$$Na_2O + SO_3 \longrightarrow Na_2SO_4$$

According to the Lewis definition this too is a neutralization reaction between a Lewis base (oxide ion) and a Lewis acid (sulfur trioxide),

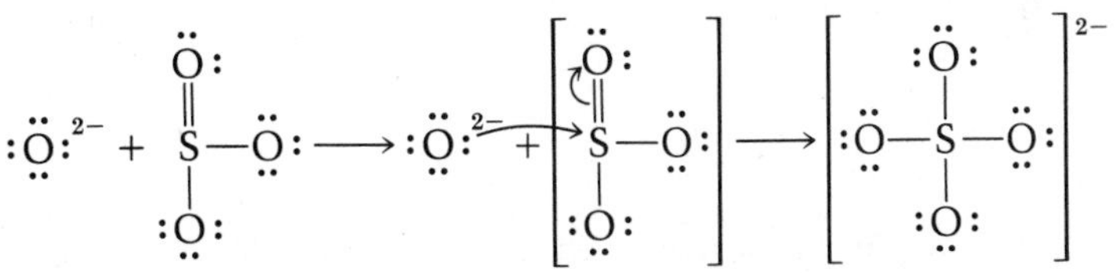

**base** **acid** **sulfate ion**

In this case we find that some electronic rearrangement must take place as the oxygen becomes attached to the sulfur. Nevertheless, the overall change can be viewed as a neutralization reaction.

Reactions of this type, between an oxide such as CaO and $SO_2$ or $SO_3$, are important for removal of sulfur oxides from the gases produced by combustion of high-sulfur fuels. For example,

$$CaO\ (s) + SO_2\ (g) \longrightarrow CaSO_3\ (s)$$

A reaction quite analogous to that just described has been used on spacecraft to remove carbon dioxide from the air breathed by the astronauts. In this case carbon dioxide is reacted with LiOH,

$$CO_2\,(g) + LiOH\,(s) \longrightarrow LiHCO_3\,(s)$$

**(lithium bicarbonate)**

(Lithium hydroxide is used because of the very low atomic weight of Li. This results in many moles of LiOH per pound.) This reaction too can be viewed as a Lewis acid-base reaction:

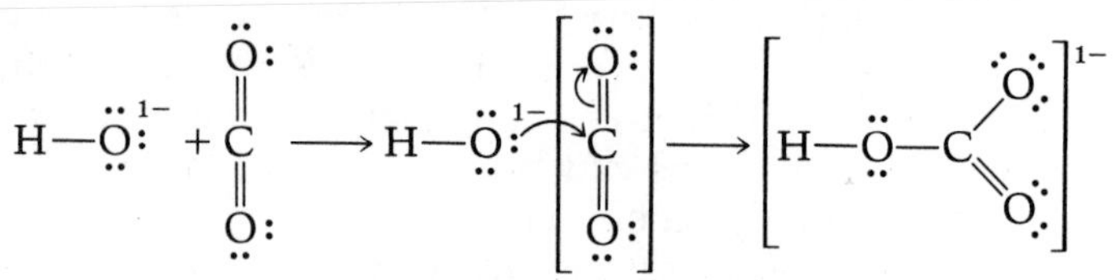

**bicarbonate ion**

The reaction between $Na_2O$ and $SO_3$ illustrates the limitations of the Brønsted-Lowry concept. Since no protons are involved in the reaction, it would never be classified as an acid-base reaction under the Brønsted-Lowry definition.

Thus far we have looked only at simple acid-base neutralizations. The acid-base reactions discussed in Sections 13.2 and 13.3 can also be treated from the Lewis point of view. In the Brønsted-Lowry theory these reactions were looked upon as competitions in which the strongest acid prevails by losing its proton. Under the Lewis definition these reactions are looked upon as constituting the displacement of one base (the weaker one) by another. Referring to our reaction of HCl with $H_2O$, for example,

$$HCl + H_2O \longrightarrow H_3O^+ + Cl^-$$

the Lewis theory interprets the change as the result of replacing the $Cl^-$ ion in HCl by the stronger base, $H_2O$.

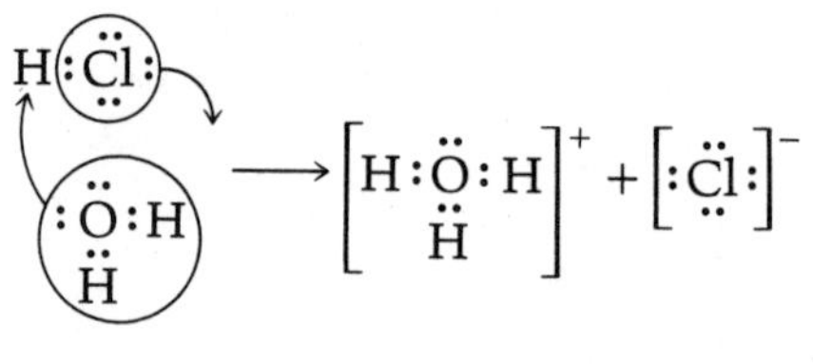

In other words, the stronger base, $H_2O$, pushes out the weaker one, $Cl^-$. Here we interpret the acid to be the $H^+$ ion rather than the entire HCl molecule.

By its definition a Lewis base is a substance that in its reactions seeks a nucleus with which it can share a pair of electrons; it is thus said to be a **nucleophile** (nucleus-loving). A base displacement reaction therefore is called a **nucleophilic displacement,** because one Lewis base (nucleophile) pushes out another. From this point of view a very large number of chemical reactions can be looked on simply as acid-base reactions in which a stronger base displaces a weaker one.

There are also acid-base reactions in which one Lewis acid displaces another. For example, consider the reaction,

$$AlCl_3 + COCl_2 \longrightarrow COCl^+ + AlCl_4^-$$

Using electron dot formulas we can analyze this reaction as follows:

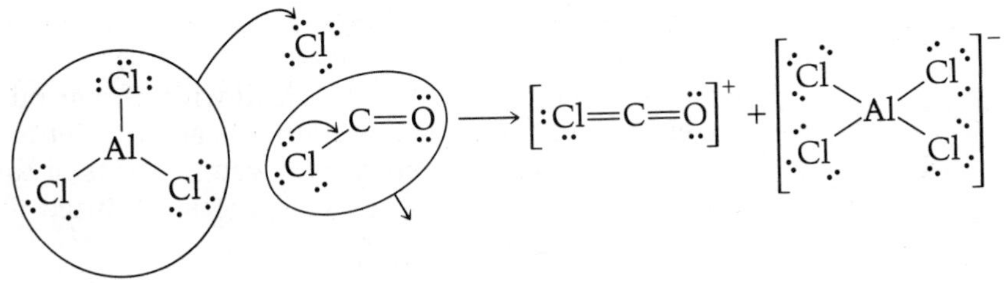

Notice that we view this reaction to take place by the displacement of the acid, $COCl^+$, by the stronger acid $AlCl_3$. In other words, we imagine the molecule, $COCl_2$, to be the product of the Lewis acid $COCl^+$ and the Lewis base $Cl^-$, and it is the base that changes partners when the displacement reaction occurs:

$$AlCl_3 + COCl_2 \longrightarrow [Cl_3Al \cdots Cl^- \cdots COCl^+] \longrightarrow AlCl_4^- + COCl^+$$

Since Lewis acids seek substances having electron pairs to which they can become bound, Lewis acids are said to be **electrophilic** (electron-loving) in character and an acid displacement reaction is said to be an **electrophilic displacement.**

With a little practice it becomes relatively simple to distinguish between nucleophilic and electrophilic displacement processes. Often, in a nucleophilic displacement one of the products is derived from a reactant species by the loss of one or more atoms and an *increase in negative charge.* For instance, in the reaction of $H_2O$ with HCl, the product $Cl^-$ is derived from the reactant, HCl, by the loss of a proton. The product species, $Cl^-$, has a charge of $-1$, while the reactant from which it is formed is neutral. In this example, the chloride ion is the base that has been displaced.

By comparison, in an electrophilic displacement reaction it frequently happens that one of the products is derived from a reactant by the loss of one or more atoms and an *increase in positive charge.* Thus $COCl^+$ is produced from $COCl_2$ by the loss of $Cl^-$. The species $COCl^+$ carries a charge of $+1$, while the $COCl_2$ molecule from which it is formed is neutral. In this case our analysis has identified $COCl^+$ as the acid that has been displaced. We can also conclude that the $Cl^-$ ion is the base that is changing partners.

The strengths of Lewis acids and bases can be compared by examining their tendency to enter into the formation of a coordinate covalent bond. When the bond is formed, electron density on the base is drawn toward the electron-poor atom of the acid. A strong base, therefore, contains an atom whose electron cloud is easily deformed, or polarized. In Section 8.2 we saw that large atoms are more easily polarized than small atoms. Thus $(CH_3)_2S$ would be expected to be a stronger Lewis base than $(CH_3)_2O$ because S is larger and more easily polarized than O. On the other hand, $(CF_3)_2S$ would be a weaker base than $(CH_3)_2S$ because the very electronegative fluorine atoms would make it more difficult to draw electron density from the S in $(CF_3)_2S$ than from the S in $(CH_3)_2S$.

Lewis acid strength is determined by the electron-attracting power of the electron-poor atom of the acid. In general, small atoms are better at attracting electrons than large atoms. The valence shell of a small atom is closer to its nucleus, and other electrons approaching this valence shell are strongly attracted. We would therefore expect $BCl_3$ to be a stronger acid than $AlCl_3$. Also, highly positively charged ions are better Lewis acids than positive ions of low charge. Thus $Fe^{3+}$ would be a stronger Lewis acid than $Fe^{2+}$.

## 13.6 THE SOLVENT SYSTEM APPROACH TO ACIDS AND BASES

Because water is so plentiful and is such a good solvent for so many substances, much of the solution chemistry that we meet in the laboratory uses water as the solvent. Chapter 5 dealt with the various kinds of reactions that are encountered in aqueous solutions and the definitions of acids and bases that were presented there were given in terms of the ions, $H_3O^+$ and $OH^-$. As mentioned before, water undergoes a limited degree of autoionization that can be described by the equation,

$$H_2O + H_2O \rightleftharpoons H_3O^+ + OH^-$$

Thus we see that an acid in water is a substance that yields, in some way, the cation characteristic of the solvent. Similarly, a base is a substance that gives the anion derived from the solvent. These last two observations form the basis of the **solvent system** approach to dealing with acid-base reactions. This approach recognizes that water is really not unique in its solvent properties and that many reactions in other solvents can be viewed as analogs of reactions that take place in aqueous media.

**LIQUID AMMONIA AS A SOLVENT.** Probably the most thoroughly studied solvent besides water is liquid ammonia. At atmospheric pressure ammonia exists as a gas at room temperature; it is therefore necessary to cool the substance to its boiling point, −33°C, or below, in order to work with the liquid, which leads to some minor experimental difficulties.

Liquid ammonia, as a solvent, has many properties that are similar to water. Like water, it undergoes a slight degree of autoionization. When it does so, it produces ammonium ion and amide ion, $NH_2^-$.

$$NH_3 + NH_3 \rightleftharpoons NH_4^+ + NH_2^-$$

By analogy with water we would predict that an acid in liquid $NH_3$ is any substance capable of forming ammonium ion. Thus any ammonium salt, such as $NH_4Cl$ for example, should show acid properties. In fact, in liquid ammonia $NH_4Cl$ is exactly analogous to the product of the reaction of HCl with $H_2O$ in aqueous solution, that is, $(H_3O)Cl$.

Following the solvent system definition, a base in liquid ammonia is any substance capable of providing amide ion—for example, $KNH_2$, potassium amide.

In water the neutralization of an acid and a base occurs by the reaction,

$$H_3O^+ + OH^- \longrightarrow 2H_2O$$

Thus the cation and anion of the solvent combine in neutralization to produce the solvent. In liquid ammonia this corresponds to the reaction,

$$NH_4^+ + NH_2^- \longrightarrow 2NH_3$$

In fact, in *any* solvent the neutralization reaction is simply the reverse of the autoionization reaction. When solutions of $NH_4Cl$ and $KNH_2$ in liquid ammonia are mixed, we have the overall reaction,

$$NH_4Cl + KNH_2 \longrightarrow KCl + 2NH_3$$

This is analogous to the reaction, in water, between HCl and KOH,

$$HCl + KOH \longrightarrow KCl + H_2O$$

or

$$(H_3O)Cl + KOH \longrightarrow KCl + 2H_2O$$

A further similarity between the acid-base phenomena in water and liquid ammonia is revealed by the behavior of certain organic dye molecules called indicators. For example, in water the indicator phenolphthalein is pink in basic solutions and colorless in acid solutions. This fact is employed in acid-base titrations, where there is a sharp color change when neutralization is achieved. This same indicator can also be used for performing titrations with liquid ammonia as a solvent. In basic solutions containing an excess of $NH_2^-$ ion the phenolphthalein is pink, while in acid solutions it is colorless.

A further similarity between acid-base behavior in these two solvents is provided by the amphoteric behavior of certain metals. For example, if an aqueous solution of $ZnI_2$ is treated with KOH, a precipitate of $Zn(OH)_2$ is formed that redissolves upon further addition of base to produce $K_2[Zn(OH)_4]$, as indicated below.

$$ZnI_2 + 2KOH \longrightarrow Zn(OH)_2\,(s) + 2KI$$
$$Zn(OH)_2 + 2KOH \longrightarrow K_2[Zn(OH)_4]$$

In liquid ammonia we find precisely the same behavior:

$$ZnI_2 + 2KNH_2 \longrightarrow Zn(NH_2)_2\,(s) + 2KI$$
$$Zn(NH_2)_2 + 2KNH_2 \longrightarrow K_2[Zn(NH_2)_4]$$

In each of these cases the addition of acid ($H_3O^+$ in water or $NH_4^+$ in ammonia) causes the reprecipitation of the zinc hydroxide, or amide; further addition of acid will cause these precipitates to dissolve, thereby regenerating the solvated zinc ion.

The similarities between the neutralization reactions in water and ammonia have also suggested other chemical parallels. By analogy we have established that $H_3O^+$ and $NH_4^+$ occupy corresponding positions in the water (aquo) and ammonia (ammono) systems. Similarly, $OH^-$ and $NH_2^-$ stand opposite one another. This leads us to postulate further that the oxide ion, $O^{2-}$, in the aquo system should be equivalent to the *imide ion*, $NH^{2-}$, or the *nitride ion*, $N^{3-}$, in the ammono system. In fact, extension of this reasoning leads us to a whole series of compounds that are analogous to each other in the two solvent systems. Some examples are given in Table 13.1.

In addition to acid-base neutralization there are many other parallels in the reactions that take place in water and liquid ammonia. For instance, we know that in water metal oxides react with the solvent to produce hydroxides,

$$Li_2O + H_2O \longrightarrow 2LiOH$$

In liquid ammonia we find the similar reaction,

$$Li_3N + 2NH_3 \longrightarrow 3LiNH_2$$

Another example is the reaction of a metal hydride with the solvent.

$$NaH + H_2O \longrightarrow H_2 + Na^+ + OH^- \qquad \text{(aquo system)}$$
$$NaH + NH_3 \longrightarrow H_2 + Na^+ + NH_2^- \qquad \text{(ammono system)}$$

Still another similarity is the reaction of an active metal with an acid to yield hydrogen plus a salt.

$$Ca + 2(H_3O)Cl \longrightarrow CaCl_2 + H_2 + 2H_2O \qquad \text{(aquo system)}$$
$$Ca + 2NH_4Cl \longrightarrow CaCl_2 + H_2 + 2NH_3 \qquad \text{(ammono system)}$$

Table 13.1
*Analogous compounds in the aquo and ammono systems*

| Aquo Compound | Ammono Compound |
|---|---|
| $H_2O$ | $NH_3$ |
| $(H_3O)Cl$ or $HCl$ | $NH_4Cl$ |
| $KOH$ | $KNH_2$ |
| $Li_2O$ | $Li_3N$ |
| $P_2O_5$ (actually $P_4O_{10}$) | $P_3N_5$ |
| $PO(OH)_3$ or $H_3PO_4$ | $[PN(NH_2)_2]_3$<br>$P(NH)(NH_2)_3$ |
| $CH_3OH$ | $CH_3NH_2$ |
| $C_2H_5OH$ | $C_2H_5NH_2$ |
| $(CH_3)_2O$ | $(CH_3)_3N$, $(CH_3)_2NH$ |
| $H_2O_2$ | $N_2H_4$ |
| $HONO_2$ or $HNO_3$ | $HNN_2$ |
| $HOCl$ | $H_2NCl$ |
| $CO(OH)_2$ or $H_2CO_3$ | $C(NH)(NH_2)_2$ |
| $CH_3COOC_2H_5$ | $CH_3C(NH)(NHC_2H_5)$ |
| $CH_3COOH$ | $CH_3C(NH)(NH_2)$ |
| $Cu(H_2O)_4{}^{2+}$ | $Cu(NH_3)_4{}^{2+}$ |
| $Si(OH)_4$ | $Si(NH_2)_4$ |
| $B(OH)_3$ | $B(NH_2)_3$ |

In Chapter 5 we saw that metathesis reactions provide a convenient route for the synthesis of certain compounds. In nonaqueous solvents, too, we can have these kinds of chemical changes. Because of different solubility relationships in the aqueous and nonaqueous media, however, it is sometimes possible to prepare compounds by metathesis in a solvent such as ammonia that cannot be made in the same way in water. For example, in liquid ammonia, $Ba(NO_3)_2$ and AgCl are both soluble. When their solutions are mixed, a white precipitate of $BaCl_2$ is produced.

$$Ba(NO_3)_2 + 2AgCl \xrightarrow{NH_3\,(l)} BaCl_2\,(s) + 2AgNO_3$$

This is, of course, exactly the reverse of the reaction that takes place in aqueous solution, where $BaCl_2$ and $AgNO_3$ are soluble and form a precipitate of AgCl when their aqueous solutions are mixed.

## 13.7 SUMMARY

The examples in the preceding section demonstrate that the solvent system approach to acids and bases has some useful and appealing features. One of these expands our thinking to reactions in solvents other than water. However, this view of acids and bases suffers from some of the same limitations as the Arrhenius theory. For example, it does not allow us to consider, as acid-base interactions, reactions that take place in the absence of a solvent.

The Brønsted-Lowry approach frees us from the restriction of having to have a solvent present, although it limits us to dealing with systems in which there is proton transfer. Clearly, the most general approach that we have considered is the Lewis theory. All of the acid-base reactions that we have examined under the various headings in this chapter can be interpreted from the Lewis point of view.

The choice of which definition of acids and bases one wishes to use in a particular instance depends largely on the sort of chemistry that is studied.

For example, in a practical sense the Arrhenius definition is perfectly satisfactory for dealing with the reactions in aqueous solution that you will encounter in the laboratory.

## INDEX TO QUESTIONS

General 1, 31

Arrhenius acids and bases 2, 3, 4

Brønsted Lowry acids and bases 5, 6, 7

Strengths of acids and bases 9, 10, 11, 12, 13, 14, 15

Factors influencing acid strength 16, 17, 18, 19

Lewis acids and bases 20, 21, 22, 23, 24, 25, 26, 27, 28, 29

Solvent system approach 8, 30, 32, 33, 34

## REVIEW QUESTIONS

**13.1** Give three properties that are characteristic of acids and bases in general.

**13.2** What is the Arrhenius definition of an acid and a base?

**13.3** Identify each of the following as Arrhenius acids or bases. For each, write a chemical equation to show its reaction with water. If necessary, refer to Table 5.3.

(a) $P_4O_{10}$ (d) HBr (f) $Al(H_2O)_6^{3+}$
(b) CaO (e) $H_2O$ (g) $Ba(OH)_2$
(c) $NH_3OH^+$

**13.4** Write the net ionic equation for the neutralization of an acid and a base in aqueous solution.

**13.5** Define acid anhydride, basic anhydride. What is the Brønsted-Lowry definition of an acid and a base? Why is it less restrictive than the Arrhenius concept?

**13.6** Identify the two acid-base conjugate pairs in each of the following reactions:

(a) $C_2H_3O_2^- + H_2O \rightleftharpoons OH^- + HC_2H_3O_2$
(b) $HF + NH_3 \rightleftharpoons NH_4^+ + F^-$
(c) $Zn(OH)_2 + 2OH^- \rightleftharpoons ZnO_2^{2-} + 2H_2O$
(d) $Al(H_2O)_6^{3+} + OH^- \rightleftharpoons Al(H_2O)_5OH^{2+} + H_2O$
(e) $N_2H_4 + H_2O \rightleftharpoons N_2H_5^+ + OH^-$
(f) $NH_2OH + HCl \rightleftharpoons NH_3OH^+ + Cl^-$
(g) $O^{2-} + H_2O \rightleftharpoons 2OH^-$
(h) $H^- + H_2O \rightleftharpoons H_2 + OH^-$
(i) $NH_2^- + N_2H_4 \rightleftharpoons NH_3 + N_2H_3^-$
(j) $HNO_3 + H_2SO_4 \rightleftharpoons H_3SO_4^+ + NO_3^-$

**13.7** Identify the acid-base conjugate pairs in each of the following reactions:

(a) $HClO_4 + N_2H_4 \rightleftharpoons N_2H_5^+ + ClO_4^-$
(b) $HSO_3^- + H_3PO_3 \rightleftharpoons H_2SO_3 + H_2PO_3^-$
(c) $C_5H_5NH^+ + (CH_3)_3N \rightleftharpoons C_5H_5N + (CH_3)_3NH^+$
(d) $CO_3^{2-} + H_2O \rightleftharpoons HCO_3^- + OH^-$
(e) $HCHO_2 + C_7H_5O_2^- \rightleftharpoons C_7H_5O_2H + CHO_2^-$
(f) $H_2C_2O_4 + CH_3NH_2 \rightleftharpoons HC_2O_4^- + CH_3NH_3^+$
(g) $H_2CO_3 + H_2O \rightleftharpoons HCO_3^- + H_3O^+$
(h) $Zn(OH)_4^{2-} + 2H_3O^+ \rightleftharpoons Zn(OH)_2 + 4H_2O$
(i) $NO_2^- + N_2H_5^+ \rightleftharpoons HNO_2 + N_2H_4$
(j) $HCN + H_2SO_4 \rightleftharpoons H_2CN^+ + HSO_4^-$

**13.8** Write autoionization reactions for the following:

(a) $H_2O$ (*l*) (b) $NH_3$ (*l*) (c) HCN (*l*)

**13.9** From Figure 13.1, place the following reactions in order of increasing tendency to proceed toward completion:

(a) $H_2O + NH_3 \rightleftharpoons NH_4^+ + OH^-$
(b) $HClO_4 + NH_2^- \rightleftharpoons ClO_4^- + NH_3$
(c) $H_2O + NO_2^- \rightleftharpoons HNO_2 + OH^-$
(d) $NH_3 + Cl^- \rightleftharpoons NH_2^- + HCl$

**13.10** Use Figure 13.1 to place the following reactions in order of increasing tendency to proceed toward completion:

(a) $OCl^- + HCl \rightleftharpoons HOCl + Cl^-$
(b) $HF + C_2H_3O_2^- \rightleftharpoons HC_2H_3O_2 + F^-$
(c) $NH_4^+ + ClO_4^- \rightleftharpoons NH_3 + HClO_4$
(d) $HNO_2 + F^- \rightleftharpoons HF + NO_2^-$

**13.11** Hydrogen sulfide is a stronger acid than phosphine, $PH_3$. What may we conclude about the strengths of their conjugate bases, $HS^-$ and $PH_2^-$?

**13.12** Given the following equilibria and equilibrium constants, arrange the acids in order of increasing strength:

(a) $HOCl + H_2O \rightleftharpoons H_3O^+ + OCl^-$

$K = 3.2 \times 10^{-8}$

(b) $NH_4^+ + H_2O \rightleftharpoons H_3O^+ + NH_3$
$K = 5.6 \times 10^{-10}$

(c) $HC_2H_3O_2 + H_2O \rightleftharpoons H_3O^+ + C_2H_3O_2^-$
$K = 1.8 \times 10^{-5}$

(d) $H_2CO_3 + H_2O \rightleftharpoons H_3O^+ + HCO_3^-$
$K = 4.2 \times 10^{-7}$

(e) $HSO_4^- + H_2O \rightleftharpoons H_3O^+ + SO_4^{2-}$
$K = 1.3 \times 10^{-2}$

**13.13** Arrange the conjugate bases in Question 13.12 in order of increasing base strength.

**13.14** What is the leveling effect? How would it apply to the strong acids, HCl and HBr? Suggest a substance that might serve as a differentiating solvent for these two acids.

**13.15** The reaction, $H_2S + H_2O \rightleftharpoons HS^- + H_3O^+$, has an equilibrium constant equal to $1.1 \times 10^{-7}$. Can you suggest a solvent where a similar reaction would have a smaller *K*? Can you suggest a solvent where a similar reaction would have a larger *K*?

**13.16** Predict the order of acid strength for each of the following pairs of protonic acids.
(a) $H_2S$ and $H_2Se$
(b) $H_2Se$ and HBr
(c) $PH_3$ and $NH_3$

(d) 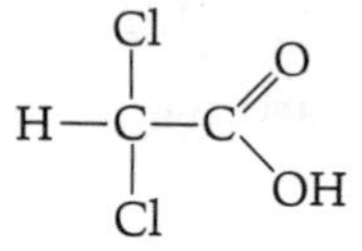 and 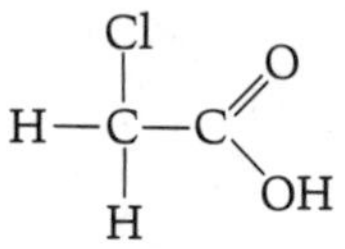

(e) 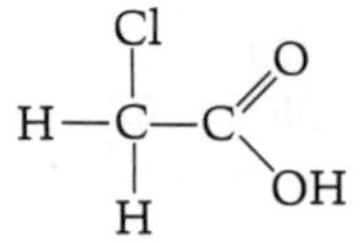 and 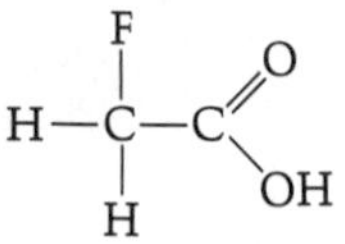

(f) 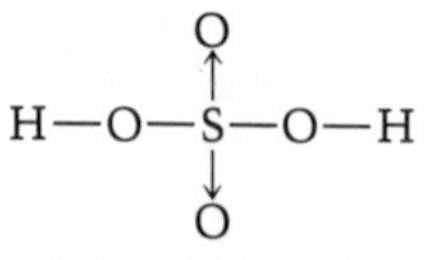 and 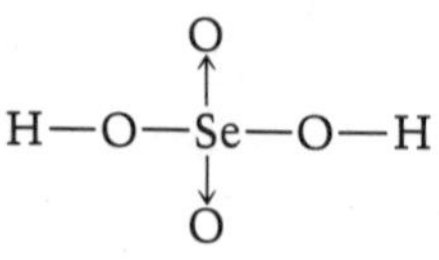

(g) $HClO_3$ and $HBrO_3$
(h) HOBr and $HBrO_3$

**13.17** Why is $HNO_3$ a stronger acid than $HNO_2$?

**13.18** Why is HCl a stronger acid than HF?

**13.19** Why is HCl a stronger acid than $H_2S$?

**13.20** What is the Lewis definition of an acid and a base?

**13.21** Boron trichloride, $BCl_3$, reacts with diethyl ether, $(C_2H_5)_2O$, to form an *addition compound,* which we can write as $Cl_3B \leftarrow O(C_2H_5)_2$. Use electron dot formulas to interpret this reaction as a Lewis acid-base neutralization.

**13.22** Indicate whether the following would be expected to serve as either a Lewis acid or a Lewis base:

| | | |
|---|---|---|
| (a) $AlCl_3$ | (e) $NO^+$ | (i) $(CH_3)_2S$ |
| (b) $OH^-$ | (f) $CO_2$ | (j) $SbF_5$ |
| (c) $Br^-$ | (g) $NH_3$ | |
| (d) $H_2O$ | (h) $Fe^{3+}$ | |

**13.23** Which would be expected to be a stronger Lewis base, $NH_3$ or $NF_3$? Explain.

**13.24** Boric acid, $B(OH)_3$, which has long been used for medicinal purposes, is a weak acid. However, it does not release protons by breaking of an O—H bond in the $B(OH)_3$ molecule. Instead, it functions as a Lewis acid by reacting with a water molecule,

$$B(OH)_3 + H_2O \rightleftharpoons B(OH)_3(H_2O)$$
$$B(OH)_3(H_2O) + H_2O \rightleftharpoons B(OH)_4^- + H_3O^+$$

Use electron dot structures to diagram this reaction. Why is $B(OH)_3$ a Lewis acid? Why is the $B(OH)_3(H_2O)$ acidic?

**13.25** Interpret each of the reactions in Question 13.6 as a Lewis acid-base displacement reaction.

**13.26** Interpret each of the reactions in Question 13.7 as a Lewis acid-base displacement reaction.

**13.27** Explain why the reaction of $CO_2$ with $H_2O$ to produce $H_2CO_3$, which we can also write as $CO(OH)_2$, can be viewed as a Lewis acid-base neutralization.

**13.28** Define nucleophile, electrophile.

**13.29** Analyze the following as either electrophilic or nucleophilic displacement reactions:
(a) $N_2O_4 + BF_3 \rightarrow NO_2^+ + NO_2BF_3^-$
(b) $SiCl_4 + 4H_2O \rightarrow Si(OH)_4 + 4HCl$
(c) $(CH_3)_3CCl + OH^- \rightarrow (CH_3)_3COH + Cl^-$
(d) $SnCl_4 + 2SeOCl_2 \rightarrow SnCl_6^{2-} + 2SeOCl^+$
(e) $Co(NH_3)_6^{3+} + Cl^- \rightarrow Co(NH_3)_5Cl^{2+} + NH_3$
(f) $Br_2 + FeBr_3 \rightarrow Br^+ + FeBr_4^-$

**13.30** What is the basic concept behind the solvent system approach to acids and bases?

**13.31** How do the Arrhenius, Brønsted-Lowry, Lewis, and solvent system concepts of acids and bases differ in terms of their general applicability?

**13.32** Concentrated sulfuric acid undergoes the autoionization reaction,

$$2H_2SO_4 \rightleftharpoons H_3SO_4^+ + HSO_4^-$$

In this solvent, acetic acid behaves as a *base* and perchloric acid behaves as an *acid.* Write chemical equations to show the following:

(a) The reaction of $HC_2H_3O_2$ with the solvent, $H_2SO_4$
(b) The reaction of $HClO_4$ with the solvent
(c) The neutralization reaction that occurs when solutions of $HC_2H_3O_2$ and $HClO_4$ in $H_2SO_4$ are mixed

**13.33** From the data in Table 13.1, write equations for reactions in liquid ammonia that are analogous to the following reactions that take place in aqueous solution:

(a) $P_4O_{10} + 6H_2O \rightarrow 4H_3PO_4$
(b) $Cl_2 + H_2O \rightarrow HCl + HOCl$
(c) $CH_3COOC_2H_5 + H_2O \xrightarrow{H_3O^+} CH_3COOH + C_2H_5OH$
(d) $Zn + 2HCl \rightarrow ZnCl_2 + H_2$
(e) $H_2CO_3 + 2OH^- \rightarrow CO_3^{2-} + 2H_2O$
(f) $SiCl_4 + 4H_2O \rightarrow Si(OH)_4 + 4HCl$
(g) $Zn + 2OH^- + 2H_2O \rightarrow Zn(OH)_4^{2-} + H_2$
(h) $Cu^{2+} + 4H_2O \rightarrow Cu(H_2O)_4^{2+}$
(i) $BCl_3 + 3H_2O \rightarrow 3HCl + B(OH)_3$

**13.34** Liquid hydrogen cyanide can be thought to undergo the autoionization,

$$2HCN \rightleftharpoons H_2CN^+ + CN^-$$

(a) Would KCN be considered an acid or a base in this solvent?
(b) $H_2SO_4$ is an acid in HCN (*l*). Write the equation for the ionization of $H_2SO_4$ in this solvent.
(c) $(CH_3)_3N$ is a base in HCN (*l*). Write an equation for the reaction of $(CH_3)_3N$ with the solvent. Using electron dot formulas, show how this reaction takes place.
(d) What is the ionic equation for the neutralization of $H_2SO_4$ by $(CH_3)_3N$ in liquid HCN?
(e) What is the net ionic equation for the neutralization reaction in this solvent?

# 14 ACID-BASE EQUILIBRIA IN AQUEOUS SOLUTION

In Chapter 5 we saw that many of the reactions that are of concern to us take place in aqueous solution. By now we also realize that chemical changes do not, in general, proceed entirely to completion, but instead approach a state of dynamic equilibrium. In Chapters 14 and 15 we will examine in greater detail, and on a quantitative basis, many of the ionic equilibria that can occur in aqueous solution. We will begin, in this chapter, with equilibria and reactions of acids and bases. This is very important because of the amphiprotic nature of water itself. Many compounds of biological interest, which occur in the aqueous environment of living systems, show acid-base properties.

## 14.1 IONIZATION OF WATER, pH

In the last chapter we saw that some solvents can be considered to undergo autoionization. Recall, for example, that the autoionization of pure water is written as

$$H_2O + H_2O \rightleftharpoons H_3O^+ + OH^-$$

Since this is an equilibrium, we can write an equilibrium expression. Following the concepts developed in Chapter 12, this can be represented as

$$K = \frac{[H_3O^+][OH^-]}{[H_2O][H_2O]}$$

The molar concentration of water, which appears in the denominator of this expression, is very nearly constant ($\approx 55.6\ M$) in both pure water and in dilute aqueous solutions. Therefore, $[H_2O]^2$ can be included with the equilibrium constant, $K$, on the left side of the above equation. We would then write

$$K \cdot [H_2O]^2 = [H_3O^+][OH^-]$$

The left side of this expression is the product of two constants which, of course, is also equal to a constant that we shall call $K_w$.

$$K_w = K[H_2O]^2$$

Our equilibrium condition therefore becomes

$$K_w = [H_3O^+][OH^-]$$

Since $[H_3O^+][OH^-]$ is the product of ionic concentrations, $K_w$ is called the

**ion product constant** for water, or frequently simply the **ionization constant** or **dissociation constant** of water.

The equation for the autoionization of water is often simplified somewhat by omitting the solvent from the expression. The dissociation reaction for water then becomes

$$H_2O \rightleftharpoons H^+ + OH^-$$

and the simplified expression for the dissociation constant of water is written simply as

$$K_w = [H^+][OH^-] \tag{14.1}$$

*The equilibrium dissociation of $H_2O$ is present in any aqueous solution, and Equation 14.1 must always be fulfilled regardless of what other equilibria may also be taking place in solution.*

The ionization constant of water at 25°C has been found to have a value of $1.0 \times 10^{-14}$ and can be used to calculate the molar concentrations of both the $H^+$ and $OH^-$ ions in pure water. From the stoichiometry of the dissociation, we see that whenever 1 mole of $H^+$ is formed, 1 mole of $OH^-$ is also produced. This means that, at equilibrium, $[H^+] = [OH^-]$. If we let $x$ equal the hydrogen ion concentration, then

$$x = [H^+] = [OH^-]$$

Substituting into Equation 14.1 gives

$$K_w = x \cdot x = x^2$$

or, because $K_w = 1.0 \times 10^{-14}$,

$$x^2 = 1.0 \times 10^{-14}$$

Taking the square root yields

$$x = 1.0 \times 10^{-7}$$

which means that the concentrations of hydrogen ion and hydroxide ion in pure water are

$$[H^+] = [OH^-] = 1.0 \times 10^{-7}\ M$$

Whenever the hydrogen ion concentration equals the hydroxide ion concentration, as it does in pure water, the solution is said to be *neutral.* An acid is a substance that makes the $H^+$ concentration greater than the $OH^-$ concentration; conversely, a base makes the $OH^-$ concentration greater than the $H^+$ concentration. However, remember that there is always *some* $OH^-$ present in an acidic solution, just as there is always *some* $H^+$ present even if the solution is basic. At all times, Equation 14.1 is obeyed if the solution is at equilibrium.

In an aqueous solution of an acid we will often want to know what the $H^+$ concentration is. In these cases it is almost always safe to assume that essentially all of the $H^+$ in the solution comes from the dissolved acid. In other words, it is usually safe to assume that the dissociation of water contributes a negligible amount of $H^+$ to the solution. This is because the presence of $H^+$ from an acid (for example, HCl) shifts the equilibrium

$$H_2O \rightleftharpoons H^+ + OH^- \tag{14.2}$$

to the left and the amount of water dissociated in a solution of an acid is even less than in pure water. Similarly, the $OH^-$ concentration in a solution of a base can be calculated just from the concentration of the solute. The $OH^-$ contributed by the dissociation of water is negligible. Example 14.1 illustrates this point.

**EXAMPLE 14.1** (a) What is the $OH^-$ concentration in a 0.001 $M$ HCl solution? (b) What is the $H^+$ concentration derived from the dissociation of the solvent?

**SOLUTION** (a) At equilibrium we must have

$$[H^+][OH^-] = 1.0 \times 10^{-14}$$

HCl is a strong acid and is essentially 100% dissociated.

$$HCl \longrightarrow H^+ + Cl^-$$

Therefore, 0.001 mole of HCl per liter gives 0.001 mole of $H^+$ per liter. Solving for $[OH^-]$,

$$[OH^-] = \frac{1.0 \times 10^{-14}}{[H^+]}$$

and

$$[OH^-] = \frac{1.0 \times 10^{-14}}{1 \times 10^{-3}} = 1 \times 10^{-11}\ M$$

(b) The hydroxide ion in part (a) comes entirely from the dissociation of water. The amount of $H^+$ derived from $H_2O$ must therefore *also* be $1 \times 10^{-11}$ $M$, as can be seen from the stoichiometry of Equation 14.2. Note that this value ($1 \times 10^{-11}\ M$) is indeed negligible compared to the $H^+$ concentration produced by the HCl ($1 \times 10^{-3}\ M$).

Hydrogen ion and hydroxide ion enter into many equilibria in addition to the dissociation of water; therefore it is frequently necessary to specify their concentrations in aqueous solutions. These concentrations may range from relatively high values to very small ones (for example, 10 $M$ to $10^{-14}$ $M$), and a logarithmic notation has been devised to simplify the expression of these quantities. In general, for some quantity $X$,

$$pX = \log \frac{1}{X} = -\log X \qquad (14.3)$$

For example, if we wish to denote the hydrogen ion concentration in a solution, we speak of pH, defined as

$$pH = \log \frac{1}{[H^+]} = -\log[H^+]$$

In a solution where the hydrogen ion concentration is $10^{-3}\ M$, we therefore have[1]

$$pH = -\log(10^{-3}) = -(-3.0)$$

or

$$pH = 3.0$$

[1] A discussion on the use of logarithms can be found in Appendix C.

Similarly, if the hydrogen ion concentration is $10^{-8}\,M$, the pH of the solution is 8.0.

Following the same approach for the hydroxide ion concentration, we can define the pOH of a solution as

$$\text{pOH} = -\log[\text{OH}^-]$$

From the equilibrium expression for the dissociation of water,

$$\log K_w = \log[\text{H}^+] + \log[\text{OH}^-]$$

Multiplying through by $-1$ gives

$$(-\log K_w) = (-\log[\text{H}^+]) + (-\log[\text{OH}^-])$$

or simply

$$\text{p}K_w = \text{pH} + \text{pOH}$$

Since $K_w = 1.0 \times 10^{-14}$, $\text{p}K_w = 14.0$, therefore we find that

$$\text{pH} + \text{pOH} = 14.0$$

In a neutral solution, $[\text{H}^+] = [\text{OH}^-] = 10^{-7}\,M$, and pH = pOH = 7.0, so that in a neutral solution we say the pH = 7.0. In an acidic solution the hydrogen ion concentration is greater than $10^{-7}\,M$ (for example, $10^{-3}\,M$) and the pH is less than 7.0. By the same token, in basic solutions the $[\text{H}^+]$ is less than $10^{-7}\,M$ (for example, $10^{-10}\,M$) and the pH is greater than 7.0. This is summarized below.

| | $[\text{H}^+]$ | $[\text{OH}^-]$ | pH | pOH |
|---|---|---|---|---|
| Acidic solution | $>10^{-7}$ | $<10^{-7}$ | $<7$ | $>7$ |
| Neutral solution | $10^{-7}$ | $10^{-7}$ | 7 | 7 |
| Basic solution | $<10^{-7}$ | $>10^{-7}$ | $>7$ | $<7$ |

Many natural substances are either acidic or basic. Citrus fruits, for example, are sour because they contain citric acid. Milk of magnesia is bitter, on the other hand, because it is basic. The degree of acidity or basicity of such substances is conveniently expressed by giving their pH. Figure 14.1 illustrates the pH of a number of common substances.

Let us now look at a few sample problems dealing with typical calculations involving pH.

Figure 14.1
*The pH of some common substances.*

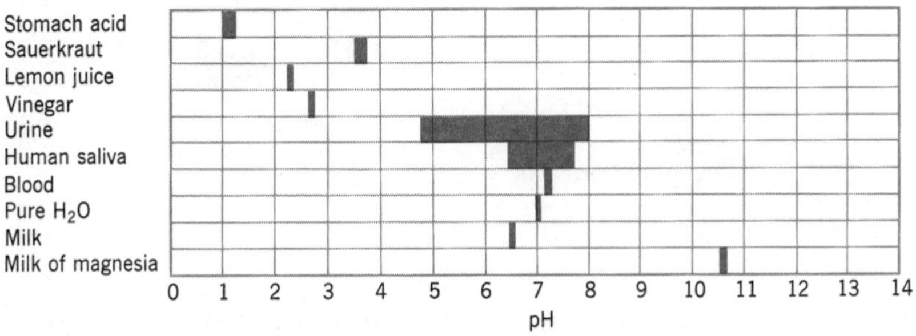

EXAMPLE 14.2 What is the pH of a 0.0020 $M$ HCl solution?

SOLUTION Since HCl is a strong acid, we can write

$$[H^+] = 0.0020\ M = 2.0 \times 10^{-3}\ M$$

We know that

$$pH = -\log[H^+]$$

and for this problem,

$$\begin{aligned} pH &= -\log(2.0 \times 10^{-3}) \\ &= -(\log 2.0 + \log 10^{-3}) \\ &= -[0.30 + (-3)] \\ &= -(-2.7) \\ &= 2.7 \end{aligned}$$

EXAMPLE 14.3 What is the pH of a $5.0 \times 10^{-4}\ M$ NaOH solution?

SOLUTION In this example the $OH^-$ concentration is given and we are asked to compute the pH. We can solve this problem in either of two ways; (1) knowing $K_w$ for water and $[OH^-]$ for this solution, we can calculate $[H^+]$ using Equation 14.1 and then proceed as we did in Example 14.2; or (2) we can calculate pOH from the $OH^-$ concentration and subtract it from $pK_w$ to obtain pH.

*Method 1.* We know that

$$K_w = [H^+][OH^-]$$

and therefore

$$\begin{aligned} [H^+] &= \frac{1.0 \times 10^{-14}}{5.0 \times 10^{-4}} \\ &= 0.2 \times 10^{-10} \quad \text{or} \quad 2.0 = 10^{-11}\ M \end{aligned}$$

Now, proceeding as we did in Example 14.2,

$$\begin{aligned} pH &= -\log(2.0 + 10^{-11}) \\ &= -[0.3 + (-11)] \\ &= 10.7 \end{aligned}$$

*Method 2.* By definition we have

$$pOH = -\log[OH^-]$$

For this problem,

$$\begin{aligned} pOH &= -\log(5 \times 10^{-4}) \\ &= -[0.7 + (-4)] \\ &= 3.3 \end{aligned}$$

The pH would be

$$\begin{aligned} pH &= pK_w - pOH \\ &= 14.0 - 3.3 \\ &= 10.7 \end{aligned}$$

**EXAMPLE 14.4** A sample of orange juice was found to have a pH of 3.80. What were the $H^+$ and $OH^-$ concentrations in the juice?

**SOLUTION** In order to compute $[H^+]$ from pH we must reverse the procedure that we followed in Example 14.2. We know that

$$pH = -\log[H^+] = 3.80$$

or

$$\log[H^+] = -3.80$$

Let's write this as the sum of a decimal fraction plus a negative integer. Thus, $-3.80$ is the same as 0.20 plus $-4.00$.

$$\log[H^+] = 0.20 + (-4.00)$$

Since 0.20 is the logarithm of 1.6, and $-4$ is the logarithm of $10^{-4}$, we can write

$$\log[H^+] = \log 1.6 + \log 10^{-4}$$

or

$$\log[H^+] = \log(1.6 \times 10^{-4})$$

Hence

$$[H^+] = 1.6 \times 10^{-4}\, M$$

To calculate $[OH^-]$ we can divide $K_w$ by $[H^+]$.

$$[OH^-] = \frac{1.0 \times 10^{-14}}{1.6 \times 10^{-4}} = 6.3 \times 10^{-11}$$

Alternatively, we could first have computed the pOH of the solution,

$$pOH = 14.00 - 3.80 = 10.20$$

From pOH we can obtain $[OH^-]$,

$$\begin{aligned} pOH = -\log[OH^-] &= 10.20 \\ \log[OH^-] &= -10.20 = 0.80 - 11.00 \\ \log[OH^-] &= \log(6.3) + \log(10^{-11}) \\ [OH^-] &= 6.3 \times 10^{-11} \end{aligned}$$

## 14.2 DISSOCIATION OF WEAK ELECTROLYTES

As a class, weak electrolytes include weak acids and bases as well as certain salts, such as $HgCl_2$ and $CdSO_4$, that are not fully dissociated in aqueous solution. In solutions of these materials, there is an equilibrium between the undissociated species and its corresponding ions. For example, acetic acid ionizes according to the equation,

$$HC_2H_3O_2 + H_2O \rightleftharpoons H_3O^+ + C_2H_3O_2^-$$

The equilibrium expression for this reaction would be

$$K = \frac{[H_3O^+][C_2H_3O_2^-]}{[HC_2H_3O_2][H_2O]}$$

As before, the concentration of $H_2O$ is considered a constant that can be included with $K$ on the left side of the equal sign. That is,

$$K \times [H_2O] = K_a = \frac{[H_3O^+][C_2H_3O_2^-]}{[HC_2H_3O_2]}$$

where we have used $K_a$ to represent an **acid dissociation constant** or **ionization constant.** This same equilibrium expression can be obtained if we simplify the dissociation by omitting the solvent. Thus, for the dissociation of acetic acid we would write

$$HC_2H_3O_2 \rightleftharpoons H^+ + C_2H_3O_2^-$$

and the equilibrium expression would then be

$$K_a = \frac{[H^+][C_2H_3O_2^-]}{[HC_2H_3O_2]}$$

In general, then, for any weak acid, HA, the simplified dissociation reaction would be written as

$$HA \rightleftharpoons H^+ + A^-$$

and its acid dissociation constant would be given by

$$K_a = \frac{[H^+][A^-]}{[HA]}$$

This same approach can also be applied to weak bases. According to the Arrhenius theory, when a weak base ionizes in aqueous solution, it does so by picking up an $H^+$ ion from the solvent. For example, with $NH_3$ we have

$$NH_3 + H_2O \rightleftharpoons NH_4^+ + OH^-$$

If we omit the solvent, the base ionization constant, $K_b$, for this reaction is

$$K_b = \frac{[NH_4^+][OH^-]}{[NH_3]}$$

The extent to which a weak acid or base undergoes ionization, as well as the value of the ionization constant, must be determined experimentally. One way of doing this is to measure the pH of a solution prepared by dissolving a known quantity of the weak acid or base in a given volume of solution, as illustrated in Example 14.5. Dissociation constants of a number of weak acids and bases are listed in Table 14.1.

We see from Table 14.1 that the values of $K_a$ and $K_b$ for weak acids and bases are quite small, ranging from $10^{-2}$ to $10^{-10}$. Recalling that numbers this small can be simplified by applying the logarithmic notation of Equation 14.3, we can write for any $K_a$,

$$pK_a = -\log K_a$$

and for any $K_b$,

$$pK_b = -\log K_b$$

For example, the $pK_a$ of acetic acid is

$$pK_a = -\log K_a = -\log(1.8 \times 10^{-5})$$
$$= 4.74$$

Table 14.1
*Ionization constants for some weak acids and bases*

| Weak Acid | Ionization | $K_a$ | $pK_a$ |
|---|---|---|---|
| Chloroacetic acid | $HC_2H_2O_2Cl \rightleftharpoons H^+ + C_2H_2O_2Cl^-$ | $1.4 \times 10^{-3}$ | 2.85 |
| Hydrofluoric acid | $HF \rightleftharpoons H^+ + F^-$ | $6.5 \times 10^{-4}$ | 3.19 |
| Nitrous acid | $HNO_2 \rightleftharpoons H^+ + NO_2^-$ | $4.5 \times 10^{-4}$ | 3.35 |
| Formic acid | $HCHO_2 \rightleftharpoons H^+ + CHO_2^-$ | $1.8 \times 10^{-4}$ | 3.74 |
| Lactic acid | $HC_3H_5O_3 \rightleftharpoons H^+ + C_3H_5O_3^-$ | $1.38 \times 10^{-4}$ | 3.86 |
| Benzoic acid | $HC_7H_5O_2 \rightleftharpoons H^+ + C_7H_5O_2^-$ | $6.5 \times 10^{-5}$ | 4.19 |
| Acetic acid | $HC_2H_3O_2 \rightleftharpoons H^+ + C_2H_3O_2^-$ | $1.8 \times 10^{-5}$ | 4.74 |
| Butyric acid | $HC_4H_7O_2 \rightleftharpoons H^+ + C_4H_7O_2^-$ | $1.5 \times 10^{-5}$ | 4.82 |
| Nicotinic acid | $HC_6H_4NO_2 \rightleftharpoons H^+ + C_6H_4NO_2^-$ | $1.4 \times 10^{-5}$ | 4.85 |
| Propionic acid | $HC_3H_5O_2 \rightleftharpoons H^+ + C_3H_5O_2^-$ | $1.4 \times 10^{-5}$ | 4.85 |
| Barbituric acid | $HC_4H_3N_2O_3 \rightleftharpoons H^+ + C_4H_3N_2O_3^-$ | $1.0 \times 10^{-5}$ | 5.00 |
| Veronal (diethylbarbituric acid) | $HC_8H_{11}N_2O_3 \rightleftharpoons H^+ + C_8H_{11}N_2O_3^-$ | $3.7 \times 10^{-8}$ | 7.43 |
| Hypochlorous acid | $HOCl \rightleftharpoons H^+ + OCl^-$ | $3.1 \times 10^{-8}$ | 7.51 |
| Hydrocyanic acid | $HCN \rightleftharpoons H^+ + CN^-$ | $4.9 \times 10^{-10}$ | 9.31 |
| **Weak Base** | **Ionization** | $K_b$ | $pK_b$ |
| Diethylamine | $(C_2H_5)_2NH + H_2O \rightleftharpoons (C_2H_5)_2NH_2^+ + OH^-$ | $9.6 \times 10^{-4}$ | 3.02 |
| Methylamine | $CH_3NH_2 + H_2O \rightleftharpoons CH_3NH_3^+ + OH^-$ | $3.7 \times 10^{-4}$ | 3.43 |
| Ammonia | $NH_3 + H_2O \rightleftharpoons NH_4^+ + OH^-$ | $1.8 \times 10^{-5}$ | 4.74 |
| Hydrazine | $N_2H_4 + H_2O \rightleftharpoons N_2H_5^+ + OH^-$ | $1.7 \times 10^{-6}$ | 5.77 |
| Hydroxylamine | $NH_2OH + H_2O \rightleftharpoons NH_3OH^+ + OH^-$ | $1.1 \times 10^{-8}$ | 7.97 |
| Pyridine | $C_5H_5N + H_2O \rightleftharpoons C_5H_5NH^+ + OH^-$ | $1.7 \times 10^{-9}$ | 8.77 |
| Aniline | $C_6H_5NH_2 + H_2O \rightleftharpoons C_6H_5NH_3^+ + OH^-$ | $3.8 \times 10^{-10}$ | 9.42 |

and for pyridine,

$$pK_b = -\log(1.7 \times 10^{-9}) = 8.77$$

We know that the smaller the value of $K_a$ and $K_b$, the weaker will be the acid or base. Relative strengths of acids and bases can also be indicated by their $pK_a$'s and $pK_b$'s. In this case the smaller the value of $pK_a$ or $pK_b$, the *stronger* is the acid or base. Let's compare, for example, the $pK_a$'s for acetic, chloroacetic, and dichloroacetic acids. Their $pK_a$'s are

$$HC_2H_3O_2 \quad pK_a = 4.74$$
$$HC_2H_2ClO_2 \quad pK_a = 2.85$$
$$HC_2HCl_2O_2 \quad pK_a = 1.30$$

The order of increasing acidity is, therefore,

acetic < chloroacetic < dichloroacetic acid

A discussion as to why the acidity of these substances increases in this fashion was included in Section 13.4.

Let's look now at some examples showing how these equilibrium constants can be calculated.

EXAMPLE 14.5 A student prepared a 0.10 *M* acetic acid solution and experimentally measured the pH of this solution to be 2.88. Calculate the $K_a$ for acetic acid and its percent dissociation.

SOLUTION To evaluate $K_a$ we must have equilibrium concentrations to substitute into the expression,

$$K_a = \frac{[H^+][C_2H_3O_2^-]}{[HC_2H_3O_2]}$$

From pH we can obtain the $H^+$ concentration.

$$\begin{aligned} pH = -\log[H^+] &= 2.88 \\ \log[H^+] &= 0.12 - 3 \\ \log[H^+] &= \log 1.3 + \log 10^{-3} \\ [H^+] &= 1.3 \times 10^{-3}\,M \end{aligned}$$

This $[H^+]$ comes from the dissociation of $HC_2H_3O_2$,

$$HC_2H_3O_2 \rightleftharpoons H^+ + C_2H_3O_2^-$$

and, from the stoichiometry of the equation, we see that

$$[H^+] = [C_2H_3O_2^-] = 1.3 \times 10^{-3}\,M$$

The concentration of undissociated $HC_2H_3O_2$ at equilibrium is equal to the original concentration, 0.10 *M*, *minus* the number of moles per liter of acetic acid that have dissociated. At equilibrium, then, we have

| | Equilibrium Concentration |
|---|---|
| $H^+$ | $1.3 \times 10^{-3}\,M$ |
| $C_2H_3O_2^-$ | $1.3 \times 10^{-3}\,M$ |
| $HC_2H_3O_2$ | $1.0 \times 10^{-1} - 0.013 \times 10^{-1} = 1.0 \times 10^{-1}\,M$ |

Note that when we compute the acetic acid concentration to the *proper number of significant figures*, the quantity that has dissociated is negligible compared to the amount initially present. Thus,

$$0.10 - 0.0013 = (0.0987) = 0.10$$

Substituting the equilibrium concentrations into the expression for $K_a$, we have

$$\begin{aligned} K_a &= \frac{(1.3 \times 10^{-3})(1.3 \times 10^{-3})}{(1.0 \times 10^{-1})} \\ &= 1.7 \times 10^{-5} \end{aligned}$$

The percent dissociation of acetic acid in this solution is found by dividing the number of moles per liter of $HC_2H_3O_2$ that have dissociated by the quantity of acetic acid that was available initially, all multiplied by 100:

$$\begin{aligned} \text{percent dissociation} &= \frac{(\text{moles/liter } HC_2H_3O_2 \text{ dissociated})}{(\text{moles/liter } HC_2H_3O_2 \text{ available})} \times 100 \\ &= \frac{1.3 \times 10^{-3}\,M}{1.0 \times 10^{-1}\,M} \times 100 = 1.3\% \end{aligned}$$

EXAMPLE 14.6 A student prepared a 0.010 *M* $NH_3$ solution and, by a freezing-point-lowering experiment, determined that the $NH_3$ had undergone 4.2% ionization. Calculate the $K_b$ for $NH_3$.

SOLUTION Ammonia ionizes in water according to the reaction,

$$NH_3 + H_2O \rightleftharpoons NH_4^+ + OH^-$$

for which we write

$$K_b = \frac{[NH_4^+][OH^-]}{[NH_3]}$$

From the stoichiometry of the ionization we see that, at equilibrium,

$$[NH_4^+] = [OH^-]$$

Since the 0.010 *M* solution undergoes 4.2% ionization, the number of moles per liter of these ions present at equilibrium is

$$[NH_4^+] = [OH^-] = 0.042 \times 0.010\ M = 4.2 \times 10^{-4}\ M$$

The number of moles per liter of $NH_3$ at equilibrium would be

$$[NH_3] = 1.0 \times 10^{-2} - 0.04 \times 10^{-2} = 0.96 \times 10^{-2}$$

When this is rounded off to the appropriate number of significant figures, we have $[NH_3] = 1.0 \times 10^{-2}\ M$ (once again, the quantity of $NH_3$ lost by ionization is negligible). Our equilibrium concentrations are therefore

| | Equilibrium Concentration |
|---|---|
| $NH_4^+$ | $4.2 \times 10^{-4}\ M$ |
| $OH^-$ | $4.2 \times 10^{-4}\ M$ |
| $NH_3$ | $1.0 \times 10^{-2}\ M$ |

When these concentrations are substituted into the equation for $K_b$, we have

$$K_b = \frac{(4.2 \times 10^{-4})(4.2 \times 10^{-4})}{(1 \times 10^{-2})}$$

or

$$K_b = 1.8 \times 10^{-5}$$

In Examples 14.5 and 14.6 we computed *K* from a knowledge of equilibrium concentrations. We can also use our knowledge of *K* to calculate the concentrations in an equilibrium mixture. Let's look at some examples.

EXAMPLE 14.7 What are the concentrations of all of the species present in a 0.50 *M* $HC_2H_3O_2$ solution?

SOLUTION First we write the chemical equation for the equilibrium,

$$HC_2H_3O_2 \rightleftharpoons H^+ + C_2H_3O_2^-$$

From Table 14.1 we find that $K = 1.8 \times 10^{-5}$ for $HC_2H_3O_2$. Therefore,

$$\frac{[H^+][C_2H_3O_2^-]}{[HC_2H_3O_2]} = 1.8 \times 10^{-5}$$

The quantities that we must substitute into this expression must represent equilibrium concentrations. In obtaining these we will construct a table as we did in Chapter 12. The question tells us that 0.50 mole of $HC_2H_3O_2$ per liter is present initially, before dissociation. There is no $C_2H_3O_2^-$, and we will neglect the $H^+$ from the dissociation of $H_2O$. Therefore, for the purposes of the problem there is no $H^+$ initially. We know that at equilibrium some of the acetic acid in the solution will have dissociated. If we therefore let $x$ equal the number of moles per liter of $HC_2H_3O_2$ that dissociate, at equilibrium we shall have produced $x$ mole/liter of $H^+$, $x$ mole/liter of $C_2H_3O_2^-$, and we shall have lost $x$ mole/liter of $HC_2H_3O_2$. At equilibrium we thus have $x$ *M* $H^+$, $x$ *M* $C_2H_3O_2^-$, and $(0.50 - x)$ *M* $HC_2H_3O_2$.

| | Initial Concentration | Change | Equilibrium Concentration |
|---|---|---|---|
| $H^+$ | 0.0 | $+x$ | $x$ |
| $C_2H_3O_2^-$ | 0.0 | $+x$ | $x$ |
| $HC_2H_3O_2$ | 0.50 | $-x$ | $0.50 - x$ |

Substituting these values into the equilibrium expression gives us

$$\frac{(x)(x)}{(0.50 - x)} = 1.8 \times 10^{-5}$$

Without simplification this expression leads to a quadratic equation that can be solved using the quadratic formula. However, in Example 12.11 we saw that it is sometimes possible to make simplifying assumptions that greatly reduce the effort required to obtain solutions to problems of this type. Because $K$ is small, very little $HC_2H_3O_2$ will have actually undergone dissociation; hence $x$ will be small. Let us assume that $x$ will be negligible compared to 0.50; that is,

$$0.50 - x \approx 0.50$$

Our equation then becomes

$$\frac{x^2}{0.50} = 1.8 \times 10^{-5}$$

or

$$x = 3.0 \times 10^{-3}$$

If we look back on our assumption, we see that $x$ is in fact small compared to 0.50 and that, when *rounded to the proper number of significant figures,*

$$0.50 - 0.003 = 0.50$$

Therefore the equilibrium concentrations of the species involved in the dissociation of the acid are

| Species | Equilibrium Concentration (*M*) |
|---|---|
| $H^+$ | $3.0 \times 10^{-3}$ |
| $C_2H_3O_2^-$ | $3.0 \times 10^{-3}$ |
| $HC_2H_3O_2$ | 0.50 |

Since the question asks for *all* concentrations, we must also calculate $[OH^-]$ that comes from the dissociation of water. Here we use $K_w$.

$$[OH^-] = \frac{K_w}{[H^+]}$$
$$= \frac{1.0 \times 10^{-14}}{3.0 \times 10^{-3}}$$
$$= 3.3 \times 10^{-12}$$

In the last example the only source of $H^+$ and $C_2H_3O_2^-$ was from the dissociation of the weak acid. Example 14.8 shows how we would handle a problem dealing with a solution where there are two sources of one of the ions.

EXAMPLE 14.8 What are the concentrations of $H^+$, $C_2H_3O_2^-$, and $HC_2H_3O_2$ in a solution prepared by dissolving 0.10 mole of $NaC_2H_3O_2$ and 0.20 mole of $HC_2H_3O_2$ in enough water to give a total volume of 1.00 liter?

SOLUTION There is only one equilibrium here with which we must be concerned,

$$HC_2H_3O_2 \rightleftharpoons H^+ + C_2H_3O_2^-$$
$$\frac{[H^+][C_2H_3O_2^-]}{[HC_2H_3O_2]} = 1.8 \times 10^{-5}$$

When $NaC_2H_3O_2$ dissolves, it is completely dissociated. It is important to remember that almost all salts are 100% dissociated in solution. Therefore 0.10 mole/liter of $NaC_2H_3O_2$ gives 0.10 mole/liter of $Na^+$ and 0.10 mole/liter of $C_2H_3O_2^-$. We are interested only in the $C_2H_3O_2^-$; the $Na^+$ is simply a *spectator ion.* The initial concentrations that are of concern to us, then, are found in the first column of our table. Since no $H^+$ is present, some $HC_2H_3O_2$ must ionize; so let's let $x$ = the number of moles per liter of $HC_2H_3O_2$ that dissociates to give $H^+$ and $C_2H_3O_2^-$. This will increase $[H^+]$ and $[C_2H_3O_2^-]$ by $x$, and decrease $[HC_2H_3O_2]$ by $x$. The equilibrium concentrations are then found in the last column for our table.

| | Initial Concentration | Change | Final Concentration |
|---|---|---|---|
| $H^+$ | 0.0 | $+x$ | $x$ |
| $C_2H_3O_2^-$ | 0.10 | $+x$ | $0.10 + x \approx 0.10$ |
| $HC_2H_3O_2$ | 0.20 | $-x$ | $0.20 - x \approx 0.20$ |

As before, we look at $K_a$ and see that $x$ will probably be small. We will therefore assume that $0.10 + x \approx 0.10$ and $0.20 - x \approx 0.20$. Substituting into the expression for $K_a$ gives

$$\frac{(x)(0.10)}{(0.20)} = 1.8 \times 10^{-5}$$
$$x = 3.6 \times 10^{-5}$$

Note that $x$ is small compared to both 0.10 and 0.20. This justifies our assumption. Finally, the equilibrium concentrations are

$$[H^+] = 3.6 \times 10^{-5}\ M$$
$$[C_2H_3O_2^-] = 0.10\ M$$
$$[HC_2H_3O_2] = 0.20\ M$$

## 14.3 DISSOCIATION OF POLYPROTIC ACIDS

Acids containing more than one atom of hydrogen that can be lost upon dissociation are known as polyprotic acids. Some examples of polyprotic acids are $H_2SO_4$ and $H_2S$, both of which contain two ionizable hydrogens, and $H_3PO_4$, which contains three. We consider these acids to lose their hydrogens, one at a time, in a stepwise fashion. Thus we write two steps for the dissociation of sulfuric acid, each with a corresponding equation for $K_a$,

$$H_2SO_4 \rightleftharpoons H^+ + HSO_4^- \qquad K_{a_1} = \frac{[H^+][HSO_4^-]}{[H_2SO_4]}$$

$$HSO_4^- \rightleftharpoons H^+ + SO_4^{2-} \qquad K_{a_2} = \frac{[H^+][SO_4^{2-}]}{[HSO_4^-]}$$

and for $H_2S$,

$$H_2S \rightleftharpoons H^+ + HS^- \qquad K_{a_1} = \frac{[H^+][HS^-]}{[H_2S]} \qquad (14.4)$$

$$HS^- \rightleftharpoons H^+ + S^{2-} \qquad K_{a_2} = \frac{[H^+][S^{2-}]}{[HS^-]} \qquad (14.5)$$

The three steps in the dissociation of $H_3PO_4$ are written

$$H_3PO_4 \rightleftharpoons H^+ + H_2PO_4^- \qquad K_{a_1} = \frac{[H^+][H_2PO_4^-]}{[H_3PO_4]}$$

$$H_2PO_4^- \rightleftharpoons H^+ + HPO_4^{2-} \qquad K_{a_2} = \frac{[H^+][HPO_4^{2-}]}{[H_2PO_4^-]}$$

$$HPO_2^{2-} \rightleftharpoons H^+ + PO_4^{3-} \qquad K_{a_3} = \frac{[H^+][PO_4^{3-}]}{[HPO_4^{2-}]}$$

Table 14.2 contains a list of some polyprotic acids and their stepwise dissociation constants. We see from this table that for sulfuric acid the first dissociation goes very nearly to completion, while the second dissociation occurs only to a relatively limited degree. Because of the near completion of its first dissociation, sulfuric acid is considered to be a strong acid. We also see that, in general, the first dissociation step of these acids occurs with the largest value of $K_a$ and that each successive step occurs with an ever-decreasing value of $K_a$. This decreasing trend in $K_a$ is reasonable considering that it should be easiest to remove a $H^+$ ion from an uncharged species and that it becomes progressively more difficult to do so as the negative charge on the ion increases.

Because the equilibria involving polyprotic acids are more complex than those of monoprotic acids, equilibrium calculations are somewhat more complicated. The next example shows, however, that a number of generalizations can be made that simplify the approach to these kinds of problems.

EXAMPLE 14.9

Hydrogen sulfide, $H_2S$, is a gas produced by anaerobic (bacterial action in the absence of air) decomposition of organic compounds. Its disagreeable odor is responsible for the terrible smell of rotten eggs. In water, $H_2S$ is a diprotic

Table 14.2
*Stepwise dissociation of some polyprotic acids at 25°C*

| Acid | Stepwise Dissociation | Dissociation Constant for Each Step | $pK_a$ |
|---|---|---|---|
| Phosphoric | $H_3PO_4 \rightleftharpoons H^+ + H_2PO_4^-$ | $K_{a_1} = 7.5 \times 10^{-3}$ | 2.13 |
| | $H_2PO_4^- \rightleftharpoons H^+ + HPO_4^{2-}$ | $K_{a_2} = 6.2 \times 10^{-8}$ | 7.21 |
| | $HPO_4^{2-} \rightleftharpoons H^+ + PO_4^{3-}$ | $K_{a_3} = 2.2 \times 10^{-12}$ | 11.66 |
| Sulfuric | $H_2SO_4 \rightleftharpoons H^+ + HSO_4^-$ | $K_{a_1}$ = very large | <0 |
| | $HSO_4^- \rightleftharpoons H^+ + SO_4^{2-}$ | $K_{a_2} = 1.2 \times 10^{-2}$ | 1.92 |
| Sulfurous | $SO_2 + H_2O \rightleftharpoons H^+ + HSO_3^-$ | $K_{a_1} = 1.5 \times 10^{-2}$ | 1.82 |
| | $HSO_3^- \rightleftharpoons H^+ + SO_3^{2-}$ | $K_{a_2} = 1.0 \times 10^{-7}$ | 7.00 |
| Hydrosulfuric | $H_2S \rightleftharpoons H^+ + HS^-$ | $K_{a_1} = 1.1 \times 10^{-7}$ | 6.96 |
| | $HS^- \rightleftharpoons H^+ + S^{2-}$ | $K_{a_2} = 1.0 \times 10^{-14}$ | 14.00 |
| Carbonic | $CO_2 + H_2O \rightleftharpoons H^+ + HCO_3^-$ | $K_{a_1} = 4.3 \times 10^{-7}$ | 6.37 |
| | $HCO_3^- \rightleftharpoons H^+ + CO_3^{2-}$ | $K_{a_2} = 5.6 \times 10^{-11}$ | 10.26 |
| Ascorbic (vitamin C) | $H_2C_6H_6O_6 \rightleftharpoons H^+ + HC_6H_6O_6^-$ | $K_{a_1} = 7.9 \times 10^{-5}$ | 4.10 |
| | $HC_6H_6O_6^- \rightleftharpoons H^+ + C_6H_6O_6^{2-}$ | $K_{a_2} = 1.6 \times 10^{-12}$ | 11.79 |

weak acid. What are the equilibrium concentrations of $H^+$, $HS^-$, $S^{2-}$, and $H_2S$ in a saturated (0.10 *M*) aqueous solution of $H_2S$?

SOLUTION The equilibria involved are shown in Equations 14.4 and 14.5. From Table 14.2, the equilibrium constants have values: $K_{a_1} = 1.1 \times 10^{-7}$ and $K_{a_2} = 1.0 \times 10^{-14}$.

Because $K_{a_1}$ is *so much greater* than $K_{a_2}$, we can safely assume that nearly all of the hydrogen ion in the solution is derived from the first step in the dissociation. In addition, only very little of the $HS^-$ formed will undergo further dissociation. On the basis of this assumption we can calculate the $H^+$ and $HS^-$ concentrations using the expression for $K_{a_1}$ alone.

$$K_{a_1} = \frac{[H^+][HS^-]}{[H_2S]}$$

If we let *x* equal the number of moles per liter of $H_2S$ that dissociate, we obtain, from the stoichiometry of the first step, *x* mole/liter of $H^+$ and *x* mole/liter of $HS^-$. At equilibrium we will have $(0.10 - x)$ mole/liter of $H_2S$ remaining.

| | Initial Concentration | Change | Equilibrium Concentration |
|---|---|---|---|
| $H^+$ | 0.0 | $+x$ | $x$ |
| $HS^-$ | 0.0 | $+x$ | $x$ |
| $H_2S$ | 0.10 | $-x$ | $0.10 - x \approx 0.10$ |

Note that as before, because $K_{a_1}$ is very small, we may assume that *x* will be negligible compared to 0.10 and write

$$[H_2S] = 0.10 - x \approx 0.10\ M$$

Substituting these equilibrium quantities into the expression for $K_{a_1}$, we have

$$\frac{(x)(x)}{0.10} = 1.1 \times 10^{-7}$$
$$x^2 = 1.1 \times 10^{-8}$$
$$x = 1.0 \times 10^{-4}$$

The equilibrium concentrations from this first dissociation are, then,

$$[H^+] = 1.0 \times 10^{-4}\,M$$
$$[HS^-] = 1.0 \times 10^{-4}\,M$$
$$[H_2S] = 0.10 - 1.1 \times 10^{-4} = 0.10\,M$$

We see that our approximation in the $H_2S$ concentration is valid, because $1.0 \times 10^{-4}$ is in fact negligible compared to 0.10.

By employing $K_{a_2}$, we can now calculate the equilibrium concentration of $S^{2-}$,

$$HS^- \rightleftharpoons H^+ + S^{2-}$$

and

$$K_{a_2} = \frac{[H^+][S^{2-}]}{[HS^-]}$$

If we let $y$ equal the number of moles per liter of $HS^-$ that dissociate, then, from the stoichiometry of this second dissociation step, the number of moles per liter of $H^+$ and $S^{2-}$ produced would also be $y$. Thus the total hydrogen ion concentration from both the first and second dissociations will be $[H^+] = (1.0 \times 10^{-4} + y)$, and the concentration of $HS^-$ that remains at equilibrium will be $(1.0 \times 10^{-4} - y)$. For this second dissociation, then,

| | "Initial" Concentration | Change | Equilibrium Concentration |
|---|---|---|---|
| $H^+$ | $1.0 \times 10^{-4}$ | $+y$ | $1.0 \times 10^{-4} + y$ |
| $S^{2-}$ | 0.0 | $+y$ | $y$ |
| $HS^-$ | $1.0 \times 10^{-4}$ | $-y$ | $1.0 \times 10^{-4} - y$ |

These quantities may be simplified by recognizing that because $K_{a_2}$ is so very small, the amount of $HS^-$ that will dissociate will also be very small. We can therefore make the assumption that $y$ will be negligible compared to $1.0 \times 10^{-4}$. Our concentrations then become

$$[H^+] = 1.0 \times 10^{-4} + y \approx 1.0 \times 10^{-4}\,M$$
$$[S^{2-}] = y$$
$$[HS^-] = 1.0 \times 10^{-4} - y \approx 1.0 \times 10^{-4}\,M$$

and we obtain

$$K_{a_2} = \frac{(1.0 \times 10^{-4})(y)}{(1.0 \times 10^{-4})} = 1.0 \times 10^{-14}$$
$$y = 1.0 \times 10^{-14}$$

Therefore,

$$[S^{2-}] = 1.0 \times 10^{-14}\,M$$

In summary, then, the concentrations of all species present at equilibrium when a 0.10 $M$ $H_2S$ solution dissociates are

$$[H^+] = 1.0 \times 10^{-4}\, M$$
$$[HS^-] = 1.0 \times 10^{-4}\, M$$
$$[S^{2-}] = 1.0 \times 10^{-14}\, M$$
$$[H_2S] = 0.10\, M$$

Note that in any solution containing only $H_2S$ the concentration of the sulfide ion is equal to $K_{a_2}$. In fact, for any polyprotic acid where $K_{a_2} \ll K_{a_1}$, the concentration of the anion formed in the second dissociation will always equal $K_{a_2}$, provided, of course, that the acid is the only solute. For example, $K_{a_2}$ for $H_3PO_4$ has a value of $6.2 \times 10^{-8}$, and in a solution containing only $H_3PO_4$ and $H_2O$, the concentration of $HPO_4^{2-}$ is $6.2 \times 10^{-8}\, M$.

Saturated solutions of $H_2S$ are sometimes used in chemical analysis to detect the presence of certain cations by the formation of an insoluble sulfide precipitate. In these analyses the concentration of the $S^{2-}$ is critical and, therefore, must be controlled. By applying Le Chatelier's principle to the dissociation of $H_2S$, we see that any increase in the $H^+$ concentration (perhaps by the addition of a strong acid) will cause a shift in the equilibrium to the left, favoring the formation of more $H_2S$ and decreasing the concentrations of both the $S^{2-}$ and $HS^-$ species. Conversely, lowering the $H^+$ concentration will increase $[HS^-]$ and $[S^{2-}]$. Thus, we can control the concentration of the $S^{2-}$ in a saturated $H_2S$ solution by varying the $H^+$ concentration. A useful equation expressing the relationship that exists between $H^+$ and $S^{2-}$ in an $H_2S$ solution can be derived by multiplying $K_{a_1}$ by $K_{a_2}$ for $H_2S$. Thus,

$$K_a = K_{a_1} \times K_{a_2} = \frac{[H^+]\cancel{[HS^-]}}{[H_2S]} \times \frac{[H^+][S^{2-}]}{\cancel{[HS^-]}}$$

$$= \frac{[H^+]^2[S^{2-}]}{[H_2S]} = 1.1 \times 10^{-21} \qquad (14.6)$$

A word of caution about the use of Equation 14.6 is in order. *This equation is useful only in situations where two of the three concentrations are given and we wish to calculate the third*. It cannot be used to determine, for example, both the $H^+$ and $S^{2-}$ concentrations in solutions of known $H_2S$ concentrations. You can verify this for yourself by using it to calculate the concentration of $H^+$ and $S^{2-}$ that are present in a 0.10 $M$ $H_2S$ solution and then comparing your answers with those that we calculated using the two dissociation constants.

The use of this equation is illustrated in Example 14.10.

EXAMPLE 14.10 Calculate the $S^{2-}$ concentration in a saturated solution (0.10 $M$) of $H_2S$ whose pH was adjusted to 2 by the addition of HCl

SOLUTION Since this is a saturated $H_2S$ solution with a known $H^+$ concentration, we can use Equation 14.6.

$$K_a = \frac{[H^+]^2[S^{2-}]}{[H_2S]}$$

Rearranging this equation to solve for $[S^{2-}]$, we have

$$[S^{2-}] = \frac{K_a[H_2S]}{[H^+]^2}$$

From the pH of this acidic solution we calculate that $[H^+] = 1 \times 10^{-2}\,M$; since the solution is saturated with $H_2S$, we know that $[H_2S] = 0.10\,M$. Substituting these values and the value of $K_a$ into our equation, we have

$$[S^{2-}] = \frac{(1.1 \times 10^{-21})(1.0 \times 10^{-1})}{(1.0 \times 10^{-2})^2}$$

or

$$[S^{2-}] = 1.1 \times 10^{-18}$$

We see very clearly that the concentration of $S^{2-}$ is dependent on the pH of the saturated solution.

## 14.4 BUFFERS

Any solution that contains both a weak acid and a weak base has the property that when small quantities of a strong acid are added, they are neutralized by the weak base, while small quantities of a strong base are neutralized by the weak acid. Such solutions are said to be **buffers,** because they have the ability to absorb these small additions of concentrated acids and bases without giving rise to a significant change in the pH of the solution.

A buffer whose pH is less than 7 generally can be prepared by mixing a weak acid with the salt of that weak acid, for example, acetic acid and sodium acetate. A buffer whose pH is greater than 7 generally can be prepared by mixing a weak base with the salt of that weak base, for example, ammonia and ammonium chloride. When $H^+$ or $OH^-$ are added to an acetic acid-acetate buffer (pH $< 7$), the following neutralization reactions take place:

$$H^+ + C_2H_3O_2^- \longrightarrow HC_2H_3O_2$$
$$OH^- + HC_2H_3O_2 \longrightarrow H_2O + C_2H_3O_2^-$$

Similarly, for the $NH_3$, $NH_4Cl$, alkaline buffer we have

$$H^+ + NH_3 \longrightarrow NH_4^+$$

and

$$OH^- + NH_4^+ \longrightarrow H_2O + NH_3$$

In an acid buffer the $H^+$ concentration (and pH) is determined by the relative concentrations of the weak acid and its conjugate base (that is, the anion). For example, for acetic acid we have

$$K_a = \frac{[H^+][C_2H_3O_2^-]}{[HC_2H_3O_2]}$$

Solving for $[H^+]$ gives

$$[H^+] = \frac{K_a[HC_2H_3O_2]}{[C_2H_3O_2^-]} \qquad (14.7)$$

To calculate $[H^+]$, and then pH, we must know $K_a$ for the weak acid (from Table 14.1) as well as the ratio of the concentrations of the weak acid and its anion.

Since salts completely dissociate in aqueous solution, the number of moles of the anion from this source is determined by the formula of the salt and the number of moles of salt dissolved. Thus a 1.0 *M* $NaC_2H_3O_2$ solution contains 1.0 mole/liter of $C_2H_3O_2^-$, while a 1.0 *M* $Ca(C_2H_3O_2)_2$ solution contains 2.0 moles/liter of $C_2H_3O_2^-$. In our buffer there will also be an additional amount of acetate ion coming from the dissociation of $HC_2H_3O_2$. The amount of $H^+$ and $C_2H_3O_2^-$ stemming from this source is very small even in solutions containing only acetic acid, and this small amount is reduced even further in the buffer because of the presence of the large concentration of $C_2H_3O_2^-$ from the salt.[2] The total anion concentration in the buffer, then, is essentially determined by the salt concentration alone, since the contribution from the dissociation of the weak acid is negligible. As in our previous calculations on weak acids, the concentration of $HC_2H_3O_2$ will not be reduced appreciably by its dissociation and, in a mixture of 1.0 *M* $NaC_2H_3O_2$ and 1.0 *M* $HC_2H_3O_2$, for example, the concentrations of both the molecular acid and the anion are 1.0 *M*. The $H^+$ concentration in such a buffer could be found from Equation 14.7 using $K_a = 1.8 \times 10^{-5}$ (Table 14.1).

$$[H^+] = (1.8 \times 10^{-5}) \frac{1.0}{1.0}$$

$$= 1.8 \times 10^{-5}\ M$$

The pH of this solution is 4.74.

Notice that when the concentrations of the acid and anion are the same in a buffer, the $H^+$ concentration in that solution is equal to the $K_a$ of the weak acid. Thus, if a buffer was prepared by mixing 0.1 mole of formic acid ($K_a = 1.8 \times 10^{-4}$ from Table 14.1) and 0.1 mole of sodium formate into a liter of solution, the resulting $H^+$ concentration would be

$$[H^+] = 1.8 \times 10^{-4}\ M$$

and

$$pH = 3.74$$

We can also use Equation 14.7 to calculate the concentrations of acid and salt that would be needed to achieve a certain pH buffer. This is illustrated in Example 14.11.

EXAMPLE 14.11 What ratio of acetic acid to sodium acetate concentration is needed to achieve a buffer whose pH is 5.70?

SOLUTION To solve this problem we need to rearrange Equation 14.7 to solve for the ratio of concentrations. Thus,

$$\frac{[HC_2H_3O_2]}{[C_2H_3O_2^-]} = \frac{[H^+]}{K_a}$$

The $H^+$ concentration when the pH is 5.70 is

$$[H^+] = 2.0 \times 10^{-6}\ M$$

[2] This can easily be seen by applying Le Chatelier's principle to the dissociation of $HC_2H_3O_2$. The presence of acetate ion from a salt causes the dissociation equilibrium of the acid to be shifted to the left and actually suppresses the dissociation.

Therefore,

$$\frac{[HC_2H_3O_2]}{[C_2H_3O_2^-]} = \frac{2.0 \times 10^{-6}}{1.8 \times 10^{-5}} = \frac{2.0 \times 10^{-6}}{18 \times 10^{-6}}$$

or

$$\frac{[HC_2H_3O_2]}{[C_2H_3O_2^-]} = \frac{1}{9}$$

As long as this ratio is maintained, the pH of an acetic acid-sodium acetate buffer is 5.70. For example, if 0.2 mole of $HC_2H_3O_2$ and 1.8 moles of $NaC_2H_3O_2$ are placed in 1 liter of solution, the pH is 5.70.

We have seen that by adjusting the ratio of concentrations of the weak acid to that of the salt, a buffer of almost any desired pH can be achieved. For example, a buffer composed of 0.01 mole of $HC_2H_3O_2$ and 1.00 mole of $NaC_2H_3O_2$ would have a pH of 6.74. However, when as little as 0.01 mole of base is added to this buffer, all of the acetic acid is neutralized and a large change in the pH of the buffer results. Therefore, we say that *the most effective pH range for any buffer is at or near the pH where the acid and salt concentrations are equal (that is,* $pK_a$). Also, in order to be most effective, the amounts of weak acid and base used to prepare the buffer must be considerably greater than the amounts of acid or base that may later be added to the buffer.

Let us now see what happens when a small quantity of acid or base is added to an acetic acid-sodium acetate buffer whose initial pH is 4.74. In this buffer the concentrations of $HC_2H_3O_2$ and $C_2H_3O_2^-$ were the same and equal to 1.0 *M* and the concentration of $H^+$ was calculated to be $1.8 \times 10^{-5}$. Suppose that 0.2 mole of $H^+$ is added to 1 liter of this buffer. The following neutralization reaction occurs:

$$H^+ + C_2H_3O_2^- \longrightarrow HC_2H_3O_2$$

Below are the number of moles per liter of all species before and after the addition.

| Initial | Final |
|---|---|
| $[H^+] = 1.8 \times 10^{-5}\ M$ | $[H^+] = x$ |
| $[C_2H_3O_2^-] = 1.0\ M$ | $[C_2H_3O_2^-] = 1.0 - 0.2\ M = 0.8\ M$ |
| $[HC_2H_3O_2] = 1.0\ M$ | $[HC_2H_3O_2] = 1.0 + 0.2\ M = 1.2\ M$ |

Substituting these final concentrations into Equation 14.7, we have

$$[H^+] = 1.8 \times 10^{-5} \left(\frac{1.2}{0.8}\right)$$
$$= 2.7 \times 10^{-5}\ M$$

The pH of the buffer after the 0.2 mole of $H^+$ was added is 4.57. The pH of the buffer changed by only 0.17 pH units.

Suppose now that we were to add 0.2 mole of $H^+$ to 1 liter of a solution of HCl whose pH = 4.74 (that is, a $1.8 \times 10^{-5}$ *M* HCl solution). Since $Cl^-$ has virtually no tendency to react with $H^+$, the final $H^+$ concentration will be 0.20 *M* ($0.20 + 1.8 \times 10^{-5} = 0.20$), and the pH of the solution will be

0.70. The change in pH in this case is 4.04 pH units, as opposed to a change of only 0.17 pH units when the same quantity of $H^+$ is added to the buffer.

Additions of strong base are also absorbed by the buffer. When 0.2 mole of $OH^-$ is added to the 1 liter of our original buffer, it is neutralized according to the reaction,

$$OH^- + HC_2H_3O_2 \longrightarrow H_2O + C_2H_3O_2^-$$

The number of moles per liter before the addition are the same as above, but the number of moles per liter after 0.2 mole of $OH^-$ is added would be

| Concentration After Addition of 0.2 mole of $OH^-$ |
|---|
| $[H^+] = x$ |
| $[C_2H_3O_2^-] = 1.0 + 0.2 = 1.2\ M$ |
| $[HC_2H_3O_2] = 1.0 - 0.2 = 0.8\ M$ |

Substituting these values into Equation 14.7, we find that the $H^+$ concentration after the addition is $1.2 \times 10^{-5}\ M$ and the resulting pH is 4.92, a change of 0.18 pH units. Finally, note that the addition of an acid to the buffer lowers the pH, while addition of a base raises the pH.

EXAMPLE 14.12 A buffer was prepared by mixing 200 ml of a 0.6 $M$ $NH_3$ solution and 300 ml of a 0.3 $M$ $NH_4Cl$ solution. (a) What is the pH of this buffer, assuming a final volume of 500 ml? (b) What will be the pH after 0.02 mole of $H^+$ is added?

SOLUTION The number of moles of $NH_3$ added to this solution is

$$\left(0.6\ \frac{\text{mole}}{\text{liter}}\right) \times 0.2\ \text{liter} = 0.12\ \text{mole}$$

and the number of moles of $NH_4^+$ added is

$$\left(0.3\ \frac{\text{mole}}{\text{liter}}\right) \times 0.3\ \text{liter} = 0.09\ \text{mole}$$

Therefore the concentrations of these ions in the 500 ml is

$$[NH_3] = \frac{0.12\ \text{mole}}{0.50\ \text{liter}} = 0.24\ M$$

$$[NH_4^+] = \frac{0.09\ \text{mole}}{0.50\ \text{liter}} = 0.18\ M$$

(a) The $OH^-$ concentration for this alkaline buffer is found by using the $K_b$ for $NH_3$:

$$NH_3 + H_2O \rightleftharpoons NH_4^+ + OH^-$$

$$K_b = \frac{[NH_4^+][OH^-]}{[NH_3]}$$

Rearranging and solving for $[OH^-]$, we have

$$[OH^-] = K_b \frac{[NH_3]}{[NH_4^+]}$$

Substituting $K_b$ for $NH_3$ from Table 14.1 and the concentrations of $NH_3$ and

$NH_4^+$ for this buffer into this equation gives

$$[OH^-] = (1.8 \times 10^{-5}) \times \left(\frac{0.24}{0.18}\right)$$
$$= 2.4 \times 10^{-5}$$
$$pOH = 4.62$$
$$pH = 9.38$$

(b) The neutralization reaction for $H^+$ in this buffer is

$$H^+ + NH_3 \longrightarrow NH_4^+$$

We are adding 0.02 mole of $H^+$ to 500 ml, or 0.04 mole of $H^+$ per liter. Therefore, the concentrations before and after the addition of the acid are

| Initial | Final |
|---|---|
| $[OH^-] = 2.4 \times 10^{-5}\ M$<br>$[NH_3] = 0.24\ M$<br>$[NH_4^+] = 0.18\ M$ | $[OH^-] = x$<br>$[NH_3] = 0.24 - 0.04\ M = 0.20\ M$<br>$[NH_4^+] = 0.18 + 0.04\ M = 0.22\ M$ |

and the $OH^-$ concentration is

$$[OH^-] = (1.8 \times 10^{-5}) \times \left(\frac{0.20}{0.22}\right)$$
$$= 1.6 \times 10^{-5}\ M$$
$$pOH = 4.80$$

The pH is therefore 9.20, a decrease of 0.18 pH units.

Buffers find many important applications. Living systems employ buffers to maintain nearly constant pH so that biochemical reactions can follow their correct paths. For example, blood contains, among other things, a $H_2CO_3/HCO_3^-$ buffer system that maintains the pH at 7.4.

In the laboratory many inorganic and organic chemical reactions are performed in buffered solutions to minimize any adverse effects caused by acids or bases that might be consumed or produced during reaction.

## 14.5 HYDROLYSIS

It was pointed out in Section 5.6 that salts are produced during an acid-base neutralization reaction. For example, NaCl is considered to be the salt of the strong acid HCl and the strong base NaOH. In a similar fashion we say that $NaC_2H_3O_2$ is the salt of a weak acid ($HC_2H_3O_2$) and a strong base (NaOH), and $NH_4Cl$ is the salt of a strong acid (HCl) and a weak base ($NH_3$). A salt such as $NH_4C_2H_3O_2$ is the salt of a weak acid and a weak base. When these salts are added to water, the pH of the resulting solution is found experimentally to be dependent on the type of salt dissolved. For example, the pH of an aqueous solution of a salt of a strong acid and a strong base is always very close to 7, while the pH for a salt of a weak acid and a strong base is greater than 7. The pH that results when each type of salt is dissolved in water is summarized below.

| Type of Salt | pH of Aqueous Solution |
|---|---|
| Strong acid–strong base | 7 |
| Weak acid–strong base | >7 |
| Strong acid–weak base | <7 |
| Weak acid–weak base | Depends on salt |

How might we account for these differences in pH?

When a salt dissolves in water, it dissociates fully to produce cations and anions that may subsequently react chemically with the solvent in a process called **hydrolysis.** For example, the cation of a salt undergoes the reaction,

$$M^+ + H_2O \rightleftharpoons MOH + H^+$$

while an anion reacts according to the equation,

$$X^- + H_2O \rightleftharpoons HX + OH^-$$

Since the $H^+$ and $OH^-$ ions produced in these reactions influence the pH of the salt solution, the extent to which the hydrolysis reactions take place determines whether the pH will be greater than, less than, or equal to 7.

Consider, for example, NaCl, the salt of a strong acid and a strong base. If this salt were to hydrolyze, the products would be NaOH and HCl. Both of these are strong electrolytes and are completely dissociated. The result is that there will be the same amount of $H^+$ and $OH^-$ in the solution, a condition that is fulfilled only when their concentrations are each $10^{-7}\,M$. Consequently, the pH of the NaCl solution is 7. Since this is the same as pure water, the net effect is that no hydrolysis actually takes place at all. In general, then, we may conclude that *anions and cations of strong acids and bases, respectively, do not undergo hydrolysis, and the salts derived from strong acids and bases yield neutral solutions.* This, however, is not the case for other types of salts.

**SALTS OF WEAK ACIDS AND STRONG BASES: ANION HYDROLYSIS.** For this type of salt we are concerned only with anion hydrolysis because, in light of the previous discussion, cations of strong bases do not undergo hydrolysis. With $NaC_2H_3O_2$ as our example of these salts, we would write for the hydrolysis reaction of the anion

$$C_2H_3O_2^- + H_2O \rightleftharpoons HC_2H_3O_2 + OH^-$$

The acetate ion, which is the conjugate base of acetic acid, is sufficiently basic to pull some protons away from water molecules, so that when equilibrium is established excess $OH^-$ is present in the solution. As a result the solution is basic. We can write an equilibrium constant (we'll call it $K'_h$) in the usual fashion.

$$K'_h = \frac{[HC_2H_3O_2][OH^-]}{[C_2H_3O_2^-][H_2O]}$$

As usual, the concentration of $H_2O$ may be included with the equilibrium constant $K'_h$, and we obtain

$$K'_h \times [H_2O] = K_h = \frac{[HC_2H_3O_2][OH^-]}{[C_2H_3O_2^-]}$$

where $K_h$ is called the **hydrolysis constant.**

This same equation can be derived simply by dividing the equation for $K_w$ by the $K_a$ of the weak acid (acetic acid in this case). Thus,

$$\frac{K_w}{K_a} = \frac{[H^+][OH^-]}{[H^+][C_2H_3O_2^-]/[HC_2H_3O_2]}$$

or

$$\frac{K_w}{K_a} = \frac{[HC_2H_3O_2][OH^-]}{[C_2H_3O_2^-]}$$

We see, then, for the salt of a weak acid and a strong base that the hydrolysis constant is equal to

$$K_h = \frac{K_w}{K_a}$$

Thus we can calculate the hydrolysis constant of the anion in this type of salt from a knowledge of $K_w$ and $K_a$ of the weak acid from which the anion is formed.

From $K_h$ and the concentration of the anion in the salt solution, we may then calculate the concentration of $OH^-$ and eventually determine the pH of the salt solution, as shown in Example 14.13.

EXAMPLE 14.13 Calculate the pH of a 0.10 $M$ $NaC_2H_3O_2$ solution.

SOLUTION Since this is the salt of a weak acid and a strong base, only the anion undergoes hydrolysis and that reaction is

$$C_2H_3O_2^- + H_2O \rightleftharpoons HC_2H_3O_2 + OH^-$$

for which we can write the equilibrium expression,

$$K_h = \frac{[HC_2H_3O_2][OH^-]}{[C_2H_3O_2^-]}$$

Before this equation becomes useful, we must first calculate $K_h$. We can do this with the equation,

$$K_h = \frac{K_w}{K_a}$$

Substituting the values of $K_w$ and $K_a$ (from Table 14.1) into this equation, we obtain

$$K_h = \frac{1.0 \times 10^{-14}}{1.8 \times 10^{-5}} = 5.6 \times 10^{-10}$$

From the magnitude of $K_h$, we see that the position of equilibrium in this hydrolysis lies far to the left in favor of $C_2H_3O_2^-$. Therefore, if we let $x$ equal the number of moles per liter of $C_2H_3O_2^-$ that undergo hydrolysis, we can set up our table of concentrations.

| | Initial Concentration | Change | Equilibrium Concentration |
|---|---|---|---|
| $HC_2H_3O_2$ | 0.0 | $+x$ | $x$ |
| $OH^-$ | 0.0 | $+x$ | $x$ |
| $C_2H_3O_2^-$ | 0.10 | $-x$ | $0.10 - x \approx 0.10$ |

Substituting these values along with $K_h$ into the hydrolysis equation, we have

$$K_h = \frac{(x)(x)}{0.10} = 5.6 \times 10^{-10}$$
$$x^2 = 5.6 \times 10^{-11}$$
$$x = 7.5 \times 10^{-6}\,M$$

Thus,

$$[OH^-] = 7.5 \times 10^{-6}\,M$$
$$pOH = 5.12$$

and

$$pH = 8.88$$

Thus the pH of this solution indicates that it is basic.

**SALTS OF STRONG ACIDS AND WEAK BASES: CATION HYDROLYSIS.** From our previous discussion we know that only the cation in this type of salt undergoes hydrolysis. For example, the hydrolysis reaction that takes place in an aqueous solution of $NH_4Cl$ is

$$NH_4^+ + H_2O \rightleftharpoons H_3O^+ + NH_3$$

As a result of the hydrolysis of the cation some of the $H_2O$ molecules are converted into $H_3O^+$ which, of course, make the solution acidic. The equation for this hydrolysis equilibrium is written as

$$K_h = \frac{[H_3O^+][NH_3]}{[NH_4^+]}$$

which can also be derived by dividing $K_w$ by $K_b$ of the weak base, $NH_3$. Therefore,

$$K_h = \frac{K_w}{K_b} = \frac{[H_3O^+][\cancel{OH^-}]}{[NH_4^+][\cancel{OH^-}]/[NH_3]} = \frac{[H_3O^+][NH_3]}{[NH_4^+]}$$

For this type of salt, then, in order to calculate the pH, we must know $K_w$, $K_b$, and the concentration of the salt, as illustrated in Example 14.14.

EXAMPLE 14.14 What is the pH of a 0.10 *M* $N_2H_5Cl$ solution?

SOLUTION We first must recognize that this is the salt of a strong acid (HCl) and a weak base ($N_2H_4$); therefore, only the cation undergoes hydrolysis. The hydrolysis reaction is

$$N_2H_5^+ + H_2O \rightleftharpoons H_3O^+ + N_2H_4$$

for which we can write for $K_h$,

$$K_h = \frac{[H_3O^+][N_2H_4]}{[N_2H_5^+]}$$

For this salt we know that

$$K_h = \frac{K_w}{K_b}$$

so that, using $K_b$ from Table 14.1,

$$K_h = \frac{1 \times 10^{-14}}{1.7 \times 10^{-6}} = 5.9 \times 10^{-9}$$

Once again, because of the size of $K_h$, the hydrolysis equilibrium lies mainly to the left. If $x$ equals the number of moles per liter of $N_2H_5^+$ that undergo hydrolysis, then

| | Initial Concentration | Change | Equilibrium Concentration |
|---|---|---|---|
| $N_2H_4$ | 0.0 | $+x$ | $x$ |
| $H_3O^+$ | 0.0 | $+x$ | $x$ |
| $N_2H_5^+$ | 0.10 | $-x$ | $0.10 - x \approx 0.10$ |

Substituting equilibrium concentrations into the expression for $K_h$ gives

$$\frac{(x)(x)}{0.10} = 5.9 \times 10^{-9}$$

$$x^2 = 5.9 \times 10^{-10}$$

$$x = 2.4 \times 10^{-5}$$

Therefore,

$$[H_3O^+] = 2.4 \times 10^{-5}$$

and

$$pH = 4.62$$

**SALTS OF WEAK ACIDS AND WEAK BASES: CATION AND ANION HYDROLYSIS.** Solutions of this type of salt can either be acidic, neutral, or basic, because both the cation and the anion of the salt undergo hydrolysis. The pH of such a salt solution is determined by the relative extent of the hydrolysis reactions of each ion. By applying what we have learned from the last two types of salts to these salts, we should be able to predict, at least qualitatively, the pH of their aqueous solutions. If the $K_a$ of the weak acid and the $K_b$ of the weak base of the salt are identical, then the extent of cation and anion hydrolysis is exactly the same (the $K_h$ for the cation is exactly equal to the $K_h$ for the anion) and the solution will be neutral. For example, in the case of $NH_4C_2H_3O_2$, where the $K_b$ of $NH_3$ is $1.8 \times 10^{-5}$ and the $K_a$ of $HC_2H_3O_2$ is $1.8 \times 10^{-5}$, the value of $K_h$ for both ions is $5.6 \times 10^{-10}$, and an aqueous solution of this salt, regardless of the concentration, is neutral. On the other hand, we would predict that an aqueous solution of $NH_4CN$ would be basic because of the relative values of the $K_h$'s of the cation and anion. For $NH_4^+$ we have

$$K_h = \frac{K_w}{K_b} = \frac{1.0 \times 10^{-14}}{1.8 \times 10^{-5}} = 5.6 \times 10^{-10}$$

and for $CN^-$ we have

$$K_h = \frac{K_w}{K_a} = \frac{1.0 \times 10^{-14}}{4.9 \times 10^{-10}} = 2.0 \times 10^{-5}$$

Since the $CN^-$ undergoes more extensive hydrolysis, this means that the equilibrium

$$CN^- + H_2O \rightleftharpoons HCN + OH^-$$

goes further to completion than does the reaction,

$$NH_4^+ + H_2O \rightleftharpoons H_3O^+ + NH_3$$

and the excess $OH^-$ ions make the solution basic.

In a similar fashion, we predict that an aqueous solution of ammonium formate, $NH_4CHO_2$, is acidic because of the relative values of the $K_h$'s for the cation and anion. The $K_h$ for $NH_4^+$ from above is $5.6 \times 10^{-10}$, while the $K_h$ for the formate ion, $CHO_2^-$, is

$$K_h = \frac{K_w}{K_a} = \frac{1.0 \times 10^{-14}}{1.8 \times 10^{-4}}$$
$$= 5.6 \times 10^{-11}$$

Thus the hydrolysis reaction,

$$NH_4^+ + H_2O \rightleftharpoons H_3O^+ + NH_3$$

occurs to a slightly greater extent than does the reaction,

$$CHO_2^- + H_2O \rightleftharpoons HCHO_2 + OH^-$$

Therefore, there is a small excess of $H_3O^+$ ions and solutions of this salt are acidic.

**HYDROLYSIS OF SALTS OF POLYPROTIC ACIDS.** An example of this type of salt is $Na_2S$, the salt of a weak acid, $H_2S$, and a strong base, NaOH. Since only the anion undergoes hydrolysis, the equilibrium reaction for this salt is

$$S^{2-} + H_2O \rightleftharpoons HS^- + OH^-$$

We see that in this reaction another anion, $HS^-$, that can also undergo hydrolysis is produced. Its equilibrium reaction is

$$HS^- + H_2O \rightleftharpoons H_2S + OH^-$$

The hydrolysis constant, $K_{h_1}$, for the first reaction is

$$K_{h_1} = \frac{K_w}{K_{a_2}} = \frac{[HS^-][OH^-]}{[S^{2-}]}$$

where $K_{a_2}$ is the acid dissociation constant for the weak acid, $HS^-$. The equilibrium constant for the second step in the hydrolysis is

$$K_{h_2} = \frac{K_w}{K_{a_1}} = \frac{[H_2S][OH^-]}{[HS^-]}$$

where $K_{a_1}$ in this case is the dissociation constant for the weak acid, $H_2S$. Substituting the values of the $K_a$'s from Table 14.2 into these equations, we obtain

$$K_{h_1} = \frac{1.0 \times 10^{-14}}{1.0 \times 10^{-14}} = 1.0$$

and

$$K_{h_2} = \frac{1.0 \times 10^{-14}}{1.1 \times 10^{-7}} = 9.1 \times 10^{-8}$$

The relative magnitudes of these two equilibrium constants indicate to us that the second hydrolysis reaction occurs to a negligible extent compared to the first; therefore, only $K_{h_1}$ need be used to determine the pH of an aqueous solution of this salt, as shown in our next example.

EXAMPLE 14.15 What is the pH of a 0.20 *M* solution of $Na_2S$?

SOLUTION From our previous discussion we know that only the first hydrolysis reaction

$$S^{2-} + H_2O \rightleftharpoons HS^- + OH^-$$

is important in determining the pH of this solution.

The equation for the hydrolysis constant is

$$K_{h_1} = \frac{[HS^-][OH^-]}{[S^{2-}]} = 1.0$$

As usual, we let $x$ equal the number of moles per liter of $S^{2-}$ that hydrolyze. Then,

| | Initial Concentration | Change | Equilibrium Concentration |
|---|---|---|---|
| $HS^-$ | 0.0 | $+x$ | $x$ |
| $OH^-$ | 0.0 | $+x$ | $x$ |
| $S^{2-}$ | 0.20 | $-x$ | $0.20 - x$ |

Our first reaction might be to neglect $x$ as we have done in problems before. However, if we do so now and solve for $x$, we obtain $x = 0.45$. This is clearly impossible, because the equilibrium $S^{2-}$ concentration becomes $-0.25$ *M*. Negative concentrations are absurd; we can't have less than nothing! Therefore we cannot neglect $x$ and must solve the problem using the quadratic formula (Appendix C).

Substituting values into the equilibrium expression gives

$$\frac{(x)(x)}{(0.20 - x)} = 1.0$$

Multiplying both sides by $(0.20 - x)$ gives

$$x^2 = (0.20 - x)1.0$$

which can be rearranged as

$$x^2 + x - 0.2 = 0$$

The quadratic formula then gives us

$$x = \frac{-1 \pm \sqrt{(1)^2 - 4(1)(-0.2)}}{(2)(1)}$$

$$x = \frac{-1 \pm \sqrt{1.8}}{2} = \frac{-1 \pm 1.34}{2}$$

Notice that two values of $x$ are obtained.

$$x = \frac{-2.34}{2} = -1.17\ M$$

$$x = \frac{0.34}{2} = 0.17\ M$$

The first value of $x$ is absurd. It has no physical meaning because it tells us that the concentration of $HS^-$ and $OH^-$ are negative. As we said before, we cannot have less than nothing. The second value of $x$ is meaningful, and we conclude that

$$x = 0.17\ M$$

and consequently

$$[OH^-] = 0.17\ M$$

from which we obtain

$$pOH = 0.77$$

The pH of the solution, therefore, is 13.23.

## 14.6 ACID-BASE TITRATION: THE EQUIVALENCE POINT

In Chapter 5 we saw that a titration is a useful and accurate way of determining the concentrations of acids and bases, provided that the equivalence point can be detected. The equivalence point, you should remember, occurs when equal numbers of equivalents of acid and base have been combined. In this section we will see how the pH of a solution changes during the course of typical acid-base titrations.

**STRONG ACID-STRONG BASE.** A typical example of a titration of a strong acid with a strong base occurs when 25.00 ml of 0.10 *M* HCl is titrated with 0.10 *M* NaOH. We can mathematically determine the pH throughout the titration by calculating the $H^+$ concentration present in the flask each time a quantity of NaOH is added to the HCl. For example, the number of moles of $H^+$ present in the 25 ml of a 0.10 *M* HCl solution is

$$\left(\frac{0.10 \text{ mole}}{1000 \text{ ml}}\right) \times 25 \text{ ml} = 2.5 \times 10^{-3} \text{ mole of } H^+$$

When 10 ml of the 0.1 *M* NaOH are added, we in fact have added

$$\left(\frac{0.10 \text{ mole}}{1000 \text{ ml}}\right) \times 10 \text{ ml} = 1.0 \times 10^{-3} \text{ mole of } OH^-$$

The neutralization reaction,

$$H^+ + OH^- \longrightarrow H_2O$$

occurs, and the amount of $H^+$ remaining is

$$(2.5 \times 10^{-3}) - (1.0 \times 10^{-3}) = 1.5 \times 10^{-3} \text{ mole of } H^+$$

The molar concentration of $H^+$ is now

$$[H^+] = \frac{1.5 \times 10^{-3} \text{ mole}}{0.035 \text{ liter}} = 4.3 \times 10^{-2}\ M$$

Table 14.3
*Titration of 25 ml of 0.10 M HCl with a 0.10 M NaOH solution*

| Volume of HCl | Volume of NaOH | Volume Total | Moles of $H^+$ | Moles of $OH^-$ | Molarity of Ion in Excess | pH |
|---|---|---|---|---|---|---|
| 25.00 | 0.00 | 25.00 | $2.5 \times 10^{-3}$ | 0 | 0.10 ($H^+$) | 1.0 |
| 25.00 | 10.00 | 35.00 | $2.5 \times 10^{-3}$ | $1.0 \times 10^{-3}$ | $4.3 \times 10^{-2}$ ($H^+$) | 1.4 |
| 25.00 | 24.99 | 49.99 | $2.5 \times 10^{-3}$ | $2.499 \times 10^{-3}$ | $2.0 \times 10^{-5}$ ($H^+$) | 4.7 |
| 25.00 | 25.00 | 50.00 | $2.5 \times 10^{-3}$ | $2.50 \times 10^{-3}$ | 0 | 7.0 |
| 25.00 | 25.01 | 50.01 | $2.5 \times 10^{-3}$ | $2.501 \times 10^{-3}$ | $2.0 \times 10^{-5}$ ($OH^-$) | 9.3 |
| 25.00 | 26.00 | 51.00 | $2.5 \times 10^{-3}$ | $2.60 \times 10^{-3}$ | $2.0 \times 10^{-3}$ ($OH^-$) | 11.3 |
| 25.00 | 50.00 | 75.00 | $2.5 \times 10^{-3}$ | $5.0 \times 10^{-3}$ | $3.3 \times 10^{-2}$ ($OH^-$) | 12.5 |

and the pH is calculated to be 1.4. The concentration of $H^+$ after further additions of NaOH have occurred are summarized in Table 14.3.

Our calculations show that the pH starts increasing slowly at first, then rises rapidly near the equivalence point, and then finally levels off after the equivalence point is reached.

If a graph is drawn of pH versus the volume of base added, we obtain the plot shown in Figure 14.2. The equivalence point occurs, in this case, at a pH of 7. At the equivalence point the solution is neutral because neither of the ions of the salt left in solution (NaCl) undergoes hydrolysis. This is not true, however, for the remaining two types of titrations that we shall consider.

**WEAK ACID-STRONG BASE.** An example quite typical of this type of titration is the titration of 25 ml of a 0.10 *M* $HC_2H_3O_2$ with a 0.10 *M* NaOH solution. Initially, when the acetic acid is the only species present, the $[H^+]$ and pH are calculated in the usual fashion from the $K_a$, and we find that the pH is 2.9. When we begin to add the strong base, NaOH, to this solution, in effect we are making a buffer. For example, when 10 ml of 0.10 *M* NaOH ($1.0 \times 10^{-3}$ mole) are added to the 25 ml of the 0.10 *M* $HC_2H_3O_2$ ($2.5 \times 10^{-3}$ mole), the following net chemical reaction takes place:

$$HC_2H_3O_2 + OH^- \longrightarrow H_2O + C_2H_3O_2^-$$

In this reaction all of the NaOH is neutralized as well as $1.0 \times 10^{-3}$ mole of $HC_2H_3O_2$. In the total volume of 35 ml, we have $(2.5 \times 10^{-3}) - (1.0 \times 10^{-3}) = 1.5 \times 10^{-3}$ mole of unreacted $HC_2H_3O_2$ as well as the $1.0 \times 10^{-3}$

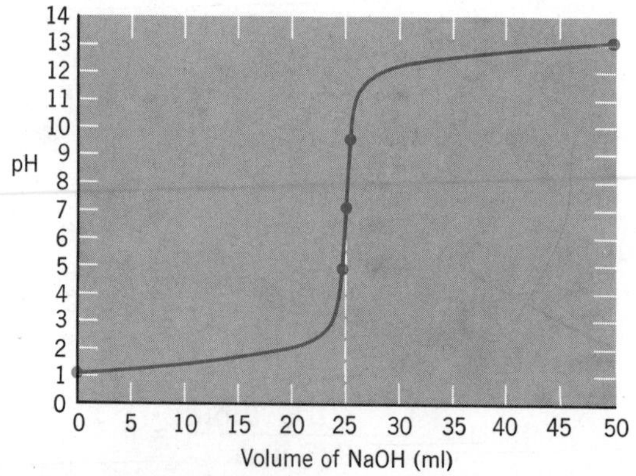


Figure 14.2
*Titration of 0.1 M HCl with 0.1 M NaOH.*

mole of acetate ion produced in the neutralization. The concentration of $HC_2H_3O_2$ and $C_2H_3O_2^-$ are, therefore,

$$[HC_2H_3O_2] = \frac{1.5 \times 10^{-3} \text{ mole}}{0.035 \text{ liter}} = 4.3 \times 10^{-2}\ M$$

$$[C_2H_3O_2^-] = \frac{1.0 \times 10^{-3} \text{ mole}}{0.035 \text{ liter}} = 2.9 \times 10^{-2}\ M$$

If we solve the $K_a$ expression for acetic acid for the $H^+$ concentration and substitute these values for $[HC_2H_3O_2]$ and $[C_2H_3O_2^-]$, we obtain

$$[H^+] = K_a \times \frac{[HC_2H_3O_2]}{[C_2H_3O_2^-]}$$

$$= 1.8 \times 10^{-5} \left(\frac{4.3 \times 10^{-2}}{2.9 \times 10^{-2}}\right)$$

$$= 2.7 \times 10^{-5}\ M$$

Therefore,

$$pH = 4.57$$

From the time of the first addition of base until the equivalence point is reached, the solution contains both acetic acid and acetate ion, and the pH may be computed in this fashion.

When a total of 25 ml of NaOH are added, all of the acetic acid is "neutralized," and we have produced $2.5 \times 10^{-3}$ mole of $NaC_2H_3O_2$ in 50 ml of solution. The resulting 0.050 *M* $NaC_2H_3O_2$ solution undergoes hydrolysis because it contains the anion of a weak acid. We have seen that for this solute the equilibrium is

$$C_2H_3O_2^- + H_2O \rightleftharpoons HC_2H_3O_2 + OH^-$$

From the last section we know that

$$K_h = \frac{[HC_2H_3O_2][OH^-]}{[C_2H_3O_2^-]} = \frac{K_w}{K_a} = 5.6 \times 10^{-10}$$

We can calculate the $OH^-$ concentration in this solution as we did previously by letting $x$ equal the number of moles per liter of $C_2H_3O_2^-$ that hydrolyze. This allows us to construct our table.

| | Initial Concentration | Change | Equilibrium Concentration |
|---|---|---|---|
| $HC_2H_3O_2$ | 0.0 | $+x$ | $x$ |
| $OH^-$ | 0.0 | $+x$ | $x$ |
| $C_2H_3O_2^-$ | 0.050 | $-x$ | $0.050 - x \approx 0.050$ |

Substituting into the $K_h$ expression gives

$$K_h = \frac{(x)(x)}{0.050} = 5.6 \times 10^{-10}$$

$$x^2 = 2.8 \times 10^{-11}$$

$$x = 5.3 \times 10^{-6}$$

Table 14.4
*Titration of 25 ml of 0.10 M $HC_2H_3O_2$ with 0.10 M NaOH*

| Ml of Base Added | Concentration of Species in Excess | pH |
|---|---|---|
| 0.0 | $1.3 \times 10^{-3}$ $(H^+)$ | 2.9 |
| 10.0 | $2.5 \times 10^{-5}$ $(H^+)$ | 4.6 |
| 24.99 | $7.2 \times 10^{-9}$ $(H^+)$ | 8.1 |
| 25.0 | $5.3 \times 10^{-6}$ $(OH^-)$ | 8.7 |
| 25.01 | $2.0 \times 10^{-5}$ $(OH^-)$ | 9.3 |
| 26.0 | $1.9 \times 10^{-3}$ $(OH^-)$ | 11.3 |

This means that $[OH^-] = 5.3 \times 10^{-6}$, from which we obtain

$$pOH = 5.28$$

and finally,

$$pH = 8.72$$

Thus the pH at which the equivalence point occurs is greater than 7. We find that this is true for any weak acid-strong base titration.

Thus far we have discussed only the first half of the titration (see Table 14.4). What takes place beyond the equivalence point? As soon as all of the weak acid has been neutralized, any further addition of NaOH suppresses the hydrolysis of the anion and the pH is then solely dependent on the concentration of $OH^-$ coming from the added NaOH. Thus we generate the last half of Table 14.4 in the same manner as we did Table 14.3 in the HCl/NaOH titration.

A graph of these data is shown in Figure 14.3, where we have plotted pH versus volume of base added. From both Table 14.4 and Figure 14.3, we can see that the change in pH near the equivalence point is not as drastic as in the case of the HCl/NaOH titration. This less rapid change near the equivalence point becomes even more pronounced for weaker acids such as HCN.

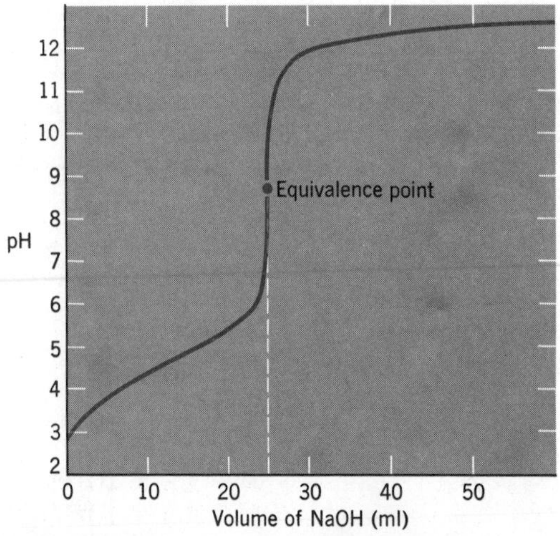


Figure 14.3
*Titration of 25 ml of 0.1 M acetic acid with 0.1 M sodium hydroxide.*

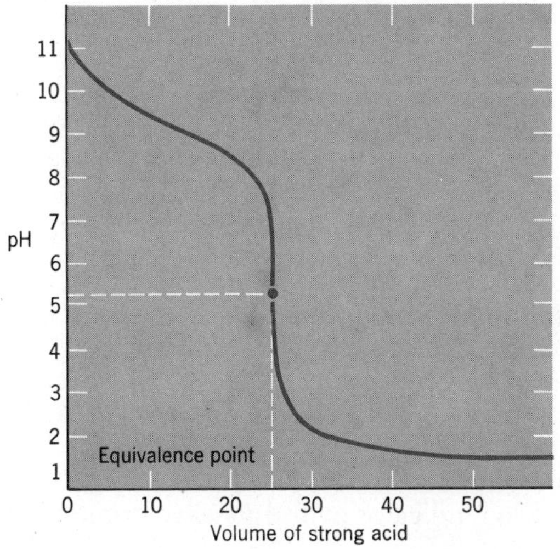


Figure 14.4
*Titration of 25 ml of 0.1 M $NH_3$ with 0.1 M HCl.*

**WEAK BASE-STRONG ACID.** When a weak base is titrated with a strong acid, the titration curve that is generated is very similar in shape to that obtained by reaction of a weak acid with a strong base. During the initial addition of acid the solution contains unreacted weak base and its salt; it therefore constitutes a buffer. At the equivalence point the solution contains the salt of the weak base, and the pH of the mixture is determined by the hydrolysis of the cation. Finally, beyond the equivalence point the pH of the solution is controlled by the excess hydrogen ion from the strong acid. The shape of the titration curve for such a titration is shown in Figure 14.4 for the titration of 25 ml of 0.10 *M* $NH_3$ with 0.10 *M* HCl. We can show that the pH at the equivalence point is less than 7 by considering the hydrolysis of the $NH_4Cl$ produced during the reaction.

From the last section we recall that the $K_h$ for $NH_4^+$ is written as

$$K_h = \frac{[H_3O^+][NH_3]}{[NH_4^+]} = \frac{K_w}{K_b} = 5.6 \times 10^{-10}$$

All of the $NH_3$ is "neutralized" in this titration when exactly 25 ml of HCl ($2.5 \times 10^{-3}$ mole) have been added. At this point, the concentration of $NH_4^+$ is

$$\frac{2.5 \times 10^{-3} \text{ mole}}{0.050 \text{ liter}} = 5.0 \times 10^{-2}\ M$$

If we let x equal the number of moles per liter of $NH_4^+$ that undergo hydrolysis

$$[H_3O^+] = x\ M$$
$$[NH_3] = x\ M$$
$$[NH_4^+] = 5.0 \times 10^{-2} - x = 5.0 \times 10^{-2}\ M$$

Substituting these concentrations into the above equation for $K_h$ gives

$$K_h = \frac{(x)(x)}{5.0 \times 10^{-2}} = 5.6 \times 10^{-10}$$

$$x^2 = 28.0 \times 10^{-12}$$
$$x = 5.3 \times 10^{-6}\ M$$
$$[H_3O^+] = 5.3 \times 10^{-6}\ M$$

and

$$pH = 5.28$$

The pH at the equivalence point of this titration is less than 7, which is typical for all weak base-strong acid titrations.

## 14.7 ACID-BASE INDICATORS

**Indicators** are often used, in very small amounts, to detect the equivalence point in an acid-base titration. They are usually weak organic acids or bases that change color on going from an acidic medium to a basic medium. Not all indicators change color at the same pH, however. The choice of indicator for a particular titration depends on the pH at which the equivalence point is expected to occur. A list of some common indicators, with their color changes and the pH ranges over which the color changes are observed, is found in Table 14.5. Let us examine briefly how these indicators work.

If we denote an indicator by the general formula HIn, we have the dissociation reaction,

$$HIn \rightleftharpoons H^+ + In^-$$

Applying Le Chatelier's principle to this equilibrium, we see that in an acid solution (excess $H^+$) the species that is present in excess is HIn. On the other hand, in basic solutions the equilibrium is shifted to the right and the predominant species is $In^-$. Therefore, HIn is said to be the "acid form" and $In^-$ the "basic form" of the indicator. The ability of HIn to function as an indicator is based on the fact that the acid and basic forms differ in color. For example, with litmus the acid form (HIn) is pink while the basic form ($In^-$) is blue.

The dissociation constant, $K_a$, for an indicator is

$$K_a = \frac{[H^+][In^-]}{[HIn]}$$

Let's solve this for the ratio $[In^-]/[HIn]$.

$$\frac{[In^-]}{[HIn]} = \frac{K_a}{[H^+]}$$

Table 14.5
*Some common indicators*

| Indicator | Color Change | pH Range in Which Color Change Occurs |
|---|---|---|
| Thymol blue | Red to yellow | 1.2–2.8 |
| Bromophenol blue | Yellow to blue | 3.0–4.6 |
| Congo red | Blue to red | 3.0–5.0 |
| Methyl orange | Red to yellow | 3.2–4.4 |
| Bromocresol green | Yellow to blue | 3.8–5.4 |
| Methyl red | Red to yellow | 4.8–6.0 |
| Bromocresol purple | Yellow to purple | 5.2–6.8 |
| Bromothymol blue | Yellow to blue | 6.0–7.6 |
| Cresol red | Yellow to red | 7.0–8.8 |
| Thymol blue | Yellow to blue | 8.0–9.6 |
| Phenolphthalein | Colorless to pink | 8.2–10.0 |
| Alizarin yellow | Yellow to red | 10.1–12.0 |

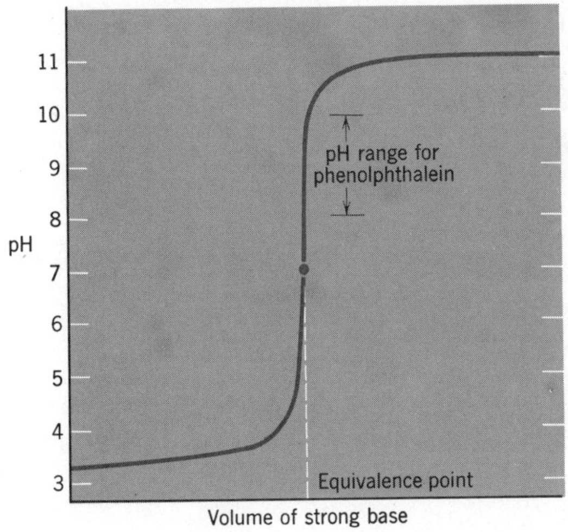


Figure 14.5
*Titration curve for the titration of a strong acid with a strong base.*

We have seen that as we pass through the equivalence point, the pH changes very rapidly. For example, in the NaOH/HCl titration described earlier, the pH changed from 4.7 to 9.3 upon the addition of only 0.02 ml of base, which corresponds to only about one-half drop of solution! This corresponds to a change in $[H^+]$ from $2 \times 10^{-5}\ M$ to $5 \times 10^{-10}\ M$. How does this affect the $[In^-]/[HIn]$ ratio?

Suppose that we were using an indicator whose $K_a = 1 \times 10^{-7}$. Then, before the equivalence point,

$$\frac{[In^-]}{[HIn]} = \frac{1 \times 10^{-7}}{2 \times 10^{-5}} = \frac{1}{200}$$

There is 200 times as much HIn as $In^-$, and the color observed is that due to HIn.

After the equivalence point,

$$\frac{[In^-]}{[HIn]} = \frac{1 \times 10^{-7}}{5 \times 10^{-10}} = \frac{200}{1}$$

Now there is 200 times as much $In^-$ as HIn, and the color that we see is due to $In^-$. Thus, as we pass through the equivalence point, there is a sudden change in the relative amounts of the acid and basic forms of the indicator, which we notice as a change in color.

If the indicator changes color at the equivalence point, the end point (when we observe the color change) occurs at the same pH as the equivalence point. Often, however, we find ourselves using an indicator whose color change takes place at a pH slightly different from that of the equivalence point. This is shown in Figure 14.5 for phenolphthalein. When the color change occurs we have actually gone slightly past the equivalence point.

In choosing our indicator, we wish to have it change color very close to the equivalence point. Phenolphthalein, for example, would be a poor choice of indicator for the titration depicted in Figure 14.4, because its color change would occur before the equivalence point. We would find that we had stopped adding acid before the equivalence point had been reached, thereby defeating the purpose of using an indicator. A better choice would be an indicator such as methyl red, where the center of the color change range occurs very near the pH at the equivalence point.

## INDEX TO QUESTIONS AND PROBLEMS (problem numbers in italics)



## REVIEW QUESTIONS

**14.1** Why can we almost always ignore the $H^+$ contributed by the dissociation of water when we calculate the $H^+$ concentration in solutions containing an acid? Under what conditions would we have to consider the $H^+$ from the dissociation of $H_2O$, even though the solute was an acid?

**14.2** How is pH defined? How is pOH defined? Why does pH + pOH = 14?

**14.3** Identify the following as representing acidic, basic, or neutral solutions:
(a) pH = 3.54 (d) pOH = 10.43
(b) pH = 8.25 (e) pOH = 2.25
(c) pOH = 7.00

**14.4** Arrange the solutions in Question 14.3 in order of increasing acidity.

**14.5** Refer to Table 14.1 and write the appropriate equilibrium constant expressions for the ionization of:
(a) benzoic acid (d) Veronal
(b) hydrazine (e) pyridine
(c) formic acid

**14.6** Write appropriate mass action expressions for $K_{a_1}$ and $K_{a_2}$ for ascorbic acid (vitamin C).

**14.7** Citric acid, which is present in many fruits and vegetables, has the formula, $H_3C_6H_5O_7$. It is a triprotic acid. Write the three equilibria for the dissociation of the acid and write the appropriate equilibrium constant expression for each step.

**14.8** What is a buffer? Explain how the following solutes function as buffers:
(a) $NaCHO_2$ and $HCHO_2$
(b) $C_5H_5N$ and $C_5H_5NHCl$
(c) $NH_4C_2H_3O_2$
(d) $NaHCO_3$

**14.9** What is hydrolysis? Without performing any computations, predict whether the following solutions will be acidic, basic, or neutral:
(a) KCl (c) $NaC_4H_7O_2$
(b) $NH_4NO_3$ (d) $C_6H_5NH_3NO_3$

**14.10** If the concentration of each solute in Question 14.9 were 0.10 *M*, which solution would be most basic? Which would be most acidic?

**14.11** Why is it necessary to consider only the first step in the hydrolysis of a salt such as $Na_2SO_3$?

**14.12** Is it possible to have a pH other than 7 at the equivalence point in an acid-base titration?

**14.13** Explain how an indicator works. Why do we want to use as little of the indicator as possible when we perform a titration?

**14.14** What indicators might be acceptable for the titration depicted in Figure 14.3? Why would we not wish to use congo red as an indicator?

**14.15** Would congo red be an acceptable indicator for the titration depicted in Figure 14.5? Explain your answer.

## REVIEW PROBLEMS

**14.16** Calculate the $H^+$ and $OH^-$ concentrations and the pH of the following solutions of strong acids and bases:
(a) 0.0010 *M* HCl
(b) 0.125 *M* $HNO_3$
(c) 0.0031 *M* NaOH
(d) 0.012 *M* $Ba(OH)_2$
(e) $2.1 \times 10^{-4}$ *M* $HClO_4$
(f) $1.3 \times 10^{-5}$ *M* HCl
(g) $8.4 \times 10^{-3}$ *M* NaOH
(h) $4.8 \times 10^{-2}$ *M* KOH

**14.17** Calculate the $H^+$ and $OH^-$ concentrations in a solution having a pH equal to:
(a) 1.30 (d) 7.80
(b) 5.73 (e) 10.94
(c) 4.00 (f) 12.61

**14.18** What is the pOH of each solution in Problem 14.17?

**14.19** A weak acid has an equilibrium constant $K_a = 3.8 \times 10^{-9}$. What is the p$K_a$ of the acid?

**14.20** A base has p$K_b$ = 3.84. What is $K_b$ of the base?

**14.21** What is the $H^+$ concentration in each of the following solutions:
(a) 0.30 *M* $HNO_2$
(b) 1.00 *M* HF
(c) 0.025 *M* HCN
(d) 0.10 *M* butyric acid
(e) 0.050 *M* barbituric acid

**14.22** Calculate the $OH^-$ concentration in the following solutions:
(a) 0.15 *M* $NH_3$
(b) 0.20 *M* $N_2H_4$
(c) 0.80 *M* $CH_3NH_2$
(d) 0.35 *M* hydroxylamine
(e) 0.010 *M* pyridine

**14.23** What is the $OH^-$ concentration in each solution in Problem 14.21?

**14.24** What is the pH of each solution in Problem 14.22?

**14.25** A 0.25 *M* solution of a monoprotic weak acid was observed to have a pH = 1.35. What is $K_a$ for this acid?

**14.26** A 0.10 *M* solution of a weak acid was found to have a pH = 5.37. What is $K_a$ for the acid?

**14.27** A weak base was found to give a solution with pH = 8.75 when its concentration was 0.10 *M*. What is $K_b$ for the base?

**14.28** What is the percent ionization of the acid in each of the following solutions:
(a) 1.0 *M* formic acid
(b) 0.010 *M* propionic acid
(c) 0.025 *M* HCN
(d) 0.35 *M* nicotinic acid
(e) 0.50 *M* HOCl
(f) 0.25 *M* $HNO_3$

**14.29** Calculate the percent ionization of each of the following acetic acid solutions. What conclusions can you draw? Can you explain on a molecular level why you obtain these results? Can you explain this using Le Chatelier's principle?
(a) 1.00 *M* $HC_2H_3O_2$
(b) 0.10 *M* $HC_2H_3O_2$
(c) 0.010 *M* $HC_2H_3O_2$

**14.30** Calculate the hydrogen ion concentration in each of the following solutions of an acid or base and its salt.
(a) 0.25 *M* $HC_2H_3O_2$, 0.15 *M* $NaC_2H_3O_2$
(b) 0.50 *M* $HCHO_2$, 0.50 *M* $NaCHO_2$
(c) 0.30 *M* $HNO_2$, 0.40 *M* $NaNO_2$
(d) 0.25 *M* $NH_3$, 0.15 *M* $NH_4Cl$
(e) 0.30 *M* $N_2H_4$, 0.50 *M* $N_2H_5NO_3$

**14.31** The pH of a 0.012 *M* solution of a weak base, BOH, was experimentally determined to be 11.40. Calculate $K_b$ for the base.

**14.32** How many grams of HCl gas would have to be dissolved in 500 ml of 1.0 *M* $NaC_2H_3O_2$ to give a solution having a pH = 4.74?

**14.33** Calculate the pH obtained by dissolving a 500-mg tablet of vitamin C in 250 ml (approx. 8 oz) of $H_2O$.

**14.34** In the stomach the fluids have a pH ≈ 1.0 due to the strong acid, HCl. What fraction of the vitamin C in a 500-mg tablet will be dissociated if the volume of fluid in the stomach is 200 ml?

**14.35** If 10 mg of sodium barbituate is swallowed, what fraction is converted to barbituric acid if the pH of the stomach is 1.0 and there are 250 ml of fluid in the stomach?

**14.36** Nicotinic acid is another name for the important vitamin, niacin. What is the pH of a 0.010 *M* solution of nicotinic acid?

**14.37** What is the molarity of a solution of acetic acid whose pH is 2.5?

**14.38** What molar concentration of hydrazine, $N_2H_4$, yields a solution whose pH = 10.64?

**14.39** A 0.010 *M* solution of a weak acid, HA, is found to have a pH of 4.55. What is the value of $K_a$ for this acid?

**14.40** Calculate the concentrations of all species present in a 1.0 *M* $H_3PO_4$ solution.

**14.41** Selenious acid, $H_2SeO_3$, has $K_{a_1} = 3 \times 10^{-3}$ and $K_{a_2} = 5 \times 10^{-8}$. What is the pH of a 0.50 *M* solution of $H_2SeO_3$? What are the equilibrium concentrations of $H_2SeO_3$, $HSeO_3^-$, and $SeO_3^{2-}$?

**14.42** What is the $HCO_3^-$ concentration in a 0.10 *M* solution of $H_2CO_3$ whose pH = 3.00? What is the $CO_3^{2-}$ concentration in this solution?

**14.43** What must the $H^+$ concentration of a saturated $H_2S$ solution be in order to give a sulfide ion concentration of $8.4 \times 10^{-15}$ *M*?

**14.44** What is the sulfide ion concentration in a saturated $H_2S$ solution whose pH is 4.60?

**14.45** What ratio of lactic acid to sodium lactate is required to give a solution having a pH = 4.25?

**14.46** Calculate the pH of each of the following buffers prepared by placing, in 1.0 liter of solution,
(a) 0.10 mole of $NH_3$ and 0.10 mole of $NH_4Cl$
(b) 0.20 mole of $HC_2H_3O_2$ and 0.40 mole of $NaC_2H_3O_2$
(c) 0.15 mole of $N_2H_4$ and 0.10 mole of $N_2H_5Cl$
(d) 0.20 mole of HCl and 0.30 mole of NaCl

**14.47** How many grams of $NaC_2H_3O_2$ must be added to 1.00 mole of $HC_2H_3O_2$ in order to prepare 1.00 liter of a buffer whose pH equals 5.15?

**14.48** What must the ratio of $NH_3$ to $NH_4^+$ be to have a buffer with a pH of 10.0?

***14.49** How many moles of HCl must be added to 1.0 liter of a mixture containing 0.010 *M* $HC_2H_3O_2$ and 0.010 *M* $NaC_2H_3O_2$ in order to give a solution whose pH = 3.0?

**14.50** How much would the pH change if 0.10 mole of HCl were added to 1.0 liter of a formic acid-sodium formate buffer containing 0.45 mole of $HCHO_2$ and 0.55 mole of $NaCHO_2$?

**14.51** How much would the pH change if 0.20 mole of NaOH were added to the buffer in Question 14.50?

**14.52** Determine the pH of each of the following salt solutions:
(a) $1.0 \times 10^{-3}$ *M* $NaC_2H_3O_2$
(b) 0.125 *M* $NH_4Cl$
(c) 0.10 *M* $Na_2CO_3$
(d) 0.10 *M* NaCN
(e) 0.20 *M* $NH_3OHCl$

**14.53** What is the percent hydrolysis of a 0.10 *M* solution of pyridinium chloride, $C_5H_5NHCl$?

**14.54** A 0.10 *M* solution of the sodium salt of a weak acid has a pH of 9.35. What is the $K_a$ of the weak acid?

**14.55** Liquid chlorine bleach is really nothing more than a dilute solution of NaOCl, usually about 5% NaOCl by weight. A particular sample of bleach was found to contain 0.67 mole/liter NaOCl. Calculate the pH of the solution.

**14.56** Veronal, a barbiturate drug, is generally administered as its sodium salt. What is the pH of a solution of $NaC_8H_{11}N_2O_3$ that contains 10 mg of the drug in 250 ml of solution?

**14.57** What would be the concentration of barbituric acid in a 0.0010 *M* solution of sodium barbiturate?

**14.58** What would be the pH at the equivalence point if 25.0 ml of 0.010 *M* barbituric acid is titrated with 0.020 *M* NaOH?

***14.59** When 50 ml of 0.20 *M* HF is titrated with 0.10 *M* NaOH, what is the pH
(a) After 5.0 ml of base has been added?
(b) When half of the HF has been neutralized?
(c) At the equivalence point?

**14.60** Sodium benzoate is often used as a preservative in packaged food products. What would be the pH of a 0.020 *M* solution of sodium benzoate?

**14.61** Plot a curve showing the pH of a solution of 100 ml of 0.10 *M* butyric acid that is gradually neutralized by the addition of solid NaOH. Do this by calculating the pH after addition of 0, 0.001, 0.005, 0.009, 0.010, and 0.011 mole of NaOH. Assume no change in volume. What is the pH at the equivalence point? What indicator in Table 14.5 could be used for this titration?

**14.62** Using the data in Tables 14.1 and 14.5, choose an indicator that is suitable for the titration of:
(a) Acetic acid with sodium hydroxide
(b) Aniline with hydrochloric acid

**14.63** An indicator, HIn, has an ionization constant, $K_a$, equal to $1 \times 10^{-5}$. If the molecular form of the indicator is yellow and the $In^-$ ion is green, what is the color of a solution containing this indicator when its pH is 7.0?

***14.64** Calculate the pH of a $1 \times 10^{-7}$ *M* solution of HCl (*Hint:* The pH does *not* equal 7.)

***14.65** Calculate the $H^+$ concentration in 0.0010 *M* $HC_2H_3O_2$.

***14.66** Calculate the concentration of all species in 0.010 *M* formic acid solution.

***14.67** Calculate the pH of 0.50 *M* $NaHCO_3$. How much will the pH change if 0.05 mole/liter of HCl is added?

***14.68** How many mililiters of 6.0 *M* HCl are required to be added to 100 ml of 0.10 *M* $NaC_2H_3O_2$ to give a solution having a pH = 4.25?

***14.69** A sample of arterial blood was found to contain $2.6 \times 10^{-2}$ mole of dissolved $CO_2$ per liter. The pH of the sample was 7.43. If it is assumed that in solution the $CO_2$ forms $H_2CO_3$, what is the $HCO_3^-$ concentration in this blood sample?

***14.70** Calculate the pH of 0.10 *M* $NH_4NO_2$.

***14.71** Determine the shape of the titration curve when 100 ml of 0.20 *M* $H_2CO_3$ is titrated with 0.10 *M* NaOH. Determine the pH at each equivalence point.

# 15 SOLUBILITY AND COMPLEX ION EQUILIBRIA

In Chapter 14 we studied ionic equilibria involving acids and bases. These are not, however, the only dynamic equilibria that can take place in aqueous solution. In this chapter we will turn our attention to the equilibria involved with salts that have very low solubilities—those that we considered to be insoluble in water in our discussions in Chapter 5. We will also look at equilibria involving species that we call "complex ions"—ions composed of a metal atom surrounded by a number of anions, or neutral molecules, to which the metal is bound.

## 15.1 SOLUBILITY PRODUCT

It was pointed out in Chapter 5 that, according to our list of solubility rules, some salts are considered soluble in water while others are said to be quite insoluble. However, even the most insoluble salts are soluble in water to at least some degree, and nearly all salts completely dissociate when they dissolve in aqueous solutions. Two examples of salts that do not dissociate completely upon dissolving are $HgCl_2$ and $CdSO_4$. Our discussion will not include this type of solid.

When a saturated solution of a salt is prepared, a dynamic equilibrium is established between the dissociated ions and the insoluble solid at the bottom of the container. For example, in a saturated solution of silver chloride we have the equilibrium

$$AgCl\ (s) \rightleftharpoons Ag^+\ (aq) + Cl^-\ (aq)$$

for which we can write

$$K = \frac{[Ag^+][Cl^-]}{[AgCl\ (s)]}$$

In Section 12.6 we saw that the *concentration* of a pure solid is independent of the amount of solid present. In other words, the concentration of the solid is a constant and can therefore be included with the constant $K$, so that

$$K[AgCl\ (s)] = K_{sp} = [Ag^+][Cl^-]$$

The equilibrium constant $K$ multiplied by the concentration of solid AgCl is still another constant called the **solubility product constant,** given the label $K_{sp}$. For example, we can obtain the expression for the $K_{sp}$ of silver acetate from its solubility equilibrium,

$$AgC_2H_3O_2\ (s) \rightleftharpoons Ag^+\ (aq) + C_2H_3O_2^-\ (aq)$$

The equilibrium condition is therefore,

$$K_{sp} = [Ag^+][C_2H_3O_2^-]$$

In the case of an insoluble solid such as $Mg(OH)_2$, the coefficients in the dissociation equilibrium are not all equal to one:

$$Mg(OH)_2\ (s) \rightleftharpoons Mg^{2+}\ (aq) + 2OH^-\ (aq)$$

The $K_{sp}$ for $Mg(OH)_2$ is then given by

$$K_{sp} = [Mg^{2+}][OH^-]^2$$

Thus the solubility product constant is equal to the product of the concentration of the ions produced in a saturated solution, each raised to a power equal to its coefficient in the balanced equation. A list of some ionic solids and their $K_{sp}$'s at temperatures ranging between 18 and 25°C is given in Table 15.1.

Table 15.1
*Solubility product constants*

| Compound | $K_{sp}$ | Compound | $K_{sp}$ |
|---|---|---|---|
| $Al(OH)_3$ | $2 \times 10^{-33}$ | PbS | $7 \times 10^{-27}$ |
| $BaCO_3$ | $8.1 \times 10^{-9}$ | $Mg(OH)_2$ | $1.2 \times 10^{-11}$ |
| $BaCrO_4$ | $2.4 \times 10^{-10}$ | $MgC_2O_4$ | $8.6 \times 10^{-5}$ |
| $BaF_2$ | $1.7 \times 10^{-6}$ | $Mn(OH)_2$ | $4.5 \times 10^{-14}$ |
| $BaSO_4$ | $1.5 \times 10^{-9}$ | MnS | $7 \times 10^{-16}$ |
| CdS | $3.6 \times 10^{-29}$ | $Hg_2Cl_2$ | $2 \times 10^{-18}$ |
| $CaCO_3$ | $9 \times 10^{-9}$ | HgS | $1.6 \times 10^{-54}$ |
| $CaF_2$ | $1.7 \times 10^{-10}$ | NiS | $2 \times 10^{-21}$ |
| $CaSO_4$ | $2 \times 10^{-4}$ | $AgC_2H_3O_2$ | $2.3 \times 10^{-3}$ |
| CoS | $3 \times 10^{-26}$ | $Ag_2CO_3$ | $8.2 \times 10^{-12}$ |
| CuS | $8.5 \times 10^{-36}$ | AgCl | $1.7 \times 10^{-10}$ |
| $Cu_2S$ | $2 \times 10^{-47}$ | AgBr | $5 \times 10^{-13}$ |
| $Fe(OH)_2$ | $2 \times 10^{-15}$ | AgI | $8.5 \times 10^{-17}$ |
| $Fe(OH)_3$ | $1.1 \times 10^{-36}$ | $Ag_2CrO_4$ | $1.9 \times 10^{-12}$ |
| $FeC_2O_4$ | $2.1 \times 10^{-7}$ | AgCN | $1.6 \times 10^{-14}$ |
| FeS | $3.7 \times 10^{-19}$ | $Ag_2S$ | $2 \times 10^{-49}$ |
| $PbCl_2$ | $1.6 \times 10^{-5}$ | $Sn(OH)_2$ | $5 \times 10^{-26}$ |
| $PbCrO_4$ | $1.8 \times 10^{-14}$ | SnS | $1 \times 10^{-26}$ |
| $PbC_2O_4$ | $2.7 \times 10^{-11}$ | $Zn(OH)_2$ | $4.5 \times 10^{-17}$ |
| $PbSO_4$ | $2 \times 10^{-8}$ | ZnS | $1.2 \times 10^{-23}$ |

There are a variety of different kinds of calculations that can be performed having to do with solubility equilibria. They can be summarized as:

1. Calculating $K_{sp}$ from solubility data
2. Calculating solubility from $K_{sp}$
3. Problems dealing with precipitation

We will begin (as you might expect) with the first type of calculation.

EXAMPLE 15.1 It was experimentally determined that at 25°C the solubility of $BaSO_4$ is 0.0091 g/liter. What is the value of $K_{sp}$ for barium sulfate?

SOLUTION From the solubility we can calculate the number of moles of $BaSO_4$ that are dissolved in 1 liter of solution.

$$\left(0.0091 \frac{\text{g}}{\text{liter}}\right) \times \left(\frac{1 \text{ mole}}{233 \text{ g}}\right) = 3.9 \times 10^{-5} \frac{\text{mole}}{\text{liter}}$$

The solubility equilibrium for $BaSO_4$ is

$$BaSO_4\ (s) \rightleftharpoons Ba^{2+}\ (aq) + SO_4^{2-}\ (aq)$$

so that for every mole of $BaSO_4$ that dissolves, 1 mole of $Ba^{2+}$ and 1 mole of $SO_4^{2-}$ are produced. Therefore the molar concentrations of $Ba^{2+}$ and $SO_4^{2-}$ in this saturated solution at 25°C are

$$[Ba^{2+}] = 3.9 \times 10^{-5}\ M$$
$$[SO_4^{2-}] = 3.9 \times 10^{-5}\ M$$

and the $K_{sp}$ would be

$$\begin{aligned} K_{sp} &= [Ba^{2+}][SO_4^{2-}] \\ &= (3.9 \times 10^{-5})(3.9 \times 10^{-5}) \\ &= 1.5 \times 10^{-9} \end{aligned}$$

EXAMPLE 15.2 The solubility of lead iodate, $Pb(IO_3)_2$, is $4.0 \times 10^{-5}$ mole per liter at 25°C. What is $K_{sp}$ for this salt?

SOLUTION First we write the chemical reaction and $K_{sp}$ expression.

$$Pb(IO_3)_2(s) \rightleftharpoons Pb^{2+} + 2IO_3^-$$
$$K_{sp} = [Pb^{2+}][IO_3^-]^2$$

When the $Pb(IO_3)_2$ dissolves, we get 1 mole of $Pb^{2+}$ and 2 moles of $IO_3^-$ for each mole of $Pb(IO_3)_2$. Therefore, when $4.0 \times 10^{-5}$ mole of $Pb(IO_3)_2$ is dissolved in 1 liter, we obtain

$$[Pb^{2+}] = 4.0 \times 10^{-5}\ M$$
$$[IO_3^-] = 2(4.0 \times 10^{-5}) = 8.0 \times 10^{-5}\ M$$

These quantities are now substituted into the $K_{sp}$ expression.

$$\begin{aligned} K_{sp} &= (4.0 \times 10^{-5})(8.0 \times 10^{-5})^2 \\ &= 2.6 \times 10^{-13} \end{aligned}$$

Let us now look at how we can determine solubility from a known value of $K_{sp}$—the second type of problem on our list.

EXAMPLE 15.3 What is the molar solubility of AgCl in water at 25°C?

SOLUTION We are concerned with the equilibrium,

$$AgCl\ (s) \rightleftharpoons Ag^+ + Cl^-$$

for which

$$K_{sp} = [Ag^+][Cl^-] = 1.7 \times 10^{-10}$$

If we let x equal the number of moles per liter of AgCl dissolved, then, because AgCl is completely dissociated in solution, the concentrations of $Ag^+$

and $Cl^-$ are

| | Initial Concentration | Change | Equilibrium Concentration |
|---|---|---|---|
| $Ag^+$ | 0.0 | $+x$ | $x$ |
| $Cl^-$ | 0.0 | $+x$ | $x$ |

Substituting the equilibrium quantities into the $K_{sp}$ expression,

$$K_{sp} = (x)(x) = 1.7 \times 10^{-10}$$
$$x^2 = 1.7 \times 10^{-10}$$
$$x = 1.3 \times 10^{-5}$$

Therefore, the molar solubility (the solubility expressed in moles per liter) of AgCl in water is $1.3 \times 10^{-5}$ *M*.

EXAMPLE 15.4 What are the concentrations of $Ag^+$ and $CrO_4^{2-}$ in a saturated solution of $Ag_2CrO_4$ at 25°C?

SOLUTION $Ag_2CrO_4$ dissolves in water according to the equilibrium,

$$Ag_2CrO_4\,(s) \rightleftharpoons 2Ag^+ + CrO_4^{2-}$$

Thus there are 2 moles of $Ag^+$ produced for every 1 mole of $CrO_4^{2-}$. If $x$ is the number of moles per liter of $Ag_2CrO_4$ that dissolve, then

| | Initial Concentration | Change | Equilibrium Concentration |
|---|---|---|---|
| $Ag^+$ | 0.0 | $+2x$ | $2x$ |
| $CrO_4^{2-}$ | 0.0 | $+x$ | $x$ |

and the $K_{sp}$ expression is

$$K_{sp} = [Ag^+]^2[CrO_4^{2-}] = (2x)^2(x)$$

Substituting the value of $K_{sp}$ (from Table 15.1) into this equation and solving for $x$, we have

$$K_{sp} = (x)(2x)^2 = 1.9 \times 10^{-12}$$
$$x(4x^2) = 4x^3 = 1.9 \times 10^{-12}$$
$$x^3 = 0.48 \times 10^{-12}$$

and

$$x = 7.8 \times 10^{-5}$$

Therefore,

$$[Ag^+] = 2(7.8 \times 10^{-5}) = 1.6 \times 10^{-4}\ M$$
$$[CrO_4^{2-}] = 7.8 \times 10^{-5}\ M$$

We now turn our attention to determining when a precipitate can form in a solution of two salts. You should recall from earlier discussions that a saturated solution is one in which the undissolved solute is in dynamic equilibrium with the solution. This is precisely the situation to which we apply $K_{sp}$. In other words, a saturated solution exists *only* when the **ion product,** that is, *the product of the concentrations of the dissolved ions each raised to its proper power,* is exactly equal to $K_{sp}$. When the ion product is less than $K_{sp}$, the solution is unsaturated, because more salt would have to dissolve in order to raise the ion concentrations to the point where the ion product equals $K_{sp}$. On the other hand, when the ion product exceeds $K_{sp}$, a supersaturated solution exists because some of the salt would have to precipitate in order to lower the ion concentrations until the ion product is equal to $K_{sp}$ once again.

In a solution a precipitate will be formed only when the mixture is supersaturated. Consequently, we may use the value of the ion product in a solution to tell us whether or not precipitation will occur. In summary, we find that

$$\left.\begin{array}{l}\text{Ion product} < K_{sp} \\ \text{Ion product} = K_{sp}\end{array}\right\} \quad \text{no precipitate will form}$$

$$\text{Ion product} > K_{sp} \quad \text{precipitation will occur}$$

**EXAMPLE 15.5** Will a precipitate of $PbSO_4$ form when 100 ml of a 0.0030 *M* $Pb(NO_3)_2$ solution is mixed with 400 ml of 0.040 *M* $Na_2SO_4$?

**SOLUTION** For a saturated solution of $PbSO_4$ we would have

$$K_{sp} = [Pb^{2+}][SO_4^{2-}]$$

To determine whether a precipitate will form, we must calculate the concentration of $Pb^{2+}$ and $SO_4^{2-}$ in our total volume of 500 ml. The 100 ml of the 0.0030 *M* $Pb(NO_3)_2$ solution contains

$$(0.1 \text{ liter}) \times \left(0.0030 \frac{\text{mole}}{\text{liter}}\right) = 0.00030 \text{ mole of } Pb^{2+}$$

(This solution also contains 0.0006 mole of $NO_3^-$, but this species is unimportant in this calculation.)

The 400 ml of $Na_2SO_4$ contains

$$(0.4 \text{ liter}) \times \left(0.040 \frac{\text{mole}}{\text{liter}}\right) = 0.016 \text{ mole of } SO_4^{2-}$$

(This solution also contains 0.032 mole of $Na^+$, but it too is unimportant in this problem.)

The concentration of the $Pb^{2+}$ in the 500 ml is then

$$\frac{0.00030 \text{ mole}}{0.5 \text{ liter}} = 0.00060 = 6.0 \times 10^{-4}\ M$$

and the concentration of $SO_4^{2-}$ in the 500 ml is

$$\frac{0.016 \text{ mole}}{0.5 \text{ liter}} = 0.032 = 3.2 \times 10^{-2}\ M$$

The ion product in the final solution is therefore

$$[Pb^{2+}][SO_4^{2-}] = (6.0 \times 10^{-4})(3.2 \times 10^{-2}) = 1.9 \times 10^{-5}$$

When we compare the ion product to the $K_{sp}$ for $PbSO_4$ ($2 \times 10^{-8}$), we find that the ion product is greater than $K_{sp}$ and therefore a precipitate will form.

From the solubility rules presented in Chapter 5 we know that it is possible to separate certain ions from each other when they are present together in solution. For instance, the addition of chloride ion to a solution containing both $Na^+$ and $Ag^+$ yields a precipitate of AgCl, thereby removing most of the $Ag^+$ from the mixture. In this case one possible product, NaCl, is soluble while the other, AgCl, is quite insoluble.

Even when both products are "insoluble," it is still frequently possible to achieve some degree of separation. Consider, for example, the salts $CaSO_4$ and $BaSO_4$. Although both have very low solubilities, as evidenced by their respective $K_{sp}$'s, we compute $CaSO_4$ to be about 1000 times more soluble, on a mole basis, than $BaSO_4$. As a result, if we had a solution containing equal concentrations of $Ca^{2+}$ and $Ba^{2+}$, we would find that as the $SO_4^{2-}$ concentration was increased in the solution, $BaSO_4$ would precipitate first. Conceivably, then, one could separate $Ca^{2+}$ and $Ba^{2+}$ by appropriately adjusting the $SO_4^{2-}$ concentration so that the $Ca^{2+}$ would remain in solution while nearly all of the $Ba^{2+}$ would be removed as $BaSO_4$. This general concept is used often in the separation of ions in qualitative analysis.

When the anion employed in a separation is derived from a weak acid, it is possible to control its concentration by appropriately adjusting the hydrogen ion concentration. This is illustrated for the selective precipitation of metal sulfides in Example 15.6.

EXAMPLE 15.6 A solution containing 0.1 *M* $Sn^{2+}$ and 0.1 *M* $Zn^{2+}$ is saturated with $H_2S$ ($[H_2S] = 0.1\ M$). What values of the hydrogen ion concentration will allow only one of these ions to be precipitated as its sulfide?

SOLUTION From Table 15.1 we have

$$\begin{array}{ll} \text{SnS} & K_{sp} = 1 \times 10^{-26} \\ \text{ZnS} & K_{sp} = 1.2 \times 10^{-23} \end{array}$$

In this problem the sulfide ion concentration must be controlled so that one ion will precipitate while the other remains in solution. Let us, therefore, calculate for each salt the value of $[S^{2-}]$ that will make the ion product equal to $K_{sp}$. For tin we have

$$K_{sp} = [Sn^{2+}][S^{2-}] = 1 \times 10^{-26}$$

Substituting the $Sn^{2+}$ concentration into the expression gives

$$(0.1)[S^{2-}] = 1 \times 10^{-26}$$
$$[S^{2-}] = 1 \times 10^{-25}\ M$$

In a similar fashion for zinc, we obtain

$$[S^{2-}] = 1.2 \times 10^{-22}\ M$$

These numbers tell us that if the sulfide ion concentration is *greater* than $1 \times 10^{-25}$ but *less than or equal to* $1.2 \times 10^{-22}$, only SnS will precipitate.

We saw in Equation 14.6 that the sulfide ion concentration is directly related to the hydrogen ion concentration; that is,

$$\frac{[H^+]^2[S^{2-}]}{[H_2S]} = K_{a_1}K_{a_2} = 1.1 \times 10^{-21}$$

We can use this expression to calculate the $[H^+]$ that gives us our desired $[S^{2-}]$.

For the lower limit, $[S^{2-}] = 1 \times 10^{-25}$. Since the solution is saturated with $H_2S$, we have $[H_2S] = 0.1\ M$. Substituting gives

$$\frac{[H^+]^2(1 \times 10^{-25})}{(0.1)} = 1.1 \times 10^{-21}$$

$$[H^+]^2 = \frac{(1.1 \times 10^{-21})(0.1)}{1 \times 10^{-25}}$$

$$= 1.1 \times 10^3$$

$$[H^+] = 3.3 \times 10^1 = 33\ M$$

This calculation implies that in order to prevent SnS from precipitating, the $H^+$ concentration must be 33 *M*. This concentration is impossible to achieve; therefore, SnS *must* precipitate.

To prevent ZnS from forming, the $S^{2-}$ concentration cannot be larger than $1.2 \times 10^{-22}\ M$. Using this value for $[S^{2-}]$ in Equation 14.6, we have

$$\frac{[H^+]^2(1.2 \times 10^{-22})}{0.1} = 1.1 \times 10^{-21}$$

$$[H^+]^2 = 0.92$$

$$[H^+] = 0.96\ M$$

Thus, when $[H^+] = 0.96\ M$, the $S^{2-}$ concentration will be $1.2 \times 10^{-22}\ M$, the highest value it can have without causing the $Zn^{2+}$ to precipitate. A hydrogen ion concentration *greater* than 0.96 *M* will produce a sulfide ion concentration *less* than $1.2 \times 10^{-22}\ M$. In summary, then, in order to achieve a separation,

$$[H^+] < 33\ M$$

and

$$[H^+] \geqslant 0.96\ M$$

## 15.2 COMMON ION EFFECT AND SOLUBILITY

In the last section we discussed the solubility of salts when only one salt was present in the solution. We now turn our attention to the effect of other dissolved ions on the solubility of an insoluble salt. If one of these foreign ions is the same as one of the ions produced by the salt, it is called a **common ion.** For example, when NaCl is added to a solution of AgCl, the common ion is $Cl^-$. The equilibrium that exists between solid AgCl and its ions can be represented as

$$AgCl\ (s) \rightleftharpoons Ag^+\ (aq) + Cl^-\ (aq)$$

and the $K_{sp}$, from Table 15.1, is $1.7 \times 10^{-10}$. When additional $Cl^-$ ions are added to this system in the form of NaCl, for example, the concentrations of $Ag^+$ and $Cl^-$ must, according to Le Chatelier's principle, shift so as to minimize the effect of this added ion.[1] In this case some of the excess $Cl^-$ ions are picked up by the $Ag^+$, and more solid AgCl is produced. Thus, in order to preserve the value of the $K_{sp}$, the $Ag^+$ concentration must decrease whenever the $Cl^-$ concentration increases and vice versa. The net result is that the salt becomes less soluble with the addition of a common ion. This phe-

[1] A similar effect was seen in the section on buffers in Chapter 14 when a common ion was added to the equilibrium dissociation of a weak acid.

nomenon, called the **common ion effect**, can be examined quantitatively as shown by the next two examples.

EXAMPLE 15.7 What is the solubility of AgCl in a 0.010 $M$ solution of NaCl?

SOLUTION For this salt we have

$$K_{sp} = [Ag^+][Cl^-] = 1.7 \times 10^{-10}$$

Before any AgCl dissolves we have an initial $Cl^-$ concentration of 0.010 $M$. We can ignore the $Na^+$ because it is not involved in the equilibrium. We now let $x$ equal the number of moles per liter of AgCl that dissolve. This increases both $[Cl^-]$ and $[Ag^+]$ by $x$. Thus,

| | Initial Concen-tration | Change | Equilibrium Concentration |
|---|---|---|---|
| $Ag^+$ | 0.0 | $+x$ | $x$ |
| $Cl^-$ | 0.010 | $+x$ | $0.010 + x \approx 0.010$ |

Note that we have assumed that we can neglect $x$ in computing the equilibrium $Cl^-$ concentration. We make this assumption because the value of $K_{sp}$ is very small. In doing so we greatly simplify the algebra. Substituting the equilibrium concentrations into the expression for $K_{sp}$ gives

$$(x)(0.010) = 1.7 \times 10^{-10}$$

or

$$x = 1.7 \times 10^{-8}\ M$$

Thus, because of the way we defined $x$, the molar solubility of AgCl is $1.7 \times 10^{-8}\ M$. We might compare this to the molar solubility of AgCl in pure water, which we found in Example 15.3 to be $1.3 \times 10^{-5}\ M$. The solubility of AgCl is indeed much less in a solution containing a common ion.

EXAMPLE 15.8 What is the solubility of $Mg(OH)_2$ in 0.10 $M$ NaOH?

SOLUTION The $K_{sp}$ for $Mg(OH)_2$ is

$$K_{sp} = [Mg^{2+}][OH^-]^2$$

Before any $Mg(OH)_2$ dissolves we have $[Mg^{2+}] = 0.0\ M$ and $[OH^-] = 0.10\ M$. We then let $x$ equal the number of moles per liter of $Mg(OH)_2$ that go into solution. This increases the $[Mg^{2+}]$ by $x$ and the $[OH^-]$ by $2x$.

| | Initial Concen-tration | Change | Equilibrium Concentration |
|---|---|---|---|
| $Mg^{2+}$ | 0.0 | $+x$ | $x$ |
| $OH^-$ | 0.10 | $+2x$ | $0.10 + 2x \approx 0.10$ |

Again, we simplify the algebra by assuming that $2x$ is negligible compared to 0.10. Substituting the equilibrium concentrations and the $K_{sp}$ for $Mg(OH)_2$ from Table 15.1 into the solubility product expression gives

$$(x)(0.10)^2 = 1.2 \times 10^{-11}$$

or

$$x = \frac{1.2 \times 10^{-11}}{(0.10)^2}$$
$$= 1.2 \times 10^{-9}\ M$$

Thus $1.2 \times 10^{-9}$ mole per liter of $Mg(OH)_2$ dissolves in a 0.10 $M$ solution of NaOH.

## 15.3 COMPLEX IONS

Many metal ions, particularly those of the transition elements, are able to combine with one or more other molecules or ions to produce more complex species that are called **complex ions.** The substances that combine with the metal ion are called **ligands** and are usually Lewis bases. They can be either (a) neutral molecules such as $H_2O$ and $NH_3$, (b) monatomic anions such as $Cl^-$ and $Br^-$, or (c) polyatomic anions such as $CN^-$ and $C_2O_4^{2-}$. One example of a complex ion that we saw in Chapter 13 is $Al(H_2O)_6^{3+}$. Another example, containing fewer ligands, is formed when $NH_3$ is added to a solution containing $Ag^+$. Its formula is $Ag(NH_3)_2^+$. The charge on a complex ion such as this is the algebraic sum of the charges of the metal ion and the ligands. Thus $Ag^+$ also forms a complex ion with $CN^-$ having the formula $Ag(CN)_2^-$.

We will discuss the details of structure and bonding of complex ions in Chapter 20. For now we will focus our attention on their dissociation equilibria and the effect that their formation has on the solubility of salts.

There are two ways of dealing with the equilibria involving complex ions. One way is to consider their dissociation equilibria. For example, the overall reaction for the equilibrium dissociation of $Ag(NH_3)_2^+$ can be written as

$$Ag(NH_3)_2^+ \rightleftharpoons Ag^+ + 2NH_3$$

Table 15.2
*Instability constants and formation constants at 25°C*

| Complex Ion | $K_{inst}$ | $K_{form}$ |
|---|---|---|
| $AlF_6^{3-}$ | $1.5 \times 10^{-20}$ | $6.7 \times 10^{19}$ |
| $Cd(CN)_4^{2-}$ | $1.3 \times 10^{-17}$ | $7.7 \times 10^{16}$ |
| $Co(NH_3)_6^{2+}$ | $1.3 \times 10^{-5}$ | $7.7 \times 10^{4}$ |
| $Co(NH_3)_6^{3+}$ | $2.0 \times 10^{-34}$ | $5.0 \times 10^{33}$ |
| $Cu(NH_3)_4^{2+}$ | $2.1 \times 10^{-13}$ | $4.8 \times 10^{12}$ |
| $Cu(CN)_2^-$ | $1.0 \times 10^{-16}$ | $1.0 \times 10^{16}$ |
| $Fe(CN)_6^{4-}$ | $1.0 \times 10^{-35}$ | $1.0 \times 10^{35}$ |
| $Fe(CN)_6^{3-}$ | $1.1 \times 10^{-42}$ | $9.1 \times 10^{41}$ |
| $Ni(NH_3)_4^{2+}$ | $1.1 \times 10^{-8}$ | $9.1 \times 10^{7}$ |
| $Ni(NH_3)_6^{2+}$ | $2.0 \times 10^{-9}$ | $5.0 \times 10^{8}$ |
| $Ag(NH_3)_2^+$ | $6.0 \times 10^{-8}$ | $1.7 \times 10^{7}$ |
| $Ag(CN)_2^-$ | $1.9 \times 10^{-19}$ | $5.3 \times 10^{18}$ |
| $Zn(OH)_4^{2-}$ | $3.6 \times 10^{-16}$ | $2.8 \times 10^{15}$ |

The equilibrium constant for this reaction is called an **instability constant.** This is because the larger the value of $K_{inst}$, the less stable is the complex as reflected by its tendency to dissociate. For the $Ag(NH_3)_2^+$ ion the equilibrium expression is

$$K_{inst} = \frac{[Ag^+][NH_3]^2}{[Ag(NH_3)_2^+]}$$

The value of the instability constant for this complex has been found to be $6.0 \times 10^{-8}$. We can see by the size of this constant that this particular complex is quite stable and will readily form whenever $Ag^+$ and $NH_3$ are added to the same solution. Other examples of complex ions and their instability constants can be seen in Table 15.2.

An alternative way of writing the equilibrium for a complex ion is as an equation representing its formation. For example,

$$Ag^+ + 2NH_3 \rightleftharpoons Ag(NH_3)_2^+$$

The equilibrium expression, of course, is simply the reciprocal of the $K_{inst}$ expression. In this case the equilibrium constant (which equals the reciprocal of $K_{inst}$) is called a **formation constant** or **stability constant.** These are also shown in Table 15.2.

$$K_{form} = \frac{[Ag(NH_3)_2^+]}{[Ag^+][NH_3]^2}$$

$$K_{form} = \frac{1}{K_{inst}}$$

In the chemical literature the equilibrium constants for complex ions are sometimes tabulated as instability constants and at other times as formation constants or stability constants. You should know the difference between them.

## 15.4 COMPLEX IONS AND SOLUBILITY

When a complex ion is formed in a solution of an insoluble salt, it reduces the concentration of free metal ion. As a result, more solid must dissolve in order to replenish the amount of metal ion lost, until that concentration required by the $K_{sp}$ of the salt is achieved. Thus the solubility of an insoluble salt generally increases when complex ions are formed. To see this more clearly, let us see what effect adding $NH_3$ has on a saturated solution of AgCl. Before any $NH_3$ is added, we have the equilibrium,

$$AgCl\ (s) \rightleftharpoons Ag^+ + Cl^-$$

Because $NH_3$ forms such a stable complex with the free silver ion, when $NH_3$ is added to this system a second equilibrium is established, namely,

$$Ag^+ + 2NH_3 \rightleftharpoons Ag(NH_3)_2^+$$

The creation of this new equilibrium upsets the first by removing some of the $Ag^+$, thereby causing the first equilibrium to shift to the right. As a result, some of the solid AgCl dissolves.

We can express the two equilibrium reactions of our example as one overall reaction obtained by simply adding together the two equilibrium reactions.

$$\begin{array}{r} AgCl\ (s) \rightleftharpoons Ag^+ + Cl^- \\ Ag^+ + 2NH_3 \rightleftharpoons Ag(NH_3)_3^+ \\ \hline AgCl\ (s) + 2NH_3 \rightleftharpoons Ag(NH_3)_2^+ + Cl^- \end{array}$$

The equilibrium constant for this overall reaction is

$$K_{eq} = \frac{[Ag(NH_3)_2^+][Cl^-]}{[NH_3]^2}$$

We can obtain this same expression by multiplying the $K_{sp}$ of AgCl by the $K_{form}$ of the complex ion.[2] Thus,

$$K_{sp} \times K_{form} = [Ag^+][Cl^-] \times \frac{[Ag(NH_3)_2^+]}{[Ag^+][NH_3]^2} = K_{eq}$$

Therefore, with a knowledge of $K_{sp}$ of the salt, $K_{form}$ of the complex ion (or $K_{inst}$), and the concentration of $NH_3$, it is possible to calculate the concentrations of $Ag^+$ and $Cl^-$ present at equilibrium and thus determine the solubility of AgCl in $NH_3$, as shown by the next example.

**EXAMPLE 15.9** What is the molar solubility of AgCl in 1 liter of 1 *M* $NH_3$ at 25°C?

**SOLUTION** As we have seen, the overall equilibrium reaction for this problem is

$$AgCl\,(s) + 2NH_3 \rightleftharpoons Ag(NH_3)_2^+ + Cl^-$$

for which we write

$$K_{eq} = \frac{[Ag(NH_3)_2^+][Cl^-]}{[NH_3]^2}$$

where

$$K_{eq} = K_{sp} \times K_{form} = (1.7 \times 10^{-10}) \times (1.7 \times 10^{7}) = 2.9 \times 10^{-3}$$

If we let $x$ equal the number of moles per liter of AgCl that dissolves, then we have the following initial and equilibrium concentrations:

| | Initial Concentration | Change | Equilibrium Concentration |
|---|---|---|---|
| $NH_3$ | 1 *M* | $-2x$ | $(1 - 2x)$ |
| $Ag(NH_3)_2^+$ | 0.0 | $+x$ | $x$ |
| $Cl^-$ | 0.0 | $+x$ | $x$ |

Substituting the concentrations at equilibrium into the $K_{eq}$ equation, we have

$$K_{eq} = \frac{(x)(x)}{(1-2x)^2} = \frac{x^2}{(1-2x)^2} = 2.9 \times 10^{-3}$$

Taking the square root of both sides, we have

$$\frac{x}{1-2x} = 5.4 \times 10^{-2}$$

[2] In general, if some equilibrium equation is obtained as the sum of two or more equations, the $K_{eq}$ for the final equation is equal to the *product* of the $K$'s of the equations that were added together.

from which we obtain

$$x = 0.049$$

Therefore, we find that 0.049 mole of AgCl will dissolve in 1 liter of 1 $M$ $NH_3$.

In Example 15.9 we assumed that when the AgCl dissolves in the ammonia solution, essentially all of the $Ag^+$ becomes complexed by $NH_3$. In other words, we said that the chloride ion concentration was equal to the concentration of $Ag(NH_3)_2^+$. Note that this assumption is valid only if $K_{form}$ is very large, indicating that the complex is very stable.

EXAMPLE 15.10 How many moles of solid NaOH must be added to 1.0 liter of $H_2O$ in order to dissolve 0.10 mole of $Zn(OH)_2$ according to the reaction,

$$Zn(OH)_2 + 2OH^- \rightleftharpoons Zn(OH)_4^{2-}$$

SOLUTION The applicable equilibrium constants are

$$\begin{array}{ll} Zn(OH)_2 & K_{sp} = 4.5 \times 10^{-17} \\ Zn(OH)_4^{2-} & K_{inst} = 3.6 \times 10^{-16} \end{array}$$

The two equilibria involved in this system are

$$Zn(OH)_2\ (s) \rightleftharpoons Zn^{2+} + 2OH^- \qquad K_{sp} = 4.5 \times 10^{-17}$$

$$Zn^{2+} + 4OH^- \rightleftharpoons Zn(OH)_4^{2-} \qquad K_{form} = \frac{1}{3.6 \times 10^{-16}} = 2.8 \times 10^{15}$$

As before, the overall reaction can be written as the sum of these two equilibria,

$$Zn(OH)_2\ (s) + 2OH^- \rightleftharpoons Zn(OH)_4^{2-}$$

for which

$$\begin{aligned} K_{eq} &= K_{sp} \times K_{form} = (4.5 \times 10^{-17})(2.8 \times 10^{15}) \\ &= 1.3 \times 10^{-1} \end{aligned}$$

Therefore,

$$\frac{[Zn(OH)_4^{2-}]}{[OH^-]^2} = 1.3 \times 10^{-1}$$

In this problem we know that 0.10 mole of Zn goes into solution where it is present as either free $Zn^{2+}$ or $Zn(OH)_4^{2-}$. Because $K_{form}$ is so very large, essentially all of the Zn will be present as the complex ion; therefore we can write

$$[Zn(OH)_4^{2-}] = 0.10\ M$$

Substituting this into the equilibrium expression gives

$$1.3 \times 10^{-1} = \frac{0.10}{[OH^-]^2}$$

and

$$[OH^-]^2 = \frac{0.10}{1.3 \times 10^{-1}} = 7.7 \times 10^{-1}$$

or

$$[OH^-] = 0.88\ M$$

This corresponds to the equilibrium concentration of free $OH^-$. In this solution, however, we also have 0.10 mole of $Zn(OH)_4^{2-}$, which contains an additional 0.40 mole of $OH^-$, 0.20 mole of which was contained in the original 0.10 mole of $Zn(OH)_2$ that dissolved. Consequently, the total number of moles of NaOH that must be added to the water is 0.88 + 0.20 = 1.08 moles.

## INDEX TO QUESTIONS AND PROBLEMS (problem numbers in italics)

Solubility product 1, 2, 3

$K_{sp}$ and solubility calculations *11, 12, 13, 14, 15, 16, 17, 18, 19, 20, 21, 22, 23, 45*

$K_{sp}$ and precipitation 4, *30, 31, 32, 33, 37, 38, 42*

Selective precipitation *34, 35, 36, 49*

Common ion effect 8, *24, 25, 26, 27, 28, 29, 41, 43, 44, 46, 47, 48*

Complex ions 5, 6, 7, 9, 10

Complex ions and solubility *39, 40*

## REVIEW QUESTIONS

**15.1** Why can the concentration of the solid be left out of the equilibrium expression for the solubility of a salt?

**15.2** Write the $K_{sp}$ expression for each of the following substances:
(a) $Ag_2S$ (d) $MgC_2O_4$
(b) $CaF_2$ (e) $Bi_2S_3$
(c) $Fe(OH)_3$ (f) $BaCO_3$

**15.3** Write the $K_{sp}$ expression for these salts:
(a) $PbF_2$ (d) $Li_2CO_3$
(b) $Cu_2S$ (e) $Ca(IO_3)_2$
(c) $Fe_3(PO_4)_2$ (f) $Ag_2Cr_2O_7$

**15.4** What condition must be met in order to have a precipitate form in a solution?

**15.5** What is a complex ion? What is a ligand? What kinds of substances are found as ligands?

**15.6** On the basis of Le Chatelier's principle, explain why the addition of solid $NH_4Cl$ to a beaker containing solid $Mg(OH)_2$ in contact with water causes the $Mg(OH)_2$ to dissolve.

**15.7** Silver forms a relatively stable complex ion, $AgI_2^-$. When a solution containing this ion is diluted with water, AgI precipitates. Explain why this happens in terms of the equilibria that are involved.

**15.8** What is the common ion effect?

**15.9** Write equilibrium expressions corresponding to $K_{form}$ for the complex ions:
(a) $AgCl_2^-$
(b) $Ag(S_2O_3)_2^{3-}$
(c) $Zn(NH_3)_4^{2+}$

**15.10** Write equilibrium expressions corresponding to $K_{inst}$ for the complex ions:
(a) $Fe(CN)_6^{4-}$
(b) $CuCl_4^{2-}$
(c) $Ni(NH_3)_6^{2+}$

## REVIEW PROBLEMS

**15.11** The solubility of CuCl in water is $1.0 \times 10^{-3}$ mole/liter. What is its value of $K_{sp}$?

**15.12** The molar solubility of $PbCO_3$ is $1.8 \times 10^{-7}$ mole/liter. What is $K_{sp}$ for $PbCO_3$?

**15.13** The solubility of barium oxalate, $BaC_2O_4$, is 0.0781 g/liter. Calculate $K_{sp}$ for $BaC_2O_4$.

**15.14** The molar solubility of $CaCrO_4$ is $1.0 \times 10^{-2}$ mole/liter. What is $K_{sp}$ for $CaCrO_4$?

**15.15** The solubility of lead iodide, $PbI_2$, is $1.4 \times 10^{-3}$ mole/liter. Calculate its $K_{sp}$.

**15.16** A student determined that 0.0981 g of $PbF_2$ was dissolved in 200 ml of saturated $PbF_2$ solution. What is $K_{sp}$ for $PbF_2$?

**15.17** The solubility of $MgF_2$ is $7.6 \times 10^{-2}$ g/liter. Calculate $K_{sp}$ for this salt.

**15.18** The solubility of $Bi_2S_3$ is $2.5 \times 10^{-12}$ g/liter. What is $K_{sp}$ for $Bi_2S_3$?

**15.19** The pH of a saturated solution of $Ni(OH)_2$ is 8.83. Calculate $K_{sp}$ for $Ni(OH)_2$.

**15.20** Using the data in Table 15.1, calculate the molar solubility of each of the following:
(a) PbS
(b) $Fe(OH)_2$
(c) $BaSO_4$
(d) $Hg_2Cl_2$ (which yields $Hg_2{}^{2+}$ and $2Cl^-$)
(e) $Al(OH)_3$
(f) $MgC_2O_4$

**15.21** Milk of magnesia is a suspension of solid $Mg(OH)_2$ in water. Calculate the pH of the aqueous phase, assuming that it is saturated with $Mg(OH)_2$.

**15.22** How many grams of $CaSO_4$ will dissolve in 600 ml of water?

**15.23** What volume of saturated HgS solution contains a single $Hg^{2+}$ ion?

**15.24** What is the molar solubility of $CaCO_3$ in 0.50 *M* $Na_2CO_3$?

**15.25** What is the molar solubility of AgCl in 0.020 *M* $AlCl_3$? Assume that $AlCl_3$ gives $Al^{3+}$ and $Cl^-$ in solution.

* **15.26** What is the molar solubility of $PbCl_2$ in 0.020 *M* $AlCl_3$? Assume that $AlCl_3$ gives $Al^{3+}$ and $Cl^-$ in solution.

**15.27** How many moles of $Ag_2CrO_4$ will dissolve in 1 liter of 0.10 *M* $AgNO_3$?

**15.28** What is the molar solubility of $CaF_2$ in 0.010 *M* NaF?

**15.29** How many grams of NaF must be added to 1 liter of solution to reduce the molar solubility of $BaF_2$ to $6.8 \times 10^{-4}$ mole/liter?

**15.30** Would a precipitate form in the following solutions?
(a) $5 \times 10^{-2}$ mole of $AgNO_3$ and $1.0 \times 10^{-3}$ mole of $NaC_2H_3O_2$ dissolved in 1.0 liter of solution
(b) $1.0 \times 10^{-2}$ mole of $Ba(NO_3)_2$ and $2.0 \times 10^{-2}$ mole of NaF dissolved in 1.0 liter of solution
(c) 500 ml of $1.4 \times 10^{-2}$ M $CaCl_2$ and 250 ml of 0.25 *M* $Na_2SO_4$ mixed to give a final volume of 750 ml

**15.31** What is the minimum pH necessary to cause a precipitate of $Fe(OH)_2$ to form in a 0.010 *M* $FeCl_2$ solution?

**15.32** A solution is prepared by mixing 100 ml of 0.20 *M* $AgNO_3$ with 100 ml of 0.10 *M* HCl. What are the concentrations of all species present in the solution when equilibrium is reached?

**15.33** Will a precipitate form in a solution containing:
(a) 0.025 *M* $CaCl_2$ and 0.0050 *M* $Na_2CO_3$
(b) 0.010 *M* $Pb(NO_3)_2$ and 0.030 *M* $CaCl_2$
(c) $1.5 \times 10^{-3}$ *M* $FeCl_2$ and $2.2 \times 10^{-3}$ *M* $Na_2C_2O_4$

**15.34** Which will precipitate first when $Na_2CrO_4$ (*s*) is gradually added to a solution containing 0.010 *M* $Pb^{2+}$ and 0.010 *M* $Ba^{2+}$? What will be the concentration of the ion precipitated first when the other ion just begins to form a precipitate?

**15.35** A solution is known to contain 0.010 *M* $Pb^{2+}$ and 0.010 *M* $Ni^{2+}$. How must the pH be adjusted to achieve the maximum separation when the solution is saturated with $H_2S$?

**15.36** A solution containing 0.10 *M* $Zn^{2+}$ and 0.10 *M* $Fe^{2+}$ is saturated with $H_2S$. What must the $H^+$ concentration be to separate these ions by selectively precipitating ZnS? What is the smallest $Zn^{2+}$ concentration that can be achieved without precipitating any of the $Fe^{2+}$ as FeS?

**15.37** What would the $H^+$ concentration have to be in order to prevent the precipitation of HgS when a 0.0010 *M* $Hg(NO_3)_2$ solution is saturated with $H_2S$? Can you explain why HgS is insoluble in concentrated (12 *M*) HCl?

**15.38** Show that ZnS is soluble in concentrated (12 *M*) HCl.

**15.39** Use the data in Table 15.2 to determine the molar solubility of AgI in 0.010 *M* KCN solution.

**15.40** The solubility of $Zn(OH)_2$ in 1 *M* $NH_3$ is $5.7 \times 10^{-3}$ mole/liter. Determine the value of the instability constant of the complex ion, $Zn(NH_3)_4^{2+}$. Ignore the reaction, $NH_3 + H_2O \rightleftharpoons NH_4^+ + OH^-$.

***15.41** What is the molar solubility of $Mg(OH)_2$ in 0.10 *M* $NH_3$ solution? Remember that $NH_3$ is a weak base.

***15.42** Will a precipitate form in a solution formed by dissolving 1.0 mole of $AgNO_3$ and 1.0 mole $HC_2H_3O_2$ in 1.0 liter of solution?

***15.43** How many moles of HCl must be added to 1.0 liter of water to dissolve completely 0.20 mole of FeS? Remember that a saturated $H_2S$ solution is 0.1 *M*.

***15.44** How many moles of solid $NH_4Cl$ must be added to 1.0 liter of water in order to dissolve 0.10 mole of solid $Mg(OH)_2$? *Hint:* Consider the simultaneous equilibria:

$$Mg(OH)_2 \rightleftharpoons Mg^{2+} + 2OH^-$$
$$NH_3 + H_2O \rightleftharpoons NH_4^+ + OH^-$$

***15.45** Plaster is composed of $CaSO_4$. Suppose that there was a leak above a ceiling through which water was seeping at the rate of 2.0 liters/day. If the plaster in the ceiling is 1.50 cm thick, how long would it take to dissolve a circular hole 1 cm in diameter? Assume that the density of the plaster is 0.97 g/ml.

***15.46** 25.0 ml of 0.10 *M* HCl is added to 1.000 liter of saturated $Mg(OH)_2$ in contact with more than enough $Mg(OH)_2$ (*s*) to react with all of the HCl. After reaction has ceased, what will be the $Mg^{2+}$ concentration? What will be the pH of the solution?

***15.47** 2.20 g of NaOH (*s*) are added to 250 ml of 0.10 *M* $FeCl_2$ solution. What weight of $Fe(OH)_2$ will be formed? What will be the concentration of $Fe^{2+}$ in the final solution?

***15.48** 1.75 g of NaOH (*s*) are added to 250 ml of 0.10 *M* $NiCl_2$ solution. What weight of $Ni(OH)_2$ will be formed? What will be the pH of the final solution? For $Ni(OH)_2$, $K_{sp} = 1.6 \times 10^{-14}$.

***15.49** Solid $Mn(OH)_2$ is added to a solution of 0.100 *M* $FeCl_2$. After reaction, what will be the concentrations of $Mn^{2+}$ and $Fe^{2+}$ in the solution? What will be the pH of the solution?

# 16 ELECTROCHEMISTRY

Electrochemistry is concerned with the conversion of electrical energy into chemical energy in **electrolytic cells,** as well as with the conversion of chemical energy into electrical energy in **galvanic** or **voltaic cells.** In an electrolytic cell a process called electrolysis takes place in which the passage of electricity through a solution provides sufficient energy to cause an otherwise nonspontaneous oxidation-reduction reaction to take place. A galvanic cell, on the other hand, provides a source of electricity that results from a spontaneous oxidation-reduction reaction taking place in solution.

Electrochemical processes have a practical importance in chemistry and in everyday life. Electrolytic cells can provide us with information concerning the chemical environment as well as the energy that is required for many important oxidation-reduction reactions to take place. For many years now, galvanic cells, such as the dry cell, have powered our flashlights, radios, and children's toys, while cells such as the lead storage battery have achieved widespread applications, especially in the automotive industry. More recently, fuel cells, in which the energy available from the combustion of fuels is converted directly into electricity, are finding many uses, especially in space vehicles. Electrochemical know-how has aided scientists in producing modern equipment for pollution analysis and biomedical research. With the aid of tiny electrochemical probes, scientists are beginning to study the chemical reactions taking place in living cells.

All of these processes will be discussed in this chapter. However, before we begin let us first understand, qualitatively, how electrolytic solutions conduct electricity.

## 16.1 METALLIC AND ELECTROLYTIC CONDUCTION

In order for a substance to be classified as a conductor of electricity, it must be able to allow electrical charges within it to be moved from one point to another for the purpose of completing an electrical circuit. From our discussion of solids in Chapter 7, we know that most metals are conductors of electricity because of the relatively free movement of their *electrons* throughout the metallic lattice. This conduction is simply called **metallic conduction.** We also know from Chapter 5 that solutions containing electrolytes have the ability to conduct electricity. In this case, however, there are no "free" electrons to carry the current. How, then, do these solutions conduct?

We can determine whether or not a solution is a conductor of electricity by using an apparatus similar to that shown in Figure 5.2 (p. 129). When the

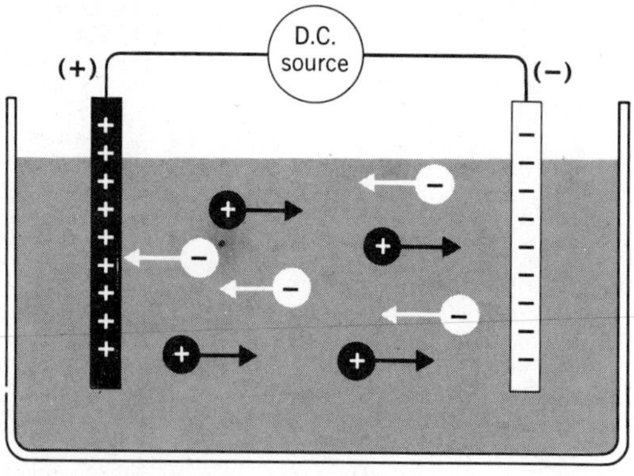


Figure 16.1
*Ion flow in an electrolytic cell.*

two electrodes are connected to a source of electricity and are dipped into a solution, we observe whether or not the bulb of the apparatus lights. The bulb will burn brightly with a strong electrolyte, somewhat more dimly with a weak electrolyte, and not at all with a nonelectrolyte. We find, therefore, that the presence of ions is necessary if a liquid is to conduct electricity, a situation that is fulfilled only by solutions of electrolytes and molten salts.

If the source of electricity to the two electrodes is a direct current, as shown in Figure 16.1, each ion in the liquid is attracted to the electrode of opposite charge. Thus, when the voltage is applied, the positive ions migrate toward the negative electrode and the negative ions move toward the positive electrode. This movement of ionic charges through the liquid, brought about by the application of electricity, is called **electrolytic conduction.**

When the ions of the liquid come in contact with the electrodes, chemical reactions take place. At the positive electrode (where there is a deficiency of electrons), the negative ions deposit electrons and are therefore oxidized. At the negative electrode (which has an excess of electrons), the positive ions pick up electrons and are reduced. Thus, during electrolytic conduction, oxidation is occurring at the positive electrode and reduction is taking place at the negative electrode. The liquid will continue to conduct electricity only as long as the oxidation-reduction reactions occurring at the electrodes continue.

The electrons that are deposited during the oxidation reaction are pumped out of the electrode by the voltage source and transferred to the negative electrode. During electrolytic conduction, then, we have electrons flowing through the exterior wire and ions flowing through the solution. This situation is illustrated in Figure 16.2*a*.

The ionic movement, as well as the reactions at the electrodes, must take place so that electrical neutrality is maintained. This means that even in the most minute part of the liquid, whenever a negative ion moves away, a positive ion must also leave, or another negative ion must immediately take its place (Figure 16.2*b*). In this way every portion of the liquid is electrically neutral at all times. During the reactions at the electrodes, electrical neutrality is assured by having equal numbers of electrons deposited and picked up. For example, whenever one electron is deposited at the positive electrode, one electron must simultaneously be picked up at the negative electrode. It is the chemical consequences of these last two processes that we focus our attention on next.

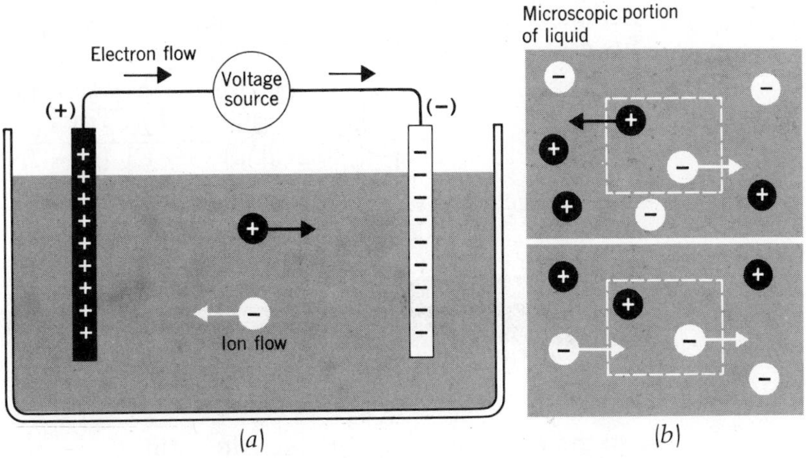


Figure 16.2
*Electrolytic conduction. (a) Electrolytic cell. (b) Maintaining electrical neutrality on a microscopic scale.*

## 16.2 ELECTROLYSIS

The chemical reactions that occur at the electrodes during electrolytic conduction constitute electrolysis. When liquid (molten) sodium chloride, for example, is electrolyzed, we find that the $Na^+$ ions move toward the negative electrode and the $Cl^-$ ions move toward the positive electrode (Figure 16.3). The reactions that take place at the electrodes are

| | | |
|---|---|---|
| Positive electrode | $2Cl^- \longrightarrow Cl_2 + 2e^-$ | oxidation |
| Negative electrode | $Na^+ + e^- \longrightarrow Na$ | reduction |

In electrochemistry we assign the terms **cathode** and **anode** according to the chemical reaction that is taking place at the electrode. *Reduction always takes place at the cathode and oxidation always takes place at the anode.* Thus, in our electrolysis reactions above, we label the negative electrode the cathode and the positive electrode the anode.

The net chemical change that takes place in the electrolytic cell is called the **cell reaction.** It is obtained by adding together the anode and cathode reactions in such a way that the same number of electrons are gained and lost. This is the same procedure that we used in the ion electron

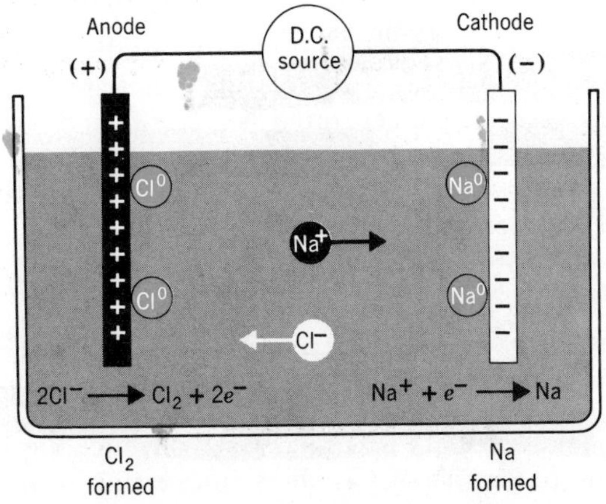


Figure 16.3
*Electrolysis of molten NaCl.*

method of balancing oxidation-reduction reactions in Chapter 5. Thus, in this case we must multiply the reduction half-reaction by 2 to get

$$\begin{array}{rl} 2Cl^- (l) \longrightarrow & Cl_2 (g) + 2e^- \\ 2Na^+ (l) + 2e^- \longrightarrow & 2Na (l) \\ \hline 2Na^+ (l) + 2Cl^- (l) \longrightarrow & Cl_2 (g) + 2Na (l) \end{array}$$

In this electrolytic cell, then, sodium is formed at the cathode and chlorine gas at the anode. This is one of the major sources of pure sodium metal and chlorine gas in the United States.

The electrolysis of aqueous solutions of electrolytes is somewhat more complex because of the ability of water to be oxidized as well as reduced. The oxidation reaction for water is

$$2H_2O (l) \longrightarrow O_2 (g) + 4H^+ (aq) + 4e^- \quad (16.1)$$

and the reduction reaction takes the form

$$2H_2O (l) + 2e^- \longrightarrow H_2 (g) + 2OH^- (aq) \quad (16.2)$$

In acidic solutions another reaction that may take place is the reduction of $H^+$, which is

$$2H^+ (aq) + 2e^- \longrightarrow H_2 (g) \quad (16.3)$$

Reaction 16.3 is not, however, a major reaction in most dilute aqueous solutions that we shall consider.

In aqueous solution, then, we have the possible oxidation and reduction of the solvent in addition to the possible oxidation and reduction of the ions of the solute. Whether the solute anion or water is going to be oxidized, or whether the solute cation or water is going to be reduced, depends on the relative ease of the two competing reactions, as we shall see in the next few examples.

**ELECTROLYSIS OF AQUEOUS NaCl.** In the electrolysis of aqueous NaCl, the following two anode (oxidation) reactions are possible:

(1) $2Cl^- (aq) \longrightarrow Cl_2 (g) + 2e^-$
(2) $2H_2O (l) \longrightarrow O_2 (g) + 4H^+ (aq) + 4e^-$

and the following two cathode (reduction) reactions are possible:

(3) $Na^+ (aq) + e^- \longrightarrow Na (s)$
(4) $2H_2O (l) + 2e^- \longrightarrow H_2 (g) + 2OH^- (aq)$

We can, of course, determine experimentally the outcome of this electrolysis by simply examining the products that are formed at the electrodes. Here we find that, in concentrated NaCl solutions (brine), chlorine gas is produced at the anode and hydrogen gas at the cathode. Therefore, during the electrolysis of such an aqueous solution of NaCl, the two half-reactions and the cell reaction are

$$\begin{array}{rl} 2Cl^- (aq) \longrightarrow & Cl_2 (g) + 2e^- \\ 2H_2O (l) + 2e^- \longrightarrow & H_2 (g) + 2OH^- (aq) \\ \hline 2H_2O (l) + 2Cl^- (aq) \longrightarrow & Cl_2 (g) + H_2 (g) + 2OH^- (aq) \end{array}$$

In this case, then, the $Na^+$ is more difficult to reduce than the $H_2O$ and the $Cl^-$ is more easily oxidized than $H_2O$. Hence we find the reduction of $H_2O$ with the oxidation of $Cl^-$. It is interesting to note that, in industry, mercury has often been used as the electrode material for this reaction and, as a result, has been a source of mercury water pollution.

**ELECTROLYSIS OF AQUEOUS $CuSO_4$.** As in our last example, we have two possible oxidation and two possible reduction reactions for the electrolysis of $CuSO_4$. These reactions are

Oxidation (1) $2SO_4^{2-}\ (aq) \longrightarrow S_2O_8^{2-}\ (aq) + 2e^-$
(2) $2H_2O\ (l) \longrightarrow O_2\ (g) + 4H^+\ (aq) + 4e^-$

and

Reduction (3) $Cu^{2+}\ (aq) + 2e^- \longrightarrow Cu\ (s)$
(4) $2H_2O\ (l) + 2e^- \longrightarrow H_2\ (g) + 2OH^-\ (aq)$

During this electrolysis we find experimentally that oxygen is produced at the anode and copper metal is deposited on the cathode. Therefore, we would write for the electrolysis of aqueous $CuSO_4$:

$$\begin{array}{rl} 2H_2O\ (l) \longrightarrow & O_2\ (g) + 4H^+\ (aq) + 4e^- \\ 2Cu^{2+}\ (aq) + 4e^- \longrightarrow & 2Cu\ (s) \\ \hline 2H_2O\ (l) + 2Cu^{2+}\ (aq) \longrightarrow & O_2\ (g) + 4H^+\ (aq) + 2Cu\ (s) \end{array}$$

Note that we multiplied the equation for the reduction of $Cu^{2+}$ by 2 in order to have equal numbers of electrons lost and gained.

In the electrolysis of aqueous $CuSO_4$, the $H_2O$ is more easily oxidized than the $SO_4^{2-}$ and the $Cu^{2+}$ is more readily reduced than the $H_2O$.

**ELECTROLYSIS OF AQUEOUS $CuCl_2$.** We should be able to apply what we have learned about the electrolysis of aqueous solutions to the electrolysis of $CuCl_2$. We would expect that the two species that could be oxidized are water and chloride ion, and that the two species that could be reduced are water and $Cu^{2+}$. Since we already know that $Cl^-$ is more easily oxidized than $H_2O$ and that $Cu^{2+}$ is more readily reduced than $H_2O$, we expect the reaction,

$$\begin{array}{rl} 2Cl^-\ (aq) \longrightarrow & Cl_2\ (g) + 2e^- \\ Cu^{2+}\ (aq) + 2e^- \longrightarrow & Cu\ (s) \\ \hline Cu^{2+}\ (aq) + 2Cl^-\ (aq) \longrightarrow & Cl_2\ (g) + Cu\ (s) \end{array}$$

This is exactly what is found experimentally.

**ELECTROLYSIS OF AQUEOUS $Na_2SO_4$.** Once again we call upon what we have learned previously in our discussion of electrolysis. We know that water is more easily oxidized than $SO_4^{2-}$ and that water is more readily reduced than $Na^+$. Therefore, in this solution $H_2O$ is both oxidized and reduced, giving us

$$\begin{array}{rl} 2H_2O\ (l) \longrightarrow & O_2\ (g) + 4H^+\ (aq) + 4e^- \\ 4H_2O\ (l) + 4e^- \longrightarrow & 2H_2\ (g) + 4OH^-\ (aq) \\ \hline 6H_2O\ (l) \longrightarrow & O_2\ (g) + 2H_2\ (g) + 4OH^-\ (aq) + 4H^+\ (aq) \end{array}$$

Note that we had to multiply the reduction reaction by 2 to achieve the same number of electrons as in the oxidation reaction.

The overall reaction, as written above, can be simplified further by recalling that $H^+$ and $OH^-$ will react to give $H_2O$.

$$4OH^- + 4H^+ \longrightarrow 4H_2O$$

Thus the true overall reaction for the electrolysis of a stirred aqueous $Na_2SO_4$ reaction is

$$2H_2O\ (l) \longrightarrow O_2\ (g) + 2H_2\ (g)$$

which is simply the reaction for the electrolysis of $H_2O$. Sodium sulfate does not participate in this electrolysis in the sense that it is not consumed at the electrodes; yet we would find experimentally that it (or some other similar salt) is needed if electrolysis of $H_2O$ is to occur. What, then, is the role of the $Na_2SO_4$? The $Na_2SO_4$ is needed in order to maintain electrical neutrality. During the oxidation of $H_2O$, $H^+$ ions are produced in the immediate vicinity of the anode. A negative ion must also be present in that vicinity to neutralize the positive charges. This is fulfilled by the $SO_4^{2-}$ ion. Likewise, at the cathode, where $OH^-$ ions are produced, there must be a positive ion present to neutralize the charges on the $OH^-$, and keep the solution electrically neutral.

## 16.3 PRACTICAL APPLICATIONS OF ELECTROLYSIS

Let's now examine some practical applications of the electrolysis process. We have already seen one such application in the production of pure metallic sodium by the electrolysis of molten sodium chloride. The commercial production of other metals such as aluminum, magnesium, and copper also employs electrolysis.

**ALUMINUM.** As you are undoubtedly aware, aluminum finds many important uses as a structural metal because of its strength and light weight. Its commercial availability has been made possible through the application of electrochemical reduction.

If we were to electrolyze an aqueous solution of an aluminum salt, such as $AlCl_3$, we would find that $H_2O$ is more easily reduced than the $Al^{3+}$. Therefore an aqueous solution of an aluminum salt cannot be used to produce the metal. A 22-year-old graduate of Oberlin College, Charles Hall, invented a process, in 1886, whereby molten $Al_2O_3$ is used. He prepared a mixture of $Al_2O_3$ with cryolite, $Na_3AlF_6$, and electrolyzed it in the molten state. The cryolite, he found, reduced the melting temperature from 2000°C for $Al_2O_3$ to 1000°C for the mixture. A diagram of the electrolysis cell is shown in Figure 16.4. The vessel holding the melted mixture is made of iron lined with carbon and serves as the cathode. Carbon rods that serve as the anode are inserted into the melt. As the oxidation-reduction reactions proceed, pure aluminum is produced at the cathode and sinks to the bottom of the vessel. The reactions at the electrodes are

| | |
|---|---|
| Anode | $3O^{2-}\ (l) \longrightarrow \frac{3}{2}O_2\ (g) + 6e^-$ |
| Cathode | $2Al^{3+}\ (l) + 6e^- \longrightarrow 2Al\ (l)$ |
| | $2Al^{3+}\ (l) + 3O^{2-}\ (l) \longrightarrow \frac{3}{2}O_2\ (g) + 2Al\ (l)$ |

Today other materials are used in place of the cryolite. These materials permit operation at still lower temperatures and are less dense than the cryolite used by Hall. This lower density of the electrolyte mix permits easier separation of the molten aluminum.

**MAGNESIUM.** Magnesium, another structural metal that is important because of its light weight, occurs to an appreciable extent in sea water. Magnesium ions are precipitated from sea water as the hydroxide and the $Mg(OH)_2$ is then converted to the chloride by treatment with hydrochloric acid. After evaporation of the water the $MgCl_2$ is melted and electrolyzed,

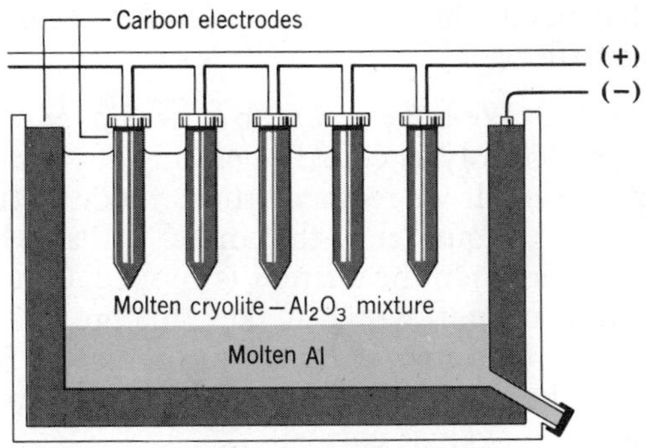


Figure 16.4
*Production of aluminum by the Hall process.*



magnesium being produced at the cathode and chlorine being evolved at the anode. The overall net reaction is simply

$$MgCl_2\ (l) \longrightarrow Mg\ (l) + Cl_2\ (g)$$

**COPPER.** An interesting application of electrolysis is the refining, or purification, of copper metal. When first separated from its ore, copper metal is about 99% pure, with iron, zinc, silver, gold, and platinum as major impurities. In the refining process the impure copper is used as the anode in an electrolytic cell containing aqueous copper sulfate as the electrolyte. The cathode of the cell is constructed of high-purity copper (Figure 16.5).

When electrolysis is carried out, the voltage across the cell is adjusted so that only copper and other more active metals, such as iron or zinc, are able to dissolve at the anode. The silver, gold, and platinum do not dissolve and simply fall off and settle to the bottom of the electrolysis cell. At the cathode only the most easily reduced species, $Cu^{2+}$, is caused to pick up electrons; hence, only copper is deposited.

The net result of the operation of this cell is that copper is transferred from the anode to the cathode while the Fe and Zn impurities remain in solution as $Fe^{2+}$ and $Zn^{2+}$. Afterwards the silver, gold, and platinum "sludge" is removed from the apparatus and sold for enough money to pay for the cost of the electricity required in the electrolysis. As a result, the purification of copper (about 99.95% pure) costs nearly nothing! However, the total produc-

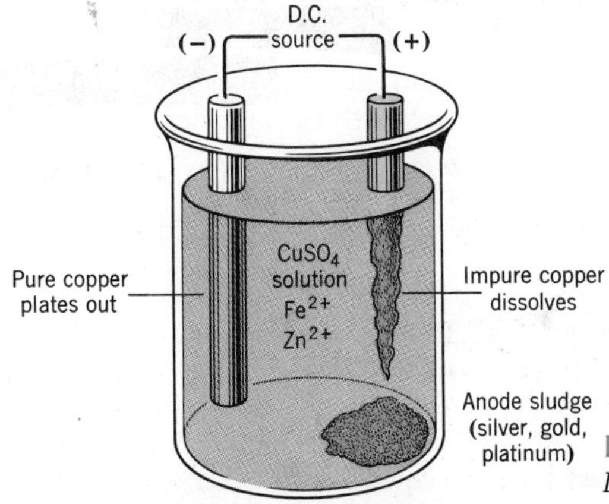


Figure 16.5
*Purification of copper by electrolysis.*

tion cost is still considerable, because it includes the mining of the crude ore and its initial purification.

**ELECTROPLATING.** We have just seen how copper can be "plated out" on an electrode in an electrolysis cell. The plating out of a metal in this fashion is called **electroplating.** If we replaced the cathode in the cell in Figure 16.5 with another metal, the surface of that metal will also become covered with a layer of pure copper when the current is applied. Other metals can be electroplated as well as copper, which makes this process of great commercial importance. In the manufacturing of automobiles, for example, various parts, such as steel bumpers, are electroplated with chromium for beauty as well as for protection against corrosion.

## 16.4 QUANTITATIVE ASPECTS OF ELECTROLYSIS

Michael Faraday was the first to describe, in a quantitative fashion, the relationship that exists between the amount of current used and the extent of the chemical change that takes place at the electrodes during electrolysis. The extent of an electrolysis reaction is, of course, related to the number of moles of electrons lost or gained in the oxidation-reduction reactions, respectively. For example, in the reaction for the reduction of silver ion to silver metal,

$$Ag^+ (aq) + e^- \longrightarrow Ag (s)$$

1 mole of electrons reacts with 1 mole of silver ions to give 1 mole, or 107.87 g, of solid silver (the amount of silver deposited can be determined experimentally by weighing the cathode before and after the current is supplied to the cell). Thus, in this case, when 107.87 g of silver are deposited on the cathode, we know that 1 mole of electrons must have passed through the cell.

The amount of electricity that must be supplied to a cell in order for *1 mole* of electrons to undergo reaction, in the above case as well as in any other oxidation-reduction reaction, is called a **faraday** ($\mathcal{F}$). In the above example, then, 1 $\mathcal{F}$ was supplied to produce the 107.87 g of silver, and it would take 2 $\mathcal{F}$ to produce 215.74 g of silver, and so on. In other words, the faraday is just another way of saying "a mole of electrons."

$$1\ (\mathcal{F}) \equiv 1 \text{ mole of electrons}$$

Another important unit is the **coulomb** (coul). One coulomb is the amount of charge that moves past any given point in a circuit when a current of 1 **ampere** (amp) is supplied for 1 sec. Thus,

$$1 \text{ coul} = 1 \text{ amp sec}$$

Experimentally, it is found that 1 $\mathcal{F}$ is equivalent to 96,487 coulombs, or 96,500 when rounded off to three significant figures. Thus,

$$1\ \mathcal{F} \sim 96{,}500 \text{ coul} \sim 1 \text{ mole of electrons}$$

Let's look at some examples of how these concepts can be applied.

EXAMPLE 16.1 In an electrolytic cell like that in Figure 16.2, how many grams of Cu will be deposited from a solution of $CuSO_4$ by a current of 1.5 amp flowing for 2.0 hr?

SOLUTION First, we must have an equation representing the reaction that takes place.

For the copper we would write

$$Cu^{2+}\,(aq) + 2e^- \longrightarrow Cu\,(s)$$

We've written a reduction because Cu is being deposited from the solution which contains $Cu^{2+}$. The equation is necessary because it gives the important relationship,

$$1 \text{ mole Cu} \sim 2 \text{ moles of electrons}$$

or simply

$$1 \text{ mole Cu} \sim 2\ \mathcal{F}$$

Our procedure now is to use the current and time to calculate the number of coulombs provided. Then we calculate the number of faradays supplied. Finally we compute the number of moles of Cu, followed by the number of grams of Cu.

$$1.5 \text{ amp} \times 2.0\ \cancel{\text{hr}} \times \left(\frac{3600 \text{ sec}}{1\ \cancel{\text{hr}}}\right) \sim 10{,}800 \text{ amp sec}$$

$$10{,}800\ \cancel{\text{amp sec}} \times \left(\frac{1 \text{ coul}}{1\ \cancel{\text{amp sec}}}\right) \sim 10{,}800 \text{ coul}$$

$$10{,}800\ \cancel{\text{coul}} \times \left(\frac{1\ \mathcal{F}}{96{,}500\ \cancel{\text{coul}}}\right) \sim 0.11\ \mathcal{F}$$

$$0.11\ \cancel{\mathcal{F}} \times \left(\frac{1\ \cancel{\text{mole Cu}}}{2\ \cancel{\mathcal{F}}}\right) \times \left(\frac{63.55 \text{ g Cu}}{1\ \cancel{\text{mole Cu}}}\right) \sim 3.5 \text{ g Cu}$$

EXAMPLE 16.2 How long would it take to produce 25.0 g of Cr from a solution of $CrCl_3$ by a current of 2.75 amp?

SOLUTION Again, we begin by writing the equation for the reaction,

$$Cr^{3+}\,(aq) + 3e^- \longrightarrow Cr\,(s)$$

from which we can say that

$$1 \text{ mole Cr} \sim 3\ \mathcal{F}$$

First we calculate the number of faradays required.

$$25.0\ \cancel{\text{g Cr}} \times \left(\frac{1\ \cancel{\text{mole Cr}}}{52.0\ \cancel{\text{g Cr}}}\right) \times \left(\frac{3\ \mathcal{F}}{1\ \cancel{\text{mole Cr}}}\right) \sim 1.44\ \mathcal{F}$$

Next we calculate the number of coulombs required.

$$1.44\ \cancel{\mathcal{F}} \times \left(\frac{96{,}500 \text{ coul}}{1\ \cancel{\mathcal{F}}}\right) \sim 139{,}000 \text{ coul}$$

Since 1 coulomb is equal to 1 amp sec, we have

$$139{,}000\ \cancel{\text{coul}} \times \left(\frac{1\ \cancel{\text{amp}} \text{ sec}}{1\ \cancel{\text{coul}}}\right) \times \left(\frac{1}{2.75\ \cancel{\text{amp}}}\right) \sim 50{,}500 \text{ sec}$$

If we convert this to hours, we get

$$50{,}500\ \cancel{\text{sec}} \times \left(\frac{1 \text{ hr}}{3600\ \cancel{\text{sec}}}\right) \sim 14.0 \text{ hr}$$

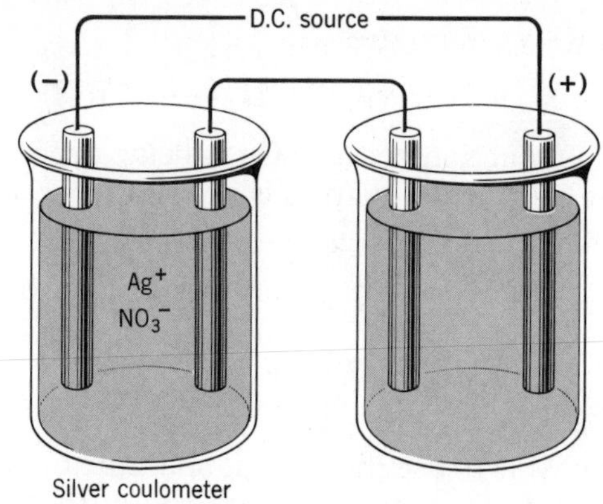


Figure 16.6
*The use of a coulometer.*

We can determine experimentally the weight of a substance that has been deposited on an electrode during electrolysis by weighing the electrode before and after the current has been supplied. The apparatus used in experiments of this kind is called a **coulometer.** In Figure 16.6 we see two such coulometers connected in series so that the same current, and thus the same number of faradays, passes through both cells. With the aid of this apparatus it is possible to use a known oxidation-reduction reaction in one cell to provide an experimental measure of the equivalent weight of an unknown in the other cell. This type of analysis is illustrated in the next example.

EXAMPLE 16.3

In the left cell of Figure 16.6 we place a solution containing $Ag^+$ ions, and in the right cell a solution whose metal ion is unknown (X). The same current is passed through both cells for the same amount of time. When the current is turned off and the electrodes are rinsed, dried, and weighed, it is found that 3.50 g of silver were deposited during the same period of time that 2.50 g of element X were deposited. What is the equivalent weight of element X?

SOLUTION

This rather long-winded problem has a relatively simple solution. Since the current and time are the same for both cells in the series circuit, the same number of faradays was passed through both. We can calculate the number of faradays supplied by using the information from the $Ag^+$ cell.

From our earlier discussion in this section, we know that the reduction reaction for $Ag^+$ is

$$Ag^+\,(aq) + e^- \longrightarrow Ag\,(s)$$

Therefore,

$$1 \text{ mole Ag} \sim 1\ \mathscr{F}$$

or

$$107.87 \text{ g Ag} \sim 1\ \mathscr{F}$$

The faraday equivalent of 3.50 g of silver is then

$$3.50\ \cancel{g} \times \left(\frac{1\ \mathscr{F}}{107.87\ \cancel{g}}\right) = 0.0324\ \mathscr{F}$$

This means that 0.0324 $\mathscr{F}$ of electricity was passed through both cells. In the cell containing the unknown, then, 0.0324 $\mathscr{F}$ deposited 2.50 g of the substance. Thus, for the unknown,

$$0.0324\ \mathscr{F} \sim 2.50\ \text{g X}$$

and 1 $\mathscr{F}$ is equivalent to

$$1\ \cancel{\mathscr{F}} \times \left(\frac{2.50\ \text{g X}}{0.0324\ \cancel{\mathscr{F}}}\right) \sim 77.2\ \text{g X}$$

Recall that the equivalent weight of a substance undergoing a redox reaction is equal to that weight of the substance that gains or loses 1 mole of electrons (that is, 1 $\mathscr{F}$). The equivalent weight of the unknown metal is therefore 77.2 g.

## 16.5 GALVANIC CELLS

In our previous discussion of electrolytic cells, chemical processes occurred because a voltage placed across electrodes forced an otherwise nonspontaneous chemical reaction to take place. We now turn to the opposite situation, in which electron flow is produced as a result of spontaneous oxidation-reduction reactions. An example of a spontaneous oxidation-reduction reaction taking place in a solution can be seen simply by placing a piece of metallic zinc into a solution of $CuSO_4$. A brownish, spongelike layer begins to form on the piece of zinc and, at the same time, the blue color of the $CuSO_4$ begins to disappear. The brownish substance forming on the zinc is metallic copper, and we write the two half-reactions that occur as

$$Cu^{2+}\,(aq) + 2e^- \longrightarrow Cu\,(s)$$
$$Zn\,(s) \longrightarrow Zn^{2+}\,(aq) + 2e^-$$

We see from these reactions that the $Cu^{2+}$ ions are spontaneously removed from the solution and are replaced by the colorless $Zn^{2+}$ ions. Thus the blue color of the solution disappears as more and more $Zn^{2+}$ ions are formed.

As long as these spontaneous reactions take place at the surface of the zinc, no useful flow of electrons can be obtained. The reaction simply generates heat. We can, however, take advantage of the electron flow that accompanies these oxidation-reduction reactions if we separate the copper solution from the zinc by placing them in a cell similar to that shown in Figure 16.7*a* or 16.7*b*. In an apparatus of this design, the electrons produced by the oxidation of the zinc must travel through the wire and into the electrode in the $CuSO_4$ solution. The electrons are then picked up by the $Cu^{2+}$ ions and reduction takes place. The electrons flowing through the external wire provide a source of electricity.

Although the zinc and copper have to be separated to obtain a useful flow of electrons, complete isolation of the two species would lead to an electrical imbalance at the electrodes, and the electron flow would soon cease. We can see how electrical imbalance would occur if we imagine that the two half-cells were completely isolated from each other and the oxidation-reduction reactions still continued to take place. On the left side of this hypothetical setup, $Zn^{2+}$ ions entering the solution would give the solution an overall positive charge. The positive charge in the solution would prevent additional $Zn^{2+}$ from entering. On the right we would find that when $Cu^{2+}$ leaves the solution the $SO_4^{2-}$ ions left behind would give the solution a negative charge. This would prevent further removal of $Cu^{2+}$.

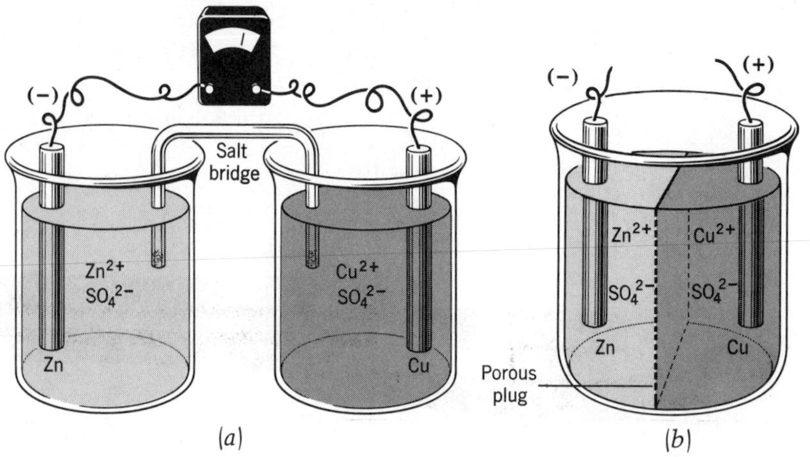


Figure 16.7
*Galvanic cell.*

Continuous flow of current, accompanied by continuous chemical activity, can take place only if electrical neutrality can be maintained in the solution. If the ions are allowed to travel from one compartment to the other, electrical neutrality can be maintained. The flow of ions, however, should not be so rapid that the half-reactions occur at the surface of the zinc electrode. The salt bridge in Figure 16.7*a* and the porous partition in Figure 16.7*b* allow for the slow mixing of the ions in the two solutions. A salt bridge is usually a tube filled with an electrolyte such as $KNO_3$ or KCl in gelatin. Cations from the salt bridge can move into one compartment to compensate for the excess negative charge, while the anions from the salt bridge diffuse into the other compartment to neutralize the excess positive charge. The porous partition in Figure 16.7*b* serves the same purpose as the salt bridge. With either the salt bridge or the porous partition in place, there is a continuous electron flow through the external wire and ion flow through the solution as a result of the spontaneous oxidation-reduction reactions taking place in the galvanic cell.

**THE SIGNS OF THE ELECTRODES IN GALVANIC CELLS.** Earlier we defined the anode in electrochemistry as the electrode where oxidation takes place and the cathode as the one where reduction occurs. In the galvanic cell just described, oxidation takes place in the zinc compartment, so that the zinc bar would be the anode and the copper electrode would be the cathode. Since electrons are released (oxidation reaction) at the zinc anode and removed (reduction reaction) at the copper cathode, the zinc bar possesses an excess of electrons compared to the copper. As a result, the zinc bar is the negative electrode and the copper is the positive electrode. This means that in galvanic cells the anode is negative and the cathode is positive, quite the opposite of what we found to be true in electrolytic cells.[1]

[1] This labeling of electrodes in galvanic cells is, however, consistent with electrolytic cells when we consider the movement of the ions within the solution. The $Zn^{2+}$ ions produced at the anode and the $SO_4^{2-}$ ions freed at the cathode must mingle with each other if electrical neutrality is to prevail in the solution. To accomplish this, some of the $Zn^{2+}$ ions must move toward the cathode and some of the $SO_4^{2-}$ ions must move toward the anode. Thus we have cations moving toward the cathode and anions moving toward the anode, which is precisely the same situation as in electrolytic cells.

## 16.6 CELL POTENTIALS

The electric current obtained from a galvanic cell is a result of electrons being pushed or forced to flow from the negative electrode, through an external wire, to the positive electrode. The "force" with which these electrons move through the wire is called the **electromotive force,** or **emf,** and is measured in **volts** (V). Actually, the volt is a measure of the energy that is capable of being extracted from the flowing electric charge. If the emf is 1 V, the passage of 1 coulomb is able to accomplish 1 joule of work.

$$1\ \mathrm{V} = \frac{1\ \mathrm{J}}{\mathrm{coul}} \qquad (16.4)$$

The emf of any galvanic cell depends on the nature of the chemical reactions taking place within the cell, the concentrations of the species undergoing oxidation and reduction, and the temperature of the cell, which we shall take to be 25°C unless otherwise noted. The greater the tendency or potential of the two half-reactions to occur spontaneously, the greater will be the emf of the cell. Thus the emf of the cell can also be called the **cell potential,** designated by the symbol $\mathscr{E}_{\text{cell}}$.

Since the cell potential depends on the concentration of the species in the cell, we define a reference potential, which is called the standard potential of the cell, $\mathscr{E}^0_{\text{cell}}$, as that emf obtained when all species in solution are at unit concentration and when any gases involved in the cell reaction are at a pressure of 1 atm.[2] Whenever we write a half-reaction or the overall cell reaction for any galvanic cell, we must specify concentrations. Thus the half-reactions and the overall reaction for the Zn/Cu cell in the last section would be written

$$\begin{array}{rcl} \mathrm{Zn}\ (s) & \longrightarrow & \mathrm{Zn^{2+}}\ (1\ M) + 2e^- \\ \mathrm{Cu^{2+}}\ (1\ M) + 2e^- & \longrightarrow & \mathrm{Cu}\ (s) \\ \hline \mathrm{Zn}\ (s) + \mathrm{Cu^{2+}}\ (1\ M) & \longrightarrow & \mathrm{Zn^{2+}}\ (1\ M) + \mathrm{Cu}\ (s) \end{array}$$

To measure the cell potential accurately, care must be taken to avoid drawing current from the cell. This is because some of the cell's voltage is required to overcome the internal resistance of the cell when current is drawn. The remaining voltage that can be measured under these conditions is less than the maximum.

One device that can usually be used to measure the emf of a cell is called a **potentiometer.** In this instrument the potential generated by the cell is balanced by an opposing potential from within the potentiometer. When the two opposing potentials are equal, no current flows and the cell potential is equal to the opposing emf, which can be read directly from the potentiometer. The voltage that is measured in this way is the maximum emf of the cell. Today modern advances in electronics have led to a variety of other instruments that are able to measure quickly and simply the emf of a cell without drawing significant amounts of current. 2

## 16.7 REDUCTION POTENTIALS

A very important and useful concept can be developed if we attempt to answer the question, what is the origin of the cell potential? To answer this question we shall use the Zn/Cu cell we just described above. In this cell we

[2] In footnote 1 in Chapter 12, it was mentioned that activities ("effective" concentrations and pressures) should be used in the mass action expression when computing $\Delta G$. This applies also to the effect of concentration on $\mathscr{E}$. The standard potential is obtained when all species are at unit activity. Only a small error is introduced, however, by using actual concentrations when solutions are relatively dilute.

have a solution containing $Zn^{2+}$ ions about one electrode and a solution containing $Cu^{2+}$ ions around the other. Each of these ions has a certain tendency to acquire electrons from its respective electrode and become reduced. In other words, there is associated with a reduction half-reaction, such as

$$Zn^{2+}\,(aq) + 2e^- \longrightarrow Zn\,(s)$$

or

$$Cu^{2+}\,(aq) + 2e^- \longrightarrow Cu\,(s)$$

in our cell, a certain intrinsic tendency to proceed from left to right that we can describe by its **reduction potential.** The larger the reduction potential for any half-reaction, the greater is its ability to undergo reduction.

When the cell reaction takes place, then, what we are actually observing is a kind of "tug-of-war." Each of the species in solution attempts to pull electrons from its electrode so as to become reduced. The species with the greatest ability to acquire electrons (the substance with the largest reduction potential) wins the tug-of-war and does undergo reduction. The loser, on the other hand, must supply the electrons to the winner and that substance is therefore oxidized.

The potential measured for the cell represents the difference in the abilities of the two ions to become reduced; that is, the cell emf corresponds to the *difference* between the reduction potentials of the two half-reactions. If we represent the standard reduction potentials for $Cu^{2+}$ and $Zn^{2+}$ as $\mathscr{E}^0_{Cu}$ and $\mathscr{E}^0_{Zn}$, respectively, then the cell potential for the Zn/Cu cell can be written as

$$\mathscr{E}^0_{cell} = \mathscr{E}^0_{Cu} - \mathscr{E}^0_{Zn}$$

Experimentally, we can only measure *positive* cell potentials. *Thus, in order to obtain the overall standard cell potential for this cell, or for any other spontaneous cell reaction, we always subtract the smaller reduction potential from the larger one.*

Since, experimentally, only overall cell potentials can be measured, we are only capable of obtaining differences between the reduction potentials for any two half-reactions. How, then, can we obtain the reduction potential for any specific half-reaction? Clearly, if the cell potential and the $\mathscr{E}^0$ for one of the half-reactions are known, the $\mathscr{E}^0$ for the other half-reaction can be calculated. What has been done, therefore, is to choose a half-reaction arbitrarily and assign to it a standard reduction potential of zero volts. All other half-reactions can then be compared to this standard and a set of relative values of $\mathscr{E}^0$ obtained.

The electrode chosen to be the standard is the hydrogen electrode, shown in Figure 16.8*a*. The hydrogen electrode consists of a platinum wire encased in a glass sleeve with hydrogen gas passing through it at a pressure of 1 atm. The platinum wire is attached to a platinum foil that is coated with a black velvet-looking layer of finely divided platinum, which serves as a catalyst for the reaction,

$$2H^+\,(aq) + 2e^- \rightleftharpoons H_2\,(g)$$

This assembly is then immersed in an acid solution whose hydrogen ion concentration is 1 *M*. When the hydrogen electrode and the copper electrode, for example, are placed together in a cell, we have the galvanic cell shown in Figure 16.8*b*.

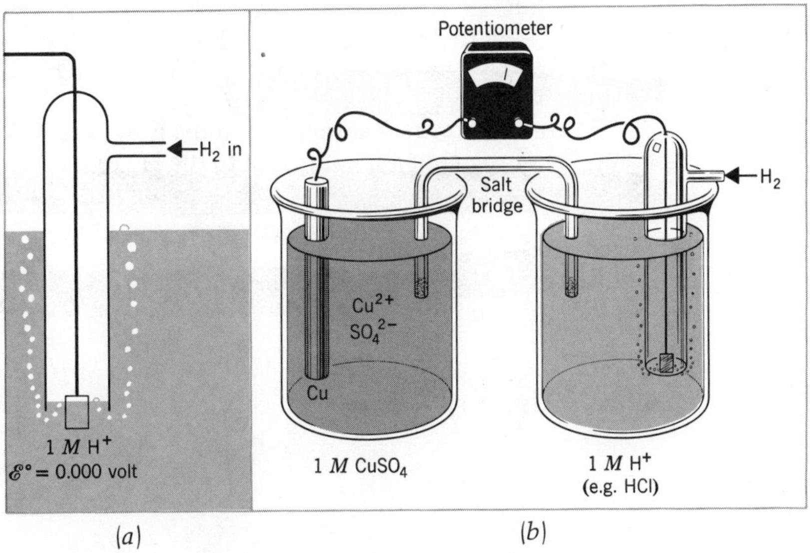


Figure 16.8
*(a) The hydrogen electrode. (b) The hydrogen electrode used in a galvanic cell.*

If the reduction potential of the species in the other half of the cell is greater than that for the hydrogen electrode (that is, if it has a positive value), the hydrogen electrode is forced to undergo oxidation. The corresponding half-reaction for the oxidation of the hydrogen electrode is

$$H_2\,(g) \longrightarrow 2H^+\,(aq) + 2e^-$$

If, on the other hand, the reduction potential of the other half-reaction is less than 0.000 V (this species would have a negative reduction potential), the hydrogen electrode undergoes reduction,

$$2H^+\,(aq) + 2e^- \longrightarrow H_2\,(g)$$

and causes the other species to become oxidized.

To illustrate this idea, let's take a closer look at the $Cu/H_2$ cell shown in Figure 16.8*b*. The potential for this cell, as measured with a potentiometer, turns out to be +0.34 V. We now must determine whether the copper is easier or more difficult to reduce than the hydrogen, in order to obtain the proper form for the $\mathscr{E}^0_{cell}$ equation. In this cell we find that copper is plated out on its electrode and, therefore the $Cu^{2+}$ ion is being reduced. This means that copper has a higher reduction potential than $H_2$ and therefore the correct form for the $\mathscr{E}^0_{cell}$ equation for the $Cu/H_2$ cell is

$$\mathscr{E}^0_{cell} = \mathscr{E}^0_{Cu} - \mathscr{E}^0_{H_2}$$

Since $\mathscr{E}^0_{H_2}$ is 0.000 V, then

$$\mathscr{E}^0_{cell} = \mathscr{E}^0_{Cu} = +0.34 \text{ V}$$

Therefore, the standard reduction potential for the reduction of copper,

$$Cu^{2+}\,(aq) + 2e^- \longrightarrow Cu\,(s)$$

is +0.34 V.

Now let's examine a cell in which hydrogen is more easily reduced than the species at the other electrode. For this example we shall replace the copper electrode in Figure 16.8*b* with a zinc electrode. The $\mathscr{E}^0_{\text{cell}}$ for this $Zn/H_2$ cell is measured to be +0.76 V.

Following our first example, we must now determine the relative ease of reduction of the zinc ion compared to the $H^+$ ion. In this cell, by the gradual disappearance of the zinc electrode, we conclude that the zinc is oxidized,

$$Zn\ (s) \longrightarrow Zn^{2+}\ (aq) + 2e^-$$

Therefore $Zn^{2+}$ must be more difficult to reduce than $H^+$. The $\mathscr{E}^0_{\text{cell}}$ for this cell would then correspond to

$$\mathscr{E}^0_{\text{cell}} = \mathscr{E}^0_{H_2} - \mathscr{E}^0_{Zn}$$

Since $\mathscr{E}^0_{H_2}$ is zero, we have

$$\mathscr{E}^0_{\text{cell}} = 0 - \mathscr{E}^0_{Zn}$$

or

$$\mathscr{E}^0_{\text{cell}} = -\mathscr{E}^0_{Zn} = +0.76\text{ V}$$

Therefore,

$$\mathscr{E}^0_{Zn} = -0.76\text{ V}$$

We see that the reduction potential for the zinc electrode in this cell is a negative quantity, which is consistent with our discussion above on the sign convention of the electrode for those species more difficult to reduce than hydrogen.

With a knowledge of the reduction potentials for the zinc and copper electrodes, we can now *predict* the cell potential for the Zn/Cu cell. This can be done even if we had no previous knowledge of the species undergoing oxidation or reduction. We know from the sizes of the reduction potentials of Cu and Zn, +0.34 and −0.76 V, respectively, that the only way we can subtract these two values from each other and obtain a positive $\mathscr{E}^0_{\text{cell}}$ is if we subtract the −0.76 V from the +0.34 V. Thus,

$$\mathscr{E}^0_{Cu-Zn} = \mathscr{E}^0_{Cu} - \mathscr{E}^0_{Zn} = +0.34 - (-0.76)$$

and we find that

$$\mathscr{E}^0_{Cu-Zn} = 1.10\text{ V}$$

This is precisely the value that is obtained experimentally when we measure the potential of this cell.

We are now in a position to determine the reduction potentials for many different half-reactions, because all we must do is construct galvanic cells in which the reduction potential of one half-cell is known, relative to the hydrogen electrode. Some standard reduction potentials, $\mathscr{E}^0$, determined in this manner are given in Table 16.1. In this table the reduction potential of the hydrogen electrode is placed in the middle, with the species more difficult to reduce than hydrogen listed above it and those more easily reduced placed below it. Such a table of reduction potentials serves many useful purposes.

1. From a table of reduction potentials we can, at a glance, pick out substances that are good oxidizing agents and those that are good reducing agents. Any species that appears on the left of the double arrow serves as an oxidizing agent if it undergoes reduction during

Table 16-1
*Standard reduction potentials at 25°C*

| Half-reaction | $\mathcal{E}^0$ (volts) |
|---|---|
| $Li^+ + e^- \rightleftharpoons Li$ | −3.05 |
| $K^+ + e^- \rightleftharpoons K$ | −2.92 |
| $Ba^{2+} + 2e^- \rightleftharpoons Ba$ | −2.90 |
| $Ca^{2+} + 2e^- \rightleftharpoons Ca$ | −2.76 |
| $Na^+ + e^- \rightleftharpoons Na$ | −2.71 |
| $Mg^{2+} + 2e^- \rightleftharpoons Mg$ | −2.38 |
| $Al^{3+} + 3e^- \rightleftharpoons Al$ | −1.67 |
| $Mn^{2+} + 2e^- \rightleftharpoons Mn$ | −1.03 |
| $2H_2O + 2e^- \rightleftharpoons H_2 + 2OH^-$ | −0.83 |
| $Zn^{2+} + 2e^- \rightleftharpoons Zn$ | −0.76 |
| $Cr^{3+} + 3e^- \rightleftharpoons Cr$ | −0.74 |
| $Fe^{2+} + 2e^- \rightleftharpoons Fe$ | −0.44 |
| $PbSO_4 + 2e^- \rightleftharpoons Pb + SO_4^{2-}$ | −0.36 |
| $Ni^{2+} + 2e^- \rightleftharpoons Ni$ | −0.25 |
| $Sn^{2+} + 2e^- \rightleftharpoons Sn$ | −0.14 |
| $Pb^{2+} + 2e^- \rightleftharpoons Pb$ | −0.13 |
| $Fe^{3+} + 3e^- \rightleftharpoons Fe$ | −0.04 |
| $2H^+ + 2e^- \rightleftharpoons H_2$ | 0.00 |
| $AgCl + e^- \rightleftharpoons Ag + Cl^-$ | 0.22 |
| $Hg_2Cl_2 + 2e^- \rightleftharpoons 2Hg + 2Cl^-$ | 0.27 |
| $Cu^{2+} + 2e^- \rightleftharpoons Cu$ | 0.34 |
| $Cu^+ + e^- \rightleftharpoons Cu$ | 0.52 |
| $I_2\,(aq) + 2e^- \rightleftharpoons 2I^-$ | 0.54 |
| $Fe^{3+} + e^- \rightleftharpoons Fe^{2+}$ | 0.77 |
| $Ag^+ + e^- \rightleftharpoons Ag$ | 0.80 |
| $Br_2\,(aq) + 2e^- \rightleftharpoons 2Br^-$ | 1.09 |
| $O_2 + 4H^+ + 4e^- \rightleftharpoons 2H_2O$ | 1.23 |
| $MnO_2 + 4H^+ + 2e^- \rightleftharpoons Mn^{2+} + 2H_2O$ | 1.28 |
| $Cr_2O_7^{2-} + 14H^+ + 6e^- \rightleftharpoons 2Cr^{3+} + 7H_2O$ | 1.33 |
| $Cl_2\,(g) + 2e^- \rightleftharpoons 2Cl^-$ | 1.36 |
| $2ClO_3^- + 12H^+ + 10e^- \rightleftharpoons Cl_2 + 6H_2O$ | 1.47 |
| $8H^+ + MnO_4^- + 5e^- \rightleftharpoons Mn^{2+} + 4H_2O$ | 1.49 |
| $PbO_2 + SO_4^{2-} + 4H^+ + 2e^- \rightleftharpoons PbSO_4 + 2H_2O$ | 1.69 |
| $H_2O_2 + 2H^+ + 2e^- \rightleftharpoons 2H_2O$ | 1.78 |
| $S_2O_8^{2-} + 2e^- \rightleftharpoons 2SO_4^{2-}$ | 2.00 |
| $F_2 + 2e^- \rightleftharpoons 2F^-$ | 2.87 |

the course of a chemical reaction. Since the substances at the bottom of the table are more easily reduced than those at the top, their ability to serve as oxidizing agents increases as we proceed down the table. Thus we could conclude, from their positions on this table, that $H^+$ is a better oxidizing agent than $Zn^{2+}$, and that $F_2$ is a better oxidizing agent than $Cl_2$. In brief, then, *good oxidizing agents are those species on the left of the double arrow at the bottom of the table.*

Each of the half-reactions listed in Table 16.1 is reversible. We saw, for example, in the last section, that $H_2$ is oxidized to $H^+$ when placed in a cell with copper, and that $H^+$ is reduced to $H_2$ when placed against zinc. When the reactions in Table 16.1 are forced to proceed from right to left, that is, when they are caused to be the oxidation step in the overall reaction, then the species appearing at the right in Table 16.1 are functioning as reducing agents by being oxidized. *All of the substances appearing on the right side of the reac-*

*tions in Table 16.1, then, could behave as reducing agents, with those species at the top right of the table, such as Li, being the best and those at the bottom right, such as $F^-$, being the poorest.*

2. From Table 16.1 we can find rather quickly which combinations of reactants lead to spontaneous oxidation-reduction reactions (when the concentrations of the reactants and products are 1 *M* and the partial pressures of any gases involved are 1 atm). It can also be determined whether or not a given reaction, as written, will proceed spontaneously in the forward direction.

Let's consider the reaction between Zn and $Cr^{3+}$. From the table we have the half-reactions,

$$Zn^{2+}\,(aq) + 2e^- \rightleftharpoons Zn\,(s) \qquad \mathscr{E}^0_{Zn} = -0.76\ V$$
$$Cr^{3+}\,(aq) + 3e^- \rightleftharpoons Cr\,(s) \qquad \mathscr{E}^0_{Cr} = -0.74\ V$$

For a spontaneous reaction to occur, the $\mathscr{E}^0_{cell}$ must be positive. We must therefore subtract $\mathscr{E}^0_{Zn}$ from $\mathscr{E}^0_{Cr}$.

$$\begin{aligned}\mathscr{E}^0_{cell} &= \mathscr{E}^0_{Cr} - \mathscr{E}^0_{Zn}\\ &= -0.74 - (-0.76)\\ &= +0.02\ V\end{aligned}$$

Now, we ask, what is the cell reaction? This can be obtained in the following way.

If we reverse the half-reaction for zinc, we can write,

$$Zn\,(s) \rightleftharpoons Zn^{2+}\,(aq) + 2e^- \qquad \mathscr{E}^0 = +0.76\ V$$

In writing the reverse reaction we have changed the sign of the $\mathscr{E}^0$ and have changed it from a reduction potential to an oxidation potential. In general, if we rewrite a half-reaction as an oxidation, its **oxidation potential** is equal to the negative of the reduction potential.

To obtain an overall redox equation, we add half-reactions corresponding to the oxidation and reduction that takes place. We can also obtain the $\mathscr{E}^0_{cell}$ by *adding* an oxidation potential and a reduction potential. In the example we have chosen, we have

$$\begin{array}{rlrl} 3(Zn\,(s) \longrightarrow & Zn^{2+}(aq) + 2e^-) & \mathscr{E}^0 &= +0.76\ V\\ 2(Cr^{3+}\,(aq) + 3e^- \longrightarrow & Cr\,(s)) & \mathscr{E}^0 &= -0.74\ V\\ \hline 3Zn\,(s) + 2Cr^{3+}\,(aq) \longrightarrow & 3Zn^{2+}\,(aq) + 2Cr\,(s) & \mathscr{E}^0 &= (+0.76) + (-0.74)\\ & & \mathscr{E}^0_{cell} &= +0.02\ V \end{array}$$

There are two further points to note here. First, observe that the $\mathscr{E}^0_{cell}$ has the same value as before. Adding a set of oxidation and reduction potentials produces the same effect as subtracting one reduction potential from another. Second, to obtain the overall redox equation the half-reactions had to be multiplied by appropriate factors so that the electrons would cancel when they were added. *The oxidation and reduction potentials, however, are not multiplied by these factors before they are added!* This is because they are intensive quantities and thus are independent of the number of moles of reactants and products involved.

We have also said we can determine whether a reaction, as written, will occur spontaneously. Let's consider the possible reaction,

$$Fe^{2+}\ (aq) + Ni\ (s) \longrightarrow Fe\ (s) + Ni^{2+}\ (aq)$$

If the reaction is spontaneous, its $\mathscr{E}^0_{cell}$ must be positive. To find if this is so, let's divide the overall equation into its half-reactions.

$$\begin{aligned} Fe^{2+}\ (aq) + 2e^- &\longrightarrow Fe\ (s) \\ Ni\ (s) &\longrightarrow Ni^{2+}\ (aq) + 2e^- \end{aligned}$$

The first equation is a reduction; its reduction potential is $\mathscr{E}^0 = -0.44$ V (from Table 16.1). The second equation is an oxidation; its oxidation potential is the negative of the reduction potential found in Table 16.1. The oxidation potential of Ni, then, is $\mathscr{E}^0 = +0.25$ V. The $\mathscr{E}^0_{cell}$ is the sum of this pair of oxidation and reduction potentials.

$$\begin{aligned} \mathscr{E}^0_{cell} &= -0.44 + 0.25 \\ &= -0.19 \text{ V} \end{aligned}$$

Since $\mathscr{E}^0_{cell}$ is negative, we conclude that the reaction of $Fe^{2+}$ with Ni is *not* spontaneous. In fact, it is the reverse reaction that is spontaneous. Therefore, we see that to obtain the spontaneous cell reaction we must reverse the direction of the half-reaction having the smallest (or most negative) reduction potential. At the same time we obtain the oxidation potential for this half-reaction by changing the sign of the reduction potential.

EXAMPLE 16.4 What will be the spontaneous reaction between the following set of half-reactions? What is the value of $\mathscr{E}^0_{cell}$?

$$(1)\ Cr^{3+}\ (aq) + 3e^- \rightleftharpoons Cr\ (s)$$
$$(2)\ MnO_2\ (s) + 4H^+\ (aq) + 2e^- \rightleftharpoons Mn^{2+}\ (aq) + 2H_2O\ (l)$$

SOLUTION From Table 16.1, reaction (1) has $\mathscr{E}^0 = -0.74$ V and reaction (2) has $\mathscr{E}^0 = +1.28$ V. These values are both reduction potentials. To obtain the overall reaction, we reverse the half-reaction having the most negative reduction potential (that is, reaction 1) and obtain its oxidation potential by changing the sign of the reduction potential. Thus,

$$\begin{aligned} 2(Cr\ (s) &\longrightarrow Cr^{3+}\ (aq) + 3e^-) \\ 3(MnO_2(s) + 4H^+(aq) + 2e^- &\longrightarrow Mn^{2+}\ (aq) + 2H_2O\ (l)) \\ \hline 2Cr\ (s) + 3MnO_2(s) + 12H^+\ (aq) &\longrightarrow 2Cr^{3+}\ (aq) + 3Mn^{2+}\ (aq) + 6H_2O\ (l) \end{aligned}$$

| | | |
|---|---|---|
| Oxidation of Cr | $\mathscr{E}^0_{Cr}$ | $= +0.74$ V |
| Reduction of $MnO_2$ | $\mathscr{E}^0_{MnO_2}$ | $= +1.28$ V |
| Cell reaction | $\mathscr{E}^0_{cell}$ | $= (+0.74 + 1.28)$ |
| | $\mathscr{E}^0_{cell}$ | $= 2.02$ V |

Again, note that we *do not* multiply the $\mathscr{E}^0$ values by the factors that we use to make the electrons cancel when combining the half-reactions.

3. In the process of combining half-reactions from Table 16.1, we see that some of the reactants in the spontaneous reaction appear on the left side of one half-reaction while the rest of the reactants are found

on the right side of another half-reaction. In Example 16.4, for instance, we have $MnO_2 + 4H^+$ from the left side of one half-reaction and Cr (*s*) from the right side of the other. The order of these reactions in Table 16.1 is

$$Cr^{3+}\ (aq) + 3e^- \rightleftharpoons Cr\ (s) \qquad \mathcal{E}^0 = -0.74\ V$$
$$MnO_2\ (s) + 4H^+\ (aq) + 2e^- \rightleftharpoons Mn^{2+}\ (aq) + 2H_2O\ (l) \qquad \mathcal{E}^0 = 1.28\ V$$

and we see that the reactants in the overall spontaneous reaction are those substances related by the diagonal line (colored arrow) running from lower left to upper right. As a general statement, then, *we can say that when comparing reactants and products having unit concentrations any species on the left of a given half-reaction will react spontaneously with a substance that is found on the right of a half-reaction located above it in Table 16.1.* We could use this rule of thumb, for example, to tell us that $Br_2$ will react spontaneously with $I^-$ to produce $Br^-$ and $I_2$, while $Br_2$ will *not* react spontaneously with $Cl^-$. Our rule, therefore, permits us to determine the course of a reaction without having to worry about subtracting electrode potentials in the proper sequence.

4. A point worth noting is that a collection of half-reactions, such as that found in Table 16.1, enables us to predict the outcome of many chemical reactions when we know only a relatively few half-reactions and their corresponding reduction potentials. From the 36 half-reactions listed in Table 16.1, for example, we can predict the results of 630 different chemical reactions! A table of this type, therefore, provides us with a very compact way of storing chemical information.
5. With a knowledge of the standard reduction potentials listed in Table 16.1, we account for the course of electrolysis reactions. For example, we know from experiment that we can produce copper by electrolyzing an aqueous solution containing $Cu^{2+}$, but that we cannot obtain aluminum in this same fashion. From Table 16.1 we see that the reduction potential of copper is +0.34 V and that for $H_2O$ it is −0.83 V. Thus copper ion is more readily reduced than $H_2O$ and will plate out on the electrode according to the half-reaction,

$$Cu^{2+}\ (aq) + 2e^- \longrightarrow Cu\ (s)$$

In the case of aluminum, however, we find that the reduction potential for $Al^{3+}$ is −1.66 V, which makes it more difficult to reduce than water. This means that when an aqueous solution containing $Al^{3+}$ ions is electrolyzed, the $H_2O$ will preferentially be reduced.

## 16.8 SPONTANEITY OF OXIDATION-REDUCTION REACTIONS

It was pointed out in Section 10.10 that the thermodynamic criterion for spontaneity of a chemical reaction is that the change in free energy, $\Delta G$, for the reaction has to be a negative quantity. In Section 10.11 we saw that $\Delta G$ also represents the maximum amount of useful work obtainable from a chemical reaction. The relationship between $\Delta G$ and maximum work ($W_{max}$) for any system takes the form,

$$\Delta G = -W_{max}$$

What is $W_{max}$ for an electrochemical cell?

The work derived from an electrochemical cell is, perhaps, not unlike that obtained from a waterwheel, shown in Figure 16.9. The amount of work that can be obtained from this waterwheel depends on two things: (1) the volume of water flowing over the blades of the wheel, and (2) the energy given to the wheel per unit volume of water as it drops to the lower level of the stream:

$$\text{work} = (\text{volume of water}) \times \left(\frac{\text{energy released}}{\text{unit volume}}\right)$$

Similarly, the work done by an electrochemical cell is dependent on (1) the number of coulombs that flow and (2) the energy available per coulomb:

$$\text{work} = (\text{number of coulombs}) \times \left(\frac{\text{energy available}}{\text{coulomb}}\right)$$

The number of coulombs that flow is equal to the number of moles of electrons that are involved in the redox reaction, $n$, multiplied by the faraday (which is the number of coulombs per mole of electrons):

$$\text{number of coulombs} = n\mathcal{F}$$

The value of $n$ depends on the nature of the half-reactions taking place in the cell and can be derived once the specific reactions are known. For example, in the Zn/Cu cell there are two electrons involved in both of the half-reactions; therefore, $n$ for this cell is 2.

The energy available per coulomb is simply the emf of the cell, because the volt is equal to the energy per coulomb (Equation 16.4):

$$\frac{\text{energy available}}{\text{coulomb}} = \text{emf}$$

When the emf is a maximum, the work derived from the cell is also a maximum. In Section 6 of this chapter we saw that the maximum emf, $\mathcal{E}_{\text{cell}}$, is that voltage obtained when no current is flowing through the cell.

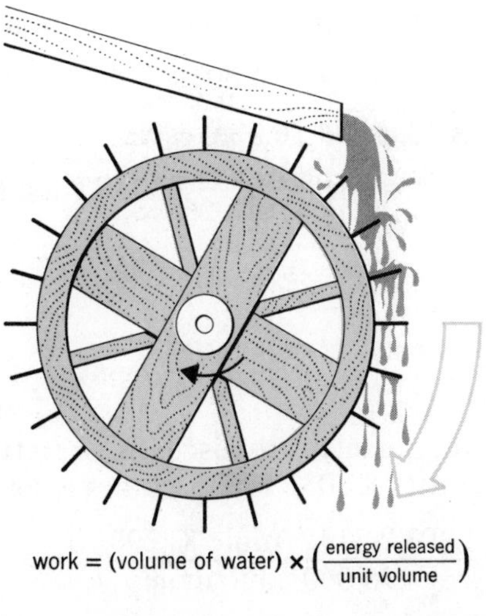

work = (volume of water) × (energy released / unit volume)


Figure 16.9
*Work obtained from a waterwheel.*

Thus the equation for maximum work for an electrochemical cell is

$$W_{max} = n \times \mathcal{F} \times \mathcal{E}$$

$$\text{joules} = (\text{moles of electrons}) \times \left(\frac{\text{coulombs}}{\text{mole}}\right) \times \left(\frac{\text{joules}}{\text{coulomb}}\right)$$

Since $\Delta G = -W_{max}$, then

$$\Delta G = -n\mathcal{F}\mathcal{E}_{cell} \qquad (16.5)$$

When all species are at unit concentration as identified by $\mathcal{E}^0$, then $\Delta G$ becomes the standard free-energy change for the reaction and is given the symbol $\Delta G^0$. Thus, the above equation becomes

$$\Delta G^0 = -n\mathcal{F}\mathcal{E}^0_{cell} \qquad (16.6)$$

With the aid of Equation 16.6 we can calculate the standard free-energy change for an oxidation-reduction reaction from a knowledge of its standard cell potential. Consider, for example, the Zn/Cu cell.

$$\begin{array}{rcl} Zn\,(s) & \longrightarrow & Zn^{2+}\,(aq) + 2e^- \\ Cu^{2+}\,(aq) + 2e^- & \longrightarrow & Cu\,(s) \\ \hline Zn\,(s) + Cu^{2+}\,(aq) & \longrightarrow & Zn^{2+}\,(aq) + Cu\,(s) \end{array}$$

The $n$ for this reaction is 2, because two electrons are transferred, $\mathcal{F}$ = 96,500 coul/mole of electrons, and $\mathcal{E}^0_{cell}$, either derived from Table 16.1 or determined experimentally, is +1.10 V; therefore,

$$\Delta G^0 = -2 \text{ moles } e^- \times \left(\frac{96{,}500 \text{ coul}}{\text{mole e}^-}\right) \times \left(\frac{1.10 \text{ J}}{\text{coul}}\right)$$
$$= -212{,}000 \text{ J}$$

This relationship between the standard cell potential and $\Delta G^0$ is an extremely important one, because it ties together two different aspects of spontaneity while at the same time giving us a readily accessible pathway for calculating standard free-energy changes. Experimentally, we have only to measure the standard cell emf, from which we can then compute the value of $\Delta G^0$. But still more important, perhaps, is that through this equation we can derive even more useful thermodynamic quantities.

## 16.9 THERMODYNAMIC EQUILIBRIUM CONSTANTS

In Section 12.4 we saw that for reactions in solution,

$$\Delta G^0 = -2.303RT \log K_c \qquad (16.7)$$

Combining Equations 16.6 and 16.7 gives us

$$\Delta G^0 = -n\mathcal{F}\mathcal{E}^0 = -2.303RT \log K_c$$

or simply

$$n\mathcal{F}\mathcal{E}^0 = 2.303RT \log K_c$$

Solving for $\mathcal{E}^0$, we have

$$\mathcal{E}^0 = \frac{2.303RT}{n\mathcal{F}} \log K_c \qquad (16.8)$$

If we choose to restrict ourselves to discussing reactions that take place at 25°C (298 K), the quantity $2.303RT/\mathcal{F}$ becomes a constant,

$$\frac{2.303RT}{\mathcal{F}} = \frac{2.303(8.314 \text{ J/mole K})(298 \text{ K})}{96{,}500 \text{ coul/mole}} = 0.0592\text{J/coul}$$

Since 1 V = 1 J/coul,

$$\frac{2.303RT}{\mathscr{F}} = 0.0592 \text{ V}$$

Thus, at 25°C, Equation 16.8 becomes

$$\mathscr{E}^0 = \frac{0.0592}{n} \log K_c$$

Solving for log $K_c$ gives

$$\log K_c = \frac{n\mathscr{E}^0}{0.0592} \tag{16.9}$$

We see, therefore, that from a knowledge of the standard cell potential, the equilibrium constant for the cell reaction can be calculated. For the Zn/Cu cell, we have

$$\log K_c = \frac{n\mathscr{E}^0}{0.0592} = \frac{2(+1.10)}{0.0592} = 37.1$$

Hence,

$$K_c \approx 1 \times 10^{37}$$

From the magnitude of this equilibrium constant, we could certainly say that the spontaneous Zn/Cu cell reaction will go very nearly to completion.

---

EXAMPLE 16.5 Using Table 16.1, determine whether or not the overall oxidation-reduction reaction,

$$\text{Sn }(s) + \text{Ni}^{2+} \longrightarrow \text{Sn}^{2+} + \text{Ni }(s)$$

is spontaneous and calculate its equilibrium constant.

SOLUTION The two half-reactions of this overall reaction are

$$\text{(oxidation)} \qquad \text{Sn }(s) \longrightarrow \text{Sn}^{2+}\ (aq) + 2e^- \qquad \mathscr{E}^0 = +0.14 \text{ V}$$

and

$$\text{(reduction)} \qquad \text{Ni}^{2+}\ (aq) + 2e^- \longrightarrow \text{Ni }(s) \qquad \mathscr{E}^0 = -0.25 \text{ V}$$

From the way the overall reaction is written, the $\mathscr{E}^0$ for the cell is

$$\begin{aligned}\mathscr{E}_{\text{cell}} &= +0.14 + (-0.25)\\ &= -0.11 \text{ V}\end{aligned}$$

This means that under *standard conditions* (unit concentrations), the reaction in this direction is nonspontaneous. We can still calculate the equilibrium constant in the same fashion as outlined above. Since two electrons are transferred during the reaction,

$$\log K_c = \frac{2(-0.11)}{0.0592} = -3.7$$

Taking the antilogarithm gives

$$K_c = 2 \times 10^{-4}$$

From the size of this equilibrium constant, we can say that this reaction will not occur to an appreciable extent in the forward direction.

## 16.10 CONCENTRATION EFFECT ON CELL POTENTIAL

Thus far we have limited our discussion to those cells containing reactants at unit concentration. In the laboratory, however, we usually do not restrict ourselves to only this one set of conditions, and it is found that the cell emf, and in fact the direction of the cell reaction, can be controlled by the concentrations of the species taking part in the reaction. Let us examine this now from a quantitative point of view.

An equation that summarizes how the free energy of the reactants and products of a given reaction varies with temperature and concentration was given in Section 12.4. For the generalized reaction,

$$aA + bB \longrightarrow eE + fF$$

this equation takes the form,

$$\Delta G = \Delta G^0 + 2.303RT \log\left(\frac{[E]^e[F]^f}{[A]^a[B]^b}\right)$$

Equations 16.5 and 16.6 (in Section 16.8) show the relationship between $\Delta G$ and $\mathcal{E}$, and $\Delta G^0$ and $\mathcal{E}^0$, respectively, for an oxidation-reduction reaction. Substituting these expressions for $\Delta G$ and $\Delta G^0$ into the above equation, we have

$$-n\mathcal{F}\mathcal{E} = -n\mathcal{F}\mathcal{E}^0 + 2.303RT \log\left(\frac{[E]^e[F]^f}{[A]^a[B]^b}\right)$$

which can be rearranged to give

$$\mathcal{E} = \mathcal{E}^0 - \frac{2.303RT}{n\mathcal{F}} \log\left(\frac{[E]^e[F]^f}{[A]^a[B]^b}\right) \qquad (16.10)$$

This equation, first developed by Walter Nernst in 1889, now bears his name and is called the **Nernst equation.**

At 25°C we have seen that the numerical value of $2.303RT/\mathcal{F}$ is 0.0592. Therefore, at 25°C the Nernst equation becomes

$$\mathcal{E} = \mathcal{E}^0 - \frac{0.0592}{n} \log\left(\frac{[E]^e[F]^f}{[A]^a[B]^b}\right) \qquad (16.11)$$

We can see from Equation 16.11 that when all ionic species are present at unit concentration, the log term becomes zero (log 1 = 0) and the emf of the cell becomes $\mathcal{E}^0$; that is, at unit concentration $\mathcal{E} = \mathcal{E}^0$. This, of course, must be true in light of our basic definition of $\mathcal{E}^0$. When the species in a cell are not present at unit concentration, $\mathcal{E}$ is generally not equal to $\mathcal{E}^0$ and the Nernst equation must be employed to calculate $\mathcal{E}$. For example, in the case of the Zn/Cu cell, whose cell reaction is

$$Zn\ (s) + Cu^{2+}\ (aq) \longrightarrow Cu\ (s) + Zn^{2+}\ (aq)$$

the Nernst equation takes the form

$$\mathcal{E} = \mathcal{E}^0 - \frac{0.0592}{n} \log \frac{[Zn^{2+}]}{[Cu^{2+}]}$$

Note that as usual we omit the concentrations of pure solids from the mass action expression.[3] Since two electrons are transferred in the reaction, n = 2 and

$$\mathscr{E} = \mathscr{E}^0 - 0.0296 \log \frac{[Zn^{2+}]}{[Cu^{2+}]}$$

Thus we see that $\mathscr{E}$ can be calculated for any particular cell if the $Zn^{2+}$ and $Cu^{2+}$ concentrations are known. This use of the Nernst equation is illustrated in the next example.

EXAMPLE 16.6 Calculate the emf of the Zn/Cu cell under the following conditions:

$$Zn\ (s) + Cu^{2+}\ (0.020\ M) \longrightarrow Cu\ (s) + Zn^{2+}\ (0.40\ M)$$

SOLUTION We have just seen that for this system the Nernst equation is

$$\mathscr{E} = \mathscr{E}^0 - 0.0296 \log \frac{[Zn^{2+}]}{[Cu^{2+}]}$$

We can also calculate $\mathscr{E}^0_{cell} = +1.10$ V for the equation as written. Substituting this $\mathscr{E}^0$ and the concentrations of the $Zn^{2+}$ and $Cu^{2+}$ into the Nernst equation, we have

$$\begin{aligned}\mathscr{E} &= 1.10 - 0.0296 \log \frac{(0.40)}{(0.020)} \\ &= 1.10 - 0.0385 \\ &= 1.06\ \text{V}\end{aligned}$$

Thus we see that under these conditions of concentration the voltage obtained from this cell is slightly less than that obtained at unit concentration.

**CONCENTRATION CELLS.** Just as the cell emf is dependent on the concentration of the ions involved in the half-reaction, so we find that the reduction potential of the individual half-reactions is also determined by the concentration of the ions involved. This effect of the concentration on the reduction potential can also be given by the Nernst equation. For example, if we consider the half-reaction,

$$Zn^{2+}\ (aq) + 2e^- \rightleftharpoons Zn\ (s)$$

the Nernst equation at 25°C takes the form,

$$\mathscr{E} = \mathscr{E}^0 - \frac{0.0592}{2} \log \frac{1}{[Zn^{2+}]}$$

As usual, we have omitted the concentration of the solid from the mass action expression.

Because the reduction potential of an electrode depends on the concentrations of the ions in solution, it is possible to construct a cell in which the cathode and anode compartments contain the same electrode materials but different concentrations of the ions. Such a cell is called a **concentration cell** and is illustrated in Figure 16.10.

[3] The Nernst equation applies exactly only if we use activities. The activity of any pure solid or liquid is equal to 1. Errors introduced by using the concentrations of the ions instead of their activities are small, as mentioned before, provided that the solutions are relatively dilute.

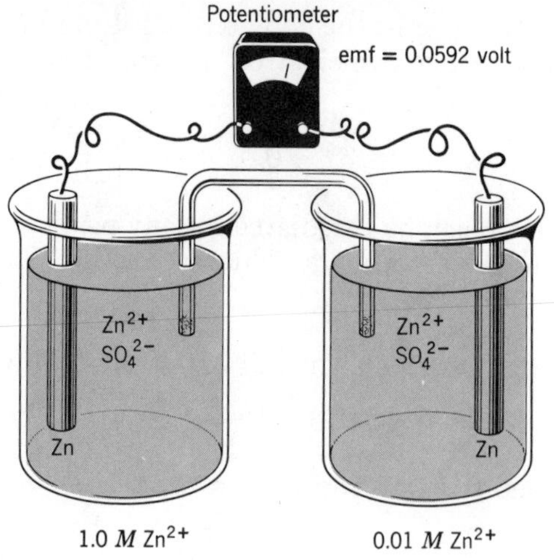


Figure 16.10
*Concentration cell.*

In Figure 16.10 we have a cell composed of two zinc electrodes placed in separate solutions of $ZnSO_4$ whose $Zn^{2+}$ concentrations are different and are separated by a salt bridge. The concentration of the $Zn^{2+}$ on the left (1.0 *M*) is 100 times greater than the $Zn^{2+}$ concentration in the right compartment, and when the circuit is completed a spontaneous reaction takes place in a direction that tends to make the two $Zn^{2+}$ concentrations become equal. Thus in the more concentrated side $Zn^{2+}$ ions disappear forming Zn (*s*), in order to decrease the $Zn^{2+}$ concentration, and in the more dilute side, more $Zn^{2+}$ will be produced. Thus we have, in the more concentrated compartment,

$$Zn^{2+}\ (1\ M) + 2e^- \longrightarrow Zn\ (s) \qquad \text{(reduction)}$$

and on the more dilute side,

$$Zn\ (s) \longrightarrow Zn^{2+}\ (0.01\ M) + 2e^- \qquad \text{(oxidation)}$$

From our earlier discussions we know that the potential of a cell is found by subtracting the reduction potential of the half-cell in which oxidation occurs from the reduction potential of the half-cell that undergoes reduction. For this concentration cell, then,

$$\mathscr{E}_{\text{cell}} = \mathscr{E}_{\text{conc}} - \mathscr{E}_{\text{dil}}$$

where $\mathscr{E}_{\text{conc}}$ and $\mathscr{E}_{\text{dil}}$ are the electrode potentials of the concentrated and dilute half cells, respectively. These are given as

$$\mathscr{E}_{\text{conc}} = \mathscr{E}^0_{\text{Zn}} - \frac{0.0592}{2} \log \frac{1}{[Zn^{2+}]_{\text{conc}}}$$

and

$$\mathscr{E}_{\text{dil}} = \mathscr{E}^0_{\text{Zn}} - \frac{0.0592}{2} \log \frac{1}{[Zn^{2+}]_{\text{dil}}}$$

Therefore,

$$\mathscr{E}_{\text{cell}} = (\mathscr{E}^0_{\text{Zn}} - \mathscr{E}^0_{\text{Zn}}) - \frac{0.0592}{2} \log \frac{1}{[Zn^{2+}]_{\text{conc}}} - \log \frac{1}{[Zn^{2+}]_{\text{dil}}}$$

or

$$\mathscr{E}_{cell} = -\frac{0.0592}{2} \log \frac{[Zn^{2+}]_{dil}}{[Zn^{2+}]_{conc}}$$

Substituting the concentrations of $Zn^{2+}$ into this expression allows us to compute the cell potential.

$$\mathscr{E}_{cell} = -\frac{0.0592}{2} \log \frac{(0.01)}{(1)}$$
$$= 0.0592 \text{ V}$$

In general, for any concentration cell we could write

$$\mathscr{E}_{cell} = -\frac{0.0592}{n} \log \frac{[M^{n+}]_{dil}}{[M^{n+}]_{conc}}$$

The voltage obtained from this type of cell is usually small and will continually decrease as the concentrations in the two compartments approach each other. The voltage becomes zero when the concentration of the ions in each compartment of the concentration cell are the same.

**SOLUBILITY PRODUCT CONSTANT.** The Nernst equation can also be useful in determining the solubility product constant of an insoluble salt. To find the $K_{sp}$ of $PbSO_4$, for example, an experiment might be designed in the following fashion: a galvanic cell is prepared consisting of $Pb/Pb^{2+}$ versus an $Sn/Sn^{2+}$ electrode with a salt bridge connecting them. In the tin compartment the $Sn^{2+}$ concentration is held constant at 1 *M*. In the lead compartment $SO_4^{2-}$ is added to precipitate $PbSO_4$ and thereby establish the equilibrium,

$$PbSO_4\ (s) \rightleftharpoons Pb^{2+}\ (aq) + SO_4^{2-}\ (aq)$$

The $SO_4^{2-}$ concentration in the lead compartment is then adjusted until it is 1 *M*, and the emf of the cell is found to be +0.22 V. It is also observed that the Pb electrode is negative with respect to the Sn electrode, thereby indicating that the Pb is undergoing oxidation while the $Sn^{2+}$ is reduced. The cell reaction must therefore be

$$Pb\ (s) + Sn^{2+}\ (1\ M) \longrightarrow Pb^{2+}\ (?) + Sn\ (s)$$

and the calculated $\mathscr{E}^0$ is

$$\mathscr{E}^0 = \mathscr{E}_{Sn} - \mathscr{E}_{Pb}$$
$$= (-0.14\text{ V}) - (-0.13\text{ V})$$
$$= -0.01\text{ V}$$

(Note that *if* the $Sn^{2+}$ and $Pb^{2+}$ concentrations were both 1 *M*, the reaction would be spontaneous from right to left, rather than from left to right.)

We can calculate the concentration of $Pb^{2+}$ by using the Nernst equation, which takes the form,

$$\mathscr{E} = \mathscr{E}^0 - \frac{0.0592}{n} \log \frac{[Pb^{2+}]}{[Sn^{2+}]}$$

for this cell reaction. We know $\mathscr{E}$, $\mathscr{E}^0$, and $[Sn^{2+}]$ and, because there are two electrons transferred in this reaction, the above equation becomes, after substitution,

$$0.22\text{ V} = -0.01\text{ V} - \frac{0.0592}{2} \log \frac{[Pb^{2+}]}{(1)}$$

Solving for $\log[Pb^{2+}]$ gives us

$$-0.22\ V = 0.01\ V + 0.0296 \log[Pb^{2+}]$$

or

$$\log[Pb^{2+}] = \frac{-0.22\ V - 0.01\ V}{0.0296\ V}$$

and

$$\log[Pb^{2+}] = -7.8$$

Taking the antilogarithm, we find, then, that the concentration of $Pb^{2+}$ in this cell is

$$[Pb^{2+}] = 2 \times 10^{-8}\ M$$

The expression for $K_{sp}$ of $PbSO_4$ is

$$K_{sp} = [Pb^{2+}][SO_4^{2-}]$$

Since

$$[Pb^{2+}] = 2 \times 10^{-8}$$
$$[SO_4^{2-}] = 1\ M$$

then

$$K_{sp} = (2 \times 10^{-8})(1) = 2 \times 10^{-8}$$

which is the value that was given in Table 15.1.

**DETERMINATION OF pH.** An extremely important application of the Nernst equation is that it can be used to calculate the concentration of a single ionic species by measuring experimentally the potential of a carefully designed cell. We have already seen one example of this in the determination of $K_{sp}$. If we were to use the $Cu/H_2$ cell, discussed in Section 16.7, we could determine the $H^+$ concentration of a solution and then calculate its pH. The cell reaction for the $Cu/H_2$ is

$$Cu^{2+}\ (aq) + H_2\ (g) \longrightarrow Cu\ (s) + 2H^+\ (aq)$$

and the corresponding form of the Nernst equation at 25°C is

$$\mathscr{E} = \mathscr{E}^0 - \frac{0.0592}{n} \log \frac{[H^+]^2}{[Cu^{2+}]p_{H_2}}$$

If the concentration of $Cu^{2+}$ is 1 $M$ and the pressure of $H_2$ is 1 atm, this equation reduces to

$$\mathscr{E} = \mathscr{E}^0 - \frac{0.0592}{n} \log[H^+]^2$$

which is the same as

$$\mathscr{E} = \mathscr{E}^0 - \frac{(0.0592)(2)}{n} \log[H^+]$$

Let's rewrite this equation as

$$\mathscr{E} = \mathscr{E}^0 + \frac{(0.0592)(2)}{n} (-\log[H^+])$$

We see that because $\mathscr{E}^0$ and the quantity $0.0592(2)/n$ are both constant for a specific reaction, then

$$\mathscr{E} \propto -\log[H^+]$$

By definition, pH = $-\log[H^+]$ and, consequently, we have

$$\mathscr{E} \propto \text{pH}$$

Thus, by measuring the emf of a galvanic cell containing a reference electrode (such as the Cu, $Cu^{2+}$ electrode here) and the hydrogen electrode, the pH of a solution can be calculated. One such application is shown in the next example.

EXAMPLE 16.7 A galvanic cell consisting of a Cu versus a hydrogen electrode was used to determine the pH of an unknown solution. The unknown was placed in the hydrogen electrode compartment and the pressure of the hydrogen gas was controlled at 1 atm. The concentration of $Cu^{2+}$ was 1 *M* and the emf of the cell at 25°C was determined to be +0.48 V. Calculate the pH of this unknown solution.

SOLUTION The cell reaction for the Cu/$H_2$ cell is

$$Cu^{2+}(1\ M) + H_2\ (g)(1\ \text{atm}) \longrightarrow Cu\ (s) + 2H^+\ (?M)$$

for which we write the Nernst equation as

$$\mathscr{E} = \mathscr{E}^0 - \frac{0.0592}{n} \log \frac{[H^+]^2}{[Cu^{2+}]p_{H_2}}$$

or, because $[Cu^{2+}] = 1\ M$ and $p_{H_2} = 1$ atm,

$$\mathscr{E} = \mathscr{E}^0 - \frac{(0.0592)(2)}{n} \log[H^+]$$

The $\mathscr{E}^0$ for this cell is +0.34 V; the value of *n* is 2. Substituting these values as well as the measured value of $\mathscr{E}$ into the equation, we have

$$+0.48 = +0.34 - 0.0592 \log[H^+]$$

and hence

$$-\log[H^+] = \frac{0.48 - 0.34}{0.0592} = 2.4$$

Therefore the pH of this solution is 2.4.

## 16.11 ION-SELECTIVE ELECTRODES

Bmit

The last example represents only a single case where, with the proper choice of electrodes, the concentration of a single ionic species can be selectively measured. Through many years of research in this particular area of electrochemistry, scientists have developed many practical electrodes whose emf depends on the concentration of only one species. Such electrodes are called **ion-selective electrodes** and are used in conjunction with a reference electrode, whose potential always remains constant and is of a known value. Thus, when an ion-selective electrode is placed in a solution with a reference electrode, only the ion-selective electrode changes in emf, and the measured voltage can immediately be used to calculate the concentration of the species being determined. Such electrodes have been found to be of great importance in such areas as chemical analysis, pollution analysis, clinical measurements, oceanography, and geology.

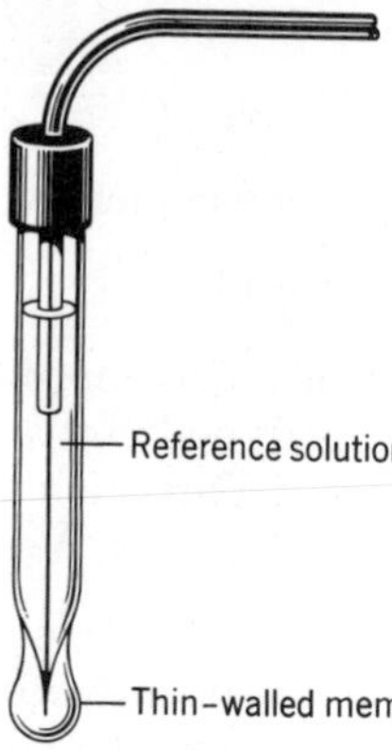


Figure 16.11
*Ion-selective electrode.*

One type of ion-selective electrode, shown in Figure 16.11, consists of a very thin-walled membrane that is sealed onto one end of a hollow tube. Inside the tube in contact with the membrane is a reference solution and immersed into this solution is a wire. The wire extends from the reference solution through the tube and out the top, to make electrical contact with the outside circuitry. The material used to make the membrane as well as the composition of the reference solution depends on the species that is to be measured. Some of the cations and anions whose concentration can be determined by these electrodes are listed in Table 16.2. Let us now discuss a few of these electrodes.

The **glass electrode** is one ion-selective electrode. The membrane in this electrode is made of an extremely thin piece of glass, and the reference solution inside is a dilute HCl solution whose $H^+$ concentration is known and which remains constant. The wire electrode is a silver wire coated with silver chloride. The emf of the glass electrode is sensitive to the relative concentrations of $H^+$ inside and outside across the thin glass membrane. Since the $H^+$ concentration inside is constant, the emf of the electrode, in effect, is determined by the concentration of $H^+$ in the solution in contact with the membrane on the outside.

The glass electrode can be made selective to various other monovalent cations such as $Na^+$, $K^+$, and $NH_4^+$ by suitable changes in the composition of the glass. Still other ions can be detected by electrodes if the glass membrane is replaced by a solid crystal. For example, when a solid crystal such as

Table 16.2
*Ions whose concentration can be determined by an ion-selective electrode*

| Cations | Anions |
|---|---|
| Cadmium ($Cd^{2+}$) | Bromide ($Br^-$) |
| Calcium ($Ca^{2+}$) | Chloride ($Cl^-$) |
| Copper ($Cu^{2+}$) | Cyanide ($CN^-$) |
| Hydrogen ($H_3O^+$) | Fluoride ($F^-$) |
| Lead ($Pb^{2+}$) | Nitrate ($NO_3^-$) |
| Mercury ($Hg^{2+}$) | Perchlorate ($ClO_4^-$) |
| Potassium ($K^+$) | Sulfide ($S^{2-}$) |
| Silver ($Ag^+$) | Thiocyanate ($SCN^-$) |
| Sodium ($Na^+$) | |

$LaF_3$ is used, fluoride ion concentrations can be detected, and when $Ag_2S$ is used, silver and sulfide ion concentrations can be determined. Other ions such as chloride, cyanide, and lead can be determined using a membrane made of silver mixed with silver sulfide.

Ion-selective electrodes have also become very important and useful in the study of biological processes. One of these, called an **enzyme-substrate electrode,** employs a glass electrode sensitive to ammonium ion coated with a thin layer of a gel containing an enzyme, as shown in Figure 16.12. Enzymes are large organic molecules that catalyze very specific chemical reactions in biochemical systems and if, for example, the enzyme in the gel is urease, the decomposition of urea to produce ammonium ion will occur when the solution around the electrode contains urea. This ammonium ion is detected by the electrode and, in effect, the electrode becomes sensitive to the presence of urea in the solution being tested.

The application of electrochemistry to biochemical research is still in its infancy. An illustration of how this field of research is progressing is the relatively recent development of miniature ion-selective electrodes that can actually be placed into a living cell to monitor changes in the concentrations of ions during the life process.

The field of medicine is also putting ion-selective electrodes to use. Miniaturized glass electrodes have been developed that fit within a small hypodermic syringe and that can monitor pH in capillary blood vessels. It is anticipated that they will be useful in monitoring respiratory distress (which shows up as a blood pH change) in the human fetus.

## 16.12 SOME PRACTICAL GALVANIC CELLS

As the final section in this chapter, let us discuss some galvanic cells that play an important part in our lives by providing us with electrical power.

**DRY CELL.** This type of cell is used in flashlights, portable radios, toys, and the like. A cutaway diagram of a typical dry cell is shown in Figure 16.13. Dry cells have an exterior layer of either cardboard or metal which serves only as a seal against the atmosphere. Inside this outer shell is a zinc cup that serves as the anode. The zinc cup is filled with a moist paste consisting of ammonium chloride, manganese dioxide, and finely divided carbon. Immersed in this paste is a graphite rod, which serves as the cathode. The

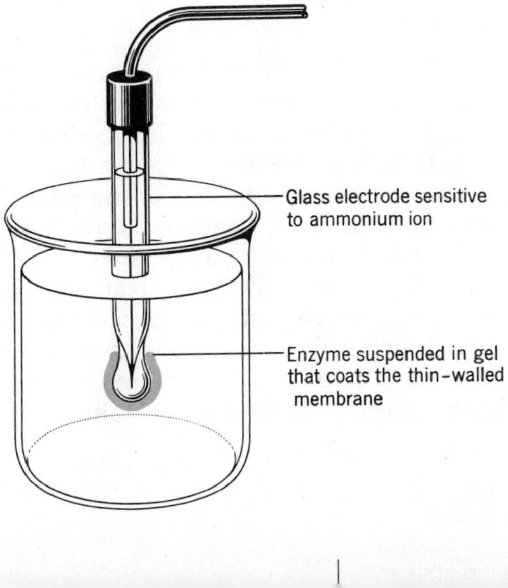


Figure 16.12
*Enzyme-substrate electrode.*

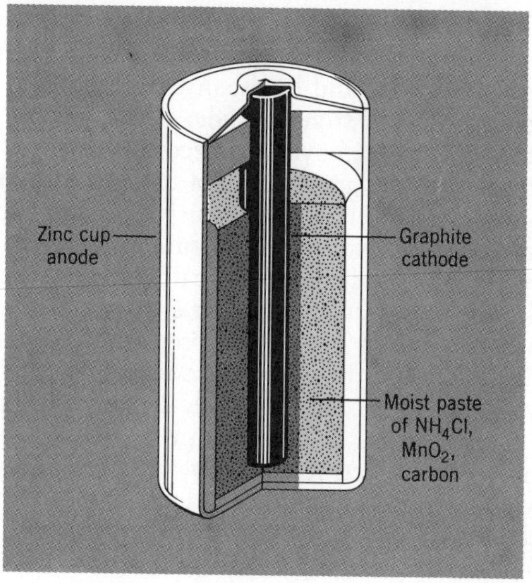


Figure 16.13
*The dry cell.*

chemical reactions that take place when the circuit is completed are actually quite complex and, in fact, are not completely understood. The following, however, is perhaps a reasonable estimate of what occurs.

At the anode zinc is oxidized,

$$Zn\,(s) \longrightarrow Zn^{2+}\,(aq) + 2e^- \qquad \text{(anode)}$$

while at the carbon cathode the $MnO_2/NH_4Cl$ mixture undergoes reduction to give a complex mixture of products. One of these reactions appears to be

$$2MnO_2\,(s) + 2NH_4^+\,(aq) + 2e^- \longrightarrow Mn_2O_3\,(s) + 2NH_3\,(aq) + H_2O\,(l) \qquad \text{(cathode)}$$

The ammonia produced at the cathode reacts with part of the $Zn^{2+}$ formed at the anode to give the complex ion, $Zn(NH_3)_4^{2+}$. Because of the complex nature of the dry cell, no simple overall cell reaction can be written.

Dry cells cannot be effectively recharged and, therefore, have a relatively short lifetime (as compared to the rechargeable lead storage and nickel-cadmium batteries, for example).

**LEAD STORAGE BATTERY.** The common automobile battery is a lead storage battery that usually delivers either 6 or 12 V, depending on the number of cells used in its construction. The inside of the battery consists of a number of galvanic cells connected to each other in series (Figure 16.14).

To increase the current output, each of the individual cells contains a number of lead anodes connected together, plus a number of cathodes, composed of $PbO_2$, also joined together. These electrodes are immersed in an electrolyte composed of dilute sulfuric acid (actually about 30% by weight in a fully charged cell). A single lead storage cell delivers 2 V, so that a 12-V battery contains six such cells connected in series.

When the external circuit is complete and the battery is in operation, the following oxidation-reduction reactions take place:

$$\text{(anode)} \quad Pb\,(s) + SO_4^{2-}\,(aq) \longrightarrow PbSO_4\,(s) + 2e^-$$
$$\text{(cathode)} \quad PbO_2\,(s) + 4H^+\,(aq) + SO_4^{2-}\,(aq) + 2e^- \longrightarrow PbSO_4\,(s) + 2H_2O\,(l)$$

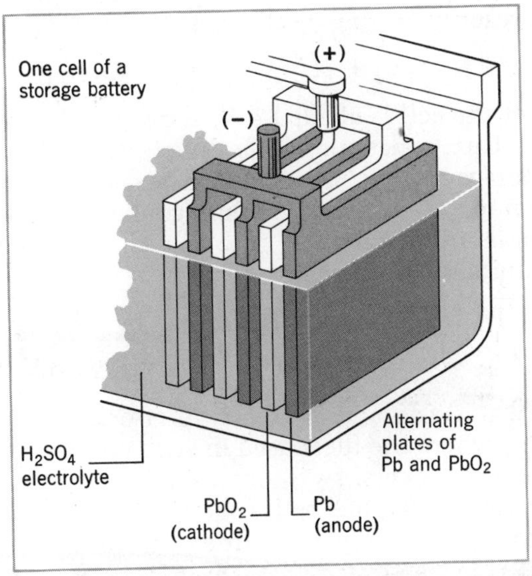


Figure 16.14
*Lead storage battery.*

and the overall reaction is

$$Pb\ (s) + PbO_2\ (s) + 4H^+\ (aq) + 2SO_4^{2-}\ (aq) \longrightarrow 2PbSO_4\ (s) + 2H_2O\ (l)$$

These batteries have the advantage that the electrode reactions can be reversed by placing across the electrodes a voltage that is slightly larger than that which the battery can deliver. The recharging operation is performed in such a way that the negative external voltage is applied to the negative pole and the positive voltage to the positive pole. In doing this, some of the $H_2SO_4$ that is used up while the battery is in operation is restored. This is accomplished by the generator or alternator of the car, or, if the battery is really run down, with the aid of a battery charger.

A convenient method of estimating the degree to which the battery has been discharged is by checking the density (or specific gravity) of the electrolyte. If the battery is in a weakened state, the electrolyte will be mostly water (the product of our overall reaction) and have a density somewhere near 1 g/ml. If, however, the battery is in good operating order, with a full charge, the density of the electrolyte will be somewhat higher than 1 g/ml (the density of concentrated sulfuric acid is 1.8 g/ml). The mechanic in a garage can perform this test with the aid of a hydrometer, a device having a float that sinks to a depth that is a function of the density of the liquid in which it is immersed.

**NICKEL-CADMIUM CELL.** A storage cell that has acquired widespread use in the last few years is the "nicad," or nickel-cadmium battery. The anode in the cell is composed of cadmium, which undergoes oxidation in an alkaline (basic) electrolyte.

anode reaction: $$Cd\ (s) + 2OH^-\ (aq) \longrightarrow Cd(OH)_2\ (s) + 2e^-$$

The cathode is composed of $NiO_2$, which undergoes reduction.

cathode reaction: $$NiO_2\ (s) + 2H_2O\ (l) + 2e^- \longrightarrow Ni(OH)_2\ (s) + 2OH^-\ (aq)$$

The net cell reaction during discharge is therefore

$$Cd\ (s) + NiO_2\ (s) + 2H_2O\ (l) \longrightarrow Cd(OH)_2\ (s) + Ni(OH)_2\ (s)$$

The voltage of the cell is about 1.4 V, somewhat less than the dry cell.

The nicad battery has some appealing features. First, it has a longer life than a lead storage battery. Second, it can be packaged in a sealed unit, much like the common dry cell. These advantages have made the nicad the choice among manufacturers of devices such as rechargeable calculators and electronic flash units in photography.

**FUEL CELLS.** Fuel cells are another means by which chemical energy may be converted into electrical energy. When gaseous fuels, such as $H_2$ and $O_2$, are allowed to undergo reaction in a carefully designed environment, electrical energy can be obtained. This type of cell finds great importance in space vehicles, where the fuels used in such cells can be the same as those used to power the rockets.

A diagram of a $H_2/O_2$ fuel cell is shown in Figure 16.15. In this cell there are three compartments separated from one another by porous electrodes. The hydrogen gas is fed into one compartment and the oxygen gas is fed into another. These gases then diffuse (not bubble) slowly through the electrodes and react with an electrolyte that is in the center compartment. The electrodes are made of a conducting material, such as carbon, with a sprinkling of platinum to act as a catalyst, and the electrolyte is an aqueous solution of a base.

At the cathode the oxygen undergoes reduction, producing $OH^-$ ions, which can be expressed as

$$O_2\ (g) + 2H_2O\ (l) + 4e^- \longrightarrow 4OH^-\ (aq)$$

These $OH^-$ ions travel to the anode where they undergo reaction with $H_2$:

$$H_2\ (g) + 2OH^-\ (aq) \longrightarrow 2H_2O\ (l) + 2e^-$$

The net reaction in the cell is

$$2H_2\ (g) + O_2\ (g) \longrightarrow 2H_2O\ (l)$$

The fuel cell is operated at a high temperature so that the water that is formed as a product of the cell reaction evaporates and may be condensed

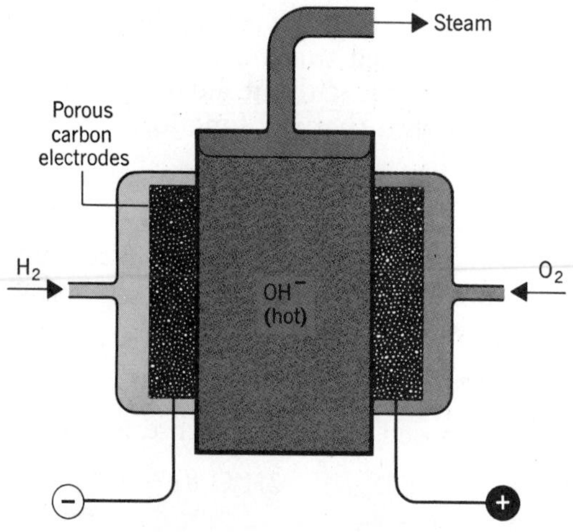


Figure 16.15
*Hydrogen-oxygen fuel cell.*

and used as drinking water for an astronaut. A number of these cells are usually placed together so that several kilowatts of power can be produced.

Fuel cells offer several advantages over other sources of energy. Unlike the dry cell or storage battery, the cathode and anode reactants may be continually supplied so that, in principle, energy can be withdrawn indefinitely from a fuel cell as long as the outside supply of fuel is maintained. Another advantage of the fuel cell is that the energy is extracted from the reactants under more nearly reversible conditions. Therefore, the thermodynamic efficiency of the reaction, in terms of producing useful work, is higher than when the reactants such as $H_2$ and $O_2$ are burned to produce heat that must be subsequently harnessed to produce work. These two advantages suggest that the development of fuel cells will probably continue at an accelerated pace in the future, particularly in light of recent energy shortages caused by greater demands for energy.



## INDEX TO QUESTIONS AND PROBLEMS (problem numbers in italics)



## REVIEW QUESTIONS

**16.1** Distinguish between: (a) electrolytic and galvanic cells, (b) metallic and electrolytic conduction, (c) oxidation and reduction.

**16.2** Why must oxidation-reduction occur in order to maintain a steady flow of electricity during electrolytic conduction?

**16.3** How do we define anode and cathode?

**16.4** Write equations for the half-reactions for the oxidation and reduction of water.

**16.5** From the reactions discussed in Section 16.2, predict the products that you would obtain in the electrolysis of an aqueous solution of $H_2SO_4$.

**16.6** What is the function of an electrolyte such as $Na_2SO_4$ or $H_2SO_4$ during the electrolysis of water? Why can't we carry out electrolysis on pure $H_2O$?

**16.7** Why is cryolite mixed with the $Al_2O_3$ prior to its electrolysis to produce Al?

**16.8** Why can't Al be produced by electrolysis of an aqueous solution containing a salt such as $Al_2(SO_4)_3$?

**16.9** Write a series of chemical equations representing the reactions involved with the recovery of Mg from sea water.

**16.10** Describe the electrolytic purification of metallic copper. Why is the process economically feasible?

**16.11** What is a faraday?

**16.12** How does a coulometer work? What advantages are there to the use of a coulometer?

**16.13** Compare the signs of the anode and cathode in galvanic and electrolytic cells.

**16.14** In Section 16.5 we saw that electrical energy can be extracted from a Zn/Cu cell. If $Cu^{2+}$ is brought into contact with metallic Zn, it is reduced to Cu while the Zn is oxidized, without the generation of electricity. In this case, what happens to the energy that is *not* being extracted as electrical energy?

**16.15** What is the function of a salt bridge in a galvanic cell?

**16.16** What is a volt? What is an ampere?

**16.17** Without computing $\mathscr{E}^0$, determine what reactions will occur spontaneously among the following sets of reactants in aqueous solution.
(a) Al (*s*), Ni (*s*), $NiSO_4$ (*aq*), $Al_2(SO_4)_3$ (*aq*)
(b) $PbO_2$ (*s*), $K_2Cr_2O_7$ (*aq*), $H_2SO_4$ (*aq*), $PbSO_4$ (*s*), $Cr_2(SO_4)_3$ (*aq*)
(c) Ag (*s*), $AgNO_3$ (*aq*), Pb (*s*), $Pb(NO_3)_2$ (*aq*)
(d) $MnO_2$ (*s*), HCl (*aq*), $Cl_2$ (*g*), $MnCl_2$ (*aq*)
(e) Mn (*s*), HCl (*aq*), $MnCl_2$ (*aq*), $H_2$ (*g*)

**16.18** Without computing $\mathscr{E}^0$, determine whether the following reactions will occur spontaneously.
(a) $2Fe^{3+} + Sn \rightarrow 2Fe^{2+} + Sn^{2+}$
(b) $Cu + 2H^+ \rightarrow Cu^{2+} + H_2$
(c) $3Mg^{2+} + 2Al \rightarrow 3Mg + 2Al^{3+}$
(d) $Mn + Zn^{2+} \rightarrow Mn^{2+} + Zn$
(e) $PbO_2 + SO_4^{2-} + 4H^+ + 2Hg + 2Cl^- \rightarrow Hg_2Cl_2 + PbSO_4 + 2H_2O$

**16.19** Without computing $\mathscr{E}^0$, determine whether the following reactions will occur spontaneously.
(a) $Ca^{2+} + Mg \rightarrow Ca + Mg^{2+}$
(b) $Pb^{2+} + 2Cl^- \rightarrow Pb + Cl_2$
(c) $2Cl^- + S_2O_8^{2-} \rightarrow Cl_2 + 2SO_4^{2-}$
(d) $6Mn^{2+} + 5Cr_2O_7^{2-} + 22H^+ \rightarrow 6MnO_4^- + 10Cr^{3+} + 11H_2O$
(e) $O_2 + 4Cl^- + 4H^+ \rightarrow 2H_2O + 2Cl_2$

**16.20** Which is the better oxidizing agent?
(a) $Li^+$ or $Ca^{2+}$
(b) $Cl_2$ or $F_2$
(c) $H_2O$ or $Al^{3+}$
(d) $S_2O_8^{2-}$ or $Cl_2$
(e) $Br_2$ or $H_2O$

**16.21** Which is the better oxidizing agent?
(a) $Cl_2$ or $ClO_3^-$
(b) $O_2$ or $Cr_2O_7^{2-}$
(c) $MnO_4^-$ or $Cr_2O_7^{2-}$
(d) $PbO_2$ or $Hg_2Cl_2$

**16.22** Which is the better reducing agent?
(a) Ni or Fe
(b) $H_2$ or Mg
(c) $Br^-$ or $I^-$
(d) $SO_4^{2-}$ or $F^-$
(e) Sn or Mn

**16.23** Which is the better reducing agent?
(a) Na or Cr
(b) $PbSO_4$ or $Cl_2$
(c) Ag or Cu
(d) $I^-$ or Sn
(e) $H_2$ or $H_2O$

**16.24** What is a concentration cell?

**16.25** Describe the anode and cathode reactions in the dry cell.

**16.26** How can one identify experimentally which electrode in a galvanic cell is the anode and which is the cathode?

**16.27** What are the reactions that take place during the discharge of the lead storage battery? What reactions occur when this battery is being charged?

**16.28** Write the anode, cathode, and overall cell reaction for the discharge of the nickel-cadmium battery.

**16.29** What is a fuel cell? What advantages does a fuel cell offer over the lead storage battery?

**16.30** What possible advantages do fuel cells offer over current electrical power plants?

## REVIEW PROBLEMS

**16.31** How many faradays would be required to reduce 1 mole of each of the following to the indicated product?
(a) $Cu^{2+}$ to $Cu^0$
(b) $Fe^{3+}$ to $Fe^{2+}$
(c) $MnO_4^-$ to $Mn^{2+}$
(d) $F_2$ to $2F^-$
(e) $NO_3^-$ to $NH_3$

**16.32** Calculate the number of electrons that corresponds to 1 coulomb of charge.

**16.33** How many faradays would be required to oxidize 1 mole of each of the following to give the indicated product?
(a) $Cu^+$ to $Cu^{2+}$
(b) Pb to $PbO_2$
(c) $Cl_2$ to $2ClO_3^-$
(d) $O_2$ to $H_2O_2$ (hydrogen peroxide)
(e) $NH_3$ to $NO_3^-$

**16.34** How many faradays are given by
(a) 8950 coul
(b) A current of 1.5 amp for 30 sec
(c) A current of 14.7 amp for 10 min

**16.35** State how many minutes it would take to
(a) Deliver 10,500 coul using a current of 25 amp
(b) Deliver 0.65 $\mathcal{F}$ using a current of 15 amp
(c) Reduce 0.20 mole of $Cu^{2+}$ to Cu using a current of 10 amp

**16.36** State how many minutes it would take to
(a) Deliver 84,200 coul using a current of 6.30 amp
(b) Deliver 1.25 $\mathcal{F}$ using a current of 8.40 amp
(c) Produce 0.50 mole of Al from molten $AlCl_3$ using a current of 18.3 amp

**16.37** How many faradays of electricity are required to produce the following?
(a) 10 ml of $O_2$ (at STP) from aqueous $Na_2SO_4$
(b) 10 g of Al from molten $Al_2O_3$ (in cryolite)
(c) 5 g of Na from molten NaCl
(d) 5 g of Mg from molten $MgCl_2$

**16.38** How many grams of Na and $Cl_2$ would be produced if a current of 25 amp was applied for 8 hr to the cell shown in Figure 16.3?

**16.39** How many grams of $O_2$ and $H_2$ are produced in 1 hr when water is electrolyzed at a current of 0.5 amp? What would be the volume, at STP, of $O_2$ and $H_2$?

**16.40** What weight of copper could be purified by a current of 100 amp for 8 hr? Refer to Figure 16.5.

**16.41** What weight of silver could be plated out on a serving tray by electrolysis of a solution containing Ag in the +1 oxidation state for a period of 8.00 hr at a current of 8.46 amp? What area would this cover, assuming that the density of Ag is 10.5 g/cm³ and the thickness of the plate is 0.010 in. (0.0254 cm)?

**16.42** How many seconds would it take to deposit 21.4 g of Ag from a solution of $AgNO_3$ by a current of 10.0 amp?

**16.43** How long would it take to deposit 35.3 g of Cr from a solution of $CrCl_3$ at a current of 6.00 amp?

**16.44** How long would it take to plate out 5 g of copper from a solution of $CuSO_4$ at a current of 5 amp?

**16.45** What current is required to deposit 0.225 g of Ni from a solution of $NiSO_4$ in 10 min?

**16.46** What current is required to produce 1.33 g of $Cl_2$ from a solution of NaCl in 45.0 min?

**16.47** In an experiment two coulometers were connected in series, one containing $CuSO_4$, the other an unknown salt. It was found that 1.25 g of copper were plated out during the same period of time as 3.42 g of the unknown metal.
(a) How many faradays passed through this coulometer?
(b) If the oxidation state of the unknown metal was +2, what was the atomic weight of the unknown?

**16.48** Two coulometers were connected in series so that the same current passes through each of them. In an experiment, 0.125 mole of Cu was deposited from a solution of $CuSO_4$ in one of the coulometers. How many moles of Cr were deposited at the same time from a $Cr_2(SO_4)_3$ solution in the other?

**16.49** What current is required to produce 50.0 ml of $O_2$, measured at STP, by the electrolysis of $H_2O$ for a period of 3.00 hr?

***16.50** A current of 0.25 amp is passed through 400 ml of a 0.250 *M* solution of NaCl for 35 min. What will be the pH of the solution after the current is turned off?

**16.51** Given the following sets of half-reactions, write the net cell reaction and calculate $\mathcal{E}^0$ for the spontaneous changes that will occur.
(a) $Hg_2Cl_2 + 2e^- \rightleftharpoons 2Hg + 2Cl^-$
$PbSO_4 + 2e^- \rightleftharpoons Pb + SO_4^{2-}$
(b) $AgCl + e^- \rightleftharpoons Ag + Cl^-$
$Cu^{2+} + 2e^- \rightleftharpoons Cu$
(c) $Mn^{2+} + 2e^- \rightleftharpoons Mn$
$Cl_2\ (g) + 2e^- \rightleftharpoons 2Cl^-$
(d) $Al^{3+} + 3e^- \rightleftharpoons Al$
$Br_2\ (aq) + 2e^- \rightleftharpoons 2Br^-$

**16.52** Determine the value of $\mathcal{E}^0$ for each of the spontaneous reactions in Question 16.17.

**16.53** Determine the value of $\mathcal{E}^0$ for each of the reactions as written from left to right in Question 16.19.

16.54 Calculate the equilibrium constants for the following cell reactions:
(a) $Ni\ (s) + Sn^{2+}\ (aq) \rightleftharpoons Ni^{2+}\ (aq) + Sn\ (s)$
(b) $Cl_2\ (g) + 2Br^-\ (aq) \rightleftharpoons$
$Br_2\ (aq) + 2Cl^-\ (aq)$
(c) $Fe^{2+}\ (aq) + Ag^+\ (aq) \rightleftharpoons$
$Ag\ (s) + Fe^{3+}\ (aq)$

16.55 Calculate the equilibrium constants for the reactions in Question 16.18.

16.56 Calculate the equilibrium constants for the reactions in Question 16.19.

16.57 Calculate $\Delta G^0_{298}$ for each reaction in Question 16.18.

16.58 Calculate $\Delta G^0_{298}$ for each reaction in Question 16.19.

16.59 Write the Nernst equation, calculate $\mathscr{E}^0$ and $\mathscr{E}$ for the following reactions:
(a) $Cu^{2+}\ (0.1\ M) + Zn\ (s) \rightarrow$
$Cu\ (s) + Zn^{2+}\ (1.0\ M)$
(b) $Sn^{2+}\ (0.5\ M) + Ni\ (s) \rightarrow$
$Sn\ (s) + Ni^{2+}\ (0.01\ M)$
(c) $F_2\ (g,\ 1\ atm) + 2Li\ (s) \rightarrow$
$2Li^+\ (1\ M) + 2F^-\ (0.5\ M)$
(d) $Zn\ (s) + 2H^+\ (0.01\ M) \rightarrow$
$Zn^{2+}\ (1\ M) + H_2\ (1\ atm)$
(e) $2H^+\ (1.0\ M) + Fe\ (s) \rightarrow$
$H_2\ (1\ atm) + Fe^{2+}\ (0.2\ M)$

16.60 Calculate $\mathscr{E}^0$, $\mathscr{E}$, and $\Delta G$ for the following cell reactions (not balanced):
(a) $Al\ (s) + Ni^{2+}\ (0.80\ M) \rightarrow$
$Al^{3+}\ (0.020\ M) + Ni\ (s)$
(b) $Ni\ (s) + Sn^{2+}\ (1.10\ M) \rightarrow$
$Sn\ (s) + Ni^{2+}\ (0.010\ M)$
(c) $Cu^+\ (0.050\ M) + Zn\ (s) \rightarrow$
$Cu\ (s) + Zn^{2+}\ (0.010\ M)$

16.61 Calculate the cell potential for the following:
(a) $Sn\ (s) + Pb^{2+}\ (0.050\ M) \rightarrow$
$Sn^{2+}\ (1.50\ M) + Pb\ (s)$
(b) $3Zn\ (s) + 2Cr^{3+}\ (0.010\ M) \rightarrow$
$3Zn^{2+}\ (0.020\ M) + Cr\ (s)$
(c) $PbO_2\ (s) + SO_4^{2-}\ (0.010\ M)$
$+ 4H^+\ (0.10\ M) + Cu\ (s) \rightarrow$
$PbSO_4\ (s) + 2H_2O + Cu^{2+}\ (0.0010\ M)$

16.62 Calculate the potential generated by a concentration cell consisting of a pair of iron electrodes dipping into two solutions, one containing 0.10 $M$ $Fe^{2+}$ and the other containing 0.0010 $M$ $Fe^{2+}$.

16.63 Calculate the potential of a concentration cell containing 0.0020 $M$ $Cr^{3+}$ in one compartment and 0.10 $M$ $Cr^{3+}$ in the other compartment with Cr (s) electrodes dipping into each solution.

16.64 The solubility product constant of AgBr is $5 \times 10^{-13}$. What will be the potential of a cell constructed using the $H_2$ electrode ($[H^+] = 1.0\ M$, $p_{H_2} = 1$ atm) versus a half-cell containing a silver wire coated with AgBr immersed in 0.010 $M$ HBr?

16.65 What is the reduction potential of a half-cell composed of a copper wire dipping into $2 \times 10^{-4}\ M$ $CuSO_4$?

16.66 A cell was constructed using the standard hydrogen electrode ($[H^+] = 1.0\ M$, $p_{H_2} = 1$ atm) in one compartment and a lead electrode in a 0.10 $M$ $K_2CrO_4$ solution in contact with undissolved $PbCrO_4$. The potential of the cell was measured to be 0.51 V with the Pb electrode as the anode. Determine the $K_{sp}$ of $PbCrO_4$ from these data.

*16.67 The standard reduction potential for $Ag^+$ is 0.80 V. Compute the standard reduction potential for the half-reaction,

$$Ag_2S\ (s) + 2e^- \rightleftharpoons 2Ag\ (s) + S^{2-}\ (aq)$$

in a solution buffered to a pH of 3.00.

*16.68 A student set up an electrolysis apparatus and passed a current of 1.22 amp through a 3 $M$ $H_2SO_4$ solution for 30.0 min. He collected the $H_2$ evolved and found that it occupied a volume, over water at 27°C, of 288 ml at a total pressure of 767 torr. Use these data to calculate the charge on the electron, expressed in the units, coulombs.

*16.69 How long will a 25-watt light bulb burn if it is powered by a lead storage battery that has available 25.0 g of Pb that can react as an anode. Assume a constant voltage of 1.5 V. (1 watt = 1 J/sec)

*16.70 What current would be required to deposit 1 $m^2$ of chrome plate having a thickness of 0.050 mm in 25 min from a solution containing $Cr_2(SO_4)_3$? The density of Cr is 7.19 g/ml.

*16.71 What weights of $H_2$ and $O_2$ would have to react each second in a fuel cell at 110°C to provide 1.0 kilowatts (kW) of power, assuming a thermodynamic efficiency of 70%. (*Hint:* Use the data in Chapter 10 to compute $\Delta G^0$ for the reaction, $H_2\ (g)$ +

$\frac{1}{2}O_2\ (g) \rightarrow H_2O\ (g)$ at 110°C. 1 watt = 1 J/sec.)

*16.72 A hydrogen electrode is immersed in a 0.10 $M$ solution of acetic acid. This electrode is connected to another consisting of an iron nail dipping into 0.10 $M$ $FeCl_2$. What will be the measured emf of this cell? Assume $p_{H_2}$ = 1 atm.

*16.73 How much work is able to be accomplished by a 5-min flow of electricity having a voltage of 110 V and a current of 1.0 amp?

*16.74 Assuming that the typical electrical generating plant has an efficiency of only about 30%, what volume of fuel having an average formula of $C_{12}H_{26}$ must be burned, giving $H_2O(g)$ and $CO_2(g)$, to produce 1 kilowatt hour (kWh) of electricity? Assume $\Delta H_f^0$ of $C_{12}H_{26}(l)$ = 291 kJ/mole and a density of 0.74 g/ml. 1 watt = J/sec.

*16.75 How long would it take to remove all of the Cr from 500 ml of 0.270 $M$ $Cr_2(SO_4)_3$ by a current of 3.00 amp?

*16.76 Calculate the value of $\Delta G$ for a system containing the following species: $Mn^{2+}$ (0.10 $M$), $Cr_2O_7^{2-}$ (0.010 $M$), $MnO_4^-$ (0.0010 $M$), $Cr^{3+}$ (0.0010 $M$). The pH of the solution is 6.00. The reaction that you should consider is

$$6Mn^{2+}\ (aq) + 5Cr_2O_7^{2-}\ (aq) + 22H^+\ (aq) \rightleftharpoons 6MnO_4^-\ (aq) + 10Cr^{3+}\ (aq) + 11H_2O\ (l)$$

Which direction will this reaction proceed to get to equilibrium from the starting conditions given above?

*16.77 A Ag/AgCl electrode dipping into 1 $M$ HCl has a standard reduction potential of +0.22 V [$AgCl\ (s) + e^- \rightleftharpoons Ag\ (s) + Cl^-\ (aq)$]. A second Ag/AgCl electrode is dipped into a solution containing $Cl^-$ at an unknown concentration. The cell generates a potential of 0.0435 V, with the electrode in the unknown solution serving as the anode. What is the $Cl^-$ concentration in the unknown?

*16.78 A student set up a galvanic cell to measure the $K_{sp}$ of CuS. On one side of the cell she had a copper electrode dipping into a 0.10 $M$ $Cu^{2+}$ solution and on the other side a zinc electrode in a $Zn^{2+}$ solution. The $Zn^{2+}$ concentration was held constant at 1 $M$ and the $Cu^{2+}$ brought to a minimum by saturating the $Cu^{2+}$ solution with $H_2S$. The emf of the cell was read as +0.67V, with the Cu electrode serving as the cathode. Calculate the $Cu^{2+}$ concentration and the $K_{sp}$ of CuS. Compare your answer to the $K_{sp}$ reported in Table 15.1. In a saturated solution the concentration of $H_2S$ is 0.10 $M$. The solution in which the CuS was formed was not buffered.

# 17 COVALENT BONDING AND MOLECULAR STRUCTURE

In Chapter 4 we found that chemical bonds can be broadly classified into two main categories: ionic bonds and covalent bonds. The ionic bond arises as a purely electrostatic attraction between oppositely charged particles and is therefore nondirectional, in the sense that the arrangement of ions in a cluster is determined simply by the balancing of attractive and repulsive forces between the ions, and not by the electronic structures of the ions. It is for this reason that the structures of ionic compounds are determined almost exclusively by packing considerations (Chapter 7). The covalent bond, on the other hand, has very definite directional properties, and covalently bound substances, such as molecules or polyatomic ions, have characteristic shapes that are usually retained when these substances undergo physical changes such as melting or vaporization.

The simple picture of the covalent bond as a pair of dots shared between two atoms is clearly not sufficient to explain molecular structure. In this chapter we shall examine some of the theories that have been developed to account for covalent bonding and the shapes of molecules. It should be kept in mind throughout this discussion that each theory represents an attempt to describe the *same* physical phenomenon. None of the theories is perfect—otherwise we would only have to consider one of them. Each has its usefulness, and each has its weaknesses. The theory that chemists apply in a particular circumstance depends largely on what aspect of the covalent bond they are attempting to explain and, to some extent, their own feelings about the validity of the various theories.

## 17.1 VALENCE BOND THEORY

There are two important approaches to chemical bonding that are based on the results of quantum mechanics. One of these, called the **valence bond theory,** permits us to retain our picture of individual atoms coming together to form a covalent bond. The other, called **molecular orbital theory,** views a molecule as a set of positive nuclei with orbitals that extend over the entire molecule. The electrons that populate these **molecular orbitals** do not belong to any individual atoms but, instead, to the molecule as a whole. We will look at the molecular orbital theory in more detail in Section 17.5.

The basic postulate of the valence bond theory is that when two atoms come together to form a covalent bond, an atomic orbital of one atom overlaps with an atomic orbital of the other, and the pair of electrons that we have come to associate with a covalent bond is shared between the two atoms in the region where the orbitals overlap. It is also postulated that the

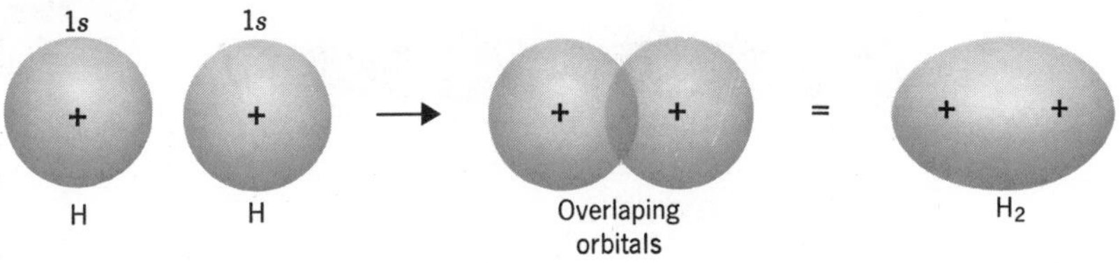


Figure 17.1
*Formation of $H_2$ by overlap of 1s orbitals.*

strength of the covalent bond is proportional to the *extent* of overlap of the atomic orbitals. As a consequence, the atoms in a molecule tend to position themselves so that there is a maximum amount of orbital overlap.

Let's see how these postulates can be applied to some familiar compounds. The simplest of these is the hydrogen molecule that is formed from two hydrogen atoms, each having a single electron in a 1*s* orbital. According to valence bond theory, we would view the H—H bond as resulting from the overlap of the two 1*s* orbitals, as shown in Figure 17.1[1]

In the HF molecule we have a somewhat different state of affairs. Fluorine has the valence-shell electron configuration

$$\text{F}\quad \underset{2s}{\underline{\uparrow\downarrow}}\quad \underset{2p}{\underline{\uparrow\downarrow}\ \underline{\uparrow\downarrow}\ \underline{\uparrow}}$$

where we find one of the 2*p* orbitals occupied by a single electron. It is with this partially occupied 2*p* orbital that the hydrogen 1*s* orbital overlaps, as illustrated by Figure 17.2. In this case the hydrogen electron and the fluorine electron can pair up and be shared between the two nuclei. Note that the 1*s* orbital of the hydrogen atom does not overlap with an already filled atomic orbital on fluorine since there would then be three electrons in the bond (two from the fluorine 2*p* orbital and one from the hydrogen 1*s* orbital), a sit-

Figure 17.2
*Formation of HF by overlap of partially filled fluorine 2p orbital with 1s orbital of hydrogen. Heavy shading indicates filled orbital, light shading indicates partially filled orbital. Atomic-Molecular Orbital Models by Science Related Materials, Inc., Janesville, Wisconsin.*

[1] In this illustration, as well as in others throughout this chapter, it is important to keep in mind that we are looking at schematic representations of orbital wave functions.

uation that is not permitted in this theory. *Only two electrons may be shared in one set of overlapping orbitals.*

Suppose that we now consider the molecule, $H_2O$. Here we have two hydrogen atoms bound to a single oxygen atom. The outer-shell electronic structure of oxygen,

$$\text{O}\quad \underset{2s}{\underline{\uparrow\downarrow}}\quad \underset{2p}{\underline{\uparrow\downarrow}\ \underline{\uparrow}\ \underline{\uparrow}}$$

tells us that there are two unpaired electrons in *p* orbitals, and we predict that the two hydrogen atoms, with their electrons in 1*s* orbitals, will bond to the oxygen by means of overlap of their 1*s* orbitals with these partially filled oxygen *p* orbitals. This is shown in Figure 17.3. Since the *p* orbitals are oriented at 90° to one another, we expect the H—O—H bond angle in water also to be 90°. Actually this angle is 104.5°. One explanation for this discrepancy (we shall see another one later) is that since the O—H bonds are highly polar, the H atoms carry a substantial positive charge and hence repel one another. This factor tends to increase the H—O—H angle. However, since the best overlap between the hydrogen 1*s* orbitals and the oxygen 2*p* orbitals occurs at an angle of 90°, the H—O—H angle cannot increase too much without there being a considerable loss in bond strength. There are thus two factors working in opposition to each other, one tending to increase the bond angle and one tending to reduce it to 90°, and it appears that a balance is obtained when the angle is 104.5°. Qualitatively, the valence bond theory can account for the geometry of the water molecule. We can also apply the theory to the ammonia molecule with reasonable success. Nitrogen, being in Group VA, has three unpaired electrons in its *p* subshell. We would expect that the three hydrogen atoms in $NH_3$ would bond to the nitrogen such that they would tend to lie along the *x*, *y*, and *z* axes at 90° to one another (Figure 17.4*a*). As in the water molecule, the H—N—H angles are larger than 90°, having values in this case of 107°; as with $H_2O$, we might attempt to explain this angle in terms of repulsion between the protons. In any event, we obtain a picture of the $NH_3$ molecule that has a geometry referred to as pyra-

Figure 17.3
*Bonding in $H_2O$. Overlap of two half-filled oxygen* 2p *orbitals with the hydrogen* 1s *orbitals. Atomic-Molecular Orbital Models by Science Related Materials, Inc., Janesville, Wisconsin.*

(a)

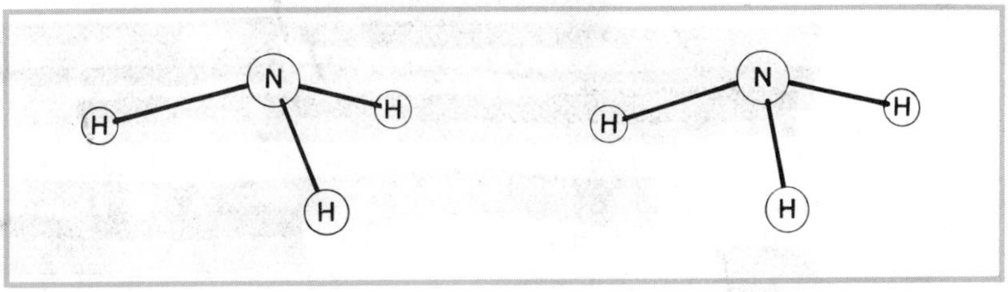


(b)

Figure 17.4

*Bonding in $NH_3$ gives pyramidal molecule. (a) Overlap of 2*p *orbitals of nitrogen with 1*s *orbitals of hydrogen. (b) Pyramidal shape of the $NH_3$ molecule. Atomic-Molecular Orbital Models by Science Related Materials, Inc., Janesville, Wisconsin.*

midal (i.e., pyramidlike), with the nitrogen atom at the apex of the pyramid and the three hydrogen atoms at the corners of the base, as shown in Figure 17.4*b*.

## 17.2 HYBRID ORBITALS

The very simple picture of the overlap of half-filled atomic orbitals we have just developed cannot be used to account for all molecular structures. For example, with carbon we would initially expect only two bonds to be formed with hydrogen since the valence shell of carbon contains only two unpaired electrons.

$$\text{C} \quad \underset{2s}{\underline{\uparrow\downarrow}} \quad \underbrace{\underline{\uparrow}\ \underline{\uparrow}\ \underline{\phantom{\uparrow}}}_{2p}$$

The species $CH_2$, however, does not exist as a stable molecule. Instead, the simplest compound between carbon and hydrogen is methane, which has the formula $CH_4$. Attempting to explain the structure of this molecule by spreading the electrons out to give

$$\underset{2s}{\underline{\uparrow}} \quad \underbrace{\underline{\uparrow}\ \underline{\uparrow}\ \underline{\uparrow}}_{2p}$$

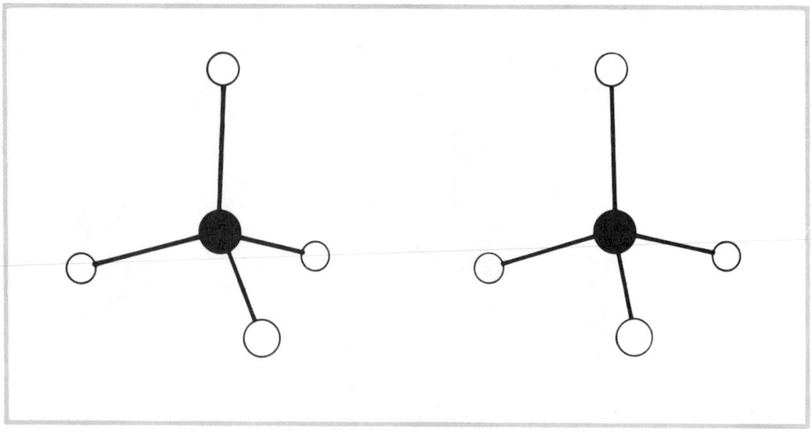

Figure 17.5
*The structure of methane. Solid sphere—carbon.*

suggests that three of the C—H bonds will be formed by overlap of hydrogen 1*s* orbitals with carbon 2*p* orbitals, while the remaining bond would be the result of the overlap of the carbon 2*s* orbital with a hydrogen 1*s* orbital. This fourth C—H bond should certainly be different from the other three bonds, because it is formed from different kinds of orbitals. It has been found experimentally, however, that *all* four C—H bonds are identical and that the molecule has a structure in which the carbon atoms lies at the center of a tetrahedron with the hydrogen atoms located at the four corners (Figure 17.5). How does valence bond theory explain the structure of this molecule?

The solution to this apparent dilemma is found in the mathematics of quantum mechanics. In that theory (see Sections 3.13 and 3.17), the solution of Schrödinger's wave equation provides us with a series of wave functions, $\psi$, each of which describes a different atomic orbital. It is the property of these mathematical functions that when they are squared, they enable us to calculate the probability of locating the electron at some point in space around the nucleus and, in fact, the spheres and figure-eights that we have been drawing roughly correspond to pictorial representations of the probability distributions predicted by the wave functions for *s* and *p* orbitals, respectively.

What is important to us here is that it is possible to combine these wave functions by appropriately adding or subtracting them to give new functions that are referred to as **hybrid orbitals.** In other words, two or more atomic orbitals are mixed together to produce a new set of orbitals and, invariably, these hybrid orbitals possess different directional properties than the pure atomic orbitals from which they are created. For example, Figure 17.6 illustrates the result of the combination of a 2*s* and a 2*p* orbital to provide a new set of two **sp hybrid orbitals.** In this drawing you will notice that we have indicated that the wave function for a *p* orbital has positive numerical values in some regions about the nucleus and negative values in others.[2] The *s* orbital, on the other hand, has the same algebraic sign everywhere. Therefore,

[2] Certain wave functions (for example, *p* orbital wave functions) can have either a positive or negative algebraic sign, depending on which region around the nucleus is examined. The square of a wave function, $\psi^2$, of course, must always be positive (a negative times a negative equals a positive).

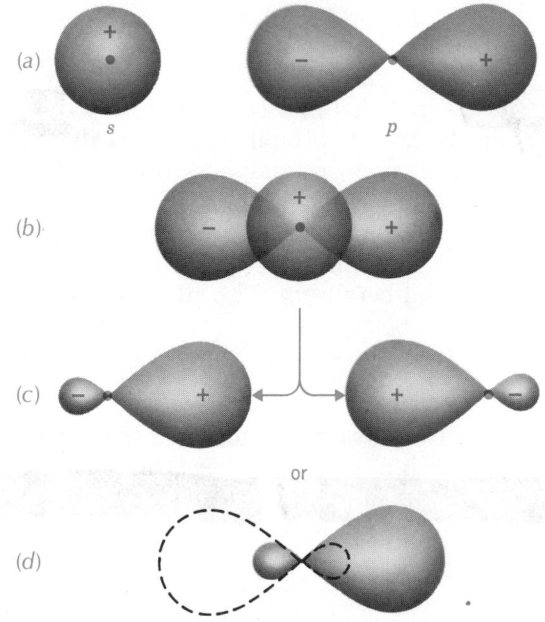


Figure 17.6
*Formation of two* sp *hybrid orbitals from an* s *and a* p *orbital. (a)* s *and* p *orbitals drawn separately. (b)* s *and* p *orbitals before hybridization. (c) Two* sp *hybrid orbitals are formed (drawn separately). (d) The two* sp *hybrid orbitals drawn together to show their directional properties. Note that one orbital points to the left, the other to the right.*

when these wave functions are alternatively added and subtracted, the new orbitals that result become larger in those regions where both functions have the same sign and smaller in regions where they are of opposite sign.

Hybrid orbitals formed in the manner just described possess some interesting properties. We see for each orbital that one lobe is much larger than the other and because of this a hybrid orbital can overlap well in only one direction—the direction in which the orbital protrudes the most. A hybrid orbital is therefore very strongly directional in its ability to enter into covalent bond formation. Furthermore, because hybrid orbitals extend out farther from the nucleus than do unhybridized orbitals, they are able to overlap more effectively with orbitals of other atoms. Consequently, bonds formed from hybrid orbitals tend to be stronger than those formed from ordinary atomic orbitals.

Thus far we have examined what occurs when one *s* and one *p* orbital are mixed together. Other combinations of orbitals are also possible, with the number of orbitals in a hybrid set, as well as their orientations, being determined by which atomic orbitals are combined. Table 17.1 contains a

Table 17.1
*Hybrid orbitals*

| Hybrid Orbitals | Number of Orbitals | Orientation |
|---|---|---|
| $sp$ | 2 | Linear |
| $sp^2$ | 3 | Planar triangle |
| $sp^3$ | 4 | Tetrahedral |
| $dsp^2$ | 4 | Square planar |
| $dsp^3$ | 5 | Trigonal bipyramidal |
| $d^2sp^3$ | 6 | Octahedral |

listing of the sets of hybrid orbitals that can be used to explain most of the molecular structures that we shall encounter in this book. Their directional properties are illustrated in Figure 17.7. Notice that the number of each kind of atomic orbital included in a combination is specified by an appropriate superscript on the atomic orbital type. Thus the $d^2sp^3$ hybrids are formed from two *d* orbitals, one *s* orbital, and three *p* orbitals.

Let us see now how we can use the information contained in Table 17.1 and Figure 17.7 to account for the structures of some typical molecules. We might begin with the substance $BeH_2$, in which there are two H atoms bound to the central beryllium. The electronic structure of Be is

$$\text{Be}\quad \underset{1s}{\underline{\uparrow\downarrow}}\quad \underset{2s}{\underline{\uparrow\downarrow}}\quad \underset{2p}{\underline{\phantom{x}}\ \underline{\phantom{x}}\ \underline{\phantom{x}}}$$

In order to form two covalent bonds with H atoms, the Be atom must provide two half-filled (i.e., singly occupied) orbitals. This can be accomplished by creating a pair of *sp* hybrids and placing one electron in each of them.

$$\text{Be}\quad \underset{1s}{\underline{\uparrow\downarrow}}\quad \underbrace{\underline{\uparrow}\ \underline{\uparrow}}_{sp}\quad \underbrace{\underline{\phantom{x}}\ \underline{\phantom{x}}}_{\text{unhybridized } 2p \text{ orbitals}}$$

The two H atoms can then bond to the beryllium atom by overlap of their respective singly occupied *s* orbitals[3] with the singly occupied Be *sp* hybrids as shown in Figure 17.8. Because of the orientation of the *sp* hybrid orbitals the H atoms are forced to lie on opposite sides of the Be, and we predict that a linear H—Be—H molecule would result. It is interesting (and perhaps comforting) to find experimentally that $BeH_2$ is indeed a linear molecule.

Let us now return to our problem of the structure of $CH_4$. If we use hybrid orbitals on the carbon atom, we find that in order to provide four orbitals with which hydrogen 1*s* orbitals can overlap, we must use a set of $sp^3$ hybrids.

$$\text{C}\quad \underset{2s}{\underline{\uparrow\downarrow}}\quad \underset{2p}{\underline{\uparrow}\ \underline{\uparrow}\ \underline{\phantom{x}}}\qquad \text{(unhybridized)}$$

gives

$$\underset{sp^3}{\underline{\uparrow}\ \underline{\uparrow}\ \underline{\uparrow}\ \underline{\uparrow}}\qquad \text{(hybridized)}$$

From Table 17.1 we see that these orbitals point toward the vertices of a tetrahedron. Hence, when the four hydrogen atoms are attached to the carbon by orbital overlap with these $sp^3$ hybrids, a tetrahedral molecule results, as shown in Figure 17.9. Again we find the predicted structure in agreement with that found by experiment.

In our earlier discussion we viewed the structure of $H_2O$ and $NH_3$ as resulting from the use of partially filled *p* atomic orbitals on the oxygen and nitrogen atoms, respectively. An alternative view of the bonding in these molecules employs $sp^3$ hybrid orbitals on the central atom. In the tetrahedral set of hybrids the orbitals are oriented at angles of 109.5° to each other. The bond angles in water (104.5°) and ammonia (107°) are not too different from the tetrahedral angle and, using water as an example, we might con-

[3] The electronic configuration of H is $1s^1$.

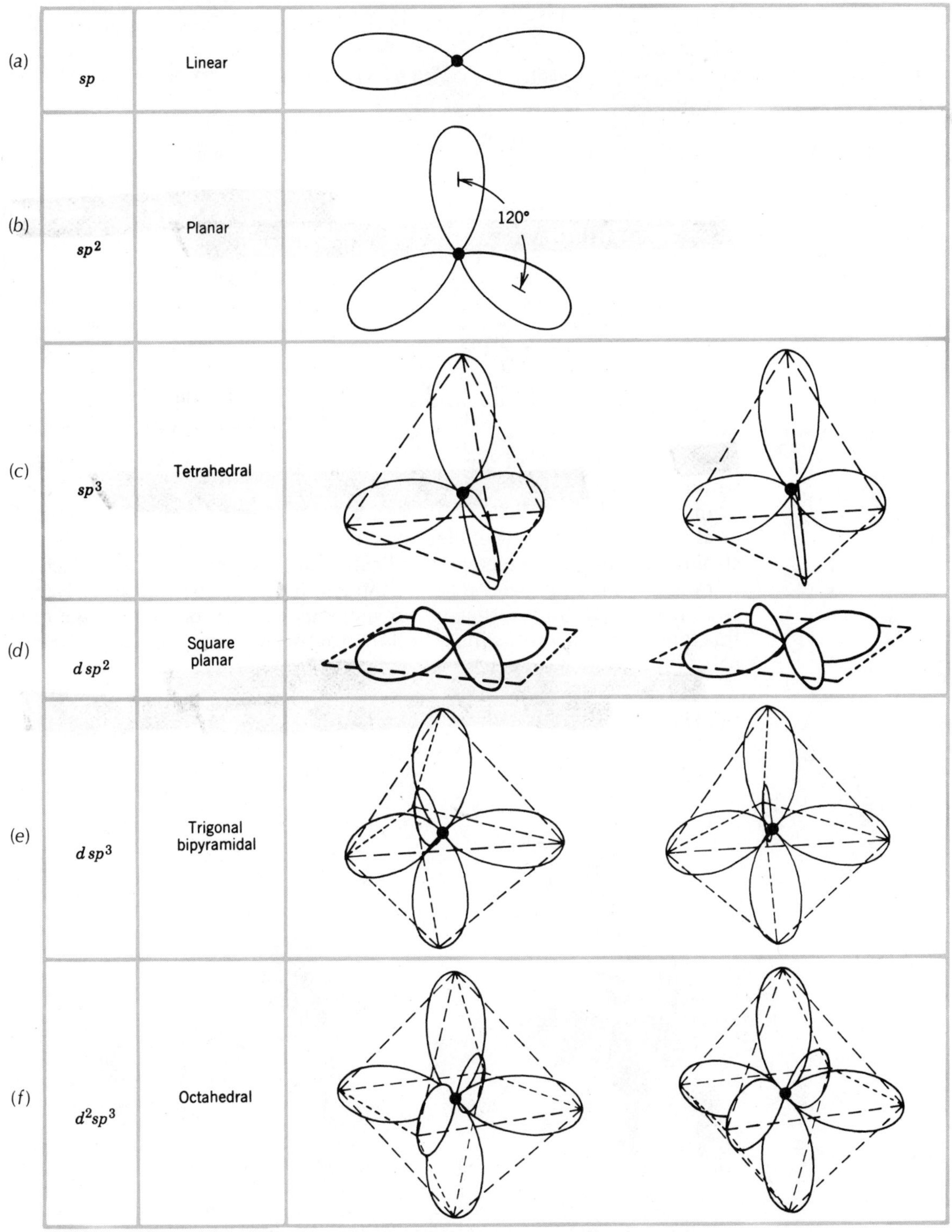


Figure 17.7
*Directional properties of hybrid orbitals. The minor lobes have been omitted for the sake of clarity. (a)* sp, *linear. (b)* $sp^2$, *planar. (c)* $sp^3$, *tetrahedral. (d)* $dsp^2$, *square planar. (e)* $dsp^3$, *trigonal bipyramidal. (f)* $d^2sp^3$ *octahedral.*

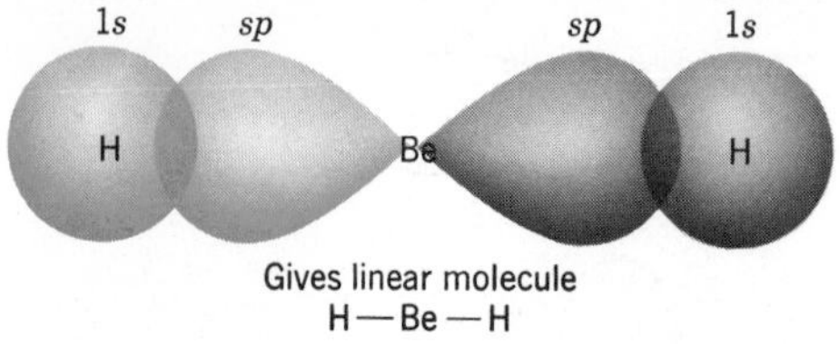


Figure 17.8
*The bonding in $BeH_2$.*

sider the molecule to result from the overlap of hydrogen 1*s* orbitals with two partially occupied $sp^3$ orbitals on the oxygen atom.

| | | |
|---|---|---|
| O | $\underline{\uparrow\downarrow}$ (2*s*) $\underline{\uparrow\downarrow}\ \underline{\uparrow}\ \underline{\uparrow}$ (2*p*) | unhybridized |
| | $\underline{\uparrow\downarrow}\ \underline{\uparrow\downarrow}\ \underline{\uparrow}\ \underline{\uparrow}$ ($sp^3$) | hybridized |
| | $\underline{\uparrow\downarrow}\ \underline{\uparrow\downarrow}\ \underline{\uparrow\cdot}\ \underline{\uparrow\cdot}$ ($sp^3$) | $H_2O$ molecule (dots = H electrons) |

Notice that only two of the hybrid orbitals are involved in bond formation while the other two contain nonbonded "lone pairs" of electrons. In the case of ammonia, three of the $sp^3$ orbitals are used in bonding while the fourth orbital contains a lone pair of electrons (Figure 17.10). There is, in fact, rather strong experimental evidence to indicate that this lone pair does indeed project out from the nitrogen atom as implied in this picture of the $NH_3$ molecule. It is worth noting that in our previous description of $NH_3$ we found this lone pair of electrons in an *s* orbital that would have spread the electron pair symmetrically about the nucleus.

In the case of $H_2O$ and $NH_3$, the H—X—H bond angles (104.5° and 107°, respectively) are less than the tetrahedral angle of 109° that is observed in

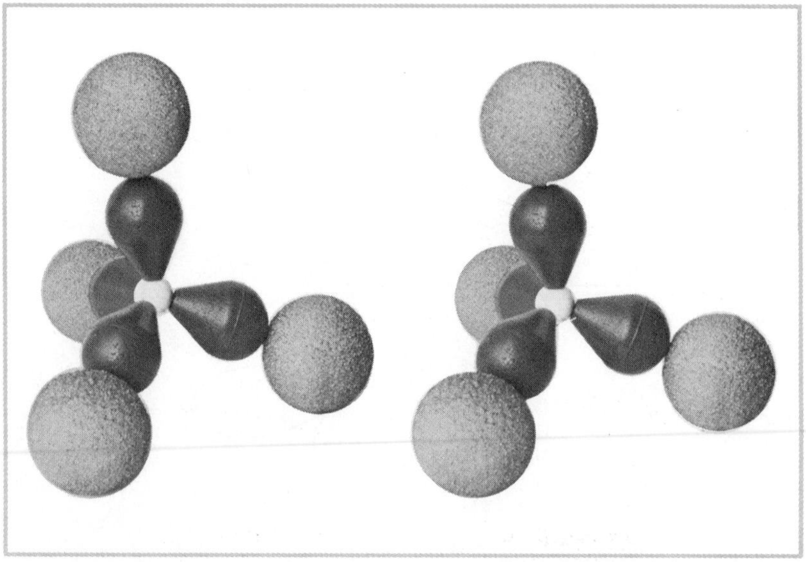

Figure 17.9
*The formation of methane by overlap of hydrogen 1s orbitals with carbon sp³ hybrids. Atomic-Molecular Orbital Models by Science Related Materials, Inc., Janesville, Wisconsin.*

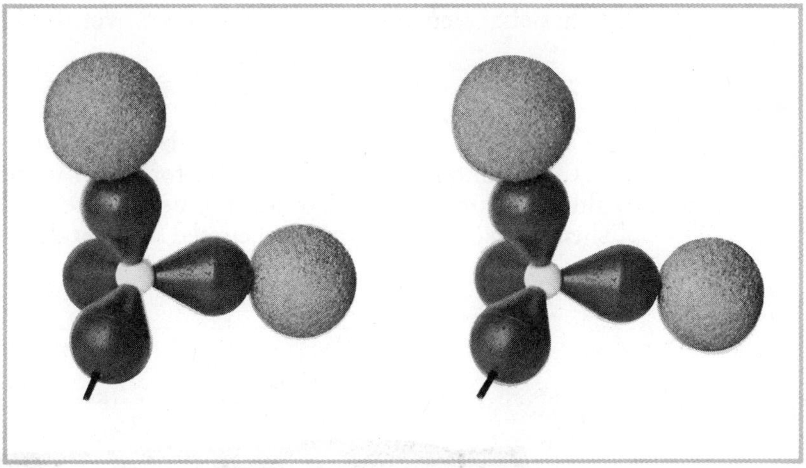

(a)

(b)

Figure 17.10

*The use of* $sp^3$ *hybrids for bonding in* $H_2O$ *and* $NH_3$ *(heavy shading indicates a filled hybrid orbital; light shading indicates a half-filled orbital). Spheres represent hydrogen 1*s *orbitals. Atomic-Molecular Orbital Models by Science Related Materials, Inc., Janesville, Wisconsin.*

the molecule, $CH_4$. One way to account for this is through the influence of the lone-pair electrons present in hybrid orbitals of the central atom. A pair of electrons in a bond is attracted to two nuclei and, therefore, might be expected to occupy a smaller effective volume than a pair of electrons in a non-bonded orbital, which experience the attraction of only one nucleus. The lone-pair electrons, then, because of their greater space requirement, tend to crowd together the electron pairs located in the bonds and hence reduce the bond angle to something less than 109°. On this basis we anticipate a greater reduction in bond angle for water than for ammonia, since the former has two lone pairs while ammonia has only one.

As another example, consider the molecule $SF_6$. Sulfur, being in Group

VIA, has six valence electrons distributed over the 3*s* and 3*p* subshells,

S ⇅ (3*s*) ⇅ ↑ ↑ (3*p*) _ _ _ _ _ (3*d*)

Here we have shown the empty 3*d* subshell as well as the 3*s* and 3*p* subshells that contain electrons. In order for sulfur to form six covalent bonds to fluorine, six half-filled orbitals must be created. This can be accomplished by using two of the unoccupied 3*d* orbitals and a $d^2sp^3$ hybrid set is formed.

S ↑ ↑ ↑ ↑ ↑ ↑ ($d^2sp^3$) _ _ _ (3*d* (unhybridized))

$SF_6$ ↑· ↑· ↑· ↑· ↑· ↑· ($d^2sp^3$) _ _ _ (3*d*) (dots = F electrons)

Since the $d^2sp^3$ orbitals point toward the corners of an octahedron, $SF_6$ is expected to have an octahedral geometry, which it does.

Finally, let's consider the $SF_4$ molecule. Again, we have sulfur in Group VIA

S ⇅ (3*s*) ⇅ ↑ ↑ (3*p*) _ _ _ _ _ (3*d*)

To form four bonds to fluorine we need four half-filled hybrid orbitals. This can be accomplished by promoting an electron from the 3*p* to the 3*d*. The hybrid orbitals that are formed are $dsp^3$—note that the *s* subshell becomes incorporated into the hybrid set even though at first glance it would not have appeared to be necessary to use it. Also note that one of the hybrid orbitals is occupied by a pair of electrons.

S ⇅ ↑ ↑ ↑ ↑ ($dsp^3$) _ _ _ _ (3*d* (unhybridized))

When the four S—F bonds are formed we obtain

S ⇅ ↑· ↑· ↑· ↑· ($dsp^3$) _ _ _ _ (3*d*)

The shape of the $SF_4$ molecule is shown in Figure 17.11. Four of the hybrid

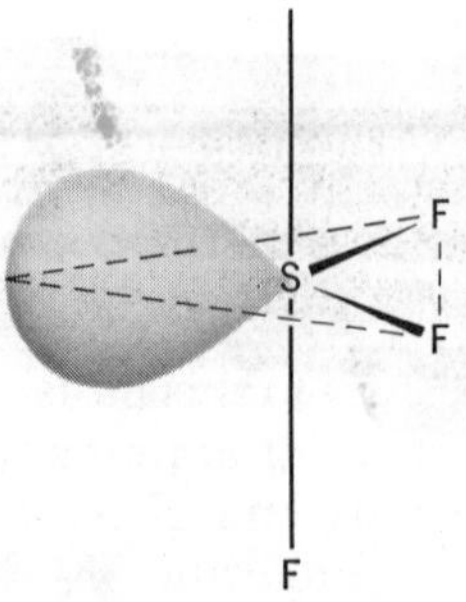


Figure 17.11
*The structure of $SF_4$. Note the lone pair of electrons in the* $dsp^3$ *hybrid orbital.*

orbitals are used to form bonds to fluorine; the other remains occupied by the unbonded *lone pair* of electrons.

Before moving on, a word should be said about the coordinate covalent bond. An example of this, you remember, is provided by the ammonium ion.

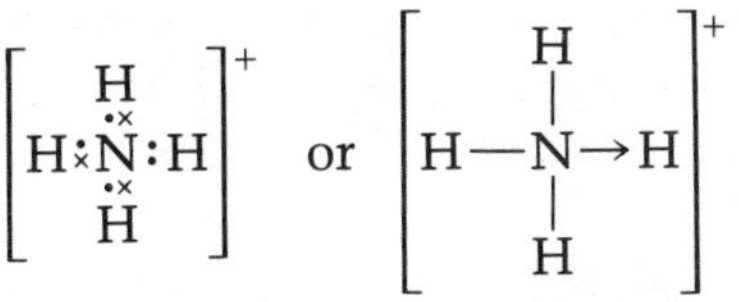

In terms of valence bond theory we can imagine the coordinate covalent bond in this ion to be formed by the overlap of an *empty* 1*s* orbital centered on a proton with the completely filled lone-pair orbital on the nitrogen of an ammonia molecule. The electron pair is then shared in the region of orbital overlap. Once the bond is formed it is a full-fledged covalent bond whose properties do not depend on its origin. Consequently, the four N—H bonds in $NH_4^+$ are identical and the ion, although tetrahedral, is usually represented as simply

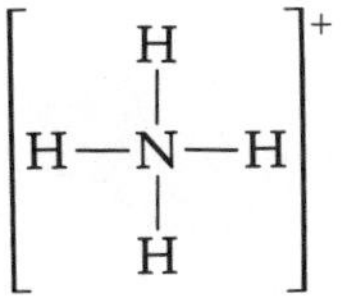

This same argument can be extended to other coordinate covalent bonds as well.

## 17.3 MULTIPLE BONDS

Double and triple bonds occur when two and three pairs of electrons, respectively, are shared between two atoms. As examples, we have seen the molecules ethylene, $C_2H_4$, and acetylene, $C_2H_2$.

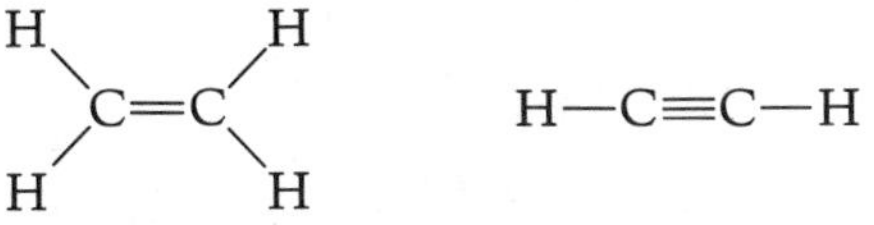

**ethylene** **acetylene**

The bonding in ethylene is usually interpreted in the following way. In order to form bonds to three other atoms (two hydrogens and one carbon), each carbon atom employs a set of $sp^2$ hybrids.

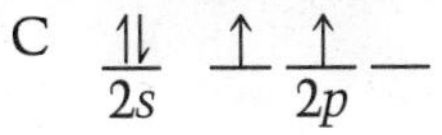

gives

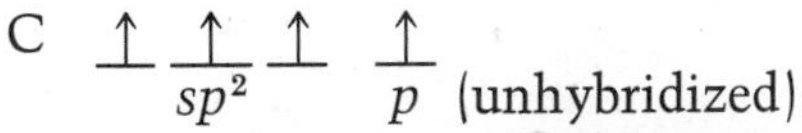

Two of these hybrid orbitals are used for overlap with hydrogen 1*s* orbitals while the third $sp^2$ orbital overlaps with a similar orbital on the other carbon atom, as shown in Figure 17.12*a*. This, then, accounts for all of the C—H bonds in $C_2H_4$ as well as *one* of the electron pairs shared between the two carbons.

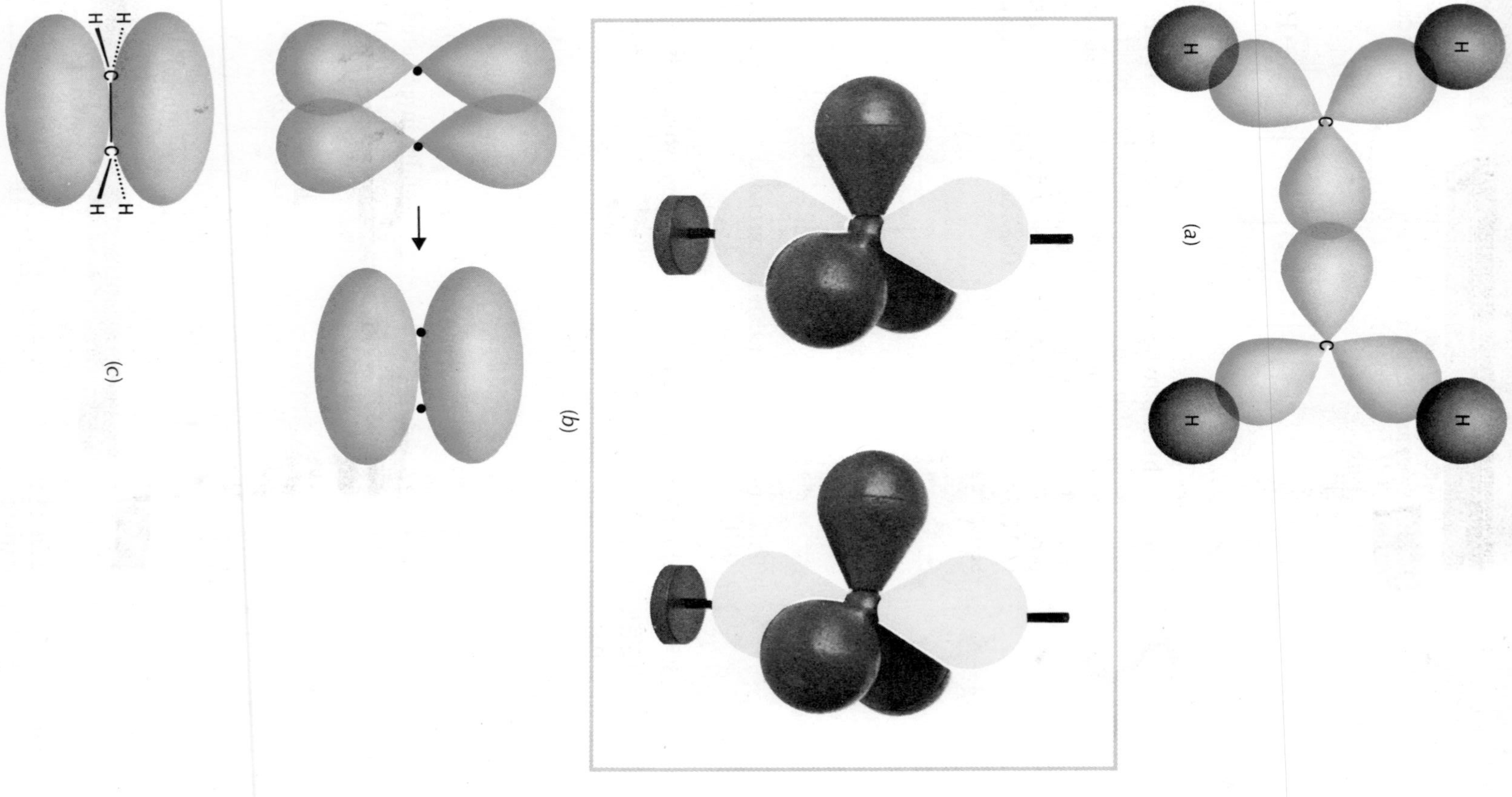
H
C
(a)
(b)
(c)

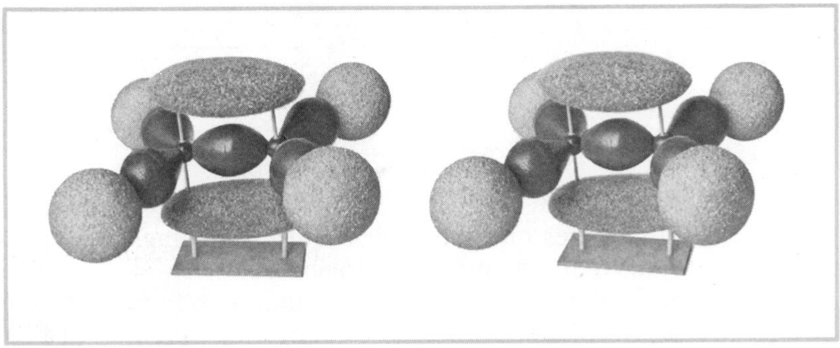

(d)

Figure 17.12
*The bonding in ethylene, $C_2H_4$. (a) Overlap of hydrogen 1s orbitals with* $sp^2$ *hybrid orbitals on carbon. The carbon atoms are also bound by overlap of* $sp^2$ *hybrid orbitals. (b) Unhybridized* p *orbital is perpendicular to plane of* $sp^2$ *hybrid orbitals. (c) Formation of π bonds by sidways overlap of* p *orbitals. (d) Photograph of a model depicting the σ and π bonds in ethylene. Atomic-Molecular Orbital Models by Science Related Materials, Inc., Janesville, Wisconsin.*

Because of the way the $sp^2$ orbitals are created, each carbon atom also has an unhybridized *p* orbital that is perpendicular to the plane of the $sp^2$ orbitals and that projects above and below the plane of these hybrids (Figure 17.12*b*). When the two carbon atoms are joined together, these *p* orbitals approach each other sideways and, in addition to the bond formed from the overlap of $sp^2$ orbitals, a second bond is formed in which the electron cloud is concentrated above and below the carbon–carbon axis. This we see illustrated in Figure 17.12*c*.

In terms of this interpretation, the double bond in ethylene consists of two distinctly different kinds of bonds, and to differentiate between them a specific notation is employed. A bond that concentrates electron density along the line joining the bound nuclei is called a **σ bond** (**sigma bond**). The overlap of the $sp^2$ orbitals of adjacent carbons therefore gives rise to a σ bond. The bond that is formed by the sideways overlap of two *p* orbitals, and that provides electron density above and below the line connecting the bound nuclei, is called a **π bond** (**pi bond**).[4] Thus in ethylene we find the double bond to consist of one σ bond and one π bond. Notice that in this double bond the two electron pairs manage to avoid one another by occupying different regions in space.

Another point to note is that bonds formed by the overlap of the hydrogen 1*s* orbitals with carbon $sp^2$ hybrid orbitals (Figure 17.12*a*) also concentrate electron density along a line joining bound atoms. Therefore, these C—H bonds would also be termed σ bonds.

In acetylene each carbon is bound to only two other atoms, a hydrogen and a carbon atom. Two orbitals are needed for this purpose, and a pair of *sp* hybrid orbitals are used.

[4] In later chapters we refer to a π bond formed by the sideways overlap of two *p* orbitals as a *pπ-pπ* bond.

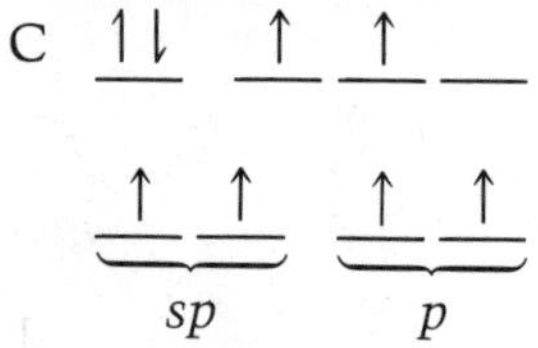


This leaves on each carbon atom two singly occupied unhybridized *p* orbitals that are mutually perpendicular as well as perpendicular to the *sp* hybrids. When the carbon atoms join by way of $\sigma$ bond formation between an *sp* hybrid orbital on each carbon, the *p* orbitals can also overlap to yield two $\pi$ bonds that surround the axis between the carbon nuclei (Figure 17.13). A triple bond therefore consists of one $\sigma$ and two $\pi$ bonds. The two $\pi$ bonds in acetylene (or in any triple bond) give a total electron distribution that is cylindrical about the bond axis. This is shown in Figure 17.13*b*.

In arriving at the structure of a molecule, such as ethylene or acetylene, for instance, the shape of the molecular framework is determined by the $\sigma$ bonds that arise from overlap of the hybrid orbitals. Double and triple bonds in a structure result from additional $\pi$ bonds. In summary, then, we find the following:

single bond—one $\sigma$ bond
double bond—one $\sigma$, one $\pi$ bond
triple bond—one $\sigma$, two $\pi$ bonds

## 17.4 RESONANCE

In Chapter 4 we saw that there are instances in which we cannot draw a single satisfactory Lewis dot formula for a molecule or ion. Some examples, you might remember, are $SO_2$, $SO_3$, and $NO_2^-$. Sulfur dioxide, for instance, was drawn as

and it was stated that the actual electronic structure of this molecule corresponds to a resonance hybrid of these two structures.

Electron dot formulas, as we have drawn them, closely correspond to the valence bond pictures that were developed in the preceding sections.

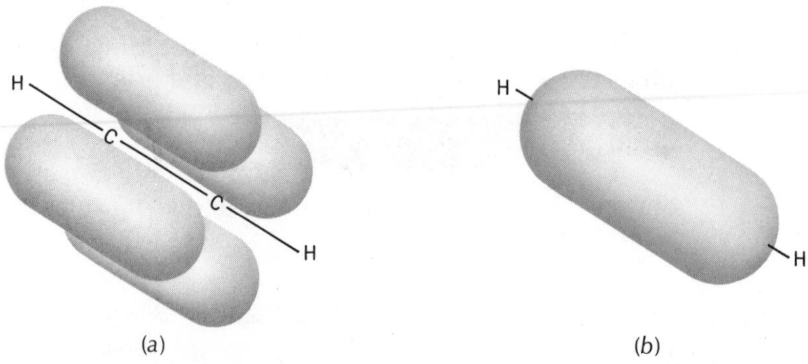


Figure 17.13
*The triple bond in acetylene consists of one $\sigma$ bond and two $\pi$ bonds. (a) Two $\pi$ bonds. (b) Cylindrical electron distribution about bond axis.*

Each pair of dots drawn between two atoms denote a pair of electrons shared in a region where atomic orbitals of the bonded atoms overlap. When we draw one of the resonance structures of $SO_2$, we are therefore referring to a bonding picture in which one S—O bond consists of a single $\sigma$ bond while the other is composed on one $\sigma$ and one $\pi$ bond.

When the valence bond theory was developed, it was recognized that there were numerous instances where a single valence bond structure was inadequate in accounting for the molecular structure. Consequently, the concept of resonance evolved. The inability in these cases to draw a single picture to describe the electron density in the molecule is one drawback of valence bond theory. Nevertheless, the correspondence between the valence bond structures that are based on orbital overlap and the simple electron dot formulas make the valence bond concept a very useful one.

## 17.5 MOLECULAR ORBITAL THEORY

In our discussion of atomic structure in Chapter 3 we saw that around an atomic nucleus there exists a set of atomic orbitals. The electronic structure of a particular atom was derived by feeding the appropriate number of electrons into this set of atomic orbitals such that (1) no more than two electrons populated a single orbital and (2) each electron was placed into the lowest energy orbital available.

**Molecular orbital theory** proceeds in much this same way. According to this theory, there exists in a molecule a certain arrangement of atomic nuclei and, spread out over these nuclei, there is a set of **molecular orbitals.** The electronic structure of the molecule is obtained by feeding the appropriate number of electrons into these molecular orbitals following the same rules that apply to the filling of atomic orbitals.

No one is quite sure what shapes molecular orbitals have in any particular molecule or ion. What *appears* to be an approximately correct picture is obtained by combining the atomic orbitals that reside on the nuclei making up the molecule. These combinations are achieved by either adding or subtracting the wave functions corresponding to the atomic orbitals that overlap one another. This is shown in Figure 17.14 for the 1*s* orbitals on two identical nuclei. Notice that when the two wave functions are *added*, the resulting molecular orbital has a shape that concentrates electron density between the two nuclei. Electrons placed in such a molecular orbital tend to

Figure 17.14

*The combination of atomic 1s orbitals to give bonding and antibonding molecular orbitals.*

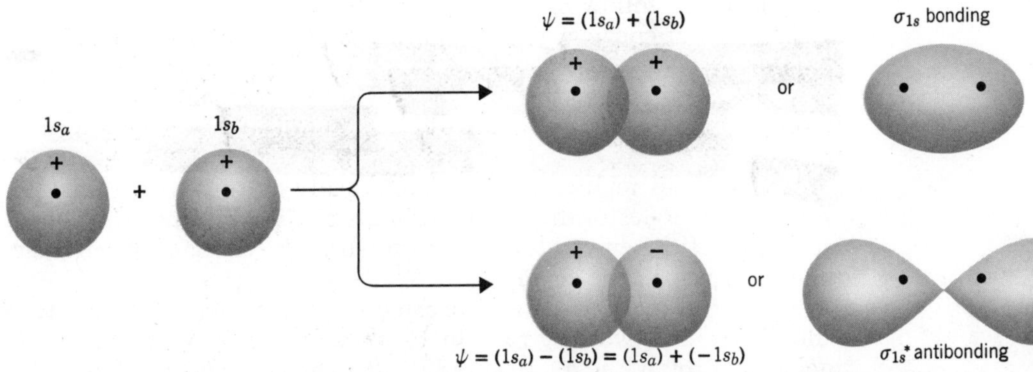

hold the nuclei together and tend to yield a stable molecule. For that reason this orbital is called a **bonding molecular orbital.** Because the electron density in the orbital is centered along the line joining the atomic nuclei, it is a $\sigma$ type of orbital; since it is derived in this case from two 1s atomic orbitals, we refer to it as the $\boldsymbol{\sigma_{1s}}$ **molecular orbital.**

You will also observe in Figure 17.14 that a second molecular orbital is obtained by subtracting one atomic orbital wave function from the other. In this instance a molecular orbital is produced that places the maximum electron density outside of the region between the two nuclei. If the electrons of a molecule are placed into this molecular orbital, they do not help to cement the nuclei together and, in fact, the unshielded nuclei repel one another. Consequently, electrons placed into this molecular orbital lead to a destabilization of the molecule and, as a result, the orbital is said to be **antibonding.** This antibonding orbital also has its greatest electron density along the line that passes through the two nuclei and is thus a $\sigma$-type orbital. Its antibonding character is denoted by an asterisk superscript; thus it is called the $\boldsymbol{\sigma^{\star}_{1s}}$ **molecular orbital.** As you might expect, we can also draw similar pictures for the combination of any pair of *s* orbitals; therefore, in a diatomic molecule we also have $\sigma_{2s}$, $\sigma^{\star}_{2s}$, $\sigma_{3s}$, $\sigma^{\star}_{3s}$, . . . , molecular orbitals.

In a molecule the *p* orbitals are also capable of interacting to produce bonding and antibonding molecular orbitals, as illustrated in Figure 17.15. Here we have arbitrarily chosen to denote the internuclear axis as the *z* axis of our coordinate system so that the *p* orbitals that point toward one another correspond to $p_z$ orbitals. Again we find that one combination of orbitals gives a bonding molecular orbital, with electron density placed between the two nuclei, while the second combination places most of the electron density outside of the region between the nuclei. The $p_z$ orbitals, like *s* orbitals, form $\sigma$-type molecular orbitals and for $2p_z$ orbitals, then, there would result $\sigma_{2p_z}$ and $\sigma^{\star}_{2p_z}$ molecular orbitals.

Having chosen the *z* axis as the internuclear axis, we find that the $p_x$ and $p_y$ orbitals on the two nuclei of our molecule are forced to overlap in a sideways fashion to produce $\pi$ and $\pi^{\star}$ molecular orbitals (Figure 17.15). Also keep in mind that the $\pi_{p_x}$ and $\pi^{\star}_{p_x}$ orbitals are the same as the $\pi_{p_y}$ and $\pi^{\star}_{p_y}$ orbitals, respectively, with the exception that they are situated at 90° to each other when viewed down the molecular axis.

We have now examined for a diatomic molecule the shapes of the molecular orbitals that can be considered to arise as a consequence of the overlap of atomic orbitals. To discuss the electronic structure of a diatomic molecule, however, we must know the relative energies of these orbitals. Once this has been established, we can then proceed with filling the orbitals with electrons, following the rules that were mentioned earlier.

Let us first consider the $\sigma_{1s}$ and $\sigma^{\star}_{1s}$ orbitals. Electrons placed into the bonding orbital lead to stable bond formation and, therefore, to an energy lower than that of two separate atoms. On the other hand, electrons placed into the antibonding orbital lead to a destabilization of the molecule and thus to a state higher in energy than the atoms from which the molecule is formed. We can represent this schematically as shown in Figure 17.16*a*, where the energies of the atomic orbitals of the separate atoms appear on either side of the energy-level diagram while the energies of the molecular orbitals appear in the center.

Using this simple diagram we can examine the bonding in the $H_2$ molecule. There are two electrons in $H_2$ that we place in the lowest-energy molecular orbital, the $\sigma_{1s}$, (Figure 17.16*b*). The electron distribution in $H_2$ is

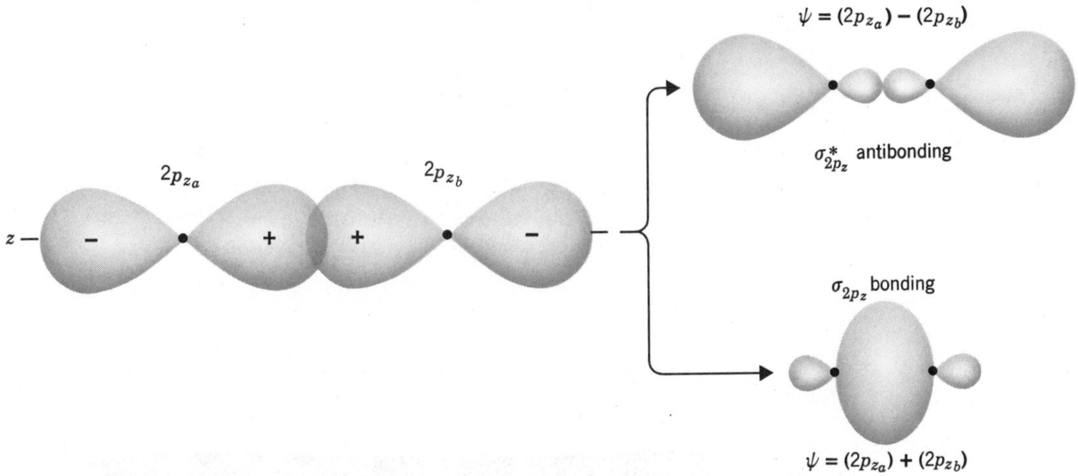


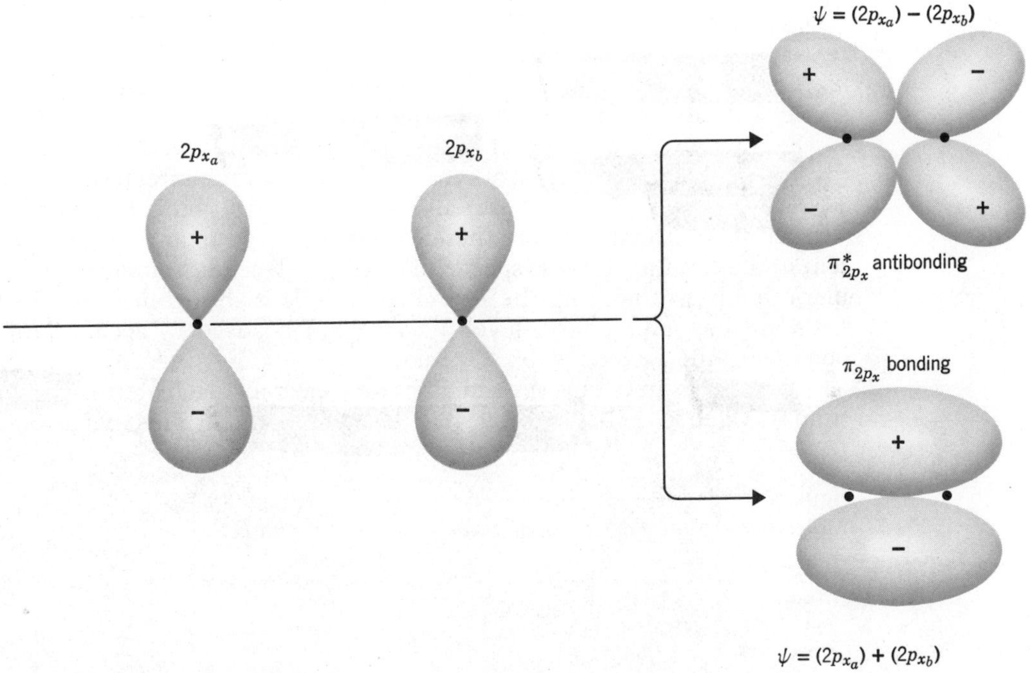


Figure 17.15
*Formation of molecular orbitals from atomic* p *orbitals.*

therefore that described by the shape of the $\sigma_{1s}$ orbital. Notice that this picture is the same as that developed in the valence bond view of $H_2$. This should not be too surprising since both theories are attempting to describe the same molecular species.

Before moving on, let us also see why the molecule $He_2$ does *not* exist. The species $He_2$ would have four electrons, two of which would be placed into the $\sigma_{1s}$ orbital and a second pair that would be forced to occupy the $\sigma_{1s}^{*}$

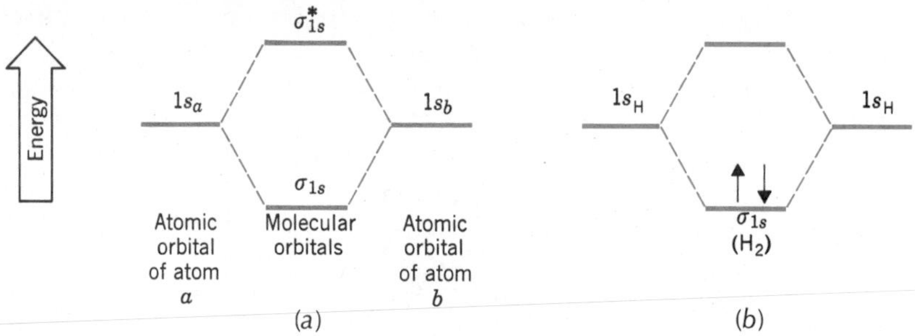


Figure 17.16
*(a) The energies of the bonding and antibonding $\sigma_{1s}$ molecular orbitals. (b) Bonding in $H_2$.*

orbital. The pair of electrons in the antibonding orbital would cancel out the stabilizing influence of the bonding pair. As a result, the net bond order, which we can define as

$$\begin{pmatrix}\text{Net bond}\\ \text{order}\end{pmatrix} = \frac{(\text{no. of } e^- \text{ in bonding MOs}) - (\text{no. of } e^- \text{ in antibonding MOs})}{2}$$

has a value of zero for $He_2$. Since the bond order in $He_2$ is zero, $He_2$ is not a stable molecule and is not observed to exist under normal conditions.

For diatomic molecules of second period elements we really only need to consider molecular orbitals that are derived from the interaction of the 2*s* and 2*p* orbitals. The 1*s* orbitals are essentially buried beneath the valence-shell orbitals and are therefore not involved to any appreciable extent in the bonding in these species. The energy-level diagram for the molecular orbitals created from the 2*s* and 2*p* orbitals is shown in Figure 17.17*a*. Let's see how this energy-level diagram can be used to account for the bonding in the molecules $N_2$, $O_2$, and $F_2$.

Nitrogen is in Group VA and, therefore, each nitrogen atom contributes from its valence shell five electrons to the $N_2$ molecule. This means that we

Figure 17.17
*Molecular orbital electron configuration in $N_2$, $O_2$ and $F_2$.*

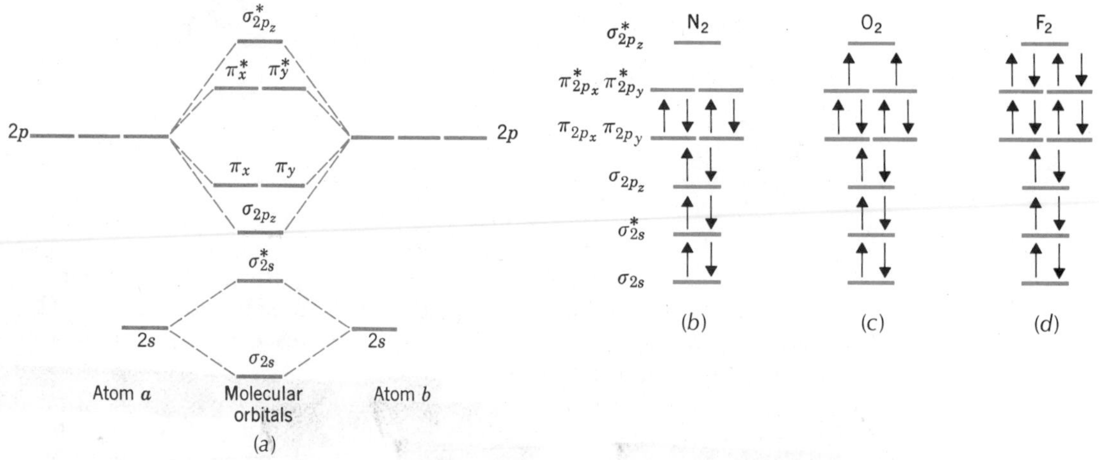

must place ten electrons into our set of molecular orbitals. As shown in Figure 17.17*b*, two electrons enter the $\sigma_{2s}$, two go into the $\sigma^{\star}_{2s}$, two more into the $\sigma_{2p_z}$ and, finally, two into each of the bonding $\pi$ orbitals, $\pi_{2p_x}$ and $\pi_{2p_y}$. As before, the two $\sigma^{\star}_{2s}$ antibonding electrons cancel the effect of the $\sigma_{2s}$ bonding electrons, leaving us with a net total of six bonding electrons (two each in the $\sigma_{2p_z}$, $\pi_{2p_x}$, and $\pi_{2p_y}$ orbitals). If, as usual, we take two electrons to represent a "bond," we find that $N_2$ is held together by a triple bond that is composed of one $\sigma$ and two $\pi$ bonds. As with $H_2$, we arrive at the same resultant description of the bonding in $N_2$ with both the valence bond and molecular orbital theories.

The real mark of success for molecular orbital theory is seen in its description of the $O_2$ molecule. This species is found experimentally to be paramagnetic with two unpaired electrons. An attempt to derive a valence bond picture for $O_2$, however, gives us

where, to satisfy the octet rule, all of the electrons appear in pairs.

The molecular orbital description of $O_2$ is seen in Figure 17.17*c*. The first 10 of the 12 valence electrons populate all of the same molecular orbitals as in $N_2$. The final two electrons must then be placed in the $\pi^{\star}_{2p_x}$ and $\pi^{\star}_{2p_y}$ antibonding orbitals; since these two orbitals are of the same energy, the electrons spread themselves out with their spins in the same direction (Hund's rule, see Section 3.15). These two antibonding $\pi$ electrons also cancel one of the pair of $\pi$-bonding electrons, so that in the final analysis we see $O_2$ to be held together by a *net* double bond (one $\sigma$ and one net $\pi$ bond) and we further note that the molecule is predicted to have two unpaired electrons, in precise agreement with experiment.

Finally, with $F_2$ (which contains two more electrons than $O_2$), we find that the two $\pi^{\star}$ antibonding orbitals are filled (Figure 17.17*d*). This leaves one net single bond, and once again the valence bond and molecular orbital theories give the same result.

The success of molecular orbital theory is not restricted merely to diatomic molecules. In more complex molecules, however, the energy-level diagrams are more difficult to predict, and we shall not attempt to extend the theory much further. One useful concept in molecular orbital theory that we can look at further, however, is the idea that molecular orbitals may extend over more than two nuclei. It is this aspect of molecular orbital theory that allows one to avoid the concept of resonance.

Consider, for example, the molecule $SO_3$. From experiment we know this to be a planar molecule (all four atoms lie in the same plane) with all three S—O bonds the same. This structure can be explained if we assume that the sulfur employs a set of $sp^2$ hybrid orbitals to form $\sigma$ bonds with the three oxygen atoms. This leaves one unhybridized $p$ orbital on the sulfur that can overlap *simultaneously* with $p$ orbitals on the three oxygen atoms, as shown in Figure 17.18. The result is the creation of a molecular orbital that extends over all four nuclei such that the electron densities in the S—O bonds are all the same. Obviously, there is no need to draw more than one bonding picture for the molecule; molecular orbital theory thus is able to explain the bonding in $SO_3$ satisfactorily without resorting, as valence bond theory does, to the rather awkward concept of resonance.

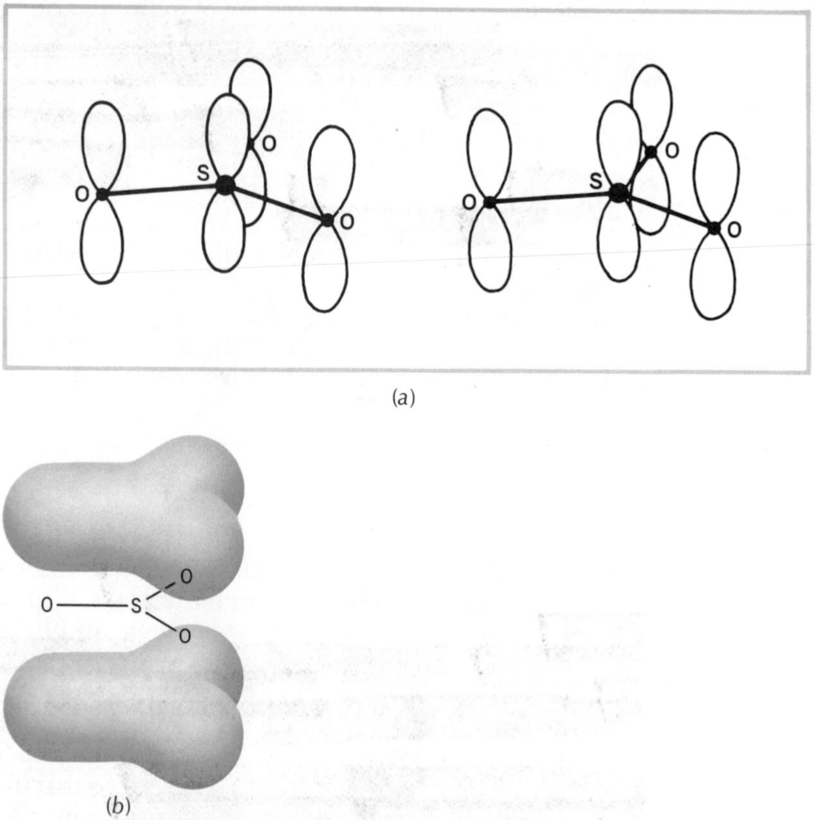


Figure 17.18
*Simultaneous overlap of atomic* p *orbitals in the* $SO_3$ *molecule. (a)* p *orbitals on the sulfur and oxygen atoms. (b) Delocalized* $\pi$ *molecular orbital.*

## 17.6 ELECTRON-PAIR REPULSION THEORY OF MOLECULAR STRUCTURE

One of the primary goals of chemical bonding theory is to explain and, hopefully, to predict molecular structure. We have seen, for example, how valence bond theory rationalizes molecular structure through the use of hybrid orbitals. Another theory that is exceedingly simple in its concept and remarkably successful in its ability to predict molecular geometry accurately is called the **electron-pair repulsion theory.** In applying it, it is not necessary to employ the notion of atomic orbitals at all. We shall see, instead, that if an electron dot structure can be drawn for a molecule, its general shape can be predicted.

The electron-pair repulsion theory proposes that the geometric arrangement of atoms, or groups of atoms (which we shall generally refer to as ligands) about some central atom is determined *solely* by the repulsion between the electron pairs present in the valence shell of the central atom. Consider, for example, the molecule $BeH_2$. The electron dot structure would be given as

$$\mathrm{H} \overset{\times}{\bullet} \mathrm{Be} \overset{\times}{\bullet} \mathrm{H}$$

where the crosses are Be electrons and the dots are hydrogen electrons. This particular molecule violates the octet rule, and there are only two pairs of electrons located in the valence shell of Be. These electron pairs would like to be as far apart as possible so that the repulsion between them is at a minimum. When there are two electron pairs in the valence shell, this minimal repulsion occurs when the electron pairs are located on opposite sides of the nucleus, so that we have

:—Be—:

The hydrogen atoms are attached to the Be through these electron pairs and the molecule should therefore have the *linear* structure,

H—Be—H

As we have seen before, this is the structure of $BeH_2$.

We can also extend this reasoning to situations involving double or triple bonds. For instance, the $CO_2$ molecule has the dot structure

:Ö::C::Ö:

where we see that there are double bonds between C and O. Both pairs of electrons in a double bond are confined to the same general region in the valence shell of an atom, so that in terms of their effect on determining molecular geometry, a group of four electrons in a double bond behaves much like a group of two electrons in a single bond. In the valence shell of carbon, then, we have *two* groups of four, and they will locate themselves on opposite sides of the C nucleus so that the repulsion between the two groups is at a minimum:

::—C—::

As before, the ligands (in this case, oxygen) are attached to the central atom through these electron pairs and we again have a *linear* structure.

O=C=O

**MORE THAN TWO PAIRS (OR GROUPS OF PAIRS) OF ELECTRONS.** When there are more than two pairs (or groups of pairs)of electrons in the valence shell we find other geometric arrangements as shown in Figure 17.19. Electron pairs arranged in the valence shell in this manner lead to minimum repulsions. Let us see how we can use these electron-pair arrangements to predict molecular structure.

**THREE GROUPS OF ELECTRONS IN THE VALENCE SHELL.** The molecule $BCl_3$ has the dot structure

:C̈l:<br>
:C̈l ˣ B̈ ˣ C̈l:

Thus there are three electron pairs around boron. We therefore expect the three chlorine atoms to be arranged about the B atom in a planar triangle. This is indeed the structure of $BCl_3$.

Let us now consider the molecule $SO_2$. The electron dot structure for one of the two resonance structures is

:Ö:S̈::Ö:

| Number of Electron Pairs | Geometric Arrangement of Electron Pairs | |
|---|---|---|
| 2 | Linear | 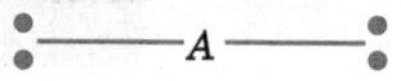 |
| 3 | Planar triangle | 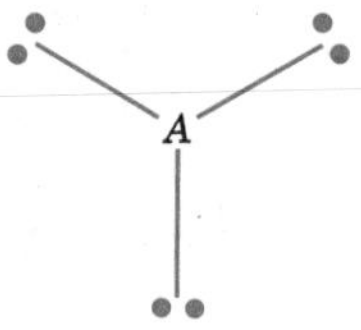 |
| 4 | Tetrahedral | 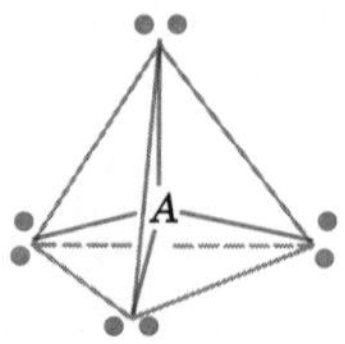 |
| 5 | Trigonal bipyramidal | 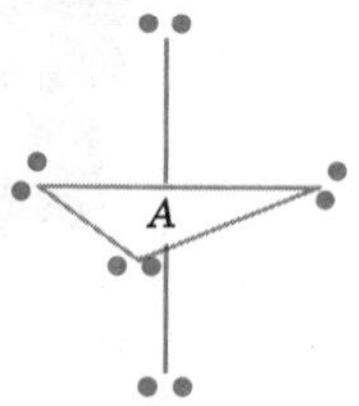 |
| 6 | Octahedral (eight sides– 6 corners) | 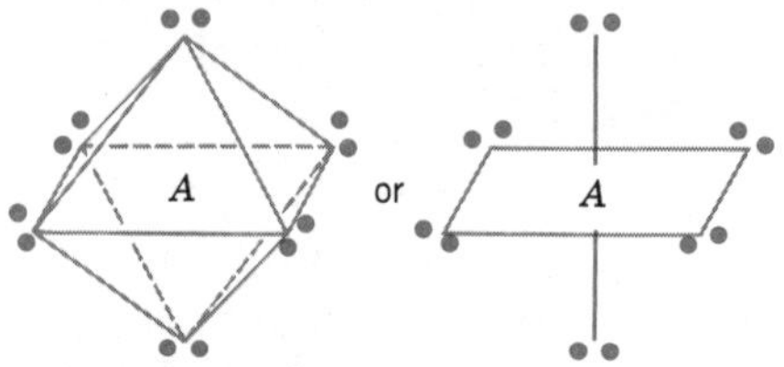 |

Figure 17.19
*Positions of electron pairs that lead to minimum electrostatic repulsions.*

About the sulfur there are again three groups of electrons, two groups each with one pair, and one group with two pairs (the double bond). These groups of electrons are situated at the corners of a triangle with the sulfur in the center.

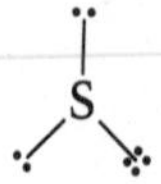

Attaching the oxygen atoms, one to a single pair and one to the double pair, we have

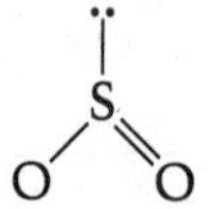

The $SO_2$ molecule is thus *not* linear, but instead has an angular, or bent, shape.[5] The nonlinear shape is caused in this case by the presence of the nonbonded, *lone pair* of electrons on the sulfur. We see, therefore, that lone pairs of electrons in the valence shell of an atom influence the molecular geometry.

In summary, when there are three groups of electrons about an atom, they are arranged at the corners of a triangle. If all of them are bonded to ligands, we have a molecule, $AX_3$, having a planar triangular shape. If only two groups are bonded, leaving one lone pair, we have a species $AX_2E$ (where we use $E$ to represent the lone pair) in which the atomic nuclei are situated so as to give an angular structure.

**FOUR GROUPS OF ELECTRONS IN THE VALENCE SHELL.** When there are four groups of electrons in the valence shell of an atom, they will situate themselves at the vertices of a tetrahedron (Figure 17.19). As shown in Figure 17.20, there are three kinds of molecules of interest that can be formed with four groups of electrons: $AX_4$, $AX_3E$, and $AX_2E_2$ (note that there is *no* uncertainty about the geometry of any $AXE_n$ molecule; $A$ and $X$ must be in a straight line).

The molecule, $AX_4$, of which $CH_4$ is an example, has a tetrahedral shape with ligands bonded through all four electron pairs. The species $AX_3E$ (for example, $NH_3$), on the other hand, contains *one* lone pair and has a geometry that we refer to as pyramidal (shaped like a pyramid). Finally, the molecule $AX_2E_2$ (for example, $H_2O$) has two lone pairs and has an angular structure. We see, then, that with four groups of electrons there are three different possible molecular shapes.

**FIVE ELECTRON PAIRS.** With five electron pairs in the valence shell of the central atom, there are four types of molecules to be considered: $AX_5$, $AX_4E$, $AX_3E_2$, and $AX_2E_3$. These are illustrated in Figure 17.21. *Note that when there are five electron pairs about an atom, any lone pairs that are present occur in the triangular plane.*

**SIX ELECTRON PAIRS.** Of the five different possibilities, only three molecule types are known: $AX_6$, $AX_5E$, and $AX_4E_2$. The molecular geometries that are associated with these formulas are shown in Figure 17.22.

**SUMMARY.** The theory that has been presented above can be applied with great success to a very large number of molecules and ions formed by the representative elements. The key to using this theory lies in your ability to write an electron dot structure for the molecule. Once this has been done, you merely count up the number of groups of electrons (either single or multiple pairs) and decide what geometry the electron pairs will assume: linear for two groups, triangular for three, and so on. The appropriate number of ligand atoms are then attached through the electron pairs, and you are now in a position to predict the structure of the molecule, that is, the structure assumed by the central atom and the ligands (when we speak of molecular structure, we refer to the positions of the atoms, *not* to the location of the electrons, even though we use the latter to predict the former). A summary of

[5] In arriving at the predicted structure of a molecule or ion for which two or more resonance structures may be drawn it is only necessary to examine one of the different reasonance formulas.

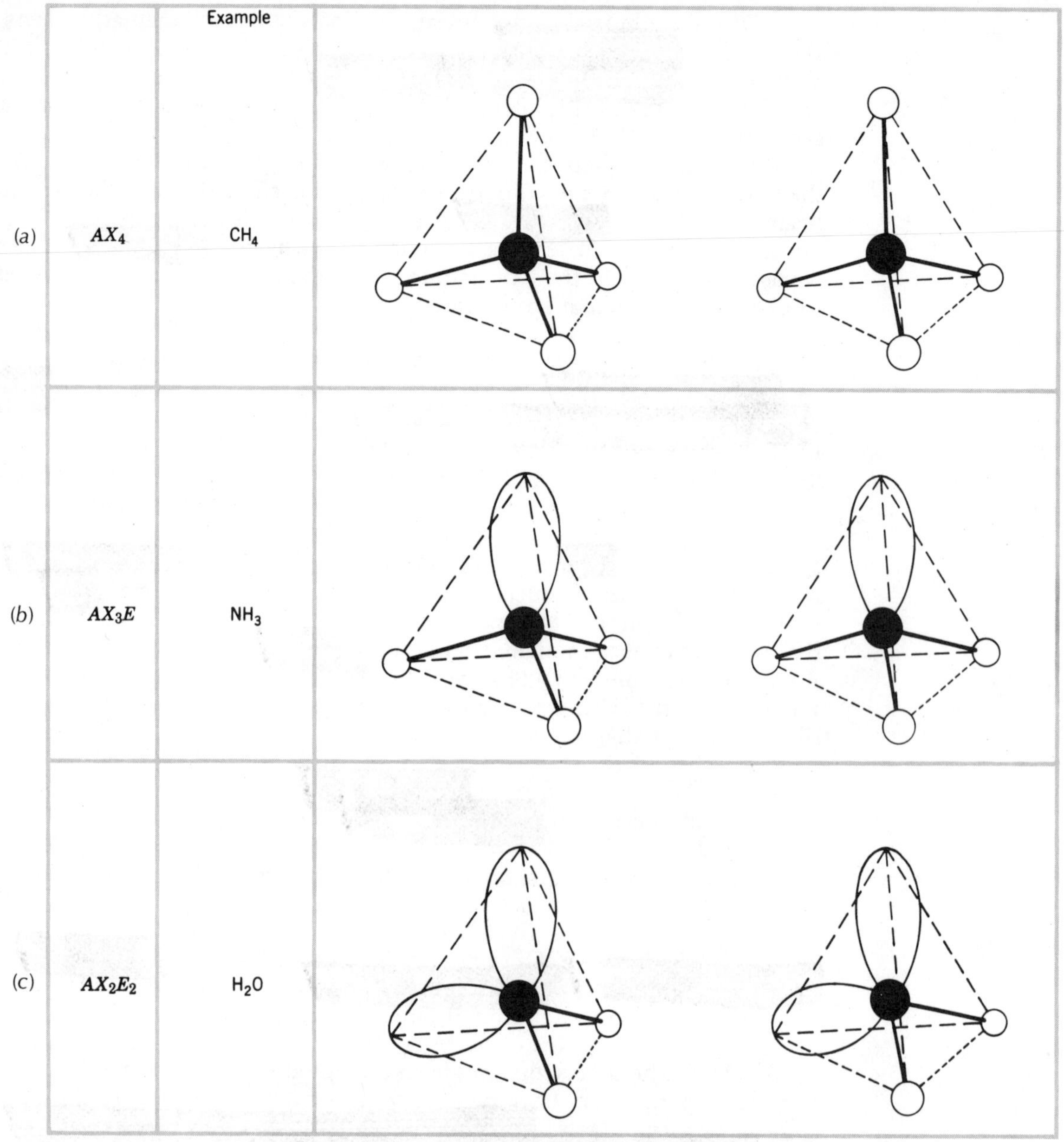


Figure 17.20
*Geometries of molecules in which the central atom has four pairs of electrons. (a) $AX_4$ (for example, $CH_4$). (b) $AX_3E$ (for example, $NH_3$). (c) $AX_2E_2$ (for example, $H_2O$).*

the molecular shapes found for different molecule types is given in Table 17.2.

By now you have undoubtedly noticed that both the valence bond theory, through the use of hybrid orbitals, and the electron-pair repulsion theory lead to very similar, if not identical, results in the prediction of

| | Molecule | Example | Structure |
|---|---|---|---|
| (a) | $AX_5$ | $PCl_5$ 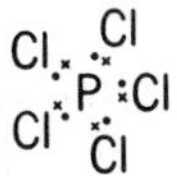 | 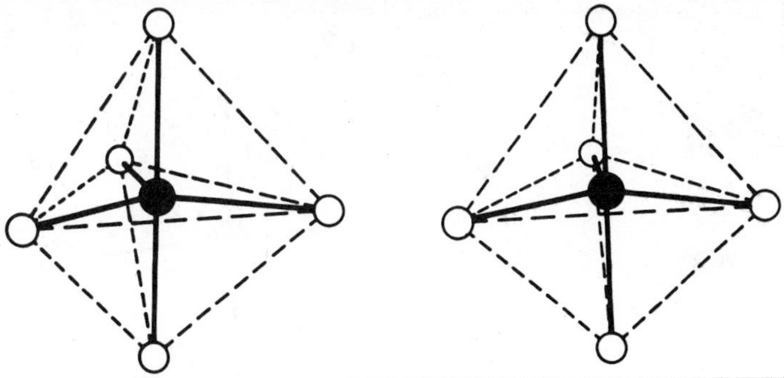 |
| (b) | $AX_4E$ | $TeCl_4$ 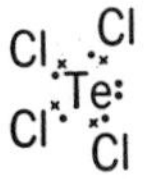 | 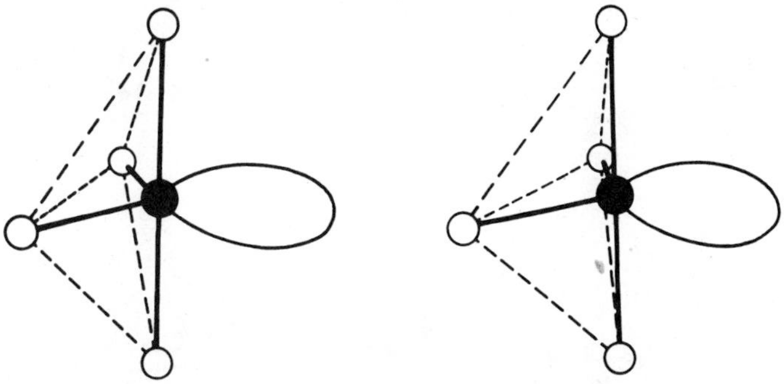 |
| (c) | $AX_3E_2$ | $ClF_3$  | 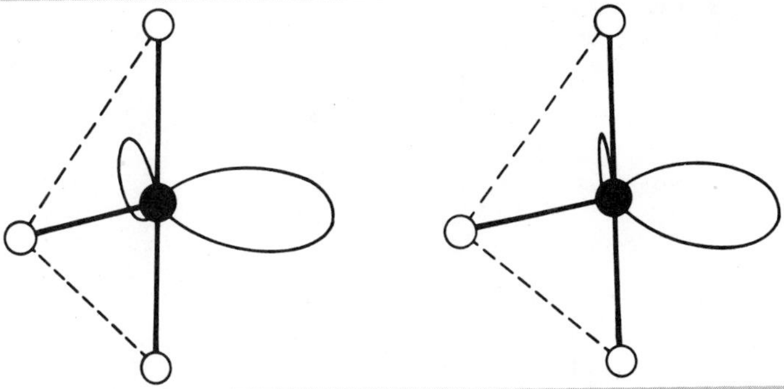 |
| (d) | $AX_2E_3$ | $I_3^-$ 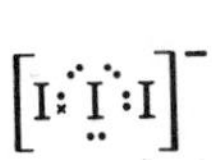 | 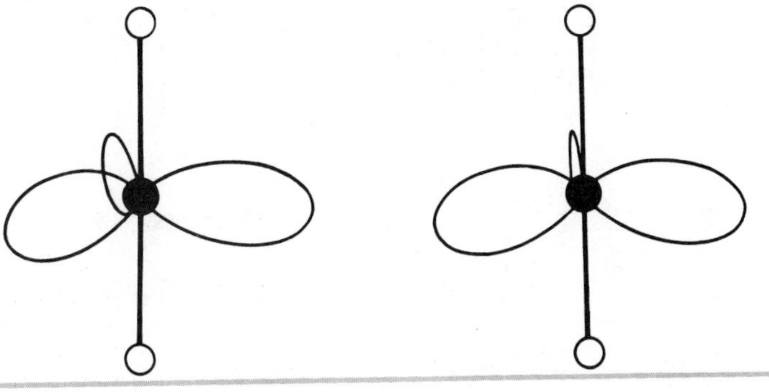 |

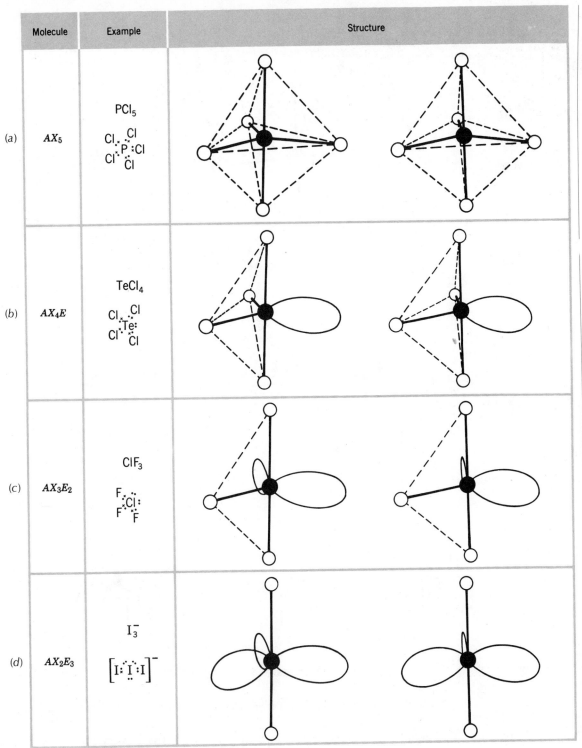

Figure 17.21
*Molecular structures that result when the central atom has five electron-pair groups.*

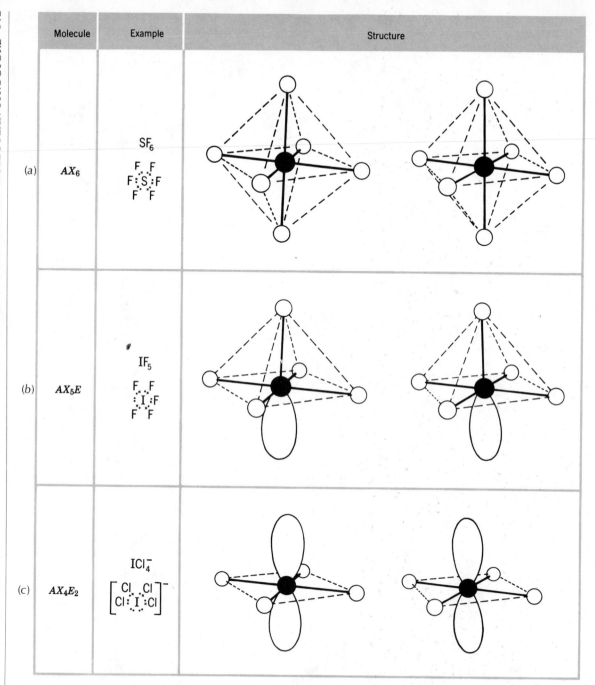

| | Molecule | Example | Structure |
|---|---|---|---|
| (a) | $AX_6$ | $SF_6$  | 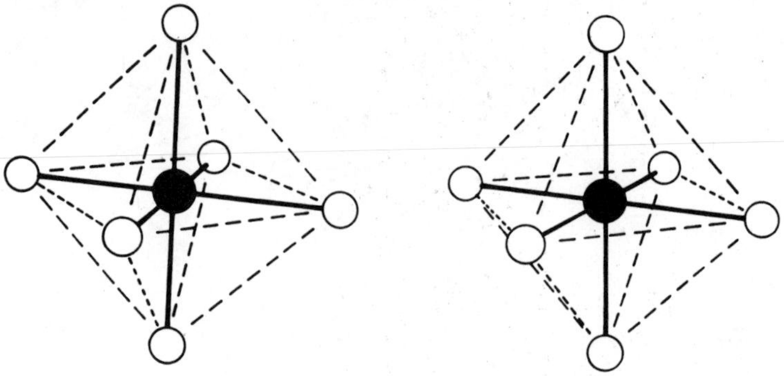 |
| (b) | $AX_5E$ | $IF_5$ 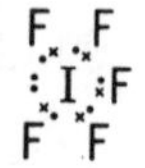 | 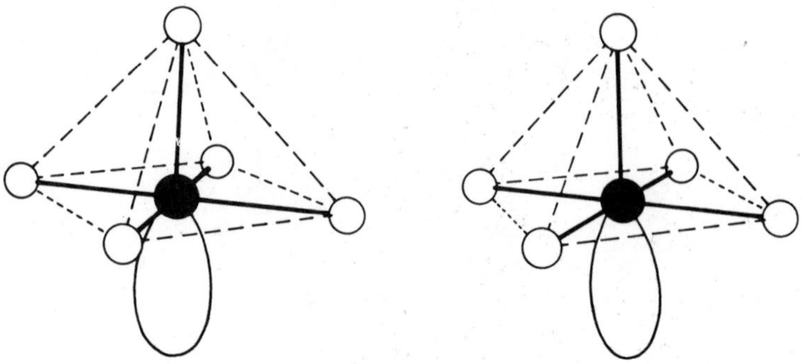 |
| (c) | $AX_4E_2$ | $ICl_4^-$ 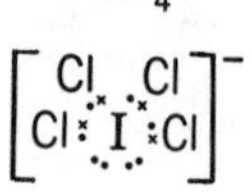 | 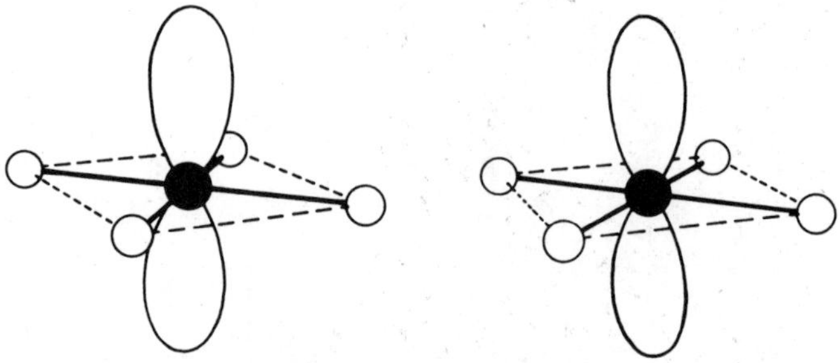 |

Figure 17.22
*Molecular structures that result when the central atom has six electron-pair groups.*

molecular geometry. These two theories are actually very much the same, with the electron-pair repulsion theory yielding structures that are the same as those we would obtain from the valence bond theory if we assumed complete hybridization in every case ($sp^3$ hybrids in $H_2O$, $sp$ hybrids on Be in $BeH_2$, $d^2sp^3$ hybrids on S in $SF_6$, etc.).

While the electron pair-repulsion theory is quite good for predicting molecular geometry, it tells us nothing about the relative stabilities of bonds or why a bond is formed at all. For these aspects of the chemical bond we must rely on valence bond and molecular orbital theory.

Table 17.2
*Summary of molecular shapes*

| Molecule Type | Shape |
|---|---|
| $AX_2$ | Linear |
| $AX_3$ | Planar triangular |
| $AX_2E$ | Angular (bent) |
| $AX_4$ | Tetrahedral |
| $AX_3E$ | Pyramidal |
| $AX_2E_2$ | Angular (bent) |
| $AX_5$ | Trigonal bipyramidal |
| $AX_4E$ | Distorted tetrahedral |
| $AX_3E_2$ | T-shaped |
| $AX_2E_3$ | Linear |
| $AX_6$ | Octahedral |
| $AX_5E$ | Square pyramidal |
| $AX_4E_2$ | Square planar |

## INDEX TO QUESTIONS

Valence bond theory 1, 2, 3, 4, 5, 6

Hybrid orbitals 7, 8, 9, 10, 11, 12, 13, 15, 16, 31, 35

Coordinate covalent bonds 14

Multiple bonds 17, 20

Resonance 18, 28

Molecular orbitals 19, 21, 22, 23, 24, 25, 26, 27

Electron-pair repulsion theory 29, 30, 31, 32, 33, 34

## REVIEW QUESTIONS

**17.1** What is the basic concept on which the valence bond theory is based?

**17.2** Use valence bond theory to explain the bonding in the $Cl_2$ molecule.

**17.3** From valence bond theory, predict the geometry of the $SnCl_2$ molecule.

**17.4** What orbitals would you expect to be involved in bonding in the $AsH_3$ molecule? The H—As—H bond angles are very close to 90°.

**17.5** How can you account for the fact that the H—S—H bond angle in $H_2S$ is approximately 92°?

**17.6** Diagram the outer-shell electronic structures of P and F. Indicate how bonding occurs between P and F to give $PF_3$. What molecular shape would you expect?

**17.7** What is a hybrid orbital?

**17.8** Why is it necessary to employ hybrid orbitals when attempting to account for the structure of methane, $CH_4$?

**17.9** On the basis of the electronic structure of the central atom, suggest what kind of hybrid orbitals would be involved in the bonding in each of the following:

| | | |
|---|---|---|
| (a) $BCl_3$ | (d) $AlCl_6^{3-}$ | (g) $PCl_3$ |
| (b) $NH_4^+$ | (e) $BeCl_2$ | (h) $TeF_4$ |
| (c) $PCl_5$ | (f) $SbCl_6^-$ | (i) $ClO_4^-$ |

**17.10** From a knowledge of the hybrid orbitals used for bonding, predict the structures of each of the species in Question 17.9.

**17.11** Diagram the bonding in $SbCl_5$. What kind of hybrid orbitals are involved in the bonding?

**17.12** It is possible for $SiF_4$ to react with $F^-$ to give $SiF_6^{2-}$ but it is not possible for $CF_4$ to form $CF_6^{2-}$. Why?

**17.13** We have discussed the reaction, $BCl_3 + NH_3 \rightarrow Cl_3BNH_3$ earlier (Chapter 13). What kind of hybrid orbitals are used by B and N before and after reaction? How does the geometry change about B and N as the reaction occurs?

**17.14** Which of the species in Question 17.9 have one or more bonds that would be considered to have been formed by way of coordinate covalent bonding? How, if at all, does a coordinate covalent bond differ from a normal covalent bond once it has been formed?

**17.15** What angles exist between the orbitals in
(a) $sp^3$ hybrids
(b) $sp^2$ hybrids
(c) $sp$ hybrids
(d) $d^2sp^3$ hybrids

**17.16** $SnCl_4$ is a volatile liquid composed of individual $SnCl_4$ molecules. Describe the bonding that is expected in this molecule.

**17.17** Describe a $\sigma$ bond; a $\pi$ bond. What constitutes a double bond? A triple bond?

**17.18** Draw the resonance structures for (a) $NO_3^-$ and (b) $NO_2^-$.

**17.19** How does molecular orbital theory view the formation of a molecule? How does molecular orbital theory differ from valence bond theory?

**17.20** Describe, in detail, the bonding in the $N_2$ molecule.

**17.21** Predict the relative stabilities of the species $N_2^+$, $N_2$, $N_2^-$. From the discussion in Section 4.7 (p. 111), how would you expect the bond lengths in these species to compare?

**17.22** Predict the relative stabilities of the species $O_2^+$, $O_2$, $O_2^-$. From the discussion in Section 4.7 (p. 111), how would you expect the bond lengths in these species to compare?

**17.23** What is the difference between a bonding and an antibonding molecular orbital? How do their energies compare?

**17.24** Use Figure 17.17*a* to draw molecular orbital energy-level diagrams for $Li_2$, $Be_2$, $B_2$, and $C_2$. Which of these should not exist, which should be paramagnetic?

**17.25** What can you predict about the stabilities of the species in Question 17.24 when one electron is (a) removed from each, (b) added to each?

**17.26** How does molecular orbital theory avoid the concept of resonance?

**17.27** The species $H_2^+$ and $He_2^+$ have been observed. Use molecular orbital theory to account for their existence.

**17.28** Give the valence bond and molecular orbital descriptions for the following species that can be drawn as two or more resonance structures: (a) $SO_2$ (b) $NO_3^-$
(c)

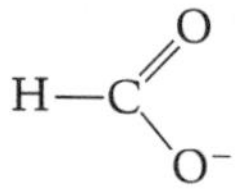


**17.29** What is the basic postulate of the electron-pair repulsion theory?

**17.30** Use the electron-pair repulsion theory to predict the geometry of each of the following:

| | | |
|---|---|---|
| (a) $NF_3$ | (g) $BrF_5$ | (l) $SnCl_4$ |
| (b) $PH_4^+$ | (h) $CCl_4$ | (m) $SO_3$ |
| (c) $CO_3^{2-}$ | (i) $AlCl_6^{3-}$ | (n) $BF_4^-$ |
| (d) $NO_2$ | (j) $SbCl_5$ | (o) $OF_2$ |
| (e) $SeCl_4$ | (k) $SnCl_2$ | (p) $XeF_4$ |
| (f) $ICl_2^-$ | | |

**17.31** Use your predictions from Question 17.30 to suggest the type of hybrid orbitals that would be used in valence bond theory to account for these geometries.

**17.32** Use the electron-pair repulsion theory to predict for each of the following: (a) the geometric arrangement of electron pairs around the central atom (written first in the formulas), (b) the molecular shape.

| | | |
|---|---|---|
| (a) $ClO_2^-$ | (g) $CH_3^+$ | (l) $CH_2O$ |
| (b) $SCl_2$ | (h) $ICl_3$ | (m) $ClO_3^-$ |
| (c) $SbCl_6^-$ | (i) $NO_3^-$ | (n) $SiF_6^{2-}$ |
| (d) $PCl_4^+$ | (j) $AsH_3$ | (o) $PCl_3$ |
| (e) $IF_4^-$ | (k) $POCl_3$ | (p) $SOCl_2$ |
| (f) $PO_4^{3-}$ | | |

**17.33** Describe the changes in molecular geometry that take place during the following reactions:
(a) $BF_3 + F^- \rightarrow BF_4^-$
(b) $PCl_5 + Cl^- \rightarrow PCl_6^-$

(c) $ICl_3 + Cl^- \rightarrow ICl_4^-$
(d) $SF_2 + F_2 \rightarrow SF_4$
(e) $C_2H_2 + H_2 \rightarrow C_2H_4$
(f)

$$\mathrm{Cl{-}\overset{\overset{\displaystyle O}{\|}}{C}{-}Cl \longrightarrow COCl^+ + Cl^-}$$

**17.34** On p. 495 it was suggested that lone electron pairs tend to repel rather strongly electron pairs between bonded atoms; in other words, lone-pair/bond-pair repulsions are greater than bond-pair/bond-pair repulsions. This will lead to structural distortions of some of the idealized molecular geometries pictured in Figures 17.20, 17.21, and 17.22. Predict the nature of these distortions and sketch the shapes of the resulting molecules.

**17.35** Hybrid orbitals are not symmetrical about the nucleus. They concentrate electron density on the side of the nucleus where the orbital is "large." Lone electron pairs in hybrid orbitals therefore are expected to contribute to the dipole moment of the molecule. It is observed experimentally that $NF_3$ is nearly a nonpolar molecule; $NH_3$ is very polar. The electronegativity difference between N and F is nearly the same as that between N and H (see Table 4.6). How does this support the view that in both $NF_3$ and $NH_3$ the nitrogen uses $sp^3$ hybrid orbitals?

# 18 CHEMISTRY OF THE REPRESENTATIVE ELEMENTS: PART I, THE METALS

In previous chapters you learned many of the concepts that chemists have applied in developing their understanding of how the elements react with each other and the kinds of compounds they produce. These earlier discussions were focused primarily on the concepts themselves, with examples of chemical behavior being used to reinforce and justify them. With these concepts available to us now, we shall direct our emphasis in the opposite direction and begin to examine, in more or less a systematic way, some of the physical and chemical properties of the elements and their compounds. For example, we shall look at what kinds of reactions the elements and their compounds undergo, what types of compounds are formed when they react and, in many cases, the types of structures that result. Chemistry discussed from this point of view is often called **descriptive chemistry.**

In the following pages quite a lot of factual information will be presented. You are not expected to remember all of it. Our discussion of descriptive chemistry is aimed at providing you with an awareness of certain important kinds of chemical reactions and compounds. Where they exist, trends in chemical and physical behavior will be pointed out and structural similarities or differences described. A knowledge of these trends in behavior will help you tie together some of the bits and pieces of factual knowledge that you may need at a later time.

## 18.1 METALS, NONMETALS, AND METALLOIDS

It is typically human of us to attempt to classify and compartmentallize knowledge in an effort to make it easier to assimilate and understand. In the discussion of the chemistry of the elements, however, this is a somewhat risky undertaking, since many of the trends in chemical properties run across boundaries that we may attempt to set up. Nevertheless, it is sometimes useful to classify the elements according to at least some of their properties.

In Section 3.10 we stated that the elements could be divided into three main categories: metals, nonmetals, and metalloids, the latter having properties that lie between the other two. We shall follow this general classification in our present discussion, although it should be kept in mind that no

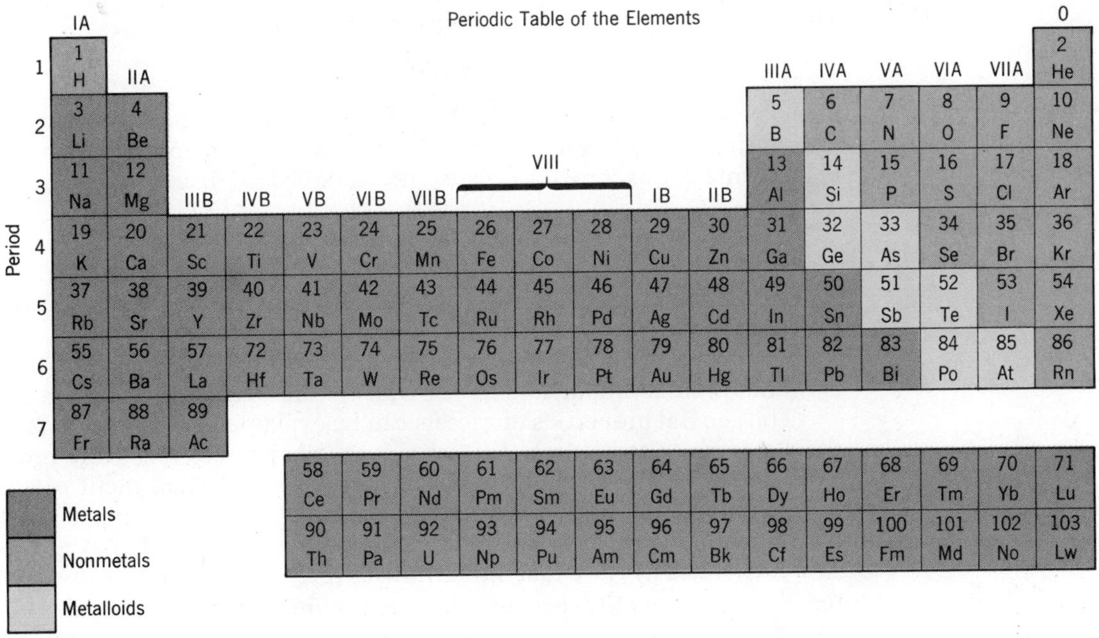


Figure 18.1
*Distribution of metals, metalloids, and nonmetals in the periodic table.*

sharp boundaries exist. In fact, many aspects of behavior extend over more than one category.

Figure 18.1 illustrates the rough breakdown of the elements into our three classes. As you can see, most of the elements exhibit metallic properties and, furthermore, most of the metals belong to the transition and inner transition (lanthanide and actinide) series. This chapter will be devoted largely to the representative (A-group) metals. The representative metalloids and nonmetals will be examined in some detail in Chapter 19, while the transition and inner transition elements will be treated separately in Chapter 20. Again, as we might expect, some characteristics that are exhibited by the representative elements are also observed for the transition elements, and vice versa.

Before we examine the chemistry of the elements in any detail, let's first take a rather brief qualitative look at what kinds of properties will identify an element as a metal, a nonmetal, or a metalloid and what sort of trends in metallic and nonmetallic character exist throughout the periodic table.

Metals have many physical properties with which we all are very familiar: for example, their electrical and thermal conductivities as well as their characteristic luster. Ductility (the ability of a substance to be drawn into wire) and malleability (its ability to be hammered into thin sheets) are two other physical properties that we associate with metals.

We can also recognize metals by their chemical properties. In earlier chapters we found that metals form oxides that are basic in nature, reacting with acids to give salts and producing hydroxide ion when they dissolve in water. Thus sodium and calcium could be identified as metals because of the reaction of their oxides with water.

$$Na_2O + H_2O \longrightarrow 2NaOH$$
$$CaO + H_2O \longrightarrow Ca(OH)_2$$

They also react directly with acids; for example,

$$CaO + 2HCl \longrightarrow CaCl_2 + H_2O$$

Some metal oxides are insoluble and are therefore unable to react with water. The do, however, react with acids. For instance, iron would be classified chemically as a metal because its oxide, $Fe_2O_3$, reacts with $H_2SO_4$.

$$Fe_2O_3 + 3H_2SO_4 \longrightarrow Fe_2(SO_4)_3 + 3H_2O$$

This reaction, in fact, is often employed to remove rust from iron or steel prior to its being coated by zinc (galvanizing) or tin. This acid treatment is called **pickling.**

Another characteristic of metals, which you may have already noticed, is that in their compounds they generally exhibit positive oxidation states.[1]

The general properties of metals can be explained quite well in terms of their electronic structures. As a group, metals possess relatively few electrons in their valence shells and, consequently, the attainment of a noble gas electron configuration is often impossible by way of sharing of electrons in ordinary covalent bonds. In addition, the metals generally have low ionization energies (IE) and electron affinities (EA). They therefore do not react readily with each other because they are not inclined to form covalent bonds and also because they have little desire to pick up electrons from another metal atom. They do form compounds with nonmetals, however, in which they tend to lose electrons and acquire a positive charge and, as we have seen before, ionic bonding is an important factor to consider in the chemistry of metals.

In the free state, metals exist in a unique type of lattice in which they tend to lose their valence electrons to the lattice as a whole. As described in Chapter 7, the result is an array of positive metal ions imbedded in a sea of very mobile electrons. Since these electrons are able to move easily and rapidly through the solid, they are able to transport thermal energy quickly and metals conduct heat well. Of course their electrical conductivity is also explained by this model. The important point here is that *the metallic lattice is a consequence of the small number of valence electrons and the low ionization energies of the metal atoms.*

By contrast, nonmetals are characteristically very poor conductors of electricity and, in fact, are classed as insulators. Many of them, such as $H_2$, $N_2$, $O_2$, and the Group 0 elements, are gases at room temperature. Those that are solids conduct heat poorly and tend to be brittle. Chemically we find that the oxides of the nonmetals are acidic. Many of them react with water to yield hydronium ion, and they react with bases to form salts containing oxoanions. The reaction of carbon dioxide with LiOH is a typical example.

$$2LiOH + CO_2 \longrightarrow Li_2CO_3 + H_2O$$

This reaction, as mentioned previously, has been used to remove $CO_2$ from the air breathed by astronauts in space vehicles.

From their positions in the periodic table we see that nonmetals generally require only a small number of electrons to reach a noble gas structure.

[1] There are compounds of certain elements in which the metal is assigned a zero or even a negative oxidation number; nickel carbonyl, $Ni(CO)_4$, for example, contains a nickel atom bonded to four neutral CO molecules and is an example of Ni in the zero oxidation state. In $NaCo(CO)_4$ the cobalt atom is assigned an oxidation number of $-1$.

Consequently, in their elemental state they combine with themselves to produce covalently bonded molecules containing two or more atoms. We shall examine these in more detail in Chapter 19. The important point is that in the process of achieving a noble gas structure all of the electrons usually become paired and are localized on individual molecules. As a result, no electrons remain available for electrical conduction and an insulator results. This absence of mobile electrons also means that the only mechanism for heat transfer through a nonmetal is by the vibrational motions of the atoms. Since this is much slower than electronic motion, nonmetals conduct heat poorly.

Chemically, nonmetals combine with themselves, other nonmetals, and with metals. They are characterized by high ionization energies and electron affinities and, therefore, by high electronegativities. In their binary compounds with metals, for instance, NaCl or $Al_2O_3$, they assume negative oxidation states. In those compounds containing more than one nonmetal, the usual rules for assigning oxidation numbers apply; the more electronegative element is assigned the negative oxidation state. Thus, in $CO_2$, because oxygen is more electronegative than carbon, we say that the oxidation number of carbon is positive (by our rules we obtain +4 for the oxidation state of carbon).

The metalloids are elements with outer-electron shells that are approximately half-filled, and with electronegativities lying between the metals and the nonmetals. As a class, they have electrical conductivities that are much smaller than those of metals but still much larger than those of nonmetals, and these semiconductor properties (discussed in Section 7.9) have made them extremely useful in electronic devices (e.g., in diodes and transistors).

Chemically, metalloids exhibit both positive and negative oxidation states. They combine with nonmetals and, in some cases, also with metals. Their oxides tend to be amphoteric, although their acid properties outweigh the basic ones. In fact, in many ways the metalloids behave chemically more as nonmetals than as metals.

## 18.2 TRENDS IN METALLIC BEHAVIOR

The trends in the metallic character of the elements can be examined by following the variations in electronegativity throughout the periodic table, since it is this property, which reflects both ionization energy and electron affinity, that is primarily responsible for metallic or nonmetallic behavior.

Within a period, ionization energy and electron affinity both increase with increasing atomic number; hence the electronegativity also increases. Consequently, as we proceed from left to right across a period, the metallic character of the elements gradually decreases and their nonmetallic character, of course, increases. In period 3, for example, Na, Mg, and Al are typical metals. In this sequence, however, the oxides become less basic (more acidic), so that we find $Al_2O_3$ to be soluble in strong bases. $Na_2O$ and MgO, on the other hand, do not react with base. Continuing to the right, silicon is classed as a metalloid and exhibits typical semiconductor properties. The remaining elements in the period, P, S, Cl, and Ar, are nonmetals that, except for Ar (which has no tendency to react) form characteristically acidic oxides.

As we descend a group, ionization energy, electron affinity, and electronegativity decrease. The metallic character of the elements therefore increase accordingly. The Group IVA elements clearly illustrate this trend.

Carbon, at the head of the group, is nonmetallic in virtually every way.[2] Below carbon are the metalloids, silicon, and germanium, and below these the metals tin and lead.

Tin is unusual because at high temperatures it forms crystals that are metallic in appearance. At low temperatures (below 13.2°C) these very gradually change to a nonmetallic, powdery form in which tin has a diamond-type lattice. Tin articles left for long periods in the cold therefore gradually crumble. At one time it was thought that some pest attacked the tin, and this disintegration of tin articles was called "tin disease."

Lead, at the bottom of the group, demonstrates only metallic properties in the elemental state.

This trend toward increasing metallic character as we proceed down within a group appears in Group VA too. The pair of elements at the top, nitrogen and phosphorus, are nonmetals, arsenic and antimony behave as metalloids and bismuth, at the bottom of the group, shows mainly metallic characteristics.

Since the metallic nature of the elements increases from right to left within a period, and from top to bottom in a group, the most metallic elements occur in the lower left corner of the periodic table. Similarly, we expect the most nometallic elements to be found in the upper right corner of the table. These trends in properties also permit us to understand why the metalloids occur as a band running diagonally from upper left to lower right. In fact, if we omit the transition elements and draw our periodic table with only the representative elements, we find the division rather even (Figure 18.2).

## 18.3 PREPARATION OF METALS

Most metals, including those of both the representative and transition elements, always occur in the combined state. For example, the oceans provide a huge storehouse of minerals in which the metals occur primarily as soluble sulfates and halides. The major source of magnesium, for instance, is from the oceans, and in the future greater attention will no doubt be focused on the oceans as a source of raw materials as supplies of ore deposits on land

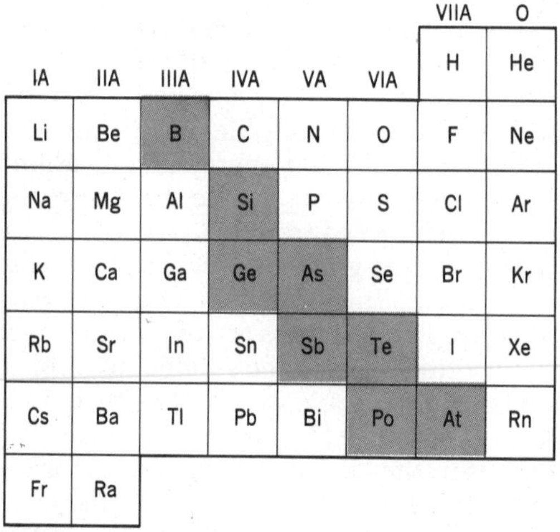
VIIA
O
H
He
IA
IIA
IIIA
IVA
VA
VIA
Li
Be
B
C
N
O
F
Ne
Na
Mg
Al
Si
P
S
Cl
Ar
K
Ca
Ga
Ge
As
Se
Br
Kr
Rb
Sr
In
Sn
Sb
Te
I
Xe
Cs
Ba
Tl
Pb
Bi
Po
At
Rn
Fr
Ra

Figure 18.2
*The representative elements. Metalloids are indicated by gray squares.*

[2] One form of carbon, graphite, does conduct electricity. This, however, is more an accident of the bonding in graphite than evidence for metallic behavior. The structure of graphite and the mechanism of its electrical conductivity is discussed in Chapter 19.

are depleted. At the present time there is already a great deal of interest in mining the "manganese nodules" that seem to line the ocean floor. The oceans also provide a source of important nonmetals such as chlorine, bromine, and iodine.

On land some metals occur as deposits of their carbonates. Limestone, for example, is primarily $CaCO_3$. A mixed $CaCO_3$, $MgCO_3$ limestone, called dolomite, is often ground up and used on farms and lawns to decrease the acidity of the soil (recall that acids react with carbonates) and to provide a source of Mg, which is needed in the production of chlorophyll.

Oxides are also important sources of metals. Two important examples are aluminum ($Al_2O_3$) and iron ($Fe_2O_3$). Sulfides are the primary sources of lead (PbS) and copper (as $Cu_2S$). Regardless of the type of ore, however, metals almost always exist in positive oxidation states, and in order to produce the free element a chemical reduction must be brought about. The nature of this reduction process depends on the ease with which the metal can be reduced.

Some metals are so easily reduced that many of their compounds can be decomposed just by heating them at relatively low temperature. Priestley, for example, in his experiments on oxygen, produced metallic mercury and oxygen from mercuric oxide by simply heating it with sunlight focused on the HgO by means of a magnifying glass. In this case HgO decomposes quite spontaneously at elevated temperatures according to the equation

$$2HgO\ (s) \longrightarrow 2Hg\ (g) + O_2\ (g)$$

The practicality of using a thermal decomposition reaction of this type to produce a free metal depends on the extent to which the reaction proceeds to completion at a given temperature. In Chapter 10 we saw that at 25°C the position of equilibrium in a reaction is governed by $\Delta G^0$. If we take $\Delta G'$ to be the equivalent of $\Delta G^0$, but at some other temperature, we have the relationship,

$$\Delta G' = \Delta H' - T\,\Delta S'$$

where $\Delta H'$ and $\Delta S'$ are the heat and entropy changes that accompany the reaction. For most systems $\Delta H$ and $\Delta S$ do not change much with temperature so that $\Delta H'$ and $\Delta S'$ can reasonably be approximated by $\Delta H^0$ and $\Delta S^0$.

The thermal decomposition reaction will be feasible when $\Delta G'$ is negative, since under these conditions an appreciable amount of product will be formed. We must now look at the magnitudes of $\Delta H'$ and $\Delta S'$, because they control the sign and magnitude of $\Delta G'$. Since a gas ($O_2$) and sometimes the metal vapor is produced in the decomposition, the process occurs with a sizable increase in entropy; hence $\Delta S'$ will be positive.

The enthalpy change for the decomposition, $\Delta H'$, is simply the negative of the heat of formation of the oxide and, since $\Delta H_f'$ is generally negative for metal oxides, $\Delta H'$ for the decomposition reaction will be positive. As a result, the sign of $\Delta G'$ is determined by the difference between two positive quantities, $\Delta H'$ and $T\,\Delta S'$.

If the metal oxide has a high negative heat of formation (if a great deal of energy is evolved when the oxide is formed), then $\Delta H'$ for the decomposition will have a large positive value. Consequently, the difference $\Delta H' - T\,\Delta S'$ will be negative *only* at very high temperatures, where $T\,\Delta S'$ is larger than $\Delta H'$. We express this by saying that the metal oxide is very stable with respect to thermal decomposition. On the other hand, if the $\Delta H_f'$ of the metal oxide is relatively small, as with HgO and certain other oxides (for example, $Ag_2O$, CuO, and $Au_2O_3$), $\Delta H'$ for the decomposition reaction is a small positive quantity and $\Delta G'$ for the reaction becomes negative at relatively low temperatures. These oxides, therefore, are said to have relatively low thermal stabilities.

EXAMPLE 18.1 Above what temperature would the decomposition of $Ag_2O$ be expected to proceed to an appreciable extent toward completion? At 25°C, $\Delta H_f^0$ for $Ag_2O$ is $-30.5$ kJ/mole, $\Delta S_f^0 = -66.1$ J/mole K.

SOLUTION Since the decomposition of $Ag_2O$,

$$Ag_2O\ (s) \longrightarrow 2Ag\ (s) + \tfrac{1}{2}O_2\ (g)$$

is the reverse of formation, we have for this reaction

$$\Delta H^0 = -\Delta H_f^0 = +30.5 \text{ kJ/mole}$$
$$\Delta S^0 = -\Delta S_f^0 = +66.1 \text{ J/mole K}$$

Let's calculate the temperature at which $\Delta G' = 0$. We will assume, as stated in the text, that $\Delta H'$ and $\Delta S'$ are approximately independent of temperature so that we can use $\Delta H^0$ and $\Delta S^0$ in the equation for $\Delta G'$.

$$\Delta G' = \Delta H^0 - T\,\Delta S^0$$

When $\Delta G' = 0$

$$0 = \Delta H^0 - T\,\Delta S^0$$

Solving for $T$,

$$T = \frac{\Delta H^0}{\Delta S^0} = \frac{30{,}500 \text{ J/mole}}{66.1 \text{ J/mole K}} = 461 \text{ K}$$

Because $\Delta H^0$ and $\Delta S^0$ are both positive, $\Delta G'$ will become negative at temperatures above 461 K (188°C). This means that above 461 K the reaction should become feasible, with much of the $Ag_2O$ undergoing decomposition.

EXAMPLE 18.2 Above what temperature would $\Delta G'$ be negative for the reaction,

$$Au_2O_3\ (s) \longrightarrow 2Au\ (s) + \tfrac{3}{2}O_2\ (g)$$

At 25°C, $\Delta H_f^0 = +80.8$ kJ/mole for $Au_2O_3$. Also for $Au_2O_3$, $S^0 = 125$ J/mole K; for Au, $S^0 = 47.7$ J/mole K; for $O_2$, $S^0 = 205$ J/mole K.

SOLUTION First, let's calculate $\Delta H^0$ and $\Delta S^0$ for the decomposition reaction. $\Delta H^0$ is simply the negative of $\Delta H_f^0$, that is, $\Delta H^0 = -80.8$ kJ for the decomposition of 1 mole of $Au_2O_3$. The value of $\Delta S^0$ is

$$\Delta S^0 = (2S^0_{Au} + \tfrac{3}{2}S^0_{O_2}) - (S^0_{Au_2O_3})$$
$$= (2 \text{ moles}) \times \left(\frac{47.7 \text{ J}}{\text{mole K}}\right) + \tfrac{3}{2} \text{ mole} \times \left(\frac{205 \text{ J}}{\text{mole K}}\right) - 1 \text{ mole} \left(\frac{125 \text{ J}}{\text{mole K}}\right)$$
$$= 278 \text{ J/K} = 0.278 \text{ kJ/K}$$

Again, we obtain $\Delta G'$ by assuming that $\Delta H'$ and $\Delta S'$ are the same as $\Delta H^0$ and $\Delta S^0$. Therefore,

$$\Delta G' = \Delta H^0 - T\,\Delta S^0$$

Substituting,

$$\Delta G' = -80.8 \text{ kJ} - T(0.278 \text{ kJ/K})$$

Note that regardless of the temperature $\Delta G'$ will be negative, because the absolute temperature is always a positive quantity. What this tells us is that $Au_2O_3$ is unstable with respect to decomposition at any temperature. It exists only because at low temperatures the rate of decomposition is very slow.



Except in a few cases, thermal decomposition is not a practical way of producing the free metals. Instead, their compounds are reacted with some substance that is a better reducing agent than the metal being sought. One of the most common agents used for the reduction of metal oxides is carbon. Tin and lead, for example, can be produced by heating their oxides with carbon.

$$2SnO + C \xrightarrow{\text{heat}} 2Sn + CO_2$$

$$2PbO + C \xrightarrow{\text{heat}} 2Pb + CO_2$$

Carbon is used in large quantities in commercial metallurgy because of its abundance and low cost. Its importance in the reduction of iron ore and in steel making will be examined in Chapter 20.

Hydrogen is another reducing agent that can be used to liberate metals of moderate chemical activity from their compounds. For instance, tin and lead oxides will also be reduced when heated under a stream of $H_2$.

$$SnO + H_2 \xrightarrow{\text{heat}} Sn + H_2O$$

$$PbO + H_2 \xrightarrow{\text{heat}} Pb + H_2O$$

The use of a more active metal to carry out the reduction is also possible. In Chapter 16 we saw that a galvanic cell could be established between two different metals, for example, Zn and Cu. In that cell the more active reducing agent, Zn, causes the $Cu^{2+}$ to be reduced. Aluminum was first prepared in 1825 by the reaction of aluminum chloride with the more active metal, potassium.

$$AlCl_3 + 3K \longrightarrow 3KCl + Al$$

As a practical source of metals, the reduction of compounds with other elements that are better reducing agents suffers from a serious limitation, specifically, the availability (and cost) of the reducing agent. Each possible reducing agent must then be generated by reacting one of its compounds with a still better reducing agent. Ultimately, of course, there must be some "best" reducing agent. How could this substance be prepared if there were no better reducing agent available that could be used to reduce its compounds?

The solution to this dilemma is electrolysis where, by applying a suitable potential, virtually any oxidation-reduction process can be brought about. Consequently, metals that themselves are very powerful reducing agents are nearly always prepared by electrolysis. Among the representative elements these include the very active elements in Groups IA and IIA (in Chapter 16 the electrolysis of molten NaCl was described).

The production of magnesium from sea water is accomplished by first treating the sea water with CaO, which precipitates $Mg(OH)_2$.

$$CaO\ (s) + H_2O \longrightarrow Ca^{2+}\ (aq) + 2OH^-\ (aq)$$
$$Mg^{2+}\ (aq) + 2OH^-\ (aq) \longrightarrow Mg(OH)_2\ (s)$$

Often the CaO used in this process is obtained by heating sea shells, which are composed of $CaCO_3$.

$$CaCO_3\ (s) \longrightarrow CaO\ (s) + CO_2\ (g)$$

After the $Mg(OH)_2$ is filtered, it is reacted with HCl and the resulting $MgCl_2$ solution is evaporated. The crystalline $MgCl_2$ is melted and electrolyzed to give molten Mg and $Cl_2$.

Although less active than the alkali and alkaline earth metals, the metallic elements in Group IIIA are generally produced by electrolysis too. When a molten salt is used for the electrolysis, a halide is generally employed because of their usually lower melting points. The production of aluminum by the Hall process described in Chapter 16 is an exception. Here, you recall, molten $Na_3AlF_6$ is used as a solvent for $Al_2O_3$.

## 18.4 CHEMICAL PROPERTIES AND TYPICAL COMPOUNDS

In the following discussion of the chemistry of the A-group metals we shall not attempt to provide an exhaustive review of all of the many kinds of compounds that they form. Instead, we shall look at some of the similarities and differences in their chemical behavior.

The elements within any given group are expected to exhibit similar properties and this is particularly evident among the elements in Groups IA and IIA. For example, the alkali metals in Group IA (we exclude hydrogen from this discussion because of its clearly nonmetallic behavior) are all extremely reactive elements that are capable of reducing water to produce hydrogen and the metal hydroxide.

$$2M + 2H_2O \longrightarrow H_2 + 2MOH$$

They yield only compounds in which their oxidation state is +1, and many of their compounds have similar solubilities in water.

The reactivity of the alkali metals is high toward all of the nonmetals. Their behavior toward oxygen is especially interesting, however. Only lithium reacts directly with oxygen to produce the normal oxide, $Li_2O$. Sodium, when it reacts with $O_2$, gives a pale yellow peroxide.

$$2Na + O_2 \longrightarrow Na_2O_2$$

This compound contains the peroxide ion, $O_2{}^{2-}$

$$\left[:\ddot{\underset{\cdot\cdot}{O}}:\ddot{\underset{\cdot\cdot}{O}}:\right]^{2-}$$

Sodium peroxide is often used as a bleaching agent because it releases $H_2O_2$ by hydrolysis.

$$Na_2O_2 + 2H_2O \longrightarrow 2Na^+ + 2OH^- + H_2O_2$$

The remaining alkali metals, K, Rb, and Cs, combine with molecular oxygen to form rather deeply colored yellow-orange superoxides. For example,

$$K + O_2 \longrightarrow KO_2$$

The superoxide ion, $O_2^-$, can be considered an oxygen molecule that has gained a single electron. This electron pairs with one of the two unpaired electrons in $O_2$ so that the $O_2^-$ ion still has one unpaired electron and is paramagnetic.

Potassium superoxide, $KO_2$, has found uses in recirculating breathing equipment. Air containing $CO_2$ is circulated over solid $KO_2$ with which it reacts, generating $O_2$.

$$4KO_2\,(s) + 2CO_2\,(g) \longrightarrow 2K_2CO_3\,(s) + 3\,O_2\,(g)$$

This prevents an undesirable buildup of $CO_2$ and regenerates $O_2$ that is consumed in respiration.

The superoxides also react with water to yield hydrogen peroxide; however, they also yield molecular oxygen as one of the products.

$$2MO_2 + 2H_2O \longrightarrow 2M^+ + 2OH^- + H_2O_2 + O_2$$

Like the alkali metals, the alkaline earths exhibit very marked group similarities. For example, they all show only a +2 oxidation state and are also quite reactive, although less so than their neighbors in Group IA. However, in both of these groups the greatest group similarities are exhibited by the elements in period 3 and below. Thus, in Group IA, the elements Na and K are more nearly alike than are Li and Na. Among the alkaline earths we again find a closer similarity between the elements Mg and Ca than between Be and Mg. In fact, in general the period 2 elements reveal a somewhat unique chemistry among both the metals and the nonmetals.

In looking for similarities in the chemistry of the A-group metals, it is interesting to find that a *diagonal relationship exists between the chemistry of the first member of a group and that of the second member of the following group.* Thus lithium is, in some ways, more like magnesium than like sodium. For example, many lithium and magnesium salts have similar solubilities that differ markedly from those of the other Group IA elements. The salts $MgF_2$ and LiF, for instance, are insoluble in water, while the corresponding salts NaF, KF, and so on, are soluble. In addition, the lithium halides (LiCl, LiBr, LiI) and $LiClO_4$ are appreciably soluble in relatively nonpolar solvents such as alcohol, as are the corresponding magnesium salts. On the other hand, sodium salts tend to be quite insoluble in alcohol.

Still another point of similarity between Li and Mg is their reactions with molecular nitrogen. Both react readily at elevated temperatures to produce nitrides.

$$6Li + N_2 \longrightarrow 2Li_3N$$
$$3Mg + N_2 \longrightarrow Mg_3N_2$$

The other Group IA metals are unreactive toward $N_2$.

The similarities in chemical behavior of lithium and magnesium are generally interpreted in terms of their ratios of ionic charge to ionic radius. This ratio is called the **ionic potential** and is usually symbolized by the Greek letter phi ($\phi$). Thus we have

$$\phi = \frac{q}{r}$$

where $q$ is the charge and $r$ is the ionic radius.

As it happens, both $Li^+$ and $Mg^{2+}$ have similar ionic potentials. The $Mg^{2+}$ ion is larger than $Li^+$, but it also has a higher charge. These two factors combine to make their ionic potentials nearly the same. Because of this each ion (i.e., both $Li^+$ and $Mg^{2+}$) behaves in very much the same fashion when placed into a chemical environment where ionic interactions are important.

The diagonal relationship between Li and Mg is repeated for the elements Be and Al and, in this latter case, it is even more striking. Once again we have a pair of ions[3] that have very similar ionic potentials. The ionic radii of $Be^{2+}$ and $Al^{3+}$ are 0.35 and 0.51Å, respectively, from which their ionic potentials are 5.88 and 5.71, quite close indeed. (For comparison, magnesium, which is found in Group IIA just below Be, has an ionic potential of 3.03). As a result, the chemistries of Be and Al are remarkably alike. For example:

1. Both Be and Al react with oxygen to form an oxide coating that protects the metal beneath from further reaction. Consequently, even though they have highly negative reduction potentials suggesting that they should be readily oxidized ($\mathscr{E}^0_{Be} = -1.85$ V, $\mathscr{E}^0_{Al} = -1.66$ V), both dissolve only slowly in acids.
2. The oxides of Be and Al, BeO, and $Al_2O_3$, are extremely high-melting (BeO, melting point = 2530°C; $Al_2O_3$, melting point = 2045°C) and very hard. Presumably the high concentration of positive charge on these very small cations leads to very strong lattice forces and hence to an unusually high lattice energy that must be overcome in order to melt the oxides. Aluminum oxide is interesting because the presence of trace impurities imparts brillant colors to its gem-quality crystals. When the impurity is $Cr^{3+}$, ruby results; when the impurities are $Fe^{2+}$, $Fe^{3+}$, and $Ti^{4+}$, the gem blue sapphire is produced. Both of these gems are currently manufactured synthetically in large quantities.
3. Beryllium and aluminum both dissolve in strong base with the evolution of hydrogen while magnesium, just below Be in Group IIA, does not dissolve in base. The chemical equations for these reactions can be represented as

$$Be + 2OH^- \longrightarrow BeO_2^{2-} + H_2$$
$$2Al + 2H_2O + 2OH^- \longrightarrow 2AlO_2^- + 3H_2$$

   The formulas given for these ions, $BeO_2^{2-}$ (beryllate) and $AlO_2^-$ (aluminate), are really oversimplifications since the actual species that exist in solution are more complex, *probably* $Be(OH)_4^{2-}$ and $Al(OH)_4(H_2O)_2^-$. Note the stoichiometric equivalence,

$$Be(OH)_4^{2-} \quad \text{and} \quad BeO_2^{2-} + 2H_2O$$
$$Al(OH)_4(H_2O)_2^- \quad \text{and} \quad AlO_2^- + 4H_2O$$

   We see here that it is frequently very difficult to ascertain the actual chemical identity of species in solution because of possible interactions with the solvent.
4. Beryllium and aluminum halides are covalent. As we shall see later in this chapter, the high ionic potential of the $Be^{2+}$ and $Al^{3+}$ ions

[3] Actually, in all known cases, there is no evidence for the existence of a simple $Be^{2+}$ ion. Instead, some degree of covalent bonding is involved in all Be compounds.

leads to a substantial degree of covalent bonding in their halides, for instance, $BeCl_2$ and $AlCl_3$. Magnesium chloride and the other alkaline earth chlorides, on the other hand, are essentially ionic in character.

The structures of these covalent halides are also somewhat similar. The simple species, $BeCl_2$ and $AlCl_3$, are electron-deficient, having only four and six electrons, respectively, in the valence shell of the metal atom. As a result, both function as Lewis acids by combining with electron-pair donors (bases). In the absence of a suitable base, however, their strong desire to achieve an octet is satisfied by forming coordinate covalent bonds to chlorine atoms of a neighboring metal halide molecule. In solid $BeCl_2$, for example, two such coordinate covalent bonds are formed to each Be atom and a long linear chainlike structure is formed. This is illustrated by the formula

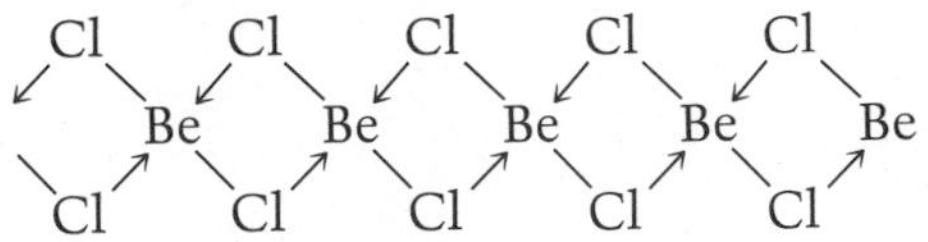

The chlorine atoms in this structure serve as bridges between adjacent Be atoms by making their lone pairs of electrons available for coordinate covalent bonding. The chain is not planar, of course, because each Be in the structure is surrounded by four electron pairs. In Chapter 17 we found that this requires that the geometry around each Be atom be tetrahedral. In other words, each Be atom in the chain lies at the center of a tetrahedron with the bridging chlorine atoms at the apexes, as shown stereoscopically in Figure 18.3.

In the simple aluminum halide species the aluminum atom is covalently bonded to three halogen atoms, and in order to complete its octet the aluminum needs to form only one coordinate covalent bond. As a result, **dimeric** species having the formula $Al_2X_6$ and containing two $AlX_3$ molecules are formed. This is illustrated for $Al_2Cl_6$ as

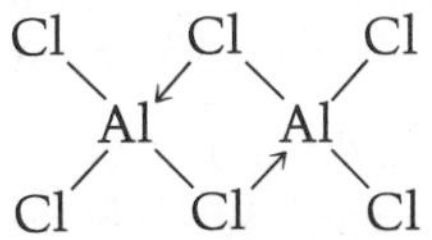

Once again the metal atom has four electron pairs in its valence shell and is surrounded tetrahedrally by halogen atoms as shown in Figure 18.4. As a final note on this discussion, it is interesting that despite the close similarities found between Be and Al in many of their chemical properties, the physiological effects of these metals are quite different. The body is able to tolerate rather large amounts of aluminum; however, even very small quantities of beryllium, particularly when inhaled as its oxide, are quite lethal. The high toxicity of Be compounds, in fact, caused a major manufacturer of fluorescent lamps to terminate the use of beryllium containing phosphors in 1949 when it was discovered that many beryllium workers had contracted chronic berylliosis from inhaling dust containing BeO. Manufacturing plants that now work with beryllium

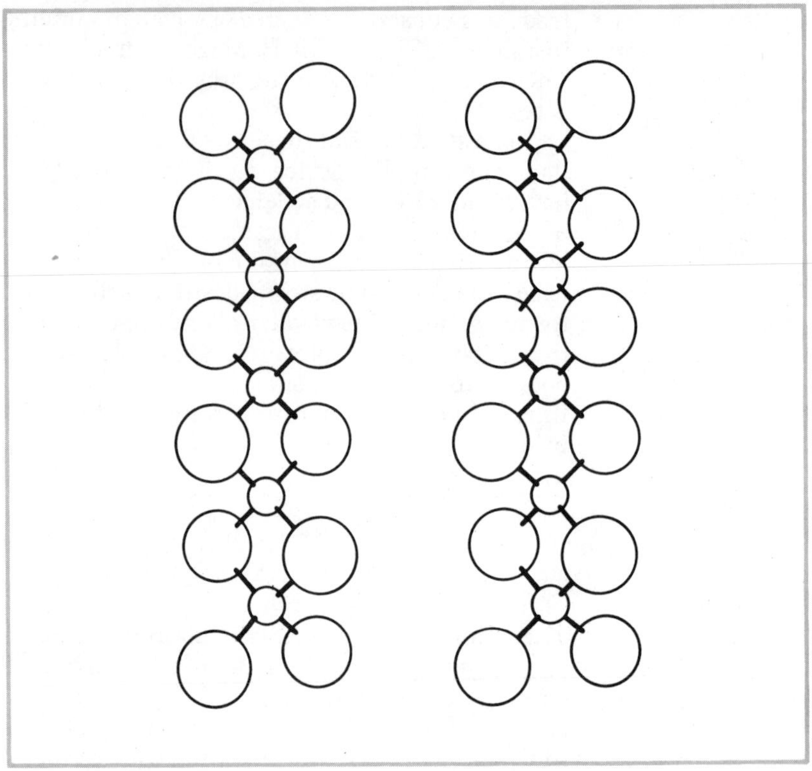

Figure 18.3
*Structure of* $(BeCl_2)_x$ *(small spheres, Be; large spheres, Cl).*

compounds take great care to ensure that their employees and the surrounding neighborhood are not exposed to beryllium in any form.

## 18.5 OXIDATION STATES

As we know, the oxidation states that an element exhibits are governed by its electronic structure. Here we wish to examine the oxidation states that characterize a particular group and the trends in their ease of formation.

**GROUP IA.** The elements of Group IA are characterized by a single electron in an *s* orbital located outside a filled, noble gas core. Loss of this one elec-

Figure 18.4
*Structure of* $Al_2X_6$ *(small spheres, Al; large spheres, X).*

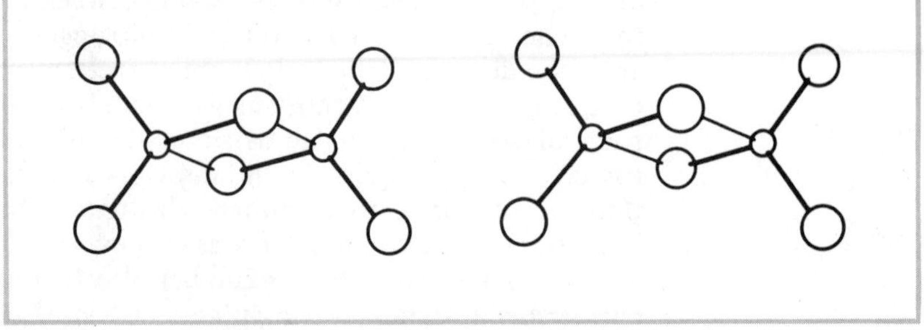

Table 18.1
*Some properties of the alkali metals*

| Element | Ionization Energies (kcal/mole) First | Second | $\mathscr{E}^0$ (V) | $M^+$ Radius (Å) |
|---|---|---|---|---|
| Lithium | 123.6 | 1735 | −3.04 | 0.68 |
| Sodium | 118.0 | 1084 | −2.71 | 0.98 |
| Potassium | 99.5 | 730 | −2.92 | 1.33 |
| Rubidium | 95.9 | 631 | −2.93 | 1.48 |
| Cesium | 89.2 | 539 | −2.92 | 1.67 |

tron gives the characteristic +1 oxidation state for the group. Since the single electron is loosely held, it is easily lost and the alkali metals are readily oxidized, as can be seen from their highly negative reduction potentials and low ionization energies in Table 18.1.

A close examination of these two quantities, however, reveals an apparent contradiction. Note that the ionization energy (IE) decreases as we proceed down within the group, suggesting that it becomes progressively easier to strip an electron from the atom as we go from Li to Cs. In Chapter 3 we saw that this is, in fact, expected. We would also anticipate that the reduction potentials should become more negative as the IE becomes smaller since the elements should become more easily oxidized. This trend is indeed followed from Na downward; however, Li has an $\mathscr{E}^0$ that is more negative than Na (or any of the other alkali metals for that matter). Why is this so?

The ionization energy, remember, is a measure of the ease with which a *gaseous* atom loses electrons to produce a gaseous cation. The reduction potential, on the other hand, is concerned with the loss of electrons by the *solid* metal to form the corresponding cation in *aqueous solution* where it is hydrated by the water molecules surrounding it. This latter process is more complex than simply removing an electron from the isolated metal atom. To understand the trends in the reduction potentials, we must break down the overall reaction into several steps. If we concentrate on the enthalpy changes involved in the reaction, we can construct the diagram in Figure 18.5. We see that the net enthalpy change is the sum of three energy terms. Two of these are endothermic, the sublimation energy, $\Delta H_{subl}$, which is the energy needed to convert the solid into gaseous atoms, and the ionization energy that we have already examined. The third quantity, called the hydration energy, is strongly exothermic. It corresponds to the energy *released* when the cation is placed into the solvent cage where it is surrounded by the water dipoles oriented in such a way that their negative ends are directed at the positive ion (Figure 18.6).

Among the alkali metals the sublimation energy remains approximately constant as we descend the group while the ionization energy decreases. To reach the peak on the energy diagram, then, we require the greatest amount of energy for Li and the least amount for Cs. However, because of its small size and high ionic potential, upon hydration $Li^+$ interacts much more strongly with the water dipoles than do any of the other Group IA ions. As a result, the hydration energy of $Li^+$ is unusually large, being much greater

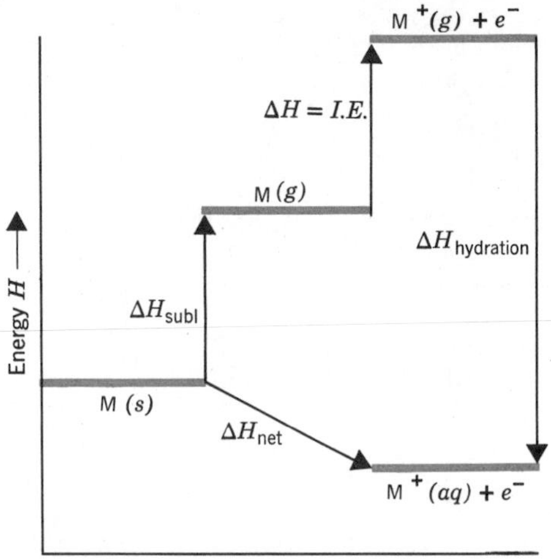


Figure 18.5
*Enthalpy diagram for the reaction:* M (s) → $M^+$(aq) + $e^-$.

than for the other $M^+$ ions in the group and, consequently, the net overall enthalpy change is most exothermic for Li. This in turn causes Li to be more easily oxidized than any other alkali metal. In other words, the extraordinarily high hydration energy of the small $Li^+$ ion more than compensates for its relatively high ionization energy and causes Li to have an unexpectedly high negative $\mathscr{E}^0$.

Associated with the easy loss of electrons from the alkali metals is their interesting behavior in liquid ammonia. We have already seen that these metals are capable of reducing water to liberate $H_2$. Ammonia is not as easily reduced as water and, when placed into this solvent, alkali metals dissolve without reaction to form deep blue solutions. It is generally agreed that this color, which is identical for liquid ammonia solutions of all of the alkali metals (as well as for Ca, Sr, and Ba from Group IIA), is a result of the presence of free electrons that have become solvated by ammonia molecules. Apparently, when the metal dissolves in $NH_3$, it loses its valence electron to become a cation. This electron becomes surrounded by $NH_3$ molecules arranged so that the positive ends of their dipoles are directed at the negatively charged electron, as shown in Figure 18.7, thereby stabilizing it through solvation. Solutions containing alkali metals in liquid ammonia are, as we would expect from the presence of readily available electrons, excellent reducing agents.

**GROUP IIA.** The elements in Group IIA each have a filled *s* subshell outside a noble gas core. When they react they always lose both of these electrons to

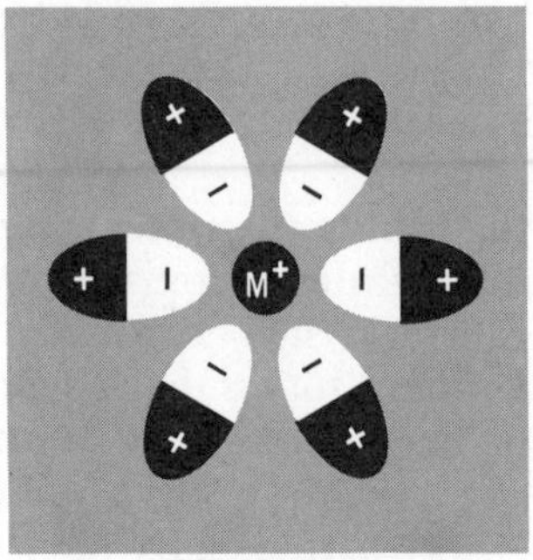


Figure 18.6
*Solvation of a cation by water dipoles.*

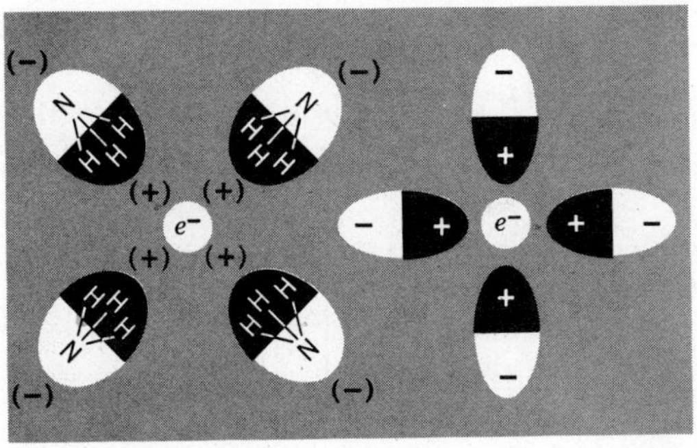


Figure 18.7
*Solvated electron in liquid ammonia.*

produce a +2 oxidation state (for example, $Mg^{2+}$, $Ca^{2+}$). In general, the Group IIA elements are easily oxidized, and we see that the reduction potentials of the alkaline earths are all highly negative (Table 18.2), although they are generally more positive than those of their neighbors in Group IA. This is a reflection of the fact that, within a given period, the alkaline earth metals have larger ionization energies than the alkali metals and, hence, lose their electrons with more difficulty.

Within the group, Ca and the heavier elements are able to reduce water at room temperature according to the reaction

$$M + 2H_2O \longrightarrow M(OH)_2 + H_2$$

Magnesium, while unable to react with cold water, will liberate hydrogen from boiling water or steam. Beryllium, on the other hand, will not react with water at all.

An interesting question to consider is why the alkaline earth metals do not form compounds in which they exhibit a +1 state, even though much more energy must be expended to remove two electrons from the atom instead of only one. For example, why do we not observe the formation of CaCl but, instead, only $CaCl_2$?

To answer the question, we must once again look further than simply the ionization energies of the isolated $Ca^0$ atom and $Ca^+$ cation (140.9 and 273.8 kcal/mole, respectively). These quantities tell us that in the gas phase $Ca^+$ is, indeed, preferred energetically over $Ca^{2+}$. In the solid or in solution, however, there are additional factors to consider.

In aqueous solution the stabilizing influence of the hydration energy is dominant. Here we find that the +2 ion, because of its smaller size and

Table 18.2
*Some properties of the alkaline earth metals*

| Element | Ionization Energies (kcal/mole) First | Second | Third | $\mathscr{E}^0$ (V) | $M^{2+}$ Radius (Å) |
|---|---|---|---|---|---|
| Beryllium | 213.9 | 417.7 | 3529 | −1.70 | 0.30 |
| Magnesium | 175.4 | 344.8 | 1838 | −2.34 | 0.65 |
| Calcium | 140.9 | 273.8 | 1181 | −2.87 | 0.94 |
| Strontium | 130.6 | 253.1 | 1005 | −2.89 | 1.10 |
| Barium | 119.6 | 229.3 | 818.7 | −2.90 | 1.29 |
| Radium | 122 | 234 | — | −2.92 | 1.52 |

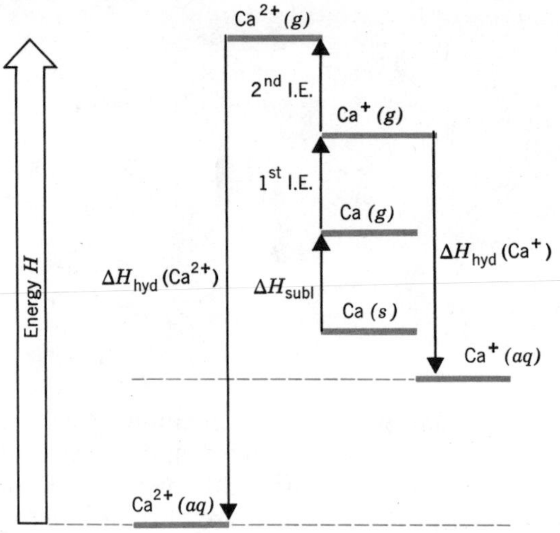


Figure 18.8
*Relative enthalpy changes for the oxidation of $Ca^0$ (s) to $Ca^+$ (aq) and $Ca^{2+}$ (aq).*

higher charge, interacts much more strongly with the solvent than does the +1 ion. It therefore has a much larger hydration energy. This additional hydration energy for the $Ca^{2+}$ ion, for example, more than compensates for the additional energy required to remove the second electron from the calcium atom. These energy changes are illustrated in Figure 18.8, where we see that even though it takes more energy to create the $Ca^{2+}$ ion than to form the $Ca^+$ ion, the much larger hydration energy of $Ca^{2+}$ causes the $Ca^{2+}$ (*aq*) species to be lower in energy (more stable) than the corresponding $Ca^+$ (*aq*) ion.

A similar argument can be made for the relative stabilities of solids containing $Ca^+$ or $Ca^{2+}$. In this case we find that it is the very large lattice energies for the $M^{2+}$ salts, compared to $M^+$ compounds, that is able to compensate for the extra energy needed to remove the second electron from the alkaline earth atom.

A question we might ask at this point is, if the additional hydration energy for $Ca^{2+}$ over $Ca^+$ can serve to stabilize the more highly charged ion, why do we not observe $Ca^{3+}$ salts? To form $Ca^{3+}$ compounds we would have to supply an additional amount of energy equal to the third ionization energy of calcium. The reaction

$$Ca^{2+} \longrightarrow Ca^{3+} + e^-$$

requires breaking into the closed-shell argon core of the $Ca^{2+}$ ion, a process that requires the input of a huge amount of energy. For Ca, the third IE has a value of 1181 kcal/mole! This extremely large investment cannot be recovered by the hydration of the $Ca^{3+}$, even though $\Delta H_{hyd}$ is larger for $Ca^{3+}$ than for $Ca^{2+}$. As a result the $Ca^{2+}$ (*aq*) ion is more stable (of lower energy) than the $Ca^{3+}$ (*aq*) species. These same arguments also apply to the other members of Group IIA, so that only a +2 state is observed.

**GROUP IIIA.** In Group IIIA we begin to see, for the first time, more than one stable oxidation state, at least among the heavier elements. While aluminum shows only a +3 oxidation state, corresponding to the loss of all three of its outer electrons, gallium, indium, and thallium exist in both a +1 state (brought about by the loss of the single outer-shell *p* electron) and a +3 state (which arises, as with Al, from the loss of all three valence electrons).

It is found that the relative stability of the lower oxidation state increases with increasing atomic number within the group. For instance, $Ga^{3+}$ is more stable than $Ga^+$, while $Tl^+$ is more stable than $Tl^{3+}$. This trend in the stabilities of the high and low oxidation states persists in later groups, as we shall see, and results, it appears, from a decreasing stability of M—X bonds with increasing size of the metal atom. As a result, insufficient energy is recovered in bond formation to make up for the energy that must be expended to remove additional electrons.

**GROUP IVA.** In Group IVA we have only two elements that can be classed as true metals, Sn and Pb. Their outer-shell electron configurations correspond to $s^2p^2$ and two oxidation states, +2 and +4, are found. As in Group IIIA, the relative stability of the +2 state increases going down within the group. Thus, $Sn^{2+}$ solutions are mild reducing agents owing to the fairly easy oxidation of Sn to the +4 state. By contrast, $Pb^{2+}$ in aqueous media has virtually no tendency to become oxidized to the +4 state. In fact, $Pb^{IV}$ compounds[4] tend to be rather good oxidizing agents in acid solution, thereby implying a strong tendency of $Pb^{IV}$ to be converted to the more stable $Pb^{II}$. In Chapter 16 we saw that this is put to practical use in the lead storage battery, where $PbO_2$ serves as the cathode while Pb serves as the anode. In the sulfuric acid electrolyte the net cell reaction is

$$Pb + PbO_2 + 2H_2SO_4 \longrightarrow 2PbSO_4 + 2H_2O$$

**GROUP VA.** Only one true metal, bismuth, is found in Group VA. It has an outer-shell configuration of $6s^26p^3$ and, like the elements at the bottom of the preceding two groups, it too forms compounds in two oxidation states. These consist of a +3 state, from the loss of the three *p* electrons, and a +5 state that corresponds to the loss (at least in principle) of all five valence electrons. In actual fact, no $Bi^{5+}$ ion exists, and the assignment of a +5 oxidation state to Bi is really a result of the rules that we use in computing oxidation numbers.

As with other representative elements at the bottom of a group, the lower oxidation state of Bi is more stable than the higher one. For example, the oxide $Bi_2O_3$ forms readily, but the formation of the higher oxide, $Bi_2O_5$, requires severe oxidizing conditions. Because of the difficulty encountered in preparing them, few compounds containing Bi in the +5 state are known. Those that do exist, for example $NaBiO_3$, have a strong tendency to acquire electrons and are very powerful oxidizing agents.

## 18.6 THE COVALENT/IONIC NATURE OF METAL COMPOUNDS

Frequently we tend to think of metal compounds with nonmetals as essentially ionic in character and, in many instances, this is indeed true. However, we should remember that no bond is purely ionic and, most important, there is a gradual transition between ionic and covalent bonding. What we wish to examine in this section are some of the factors that allow us to predict trends in ionic (or covalent) bonding within a series of compounds.

The basic concept that we shall expand upon is that the electron cloud surrounding an atom or ion is somewhat soft and "mushy" and can be dis-

[4] We shall see later in this chapter that highly charged ions tend to become involved in a significant degree of covalent bonding. It is doubtful whether a true $Pb^{4+}$ ion can actually exist in chemical compounds and therefore the formal oxidation state of the lead, in roman numerals, is used instead of an ionic charge.

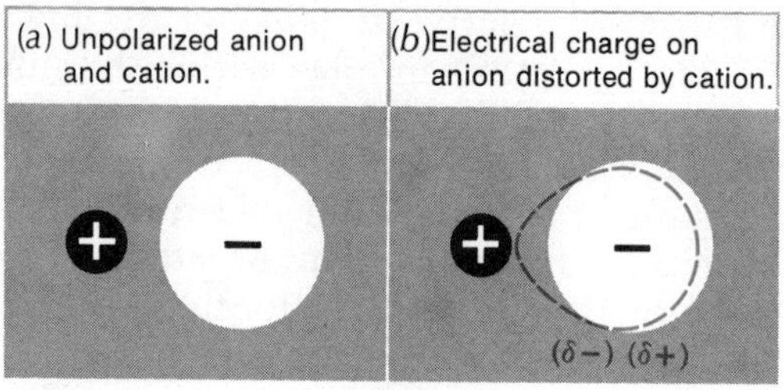


Figure 18.9
*Polarization of an anion by the positive charge on the cation. (a) Unpolarized anion and cation. (b) Electrical charge on anion distorted by cation.*

torted when in the presence of another ion. Consider, for example, the two ions in Figure 18.9*a*. The positive ion here is depicted as smaller than the negative one, reflecting the fact that cations are generally smaller than anions. Because of its positive charge, the cation tends to draw the electron cloud of the anion toward itself. This gives an electron distribution somewhat like that in Figure 18.9*b*, in which the anion has been distorted so that in addition to being an ion, it is now something of a dipole as well. We say, therefore, that the electron cloud of the anion has been **polarized** by the cation.

When the anion becomes polarized, electron density is drawn into the region between the two nuclei. Since a covalent bond consists of electrons shared *between* nuclei, the polarization of the anion results in the partial formation of a covalent bond. The extent to which this polarization occurs determines the degree of covalent character in the ionic bond. The limit of this, of course, is where the electrons of the anion are pulled toward the cation so much that a nonpolar covalent bond is formed.

At this point we must ask, what factors influence the degree to which anion polarization occurs? As we might expect, one of these is the ionic potential of the cation, that is, the ratio of its charge to its radius. We have said before that this quantity is a measure of the "concentration" of charge, and cations in which the charge is highly concentrated are better able to distort a neighboring electron cloud than cations in which the positive charge is thinly spread and diffuse.

The ionic potential increases with decreasing size of the cation and hence, *small ions are more effective at producing anion polarization than are large ones.* The ionic potential also increases as the charge on the cation gets larger, so that *for a given size, multiply charged ions produce a greater degree of polarization than do singly charged ions.* We can see how these ideas apply if we examine some examples.

As we proceed down within a group, the ions of a given charge become progressively larger and their ionic potentials therefore decrease. In Group IA, for instance, we find the lithium ion to be very small, $Na^+$ somewhat larger, $K^+$ larger still, and so on. Since $Li^+$, because of its size, has the largest ionic potential in this series of ions, it should also form compounds with the greatest degree of covalent character. As it happens, among the alkali metals Li is in fact the only element that forms compounds with organic molecules (called organolithium compounds) that have properties usually associated with covalently bonded substances, such as low melting points and appreciable solubilities in nonpolar solvents. The corresponding compounds of the other alkali metals, however, are predominantly ionic.

In Group IIA we would expect the degree of covalent character to be greater because of the greater charge on the ions of the group. It has already been mentioned that Be compounds are all covalent to at least some degree and that there is little evidence for the existence of a true $Be^{2+}$ ion. In fact, when electrolysis of $BeCl_2$ is carried out, NaCl must be added as an electrolyte; otherwise the molten $BeCl_2$ does not conduct electricity. Magnesium compounds, as pointed out earlier, are in many ways similar to Li compounds, and Mg forms organomagnesium compounds which, like organolithium compounds, exhibit properties characteristic of covalent substances. However, Mg differs from Be in that there are many compounds of Mg, such as the oxide, for example, that are distinctly ionic. Finally, the metals below Mg in Group IIA, because of their larger size and smaller ionic potentials, form compounds that are essentially ionic. In Group IIA, as in the first group, we see that as we descend the group there is a trend toward decreasing covalent character in the bonds.

The effect of charge is also seen for compounds such as $SnCl_4$ and $PbCl_4$. These, at first glance, appear to be salts of the cations $Sn^{4+}$ and $Pb^{4+}$; however, each is quite covalent. For example, both are *liquids* with low melting points (melting point of $SnCl_4$ = −33°C, $PbCl_4$ = −15°C). In fact, we might compare them with the other Group IVA tetrahalides $CCl_4$ (melting point = −23°C), $SiCl_4$ (melting point = −70°C) and $GeCl_4$ (melting point = −49.5°C). The very high ionic potential of a +4 ion thus appears to polarize an anion such as $Cl^-$ to the extent that covalently bonded molecules result.

In addition to vertical trends in covalent character, we can also look at variations in the nature of the bonding as we move from left to right within a period. Here, as you know, the cations become progressively smaller and more highly charged, for example, $Li^+$, $Be^{2+}$, and $B^{3+}$, and so on. Consequently, the ionic potential increases rapidly and the degree of covalent bonding in their compounds does so too. Thus LiCl is predominantly ionic, $BeCl_2$ is covalent, and so are the remaining halides as we continue across period 2 (i.e., $BCl_3$, $CCl_4$, etc.).

In period 3 the cations are larger than those in the preceding period, with the result that the covalent character of the bonds shows up later in the period. For example, NaCl and $MgCl_2$ are both essentially ionic, and it is not until Group IIIA, with $AlCl_3$, that the halides become largely covalent. We might predict, however, that with a given anion the bonds to $Mg^{2+}$ would be less ionic than those to $Na^+$. Recall, for instance, that covalent organomagnesium compounds exist while the corresponding compounds with sodium are ionic.

The type of outer-shell electron configuration possessed by the cation also affects its ability to polarize an anion. Cations with pseudonoble gas configurations of 18 electrons ($ns^2np^6nd^{10}$), such as those formed from the metals immediately following a transition series, appear to have higher ionic potentials than similarly sized ions having the same net charge but with only an octet of electrons. This is because the *d* electrons that are added to the atom as the transition series is crossed do not completely shield the nuclear charge, which has also increased. The net effect is that a cation with a pseudonoble gas structure presents to a neighboring anion an *effective* positive charge that is actually somewhat greater than its net ionic charge. *As a result, these cations behave as if their effective ionic potentials were higher than those calculated from their net ionic charges and, consequently, they are more effective at distorting the electron cloud of the anion.*

Table 18.3
*Melting points of some metal chlorides*

| Pre-transition Elements | | | | Post-transition Elements | | | |
|---|---|---|---|---|---|---|---|
| M—Cl Distance (Å) | | Compound | Melting Point (°C) | M—Cl Distance (Å) | | Compound | Melting Point (°C) |
| Na—Cl | 2.81 | NaCl | 800 | Ag—Cl | 2.77 | AgCl | 455 |
| Rb—Cl | 3.29 | RbCl | 715 | Tl—Cl | 3.31 | TlCl | 430 |
| Ca—Cl | 2.74 | $CaCl_2$ | 772 | Cd—Cl | 2.76 | $CdCl_2$ | 568 |
| Sr—Cl | 3.02 | $SrCl_2$ | 873 | Pb—Cl | 3.02 | $PbCl_2$ | 501 |
| | | | | Hg—Cl | 2.91 | $HgCl_2$ | 276 |
| Mg—Cl | 2.46 | $MgCl_2$ | 708 | Zn—Cl | 2.53 | $ZnCl_2$ | 283 |
| | | | | Sn—Cl | 2.42 | $SnCl_2$ | 246 |

This effect is illustrated in Table 18.3 where the melting points of several metal chlorides are presented. We associate high melting points with a high degree of ionic bonding where the attractive forces between particles within the solid are very strong. On the other hand, low melting points are characteristic of substances that contain covalently bonded molecules in which the attractive forces between neighboring molecules are relatively weak. As a rough guide, then, we can use melting points of compounds of similar composition as a measure of their degree of covalent bonding, and we expect their melting points to decrease as the bonding becomes progressively more covalent.[5]

In Table 18.3 we compare salts having similar metal-chlorine bond distances. The salts on the left of the table have cations with an octet in their outer shell while those on the right contain cations with either a pseudonoble gas $ns^2np^6nd^{10}$ structure ($Ag^+$, $Zn^{2+}$, $Cd^{2+}$, $Hg^{2+}$) or a pseudonoble gas structure plus two electrons ($Tl^+$, $Pb^{2+}$, $Sn^{2+}$). The effectiveness of these post-transition metal ions at polarizing the chloride ion is revealed by the consistently lower melting points of their salts. Since on each line we have cations of essentially the same effective size (as shown by nearly the same M—Cl distances), the difference in covalence must be related to their effective ionic potentials, being less for elements that precede a transition series than for elements at the end of the series or for those in Groups IIIA and IVA that follow the transition elements.

The covalent character of a metal-nonmetal bond is also influenced by the nature of the anion. For a given cation the bond becomes less ionic as the anion becomes more easily deformed; therefore we must examine trends in anion polarizability. In general, it is found that *for a given charge, anions become more readily distorted as they become larger.* In a large anion the outer electrons are further from the nucleus and are spread more thinly. Consequently, they are able to be influenced to a greater degree by a cation. Thus, among the halides, $F^-$, $Cl^-$, $Br^-$, and $I^-$, we expect to find the greatest amount of covalent character in compounds in which a metal is combined

[5] This generalization breaks down for polymeric covalent structures such as those found in quartz or diamond. In this case, because of the size of the giant molecules, very high melting points are found even though the bonding is covalent.

Table 18.4
*Colors of silver halides*

| Compound | Color | Anion Radius (Å) |
|---|---|---|
| AgF | White | 1.36 |
| AgCl | White | 1.81 |
| AgBr | Cream | 1.95 |
| AgI | Yellow | 2.16 |

with $I^-$. This is indeed the case, as demonstrated, for example, by the colors of the silver halides (Table 18.4).

Ionic compounds generally do not absorb light in the visible region of the spectrum. Therefore, they appear white, or colorless, when bathed in white light (e.g., NaCl is a colorless crystalline substance). The absorption that does take place with ionic compounds occurs in the shorter-wavelength (higher-frequency, higher-energy) ultraviolet region. The energy absorbed by the compound is used to shift an electron from the anion to the cation, for example,

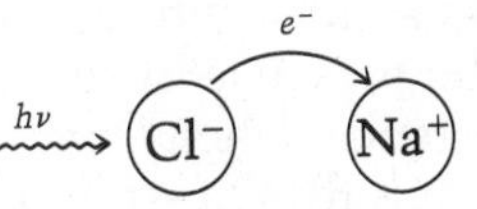


The absorption of a band of wavelengths from the ultraviolet "rainbow" produces a **charge-transfer absorption band.**

As the bond between the metal and nonmetal becomes more covalent (that is, as electron density shifts away from the anion in the direction of the cation), less energy is required to produce the charge transfer. As a result, the absorption band shifts toward the lower-energy visible region of the spectrum, where the removal of some colors from white light gives rise to reflected colors that represent the remainder of the visible spectrum.

The depth of color in compounds (particularly those of the nontransition elements) can often be taken to be a measure of the degree of covalent character in the metal-nonmetal bonds. Among the silver halides we see that as the anion grows in size, the compounds become progressively deeper in color indicating a progressive rise in covalent character in the Ag—*X* bonds.[6]

We also find a similar relationship between ionic character and anion size if we compare metal oxides ($r_{O^{2-}} = 1.40$ Å) and sulfides ($r_{S^{2-}} = 1.84$ Å). Both of these ions are colorless as evidenced by the fact that both $Na_2O$ and $Na_2S$ are colorless. However, with aluminum we find $Al_2O_3$ to be white while $Al_2S_3$ is yellow, suggesting a greater degree of ionic bonding in the oxide. This is further supported by comparing the melting points of these two compounds; $Al_2O_3$ melts at 2045°C, while $Al_2S_3$ melts at a much lower temperature, 1100°C. Additional examples are provided in Table 18.5.

*For anions of a given size, polarization also increases with an increase in the charge on the anion.* We can compare, for example, compounds con-

[6] The salts NaF, NaCl, NaBr, and NaI, which are predominantly ionic, are colorless. This indicates that the halide ions themselves are colorless. The color of the Ag*X* compounds is therefore a reflection of covalent bonding.

Table 18.5
*Properties of some metal oxides and sulfides*

| Oxides | | | Sulfides | | |
|---|---|---|---|---|---|
| | Color | Melting Point (°C) | | Color | Melting Point (°C) |
| $Al_2O_3$ | White | 2045 | $Al_2S_3$ | Yellow | 1100 |
| $Ga_2O_3$ | White | 1900 | $Ga_2S_3$ | Yellow | 1255 |
| $Sb_2O_3$ | White | 656 | $Sb_2S_3$ | Yellow-red | 550 |
| $Bi_2O_3$ | Yellow | 860 | $Bi_2S_3$ | Brown-black | 685 decomp. |
| $SnO_2$ | White | 1127 | $SnS_2$ | Yellow | 882 |
| ZnO | White | 1975 | ZnS | White | 1850 |

taining chloride ($r$ = 1.81 Å) and sulfide ($r$ = 1.84 Å) as shown in Table 18.6. In each case the sulfide has a deeper color than the corresponding chloride salt, indicating that the compounds containing the more highly charged sulfide ion are more covalent. We also see that, in general, most metal sulfides, except those of the alkali and alkaline earth metals, are deeply colored and possess a quite substantial degree of covalent bonding.

The presence of these colored compounds can often be seen about us. For instance, when silver tarnishes, it reacts with traces of $H_2S$ in the air to give a dull film of $Ag_2S$. This hydrogen sulfide, produced generally from decomposing organic matter, also darkens lead-based paint. The pigment, $Pb_3(OH)_2(CO_3)_2$, called *white lead,* reacts with $H_2S$ to form black PbS.

The fact that many of these compounds possess rather striking colors has also been put to use throughout history. For example, the brilliant yellow of natural CdS and the "vermilion" red of HgS have led them to be used as pigments for the oil paints used by artists. Recently, several sticks of black PbS that had been used as a type of mascara were recovered from an ancient Egyptian burial ground.

Table 18.6
*Colors of metal chlorides and sulfides*

| Chloride ($r$ = 1.81 Å) | | Sulfide ($r$ = 1.84 Å) | |
|---|---|---|---|
| AgCl | White | $Ag_2S$ | Black |
| CuCl | White | $Cu_2S$ | Black |
| AuCl | Yellow | $Au_2S$ | Brown-black |
| $CdCl_2$ | White | CdS | Yellow |
| $HgCl_2$ | White | HgS | Black or red, depending on crystal structure |
| $PbCl_2$ | White | PbS | Black |
| $SnCl_2$ | White | SnS | Black |
| $AlCl_3$ | White | $Al_2S_3$ | Yellow |
| $GaCl_3$ | White | $Ga_2S_3$ | Yellow |
| $BiCl_3$ | White | $Bi_2S_3$ | Brown-black |

## 18.7 HYDROLYSIS

The concepts developed in the last section, you may have noticed, are very similar to those presented in Chapter 13, where we discussed the relative degrees of acidities of Brønsted acids. In fact, they are essentially identical, and we can use them to understand trends in metal ion hydrolysis.

Let us consider, for example, the hydrolysis of a metal ion, $M^{n+}$. In water, as you may recall, the ion will be surrounded by water dipoles oriented with their negative ends toward the cation. The positive charge of the ion will distort the electron cloud surrounding the oxygen nucleus of a neighboring water molecule, inducing a certain degree of covalent bonding to the $M^{n+}$—$OH_2$ bond. The electron density drawn into the M—O bond, however, tends to come from the O—H bonds of the $H_2O$. As a result, the hydrogen atoms become more positively charged and therefore become more easily removed as $H^+$. We see, then, that the extent of hydrolysis depends on the amount of charge removed from the O—H bonds and transferred to the M—O bonds. However, the degree of covalent character produced in the M—O bonds is determined, in turn, by the ionic potential of the cation. Consequently, we expect that the tendency for the $M(H_2O)_x{}^{n+}$ species to lose protons (i.e., the tendency for the cation by hydrolyze) will be greatest for ions having large ionic potentials.

This is, in fact, precisely what is found. The ions of Groups IA and IIA, for the most part, have virtually no tendency to hydrolyze. Beryllium is an exception because of its small size and high charge and, hence, high ionic potential. Aqueous solutions of Be salts are therefore acidic because of the reaction

$$Be(H_2O)_4{}^{2+} + H_2O \longrightarrow Be(H_2O)_3(OH)^+ + H_3O^+$$

Of the ions of the metals in Groups IIIA, IVA, and VA, almost all exhibit a substantial degree of hydrolysis. In our earlier discussions (Chapter 13) we saw the hydrolysis of $Al^{3+}$, which probably exists as $Al(H_2O)_6{}^{3+}$ in water. The elements below Al, in their +3 states, have a pseudonoble gas structure and, consequently, have high ionic potentials just as $Al^{3+}$ does. As a result, the $Ga^{3+}$, $In^{3+}$ and $Tl^{3+}$ ions tend to be rather extensively hydrolyzed too. Because of their much lower ionic potentials, the +1 states of these three metals are expected to hydrolyze to a much lesser degree. This is true for $Tl^+$, compared to $Tl^{3+}$; since the other two, $Ga^+$ and $In^+$, do not exist in aqueous solution we can only speculate about their behavior.

The aqueous chemistry of the Group IVA metals too, is influenced by hydrolysis. Tin(II) solutions, for example, are extensively hydrolyzed, and $Pb^{2+}$, although larger in size and therefore of lower ionic potential than $Sn^{2+}$, is also hydrolyzed to a degree. In the higher oxidation state characteristic of the metals of this group we might expect even greater hydrolysis and, indeed, this occurs.

Moving to the right again, we find that $Bi^{3+}$, because of its high charge, is very extensively hydrolyzed. In this case, in fact, there is evidence for the species $BiO^+$ (called the bismuthyl ion) produced by the reaction

$$Bi^{3+} + H_2O \longrightarrow BiO^+ + 2H^+$$

Thus the $BiO^+$ ion can be looked upon as the product of very severe hydrolysis in which two protons of a water molecule are lost entirely. We shall encounter ions of this type again in our discussion of the transition elements when we meet ions of high charge (and hence high ionic potential).

## INDEX TO QUESTIONS AND PROBLEMS (problem numbers in italics)



## REVIEW QUESTIONS

**18.1** On a piece of paper, sketch the shape of the periodic table and roughly mark off those regions where we find the metals, the nonmetals, and the metalloids.

**18.2** Write the symbol for (a) a representative element, (b) an alkali metal, (c) an alkaline earth metal, (d) an inner transition element, (e) a halogen, (f) a noble gas.

**18.3** What are three physical properties that are generally characteristic of metals? What chemical characteristics do metals possess in common?

**18.4** How is iron "pickled"? What is the purpose of this treatment?

**18.5** Why is LiOH used to trap $CO_2$ in space vehicles rather than NaOH, which is much less expensive?

**18.6** Sulfur dioxide can be removed from exhaust gases by passing the hot gases over CaO. Write a chemical equation for the reaction of $SO_2$ with CaO.

**18.7** Give the outer-shell electron configuration of each of the following: (a) Ca, (b) Tl, (c) Bi, (d) Cs, (e) Sn.

**18.8** Why are metals better conductors of heat than nonmetals?

**18.9** Why, under ordinary conditions, do metals not form simple molecular species with each other by sharing electrons in covalent bonds?

**18.10** How does the band theory of solids (Chapter 7) differentiate among conductors, semiconductors, and insulators?

**18.11** How does the metallic character of the elements depend on electronegativity? What vertical and horizontal trends in metallic character exist in the periodic table? Illustrate these trends for the elements in the second period and in Group IVA.

**18.12** In each pair below, choose the element expected to have the more metallic character.
(a) Li or Be
(b) B or Al
(c) Al or Cs
(d) Sn or P
(e) Ga or I

**18.13** What is meant by amphoteric? Write chemical equations to illustrate the amphoteric behavior of beryllium and aluminum.

**18.14** Write chemical equations showing the chemical reactions involved in separating Mg from sea water.

**18.15** What is "tin disease"?

**18.16** Using Table 10.1, predict which of the oxides, CuO or ZnO, should be most stable with respect to thermal decomposition to the free elements.

**18.17** Write chemical reactions illustrating the chemical reduction of a metal compound using (a) carbon, (b) hydrogen, (c) sodium, (d) electrolysis.

**18.18** Why is carbon a preferred reducing agent in commercial metallurgy?

**18.19** Why are halide salts often used when electrolysis of a molten salt is carried out?

**18.20** Write chemical equations for the reactions that occur when the alkali metals are exposed to molecular oxygen.

**18.21** Write chemical equations for the hydrolysis of (a) an oxide, (b) a peroxide, (c) a superoxide.

**18.22** Why is $KO_2$ used in recirculating breathing equipment?

**18.23** Construct the molecular orbital energy-level diagram for the superoxide ion, $O_2^-$, and the peroxide ion, $O_2^{-2}$. How should the bond lengths compare for these species?

18.24 List three ways that lithium and magnesium are similar to one another.

18.25 To what do we attribute the diagonal relationships that exist between Li and Mg, and between Be and Al?

18.26 Why must NaCl be added to molten $BeCl_2$ in order to produce Be by electrolysis?

18.27 Compare the structures of $BeCl_2$ and $AlCl_3$. Why doesn't $AlCl_3$ form a linear chain polymer as $BeCl_2$ does? What type of hybrid orbitals would be used by the Be and Al in these species?

18.28 To what do we attribute the unusually high negative reduction potential of lithium? Why do we not observe $Li^{2+}$ in compounds?

18.29 What is the active reducing agent in solutions of alkali metals in liquid ammonia? What prevents the formation of stable solutions of alkali metals in water?

18.30 Why do the alkaline earth metals only form compounds in which they exhibit a +2 oxidation state? On this basis, and using the data in Table 3.7 on p. 90, can you suggest why Al exhibits only a +3 oxidation state and not a +4 state?

18.31 Among the representative elements, why are the lower oxidation states preferred by elements at the bottom of a given group in the periodic table?

18.32 Predict the better oxidizing agent in each of the following pairs.
(a) $Bi_2O_3$ or $Bi_2O_5$
(b) $SnO_2$ or SnO
(c) $PbO_2$ or PbO
(d) TlCl or $TlCl_3$

18.33 Predict which of the following exhibit the greater degree of covalent bonding.
(a) $Bi_2O_3$ or $Bi_2O_5$
(b) PbO or PbS
(c) CaO ($r_{Ca^{2+}} = 0.99$ Å) or SnO ($r_{Sn^{2+}} = 0.93$ Å)
(d) $Na_2S$ or MgS
(e) LiCl ($r_{Cl^-} = 1.81$ Å) or $Li_2S$ ($r_{S^{2-}} = 1.84$ Å)

18.34 Predict which of the following should be more ionic.
(a) SnO or SnS
(b) SnS or $SnS_2$
(c) SnS or PbS
(d) $BCl_3$ or $BeCl_2$

18.35 Which of the following compounds should have the higher melting point?
(a) $CaCl_2$ or $SnCl_2$ ($r_{Ca^{2+}} = 0.99$ Å, $r_{Sn^{2+}} = 0.93$ Å)
(b) $BeCl_2$ or $BeF_2$
(c) $BeCl_2$ or $MgCl_2$
(d) $SnCl_2$ and $SnCl_4$

18.36 For each pair, predict which compound should be more deeply colored.
(a) $HgCl_2$ ($r_{Cl^-} = 1.81$ Å) or HgS ($r_{S^{2-}} = 1.84$ Å)
(b) $Ag_2O$ or $Ag_2S$
(c) SnO or PbO
(d) SrS ($r_{Sr^{2+}} = 1.12$ Å) or PbS ($r_{Pb^{2+}} = 1.20$ Å)

18.37 Which compound in each pair in Question 18.36 would have the lower melting point?

18.38 For each of the pairs below, pick the ion that should undergo the greater degree of hydrolysis. Explain your choice in each case.
(a) $Li^+$ or $Be^{2+}$
(b) $Be^{2+}$ or $Ca^{2+}$
(c) $B^{3+}$ or $Al^{3+}$
(d) $Tl^+$ or $Tl^{3+}$
(e) $Sn^{4+}$ or $Sn^{2+}$

18.39 Choose the ion in each pair below that should give the lower pH when its concentration is 1 *M*.
(a) $Be^{2+}$ or $Mg^{2+}$
(b) $Be^{2+}$ or $B^{3+}$
(c) $Sn^{2+}$ ($r = 0.93$ Å) or $Ca^{2+}$ ($r = 0.99$ Å)
(d) $Sr^{2+}$ ($r = 1.12$ Å) or $Pb^{2+}$ ($r = 1.20$ Å)
(e) $Sn^{2+}$ or $Pb^{2+}$

18.40 What is responsible for the color observed in compounds like $SnS_2$ and PbS?

18.41 When a solution of $BiCl_3$ in concentrated HCl is diluted with water, a precipitate having the formula BiOCl is formed. Explain this reaction in terms of hydrolysis of $Bi^{3+}$.

18.42 Construct a table showing the oxidation states found for each of the representative metals. Look up the reduction potentials for each of these species (i.e., for the reaction $M^{n+} + ne^- \rightarrow M$). What vertical and horizontal trends exist in the standard reduction potentials? What atomic properties do these trends correlate with?

REVIEW PROBLEMS

**18.43** Given the following thermodynamic data, calculate the hydration energy for the $Na^+$ ion

$$\Delta H_f^0 \text{ of } Na^+ (aq) = -57.28 \text{ kcal/mole}$$
$$\Delta H_{atom}^0 \text{ of Na} = 25.98 \text{ kcal/mole}$$
$$\text{IE of Na} = 118.0 \text{ kcal/mole}$$

**18.44** Using the data for the atomization energy of Na (Problem 18.43) and the first and second ionization energies for Na in Table 18.1, compute the value of the hydration energy required to produce a negative $\Delta H_f$ for $Na^{2+}$ $(aq)$.

**18.45** From the data in Tables 10.1 and 10.4, calculate the temperature above which the thermal decomposition of ZnO should become feasible.

**18.46** Given the data below, determine the temperature at which $K_P = 1$ for the reaction,

$$CuO\ (s) \longrightarrow Cu\ (s) + \tfrac{1}{2}O_2\ (g)$$

For CuO $(s)$, $\Delta H_f^0 = -155$ kJ/mole. Absolute entropies; CuO $(s)$, 43.5 J/mole K; Cu $(s)$, 33.3 J/mole K; $O_2$ $(g)$, 205.0 J/mole K.

***18.47** Calculate $K_P$ at 100, 500, and at 2000°C for the reaction,

$$MoO_3\ (s) \longrightarrow Mo\ (s) + \tfrac{3}{2}O_2\ (g)$$

given the following data:

| | $\Delta H_f^0$ (kcal/mole) | $S^0$ (cal/mole K) |
|---|---|---|
| $MoO_3$ $(s)$ | −180.3 | 18.68 |
| Mo $(s)$ | 0.0 | 6.83 |
| $O_2$ $(g)$ | 0.0 | 49.00 |

***18.48** The standard reduction potential of potassium is −2.92 V. $\Delta H_{hyd}^0$ of $K^+$ in 1 $M$ aqueous solution is 759 kJ/mole. The atomization energy of K is 90.0 kJ/mole and the ionization energy of K $(g)$ is 418 kJ/mole. Calculate $\Delta S^0$ for the process,

$$K\ (s) \longrightarrow K^+\ (1\ M) + e^-$$

# 19 CHEMISTRY OF THE REPRESENTATIVE ELEMENTS: PART II, THE METALLOIDS AND NONMETALS

In this chapter we will conclude our discussion of the representative elements by considering the chemistries of the metalloids and nonmetals. These two classes of elements are conveniently treated together because in many ways their chemical properties are similar, owing to the covalent nature of many (if not most) of their compounds.

As in Chapter 18, we will continue to concentrate on points of similarity and trends in behavior that allow us to correlate more efficiently the factual descriptive information. In this vein we shall discuss, in addition to the free elements themselves, three classes of compounds formed by the metalloids and nonmetals, specifically, their compounds with hydrogen, oxygen, and the halogens.

Among the various aspects of the chemistry of these elements that we shall encounter in this chapter, one of the most fascinating is the variety of structures exhibited by both the free elements and their compounds. This is particularly pronounced for these elements because of the predominance of covalence among their compounds. Covalent bonds, you recall, are strongly directional in nature, and we shall come to see that the basic structural units formed by atoms covalently bound to one another can be combined in various ways to give quite complex overall structures. What you should note, however, is that even the most complex structures can be understood by considering them in terms of the *simple* structural units of which they are composed.

## 19.1 THE FREE ELEMENTS

There are two aspects of the elemental forms of the metalloids and nonmetals that we shall examine. The first of these, discussed in this section, is the way in which these elements may be obtained in their free state. The second, which we shall attend to in the next section, is the variety of structural forms that they exhibit.

In the last chapter we made the rather broad generalization that metals, in their compounds, usually exist only in positive oxidation states. As it happens, this statement can be extended for the most part also to the metalloids, since most of their compounds contain the metalloid combined with a nonmetal, either in a molecular structure such as $SiO_2$ or in an oxoanion such as is found in the silicates. In these combinations the metalloids have a lower electronegativity than the nonmetal and, consequently, exist in positive oxidation states. As a result, the free elements are produced by the reduction of metalloid compounds, usually using either carbon or hydrogen as a chemical reducing agent. For example, boron is obtained by passing a mixture of $BCl_3$ vapor and hydrogen gas over a hot wire, upon which occurs the reaction

$$2BCl_3 + 3H_2 \longrightarrow 2B + 6HCl$$

On the other hand, elemental silicon is produced by heating $SiO_2$ with carbon in an electric furnace, where the reaction

$$SiO_2 + C \longrightarrow Si + CO_2$$

takes place spontaneously once the temperature exceeds approximately 3000°C (below this temperature the reverse reaction is actually favored).

The remaining metalloids may be obtained from their oxides by heating them with either carbon or hydrogen; for example,

$$GeO_2 + C \longrightarrow Ge + CO_2$$
$$GeO_2 + 2H_2 \longrightarrow Ge + 2H_2O$$

Similarly, we also have

$$2As_2O_3 + 3C \longrightarrow 4As + 3CO_2$$
$$2Sb_2O_3 + 3C \longrightarrow 4Sb + 3CO_2$$

and

$$As_2O_3 + 3H_2 \longrightarrow 2As + 3H_2O$$
$$Sb_2O_3 + 3H_2 \longrightarrow 2Sb + 3H_2O$$

In very pure form, silicon and germanium have found widespread application in the electronics industry, where they are used in transistors and photoconduction devices. Sophisticated miniaturized circuits deposited on tiny wafers of silicon have made possible intricate small computers to guide spacecraft and missiles, perform computations in hand-held calculators, and operate automatic-exposure cameras.

Unlike the metals and metalloids, it is difficult to make general statements about the preparation of the nonmetals. Some of them, such as the noble gases, are always found uncombined in nature. Others, while present in many naturally occurring compounds, also are found extensively in the free state as well. For instance, our atmosphere is composed primarily of elemental nitrogen, $N_2$ (about 80%), and oxygen, $O_2$ (about 20%). While both nitrogen and oxygen are found in a vast number of compounds, certainly their most economical source is simply the air itself. The atmosphere is also the major source of the noble gases, even though they are present only in very small quantities. Of the noble gases, only helium and radon are not obtained primarily from the atmosphere. Helium is found in gaseous deposits beneath the earth's crust where it has collected after being produced by the capture of electrons by alpha particles (He nuclei) that are formed during the radioactive decay of elements such as uranium. Radon itself is radioactive and is produced by the radioactive decay of still heavier elements. Since radon spontaneously decomposes into other elements, it occurs only in minute quantities in nature.

Sulfur and carbon are two other elements that occur naturally in both the combined and free states. There are, for instance, many naturally occurring sulfates (for example, $BaSO_4$, $CaSO_4 \cdot 2H_2O$) and sulfides ($FeS_2$, CuS, HgS, PbS, ZnS). In the free state sulfur has been found in large underground deposits from which it is mined using a rather clever method developed by an engineer, Herman Frasch, in 1890. The process, which has come to bear his name, involves forcing superheated water under pressure into the sulfur deposit, causing the sulfur to melt. Once molten, the sulfur-water mixture is then foamed to the surface using compressed air. Huge quantities of this element, mined by the Frasch process, are used annually to produce such important industrial chemicals as sulfuric acid.

Turning to carbon, we find that most of the naturally occurring compounds are carbonates, for example, limestone ($CaCO_3$). In the free state carbon is found in two forms, diamond and graphite.

Since nonmetals combine with each other as well as with metals, no strict generalizations can be made concerning their recovery from compounds. When combined with a metal, the nonmetal is found in a negative oxidation state and, therefore, in order to generate the free element an oxidation must be brought about. For example, the halogens $Cl_2$, $Br_2$, and $I_2$ can be conveniently prepared in the laboratory in this manner by reacting one of their salts with an oxidizing agent such $MnO_2$ in acid solution, as in the equation

$$2X^- + MnO_2 + 4H^+ \longrightarrow X_2 + Mn^{2+} + 2H_2O$$

where X = Cl, Br, or I.

Chlorine is a very important industrial chemical, and vast quantities (approximately 10 million tons annually) are produced by electrolysis of NaCl, both aqueous and molten. Chlorine is used in large amounts in water treatment and in the production of vinyl chloride, which is used to manufacture vinyl plastics.

The halogens themselves can also serve as oxidizing agents in replacement reactions. Since the tendency to acquire electrons decreases as we proceed downward in a group, the ability of the halogen to serve as an oxidizing agent decreases too. This is seen in their reduction potentials (Table 19.1), which decrease from fluorine to iodine. As a result, a given halogen is a better oxidizing agent than the other halogens below it in Group VIIA and is able to displace them from their binary compounds with metals. Thus $F_2$

Table 19.1
*Reduction potentials of the halogens*

| Reaction | $\mathscr{E}^0$ (V) |
|---|---|
| $F_2 + 2e^- \rightleftharpoons 2F^-$ | 2.87 |
| $Cl_2 + 2e^- \rightleftharpoons 2Cl^-$ | 1.36 |
| $Br_2 + 2e^- \rightleftharpoons 2Br^-$ | 1.09 |
| $I_2 + 2e^- \rightleftharpoons 2I^-$ | 0.54 |

will displace $Cl^-$, $Br^-$, and $I^-$, while $Cl_2$ will displace only $Br^-$ and $I^-$ but not $F^-$, and so on. This is illustrated by these typical reactions.

$$F_2 + \begin{Bmatrix} 2NaCl \\ 2NaBr \\ 2NaI \end{Bmatrix} \longrightarrow 2NaF + \begin{Bmatrix} Cl_2 \\ Br_2 \\ I_2 \end{Bmatrix}$$

$$Cl_2 + \begin{Bmatrix} 2NaBr \\ 2NaI \end{Bmatrix} \longrightarrow 2NaCl + \begin{Bmatrix} Br_2 \\ I_2 \end{Bmatrix}$$

$$Cl_2 + NaF \longrightarrow \text{no reaction}$$

$$Br_2 + 2NaI \longrightarrow 2NaBr + I_2$$

$$Br_2 + \begin{Bmatrix} NaF \\ NaCl \end{Bmatrix} \longrightarrow \text{no reaction}$$

The relative oxidizing power of the halogens is used in the commercial preparation of $Br_2$. Bromine is isolated from sea water and brine solutions pumped from deep wells by passing $Cl_2$ followed by air, through the liquid. The $Cl_2$ oxidizes the $Br^-$ to $Br_2$ and the air sweeps the volatile $Br_2$ from the solution. Bromine is used primarily to make ethylene bromide, $C_2H_4Br_2$, which is added to "antiknock" gasoline along with tetraethyllead, $Pb(C_2H_5)_4$. Upon combustion the Br combines with the Pb to form volatile $PbBr_2$ which escapes with the exhaust. Without the ethylene bromide lead deposits would form within the engine.

Fluorine, because of its position as the most powerful chemical oxidizing agent, can only be obtained by electrolytic oxidation. This process must be carried out in the absence of water, since water is more easily oxidized than the fluoride ion and, if $H_2O$ is present, the reaction

$$2H_2O \longrightarrow O_2 + 4H^+ + 4e^- \qquad \mathscr{E}^0 = -1.23\ V$$

will occur in preference to

$$2F^- \longrightarrow F_2 + 2e^- \qquad \mathscr{E}^0 = -2.87\ V$$

In practice, a molten mixture of KF and HF, which has a lower melting point than KF alone, is electrolyzed, producing $H_2$ at the cathode and $F_2$ at the anode.

Nonmetals can also be extracted from their compounds by reduction if the nonmetal happens to exist in a positive oxidation state. We have seen how this applies to the metalloids, and the general procedure employed there can also be used for the nonmetals. For instance, elemental phosphorus is produced from a phosphate such as $Ca_3(PO_4)_2$, where it is found in the +5 state. In this case, the $Ca_3(PO_4)_2$ is heated with a mixture of carbon and $SiO_2$ (sand).

$$Ca_3(PO_4)_2 + 3SiO_2 + 5C \longrightarrow 3CaSiO_3 + 5CO + 2P$$

In this reaction the $SiO_2$ is present to combine with the calcium. Since the reduction of $SiO_2$ by carbon requires extremely high temperatures, only the phosphorus is reduced to the element.

## 19.2 MOLECULAR STRUCTURE OF THE NONMETALS AND METALLOIDS

We have seen that when atoms react they have a strong tendency to acquire a noble gas electron configuration, either by mutual sharing of electrons or by electron transfer. Except for the noble gases themselves, the nonmetals and metalloids have incomplete valence shells and, therefore, their individual atoms tend to combine until a noble gas structure is achieved. For example, hydrogen atoms unite to form $H_2$ so that each H atom acquires the He configuration

$$\mathrm{H}\cdot + \mathrm{H}\cdot \longrightarrow \mathrm{H}:\mathrm{H}$$

In a similar fashion, the atoms of the other nonmetals and metalloids combine to give structures containing two or more atoms.

One of the controlling factors in determining the complexity of the molecular structures that the nonmetals and metalloids exhibit is the ability of the second-period elements, C, N, and O, to enter into multiple bonding by way of overlap of adjacent *p* orbitals ($p\pi$-$p\pi$ double and triple bonds) as described in Chapter 17. As it turns out, *there is very little tendency for elements in the third and succeeding periods to form multiple bonds of this type.* Presumably this is because their larger size prevents them from approaching each other too closely. Consequently, for these heavier elements, the sideways overlap of *p* orbitals, required for $p\pi$-$p\pi$ multiple bonding, is not very effective, and the formation of two (or three) separate single bonds tends to be preferred energetically over the formation of one double (or triple) bond. The net result, regardless of how we justify it, is that *elements of the second period are able to form multiple bonds fairly readily, while the elements below them in the following periods have a tendency to prefer single bonds.* This phenomenon is particularly striking when we examine the structures of the elements in their free state.

We have seen that some of the elements in the second period form stable diatomic molecules, for example, $N_2$, $O_2$, and $F_2$. In these three cases the valence shells of the atoms are completed by sharing three, two, and one electron, respectively, with nitrogen and oxygen participating in $p\pi$-$p\pi$ bonding. It is because these elements are capable of achieving a stable electron configuration by sharing electrons with a *single* neighbor that they are able to form simple diatomic molecules in their elemental state.

Oxygen, in addition to forming the stable species $O_2$, also can exist in another exceedingly reactive molecular form, $O_3$, called **ozone,** the structure of which may be represented as a resonance hybrid

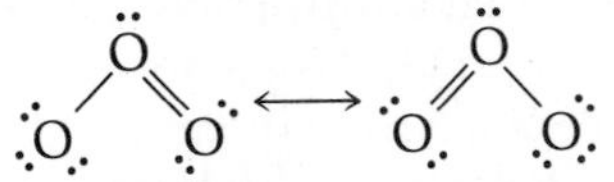

This unstable molecule (as evidenced by its endothermic heat of formation from $O_2$, $\Delta H_f = +34$ kcal/mole) can be generated by the passage of an electric discharge through ordinary $O_2$, and its pungent odor can often be detected in the vicinity of electrical equipment. It is also formed in limited quantities in the upper atmosphere by the action of ultraviolet radiation from the sun on $O_2$. Its presence in the upper atmosphere shields the earth and its creatures from exposure to intense, and harmful, ultraviolet light. In recent years alarm has grown that chlorofluorocarbon propellants such as $CCl_2F_2$ in aerosol products are reacting with ozone and depleting this shield. In fact, steps are underway that will ultimately ban this use of chlorofluorocarbons.

Ozone is currently believed to be one of the major constituents of photochemical smog, a type of air pollution that has accompanied the increased use of the internal combustion engine.

When an automobile engine operates, it gives off a mixture of gases in its exhaust, including unburned hydrocarbons from the fuel and small amounts of NO produced from the oxidation of atmospheric $N_2$. The series of reactions that produce the smog are thought to include the following:

1. Oxidation of NO by atmospheric oxygen

$$2NO + O_2 \longrightarrow 2NO_2$$

The brown color of $NO_2$ is often seen and the irritating effect of $NO_2$ on the nasal passages and lungs is often felt in the pollution haze that settles over large cities.

2. A photochemical reaction in which a molecule of $NO_2$ absorbs a photon (energy $= h\nu$)

$$NO_2 + h\nu \longrightarrow NO + O$$

3. Reaction of atomic oxygen with molecular oxygen to give ozone

$$O + O_2 \longrightarrow O_3$$

4. A host of other reactions in which hydrocarbons react with $O_3$ and nitrogen oxides to produce a range of unpleasant products

A characteristic property of photochemical smog is its oxidizing properties, which result primarily from high concentrations of $O_3$.

The oxidizing power of ozone is being investigated as an alternative method of purifying drinking water instead of chlorination. It has been found that $Cl_2$ in drinking water is able to form chlorine compounds with some of the organic compounds that are also present in small amounts. It is feared that long-range toxic effects may occur with these chlorinated organics, a situation that can be avoided if ozone is used to kill bacteria instead of chlorine.

The existence of an element in more than one form, either as the result of differences in molecular structure, as with $O_2$ and $O_3$, or as a consequence of differences in the packing of atoms or molecules in the solid, is a phenomenon called **allotropism.** Oxygen is only one of several nonmetals that exist in different allotropic forms, although the phenomenon is not limited to the nonmetals since tin, an element with mostly metallic properties, exhibits a metallic lattice at high temperature and a nonmetallic lattice at low temperature.

If we next turn our attention to carbon, also a second-period element, we see that it must share four electrons to complete its octet. Now there is just no way for carbon to form a "quadruple bond," and a simple $C_2$ species is not stable under ordinary conditions. Instead, carbon tends to complete its octet in either of two ways so that two allotropic forms of elemental carbon are found. One of these is diamond. In diamond (see Chapter 7) each carbon atom is covalently bonded to four others located at the corners of a tetrahedron. Each of those atoms, in turn, is bonded to three more, and so on, as illustrated in Figure 19.1. In this fashion a three-dimensional network is created so that a diamond crystal (for example, a gem-quality diamond) consists of a huge number of carbon atoms covalently bonded together in

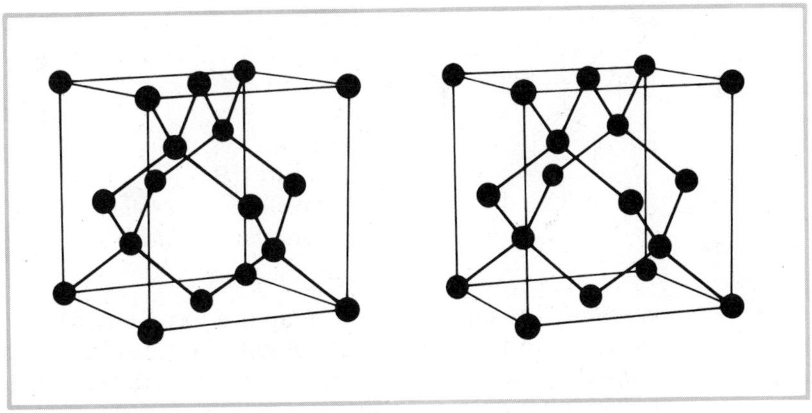

Figure 19.1
*Structure of diamond.*

one gigantic molecule. Since breaking a diamond crystal involves rupturing a very large number of covalent bonds, diamond is very hard.

In the second allotrope of carbon, graphite, the atoms are arranged in the form of hexagonal rings fused together in large planar sheets, perhaps somewhat reminiscent of chicken wire (Figure 19.2). Each carbon atom is surrounded by three nearest neighbors located at angles of 120° from one another, and the molecular framework is therefore based on $\sigma$ bonds produced by overlap of $sp^2$ hybrid orbitals on the carbon atoms. On each of the carbon atoms throughout the entire structure there remains an unhybridized *p* orbital, each containing one electron, and these *p* orbitals are situated ideally for $p\pi$-$p\pi$ overlap. The result, then, is a huge delocalized $\pi$ electron cloud extending across the graphite sheet above and below the plane of the carbon atoms.[1] The free movement of electrons in this $\pi$ cloud accounts for the electrical conductivity of graphite. An electron can be pumped into the cloud at one end of the sheet and another removed from the other end to give a net transfer of electrons through the solid.

In the total graphite structure the planes of carbon atoms are stacked in layers so that each carbon atom lies above another in every second layer, as shown in Figure 19.4. Within any given layer, adjacent carbon atoms are fairly close together (1.41 Å) while the spacing between successive planes is much greater (3.35 Å). The different planes of carbon atoms are not held together by covalent bonds but instead, by very much weaker London forces. As a result, the layers are able to slide over one another with relative

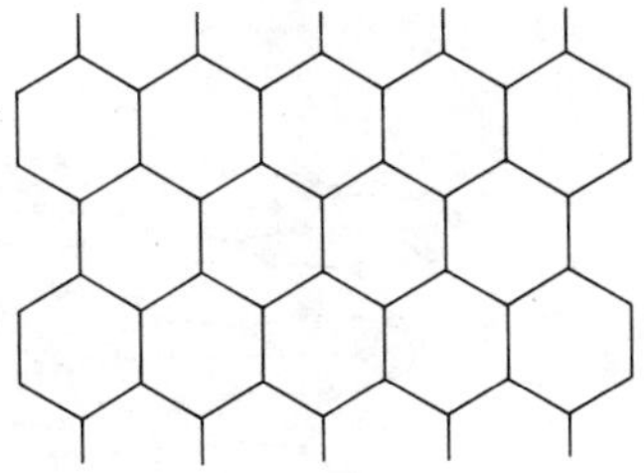

Figure 19.2
*Sigma-bond framework of a graphite sheet.*

[1] In the valence bond approach we imagine that adjacent carbon *p* orbitals overlap to form simple $\pi$ bonds and that the total bonding picture can be viewed as a composite of resonance structures such as that shown in Figure 19.3 on page 550.

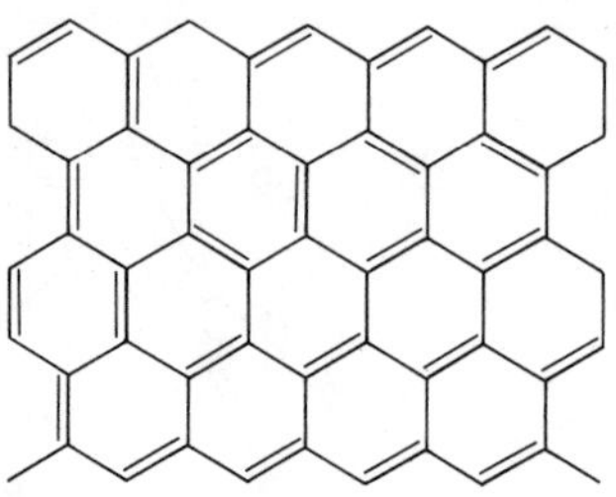

Figure 19.3
*One of the resonance structures for graphite.*

ease and, as you may know, graphite has many applications as a dry lubricant. As every school child is taught, the "lead" in a "lead pencil" is actually graphite, and when one writes with a pencil, the layers of graphite slide off one another onto the paper.

In graphite carbon exhibits multiple bonding, as do nitrogen and oxygen in their molecular forms. When we move now to the third and successive periods, a different state of affairs exists. Here we have atoms that prefer to form only single bonds with other atoms and that have relatively little tendency to participate in $p\pi$-$p\pi$ multiple bonds. The molecular structures of the free elements reflect this.

Chlorine, because it only needs to form a single covalent bond to complete its octet, exists as diatomic molecules. Bromine and iodine form diatomic $Br_2$ and $I_2$ for the same reason. The structures of the remaining nonmetals, however, are considerably more complex. With sulfur, for instance, we find that each atom must share two electrons to fill its valence shell, and it does so by forming two single bonds to *two different* sulfur atoms. These, in turn, must also be bonded to two separate S atoms and a —S—S—S—S— sequence is produced. Actually, in its most stable form the sulfur atoms are arranged in puckered, eight-membered $S_8$ rings having a crownlike structure, illustrated in Figure 19.5. Selenium too forms $Se_8$ rings in one of its allotropic forms. Selenium and tellurium also exist in a gray form in which there are long $Se_x$ and $Te_x$ chains, respectively.

Elemental sulfur exhibits rather interesting behavior when it is heated. At room temperature the most stable allotrope of sulfur contains the $S_8$ rings packed into a rhombic crystal structure. If the sulfur is melted and allowed to cool and solidify slowly, a second allotrope is produced in which the $S_8$ rings are stacked in a monoclinic crystal structure. The monoclinic form, however, is stable only above 95.5°C and, when allowed to stand at

Figure 19.4
*Total graphite structure.*

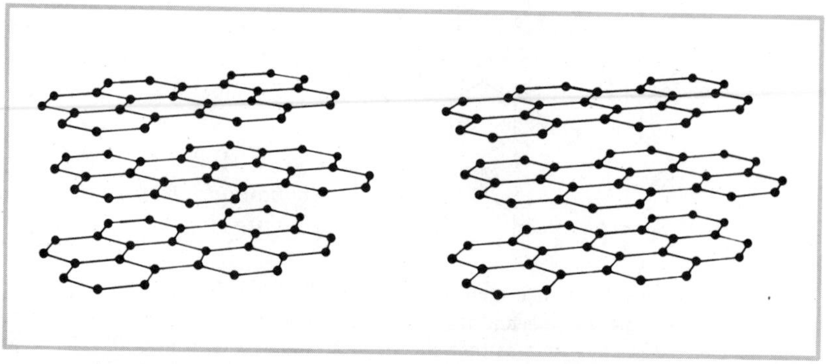

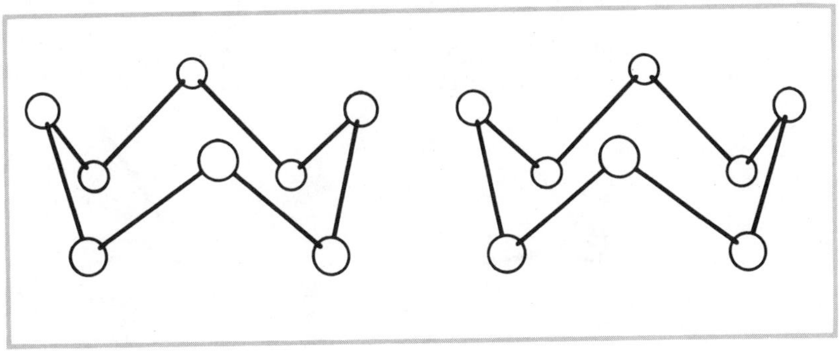

Figure 19.5
*The structure of the $S_8$ ring.*

room temperature, gradually reverts to the rhombic modification. The reverse transformation (rhombic → monoclinic) can be brought about too if the rhombic form is held at a temperature above 95.5°C, although the rate of transformation is slow. Usually when sulfur is heated, the rhombic form melts before it has an opportunity to revert to the monoclinic.

Liquid sulfur, at its melting point, is yellow, relatively nonviscous, and is composed primarily of $S_8$ rings in the random orientations that characterize a liquid. As the liquid is heated to higher temperatures, it begins to thicken and darken, becoming a dark red, molasseslike substance. At still higher temperatures it thins out again and becomes lighter in color, and finally it boils at 445°C. This rather unusual behavior is explained in the following way.

As the temperature of the liquid sulfur is raised, thermal motion, transmitted into molecular vibrations, begins to break S—S bonds, and the $S_8$ rings begin to open, producing $S_8$ chains in which the end S atoms each have one unpaired electron.

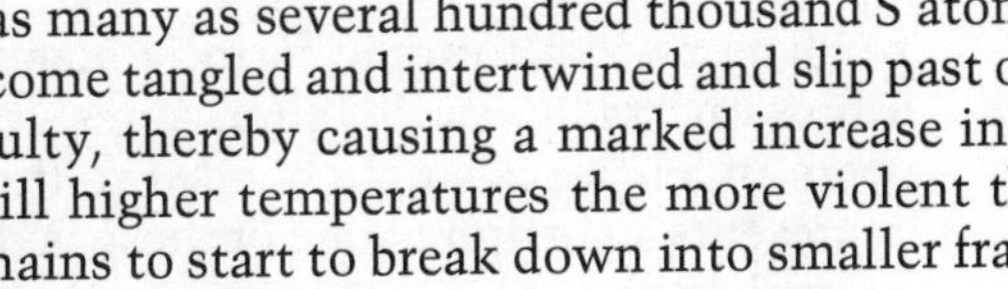

When one terminal S atom encounters another, a single bond can be created by the pairing of electrons, and the $S_8$ chains begin to couple together to produce first $S_{16}$, then $S_{24}$, $S_{32}$, $S_{40}$, and so on, until extremely long chains possessing perhaps as many as several hundred thousand S atoms are produced. These chains become tangled and intertwined and slip past one another only with great difficulty, thereby causing a marked increase in the viscosity of the liquid. At still higher temperatures the more violent thermal motions cause the long chains to start to break down into smaller fragments, and the liquid becomes mobile again.

It's interesting to note that if the thickened liquid sulfur is cooled rapidly, by being poured into cold water, for example, the sulfur atoms do not have an opportunity to be transformed into $S_8$ rings and, as a result, a supercooled liquid called **amorphous sulfur** (also referred to as plastic sulfur) is produced that has many of the elastic properties of rubber. When allowed to stand, the $S_x$ chains of the amorphous sulfur gradually revert to the more thermodynamically stable $S_8$ rings of the rhombic form.

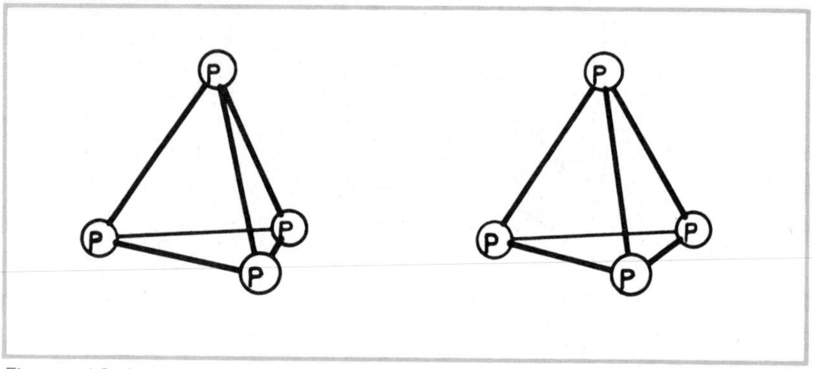


Figure 19.6
*The structure of white phosphorus, $P_4$.*

Let's move now to the left and examine the element phosphorus. Being in Group VA, an atom of phosphorus has three unpaired electrons,

$$\cdot \overset{\cdot}{\underset{\cdot\cdot}{\mathrm{P}}} \cdot$$

Therefore, to achieve a noble gas structure it must acquire three more. Since there is little tendency for this element to form multiple bonds, as nitrogen does when it forms $N_2$, the octet is completed by the formation of three single covalent bonds to three different phosphorus atoms.

The simplest elemental form of phosphorus is a waxy solid called white phosphorus consisting of $P_4$ molecules in which each P atom lies at the corner of a tetrahedron, as illustrated in Figure 19.6. In this structure we see that each P atom is nicely bound to three others. This particular allotrope of phosphorus is very reactive because of the highly strained P—P—P bond angle of 60°. White phosphorus ignites spontaneously in air to produce the oxide and, for this reason, it is used in military incendiary devices. You've probably seen movies in which exploding phosphorus shells produce arching showers of smoking particles.

A second allotrope that is much less reactive is red phosphorus and, at the present time, its structure is unknown, although it has been suggested that it contains $P_4$ tetrahedra linked at the corners. Red phosphorus is used in explosives and mixed with fine sand on the striking surface used to light matches. In the latter case, friction caused by the match being drawn across the surface ignites the phosphorus, which in turn ignites the ingredients in the match head.

The third allotropic form is black phosphorus, formed by heating the white variety at very high pressures. It has a layer structure in which each P atom in a layer is singly bonded to three others (Figure 19.7). This drawing represents a portion of one of these layers (each consisting of two planes of P atoms). In the solid the layers are stacked upon each other with only weak London forces between them. Consequently, black phosphorus looks and behaves physically much like graphite. Like red phosphorus, it is quite unreactive.

The elements below phosphorus, arsenic, and antimony are also able to form somewhat unstable yellow allotropic forms containing $As_4$ and $Sb_4$ molecules. The most stable modifications have a metallic appearance with structures similar to black phosphorus.

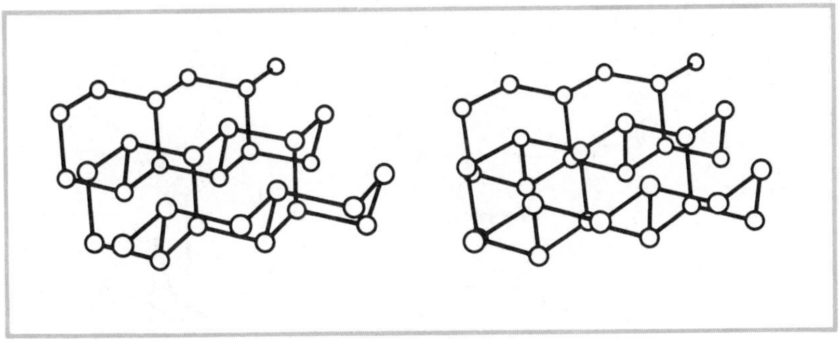

Figure 19.7
*The structure of black phosphorus.*

Finally, we look at the heavier elements in Group IVA, silicon, and germanium. These elements each must acquire, through sharing, four electrons, and they do so by the formation of single bonds. As a result, Si and Ge (as well as the low-temperature nonmetallic form of tin) have the diamond structure shown in Figure 19.1. Since silicon and germanium do not appear able to enter into $p\pi$-$p\pi$ bonding, graphitelike structures do not exist.

There is still one element whose structure we have not considered, namely boron. This period 2 element, found in Group IIIA, is quite unlike any of the others, since there is no simple way for it to complete its valence shell. Sharing electrons in such a way as to give three ordinary single bonds still would leave each boron atom with only six electrons around it; as a result, there is no easy way to understand its structure.

Boron exists in several different crystalline forms, each of which is characterized by clusters of 12 boron atoms located at the vertices of an icosahedron (a 20-sided geometric figure) as shown in Figure 19.8*a*. Each boron atom within a given cluster is equidistant from five others and, in the solid, each of these is also joined to yet another boron atom outside the cluster (Figure 19.8*b*). The electrons available for bonding are therefore delocalized to a large extent over many boron atoms.

The linking together of $B_{12}$ units produces a large three-dimensional network solid that is very difficult to break down. Boron, therefore, is very hard (it is the second hardest element) and has a very high melting point (about 2200°C).

## 19.3 OXIDATION NUMBERS

The oxidation-number concept is useful in coordinating the descriptive chemistry of the metals because most compounds of the metals are appreciably ionic and the metal atom therefore carries a positive charge that is at least approximated by the oxidation number assigned to it. With many (if not most) of the compounds of the nonmetals and metalloids,[2] however, this is not the case, particularly in those instances where the element, because of the rules of assigning oxidation numbers, exists in a high positive oxidation state. Thus sulfur in $SO_4^{2-}$ certainly does not carry a +6 charge, as its oxidation number would imply. As a result, oxidation numbers are only of somewhat limited usefulness in discussing and comparing the chemistries of the

[2] For the remainder of this chapter we shall not make the formal distinction between metalloids and nonmetals. The term nonmetal will be used to refer to any element that is clearly not a metal.

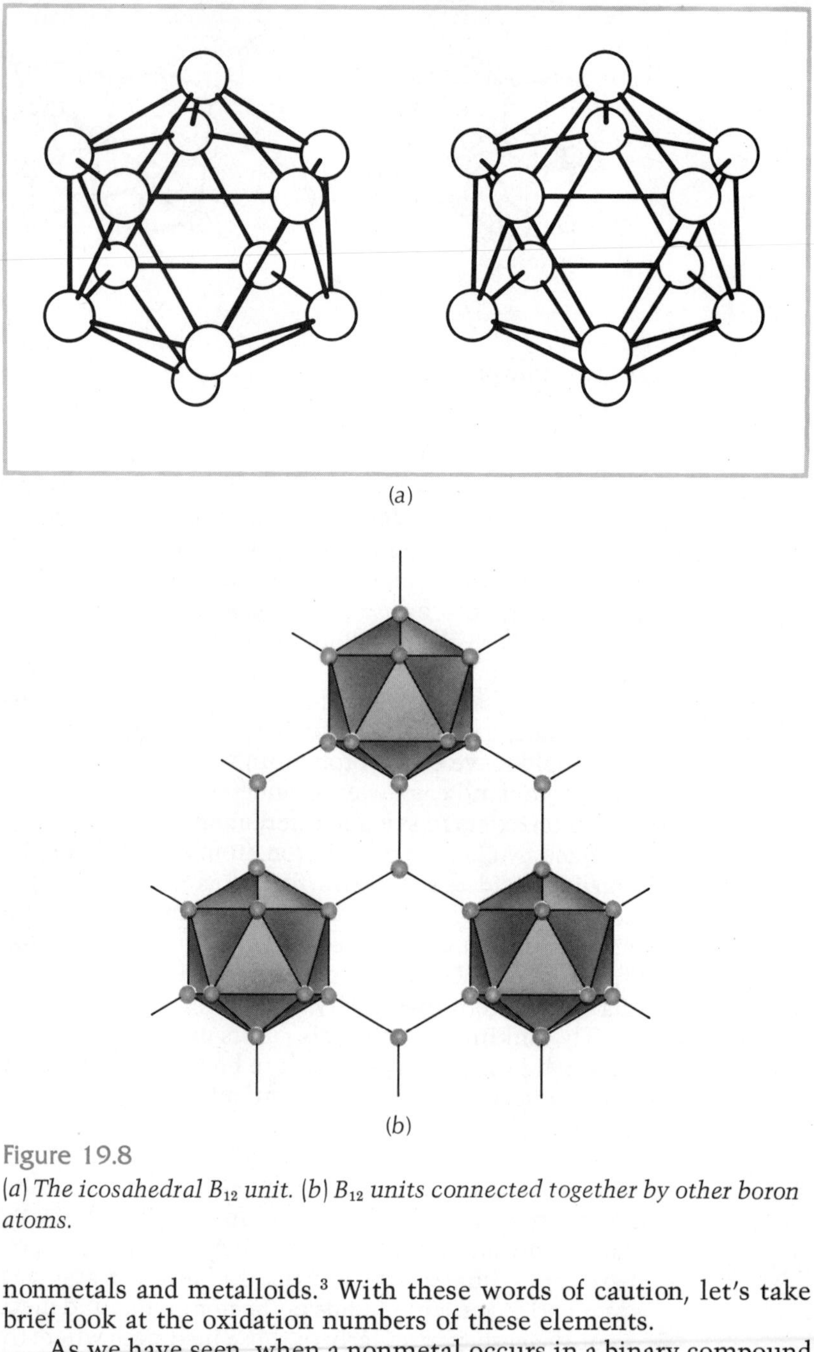

(a)

(b)

Figure 19.8
*(a) The icosahedral $B_{12}$ unit. (b) $B_{12}$ units connected together by other boron atoms.*

nonmetals and metalloids.[3] With these words of caution, let's take a very brief look at the oxidation numbers of these elements.

As we have seen, when a nonmetal occurs in a binary compound with a metal, such as NaCl or $Mg_2Si$, the oxidation number of the nonmetal is negative. These compounds are largely ionic in character, and the oxidation number of the nonmetal corresponds to the charge on the anion. We saw in Chapter 4 that this anionic charge is determined by the number of electrons that must be acquired by the element in order to achieve a noble gas struc-

[3] They still, of course, retain their usefulness when we wish to discuss redox reactions involving their compounds.

ture. As a result, we find that the halogens pick up one electron to form anions with a −1 charge (for example, $F^-$, $Cl^-$, $Br^-$, $I^-$). In like manner the elements in Group VIA acquire two electrons and form anions with a charge of −2 (for example, $O^{2-}$, $S^{2-}$), while those in Group VA gain three electrons to produce anions with a −3 charge (for example, $N^{3-}$ in $Li_3N$), and so forth.

When a nonmetal is combined with another nonmetal, its oxidation state is determined by whether the other nonmetal has a higher or lower electronegativity. The basic rules of assigning oxidation numbers thus apply and, because of the myriad of ways that the nonmetals combine with each other, it is a property of nonmetals that they exhibit multiple oxidation states. Nitrogen, for example, occurs in every integral oxidation state between −3 and +5, as seen in Table 19.2. Fractional oxidation numbers also occur among the nonmetals, as we saw in Chapter 4 where the $S_4O_6^{2-}$ ion, containing sulfur with an oxidation number of $+\frac{5}{2}$, was used as an example. In every case that we shall consider, the highest positive oxidation state equals the group number. With the halogens (Group VIIA) it is +7, as found in the $ClO_4^-$ ion, while the Group VIA elements show a maximum oxidation number of +6 (S in $SO_4^{2-}$). In a similar fashion, for Group VA the maximum is +5 (N in $NO_3^-$) and in Group IVA it is +4 (Si in $SiO_2$).

## 19.4 NONMETAL HYDRIDES

Hydrogen is quite a special element, possessing only one electron in its 1*s* orbital. It can achieve the noble gas structure of helium either by acquiring an additional electron to give the hydride ion, $H^-$, as it does when it reacts with the very active metals, or it can complete its valence shell through covalent bond formation. Because of its relatively high electronegativity, it forms covalent bonds with the nonmetals. As a class, binary compounds with hydrogen are called hydrides, whether the hydrogen exists in a positive or negative oxidation state.

Except for the very special case of boron, which we shall treat in some detail later, and the hydrogen difluoride ion $HF_2^-$, in which a H atom lies equidistant between two F atoms (that is, [F—H—F]$^-$), hydrogen is capable of binding to only one other atom by way of electron sharing, because it seeks only one additional electron. As a result, the structural chemistry of the simple hydrides is rather straightforward and, perhaps, even somewhat

Table 19.2
*Oxidation states of nitrogen*

| Oxidation State | Example | Preparative Reaction |
|---|---|---|
| −3 | $NH_3$ (ammonia) | $N_2 + 3H_2 \longrightarrow \mathbf{2HN_3}$ |
| −2 | $N_2H_4$ (hydrazine) | $2NH_3 + NaOCl \longrightarrow \mathbf{N_2H_4} + NaCl + H_2O$ |
| −1 | $NH_2OH$ (hydroxylamine) | $NaNO_2 + NaHSO_3 + SO_2 + H_2O \longrightarrow 2NaHSO_4 + \mathbf{NH_2OH}$ |
| 0 | $N_2$ | $NH_4NO_2\ (s) \xrightarrow{\Delta} \mathbf{N_2}\ (g) + 2H_2O\ (g)$ |
| +1 | $N_2O$ (nitrous oxide) | $NH_4NO_3\ (s) \xrightarrow{\Delta} \mathbf{N_2O}\ (g) + 2H_2O\ (g)$ |
| +2 | NO (nitric oxide) | $4NH_3 + 5O_2 \longrightarrow \mathbf{4NO} + 6H_2O$ |
| +3 | $N_2O_3$ (dinitrogen trioxide) | $NO + NO_2 \xrightarrow{-20°C} \mathbf{N_2O_3}$ |
| +4 | $NO_2$ (nitrogen dioxide) | $2NO + O_2 \longrightarrow \mathbf{2NO_2} \xrightarrow{cool} \mathbf{N_2O_4}$ |
| +5 | $HNO_3$ (nitric acid) | $3NO_2 + H_2O \longrightarrow \mathbf{2HNO_3} + NO$ |

Table 19.3
*Catenation among nonmetal hydrides*

| | | | | | |
|---|---|---|---|---|---|
| Group IVA | $CH_4$<br>$C_2H_6$<br>$C_3H_8$<br>.<br>.<br>.<br>$C_nH_{2n+2}$<br>+<br>many others | $SiH_4$<br>$Si_2H_6$<br>$Si_3H_8$<br>.<br>.<br>.<br>$Si_6H_{14}$ | $GeH_4$<br>$Ge_2H_6$<br>$Ge_3H_8$ | $SnH_4$<br>$Sn_2H_6$ | |
| Group VA | $NH_3$<br>$N_2H_4$ | $PH_3$<br>$P_2H_4$ | $AsH_3$ | $SbH_3$ | $BiH_3$ |
| Group VIA | $H_2O$<br>$H_2O_2$ | $H_2S$<br>$H_2S_2$<br>$H_2S_n$ $(n = 1 - 6)$ | $H_2Se$ | $H_2Te$ | $H_2Po$ |

mundane. With the halogens it forms compounds with the general formula HX (e.g., HF, HCl). Similarly, we find, quite expectedly, that the Group VIA elements form molecules of general formula $H_2X$ (e.g., $H_2O$, $H_2S$), the elements of Group VA form $H_3X$ (usually written $XH_3$, for example, $NH_3$, $PH_3$), and those in Group IVA form $H_4X$ (or $XH_4$, for example, $CH_4$, $SiH_4$). The geometries of these molecules are all readily predicted by the electron-pair repulsion theory discussed in Section 17.6.

In addition to these simple hydrides containing a single atom of the nonmetal, there are others that possess two or more nonmetal atoms. Some examples are given in Table 19.3. All of these compounds are characterized by nonmetal atoms of the same element linked directly to one another, a phenomenon called **catenation.** Thus hydrogen peroxide, $H_2O_2$, has the Lewis structure

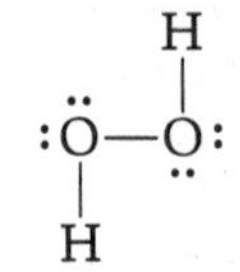

Similarly, we have others such as[4]

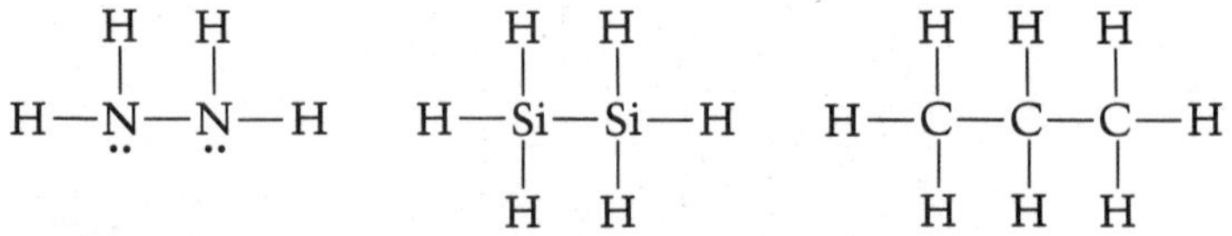

**hydrazine** **disilane** **propane**

The ability of nonmetals to form compounds in which they bond to other like atoms varies greatly. You will notice, for example, that in Group VIA only oxygen and sulfur form such compounds. In Group VA we find that both nitrogen and phosphorus catenate, but the chain length seems to

[4] The halogens do not exhibit this property except in the simple elemental state such as $F_2$, $Cl_2$, etc., and in the ion, $I_3^-$.

be limited to two atoms. When we proceed to Group IVA, all of the elements, down to and including tin, exhibit this property and here we find chains containing three, four, and even more atoms. We also see that the tendency toward catenation generally decreases downward in a group, as evidenced by the trend toward shorter chains demonstrated by the heavier elements in Group IVA, Ge and Sn.

Of all of the elements, carbon has the greatest capacity to form bonds to itself. In fact, the broad area of organic chemistry is concerned entirely with hydrocarbons and compounds that are derived from them by substituting other elements for hydrogen. Organic compounds, then, are compounds in which the molecular framework consists primarily of carbon-carbon chains. The unique ability of carbon to form such diverse compounds containing these long, stable carbon chains is undoubtedly the reason why life has evolved around the element carbon instead of around another element such as silicon.

Catenation is a property that is not restricted to the nonmetal hydrides alone; it occurs in other compounds as well. Sulfur, for example, has a rather marked tendency to form bonds to other sulfur atoms, as we saw in our discussion of the free element. This carries over to its compounds too. For instance, if an aqueous solution containing $S^{2-}$ is heated with elemental sulfur, a series of polysulfide ions, $S_2^{2-}$, $S_3^{2-}$, . . . , $S_x^{2-}$, are formed. We might illustrate this kind of reaction as

Addition of strong acid to these solutions produces the corresponding hydrides ($H_2S_2$, $H_2S_3$, . . . , $H_2S_x$).

Another similar reaction occurs when an aqueous solution containing sulfite ion is heated with sulfur. This reaction produces **thiosulfate** ion,[5] $S_2O_3^{2-}$,

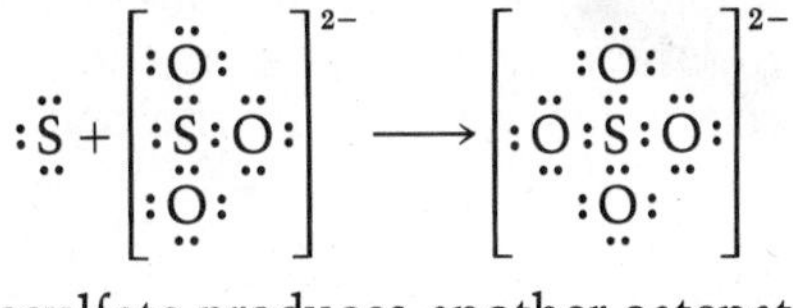

Oxidation of thiosulfate produces another catenated sulfur species, **tetrathionate** ion, $S_4O_6^{2-}$, having the structure

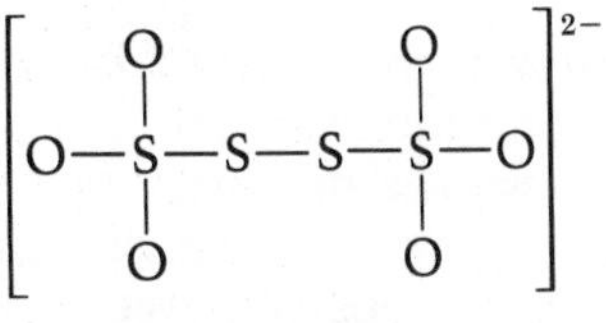

Other ions of similar structure are **dithionate** ($S_2O_6^{2-}$), **trithionate** ($S_3O_6^{2-}$), **pentathionate** ($S_5O_6^{2-}$), and **hexathionate** ($S_6O_6^{2-}$). Their structures are, respectively,

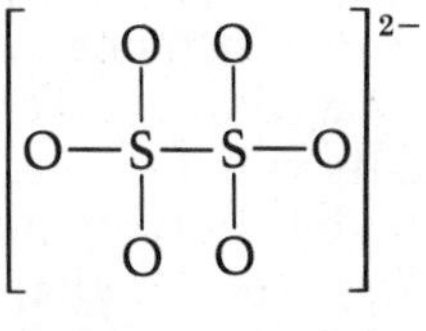

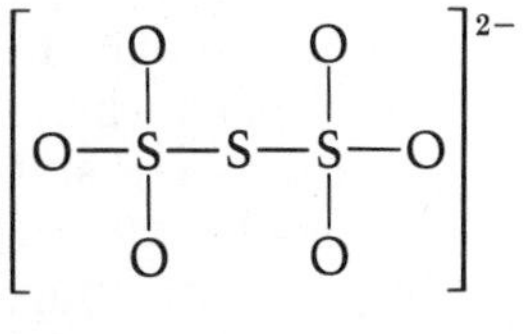

[5] The prefix thio in this case implies substitution of sulfur for oxygen. Thus the thiosulfate ion is a sulfate ion in which one oxygen atom has been replaced by a sulfur atom.

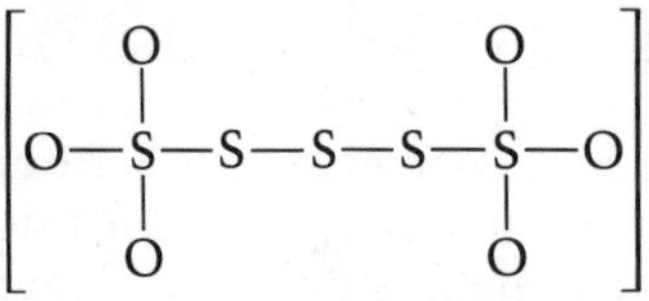 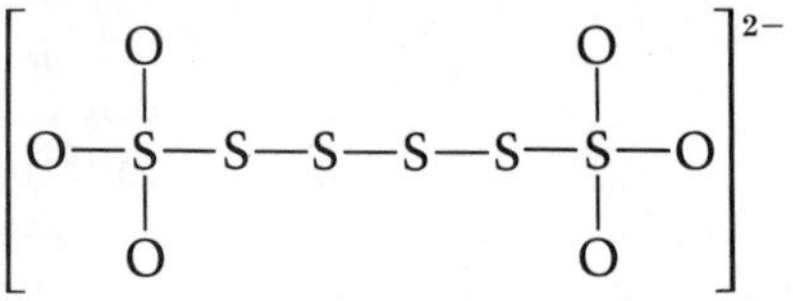

In addition to sulfur, some other common nonmetals also form catenated species that are not hydrides. Some examples are the following:

1. *Carbon.* For example, oxalic acid:

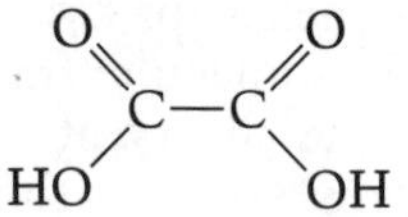

2. *Nitrogen.* For example, azides containing the ion $N_3^-$ (derived from hydrazoic acid, $HN_3$):

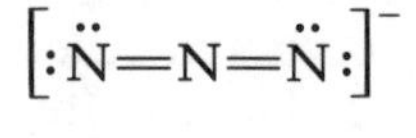

3. *Oxygen.* For example, peroxides such as peroxydisulfate ion, $S_2O_8^{2-}$, which may be considered to be derived, at least in a formal way, from $H_2O_2$,

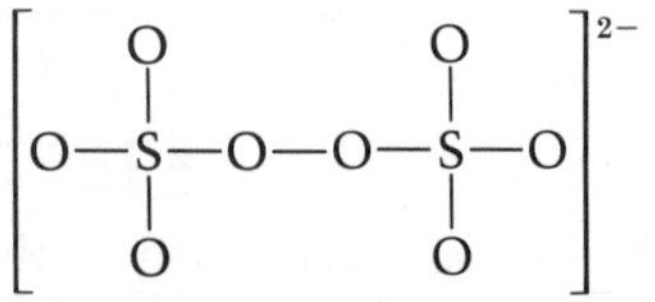

## 19.5 PREPARATION OF THE HYDRIDES

Nonmetal hydrides are produced as products of many different chemical reactions; however, we shall consider only two general methods of preparation here. One of these is the direct combination of the elements, as illustrated, for example, by the reaction of hydrogen with either chlorine

$$H_2 + Cl_2 \longrightarrow 2HCl$$

or with oxygen

$$2H_2 + O_2 \longrightarrow 2H_2O$$

However, this method is not applicable to all of the hydrides, as we can see by examining some of their thermodynamic properties shown in Table 19.4. Here we see that only the hydrides of the more active nonmetals possess negative free energies of formation. Those lying below the heavy line in the table have positive free energies of formation and from a practical standpoint cannot be prepared directly from the free elements. Instead an indirect procedure must be employed.

The rates of reaction toward hydrogen vary substantially among the nonmetals. In period 2, for instance, fluorine reacts immediately with hydrogen when they are placed in contact. On the other hand, $H_2$ and $O_2$ mixtures are stable virtually indefinitely, unless the reaction is initiated in some way, for example, by applying heat or introducing a catalyst.

Nitrogen is even less reactive than oxygen, not only toward hydrogen, but toward nearly all other chemical reagents as well. Presumably this is because of the high stability of the $N_2$ molecule that arises as a consequence of

Table 19.4
*Standard enthalpies and free energies of formation of nonmetal hydrides*

| | $XH_n$<br>$\Delta G_f^0$(kJ/mole)<br>$\Delta H_f^0$(kJ/mole) | | | |
|---|---|---|---|---|
| $BH_3$<br>Not stable,<br>simplest hydride is<br>$B_2H_6$ | $CH_4$<br>−74.9<br>−50.6 | $NH_3$<br>−46.0<br>−16 | $H_2O$<br>−242<br>−228 | HF<br>−271<br>−273 |
| | $SiH_4$<br>+34<br>+56.9 | $PH_3$<br>+5.4<br>+13 | $H_2S$<br>−21<br>−33 | HCl<br>−92.5<br>−95.4 |
| | $GeH_4$<br>(positive) | $AsH_3$<br>+66.5<br>+69.0 | $H_2Se$<br>+30<br>+16 | HBr<br>−36<br>−53.6 |
| | | $SbH_3$<br>(?) | $H_2Te$<br>+154<br>+138 | HI<br>+26<br>+2 |

its strong triple bond (the bond energy of $N_2$ is 946 kJ/mole, compared to 502 and 159 kJ/mole for $O_2$ and $F_2$, respectively).

The production of ammonia by reaction of $N_2$ and $H_2$ is undoubtedly one of the most important industrial chemical reactions, since virtually all useful nitrogen compounds, such as chemical fertilizers, for example, may be prepared from $NH_3$ in one way or another.

The Haber process, developed in Germany during World War I, employs the reaction

$$N_2\,(g) + 3H_2\,(g) \rightleftharpoons 2NH_3\,(g) \qquad \Delta H^0 = -92\text{ kJ}$$

As we can see, the reaction is exothermic and we predict, on the basis of the arguments presented in Chapter 12, that the greatest yield of $NH_3$ would be achieved if the reaction were permitted to come to equilibrium at low temperature and high pressure. However, the reaction takes place very slowly at ordinary temperatures, even in the presence of a catalyst (iron containing a small amount of oxide serves as a heterogeneous catalyst). Consequently, a high temperature is used, even though the quantity of $NH_3$ produced is somewhat reduced. The actual conditions that are employed are a pressure of approximately 1000 atm and a temperature of 400 to 500°C.

The second method of preparation of nonmetal hydrides involves the addition of protons, from a Brønsted acid, to the conjugate base of a nonmetal hydride, a reaction that we might depict as

$$X^{n-} + nHA \longrightarrow H_nX + nA^-$$

where $X^{n-}$ is the conjugate base of the hydride $H_nX$, and HA is the Brønsted acid. Let's look at some examples.

The hydrogen halides are commonly prepared in the laboratory by treating a halide salt with a nonvolatile acid such as sulfuric or phosphoric acid.

$$NaCl\,(s) + H_2SO_4\,(l) \longrightarrow HCl\,(g) + NaHSO_4\,(s)$$
$$NaCl\,(s) + H_3PO_4\,(l) \longrightarrow HCl\,(g) + NaH_2PO_4\,(s)$$

In these examples HCl is removed as a gas, which causes the reaction to proceed to completion.

With the heavier halogens, Br and I, sulfuric acid cannot be used because it is a sufficiently strong oxidizing agent to oxidize the halide ion to the free halogen. For example, when treated with $H_2SO_4$, $I^-$ reacts as follows:

$$2I^- + HSO_4^- + 3H^+ \longrightarrow I_2 + SO_2 + 2H_2O$$

Phosphoric acid, being a much weaker oxidizing agent than $H_2SO_4$, simply supplies protons to $I^-$, and HI can therefore be produced in a reaction analogous to the production of HCl above, that is,

$$NaI\ (s) + H_3PO_4\ (l) \longrightarrow HI\ (g) + NaH_2PO_4\ (s)$$

As we proceed from right to left across a period (for example, from fluorine toward carbon) we have seen that the acid strength of the $H_nX$ compounds decrease. Thus HF is a stronger acid than $H_2O$ which, in turn, is stronger than $NH_3$, and so forth. This means that the strengths of their corresponding conjugate bases *increase* from right to left ($C^{4-} > N^{3-} > O^{2-} > F^-$). As a result, the strength of the Brønsted acid required to react with the anion of the nonmetal to produce the hydride decreases. For example, the production of HF, whose conjugate base, $F^-$, is weak, requires a strong acid such as $H_2SO_4$. Oxide ion, on the other hand, is a much stronger base than $F^-$ and when treated with even a relatively weak source of protons, oxide ion gobbles them up to produce water.

$$O^{2-} + 2H^+ \longrightarrow H_2O$$

This is a reaction that we have seen before in Chapters 5 and 13.

Nitride ion, $N^{3-}$, is expected to be even a stronger base than $O^{2-}$. Therefore it is not surprising to find that $Mg_3N_2$ reacts with the weak acid, $H_2O$, to produce $NH_3$ in a reaction that we can interpret as a hydrolysis of the $N^{3-}$ ion.

$$Mg_3N_2 + 6H_2O \longrightarrow 3Mg(OH)_2 + 2NH_3$$

Metal carbides, which can be prepared by heating an active metal with carbon, also react with water in the same fashion. Aluminum carbide, for instance, which contains $C^{4-}$ ions, hydrolyzes according to the reaction,

$$Al_4C_3 + 12H_2O \longrightarrow 4Al(OH)_3 + 3CH_4$$

There are also carbides in which the carbon atoms exist in discrete pairs which we might write as $C_2^{2-}$, such as in $CaC_2$. This ion has the structure

$$\left[:C:::C:\right]^{2-}$$

and upon hydrolysis yields acetylene, $C_2H_2$,

$$CaC_2 + 2H_2O \longrightarrow Ca(OH)_2 + C_2H_2$$

Acetylene used in welding torches is prepared in this way.

This general method of preparation also extends to the third, fourth, and fifth periods, too, with the same trends in the strength of the Brønsted acid required to liberate the hydride. In period 3 we have these anions:

| Group | IV | V | VI | VII |
|---|---|---|---|---|
| Anion | $Si^{4-}$ | $P^{3-}$ | $S^{2-}$ | $Cl^-$ |

We again expect the anions to become increasingly basic as we move from right to left (from $Cl^-$ to $Si^{4-}$); therefore the strength of the Brønsted acid needed to protonate the anion decreases. To form HCl from NaCl, a strong acid is required. Sulfide ion, on the other hand, is sufficiently basic to be highly hydrolyzed in aqueous solution (as we found in Chapter 14), and solutions containing a soluble sulfide such as $Na_2S$ always have a strong odor of $H_2S$ because of the reaction,[6]

$$S^{2-} + 2H_2O \rightleftharpoons H_2S + 2OH^-$$

Many insoluble metal sulfides dissolve in acids with the evolution of $H_2S$.

Phosphides, like sulfides, also hydrolyze on contact with water. However, because the $P^{3-}$ ion is more basic than the $S^{2-}$ ion, the hydrolysis proceeds essentially to completion. Thus aluminum phosphide, AlP, reacts with water to produce phosphine, $PH_3$.

$$AlP + 3H_2O \longrightarrow Al(OH)_3 + PH_3$$

Moving left to Group IVA, we again find that a hydrolysis reaction serves to prepare silicon hydrides. A metal silicide such as $Mg_2Si$ (which can be formed by simply heating Mg and Si together) reacts with water to generate a mixture of silanes; $SiH_4$, $Si_2H_6$, $Si_3H_8$, etc., up to $Si_6H_{14}$.

The heavier nonmetals behave in much the same fashion as those above them. Thus $H_2Se$ and $H_2Te$, like $H_2S$, can be prepared by adding an acid to a metal selenide or telluride. Arsine, $AsH_3$, like phosphine, $PH_3$, is made by the hydrolysis of a metal arsenide such as $Na_3As$ or AlAs, and the germanes, $GeH_4$, $Ge_2H_6$, and $Ge_3H_8$ are produced by the action of dilute HCl on $Mg_2Ge$.

## 19.6 BORON HYDRIDES

We now come to a very interesting and unique series of hydrogen compounds, the boron hydrides. From its electronic structure we might expect boron to have chemical properties somewhat similar to those of aluminum (a metal), which is just below it in Group IIIA. The extremely small size of the $B^{3+}$ "ion," and its correspondingly high ionic potential, however, prevent it from forming ionic compounds and bonds formed to boron are always covalent.

Because of its three valence electrons, we might predict that the simplest boron hydride would be $BH_3$,

$$\begin{array}{c} H \\ \cdot\cdot \\ H:B:H \end{array}$$

While this compound has been observed, it is not stable. Instead, the simplest stable molecule has the formula $B_2H_6$, the structure of which has been determined to be that shown in Figure 19.9. If you look carefully at this structure, you will see that it appears to contain a total of eight B—H bonds. The $B_2H_6$ molecule, however, possesses only 12 valence electrons—sufficient to form only six ordinary single bonds.

In the diborane molecule, $B_2H_6$, there are two different kinds of bonds. Four of the bonds are ordinary B—H bonds, in the sense that they are formed from the overlap of an $sp^3$ orbital on boron with a 1*s* orbital on hydrogen. At two electrons apiece, these account for eight of the 12 valence electrons.

The other bonds correspond to those in which hydrogen atoms serve as bridges between the two boron atoms. We might view each of these bonds as

[6] Actually there is a two-step equilibrium involving both $HS^-$ and $H_2S$. See Chapter 14.

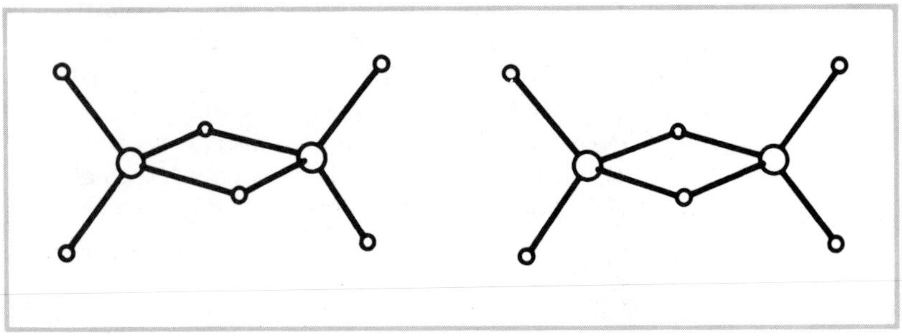

Figure 19.9
*The diborane molecule, $B_2H_6$. Large atom, boron; small atom, H.*

arising from the simultaneous overlap of one $sp^3$ hybrid orbital from each B atom, with the 1*s* orbital of the bridging hydrogen atom, as shown in Figure 19.10. This picture leads to two bonds, called **three-center bonds,** that each extend over three nuclei and that each contain a single pair of electrons. This produces a rather odd situation where three atoms in a three-center bond are held together by only two electrons.

We might note that valence bond theory, which always considers pairs of electrons shared between two atoms, cannot easily handle this, and other, boron hydrides. In fact, boron compounds containing three-center bonds are often spoken of as being electron-deficient because there does not appear to be, on the basis of our usual view of bonding, sufficient electrons to account for all of the bonds. Molecular orbital theory, on the other hand, is not embarrassed by this molecule since we simply have a case of a delocalized molecular orbital extending over three nuclei, not too much different in principle than the delocalized $\pi$ bonds in the $SO_3$ molecule discussed in Chapter 17.

## 19.7 GEOMETRIC STRUCTURES OF THE NONMETAL HYDRIDES

In general, we have seen that nonmetal hydrides are formed in such a way that the nonmetal achieves a noble gas configuration consisting of four electron pairs in its valence shell. We have seen before that these electron pairs tend to situate themselves at the vertices of a tetrahedron and therefore the molecular structures that we find for most of the hydrides are based primarily on the tetrahedron.

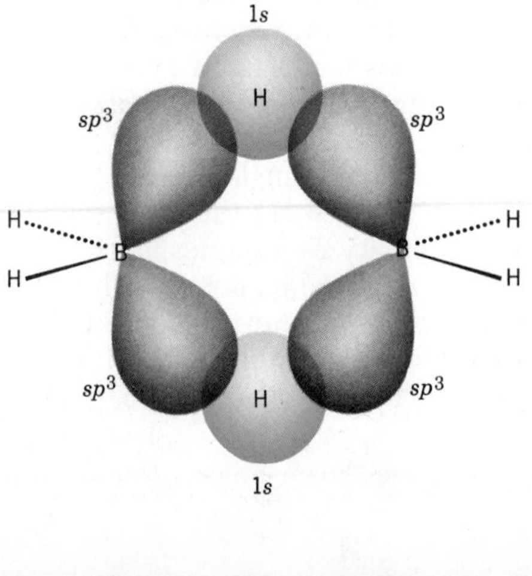


Figure 19.10
*Three-center bonding in $B_2H_6$.*

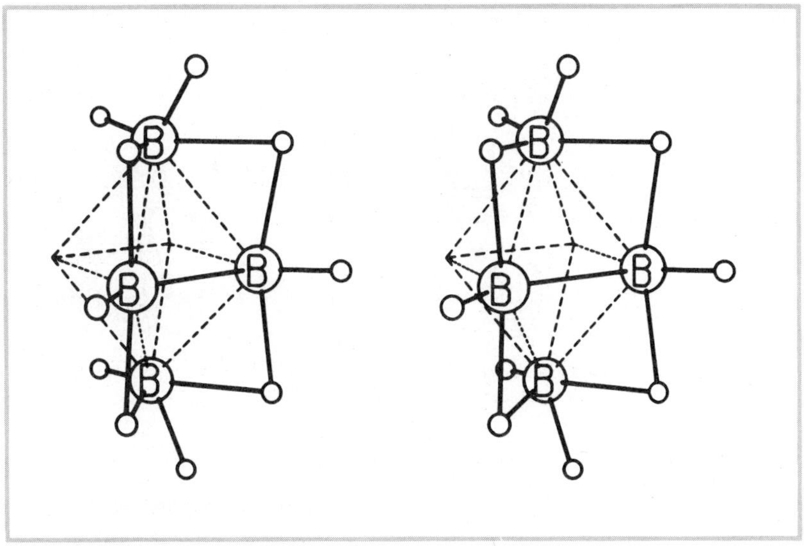


Figure 19.11
*Structure of $B_4H_{10}$.*

The boron hydrides offer us some interesting variations in geometry. With $B_2H_6$ there is an approximately tetrahedral arrangement of hydrogen atoms about each boron (Figure 19.9), and we can say that the total molecular structure is derived by the sharing of an edge between two tetrahedra.

There are, in addition to $B_2H_6$, more complex boron hydrides. Two of these are $B_4H_{10}$ and $B_5H_9$ whose structures are shown in Figures 19.11 and 19.12, respectively. In each case, notice that the boron atoms are found to occupy positions that correspond, at least roughly, to most (although not all) of the vertices of another familiar geometric figure, the octahedron.

Still other boron hydrides have structures in which the boron atoms sit at some of the vertices of an icosahedron. This geometric figure, you might recall, describes the structure of elemental boron (Figure 19.8). Examples of these more complex boron hydrides, including the $B_{12}H_{12}^{2-}$ ion, are shown in Figure 19.13.

Figure 19.12
*Structure of $B_5H_9$.*

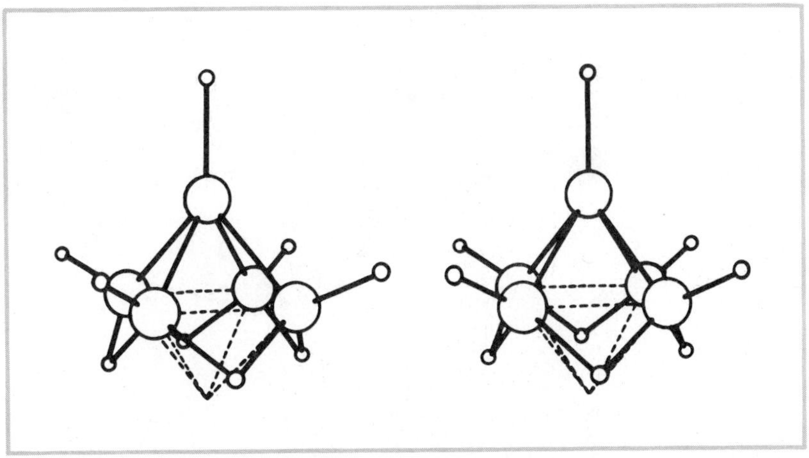

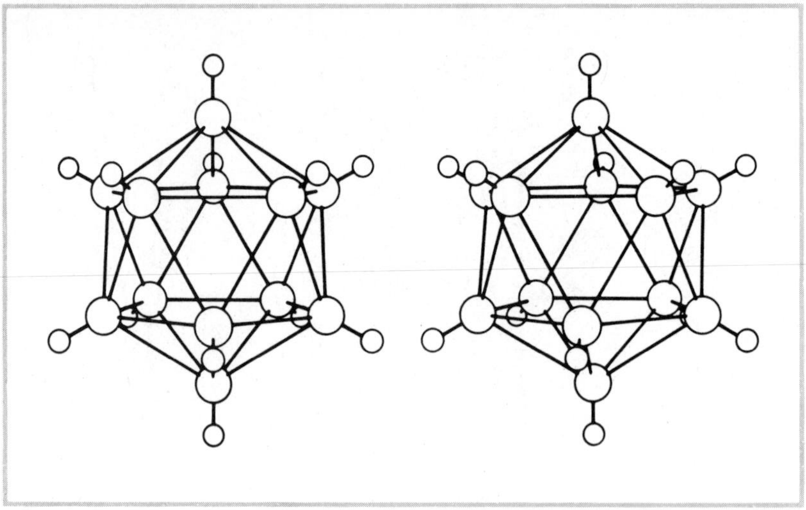

(a)

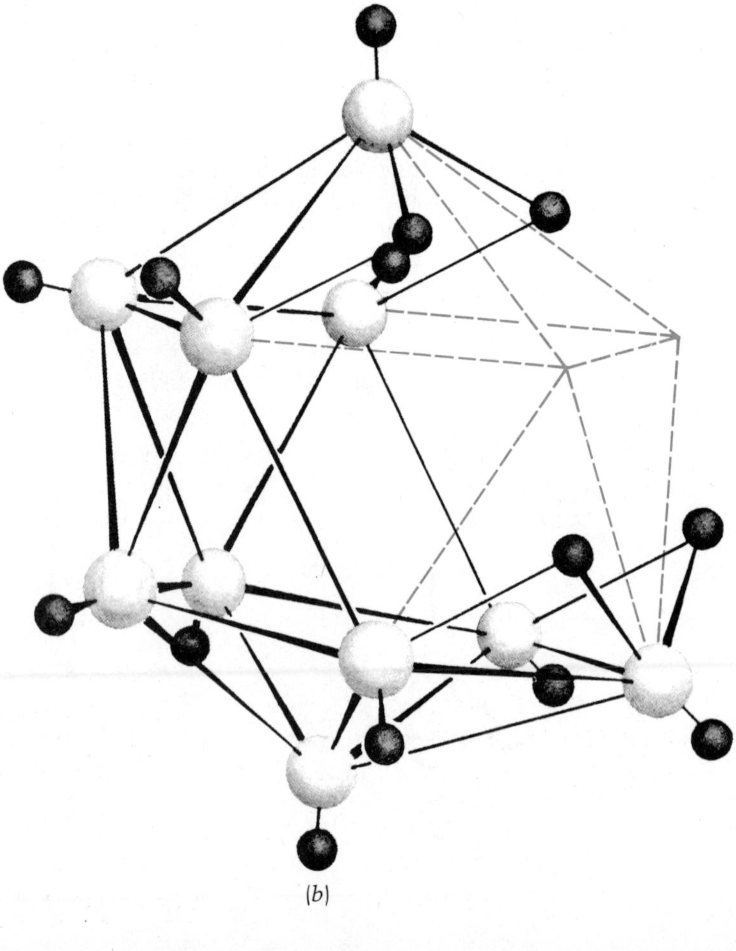

(b)

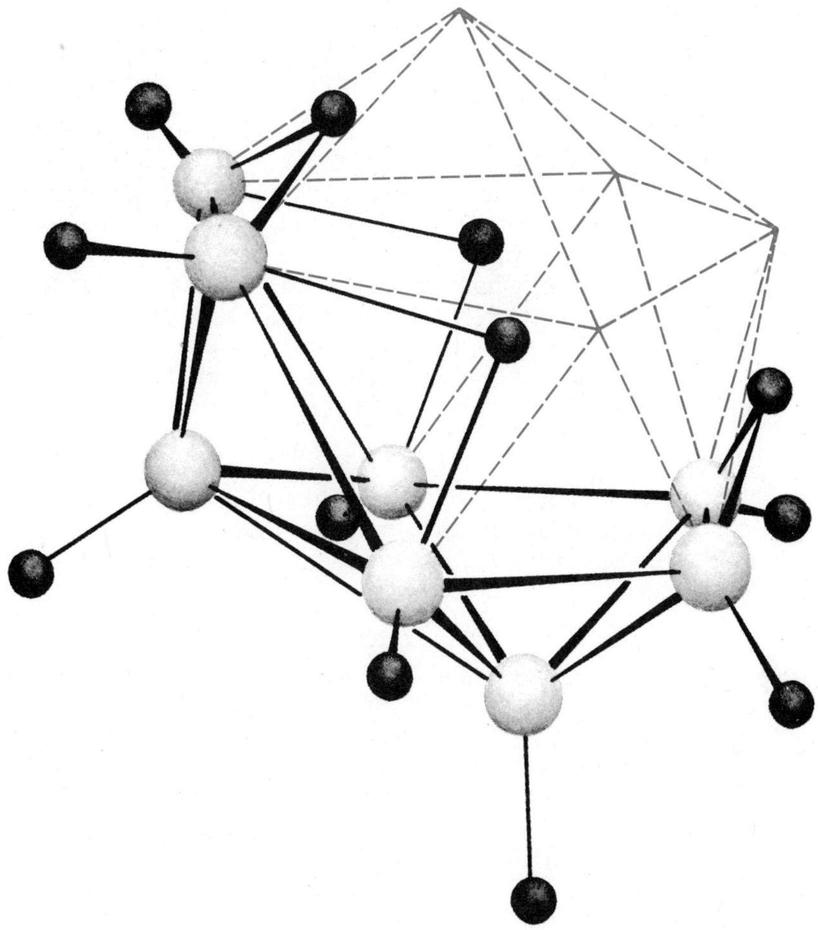

(c)

Figure 19.13

*The structures of (a) $B_{12}H_{12}^{2-}$ (large spheres = boron), (b) $B_{10}H_{14}$, and (c) $B_8H_{12}$. Note the similarity to the $B_{12}$ unit. (From Earl L. Muetterties,* The Chemistry of Borons and Its Compounds, *Copyright 1967, John Wiley & Sons, New York. Reprinted by permission.)*

## 19.8 OXYGEN COMPOUNDS OF THE NONMETALS

The oxides and those compounds derived from them, the oxoacids (acids containing oxygen) and oxoanions (anions containing oxygen), are among the most important compounds of the nonmetals. Here we find many of the common reagents encountered in the laboratory, for example, acids such as $H_2SO_4$ and $HNO_3$ and their salts, the sulfates and nitrates. In addition, most minerals are either oxides, such as $SiO_2$ (silica) and $Fe_2O_3$ (hematite), or contain oxoanions of nonmetals, for example, the carbonates (limestone) and silicates (asbestos, mica, and others).

We have already seen that nonmetal oxides exhibit acidic properties and that a relatively simple relationship exists between an oxide and the acid derived from it by reaction with water. Thus we have many reactions in which water combines with an oxide; for example,

$$SO_2 + H_2O \longrightarrow H_2SO_3$$
$$CO_2 + H_2O \longrightarrow H_2CO_3$$

You also will recall that neutralization of these acids gives their corresponding anions.

This relationship exists for many oxides, oxoacids, and oxoanions; however, there are also many cases in which the relationship is more a formality than an actual fact. For example, some oxides such as $SiO_2$ are insoluble and therefore do not react to give acids. In other instances the oxide cannot react with water to give an acid in which the nonmetal retains the same oxidation state. Nitric oxide, NO, is unreactive toward water while nitrogen dioxide **disproportionates** (undergoes an oxidation-reduction in which the same reactant is both oxidized *and* reduced).

$$3NO_2 + H_2O \longrightarrow 2HNO_3 + NO$$

In still other cases the parent acid of an anion either is unknown or cannot be obtained in the pure state. Carbonic acid, $H_2CO_3$, for example, decomposes as its aqueous solution is concentrated by evaporation of the solvent.

## 19.9 PREPARATION OF NONMETAL OXIDES

Table 19.5 contains a list of many of the oxides of the nonmetals. Here, once again, we find that there is no single preparative procedure that can be employed in every case because of the rather wide variety of properties of the nonmetals. We can, however, find some generalizations that apply in a reasonably large number of instances.

One method that can be employed in most cases is simply the direct union of the elements, as typified by the reactions:

$$S + O_2 \longrightarrow SO_2$$
$$C + O_2 \longrightarrow CO_2 \quad \text{(excess oxygen)}$$
$$C + \tfrac{1}{2}O_2 \longrightarrow CO \quad \text{(limited supply of oxygen)}$$
$$H_2 + \tfrac{1}{2}O_2 \longrightarrow H_2O$$

Table 19.5
*Typical oxides of the nonmetals*

| | | | | |
|---|---|---|---|---|
| Group III | $B_2O_3$ | | | |
| Group IV | CO<br>$CO_2$ | $SiO_2$ | $GeO_2$ | |
| Group V | $N_2O$<br>NO<br>$N_2O_3$<br>$NO_2$; $(N_2O_4)$<br>$N_2O_5$ | $P_4O_6$<br>$P_4O_{10}$ | $As_4O_6$<br>$As_2O_5$[a] | $Sb_4O_6$<br>$Sb_2O_5$[a] |
| Group VI | $O_2$<br>$O_3$ | $SO_2$<br>$SO_3$ | $SeO_2$<br>$SeO_3$ | $TeO_2$<br>$TeO_3$ |
| Group VII | $OF_2$<br>$O_2F_2$ | $Cl_2O$<br>$ClO_2$<br>$Cl_2O_7$ | $Br_2O$<br>$BrO_2$ | $I_2O_5$<br>$I_2O_7$ |

[a] Molecular structure unknown.

Table 19.6
*Thermodynamic properties of some nitrogen oxides*

| Oxide | $\Delta H_f^0$(kJ/mole) | $\Delta G_f^0$(kJ/mole) |
|---|---|---|
| $N_2O$ (*g*) | +81.5 | +104 |
| NO (*g*) | 90.4 | 86.8 |
| $NO_2$ (*g*) | 38 | 51.9 |
| $N_2O_4$ (*g*) | 9.7 | 98.3 |
| $N_2O_5$ (*g*) | 11 | 115 |

Not all oxides can be prepared in this manner; however. For example, in Table 19.6 we see that many of the oxides of nitrogen have positive free energies of formation and, therefore, from what we know of thermodynamics, they cannot be synthesized directly from the elements.[7] In these instances, and in others too, indirect methods of preparation are employed.

The indirect procedures are, expectedly, many in number. However, once more, a few generalizations can be made. In some cases an oxide can be prepared from a lower oxide by further reaction with oxygen. The synthesis of $SO_3$, for example, consists of catalytic oxidation of $SO_2$,

$$2SO_2 + O_2 \longrightarrow 2SO_3$$

This very thermodynamically favorable reaction ($\Delta G^0 = -140$ kJ) is slow under ordinary conditions but proceeds rapidly in the presence of a catalyst. Recall (p. 346) that this reaction is promoted in catalytic mufflers originally designed to speed up the conversion of CO to $CO_2$.

$$2CO + O_2 \longrightarrow 2CO_2$$

The combustion of CO is an important industrial reaction because CO is often used as a fuel. When white-hot carbon is treated with steam an endothermic reaction takes place producing a mixture of CO and $H_2$, both of which can burn. This mixture is called water gas.

$$C\ (s) + H_2O\ (g) \longrightarrow CO\ (g) + H_2\ (g)$$

Another technique that serves to produce oxides is the combustion of nonmetal hydrides. Methane, the chief constituent of natural gas, and other hydrocarbons, burn to produce $CO_2$ and $H_2O$ when an excess of $O_2$ is present.

$$\underset{\textbf{methane}}{CH_4} + 2O_2 \longrightarrow CO_2 + 2H_2O$$

$$\underset{\substack{\textbf{octane}\\\textbf{(gasoline)}}}{2C_8H_{18}} + 25O_2 \longrightarrow 16CO_2 + 18H_2O$$

When insufficient $O_2$ is available, as in an automobile engine, CO may be produced instead of $CO_2$.

[7] Since they also have positive $\Delta S_f^0$, it should be possible to prepare them at very high temperatures where $\Delta H - T\,\Delta S$, and hence $\Delta G$, is negative. This is the case with NO, which is formed in small amounts near 3000°C. The high-temperature production of NO in motor vehicle engines is a major source of urban air pollution.

A reaction of this general type, which is of great commercial importance, is the oxidation of ammonia. In this case a platinum catalyst is used and the reaction is

$$4NH_3 + 5O_2 \longrightarrow 4NO + 6H_2O$$

The NO formed in this reaction is readily oxidized further to produce $NO_2$,

$$2NO + O_2 \longrightarrow 2NO_2$$

which, as noted earlier, disproportionates when dissolved in water to yield $HNO_3$ and NO:

$$3NO_2 + H_2O \longrightarrow 2HNO_3 + NO$$

The commercial application of this sequence of reactions accounts for the major source of nitric acid and nitrates used in the manufacture of explosives, fertilizers, plastics, and many other useful substances. In fact, the development of this process in Germany by Wilhelm Ostwald, accompanied by the successful preparation of $NH_3$ from $N_2$ and $H_2$ by Haber, are said to have prolonged World War I, since the Allied blockade of Germany was unable to halt the German manufacture of munitions which had depended, prior to these processes, on the importation of nitrates from other countries.

Finally, another indirect method of obtaining nonmetal oxides makes use of oxidation-reduction reactions. For instance, when nitric acid serves as an oxidizing agent, the nitrate ion is reduced and, depending on conditions, nitrogen in any oxidation state can be produced. When concentrated nitric acid is used, the reduction product is frequently $NO_2$ while dilute solutions of $HNO_3$ often yield NO as the reduction product.

$$4HNO_3 + Cu \longrightarrow Cu(NO_3)_2 + 2NO_2 + 2H_2O \qquad \text{(concentrated)}$$
$$8HNO_3 + 3Cu \longrightarrow 3Cu(NO_3)_2 + 2NO + 4H_2O \qquad \text{(dilute)}$$

Similarly, hot concentrated sulfuric acid is a fairly potent oxidizing agent, the reduction product usually being $SO_2$; for example,

$$Cu + 2H_2SO_4 \longrightarrow CuSO_4 + SO_2 + 2H_2O$$

## 19.10 THE STRUCTURE OF NONMETAL OXIDES

The structures that are found for the nonmetal oxides are once again determined significantly by the ability of the nonmetal to form multiple bonds, in this case to oxygen atoms. In our earlier discussion of the structures of the free elements we saw that period 2 elements have a substantial tendency to enter into such bonding; therefore the oxides of the period 2 elements are simple monomeric species. Carbon, for example, forms CO and $CO_2$, which we can write as

$$:C{\equiv}O: \quad \text{and} \quad :\ddot{O}{=}C{=}\ddot{O}:$$

respectively.

With nitrogen there are NO and $NO_2$. Both of these have an odd number of electrons and therefore are not able to satisfy the octet rule, since at least one atom must be left with an odd number of electrons. The structure of the NO molecule is usually represented as a resonance hybrid of the dot structures

$$\cdot\ddot{N}{=}\ddot{O}: \quad \text{and} \quad :\ddot{N}{=}\ddot{O}\cdot$$

Molecular orbital theory gives a better picture of this molecule, which we might compare to the species $O_2^+$. These two (NO and $O_2^+$) are isoelectronic, that is, they have the same number of electrons; hence we might expect that electronically they would not be too different. The MO energy-level diagram for $O_2^+$, obtained by removing one of the antibonding electrons from $O_2$ (Figure 17.17), is shown in Figure 19.14*a*. The corresponding energy diagram for NO is found in Figure 19.14*b*.

The molecule $NO_2$ is another interesting species. Using electron dot formulas we can indicate its structure by two resonance forms, neither of which is capable of fully satisfying the octet rule.

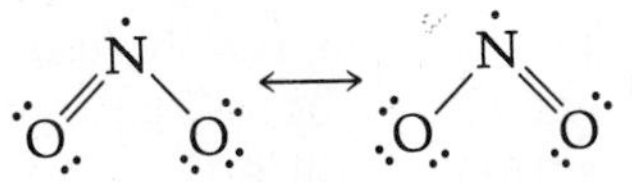

In each of these structures an unpaired electron resides on the nitrogen atom and in samples of $NO_2$, there is a dimerization equilibrium,

$$\underset{\text{(brown)}}{2NO_2} \rightleftharpoons \underset{\text{(colorless)}}{N_2O_4}$$

in which the $N_2O_4$ (dinitrogen tetroxide) molecule has the structure

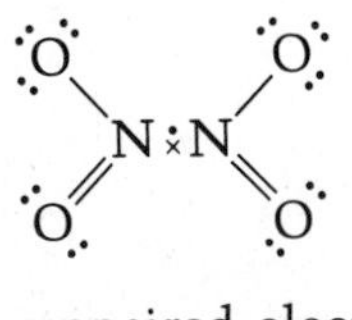

**(one of four resonance forms)**

In $N_2O_4$ the unpaired electrons of two $NO_2$ molecules have paired to form a covalent bond between the two nitrogen atoms. Liquid $N_2O_4$ is an effective oxidizing agent and has been used as the oxidant in liquid-fuel rocket engines.

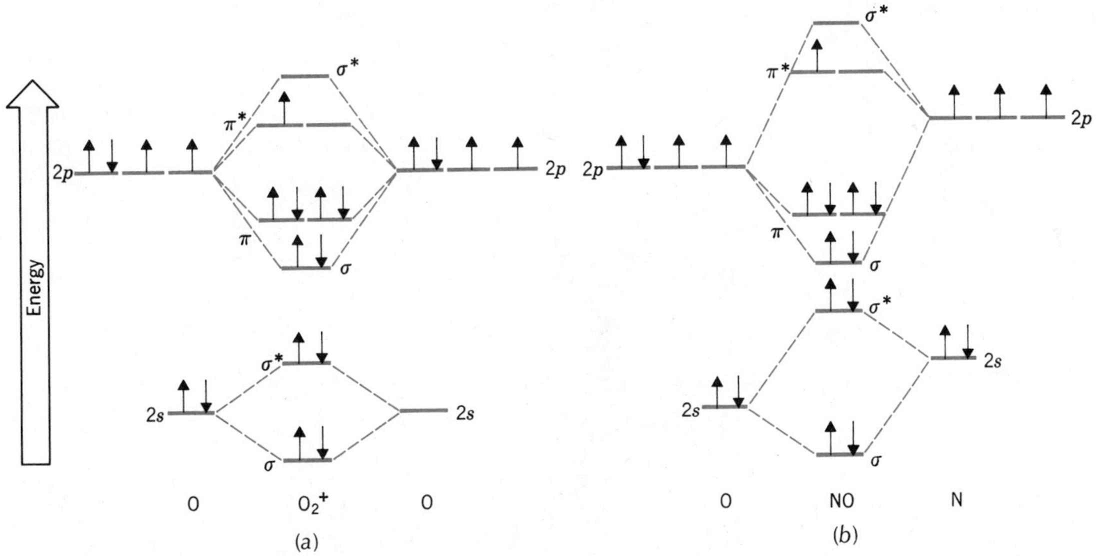


Figure 19.14
*Molecular orbital energy-level diagrams for $O_2^+$ and NO. (a) $O_2^+$. (b) NO. Note that the energies of the atomic orbitals on the isolated atoms are not the same.*

In period 3 there is considerably less tendency for the nonmetals to enter into multiple bonding. Sulfur, though, does appear able to do so, as evidenced by the oxides $SO_2$ and $SO_3$. Recall that these can be represented by resonance structures such as

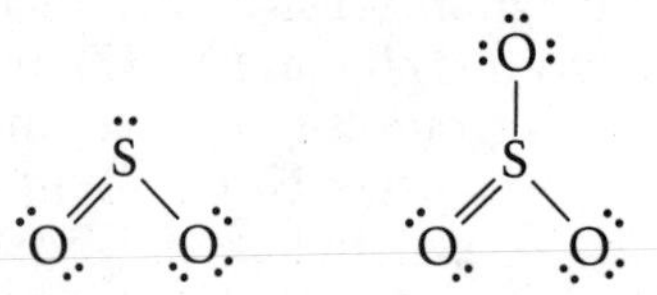

**(two structures) (three structures)**

With phosphorus there is even less tendency toward multiple bonding, and when phosphorus reacts with oxygen two oxides are formed, depending on reaction conditions,

$$P_4 + 3O_2 \longrightarrow P_4O_6$$
$$P_4 + 5O_2 \longrightarrow P_4O_{10}$$

The structures of these oxides bear a remarkably simple relationship to the elemental $P_4$ unit which, as we saw in Figure 19.6, is tetrahedral in shape. In Figure 19.15*a* is shown the structure of $P_4O_6$ in which we find an

Figure 19.15
*The oxides of phosphorus. (a) $P_4O_6$. (b) $P_4O_{10}$.*

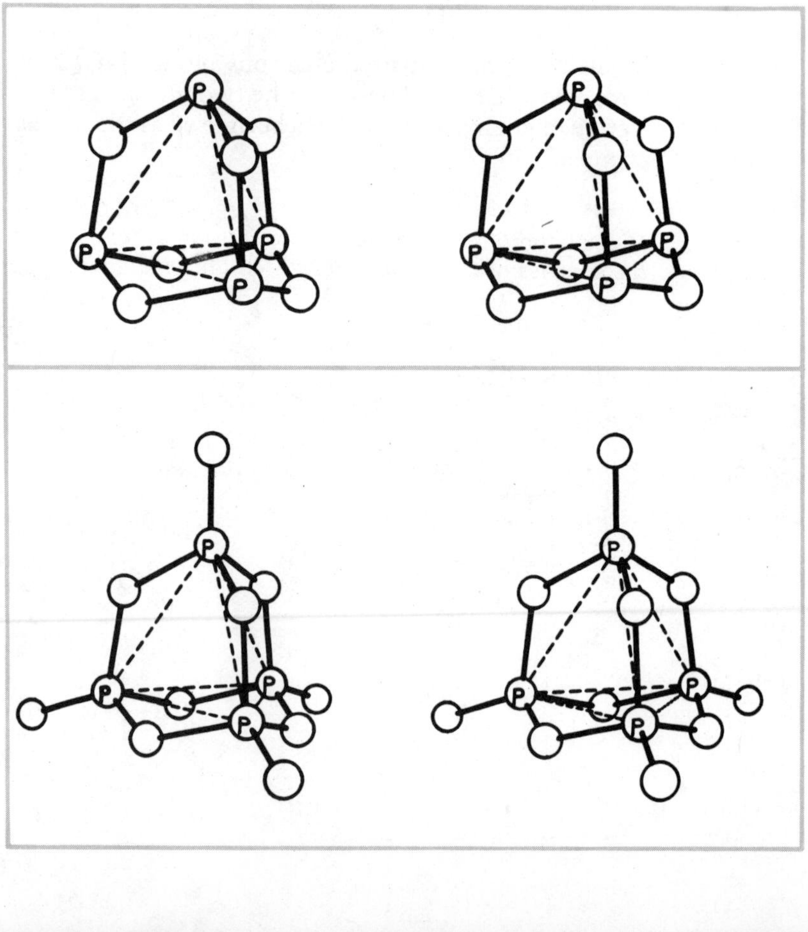

(a)

(b)

oxygen placed along each of the six edges of the $P_4$ tetrahedron. In the $P_4O_{10}$ molecule, Figure 19.15*b*, we find an additional oxygen atom attached to each of the P atoms at the vertices of the tetrahedron.

In the $P_4O_6$ molecule we can imagine that each of the three unpaired electrons of a phosphorus atom,

$$\cdot\underset{\cdot\cdot}{\dot{\mathrm{P}}}\cdot$$

is shared with one electron of an oxygen atom in a P—O—P bridge,

$$:\underset{\bullet}{\dot{\mathrm{P}}}\overset{\times}{\underset{\times}{}}\overset{\times\times}{\underset{\times\times}{\mathrm{O}}}\overset{\times}{\underset{\times}{}}\underset{\bullet}{\dot{\mathrm{P}}}:$$

with each P atom participating in three such bridges. Each P atom in $P_4O_6$ therefore has an unshared lone pair of electrons. In $P_4O_{10}$, four additional oxygen atoms become attached to these P atoms by coordinate covalent bonds so that in $P_4O_{10}$ each P atom is surrounded by four oxygen atoms in an approximately tetrahedral arrangement.

Silicon, in Group IVA, forms only one oxide, $SiO_2$. Silicon dioxide, while having the same empirical formula as carbon dioxide, is nevertheless quite different structurally since silicon has virtually no tendency to form Si—O multiple bonds. In quartz, and other forms of $SiO_2$ (for example, sand), each Si atom is surrounded tetrahedrally by four oxygen atoms, each of which is bound to another Si atom. The resulting structure, a portion of which is shown in Figure 19.16, is quite complex, extending in a network fashion in three dimensions.

A few features of this structure are worth noting. The overall hexagonal nature of quartz can be seen from Figure 19.17, which is the same as Figure 19.16 but with additional lines added for emphasis. Another interesting point is that the quartz structure is built up of —Si—O—Si—O—Si— spiral chains, one of which is outlined by the circle in Figure 19.17. In Figure 19.16 you are looking down the axis of these spiral chains; a side view of one of

Figure 19.16
*The structure of quartz. Solid spheres are silicon; open spheres are oxygen.*

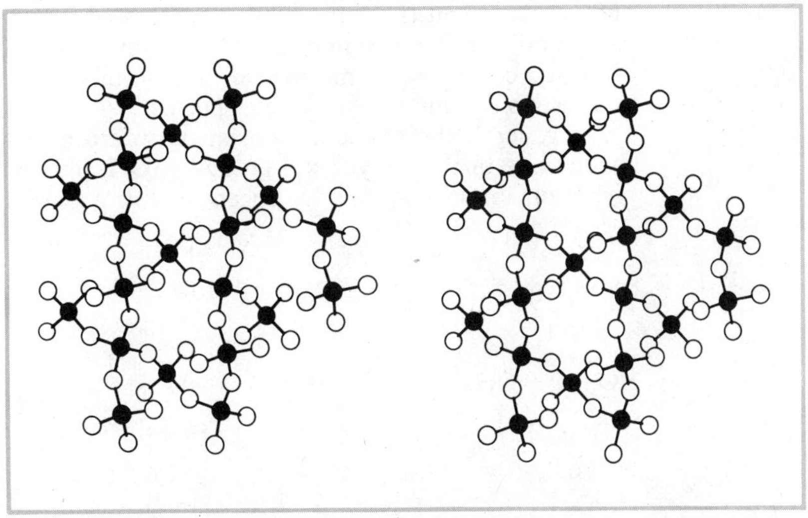

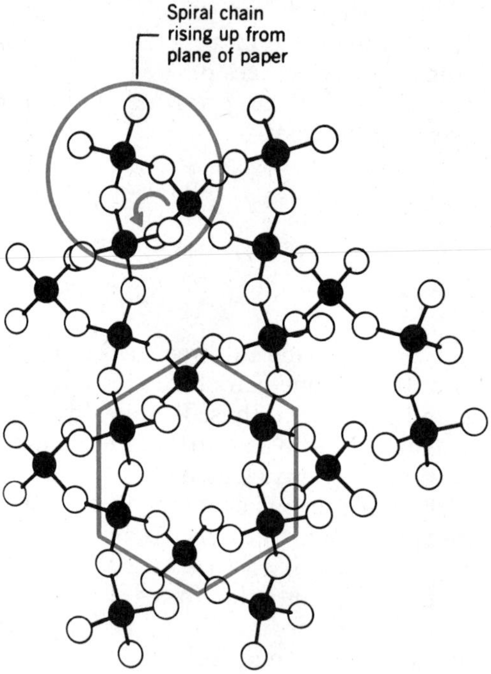


Figure 19.17
*The structure of quartz showing the hexagonal nature of the crystal.*

these chains containing three repetitions of the spiral is seen in Figure 19.18.

The existence of these spiral chains of $SiO_4$ tetrahedra in quartz impart an interesting property to the overall structure. Quartz crystals occur in two varieties that are similar to one another in the same way that your left and right hands are related, that is, as mirror images of each other. In quartz this is because of the two possible directions of rotation of the spiral chains, in one case clockwise and in the other counterclockwise, as shown in Figure 19.19.

## 19.11 SIMPLE OXOACIDS AND OXOANIONS

The oxoacids and their anions can be divided into two categories, simple monomeric acids and anions containing one atom of the nonmetal (Table 19.7), and complex polymeric acids and anions. We shall devote our attention in this section to the simple species.

In general, the oxoacids consist of an atom of nonmetal to which is bonded one or more **hydroxyl groups,** —OH, plus, perhaps, additional oxygen atoms not bonded to hydrogen, to give a generalized formula

$$XO_m(OH)_n$$

Some typical examples are given below.

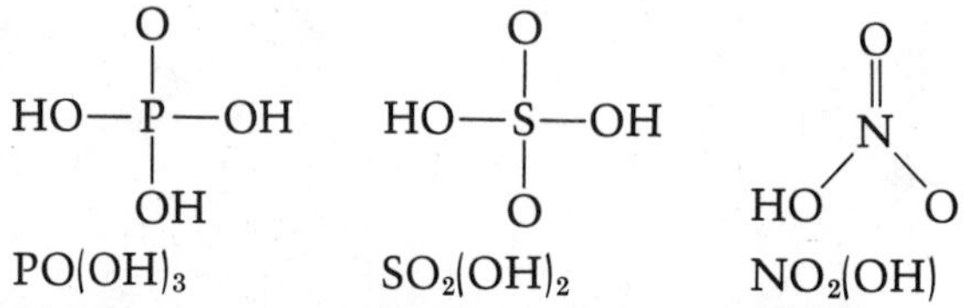


**(two resonance forms)**

| | | |
|---|---|---|
| $PO(OH)_3$ | $SO_2(OH)_2$ | $NO_2(OH)$ |
| $H_3PO_4$ | $H_2SO_4$ | $HNO_3$ |

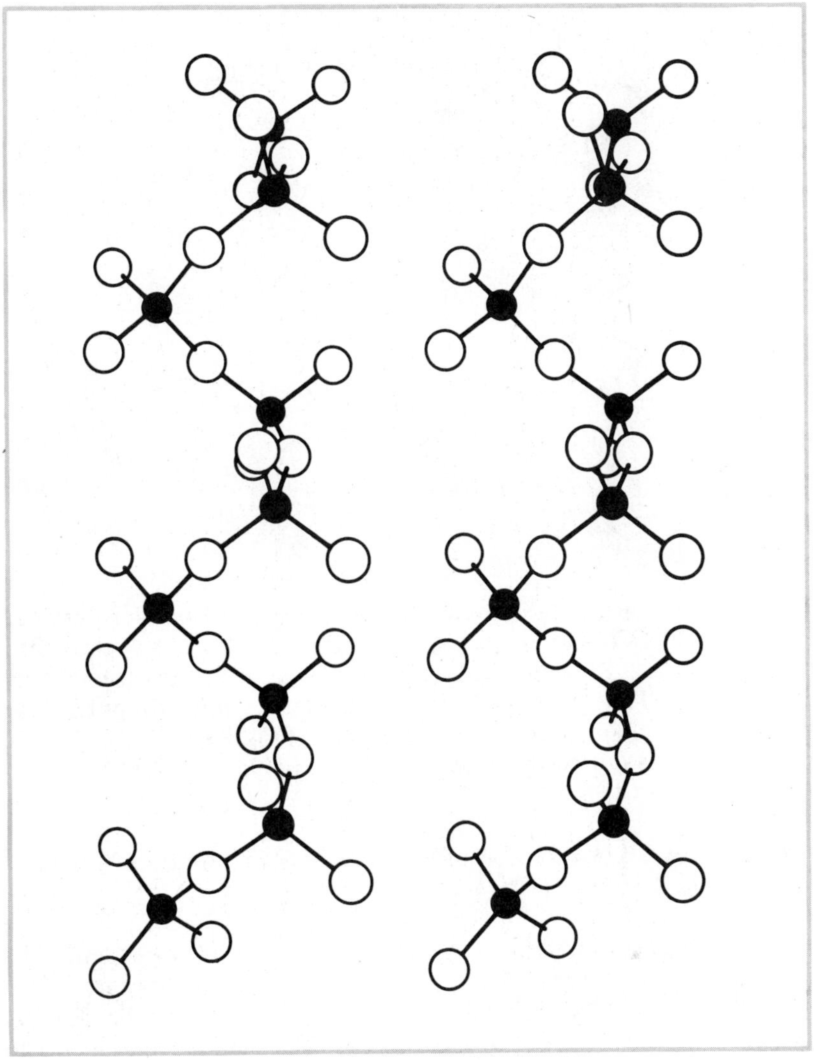

Figure 19.18
*Spiral chains of $SiO_4$ tetrahedra.*

In Chapter 13 we saw that polarization of the O—H bond of the hydroxyl group by the atom of the nonmetal permits the H atom to be removed, more or less readily, as an $H^+$ ion. To reflect this, the formulas of these acids are written with the hydrogens first, followed by the formula of the anion that remains when the protons are removed by neutralization.

The geometric structures of the monomeric acids and anions, like those of the simple oxides, are readily predicted using the electron-pair repulsion theory. Rather than discuss them at length here, try Question 19.48 at the end of the chapter.

For the preparation of these substances, once again there are no specific reactions applicable to all cases. We have seen that the oxides often react with water to form acids and that the oxoanions can be obtained from them by neutralization. In cases where more than one oxide is formed by the non-

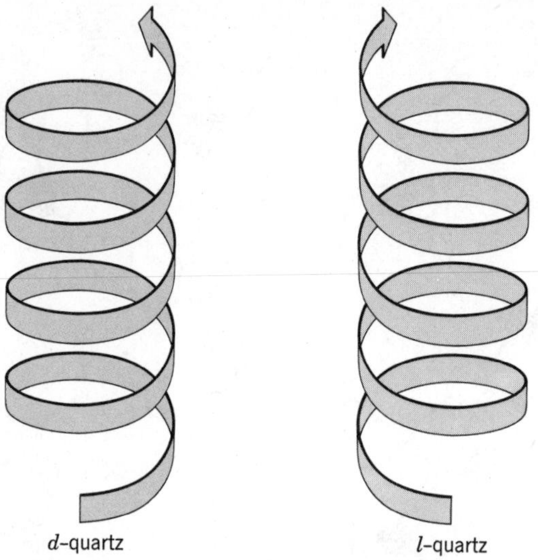


Figure 19.19
*Left- and right-handed helices in quartz. The spiral chains of $SiO_2$ tetrahedra can twist in either of two directions.*

metal, different acids can be produced by hydrolysis. Thus we have seen that $SO_2$ and $SO_3$ produce $H_2SO_3$ and $H_2SO_4$, respectively.

The industrial reaction of greatest importance today is the production of $H_2SO_4$ from $SO_3$. In actual practice, sulfur is burned to produce $SO_2$ which is then catalytically oxidized to $SO_3$. Pure $SO_3$ dissolves in water very slowly; so instead it is dissolved in $H_2SO_4$ to give $H_2S_2O_7$.

$$H_2SO_4 + SO_3 \longrightarrow H_2S_2O_7$$

Upon addition of $H_2O$, the $H_2S_2O_7$ produces $H_2SO_4$.

$$H_2S_2O_7 + H_2O \longrightarrow 2H_2SO_4$$

About 30 million tons of $H_2SO_4$ are produced annually. Most of it is used to make phosphoric acid and phosphate fertilizers.

With phosphorus, hydrolysis of the oxides $P_4O_6$ and $P_4O_{10}$ yield $H_3PO_3$ and $H_3PO_4$. These latter two acids are interesting because only $H_3PO_4$ is able to provide three protons for neutralization and thus produce salts containing the $PO_4^{3-}$ ion. Complete neutralization of $H_3PO_3$, on the other hand, gives the ion $HPO_3^{2-}$. The differences in behavior between these two acids can be traced to their structures.

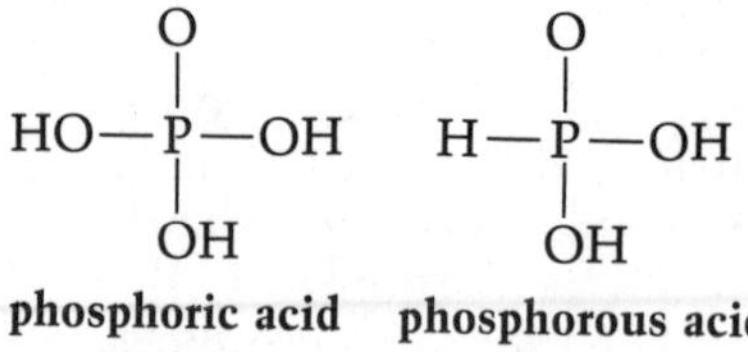

**phosphoric acid** $H_3PO_4$ **phosphorous acid** $H_3PO_3$

In each case phosphorus is bonded to four atoms. In $H_3PO_3$, however, one of the hydrogens is covalently bonded directly to the phosphorus atom and is not able to be removed as a proton. Consequently, $H_3PO_3$ is only a diprotic acid.

Phosphoric acid is another very important industrial chemical. It is made by dissolving $P_4O_{10}$ in water, as described earlier, as well as by reacting

Table 19.7
*Monomeric oxoacids and oxoanions of the nonmetals*

| | | | | |
|---|---|---|---|---|
| Group IIIA | $H_3BO_3$ (no simple borates) | | | |
| Group IVA | $H_2CO_3$ ($CO_3^{2-}$) | $H_4SiO_4$[a] ($SiO_4^{4-}$) | $H_4GeO_4$[a] ($GeO_4^{4-}$) | |
| Group VA | $HNO_2$ ($NO_2^-$) $HNO_3$ ($NO_3^-$) | $H_3PO_3$ ($HPO_3^{2-}$) $H_3PO_4$ ($PO_4^{3-}$) | $H_3AsO_4$ ($AsO_4^{3-}$) | |
| Group VIA | | $H_2SO_3$ ($SO_3^{2-}$) $H_2SO_4$ ($SO_4^{2-}$) | $H_2SeO_3$ ($SeO_3^{2-}$) $H_2SeO_4$ ($SeO_4^{2-}$) | $H_2TeO_3$[a] ($TeO_3^{2-}$) $Te(OH)_6$ ($TeO(OH)_5^-$) |
| Group VIIA | HOF | HOCl ($OCl^-$) $HClO_2$ ($ClO_2^-$) $HClO_3$ ($ClO_3^-$) $HClO_4$ ($ClO_4^-$) | HOBr ($OBr^-$) $HBrO_2$ ($BrO_2^-$) $HBrO_3$ ($BrO_3^-$) $HBrO_4$ ($BrO_4^-$) | HOI ($OI^-$) $HIO_3$ ($IO_3^-$) $H_5IO_6$ ($H_2IO_6^{3-}$) $HIO_4$ ($IO_4^-$) |

[a] Not observed.

phosphate rock, composed of $Ca_3(PO_4)_2$, with concentrated sulfuric acid in a metathesis reaction,

$$Ca_3(PO_4)_2\ (s) + 3H_2SO_4\ (l) \longrightarrow 3CaSO_4\ (s) + 2H_3PO_4\ (l)$$

Phosphoric acid is used in manufacturing detergents, fertilizers, cosmetics, and soft drinks.

Oxoacids and their anions may also be prepared by oxidation-reduction reactions. We have seen, for instance, the reaction of $NO_2$ with water.

$$3NO_2 + H_2O \longrightarrow 2HNO_3 + NO$$

Another reaction that is at least somewhat similar occurs when an equimolar mixture of NO and $NO_2$ is dissolved in water.

$$NO + NO_2 + H_2O \longrightarrow 2HNO_2$$

Here we have nitrogen(II) oxide and nitrogen(IV) oxide which, taken together, are equivalent to $N_2O_3$, the acid anhydride of $HNO_2$. In fact, if this equimolar mixture of oxides is condensed to a liquid, or frozen to a solid, molecules having the formula $N_2O_3$ are created. A nitrite salt can also be prepared from this $NO/NO_2$ mixture by reacting it with a base.

$$NO + NO_2 + 2OH^- \longrightarrow 2NO_2^- + H_2O$$

Other oxoacids and anions that may be obtained by oxidation-reduction reactions are those containing the halogens. For example, with the exception of fluorine, the halogens react with water in a disproportionation reaction.

$$X_2 + H_2O \longrightarrow HOX + H^+ + X^-$$

Thus $Cl_2$ gives hypochlorous acid, HOCl, and hydrochloric acid, HCl; $Br_2$ yields HOBr and HBr, and so forth. In the presence of a base we expect the corresponding acids to be neutralized and the oxoanions to be formed, thus driving these reactions toward completion. With chlorine, for example, we have

$$Cl_2 + 2OH^- \longrightarrow OCl^- + Cl^- + H_2O$$

and, in general, there is the reaction

$$X_2 + 2OH^- \longrightarrow OX^- + X^- + H_2O$$

However, here we have an interesting case of kinetics influencing the ultimate products of a chemical reaction.

The hypohalite ions themselves ($OCl^-$, $OBr^-$, $OI^-$) have a tendency to disproportionate to give the halate ($XO_3^-$) and halide ($X^-$).

$$3OX^- \rightleftharpoons XO_3^- + 2X^-$$

The equilibrium constants for this reaction with the different halogens are

$$3OCl^- \rightleftharpoons ClO_3^- + 2Cl^- \qquad K \approx 10^{27}$$
$$3OBr^- \rightleftharpoons BrO_3^- + 2Br^- \qquad K \approx 10^{15}$$
$$3OI^- \rightleftharpoons IO_3^- + 2I^- \qquad K \approx 10^{20}$$

The large values for these *K*'s suggest that all of the $OX^-$ ions should be transformed rather completely to the corresponding $XO_3^-$ ions. Solutions of hypochlorite ion, however, are reasonably stable when kept cool, although disproportionation to $ClO_3^-$ does occur if a solution of hypochlorite is heated. The stability of cold solutions containing $OCl^-$ is further illustrated by the fact that commercial liquid bleach consists of a 5% solution of NaOCl. Hypobromite ion, on the other hand, rapidly disproportionates at room temperature while $OI^-$ reacts so rapidly that it is never even observed at all when $I_2$ is dissolved in base. We see, then, that it is slow kinetics that accounts for the "stability" of solutions of $OCl^-$ rather than simply thermodynamics. If all of the $OX^-$ ions were to disproportionate as rapidly as $OI^-$, the reaction of the elemental halogens with base would lead directly to the production of the $XO_3^-$ ions.

There are other disproportionation reactions that yield oxoanions that can be caused to occur in the solid state by the application of heat. For example, strong heating of a nitrate such as $NaNO_3$ can drive off oxygen to give the nitrite,

$$2NaNO_3 \longrightarrow 2NaNO_2 + O_2$$

Similarly, $KClO_3$, when heated at moderate temperatures in the *absence* of a catalyst such as $MnO_2$, disproportionates to give the perchlorate,

$$4KClO_3 \longrightarrow 3KClO_4 + KCl$$

(In the presence of a catalyst, or when heated to high temperatures, decomposition to KCl and $O_2$ occurs; $2KClO_3 \rightarrow 2KCl + 3O_2$.)

It was mentioned earlier that there are some acids that do not exist in the pure liquid state but, instead, can be obtained only in solution. Nitrous acid, $HNO_2$, and chloric acid, $HClO_3$, are examples. On the basis of the equilibria involved, we would expect that any weak acid could be produced in solution by adding a strong acid such as HCl or $H_2SO_4$ to a salt of the weak acid. Nitrous acid can be conveniently prepared by acidifying a nitrite salt.

$$NO_2^- + H^+ \rightleftharpoons HNO_2$$

The production of $HNO_2$ by this reaction has raised some fears that the use of nitrites to preserve the red color of meat products such as frankfurters may create the potential for cancer. Nitrous acid is known to react with organic compounds called amines to give products called nitrosoamines, which have been proven to be potent carcinogens. In the acid environment of the stomach nitrites are converted to $HNO_2$, and the proteins in the meat contain the components of amines. The possibility of reaction to form nitrosoamines therefore exists in the stomach.

Chloric acid, $HClO_3$, is strong and cannot be generated in the same manner as $HNO_2$. The solubility rules that you learned (hopefully) in Chapter 5 can, however, point to a preparative method, at least in this case. Mixing solutions of barium chlorate and sulfuric acid yields a precipitate of $BaSO_4$ and leaves $HClO_3$ in solution.

$$Ba(ClO_3)_2\,(aq) + H_2SO_4\,(aq) \longrightarrow BaSO_4\,(s) + 2HClO_3\,(aq)$$

As a general method of preparation of a strong acid, however, this obviously requires the use of a barium salt that is more soluble than $BaSO_4$.

## 19.12 POLYMERIC OXOACIDS AND OXOANIONS

Polymeric oxoacids and oxoanions consist of two or more nonmetal atoms linked together in some fashion via oxygen bridges. They occur primarily with the elements in the third and succeeding periods, although boron, a second-period element, also forms complex borates (which we shall not discuss here).

In the polymeric species that we will now examine, the basic structural unit is the $XO_4$ tetrahedron. Polymeric acids or ions are constructed by linking $XO_4$ tetrahedra together through commonly shared corners, resulting in structures that range from simple to complex. However, we shall see that relatively simple, straightforward structural relationships exist between the polymeric acids and anions and the simple oxoacids discussed in the last section.

The oxoacids formed by the nonmetals of period 3, in their highest oxidation states, can be represented by the formulas

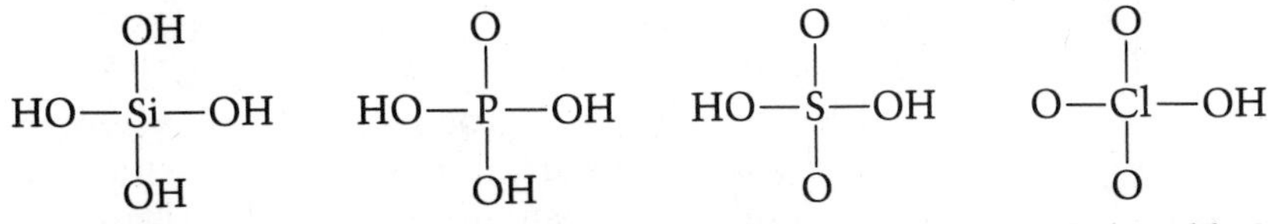

**orthosilicic acid (never isolated)** **orthophosphoric acid** **(ortho)sulfuric acid** **(ortho)perchloric acid**

These correspond to the "ortho" acids, although in the latter two the ortho prefix is omitted, as indicated by the parentheses.

As noted earlier, the hydrogen atoms of the —OH groups are acidic and may be removed by reaction with a base to yield the series of anions,

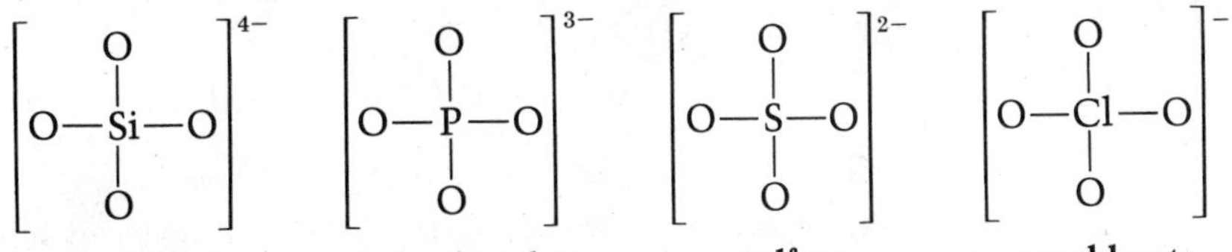

**orthosilicate** **orthophosphate** **sulfate** **perchlorate**

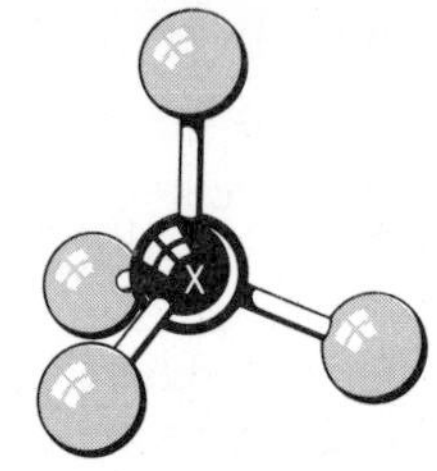

Figure 19.20
*Tetrahedral "ortho" anions,* $XO_4^{n-}$.

Notice that the number of negative charges on the anion is equal, in each case, to the number of —OH groups attached to the central atom in the parent acid. In addition, in each of these anions the central atom is surrounded by four electron pairs (in the X—O bonds) and each therefore has a tetrahedral structure that we can depict as shown in Figure 19.20.

The polymeric oxoacids and anions are related to these simple species in a rather direct way, because they are formed, if not always in practice, at least in principle, by joining these simple tetrahedral units together into more complex structures. We may view this as being accomplished by the removal of the constituents of a water molecule from two —OH groups on adjacent molecules of the acid. We can illustrate this in a general way as

$$-\overset{|}{\underset{|}{X}}-\boxed{O-H \quad H}-O-\overset{|}{\underset{|}{X}}- \longrightarrow -\overset{|}{\underset{|}{X}}-O-\overset{|}{\underset{|}{X}}- + H_2O$$

where the resulting product contains an oxygen bridge that is characteristic of these polymeric species in general.

Another important point to notice is that this bridging oxygen atom had its origin in an —OH group bonded to the central atom. Therefore, the maximum complexity of the polymeric acid (or anion) is determined by the number of —OH groups that are bound to the nonmetal in the simple monomeric acid. Thus we expect that silicon can form a maximum of four oxygen bridges to other Si atoms, phosphorus can form three, sulfur two, and chlorine only one.

Let us now begin to examine some of these polymers. When one molecule of water is eliminated from between two molecules of acid, the species $H_nX_2O_7$ are formed (Table 19.8). These constitute the so-called **pyro** acids,[8] and their neutralization produces the corresponding oxoanions. In this table we have also included the oxide, $Cl_2O_7$, which is essentially structurally identical to the pyroanions (Figure 19.21). In fact, $Cl_2O_7$ is prepared by reacting $HClO_4$ with a powerful dehydrating agent such as $P_4O_{10}$. In addition, $Cl_2O_7$ is the acid anhydride of $HClO_4$ and reacts with water to form this acid.

The prefix "pyro" that is used with these acids and anions is derived from the fact that they may be formed by heating either the free acids, or one of their acid salts, to drive off the $H_2O$ molecule from between two tetrahedral units. The pyrosulfate ion, for example, can be obtained by heating a salt such as $KHSO_4$.

$$2KHSO_4 \longrightarrow K_2S_2O_7 + H_2O$$

[8] The names for these acids are derived from the Greek, *pyros,* meaning fire. Often they may be formed by driving water out of the corresponding ortho acid or one of its salts.

Table 19.8
*"Pyro" acids and anions*

| Silicon | Phosphorus | Sulfur | Chlorine |
|---|---|---|---|
| 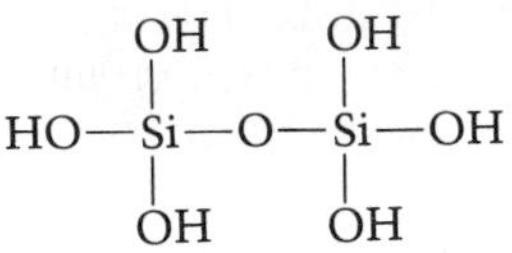 <div align="center">OH   OH<br>\|     \|<br>HO—Si—O—Si—OH<br>\|     \|<br>OH   OH<br>**pyrosilicic acid**<br>$H_6Si_2O_7$</div> | 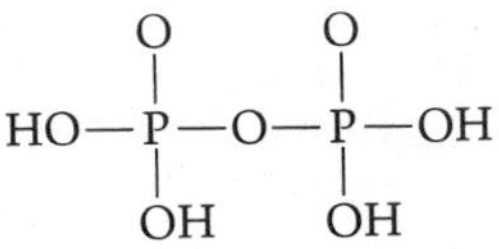 <div align="center">O    O<br>\|\|   \|\|<br>HO—P—O—P—OH<br>\|     \|<br>OH   OH<br>**pyrophosphoric acid**<br>$H_4P_2O_7$</div> | 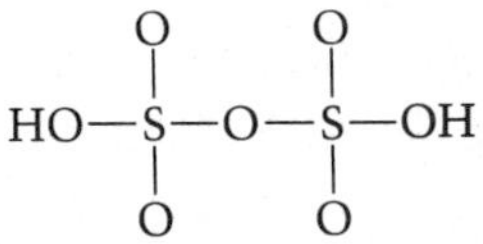 <div align="center">O    O<br>\|\|   \|\|<br>HO—S—O—S—OH<br>\|\|   \|\|<br>O    O<br>**pyrosulfuric acid**<br>$H_2S_2O_7$</div> | 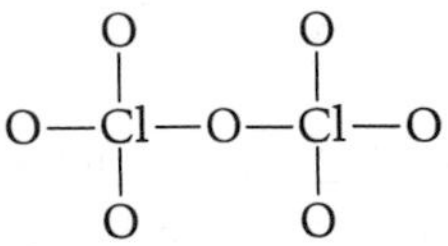 <div align="center">O    O<br>\|\|   \|\|<br>O—Cl—O—Cl—O<br>\|\|   \|\|<br>O    O<br>**chlorine heptoxide**<br>$Cl_2O_7$<br>$(H_0Cl_2O_7)$</div> |
| 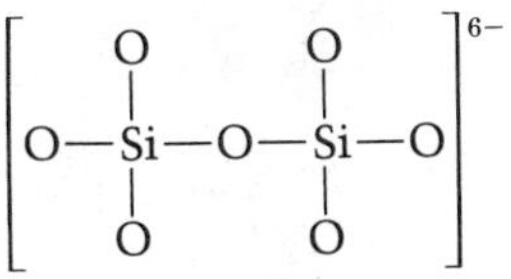 <div align="center">$\begin{bmatrix} O\quad\ O \\ \| \quad\ \| \\ O{-}Si{-}O{-}Si{-}O \\ \|\quad\ \| \\ O\quad\ O \end{bmatrix}^{6-}$<br>**pyrosilicate**<br>$Si_2O_7{}^{6-}$</div> | 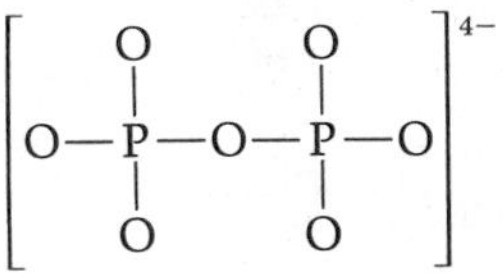 <div align="center">$\begin{bmatrix} O\quad\ O \\ \| \quad\ \| \\ O{-}P{-}O{-}P{-}O \\ \|\quad\ \| \\ O\quad\ O \end{bmatrix}^{4-}$<br>**pyrophosphate**<br>$P_2O_7{}^{4-}$</div> | 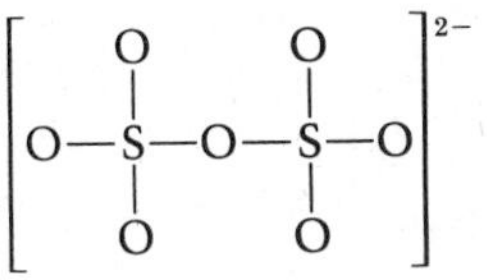 <div align="center">$\begin{bmatrix} O\quad\ O \\ \| \quad\ \| \\ O{-}S{-}O{-}S{-}O \\ \|\quad\ \| \\ O\quad\ O \end{bmatrix}^{2-}$<br>**pyrosulfate**<br>$S_2O_7{}^{2-}$</div> | |

The acid, $H_2S_2O_7$, on the other hand, is formed by dissolving $SO_3$ in concentrated $H_2SO_4$,

$$SO_3 + H_2SO_4 \longrightarrow H_2S_2O_7$$

We saw this reaction in our earlier discussion on the preparation of $H_2SO_4$. Pyrophosphoric acid is produced when $H_3PO_4$ is heated.

$$2H_3PO_4 \longrightarrow H_4P_2O_7 + H_2O$$

The pyrosilicate ion is observed only in certain minerals, for example, thortveitite, $Sc_2Si_2O_7$. The acid, $H_2Si_2O_7$, cannot be isolated.

The ease with which water may be removed from the acid to cause polymerization is not the same for all of the period 3 nonmetals but, instead, increases from Cl to Si. To prepare $Cl_2O_7$ from $HClO_4$ requires a very power-

Figure 19.21
*Structure of the "pyro" anions, $X_2O_7{}^{m-}$. Solid sphere = X.*

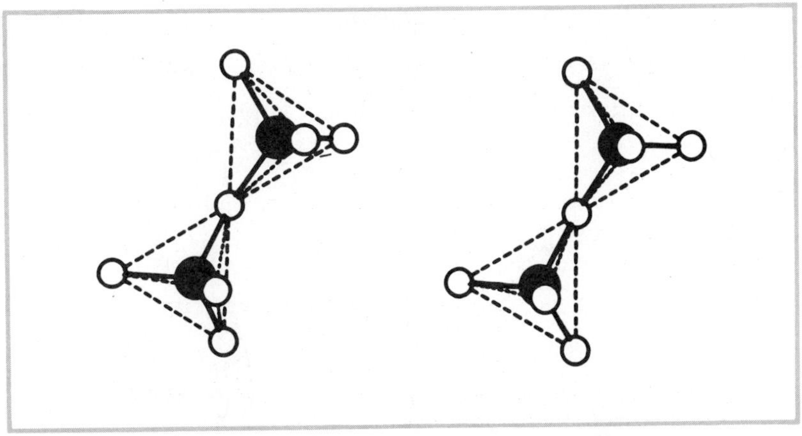

ful dehydrating agent. As we move to the left, conditions become less severe until, with silicon, the acidification of a solution containing a soluble silicate leads to the immediate precipitation of hydrated silica, $SiO_2(H_2O)_x$, the end result of large-scale spontaneous dehydration-polymerization. With the exception of silicon, the dehydration process that leads to polymerization can be reversed by the addition of water to the polymeric species. Pyrosulfuric acid, for example, reacts with water to form $H_2SO_4$.

$$H_2S_2O_7 + H_2O \longrightarrow 2H_2SO_4$$

Except for chlorine, all of the dimeric "pyro" acids discussed above have additional —OH groups that can participate in further bridging, again by elimination of $H_2O$, to give the **meta** series of acids and anions. The sharing of two —OH groups of the original ortho acid thus can give a chainlike structure containing a very large number of repeating units as shown below.

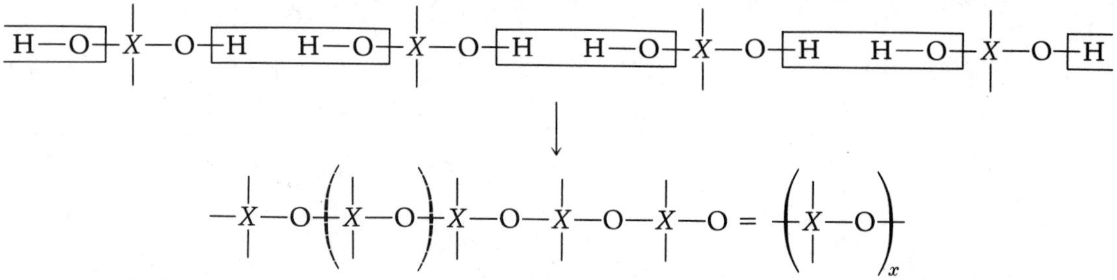

The "meta" anions derived from the acids all consist of $XO_4$ tetrahedra linked through shared corners and possess the empirical formula $XO_3^{n-}$. The general structure is illustrated in Figure 19.22.

For silicon the free acid is unknown, while the metasilicate ion, $SiO_3^{2-}$, occurs in several minerals [for example, spodumene, $LiAl(SiO_3)_2$, a major source of Li]. Not surprisingly, these are observed to have a fiberlike appearance.

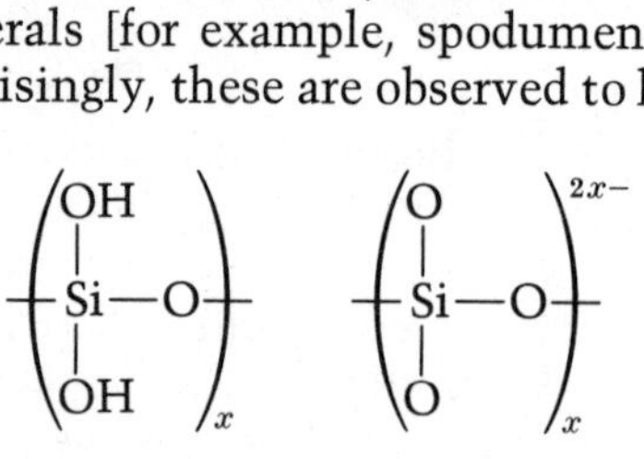

**metasilicic acid (not isolated)** **metasilicate ion**

With phosphorus we have a similar situation. Heating $NaH_2PO_4$ drives off water to yield $(NaPO_3)_n$,

$$nNaH_2PO_4 \longrightarrow (NaPO_3)_n + nH_2O$$

where we again have a linear polyanion

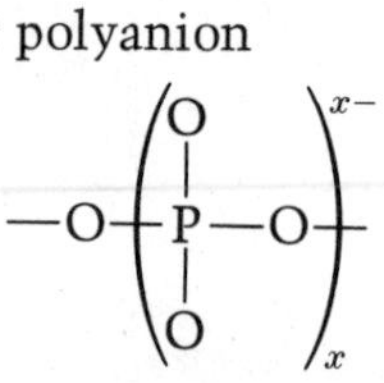

which we can consider to be derived from metaphosphoric acid.

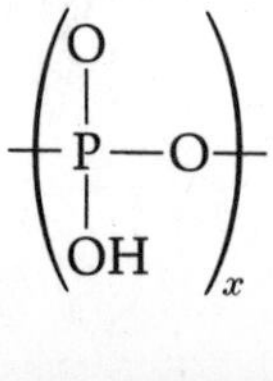

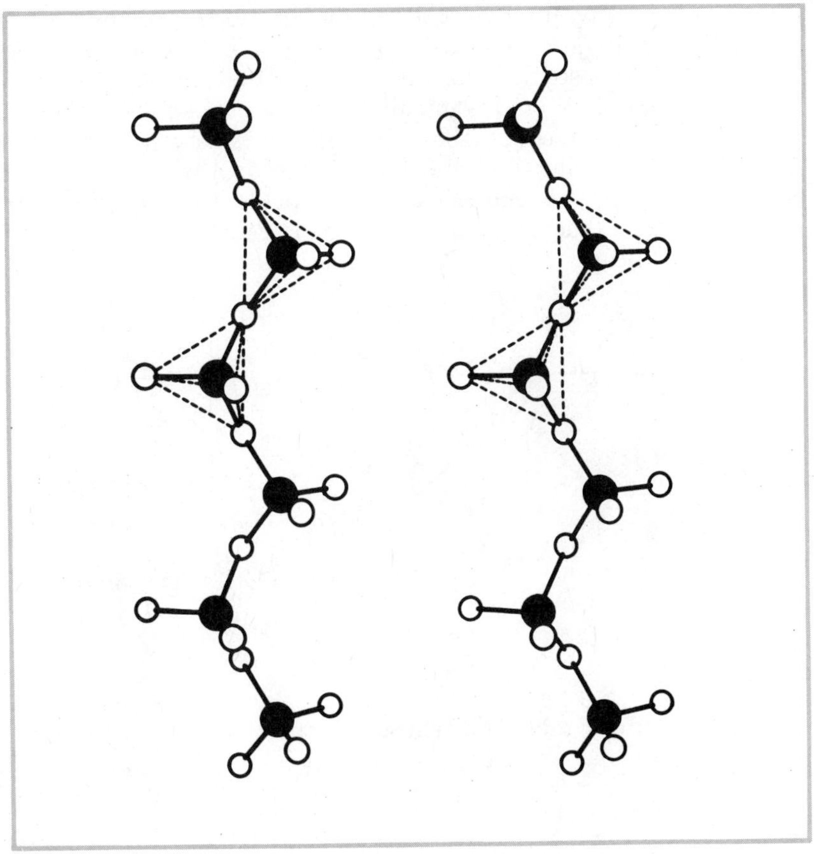

Figure 19.22
*Simple linear* $(XO_3^{n-})_x$ *formed by sharing two corners on every* $XO_4$ *tetrahedron. (Solid spheres = X.)*

Note that only one of the nonbridging oxygens in metaphosphoric acid belongs to an OH group while in metasilicic acid both do. As a result, the metaphosphate ion carries a charge of −1, while the $SiO_3$ unit bears a −2 charge.

A metaphosphate that is a component of many solid detergent mixtures is sodium tripolyphosphate, $Na_5P_3O_{10}$, which contains the ion,

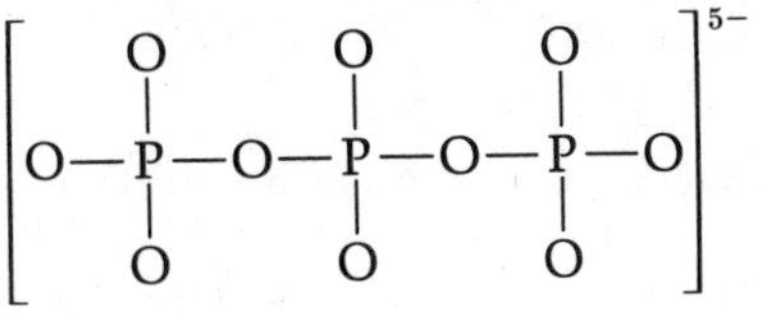

The presence of this ion in the wash water has several beneficial effects. For example, it forms soluble complexes with iron and manganese ions, which can stain clothing, and also helps to keep dirt particles suspended so that they can be readily washed away.

In liquid detergents sodium or potassium pyrophosphate ($P_2O_7^{4-}$) is used because it hydrolyzes to $PO_4^{3-}$ less rapidly than the $P_3O_{10}^{5-}$ ion.

Unfortunately, phosphates in detergents also have a damaging effect on the environment. They collect in lakes where they serve as nutrients for excessive growths of algae, far beyond the amount that can be consumed by fish and other marine life. Eventually the algae die. The bacteria that decompose them consume most of the oxygen in the water, and other marine life (fish, etc.) is unable to survive. In some localities this problem has become so severe that phosphate detergents have been banned entirely.

Earlier we saw that the removal of an $H_2O$ molecule from a pair of $HClO_4$ molecules leads to the oxide $Cl_2O_7$. By the same token we might expect that the removal of two molecules of water from $H_2SO_4$ might lead to a linear chainlike $SO_3$ species.

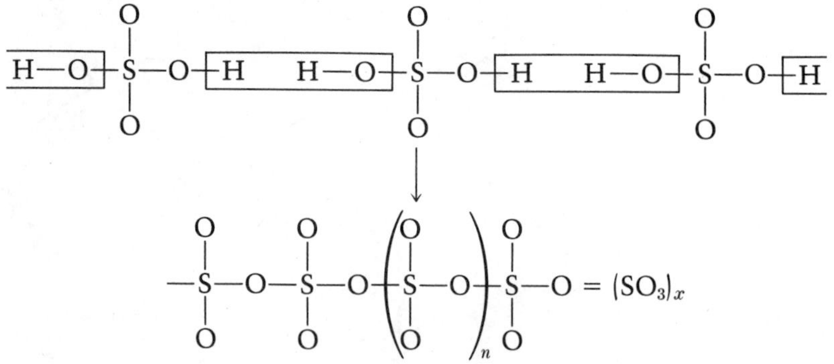

Such a substance does actually exist. We have spoken earlier of $SO_3$ as a simple molecule whose structure can be represented by resonance forms such as

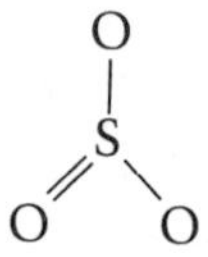

However, when a trace of $H_2O$ is added to pure $SO_3$ a polymerization takes place to produce a fibrous solid.

We can view this reaction in the following way. An $H^+$ from the $H_2O$ adds itself to the oxygen of one $SO_3$ molecule.

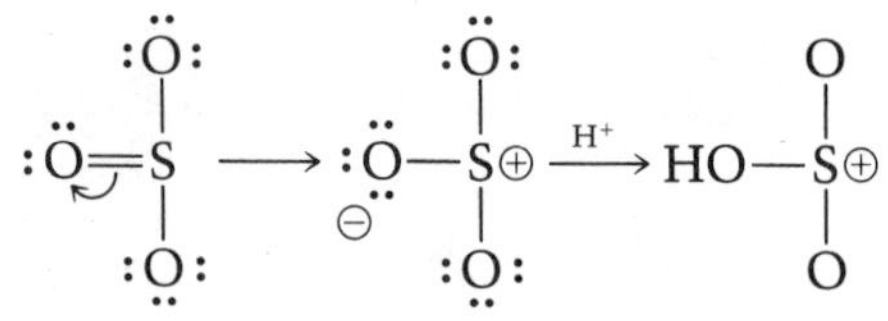

This species then adds another $SO_3$ unit,

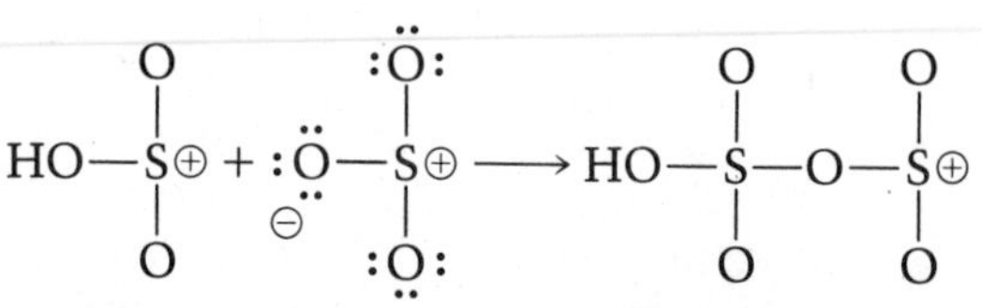

a process that is repeated over and over to produce a long chain of $SO_3$ molecules. Finally, the positive charge on the sulfur atom at the end of the chain is neutralized by the addition of the $OH^-$ left over from the water molecule that had lost its proton. The end result is a species,

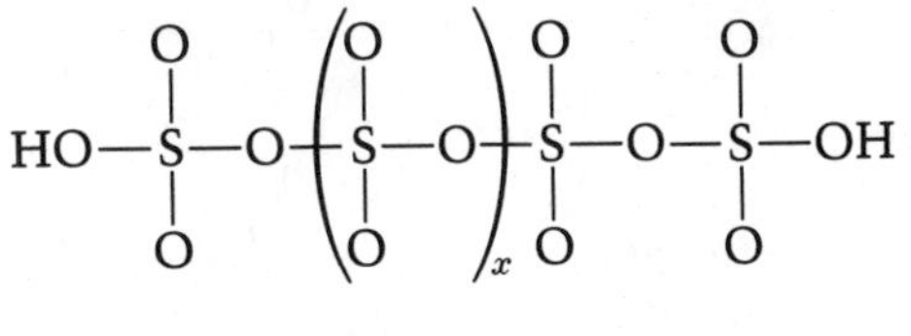

If the chain is extremely long, its empirical formula essentially reduces to simply $SO_3$ because the $H^+$ and $OH^-$ make up an insignificant part of the whole molecule.

In addition to infinite chain anions, the formation of two bridges to one atom can also lead to cyclic structures. For example, heating $(NH_4)_2HPO_4$ leads to the reaction

$$3(NH_4)_2HPO_4 \longrightarrow H_3P_3O_9 + 6NH_3 + 3H_2O$$

in which $H_3P_3O_9$ is cyclotrimetaphosphoric acid (quite a mouthful, isn't it?),

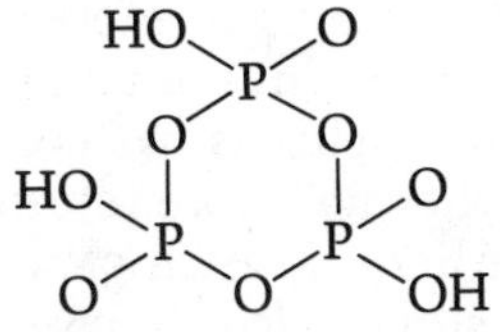

This is a triprotic acid and yields salts such as $Na_3P_3O_9$ containing the cyclotrimetaphosphate ion

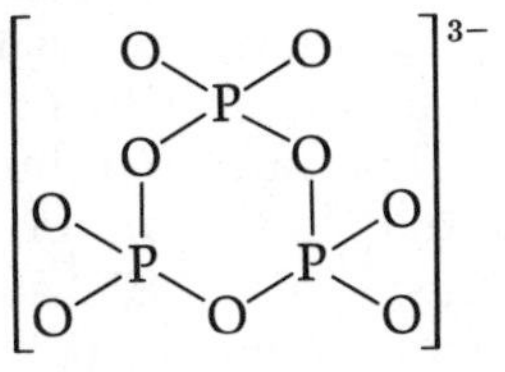

In a similar fashion, discrete polymeric anions are also found with silicon. For example, there is the $Si_6O_{18}^{12-}$ ion found in the mineral beryl, $Be_3Al_2(Si_6O_{18})$. This ion is pictured in Figure 19.23. Note that once again we have a series of $SiO_4$ tetrahedra in which each share two corners and where the empirical formula of the *anion* is $SiO_3^{2-}$. Beryl is the only important mineral source of beryllium; a very intensely green transparent form of beryl is better known as emerald.

The formation of cyclic species even occurs with $SO_3$, which solidifies in a form containing rings composed of three $SO_3$ units,

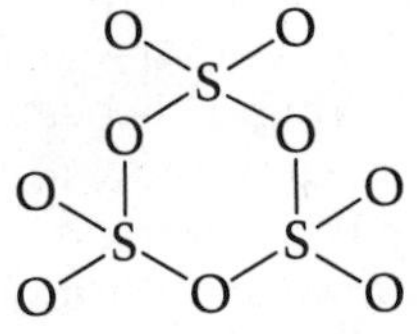

entirely analogous to the $P_3O_9^{3-}$ ion.

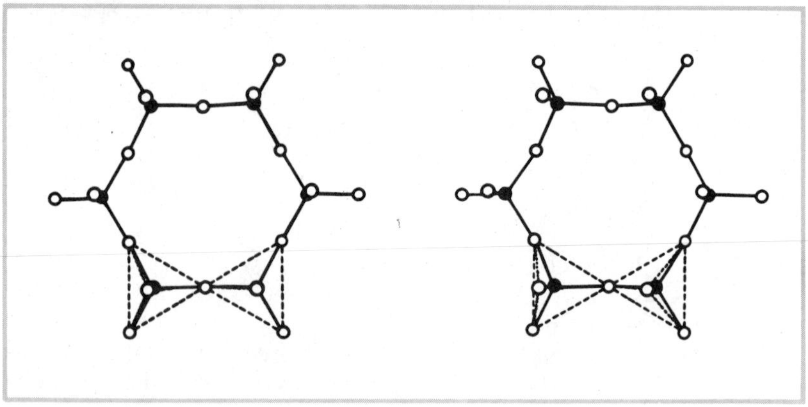

Figure 19.23
*Cyclic structure of the $Si_6O_{18}^{12-}$ anion. Notice that there are 12 nonbridging oxygen atoms that each carry a single negative charge. (Solid spheres = X.)*

The sharing of more than two corners of the tetrahedron, which produces either three or four oxygen bridges, leads to even more complex structures. Phosphorus has only three oxygen atoms that can form bridges and when they do so, we return to the oxide, $P_4O_{10}$ (Figure 19.15). It was pointed out earlier that each of the phosphorus atoms in this molecule is tetrahedrally surrounded by oxygen atoms. The structure of $P_4O_{10}$ is thus related to $H_3PO_4$ by the elimination of water from the three —OH groups, so that four $PO_4$ tetrahedra can each share three of their corners with other tetrahedra.

Silicon, with four oxygen atoms that can function as bridges, forms the most complex series of anions. In addition to the simple strands of $SiO_4$ tetrahedra in the infinite chain anion $SiO_3^{2-}$ (Figure 19.22), double strands, formed by the sharing of three corners by *every other* $SiO_4$ unit, are also found. This time, an infinite *double* chain results, a small segment of which is illustrated in Figure 19.24. In this, as well as in the other silicate anions, each unshared oxygen atom carries a negative charge and the repeating unit along the chain is $Si_4O_{11}^{6-}$. This anion is found in asbestos, $[Ca_2Mg_5(Si_4O_{11})_2(OH)_2]_x$, and, as we might expect, asbestos has a fiberlike nature because of the presence of the long $(Si_4O_{11}^{6-})_x$ chains which line up more or less parallel to one another.

Asbestos has many useful properties. Its fibrous nature makes it excellent as a reinforcing filler in brake and clutch linings. Because it is fireproof it is spun and woven into fabrics used in fireproof curtains and ironing board covers. It is used in insulation and sealers in cars, trucks, and planes. Once again, however, nature takes as well as gives. Asbestos fibers breathed into the lungs have been implicated in many cases of lung cancer. There is also evidence that if consumed in food it can produce stomach cancer.

Still another type of silicate is formed if *each* $SiO_4$ tetrahedron shares three of its corners so that each Si is attached to three others by oxygen bridges. When this occurs a planar sheet of $SiO_4$ units results. A portion of one of these is shown in Figure 19.25 with the repeating unit, $Si_2O_5^{2-}$, outlined by the rectangle.

A number of different minerals are known to contain these $(Si_2O_5^{2-})_x$ sheets. They differ in the way the silicate layers are stacked, and the nature of the cations and other anions that are also present in the structure; how-

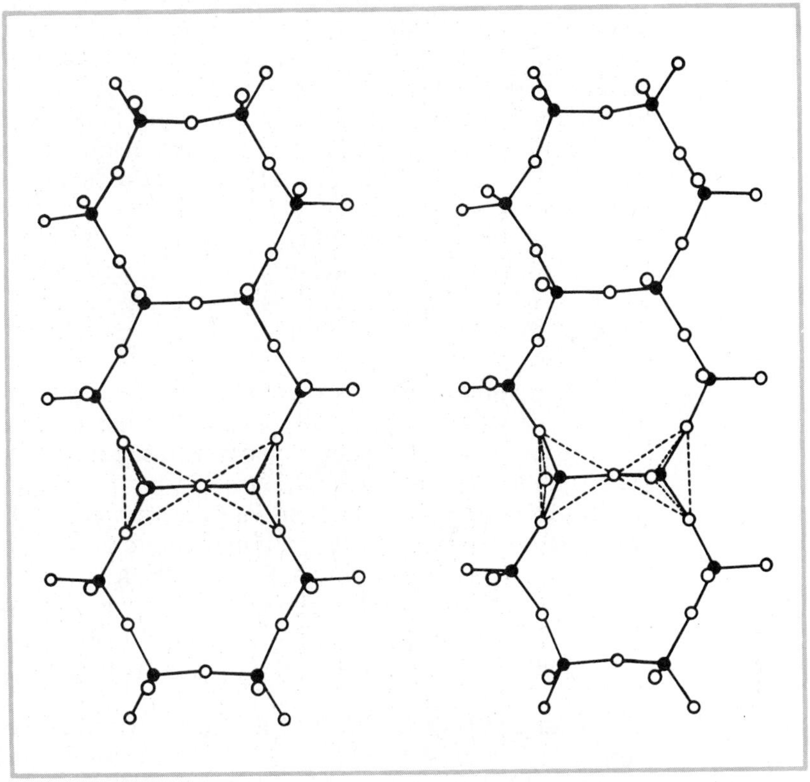

Figure 19.24
*Linear double chain in which every other $SiO_4$ tetrahedron shares three corners. (Solid spheres = Si.)*

Figure 19.25
*Planar sheet silicate formed by sharing three corners on every $SiO_4$ tetrahedron.*

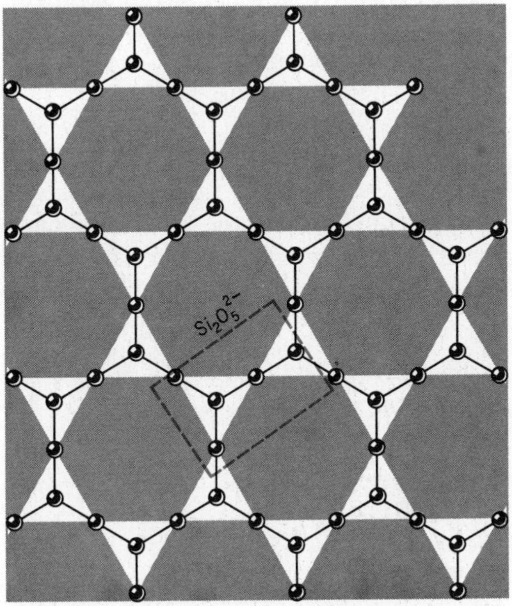

ever, all have certain similarities to each other. Some examples are talc (used in bath powder) and soapstone, in which the $(Si_2O_5^{2-})_x$ sheets are packed together with cations in such a way that there is a minimum of attractive forces between successive layers. These layers therefore slide over each other easily and both of these minerals feel slippery.

In other related minerals there is substitution of another element for Si. In mica, for instance every fourth Si atom is replaced by an $Al^{3+}$ ion. The properties of mica are therefore different from talc; however, the layer structure is still apparent. A mica-type material with which you may be familiar is *vermiculite,* used in place of soil in propagating house plants and as a cushioning filler in packaging items for shipment. This solid flakes in thin flat layers characteristic of mica.

When Si finally shares all four of its oxygen atoms with other Si atoms, a three-dimensional framework occurs that has the empirical formula $SiO_2$. This is found in the mineral quartz whose structure was discussed earlier (see Figures 19.16 through 19.19).

In concluding this section we note that there are great similarities of structure among the oxides, polymeric anions, and even the simple anions of the third-period nonmetals. Some of these relationships are summarized in Table 19.9. In all of these species, as we hope you have noticed, the simple tetrahedron is found, and even the most complex structures are simply obtained by combining tetrahedra in many different ways.

## 19.13 HALOGEN COMPOUNDS OF THE NONMETALS

The halogens (F, Cl, Br, and I) as a class form a large number of different compounds with the nonmetals. We shall look at some of their binary compounds in this section, with an eye toward those factors that control the kinds of compounds that are formed as well as one aspect of their reactivity, their tendency to react with water.

In Table 19.10 you will find a list (although not an exhaustive one) of many of the compounds that are formed between the nonmetals and the halogens. The structures of the substances found in this table can, without exception, be predicted on the basis of the electron-pair repulsion theory.

In Table 19.10 we see that most of the nonmetals form more than one compound with a given halogen. The number of halogen atoms that may become bound to any particular nonmetal can be related to two factors. One of

Table 19.9
*Summary of oxoanions of the nonmetals*

| Anions | Silicon | Phosphorus | Sulfur | Chlorine |
|---|---|---|---|---|
| Simple anions (no X—O—X bridges) | $SiO_4^{4-}$ | $PO_4^{3-}$ | $SO_4^{2-}$ | $ClO_4^-$ |
| "Pyro" anions (one X—O—X bridge) | $Si_2O_7^{6-}$ | $P_2O_7^{4-}$ | $S_2O_7^{2-}$ | $Cl_2O_7$ |
| "Meta" anions (two X—O—X bridges) | $(SiO_3)_x^{2x-}$ | $(PO_3)_x^{x-}$ | $(SO_3)_x$ | |
| Two-dimensional sheets (three X—O—X bridges) | $(Si_2O_5)_2^{2x-}$ | $P_4O_{10}$ | | |
| Three-dimensional network (four X—O—X bridges) | $(SiO_2)_x$ | | | |

Table 19.10
*Halogen compounds of the nonmetals*

| | | | | |
|---|---|---|---|---|
| Group IIIA | $BX_3$ (X = F, Cl, Br, I)<br>$BF_4^-$ | | | |
| Group IVA | $CX_4$ (X = F, Cl, Br, I) | $SiF_4$<br>$SiF_6^{2-}$<br>$SiCl_4$ | $GeF_4$<br>$GeF_6^{2-}$<br>$GeCl_4$ | |
| Group VA | $NX_3$ (X = F, Cl, Br, I)<br>$N_2F_4$ | $PX_3$ (X = F, Cl, Br, I)<br>$PF_5$<br>$PCl_5$<br>$PBr_5$ | $AsF_3$<br>$AsF_5$ | $SbF_3$<br>$SbF_5$ |
| Group VIA | $OF_2$<br>($O_2F_2$)<br>$OCl_2$<br>$OBr_2$ | $SF_2$<br>$SCl_2$<br>$S_2F_2$<br>$S_2Cl_2$<br>$SF_4$<br>$SCl_4$<br>$SF_6$ | $SeF_4$<br>$SeF_6$<br>$SeCl_2$<br>$SeCl_4$<br>$SeBr_4$ | $TeF_4$<br>$TeF_6$<br>$TeCl_4$<br>$TeBr_4$<br>$TeI_4$ |
| Group VIIA | ICl<br>IBr<br>BrF<br>BrCl<br>ClF | $ClF_3$<br>$BrF_3$<br>$ICl_3$<br>$IF_3$ | $ClF_5$<br>$BrF_5$<br>$IF_5$ | $IF_7$ |

these is the electronic structures of the elements that are combined together; the other has to do with the sizes of the atoms.

Each halogen atom contains seven electrons in its valence shell and requires only one more to achieve the stable noble gas configuration. As a result, there is little tendency for them to form $p\pi$-$p\pi$ multiple bonds with other nonmetals. Furthermore, the halogens ordinarily do not accept electrons in the formation of coordinate covalent bonds because this would mean the addition of two electrons to a valence shell that already contains seven, thereby exceeding the stable octet by one electron.[9]

On this basis, then, we can divide the halogen compounds into two groups, those that obey the octet rule and those that do not. The compositions of the compounds in the first category are determined by the number of electrons that a given nonmetal requires to reach an octet, since each atom bonded to a halogen atom furnishes one electron. For example, in Group VIIA (the halogens themselves) only one electron is needed and only one bond is formed. Thus the halogens are diatomic and substitution of one halogen for another is possible, as we see for substances such as ClF, BrF, BrCl, BrI, and ICl.

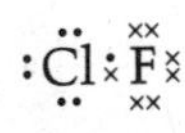

In Group VIA each element requires two electrons to reach an octet and, hence, we find that compounds such as $OF_2$ and $SCl_2$ are formed. Similarly,

[9] The halide ions (for example, $Cl^-$) do *furnish* electron pairs toward the formation of coordinate covalent bonds and are common ligands in complex ions.

in Group VA three electrons are given to the central atom by three halogens in molecules such as $NF_3$, $PF_3$, $AsF_3$, and so on. On the other hand the Group IVA elements pick up four electrons from four halogen atoms in compounds such as $CCl_4$ and $SiCl_4$.

Boron, in Group IIIA, is once again a special case. Since the boron atom has only three valence electrons, it forms only three ordinary covalent bonds to the halogens. With fluorine, however, $BF_3$ can add on an additional $F^-$ ion to form the $BF_4^-$, tetrafluoroborate, anion.

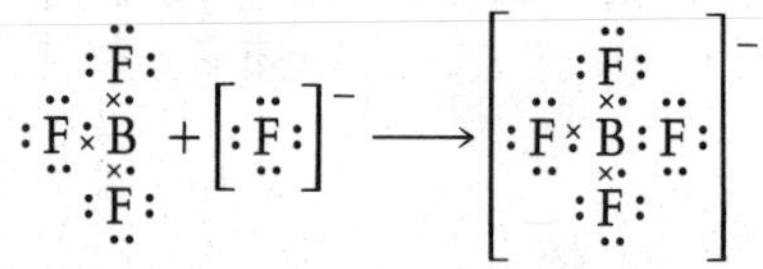

We see that the fourth B—F bond can be considered to be a coordinate covalent bond (although by now you know that we really cannot distinguish the source of the electrons once the bond has been formed). Since the boron halides are electron-deficient, in the sense that there is less than an octet of electrons in the valence shell of boron, they are all powerful Lewis acids and the formation of the $BF_4^-$ ion is a typical Lewis acid-base reaction.

In the second category of halogen compounds we have substances in which more than four pairs of electrons surround the central atom. These are limited to those nonmetals beyond the second period, because the second-period elements have a valence shell that can contain a maximum of only eight electrons corresponding to the completion of the 2*s* and 2*p* subshells. The elements below the second period, however, also have in their valence shell a low-energy set of vacant *d* orbitals as well as the *s* and *p* subshells. These *d* orbitals may be used, through hybridization, to make additional electrons available for bonding, as illustrated below for sulfur in the molecule $SF_6$.

In its ground state the electron configuration of sulfur can be represented as

S [Ne] ⇅ (3*s*) ⇅ ↑ ↑ (3*p*) _ _ _ _ _ (3*d*)

The use of two 3*d* orbitals permits the formation of a set of $d^2sp^3$ hybrid orbitals that are each singly occupied.

S [Ne] ↑ ↑ ↑ ↑ ↑ ↑ ($d^2sp^3$) _ _ _ (unhybridized 3*d* orbitals)

This provides six electrons that may each pair with one electron of a fluorine atom to form $SF_6$.

S(in $SF_6$) [Ne] ↑x ↑x ↑x ↑x ↑x ↑x ($d^2sp^3$) _ _ _ (unhybridized 3*d*)

(*x*'s represent fluorine electrons)

The structure of $SF_6$ is octahedral (Figure 19.26) as we would expect, both

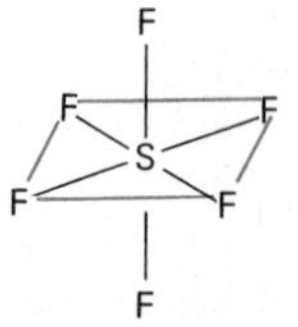


Figure 19.26
*Octahedral* $SF_6$.

from the directional properties of the $d^2sp^3$ hybrid set and from the electron-pair repulsion theory, which says that when there are six electron pairs in the valence shell of an atom, they will situate themselves at the vertices of an octahedron.

In summary, then, we can account for many of the halogen compounds with nonmetals below the second period on the basis of an expansion of the octet that occurs through the use of available *d* orbitals in hybrids that make additional valence electrons available for bonding.

The second factor that influences the number of halogen atoms that become bound to a nonmetal is the relative sizes of the different atoms. In Table 19.10 you will notice that the compounds that contain a large number of halogen atoms are formed from nonmetals found toward the bottom of the periodic table. This makes sense since these nonmetals are expected to be large and therefore able to accommodate a relatively large number of bonded atoms with a minimum of crowding. On the other hand, an element near the top of a group would be small and only a relatively few halogen atoms could be expected to be packed about it.

The Group VIA halides serve quite nicely to illustrate this point. Sulfur is able to form species such as $SF_4$ and $SF_6$ with the smallest halogen, fluorine. With chlorine, however, only $SCl_4$ exists while $SBr_4$ and $SI_4$ are not known. The much larger atom, Te, however, forms all of the tetrahalides, $TeF_4$, $TeCl_4$, $TeBr_4$, and $TeI_4$.

In Group VIIA, too, we find that size influences the number of atoms that can be bonded to a given halogen. Both chlorine and bromine form compounds having the formulas $XF_3$ and $XF_5$; however, $ClF_5$ has only recently been synthesized and is difficult to prepare. Iodine, on the other hand, is able to accommodate up to seven fluorine atoms to give $IF_7$, whose structure is given in Figure 19.27.

Figure 19.27
*Structure of $IF_7$ (pentagonal bipyramid).*

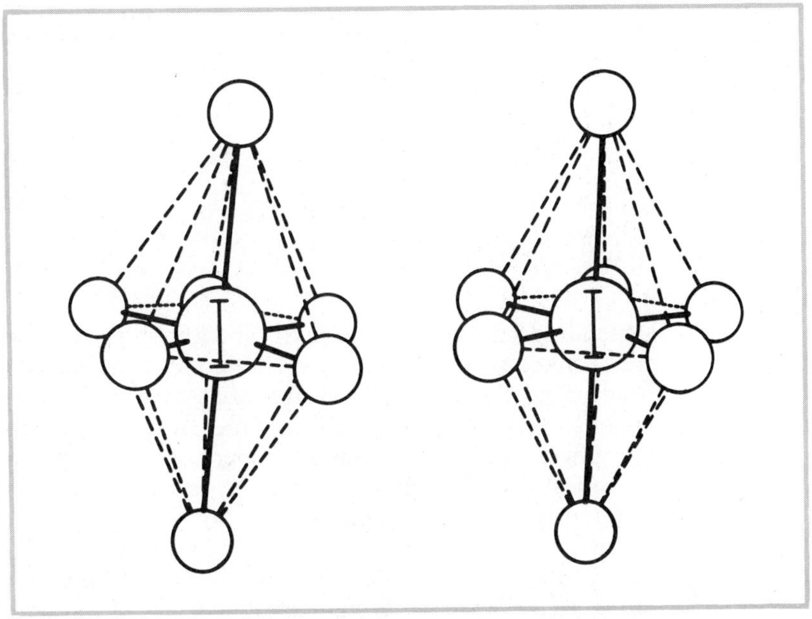

Another interesting facet of the chemistry of the nonmetal halides is their reactivity toward compounds containing an —OH group, the most familiar of which is water. Here once again we find that both thermodynamics and kinetics are involved in determining the course of reactions.

The kind of reaction on which we shall focus our attention is the hydrolysis of the nonmetal halide to produce either the oxoacid, or an oxide, plus the corresponding hydrogen halide. Some examples are the reactions of $PCl_5$, $SiCl_4$, and $SF_4$ with water.

$$PCl_5 + 4H_2O \longrightarrow H_3PO_4 + 5HCl$$
$$SiCl_4 + 2H_2O \longrightarrow SiO_2 + 4HCl$$
$$SF_4 + 2H_2O \longrightarrow SO_2 + 4HF$$

These reactions occur very rapidly and proceed to completion with the evolution of considerable amounts of heat. In fact, it is quite common for many of the halogen compounds of the elements below period 2 to react very rapidly in this same way. For example, the tin(IV) and lead(IV) halides, which we saw were covalent in Chapter 18, also hydrolyze in this manner with the formation of a mixture of species including complexes of $Sn^{4+}$ with the halide ion.

A very important class of compounds called **silicones** are produced by hydrolysis of compounds such as $(CH_3)_2SiCl_2$. The formation of silicone polymers can be thought of as proceeding by formation of a hydroxy intermediate which then eliminates water in much the same way that $Si(OH)_4$ polymerizes.

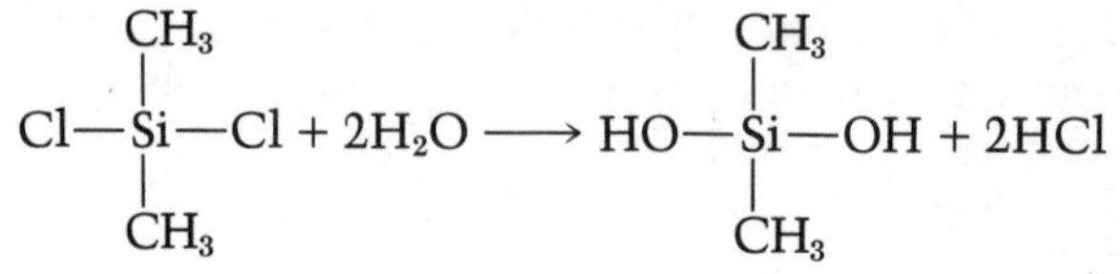

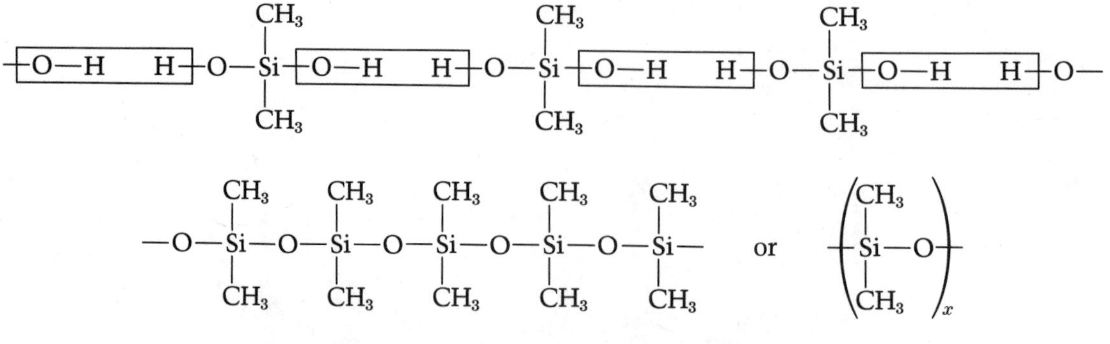

Depending on the chain length, and the degree to which chains may be cross-linked to one another, the silicones may be oils, greases, or rubbery solids. They are useful in waterproofing garments. In low-temperature applications the oils remain fluid (hydrocarbon oils become very viscous), and the rubbers retain their elastic properties (ordinary rubber becomes brittle). Silicones are also unaffected by hydrocarbon solvents and greases that soften ordinary rubber and they are not attacked by ozone in the air, which causes ordinary rubber to crack.

There are also some nonmetal halogen compounds that are quite *unreactive* toward water, for example, $CCl_4$, $SF_6$, and $NF_3$. In these cases it is unfavorable kinetics, rather than thermodynamics, that prevents the hydrolysis from taking place.

We can compare, for example, the potential hydrolysis reactions of $CCl_4$ and $SiCl_4$. Calculations based on thermodynamics imply that both should proceed very nearly to completion.

$$SiCl_4\,(l) + 2H_2O\,(l) \longrightarrow SiO_2\,(s) + 4HCl\,(aq) \qquad \Delta G^0 = -282\text{ kJ}$$
$$CCl_4\,(l) + 2H_2O\,(l) \longrightarrow CO_2\,(g) + 4HCl\,(aq) \qquad \Delta G^0 = -377\text{ kJ}$$

In fact, we see from the values of $\Delta G^0$ that the hydrolysis of $CCl_4$ is even more "spontaneous" than $SiCl_4$. Kinetically, however, the hydrolysis of $CCl_4$ is essentially prohibited. This is attributed to the absence of a low-energy path for the hydrolysis of $CCl_4$. Attack by water on the carbon atom of $CCl_4$ is prevented by the crowding of the Cl atoms. In $SiCl_4$, on the other hand, the larger Si atom provides a greater opportunity for attack and, in addition, the presence of low-energy *3d* orbitals in the valence shell of the Si atom permits a temporary bonding of the water molecule to the Si atom prior to the expulsion of a molecule of HCl. The mechanism of this hydrolysis is believed to be

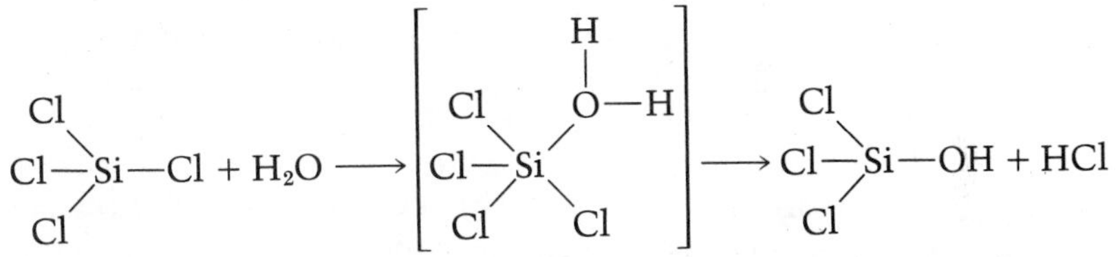

Repetition of this process eventually yields $Si(OH)_4$ (orthosilicic acid), which loses water spontaneously to give the hydrated $SiO_2$.

The stability of $SF_6$ and $NF_3$ toward hydrolysis can also be attributed to the absence of a low-energy reaction mechanism. Like $CCl_4$, $SF_6$ should also undergo hydrolysis quite spontaneously, the value of $\Delta G^0$ for the reaction being tremendous.

$$SF_6\,(g) + 4H_2O\,(l) \longrightarrow H_2SO_4\,(aq) + 6HF\,(g) \qquad \Delta G^0 = -423\text{ kJ}$$

However, the crowding of the fluorine atoms around the sulfur atom apparently prevents attack by water (even up to 500°C), as well as by most other reagents. This crowding is absent with $SF_4$, and hydrolysis by water is instantaneous.

The resistance of $NF_3$ toward attack by water cannot be attributed to interference by the fluorine atoms as in $SF_6$, since the $NF_3$ molecule is pyramidal, with the nitrogen atom being quite openly exposed to an attacking water molecule. We might compare $NF_3$ with $NCl_3$, which *does* hydrolyze (if it doesn't explode first—$NCl_3$ is extremely unstable). The mechanism for this reaction appears to involve the initial formation of a hydrogen bond from water to the lone pair on the nitrogen atom, followed by expulsion of hypochlorous acid.

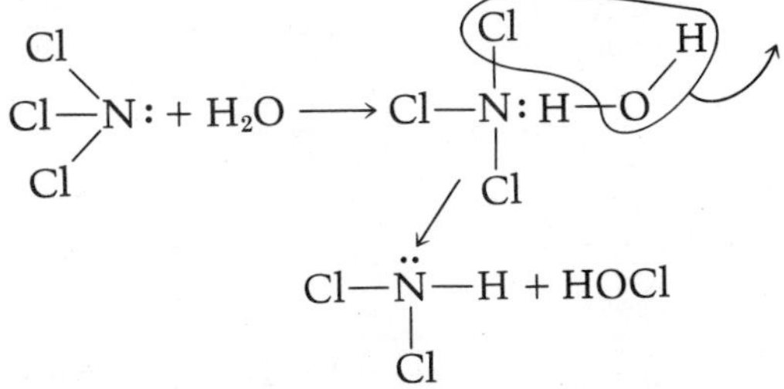

The ultimate products of the hydrolysis are ammonia and HOCl. This mechanism is not favorable for $NF_3$ because of its very low basicity. In this case the highly electronegative fluorine atoms draw electron density from the nitrogen atom. As a result, it has been suggested that the lone pair of electrons on nitrogen, in $NF_3$, may not be available to serve as a point of attachment for the $H_2O$ molecule which, as we see above, is a necessary step in the mechanism proposed for the hydrolysis of $NCl_3$.

## 19.14 NOBLE GAS COMPOUNDS

In our discussion of the nonmetals we have not mentioned compounds of the noble gases. These are rather unusual substances because, on the basis of the electronic structure of the noble gases, we would perhaps not have predicted their existence. In fact, until 1962 chemists firmly believed that these elements were totally incapable of forming compounds (other than several **clathrates,** in which the noble gas atoms are trapped in cagelike sites within a crystalline lattice). For this reason chemists had referred to them as the inert gases. Today they are spoken of as the noble gases in recognition of the fact that they do react but nevertheless possess a very low degree of chemical reactivity.

The first real chemistry of the noble gases was discovered in 1962 by Neil Bartlett at the University of British Columbia. He had found that molecular oxygen, $O_2$, reacts with $PtF_6$ to form an orange-red compound, $O_2PtF_6$, containing the ion, $O_2^+$. Since the ionization energies of $O_2$ and Xe are nearly the same (1210 and 1170 kJ/mole, respectively), he reasoned that Xe should react in the same way that $O_2$ does; when he reacted Xe with $PtF_6$ he isolated a yellow compound, containing Xe, that was formulated as $XePtF_6$.

After the initial report by Bartlett, it was not long before chemists at Argonne National Laboratory found that Xe also reacts directly with fluorine at elevated temperatures. This reaction yields a series of fluorides, $XeF_2$, $XeF_4$, and $XeF_6$. Other reactions and compounds were soon discovered and a partial list of the known Xe compounds is given in Table 19.11. The oxides and oxofluorides result from the hydrolysis of the fluorides,

$$XeF_6 + 3H_2O \longrightarrow XeO_3 + 6HF$$
$$XeF_6 + H_2O \longrightarrow XeOF_4 + 2HF$$

Table 19.11
*Some compounds of xenon*

| | Melting Point (°C) | Physical Form |
|---|---|---|
| Fluorides | | |
| $XeF_2$ | 140 | Colorless crystals |
| $XeF_4$ | 114 | Colorless crystals |
| $XeF_6$ | 47.7 | Colorless crystals |
| Oxides | | |
| $XeO_3$ | Explodes | Colorless crystals |
| $XeO_4$ | Explodes | Colorless gas |
| Salts | | |
| $XePtF_6$[a] | — | Red-orange crystals |
| $CsXeF_7$ | Decomp > 50°C | Colorless solid |
| $Cs_2XeF_8$ | Decomp > 400°C | Yellow solid |

[a] Since shown to be more complex; $Xe(PtF_6)_x$, where $x$ lies between 1 and 2.

Some of these compounds are quite unstable and tend to decompose. This is particularly true for the oxides $XeO_3$ and $XeO_4$, which explode ($XeO_3$ has $\Delta H_f^0 = +400$ kJ/mole). Others, on the other hand, appear quite stable. For example, $Cs_2XeF_8$ does not decompose even when heated to 400°C, and the fluorides have moderately high melting points, suggesting a modest degree of thermal stability.

The structure and bonding in these compounds is quite interesting. Since Xe has four pairs of electrons in its valence shell, corresponding to a completed 5*s* and 5*p* subshell, unpairing of electrons and expansion of the octet must occur to provide unpaired electrons for bonding. Let us consider $XeF_2$ and $XeF_4$.

The electronic structure of Xe can be represented as

Xe ⇅ (5*s*) ⇅ ⇅ ⇅ (5*p*) _ _ _ _ _ (5*d*)

In order to form $XeF_2$ one electron must be promoted to the 5*d* subshell, followed by hybrid orbital formation. The smallest hybrid set that will accommodate all of our electrons is $dsp^3$.

Xe ⇅ (5*s*) ⇅ ⇅ ↑ (5*p*) ↑ _ _ _ _ (5*d*)

gives

Xe ⇅ ⇅ ⇅ ↑ ↑ ($dsp^3$) _ _ _ _ (unhybridized 5*d*)

The two unpaired electrons can now be used in bonding to fluorine to give

Xe ⇅ ⇅ ⇅ ↑x ↑x ($dsp^3$) _ _ _ _ (unhybridized 5*d*)

(x's represent fluorine electrons)

In Chapter 17 we saw that the $dsp^3$ hybrids point to the vertices of a trigonal bipyramid. In terms of the electron-pair repulsion theory these five electron pairs will also be situated in this fashion, and from our rules (presented on p. 509), we expect that the three lone pairs will locate themselves in the triangular plane with the fluorine atoms above and below (Figure 19.28). The $XeF_2$ molecule is therefore linear.

In the case of $XeF_4$ we must provide four unpaired electrons for bonding to fluorine. This requires promotion of two electrons to the 5*d* subshell and the formation of $d^2sp^3$ hybrid orbitals.

Xe ⇅ (5*s*) ⇅ ↑ ↑ (5*p*) ↑ ↑ _ _ _ (5*d*)

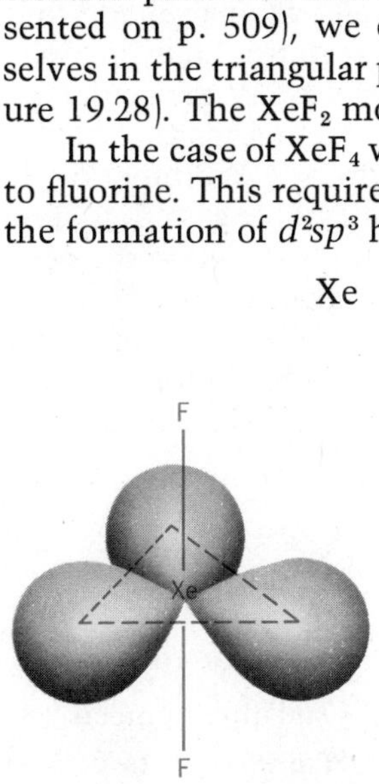


Figure 19.28
*Molecular structure of $XeF_2$.*

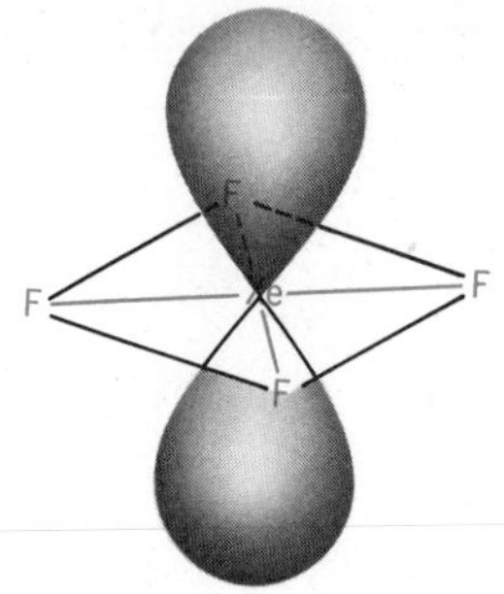


Figure 19.29
*Molecular structure of $XeF_4$.*

gives

Xe ⇅ ⇅ ↑ ↑ ↑ ↑ (${d^2sp^3}$) __ __ __ (unhybridized $5d$)

Finally, bonding with fluorine gives

Xe (in $XeF_4$) ⇅ ⇅ ↑x ↑x ↑x ↑x (${d^2sp^3}$) __ __ __ (unhybridized $5d$)

In $XeF_4$ there are six electron pairs about the Xe and both valence bond theory, with its $d^2sp^3$ hybrids, as well as the electron-pair repulsion theory predict that these are directed toward the corners of an octahedron. As we have seen before, in a similar situation, the two lone pairs occupy positions on opposite sides of a square plane containing the four ligand atoms, so that $XeF_4$ has a square planar structure (Figure 19.29).

Since the initial discovery of noble gas compounds by Bartlett, three of the noble gases have been demonstrated to form compounds, Rn, Xe, and Kr. The lighter ones, because of their much higher ionization energies, do not appear able to form chemical compounds.

Bartlett's work on Xe has taught chemists an important lesson. So firmly convinced were they that the noble gases were totally unreactive that after some initial experiments attempting to react Xe with fluorine had failed in the 1930s, no further efforts were made to explore the possibility that they were not inert. It is interesting that the noble gas compounds that have been obtained do not present any particular problem in bonding. In fact, many of the interhalogen compounds that had already been found (for example, $BrF_5$, $ICl_4^-$, $IF_7$) have the same number of electrons in the valence shell of the central atom as do the noble gas compounds. The same concepts that we applied to other compounds in which the octet was exceeded, therefore, work with the noble gas compounds too. The failure by chemists to recognize the possibility that the noble gases might react reflects a "blind spot" in their thinking that was probably founded in an overzealous acceptance of the stability and inertness of the $ns^2np^6$ octet of electrons.

## INDEX TO QUESTIONS

General 1, 23
Preparation of free elements 2, 3, 4, 5, 6, 7, 8
Structure of free elements 9, 11, 12, 13, 14, 15, 16
Allotropes 10, 11, 12, 14
Oxidation numbers 17, 18, 20, 21
Hydrides 19, 26

Catenation 24, 25, 27

Preparation of hydrides 29, 30, 31, 32, 33

Boron hydrides 34, 35

Structure of nonmetal hydrides 22

Oxygen compounds, properties 36

Preparation of nonmetal oxides 37, 38, 39, 41

Structure of nonmetal oxides 40, 45, 47

Simple oxoacids and oxoanions 42, 43, 44, 46, 48, 49, 50, 53

Polymeric oxoacids and oxoanions 28, 51, 52, 54, 55, 56, 57, 61, 62

Halogen compounds of nonmetals 50, 58, 59, 60, 61, 62, 63, 64

Noble gas compounds 65

## REVIEW QUESTIONS

**19.1** Why are the chemistries of metalloids and nonmetals conveniently discussed together? In what ways do metalloids resemble nonmetals? How do metalloids differ from nonmetals?

**19.2** Why are metalloids usually produced by chemical reduction rather than by oxidation?

**19.3** Write chemical equations for
(a) The chemical reduction of $BCl_3$ with hydrogen
(b) The production of Si from $SiO_2$ using carbon as a reducing agent
(c) The reduction of $As_2O_3$ with hydrogen

**19.4** What is the major source of $N_2$, $O_2$? Why is He found in underground deposits? Why are only small quantities of Rn observed in nature?

**19.5** Write a balanced chemical equation for the production of $Cl_2$ from HCl by reaction with $KMnO_4$ (the manganese is recovered as $MnCl_2$). What is the net ionic equation for the reaction?

**19.6** Complete the following chemical equations. If no reaction occurs, write N.R.
(a) $Cl_2 + KI$
(b) $Br_2 + CaF_2$
(c) $I_2 + MgCl_2$
(d) $F_2 + SrCl_2$

**19.7** Describe the recovery of $Br_2$ from sea water. Why is $C_2H_4Br_2$ added to leaded gasoline? Write a chemical equation for the combustion of $Pb(C_2H_5)_4$ in the presence of $C_2H_4Br_2$ (assume that C and H appear as $CO_2$ and $H_2O$ in the exhaust).

**19.8** Why can't fluorine be produced by the electrolysis of aqueous NaF? What will be the electrolysis reaction in this case?

**19.9** What appears to be the dominant factor in determining the complexity of the molecular structures of the elemental nonmetals?

**19.10** How is ozone formed in photochemical smog? Why is ozone being considered as a replacement for chlorine in killing bacteria in municiple water treatment?

**19.11** What are the two allotropic forms of carbon? How do they differ from one another in terms of structure and physical properties?

**19.12** What are the two allotropic forms of sulfur? How do they differ? Outline the physical changes that take place when sulfur is gradually heated to its boiling point and relate them to the structural changes that occur.

**19.13** Compare the structures of $N_2$ and $P_4$. What is the P—P—P bond angle in $P_4$? How does this compare to the normal bond angles that would occur with overlap of *p* orbitals? Can you suggest how this might account for the high reactivity of $P_4$?

**19.14** What are the structural differences between white and black phosphorus?

**19.15** How do we account for the electrical conductivity of graphite?

**19.16** What is the structural unit that occurs in elemental boron? Practice sketching its shape.

**19.17** What limitations are there on the usefulness of the oxidation number concept as a means of classifying compounds of the nonmetals? What is the oxidation number of sulfur in each of the following: $SO_2$, $S_2O_3^{2-}$, $SO_4^{2-}$, $S_2O_6^{2-}$, $S_5O_6^{2-}$, $S_6O_6^{2-}$? Draw the structure of each species. Among these compounds, what (if any) relationship exists between oxidation number and the number of covalent bonds formed by each sulfur atom?

**19.18** Write electron dot formulas of the compounds of nitrogen listed in Table 19.2.

**19.19** The following are arranged in order of decreasing basicity: $NH_3 > N_2H_4 > NH_2OH$. Can you explain this order in terms of the structures of the molecules?

**19.20** What are the maximum oxidation numbers of the nonmetals in each of the A Groups? Give an example for each of the nonmetals in period 3.

**19.21** What charges would be expected on each of the simple anions of the nonmetals?

**19.22** What are the molecular shapes of each of the following: $PH_3$, $H_2S$, $CH_4$, $H_2Se$, $Si_2H_6$?

**19.23** What explanation can be offered for the fact that the nonmetals in the third and succeeding periods do not appear to form stable $p\pi$-$p\pi$ bonds?

**19.24** What is meant by catenation? Which element exhibits this property in the largest number of compounds?

**19.25** The structures of the polythionate ions ($S_2O_6^{2-}$, $S_3O_6^{2-}$, $S_4O_6^{2-}$, etc.) were written on pp. 557 and 558 with the sulfur atoms in a straight line. The S—S—S angle in these compounds is actually in the neighborhood of 100° (rather than 180°). Make a series of sketches that more accurately depict the structures of these ions.

**19.26** Hydrogen forms ionic hydrides with the Group IA elements and also with the heavier elements in Group IIA. Write chemical equations for the reactions of Na and Ca with elemental $H_2$.

**19.27** Would the polysulfide ions, $S_x^{2-}$, be expected to be linear? Explain your answer.

**19.28** The peroxydiphosphate ion has the same general composition as the peroxydisulfate ion, $S_2O_8^{2-}$, with P substituted for sulfur. Draw the electron dot structure for the peroxydiphosphate ion and indicate the charge on the ion. (Hint: The charge is *not* $-2$.)

**19.29** What two general methods for the preparation of nonmetal hydrides were presented in this chapter?

**19.30** Why do the anions of the nonmetals become easier to protonate as we move from right to left across a period (e.g., from $Cl^-$ to $Si^{4-}$)?

**19.31** What is the Haber process? Why is the chemical reaction in the Haber process run at a high pressure and high temperature?

**19.32** Under the proper conditions, HI will react with $H_2SO_4$ to generate $I_2$ and $H_2S$. Write a balanced chemical equation for this reaction.

**19.33** Write balanced chemical equations for the hydrolysis of each of the following:

| | |
|---|---|
| (a) $Mg_3N_2$ | (d) $Ca_3P_2$ |
| (b) $CaC_2$ | (e) $Na_2S$ |
| (c) $Mg_2Si$ | (f) NaF |

**19.34** What is the structure of $B_2H_6$? What type of bonding exists in this molecule?

**19.35** In what way are the structures of $B_4H_{10}$ and $B_5H_9$ related? Compare the number of B—H—B three-center bonds and the number of "normal" B—H bonds in the two structures.

**19.36** What is an oxoacid? What is an oxoanion?

**19.37** List three methods of preparing nonmetal oxides and write a chemical reaction to illustrate each one.

**19.38** What is a disproportionation reaction? Give two examples.

**19.39** Complete and balance the following equations:

(a) $C_2H_6 + O_2 \rightarrow$
(b) $PH_3 + O_2 \rightarrow$
(c) $H_2S + O_2 \rightarrow$
(d) $SiH_4 + O_2 \rightarrow$
(e) $NH_3 + O_2 \rightarrow$
(f) $AsH_3 + O_2 \rightarrow$

**19.40** Write electron dot structures for the following:

| | |
|---|---|
| (a) $SO_2$ | (d) $N_2O_4$ |
| (b) $NO_2$ | (e) CO |
| (c) NO | |

**19.41** Write a chemical equation for the combustion of $C_8H_{18}$ (octane) in the presence of excess oxygen.

**19.42** Write chemical equations for the production of $HNO_3$ using $N_2$, $H_2$, $O_2$, and $H_2O$ as starting materials.

**19.43** Write chemical equations for the production of $H_2SO_4$ using $O_2$, S, and $H_2O$ as starting materials.

**19.44** What two methods are employed in the production of $H_3PO_4$? Write chemical equations for the reactions involved.

**19.45** Compare the molecular structures of $P_4$, $P_4O_6$, and $P_4O_{10}$.

**19.46** How do the structures of $H_3PO_3$ and $H_3PO_4$ compare? Write the formulas for all of the sodium salts that can be obtained by either partial or complete neutralization of these acids.

**19.47** What is the basic structural unit in the structure of quartz? Why do two kinds of quartz crystals exist? What is the relationship between them?

**19.48** What are the geometric structures of the following:

(a) $PO_4^{3-}$ (d) $S_2O_3^{2-}$
(b) $ClO_2^-$ (e) $SO_3^{2-}$
(c) $NO_3^-$ (f) $ClO_3^-$

**19.49** Give examples of three methods of preparing oxoacids or their anions.

**19.50** Why are solutions of $OCl^-$ stable while $OI^-$ disproportionates into $I^-$ and $IO_3^-$?

**19.51** Use electron dot structures to show how $H_2S_2O_7$ results from the removal of one molecule of $H_2O$ from two molecules of $H_2SO_4$.

**19.52** Pyrophosphoric acid may be obtained from its sodium salt in the following way. The sodium salt is dissolved in water and treated with lead nitrate to precipitate lead pyrophosphate which, when treated with $H_2S$, liberates the free acid. Write chemical equations for all of the reactions that take place.

**19.53** Write chemical equations for the disproportionation of $KClO_3$ when it is heated at moderate temperatures in the absence of a catalyst. What reaction occurs if a catalyst such as $MnO_2$ is present?

**19.54** Write generalized formulas for the "ortho" anions, "pyro" anions, and "meta" anions of the period 3 elements.

**19.55** Illustrate the types of structures that are obtained by joining $XO_4$ tetrahedra through the sharing of the following:

(a) One corner (c) Three corners
(b) Two corners (d) Four corners

**19.56** Describe how the properties of asbestos and talc are related to the structures of the silicate anions they contain.

**19.57** What metaphosphate ion is used in many solid detergents? What is their function? What environmental problems do they cause?

**19.58** What are the structures of the following:

(a) $PCl_3$ (e) $ClF_5$ (h) $SiF_4$
(b) $BBr_3$ (f) $BrF_4^-$ (i) $SF_6$
(c) $SF_4$ (g) $SCl_2$ (j) $I_3^-$
(d) $ClF_3$

**19.59** In $SF_6$, $d^2sp^3$ hybrid orbitals are used to make all six valence electrons of sulfur available for bonding. What type of hybrid orbitals would be used in the bonding in $SF_4$? How does the resulting molecular structure compare with that predicted by the electron-pair repulsion theory?

**19.60** How does size influence the number of halogen atoms that can combine with a given nonmetal in (a) period 2, (b) period 5?

**19.61** Use electron dot structures to show how hydrolysis of $(CH_3)_2SiCl_2$ gives rise to silicone polymers.

**19.62** If some $(CH_3)_3SiCl$ is added to $(CH_3)_2SiCl_2$ before hydrolysis, shorter polymeric chains are produced than if pure $(CH_3)_2SiCl_2$ is hydrolyzed. Explain.

**19.63** Sulfur hexafluoride is so stable toward reaction with other reagents and toward decomposition that it is used as a gaseous insulator in high-voltage generators. What accounts for this inert behavior?

**19.64** Why is $SiCl_4$ so susceptible to attack by water while $CCl_4$ is unreactive toward water?

**19.65** Write electron dot structures for $XeO_3$ and $XeO_4$. What geometric structures would you expect them to exhibit? What structure would you expect for $XeOF_2$ (Xe bonded to two fluorine atoms and one oxygen atom)?

# 20 THE TRANSITION ELEMENTS

In this chapter we consider the collection of elements, generally called the transition elements, that fit into the periodic table between Groups IIA and IIIA. We saw in Chapter 3 that these elements arise as a consequence of the gradual filling of *d* and *f* subshells and, as a rule, a transition element is usually considered to be one that possesses an incompletely filled *d* or *f* subshell in either the free state or in one of its compounds. We will also include in this discussion the elements in Group IIB, zinc, cadmium, and mercury, which are found at the extreme right of the transition elements and which complete a transition series (a horizontal row) by having their outer *d* subshells filled.

## 20.1 GENERAL PROPERTIES

To discuss their properties it is convenient to divide the transition elements into two categories. These are the **d-block elements** (or **main transition elements**), which in our condensed version of the periodic table are located in the main body of the table between Groups IIA and IIIA, and the **inner transition elements,** which correspond to the two long rows of 14 elements each that *are placed just below the table* (Figure 20.1). The *d*-block elements themselves consist of three rows that are frequently referred to as the first, second, and third transition series.

Like the representative elements, most of the *d*-block transition elements possess certain vertical similarities in chemical and physical properties and are therefore divided into groups, designated as B groups. They begin with Group IIIB on the left and proceed through Group VIIB. Next there follows a set of nine elements collectively termed Group VIII, and finally, on the right, we find Groups IB and IIB. This numbering sequence of the B groups results because the group numbers are chosen to correspond to the highest positive oxidation state that their elements normally exhibit.

The division of the periodic table into A and B groups (for example, IIIA and IIIB) suggests that there may be certain parallels between the two, and to a limited degree this is true. The similarities, however, are restricted primarily to likenesses of composition, structure, and maximum positive oxidation state, rather than chemical reactivity. Some examples of these are found in Table 20.1.

The Group VIII elements, which lie between Groups VIIB and IB, are classed differently from the other *d*-block elements because they have no counterparts among the representative elements. Within this group there are greater *horizontal similarities* than vertical ones, and the description of

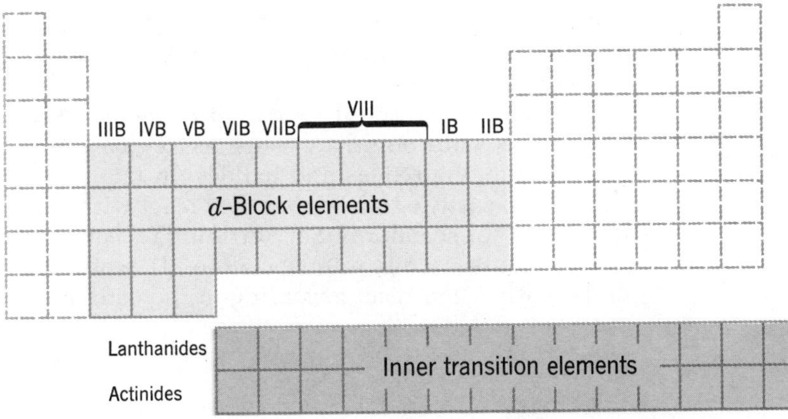


Figure 20.1
*The transition elements.*

the behavior of these elements is usually organized on the basis of horizontal groups of three elements each, called **triads.** Each triad is named after the best-known element within it. Thus we have the iron triad, the palladium triad, and the platinum triad.

As a class, the transition elements are all typical metals; they possess a characteristic metallic luster and are good conductors of heat and electricity. Silver has the highest electrical and thermal conductivity of any metal, followed closely by copper. Copper, of course, is used in vast quantities in electrical wiring. Silver is the preferred coating for mirrors because of its high reflectivity.

The chemical and physical properties of the transition elements cover a wide range and account for the wide range of uses to which they are applied. Some of them are very hard and strong and are used as structural metals, either in the pure state or as alloys. Iron is the prime example; steels of different properties are formed by incorporating it with other transition elements such as chromium, cobalt, and nickel. Even copper, which is very soft when pure, can be made very strong by forming an alloy with beryllium. Beryllium-copper alloys are used in place of steel in nonsparking tools for use in explosive atmospheres and as high-quality springs in cameras and other precision instruments.

The melting points of the transition elements also vary over a wide range. Most are high-melting. Tungsten, with a melting point of approxi-

Table 20.1
*Some similarities of chemical composition between* A *and* B *group compounds*

| Group | Compounds | Group | Compounds |
|---|---|---|---|
| IVA | $CCl_4$, $SnCl_4$, $CO_2$ | IVB | $TiCl_4$, $TiO_2$ |
| VA | $PO_4^{3-}$, $POCl_3$ | VB | $VO_4^{3-}$, $VOCl_3$ |
| VIA | $SO_4^{2-}$, $S_2O_7^{2-}$ | VIB | $CrO_4^{2-}$, $Cr_2O_7^{2-}$ |
| VIIA | $ClO_4^-$, $Cl_2O_7$ | VIIB | $MnO_4^-$, $Mn_2O_7$ |
| IA | NaCl | IB | CuCl, AgCl |
| IIA | $CaCl_2$ | IIB | $ZnCl_2$ |

mately 3400°C, is used as filaments in light bulbs. At the other extreme is mercury, which is a liquid at room temperature and is useful as a liquid in thermometers.

The chemical reactivity of the free elements varies greatly too. Most react directly with nonmetals such as oxygen and the halogens to produce the corresponding oxides and halides. In fact, some of the transition elements are so easily oxidized that they react with water to liberate hydrogen. This is true for scandium (Sc), yttrium (Y), lanthanum (La), and the lanthanide elements (atomic numbers 58 to 71), which have very negative reduction potentials and react according to the equation

$$M + 3H_2O \longrightarrow \tfrac{3}{2}H_2 + M(OH)_2$$

Other transition elements, such as Pt and Au, are very resistant to oxidation and are insoluble in both protonic acids such as HCl as well as oxidizing acids such as $HNO_3$. It is interesting, however, that these two metals do dissolve slowly in a 3:1 mixture of HCl and $HNO_3$ (called *aqua regia*).

Despite some rather marked differences in behavior, there are several characteristics that the transition elements have in common with each other:

1. **Multiple oxidation states.** With only a few exceptions, the transition elements tend to exhibit more than one oxidation state.
2. **Many of their compounds are paramagnetic.** Because the transition elements tend to have partially completed *d* or *f* subshells in both the free state and in their compounds, the metal atoms often possess unpaired electrons. These impart the property of paramagnetism.
3. **Many (if not most) of their compounds are colored.** The origin of the colors of complex ions of the transition elements will be discussed in Section 20.11.
4. **They have a strong tendency to from complex ions.** As a group, these elements form a huge number of complex ions of varying degrees of complexity. The last six sections of this chapter are devoted to a discussion of their structures and bonding.

## 20.2 ELECTRONIC STRUCTURE AND OXIDATION STATES

In Chapter 3 we saw that as we proceed from left to right across a period through the main transition elements, there is a gradual filling of the *d* subshell that lies just below the outer shell. In period 4, for example, this corresponds to the 3*d* subshell, as seen for the electronic structures of the first row elements given in Table 20.2. Each of these elements possesses a com-

Table 20.2
*Electronic structures for elements in the first transition series*

| | |
|---|---|
| Sc $[Ar]3d^1 4s^2$ | Fe $[Ar]3d^6 4s^2$ |
| Ti $[Ar]3d^2 4s^2$ | Co $[Ar]3d^7 4s^2$ |
| V $[Ar]3d^3 4s^2$ | Ni $[Ar]3d^8 4s^2$ |
| Cr $[Ar]3d^5 4s^1$ | Cu $[Ar]3d^{10} 4s^1$ |
| Mn $[Ar]3d^5 4s^2$ | Zn $[Ar]3d^{10} 4s^2$ |

Table 20.3
*Electronic structures of elements of the second and third transition series*

| Period 5 Second Transition Series | Period 6 Third Transition Series |
|---|---|
| Y [Kr]$5s^2 4d^1$ | La [Xe]$6s^2 5d^1$ |
| Zr [Kr]$5s^2 4d^2$ | Hf [Xe,$4f^{14}$]$5d^2 6s^2$ |
| Nb [Kr]$5s^1 4d^4$ | Ta [Xe,$4f^{14}$]$5d^3 6s^2$ |
| Mo [Kr]$5s^1 4d^5$ | W [Xe,$4f^{14}$]$5d^4 6s^2$ |
| Tc [Kr]$5s^1 4d^6$ | Re [Xe,$4f^{14}$]$5d^5 6s^2$ |
| Ru [Kr]$5s^1 4d^7$ | Os [Xe,$4f^{14}$]$5d^6 6s^2$ |
| Rh [Kr]$5s^1 4d^8$ | Ir [Xe,$4f^{14}$]$5d^7 6s^2$ |
| Pd [Kr]$5s^0 4d^{10}$ | Pt [Xe,$4f^{14}$]$5d^9 6s^1$ |
| Ag [Kr]$5s^1 4d^{10}$ | Au [Xe,$4f^{14}$]$5d^{10} 6s^1$ |
| Cd [Kr]$5s^2 4d^{10}$ | Hg [Xe,$4f^{14}$]$5d^{10} 6s^2$ |

pleted argon core with additional electrons in the $3d$ and $4s$ subshells. Notice once again that chromium and copper are anomalous owing to the extra stability that is associated with half-filled and filled subshells. Similar irregularities are found in the second and third transition series (Table 20.3), although other factors in addition to those having to do with half-filled and filled subshells are apparently involved and, therefore, no simple correlations can be made.

Among the inner transition elements, the lanthanides and actinides (so named because lanthanum and actinium have properties more or less typical of their respective series), there is a gradual filling of an $f$ subshell that lies *two* shells below the outer shell. Thus, in Table 20.4 we see that as we pass through the lanthanides in period 6, the $4f$ subshell is completed; while in the following period, as we pass through the actinides, the $5f$ subshell becomes populated.

Table 20.4
*Electronic structure of the lanthanide and actinide elements*

| Lanthanides | Actinides |
|---|---|
| La [Xe]$5d^1 6s^2$ | Ac [Rn]$6d^1 7s^2$ |
| Ce [Xe]$4f^2 6s^2$ | Th [Rn]$6d^2 7s^2$ |
| Pr [Xe]$4f^3 6s^2$ | Pa [Rn]$5f^2 6d^1 7s^2$ |
| Nd [Xe]$4f^4 6s^2$ | U [Rn]$5f^3 6d^1 7s^2$ |
| Pm [Xe]$4f^5 6s^2$ | Np [Rn]$5f^5 7s^2$ |
| Sm [Xe]$4f^6 6s^2$ | Pu [Rn]$5f^6 7s^2$ |
| Eu [Xe]$4f^7 6s^2$ | Am [Rn]$5f^7 7s^2$ |
| Gd [Xe]$4f^7 5d^1 6s^2$ | Cm [Rn]$5f^7 6d^1 7s^2$ |
| Td [Xe]$4f^9 6s^2$ | Bk [Rn]$5f^8 6d^1 7s^2$ |
| Dy [Xe]$4f^{10} 6s^2$ | Cf [Rn]$5f^{10} 7s^2$ |
| Ho [Xe]$4f^{11} 6s^2$ | Es [Rn]$5f^{11} 7s^2$ |
| Er [Xe]$4f^{12} 6s^2$ | Fm [Rn]$5f^{12} 7s^2$ |
| Tm [Xe]$4f^{13} 6s^2$ | Md [Rn]$5f^{13} 7s^2$ |
| Yb [Xe]$4f^{14} 6s^2$ | No [Rn]$5f^{14} 7s^2$ |
| Lu [Xe]$4f^{14} 5d^1 6s^2$ | Lr [Rn]$5f^{14} 6d^1 7s^2$ |

The chemical and physical properties of the transition elements are controlled, of course, by their electronic structures. With the *d*-block elements the outer *s* and underlying *d* subshells are of nearly equal energy and, consequently, when these elements react, the *d* electrons are able to participate in bonding. The importance of the *d* electrons in determining the chemistry of these elements accounts for their varied chemical properties, including the multiplicity of oxidation states, illustrated for the first transition series in Table 20.5.

Examination of Table 20.5 reveals several interesting points. First, we see that *except for Sc all of the first row elements show a +2 oxidation state.* You might recall that, in general, when electrons are lost by an atom, those lost first come from the subshell having the highest principal quantum number. Consequently the 4*s* electrons are lost before the 3*d*; therefore the +2 state results simply from the loss of the two outer 4*s* electrons. Exceptions to this are Cr and Cu where a 3*d* electron is also lost.

A second point is that *elements at the left of a transition series prefer the highest oxidation state and, as we proceed to the right, the stabilities of*

Table 20.5
*Oxidation states of the first transition series elements*

| | | |
|---|---|---|
| Sc | (+2) | Not known |
| | **+3** | Only oxidation state; Sc reduces $H_2O$ to $H_2(g)$ |
| Ti | +2 | Not stable in $H_2O$, reduces water |
| | +3 | Prepared by reducing Ti(IV) with Zn |
| | **+4** | Most stable oxidation state |
| V | +1 | |
| | +2 | Easily oxidized |
| | +3 | Stable |
| | **+4** | Most stable under ordinary conditions |
| | +5 | $V_2O_5$ moderate oxidizing agent |
| Cr | +2 | Very easily oxidized |
| | **+3** | Most stable |
| | +6 | $CrO_4^{2-}$, $Cr_2O_7^{2-}$; good oxidizing agent |
| Mn | **+2** | Most stable |
| | +3 | Stable in complex ions |
| | +4 | $MnO_2$; good oxidizing agent |
| | +6 | $MnO_4^{2-}$ (manganate ion) stable in basic solution only |
| | +7 | $MnO_4^-$ (permanganate ion) very powerful oxidizing agent |
| Fe | +2 | Stable but easily oxidized to +3 |
| | **+3** | Most stable |
| | +4 | |
| | +6 | Rare |
| Co | **+2** | Most stable in water |
| | +3 | $Co^{3+}$ oxidizes water, stable in complex ions |
| Ni | **+2** | Most stable |
| | +3 | Rare, powerful oxidizing agent |
| Cu | +1 | $Cu^+$ disproportionates in water, $2Cu^+ \longrightarrow Cu^{2+} + Cu$. Stable in complex ions and insoluble CuCl |
| | **+2** | Most stable in water |
| Zn | **+2** | Only oxidation state for zinc |

*the lower oxidation states increase relative to the higher ones.* Thus with scandium all three electrons ($3d^14s^2$) are lost, and only the maximum +3 state is observed. With titanium, +2, +3, and +4 states are found; however, in aqueous solution the +4 state, corresponding to the "loss" of all four outer electrons, is most stable. In fact, the +2 state is so unstable that it reduces water to produce $H_2$ and Ti(IV), indicating that an oxidizing agent as weak as $H_2O$ is capable of removing electrons from $Ti^{2+}$.

As we continue to move to the right through the first transition series, we find that up to manganese the maximum oxidation state that is observed corresponds to the group number. However, the most stable state gradually becomes the lower one, while the higher oxidation states tend to become good oxidizing agents. With chromium in Group VIA, for example, the most stable state is $Cr^{3+}$. The +2 state is very easily oxidized to +3, while the +6 state, found in the $CrO_4^{2-}$ (chromate) and $Cr_2O_7^{2-}$ (dichromate) ions, is easily reduced. For manganese, the next element to the right, the +2 oxidation state is most stable and higher states tend to be very readily reduced. Permanganate ion, $MnO_4^-$, containing Mn(VII) is a very powerful oxidizing agent.

$$MnO_4^- + 8H^+ + 5e^- \rightleftharpoons Mn^{2+} + 4H_2O \qquad \mathscr{E}^0 = +1.49$$

After manganese, the occurrence of high positive oxidation states is rare. With iron the +2 state is fairly easily oxidized to $Fe^{3+}$, although stable solutions (as well as compounds) containing $Fe^{2+}$ are readily prepared. Moving to the right to cobalt, the +2 state tends to be most stable and, in fact, in the absence of complex ion-forming agents $Co^{3+}$ oxidizes water.

$$2H_2O + 4Co^{3+} \longrightarrow O_2 + 4H^+ + 4Co^{2+}$$

For nickel the +3 state is very rare, and only the +2 state is observed under ordinary conditions. Following nickel we have copper, which forms compounds in both the +1 and +2 states, although the +1 state is easily oxidized. Finally we arrive at zinc, which can lose only two electrons to form $Zn^{2+}$, having a pseudonoble gas configuration.

The trend toward increasing stability of the lower oxidation states as we proceed across a transition series occurs for the second and third series too, as seen in Table 20.6. In the second row, for instance, the maximum oxidation state is most stable for Zr while the lowest oxidation state is most stable for Ag. In the third transition series we again see that the lower oxidation states become more prevalent and increasingly more stable as we move

**Table 20.6**
*Oxidation states of second and third transition series elements (common oxidation states in boldface type)*

| | | | |
|---|---|---|---|
| Y | **+3** | La | **+3** |
| Zr | +2, +3, **+4** | Hf | +3, **+4** |
| Nb | +2, +3, +4, **+5** | Ta | +2, +3, +4, **+5** |
| Mo | +2, +3, **+4**, **+5**, **+6**, +8 | W | +2, +3, +4, **+5**, **+6** |
| Tc | +2, +3, **+4**, +5, +6, **+7** | Re | **+3**, **+4**, +5, +6, **+7** |
| Ru | **+2**, **+3**, **+4**, +5, +6, +7, +8 | Os | +2, +3, **+4**, +5, **+6**, **+8** |
| Rh | +1, +2, **+3**, **+4**, +5, +6 | Ir | +1, +2, **+3**, **+4**, +5, +6 |
| Pd | **+2**, +3, **+4** | Pt | **+2**, +3, **+4**, +5, +6 |
| Ag | **+1**, +2, +3 | Au | **+1**, **+3** |
| Cd | **+2** | Hg | **+1**, **+2** |

to the right. Thus mercury, at the end of the series in Group IIB, forms compounds in both the +1 and +2 states. The +1 oxidation state of mercury seems surprising at first glance since it would appear to contain a single electron in an *s* orbital outside a pseudonoble gas core. However, experimental evidence indicates that the mercury(I) species is actually a dimer, $Hg_2^{2+}$. For example, mercurous compounds such as $Hg_2Cl_2$ are diamagnetic, indicating that no unpaired electrons are present. A simple $Hg^+$ ion would contain an odd number of electrons (79) and could not possibly have all of them paired. In the $Hg_2^{2+}$ ion, two $Hg^+$ ions seem to be joined by a covalent bond, thus pairing the odd electrons on each mercury atom. Other evidence for the existence of the $Hg_2^{2+}$ species relates to equilibria, as illustrated in Problem 20.53 at the end of the chapter.

Among the representative elements we found that as we proceed downward within a group the lower oxidation states become increasingly more stable compared to the higher ones. In the transition elements this trend is reversed and *in the second and third transition series it is found that the higher oxidation states are preferred.* For example, in Group VIB the $CrO_4^{2-}$ ion, with Cr in the +6 oxidation state, is a potent oxidizing agent, indicating that the chromate ion wants to pick up electrons to reduce the Cr to a lower oxidation state. In contrast to this, the $MoO_4^{2-}$ ion and other complex polynuclear molybdates (species containing more than one Mo atom) are poor oxidizing agents owing to the stability of the Mo(VI) species.

In a similar fashion, comparing the Group VIIB elements, we find that the $MnO_4^-$ ion is a very powerful oxidizing agent while the $ReO_4^-$ ion is a very weak oxidizing agent. In Group VIII, iron forms compounds in the +2 and +3 states, although there are unusual and relatively unstable compounds containing iron in oxidation states as high as +6. For example, the $FeO_4^{2-}$ ion [containing Fe(VI)] is an even stronger oxidizing agent in acid solution than the permanganate ion. Below iron, however, we find ruthenium and osmium in oxidation states as high as +8 (for example, $RuO_4$ and $OsO_4$).

In contrast to the wide range of chemical properties of the *d*-block elements, the lanthanides exhibit a remarkable sameness of properties. The 4*f* subshell, which is only partially filled for most of these elements, is buried beneath the outer 5*d* and 6*s* subshells and does not interact to an appreciable extent with the surrounding chemical environment. Consequently the chemistry of the lanthanides, like that of lanthanum itself, is predominantly that of the +3 ion, and differences in behavior depend primarily on differences in ionic size.

The actinide elements exhibit a greater variation in oxidation numbers than do the lanthanides (e.g., uranium forms compounds in the +3, +4, +5, and +6 oxidation states). This is sometimes ascribed to the notion that the 5*f* orbitals of the actinides project out further toward the periphery of the atom than do the 4*f* orbitals of the lanthanides. As a result, the 5*f* orbitals are able to become involved to a greater degree in chemical bonding; hence more complex chemistry is observed.

## 20.3 ATOMIC AND IONIC RADII

We have seen before that many trends in properties can be correlated with variations that occur in atomic and ionic radii. This is true among the transition elements as well as the representative elements.

In Chapter 3 the horizontal and vertical trends in atomic size were discussed. Let's briefly review them here. As we move across a given transition series there is only a gradual decrease in atomic radius. This is because the

Table 20.7
*Atomic Radii* (Å)

| Sc | Ti | V | Cr | Mn | Fe | Co | Ni | Cu | Zn |
|---|---|---|---|---|---|---|---|---|---|
| 1.62 | 1.47 | 1.34 | 1.27 | 1.26 | 1.26 | 1.25 | 1.24 | 1.28 | 1.38 |
| Y | Zr | Nb | Mo | Tc | Ru | Rh | Pd | Ag | Cd |
| 1.80 | 1.60 | 1.46 | 1.39 | 1.36 | 1.34 | 1.34 | 1.37 | 1.44 | 1.54 |
| La | Hf | Ta | W | Re | Os | Ir | Pt | Au | Hg |
| 1.87 | 1.58 | 1.46 | 1.39 | 1.37 | 1.35 | 1.36 | 1.38 | 1.44 | 1.57 |

3*d* electrons that are added to the atom shield the outer 4*s* electrons quite well from the increasing nuclear charge and, as a result, the effective nuclear charge experienced by the outer electrons rises only slowly. Consequently, only a small size decrease occurs (Table 20.7).

Vertically, we find a rather large increase in size among the *d*-block elements between periods 4 and 5 (from the first transition series to the second), just as we expect. However, between periods 5 and 6 there is only a very small size increase (and in some cases, none at all). This is a consequence of the **lanthanide contraction,** the gradual size decrease that occurs across the lanthanide series (from atomic numbers 58 to 71). Apparently this just cancels the size increase that would be expected as we go from period 5 to period 6. Consequently, the period 6 elements that follow the lanthanides are essentially the same size as those above them in period 5.

These variations in size have some very pronounced chemical and physical consequences. They can be correlated, for example, with variations in ionization energies (IE), as shown in Table 20.8. Here we see that the gradual horizontal size decrease that we associated with an increase in effective nuclear charge is also accompanied by an increase in IE. As we might expect, this increasing difficulty encountered in removing an outer electron from the isolated atoms is reflected in a gradual (although not altogether uniform) rise in their standard reduction potentials. In other words, as we proceed across the table from left to right, it generally becomes more difficult to oxidize the elements.

The effect of the lanthanide contraction is demonstrated in these properties as well. Among the representative elements the IE generally decreases

Table 20.8
*Ionization energy (kJ/mole)*

| Sc | Ti | V | Cr | Mn | Fe | Co | Ni | Cu | Zn |
|---|---|---|---|---|---|---|---|---|---|
| 632 | 660 | 651 | 653 | 718 | 763 | 760 | 737 | 746 | 907 |
| Y | Zr | Nb | Mo | Tc | Ru | Rh | Pd | Ag | Cd |
| 616 | 672 | 665 | 694 | 720 | 711 | 720 | 805 | 732 | 869 |
| La | Hf | Ta | W | Re | Os | Ir | Pt | Au | Hg |
| 540 | 675 | 763 | 771 | 761 | 842 | 868 | 866 | 891 | 1008 |

as we proceed down a group, paralleling the increase in size. This phenomenon is also observed among the transition elements on going from period 4 to period 5. However, from periods 5 to 6 there is an increase in nuclear charge without an accompanying increase in size and, as a result, the IE increases. This in turn manifests itself in reduction potentials that tend to be quite high for the third transition series elements, thereby accounting for their virtually inert behavior toward many oxidizing agents.

Still another consequence of the lanthanide contraction is the very high densities of the third-row transition elements. For example, from Rh to Ir we find an increase of 89 mass units (about an 87% increase in mass) with only a 0.02 Å increase in radius. Iridium therefore has 87% more mass packed into essentially the same size atom; it is not surprising, therefore, to find that Ir is about 81% more dense than Rh ($d_{Ir}$ = 22.5 g/ml, $d_{Rh}$ = 12.4 g/ml).

## 20.4 METALLURGY

Metallurgy is the process whereby a metal is extracted from its ore and brought to the point where it can be put to practical use. In Chapter 18 we discussed one aspect of this, the reduction of metal compounds to produce the free element. The desirable physical properties of many of the transition metals, such as high strength, hardness, and high melting points, make them extremely important in modern technology. Some of the methods and procedures that are used to obtain them from their ores are worth examining.

An **ore** is a substance that contains a particular desirable constituent in a high enough concentration that its extraction from the ore is economically worthwhile. Thus, for example, many minerals may contain small amounts of iron but only those that are rich in iron are used to prepare this metal. As the earth's reserves of rich ores are consumed, it will become the job of the chemist to devise new ways to obtain metals such as iron from less rich ores.

Generally, we can divide metallurgical processes into three categories:

1. **Concentration.** Ores that contain substantial amounts of impurities, such as rock, must often be treated to concentrate the metal-bearing constituent. Pretreatment of an ore is also carried out to convert some metal compounds into substances that can be more easily reduced.
2. **Reduction.** This is the topic discussed in Chapter 18. The particular procedure employed for a given metal depends on its ease of reduction to the free state.
3. **Refining.** Often, during reduction, substantial amounts of impurities become introduced into the metal. Refining is the process whereby these impurities are removed and the composition of the metal adjusted (alloys formed) to meet specific applications.

Let us now take a brief look at each of these steps as they apply to some important metals.

**CONCENTRATION.** Not all ores have to be subjected to a pretreatment step prior to reduction, although most of them must. These pretreatment procedures involve the separation of the metal-bearing component of the ore from unwanted or interfering impurities. This is particularly important for low-grade ores in which the desired metal is present only in small amounts.

As expected, different methods are applied to different ores, depending on the specific properties of the impurities and the metal compounds. We can divide these procedures into two classes: *physical separations,* in which the chemical compositions of the constituents are not altered, and *chemical separations,* which make use of the chemical properties of the different substances in the ore. Some metals, such as silver and gold, are found in deposits as the free element, and their recovery simply involves removing them from the rock and sand with which they are mixed.

One of the earliest forms of physical separation was used by the "forty-niners" in panning for gold. A mixture of sand containing (hopefully!) particles of metallic gold was placed in a shallow pan with water. The mixture was swirled about and the sand was washed away, leaving the gold dust in the bottom of the pan. The success of this procedure was based on the fact that gold is about nine times as dense as the sand and gravel impurities. As a result, the lighter impurities are more easily washed away than the more dense metal.

Another way of removing metallic gold (and silver) from its ore is to treat the mixture with metallic mercury (a liquid) in which silver and gold dissolve to form an alloy called an **amalgam.** The silver and gold are later recovered by distilling away the mercury that is reclaimed and used again. You are probably familiar with silver and gold amalgams as the material used by dentists to fill teeth.

A physical separation technique that can be applied to the sulfide ores of zinc, copper, and lead is called **flotation.** In this process, illustrated in Figure 20.2, the ore is finely ground and added to a mixture of water and oil containing suitable additives. The metal-bearing component of the ore becomes coated by the oil while the unwanted material, called the **gangue,** is wetted by the water. A stream of air is then blown through the mixture and the oil-covered mineral is carried to the surface by bubbles where it is trapped in a froth that can be removed to recover the metal compound. The gangue, on

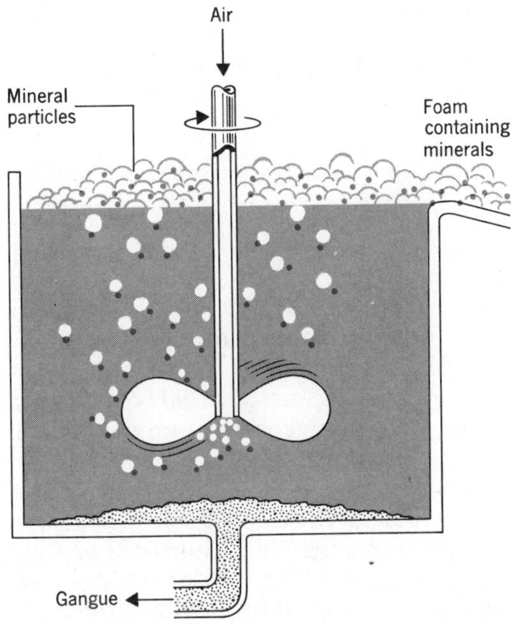


Figure 20.2
*The flotation process.*

the other hand, simply settles to the bottom of the apparatus and is later discarded.

Chemical methods of concentrating the metal-bearing component of an ore vary considerably because of the variety of chemical properties exhibited by the metals and their compounds. For example, aluminum, whose electrolytic reduction was described in Chapter 16, occurs in deposits of bauxite, a form of $Al_2O_3$. In this case the ore is concentrated by taking advantage of the amphoteric behavior of aluminum. The bauxite is treated with concentrated base, which dissolves the $Al_2O_3$ to produce aluminate ion, $AlO_2^-$.[1]

$$Al_2O_3 + 2OH^- \longrightarrow 2AlO_2^- + H_2O$$

After it is removed from the gangue, the solution is acidified, precipitating $Al(OH)_3$, which yields pure $Al_2O_3$ when heated.

$$2Al(OH)_3 \longrightarrow Al_2O_3 + 3H_2O$$

This purified aluminum oxide serves as the charge in the Hall process discussed previously.

Another chemical pretreatment, often given to a sulfide ore, is called **roasting.** Here the ore is heated in air, converting the metal sulfide to an oxide that is more conveniently reduced.

$$2PbS + 3\ O_2 \longrightarrow 2PbO + 2SO_2$$
$$2ZnS + 3\ O_2 \longrightarrow 2ZnO + 2SO_2$$

This industrial process is a severe source of air pollution unless the $SO_2$ is recovered.

**REDUCTION.** In Chapter 18 we saw that there are several ways of reducing a metal compound to produce the free element. All of these methods have found application in the production of metals commercially. Titanium, for example, is prepared by reacting $TiCl_4$ with the more active metal, magnesium. The $TiCl_4$ is produced from rutile, a fairly pure source of $TiO_2$, by reaction with chlorine gas and carbon.

$$TiO_2 + 2C + 2Cl_2 \longrightarrow TiCl_4 + 2CO$$

The titanium tetrachloride is a volatile liquid (boiling point = 136°C) and can be separated from impurities by distillation. The final reduction to the metal follows the equation

$$TiCl_4 + 2Mg \longrightarrow Ti + 2MgCl_2$$

Titanium is a very useful metal because it is considerably lighter than steel ($d_{Fe}$ = 7.86 g/ml, $d_{Ti}$ = 4.51 g/ml), yet does not lose its strength at high temperature as does aluminum. It is used in large quantities in jet engines, and replaces aluminum and steel in other aircraft applications. Its oxide, $TiO_2$, is also used in large quantities as a white pigment in paint. It is better than white lead, $Pb_3(OH)_2(CO_3)_2$, because it appears to be of low toxicity and because it doesn't darken in the presence of $H_2S$ as the lead-based pigments do.

Active metals such as magnesium are very expensive reducing agents because they themselves are difficult and costly to prepare. As a result, less expensive reducing agents are employed whenever possible. One of the cheapest of all is carbon, in the form of coke, which is produced from coal by heating it at high temperatures in the absence of air. This treatment drives

[1] The exact nature of this species is not known with certainty. See p. 526.

off the volatile components of the coal (from which other important chemicals are derived), leaving nearly pure carbon behind.

Some typical reductions of metal oxides with carbon were illustrated in Section 18.3, and many of the transition metals (for example, Zn, Cd, Fe, Co, Ni, Mo) can be prepared in this way. Undoubtedly the most important chemical reduction brought about with carbon is that of iron oxide, $Fe_2O_3$, to iron. This is accomplished in the **blast furnace,** developed in about 1300 A.D. and which in modern times takes the form shown in Figure 20.3.

The blast furnace is charged with a mixture of limestone, coke, and iron ore. The ore is generally composed primarily of $Fe_2O_3$ with impurities of $SiO_2$ (sand, ~10%) and smaller amounts of compounds containing sulfur, phosphorus, aluminum, and manganese. Heated air is forced in at the bottom of the furnace, where it reacts with carbon in a very exothermic reaction to produce carbon dioxide.

$$C + O_2 \longrightarrow CO_2 \qquad \Delta H = -394 \text{ kJ}$$

The large amount of heat generated in this region of the furnace raises the temperature to nearly 1900°C. As the hot gases rise, the $CO_2$ reacts with ad-

Figure 20.3
*The blast furnace.*

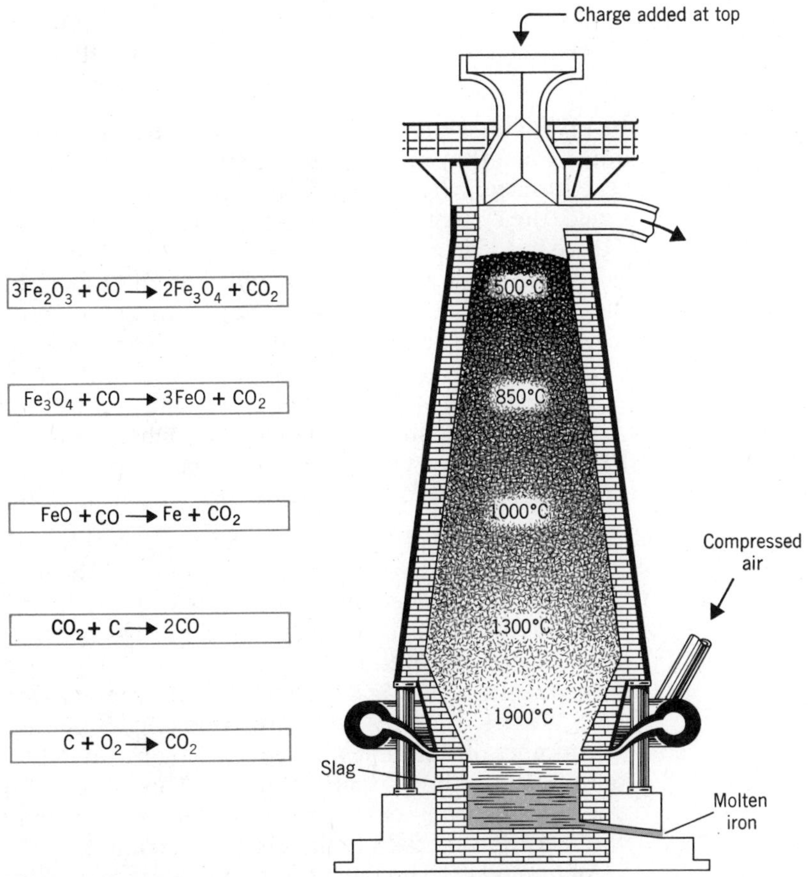

ditional carbon in an endothermic reaction to form carbon monoxide, the active reducing agent in the furnace.

$$CO_2 + C \longrightarrow 2CO \qquad \Delta H = +173 \text{ kJ}$$

The reduction of the iron oxide takes place in a series of steps. Near the top of the furnace, $Fe_2O_3$ is reduced to $Fe_3O_4$.

$$3Fe_2O_3 + CO \longrightarrow 2Fe_3O_4 + CO_2$$

Farther down, in a hotter region of the furnace, this is reduced to FeO.

$$Fe_3O_4 + CO \longrightarrow 3FeO + CO_2$$

Finally, still farther down the FeO is reduced to the metal which, at these high temperatures, is a liquid and trickles down to form a pool of molten metal at the base of the tower.

$$FeO + CO \longrightarrow Fe + CO_2$$

The function of the limestone in the furnace is to provide a basic medium with which acidic oxides, such as $SiO_2$ and $P_2O_5$, can react. At elevated temperatures limestone, $CaCO_3$, decomposes to form lime (CaO) and $CO_2$ according to the equation,

$$CaCO_3 \longrightarrow CaO + CO_2$$

The lime then reacts with the acidic oxides,

$$CaO + SiO_2 \longrightarrow CaSiO_3$$
$$3CaO + P_2O_5 \longrightarrow Ca_3(PO_4)_2$$
$$CaO + Al_2O_3 \longrightarrow Ca(AlO_2)_2$$

The products of these reactions have relatively low melting points and are liquids when they are formed. The mixture, called **slag,** also runs to the base of the furnace, where it floats atop the molten iron. As these two layers are formed, the charge in the furnace settles and additional limestone-coke-ore mixture is added at the top. In this way the blast furnace operates continuously, with fresh charge being added at the top and molten iron and slag being tapped off at the bottom. These furnaces are often run for months at a time before being shut down for routine maintenance.

The liquid iron, when it is withdrawn from the blast furnace, is called **pig iron** and consists of about 95% Fe and approximately 4% carbon, with small amounts of silicon, manganese, phosphorus, and sulfur. This somewhat impure iron is very hard and can be poured into molds as **cast iron.** The slag that comes from the furnace can be used in making cement.

**REFINING.** In the process of separating a metal from its ore, impurities are often introduced that impart undesirable properties to the final product. Consequently, it is generally necessary to purify the metal before it can be put to practical use. This purification process is called **refining.**

The specific procedure employed for refining a given metal depends on the chemical and physical properties of the metal as well as the properties of the impurities. As a result, there is no single method applicable to a very large number of different metals. We saw in Chapter 16 that copper can be economically refined electrolytically. This occurs, however, primarily because the silver and other precious metals recovered from the electrolytic cell offset the generally high cost of electricity.

An interesting process for refining nickel, called the Mond process, makes use of the relative ease of formation of a compound formed between nickel and carbon monoxide.

$$Ni + 4CO \longrightarrow Ni(CO)_4$$

Compounds of this general type are called **carbonyls:** nickel carbonyl, besides being easily formed, is also very volatile (and very poisonous). The impure nickel is therefore treated with CO at a moderately low temperature of 60°C, where the $Ni(CO)_4$ that is formed exists as a gas. This is circulated to another portion of the apparatus, where it is heated to about 200°C and decomposes to give pure nickel plus CO, which can be recycled through the process.[2]

The most important commercial refining process involves the conversion of pig iron into steel. This requires that impurities such as silicon, sulfur, and phosphorus be removed and the carbon content lowered significantly from the approximately 4% introduced into the pig iron in the blast furnace.

Modern steelmaking began with the introduction of the **Bessemer converter** in England in 1856. A batch of molten pig iron from the blast furnace, weighing about 25 tons, is transferred to a tapered cylindrical vessel containing a refractory lining (Figure 20.4). The composition of the lining is determined in part by the nature of the impurities in the iron. Since these impurities are usually silicon, phosphorus, and sulfur, whose oxides are acidic, a basic lining of dolomite (a $MgCO_3$, $CaCO_3$ mineral) is generally used. A blast of air (or oxygen) is blown through the melt from a set of small holes at the bottom of the vessel. The oxygen passing through the molten metal converts the silicon, phosphorus, and sulfur to oxides that then react with the lining to form a slag. The carbon in the pig iron is also oxidized to CO and its concentration is also reduced. The conversion of the pig iron to

Figure 20.4
*(a) Bessemer converter. (b) Open hearth furnace.*

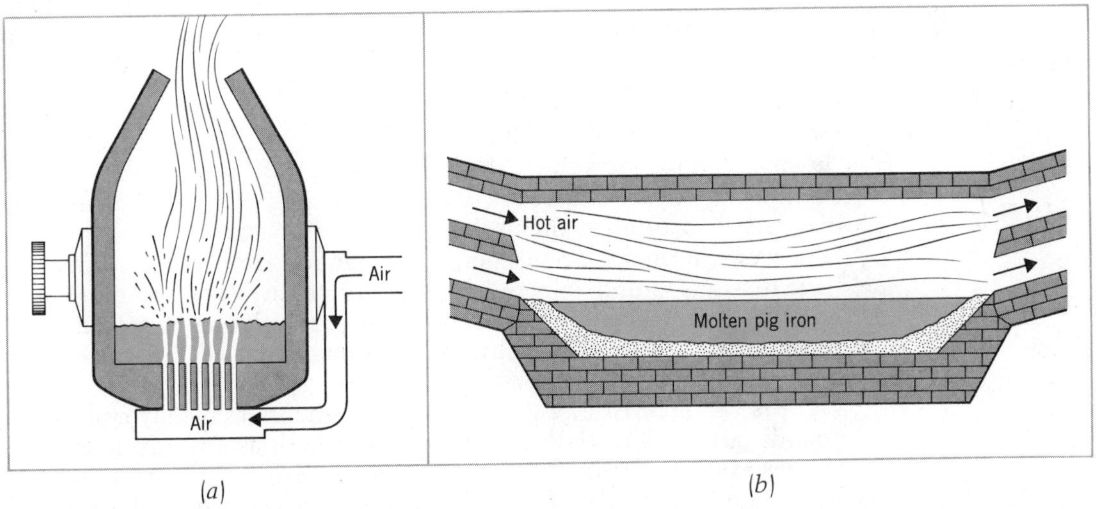


[2] For a time it was believed that extremely toxic $Ni(CO)_4$ was responsible for the so-called Legionnaires' disease that killed a group of people attending an American Legion Convention in Philadelphia in 1976. Later, however, this idea was abandoned.

steel by this process is rapid, requiring about 15 minutes, and gives rise to a spectacular display of fire and showers of sparks. The reaction is difficult to control, however, and the quality of the steel produced in the Bessemer converter can be quite variable.

A somewhat newer method that has almost replaced the Bessemer process employs an **open hearth furnace,** a large, shallow hearth usually lined with a basic oxide refractory (for example, MgO, CaO). The furnace is charged with a mixture of pig iron, $Fe_2O_3$, scrap iron, and limestone. A mixture of burning gases and hot air is played over the surface of the charge to maintain it in a molten state while a series of chemical reactions take place. Impurities in the steel are oxidized by the $Fe_2O_3$ and air. Carbon dioxide, formed by oxidation of the carbon in the pig iron, bubbles out of the mixture, keeping it stirred, while the $SiO_2$ and other acidic oxides combine with CaO (from the limestone) and the refractory lining to form a slag. This entire process takes much longer than the Bessemer process, requiring 8 to 10 hours to complete. However, the quality of the steel is much more easily controlled because chemical analyses can be constantly carried out on samples of the mixture. The increased length of time required to process a batch of steel is also offset by the fact that much larger quantities (about 200 tons) can be handled at one time. In addition, prior to being poured from the furnace, other metals (e.g., cobalt, chromium, nickel, vanadium, and tungsten) can be added to the steel to form alloys having special properties. A typical stainless steel, for instance is composed of approximately 72% iron, 19% chromium, and 9% nickel.

Modern methods of chemical analysis, making use of high-speed computers, have enabled a return to a modified form of the Bessemer process. This newer procedure, which is replacing the open hearth furnace because of its speed, involves forcing a mixture of powdered $CaCO_3$ and oxygen gas into the molten pig iron. This rapidly burns away the impurities, which form a slag. The characteristic emission spectra of the elements in the steel permit rapid chemical analysis, and additives can be incorporated into the steel in the proper proportions to give a product with the desired properties. This **basic oxygen process** takes only about 20 to 25 minutes to complete, thereby giving a very substantial savings in time (and, of course, *money*) over the open hearth process.

## 20.5 MAGNETISM

In Chapter 3 we saw that the presence of unpaired electrons in an atom or molecule impart to the substance the property called paramagnetism. The tiny electron magnets cause the atom or molecule as a whole to behave as a small magnet. When these are placed into a magnetic field, the microscopic magnets tend to align themselves with and be attracted toward the field. However, thermal motion operates to randomize the orientations of the little magnets with the net result that only a relatively small number of tiny magnets are aligned with the field at any particular instant. Consequently, an ordinary paramagnetic substance is drawn only weakly into an external magnetic field.

It is characteristic of the transition elements that they and their compounds possess a partially filled *d* or *f* subshell; as a result, many of these species exhibit the phenomenon of paramagnetism. The prediction of magnetic properties of transition metal compounds is therefore one of the requisites of a theory of bonding that is applicable to these substances. We shall explore this a little further when we discuss complex ions later in this chapter.

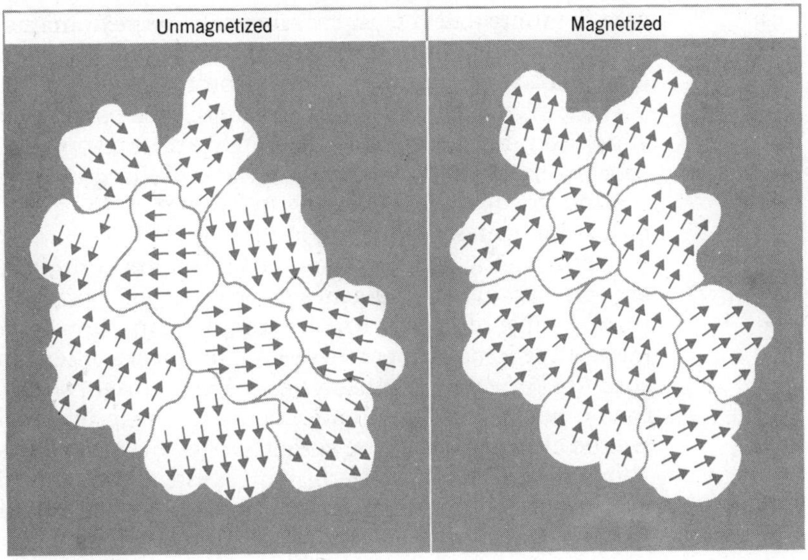


Figure 20.5
*Domains in a ferromagnetic solid.*

Related to the property of paramagnetism is the phenomenon called **ferromagnetism,** observed for the three pure elements, iron, cobalt, and nickel. Ferromagnetic materials, like paramagnetic ones, are also attracted to a magnetic field; however, the magnitude of the interaction for a ferromagnetic substance is approximately a million times stronger than with paramagnetic materials. How does this occur?

The origin of ferromagnetism is the same as paramagnetism—that is, the existence of unpaired electrons in the ferromagnetic material. In these substances it is believed that there exist regions, called **domains,** containing very large numbers of paramagnetic atoms that have their atomic magnets all lined up in the same direction, as illustrated in Figure 20.5. Ordinarily these domains are randomly oriented in the ferromagnetic solids so that even though each domain behaves as a relatively large magnet, their effects cancel. When placed in a magnetic field, the domains tend to become oriented much the same as the atomic magnets of a paramagnetic substance. In this case, however, each time one domain becomes aligned with the field, millions of tiny atomic magnets become aligned. As a result, the interaction between the ferromagnetic solid and the magnetic field is very much larger than that experienced by paramagnetic substances.

When the magnetic field is removed from a paramagnetic substance, its atomic magnets very quickly become randomly oriented and no permanent magnetism is induced. For a ferromagnetic substance, however, the domains tend to remain in the orientation in which they found themselves when the external magnetic field was present. This alignment of domains in the absence of an external field causes the substance to possess a residual magnetism; we say that it has become *permanently magnetized.* Any piece of iron (e.g., a pin) can be magnetized simply by stroking it with another permanent magnet.

A permanent magnet is not really permanent, because the magnetism may be destroyed either by heating the solid or by pounding it. In the first case the increased thermal motion causes the domains to become randomly oriented, while in the second instance violent vibrational motions cause the domains to twist and turn and become disoriented.

The phenomenon of ferromagnetism is associated only with the solid state. Iron, for example, is no longer ferromagnetic when it is melted. Instead it exhibits only paramagnetism. Melting of the solid thus appears to destroy the domains, and each individual atom in the liquid behaves more or less independently of the others nearby.

Even in the solid state not all elements containing unpaired electrons are ferromagnetic. Manganese, for example, possesses five unpaired electrons compared to only four for iron; yet pure iron is ferromagnetic but pure manganese is not. Apparently a requirement for ferromagnetism is that the spacings between paramagnetic ions be just right so that they may lock onto each other to form a domain. Nonferromagnetic metals, in which the ions are too close together, can sometimes be made ferromagnetic by forming an alloy. Such is the case with manganese, where the addition of the proper amount of copper permits the $Mn^{2+}$ ions in the metallic lattice to interact strongly and form domains, thereby producing a ferromagnetic alloy.

## 20.6 COORDINATION COMPOUNDS

The transition elements are known for their ability to form many complex compounds (for example, complex ions) in which the metal cation is surrounded by two or more ions or molecules, generally referred to as ligands. These complexes are also called **coordination compounds** because, from the point of view of the valence bond theory (the first bonding theory to be applied to them), they are considered to be held together by *coordinate* covalent bonds between the ligands and the metal. Because of the extremely large number of these coordination compounds, as well as their unusual colors, magnetic properties, structures, and chemical reactions, the study of them has become one of the major areas of inorganic chemical research.

Coordination compounds have found a number of important uses. Unexposed silver salts in photographic film and paper are removed by dissolving these salts in a solution containing thiosulfate ion, with which $Ag^+$ forms a complex ion. Complex ions are used in water softening (phosphates binding to iron and manganese ions) and as catalysts in a variety of industrial processes. The formation of complex ions has also been used to alleviate poisoning produced by beryllium and lead.

As the study of biochemistry has progressed, it has also become evident that many biologically important molecules owe their biological activity to a metal ion held in a "complex ion" within the molecule. Hemoglobin, containing iron(II) atoms, is a well-known example. The importance of metal ions in biosystems (not to mention the increased availability of research funding in biochemistry), has recently turned many inorganic chemists into bioinorganic chemists. We shall take a closer look at metal-containing biomolecules in Chapter 22.

The father of modern coordination chemistry was Alfred Werner, who received the Nobel Prize in Chemistry in 1913 for his work on these compounds. Werner was the first to recognize that metal ions could combine with other molecules or ions through more than one type of "valence" to produce relatively stable complex species. He was also the first to propose structures of complex ions that were consistent with their properties.

Metal complexes are formed with many kinds of ligands, including ions such as $Cl^-$, $CN^-$, and $NO_2^-$, as well as neutral molecules such as $H_2O$ or

$NH_3$. Nearly all ligands, however, have one thing in common; they possess a lone pair of electrons that may be shared with the metal cation in coordinate covalent bonds. In this sense the formation of a complex can be viewed as a Lewis acid-base reaction; in general, we can expect that the ligands in coordination compounds will all be Lewis bases with few exceptions.

In a complex, the ligands that are attached to the metal are considered to be in a **first coordination sphere,** and in solution they are held tightly by the metal ion, compared to other ions and molecules that might also be present nearby in the mixture. When the formula of a metal complex is written, it is usually the practice to indicate the species that are bonded to the metal in the first coordination sphere by enclosing them and the metal ion within square brackets. An example is the ion

$$[CoCl_6]^{3-}$$

These brackets are *not* to be confused with those that we used earlier when we wished to denote molar concentration.[3] Note that the charge on the complex is indicated *outside* the brackets, showing that the entire complex ion carries, in this example, a charge of $-3$.

Ligands such as $Cl^-$ or $NH_3$, which have one atom that can bond to a metal cation, are said to be **monodentate** (one "tooth") ligands. There are also many molecules and ions that are able to attach themselves to a metal ion through more than one donor atom to produce a cyclic *ring* type of arrangement. Two very common examples that have been much studied are oxalate ion, $C_2O_4^{2-}$,

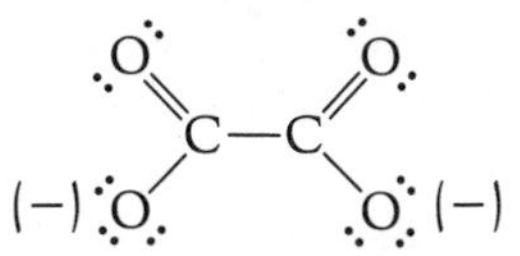

and ethylenediamine, $H_2N—CH_2—CH_2—NH_2$,

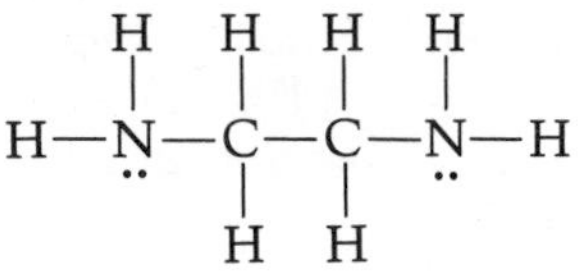

They bond to a metal ion as shown below.[4]

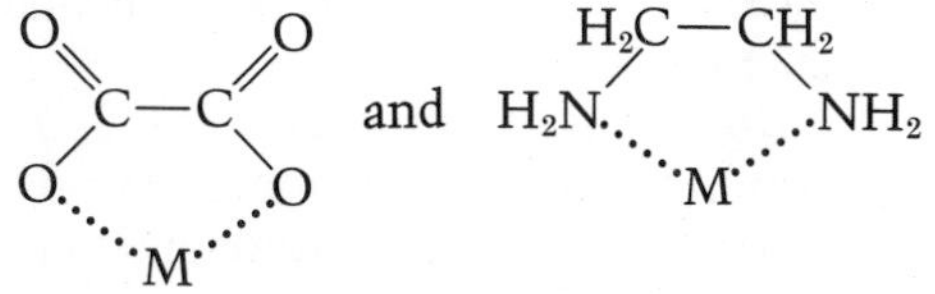

With cobalt(III), for example, these two ligands form complexes such as $[Co(C_2O_4)_3]^{3-}$ and $[Co(H_2NCH_2CH_2NH_2)_3]^{3+}$, whose structures are illustrated in Figure 20.6. Ethylenediamine is such a common ligand in coordination chemistry that in writing formulas containing it the abbreviation, **en,**

[3] In our earlier discussions of complex ion equilibria, we avoided this notation specifically to prevent such confusion.

[4] The origin of chemical terminology is sometimes rather colorful. Complexes of this general type are often called **chelates,** from the Greek *chele,* meaning claw. The ligand in this case bites the metal with two claws (donor atoms) much like a crab.

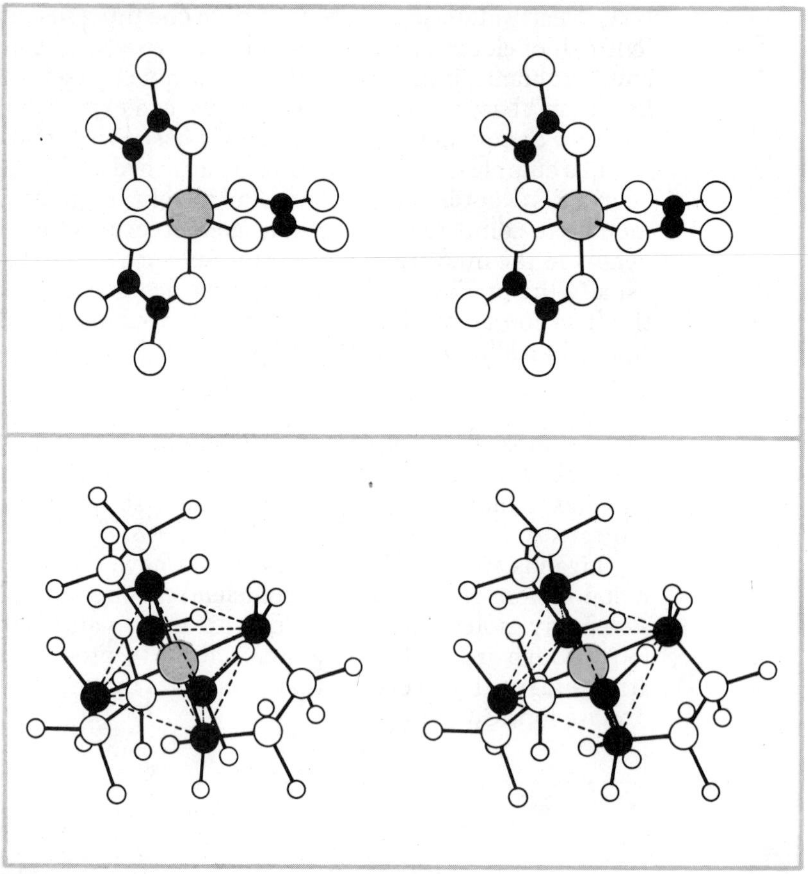

(a) (b)

Figure 20.6
*Structures of $[Co(C_2O_4)_3]^{3-}$ and $[Co(en)_3]^{3+}$. (a) $[Co(C_2O_4)_3]^{3-}$. The colored atom is cobalt, solid black atoms are carbon, and white spheres are oxygen atoms. (b) $[Co(en)_3]^{3+}$. The colored atom is cobalt, solid black atoms are nitrogen, the large white atoms are carbon, and the small white atoms are hydrogen.*

usually is used. As a result, the latter ion is generally written as simply $[Co(en)_3]^{3+}$.

Molecules or ions, such as $C_2O_4^{2-}$, or en, that have two atoms that may coordinate to a metal ion are said to be **bidentate** ligands. There are also more complex **polydentate** ligands containing three, four, or more donor atoms. Table 20.9 contains a list of some common mono- and bidentate ligands, as well as a few examples of polydentate ligands.

A particularly important polydentate ligand is ethylenediaminetetraacetic acid (EDTA). Its six coordinating atoms firmly attach themselves to free metal ions, and EDTA has been used as an antidote in lead poisoning because it binds $Pb^{2+}$, thereby preventing it from inhibiting certain important enzyme functions. EDTA also binds to iron and calcium and is used as water softeners in products such as shampoos. EDTA is added to foods to tie up metal ions that catalyze oxidation (and hence deterioration) of the food product. It has also been found that EDTA increases the storage life of whole blood by removing $Ca^{2+}$, which promotes clotting.

Table 20.9
*Common ligands found in complex ions*

| | | | | |
|---|---|---|---|---|
| *Monodentate* | $H_2O$ | Water | $Br^-$ | Bromide |
| | $NH_3$ | Ammonia | $I^-$ | Iodide |
| | $CN^-$ | Cyanide | $NO_2^-$ | Nitrite |
| | $OH^-$ | Hydroxide | $SCN^-$ | Thiocyanate |
| | $F^-$ | Fluoride | $S_2O_3^{2-}$ | Thiosulfate |
| | $Cl^-$ | Chloride | | |

*Bidentate*

**oxalate**

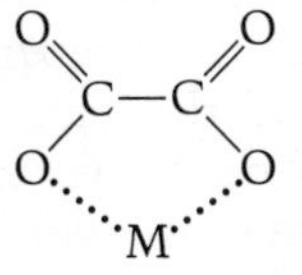

**ethylenediamine**

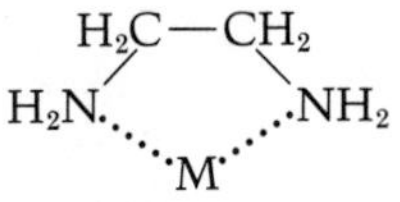

***o*-phenanthroline**

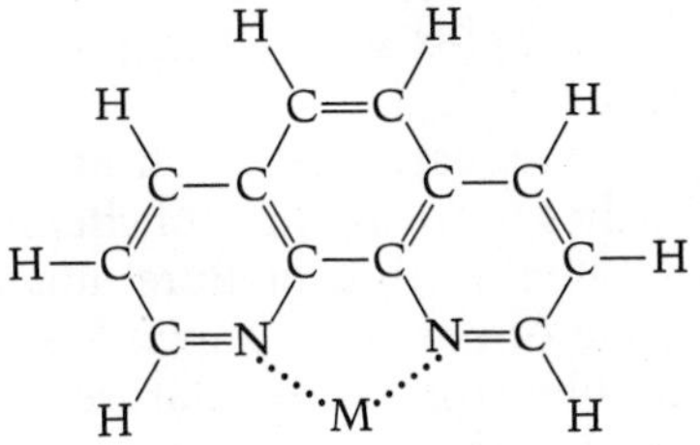

or

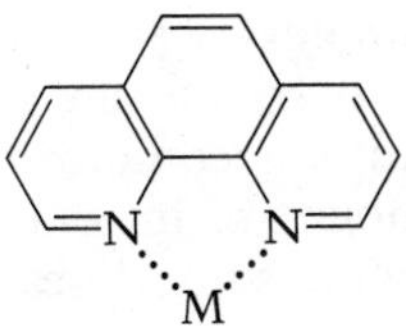

**dipyridyl**

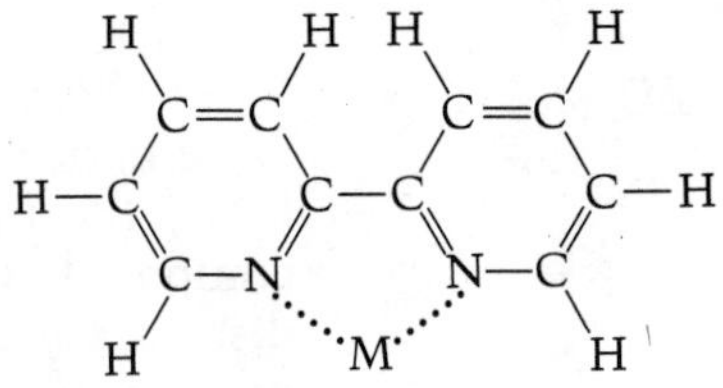

or

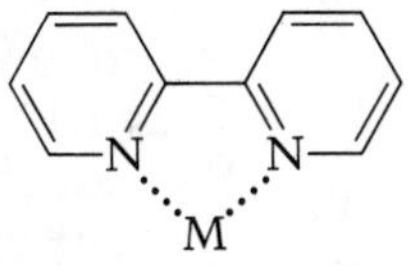

*Polydentate*

(coordinating atoms indicated with asterisks)

Diethylenetriamine (three coordinating atoms)

$H_2\overset{*}{N}-CH_2-CH_2-\overset{*}{N}H-CH_2-CH_2-\overset{*}{N}H_2$

Ethylenediaminetetraacetate (six coordinating atoms)

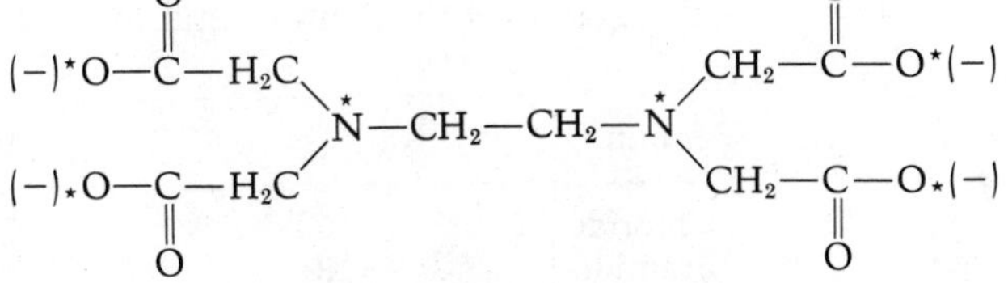

**also called EDTA**

## 20.7 COORDINATION NUMBER

The term **coordination number** (C.N.) refers to the total number of ligand atoms that are bound to a given metal ion in a complex. These atoms may be supplied by either monodentate or polydenate ligands, or both. Thus, in the three complexes, $[CoCl_6]^{3-}$, $[Co(en)_2Cl_2]^+$, and $[Co(en)_3]^{3+}$, the C.N. of cobalt is the same, since in each case there are *six* donor atoms about the $Co^{3+}$ ion.

Coordination numbers ranging from 2 to more than 8 are observed in various coordination compounds, the C.N. in any given instance being determined by the nature of the metal ion, its oxidation state, and to some extent, the ligands and the environment surrounding the complex. The most common coordination numbers are observed to be 2, 4, and 6. The basic structural types that are found for these are shown in Figure 20.7.

By far the most frequently occurring coordination number in transition metal complexes is 6, and the geometry that is observed in nearly all instances is octahedral. A simple two-dimensional way of representing the octahedral geometry is shown in Figure 20.8. The dashed rectangle represents the square plane (viewed in perspective) that joins the upper and lower pyramids in the octahedron. The six solid lines connect the center of the metal cation to the coordinated ligand atoms. This arrangement is illustrated in Figure 20.8 for the complex ion, $[CoCl_6]^{3-}$.

## 20.8 NOMENCLATURE

In an effort to communicate with each other, chemists attempt to devise systematic approaches to the naming of chemical compounds. On a periodic basis, the International Union of Pure and Applied Chemistry (IUPAC), composed of a group of chemists drawn from all over the world, meets to discuss current problems in nomenclature. In this way the systematic nomenclature of compounds continually evolves to meet our needs as new compounds and structures are discovered.

Below are some of the rules that have been developed by the IUPAC to name coordination complexes. Some of the names assigned to complexes following these rules may sound odd, and even funny. Remember, however, that we are primarily interested in formulating a name that is able to transmit the maximum amount of information in the shortest possible name. The end result is therefore sometimes difficult to pronounce.

### RULES OF NOMENCLATURE OF COORDINATION COMPOUNDS

1. **Cationic species are named before anionic species.** This is just like other cases of ionic compounds, such as NaCl, which is named as sodium chloride (cation, anion).
2. **Within a complex ion, the ligands are named first, followed by the metal ion.** This is opposite to the sequence in which they appear in the formula. For example, the complex $[Co(NH_3)_6]^{3+}$ is named by specifying the ammonia first, then the cobalt.
3. **The names of anionic ligands end in the suffix -o.**
   (a) Ligands whose names end in *ide* have this suffix replaced by *-o*.

| Anion | | Ligand |
|---|---|---|
| Chloride | $Cl^-$ | Chloro |
| Bromide | $Br^-$ | Bromo |
| Cyanide | $CN^-$ | Cyano |
| Oxide | $O^{2-}$ | Oxo |

| | | |
|---|---|---|
| C.N. = 2 | Linear |  |
| C.N. = 4 | Tetrahedral | 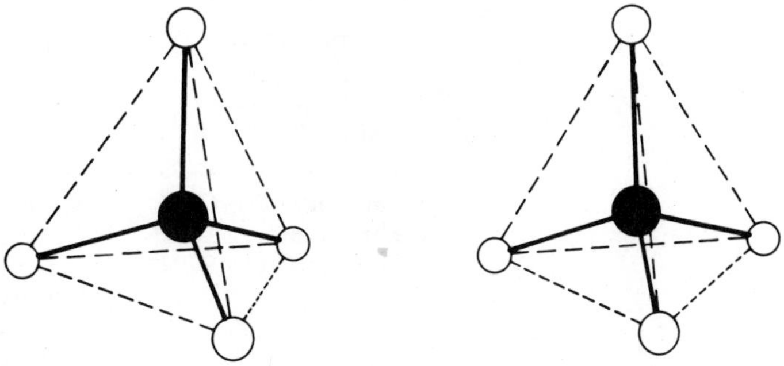 |
| | Square planar | 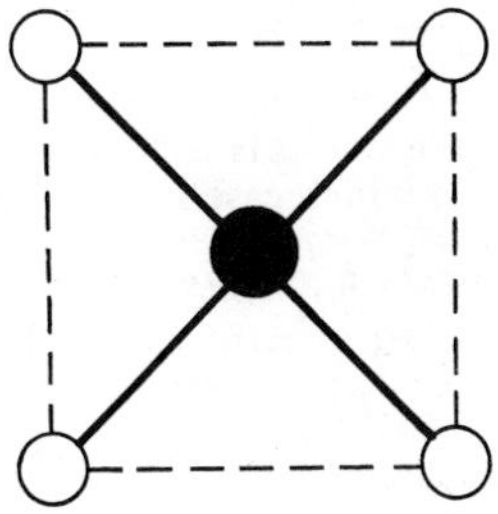 |
| C.N. = 6 | Octahedral | 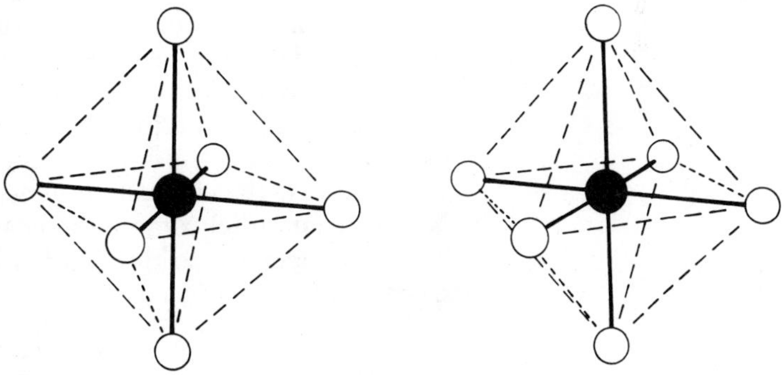 |

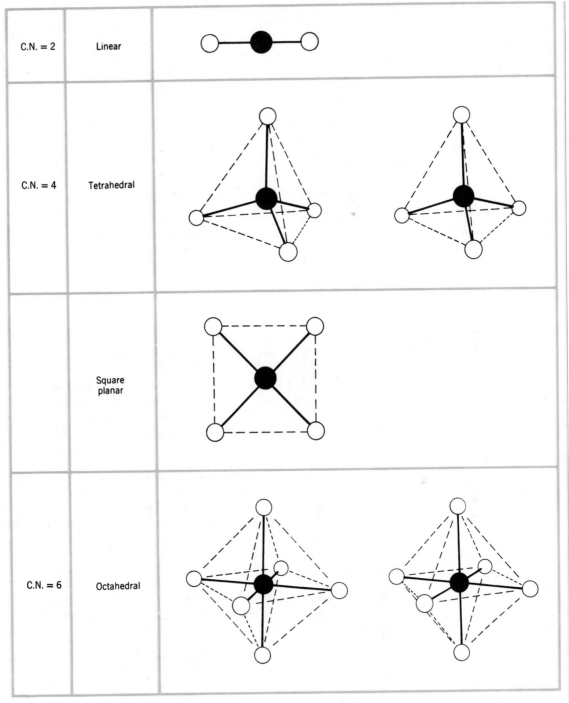

Figure 20.7
*Structure types for coordination 2, 4, and 6.*

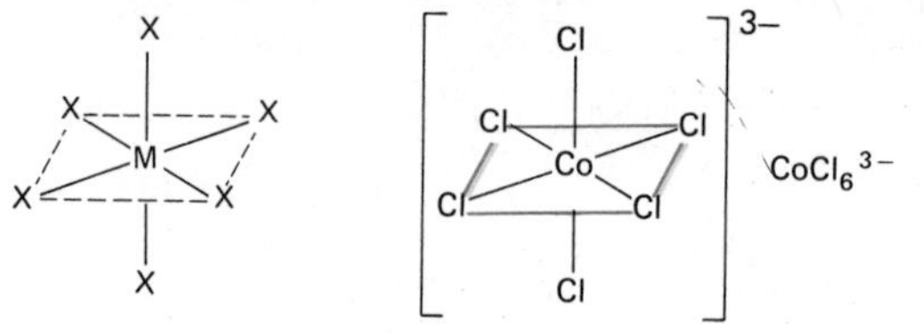


Figure 20.8
*Two-dimensional representation of octahedral coordination.*

(b) Ligands whose names end in *-ite* or *-ate* become *-ito* and *-ato*, respectively.

| Anion | | Ligand |
|---|---|---|
| Carbonate | $CO_3{}^{2-}$ | Carbonato |
| Thiosulfate | $S_2O_3{}^{2-}$ | Thiosulfato |
| Thiocyanate | $SCN^-$ | Thiocyanato (bonded through sulfur) Isothiocyanato (bonded through nitrogen) |
| Oxalate | $C_2O_4{}^{2-}$ | Oxalato |
| Nitrite (bonded through oxygen, $ONO^-$)[a] | $NO_2{}^-$ | Nitrito |

[a] An exception to this is $NO_2{}^-$ when bonded through nitrogen, in which case it is named as *nitro.*

4. **Neutral ligands are given the same names as the neutral molecule.** Thus ethylenediamine as a ligand is called ethylenediamine in the name of the complex. Two very important exceptions to this, however, are

   $H_2O$ Aquo $\qquad$ $NH_3$ Ammine (note double *m*)

5. **When there are more than one of a particular ligand, the number is specified by di = 2, tri = 3, tetra =4, penta = 5, hexa = 6, and so forth. When confusion might result, the prefixes bis = 2, tris = 3, tetrakis = 4, and so forth, are employed.** Thus the presence of two chloride ions is specified as *dichloro.* However, because ethylenediamine already contains the term *di,* two of these molecules are indicated by placing the name of the ligand in parentheses preceded by the term *bis;* that is, *bis(ethylenediamine).*
6. **Negative (anionic) complex ions always end in the suffix -ate.** This suffix is appended to the English name of the metal atom in most cases.

| Element | Metal as Named in Anionic Complex |
|---|---|
| Aluminum | Aluminate |
| Chromium | Chromate |
| Manganese | Manganate |
| Nickel | Nickelate |
| Cobalt | Cobaltate |
| Zinc | Zincate |
| Molybdenum | Molybdate |
| Tungsten | Tungstate |

For some metals, the *ate* is appended to the Latin stem.

| Element | Stem | Metal as Named in Anionic Complex |
|---|---|---|
| Iron | Ferr- | Ferrate |
| Copper | Cupr- | Cuprate |
| Lead | Plumb- | Plumbate |
| Silver | Argent- | Argentate |
| Gold | Aur- | Aurate |
| Tin | Stann- | Stannate |

In neutral or positively charged complexes the metal *always* appears with the common English name for the element.

7. **The oxidation number of the metal in the complex is written in Roman numerals within parentheses following the name of the metal.** For example,

$[Co(H_2O)_6]^{3+}$ is the hexaaquocobalt(III) ion.
$[CoCl_6]^{3-}$ is the hexachlorocobaltate(III) ion.

Note that the charge on the complex is obtained as the *algebraic sum* of the oxidation number of the metal and the charges on the ligands.

Some additional examples illustrate these rules.

| | |
|---|---|
| $[Ni(CN)_4]^{2-}$ | Tetracyanonickelate(II) ion |
| $[Co(NH_3)_4Cl_2]^+$ | Dichlorotetraamminecobalt(III) ion |
| $Na_3[Cr(NO_2)_6]$ | Sodium hexanitrochromate(III) |
| $[Ag(NH_3)_2]^+$ | Diamminesilver(I) ion |
| $[Ag(CN)_2]^-$ | Dicyanoargentate(I) ion |
| $[Co(en)_3]Cl_3$ | Tris(ethylenediamine)cobalt(III) chloride |
| $[Cr(NH_3)_3Cl_3]$ | Trichlorotriamminechromium(III) |

## 20.9 ISOMERISM AND COORDINATION COMPOUNDS

When two different compounds have the same molecular formula, but differ in the way that their atoms are arranged, they are said to be **isomers** of one another. For example, there are two compounds having the general formula:

$$Cr(NH_3)_5SO_4Br$$

One of these we should formulate as

$$[Cr(NH_3)_5SO_4]Br$$

because it yields a precipitate of AgBr when treated in aqueous solution with $AgNO_3$ but does not give a precipitate of $BaSO_4$ when treated with $Ba(NO_3)_2$. This latter observation means that the $SO_4^{2-}$ is not free in the solution and, hence, must be bound to the chromium.

The second compound is written as

$$[Cr(NH_3)_5Br]SO_4$$

and produces $BaSO_4$ when treated with $Ba(NO_3)_2$. On the other hand, addition of $AgNO_3$ to a solution of the compound does not yield AgBr.

The two compounds just described have different chemical properties and are clearly different chemical substances, even though they are composed of the same number of the same kinds of atoms. This particular type of isomerism is not uncommon among coordination compounds and is called **ionization isomerism.**

Another type of isomerism that is very important is called **stereoisomerism,** and results when a given molecule or ion can exist in more than one structural form in which the same atoms are bound to one another but find themselves oriented differently in space. To illustrate this, we will focus our attention on octahedral complexes because they represent the most common structural type.

The simplest form of stereoisomerism results when a complex has the general formula $Ma_4b_2$ in which *a* and *b* represent monodentate ligands. An example would be the ion, $[Co(NH_3)_4Cl_2]^+$. (How would you name it?) This complex can exist in two different isomeric forms, called **geometrical isomers,** as shown in Figure 20.9. As you can see, in one of these isomers the two *b* ligands are located across from one another on opposite sides of the metal ion. Such an isomer is given the designation **trans** (Latin *trans* means "across"). The other isomer has the two *b* ligands adjacent to one another and is referred to as the **cis** isomer (L. *cis* = on the same side). Thus the two isomers would be specified as

$$trans\text{-}[Co(NH_3)_4Cl_2]^+$$

and

$$cis\text{-}[Co(NH_3)_4Cl_2]^+$$

Because *cis*- and *trans*-isomers possess different structures, they are different chemical species, each with its own set of chemical and physical properties. While these properties may often be similar, the fact that they are different clearly tells us that the two structures represent truly different compounds.

Geometrical isomers also occur when there are bidentate ligands in a complex, as illustrated by the *cis* and *trans* forms of the ion $[Cr(en)_2Cl_2]^+$ shown in Figure 20.10. Once again, in the *trans* form we see that the chloride ligands are on opposite sides of the metal while in the *cis* form they are alongside one another.

A second form of stereoisomerism is called **optical isomerism.** Optical isomers, which, as we shall see, affect polarized light differently, bear the same relationship to each other as do your left and right hands; that is, they are **nonsuperimposable mirror images** of one another. To see what this means, try this simple experiment. Place your right hand in front of a mirror, with the palm toward the mirror, and hold your left hand alongside with the palm facing you (Figure 20.11). Notice that the image of your *right* hand in the mirror looks the same as your left hand and hence we can say that your left and right hands are mirror images of each other. The nonsuperimposable aspect arises because your left and right hands, while similar in appearance, do not match exactly when one is placed over the other, both with palms down; the thumbs point in different directions. This difference is perhaps seen even more clearly if you attempt to place your right hand into a left-hand glove; it doesn't fit properly. Thus your left and right hand, and optical isomers too, cannot be superimposed on each other.

An example of a complex that exists as two nonidentical mirror image isomers is the $[Co(en)_3]^{3+}$ ion. The mirror image relationship between the

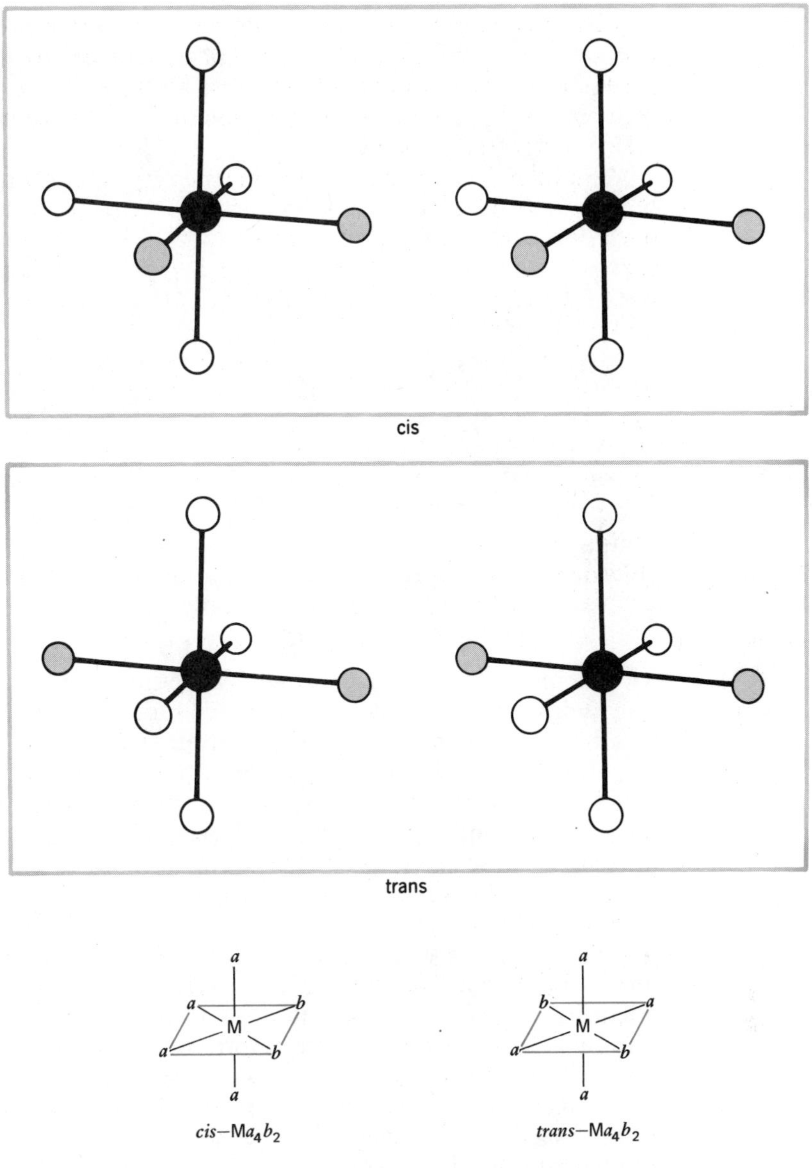


Figure 20.9
Cis-trans *isomers for complexes* $Ma_4b_2$.

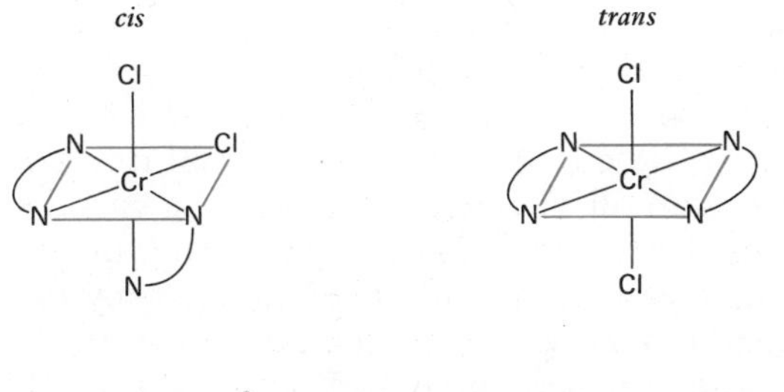


N⌒N represents the bidentate ethylenediamine ligand

Figure 20.10
*Cis-trans isomerism in* $[Cr(en)_2Cl_2]^+$.

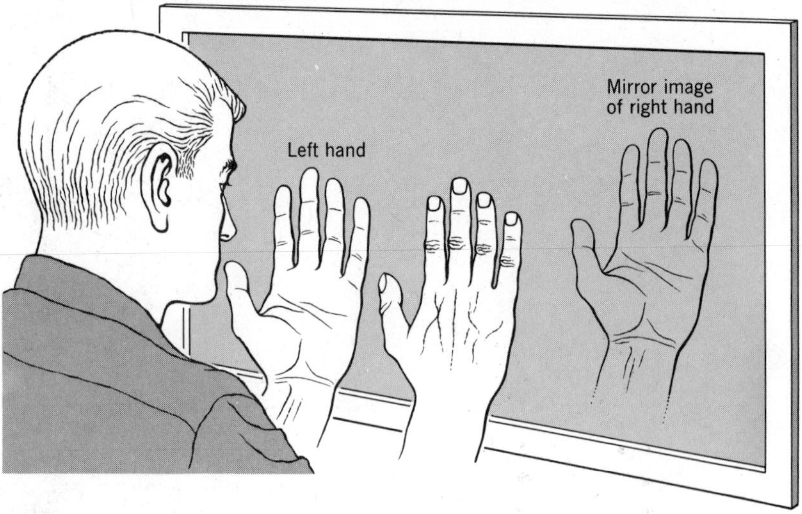


Figure 20.11
*Illustration of nonsuperimposable mirror images.*

two isomers shown in Figure 20.12 is demonstrated in Figure 20.13 (where the hydrogen atoms have been omitted for simplicity). A pair of isomers related in this manner are called **enantiomers.**

In such complexes it is the arrangement of the chelate rings that gives rise to the optical isomerism and, in general, any octahedral complex containing three bidentate ligands will exist as two optical isomers. In two dimensions these are usually represented as shown in Figure 20.14.

Optical isomerism is also important for the *cis* form of complex ions containing two bidentate ligands and two monodentate ligands; for example, *cis*-$[Co(en)_2Cl_2]^+$ (Figure 20.15*a*). The *trans* form of this complex does not exhibit optical isomerism, however, because it and its mirror image are identical. In Figure 20.15*b* the *trans* isomer is drawn with a plane cutting through it. Notice that all of the atoms on one side of this plane are oriented, relative to the plane, exactly as the atoms on the other side. Such an imaginary plane is called a mirror plane, and, it is fact that any molecule or ion that has a structure such that it possesses a mirror plane will not exist as two distinct optical isomers.

In general, the properties of optical isomers are identical except for the way in which they interact with outside influences that are able to distinguish between left and right handedness. The situation here is analogous to having a group of baseball players, some of whom are left-handed and some of whom are right-handed. Since they are all able to toss a baseball with equal ease, a baseball will not differentiate between left- and right-handed players. The same applies to a bat, since there are no left- or right-handed baseball bats. A fielder's glove, however, will fit only one hand. A glove designed to be worn on the left hand cannot be used by a player who catches the ball in his right hand. In this case, the glove differentiates between these two kinds of players because it too has a left or right handedness to it. In this same fashion, optical isomers interact in an identical way with most chemical reagents and physical probes. They do differ, however, in the way in which they react toward polarized light.

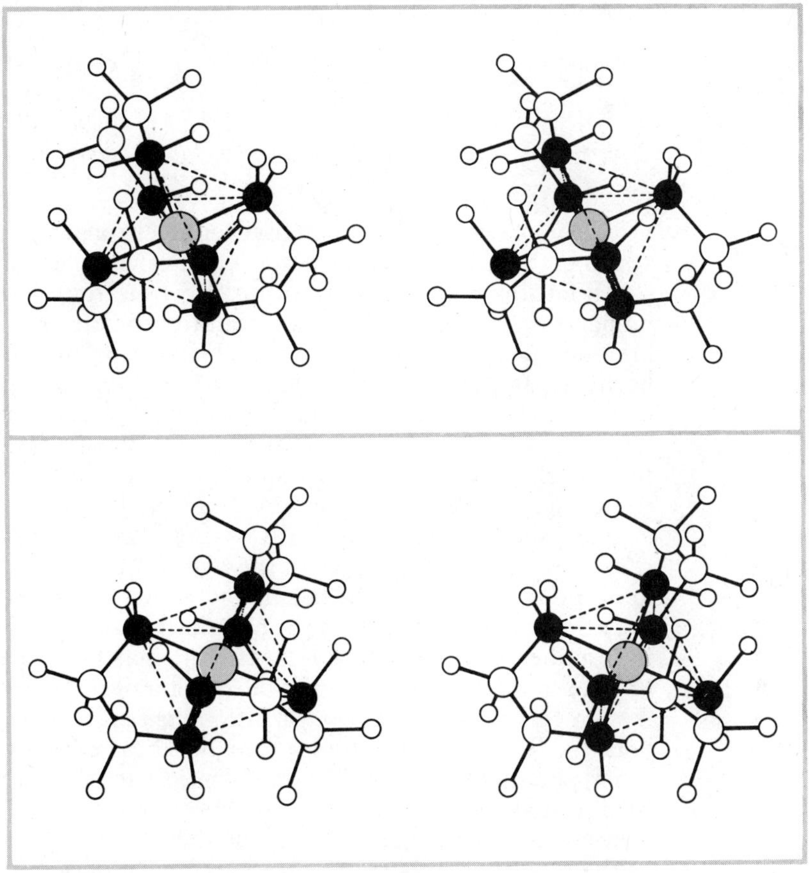

(a)

(b)

Figure 20.12
*The complex $[Co(en)_3]^{3+}$. The two structures shown are nonsuperimposable mirror images of one another.*

Figure 20.13
*Mirror image relationship between optical isomers of $[Co(en)_3]^{3+}$.*

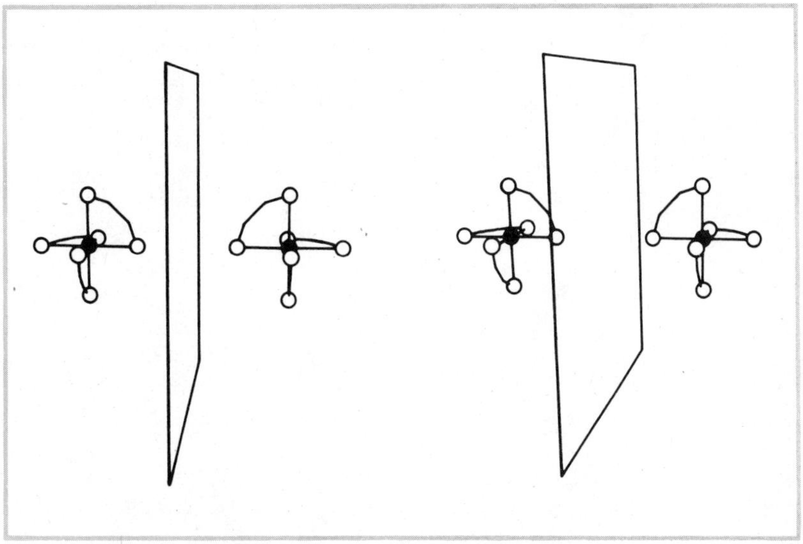

Figure 20.14
*Two-dimensional representation of optical isomerism for the complex ion* $[Co(en)_3]^{3+}$.

Light, in general, is composed of electromagnetic radiation that possesses both electric and magnetic vectors. These vectors oscillate in a sinusoidal fashion perpendicular to the direction from which the light wave is propagated (Figure 20.16). If we examine the electric vectors, all different orientations are observed in an unpolarized beam. However, when such a beam is passed through a polarizing medium, only the vibrations in one direction (plane) remain. The result is called plane polarized light. A unique feature of optical isomers is that when plane polarized light is passed through them (or their solutions) the plane of polarization is rotated through some angle, $\theta$, as shown in Figure 20.17. Substances that exhibit this property are also said to be **optically active.** One enantiomer (optical isomer) causes the light to be rotated to the right (clockwise when viewed down the axis of the oncoming light beam) and is said to be **dextrorotatory,** whereas the other enantiomer causes the polarized beam to be rotated to the left and is described as **levorotatory.** The two isomers are therefore designated as *d* or *l* depending on the direction of rotation of the polarized light.

An equal mixture of two enantiomers tends to rotate polarized light to both the left and right simultaneously. These effects therefore cancel each other, and such a mixture shows no optical activity; it is said to be **racemic.** In almost all cases, when enantiomers are produced in a chemical reaction, they are formed in equal numbers so that a racemic mixture is formed. One of the arts in chemistry is the separation of optical isomers from one another.

## 20.10 BONDING IN COORDINATION COMPOUNDS: VALENCE BOND THEORY

There are three important properties of transition metal complexes that must be explained by a bonding theory: (1) structure, (2) magnetic properties, and (3) color (nearly all coordination complexes of the transition elements are colored).

In the earliest theories of coordination complexes the metal was considered to be attached to the ligands by way of coordinate covalent bonds, and

Figure 20.15
(*a*) *Optical isomers for cis-*$[Co(en)_2Cl_2]^+$. (*b*) *trans-*$[Co(en)_2Cl_2]^+$.

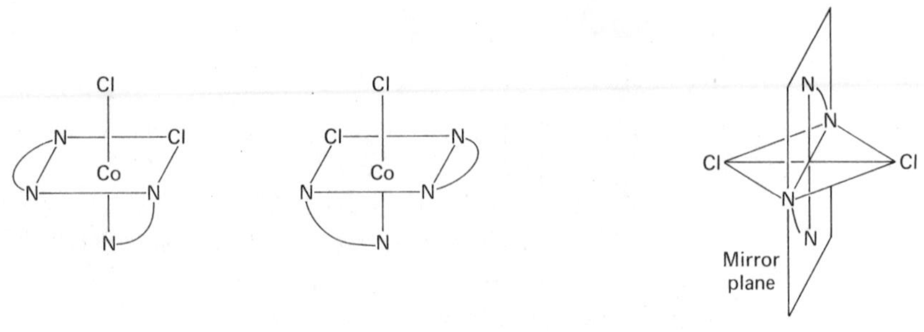


(*a*) (*b*)

(a)

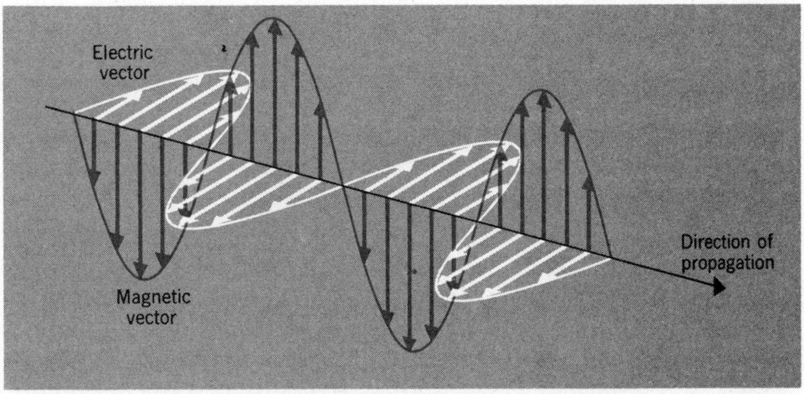


(b)

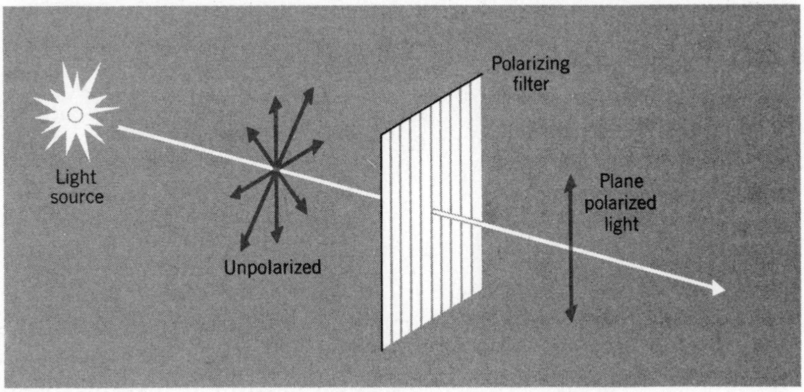


Figure 20.16

*Polarized light. (a) Electromagnetic radiation composed of electric and magnetic vectors. (b) Orientation of electric vectors in unpolarized and polarized light.*

Figure 20.17

*Rotation of plane polarized light by an optical isomer in solution. Plane of polarization is rotated by an angle $\theta$ as it passes through the solution containing the optically active compound. In this example the light is rotated to the left; the substance in the solution is said to be levorotatory.*

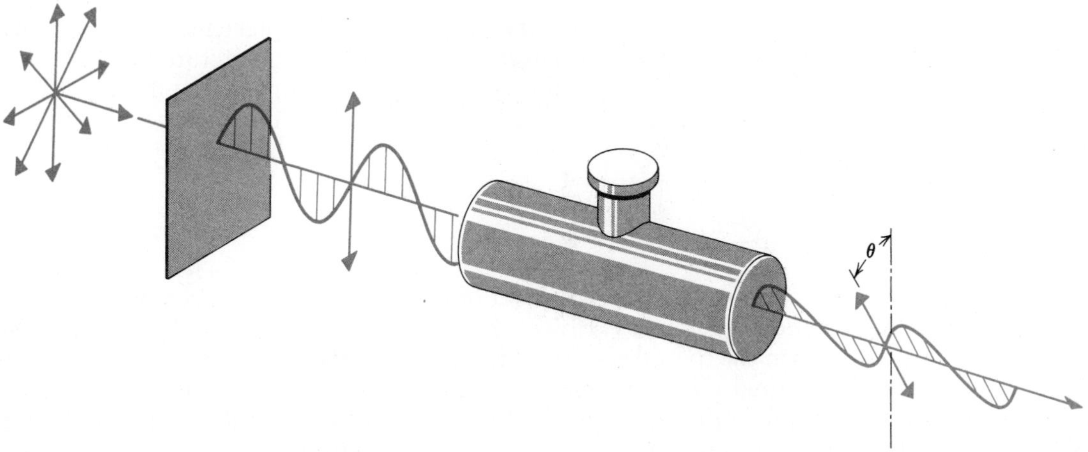

the first serious attempt to explain the structures and magnetic properties of complexes was made by applying the concepts of valence bond theory. Much of what follows is simply an extension of some of the ideas that were developed in Chapter 17.

In valence bond theory, you recall, a bond is formed by the overlap of two orbitals and the subsequent sharing of a pair of electrons between the two atoms in the region of overlap. Ligands, as a rule, do not possess unpaired electrons and, hence, the bonding in the complex must result from the overlap of ligand orbitals, containing lone pairs of electrons, with vacant orbitals on the metal ion, thereby giving rise to the coordinate covalent bond. To see how this occurs, let's consider an octahedral ion such as the blue-violet $[Cr(H_2O)_6]^{3+}$, characteristic of simple Cr(III) salts in aqueous solution. The electronic structure of a free chromium atom is

| | 3*d* | 4*s* | 4*p* | 4*d* |
|---|---|---|---|---|
| Cr [Ar] | ↑ ↑ ↑ ↑ ↑ | ↑ | _ _ _ | _ _ _ _ _ |

where we have shown the empty 4*p* and 4*d* subshells as well as those that are occupied by electrons. The central $Cr^{3+}$ ion in our complex results from the loss of three electrons (starting with the 4*s*) to give

| | 3*d* | 4*s* | 4*p* | 4*d* |
|---|---|---|---|---|
| Cr [Ar] | ↑ ↑ ↑ _ _ | _ | _ _ _ | _ _ _ _ _ |

Now, in order to obtain an octahedral geometry, we saw in Chapter 17 that a $d^2sp^3$ set of hybrid orbitals must be employed. In addition, these orbitals must be empty so that the electron pairs from the ligands can be placed into them when the bonds are formed. In this case, suitable hybrids can be constructed using the two vacant 3*d* orbitals. The 3*d* orbitals are preferred here over the 4*d* because they are lower in energy; therefore stronger bonds are formed when the 3*d* orbitals are used instead of the 4*d*. If we use dots to represent ligand electrons, we can now write the electronic structure for the complex as

| | 3*d* | 4*s* | 4*p* | 4*d* |
|---|---|---|---|---|
| $[Cr(H_2O)_6]^{3+}$ | ↑ ↑ ↑ ‥ ‥ | ‥ | ‥ ‥ ‥ | _ _ _ _ _ |

($d^2sp^3$: the two ‥ 3*d*, the ‥ 4*s* and the three ‥ 4*p* orbitals)

Experimental measurements demonstrate the presence of three unpaired electrons in this ion, as suggested by our bonding picture.

Let's consider next the emerald-green complex ion, $[Ni(H_2O)_6]^{2+}$. The electron configuration of the $Ni^{2+}$ ion (obtained in the same manner as $Cr^{3+}$ above) would be

| | 3*d* | 4*s* | 4*p* | 4*d* |
|---|---|---|---|---|
| $Ni^{2+}$ [Ar] | ↑↓ ↑↓ ↑↓ ↑ ↑ | _ | _ _ _ | _ _ _ _ _ |

Once again we must have a set of vacant $d^2sp^3$ hybrids in order to obtain the octahedral geometry. However, this time we cannot use a pair of 3*d* orbitals to form the hybrid set. At best, we could obtain only one empty 3*d* orbital if we paired all of the electrons in the 3*d* subshell. When the hybrids are created, both *d* orbitals must come from the *same* subshell and, therefore, two 4*d* orbitals must be used. The electronic structure of the complex then

becomes

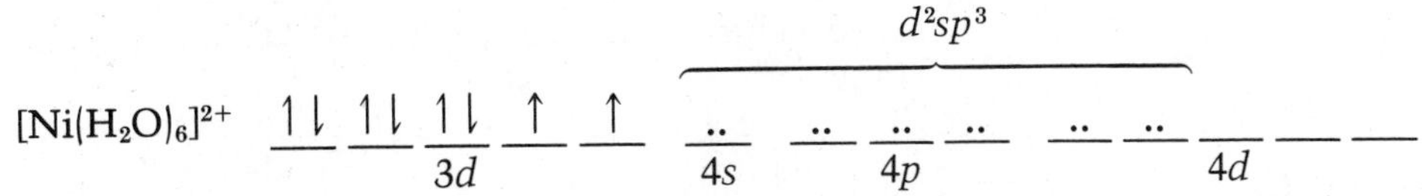


Note that the complex contains two unpaired electrons, again in agreement with experiment.

We have now seen two complex ions that can be considered to employ $d^2sp^3$ hybrids for bonding. In valence bond language, when 3*d* orbitals are used to form the hybrids, an **inner orbital complex** is said to result. On the other hand, when the 4*d* orbitals are used an **outer orbital complex** is formed.

In these last two examples there really was no choice as to which type of bonding (inner or outer orbital) would occur. Let's look at a situation, now, where we do have a choice. An example is Co(III). The electronic structure of the $Co^{3+}$ ion is

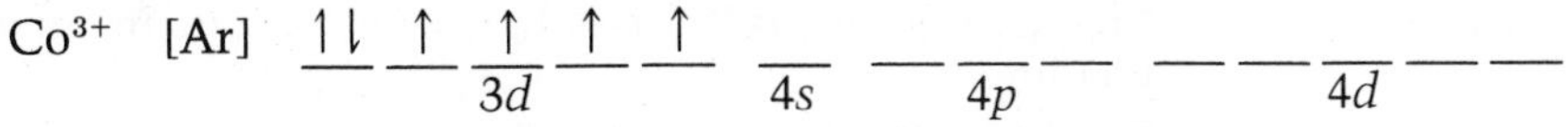


In this case the $d^2sp^3$ hybrid can be formed in either of two ways. One is to make use of two 4*d* orbitals, thereby giving an outer orbital complex. The second is to pair the electrons together to produce two vacant 3*d* orbitals that can be used in the hybrids. This gives rise to an inner orbital complex.

*Outer orbital*[5]

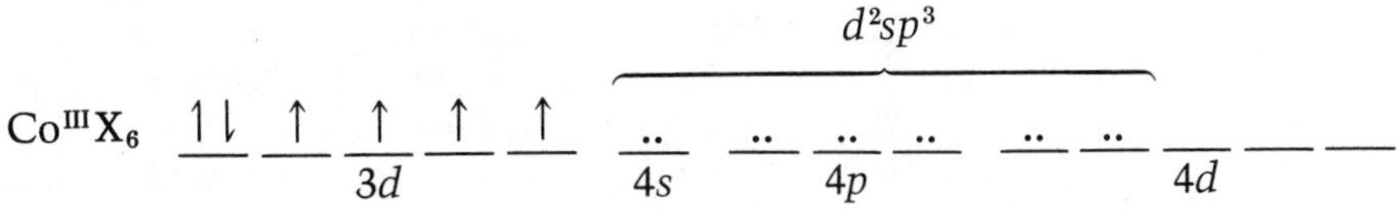


*Inner orbital*

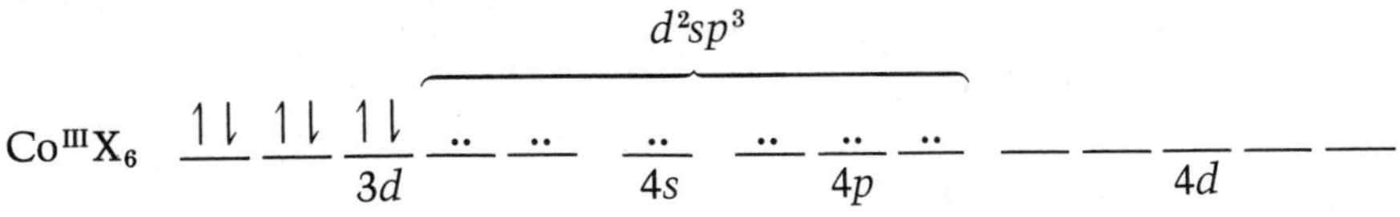


Notice that we can distinguish experimentally between these two possibilities by examining the number of unpaired electrons in the complex. The outer orbital complex has four unpaired electrons; the inner orbital complex has none. The latter complex is diamagnetic.

Both inner and outer orbital complexes are possible whenever the metal ion contains either four, five, or six *d* electrons. In each case, two empty 3*d* orbitals can be created by the pairing of electrons. Failure to pair them, however, leads to the formation of outer orbital complexes.

How do we make a choice between inner and outer orbital bonding? When there is a choice, what determines which of the two possibilities will occur? To answer these questions we must consider two opposing factors:

[5] Recall that the Roman numeral superscript here is meant to indicate the oxidation state of the metal.

1. As stated earlier, when $3d$ orbitals are used to form the hybrid orbitals, stronger metal-ligand bonds result than when $4d$ orbitals are used. This favors the pairing of electrons and the production of inner orbital complexes.
2. The pairing of electrons required to produce the necessary vacant $3d$ orbitals for inner orbital bonding also requires an input of energy. Since this energy doesn't have to be invested if the 4d orbitals are used to form the hybrids, this factor favors the production of outer orbital complexes.

The way in which these two factors come into play is illustrated in Figure 20.18. In the first drawing we see that the energy released when the bonds are formed using hybrids composed of $3d$ orbitals is so great that it more than compensates for the pairing energy, and the resulting inner orbital complex is of lower energy (more stable) than the outer orbital complex. In this case the preferred complex would be the inner orbital one. On the right in Figure 20.18 we find the other situation where the energy released upon inner orbital bond formation does not lead to an overall lower energy than that achieved with the formation of an outer orbital complex. In this case outer orbital bonding would occur in preference to inner orbital bonding.

As a general rule, for first-row transition metal ions with either a $d^4$ or $d^6$ electron configuration, most ligands tend to lead to the production of inner orbital complexes. Exceptions are the ligands $H_2O$ and $F^-$, which usually produce outer orbital complexes.

With a $3d^5$ configuration the subshell is half-filled, and we have noted earlier that a half-filled subshell in which all electrons have the same spin possesses a certain extra stability. As a result, these electron configurations are difficult to disturb and electron pairing is difficult to accomplish. Consequently, metal ions with a $d^5$ structure tend to keep their electrons spread out with parallel spins and thus tend to form outer orbital complexes with most ligands. An exception to this occurs with cyanide ion. This particular ligand forms very stable metal-ligand bonds,[6] sufficiently stable that electron pairing of the $d^5$ configuration can occur. This is illustrated by the hexacyanoferrate(III) ion (also called ferricyanide ion), $[Fe(CN)_6]^{3-}$. For iron(III) we have

$Fe^{3+}$ [Ar] ↑ ↑ ↑ ↑ ↑ (3d) __ (4s) __ __ __ (4p) __ __ __ __ __ (4d)

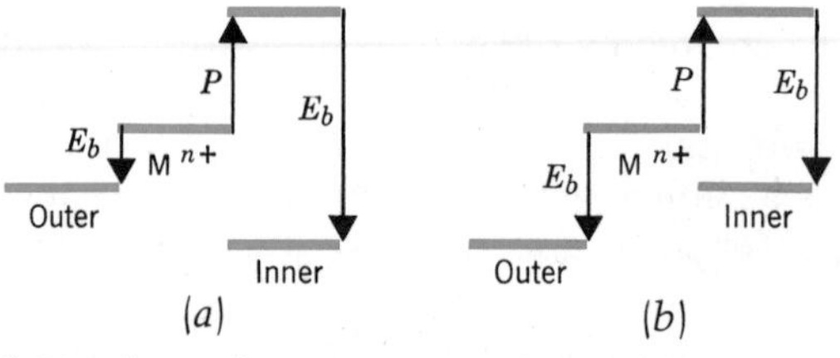


Figure 20.18
*Energy changes in the production of inner orbital and outer orbital complexes. (a) inner orbital favored;* $E_{outer} > E_{inner}$. *(b) Outer orbital favored;* $E_{outer} < E_{inner}$. *(*$E_b$ = *Energy released upon bond formation. P = pairing energy.)*

[6] Cyanides, such as KCN or HCN, are very poisonous because of the ability of $CN^-$ to irreversibly bind to iron atoms in hemoglobin and because $CN^-$ is able to inactivate certain enzymes.

and for the $[Fe(CN)_6]^{3-}$ inner orbital complex,

$[Fe(CN)_6]^{3-}$ ↑↓ ↑↓ ↑ ·· ·· (3d) ·· (4s) ·· ·· ·· (4p) — — — — — (4d); $d^2sp^3$ hybrids: the two 3d pairs (··), 4s and 4p

**OTHER GEOMETRIES.** Valence bond theory can also be applied to geometries other than octahedral. For example, the complex ion, $[Ni(CN)_4]^{2-}$, is found to have a square planar shape and is diamagnetic. In Table 17.1 we find that in order to have a square planar geometry, a *dsp*$^2$ set of hybrid orbitals must be used by the nickel ion. Once again we have nickel(II) and hence a $d^8$ electron configuration.

$Ni^{2+}$ [Ar] ↑↓ ↑↓ ↑↓ ↑ ↑ (3d) — (4s) — — — (4p)

We can create the one empty *d* orbital needed for the hybrids by pairing the electrons in the 3*d* subshell, so that for the complex we have

$[Ni(CN)_4]^{2-}$ ↑↓ ↑↓ ↑↓ ↑↓ ·· (3d) ·· (4s) ·· ·· — (4p); $dsp^2$ hybrids: the last 3d (··), 4s and two 4p

Since we have paired all of the electrons the complex should be diamagnetic, which it is.

Tetrahedral complexes such as $[CoCl_4]^{2-}$ can also be accounted for. This ion, containing cobalt(II) with a $d^7$ configuration, makes use of tetrahedral $sp^3$ hybrids and has the electronic structure

$[CoCl_4]^{2-}$ ↑↓ ↑↓ ↑ ↑ ↑ (3d) ·· (4s) ·· ·· ·· (4p); $sp^3$ hybrids: 4s and 4p

From our discussion, valence bond theory appears to be quite effective at accounting for the structures and magnetic properties of complex ions. However, there are some problems with the theory. One very serious drawback is that it does not allow us to explain why complex ions exist in such a wide profusion of colors, even where they contain the same metal ion in the same oxidation state. Table 20.10, for example, contains a list of some typi-

Table 20.10
*Colors of some octahedral cobalt(III) complexes*

| Complex | Color |
|---|---|
| $[Co(H_2O)_6]^{3+}$ | Pink |
| $[Co(NH_3)_6]^{3+}$ | Yellow-red |
| $[Co(NO_2)_6]^{3-}$ | Yellow |
| $[Co(en)_3]^{3+}$ | Yellow-orange |
| $[Co(C_2O_4)_3]^{3-}$ | Green |
| $[Co(en)(C_2O_4)_2]^{-}$ | Red-violet |
| $[Co(EDTA)]^{-}$ | Violet |
| $[Co(CN)_6]^{3-}$ | Colorless |

cal cobalt(III) complexes and their colors. There are also certain complex ions that are quite difficult to account for in a reasonable and satisfying way. The ion $[Co(NO_2)_6]^{4-}$ is one of these. This ion contains Co(II), a $d^7$ ion,

$Co^{2+}$ [Ar] ↑↓ ↑↓ ↑ ↑ ↑ (3d) __ (4s) __ __ __ (4p) __ __ __ __ __ (4d)

With water we saw that cobalt(II) forms an outer orbital complex containing three unpaired electrons. The $[Co(NO_2)_6]^{4-}$ ion, however, contains only one unpaired electron; therefore, it must be postulated that two of the three unpaired electrons in $Co^{2+}$ become paired and that the third is promoted to the 4*d* subshell in order to make two 3*d* orbitals available for inner orbital complex formation. The final result, then, looks like this:

$[Co(NO_2)_6]^{4-}$ ↑↓ ↑↓ ↑↓ ·· ·· (3d) ·· (4s) ·· ·· ·· (4p) ↑ __ __ __ __ (4d)

($d^2sp^3$: the two 3*d*, the 4*s* and the three 4*p* orbitals holding the ligand electron pairs)

Thus with valence bond theory we can account for the magnetic properties of this ion, but they certainly are not what we would have predicted. Let us now look at another bonding theory that manages to avoid some of the pitfalls of valence bond theory as applied to these transition metal complexes.

## 20.11 CRYSTAL FIELD THEORY

A second theory of bonding in transition metal complexes, that has been extensively applied over the past 20 years, is called **crystal field theory** (CFT). It was developed by physicists in the early 1930s to deal with metal ions trapped in crystalline lattices, but it was not until the early 1950s that chemists realized that it could be applied to coordination complexes in general. It differs from valence bond theory in that it views the complex as held together by purely electrostatic attactions, that is, *in its simplest form, CFT ignores covalent bonding.* The most significant aspect of the theory, however, is its concern with the effect that the ligands have on the energies of the *d* orbitals of the metal.

Generally, the ligands in a transition metal complex are either anions, or they are polar molecules. In the latter case the negative ends of the ligand dipoles point in the direction of the metal cation. Let us examine how these ligands affect the *d* orbitals. The simplest complex ion that we can consider for this purpose is the $[Ti(H_2O)_6]^{3+}$ cation, consisting of a $Ti^{3+}$ ion surrounded octahedrally by six water molecules. Titanium(III) has a single 3*d* electron,

$Ti^{3+}$ ↑ __ __ __ __ (3d) __ (4s) __ __ __ (4p)

Which of the five 3*d* orbitals will this electron prefer to occupy?

In the free gaseous $Ti^{3+}$ ion all of the 3*d* orbitals have exactly the same energy; we say that they are **degenerate.** In the presence of the ligands, however, some of this degeneracy is removed and we find that not all of the 3*d* orbitals have the same energy. To understand why, we must consider the spatial arrangement of the five 3*d* orbitals.

In Chapter 3 we discussed the shapes of the *p* orbitals; each one consists of a pair of lobes directed along a coordinate axis. The *d* orbitals are somewhat more complex, as we can see in Figure 20.19. Four of them, labeled $d_{xy}$, $d_{xz}$, $d_{yz}$, and $d_{x^2-y^2}$, have the same shape, being composed of four lobes each.

The fifth, the $d_{z^2}$, consists of two large lobes directed along the positive and negative $z$ axis plus a donut of charge in the $xy$ plane. For our purposes here, it is important to notice that two of these $d$ orbitals have lobes that are pointed along the coordinate axes (the $d_{x^2-y^2}$ and $d_{z^2}$ orbitals), while the other three (the $d_{xy}$, $d_{xz}$, and $d_{yz}$) have lobes that point between the axes at 45° angles to them.

We can construct an octahedral complex ion by placing the six ligands along the coordinate axes as shown in Figure 20.20. If we now look at the $d$ orbitals of the central ion in the complex (Figure 20.21), we see that the $d_{x^2-y^2}$ and $d_{z^2}$ orbitals have their lobes pointing directly at the ligands. The $d_{xy}$, $d_{xz}$, and $d_{yz}$ orbitals, on the other hand, point between the ligands. Let's imagine, now, placing an electron into one of these $d$ orbitals. If we place it into either the $d_{x^2-y^2}$ or $d_{z^2}$ orbital, it will be forced to spend much of its time in the vicinity of the ligands and, hence, it will tend to be strongly repelled by the negative charge of the ligands (remember that pointing toward the central ion we have either a negative charge because of an anionic ligand or the negative end of a ligand dipole). On the other hand, this repulsion will be much less if the electron is placed into either the $d_{xy}$, $d_{xz}$, or $d_{yz}$ orbitals where it can avoid the ligands.

In the $[Ti(H_2O)_6]^{3+}$ ion, then, the single $d$ electron of the $Ti^{3+}$ ion will experience the least repulsion and, hence, be of lowest energy, when it is located in the $d_{xy}$, $d_{xz}$, or $d_{yz}$ orbitals. In effect, then, the energies of the $d$ orbitals are no longer all the same. Three of them are lower in energy while the other two are higher in energy. This splitting of the energies of the $d$ orbitals of a metal ion that occurs when the ion is placed into the electrostatic field of the ligands is indicated graphically in Figure 20.22. For reasons beyond the scope of this book, for an octahedral field the upper level containing two orbitals is labeled $e_g$ and the lower level, consisting of three orbitals, is referred to as the $t_{2g}$ level. The energy difference between the $e_g$ and $t_{2g}$ levels is denoted as $\Delta$.

In the lowest energy state of the $[Ti(H_2O)_6]^{3+}$ ion, the single $3d$ electron of titanium will be located in one of the orbitals in the low-energy $t_{2g}$ level. When the ion absorbs light, the energy absorbed is able to promote this electron to the $e_g$ level, as shown in Figure 20.23. The color of the absorbed light depends on the magnitude of $\Delta$.

Recall that the energy of a photon is related to the frequency of the light, $\nu$, by the equation

$$E = h\nu \qquad (h \text{ is Planck's constant})$$

and that the wavelength is, in turn, related to frequency by

$$\lambda = \frac{c}{\nu} \qquad (c \text{ is the speed of light})$$

Thus high frequency is associated with high energy and short wavelength. As the magnitude of $\Delta$ increases, more energy is required to raise the electron from the $t_{2g}$ to the $e_g$ level. Hence, light of higher frequency (shorter wavelength) must be used. Generally, for most complex ions the wavelength of the absorbed light required to cause the promotion lies in the visible region of the spectrum. Since in the absorption process, a portion of the visible spectrum is removed from white light as it passes through (or is reflected from) the complex, the observed color of the complex is due to the wavelengths that remain. For example, the $[Ti(H_2O)_6]^{3+}$ complex has a $\Delta$ that corresponds to energies associated with light in the green region of the

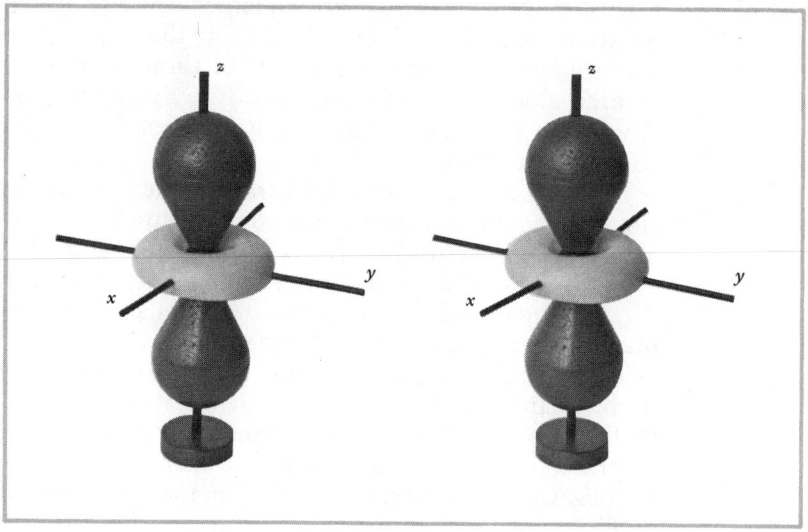


(a)

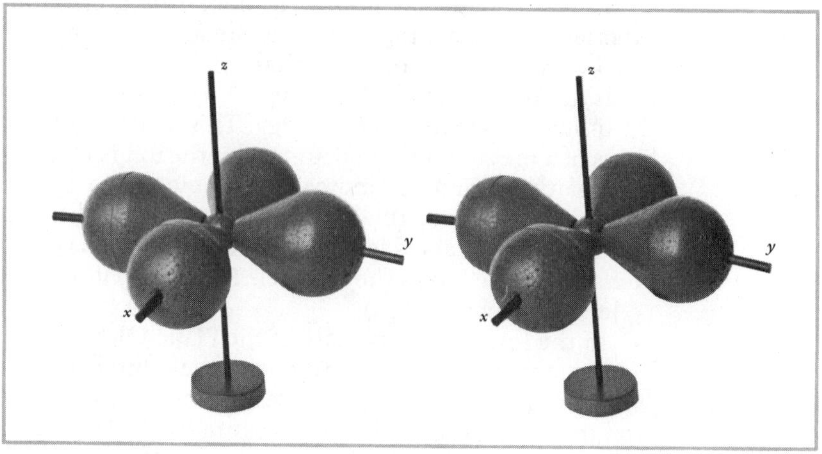


(b)

Figure 20.19
*Directional properties of the* d *orbitals. (a)* $d_{z^2}$. *(b)* $d_{x^2-y^2}$. *(c)* $d_{yz}$. *(d)* $d_{xy}$. *(e)* $d_{xz}$. *Atomic-Molecular Orbital Models by Science Related Materials, Inc., Janesville, Wisconsin.*

spectrum. When white light is passed through a solution of this complex, green is removed and the solution appears violet.

For a given metal ion, different ligands have different effects on the splitting of the *d* orbitals, that is, on $\Delta$. By examining the absorption spectra of various complexes, we can arrange the ligands in order of their ability to produce a large $\Delta$. This series is called the **spectrochemical series** and can be given in abbreviated form as

$$I^- < Br^- < Cl^- < F^- < OH^- < H_2O < NH_3 < en < NO_2^- < CN^-$$

Thus $I^-$ is poorest at splitting the energies of the $t_{2g}$ and $e_g$ orbitals, and $CN^-$ is best. What is particularly interesting is that this same series applies for essentially any metal in any oxidation state. However, whereas the order is

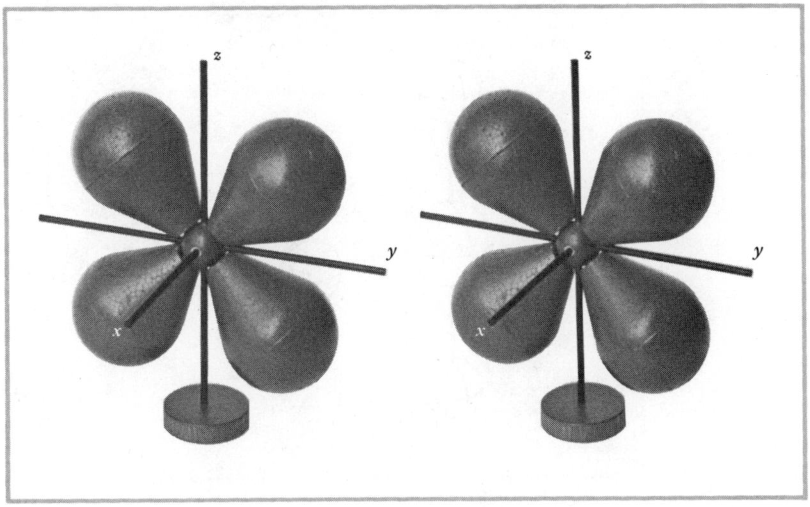


(c)

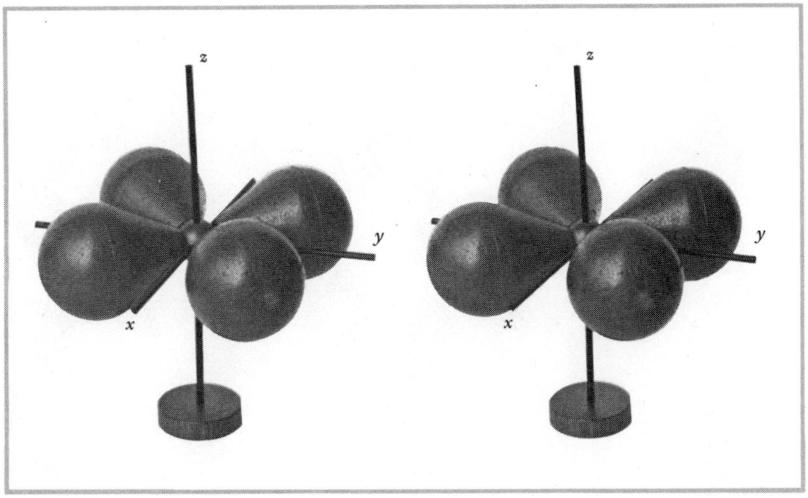


(d)

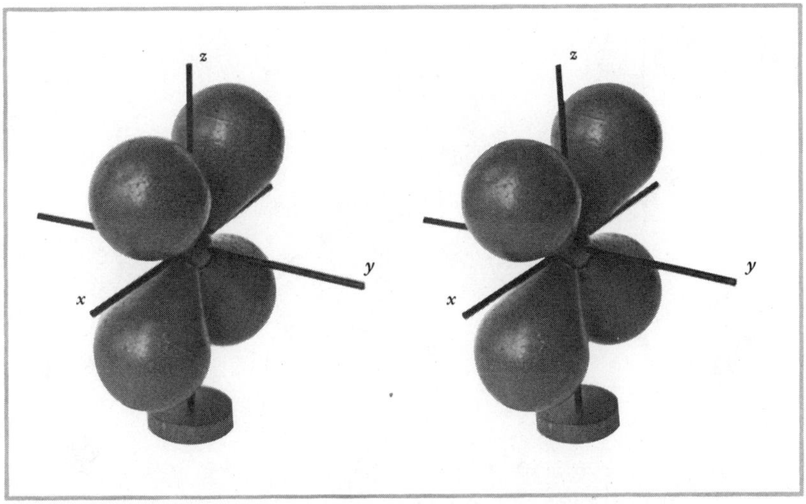


(e)

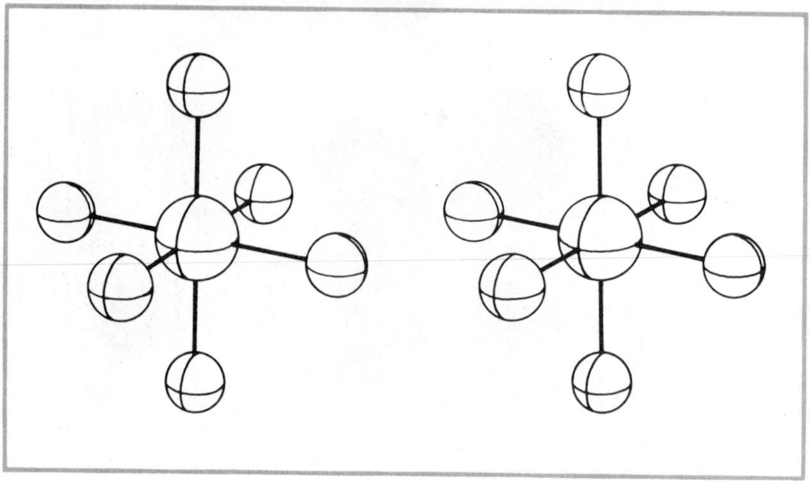

**Figure 20.20**
*Octahedral arrangement of ligands about a central metal ion.*

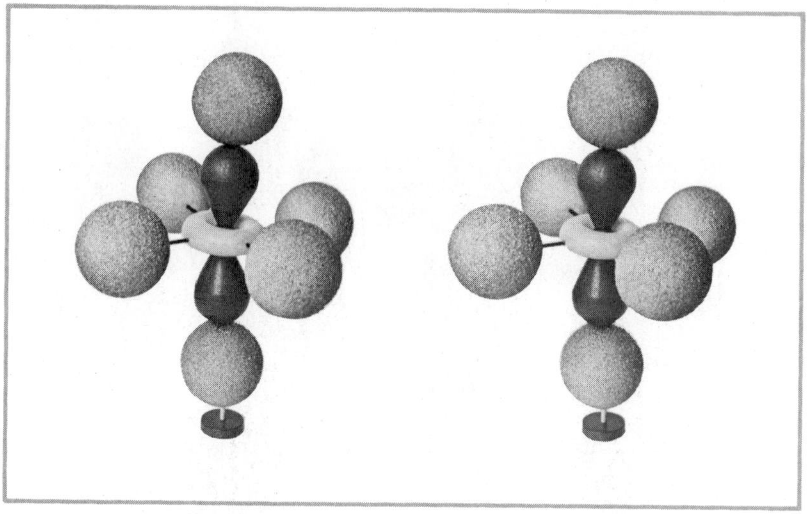

(*a*)

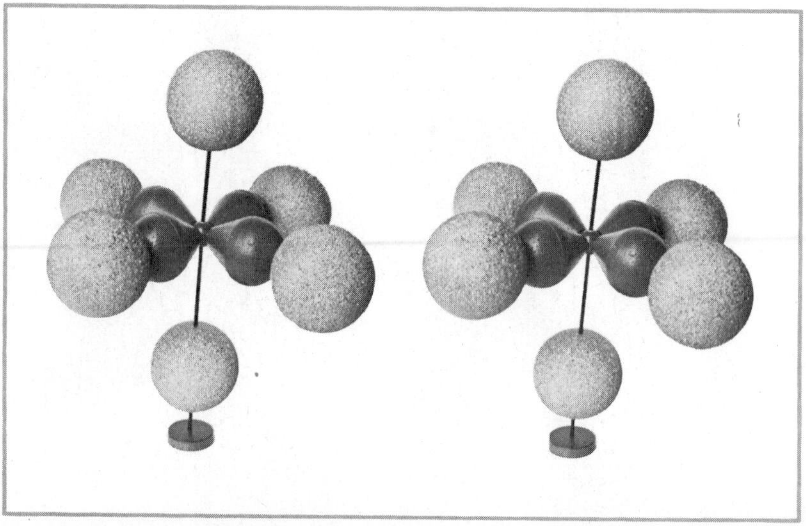

(*b*)

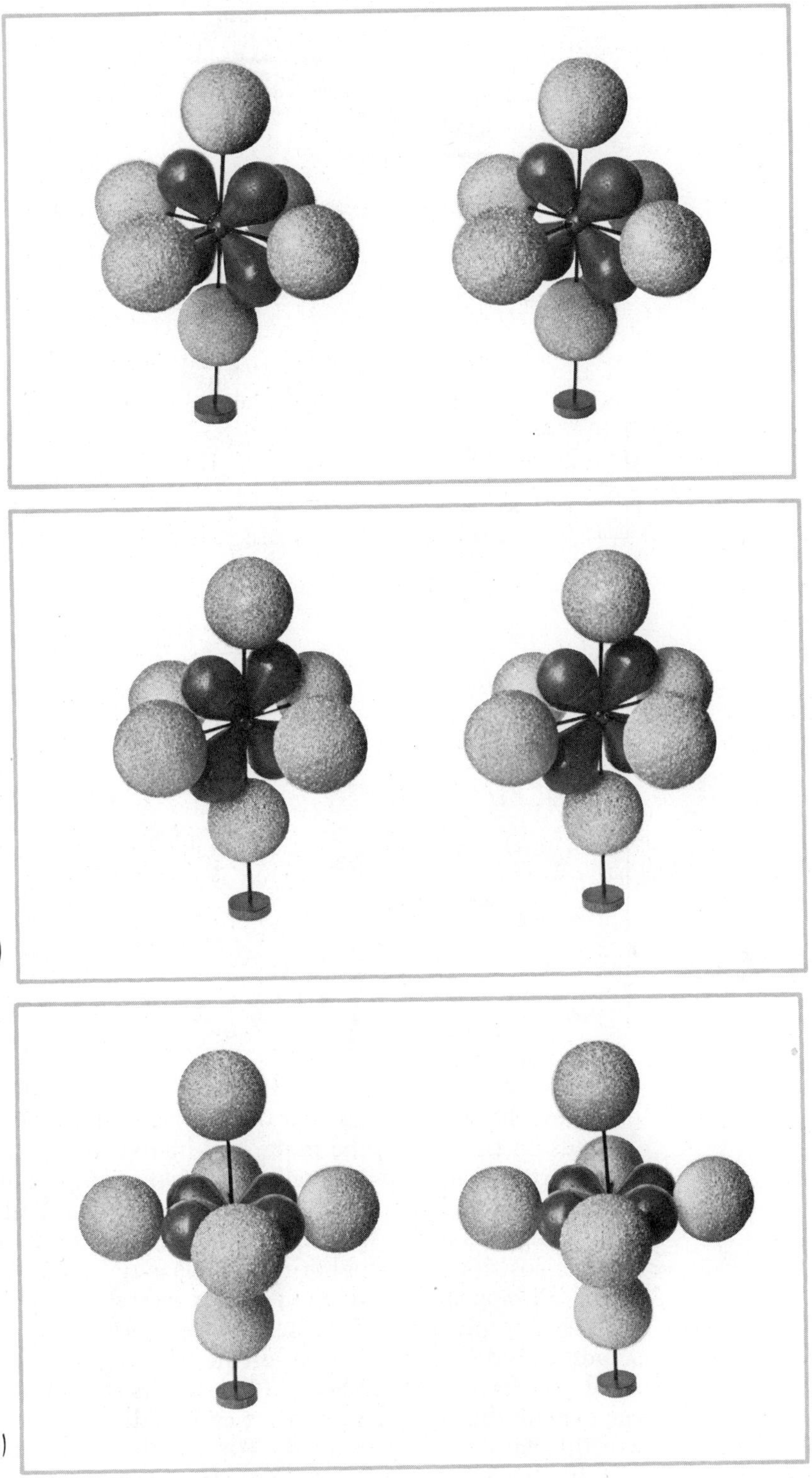

(c)

(d)

(e)

Figure 20.21

*Interaction of ligands with the* d *orbitals of the metal.* (*a*) $d_{z^2}$. (*b*) $d_{x^2-y^2}$. (*c*) $d_{yz}$. (*d*) $d_{xz}$. (*e*) $d_{xy}$. *Atomic-Molecular Orbital Models by Science Related Materials, Inc., Janesville, Wisconsin.*

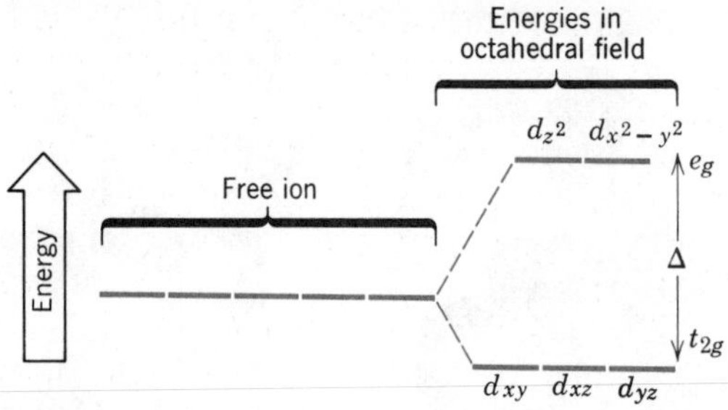


Figure 20.22
*Splitting of energies of the* d *orbitals in an octahedral crystal field.*

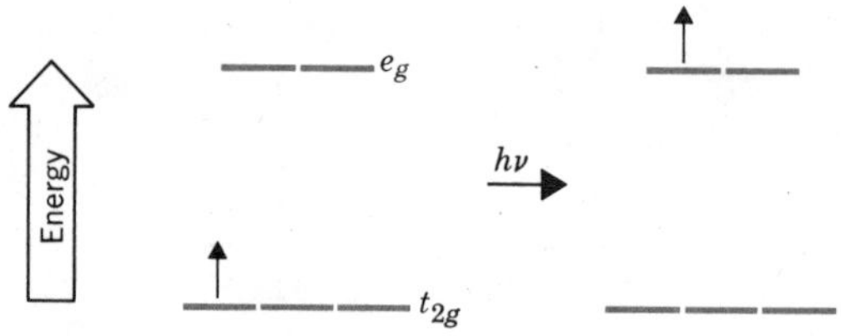


Figure 20.23
*Absorption of light by* $[Ti(H_2O)_6]^{3+}$ *ion. Absorption of light* ($h\nu$) *promotes electron from* $t_{2g}$ *to* $e_g$.

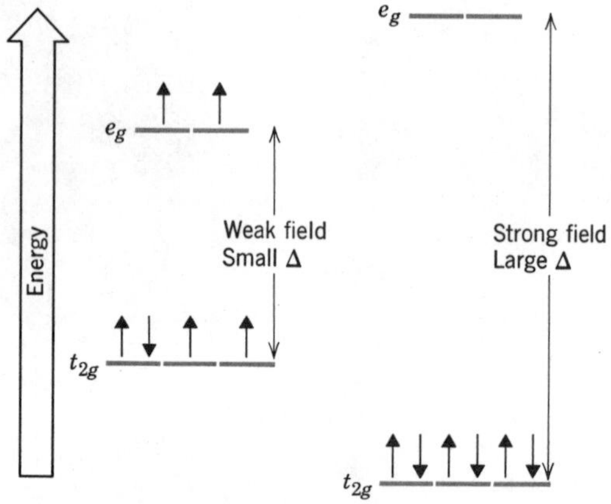


Figure 20.24
*Electron configuration of Co(III) in weak and strong crystal fields.*

the same, the actual magnitude of $\Delta$ for a given complex in a given geometry depends on the ligand, the metal and its oxidation state.

As with nearly any generalization that we might make in attempting to describe chemical properties, there are exceptions. This is true here with the order of ligands in the spectrochemical series where, in some instances, the relative positions of neighboring ligands is reversed. With cobalt(III), for example, $Cl^-$ appears to produce a greater crystal field splitting than $F^-$. Nevertheless, the spectrochemical series often serves as a useful guide in understanding (and sometimes even predicting) properties of complexes. For example, we have just seen how CFT accounts for the colors of complexes. We can explain their magnetic properties as well. Consider, for example, the cobalt(III) complexes of $F^-$ and $Cl^-$. The metal ion here contains six *d* electrons and, in a weak crystal field (one that produces a small $\Delta$), they will be unpaired as much as possible, as shown in Figure 20.24*a*, to give a complex with four unpaired electrons. This is what occurs with $F^-$ in the $[CoF_6]^{3-}$ ion.

When the ligand produces a large crystal field (and hence a large $\Delta$), we have the possibility of pairing all of the $d$ electrons in the $t_{2g}$ level (Figure 20.24*b*) to produce a diamagnetic complex. This will occur if the magnitude of $\Delta$ is greater than the energy needed to pair the electrons in a given orbital. In other words, when the pairing energy (let's call it $P$) is less than $\Delta$, more energy is required to place the electron in the $e_g$ orbital than is required to pair them and place them in the $t_{2g}$ level. This happens when the ligand is $Cl^-$.

For Co(III) complexes in general (or for that matter, any $d^6$ system), a paramagnetic complex with four unpaired electrons will occur whenever $\Delta < P$; diamagnetic complexes will be formed when $\Delta > P$. In general, we speak of these two possibilities as **high-spin** complexes (minimum pairing of electrons) and **low-spin** complexes (maximum pairing of electrons from the $e_g$ into the $t_{2g}$). Comparing them to the valence bond treatment, we find that low-spin complexes correspond to inner orbital complexes whereas high-spin complexes correspond to outer orbital complexes.

The possibility of both low-spin and high-spin complexes exists when the central metal ion contains four, five, six, or seven $d$ electrons. The electron configurations of the $t_{2g}$ and $e_g$ levels in these species are left to you as an exercise (Question 20.52). For $d^1$, $d^2$, and $d^3$ systems the electrons will naturally prefer the three low-energy $t_{2g}$ orbitals, because no pairing is required; therefore, only one type of electron configuration will be found for them. Likewise, with a $d^8$ or $d^9$ configuration, six electrons will be forced to occupy the $t_{2g}$ (thereby filling it) and the $e_g$ level will contain either two or three electrons, respectively. Once again we see that $d^8$ and $d^9$ ions will each have only one type of electron configuration in an octahedral complex.

With this as background, we note that the magnetic properties of the $[Co(NO_2)_6]^{4-}$ ion, which presented such a problem with the valence bond theory, are easily explained in terms of the CFT. Recall that this complex contains the $Co^{2+}$ ion, a $d^7$ system. From the spectrochemical series we also note that $NO_2^-$ produces a very strong crystal field; therefore, we would expect that $\Delta$ is probably quite large, larger in fact than the pairing energy. Under these circumstances there will be pairing of electrons in the $t_{2g}$ level, as we see in Figure 20.25. Since the $t_{2g}$ level can accommodate only six electrons (three pairs), the seventh electron is forced to occupy the $e_g$ level. The complex is therefore low-spin (analogous to inner orbital of the valence bond theory) and will contain a single unpaired electron, in agreement with experiment. Thus we see that there is really nothing unusual about the $[Co(NO_2)_6]^{4-}$ ion, quite opposite to what we would have concluded based on the valence bond theory.

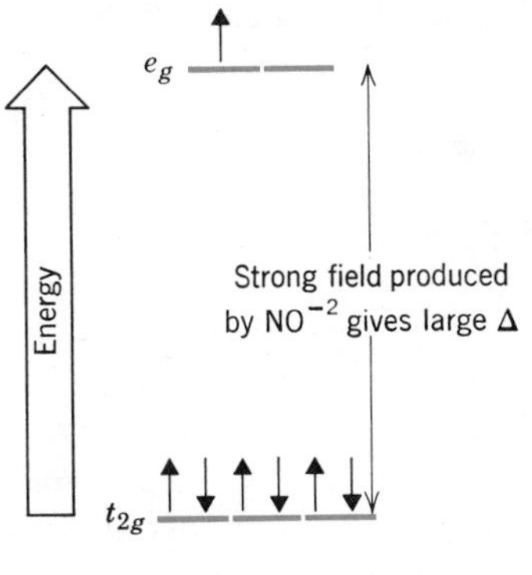


Figure 20.25
*Pairing of electrons in* $t_{2g}$ *level in* $[Co(NO_2)_6]^{4-}$.

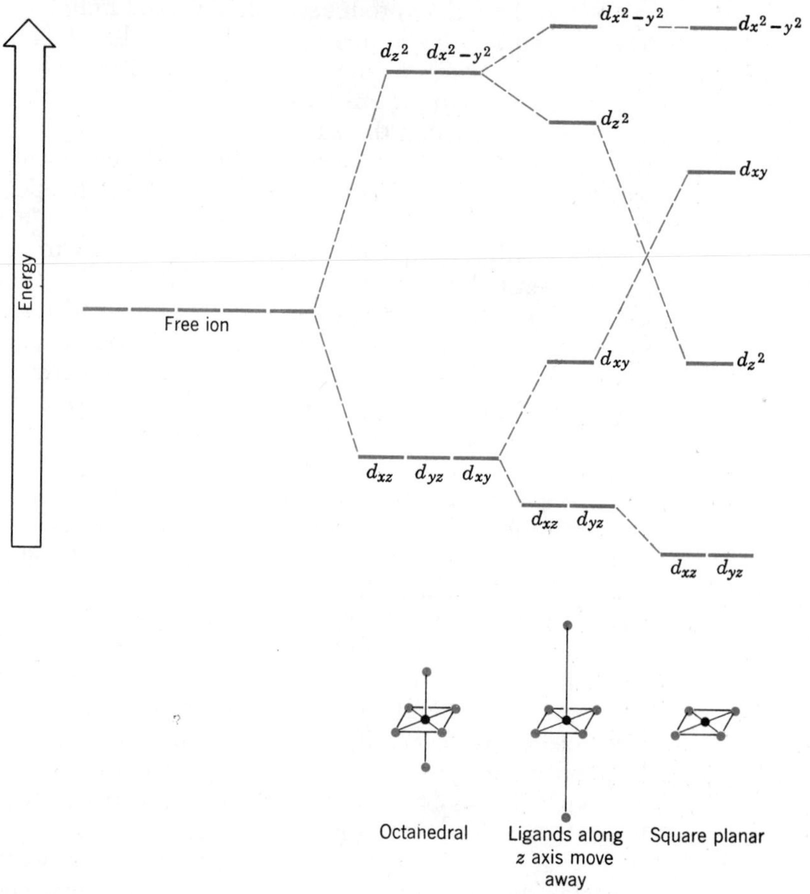


Figure 20.26
*Splitting pattern of the* d *orbitals changes as the geometry of the complex changes.*

Crystal field theory can be extended to other geometries besides octahedral, the difference being that other splitting patterns are observed. For example, a square planar complex can be thought of as being derived from an octahedral complex by removing the ligands that lie along the *z* axis. As shown in Figure 20.26, when this occurs the energies of the $d_{z^2}$, $d_{xz}$ and $d_{yz}$ orbitals decrease because an electron placed into them experiences less repulsion than in an octahedral complex. Also, by removing the ligands along the *z* axis, those along the *x* and *y* axes can move in slightly and therefore

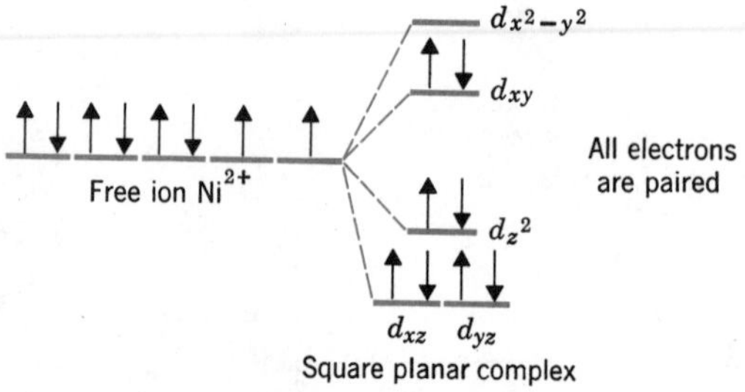


Figure 20.27
*Diamagnetic* $[Ni(CN)_4]^{2-}$, *a square planar complex.*

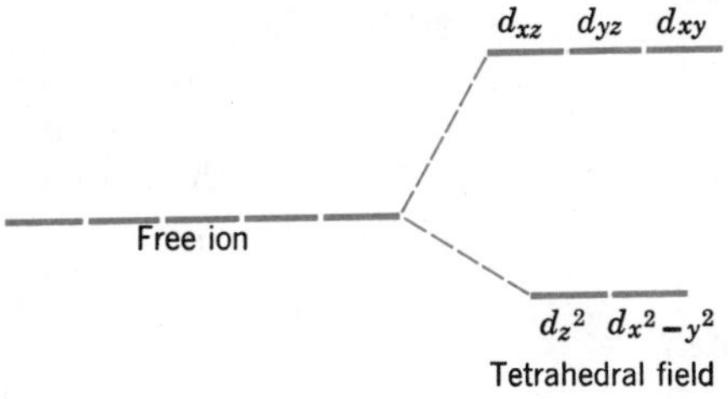


Figure 20.28
*Splitting pattern for a tetrahedral field.* $\Delta_{tet} \approx \frac{4}{9}\,\Delta_{oct}$.

the energies of the $d_{x^2-y^2}$ and $d_{xy}$ orbitals rise somewhat. In the $[Ni(CN)_4]^{2-}$ ion, the energy separation between the $d_{xy}$ and $d_{x^2-y^2}$ is large enough that the eight $d$ electrons of the $Ni^{2+}$ ion can exist as four pairs (Figure 20.27).

In tetrahedral complexes the splitting pattern of the $d$ orbitals is that shown in Figure 20.28. Notice that the order of the energy levels is exactly opposite to that found in octahedral complexes. The magnitude of $\Delta$ is also considerably smaller (actually, $\Delta_{tetr} \approx \frac{4}{9}\,\Delta_{oct}$ for the same ligands and metal ion). The small $\Delta$ observed for tetrahedral complexes is always less than the pairing energy, and tetrahedral complexes are always of the high-spin variety. Note that this agrees with the valence bond theory in which $sp^3$ hybrids were used by the metal, thereby making all of the $d$ orbitals available for spreading out the $d$ electrons.

## INDEX TO QUESTIONS AND PROBLEMS (problem numbers in italics)

General 1, 2, 3, 4, 5, 6
Electronic structure, 7, 8, 9
Oxidation states 10, 11, 12, 13, *53*
Atomic and ionic radii 14, 15, 16, *54*
Metallurgy 17, 18, 19, 20, 21, 22, 23, 24, 25
Magnetism 26
Coordination compounds, general 27, 28, 29
Structures of coordination compounds 30, 31, 32
Nomenclature 33, 34, 35, 36
Isomerism 37, 38, 39, 40, 41
Bonding: valence bond theory 42, 43, 44, 45
Bonding: crystal field theory 46, 47, 48, 49, 50, 51, 52

## REVIEW QUESTIONS

**20.1** What is a transition element? What is an inner transition element?

**20.2** What similarities exist between the Group VIIA and VIIB elements? In what way are they different?

**20.3** You no doubt are familiar with quite a few of the $d$-block metals. Make a list of those with which you are familiar and give as many applications of each as you can.

**20.4** What relationship (if any) exists between the A and B groups in the periodic table?

**20.5** What is *aqua regia*?

**20.6** Give four general properties of the transition elements.

**20.7** Write the electron configurations of each of the members of the first-row transition elements.

**20.8** Use the rules you learned in Chapter 3 to write electron configurations for the following elements. Compare them with the actual configurations given in Tables 20.2 through 20.4.

(a) Zr (b) Os (c) Pr (d) Ag (e) Mo (f) Pt (g) U (h) Tm (i) Au (j) Hg

**20.9** Write electron configurations for the following ions:

(a) $Cu^+$ (b) $Fe^{2+}$ (f) $Pt^{2+}$ (g) $Cd^{2+}$

(c) $Ti^{4+}$
(d) $V^{3+}$
(e) $Ag^{+}$
(h) $Co^{3+}$
(i) $Fe^{3+}$
(j) $Sc^{3+}$

**20.10** What accounts for the occurrence of a +2 oxidation state for a large number of transition elements?

**20.11** How do the relative stabilities of high and low oxidation states of the transition elements vary as we move from left to right across a given period?

**20.12** Compare the relative stabilities of high and low oxidation states for the elements in Group VIB.

**20.13** When solutions containing chromate ion ($CrO_4^{2-}$) are acidified, dichromate ion ($Cr_2O_7^{2-}$) is produced. Use structural formulas to indicate how this polymerization occurs. What oxoanions of sulfur are analogous to $CrO_4^{2-}$ and $Cr_2O_7^{2-}$?

**20.14** How do atomic radii vary from left to right across the first-row transition elements? How do we account for this variation?

**20.15** What is the lanthanide contraction? How does it affect the properties of the *d*-block elements in period 6?

**20.16** The chemistries of Zr and Hf are very similar; the elements are always found together in nature and are difficult to separate from one another. What explanation can be offered to account for the similar properties of Zr and Hf compounds?

**20.17** What is an *ore*?

**20.18** What are the three steps that are involved in extracting a metal from its ore and making it ready for practical use?

**20.19** Describe the process called flotation. What is meant by *roasting* as applied to metallurgy?

**20.20** What is an amalgam? What is a common application of amalgams?

**20.21** Write chemical equations for (a) the reduction of $Fe_2O_3$ in the blast furnace and (b) the production of slag from $SiO_2$ and $CaCO_3$.

**20.22** Write chemical equations for the purification of bauxite, $Al_2O_3$.

**20.23** Why is $TiO_2$ a better paint pigment than white lead? Why is titanium metal useful in the manufacture of aircraft?

**20.24** What is the Mond process?

**20.25** Why must pig iron be refined to be useful as a strong structural metal? What is a Bessemer converter; an open hearth furnace; the basic oxygen process?

**20.26** Compare the properties of paramagnetism and ferromagnetism. Why can a ferromagnetic material become permanently magnetized?

**20.27** Define ligand, first coordination sphere, coordination compound, monodentate ligand, polydentate ligand, chelate, and coordination number.

**20.28** What are some applications of coordination compounds?

**20.29** What are some uses of EDTA?

**20.30** Sketch the common structures found for complex ions with coordination number 4 and coordination number 6.

**20.31** Sketch the structure of an octahedral EDTA complex.

**20.32** Nitrilotriacetic acid, NTA (structure shown below), was used by detergent manufacturers for a while in place of phosphates because it is biodegradable and does not promote the growth of algae. However, it was found to increase the solubility of some heavy metals that are poisonous to a variety of life forms, and its use has been discontinued. Can you suggest, using appropriate chemical equations and structural formulas, how NTA dissolves metal compounds?

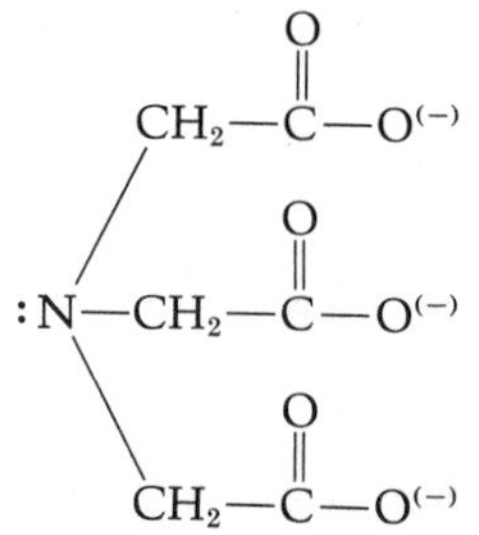

**nitrilotriacetate ion**

**20.33** Give IUPAC names for each of the following:
(a) $[Ni(NH_3)_6]^{2+}$
(b) $[CrCl_3(NH_3)_3]^0$
(c) $[Co(NO_2)_6]^{3-}$
(d) $[Mn(C_2O_4)_3]^{3-}$
(e) $MnO_4^-$

**20.34** What are the IUPAC names for the following:
(a) $[AgI_2]^-$
(b) $[Cr(NH_3)_5Cl]^{2+}$
(c) $[Co(H_2O)_4(NH_3)_2]Cl_2$

(d) $[Co(en)_2(H_2O)_2]_2(SO_4)_3$
(e) $[Cr(NH_3)_4Cl_2]Cl$

**20.35** Write chemical formulas for the following:
(a) Dicyanotetraaquoiron(III) ion
(b) Oxalatotetraamminenickel(II)
(c) Potassium hexacyanomanganate(III)
(d) Tetrachlorocuprate(II) ion
(e) Tetraoxochromate(VI) ion

**20.36** Write chemical formulas for the following:
(a) Tetrachloroaurate(III) ion
(b) Dinitrobis(ethylenediamine)iron(III) sulfate
(c) Carbonatotetraamminecobalt(III) nitrate
(d) Ethylenediaminetetraacetatoferrate(II) ion
(e) Dithiosulfatoargentate(I) ion

**20.37** What is meant by *isomer?* What are stereoisomers?

**20.38** Sketch the isomers of $[Co(NH_3)_2Cl_4]^-$. Identify *cis* and *trans* isomers. How many isomers are there for the complex, $[Co(NH_3)_3Cl_3]$? Sketch them.

**20.39** Sketch the isomers of $[Cr(en)_2Cl_2]^+$. Identify *cis* and *trans* isomers and indicate any isomers that exhibit optical isomerism.

**20.40** Draw the two optical isomers of $[Co(EDTA)]^-$.

**20.41** What are enantiomers? What is meant by *racemic?*

**20.42** What is the difference between an inner orbital complex and an outer orbital complex?

**20.43** Use valence bond theory to predict the electron configuration, the type of bonding (inner orbital or outer orbital), and the number of unpaired electrons for each of the following:
(a) $[VCl_6]^{3-}$
(b) $[Ni(NH_3)_6]^{2+}$
(c) $[Fe(NH_3)_6]^{3+}$
(d) $[Co(CN)_6]^{3-}$
(e) $[CrCl_6]^{3-}$

**20.44** Predict the number of unpaired electrons in (a) $[Cr(H_2O)_6]^{2+}$ (b) $[Cr(CN)_6]^{4-}$

**20.45** What magnetic properties would you predict for the square planar complex, $[Cu(NH_3)_4]^{2+}$?

**20.46** Sketch on appropriate coordinate axes the shapes of the five *d* orbitals.

**20.47** Diagram the crystal field splitting of the *d* orbitals, and indicate the electron population of each energy level, in the paramagnetic complex $[Mn(H_2O)_6]^{3+}$. Label the energy levels.

**20.48** How does crystal field theory account for the colors of complex ions?

**20.49** What relationship exists between $\Delta$ (the crystal field splitting) and the pairing energy in determining whether a given complex will be paramagnetic or diamagnetic?

**20.50** What are meant by high-spin and low-spin complexes? How do these compare with inner orbital and outer orbital complexes in the valence bond theory?

**20.51** Sketch the CFT splitting patterns of the *d* orbitals for
(a) square planar and
(b) tetrahedral complexes.

**20.52** Using the CFT splitting pattern for an octahedral complex, indicate the high-spin and low-spin distribution of electrons among the $t_{2g}$ and $e_g$ levels for the configurations: (a) $d^4$, (b) $d^5$, (c) $d^6$, (d) $d^7$.

## REVIEW PROBLEMS

**20.53** Saturated solutions of mercurous chloride were prepared by adding the solid to solutions having various chloride ion concentrations. The total concentration of mercury in each solution was then determined. Use the concepts that you learned having to do with equilibrium and solubility product to show that the data at the right are only consistent with mercury(I) having the formula $Hg_2^{2+}$, [and hence mercury(I) chloride being $Hg_2Cl_2$], and not with $Hg^+$ (and therefore HgCl).

| Chloride Ion Concentration | Moles of Mercury per Liter |
|---|---|
| 1.0 *M* | $2.2 \times 10^{-18}$ |
| 0.5 *M* | $8.8 \times 10^{-18}$ |
| 0.2 *M* | $5.5 \times 10^{-17}$ |
| 0.1 *M* | $2.2 \times 10^{-16}$ |

**20.54** The atomic radii of Zr and Hf are 1.60 and 1.58 Å, respectively. Zr and Hf have the same crystal structure (hexagonal closest packed). The density of Zr is 6.49 g/cm³. Calculate the approximate density of Hf.

# 21 ORGANIC CHEMISTRY

As far back as the eighteenth century chemists were able to distinguish between two types of compounds, those derived from plants and animals and those from the mineral constituents of the earth. The latter type, called inorganic substances, have received much of our attention thus far in this text. This chapter and the next are devoted to a discussion of organic compounds, many of which find their origin in the nature of living things but most of which have been synthesized in the laboratory.

One does not have to delve very far into the chemistry of life before observing that the element carbon is present in all of the molecules in life's makeup. Of all of the 100 odd elements, carbon is the only one found universally in these substances and, thus, organic chemistry has become known as the study of carbon and its compounds. Besides carbon, organic molecules contain relatively few other elements; among the most prevalent are hydrogen, oxygen, nitrogen, and to a lesser extent phosphorus and sulfur.

Until 1828 chemists believed that the only source of these organic compounds was from nature itself. It was thought that it was impossible to synthesize them in the laboratory because nature's "vital force" was missing. In 1828 Friedrich Wöhler first synthesized urea (a compound found in urine) from inorganic materials. Wöhler evaporated an aqueous solution of the inorganic salt ammonium cyanate which resulted in the production of urea:

$$\underset{\textbf{ammonium cyanate}}{NH_4OCN\ (aq)} \longrightarrow \underset{\textbf{urea}}{CO(NH_2)_2}$$

As a result of this first synthesis, many scientists began to attempt to prepare other organic materials with a good deal of success. Gradually the vital force theory was abandoned, and during the next several years a great profusion of organic compounds were made. Today there are well over two million known organic compounds (compared to about 100,000 inorganic substances), with new ones being discovered every day.

The great abundance of organic compounds is a result of carbon's ability to bond to itself to form long chains, rings, and complex combinations of both. By comparison, silicon (which is below carbon in the periodic table) is able to form only relatively short chains when bonded to itself. Silicon is not able to form chains containing thousands of atoms bonded together as carbon does as, for example, in rubber, plastics, and synthetic fibers.

In organic chemistry we are fortunate to be able to categorize systematically a gigantic number of compounds into a relatively small number of groups quite successfully. In this chapter we shall survey a number of these groups and look at some samples of organic reactions. The goal here will be to provide an overview of some of the topics that are of importance in this vast field of organic chemistry.

## 21.1 HYDROCARBONS

We begin our classification with the **hydrocarbons,** which are compounds containing only carbon and hydrogen. All of the remaining types of organic compounds can then be looked upon as being derived from the hydrocarbons. Hydrocarbons can be divided into two main categories: **aliphatic hydrocarbons,** which include straight-chain, branched-chain, and cyclic compounds, and **aromatic hydrocarbons,** which contain highly stable rings of carbon atoms. The aliphatic hydrocarbons can be further subdivided into two groups based on the multiplicity of the carbon–carbon bond: **saturated hydrocarbons** that contain only carbon–carbon single bonds; and **unsaturated hydrocarbons** that possess at least one carbon–carbon double bond or triple bond.

**SATURATED HYDROCARBONS.** There is only one type of noncyclic saturated hydrocarbon while, as we shall soon see, there are two types of noncyclic unsaturated hydrocarbons. The compounds that constitute the saturated hydrocarbons are collectively called the **alkanes** or **paraffins.** The first ten members of the straight-chain (normal) alkanes are listed in Table 21.1 in order of an increasing number of carbon atoms in the chain.

The alkanes are important as fuels and as raw materials in the synthesis of other important organic compounds. They occur in abundance in petroleum from which they may be separated to a large degree by fractional distillation. Methane, a gas at room temperature, is the "natural gas" used for cooking and as a heating fuel. Propane, which liquefies under high pressure, is used as a fuel in many rural areas where tanks of LPG (liquefied petroleum gas) are delivered to homes. Butane liquefies more easily than propane and is used in cigarette lighters. Octane, of course, has a boiling point that places it in the range of gasoline fuel. Heavier alkanes are found in kerosene, lubricating oils, and in the paraffins used to make candles.

The names of each of the members of the alkanes are composed of two parts. The first part, *meth-*, *eth-*, *prop-*, and so on in Table 21.1, reflects the

Table 21.1
*First ten members of the straight-chain alkanes*

| Formula | Name | Boiling Point (°C) at 1 atm |
|---|---|---|
| $CH_4$ | Methane | −161 |
| $C_2H_6$ | Ethane | −89 |
| $C_3H_8$ | Propane | −44 |
| $C_4H_{10}$ | Butane | −0.5 |
| $C_5H_{12}$ | Pentane | 36 |
| $C_6H_{14}$ | Hexane | 68 |
| $C_7H_{16}$ | Heptane | 98 |
| $C_8H_{18}$ | Octane | 125 |
| $C_9H_{20}$ | Nonane | 151 |
| $C_{10}H_{22}$ | Decane | 174 |

number of carbon atoms in the chain. The second part, which is the same for all the members, is *ane* after the parent name alk*ane*. Thus we have methane, an alkane with one carbon, ethane having two carbons, propane consisting of three carbons, and so on. You should become familiar with the names of all ten of these simpler alkanes, for they serve as the basis for naming many of the remaining organic compounds.

In listing the alkanes by increasing number of carbon atoms, two things become apparent. First, the molecular formula for each of the members of this series can be represented by a single general formula, $C_nH_{2n+2}$, where $n$ is the number of carbons in the molecular chain. For example, the formula for the alkane with four carbon atoms would be $C_4H_{10}$ and, according to Table 21.1, would be called butane. Second, we see that any two successive members of the series differ from each other by a single $CH_2$ group (this becomes quite apparent after ethane). Such a series, where one member differs from the next by the same repeating cluster of atoms, is called a **homologous series.** The alkanes form such a series.

In organic chemistry it is important for us to be able to write molecular formulas, which indicate the number of each of the various atoms present in a molecule, as well as structural formulas, which show the relative positions of each of the atoms in a molecule. Since carbon forms the backbone of all organic compounds, it is mainly the shape of the carbon skeleton that is responsible for the overall shape of the various molecules. Carbon normally forms four covalent bonds, and in the alkanes these are single bonds. Each of the carbon atoms is $sp^3$ hybridized, with the hybrid orbitals pointing to the corners of a tetrahedron.

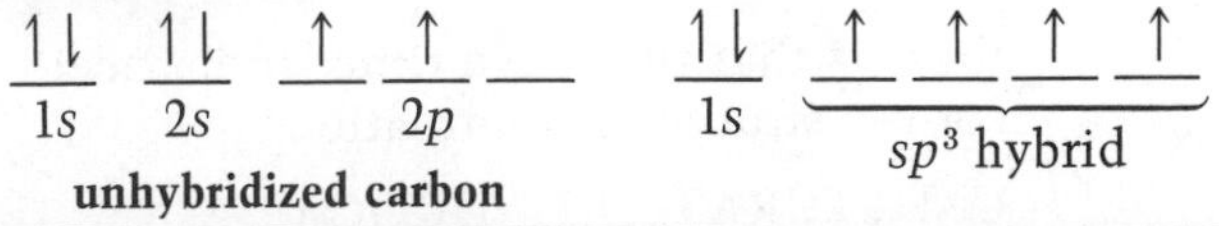


Thus in the straight- and branched-chain alkanes carbon is tetrahedrally surrounded by bonded atoms. There are several ways of representing tetrahedral carbon in two dimensions, as illustrated in Figure 21.1 for methane. The most common, and the simplest, two-dimensional representation used for most organic compounds is the structural formula shown as Figure 21.1*a*. The remaining members of the straight-chain alkanes are drawn simply by connecting together several tetrahedral carbons along with their respective hydrogens.

In Figure 21.2 we see the two- and three-dimensional representations of a four-carbon alkane, butane. Although the two-dimensional representation drawn in Figure 21.2*a* shows a flat molecule with a straight carbon chain, you should remember that molecules such as this are not planar. The tetrahedral geometry about the carbon atoms prevents them from being in a straight line, as shown in Figure 21.2*b*.

Often we find it convenient to condense structural formulas somewhat by not drawing all of the C—H bonds. For instance, the butane formula can be condensed to either

$$CH_3—CH_2—CH_2—CH_3$$

or

$$CH_3CH_2CH_2CH_3$$

We will often use condensed formulas in this chapter. The advantage of

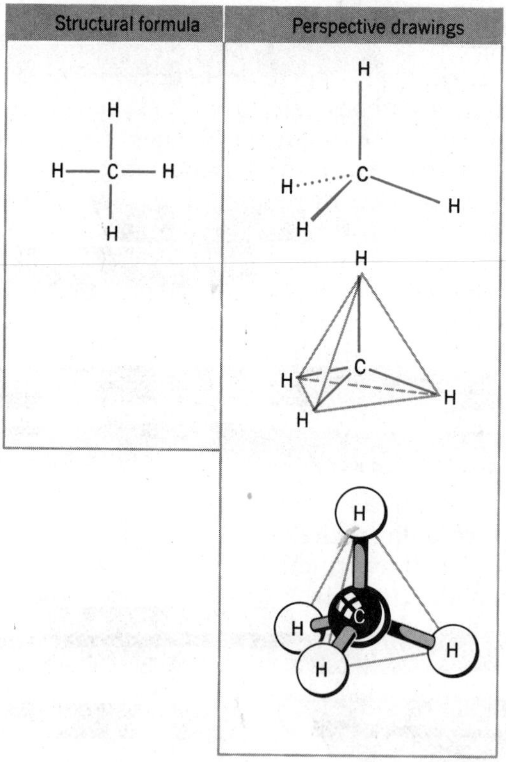


Figure 21.1
*Two-dimensional representations of tetrahedral carbon. Methane is chosen as an example. (a) Structural formula. (b) Perspective drawings.*

these is that they save time (and space) when writing them; yet they still convey structural information.

**UNSATURATED HYDROCARBONS.** The unsaturated hydrocarbons can be divided into two groups: **alkenes** or **olefins,** which contain at least one

Figure 21.2
*Two- and three-dimensional illustrations of the structure of butane. (a) Two-dimensional drawing. (b) Three-dimensional drawing. Note that carbon atoms (white spheres) are not actually in a straight line.*

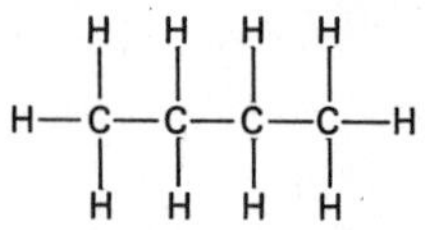

(a)

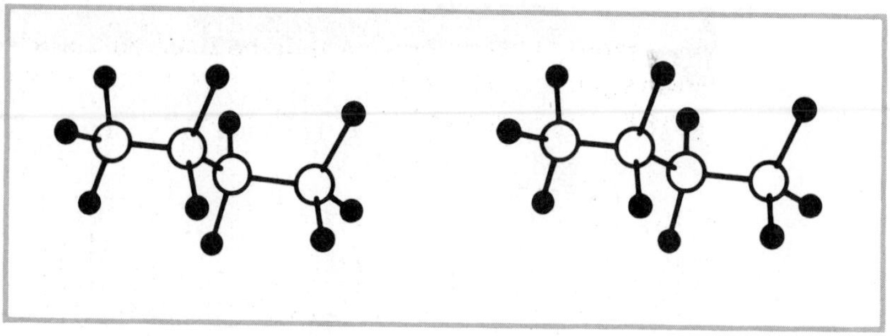

(b)

carbon–carbon double bond and **alkynes,** which contain a carbon–carbon triple bond.

Following a procedure similar to that used for the alkanes, we can introduce the alkenes by considering a series of compounds with one carbon–carbon double bond and an increasing number of carbon atoms in the chain. The simplest alkene contains two carbons:

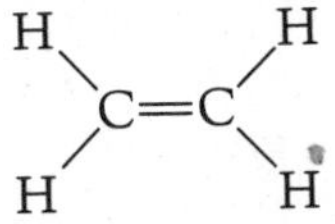


($C_2H_4$)

**ethene**
**(ethylene)**

which is then followed by the three-carbon alkene,

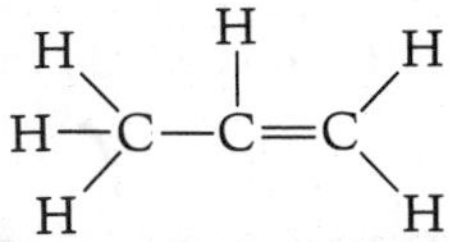


($C_3H_6$)

**propene**
**(propylene)**

Next come the alkenes, with the formulas $C_4H_8$, $C_5H_{10}$, $C_6H_{12}$, and so on. Thus we see that the alkenes, like the alkanes, form a homologous series, with a $CH_2$ group being the difference between any two successive members. We also see that, like the alkanes, the alkenes can collectively be represented by one general formula. The formula that is the same for the alkenes is $C_nH_{2n}$, where $n$ is the number of carbons in the molecule.

The names of the straight-chain alkenes can also be viewed as consisting of two parts. The first, which is the same used in the naming of the alkanes, indicates the number of carbons in the chain: *eth-*, two; *prop-*, three; *but-*, four; and so on. To this stem we add *ene,* which tells us that there is a carbon–carbon double bond present. The first two members of the alkenes (drawn above) are thus ethene and propene. These are also called by their older names, ethylene and propylene and are probably the best known of the alkenes because of their use in making the plastics, polyethylene and polypropylene. We shall discuss the nomenclature of these and other hydrocarbons in greater detail in Section 21.3.

The bonding in ethene was discussed earlier in Section 17.3 and will, therefore, only be reviewed briefly here. The carbon atoms participating in the double bond are each $sp^2$ hybridized. The overlap of the $sp^2$ hybrid orbitals between the two carbon atoms produces a $\sigma$ bond along the C—C axis. Each carbon also possesses an unhybridized pure $p$ orbital. These overlap sideways producing a $\pi$ bond having electron density concentrated above and below the bond axis. Because of the geometry of the $sp^2$ hybrids and the restriction that the unhybridized $p$ orbitals must overlap in the $\pi$ bond, a planar configuration of atoms with approximately 120° bond angles is formed.

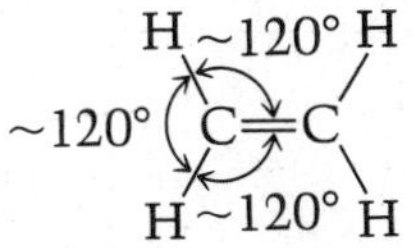

Besides the straight-chain alkenes there also exist alkenes with more than one double bond, alkenes that have other groups attached to the straight chain, alkenes with branched chains, and alkenes that are cyclic in structure.

The second type of unsaturated hydrocarbon that still remains to be discussed here is the alkynes, a series of compounds containing a carbon–carbon triple bond. The simplest, but truly one of the most important alkynes is HC≡CH, which is called ethyne or, more commonly, acetylene. This, of course, is the fuel used in welding torches. The next two members of this group are $CH_3$—C≡CH, propyne, and $CH_3$—$CH_2$—C≡CH, butyne. In naming these compounds we once again use the stems *eth-*, *prop-*, and *but-* to mean two, three, and four carbons, respectively. To this stem is added *yne* to denote the existence of the carbon–carbon triple bond.

The bonding in the carbon–carbon triple bond was also discussed earlier (Section 17.3). The hybridization used by the two carbons in the triple bond is *sp*. Thus one of the bonds is formed by an *sp-sp* overlap. The remaining two bonds are both $\pi$ bonds that are formed by $p_x$-$p_x$ and $p_y$-$p_y$ overlap.

## 21.2 ISOMERS IN ORGANIC CHEMISTRY

Beginning with hydrocarbons containing four carbon atoms, we find that besides the normal straight-chain structures, there also exist branched-chain structures bearing the same molecular formula. For example, we find that there are two compounds with the formula $C_4H_{10}$.

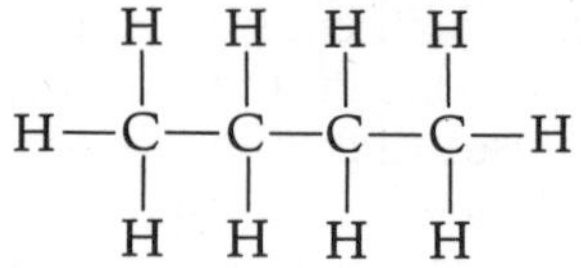

(1)
**straight chain**
**m.p. = −138.3°C**
**b.p. = −0.5°C**

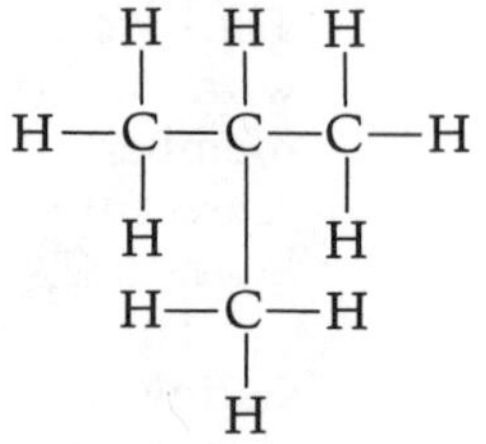

(2)
**branched chain**
**m.p. = −159°C**
**b.p. = −12°C**

Two or more compounds that have the same molecular formula but differ in the sequence in which the atoms are joined together are said to be **structural isomers.** Thus there are two structural isomers of butane, each with its own chemical and physical properties. This is an important point to keep in mind. Each structural isomer is a unique chemical compound.

Each of the remaining members of the alkane series show an even greater number of isomers. With $C_5H_{12}$, for example, we can write three structures:

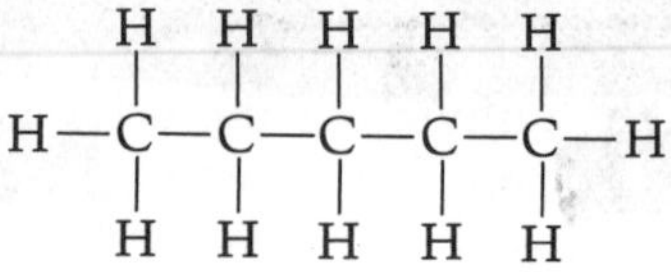

(1)

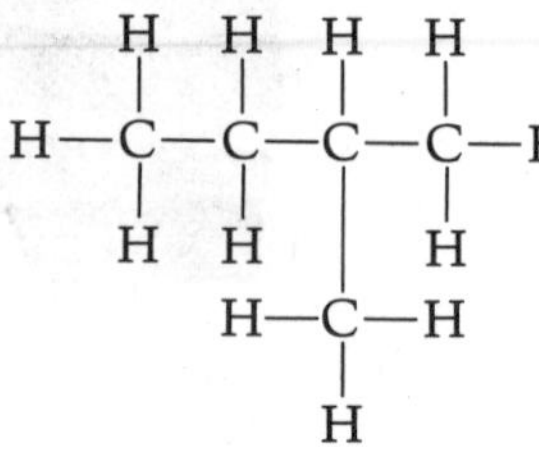

(2)

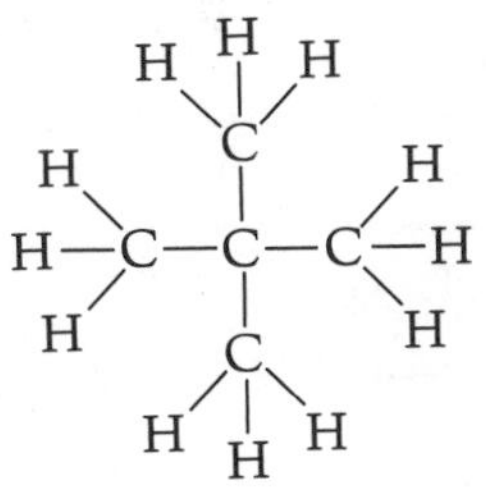

(3)

In the case of $C_6H_{14}$ we find five isomers, which are listed in Table 21.2. In this table we have left room for you to add the names of each of the isomers following our discussion of nomenclature in the next section.

Structural isomerism also exists in the alkenes. Butene ($C_4H_8$), for example, can be written as a straight chain with the double bond between the first and second carbon atoms.

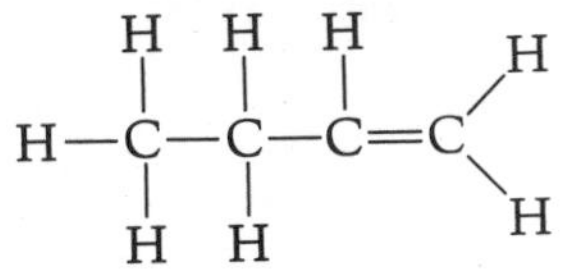

($CH_3—CH_2—CH=CH_2$)

**isomer 1**

or with the double bond between the middle two carbons,

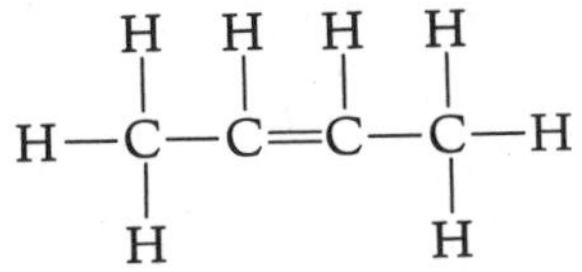

($CH_3—CH=CH—CH_3$)

**isomer 2**

or as a branched isomer in which two $CH_3$ groups are attached to one of the carbon atoms participating in the carbon–carbon double bond.

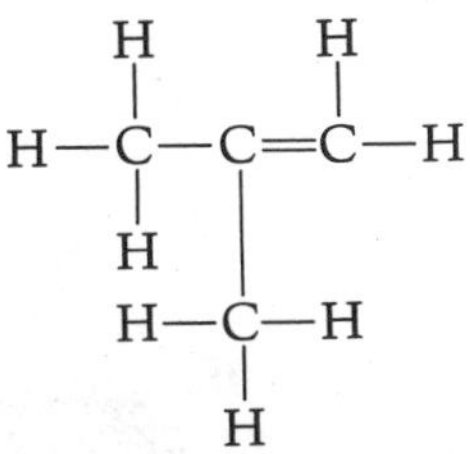

($(CH_3)_2C=CH_2$)

**isomer 3**

Note that the arrangement of atoms in the molecules is different in all three cases, while the molecular formula, $C_4H_8$, is the same. Thus in alkenes we have isomers that result from branching of the carbon chains as well as isomers that result from a difference in the relative position of the double bond.

Another type of isomerism exists in organic compounds in which the sequence of atoms in the molecule is the same but in which the relative positions of the atoms or groups of atoms are different. In general this type of isomerism is called **stereoisomerism**, and is found in two forms: geometrical isomerism and optical isomerism (see Section 20.9).

Table 21.2
*The five isomers of hexane*

| Isomer | Name (Fill in this column after reading Section 21.3) |
| --- | --- |
| 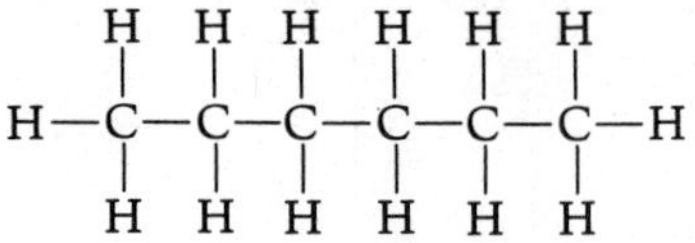 | |
| 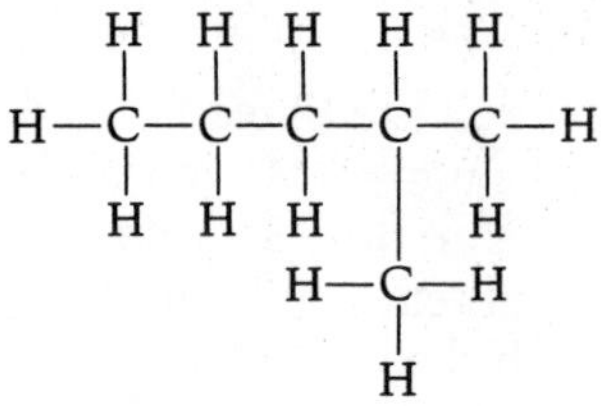 | |
| 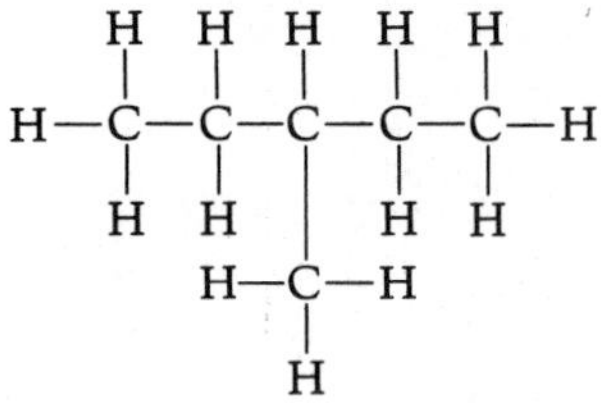 | |
| 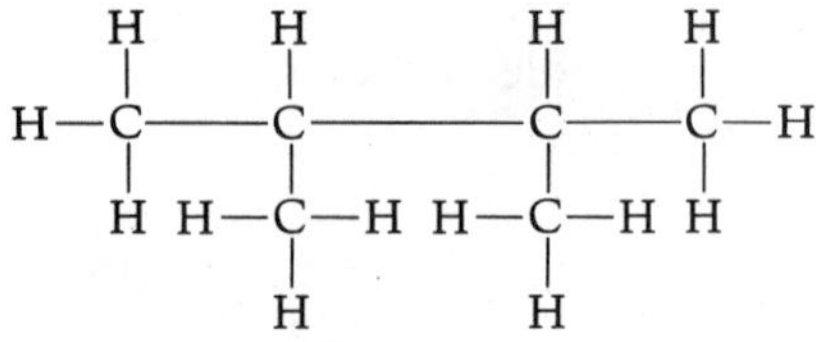 | |
| 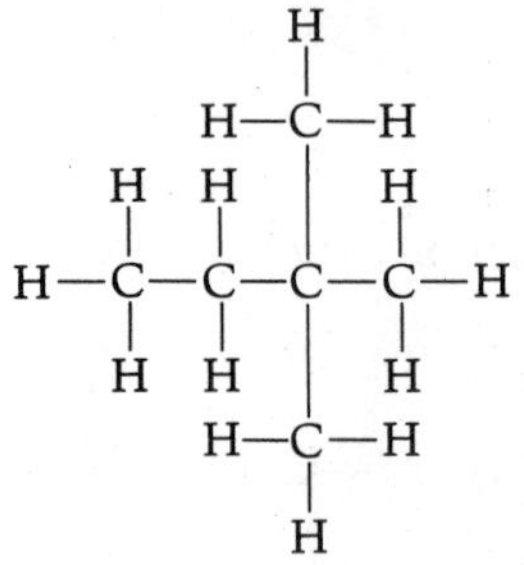 | |

**GEOMETRICAL ISOMERISM.** In organic compounds *cis* and *trans* geometrical isomers can occur in molecules that possess one or more carbon–carbon double bonds. This is illustrated by one of the isomers of butene,

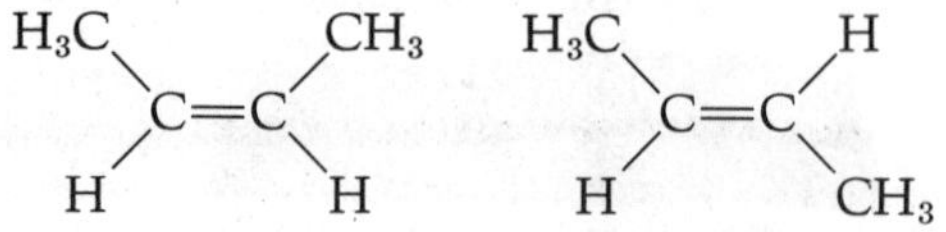

*cis* **isomer** *trans* **isomer**

These are not identical molecules because there is no free rotation of the atoms about the double bond. If there were, one of the —CH($CH_3$) groups could rotate 180° to convert *cis* into *trans* or vice versa. However, such a rotation would cause the unhybridized *p* orbitals that overlap to give the bond to become misaligned and thus destroy the $\pi$ bond. In effect, then, conversion of *cis* to *trans* in this manner would require the breaking of a bond. The energy available through molecular collision is generally not sufficient to accomplish this, and the rate of interconversion between *cis* and *trans* isomers is usually so slow that the individual isomers can be isolated.

**OPTICAL ISOMERISM.** Optical isomers, as we saw in Section 20.9, are molecules that have the same formula but possess structures that are nonsuperimposable mirror images of each other. As a rule, in organic compounds the presence of an **asymmetric carbon atom** is responsible for optical isomerism. An asymmetric carbon atom occurs when the carbon is bonded to four *different* atoms or groups of atoms; for example,

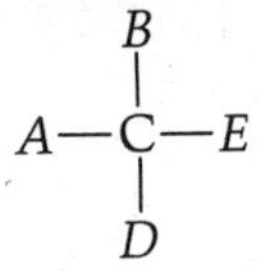


where *A, B, D,* and *E* represent different groups. For instance, the isomer of heptane with the structural formula

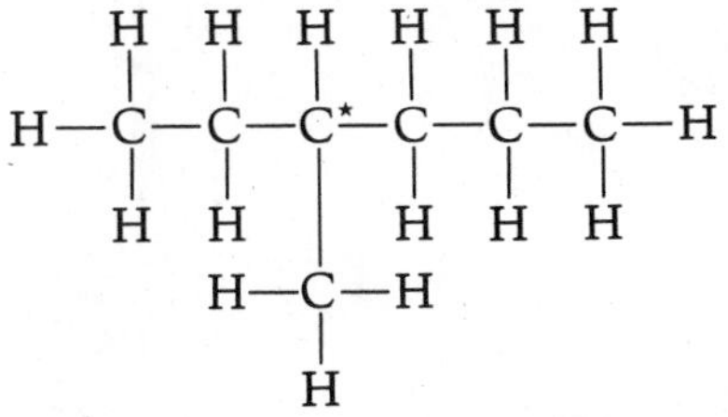

has one asymmetric carbon atom, the one marked with an asterisk. This can be seen more readily if we rewrite the structural formula for this isomer in a slightly more condensed fashion.

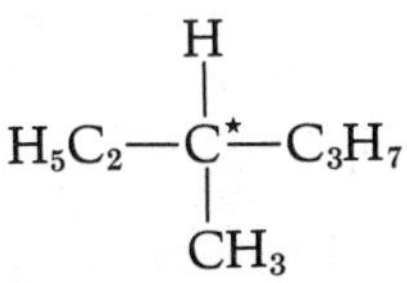


Let us see how compounds with an asymmetric carbon atom give rise to optical isomers and why molecules without one cannot. To show this, we compare the structures of two different isomers of a compound to see if one is the nonsuperimposable mirror image of the other. If this is true, then they are optical isomers.

In the case of our general molecule, the two structures shown in Figure 21.3 are mirror images of each other with nonsuperimposable structures. Note that when one is placed over the other, the *D*'s and *B*'s line up, but the *E*'s and *A*'s do not. Furthermore, we would find that, regardless of the amount of manipulation of the two structures, we could never bring about the situation where one structure would be exactly superimposable upon the other.

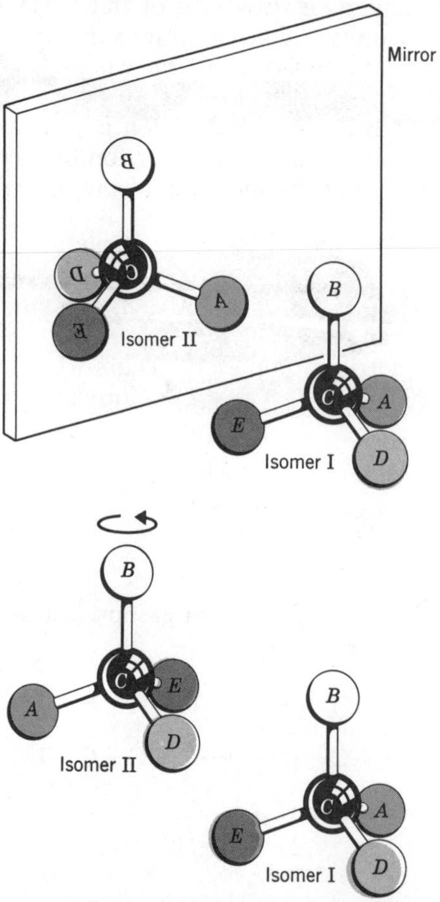


Figure 21.3
*Optical isomers. (a) Isomer II is a reflection of isomer I in the mirror; that is, it is the mirror image of isomer I. (b) Isomer II is rotated about the* B—C *bond so that atoms* D *in both I and II are in same relative position. Isomers I and II are not superimposable. Atoms* B, C, *and* D *match, but* A *and* E *do not.*

If, in each structure there are two groups that are alike, we destroy the asymmetry of the central carbon and the two structures no longer represent optical isomers; they are, in fact, identical. For example, by replacing the *E* in our general formula with another *A*, we would then have the two structures shown in Figure 21.4, which are directly superimposable on each other. Thus we see that if an asymmetric carbon atom is present, optical isomerism will occur.

## 21.3 NOMENCLATURE

We have already seen some of the basic elements of naming organic compounds in our discussion of the hydrocarbons. The complexities produced by isomerism, and the introduction of atoms other than carbon and hydrogen, create a need for a systematic procedure for naming compounds. The guidelines presented below are currently followed in the modern chemical literature. Unfortunately, many of the older, nonsystematic *common names* are still often used in less formal situations. For instance, ethyne, $C_2H_2$, is also called acetylene; ethene and propene are also known as ethylene and propylene. A student of organic chemistry must be aware of this dual nomenclature.

The systematic nomenclature of organic compounds, like that of coordination compounds, has been established by the IUPAC (p. 619). The application of the IUPAC system to the saturated and unsaturated hydrocarbons is defined by the following set of rules.

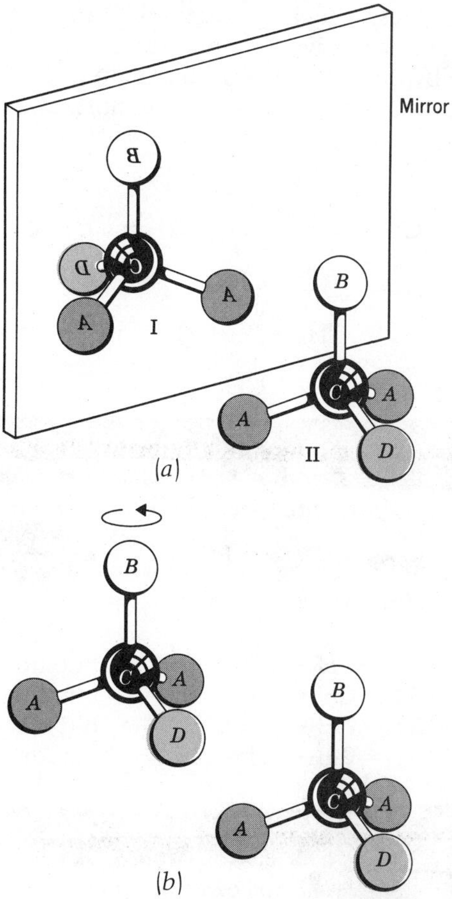

Figure 21.4
(a) Lack of optical isomers when two groups attached to carbon are the same. (b) "Isomer" I rotated about B—C bond so that atoms D are matched. These two are identical and are not isomers of each other.

## RULES OF NOMENCLATURE OF ORGANIC COMPOUNDS

1. **The longest unbroken chain of carbon atoms in a molecule serves as the parent name for any hydrocarbon or its derivative.** With the alkenes this longest chain must contain the double bond; in the alkynes the triple bond must be included. For example, each of the following compounds would be named as a derivative of pentane, because in each case the longest carbon chain consists of five carbon atoms.

$$CH_3-\underset{\underset{\displaystyle CH_3}{|}}{\overset{\overset{\displaystyle CH_3}{|}}{C}}-CH_2-CH_2-CH_3$$

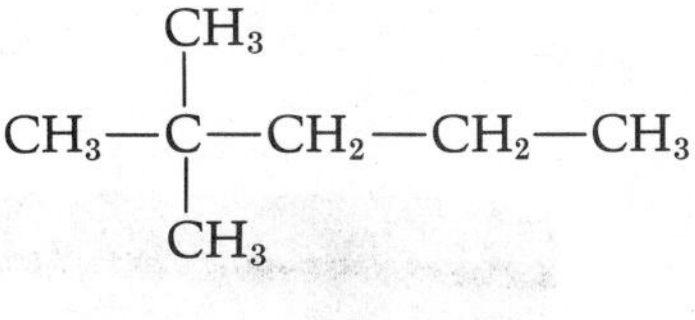

$$CH_3-CH_2-\underset{\underset{\displaystyle CH_3}{|}}{CH}-\underset{\underset{\displaystyle CH_3}{|}}{CH}-CH_3$$

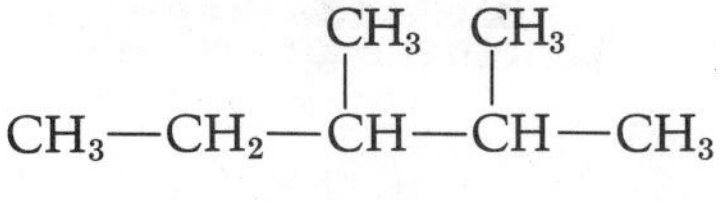

$$CH_3-\underset{\underset{\displaystyle CH_3}{|}}{\overset{\overset{\displaystyle CH_3}{|}}{C}}-CH_2-\underset{\underset{\displaystyle CH_3}{|}}{\overset{\overset{\displaystyle CH_3}{|}}{C}}-CH_3$$

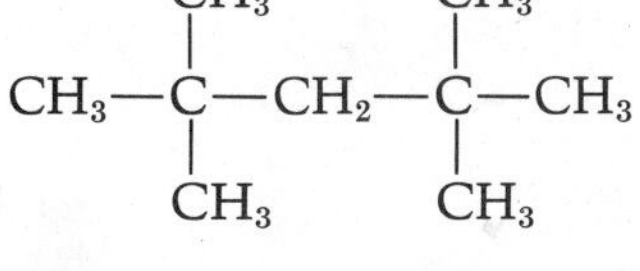

Sometimes, when a structural formula is drawn, the longest carbon chain is not written in a straight line. For instance, the compound below has an eight-carbon chain (can you find it?) and would be named as a derivative of octene (note the double bond).

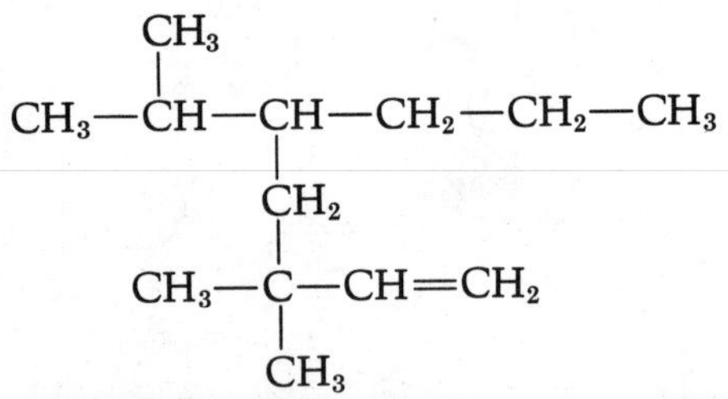

2. **The stem of the parent name indicates the number of carbon atoms in the chain and has the following endings: -ane** when only single bonds are present, **-ene** when there is a carbon–carbon double bond, and **-yne** when a triple bond is present. When two double bonds are present the ending is **-diene,** and when three are present the ending is **-triene.** Thus we would have

| | |
|---|---|
| $CH_3{-}CH_2{-}CH_3$ | butane |
| $CH_3{-}CH_2{-}CH{=}CH_2$ | butene |
| $CH_3{-}CH_2{-}C{\equiv}CH$ | butyne |
| $CH_2{=}CH{-}CH{=}CH_2$ | butadiene |

3. **Branched isomers are named as derivatives of straight-chain hydrocarbons in which one or more hydrogen atoms are replaced by hydrocarbon fragments.** For example, isomer 2 of butane (on p. 650) can be considered to be derived from propane where a $CH_3$ group has replaced one of the middle hydrogen atoms.

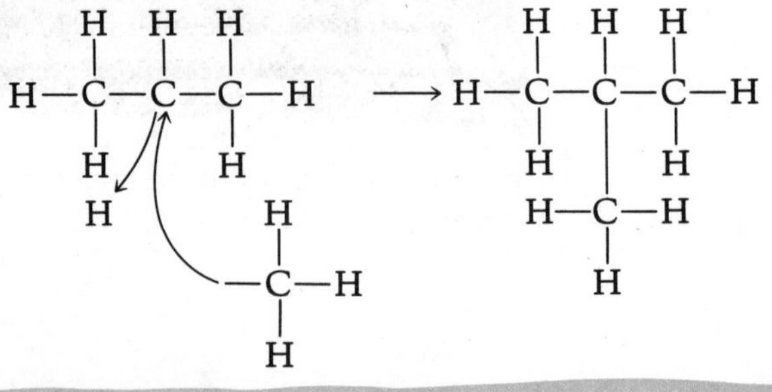

Hydrocarbon groups that are derived from members of the alkane series are called **alkyl groups.** The $CH_3$ group is derived from methane and is called a methyl group.

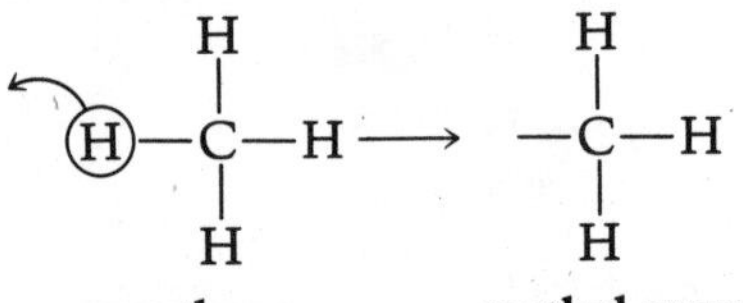

**methane** **methyl group**

Similarly, the $C_2H_5$ group is derived from ethane and is called an ethyl group.

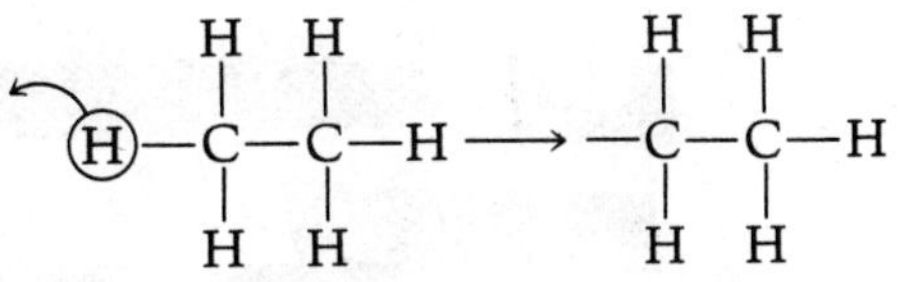

**ethane** **ethyl group**

Thus, in naming an alkyl group the ending **-yl** is added to the alkane stem that indicates the number of carbon atoms in the group. Some additional alkyl groups are listed in Table 21.3.

Referring back to the branched isomer of butane, we see that it has a three-carbon chain as its longest and a $CH_3$ group attached to the middle carbon; therefore it would be called

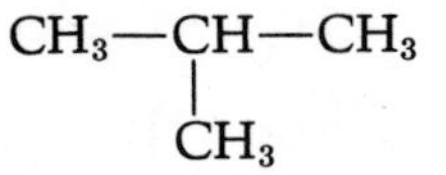

methylpropane

Likewise we would have

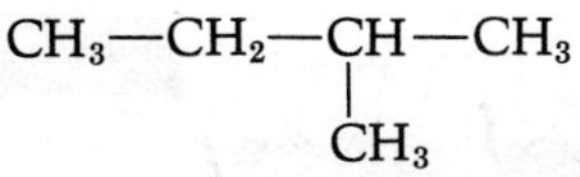

methylbutane

and

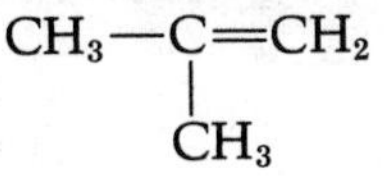

methylpropene

Table 21.3
*The names of some alkyl groups*

| Alkyl Group | Condensed Formula | Name |
|---|---|---|
| 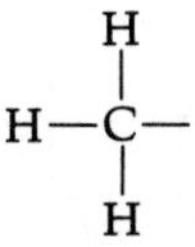 | $CH_3—$ | Methyl |
| 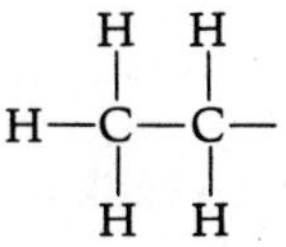 | $CH_3—CH_2—$ | Ethyl |
| 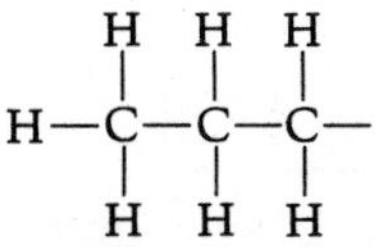 | $CH_3—CH_2—CH_2—$ | Propyl |
| 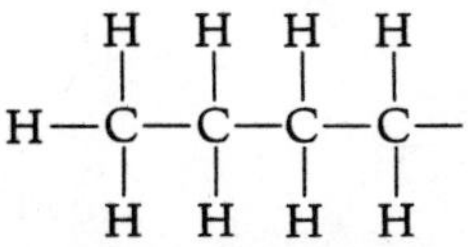 | $CH_3—CH_2—CH_2—CH_2—$ | Butyl |

4. **In order to denote the positions of alkyl groups attached to the parent chain, as well as the positions of double and triple bonds, a numbering system is used.** In alkanes the numbering starts from the end of the molecule that gives the lowest numbers to the alkyl groups. When multiple bonds are present the numbering begins from the end of the molecule that gives the lowest numbers to the multiple bonds. In the case of an alkyl group, the number identifying its position immediately precedes its name, while the numbers identifying a double or triple bond precede the name of the parent chain. Examples of this include

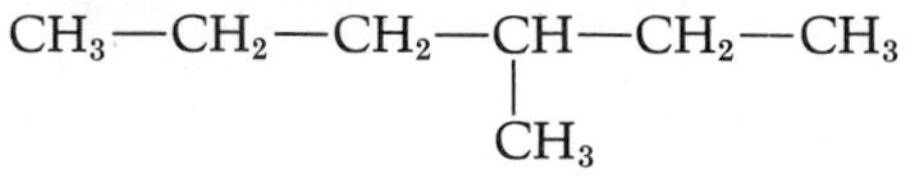 3-methylhexane

$CH_3—CH{=}CH—CH_3$ 2-butene

$CH_3—CH{=}CH—CH_2—CH{=}CH_2$ 1,4-hexadiene

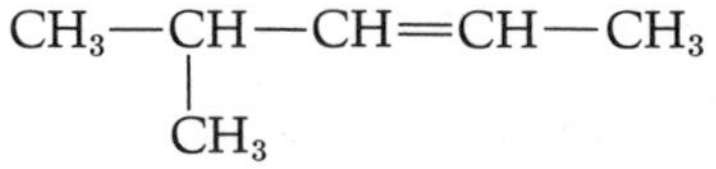 4-methyl-2-pentene

5. **In compounds were more than one alkyl group is attached to a carbon chain then: (1) when they are identical the prefixes di- (two), tri- (three), tetra- (four), and so on, immediately precede its name; and (2) if they are different, they are listed in alphabetical order along with their respective locations.** Some examples of this rule are

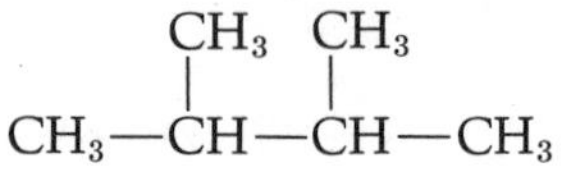 2,3-dimethylbutane

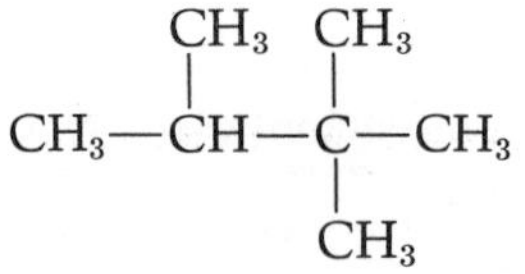 2,2,3-trimethylbutane

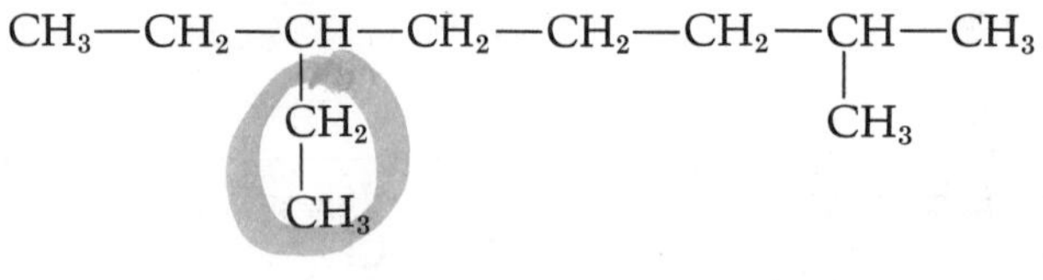 6-ethyl-2-methyloctane

As an exercise, you should now go back and fill in the names of all the isomers of hexane in Table 21.2.

EXAMPLE 21.1 (a) Give the IUPAC name for the following compounds:

(1) 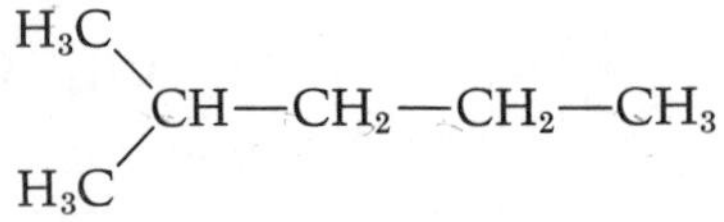

(2) 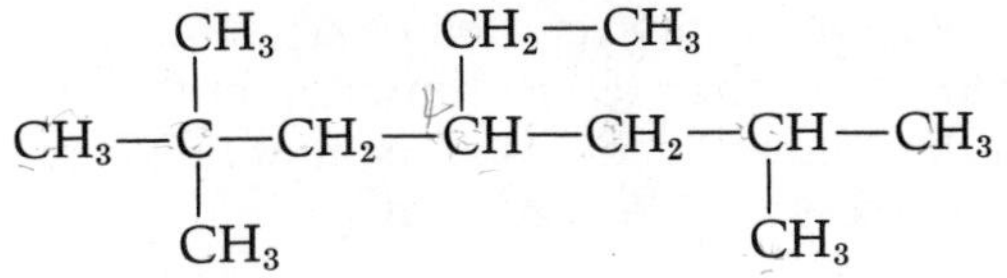

(3) 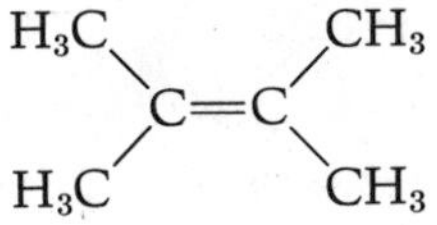

(4) 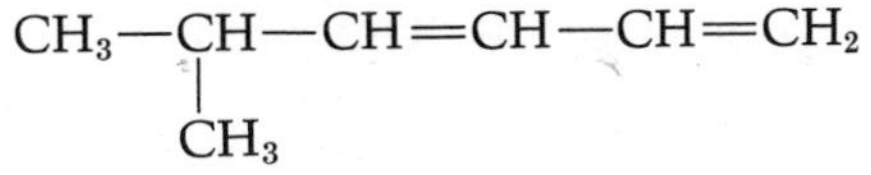

(5) $CH_3{-}CH_2{-}C{\equiv}C{-}CH_2{-}CH_3$

(b) Write structural formulas for the following:
 (1) 2,3-dimethylbutane
 (2) 2-pentyne
 (3) 2-ethyl-1-butene
 (4) 1,5-octadiene
 (5) 2-ethyl-3-methyl-1-pentene

SOLUTION (a) (1) 2-methylpentane
 (2) 4-ethyl-2,2,6-trimethylheptane
 (3) 2,3-dimethyl-2-butene
 (4) 5-methyl-1,3-hexadiene
 (5) 3-hexyne

(b) (1) 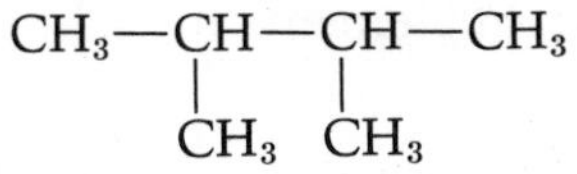

(2) $CH_3{-}CH_2{-}C{\equiv}C{-}CH_3$

(3) 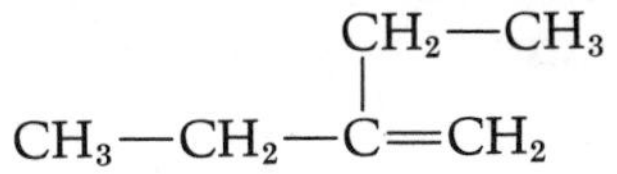

(4) $CH_3{-}CH_2{-}CH{=}CH{-}CH_2{-}CH_2{-}CH{=}CH_2$

(5) 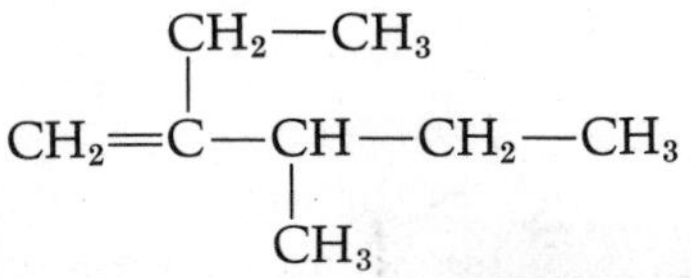

**COMMON NAMES.** As indicated earlier, quite a large number of organic compounds were known prior to the introduction of the IUPAC rules. As a result, many of them had already been named using other systems of nomenclature, and some of these names are carried over into current usage. For example, methylpropane,

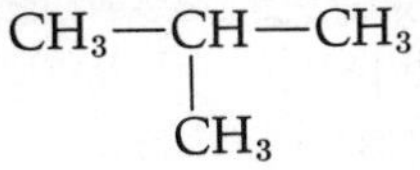

is more commonly referred to as isobutane. This name indicates that there are four carbons present (butane), but that they are not in a straight chain and that the molecule is an isomer of butane, hence *iso.* In general, alkanes that possess the arrangement

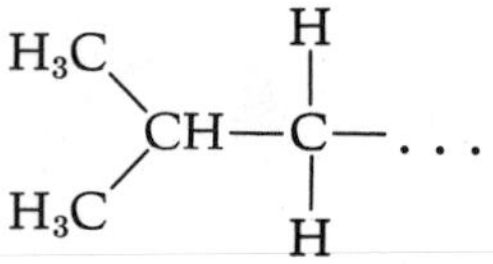

are called *iso* compounds. Thus we have

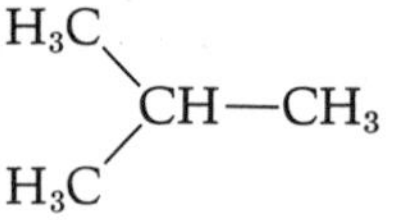

**isobutane**

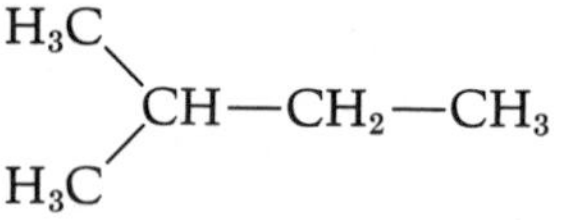

**isopentane**

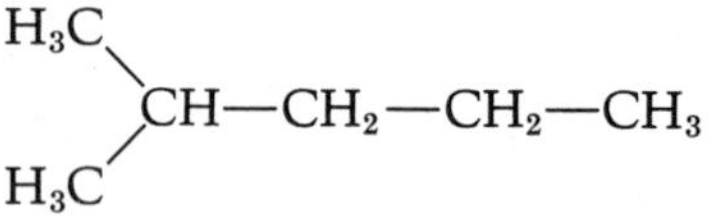

**isohexane**

In the case of the alkynes, the first member, acetylene, is so important that its name is used in the common nomenclature of all the remaining members. Thus we have

| Alkyne | Common | IUPAC |
|---|---|---|
| $CH_3—C{\equiv}CH$ | Methylacetylene | Propyne |
| $CH_3—CH_2—C{\equiv}CH$ | Ethylacetylene | 1-Butyne |
| $CH_3—CH_2—CH_2—C{\equiv}CH$ | Propylacetylene | 1-Pentyne |

We will see that the use of common names carries over to organic compounds that also contain elements in addition to carbon and hydrogen.

## 21.4 CYCLIC HYDROCARBONS

**CYCLIC ALKANES.** With alkanes containing three or more carbon atoms it is possible for one end of the molecule to become attached to the other end, forming a ring. An example of a three-carbon hydrocarbon that exists in a cyclic or ring structure is

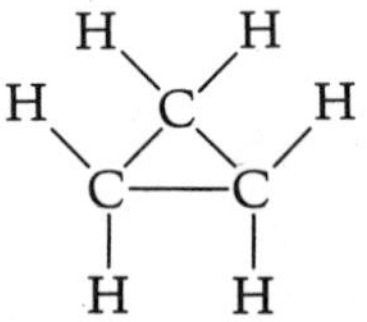

This compound, as well as the other members of this group, contains only carbon–carbon single bonds and is, therefore, saturated. These hydrocarbons are named, then, in the same manner as the straight-chain alkanes, but with the prefix *cyclo-* being added. The compound above would therefore be named cyclopropane (*cyclo* meaning ringed, *prop* for three carbon atoms, and *ane* because it is saturated). Some of the other cycloalkanes, with their molecular and structural formulas, are listed in Table 21.4. Note that the general formula for the cycloalkanes is $C_nH_{2n}$. This is the same as the general formula for the noncyclic alkenes and demonstrates that it is very dangerous to attempt to write structures from molecular formulas alone.

Table 21.4
*Some cycloalkanes*

| Structural Formula | Molecular Formula | Name |
|---|---|---|
| 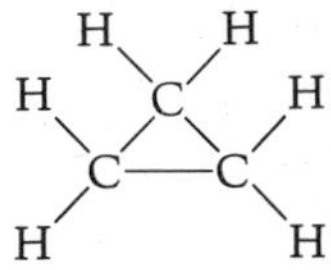 | $C_3H_6$ | Cyclopropane |
| 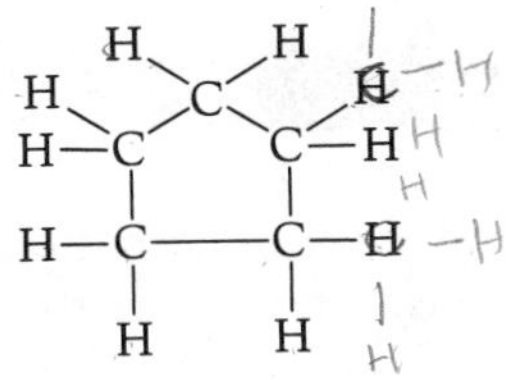 | $C_5H_{10}$ | Cyclopentane |
| 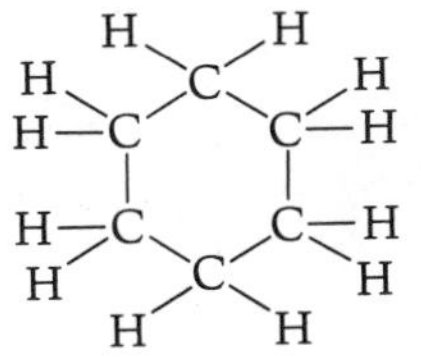 | $C_6H_{12}$ | Cyclohexane |
| 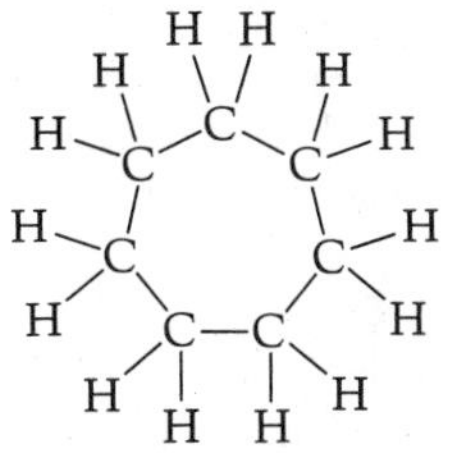 | $C_7H_{14}$ | Cycloheptane |

In cyclopropane, shown above, the C—C—C bond angle is 60°, while in cyclobutane,

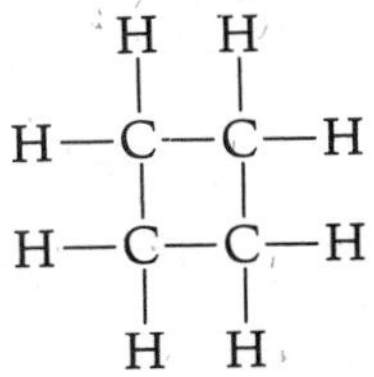

the C—C—C bond angle is about 90°. In both cases the bond angle is much less than the stable tetrahedral bond angle of 109° that is exhibited by carbon in the straight-chain alkanes. These two species then are said to be quite *strained* and, as a result, are very unstable. This manifests itself in their high degree of reactivity. Cyclopropane, for example, which is used as a very fast-acting anesthetic, forms extremely flammable mixtures with air. The straight-chain propane, on the other hand, is much less reactive.

The next two members that follow cyclobutane (i.e., cyclopentane and cyclohexane) are quite stable. In cyclopentane, which has a nonplanar pentagonal structure, the C—C—C bond angle is very nearly 109° and, as a result, we expect it to be stable. In cyclohexane we find that in order to achieve more closely the tetrahedral angle of 109°, the ring is warped or

puckered. In a planar or flat hexagon,

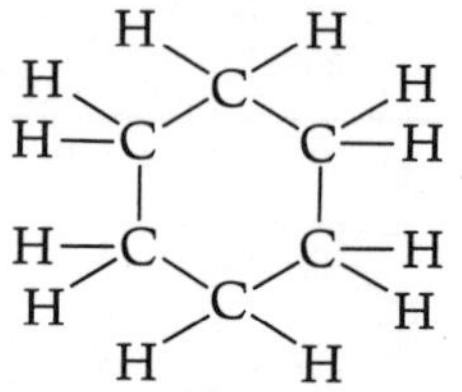

the C—C—C bond angle would have to be 120° and, as a result, the structure would be strained. Two structures of cyclohexane that are free of this "angle strain" are shown in Figure 21.5.

**CYCLIC ALKENES.** Cyclic alkenes are formed containing three or more carbon atoms and are named in a fashion similar to that for the cycloalkanes. For example, cyclopentene would be

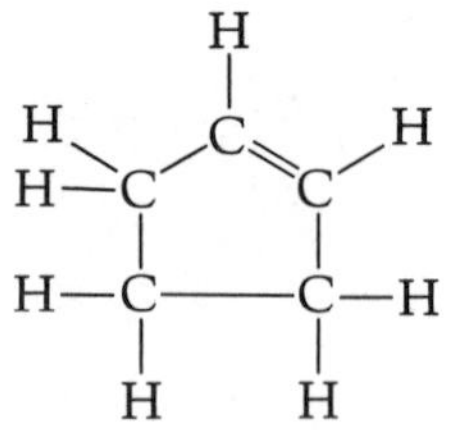

Figure 21.5
*Two structures of cyclohexane that are free of "angle strain."*

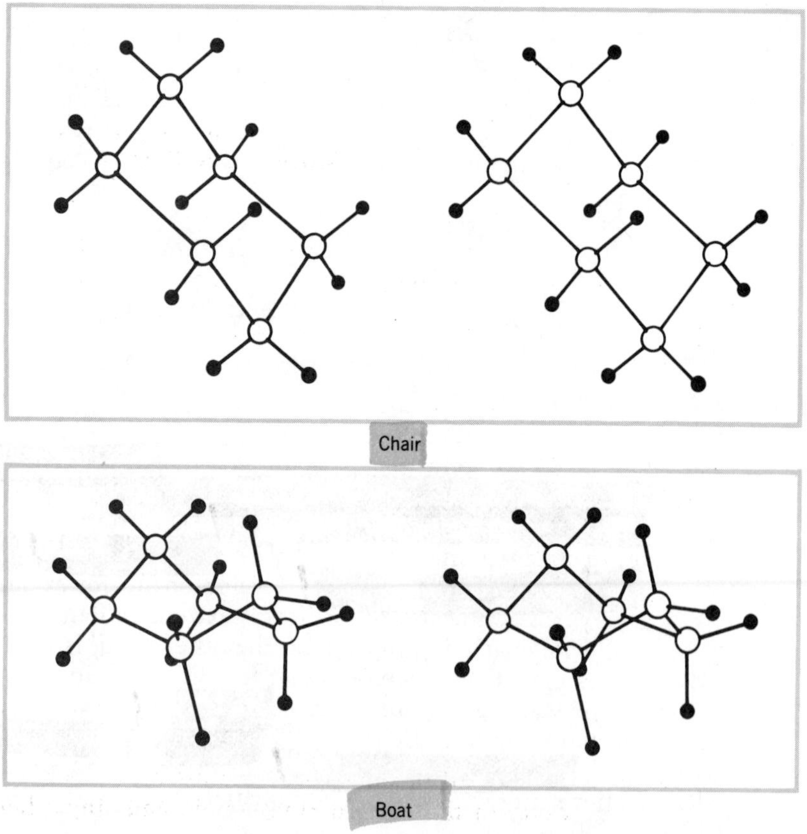

Other examples include

**1,3-cyclopentadiene**

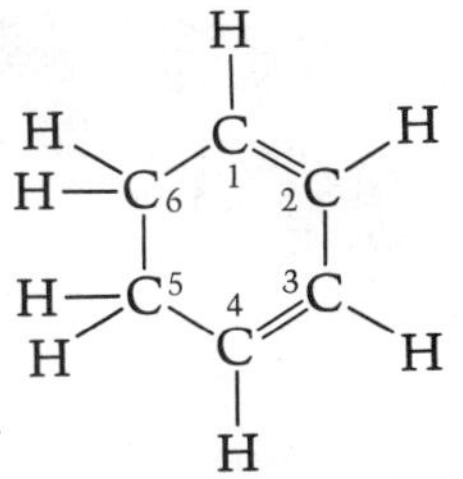

**1,3-cyclohexadiene**

In these compounds the numbering begins at the carbon with the first double bond and continues in the direction that leads to the smallest numbers for the remaining double bonds. Following this idea, the compound

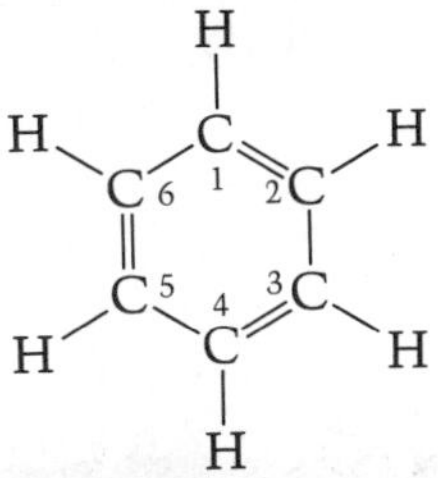

would be called 1,3,5-cyclohexatriene. However, the orientation of the orbitals that would be involved in the bonding in this molecule gives rise to some special properties. This compound is called benzene and forms the basis of another entire series of organic compounds.

## 21.5 AROMATIC HYDROCARBONS

Since benzene has properties so unlike those of the other cycloalkenes, it is placed in a separate class. Benzene, and the host of benzenelike compounds (those containing a "benzene ring"), are collectively called the *aromatic compounds* because many of them have pleasing "aromatic" odors. In spite of their pleasant odors, however, benzene and many other aromatic compounds are quite toxic.

The main structural feature in the aromatic compounds that is responsible for their distinctive chemical properties is the benzene ring. We saw in the last section that the stoichiometry of benzene corresponds to that of the cyclotriene,

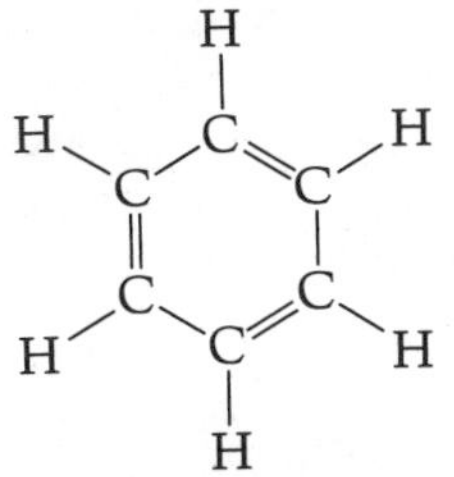

Physical and chemical evidence, however, reveals that this does not give an accurate representation of the molecule. For instance, it is found experimentally that all of the C—C bond distances in benzene are the same. If there were, in fact, alternating double and single bonds, some bond distances (C=C double bonds) would be shorter than others. In addition, the benzene molecule does not readily undergo chemical reactions typical of molecules containing double bonds.

The uniform bond distances in benzene can be accounted for by resonance. The two resonance structures (also called Kekulé structures[1]) that are usually written are

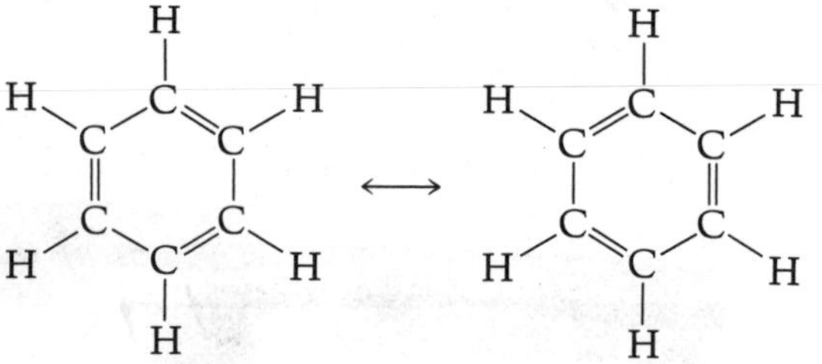

These are usually represented simply as

and

It is understood that at each vertex of the hexagon there is a carbon atom bound to a hydrogen atom.

On an atomic orbital level we see that in order to achieve a C—C—C bond angle of 120°, the carbon atoms must $sp^2$ hybridize. Thus on each of the carbons in the benzene ring we have three of the valence electrons in an $sp^2$ hybrid and a fourth in a pure $p$ orbital. The carbon skeleton showing only the hybridization can be seen in Figure 21.6. This same skeleton showing the usual dash for the $\sigma$ bonds as well as the electrons in the pure $p$ orbitals is illustrated in Figure 21.7.

According to *valence bond theory,* the two resonance structures of benzene would be formed by the overlap of pairs of adjacent $p$ orbitals, as shown in Figure 21.8.

In Section 17.5 it was pointed out that the *molecular orbital theory* can quite successfully explain the bonding in polyatomic molecules without resorting to resonance. According to this theory the six unhybridized atomic orbitals overlap to form a molecular orbital that extends over the entire molecule. In this respect the electron density, and of course the length, of those bonds involved in the production of the molecular orbital, are identical. This situation is illustrated in Figure 21.9 for benzene, where we see the $\sigma$ bonds as the solid lines between the carbons and the $\pi$ bonds as a dispersed cloud with electron density above and below the carbon ring. You may recall that electrons belonging to a molecular orbital that extends over several nuclei (such as the six $\pi$ electrons in benzene) are said to be delocalized. The benzene ring is frequently drawn as

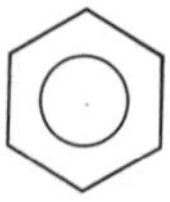

to emphasize the delocalized nature of the $\pi$ electrons.

[1] These structures are named after the German chemist, August Kekulé, who first proposed them in 1865.

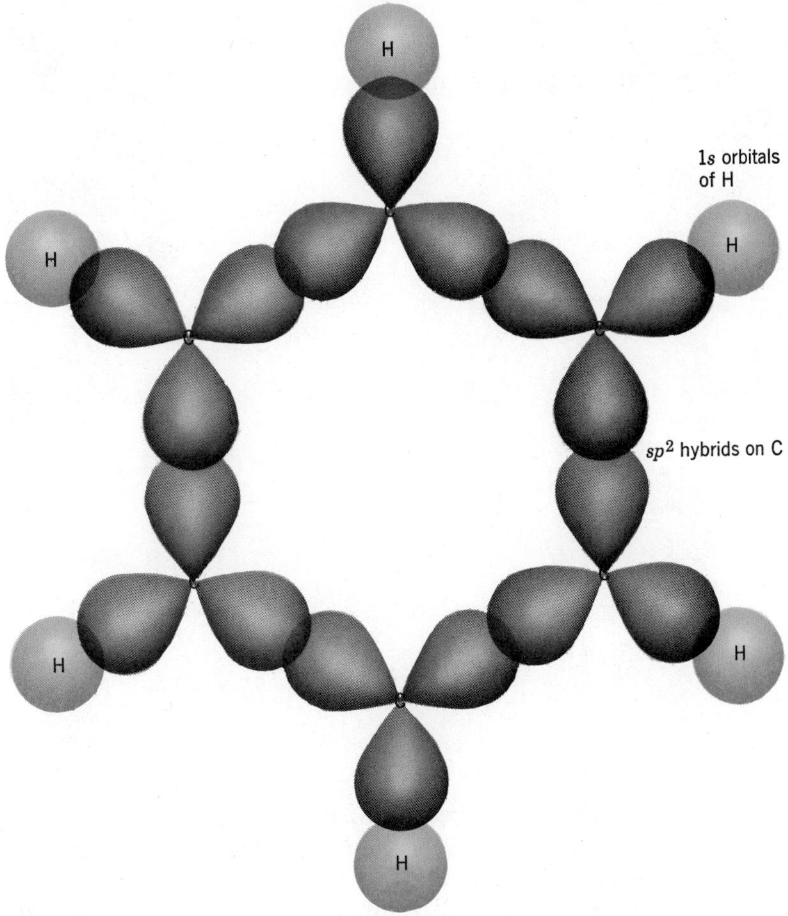


Figure 21.6
*Sigma-bond framework in benzene. Each carbon uses* $sp^2$ *hybrid orbitals.*

Figure 21.7
*Electrons in unhybridized* p *orbitals on C atoms in benzene.*

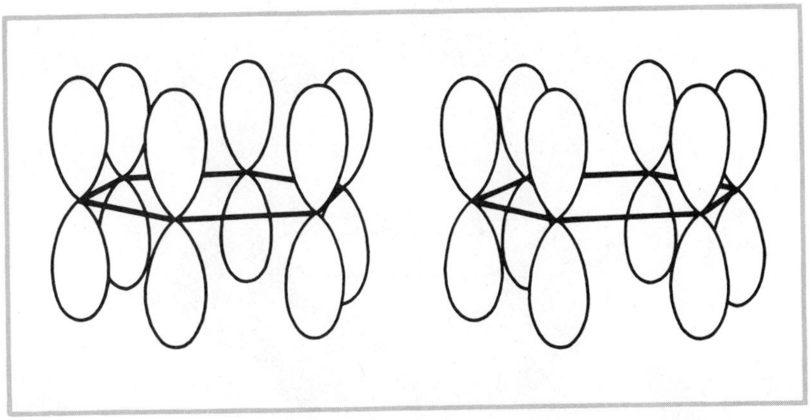

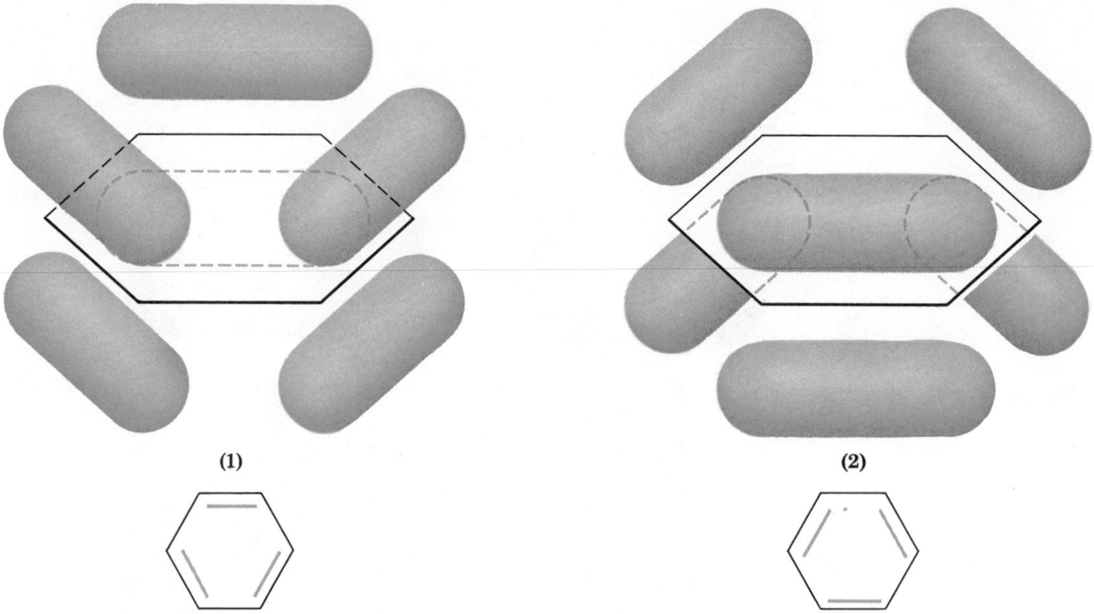


Figure 21.8
*Valence bond resonance structures of benzene.*

The delocalization of the $\pi$-electron cloud in benzene leads to a very stable ring structure. In fact, thermochemical calculations show that benzene is more stable than the hypothetical 1,3,5-cyclohexatriene by approximately 150 kJ/mole. Thus we say that the *delocalization energy* or *resonance stabilization energy* of benzene is 150 kJ/mole. A similar high resonance stabilization energy is found in all members of the aromatic family.

Aromatic compounds are important in many ways. We often find them in solvents of various kinds as well as in many plastics. The benzene ring is

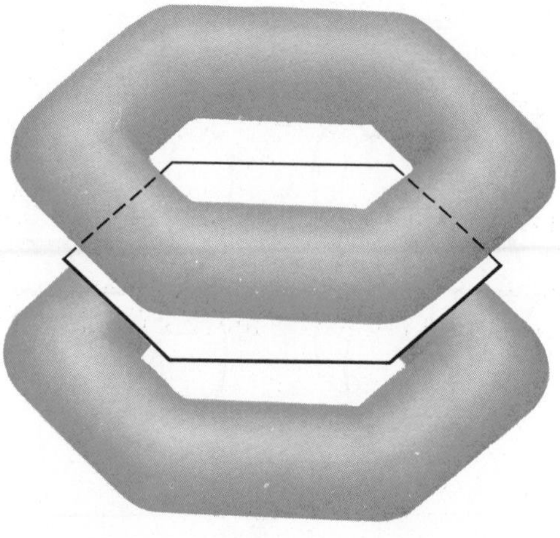

Figure 21.9
*Molecular orbital delocalized electron cloud for benzene.*

also found in many biologically important compounds such as vitamins, proteins, and hormones. Unfortunately, the human body does not possess the ability to synthesize the benzene ring and we must therefore obtain them from an outside source. Plants have the ability to synthesize some of the needed aromatics and therefore they are essential to our diets.

Not all aromatics are beneficial to us. In fact, some of them are extremely harmful. One example is the compound 1,2-benzopyrene, composed of five benzene rings fused together.

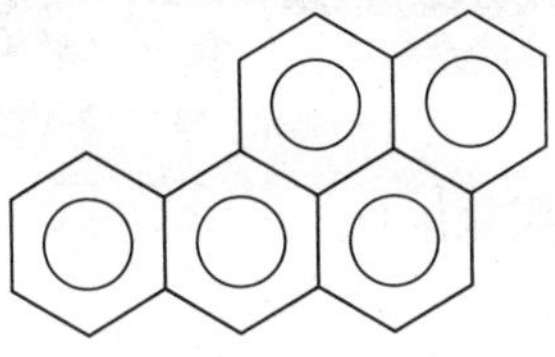

1,2-benzopyrene

It has been found in cigarette smoke, in the exhaust from gasoline engines, and even on charcoal-broiled steaks. It is one of the most potent carcinogens (cancer-producing agents) known. For example, even small amounts applied to a shaved area of a mouse produces skin cancer very nearly 100% of the time.

**NOMENCLATURE OF SOME BENZENE DERIVATIVES.** The nomenclature of the aromatic hydrocarbons follows much the same pattern as was outlined for the alkanes and alkenes. Here, *benzene is generally taken to be the parent and the attached groups are named as before.* For example, the compound,

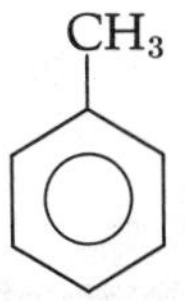


is called methylbenzene. Its common name is *toluene.* Another example is ethylbenzene,

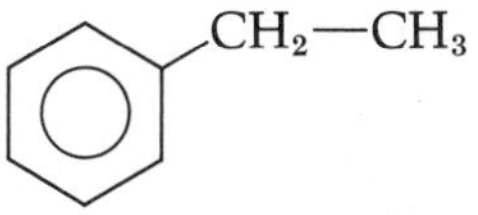


In a compound where there is more than one substituent attached to the ring, there are two systems that may be used to identify their positions. In one of these the carbon atoms in the ring are numbered 1 to 6, beginning with the carbon that is bonded to the first group and continuing in such a direction as to lead to the lowest numbers for the remaining ones. For example, the compound,

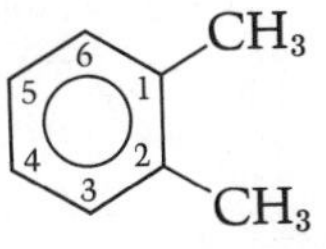

would be called 1,2-dimethylbenzene, whereas the compound,

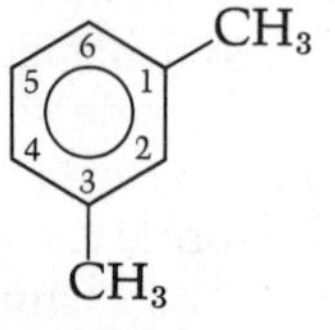

would be named 1,3-dimethylbenzene.

When the positions of the substituent groups are named, rather than numbered, the terms *ortho, meta,* and *para* are used, and the common name of the parent compound is usually employed. The terms **ortho, meta,** and **para,** respectively, label the next three consecutive positions that are adjacent to the first group. For example, the *ortho, meta,* and *para* positions on toluene are

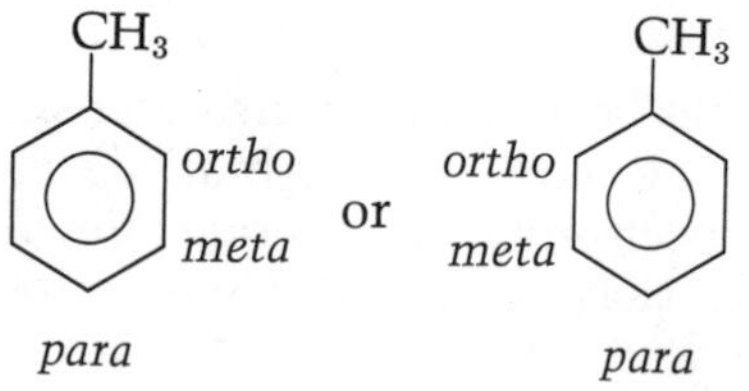


Thus the compounds

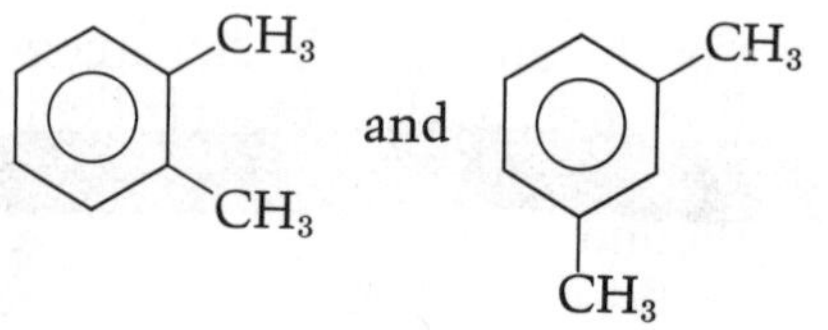

would be called *ortho*-methyltoluene and *meta*-methyltoluene, respectively. These compounds also have common names, which are *ortho*-xylene and *meta*-xylene (or simply *o*-xylene and *m*-xylene), respectively.

In some of the larger, more complex, compounds *it is sometimes easier to consider the benzene ring as the attached group and the longest carbon chain as the parent.* In this case the benzene ring ($C_6H_5$) is called **phenyl** (rhymes with kennel), and the same rules developed earlier for the alkanes and alkenes are employed. For example, we have

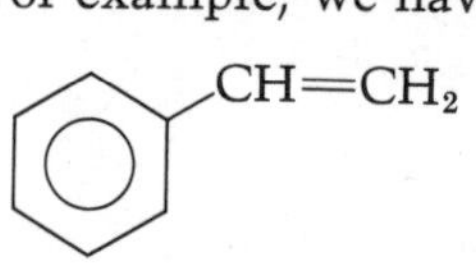

which we could call phenylethene. Its more common name is *styrene.* This compound is used to manufacture the plastic polystyrene, which is easily molded into various forms (e.g., model airplanes) and foamed to give the thermal insulating product, *styrofoam.*

Some other examples are

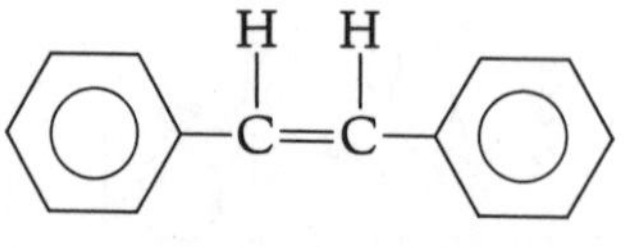

**1,2-diphenylethene**

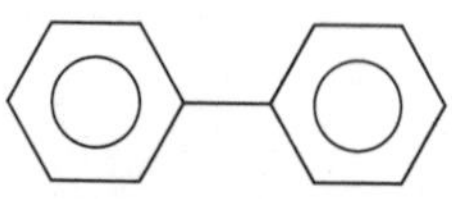

**biphenyl**

The latter compound, biphenyl, has become the villain in a very serious pollution problem. When the hydrogen atoms in biphenyl are replaced by chlorine atoms, polychlorinated biphenyls (PCBs) are produced. These have many desirable physical and chemical properties that have led to their widespread use in many consumer products. For example, they are excellent flame retardants and, as oils, have been used in many types of electric motors; for example, in refrigerators, air conditioners, clothes washers and dryers, and furnace blowers. They are also used in electrical transformers such as those seen on utility poles which feed electricity into homes. Over the past five decades large quantities of these very stable PCBs have found their way into the environment where they are just now beginning to be seen as a potential hazard of immense proportions. It has been discovered that they produce severe acne and loss of hair in humans, while in test animals, PCBs have caused birth defects, cancer, and death. Their presence in high concentrations in sediments in the Hudson River has already led to a ban on commercial fishing in this river. At the same time, PCBs are beginning to show up in fish caught in other areas too.

## 21.6 HYDROCARBON DERIVATIVES

Only a small fraction of all organic compounds contain just carbon and hydrogen; most contain other elements as well. In attempting to organize these it is most convenient to view them as being derived from a hydrocarbon by replacing one or more of the hydrogens of the parent molecule by other atoms or groups of atoms. That is why we call them hydrocarbon derivatives. Usually these attached atoms or groups bestow some characteristic property to the molecule so that any molecule with the same grouping will react chemically in a similar fashion. Groups bestowing such properties on organic compounds are called **functional groups.** The study of organic chemistry is greatly simplified, then, by examining the properties and reactions of various functional groups.

We have already seen two functional groups within the hydrocarbons: the carbon–carbon double bond and the carbon–carbon triple bond. Compounds containing these functional groups were called alkenes and alkynes, respectively. Some typical reactions of these compounds will be discussed below. Some other important functional groups, such as the halogens (X), hydroxyl group (—OH, in alcohols) and so on, are listed in Table 21.5.

For convenience, organic compounds can be viewed as composed of two parts: a hydrocarbon fragment that is generally denoted as R (aliphatic) or Ar (aromatic) plus one or more functional groups, such as those listed in Table 21.5. Thus we have R—X or Ar—X, an alkyl halide or aryl halide, respectively, R—OH, an alcohol, and so forth. Let's now take a brief look at some of these functional groups and some of the properties they impart to organic compounds.

**ALKENES AND ALKYNES.** The double and triple bonds in alkenes and alkynes show a characteristic tendency to undergo addition reactions, that is, reactions in which the components of a reactant molecule become incorporated into the alkene or alkyne. Some examples are

$$CH_3{-}CH{=}CH{-}CH_3 + HBr \longrightarrow CH_3{-}\underset{\displaystyle H}{\underset{|}{CH}}{-}\underset{\displaystyle Br}{\underset{|}{CH}}{-}CH_3$$

Table 21.5

*Some functional groups in organic compounds*

| Functional Group | Compound Class | Example | Name |
|---|---|---|---|
| $-\overset{\vert}{C}=\overset{\vert}{C}-$ | Alkenes | $CH_2{=}CH_2$ | Ethene |
| $-C\equiv C-$ | Alkynes | $CH\equiv CH$ | Ethyne |
| F, Cl, Br, I | Halides | $CH_3Cl$ | Chloromethane |
| $-OH$ | Alcohols | $CH_3OH$ | Methanol |
| $-C(=O)H$ | Aldehydes | $CH_3-\overset{O}{\overset{\Vert}{C}}-H$ or $CH_3CHO$ | Ethanal (acetaldehyde) |
| $-C(=O)-$ | Ketones | $CH_3-\overset{O}{\overset{\Vert}{C}}-CH_3$ | Propanone (acetone) |
| $-C(=O)OH$ | Carboxylic acids | $CH_3-\overset{O}{\overset{\Vert}{C}}-OH$ | Ethanoic acid (acetic acid) |
| $-\overset{\vert}{\underset{\vert}{C}}-NH_2$ | Amines | $CH_3NH_2$ | Methylamine |
| $-\underset{O}{\underset{\Vert}{C}}-N\langle$ | Amides | $CH_3-\overset{O}{\overset{\Vert}{C}}-NH_2$ | Ethanamide (acetamide) |
| $-\overset{\vert}{\underset{\vert}{C}}-O-\overset{\vert}{\underset{\vert}{C}}-$ | Ethers | $CH_3-O-CH_3$ | Dimethyl ether |
| $-\overset{\vert}{\underset{\vert}{C}}-\overset{O}{\overset{\Vert}{C}}-O-\overset{\vert}{\underset{\vert}{C}}-$ | Esters | $CH_3-\overset{O}{\overset{\Vert}{C}}-O-CH_3$ | Methyl acetate |

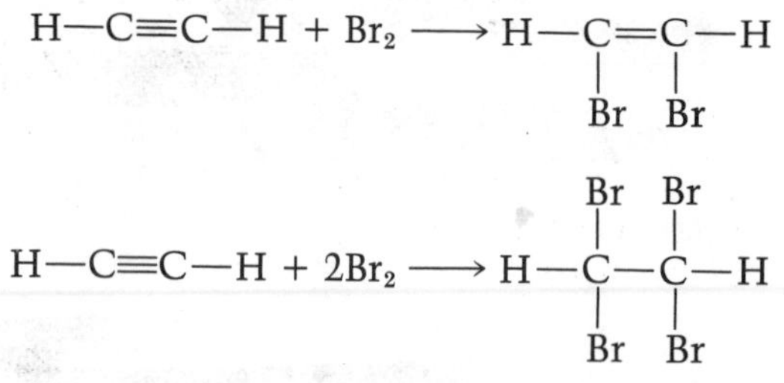

$$CH_2{=}CH_2 + H_2O \xrightarrow{H^+} CH_3CH_2OH$$

In this last case $H^+$ serves as a catalyst and is written over the arrow.

The reactions above are termed *electrophilic addition reactions* because there is evidence that they proceed by a mechanism in which the high electron density associated with the multiple bond is responsible for the initial attachment of an electron-poor fragment of the reactant. Such an electron-poor species would seek electrons and therefore would be termed an electrophile. For example, when HBr is added to ethylene the mechanism is believed to involve the initial attachment of the proton. The positive charge is transferred to the adjacent carbon atom which rapidly bonds with the $Br^-$ ion.

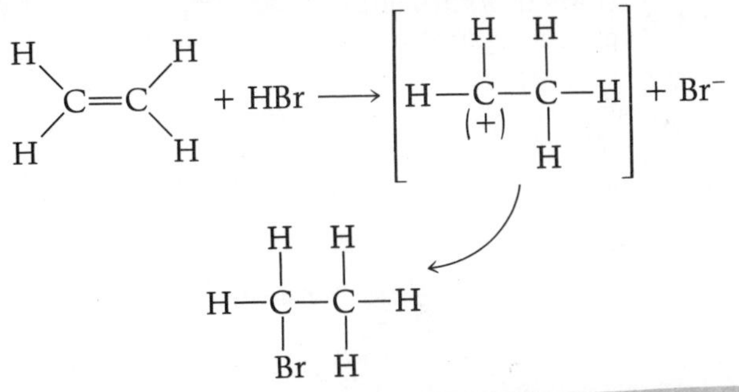

When a substance such as HBr is added to an unsymmetrical alkene there are two possibilities for reaction.

$$CH_3{-}CH{=}CH_2 + HBr \longrightarrow \underset{\textbf{2-bromopropane}}{CH_3{-}CHBr{-}CH_3}$$

$$CH_3{-}CH{=}CH_2 + HBr \longrightarrow \underset{\textbf{1-bromopropane}}{CH_3{-}CH_2{-}CH_2Br}$$

In this situation it happens that very little of the 1-bromo product is formed. Nearly 90% of the product is 2-bromopropane. The products of such reactions can be predicted by *Markovnikov's*[2] *rule,* which states that during an electrophilic addition to ordinary alkenes the hydrogen of the reactant becomes attached to that carbon of the alkene that is already bonded to the most hydrogen atoms. In propene, the end carbon has two hydrogen atoms while the middle carbon has only one. The H of the HBr therefore becomes attached to the end carbon, while the Br becomes bonded to the middle carbon. Other unsymmetrical reagents that follow this rule are $H_2O$, HCN, and the other hydrogen halides. For example, we find that the addition of water to 2-methylpropene gives the following:

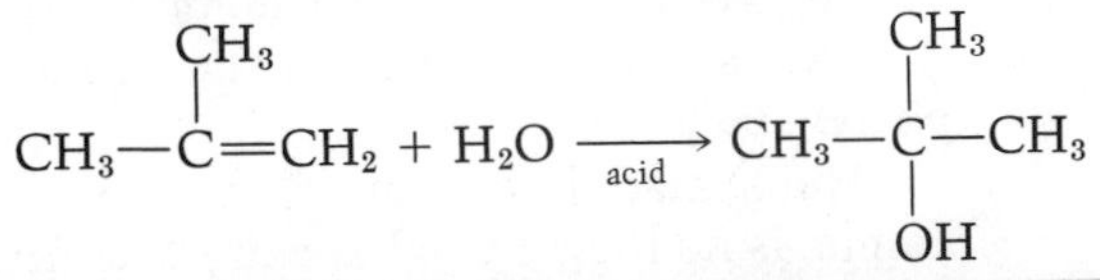

The tendency of the unsaturated hydrocarbons to undergo addition reactions can be contrasted with the tendency of the alkanes to react through substitution. For example, methane will react with $Cl_2$ to give a variety of products in which Cl replaces H.

$$CH_4 + Cl_2 \longrightarrow \left\{\begin{matrix} CH_3Cl \\ CH_2Cl_2 \\ CHCl_3 \\ CCl_4 \end{matrix}\right\} + HCl$$

[2] The publication stating Vladimir W. Markovnikov's rule was published in 1905, one year after his death.

Under similar conditions the alkenes and alkynes have little tendency to undergo substitution.

**21.7 HALOGEN DERIVATIVES**

In the halogen derivatives a halogen has replaced a hydrogen in the parent hydrocarbon. Some examples of these were seen above as products of reactions of hydrocarbons with halogens or hydrogen halides. In naming these compounds the halogen is specified as fluoro, chloro, bromo, or iodo.

$$CH_3-CH-CH_2$$

Cl    Cl

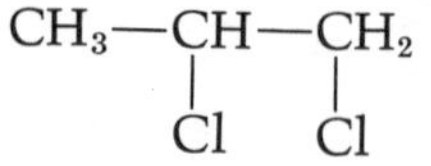

**1,2-dichloropropane**

$$Cl-\overset{\displaystyle Cl}{\underset{\displaystyle Cl}{C}}-H$$

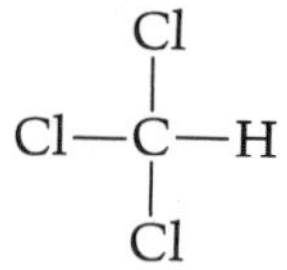

**trichloromethane (chloroform)**

$$Cl-\overset{\displaystyle Cl}{\underset{\displaystyle Cl}{C}}-Cl$$

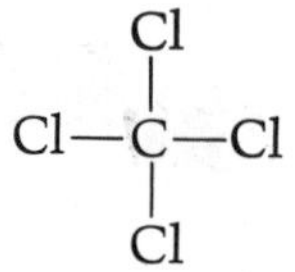

**tetrachloromethane (carbon tetrachloride)**

$$\underset{Cl}{\overset{H}{\phantom{.}}}C=C\underset{Cl}{\overset{Cl}{\phantom{.}}}$$

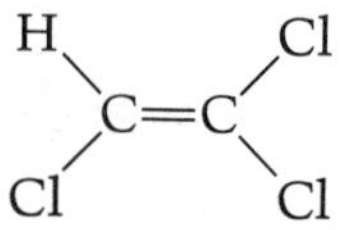

**trichloroethene**

$$F-\overset{\displaystyle F}{\underset{\displaystyle Cl}{C}}-Cl$$

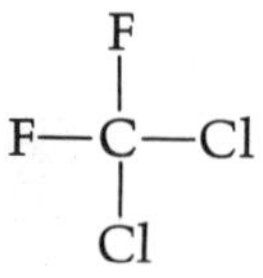

**dichlorodifluoromethane** (*Freon-12*)

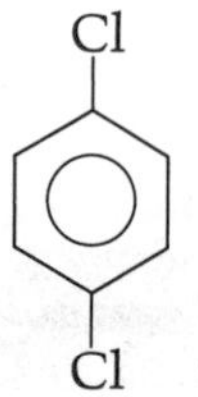

**p-dichlorobenzene**

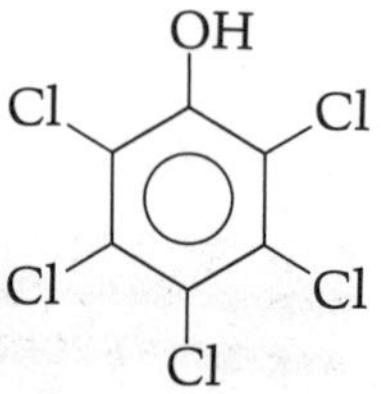

**pentachlorophenol**

Many of the halogenated hydrocarbons have important commercial applications. Trichloroethene, for example, is a common dry cleaning solvent. The *Freons* are used as refrigerants and propellants in aerosol products. As you may be aware, there is considerable controversy over whether these materials are seriously depleting the ozone shield in the earth's upper atmosphere (see p. 547).

Halogenated compounds are also toxic. Carbon tetrachloride, for example, is no longer used as a dry cleaning solvent because it is a cumulative poison. Many insecticides contain halogenated compounds. For example, p-dichlorobenzene (above) has been used in moth balls. Pentachlorophenol, also shown above, is used as a wood preservative because it is toxic to creatures that attack wood.

The alkyl halides can be prepared by addition of halogen or hydrogen halides to alkenes, as we have seen, as well as by substitution of a halogen for a hydrogen in an alkane.

$$H-\overset{\displaystyle CH_3}{\underset{\displaystyle CH_3}{C}}-CH_3 \xrightarrow[\text{light, }120°]{Br_2} Br-\overset{\displaystyle CH_3}{\underset{\displaystyle CH_3}{C}}-CH_3$$

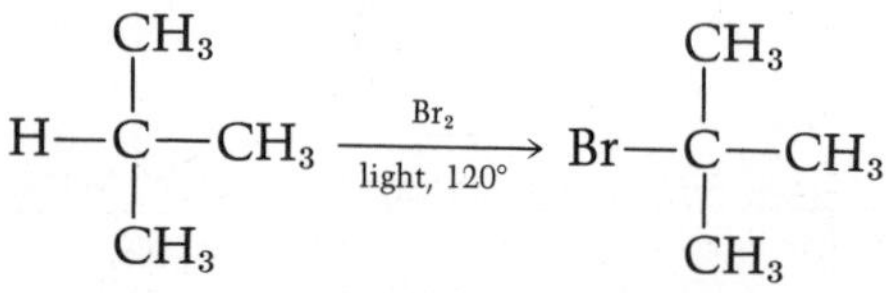

Here the Br replaces the hydrogen that is bonded least strongly to the carbon in the molecule.

Perhaps the most important method of preparing alkyl halides is from alcohols. During this reaction the OH of the alcohol is displaced by a halide ion (nucleophile). Examples of this are

$$CH_3-CH_2-CH_2-CH_2-OH + HBr \longrightarrow CH_3-CH_2-CH_2-CH_2-Br + H_2O$$

$$CH_3-\overset{\overset{\displaystyle CH_3}{|}}{\underset{\underset{\displaystyle CH_3}{|}}{C}}-OH + HCl \longrightarrow CH_3-\overset{\overset{\displaystyle CH_3}{|}}{\underset{\underset{\displaystyle CH_3}{|}}{C}}-Cl + H_2O$$

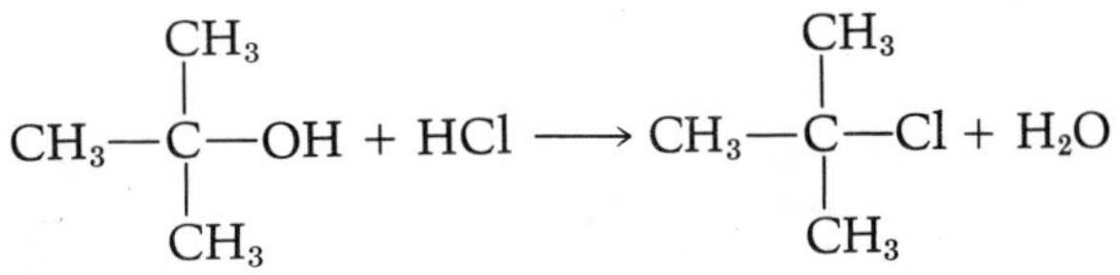

The reactions of alkyl halides can be analyzed as Lewis acid-base reactions. Because of the polar nature of the C—X bond the alkyl halides can undergo either a *nucleophilic substitution* $(S_N)$, or an *elimination (E)* reaction when they are combined with basic reagents. In general these reactions can be written as

$$S_N \qquad\qquad R-X + Z^- \longrightarrow R-Z + X^-$$

$$E \qquad R-\overset{\overset{\displaystyle H}{|}}{\underset{\underset{\displaystyle H}{|}}{C}}-\overset{\overset{\displaystyle H}{|}}{\underset{\underset{\displaystyle X}{|}}{C}}-H + Z^- \longrightarrow \underset{R}{\overset{H}{\diagdown}}C=C\underset{H}{\overset{H}{\diagup}} + HZ + X^-$$

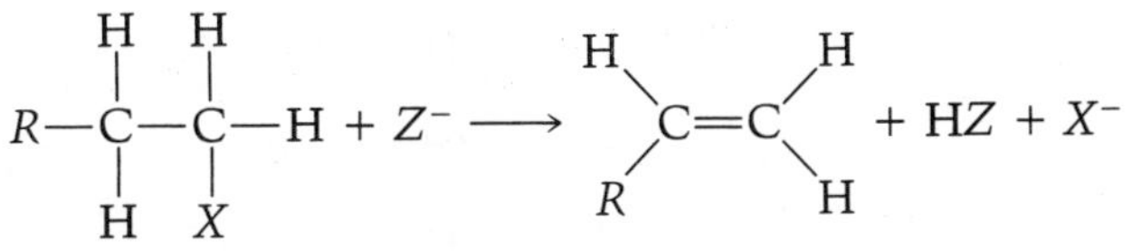

In the $S_N$ reaction the base, $Z^-$, replaces the weaker base $X^-$. Some nucleophiles that can be used in this type of reaction include $OH^-$, $H_2O$, and $NH_3$. For example, we have

$$CH_3-Br + OH^- \longrightarrow CH_3-OH + Br^-$$
$$CH_3-CH_2-Cl + H_2O \longrightarrow CH_3-CH_2-OH + HCl$$
$$CH_3-Br + NH_3 \longrightarrow CH_3-NH_2 + HBr$$

An elimination reaction is just the reverse of an addition reaction. In this case a hydrogen and a halogen on two adjacent carbons of the alkyl halide are eliminated, and a double bond between these two carbon atoms is created.

$$H-\overset{\overset{\displaystyle H}{|}}{\underset{\underset{\displaystyle H}{|}}{C}}-\overset{\overset{\displaystyle H}{|}}{\underset{\underset{\displaystyle Cl}{|}}{C}}-H + KOH \xrightarrow{\text{alcohol}} \underset{H}{\overset{H}{\diagdown}}C=C\underset{H}{\overset{H}{\diagup}} + H_2O + KCl$$

removed as HCl

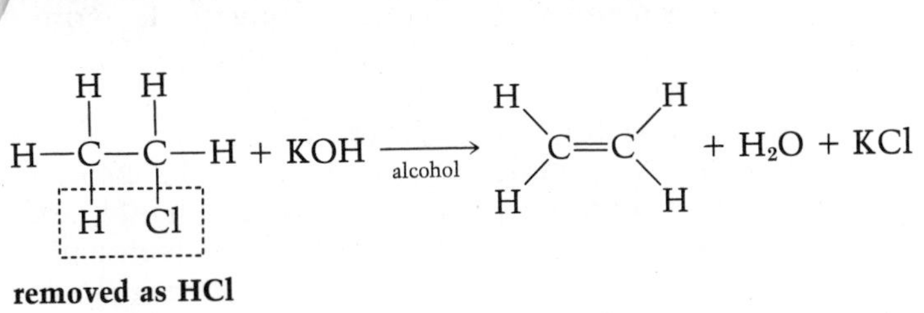

The $OH^-$ of the strong base removes a $H^+$ from the carbon next to the halogen, with a simultaneous formation of a double bond and the expulsion of the halide ion. If the base is moderately weak, there is a competition between the substitution and the elimination paths. Substitution is favored when extremely weak bases are used, whereas a very strong base favors the elimination pathway.

## 21.8 IMPORTANT OXYGEN-CONTAINING DERIVATIVES

Among these compounds we have the alcohols, aldehydes and ketones, carboxylic acids and esters, and the ethers. Many are related to each other and therefore we discuss them together in this section.

ALCOHOLS. Alcohols are characterized by the functional group, —OH. Some typical alcohols with their IUPAC and common names are

| Formula | IUPAC Name | Common Name |
|---|---|---|
| $CH_3$—OH | Methanol | Methyl alcohol |
| $CH_3$—$CH_2$—OH | Ethanol | Ethyl alcohol |
| $CH_3$—$CH_2$—$CH_2$—OH | 1-Propanol | *n*-Propyl alcohol |

As you can see, the name of an alcohol is obtained from the name of the parent alkane by replacing the *e* by *ol*. Also, when necessary, the position of the OH group is identified by number. Thus we have

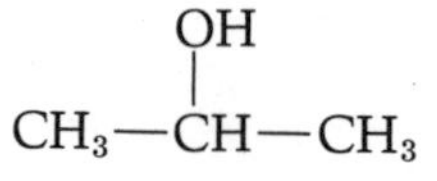
2-propanol

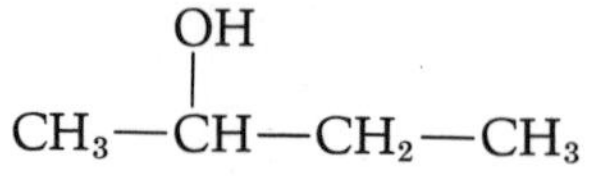
2-butanol

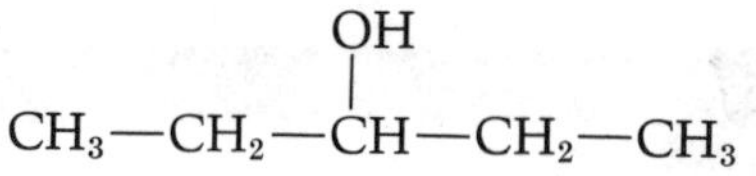
3-pentanol

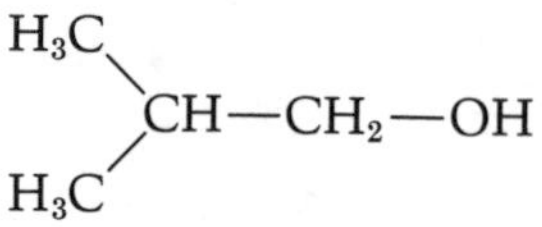
2-methyl-1-propanol

There are many important alcohols. Ethanol, of course, is found in alcoholic beverages. Compared with most other alcohols, ethanol is relatively nontoxic. When it is consumed even in small quantities, however, it causes the blood vessels to dilate, resulting in a lowering of the blood pressure followed by a general feeling of relaxation. In larger amounts ethanol causes intoxication, and excessively prolonged use can permanently damage the liver and eventually lead to death.

Methanol, also known as wood alcohol, is a deadly poison, whereas the remaining alcohols, other than ethanol, are somewhat "milder poisons." Methanol can cause blindness and eventually total loss of motor control and death. Perhaps one familiar use of methanol is in "dry gas," which is added to gasoline to cause suspended water droplets in the fuel to dissolve. In this way water that may have condensed in a fuel tank can pass harmlessly through the engine.

Methanol is prepared in large quantities from the reaction between carbon monoxide and hydrogen in the presence of a metal oxide catalyst at high temperature and pressure. The balanced equation for this preparation is

$$CO + 2H_2 \xrightarrow[\substack{3000\text{ psi} \\ 350\text{–}400°C}]{ZnO/Cr_2O_3} CH_3OH$$

The common method of preparing ethanol is through the fermentation of carbohydrates (sugars or starch). Sugars are converted into ethanol and carbon dioxide by the action of yeast in the absence of oxygen.

$$\underset{\textbf{carbohydrate}}{C_6H_{12}O_6} \xrightarrow{\text{yeast}} \underset{\textbf{ethanol}}{2C_2H_5OH} + 2CO_2$$

Another important alcohol is 2-propanol, commonly known as isopropyl alcohol, which is used widely as rubbing alcohol.

The number of groups attached to the carbon to which the OH is bonded aids us in classifying the alcohols. The C—OH grouping is called the **carbinol group,** and the carbon of this group is referred to as the **carbinol carbon.** Compounds in which there is one hydrocarbon group (R) attached to the carbinol carbon are known as **primary alcohols.** Alcohols that contain two such R groups are known as **secondary alcohols. Tertiary alcohols,** on the other hand, have three R groups bonded to the carbinol carbon.

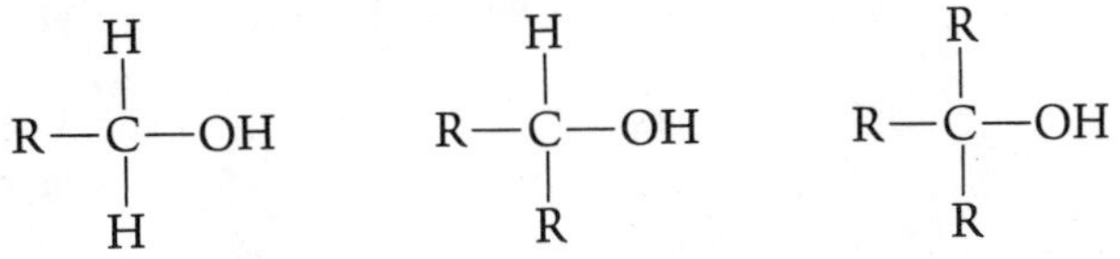

**primary alcohol** **secondary alcohol** **tertiary alcohol**

Some specific examples are

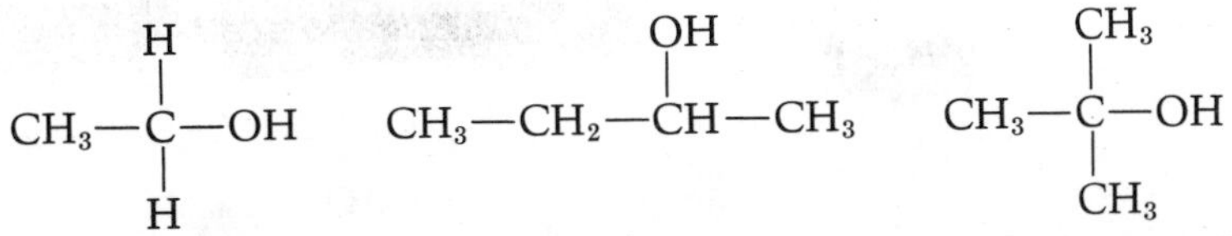

**a primary alcohol** **a secondary alcohol** **a tertiary alcohol**

Following the IUPAC nomenclature, these alcohols are called ethanol, 2-butanol, and 2-methyl-2-propanol. Their common names are ethyl alcohol, *sec*-butyl alcohol, and *tert*-butyl alcohol.

Alcohols containing more than one hydroxyl group are also possible and are called polyhydroxy alcohols. Important examples are ethylene glycol (1,2-ethanediol),

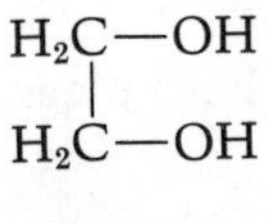

ethylene glycol

and glycerol (1,2,3-propanetriol)

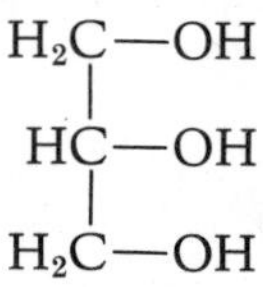

glycerol (also, glycerin)

These compounds are both soluble in water in all proportions. Ethylene glycol is used widely as an automotive antifreeze and in the manufacture of polyesters such as *Dacron* and *Mylar.* Glycerol occurs in fats, as we shall see in the next chapter. It too is used in antifreeze applications where a nontoxic antifreeze is necessary (ethylene glycol is poisonous; glycerol is not. Glycerol is also sometimes added to alcoholic beverages to promote "smoothness."

Both these compounds can be nitrated by their cautious addition to a mixture of nitric and sulfuric acid.

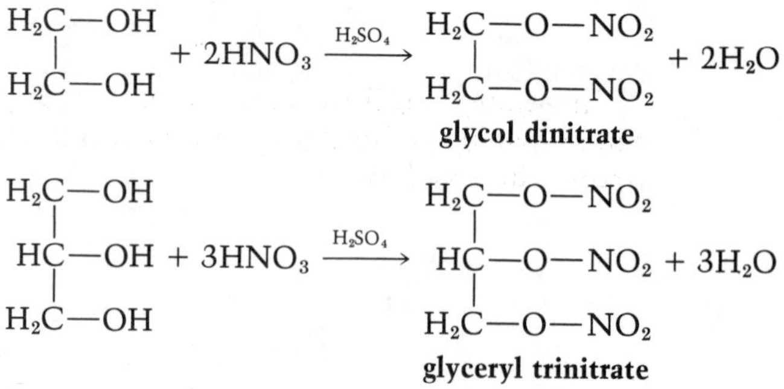


The products of these reactions must be handled with extreme caution. Glyceryl trinitrate is also called nitroglycerin and is used with glycol dinitrate in the production of dynamite.

**ALDEHYDES AND KETONES.** Aldehydes and ketones are characterized by the presence of a **carbonyl group, >C=O**. In aldehydes this occurs on an end carbon,

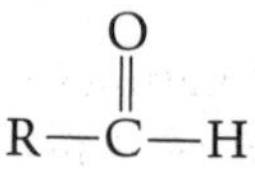

aldehyde

while in ketones it occurs on one of the middle carbon atoms,

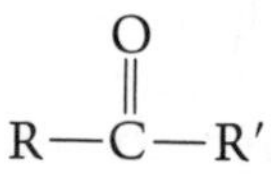

ketone

(R and R′ indicate the possibility of having different alkyl groups attached to the carbonyl group.)

Many aldehydes have pleasant odors, particularly those in which the R group is aromatic.

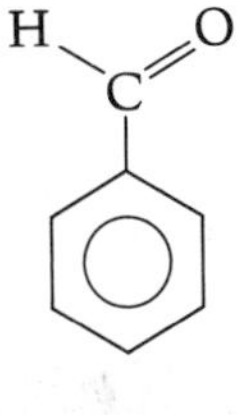

**benzaldehyde (bitter almonds)**

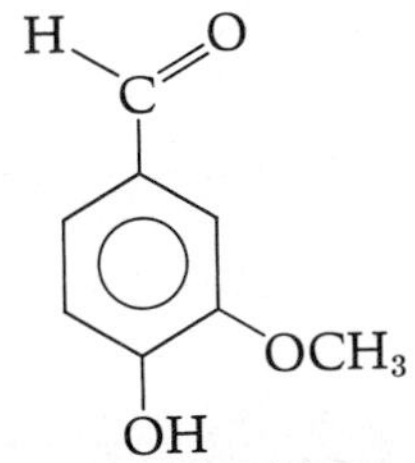

**vanillin (vanilla bean)**

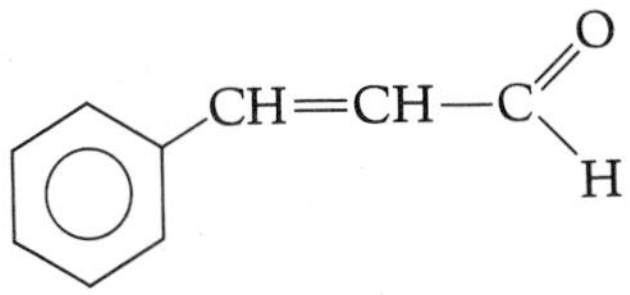

**cinnamaldehyde (cinnamon)**

Ketones often have very desirable solvent properties. Acetone, for example, is found in nail polish remover; methyl ethyl ketone is a solvent in airplane glue.

$$CH_3-\overset{O}{\overset{\|}{C}}-CH_3 \qquad CH_3-\overset{O}{\overset{\|}{C}}-CH_2-CH_3$$

**acetone** **methyl ethyl ketone**

In naming aldehydes and ketones, the *–e* of the corresponding hydrocarbon parent is dropped and replaced by **-al** for aldehydes or **-one** for ketones. Thus we have

$$CH_3-\overset{O}{\overset{\|}{C}}-H$$ ethanal (common: acetaldehyde)

The IUPAC names of the ketones, acetone and methyl ethyl ketone, would be propanone and butanone, respectively. When there are possible alternative locations of the carbonyl group, its position is indicated by number.

$$CH_3-\overset{O}{\overset{\|}{C}}-CH_2-CH_2-CH_3 \qquad CH_3-CH_2-\overset{O}{\overset{\|}{C}}-CH_2-CH_3$$

**2-pentanone** **3-pentanone**

Aldehydes and ketones are related to alcohols through oxidation and reduction. Oxidation of a primary alcohol can yield an aldehyde (although these are usually difficult to isolate because they are generally easily oxidized further).

$$CH_3-\underset{H}{\overset{H}{C}}-OH \xrightarrow{(O)} CH_3-\overset{H}{C}=O$$

**ethanol** **ethanal**

Secondary alcohols are oxidized to ketones.

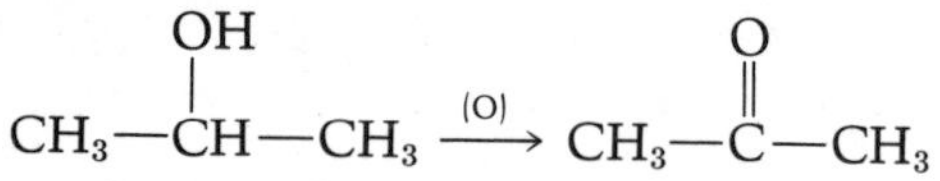

$$\underset{\textbf{2-propanol}}{\overset{\displaystyle OH}{CH_3-\overset{|}{C}H-CH_3}} \xrightarrow{(O)} \underset{\substack{\textbf{propanone} \\ \textbf{(acetone)}}}{\overset{\displaystyle O}{CH_3-\overset{\|}{C}-CH_3}}$$

The symbol (O) is used here to indicate oxidation without specifying the oxidizing agent. Oxidation of tertiary alcohols, which is considerably more difficult, breaks down the carbon chain.

The reason for the various products seems to be the number of hydrogens that are attached to the carbinol carbon. During the oxidation reaction one hydrogen is eliminated from the carbinol carbon as well as the one on the OH. With a primary alcohol, therefore, one hydrogen remains on the carbon after oxidation, whereas with a secondary alcohol none remain. Since no hydrogens are attached to the carbinol carbon in a tertiary alcohol, no reaction occurs when $H_2CrO_4$ or other similar oxidizing materials are added.

The oxidation of alcohols to aldehydes and ketones can be reversed through reduction. Often a metal hydride is used (for example, $NaBH_4$, sodium borohydride) because of the availability, at least in principle, of electron-rich $H^-$ ions.

$$\overset{\displaystyle O}{CH_3-\overset{\|}{C}-H} \xrightarrow{NaBH_4} \overset{\displaystyle H}{CH_3-\overset{|}{\underset{|}{C}}-OH}$$
$$\phantom{xxxxxxxxxxxxxxxxxxxxxxxxxxxxxxxxxxx} H$$

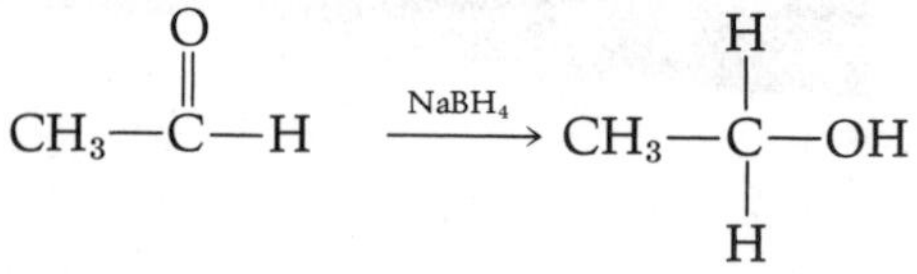

$$\overset{\displaystyle O}{CH_3-\overset{\|}{C}-CH_3} \xrightarrow{NaBH_4} \overset{\displaystyle OH}{CH_3-\overset{|}{C}H-CH_3}$$

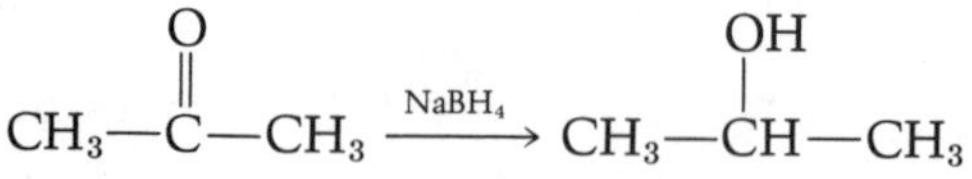

**ORGANIC ACIDS AND ESTERS.** Organic acids are characterized by the presence of the **carboxyl group,** —COOH. Structurally this is

$$\overset{\displaystyle O}{R-\overset{\|}{C}-O-H}$$

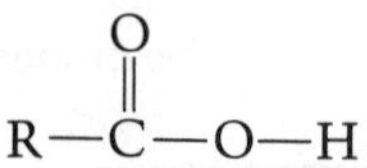

The presence of the lone oxygen bonded to the carbon that is attached to the —OH group polarizes the O—H bond and permits the H to be lost as a proton.

$$\overset{\displaystyle :\!O\!:}{R-\overset{\|}{C}-\overset{..}{\underset{..}{O}}-H} + H_2O \longrightarrow \left(\overset{\displaystyle :\!O\!:}{R-\overset{\|}{C}-\overset{..}{\underset{..}{O}}\!:}\right)^{-} + H_3O^+$$

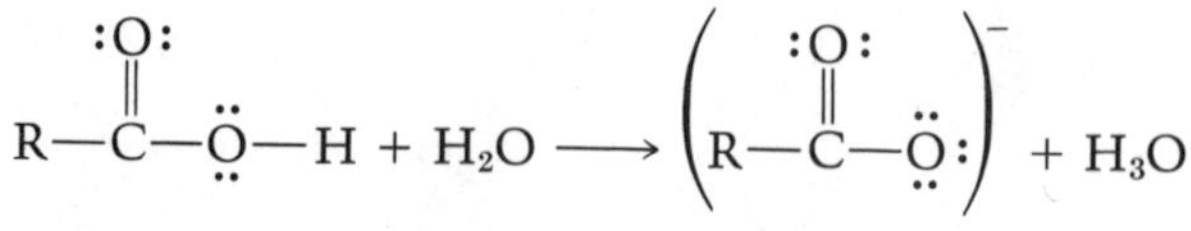

The acidity of the carbonyl group was discussed earlier in Chapter 13.

Many important organic compounds are acids or their salts. Some examples are shown at the top of the next page.

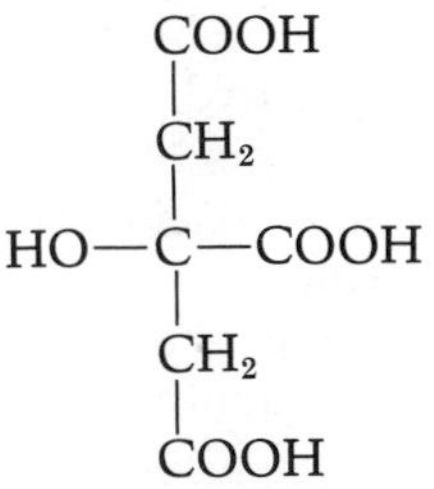

**citric acid**
**(citrus fruits)**

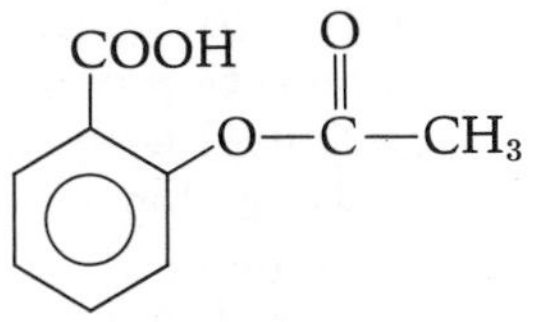

**acetylsalicylic acid**
**(aspirin)**

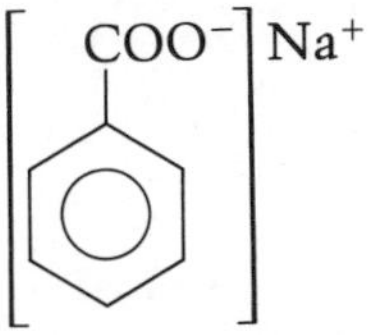

**sodium benzoate**
**(a food preservative)**

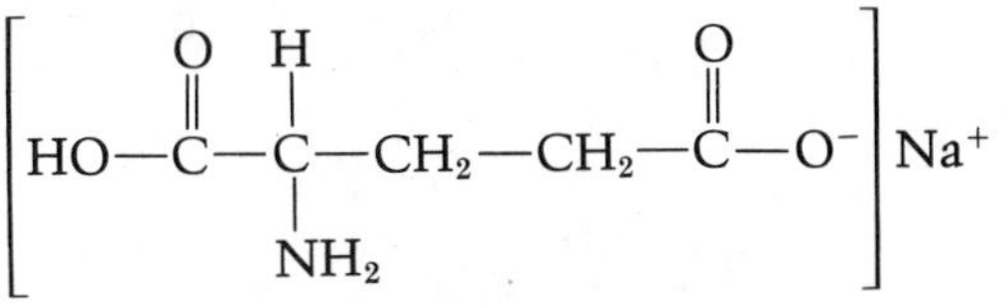

**monosodium glutamate, MSG**
**(*Áccent,* a flavor enhancer)**

Organic acids derived from hydrocarbons are named by dropping the $-e$ from the end of the name of the parent and adding **-oic acid.**

$$CH_3—\overset{\overset{O}{\|}}{C}—OH \qquad \text{ethanoic acid (acetic acid)}$$

$$CH_3—CH_2—CH_2—\overset{\overset{O}{\|}}{C}—OH \qquad \text{butanoic acid (butyric acid, from rancid butter)}$$

One method of preparing an acid is by oxidation of an aldehyde (or by thorough oxidation of a primary alcohol).

$$\underset{\textbf{ethanol}}{CH_3CH_2OH} \xrightarrow{(O)} \underset{\textbf{ethanal}}{CH_3CHO} \xrightarrow{(O)} \underset{\textbf{ethanoic acid}}{CH_3COOH}$$

**Esters,** the other compound type that is the subject of this subsection, are characterized by the presence of the functional group,

$$R—\overset{\overset{O}{\|}}{C}—O—R'$$

They are products of the acid-catalyzed elimination of water from between a carboxylic acid and an alcohol, a process called **esterification.**

$$\underset{\textbf{acetic acid}}{CH_3—\overset{\overset{O}{\|}}{C}—O—H} + \underset{\textbf{ethyl alcohol}}{H—O—CH_2CH_3} \overset{H^+}{\rightleftharpoons} \underset{\textbf{ethyl acetate}}{CH_3—\overset{\overset{O}{\|}}{C}—O—CH_2CH_3} + H_2O$$

This reaction leads to an equilibrium. The position of equilibrium can be shifted to the right by employing a dehydrating agent to remove $H_2O$ from the reaction mixture as it is formed. This type of reaction, where molecules are joined with the simultaneous elimination of another smaller molecule, is also called a **condensation reaction.**

When an alcohol is treated with an inorganic acid an inorganic ester is produced. For example, when nitric acid is *cautiously* added to ethyl alcohol,

$$CH_3-CH_2-OH + H-O-\overset{\overset{\displaystyle O}{\|}}{N}-O \longrightarrow CH_3-CH_2-O-\overset{\overset{\displaystyle O}{\|}}{N}-O + H_2O$$

the product is ethyl nitrate (quite explosive). Organic phosphate esters of the type

$$R-O-\begin{array}{c} O \\ | \\ P \\ | \\ OH \end{array}-OH$$

are very important in biological systems as are esters of the trihydroxy alcohol, glycerol.

The reaction to produce an ester is reversible and the insertion of an $H_2O$ molecule into the ester to give the acid and alcohol is termed hydrolysis. An example is the hydrolysis of methyl acetate.

$$\underset{\textbf{methyl acetate}}{CH_3-O-\overset{\overset{\displaystyle O}{\|}}{C}-CH_3} + H_2O \overset{\text{acid}}{\rightleftharpoons} \underset{\textbf{methanol}}{CH_3OH} + \underset{\textbf{acetic acid}}{CH_3-\overset{\overset{\displaystyle O}{\|}}{C}-OH}$$

This equilibrium is established rapidly, and in order to drive the reaction to the right, the alcohol can be removed by distillation.

In the base-catalyzed hydrolysis, also called **saponification,** the acid that is produced in the forward reaction is neutralized, thereby shifting the position of equilibrium to the right. For example,

$$\underset{\textbf{propyl ethanoate}}{CH_3-CH_2-CH_2-O-\overset{\overset{\displaystyle O}{\|}}{C}-CH_3} + NaOH \xrightarrow{H_2O}$$

$$CH_3-CH_2-CH_2-OH + NaO-\overset{\overset{\displaystyle O}{\|}}{C}-CH_3$$

Esters, particularly those of low molecular weight, generally have pleasant, rather agreeable odors, as seen in Table 21.6. Some of them have very desirable solvent properties and are used in paints and varnishes.

Table 21.6
*Odors of some common esters*

| Name | Formula | Odor |
|---|---|---|
| *n*-Amyl acetate | $CH_3COOCH_2(CH_2)_3CH_3$ | Banana |
| *n*-Octyl acetate | $CH_3COOCH_2(CH_2)_6CH_3$ | Orange |
| *iso*-Amyl butyrate | $CH_3(CH_2)_2COOCH(CH_3)CH_2CH_2CH_3$ | Pear |

**ETHERS.** We saw that esters are produced by a condensation reaction between an alcohol and an acid. Alcohols can also undergo self-condensation to give **ethers** in which two hydrocarbon units are joined by an oxygen bridge.

$$R{-}OH + HO{-}R \longrightarrow \underset{\textbf{ether}}{R{-}O{-}R} + H_2O$$

A specific example is

$$2CH_3{-}CH_2{-}OH \xrightarrow[H_2SO_4]{conc} \underset{\textbf{diethyl ether}}{CH_3{-}CH_2{-}O{-}CH_2{-}CH_3} + H_2O$$

The concentrated sulfuric acid is used as a dehydrating agent to help in the removal of the $H_2O$ as it is formed. Diethyl ether is used as an anesthetic. It must be used with care because it is extremely flammable. Most ethers are used primarily as solvents.

## 21.9 AMINES AND AMIDES

The **amines** are a group of compounds identified by the functional group $-NH_2$. As a result they are commonly viewed as derivatives of ammonia. Below are some typical amines.

| Formula | IUPAC Name | Common Name |
|---|---|---|
| $CH_3{-}NH_2$ | Aminomethane | Methylamine |
| $CH_3{-}CH_2{-}NH_2$ | Aminoethane | Ethylamine |
| $CH_3{-}CH_2{-}NH{-}CH_2{-}CH_3$ | Ethylaminoethane | Diethylamine |
| $(CH_3)_3C{-}NH_2$ | 2-Amino-2-methylpropane | *t*-Butylamine |
| $H_2N{-}CH_2{-}CH_2{-}NH_2$ | 1,2-Diaminoethane | Ethylenediamine |
| 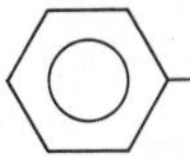 $-NH_2$ | Aminobenzene | Aniline |

Amines can be classified as being primary, secondary, or tertiary, depending on the number of R groups attached to the nitrogen.

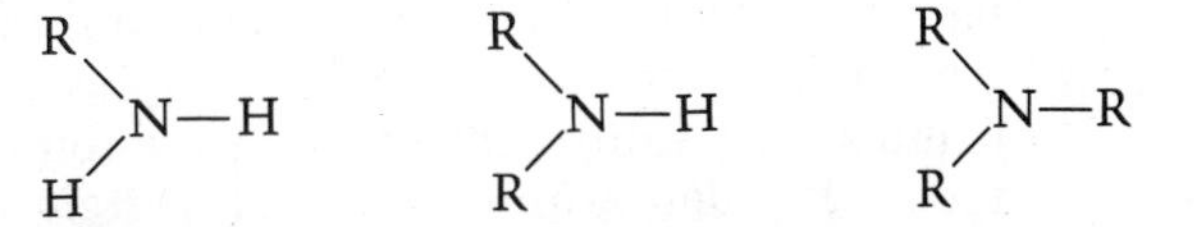

**primary amine** **secondary amine** **tertiary amine**

All the amines in our list above are primary amines except diethylamine, which is a secondary amine.

Amines also exist in which the nitrogen is a member of a ring. Examples of this type are

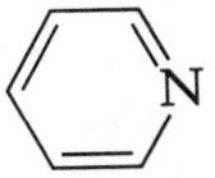

pyridine

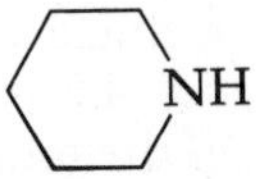

piperidine

These compounds are referred to as **heterocycles** because not all the atoms in the ring are identical.

Amines, like ammonia, are weak bases. Most of them also have very unpleasant odors. The stench of decaying protein, for instance, can be traced to compounds like those below.

$H_2N{-}CH_2{-}CH_2{-}CH_2{-}CH_2{-}NH_2$ putrescine

$H_2N{-}CH_2{-}CH_2{-}CH_2{-}CH_2{-}CH_2{-}NH_2$ cadaverine

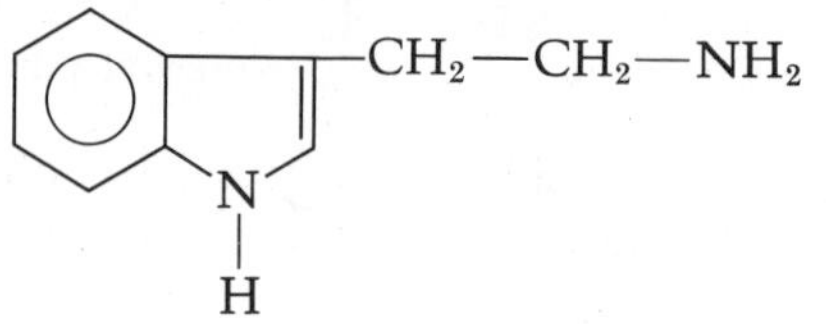

skatole (in feces)

**Amides** are identified by the functional group $-\overset{\overset{\displaystyle O}{\|}}{C}-NH_2$. Some examples of this type of compound are

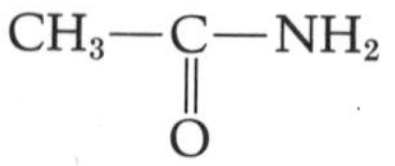

**acetamide**

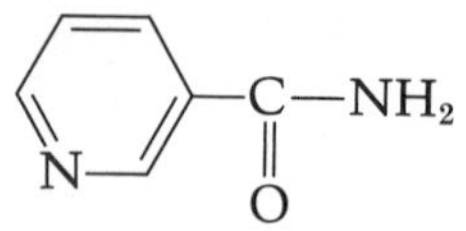

**nicotinamide**

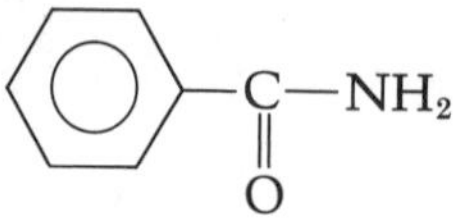

**benzamide**

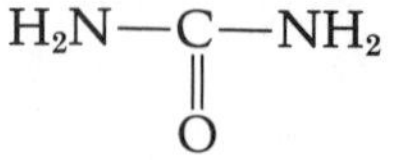

**urea**

The functional groups of the amines and amides are found in many important biological compounds that will be discussed in the next chapter. These compounds include the nucleic acids, the amino acids, thiamin, riboflavin, and biotin. The heterocyclic amines are also found as a basic unit of a group of compounds known as alkaloids. Alkaloids are rather complex compounds containing nitrogen that are found in plants. Compounds such as nicotine, codeine, morphine, and lysergic acid diethylamide (LSD) are all alkaloids.

## 21.10 POLYMERS

Polymers are very large molecules that are made by bonding together many smaller molecules that we call **monomers.** For example, polyethylene (plastic food wrap) is prepared by linking together a large number of $CH_2{=}CH_2$ monomer units to give a hydrocarbon having the general formula, $({-}CH_2{-}CH_2{-})_n$. Many naturally occurring substances are polymers, for example, rubber, starch, proteins, and the nucleic acids. Man-made polymers include such familiar materials as Bakelite, Melmac, Nylon, Dacron, Plexiglass, Teflon, and polyvinyl chloride (PVC).

Polymers can be made either by direct addition of their monomeric units (to make an **addition polymer**) or by condensation in which a small molecule is lost when the monomers are linked together (forming a **conden-**

**sation polymer**). Vinyl chloride,

$$CH_2 {=} CH$$
$$|$$
$$Cl$$

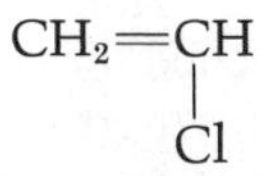

can be made to polymerize in the presence of a peroxide initiator to form polyvinyl chloride. This reaction can be seen as

$$n \ CH_2 {=} CH \xrightarrow{\text{peroxide}} \left( CH_2 {-} CH \right)_n$$
$$| \qquad\qquad\qquad\qquad |$$
$$Cl \qquad\qquad\qquad\qquad Cl$$

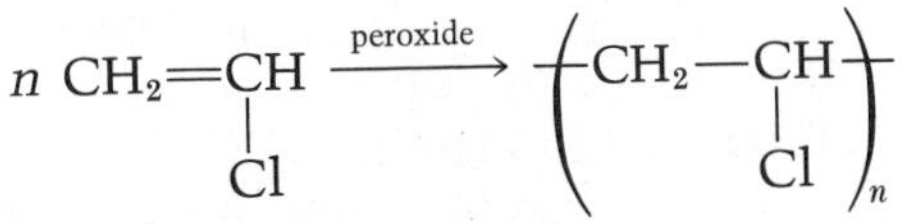

In general, the formula for polyvinyl chloride is

$$\left( CH_2 {-} CH \right)_n$$
$$|$$
$$Cl$$

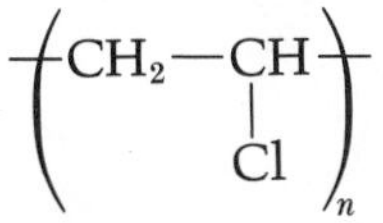

This material finds many uses, for example, in phonograph records and plastic pipe. It can also be mixed with esters that soften the polymer. The softened material finds uses in such products as plastic garden hoses, table-cloths, raincoats and "vinyl leather" products.

Condensation polymers are prepared by the reaction between two species, each of which has more than one functional group. The resulting polymer is said to be a **copolymer** because it consists of two different monomers. For example, *Nylon* is formed when a dicarboxylic acid (i.e., a carboxylic acid that contains two —COOH groups) reacts with a diamine (an amine with two —NH$_2$ groups). The overall reaction for the production of Nylon is

$$n\text{HOOC(CH}_2)_4\text{COOH} + \quad n\text{H}_2\text{N(CH}_2)_6\text{NH}_2 \xrightarrow[\text{280°C}]{-\text{H}_2\text{O}}$$

**adipic acid**        **hexamethylenediamine**

$$\left( C(CH_2)_4C {-} N(CH_2)_6N \right)_n + H_2O$$

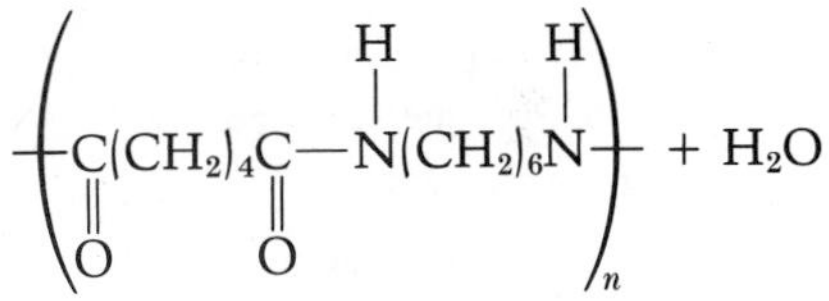

**Nylon**

*Dacron*, a **polyester**, is prepared by the reaction of methyl terephthalate (a diester) with ethylene glycol in the presence of an acid or base.

$$n\text{CH}_3\text{OOC} \langle \bigcirc \rangle \text{COOCH}_3 + n\text{HOCH}_2\text{CH}_2\text{OH} \xrightarrow[\text{base}]{\text{acid or}}$$

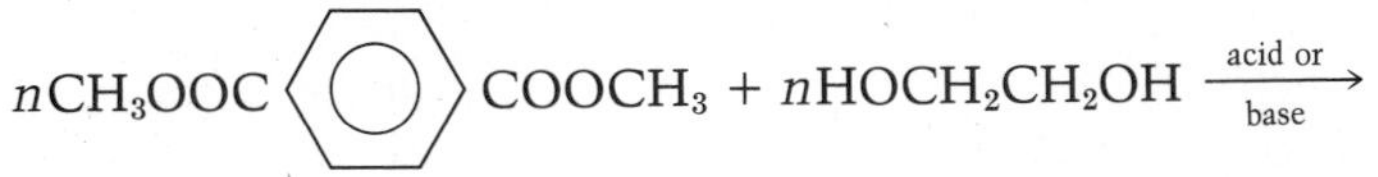

**methyl terephthalate**        **ethylene glycol**

$$\left( C \langle \bigcirc \rangle C {-} OCH_2CH_2O \right)_n + n\text{CH}_3\text{OH}$$

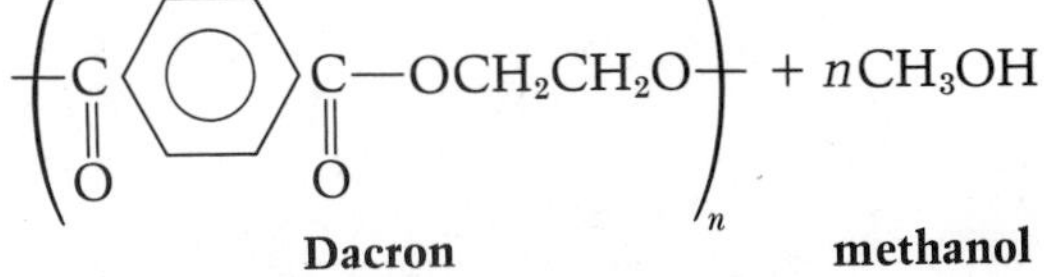

**Dacron**        **methanol**

In the production of Nylon, $H_2O$ is eliminated; and with Dacron, methanol is eliminated. Table 21.7 contains a number of important polymers, the reactants needed for their production, and whether they are addition or condensation polymers.

The formation of bonds between adjacent polymer molecules can also be brought about. In this case, the greater the degree of this cross-linking between parallel rows of polymer molecules, the stronger will be the material. Bakelite, for example, owes its strength and hardness to the three-dimensional network of covalent bonds throughout the entire polymer, as illustrated in Figure 21.10.

Natural rubber, too, can be made harder and stronger by a process known as **vulcanization.** In this reaction sulfur bridges between different chains create cross-links that lead to a tougher material.

Table 21.7
*Compositions of some common polymers*

| Monomer | Polymer | Type |
|---|---|---|
| $CH_2{=}CH_2$ **ethylene** | Polyethylene | Addition |
| $CH_2{=}CHCl$ **vinyl chloride** | Polyvinyl chloride (PVC) | Addition |
| $F_2C{=}CF_2$ **tetrafluoroethylene** | Teflon | Addition |
| 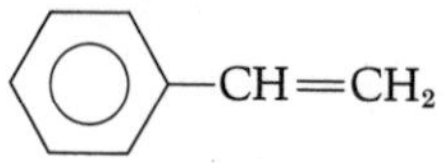 **styrene** | Polystyrene | Addition |
| $CH_2{=}C(CH_3)COOCH_3$ **methyl methacrylate** | Plexiglass | Addition |
| $HOOC{-}(CH_2)_4{-}COOH$ + $NH_2{-}(CH_2)_6NH_2$ **adipic acid** **hexamethylenediamine** | Nylon | Condensation |
| $HOOC{-}C_6H_4{-}COOH$ + $HO{-}CH_2CH_2{-}OH$ **terephthalic acid** **ethylene glycol** | Dacron | Condensation |
| $C_6H_5OH$ + HCHO **phenol** **formaldehyde** | Bakelite | Condensation |
| 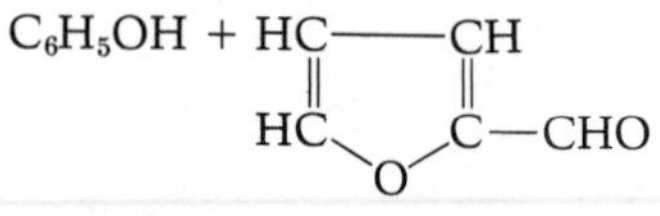 **phenol** **furfural** | Durite | Condensation |
| 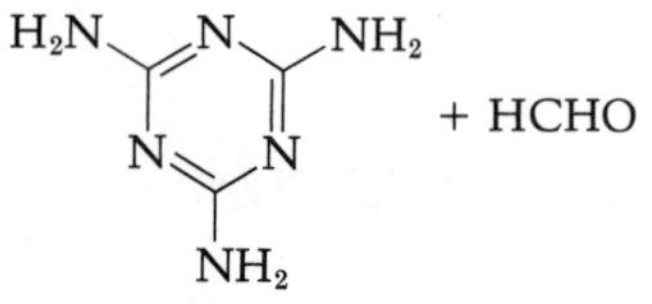 **melamine** | Melmac | Condensation |

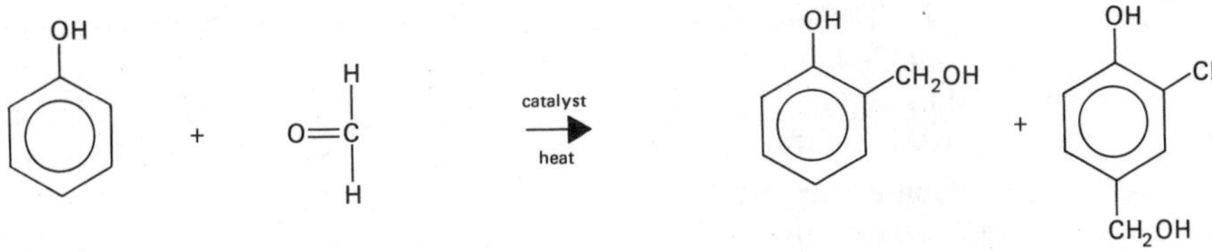


**phenol** **formaldehyde** **(salicyl alcohol)**

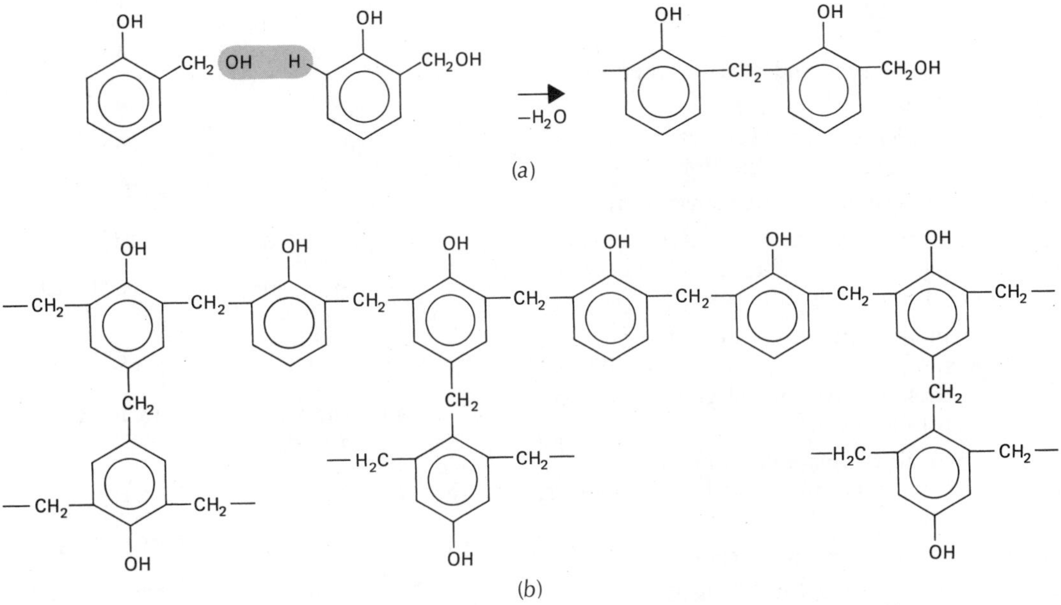


Figure 21.10
*Bakelite. (a) Polymerization of salicyl alcohol can give a linear polymer. (b) Continued polymerization, with cross-linking, gives a rigid three-dimensional structure called Bakelite.*

## INDEX TO QUESTIONS

Hydrocarbons 1, 2, 3, 4, 12
Isomerism 5, 6, 10, 11, 16, 26
Structure and bonding 7, 8, 9, 20, 21, 35
Nomenclature 10, 4, 15, 17, 18, 19, 25, 33
Familiar properties and applications 13, 22, 30, 31
Functional groups 24, 38, 39, 44
Reactions 27, 29, 36, 37, 40, 41, 42, 43, 44
Mechanisms 28, 32, 34
Polymers 46, 47, 48, 49, 50, 51

## REVIEW QUESTIONS

**21.1** What is the difference between a saturated and an unsaturated hydrocarbon?

**21.2** The straight-chain alkanes are nonpolar molecules. How do we explain the fact that their boiling points increase from $CH_4$ to $C_{10}H_{22}$?

**21.3** What would be the molecular formulas for: (a) an alkane having 30 carbon atoms, (b) an alkene having 27 carbon atoms, (c) an alkyne having 33 carbon atoms?

**21.4** What would be the molecular formula for a straight-chain hydrocarbon having 17 carbon atoms and (a) all C—C single bonds, (b) one C—C double bond, (c) one C—C triple bond, (d) three C—C double bonds, (e) two C—C triple bonds?

**21.5** How does optical isomerism arise in organic compounds? Draw the optical isomers of

$$
\begin{array}{c}
H \\
| \\
Br-C-I \\
| \\
Cl
\end{array}
$$

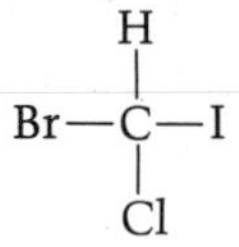

**21.6** Sketch the cis and trans isomers of 2-pentene.

**21.7** What geometry do we expect for the molecules described in Question 21.5?

**21.8** The molecule $C_2Cl_4$ is planar. Why?

**21.9** The carbon atoms in 2-butyne lie in a straight line while those in butane do not. Explain why this is so.

**21.10** Draw the structural formulas for and name the nine isomers of heptane. Which of these isomers would give rise to optical isomerism?

**21.11** Draw all the possible isomers of hexene and show geometric isomers wherever possible.

**21.12** What is a homologous series?

**21.13** Name some uses of the alkanes.

**21.14** Name the following compounds:

(a) 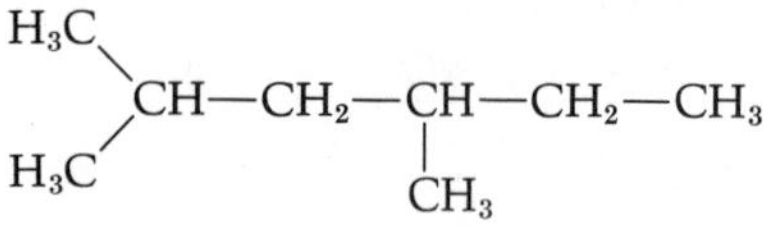
$$
\begin{array}{c}
H_3C \\
\phantom{H_3C}\searrow \\
\phantom{H_3}CH-CH_2-CH-CH_2-CH_3 \\
\phantom{H_3C}\nearrow \qquad\qquad | \\
H_3C \qquad\qquad CH_3
\end{array}
$$

(b) 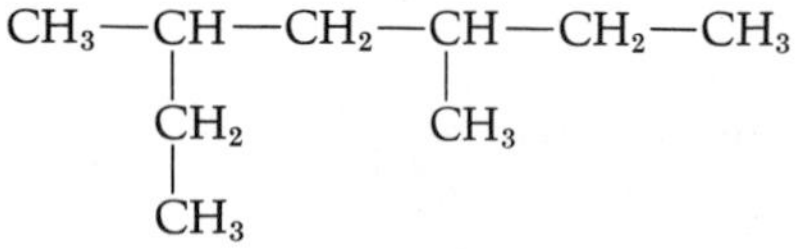
$$
\begin{array}{c}
CH_3-CH-CH_2-CH-CH_2-CH_3 \\
\quad | \qquad\qquad | \\
\quad CH_2 \qquad\quad CH_3 \\
\quad | \\
\quad CH_3
\end{array}
$$

(c) 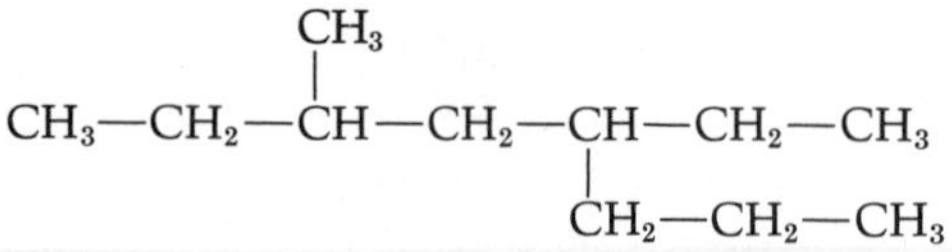
$$
\begin{array}{c}
\qquad\qquad CH_3 \\
\qquad\qquad | \\
CH_3-CH_2-CH-CH_2-CH-CH_2-CH_3 \\
\qquad\qquad\qquad\qquad\qquad | \\
\qquad\qquad\qquad\qquad CH_2-CH_2-CH_3
\end{array}
$$

(d) 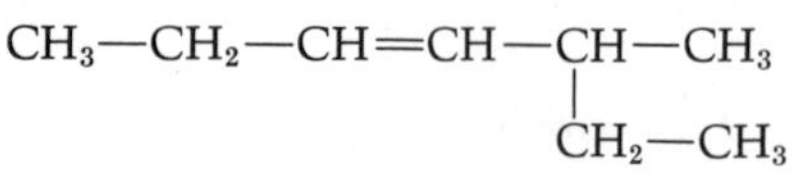
$$
CH_3-CH_2-CH=CH-CH-CH_3 \\
\qquad\qquad\qquad\qquad\quad | \\
\qquad\qquad\qquad\qquad CH_2-CH_3
$$

(e) 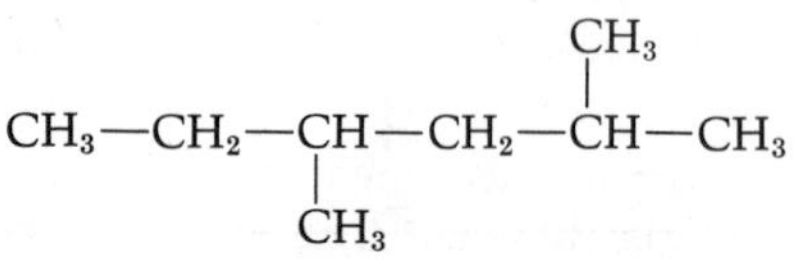
$$
CH_3-CH_2-CH-CH_2-\overset{\displaystyle CH_3}{\overset{|}{CH}}-CH_3 \\
\qquad\qquad | \\
\qquad\qquad CH_3
$$

**21.15** Name the following compounds:

(a) 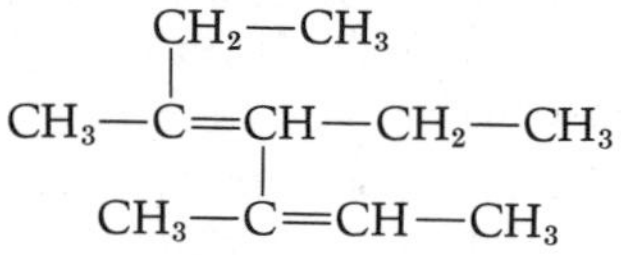
$$
\begin{array}{c}
\qquad\qquad CH_2-CH_3 \\
\qquad\qquad | \\
CH_3-C=CH-CH_2-CH_3 \\
\quad | \\
CH_3-C=CH-CH_3
\end{array}
$$

(b) 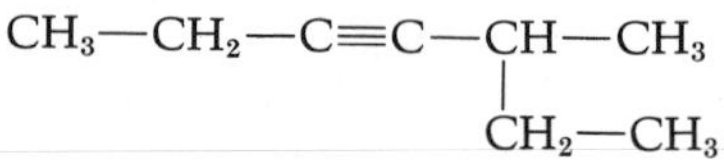
$$
CH_3-CH_2-C\equiv C-CH-CH_3 \\
\qquad\qquad\qquad\qquad | \\
\qquad\qquad\qquad\quad CH_2-CH_3
$$

(c) 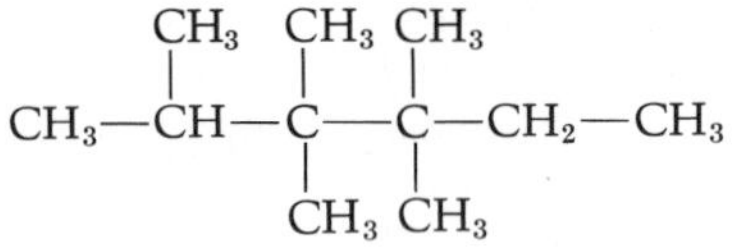
$$
\begin{array}{c}
\quad CH_3 \quad CH_3 \; CH_3 \\
\quad | \qquad | \qquad | \\
CH_3-CH-C-\!-C-CH_2-CH_3 \\
\qquad\qquad | \qquad | \\
\qquad\quad CH_3 \; CH_3
\end{array}
$$

(d) 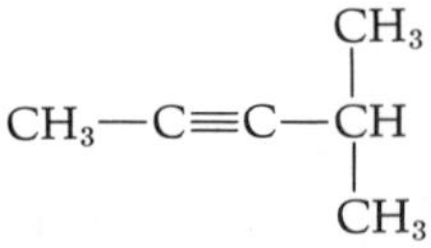
$$
\begin{array}{c}
\qquad\qquad CH_3 \\
\qquad\qquad | \\
CH_3-C\equiv C-CH \\
\qquad\qquad | \\
\qquad\qquad CH_3
\end{array}
$$

(e) 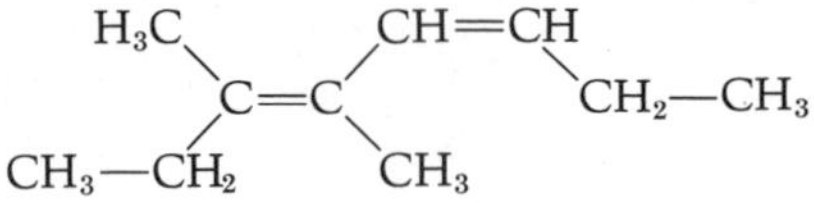
$$
\begin{array}{c}
H_3C \qquad\qquad\quad CH=CH \\
\phantom{H_3C}\searrow \qquad\qquad\nearrow \qquad\searrow \\
\qquad C=C \qquad\qquad CH_2-CH_3 \\
\phantom{CH}\nearrow \qquad\searrow \\
CH_3-CH_2 \qquad CH_3
\end{array}
$$

**21.16** Draw the remaining isomers of the compound,

$$
\begin{array}{c}
\quad CH_3 \; CH_3 \\
\quad | \qquad | \\
CH_3-C=C-CH_3
\end{array}
$$

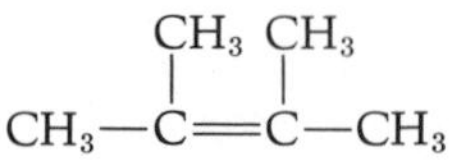

**21.17** Write structural formulas for the following:
(a) 2-methylpentane
(b) 2,3-dimethylbutane
(c) 2,3-dimethyl-2-butene
(d) 1,3,5-octatriene
(e) 3,3,4-trimethyl-1-pentyne

**21.18** What is the proper IUPAC name for 2,4-diethyl-3,3,4-trimethylpentane?

**21.19** Write the structural formula for (a) cis-1,2-dichloropropene, (b) isohexane, (c) isopropyl alcohol, (d) m-dichlorobenzene.

**21.20** What would be the C—C—C bond angles in planar cyclopropane, cyclobutane, cyclopentane, cyclohexane?

**21.21** The chair form of cyclohexane is more stable (of lower energy) than the boat form by several kilocalories per mole. By examining these two structures, can you suggest why the boat form has a higher energy?

**21.22** What is a carcinogen? In what way is the structure of 1,2-benzopyrene similar to graphite? Can you suggest why incomplete combustion of many hydrocarbons produces products similar to 1,2-benzopyrene?

**21.23** Describe the bonding in benzene. How does it compare to the bonding in graphite?

**21.24** What is a functional group? Give the structural formula of
(a) An aldehyde
(b) A ketone
(c) A carboxylic acid
(d) An amine
(e) An alcohol
(f) An ester
(g) An ether

**21.25** Write structural formulas for the following compounds:
(a) 2-methyl-1-butene
(b) 2,3-dimethyl-2-butanol
(c) 2-bromo-1-phenylpropane
(d) 3-methyl-2-pentanone
(e) 1,3,5-tribromo-2,4,6-trichlorobenzene

**21.26** Draw *all* of the structural isomers, including cyclic structures, of $C_3H_4Cl_2$.

**21.27** What type of reaction is characteristic of the alkenes and alkynes? What type of reaction is characteristic of the alkanes?

**21.28** Discuss the various steps in the mechanism of the reaction of $C_2H_4$ with HBr (p. 671) in terms of the Lewis acid-base concept.

**21.29** A **carbonium ion** is a cation in which the positive charge resides on a carbon atom. Use the results predicted by Markovnikov's rule for the reaction of HBr with 1-pentene and 2-methyl-2-butene to place the following carbonium ions in order of increasing stability. Explain your reasoning.

(a) $CH_3{-}CH_2{-}CH_2{-}CH_2{-}\overset{(+)}{C}H_2$

(a primary carbonium ion)

(b) $CH_3{-}CH_2{-}CH_2{-}\overset{(+)}{C}H{-}CH_3$

(a secondary carbonium ion)

(c) $CH_3{-}\underset{(+)}{\overset{CH_3}{\overset{|}{C}}}{-}CH_2{-}CH_3$

(a tertiary carbonium ion)

**21.30** What are some uses of halogenated hydrocarbons?

**21.31** Examine the labels of a number of common household insecticides and make a list of their active ingredients. Identify the kinds of functional groups present in these ingredients.

**21.32** Compare nucleophilic substitution and elimination reactions for bromoethane.

**21.33** Name the following compounds using IUPAC rules.

(a)

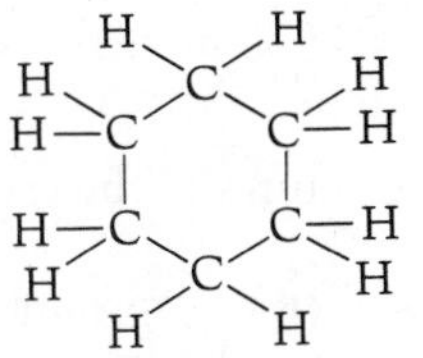

(b)

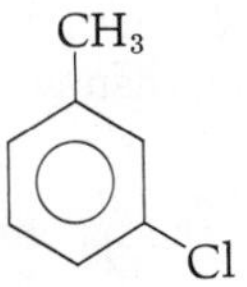

(c) $CH_3{-}CH_2{-}\overset{O}{\overset{\|}{C}}{-}CH_3$

(d) $CH_3{-}\overset{CH_3}{\overset{|}{C}H}{-}CH_2{-}NH_2$

**21.34** Using Markovnikov's rule, complete the following addition reactions. Name the reactants and products.

(a) $CH_3{-}CH_2{-}CH{=}CH_2 + HI \longrightarrow$

(b) $CH_3{-}CH{=}CH_2 + H_2O \xrightarrow[H_2SO_4]{}$

(c) $CH_3{-}CH_2{-}CH{=}C(CH_3)_2 + H_2O \xrightarrow[H_2SO_4]{}$

**21.35** Among the alcohols,

$$CH_3{-}(CH_2)_x{-}CH_2OH$$

as x increases the molar solubility in water decreases. Why does this occur?

**21.36** What products (if any) are obtained by the *mild* oxidation, using $K_2Cr_2O_7$ in acid solution, of each of the following:

(a) $CH_3{-}CH_2{-}OH$

(b) $CH_3{-}\overset{CH_3}{\overset{|}{C}H}{-}CH_2{-}OH$

(c) $CH_3{-}\underset{OH}{\underset{|}{\overset{CH_3}{\overset{|}{C}}}}{-}CH_3$

(d) $CH_3{-}\overset{\overset{OH}{|}}{C}H{-}CH_3$

(e) $CH_3{-}CH_2{-}CHO$

(f) $CH_3{-}COOH$

(g) $CH_3{-}\overset{\overset{O}{\|}}{C}{-}CH_3$

**21.37** What chemical reactions discussed in this chapter would be used to distinguish between the following:
(a) 2-propanol and 2-methyl-2-propanol
(b) 1-butanol and 2-butanol
(c) *n*-butane and 1-butene
(d) ethanal and 2-propanone

**21.38** Vanillin (p. 677) contains three types of functional groups. What are they?

**21.39** Draw the structures of the esters formed from
(a) acetic acid and 2-propanol
(b) acetic acid and 1-pentanol
(c) benzoic acid and methanol
(d) formic acid (methanoic acid) and methanol

**21.40** Write equations for the saponification of:

(a) $CH_3{-}CH_2{-}\overset{\overset{O}{\|}}{C}{-}O{-}CH_3$

(b) $CH_3{-}CH_2{-}O{-}\overset{\overset{O}{\|}}{C}{-}CH_2{-}CH_2{-}\overset{\overset{O}{\|}}{C}{-}O{-}CH_2{-}CH_3$

**21.41** The synthesis of organic compounds from simple starting materials is an important aspect of organic chemistry. From the reactions described in this chapter, describe how you could prepare
(a) dichloroethane from ethene
(b) propanoic acid from 1-propanol
(c) 2-propanol from 1-chloropropane
(d) ethyl acetate from ethanal
(e) methyl ethyl ketone from 1-bromobutane

**21.42** Predict the results of the following reactions:

(a) $CH_3{-}CH_2OH + CH_3{-}\overset{\overset{O}{\|}}{C}{-}OH \xrightarrow{H^+}$

(b) $CH_3{-}CH_2{-}CH_2{-}OH \xrightarrow[\Delta]{KMnO_4}$

(c) $CH_3{-}CH_2{-}\overset{\overset{OH}{|}}{C}H{-}CH_3 \xrightarrow{KMnO_4}$

(d) $CH_3{-}CH_2{-}CH_2{-}CH_2{-}OH \xrightarrow[conc]{H_2SO_4}$

**21.43** Write a chemical equation showing the saponification of propyl acetate. What drives this reaction toward completion?

**21.44** Compare the products obtained by reduction of an aldehyde and a ketone with hydrogen.

**21.45** Diethylamine gives a basic solution in water. Why?

**21.46** Styrene (phenylethene) forms an addition polymer. Sketch the structure of the repeating unit in polystyrene.

**21.47** What is the difference between addition polymerization and condensation polymerization?

**21.48** *Saran* is an addition copolymer of vinyl chloride and vinylidene chloride, $CH_2{=}CCl_2$. Sketch a portion of the polymer chain showing several monomer units.

**21.49** If ethylene glycol and dimethylmalonate were to form a condensation polymer by the elimination of methanol, what would be the structure of the repeating unit in the polymer chain?

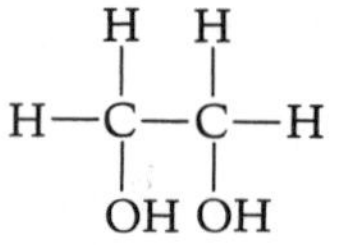

**ethylene glycol**

$CH_3{-}O{-}\overset{\overset{O}{\|}}{C}{-}CH_2{-}\overset{\overset{O}{\|}}{C}{-}O{-}CH_3$

**dimethylmalonate**

**21.50** Nylon stockings appear practically to disintegrate when hydrochloric or sulfuric acid is accidentally spilled on them. Actually, they dissolve very rapidly. Can you suggest what occurs chemically when nylon is dissolved by acid?

**21.51** What is cross-linking? What effect does it have on the physical properties of a polymer?

# 22 BIOCHEMISTRY

Without question, one of the most active areas of chemical research today is the field of biochemistry which, as its name implies, is concerned with the chemistry that takes place in living systems. The modern biochemist views a living organism as a collection of organic molecules that interact with each other and with their environment in a very unique and special way. When isolated from a living system these biomolecules are themselves lifeless. They obey all of the laws of chemistry and thermodynamics that we have examined up to now, and it is the goal of the biochemist to understand the functions and intricate interactions of these molecules that give rise to the phenomenon that we call life.

The field of biochemistry is very large and quite complex, and we certainly cannot hope to explore it fully in a single chapter. Instead, we shall be content to examine some of the types of biomolecules and their apparent functions in the operations of a living cell.

Nearly all of the compounds found in a living system are composed primarily of carbon, hydrogen, oxygen, nitrogen, and some sulfur and, for the most part, they depend on carbon for their molecular backbone. In general these molecules are very large, having molecular weights ranging up to a million or more. We shall see, however, that in many cases these **macromolecules** are constructed using a relatively small number of different simple molecules. For the purpose of discussing the various kinds of biomolecules we can place nearly all of them into one or another of four classes: **proteins, carbohydrates, lipids,** and **nucleic acids.**

Before we proceed, remember that you are not expected to memorize the names and formulas of all of the different compounds that we will discuss in this chapter. Instead, concentrate your attention on the types of molecules that are involved, the way that they combine with each other, and the general features of the structures that result.

## 22.1 PROTEINS

Proteins are very large molecules having molecular weights ranging from about 6000 to approximately 1,000,000. They constitute nearly 50% of the dry weight of cells and, depending on the individual protein, serve a variety of different functions within a living organism. Some of them are hormones, like insulin, which serve as chemical messengers that coordinate certain biochemical activities. Insulin, for example, controls the level of sugar in the bloodstream by promoting its absorption into cells. Others are enzymes that act as catalysts for biochemical reactions. We shall discuss these further in the next section. Some proteins serve to transport substances through the organism. Hemoglobin, for instance, carries oxygen in the bloodstream and delivers it to different parts of the body. There are long fibrous proteins, such as *actin* and *myosin,* that are found in muscle. Another fibrous protein, α-*keratin,* serves as the major constituent of hair, nails, and skin, while *collagen* is the prime constituent of tendons. Proteins are also found in toxins (poisonous materials) such as botulinus toxin as well as in antibodies.

Despite their wide range of functions, all proteins have something in common with one another. They are polymers made up by linking together in various combinations, a number of different simple monomeric units called **α-amino acids.**

An amino acid is a bifunctional organic molecule that contains both a carboxyl group, —COOH, as well as an amine group, —NH$_2$. In an α-amino acid the amine group is located on the carbon atom adjacent to the carboxyl group (the α-carbon atom). This gives a structure that we can generalize as

$$R-\overset{\overset{\displaystyle H}{|}}{\underset{\underset{\displaystyle NH_2}{|}}{C}}-COOH$$

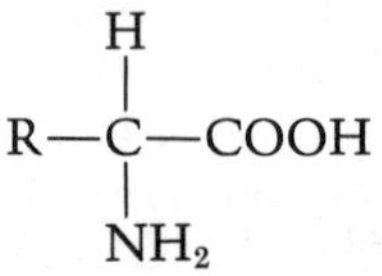

Since the —NH$_2$ group (like ammonia) is basic and the —COOH group is acidic, in neutral solution the amino acid exists in an internal ionic form called a *zwitterion* where the proton of the —COOH group is transferred to the NH$_2$ to give

$$R-\overset{\overset{\displaystyle H}{|}}{\underset{\underset{\displaystyle \overset{\ominus}{N}H_3}{|}}{C}}-COO^{\ominus}$$

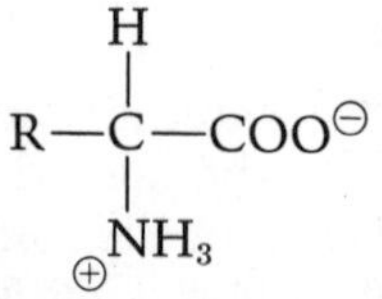

A very interesting and important fact is that in all organisms nearly all proteins are constructed using as building blocks a set of only 20 α-amino acids. These are shown in Figure 22.1. Note that except for proline, all fit the general formula above. Another point of interest is that except for glycine, all of these amino acids have four different groups attached to the α-carbon atom, which is therefore an asymmetric carbon atom. Each of these amino acids can therefore exist in two different isomeric forms (optical isomers). In proteins, however, only one isomer of each is commonly observed to occur. Apparently substitution of one isomer for another destroys the biological activity of the protein molecule.

In a protein molecule, amino acids are linked together to form a long chain. This can be viewed as the result of the elimination of a water molecule from between the —NH$_2$ group of one amino acid molecule and the —COOH group of another,

$$H_2N-\overset{\overset{\displaystyle H}{|}}{\underset{\underset{\displaystyle R_1}{|}}{C}}-\overset{\overset{\displaystyle O}{\|}}{C}-\!\boxed{O-H} \quad \boxed{H}\!-N-\overset{\overset{\displaystyle H}{|}}{\underset{\underset{\displaystyle R_2}{|}}{C}}-COOH \xrightarrow{-H_2O}$$

$$H_2N-\overset{\overset{\displaystyle H}{|}}{\underset{\underset{\displaystyle R_1}{|}}{C}}\!-\!\overset{\overset{\displaystyle O}{\|}}{C}-N\!-\overset{\overset{\displaystyle H}{|}}{\underset{\underset{\displaystyle R_2}{|}}{C}}-COOH$$

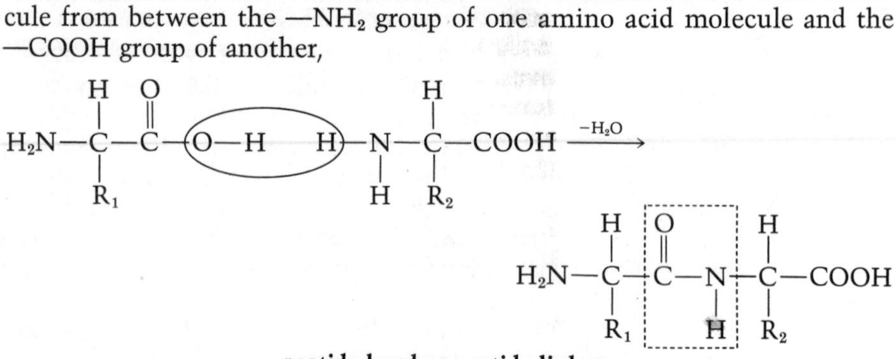

**peptide bond or peptide linkage**

The molecule that results is called a **peptide** and the group of atoms within the dotted line constitutes a **peptide bond** (amide bond), or **peptide linkage.**[1] In the particular example above, the peptide is composed of two amino acids and is said to be a **dipeptide.** Since one end of this molecule contains a carboxyl group and the other a free —$NH_2$ group, additional amino acids may be joined to give ultimately a **polypeptide,** a long chain composed of many amino acid molecules linked by peptide bonds. A segment of such a chain could be indicated as

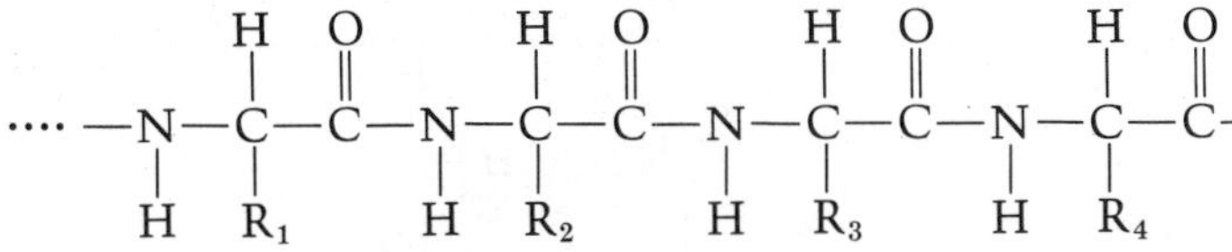

The backbone of the chain is thus the same series of atoms repeated over and over again, with only the R groups changing as we move along the chain.

Before we continue, we should note that in the formation of a protein the linking together of the different amino acids is not a random process. Each molecule of a given protein has the same sequence of amino acids along its polypeptide chain. In fact, it is this very sequence that imparts to a protein its own specific properties.

The amino acid sequence that exists in a polypeptide is called its **primary structure.** In addition to this, the polypeptide chain twists and turns and assumes a **secondary structure** that is determined by hydrogen bonding that occurs between different groups along the chain. An example of this is found in the fibrous protein, $\alpha$-keratin (the major component of hair) in which the polypeptide chains coil themselves into the **$\alpha$-helix,** shown in Figure 22.2. The hydrogen bonding, in this case, takes place between the oxygen atom in a carbonyl group ($>C{=}O$) and a hydrogen atom attached to a nitrogen atom that lies in an adjacent loop of the helix (Figure 22.3). This therefore serves to hold the chain in its coiled shape.

In globular proteins, so named because of their overall shape, coiled polypeptide chains are also folded to give a complex three-dimensional structure, referred to as its **tertiary structure.** This is shown in Figure 22.4 for the protein myoglobin, a substance that stores oxygen in muscle tissue until it is needed in metabolic oxidation. It is the presence of large amounts of myoglobin in the leg and thigh muscles of birds, for instance, that gives this meat a darker color than the breast meat. Myoglobin is also responsible for the red color of beef steak.

The tertiary structure of a protein is controlled by several different kinds of interactions that serve to hold the folded segments of the chain in place. For example, besides hydrogen bonding there are also ionic attractions that occur between a negatively charged deprotonated carboxyl group (like that found in the R group of glutamic acid) and a positively charged protonated amine group (like that found in lysine). This is shown in Figure 22.5.

The solvent is also important in determining the shape of the protein molecule. In the presence of the polar solvent water, nonpolar R groups such

[1] Perhaps an interesting point to note here is that the peptide linkages that holds the amino acids together in the polypeptide chains are the same as the linkages that hold the monomer units together in that familiar synthetic polymer, *Nylon* (p. 683).

Figure 22.1
The 20 amino acids found in most proteins.

| Nonpolar | |
|---|---|
| Alanine (ala) | 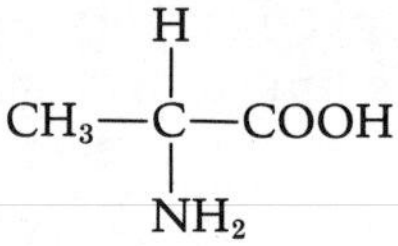 |
| Valine (val) | 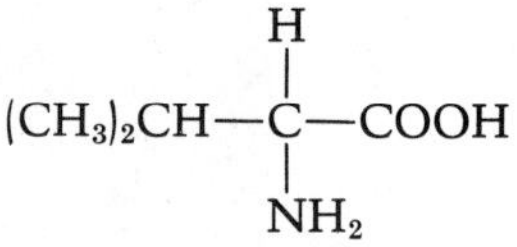 |
| Leucine (leu) | 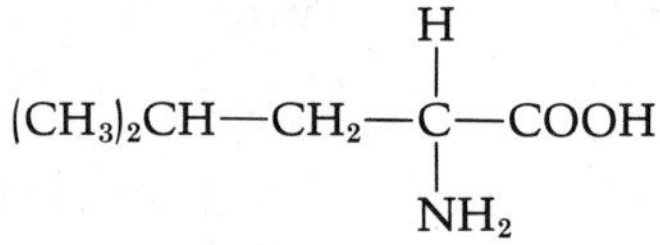 |
| Isoleucine (ile) | 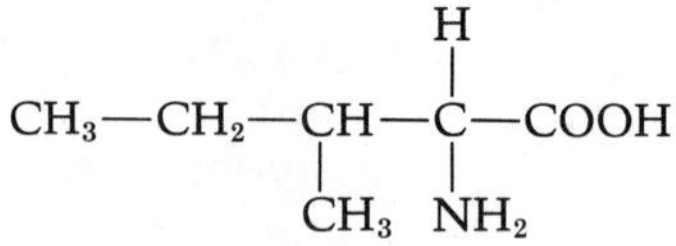 |
| Proline (pro) | 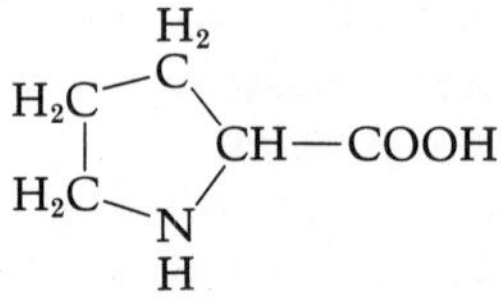 |
| Methionine (met) | 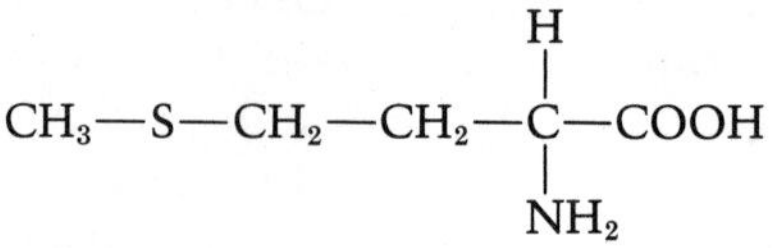 |
| Phenylalanine (phe) | 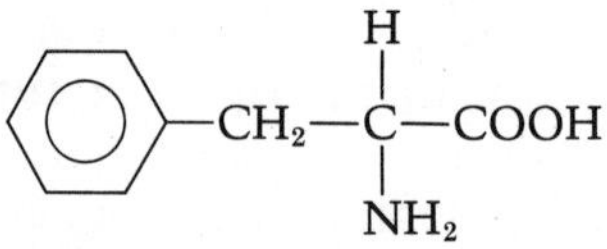 |
| Tryptophan (trp) | 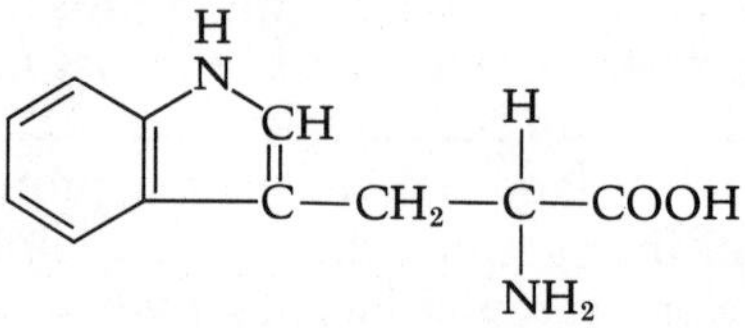 |

as the phenyl ring in phenylalanine (see Figure 22.1) are forced toward the center of the folded polypeptide chain, away from the solvent. This is the same phenomenon, you may remember, that leads to the low solubility of nonpolar substances in polar solvents. It too helps determine the tertiary structure of proteins because the polypeptide chain tends to fold in such a way that nonpolar groups do not contact the solvent.

| Polar | | | |
|---|---|---|---|
| Glycine (gly) | 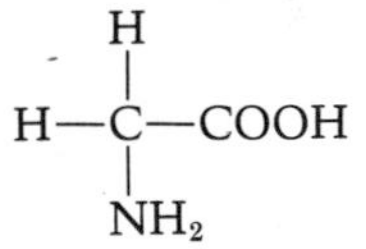 | Threonine (thr) | 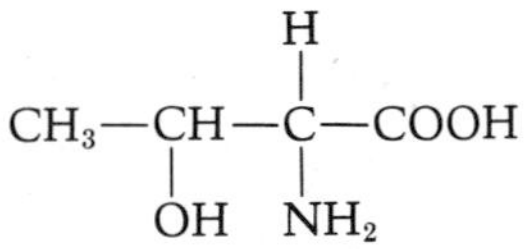 |
| Serine (ser) | 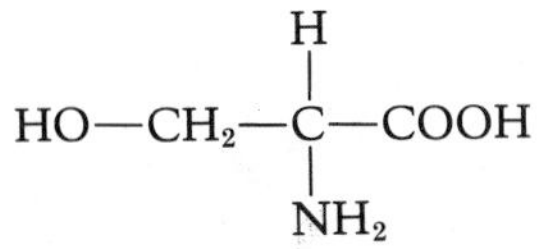 | Cysteine (cys) | 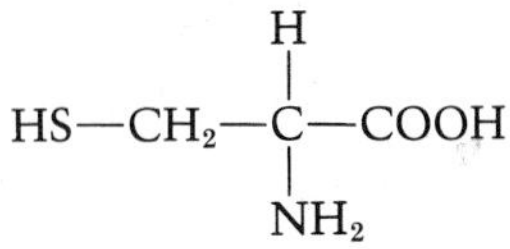 |
| Tyrosine (tyr) | 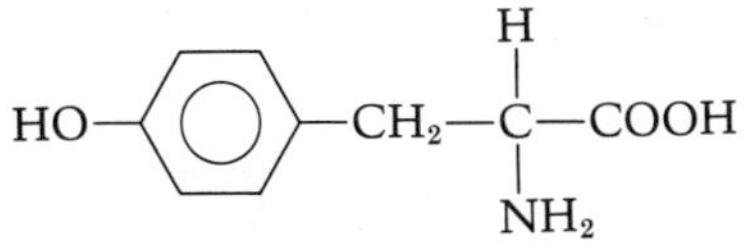 | Lysine (lys) | 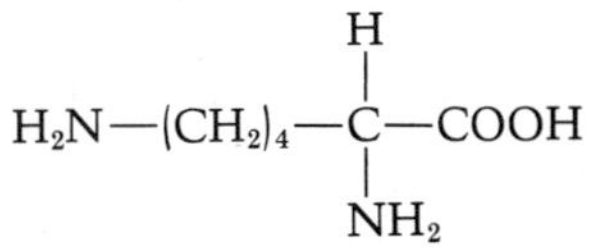 |
| Asparagine (asn) | 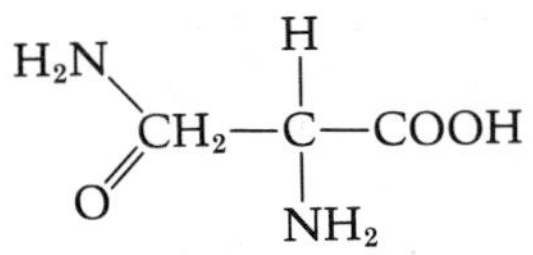 | Aspartic acid (asp) | 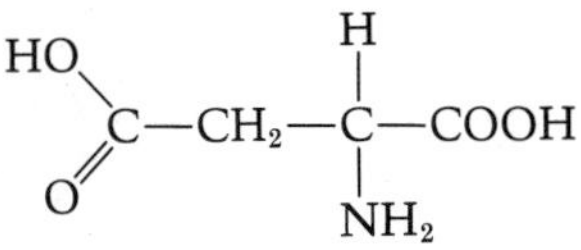 |
| Glutamine (gln) | 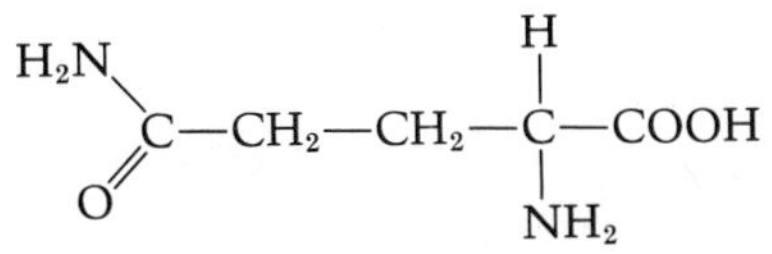 | Glutamic acid (glu) | 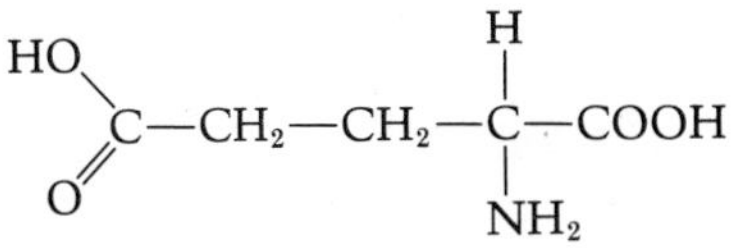 |
| Arginine (arg) | 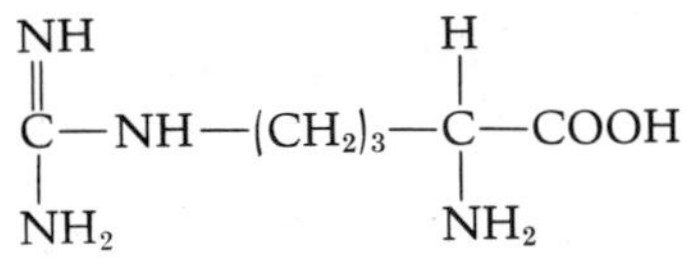 | Histidine (his) | 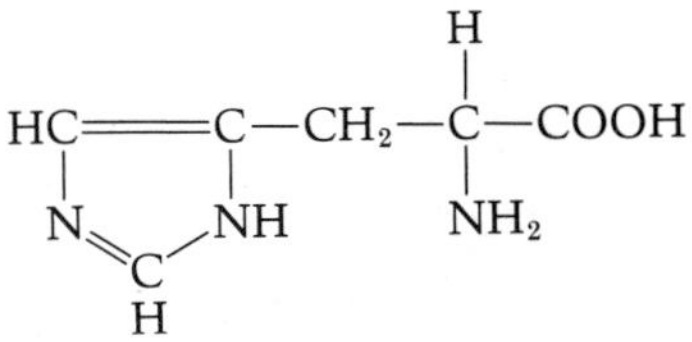 |

Still another type of interaction that maintains the folded conformation of the protein molecule is the formation of covalent bonds between cysteine molecules located at different points along the chain. This occurs by partial oxidation of the —SH (thiol) group,

$$R{-}SH + HS{-}R \xrightarrow{\text{oxidation}} H_2O + R{-}S{-}S{-}R$$

The resulting linkage is called a **disulfide bridge.**

Disulfide bridges not only help to keep the polypeptide chain folded but can also bind two such chains together. For example, beef insulin (Figure 22.6) consists of two polypeptide chains that are cross-linked at two points by these disulfide bridges.[2]

There is an interesting sidelight to the subject of the disulfide bridge in protein chemistry. The curl (or lack of curl) in hair is determined by protein conformation locked in place by disulfide bridges. The "permanent wave"

[2] The elucidation of the primary structure of this protein by Frederick Sanger won him the Nobel Prize in 1958.

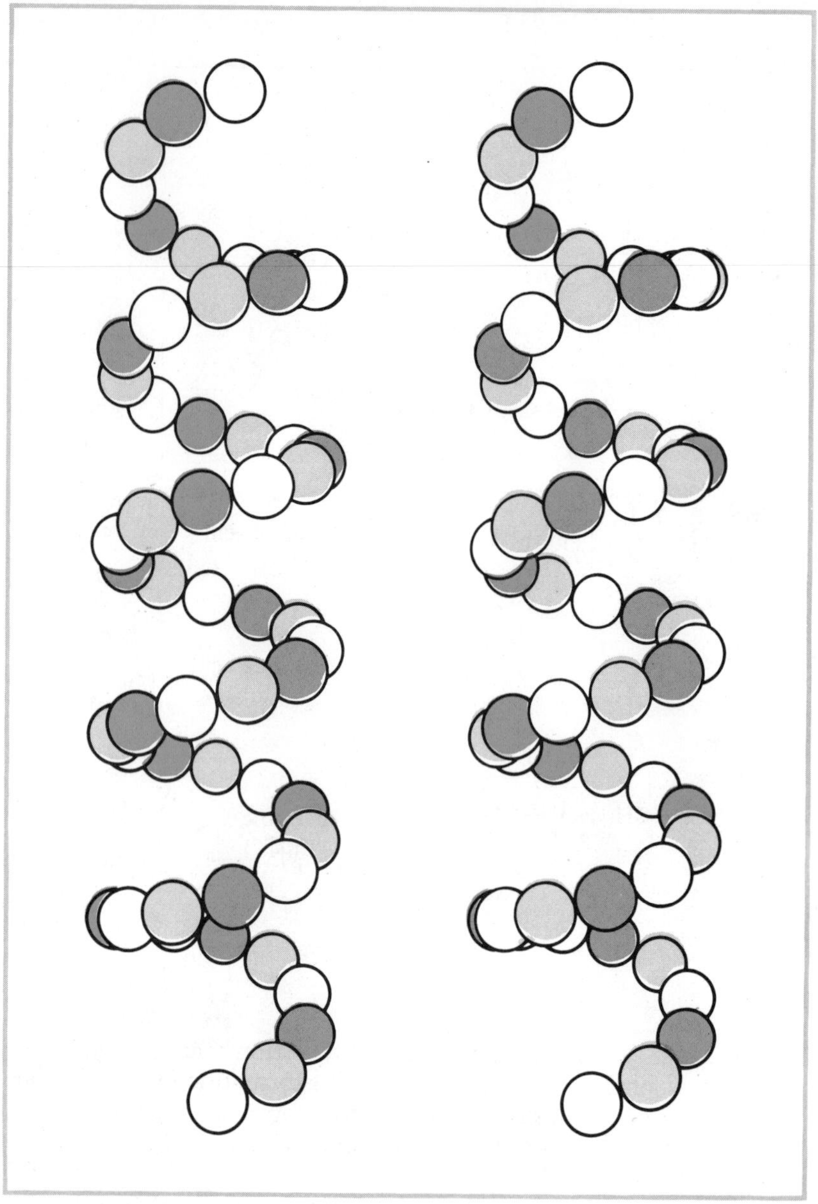

Figure 22.2

*The α-helix, composed of a repetition of amino acid units.*

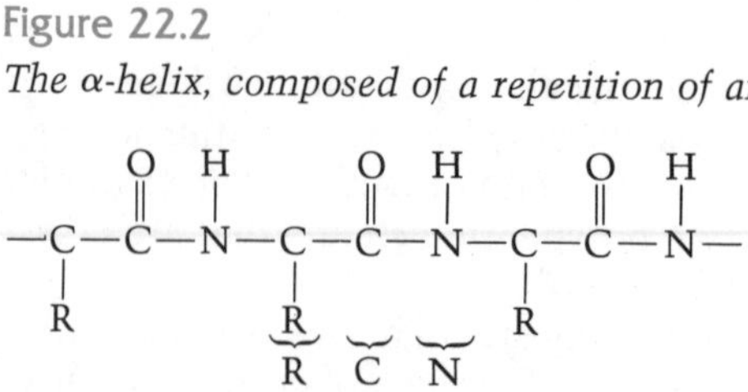


*In the illustration N is white, C is pale orange and R is dark orange.*

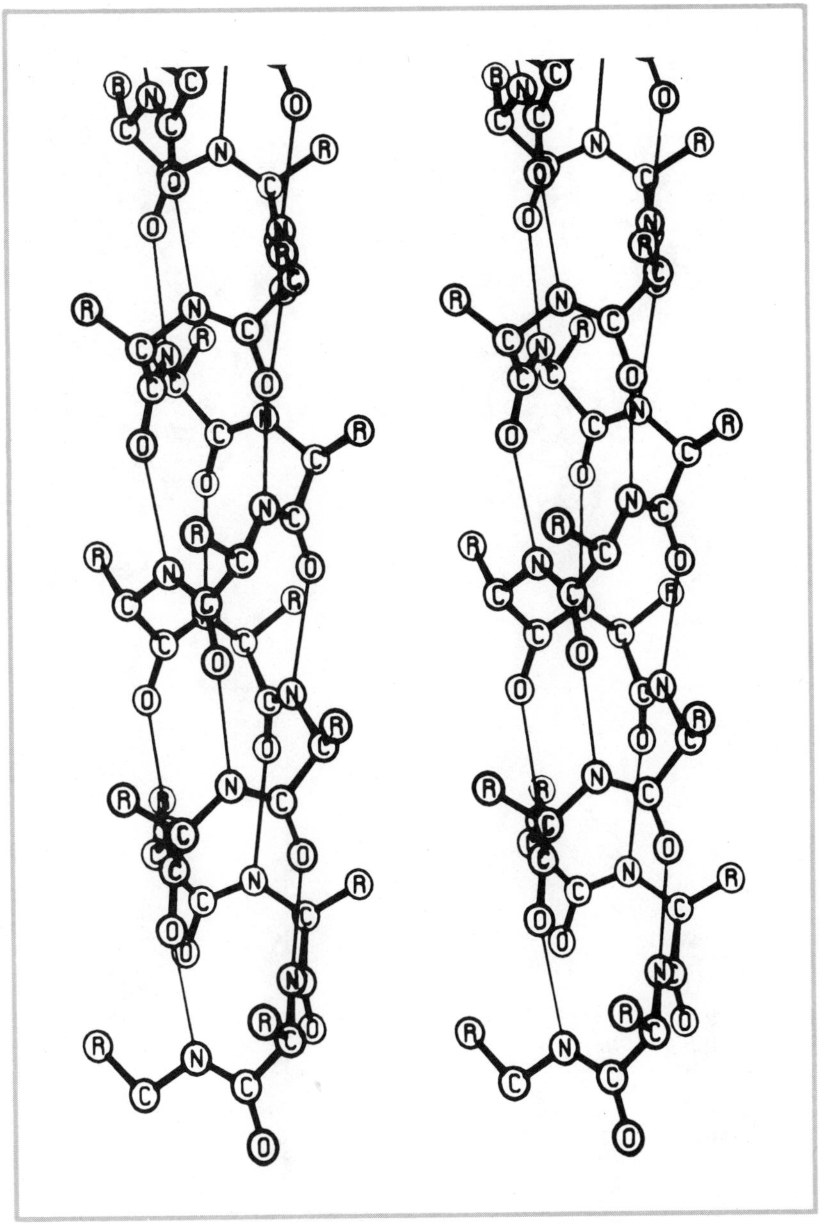


Figure 22.3
*Hydrogen bonding (light lines) in the α-helix. Courtesy of Carroll K. Johnson, Oak Ridge National Laboratory, Oak Ridge, Tennessee.*

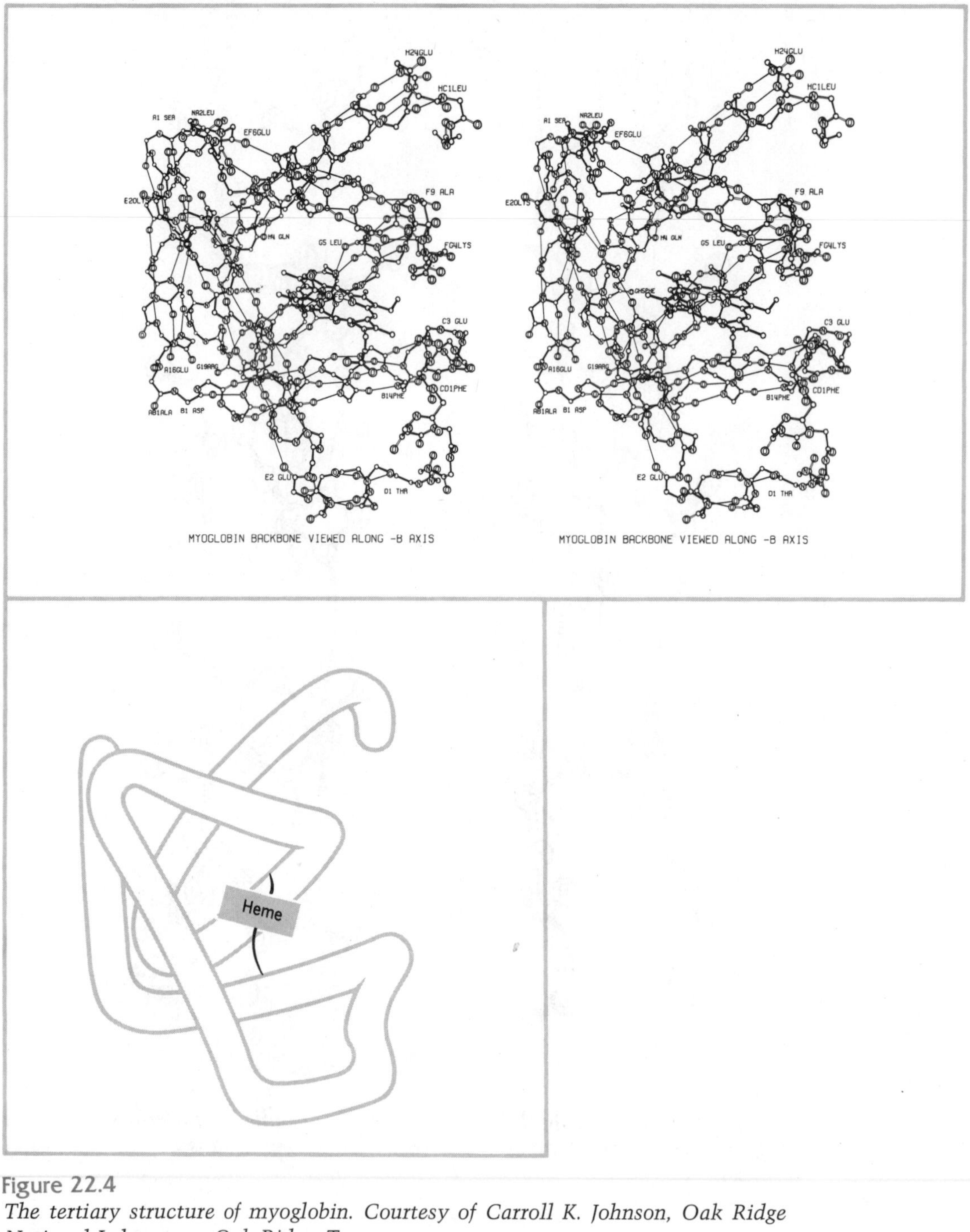


Figure 22.4
*The tertiary structure of myoglobin. Courtesy of Carroll K. Johnson, Oak Ridge National Laboratory, Oak Ridge, Tennessee.*

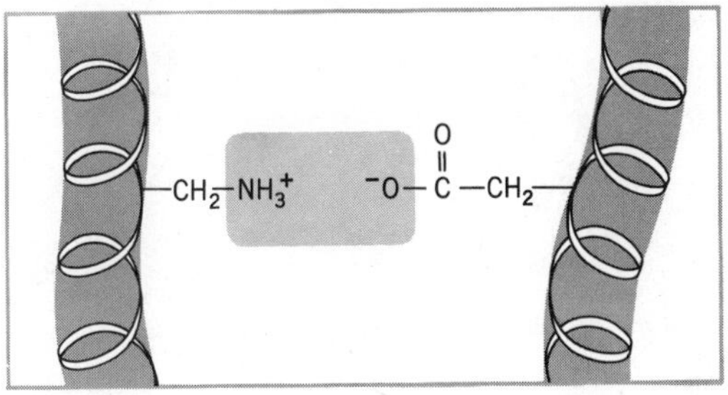


Figure 22.5
*Ionic interactions in determining the tertiary structure of proteins.*

treatment that women (and, recently, men too) use to produce curly hair involves two chemical reactions. The hair is first treated with a chemical able to break the disulfide bridges so that the protein chains in the hair are free to twist into any desired shape, as determined by the curlers. A second solution is then applied which produces a mild oxidation, thereby causing the disulfide bridges to be reestablished. These newly formed disulfide bridges "set" the hair in the newly curled shape. The only reason a permanent wave isn't really "permanent" is because new hair grows that hasn't received the setting treatment.

Proteins that contain more than one independent polypeptide chain exhibit still another degree of structural sophistication, called **quaternary structure.** This is determined by the way in which the folded chains orient themselves with respect to one another. A good example of this occurs in hemoglobin, Figure 22.7. This protein consists of four polypeptide chains; two α-chains each containing 141 amino acids and two β-chains each with 146 amino acids.

In addition to the polypeptide chains, this protein also contains groupings of atoms, called **heme groups,** that serve to bind oxygen so that it can be transported through the bloodstream and be deposited at oxygen-poor cells. As depicted in Figure 22.7, the four folded polypeptide chains with their heme groups are packed together in a roughly tetrahedral fashion.

The heme group in hemoglobin is also found in myoglobin and accounts for the red color of blood and muscle tissue. The structure of heme is

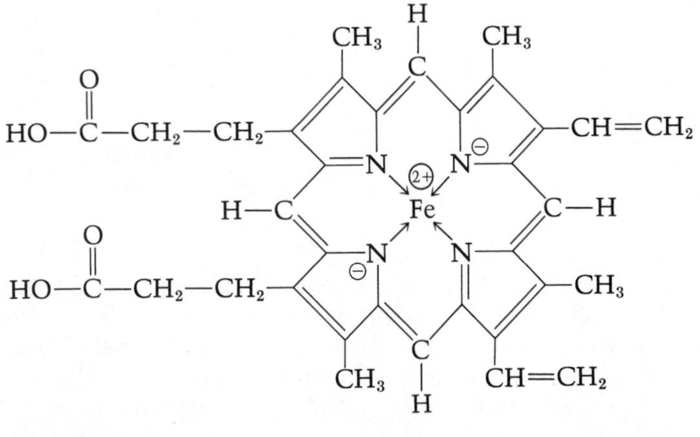

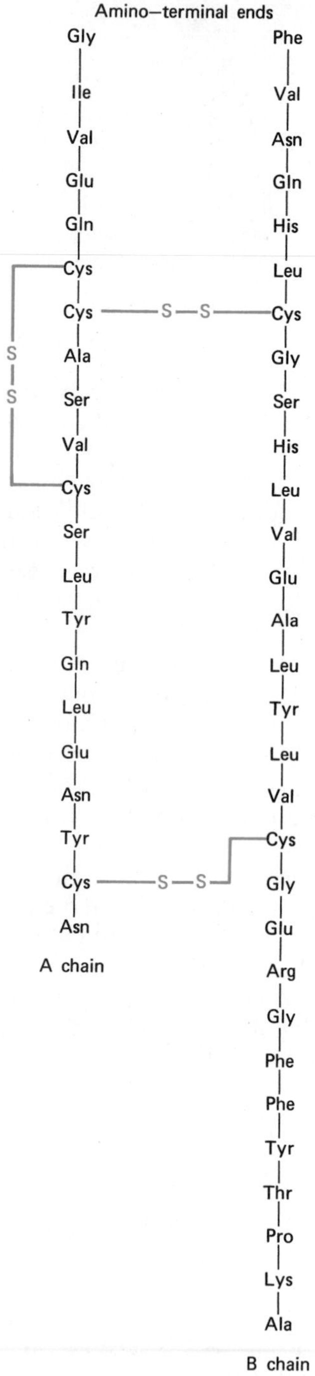


Figure 22.6
*Amino acid sequence in beef insulin.*

and is, in fact, a complex ion containing iron(II) enclosed within a square planar grouping of nitrogen atoms. In hemoglobin each heme group is attached to its polypeptide by additional coordination to the nitrogen atom of a histidine, as shown in Figure 22.8. The sixth coordination site about the iron(II) is empty and is used to bind an oxygen molecule.

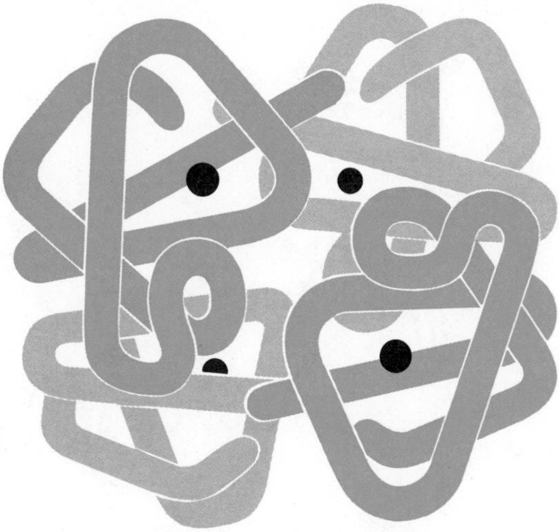

Figure 22.7
*Quaternary structure of hemoglobin. Four globular protein molecules containing heme groups are packed into hemoglobin structure. Adapted from R. E. Dickerson and I. Geis,* The Structure and Action of Proteins, *W. A. Benjamin, Inc., Menlo Park, California, 1969. Original illustration copyright 1969 by R. E. Dickerson and I. Geis.*

The basic square planar ligand structure in heme is called a **porphyrin** and forms very stable complexes with several different metal ions. For example, a structure very similar to heme containing $Mg^{2+}$ instead of $Fe^{2+}$ is found in chlorophyll, the green pigment found in plants that is used in photosynthesis. Still another porphyrin structure, containing $Co^{2+}$ in the center, exists in a substance called vitamin $B_{12}$ coenzyme (Figure 22.9). Thus, even though most of the structures of biomolecules are made up of carbon, hydrogen, nitrogen, and oxygen, some metals are also of critical importance to the well-being of a living organism.

## 22.2 ENZYMES

Enzymes are globular proteins that serve to catalyze specific biochemical reactions with what can only be judged as amazing effectiveness. In some cases reactions are speeded up, with respect to their uncatalyzed paths, by factors ranging from $10^9$ to $10^{20}$! Competing side reactions are not affected and are very slow by comparison. The result is that essentially 100% of the reactants are funneled through the same reaction path. In this way a buildup

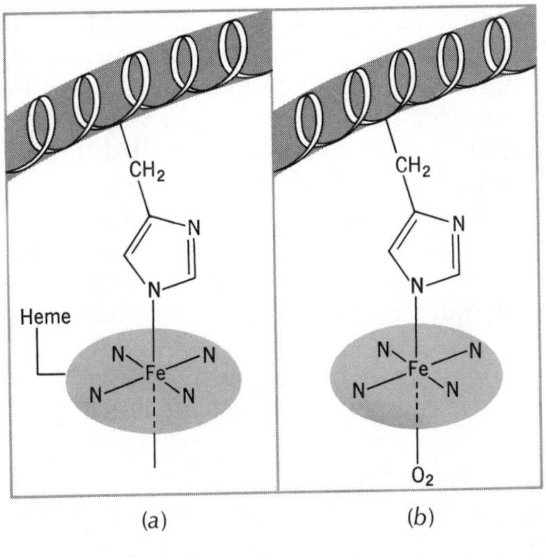


Figure 22.8
*Attachment of heme group to hemoglobin protein by coordination of histidine to the iron atom of heme. (a) Sixth coordination site is vacant. (b) When hemoglobin carries oxygen, the $O_2$ is bound to Fe of heme at the sixth coordination site.*

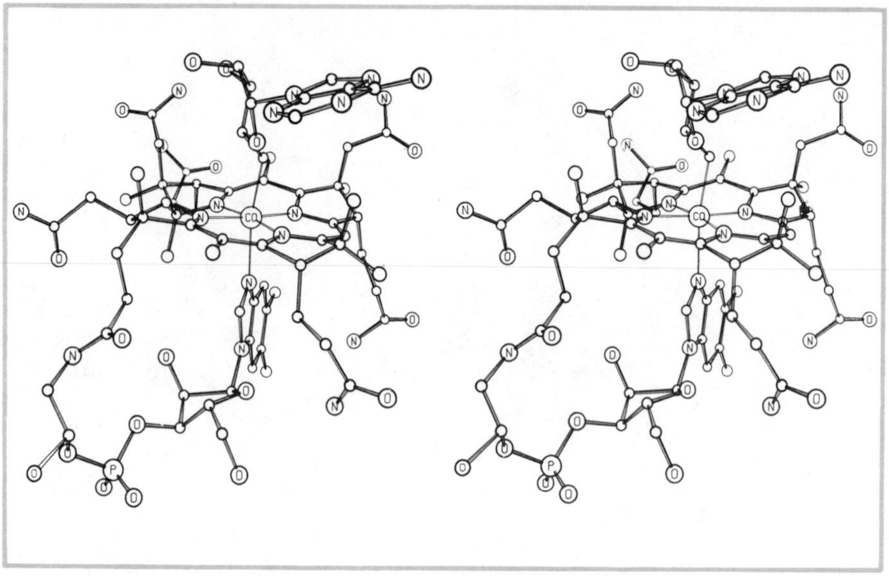

Figure 22.9
*Vitamin B-12 coenzyme. Courtesy of Carroll K. Johnson, Oak Ridge National Laboratory, Oak Ridge, Tennessee.*

of by-products that would cause a waste-disposal problem for the organism is avoided. Enzymes also provide the organism with a way of controlling the rates of the reactions that take place because biochemical reactions do not occur at appreciable rates in the absence of the catalyst. Removal, or at least temporary blockage, of a critically important enzyme "turns off" the chemistry which that particular enzyme catalyzes. Thus, in a very real sense enzymes direct the chemical reactions that take place in a living cell.

Some enzymes require an additional substance called a **coenzyme** in order to function. Many of the vitamins that we must ingest to maintain good health are precursors of coenzymes. An example is vitamin $B_{12}$, the absence of which from the diet leads to a deficiency disease known as pernicious anemia. In the body vitamin $B_{12}$ is converted to its coenzyme, the structure of which we saw in Figure 22.9. Many metals, like cobalt, are needed in small amounts to promote enzyme activity.

The mechanism by which an enzyme acts has long been the subject of intense research. It appears to depend on the ability of the enzyme to bind very selectively to a reactant molecule (called the **substrate** of the enzyme). There thus seems to be a "lock-and-key" relationship between an enzyme and its substrate, where the substrate molecule just precisely fits into (or onto) the folded globular protein. There is evidence that when this occurs, there is a slight alteration in the shape of the enzyme which strains certain key bonds in the substrate, thereby making them more susceptible to chemical attack.

Some enzymes are very specific in their activity, affecting the rate of reaction of a single compound. Others are less choosy and simply promote a certain kind of chemical reaction on a whole class of compounds having similar structures. This behavior, too, is the direct result of the lock-and-key relationship between the enzyme and its substrate. An example is the enzyme *chymotrypsin,* which accelerates the hydrolysis of the dotted bond in the compounds,

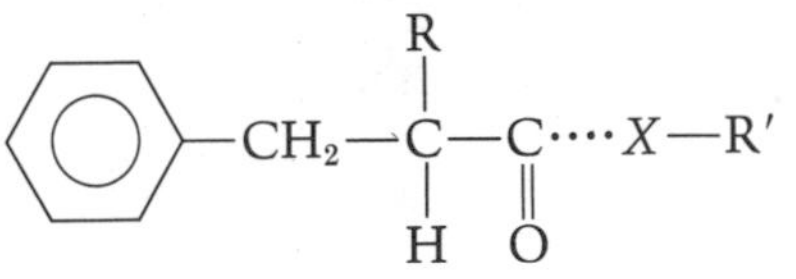

where $X$ = N or O

An example is

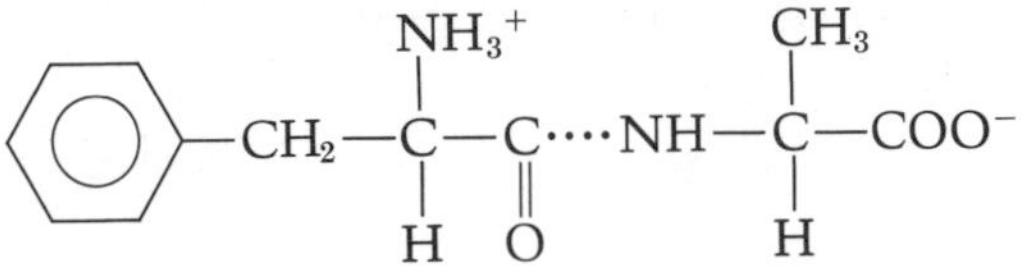

In this case it is thought that the hydrophobic benzene ring serves to position the substrate molecule on the enzyme (Figure 22.10), which then interacts with the >C=O group in a way that makes the dotted bond more susceptible to hydrolysis. Since the function of the enzyme depends on the hydrophobic tail and the proper location of the >C=O group, a family of similar compounds are affected by the enzyme.

**ENZYME INHIBITION.** When a substance other than the enzyme substrate becomes bound to the active site of an enzyme, the catalytic activity is lost and the enzyme is said to be inhibited. In some cases this inhibition is irreversible. This occurs when the inhibitor becomes permanently bound to the enzyme, by covalent bond formation, and cannot be displaced. An example

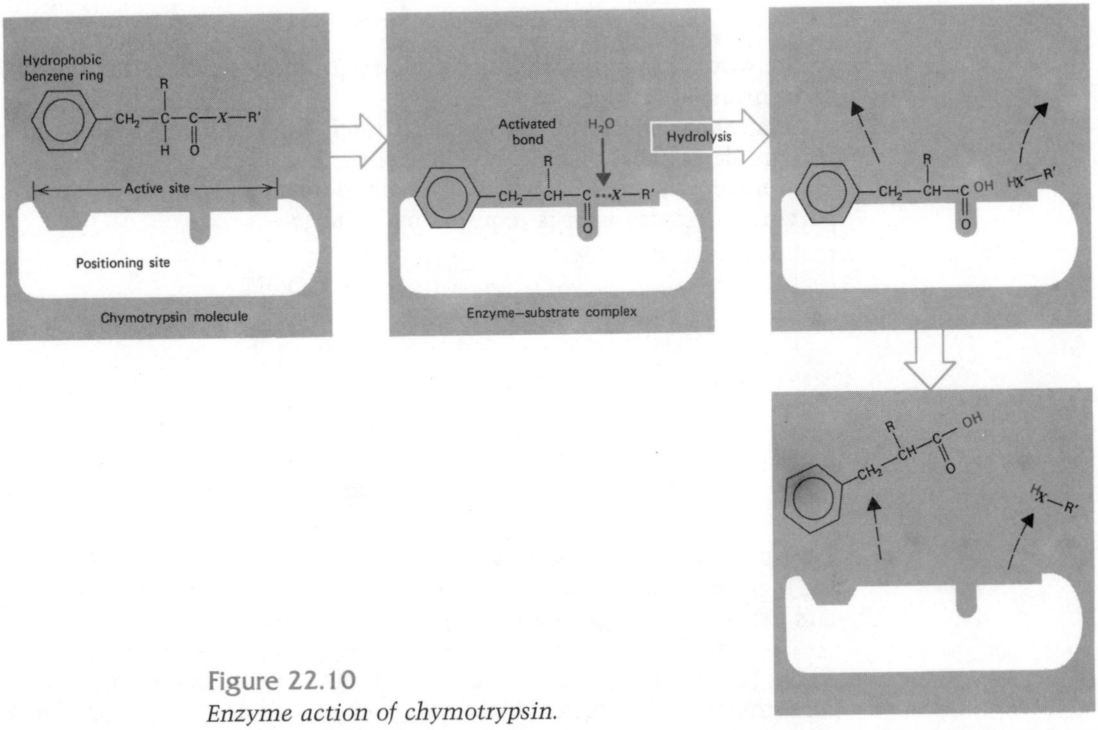


Figure 22.10
*Enzyme action of chymotrypsin.*

is diisopropylfluorophosphate,

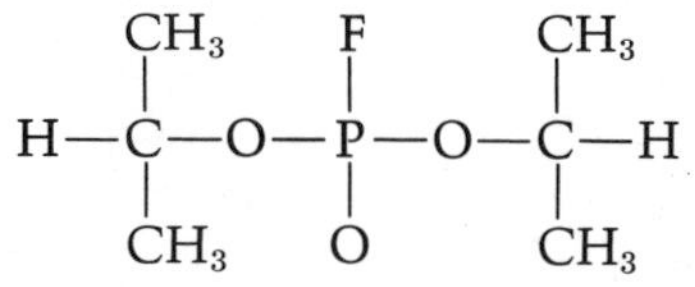

a highly toxic nerve poison and an ingredient in some nerve gases. This molecule reacts with and poisons (inhibits) an enzyme called *acetylcholine esterase,* which is required for the transport of impulses along nerve tissue.

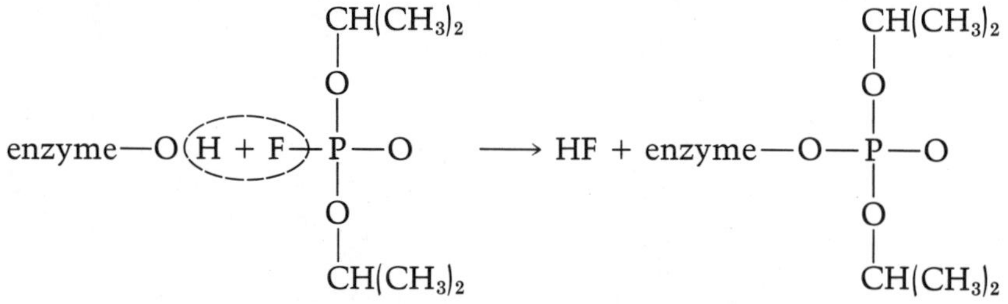

A second type of enzyme inhibition is called **competitive inhibition,** in which there is a competition between the inhibitor and the substrate for the enzyme active site. This is a system in which there are two simultaneous equilibria, one between the enzyme ($E$) and the substrate ($S$),

$$E + S \rightleftharpoons ES$$

and one between the enzyme and the inhibitor ($I$),

$$E + I \rightleftharpoons EI$$

As we would predict from Le Chatelier's principle, increasing the substrate concentration displaces the inhibitor.

An example of competitive inhibition is the action of the sulfa drug, sulfanilamide. This molecule, shown below, bears a very close similarity to *p*-aminobenzoic acid, which is acted upon by an enzyme to produce an important coenzyme that is required by bacteria.

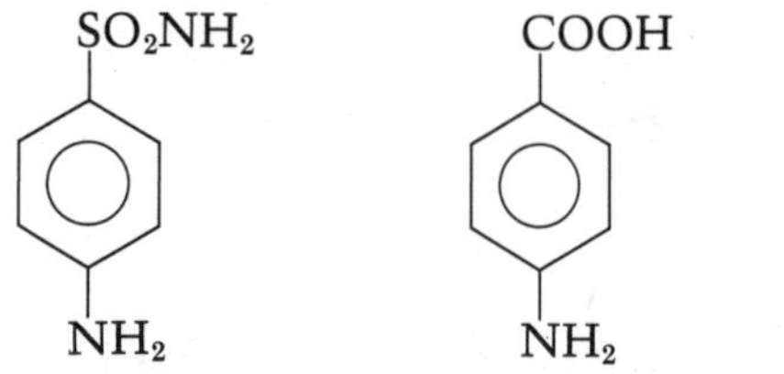

**sulfanilamide** **_p_-aminobenzoic acid**

The sulfanilamide, by occupying the active site of the enzyme that works on the *p*-aminobenzoic acid, prevents the production of the required coenzyme and hence leads to the demise of the bacterium.

As a final note, enzyme activity is also affected by temperature and pH. These factors alter the conformation and shape of the globular protein structure; therefore, changes in temperature or pH can destroy the precise fit that must exist between the enzyme and its substrate in order to obtain the desired catalytic activity.

## 22.3 CARBOHYDRATES

The carbohydrates form an important class of compounds that are used by living organisms in a variety of ways: as a source of energy, as a source of carbon to be used in the synthesis of other biomolecules, and as a structural element in cells and tissues. Historically the name carbohydrate arose as a consequence of the empirical formula exhibited by many of them, $C_n(H_2O)_m$, which suggested that they were *hydrates of carbon.* Examples are glucose, $C_6H_{12}O_6$, having the empirical formula $CH_2O$, and sucrose (ordinary cane sugar), $C_{12}H_{22}O_{11}$, with the empirical formula $C_{12}(H_2O)_{11}$. The name carbohydrate has remained with these substances even though it is now known that they do not contain intact water molecules.

Most carbohydrates, such as starch and cellulose, are very large molecules having enormous molecular weights. However, like the proteins, they are composed of many relatively simple units arranged in long polymeric chains. The simplest such units are called **monosaccharides,** and constitute the **simple sugars.** As a class, the monosaccharides are polyhydroxy aldehydes or ketones, the simplest of which is glyceraldehyde,

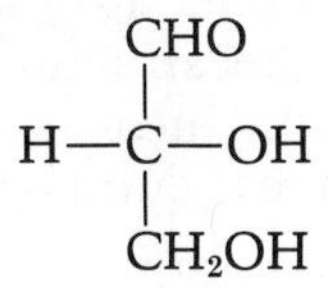

Glyceraldehyde is a **triose,** *tri* denoting three carbon atoms and *ose,* the characteristic ending used in naming the sugars (for example, gluc*ose,* sucr*ose,* and fruct*ose*).

Glyceraldehyde, like the other saccharides, contains an asymmetric carbon atom and exhibits optical isomerism. In the two-dimensional structural formulas written for the saccharides, the H and OH units that are attached to the asymmetric carbon atoms project upward from the paper while the bonds to other carbon atoms project downward. For example, the two optical isomers of glyceraldehyde are shown in Figure 22.11.

Among the saccharides it is generally observed that one optical isomer is significantly more important than the others. Glucose, for example, contains four asymmetric carbon atoms (indicated by asterisks).

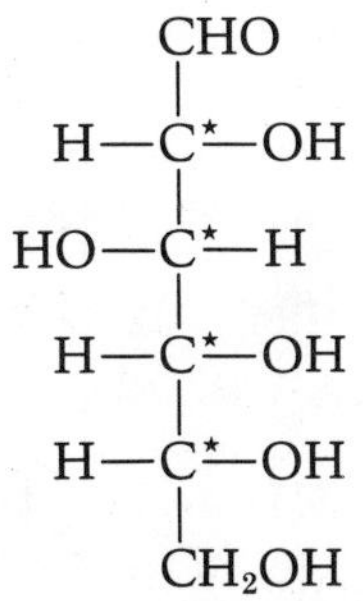

**D-glucose**

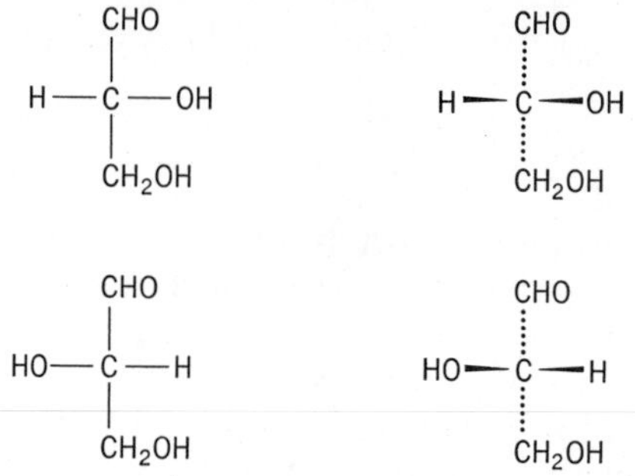

Figure 22.11
*Optical isomers of glyceraldehyde.*

There are 16 possible optical isomers of glucose, the most important one being D-glucose, shown above.

By far the most important monosaccharides are those containing five and six carbon atoms, the pentoses and hexoses, respectively. Some of the more prominent ones are found in Table 22.1. The most common hexose is glucose, whose structure is shown above. Glucose, however, like most of the other pentoses and hexoses, exists predominantly in a cyclic structure in which the molecule turns on itself as shown in Figure 22.12. When the ring is closed, the —OH group that is created from the aldehyde functional group can point either up or down (this is the —OH group on the rightmost carbon atom in the structures drawn in Figure 22.12). Two isomers are thus created, α-D-glucose and β-D-glucose. As we shall see, the orientation of this —OH group is quite significant in the polysaccharides, starch, and cellulose.

Table 22.1
*Some important monosaccharides*

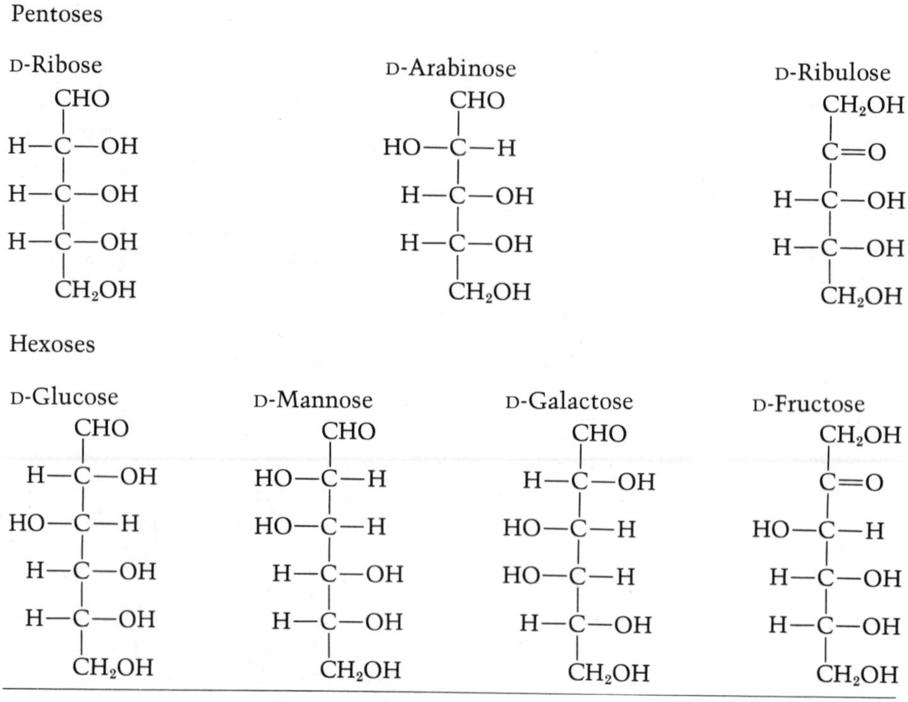

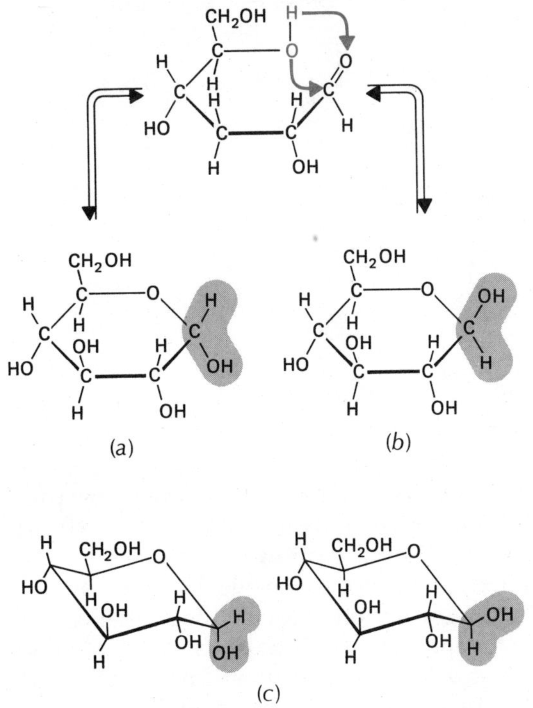


Figure 22.12
*Cyclic structures for glucose. (a) α-D-glucose. (b) β-D-glucose. (c) Puckered ring. Note orientation of H and OH on rightmost carbon.*

Another important six-carbon sugar is fructose. In its open-chain structure the molecule is a ketone.

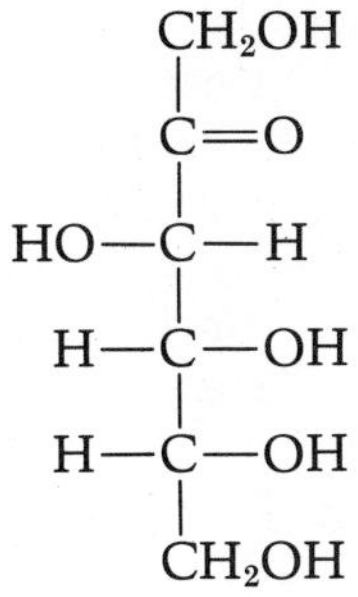


Like glucose, however, it too prefers a cyclic structure, as shown in Figure 22.13. In this case a five-membered ring is formed.

The five-membered ring also occurs in two very important pentoses, **ribose** and **deoxyribose** (Figure 22.14), sugars that are part of the backbone of RNA and DNA, respectively. We'll examine the structures of these in Section 22.5.

In the more complex sugars and the polysaccharides, monosaccharide units are condensed together by way of C—O—C bridges called **glycoside linkages.** Sucrose, for example, is a disaccharide consisting of a glucose and a fructose unit joined by eliminating $H_2O$ from an —OH group on the glucose and an —OH group on the fructose. This is illustrated in Figure 22.15. As indicated, addition of $H_2O$ to the glycoside linkage (hydrolysis) splits the sucrose molecule into the simple monosaccharides from which it is formed. This hydrolysis reaction is accelerated by the presence of dilute acid and, in

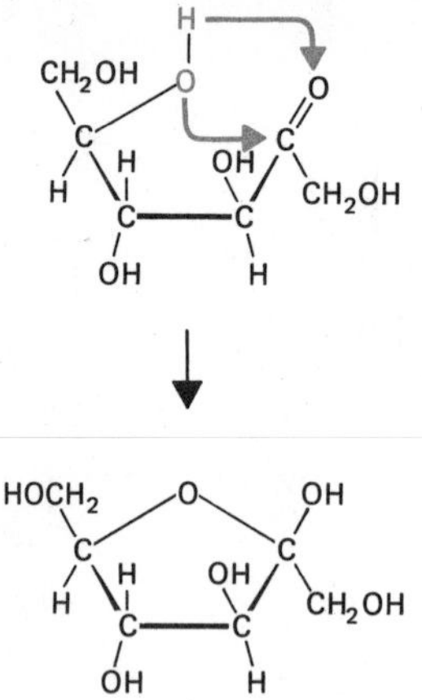

Figure 22.13
*Cyclic structure for fructose.*

general, polysaccharides can be broken down into their simple sugars by this reaction. Special enzymes in saliva start "digesting" carbohydrates in the mouth by this hydrolysis reaction.

The formation of two glycoside linkages by a single monosaccharide unit permits the formation of long polymeric chains called polysaccharides, the two most important of which are starch and cellulose. Starch (amylose) is composed of $\alpha$-D-glucose units strung together, while cellulose is composed of $\beta$-D-glucose units, as shown in Figure 22.16.

The difference between these two structures is rather subtle, but nevertheless has very profound effects. In starch the polysaccharide chains tend to coil in a helical structure with the polar —OH groups pointing outward. When placed into water these —OH groups on the starch molecule interact strongly with the polar solvent and cause the starch to be slightly water-soluble. Cellulose, on the other hand, forms linear chains that interact with each other via hydrogen bonding. This phenomenon gives wood, which is composed of approximately 50% cellulose, its structural strength.

The relatively minor structural differences between starch and cellulose also account for the fact that starch can be digested by humans but cellulose cannot. In the digestive tract the starch molecule is hydrolyzed enzymatically, which requires a certain fit between the carbohydrate molecule and the enzyme. With cellulose this necessary fit is not achieved and, hence, cellulose is unaffected. In termites, cows, and many other animals, however, cellulose is hydrolyzed and digested with the aid of bacteria in their digestive tract.

## 22.4 LIPIDS

A third class of biomolecules is made up of the **lipids,** water-insoluble substances that can be extracted from other cell components by nonpolar organic solvents (hydrocarbon solvents, carbon tetrachloride, etc.). Lipids serve mainly as storage of energy-rich fuel for use in metabolism (for example, in fats) and as a major structural element in cell membranes.

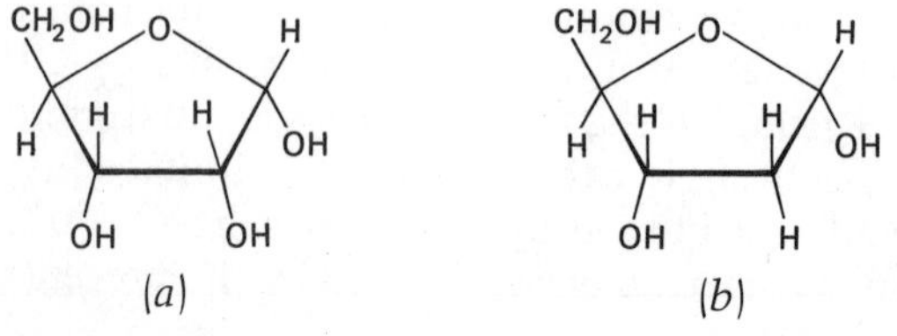

Figure 22.14
*(a) Ribose and (b) deoxyribose.*

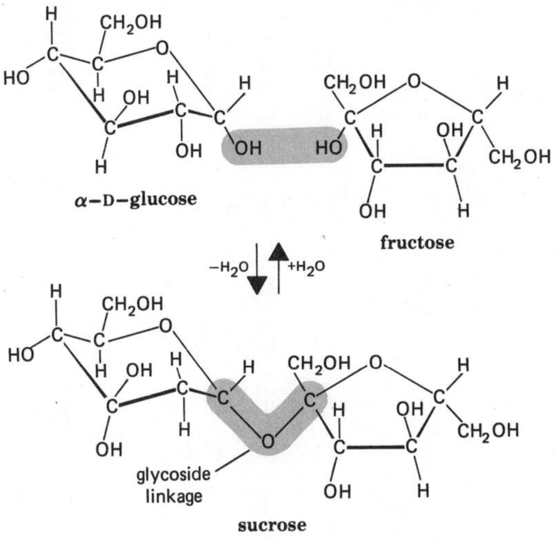


Figure 22.15
*Sucrose.*

As was true with the proteins and carbohydrates, most lipids are composed of simpler substances. The primary building blocks of the lipids are called **fatty acids,** long unbranched hydrocarbon chains, from 12 to 28 carbon atoms long, terminated at one end with the carboxyl group characteristic of organic acids. Nearly all of the naturally occurring fatty acids have an even number of carbon atoms and occur with both saturated and unsaturated chains. Some typical examples are shown in Table 22.2.

Most lipids can be classed as either **neutral lipids** or **polar lipids.** Fats, for example, are neutral lipids and are esters of the fatty acids with the alcohol, glycerol.

Figure 22.16
*Structures of the polysaccharides, starch and cellulose. (a) Amylose (starch). (b) Cellulose.*

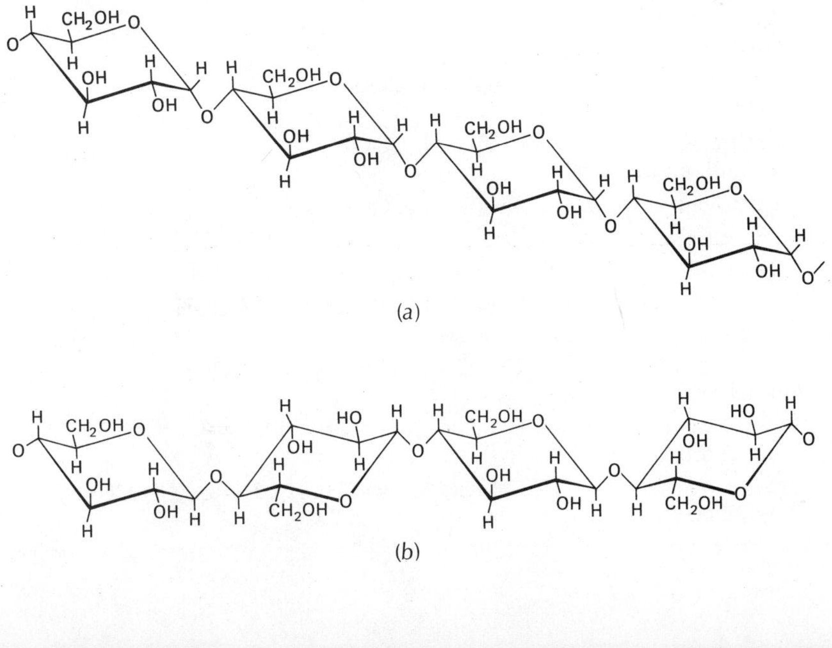

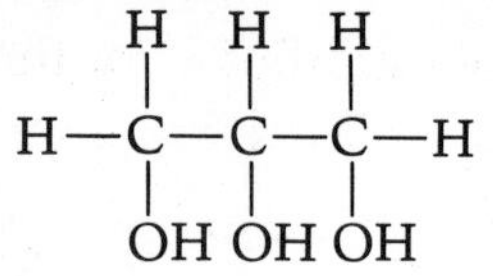

The resulting triester is called a **triglyceride.**

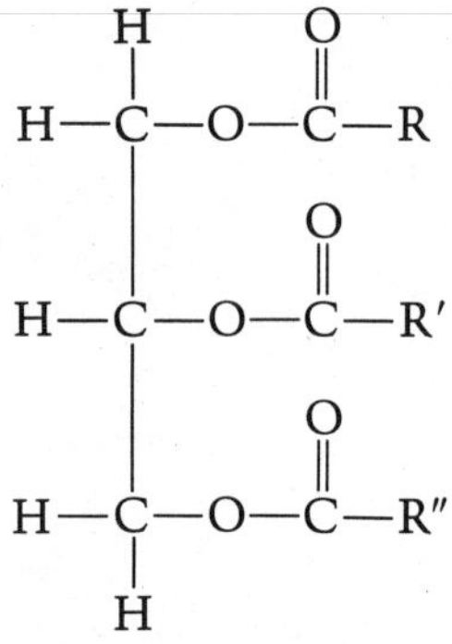

As you might expect, many different triglycerides are found to occur, as determined by the nature and location of the fatty acids attached to the glycerol molecule. Lipids containing saturated fatty acids, such as tristearin

Table 22.2
*Some naturally occurring fatty acids*

| Fatty Acid | | Melting Point (°C) |
|---|---|---|
| *Saturated* | | |
| Lauric acid (coconut or palm) kernal oil) | $CH_3(CH_2)_{10}COOH$ | 44 |
| Myristic acid (nutmeg fat) | $CH_3(CH_2)_{12}COOH$ | 54 |
| Palmitic acid (palm oil, animal fats) | $CH_3(CH_2)_{14}COOH$ | 63 |
| Stearic acid (animal fats) | $CH_3(CH_2)_{16}COOH$ | 70 |
| *Unsaturated* | | |
| Palmitoleic acid (butter fat) | $CH_3(CH_2)_5CH{=}CH(CH_2)_7COOH$ | −1 |
| Oleic acid (olive oil, animal fats) | $CH_3(CH_2)_7CH{=}CH(CH_2)_7COOH$ | 13.4 |
| Linoleic acid (linseed oil) | $CH_3(CH_2)_4CH{=}CHCH_2CH{=}CH(CH_2)_7COOH$ | −5 |
| Linolenic acid (linseed oil) | $CH_3(CH_2)CH{=}CHCH_2CH{=}CHCH_2CH{=}CH(CH_2)_7COOH$ | −11 |

(glycerol esterified with three stearic acid molecules),

$$
\begin{array}{c}
\text{H} \quad\quad \text{O} \\
| \quad\quad\quad || \\
\text{H}-\text{C}-\text{O}-\text{C}-\text{C}_{17}\text{H}_{35} \\
| \\
\quad\quad\quad \text{O} \\
\quad\quad\quad || \\
\text{H}-\text{C}-\text{O}-\text{C}-\text{C}_{17}\text{H}_{35} \\
| \\
\quad\quad\quad \text{O} \\
\quad\quad\quad || \\
\text{H}-\text{C}-\text{O}-\text{C}-\text{C}_{17}\text{H}_{35} \\
| \\
\text{H}
\end{array}
$$

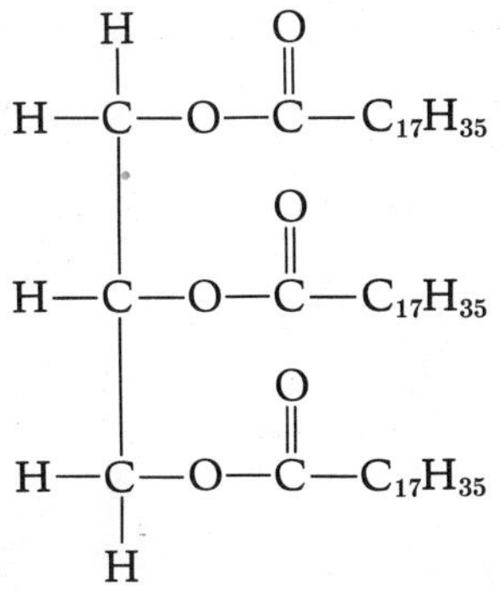

are solids, while those containing three unsaturated fatty acids are liquids at room temperature. An example of this latter type is *triolein*, in which glycerol is esterified with three oleic acid molecules. This substance is the major constituent of olive oil. The liquid unsaturated triglycerides that are found in vegetable oils, such as olive oil and corn oil, serve as the basis of oleomargarine. Addition of hydrogen to the double bonds of unsaturated vegetable oils produces saturated chains and hence solid fats.

In recent years the growth of the processed food industry has led to widespread use, as food additives, of monoglycerides and diglycerides, in which only one or two of the —OH groups of glycerol are esterified. These are added to foods as emulsifiers to improve texture and to keep oils suspended.

Fats, like other esters, can be **saponified** upon treatment with aqueous base. The products of this reaction are glycerol plus the anions of the fatty acids that were bound to the glycerol in the fat.

$$
\begin{array}{c}
\text{H}_2\text{COOCC}_{17}\text{H}_{35} \\
| \\
\text{HCOOCC}_{17}\text{H}_{35} \\
| \\
\text{H}_2\text{COOCC}_{17}\text{H}_{35}
\end{array}
\xrightarrow[\text{H}_2\text{O}]{\text{OH}^-}
\begin{array}{c}
\text{H}_2\text{COH} \\
| \\
\text{HCOH} \\
| \\
\text{H}_2\text{COH}
\end{array}
+ \ 3\text{C}_{17}\text{H}_{35}\text{COO}^-
$$

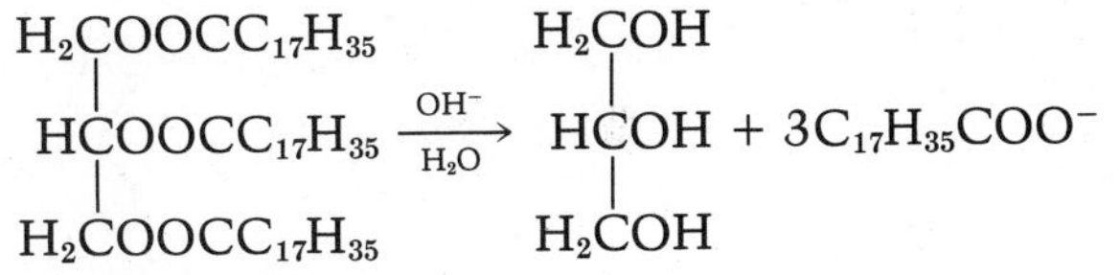

These anions constitute a soap and have rather peculiar properties that result from having a polar hydrophilic "head" and a nonpolar hydrophobic "tail."

$$
\text{CH}_3-\text{CH}_2-\ \cdots\ -\text{CH}_2-\text{CH}_2-\text{CH}_2-\text{C}\overset{\displaystyle\text{O}}{\underset{\displaystyle\text{O}^-}{<}}
$$

$$\underbrace{\qquad\qquad\qquad\qquad\qquad}_{\text{tail}}\quad\underbrace{\qquad}_{\text{head}}$$

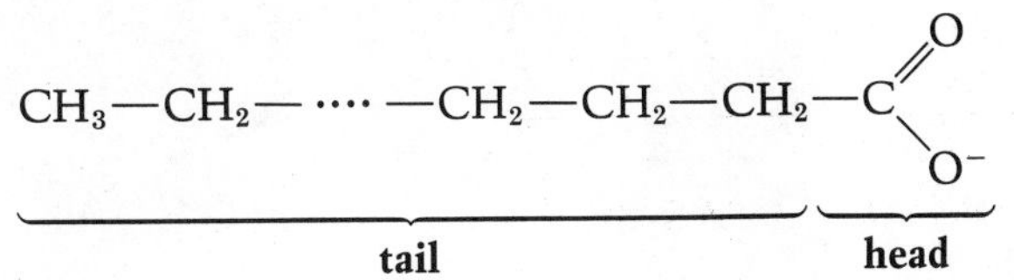

The polar end of the anion tends to be water-soluble while the other end, the nonpolar hydrocarbon tail, tends to be insoluble in water but soluble in nonpolar solvents. As a result, in water these anions group themselves together into small globules called **micelles** in which the nonpolar tails "dissolve" in each other, leaving the polar heads facing outward toward the aqueous surroundings. This is illustrated in Figure 22.17. The same properties that lead

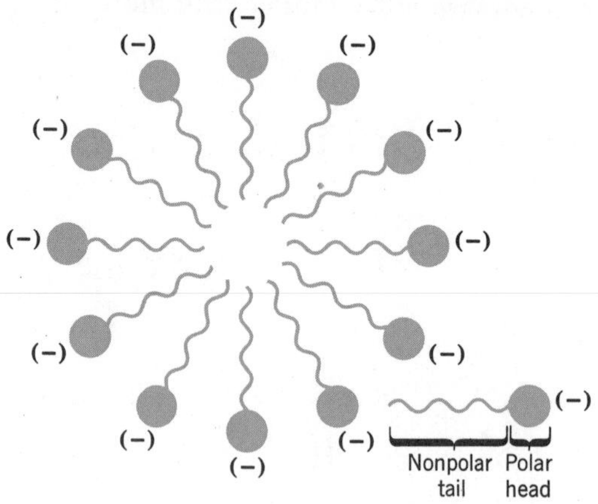


Figure 22.17
*Micelle formation with soap.*

to micelle formation are also responsible for the ability of soap to dissolve grease. In this case the nonpolar tails dissolve in the grease particle and the polar heads dissolve in water (Figure 22.18). This keeps the grease particle suspended in water so that it can be rinsed away.

Another very important class of lipids are the **phospholipids.** These are polar lipids and, like the fats, are esters of glycerol. In this case, however, only two fatty acid molecules are esterified to glycerol, at the first and second carbon atom. The remaining end position of the glycerol is esterified to a molecule of phosphoric acid, which in turn is also esterified to another alcohol. This gives a general structure,

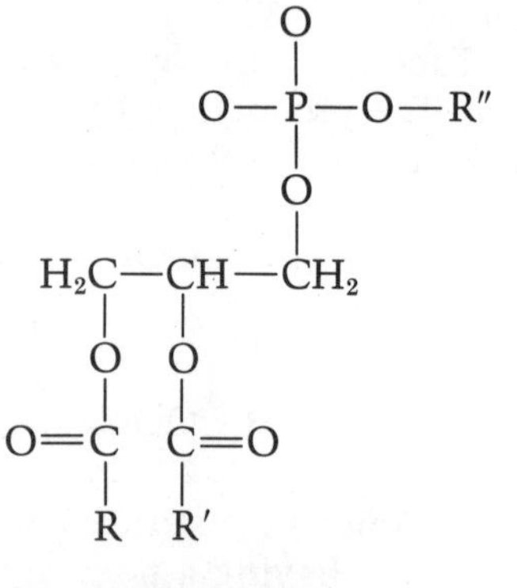

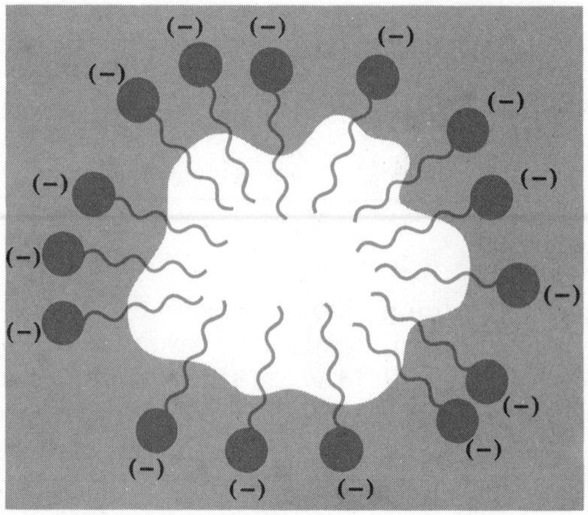


Figure 22.18
*The dissolving of grease globule by soap.*

An example is the phospholipid *phosphatidyl ethanolamine*,

$$CH_3-CH_2-CH_2-CH_2-CH_2-CH_2-CH_2-(CH_2)_8\overset{\overset{\displaystyle O}{\|}}{C}-O-\overset{\overset{\displaystyle H}{|}}{\underset{|}{C}}-H$$

$$CH_3-CH_2-CH_2-CH_2-CH_2-CH_2-CH=CH-(CH_2)_7\overset{\overset{\displaystyle O}{\|}}{C}-O-\overset{|}{\underset{|}{C}}-H$$

$$H-\overset{\overset{\displaystyle H}{|}}{\underset{|}{C}}-O-\overset{\overset{\displaystyle O}{\|}}{\underset{\displaystyle O}{P}}-O^{(-)}$$

$$\underset{NH_2^{(+)}}{\overset{\displaystyle CH_2}{\underset{\displaystyle CH_2}{|}}}$$

$\underbrace{\hspace{4cm}}_{\text{nonpolar tail}}$  $\underbrace{\hspace{2cm}}_{\text{polar head}}$

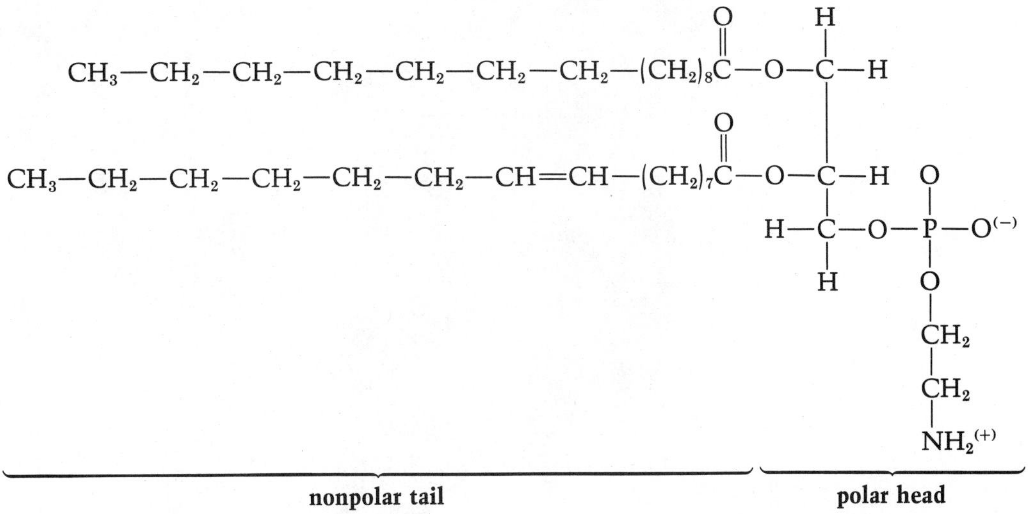

As indicated, this type of lipid contains a polar head and nonpolar tail, much the same as the anions of the fatty acids.

Cell membranes are composed of phospholipids and proteins in about equal proportion. The phospholipids in the membrane appear to be arranged in a double layer, or **bilayer** (Figure 22.19) in which the nonpolar tails face each other, thereby exposing the polar heads to the aqueous environment on either side of the membrane. As shown in Figure 22.20, the proteins found in the membrane are embedded in the mosaic formed by the lipids. Much research today is centered on the mechanism of transport of matter and energy across such membranes.

Finally, there is a third class of lipids that do not contain fatty acids and glycerol, and that do not undergo saponification when treated with a base. Included in this group are the **steroids,** complex substances that possess unusually high biological activity.

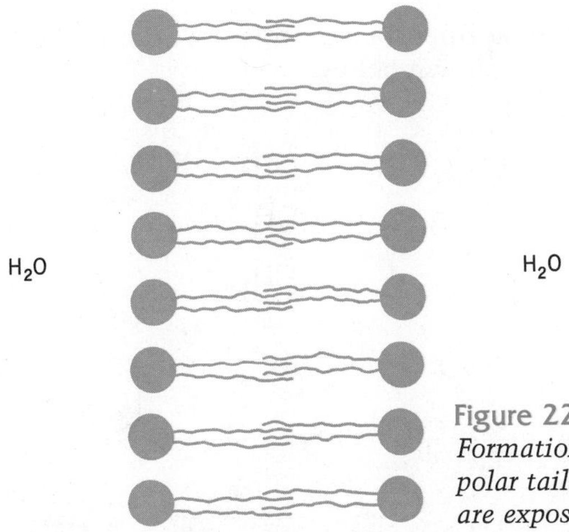

$H_2O$ $H_2O$

Figure 22.19
*Formation of bilayer by phospholipids. Nonpolar tails dissolve in each other. Polar heads are exposed to aqueous environment.*

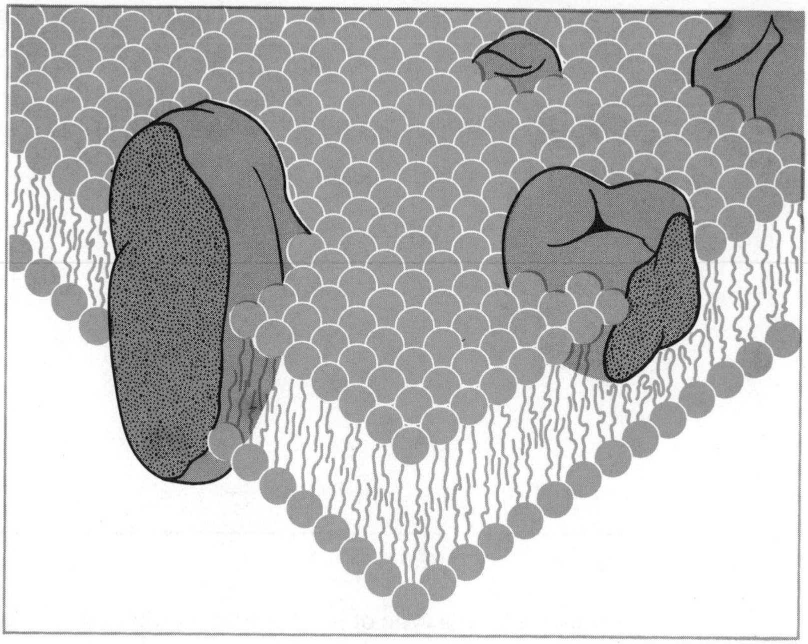

Figure 22.20
*Bilayer structure of cell membrane. Lipid mosaic model: irregularly shaped proteins float randomly in a lipid sea. From S. J. Singer,* Annals of the New York Academy of Sciences, *Vol. 195, p. 21, 1962. Used by permission of the publisher and author.*

The steroids have in common a fused ring structure.

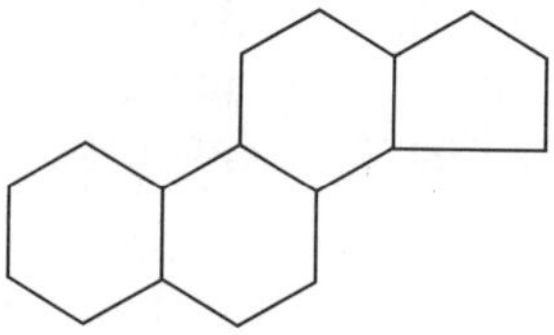

Examples of some important steroids, whose names you may have come across before, are shown below.

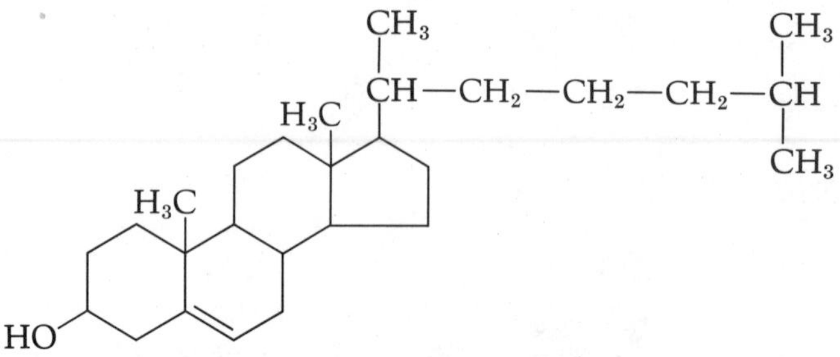


**cholesterol**

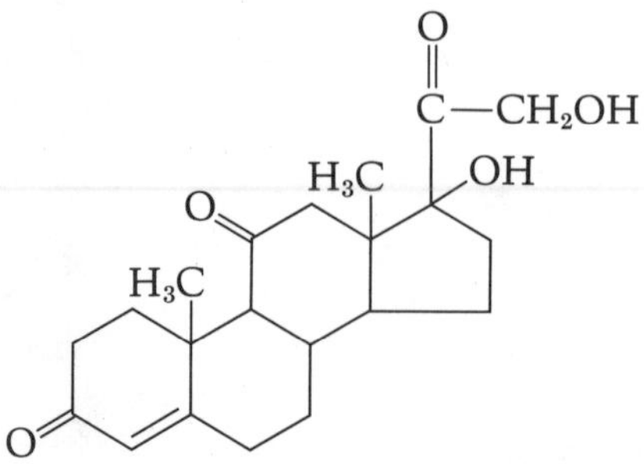


**cortisone**
**(affects protein metabolism)**

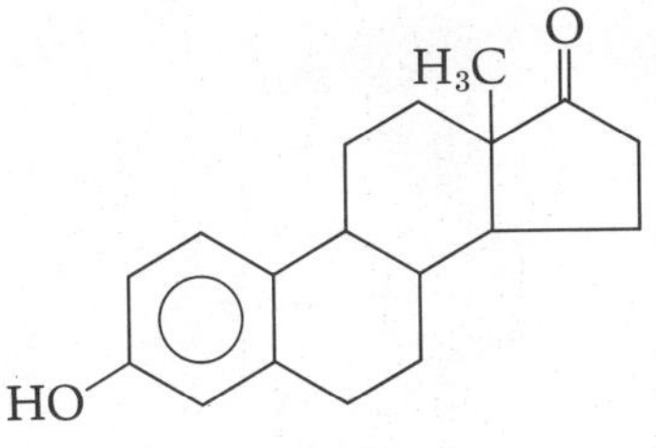


estrone

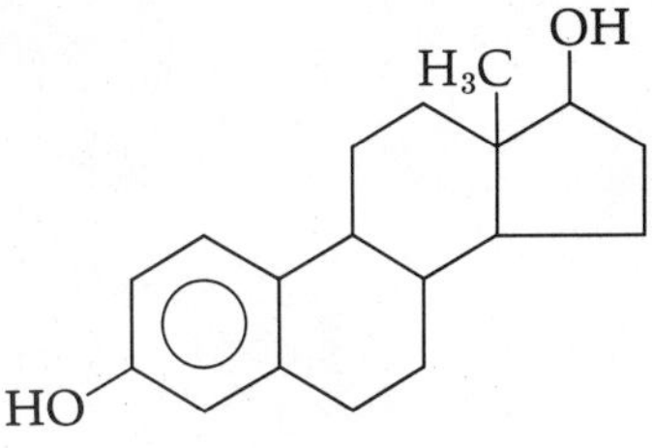


estradiol

(female sex hormones: note similarity in structures)

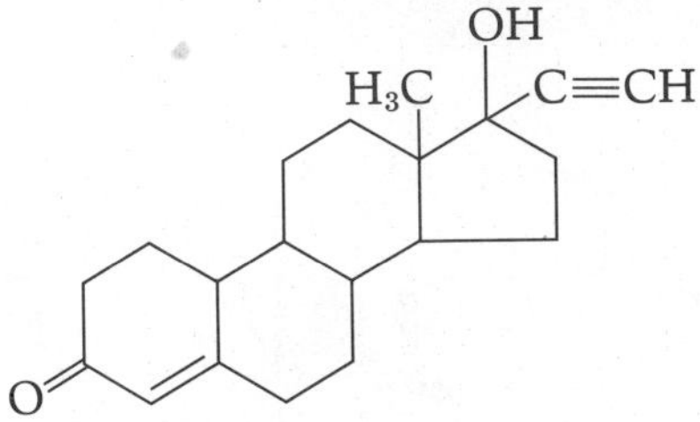


norethindrone: an
oral contraceptive
("the pill")

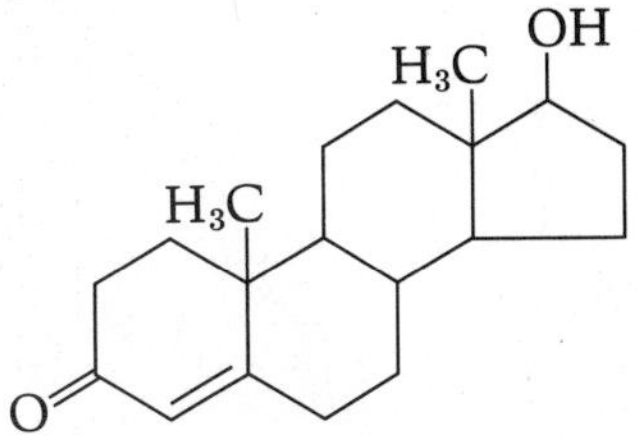


testosterone
(male sex hormone)

## 22.5 NUCLEIC ACIDS

One of the most intriguing aspects of biochemistry has been, from its very beginnings, the mechanism whereby an organism transmits its genetic information from one generation to the next during cell division. It is now believed, with a good deal of confidence, that this process is controlled by a substance found in the nucleus of the cell, the **nucleic acid, DNA** (deoxyribonucleic acid). Furthermore, DNA in conjunction with **RNA** (ribonucleic acid, another type of nucleic acid) is responsible for the synthesis of the proteins that are characteristic of a given organism.

Nucleic acids, like the proteins and carbohydrates that we looked at earlier, are polymers. The simpler units that make up the nucleic acid are called **nucleotides,** and are themselves composed of three even simpler molecules. They include the following:

1. **A nitrogenous base.** These are heterocyclic organic compounds having two or more nitrogen atoms in the ring skeleton. They are called bases because the lone pairs of electrons on the nitrogen atoms make them Lewis bases. These substances are shown in Figure 22.21.
2. **A five-carbon sugar (pentose).** In RNA this sugar is ribose, whereas in DNA the sugar is deoxyribose. These two are shown in Figure 22.22. Notice that they differ only at carbon atom number 2 in the ring.
3. **Phosphoric acid.** $H_3PO_4$, as we shall see, forms esters to —OH groups of the sugar to bind nucleotide segments together.

A molecule called a **nucleoside** is formed from these components by condensing a molecule of the base with the appropriate pentose. For example, adenine combines with ribose and deoxyribose at carbon number 1 to give the compounds shown in Figure 22.23*a*. Adenosine is an important constituent of ATP (adenosine triphosphate) and ADP (adenosine diphosphate), both of which are involved in energy-transfer processes in a cell.

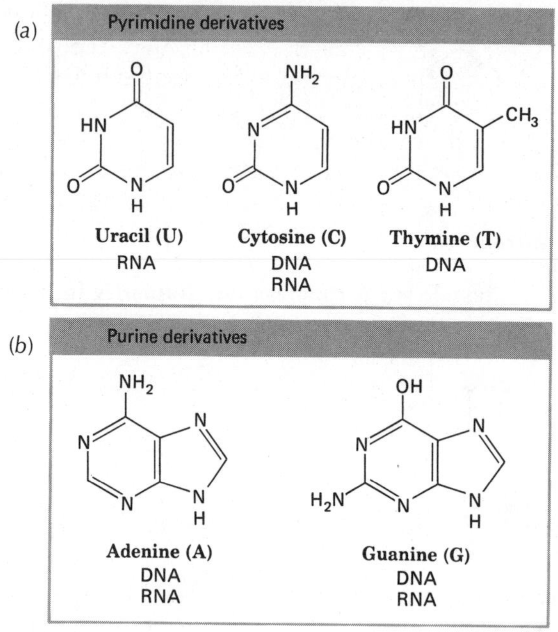


Figure 22.21
*Nitrogenous bases found in DNA and RNA. (a) Pyrimidine derivatives. (b) Purine derivatives.*

Finally, linkage of phosphoric acid to carbon atom number 5 (Figure 22.23*b*) produces a **nucleotide,** the basic building block of both DNA and RNA.

The nucleic acids are condensation polymers of the nucleotide monomers and are formed by the creation of an ester linkage from the phosphoric acid residue on one nucleotide to the hydroxy group on carbon number 3 in the pentose of the second nucleotide, as illustrated in Figure 22.24. The result is a very long polymeric chain, possessing up to a billion or so nucleotide units in DNA!

In Figure 22.21 it was indicated that the base compositions of DNA and RNA are not the same. In DNA the organic bases adenine (A), guanine (G), cytosine (C), and thymine (T) occur bound to the deoxyribose ring, whereas in RNA the bases adenine, guanine, cytosine, and uracil (U) are found. As in the proteins, the sequence of bases along the DNA or RNA chain establishes its primary structure, which controls the specific properties of the nucleic acid. For instance, the base sequence in DNA contains coded information that the cell utilizes in the synthesis of its own characteristic proteins. We shall look at this more closely in the next section.

The truly remarkable properties of DNA that account for its ability to reproduce itself exactly are dictated by its secondary structure, in which two

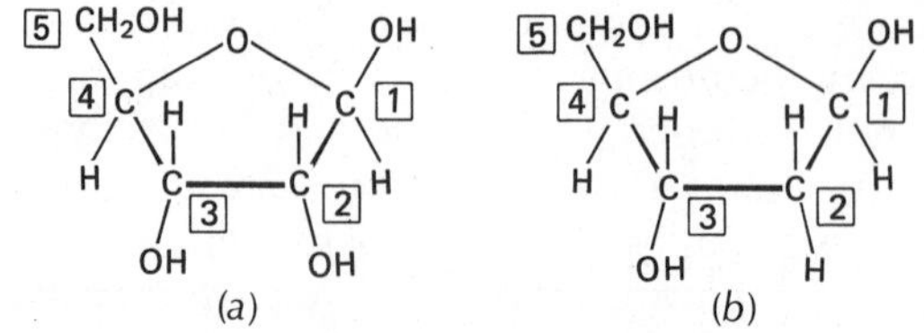


Figure 22.22
*(a) Ribose and (b) deoxyribose.*

strands of DNA intertwine into the now much celebrated **double helix** proposed by Watson and Crick, for which they received the Nobel Prize in 1962.[3] The key to the formation of the double helix, as well as the function of DNA and RNA in protein synthesis, lies in the interaction of the nitrogenous bases by way of hydrogen bonding.

Consider, for example, the bases cytosine and guanine, situated across from one another on separate DNA strands. The structure of these bases, as shown below, is such that they are ideally suited to interact with each other through the formation of hydrogen bonds.

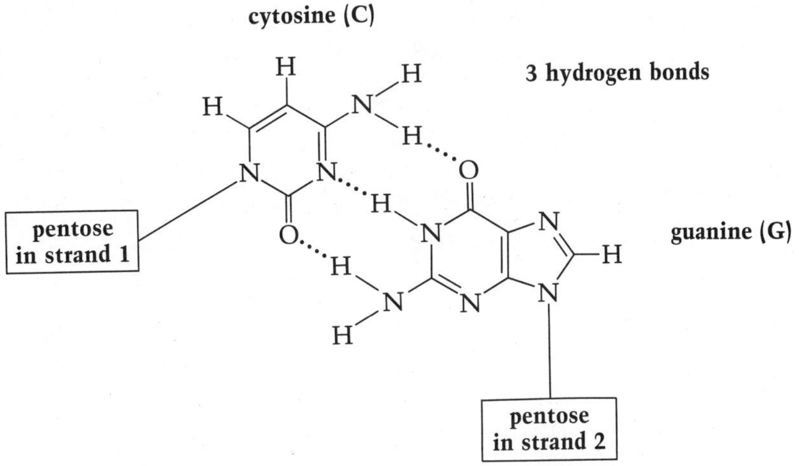


A similar relationship holds for the bases thymine and adenine.

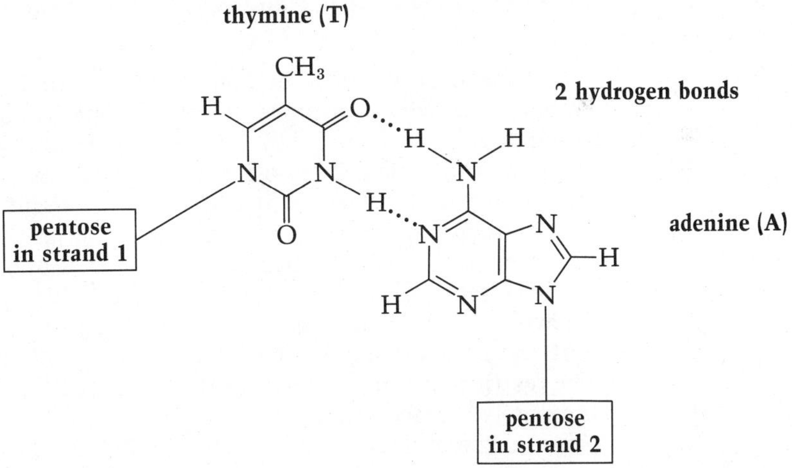


Thus cytosine and guanine fit together like a hand in a glove, as do thymine and adenine. Now, in DNA there is always the same amount of C as G. In addition, the quantities of T and A are the same. Furthermore, the total amount of C and T together is always equal to the combined total of G and A. If you think about this for a while, you will see that this suggests that C and G are paired together, as are T and A.

[3] Watson and Crick shared the 1962 Nobel Prize with Wilkins for their use of Rosalind Franklin's X-ray data in the deduction of the double-helix structure of DNA.

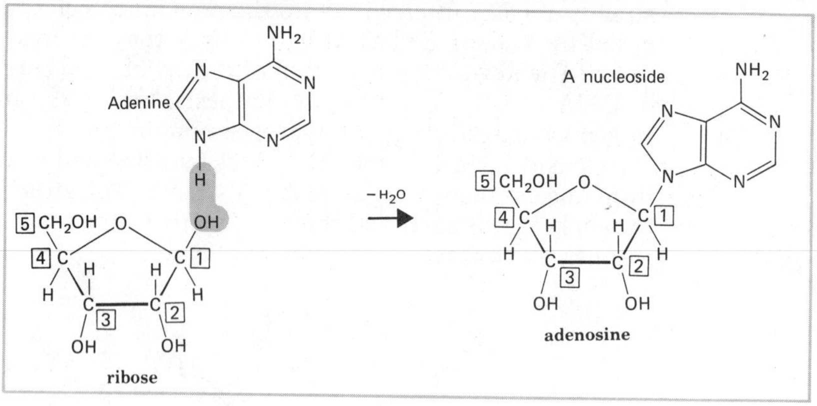


(a)

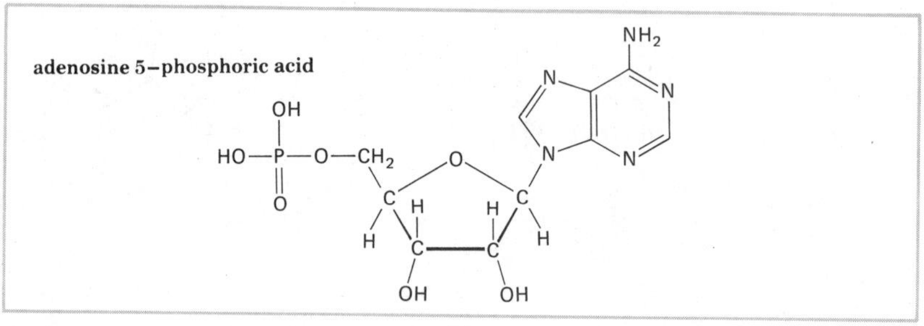


(b)

Figure 22.23

*Nucleosides and nucleotides. (a) Adenine combines with ribose to form a nucleoside. (b) Linkage of phosphoric acid to carbon 5 gives a nucleotide.*

These observations, in conjunction with X-ray diffraction data, led Watson and Crick to propose the double-helical structure of DNA, shown schematically in Figure 22.25. In this structure we see that for each G on one strand there is a C opposite it, across the axis of the helix, on the other strand. A similar relationship also holds for T and A. The two DNA strands are not identical, but rather complement one another, and it is this property that accounts for the replication of the DNA upon cell division.

It is believed that during cell division the two DNA strands begin to unravel, as shown in Figure 22.26, giving the two complementary chains that serve as templates for the construction of two new daughter chains. The restrictions on the base pairing (that is, T with A and C with G) causes the newly formed strands to be identical to the departing complementary parent chain and, as a result, a pair of DNA double helices are produced that are exact copies of the original.

## 22.6 PROTEIN SYNTHESIS

The DNA found in the nucleus of a cell serves indirectly to determine the makeup of the proteins that are synthesized at *ribosomes* located outside the nucleus. The genetic information that determines the amino acid sequence in each of the enzymes is stored in the DNA in a genetic code that is made up of the sequence of bases along a DNA strand. The transcription of this code to the site of protein synthesis, as well as the decoding and construction of the polypeptide chain of the protein, is accomplished by the other nucleic acids, the RNA.

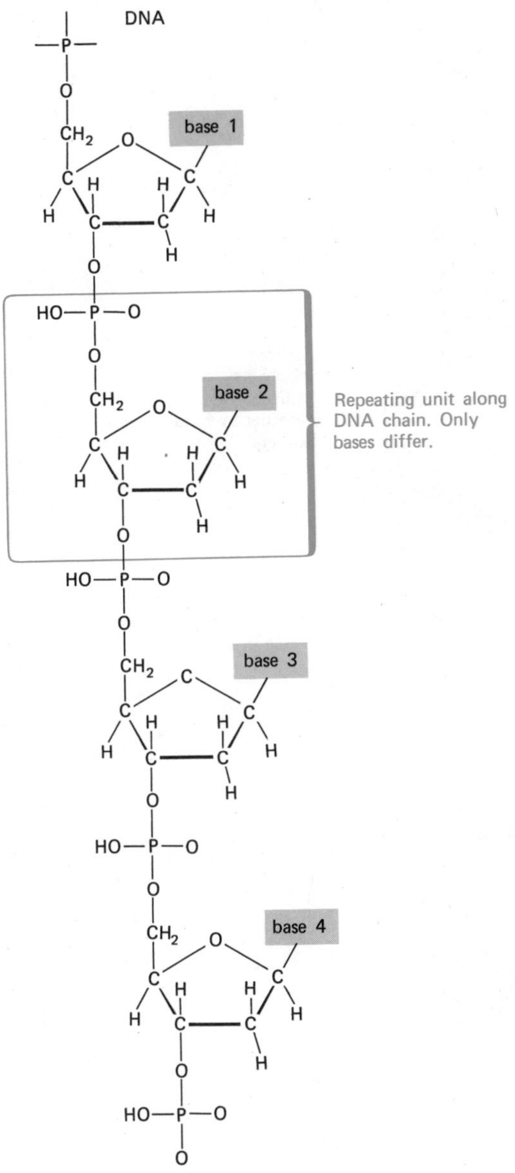


Figure 22.24
*Polymerization of nucleotides give nucleic acids.*

Unlike DNA, RNA occurs only in single strands. In addition, there are several types of RNA. One of these is called **messenger RNA, mRNA,** and serves to carry the genetic code from the DNA template within the nucleus to the ribosomes outside. Another type of RNA molecule, called **transfer RNA, tRNA,** is much smaller than mRNA. It acts as an amino acid carrier and through a decoding mechanism that we shall examine momentarily, adds the amino acid to the growing polypeptide chain at just the right place at just the right time.

The mechanism by which this process is believed to occur is not really too different from that involved in the duplication of DNA itself. It is known, for instance, that the production of mRNA takes place within the nucleus on an untwisted segment of a DNA chain. This segment corre-

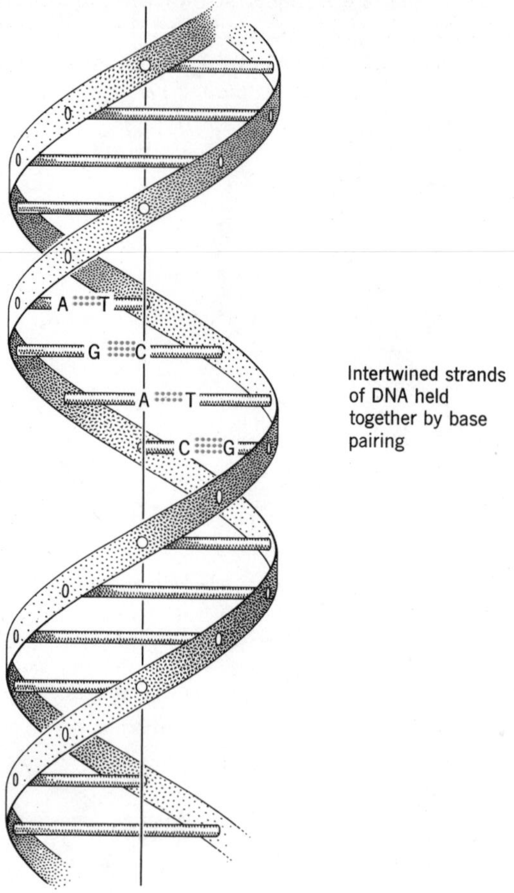


Figure 22.25
*The double helix. Intertwined strands of DNA held together by base pairing.*

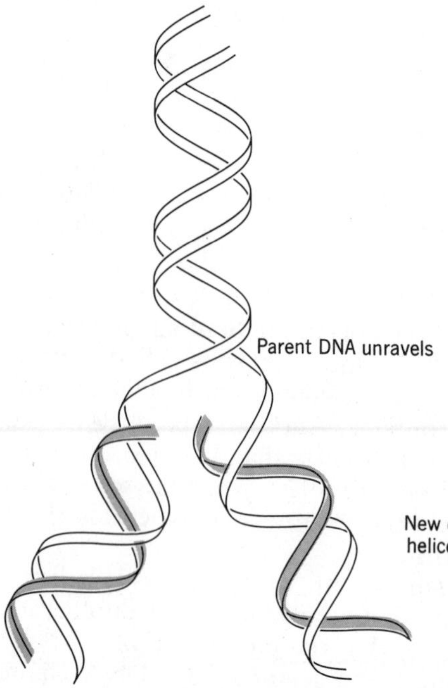


Figure 22.26
*Replication of DNA.*

sponds to the gene that is characteristic of the particular protein to be synthesized. The mRNA strand produced contains a sequence of bases that is determined, through base pairing, by the sequence of bases in the DNA; however, the pairing scheme in this case is

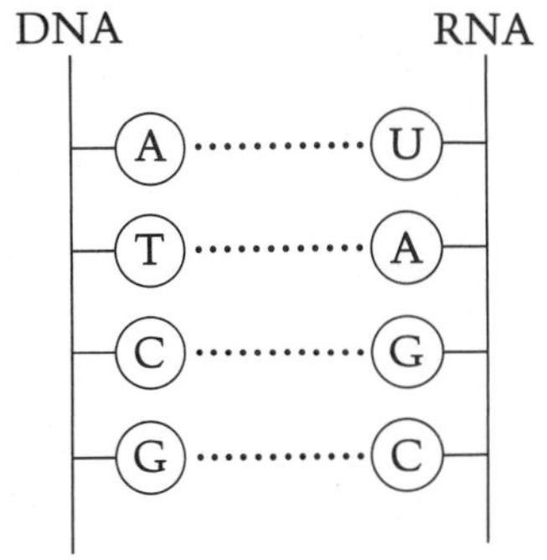


Thus uracil occurs in RNA instead of thymine. Once formed, the mRNA is transported from the nucleus to the active site of protein synthesis.

It is now known that the genetic code that directs the insertion of amino acids in the proper sequence in the growing polypeptide chain consists of sets of three bases (called **codons**). The amino acids are brought to the proper place by tRNA, which is able to decipher the code.

A molecule of tRNA contains from 75 to 90 nucleotide units containing the bases U, A, G, C, and, in addition, many minor bases. What is interesting is that although the tRNAs corresponding to different amino acids have different minor bases and base sequences, they can all be brought into a characteristic cloverleaf shape if maximum base pairing is assumed. For example, the yeast tRNA for the amino acid alanine has the structure shown in Figure 22.27. The molecule is folded in such a way as to give maximum base pairing (colored lines). On the stem of this RNA molecule, alanine becomes attached by the action of the proper enzyme. The two arms to either side of the site of attachment of the amino acid appear to be important in the interaction of the tRNA with this enzyme. The bottom loop contains the three-unit anticodon that attaches itself to the complementary condon on the mRNA chain.

The sequence of operations that leads to the synthesis of a polypeptide is summarized in Figure 22.28. In the bacterium, *E. coli,* for example, it is known that the amino acid sequence is initiated by a derivative of methionine. The tRNA containing this derivative becomes attached at the head of the mRNA through its codon-anticodon pairing. Once this is in place the next amino acid in the sequence is brought into place by its tRNA, which becomes bound to the mRNA at the second codon site. This is followed by an enzyme-induced formation of the peptide linkage and the subsequent departure of the tRNA from the first codon. Next, the third amino acid is delivered by its tRNA, which attaches itself to the third codon. Again a peptide linkage is formed and the second tRNA leaves. This process is repeated over and over along the mRNA as the polypeptide chain grows in size. The chain is finally terminated when a "nonsense" codon (one that cannot be recognized by any tRNA molecule) is encountered.

**THE GENETIC CODE.** Through a series of very clever experiments, which earned Nirenberg, Holley, and Khorana a Nobel Prize in 1968, the base sequences in the genetic code triplets were determined. These are given in

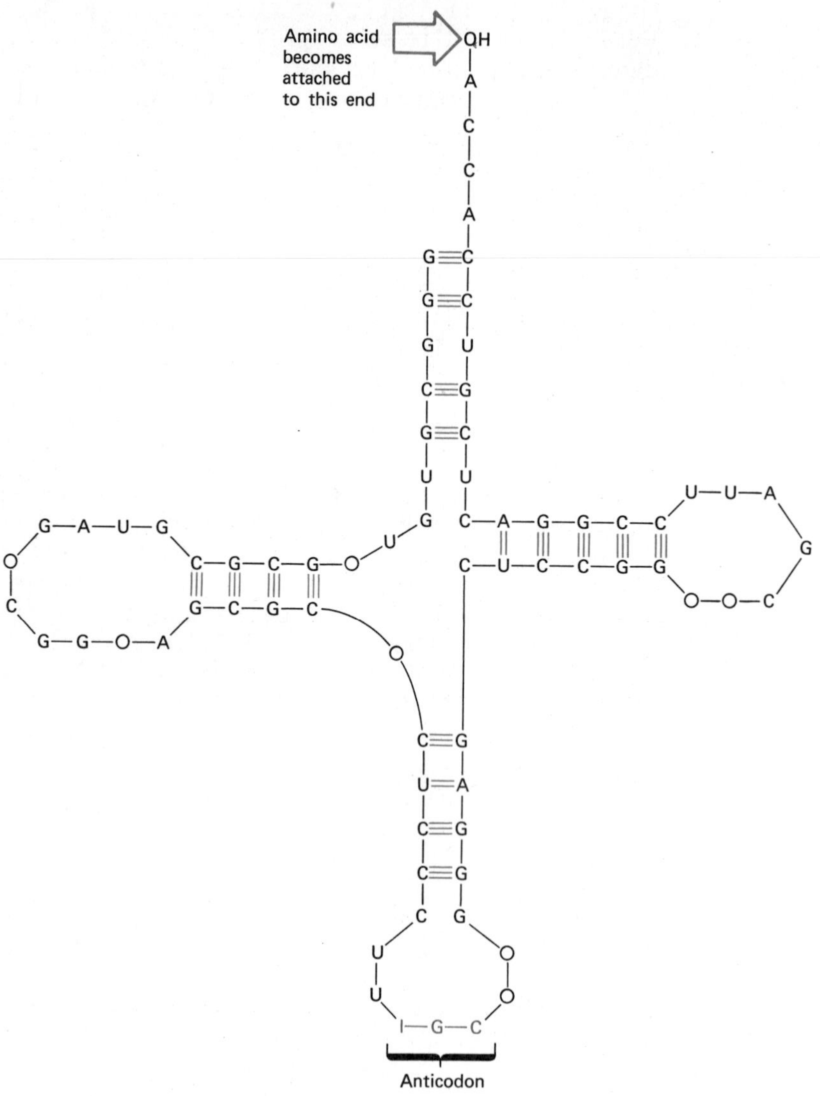


Figure 22.27
*The base sequence in a yeast alanine tRNA*

Table 22.3. Notice that most of the amino acids are specified by more than one code "word." For these, apparently, the attachment of the tRNA to the mRNA is determined primarily by the first two bases in the sequence (which are usually the same for a given amino acid). The interaction between the third base in the mRNA codon and the tRNA anticodon thus does not appear to be as critical as the first two in determining specificity.

With this code we can now see how the amino acid sequence in a polypeptide is fixed by the primary structure of the DNA. Consider, for example, a segment of DNA having the base sequence,

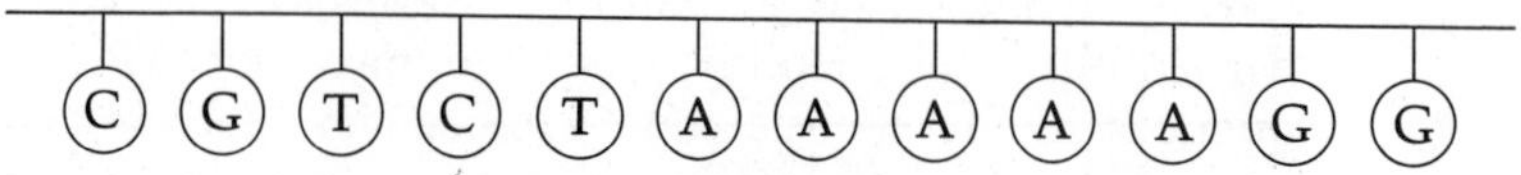

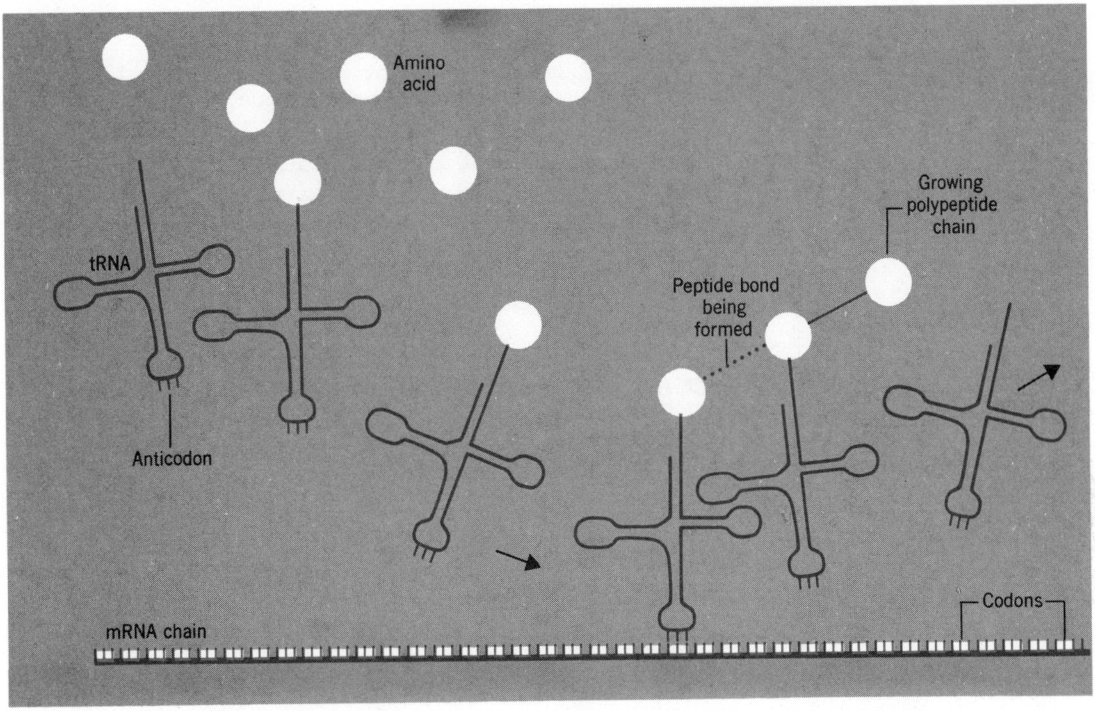


Figure 22.28
*Synthesis of polypeptide.*

The mRNA formed from it will have the base sequence,

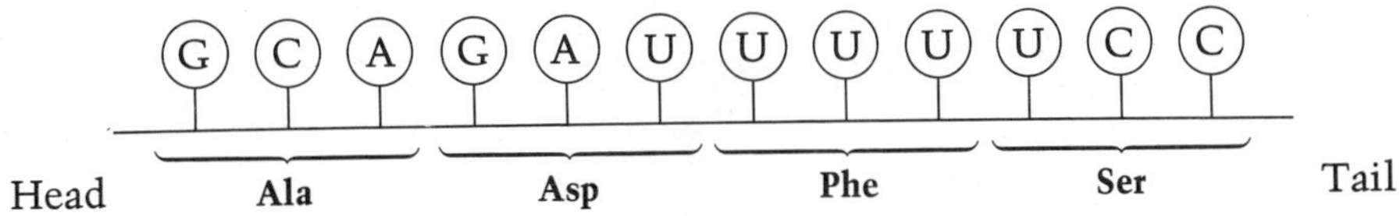


which, as you can see, is composed of the code words for the polypeptide

Ala—Asp—Phe—Ser

The triplet code presented in Table 22.3 has been shown to be applicable to such diverse species as the bacterium, *E. coli*, the tobacco plant, the guinea pig, and even man. It is widely believed that this code is universal for all species.

**MUTATIONS.** Any factor that will have the effect of altering the base sequence in a DNA molecule will, in effect, alter the mRNA transcribed from it and thereby cause a change in the amino acid sequence that is produced using that mRNA as a template. Provided that this does not prove fatal to the cell, a mutation will have occurred that will be transmitted from one generation to another as the DNA reproduces itself upon cell division.

Often these mutations prove harmful, although not immediately lethal to the organism in which they occur, and give rise to symptoms that cause them to be referred to as diseases. Since the origin of the diseases is in the genetic material of the cell, they are said to be genetic diseases. Many such

Table 22.3
*The genetic code*

| First Base in Codon | Second Base in Codon: U | C | A | G |
|---|---|---|---|---|
| U | UUU, UUC: Phe; UUA, UUG: Leu | UCU, UCC, UCA, UCG: Ser | UAU, UAC: Tyr; UAA[b]; UAG[b] | UGU, UGC: Cys; UGA[b]; UGG: Trp |
| C | CUU, CUC, CUA, CUG: Leu | CCU, CCC, CCA, CCG: Pro | CAU, CAC: His; CAA, CAG: Gln | CGU, CGC, CGA, CGG: Arg |
| A | AUU, AUC, AUA: Ile; AUG: Met[a] | ACU, ACC, ACA, ACG: Thr | AAU, AAC: Asn; AAA, AAG: Lys | AGU, AGC: Ser; AGA, AGG: Arg |
| G | GUU, GUC, GUA, GUG: Val | GCU, GCC, GCA, GCG: Ala | GAU, GAC: Asp; GAA, GAG: Glu | GGU, GGC, GGA, GGG: Gly |

[a] AUG appears to initiate an amino acid sequence with methionine.
[b] Nonsense codons: these do not code for any amino acid. They serve to terminate peptide chains.

diseases are recognized. Some of the more well known are cystic fibrosis, hemophelia, and sickle-cell anemia. Some evidence even suggests that schizophrenia may be of genetic origin.

In sickle-cell anemia, for example, the red blood cells assume a crescent shape instead of the flat disklike shape of normal cells. This is caused by a mutated DNA in the gene that is responsible for the synthesis of hemoglobin. It has been found that in sickle-cell hemoglobin a *glutamic acid* unit in one of the hemoglobin chains is replaced by *valine.* This alters the secondary and tertiary structure of the protein and reduces its ability to carry oxygen.

Sickle-cell anemia is thus a disease with its origin in the DNA of the cell nucleus and is therefore passed from one generation to the next because of its genetic nature.

INDEX TO QUESTIONS

Amino acids 1, 5, 13, 15
Proteins 2, 3, 4, 6, 7, 8, 9, 10
Enzymes 11, 12, 14
Carbohydrates 16, 17, 18, 19, 20, 21
Lipids 22, 23, 24, 25, 26, 27, 28, 29
Nucleic acids 30, 31, 32, 33
Protein synthesis 34, 35, 36, 37, 38, 40
Mutations 39, 41

## REVIEW QUESTIONS

**22.1** What is an α-amino acid? Indicate how amino acids are linked together to give a dipeptide; a polypeptide.

**22.2** Lye, NaOH, is able to dissolve proteins such as hair that are lodged in a sink drain. What chemical reaction is involved?

**22.3** Describe what is meant by the primary structure of a protein. What is meant by secondary structure, tertiary structure, and quaternary structure?

**22.4** Indicate three functions served by proteins in a living organism.

**22.5** What is a zwitterion? Why do amino acids form zwitterions?

**22.6** Describe the α-helix structure found in many polypeptides. What holds the polypeptide chain in this helical conformation?

**22.7** How do the properties of the solvent affect the tertiary structures of proteins? What interactions in addition to hydrogen bonding determine tertiary structure?

**22.8** What is meant by a quaternary structure? What proteins exhibit this kind of structural sophistication?

**22.9** How do the roles played by hemoglobin and myoglobin differ? In what way are they similar?

**22.10** What is a porphyrin? Name two biologically important porphyrin structures.

**22.11** In what sense do enzymes guide the chemistry that takes place in living organisms? What problems would a living cell encounter if it were not for the existence of enzymes?

**22.12** Describe in qualitative terms how an enzyme operates. What is meant by *enzyme substrate*? What is enzyme inhibition? How do the sulfa drugs function?

**22.13** Monosodium glutamate (MSG), which is the monosodium salt of glutamic acid, is a popular flavor enhancer. However, only the L-isomer is effective. Is this particularly surprising? Explain your answer.

**22.14** How does diisopropylflurophosphate function as a nerve gas? Many insecticides contain organic phosphates. Can you guess how they function?

**22.15** Only one of the basic set of 20 amino acids does not exhibit optical isomerism. Which one is it? Why is it not optically active?

**22.16** What is a monosaccharide? Give an example of (a) a pentose and (b) a hexose.

**22.17** How many optical isomers of fructose are there?

**22.18** Compare the structures of starch and cellulose. What is the major difference between them?

**22.19** Using the method described on p. 703, draw formulas for the remaining optical isomers of

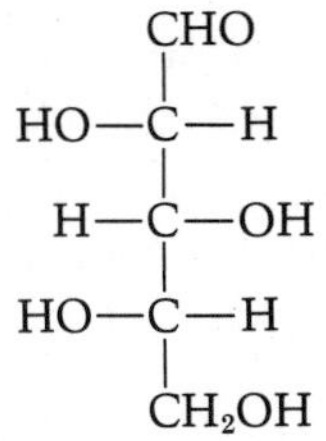


**22.20** Sucrose, cane sugar, is used in large amounts in food products of all kinds. What monosaccharides are produced by hydrolysis of a sucrose molecule?

**22.21** What functions do the carbohydrates serve in living organisms?

**22.22** What is a lipid? What functions do they serve in living systems?

**22.23** What is a fatty acid? Give an example of a triglyceride. What difference in physical properties exists between the saturated and unsaturated triglycerides?

**22.24** Write a chemical equation for the saponification of tristearin.

**22.25** Give the structure of triolein. What product is formed upon addition of $H_2$ to the double bonds in triolein?

**22.26** What is a soap? How does it form micelles? What is responsible for the ability of soap to dissolve grease?

**22.27** How does a phospholipid differ from a triglyceride?

**22.28** How are phospholipids believed to contribute to the structure of cellular membranes?

**22.29** What characteristic structural feature is found among the steroids?

**22.30** Describe the structure of a *nucleotide.* What is the difference between the nucleotide units in DNA and RNA?

**22.31** Imagine that a single DNA strand contained a segment with the composition,

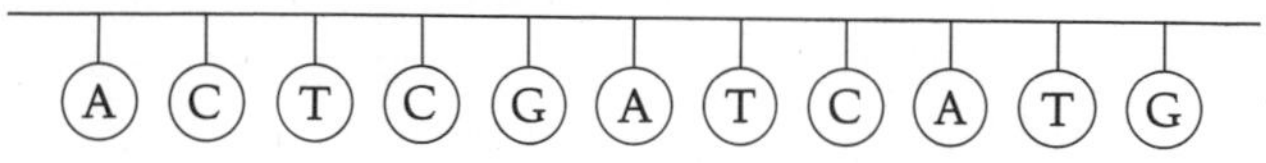


What would be the base sequence in the complementary strand?

**22.32** What holds the two DNA strands together in the double helix? Illustrate the base pairing between C and G; between T and A.

**22.33** What occurs with DNA during cell division? How is the DNA replicated?

**22.34** What is meant by the *genetic code*? Describe the functions of mRNA and tRNA.

**22.35** Give a base sequence that would have to exist on mRNA to give a polypeptide with the amino acid sequence,

Arg—Leu—Lys—Gly—Cys

**22.36** If a DNA strand contains the base sequence,

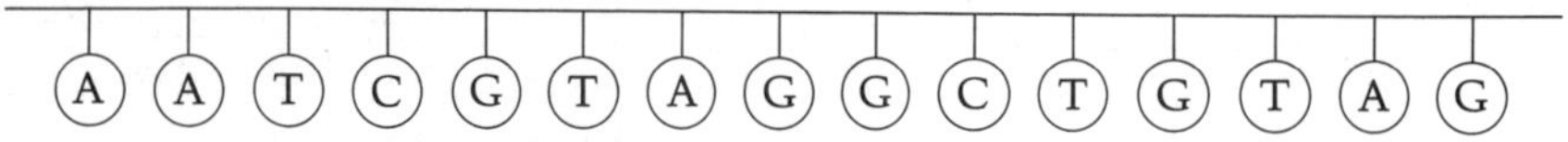


what will be the base sequence transcribed onto the mRNA?

**22.37** What amino acid sequence is specified by the base sequence (from left to right) in the following mRNA strand?

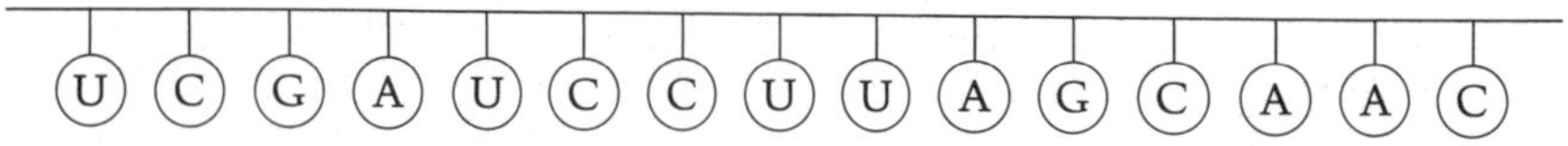


**22.38** What base sequence on a DNA strand would give a mRNA that would produce a polypeptide having the amino acid sequence,

Ala—Pro—Asp—Tyr—Ile—Gly

**22.39** Suppose that through some mutational change the first base was removed from the sequence in the DNA strand depicted in Question 22.36. What amino acid sequence would result in the polypeptide synthesized using this modified (mutant) DNA template?

**22.40** What amino acid sequence is specified by the base sequence (starting at the left) in the DNA strand in Question 22.36?

**22.41** What is a genetic disease? Must it always be fatal to the organism?

# 23 NUCLEAR CHEMISTRY

In our discussion of chemistry up to now we have paid very little attention to the atomic nuclei, other than to note that their charge determines the number of electrons in the neutral atom. There are, however, nuclear phenomena that have applications to chemistry and we will explore them in this chapter. At the same time, we shall also explore the nuclear changes that occur in radioactive elements as well as those that can cause an otherwise stable element to become radioactive.

## 23.1 SPONTANEOUS RADIOACTIVE DECAY

Henry Becquerel, a French chemist, was the first to discover that some elements, especially the heavier ones, spontaneously emit radiation and are said to be radioactive. He found, somewhat to his surprise, that when a photographic film was exposed to the salt, potassium uranyl sulfate, $K_2UO_2(SO_4)_2$, the film was blackened. He reasoned that in order to accomplish this the uranium salt must emit a powerful radiation that upon striking the film blackened it. In later experiments he also observed that the rate at which radiation was emitted from this salt was directly dependent on the amount of sample present.

Two colleagues of Becquerel, Marie and Pierre Curie, were successful in isolating two other radioactive isotopes from uranium ores. These isotopes, which are more intensely radioactive than uranium, were named polonium, Po, and radium, Ra. For their work in this field Becquerel and the two Curies were awarded the Nobel Prize for Physics in 1903.

We now know that the radiation emitted from radioactive materials is of three main types (as was first pointed out in Section 3.4): **alpha particles** ($\alpha$), **beta particles** ($\beta$), and **gamma rays** ($\gamma$). Experimentally it is found that the $\alpha$-particle carries a positive charge, the $\beta$-particle is negatively charged, and the $\gamma$-rays carry no charge. The actual charges and masses of these and other particles with which we shall be concerned are listed in Table 23.1.

When a substance spontaneously emits one or more of these types of radiation, it is said to be radioactive. The uranium in the potassium uranyl sulfate of Becquerel is radioactive and spontaneously emits $\alpha$-particles. The most abundant isotope of uranium is $^{238}_{92}U$, which upon emitting $\alpha$-particles becomes $^{234}_{90}Th$. This can be written as

$$^{238}_{92}U \longrightarrow {}^{4}_{2}He + {}^{234}_{90}Th$$

Thus we say that $^{238}U$ decays to $^{234}Th$ by the emission of $\alpha$-particles. Isotopic reactants and products in these decay reactions are also called **parent** and

Table 23.1
*Basic types of particles emitted by radioisotopes*

| Particle | Approximate Mass (amu) | Charge | Symbol | Type |
|---|---|---|---|---|
| Alpha | 4 | +2 | ${}^{4}_{2}\mathrm{He}$ | Particle |
| Beta | 0 | −1 | ${}^{0}_{-1}e$ | Particle |
| Gamma | 0 | 0 | $\gamma$ | Electromagnetic radiation |
| Neutron | 1 | 0 | ${}^{1}_{0}n$ | Particle |
| Proton | 1 | +1 | ${}^{1}_{1}p$ (${}^{1}_{1}\mathrm{H}$) | Particle |
| Positron | 0 | +1 | ${}^{0}_{1}e$ | Particle |

**daughter** isotopes, or **nuclides**, respectively. The word nuclide is used as a general term when referring to a nucleus with a specified number of protons and neutrons. Thus the daughter isotope of nuclide ${}^{238}_{92}\mathrm{U}$ is ${}^{234}_{90}\mathrm{Th}$.

The thorium isotope produced in the above decay reaction is itself radioactive and spontaneously emits $\beta$-particles. As a result, when the thorium is formed, it then decays to an isotope of protactinium, the reaction being

$$\underset{\textbf{(parent)}}{{}^{234}_{90}\mathrm{Th}} \longrightarrow \underset{(\boldsymbol{\beta})}{{}^{0}_{-1}e} + \underset{\textbf{(daughter)}}{{}^{234}_{91}\mathrm{Pa}}$$

Highly energetic $\gamma$-rays are emitted by nearly all radioactive materials, a process that occurs without a change in the mass or the charge of the isotope. As a result they are often omitted when writing decay reactions like those shown above.

In balancing radioactive decay reactions, it is important to remember that the total mass and charge of all of the species on the left of the arrow must exactly balance with the total mass and charge of those species on the right (check the two reactions above to see if we wrote them correctly). Thus, whenever an isotope decays by $\alpha$-emissions, the daughter isotope always has an atomic mass number that is four units lower than the parent and an atomic number two units lower. In the case of a $\beta$-emitter, on the other hand, the daughter has an atomic mass number that is the same as the parent and an atomic number one unit *higher* than its parent.

EXAMPLE 23.1 (a) ${}^{234}\mathrm{U}$ decays by $\alpha$-emissions. What is its daughter isotope?
(b) ${}^{214}\mathrm{Pb}$ decays to ${}^{214}\mathrm{Bi}$. By what type of radiation is this accomplished?

SOLUTION (a)

$${}^{234}_{92}\mathrm{U} \longrightarrow {}^{4}_{2}\mathrm{He} + {}^{230}_{90}X$$

The equation can first be written in this form; and then we simply identify the element with $Z = 90$. The complete equation is then written as

$${}^{234}_{92}\mathrm{U} \longrightarrow {}^{4}_{2}\mathrm{He} + {}^{230}_{90}\mathrm{Th}$$

(b) The statement of the problem means

$${}^{214}_{82}\mathrm{Pb} \longrightarrow {}^{214}_{83}\mathrm{Bi} + {}^{0}_{-1}X$$

From Table 23.1 we see that ${}^{0}_{-1}X$ corresponds to ${}^{0}_{-1}e$; therefore the decay occurs by $\beta$-emission.

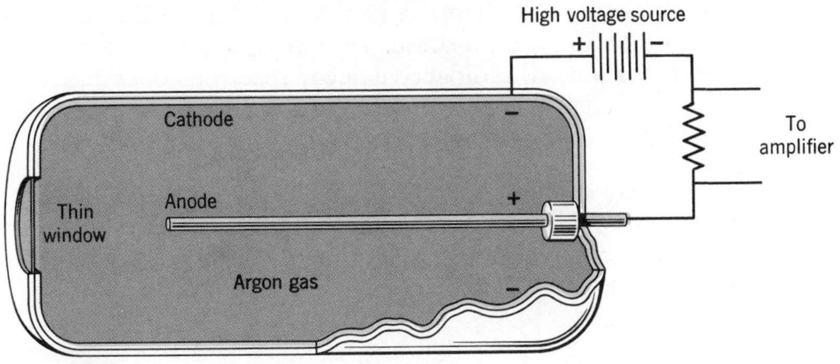


Figure 23.1

*The Geiger-Müller counter. An α- or β-particle enters the Geiger tube through the thin window shown at the left of the apparatus. As the particle passes through the gas inside the tube, it ionizes argon atoms along its path. These ions cause an electrical breakdown (discharge) between the wire and the wall of the tube, thereby producing a current pulse. This current pulse is readily amplified and counted electronically.*

We have seen above that $^{238}U$ decays to $^{234}Th$, which is also radioactive and decays to $^{234}Pa$. This isotope is also unstable and decays to $^{234}U$ (how?), which is also radioactive, and so on. This decay process continues until a stable (nonradioactive) isotope of an element is formed. This entire scheme, where one isotope decays to another, and so on, is called a **radioactive series** or **decay series.** $^{238}U$, for example, decays by some 14 steps to stable $^{206}Pb$.

**KINETICS OF RADIOACTIVE DECAY.** In Chapter 11 we saw that the rate of a chemical reaction is given by the change in the concentration of a reactant over a certain small time interval. For a radioactive substance, its concentration is directly proportional to the number of particles or rays emitted by the substance per minute. These emissions can be detected and counted by a suitable device such as the Geiger-Müller counter shown schematically in Figure 23.1.

If, for a particular radioactive isotope, we construct a graph in which the number of counts per minute, **cpm,** of particles emitted is plotted against time, we obtain a curve similar to that shown in Figure 23.2*a*. If the log of

Figure 23.2

*Kinetics of radioactive decay. (a) A graph of concentration versus time. (b) A logarithmic plot of concentration versus time. Straight line indicates first-order process.*

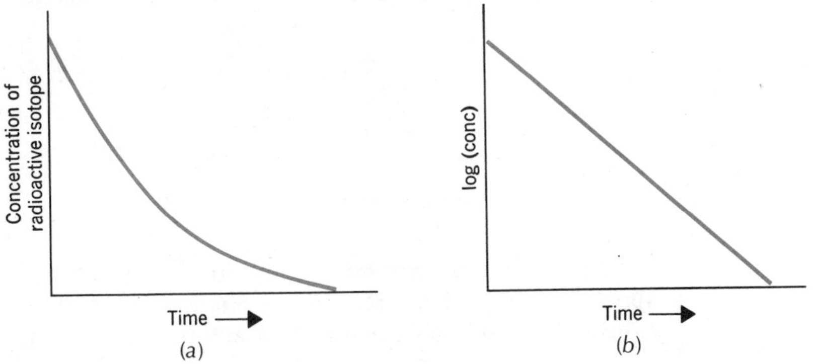

the cpm versus time is graphed, a straight line is obtained as shown in Figure 23.2*b*. Both of these plots are indicative of a first-order reaction and, as it turns out, all radioactive decay reactions obey first-order kinetics. The generalized rate law for any radioactive decay process would then take the form,

$$\text{rate} = k(\text{concentration of isotope})^1$$

By applying what we have learned in Chapter 11 to the decay process, it is possible to derive an equation that is helpful in determining rate constants for decay reactions. According to Section 11.1, for any first-order reaction, such as

$$A \longrightarrow B$$

we can express its rate law as

$$-\frac{\Delta[A]}{\Delta t} = k[A]$$

Rearranging this equation to take the form

$$-\frac{\Delta[A]}{[A]} = k(\Delta t)$$

and using calculus, we can derive the equation,

$$\log \frac{[A]_0}{[A]} = \frac{kt}{2.30} \qquad (23.1)$$

In this equation $[A]_0$ is the initial concentration of the isotope (concentration at time zero) and $[A]$ is its concentration at any time $t$. Equation 23.1 can be employed to calculate rate constants as shown in the following example.

EXAMPLE 23.2 A chemist determined that after exactly 1 week his initial 10.0 $\mu$g (1 $\mu$g = $10^{-6}$ g) of $^{222}$Rn had decayed, and he now had 2.82 $\mu$g of the radon. What is the rate constant for the $\alpha$-decay of $^{222}$Rn?

SOLUTION The rate constant can easily be obtained by using Equation 23.1, where $[A]_0$ = 10.0 g, $[A]$ = 2.82 g, and $t$ = 7.00 days. Solving Equation 23.1 for $k$, we have

$$k = \frac{2.30}{t} \log \frac{[A]_0}{[A]}$$

and substituting the values given above into this equation, we have

$$k = \left(\frac{2.30}{7.00 \text{ days}}\right) \log \left(\frac{10.0}{2.82}\right)$$

or

$$k = 0.181 \text{ days}^{-1}$$

Another very important consequence of the kinetic studies described above is that the time required for half of any radioactive isotope to decay, called its **half-life,** $t_{1/2}$, is constant and is independent of the initial concen-

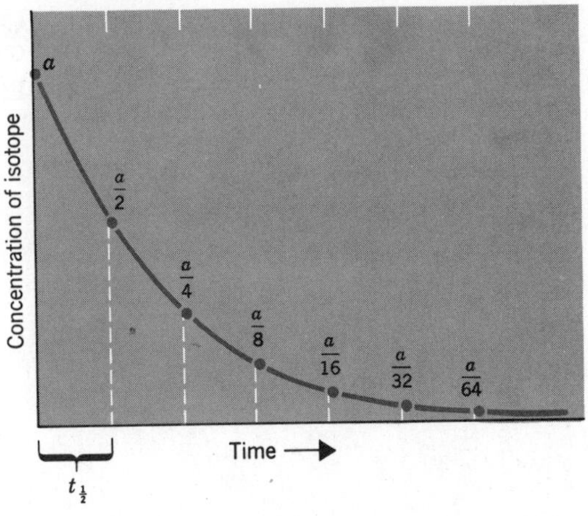


Figure 23.3
*A graph of a first-order radioactive decay illustrating the concept of half-life.*

tration of the isotope. Thus, if we were to begin with 10.0 g of a particular isotope, after one half-life only 5.0 g would remain. In a time equal to another half-life this would decay to 2.5 g, and so on. This is represented graphically in Figure 23.3. In this figure we see, in general, how an isotope decays through many half-lives. We can show that the half-life of any substance is inversely proportional to the rate constant for its decay. The actual relationship as derived takes the form (see Review Question 23.11)

$$t_{1/2} = \frac{0.693}{k} \tag{23.2}$$

From this equation we see that if the rate constant for decay is large, its half-life will be small and vice versa. Also, we see that if either $t_{1/2}$ or $k$ is known, the other can be calculated.

EXAMPLE 23.3 (a) What is the half-life of $^{222}$Rn and (b) What is the fraction of a sample of Rn that decays in 1 week?

SOLUTION (a) We can calculate the half-life of Rn directly from Equation 23.2 by substituting in the value of $k$ calculated in Example 23.2,

$$t_{1/2} = \frac{0.693}{k}$$

and $k = 0.181\ \text{days}^{-1}$ so that

$$t_{1/2} = \frac{0.693}{0.181\ \text{days}^{-1}}$$

or

$$t_{1/2} = 3.83\ \text{days}$$

(b) We first have to calculate the fraction left after decay. By this we mean the ratio $[A]/[A]_0$. Therefore we employ Equation 23.1 for this purpose and solve it for $[A]_0/[A]$.

$$\log \frac{[A]_0}{[A]} = \frac{kt}{2.30}$$

Substituting in $k = 0.181 \text{ days}^{-1}$ and $t = 7.00$ days, we have

$$\log \frac{[A]_0}{[A]} = \frac{(0.181 \text{ days}^{-1})(7.00 \text{ days})}{2.3}$$

or

$$\log \frac{[A]_0}{[A]} = 0.551$$

and

$$\frac{[A]_0}{[A]} = 3.56$$

so that

$$\frac{[A]}{[A]_0} = 0.281$$

This is the fraction left after 1 week and the fraction decayed is simply 1 − 0.281 or 0.719.

Half-lives are important for two reasons. First, they indicate the stability of a particular isotope; the longer the half-life the more stable is the material. Second, the half-life can be effectively used to measure the age of such things as rocks, bones, ancient pieces of art, and the like.

**USES OF RADIOACTIVE DECAY.** The age of rocks containing uranium can be dated by determining the ratio of $^{238}U$ to $^{206}Pb$. Remember that $^{206}Pb$ is the stable isotope to which $^{238}U$ eventually decays. Using a $^{238}U/^{206}Pb$ ratio of 1 to 0 as corresponding to zero time and a $^{238}U/^{206}Pb$ ratio of 1 to 1 as corresponding to the time for one half-life (that is, $4.5 \times 10^9$ years), the age of these rocks can be closely approximated. The oldest rocks that have been found on earth have an age of $4.55 \times 10^9$ years using this method.

Rocks that do not contain uranium are presently being dated using a potassium-argon method. This method makes use of the reaction,

$$^{40}_{19}K + {}^{0}_{-1}e \longrightarrow {}^{40}_{18}Ar \qquad t_{1/2} = 1.3 \times 10^9 \text{ years}$$

The electron in this case comes from the 1*s* orbital of the potassium and is captured by the unstable $^{40}_{19}K$ nucleus. In the dating procedure the ratio of $^{40}K$ to argon is measured, and the age of the rock is determined in the same manner as the uranium dating described above.

The age of materials that were once living, such as bones and wood, can be estimated quite accurately by measuring their ratio of $^{14}C$ to $^{12}C$. Carbon-14 is radioactive and is constantly being produced in the upper atmosphere by the bombardment of cosmic neutrons upon $^{14}_{7}N$, which is present there in large amounts. The equation for this reaction is

$$^{14}_{7}N + {}^{1}_{0}n \longrightarrow {}^{14}_{6}C + {}^{1}_{1}p$$

The carbon-14 thus produced immediately begins to decay

$$^{14}_{6}C \longrightarrow {}^{14}_{7}N + {}^{0}_{-1}e \qquad t_{1/2} = 5770 \text{ years}$$

so that a steady-state concentration (15 cpm per gram of carbon) of this radioactive nuclide is maintained in the atmosphere. The $^{14}C$ becomes incorporated into the carbon dioxide ($^{14}CO_2$) in the air where it can be taken in by plants through the process of photosynthesis. The intake of $^{14}C$ into animals is by the consumption of such plants or by the consumption of plant-eating animals. While they are alive, plants and animals consume and excrete carbon so that they also maintain a steady-state concentration of $^{14}C$ and are thus in equilibrium with their surroundings. Once they die, however, the $^{14}C$ that they possess is not replaced as it decays and hence the $^{14}C$ concentration begins to decrease. The half-life of the $^{14}C$ is 5770 years; therefore, if we find that the carbon-14 concentration in an object that had once been living has dropped to half its initial value, we could conclude that the object is 5770 years old. However, atmospheric nuclear testing has made it impossible for our current history to be dated this way by future archaeologists.

EXAMPLE 23.4 A piece of charcoal from the ruins of a settlement in Japan was found to have a $^{14}C/^{12}C$ ratio that was 0.617 times that found in living organisms. How old is this piece of charcoal?

SOLUTION The answer to this problem can be obtained once again by employing Equations 23.2 and 23.1. From Equation 23.2 we can obtain the $k$ for the decay of $^{14}C$.

$$k = \frac{0.693}{5770 \text{ years}} = 1.20 \times 10^{-4} \text{ years}^{-1}$$

Substituting this value and that for the ratio of $^{14}C/^{12}C$ into Equation 23.1, we have

$$\log \frac{[A]_0}{[A]} = \frac{kt}{2.30}$$

$$\log \left(\frac{1.000}{0.617}\right) = \log 1.62 = 0.210 = \frac{(1.20 \times 10^{-4} \text{ years}^{-1})t}{2.30}$$

or

$$t = 4030 \text{ years}$$

## 23.2 NUCLEAR TRANSFORMATIONS

Lord Rutherford, in 1919, performed the first known **nuclear transformation** of one element into another. Rutherford found that by bombarding $^{14}_{7}N$ with $\alpha$-particles he produced the isotope of oxygen, $^{17}_{8}O$. The equation for this process is

$$^{14}_{7}N + ^{4}_{2}He \longrightarrow \left[^{18}_{9}F\right] \longrightarrow ^{17}_{8}O + ^{1}_{1}H$$

The $^{18}_{9}F$ isotope of fluorine is a very unstable intermediate that rapidly decays to the products above.

In 1933 Irene Curie[1] and her husband Frederick Joliot showed that other light elements could similarly be transformed by bombardment with $\alpha$-particles. For example, $^{27}_{13}Al$ can be transformed into $^{30}_{15}P$ by this process.

$$^{27}_{13}Al + ^{4}_{2}He \longrightarrow ^{30}_{15}P + ^{1}_{0}n$$

Reactions of this type, where an $\alpha$-particle is used for bombardment and a neutron is one of the products, is known as an alpha, neutron reaction sym-

[1] Irene Joliot-Curie was the daughter of Marie and Pierre Curie, the discoverers of polonium and radium.

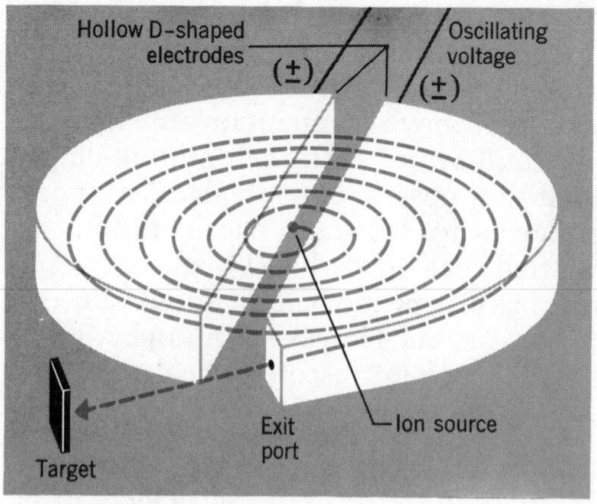


Figure 23.4
*Diagram of a cyclotron.*

bolized by $(\alpha,n)$. A shorthand notation for the above reaction is ${}^{27}_{13}Al(\alpha,n){}^{30}_{15}P$.

Since 1933 many isotopes have been produced by bombardment reactions, some where particles other than $\alpha$-particles have been used. One of the main problems with such experiments is that a positively charged nucleus is being bombarded with positively charged particles. Heavy elements, with their very highly positive nuclei, will repel particles with positive charges like the $\alpha$-particle. One way to circumvent this problem is to use neutrons as the bombarding particles.

Neutrons, which have no charge, are not repelled by the nucleus and, therefore, are excellent materials for bombardment reactions. The supply of neutrons for these reactions can be obtained from either of two sources, from a transformation reaction in which neutrons are produced or from a nuclear reactor in which fission reactions, which will be examined in Section 23.6, occur at a controlled rate. Examples of neutron-producing reactions are the ${}^{27}_{13}Al(\alpha,n){}^{30}_{15}P$ reaction mentioned above and

$$ {}^{9}_{4}Be + {}^{4}_{2}He \longrightarrow {}^{12}_{6}C + {}^{1}_{0}n $$

where the beryllium isotope undergoes an $\alpha,n$ reaction.

A second way that nuclear transformations can be brought about is through the use of particle accelerators. Particle accelerators, such as the cyclotron described in Figure 23.4, speed up particles to extremely high velocities and then direct them at target nuclei. At these speeds the positive particles are able to overcome the coulombic repulsion of the nucleus and collide with it. Accelerators have been used extensively by Dr. Glenn Seaborg and his colleagues at the University of California in producing many of the transuranium elements, that is, elements 93 to 105. In these accelerators positive ions of such isotopes as ${}^{2}_{1}H$ (deuterium), ${}^{12}_{6}C$, ${}^{13}_{6}C$, ${}^{16}_{8}O$, ${}^{14}_{7}N$, ${}^{10}_{6}B$, as well as ${}^{4}_{2}He$ have been used in producing new, man-made, elements. A listing of these bombardment reactions and the elements they produce is shown in Table 23.2. Many of these elements are extremely short-lived and, as a result, only a few atoms, especially of the high-atomic-numbered isotopes, have ever been formed.

## 23.3 NUCLEAR STABILITY

Experimentally it is observed that all of the elements with atomic numbers greater than 83 (bismuth) are radioactive and possess no known stable isotopes. On the other hand, all of the lighter elements, with the exception of

Table 23.2
*Elements produced by particle accelerators*

| |
|---|
| ${}^{238}_{92}U + {}^{4}_{2}He \longrightarrow {}^{239}_{94}Pu + 3\,{}^{1}_{0}n$ |
| ${}^{239}_{94}Pu + {}^{4}_{2}He \longrightarrow {}^{240}_{95}Am + {}^{1}_{1}H + 2\,{}^{1}_{0}n$ |
| ${}^{239}_{94}Pu + {}^{4}_{2}He \longrightarrow {}^{242}_{96}Cm + {}^{1}_{0}n$ |
| ${}^{244}_{96}Cm + {}^{4}_{2}He \longrightarrow {}^{245}_{97}Bk + {}^{1}_{1}H + 2\,{}^{1}_{0}n$ |
| ${}^{238}_{92}U + {}^{12}_{6}C \longrightarrow {}^{246}_{98}Cf + 4\,{}^{1}_{0}n$ |
| ${}^{238}_{92}U + {}^{14}_{7}N \longrightarrow {}^{247}_{99}Es + 5\,{}^{1}_{0}n$ |
| ${}^{238}_{92}U + {}^{16}_{8}O \longrightarrow {}^{249}_{100}Fm + 5\,{}^{1}_{0}n$ |
| ${}^{253}_{99}Es + {}^{4}_{2}He \longrightarrow {}^{256}_{101}Md + {}^{1}_{0}n$ |
| ${}^{246}_{96}Cm + {}^{13}_{6}C \longrightarrow {}^{254}_{102}No + 5\,{}^{1}_{0}n$ |
| ${}^{252}_{98}Cf + {}^{10}_{5}B \longrightarrow {}^{257}_{103}Lw + 5\,{}^{1}_{0}n$ |
| ${}^{249}_{98}Cf + {}^{12}_{6}C \longrightarrow {}^{257}_{104}Ku + 4\,{}^{1}_{0}n$ |

technetium ($Z = 43$) and promethium ($Z = 61$), have one or more stable, nonradioactive isotopes. In addition, radioactive isotopes undergo nuclear transformations that lead ultimately to stable nuclei. Sometimes this is accomplished by a simple one-step process, while in other cases a series of nuclear reactions occur before a stable isotope is reached. A question that naturally arises from these observations is, what factors give rise to stable or unstable nuclei?

Little is known about the nature of the forces that hold a nucleus together. Some interesting facts concerning nuclear stability emerge, however, if we examine the numbers of protons and neutrons found in stable nuclei. For example, if we make a graph of the number of neutrons versus the number of protons in different nuclei, we find that all of the stable isotopes fall in a narrow band, which we might call a **band of stability,** as shown in Figure 23.5.

In this illustration we see that at low atomic numbers stable nuclei possess approximately equal numbers of protons and neutrons. Above about $Z = 20$, however, the number of neutrons always exceeds the number of protons and the neutron-to-proton ratio gradually increases to about 1.5 at the upper end of the band of stability. Apparently, as the number of protons in the nucleus increases, there must be more and more neutrons present to help overcome the strong repulsion forces between the protons. It also seems that there is an upper limit to the number of protons that can exist in a stable nucleus, that number being reached at bismuth.

Nuclei that lie outside the band of stability are unstable and decay in a manner that tends to give them a stable neutron-to-proton ($n/p$) ratio. On

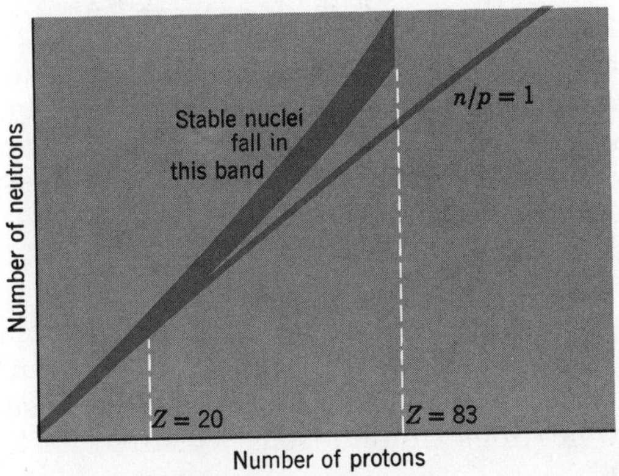


Figure 23.5
*Band of stability.*

this basis, then, we can understand why certain nuclei undergo the type of radioactive decay that they do. For instance, a nucleus that lies above the band of stability must either lose neutrons or gain protons in order to achieve stability. Thus we can understand why elements such as $^{14}C$ (which lies above the band) decay by $\beta$-emission, because this process converts a neutron into a proton ($^{1}_{1}p$).

$$^{1}_{0}n \longrightarrow {}^{1}_{1}p + {}^{0}_{-1}e$$

For $^{14}C$ we have

$$^{14}_{6}C \longrightarrow {}^{14}_{7}N + {}^{0}_{-1}e$$

Another way that an element located above the band can achieve a stable $n/p$ ratio is by emitting a neutron, although this particular mode of decay is rare. An example is the decay of $^{137}I$.

$$^{137}_{53}I \longrightarrow {}^{136}_{53}I + {}^{1}_{0}n$$

Elements located *below* the band of stability must increase their $n/p$ ratio in order to achieve stability. This is accomplished generally in one of two ways. One of these involves the emission of a positron, a particle having the same mass as the electron but with a unit positive charge. The positron is symbolized as $^{0}_{1}e$. The ejection of a positron by an unstable nucleus converts a proton into a neutron,

$$^{1}_{1}p \longrightarrow {}^{1}_{0}n + {}^{0}_{1}e$$

An example is the decay of $^{11}C$.

$$^{11}_{6}C \longrightarrow {}^{11}_{5}B + {}^{0}_{1}e$$

The second mode of decay that results in an increased $n/p$ ratio is called **electron capture.** In this case the unstable nucleus captures an electron, usually from its own 1*s* orbital. Since the captured electron most often originates in the *K* shell, the process is also called **K-capture.** The addition of this electron to the nucleus transforms a proton into a neutron.

$$^{1}_{1}p + {}^{0}_{-1}e \longrightarrow {}^{1}_{0}n$$

Two examples of decay by *K*-capture are

$$^{7}_{4}Be + {}^{0}_{-1}e \xrightarrow{K\text{-capture}} {}^{7}_{3}Li$$

and

$$^{40}_{19}K + {}^{0}_{-1}e \xrightarrow{K\text{-capture}} {}^{40}_{18}Ar$$

The vacancy created in the 1*s* subshell as a result of *K*-capture is only temporary, and electrons from higher energy levels quickly drop to fill the 1*s* orbital. Since electrons are falling from higher energy levels to lower ones, energy is emitted in the form of electromagnetic radiation (light), in this instance in the X-ray region of the spectrum.

Elements having atomic numbers higher than 83, that is, those beyond the end of the band of stability, cannot find their way to a stable $n/p$ ratio by any of the decay modes that we just discussed. In these cases the unstable nuclei must lose both protons *and* neutrons. As a result their decay usually involves emission of $\alpha$-particles, since each $\alpha$-emission removes two protons and two neutrons simultaneously. Earlier, for example, we saw this type of decay process for uranium, that is,

$$^{238}_{92}U \longrightarrow {}^{4}_{2}He + {}^{234}_{90}Th$$

Another type of nuclear transformation that is available to the heavy elements is **fission,** in which a heavy nucleus splits into several much lighter fragments, many of which may also lie outside the band of stability and hence may be radioactive. The smaller nuclei that are produced by fission, if they are unstable, are able to undergo the simpler types of decay in order to produce a stable nucleus. We shall take a closer look at nuclear fission in Section 23.6.

We may also observe that nuclei with even numbers of protons and neutrons are apparently more stable than those containing an odd number of these particles. For example, there are 157 stable isotopes in which there are an even number of both protons and neutrons, 52 isotopes having an even number of protons and an odd number of neutrons, and 50 with an even number of neutrons but an odd number of protons. By contrast, there are only five stable nuclides in which there are an odd number of both protons and neutrons.

| Protons<br>Neutrons | Even<br>Even | Even<br>Odd | Odd<br>Even | Odd<br>Odd |
|---|---|---|---|---|
| Stable nuclei | 157 | 52 | 50 | 5 |

This phenomenon suggests that in stable nuclei protons and neutrons each tend to be paired, in much the same way that electrons become paired in the outer region of the atom, and that extra stability, as evidenced by the number of stable nuclides, results when pairing takes place with both protons and neutrons. On the other hand, when pairing cannot occur, as must be true when the numbers of protons and neutrons are both odd, very few stable isotopes occur (most isotopes having odd numbers of both protons and neutrons are radioactive).

A final observation on nuclear stability is that nuclei that contain certain specific numbers of protons and neutrons possess a degree of extra stability. These so-called *magic numbers* for protons and neutrons are 2, 8, 20, 28, 50, and 82, with an additional magic number of 126 for neutrons. When nuclei contain a magic number of both protons and neutrons, they are said to be *doubly magic* and are extremely stable. Examples are $^{4}_{2}He$, $^{16}_{8}O$, $^{40}_{20}Ca$, and $^{208}_{82}Pb$.

The occurrence of these magic numbers suggests a shell structure for the nucleus somewhat akin to the shell structure exhibited by electrons. For example, we have seen that very stable (unreactive) electron configurations occur when an atom contains magic numbers of 2, 8, 18, 36, or 54 electrons, corresponding to the noble gases, He through Kr. In the nucleus, then, it seems that nuclear shells of either protons or neutrons become completed when these magic numbers are reached and that a particularly stable nucleus occurs whenever there is a completed shell of either neutrons or protons. Exceptionally stable nuclei result when the nucleus contains filled shells of protons and neutrons simultaneously.

## 23.4 EXTENSION OF THE PERIODIC TABLE

The heaviest naturally occurring element is uranium and, as we saw earlier in this chapter, elements beyond $Z = 92$ are all prepared artificially by bombarding lighter nuclei with protons, $\alpha$-particles, and the positive ions of some of the second-period elements. The discovery of these new elements quite expectedly prompted chemists to begin to think about a whole host of new elements with new and interesting properties to be studied. However, it soon became apparent that as the atomic number of the artificial element became higher, its half-life became shorter and, hence, the prospects for stable elements of very high atomic numbers became dim.

Recent calculations by many nuclear physicists, based on the nuclear shell model, now suggest that a closed nuclear shell for protons exists at $Z = 114$ and that one for neutrons occurs at 184. As a result, chemists have once again begun to speculate about the possibilities of new stable elements.

One proposed extension of the periodic table to include these heavy and superheavy elements is shown in Figure 23.6. Recall that the actinide series, which occurs as the result of the filling of the 5*f* subshell, ends at lawrencium, $Z = 103$. The next element, 104, therefore lies under hafnium if we follow the scheme for the filling of subshells developed in Chapter 3. Elements 104 to 112, therefore, would correspond to the filling of the 6*d* sub-

Figure 23.6
*Extended periodic table. From G. Seaborg,* Journal of Chemical Education, *Vol. 46, p. 626, October 1969. Used by permission.*

| | | | | | | | | | | | | | | | | | |
|---|---|---|---|---|---|---|---|---|---|---|---|---|---|---|---|---|---|
| H 1 | | | | | | | | | | | | | | | | | He 2 |
| Li 3 | Be 4 | | | | | | | | | | | B 5 | C 6 | N 7 | O 8 | F 9 | Ne 10 |
| Na 11 | Mg 12 | | | | | | | | | | | Al 13 | Si 14 | P 15 | S 16 | Cl 17 | Ar 18 |
| K 19 | Ca 20 | Sc 21 | Ti 22 | V 23 | Cr 24 | Mn 25 | Fe 26 | Co 27 | Ni 28 | Cu 29 | Zn 30 | Ga 31 | Ge 32 | As 33 | Se 34 | Br 35 | Kr 36 |
| Rb 37 | Sr 38 | Y 39 | Zr 40 | Nb 41 | Mo 42 | Tc 43 | Ru 44 | Rh 45 | Pd 46 | Ag 47 | Cd 48 | In 49 | Sn 50 | Sb 51 | Te 52 | I 53 | Xe 54 |
| Cs 55 | Ba 56 | La 57 | Hf 72 | Ta 73 | W 74 | Re 75 | Os 76 | Ir 77 | Pt 78 | Au 79 | Hg 80 | Tl 81 | Pb 82 | Bi 83 | Po 84 | At 85 | Rn 86 |
| Fr 87 | Ra 88 | Ac 89 | (104) | (105) | (106) | (107) | (108) | (109) | (110) | (111) | (112) | (113) | (114) | (115) | (116) | (117) | (118) |
| (119) | (120) | (121) | (154) | (155) | (156) | (157) | (158) | (159) | (160) | (161) | (162) | (163) | (164) | (165) | (166) | (167) | (168) |

| | | | | | | | | | | | | | | |
|---|---|---|---|---|---|---|---|---|---|---|---|---|---|---|
| Lanthanides | Ce 58 | Pr 59 | Nd 60 | Pm 61 | Sm 62 | Eu 63 | Gd 64 | Tb 65 | Dy 66 | Ho 67 | Er 68 | Tm 69 | Yb 70 | Lu 71 |
| Actinides | Th 90 | Pa 91 | U 92 | Np 93 | Pu 94 | Am 95 | Cm 96 | Bk 97 | Cf 98 | Es 99 | Fm 100 | Md 101 | No 102 | Lr 103 |
| Super-actinides | (122) | (123) | (124) | | | | | | | | | | | (153) |

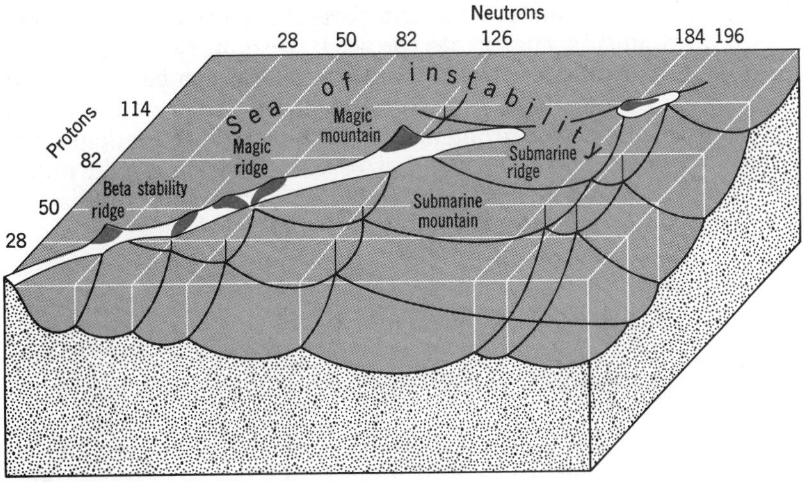


Figure 23.7
*Known and predicted regions of nuclear stability, surrounded by a sea of instability. From G. Seaborg,* Journal of Chemical Education, *Vol. 46, p. 626, October 1969. Used by permission.*

shell. Next we have six elements in the *p*-block (113 to 118), which would have their 7*p* subshell gradually filled. Elements 119 and 120, in period 8, correspond to the completion of the 8*s* subshell, and following element 121 there occurs a sequence of 32 inner transition elements, from $Z = 122$ to 153. These *superactinides* would be accounted for by the completion of first the 6*f* subshell (14 elements) followed by the filling of a 5*g* subshell (a *g* subshell would contain nine orbitals that can accommodate 18 electrons; therefore we would have 18 more elements, 136 to 153). After the superactinides we would fill the 7*d* subshell (elements 154 to 162) and then the 8*p* subshell (163 to 168).

In the search for new elements it is expected that nuclides that differ much from $Z = 114$ would be extremely unstable and would decompose by fission with very short half-lives. However, in the vicinity of $Z = 114$ it has been suggested that fission should not occur and that the half-lives of these elements with respect to $\alpha$- and $\beta$-decay should be long enough so that it should be possible to detect them and perhaps even investigate their chemical properties. There is even the possibility that these superheavy elements will not be radioactive at all.

The relative stabilities of nuclides containing differing numbers of protons and neutrons have been dramatized in a drawing (Figure 23.7) published by Dr. Glenn T. Seaborg, formerly head of the U. S. Atomic Energy Commission. Here the stable nuclei of our band of stability are shown as a long peninsula extending out into a sea of instability. Stable nuclei correspond to points above sea level, whereas submerged regions constitute unstable nuclei. Notice that nuclei with magic numbers of either protons or neutrons are shown as higher, more stable ridges, while doubly magic nuclei are shown as mountains of stability.

The superheavy elements with approximately a magic number of 114 protons and either 184 or 196 neutrons are depicted as an island of stability separated from the peninsula by a region of high nuclear instability. As a result, in order to reach this island we cannot bombard stable (or relatively stable) nuclei with light particles such as $^4_2He$, because this simply places the product nuclei into the sea of instability where they decompose rapidly before any additional mass and charge can be added. Consequently, the jump to the island must be made in one step. At the present time this presents substantial problems, because bombarding nuclei must contain a $n/p$ ratio of at least 1.6 (184/114 = 1.61). However, light nuclei such as $^{40}_{18}Ar$, while possessing sufficient protons to give the desired atomic number by a reaction such as

$$^{248}_{96}Cm + ^{40}_{18}Ar \longrightarrow ^{284}_{114}X + 4^1_0n$$

do not contain enough neutrons to place the product isotope within the island of stability. For example, $^{284}_{114}X$ contains only 284 − 114 = 170 neutrons, 14 less than the 184 that we would want in order to achieve a doubly magic nucleus. Research today is directed at obtaining suitable target and projectile nuclei that will give not only $Z = 114$ but also a sufficient number of neutrons to place the product nucleus within the bounds of the predicted island of stability.

While physicists continue their search for ways to synthesize these superheavy elements, other scientists are looking for evidence of their current or past existence in the universe. In 1975 evidence was discovered by Dr. Edward Anders at the University of Chicago's Fermi Institute that suggested the one-time existence of element 114 (or perhaps 115 or 113) in a meteorite that fell in Mexico in 1969. Other scientists at Oak Ridge National Laboratories have found crystals that may once have contained elements 116 and 126.

## 23.5 CHEMICAL APPLICATIONS

Ordinarily, nuclear changes such as those involved in radioactive decay have very little direct effect on chemical reactions, although in some cases the high-energy radiation emitted by radioactive nuclei can influence the products of reaction. This radiation is generally capable of disrupting chemical bonds. If the cleavage of a chemical bond occurs in DNA, for example, mutations of the DNA strand can be brought about. Mutational changes can also take place by reactions between DNA and the products of other cleavage reactions. For instance, free radicals are produced by the splitting of an H—O bond in water. This takes place through a series of steps, the first of which is the absorption of radiation by the water molecule and the subsequent ejection of an electron

$$H_2O + \underbrace{h\nu}_{\textbf{energy from radiation}} \longrightarrow H_2O^+ + e^-$$

The final products of the reaction are a hydrogen atom, H· (the dot indicates the unpaired electron in the radical) and a *hydroxyl radical*, ·OH. The overall change is

$$H_2O + h\nu \longrightarrow H\cdot + \cdot OH$$

Free radicals are extremely reactive because of the presence of their unpaired electrons, and their interactions with DNA can cause mutations or otherwise disrupt the replication of the DNA strands.

It is only in rare cases that the chemist makes direct use of the energy emitted in nuclear transformations. Most chemical applications of radioactive nuclides stem from their ease of identification and detection, even when they are present in very small amounts. Hence radioactive isotopes are usually employed in **tracer studies,** where they may be added in very small amounts and used to follow, or trace, the course of a chemical reaction. The range of applications of these tracer techniques is limited only by the imagination and ingenuity of the experimenter. Let us take a brief look at some examples that demonstrate the scope of these applications.

ANALYTICAL CHEMISTRY. There are many examples of analytical uses for radioactive isotopes. One of these techniques, called **isotope dilution,** can be used when it is impossible to separate completely a desired substance from a mixture. In this case, a small measured amount of the substance containing a known quantity of a radioactive isotope is *added* to the mixture. After making sure that complete mixing has occurred, a small amount of the *pure* desired substance is separated from the mixture. This sample will contain some of the added radioactive isotope, and from the proportion of the labeled isotope present in the sample the total quantity of the substance in the original mixture can be computed.

Consider, for instance, a mixture of salts of similar solubilities, such as a mixture of $KNO_3$ and NaCl. By fractional crystallization only a portion of the $KNO_3$ can be separated from the mixture. As a result, we cannot determine, in a simple fashion, how much of this salt is in the mixture.

Suppose, now, that 1.0 g of $KNO_3$ containing a small amount of radioactive $^{40}K$ is added to the salt mixture and then some $KNO_3$ (now containing K from the original mixture as well as from the added tagged $KNO_3$) is separated by fractional crystallization. If the specific activity[2] of this $KNO_3$ has dropped to 1% of the specific activity of the added $KNO_3$, then we know that only 1% of the added solid has been recovered in our $KNO_3$ sample and that the other 99% of the $KNO_3$ must have been present in the original mixture. In other words, after we had added the 1 g of labeled $KNO_3$ there was a 99-to-1 ratio of unlabeled to labeled salt. Therefore the original mixture must have contained 99 g of $KNO_3$.

Isotope dilution methods are also used when the volume of a liquid in an irregular container must be measured. A small known volume of radioactive material is added and after mixing is complete, the extent of dilution allows one to calculate backwards to find the initial liquid volume. This method has been used to measure blood volumes in living animals and the volumes of underground reservoirs of water.

Another technique that is applicable to analytical chemistry is called **neutron activation analysis.** When nonradioactive isotopes are bombarded by neutrons, heavy isotopes of these elements can be produced. The product of this reaction may lie outside the band of stability and hence be radioactive. Even if another nonradioactive nucleus is produced, however, the absorption of these neutrons generally gives nuclei that are excited and that emit $\gamma$-radiation in much the same way that an excited atom emits light when it returns to the ground state.

$$^{A}_{Z}X + ^{1}_{0}n \longrightarrow ^{A+1}_{Z}X^{*} \qquad \text{(the asterisk indicates an excited nucleus)}$$
$$^{A+1}_{Z}X^{*} \longrightarrow ^{A+1}_{Z}X + h\nu \qquad (\gamma \text{ photon})$$

Since each element has its own characteristic $\gamma$-emission spectrum, an analysis of the energies of the $\gamma$-emissions from the activated sample allows

[2] Specific activity is defined as the number of counts per minute per gram of sample.

its composition to be determined. In addition, from the intensity of the emitted $\gamma$-radiation, the concentration of each element can be computed.

This technique has some very useful advantages. First, it is nondestructive. Since the number of nuclei that must be activated to perform the analysis is small, most of the sample is unaffected. Second, as implied in the preceding sentence, the method is very sensitive and is, therefore, well suited to the analysis of trace amounts of impurities. In some cases, sensitivities of the order of $10^{-12}$ g can be achieved.

**DESCRIPTIVE CHEMISTRY.** Many of the elements having atomic numbers greater than $Z = 83$ (bismuth) have short half-lives and, hence, are not observed to occur naturally. Instead, they must be synthesized in particle accelerators; consequently, only extremely small quantities of these elements have ever been prepared. A question then arises: How can we study their chemistry if we cannot even obtain enough to be able to see them?

To arrive at the solution to this problem let us consider the element astatine. Astatine was first produced in the cyclotron by the reaction,

$$^{209}_{83}\text{Bi} + {}^{4}_{2}\text{He} \longrightarrow {}^{211}_{85}\text{At} + 2\,{}^{1}_{0}n$$

in which the $^{211}$At produced has a half-life of only about 7.5 hr. The most stable isotope, $^{210}$At, has a half-life of only 8.3 hr, so that large quantities of the element cannot be accumulated.

Since astatine occurs in Group VIIA, we expect the element to be similar in some of its properties to iodine. To verify this the astatine is added as a tracer in reactions involving iodine, and the fate of the At is followed as the iodine undergoes reactions. If in a given reaction the At occurs in the products along with the iodine, we conclude that, in this reaction, At behaves just as I does. Hence, we have discovered something about the chemical behavior of an element that we cannot even see! For instance, it is observed that, like iodine, elemental astatine is rather volatile, since it is carried with the iodine when $I_2$ is sublimed. In solution $At^-$ is carried from solution along with $I^-$ upon the addition of $Ag^+$. Thus we conclude that AgAt is insoluble just as is AgI.

**REACTION MECHANISMS.** In Chapter 11 we saw that a study of the effect of the concentrations of the reactants on the rate of a chemical reaction can often give some insight into the mechanism of the reaction. Such studies, however, seldom answer all of the questions that we might ask about the reaction mechanism. Consider, for example, the reaction of an alcohol and an organic acid to produce an ester and water,

$$\text{R—OH} + \text{HO—}\overset{\displaystyle \text{O}}{\overset{\|}{\text{C}}}\text{—R}' \longrightarrow \text{R—O—}\overset{\displaystyle \text{O}}{\overset{\|}{\text{C}}}\text{—R}' + H_2O$$

Upon the formation of the ester molecule, two hydrogen atoms and one oxygen atom are removed from the alcohol and acid to become a molecule of water. There seems little doubt about the origin of the two hydrogen atoms; however, there is a question about which one of the —OH oxygen atoms is removed and finds its way into the $H_2O$ molecule.

This question can be resolved by carrying out the reaction with a labeled oxygen (for example, $^{18}$O) incorporated into the OH group of either the alcohol or the acid. For instance, if the alcohol is labeled with $^{18}$O, it is found that all of the labeled oxygen becomes incorporated into the ester. On the other hand, if the acid contains $^{18}O$ in the OH group, all of the labeled oxygen ends up in the water with none in the ester. It is clear, therefore, that the reaction involves the removal of the OH from the acid and the H from the alcohol.

Reaction using labeled alcohol

$$R{-}O^{*}{-}H + H{-}O{-}\overset{\overset{\displaystyle O}{\|}}{C}{-}R' \longrightarrow R{-}O^{*}{-}\overset{\overset{\displaystyle O}{\|}}{C}{-}R' + H_2O$$

Reaction using labeled acid

$$R{-}O{-}H + H{-}O^{*}{-}\overset{\overset{\displaystyle O}{\|}}{C}{-}R' \longrightarrow R{-}O{-}\overset{\overset{\displaystyle O}{\|}}{C}{-}R' + H_2O^{*}$$

Many similar experiments using tagged atoms have been employed to aid in the elucidation of a large number of reaction mechanisms, including biological and biochemical processes. For instance, labeled water can be added to the root system of a plant, and its progression into the stem—and ultimately into the leaves—can be traced. Experiments using $^{14}C$-labeled $CO_2$ have been used to follow the course of carbon in photosynthesis in plants. In this case plants are exposed to $CO_2$ and, at various intervals, are killed and their cellular components separated to determine which compounds have had $^{14}C$ built into them. In this way the sequence of reactions in photosynthesis can be unraveled.

## 23.6 NUCLEAR FISSION AND FUSION

In the late 1930s in Germany, two chemists, Otto Hahn and Fritz Strassmann, and a physicist, Lise Meitner, found that when $^{235}U$ was bombarded with neutrons the unexpected products were the isotopes $^{139}Ba$ and $^{94}Kr$, as well as three neutrons:

$$^{235}_{92}U + ^{1}_{0}n \longrightarrow ^{139}_{56}Ba + ^{94}_{36}Kr + 3\,^{1}_{0}n$$

The significance of this accidental discovery was explained by Meitner and her nephew, Otto Frisch, who pointed out that fragmentation (splitting) of the $^{235}U$ was taking place, and that a very large amount of energy was emitted during the process. The splitting of an atom into two approximately equal parts is known as **fission.** The results of the fission of $^{235}U$ were tested and substantiated by Enrico Fermi at Columbia University in New York City and by physicists at Berkeley in California. Unfortunately, scientists saw military applications of the fission process, and it was Albert Einstein who alerted President Roosevelt to its possibilities. Roosevelt responded by establishing the Manhattan Project, the research efforts of which led to the two bombs that were dropped on Hiroshima and Nagasaki, and ultimately to the production of energy by controlled nuclear fission.

Since three neutrons are produced during each fission of $^{235}U$ and since neutrons are required as a reactant for each fission process, then, potentially at least, the initial reaction is capable of triggering several additional reactions, as we can see in Figure 23.8. These reactions can in turn trigger many more and so on, permitting a nuclear chain reaction to take place. This is indeed what takes place if enough pure $^{235}U$ is present. Each fission reaction causes several others to take place, with the evolution of a tremendous amount of energy.

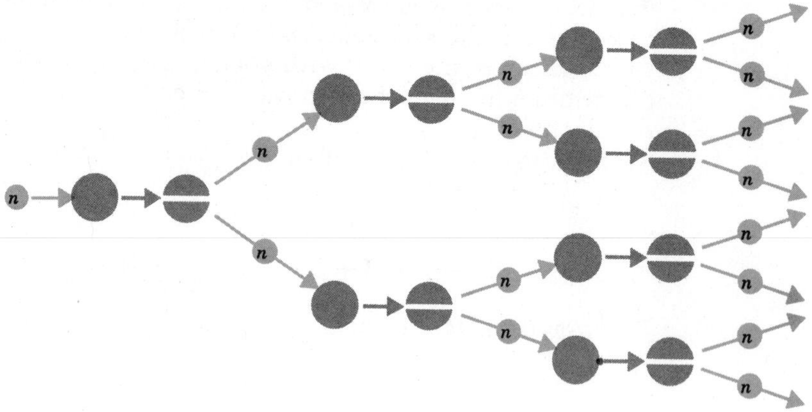


Figure 23.8
*Fission chain reaction.*

We mentioned earlier that the most abundant isotope of uranium found in its naturally occurring ores is $^{238}U$. This isotope is nonfissionable and can, in fact, prohibit the fission chain reaction of the $^{235}U$ from occurring. The $^{238}U$ absorbs the neutrons emitted during the fission reaction, thus preventing the chain from continuing. There is, then, a minimum quantity of $^{235}U$ that must be present in order for the fission reaction to sustain itself. This minimum quantity needed for fission is called the isotope's **critical mass.**

**NUCLEAR REACTORS.** In a nuclear reactor, fission reactions take place, but at a controlled rate: slow enough to avoid a chain explosion but fast enough to produce usable heat. The fission reactions are controlled by the use of control rods made of such materials as cadmium, which absorb neutrons and thus prohibit the chain from occurring too rapidly. When these rods are extended all the way into the reactor (or pile), fission occurs very slowly. The rate can be increased by withdrawing the rods. When they are pulled out, they absorb fewer and fewer neutrons and the reaction occurs faster and faster. The ideal position of the rods is at that point where the fission reaction is just able to sustain itself at a desired level.

The large amount of energy generated in the controlled nuclear fission reaction appears primarily as heat, and hence nuclear reactors must be cooled. There are, of course, many current applications of this released thermal energy in the production of electrical power. The heat removed from the reactor is used to convert water to steam, which is then used to drive turbines that generate electricity (Figure 23.9).

As with our supply of fossil fuels, there is only a limited supply of the fissionable $^{235}U$, and nuclear reactors would face a somewhat uncertain future were it not possible to generate other fissionable isotopes. In a **breeder reactor** some of the control rods are replaced with rods containing $^{238}U$. Some of the neutrons produced in the fission reaction are absorbed by the $^{238}U$ and give the reaction,

$$^{238}_{92}U + ^{1}_{0}n \longrightarrow ^{239}_{92}U$$

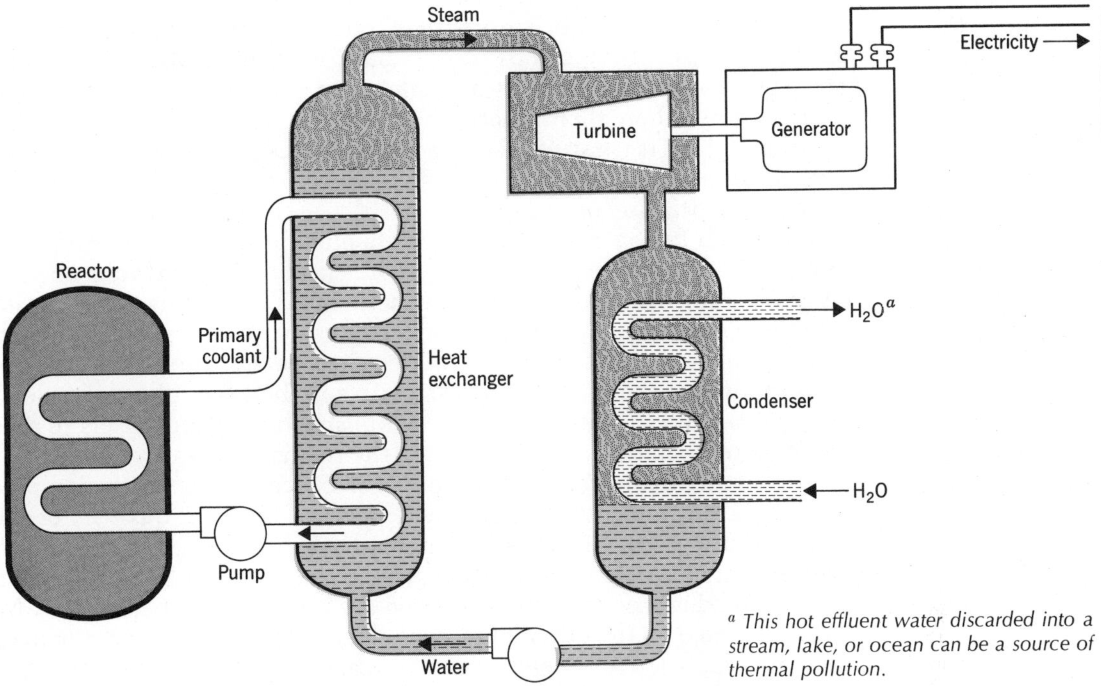


[a] *This hot effluent water discarded into a stream, lake, or ocean can be a source of thermal pollution.*

Figure 23.9
*Application of nuclear fission to the production of electricity.*

The $^{239}_{92}U$ decays rapidly to yield ultimately $^{239}_{94}Pu$ which, like $^{235}_{92}U$, is fissionable.[3]

$$^{239}_{92}U \longrightarrow {}^{239}_{93}Np + {}^{0}_{-1}e$$
$$^{239}_{93}Np \longrightarrow {}^{239}_{94}Pu + {}^{0}_{-1}e$$

Breeder reactors thus have the useful property that they produce as much or *more* fissionable isotopes as they consume.

Nuclear reactors, in general, have the undesirable effect of producing highly radioactive by-products, some with extremely long half-lives, that are nearly impossible to dispose of safely. Nevertheless, the energy crisis in recent years has placed increasing pressure on the need to develop additional sources of electrical power, including the exploitation of nuclear energy.

**NUCLEAR FUSION.** Quite the opposite of nuclear fission (fragmentation) is nuclear fusion. In this process two isotopes, usually very light ones, are brought together to form a new product. In so doing a very large amount of energy is released. The fusion reaction known as the hydrogen bomb is the reaction,

$$^{2}_{1}H + {}^{3}_{1}H \longrightarrow {}^{4}_{2}He + {}^{1}_{0}n + \text{energy}$$

This is also one of the reactions taking place on the sun and accounts for the production of a good deal of its energy.

[3] Plutonium is exceedingly poisonous. Even very tiny amounts produce cancer with near certainty.

Fusion reactions, not surprisingly, possess a high energy of activation mainly because of the electrostatic repulsion between the two nuclei that are being joined together. As a result, they occur only at extremely high temperatures where their kinetic energies are sufficient to overcome this repulsion. In fact, it is estimated that temperatures of approximately 200 million degrees Celsius are needed for fusion to occur (by comparison, the average temperature of the sun is $4 \times 10^6$ °C). The temperatures required to initiate such fusion reactions can be supplied by using an atomic (fission) bomb as a sort of "nuclear match." The energy obtained from one fusion reaction is sufficient to cause other reactions to occur; thus a chain reaction is set up, resulting in a thermonuclear explosion. In controlled fusion applications, such as the generation of electrical power, the use of an atomic bomb to initiate the fusion process is, to say the least, unacceptable. Currently work is centering on the use of multiple high-energy lasers to provide the high temperatures required to get the fusion process started.

As a potential source of commercial electrical power, the fusion process has several advantages over the fission reaction. First, the quantity of energy liberated in nuclear fusion is much greater than in fission. Another important advantage is that fusion reactions are relatively "clean" in the sense that the products of the fusion reaction are generally not radioactive. In fission reactions, on the other hand, many of the products and by-products are unstable radioactive nuclei. Fission reactors, therefore, pose a waste-disposal problem not anticipated for potential fusion reactors. As a result, scientists are attempting to establish controlled fusion reactors. One of the main obstacles to this, however, is the lack of a container that is able to hold a reaction mass that is at a temperature of $2 \times 10^8$ °C! Current research is aimed at maintaining the reacting mass of ions (called a **plasma**) suspended and enclosed within a powerful magnetic field.

## 23.7 NUCLEAR BINDING ENERGY

Nuclei are composed of protons and neutrons. Naturally we would expect that if we added up the mass of all of the protons that go into forming a nucleus, and then added in the mass of all of the neutrons, we would obtain the mass of the nucleus. Actually, however, the mass of the nucleus is always somewhat *less* than the total mass of the individual protons and neutrons. How can this be?

Within the nucleus the nuclear particles are bound together by very strong forces, the nature of which is not understood very well. Nevertheless, enormous amounts of energy have to be supplied to separate the nucleus into its component protons and neutrons. It follows that if we were to form a nucleus, this same large amount of energy would be released.

Einstein showed that mass and energy are related by his now famous equation, $E = mc^2$, where $c$ is the speed of light. The energy liberated when the nucleus is formed comes at the expense of some of the mass of the nucleons. In other words, some mass is converted to the energy that is liberated as the nucleus is formed. Consequently, the final mass of the nucleus is less than we might have expected it to be.

The energy needed to decompose the nucleus (or the energy released when it is formed) is called the **binding energy.** The difference between the actual mass of a nucleus and the sum of the masses of its individual protons and neutrons is termed the **mass defect.** Let's look at an example.

A $^4_2He$ atom is composed of two protons, two neutrons, and two electrons. These individual particles have the following masses:

| | |
|---|---|
| $p$ | 1.007277 amu |
| $n$ | 1.008665 amu |
| $e^-$ | 0.0005486 amu |

The calculated weight of a $^4_2$He atom is therefore

$$(2 \times 1.007277 \text{ amu}) + (2 \times 1.008665 \text{ amu}) + (2 \times 0.0005486 \text{ amu}) = 4.032981 \text{ amu}$$

$$\text{calculated mass } ^2_4\text{He} = 4.032981 \text{ amu}$$

The actual mass of $^4_2$He, as measured with a mass spectrometer, is 4.002603 amu. The mass defect is, then, the difference between the computed and measured mass; that is,

$$4.032981 \text{ amu} - 4.002603 \text{ amu} = 0.030378 \text{ amu}$$

Thus when a helium nucleus is formed from two protons and two neutrons, 0.030378 amu of mass is converted to energy and is released. How much energy does this represent?

Suppose that we were to form 1 mole of He atoms. The total mass lost would then be 0.030378 g. We can use Einstein's equation to calculate the energy equivalent. Using $c = 2.9979 \times 10^{10}$ cm/sec, we have

$$\begin{aligned} E &= (0.030378 \text{ g}) \times (2.9979 \times 10^{10} \text{ cm/sec})^2 \\ &= 2.730 \times 10^{19} \text{ g cm}^2/\text{sec}^2 \end{aligned}$$

Since 1 erg = 1 g cm²/sec², 1 J = 10⁷ erg, 1 kJ = 10³ J,

$$E = 2.73 \times 10^9 \text{ kJ/mole}$$

For comparison, combustion of 1 mole of $CH_4$ liberates only $8.9 \times 10^2$ kJ. The binding energy, therefore, represents a huge amount of energy.

Let us now look at the average binding energy per nucleon that occurs for various atoms. This is usually expressed in energy units of *MeV per nucleon.* Nuclear physicists generally deal in energy units of MeV. One MeV is 1 million **electron volts,** where the electron volt is the kinetic energy that an electron would acquire if it were accelerated from one electrode to another across a potential difference of 1 volt. Then,

$$1 \text{ MeV} = 10^6 \text{ eV}$$

From Einstein's equation, the energy equivalence of 1 amu can be calculated. Expressed in MeV, this is

$$1 \text{ amu} \sim 931 \text{ MeV}$$

For $^4_2$He the binding energy is therefore

$$0.030378 \text{ amu} \times \left(\frac{931 \text{ MeV}}{1 \text{ amu}}\right) = 28.3 \text{ MeV}$$

Since $^4_2$He is composed of four nucleons ($2p$, $2n$),

$$\text{average binding energy per nucleon} = \frac{28.3 \text{ MeV}}{4 \text{ nucleons}} = 7.07 \text{ MeV/nucleon}$$

The average binding energy per nucleon varies, of course, for different atoms. Figure 23.10 is a plot of this average binding energy versus mass number.

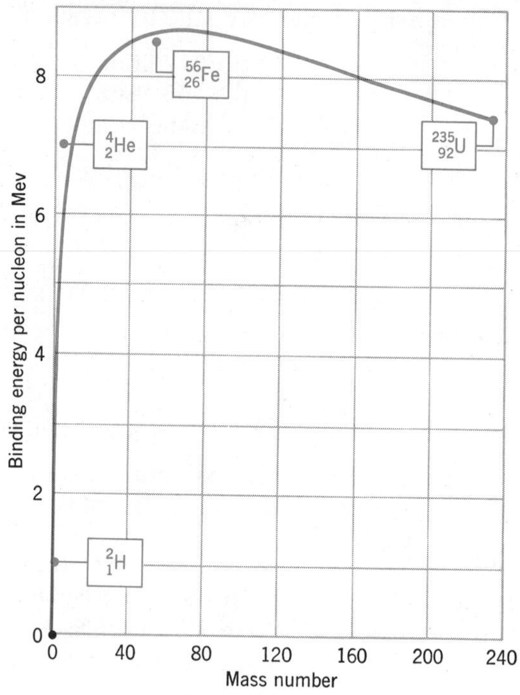


Figure 23.10
*Average nuclear binding energy per nucleon for atoms of different mass number.*

Examination of Figure 23.10 leads to several observations. We see that atoms of intermediate mass have larger binding energies than either very light atoms or very heavy ones. This means that when light nuclei are combined, as during fusion, there is a net increase in binding energy. As the heavier nucleus is formed, an amount of energy equal to this extra binding energy is released. This is the origin of the energy given off during a fusion reaction.

We can also see in Figure 23.10 that when a very heavy nucleus is split to give lighter fragments, the fission products have a greater binding energy per nucleon. Energy equal to the increased binding energy is therefore released during the fission process.

In summary, then, the energy changes in both fission and fusion are the result of changes in the binding energy experienced by the protons and neutrons as the products of these nuclear reactions are formed.

## INDEX TO QUESTIONS AND PROBLEMS (Problem numbers in italics)

General 10, 11

Decay particles 1, 2

Kinetics of radioactive decay *26, 27, 28, 29, 30, 31, 38*

Radioactive dating *32, 33*

Nuclear equations 3, 4, 5, 6, 7, 8, 9

Nuclear stability 12, 13, 14, 15, 16, 20

Extended periodic table 17, 18, 19, 21

Chemical applications 23, 24, 25, *39, 40*

Nuclear binding energy *34, 35, 36, 37, 41, 42*

## REVIEW QUESTIONS

**23.1** What are the three main types of radiation emitted by radioactive nuclei? What are their properties?

**23.2** What differences and similarities exist between $\beta$-particles and positrons?

**23.3** Complete and balance the following nuclear equations:
(a) $^{81}_{36}Kr + ^{0}_{-1}e \rightarrow ?$
(b) $^{104}_{47}Ag \rightarrow ^{0}_{1}e + ?$
(c) $^{73}_{31}Ga \rightarrow ^{0}_{-1}e + ?$
(d) $^{104}_{48}Cd \rightarrow ^{104}_{47}Ag + ?$
(e) $? + ^{0}_{-1}e \rightarrow ^{54}_{24}Cr$

**23.4** Complete and balance the following nuclear equations:
(a) $^{47}_{20}Ca \rightarrow ^{47}_{21}Sc + ?$
(b) $^{55}_{27}Co \rightarrow ^{55}_{26}Fe + ?$
(c) $^{220}_{86}Rn \rightarrow ^{116}_{84}Po + ?$
(d) $^{54}_{26}Fe + ^{1}_{0}n \rightarrow ^{1}_{1}H + ?$
(e) $^{46}_{20}Ca + ^{1}_{0}n \rightarrow ?$

**23.5** Complete and balance the following nuclear equations:
(a) $^{135}_{53}I \rightarrow ^{135}_{54}Xe + ?$
(b) $^{245}_{97}Bk \rightarrow ^{4}_{2}He + ?$
(c) $^{238}_{92}U + ^{12}_{6}C \rightarrow ^{246}_{98}Cf + ?$
(d) $^{96}_{42}Mo + ^{2}_{1}H \rightarrow ^{1}_{0}n + ?$
(e) $^{20}_{8}O \rightarrow ^{20}_{9}F + ?$

**23.6** Complete and balance the following nuclear equations:
(a) $^{35}_{17}Cl + ^{1}_{0}n \rightarrow ^{35}_{16}S + ?$
(b) $^{40}_{19}K \rightarrow ^{0}_{-1}e + ?$
(c) $^{98}_{42}Mo + ^{1}_{0}n \rightarrow ^{0}_{-1}e + ?$
(d) $^{229}_{90}Th \rightarrow ^{4}_{2}He + ?$
(e) $^{184}_{80}Hg \rightarrow ^{184}_{79}Au + ?$

**23.7** Write balanced equations for the nuclear decay reactions below.
(a) Alpha emission by $^{11}_{5}B$
(b) Beta emission by $^{90}_{38}Sr$
(c) Neutron absorption by $^{107}_{47}Ag$
(d) Neutron emission by $^{88}_{35}Br$
(e) Electron absorption by $^{116}_{51}Sb$
(f) Positron emission by $^{70}_{33}As$
(g) Proton emission by $^{41}_{19}K$

**23.8** Write nuclear equations for the following processes:
(a) $^{27}_{13}Al(\alpha,n)^{30}_{15}P$
(b) $^{209}_{83}Bi(d,n)^{210}_{84}Po$; ($d$ = deuteron, $^{2}_{1}H$)
(c) $^{15}_{7}N(p,\alpha)^{12}_{6}C$
(d) $^{12}_{6}C(p,\gamma)^{13}_{7}N$
(e) $^{14}_{7}N(\alpha,p)^{17}_{8}O$

**23.9** Write nuclear equations for the following:
(a) $^{242}_{96}Cm(\alpha,n)^{245}_{98}Cf$
(b) $^{108}_{48}Cd(n,\gamma)^{109}_{48}Cd$
(c) $^{14}_{7}N(n,p)^{14}_{6}C$
(d) $^{27}_{13}Al(d,\alpha)^{25}_{12}Mg$
(e) $^{249}_{98}Cf(^{18}_{8}O,4n)^{263}_{106}Xe$

**23.10** Describe how the Geiger-Müller counter works.

**23.11** Show that Equation 23.1 reduces to Equation 23.2 if we take $[A] = \frac{1}{2}[A]_0$.

**23.12** What is the significance of the *band of stability*? What decay processes are likely to occur for nuclides that have $n/p$ ratios that place them above the band of stability?

**23.13** Elements with atomic numbers greater than 83 generally decay by either $\alpha$-emission or fission. Why are the other forms of decay less likely for these nuclides?

**23.14** What is a *magic number*? What magic numbers occur for protons? For neutrons? What magic numbers do we observe for orbital electrons?

**23.15** In the absence of any specific information about their actual stability, rank the following nuclides in their expected order of decreasing stability.
$^{4}_{2}He$ $^{39}_{20}Ca$ $^{10}_{5}B$ $^{71}_{32}Ge$ $^{58}_{28}Ni$

**23.16** What would you anticipate for the order of increasing nuclear stability for the following nuclides?
$^{3}_{2}He$ $^{40}_{20}Ca$ $^{116}_{50}Sn$ $^{13}_{6}C$ $^{192}_{77}Ir$

**23.17** What chemical and physical properties would you predict for element number 114? If any of this element were formed at the time the universe came into being, where among earthly minerals would be a likely place to search for it?

**23.18** If element 116 is found, what would be the expected formula of (a) its sodium salt, (b) its simple hydride, (c) its oxide? Would 116 be a metal or a nonmetal?

**23.19** What would be the probable mass numbers of the most stable isotopes of element 114?

**23.20** The element $^{34}_{17}Cl$ emits $\gamma$-radiation with energies of 0.14, 1.15, 2.27, 3.22, and 4.80 MeV. How does this observation support the nuclear shell theory?

**23.21** Explain the difficulties that must be overcome if a stable element with $Z = 114$ is to be made by nuclear bombardment.

**23.22** Technetium and promethium do not possess any stable isotopes. In light of the discussion in Section 23.3, can you comment on this observation?

**23.23** Dinitrogen trioxide, $N_2O_3$, is largely dissociated into NO and $NO_2$ in the gas phase where there exists the equilibrium, $N_2O_3 \rightleftharpoons NO + NO_2$. In an effort to determine the structure of $N_2O_3$, a mixture of NO and $^*NO_2$ was prepared containing isotopically labeled N in the $NO_2$. After a period of time the mixture was analyzed and found to contain substantial amounts of both $^*NO$ and $^*NO_2$. Explain how this is consistent with the structure for $N_2O_3$ being ONONO.

**23.24** The reaction $(CH_3)_2Hg + HgI_2 \rightarrow 2CH_3HgI$ is believed to occur through a transition state with the structure

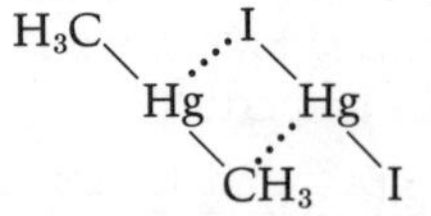

If this is so, what should be observed if $CH_3HgI$ and $^*HgI_2$ are mixed? Explain your answer.

**23.25** *Racemization* is a chemical reaction in which one optical isomer of a compound is converted into its mirror image. One possible mechanism for the racemization of octahedral complex ions containing three bidentate ligands involves the temporary loss of one of the ligands,

$$d\text{-}[M(AA)_3] \longrightarrow \begin{pmatrix} M(AA)_2 \\ + \\ AA \end{pmatrix} \longrightarrow l\text{-}[M(AA)_3]$$

This can be pictured as shown in Figure 23.11. Can you suggest a simple experiment, making use of radioisotopes, that would be able to confirm whether or not this mechanism is operative in the racemization of the $[Co(C_2O_4)_3]^{3-}$ ion?

## REVIEW PROBLEMS

**23.26** Cobalt-60 has a half-life of 5.26 years. If 1.00 g of $^{60}Co$ was allowed to decay, how much would be present after (a) one half-life, (b) three half-lives, (c) five half-lives?

**23.27** Selenium-75 has a half-life of 120.0 days. If we began with 8.00 g of $^{75}Se$, how much would remain after (a) 240 days, (b) 480 days, (c) 960 days?

**23.28** The rate constant for the decay of $^{45}Ca$ is $4.23 \times 10^{-3}$ days$^{-1}$. What is the half-life of $^{45}Ca$?

**23.29** The rate constant for the decay of $^{36}Cl$ is $2.30 \times 10^{-6}$ year$^{-1}$; what is the half-life of $^{36}Cl$?

**23.30** The half-life of $^{51}Cr$ is 27.72 days. What is the rate constant for decay of $^{51}Cr$?

**23.31** The half-life for the decay of $^{109}Cd$ is 470 days. What is the value for the rate constant for this decay?

**23.32** A sample of rock was found to contain $2.07 \times 10^{-5}$ mole of $^{40}K$ and $1.15 \times 10^{-5}$ mole of $^{40}Ar$. If we assume that all of the

Figure 23.11
*A possible mechanism for the racemization of an octahedral* $[M(AA)_3]$ *complex* (*AA = bidentate ligand*).

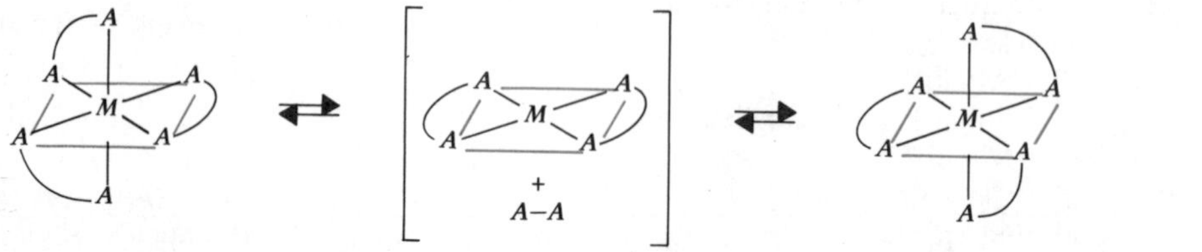

$^{40}Ar$ came from the decay of $^{40}K$, what is the age of the rock ($t_{1/2} = 1.3 \times 10^9$ years for $^{40}K$)?

**23.33** The $^{14}C$ content of an ancient piece of wood was found to be one-eighth of that in living trees. How old is this piece of wood ($t_{1/2} = 5770$ years for $^{14}C$)?

**23.34** When an electron and a positron (a positively charged electron) encounter each other, they destroy each other with the production of energy. How much energy, in joules, results from such an encounter? The rest mass of the electron is $9.1096 \times 10^{-28}$ g.

**23.35** Calculate the binding energy for the following isotopes given the following masses: $p = 1.007277$ amu, $n = 1.008665$ amu, $e = 5.4859 \times 10^{-4}$ amu.

| Isotope | Actual Atomic Mass (amu) |
|---|---|
| $^{7}_{3}Li$ | 7.01600 |
| $^{14}_{7}N$ | 14.003074 |
| $^{19}_{9}F$ | 18.99840 |

**23.36** What is the binding energy of $^{56}_{26}Fe$ (atomic mass, 55.9349 amu)? Where is this on the binding energy versus mass number curve? Why don't we have to worry about an enemy that claims they have developed the iron bomb?

**23.37** Calculate the energy liberated in the fusion reaction to produce 1 mole of helium from deuterium.

$$^{2}_{1}H + ^{2}_{1}H \longrightarrow ^{4}_{2}He$$

The accurate atomic masses are

$$^{2}_{1}H = 2.014102 \text{ amu}$$
$$^{4}_{2}He = 4.002603 \text{ amu}$$

***23.38** The following data were obtained for the decay of $^{47}Ca$.

| Time (hr) | cpm |
|---|---|
| 0.0 | 4720 |
| 8.0 | 4485 |
| 12.0 | 4372 |
| 24.0 | 4050 |
| 48.0 | 3475 |
| 72.0 | 2983 |
| 96.0 | 2560 |

Determine (a) the rate constant for the decay, (b) the half-life of $^{47}Ca$.

***23.39** A large, complex piece of apparatus has built into it a cooling system containing an unknown volume of cooling liquid. It is desired to measure the volume of the coolant without draining the lines. To the coolant was added 10 ml of methanol labeled with $^{14}C$ and having a specific activity of 580 cpm per gram. The coolant was permitted to circulate to assure complete mixing before a sample was withdrawn that was found to have a specific activity of 29 cpm per gram. Calculate the volume of coolant in the system. The density of methanol is 0.792 g/ml and the density of the coolant is 0.884 g/ml.

***23.40** A complex ion of chromium(III) with oxalate ion was prepared from $^{51}Cr$-labeled $K_2Cr_2O_7$, having a specific activity of 843 cpm/gram, and $^{14}C$-labeled oxalic acid, $H_2C_2O_4$, having a specific activity of 345 cpm/gram. Chromium-51 decays by electron capture with the emission of a $\gamma$-ray, whereas $^{14}C$ is a pure $\beta$-emitter. Because of the characteristics of the $\alpha$- and $\gamma$-detectors, each of these isotopes may be counted independently. A sample of the complex ion was observed to give a $\gamma$-count of 165 cpm and an $\alpha$-count of 83 cpm. From these data, determine the number of oxalate ions bound to each Cr(III) in the complex ion. (Hint: For the starting materials calculate the cpm per mole of Cr and oxalate, respectively.)

***23.41** Compare the energies liberated in the following fusion reactions:
(a) $2\,^{2}_{1}H \rightarrow ^{4}_{2}He$
(b) $2\,^{12}_{6}C \rightarrow ^{24}_{12}Mg$
Which reaction produces the most energy per mole of product? If each were equally feasible from an engineering standpoint, which would be preferable? Actual atomic masses: $^{2}_{1}H = 2.014102$, $^{4}_{2}He = 4.002603$, $^{24}_{12}Mg = 23.98504$.

***23.42** How many gallons of gasoline, $C_8H_{18}$, having a density of 0.703 g/ml, would have to be burned (to give $H_2O$ (*l*) + $CO_2$ (*g*)) to produce the same amount of energy as in the production of 1 mole of $^{4}_{2}He$ by the fusion reaction described in Problem 23.37? 1 gallon = 3.8 liters. $\Delta H_f^0$ of $C_8H_{18}$ (*l*) is $-208.4$ kJ/mole.

# APPENDIX A SOME COMMONLY ENCOUNTERED GEOMETRICAL SHAPES

A subject of vital concern in chemistry is the geometry, or shapes, of complex molecules and ions. For example, knowledge of molecular geometry permits the testing of theories of bonding. In some instances knowledge of molecular geometry can also account for the products of chemical reaction, and it seems almost certain that the shapes of biologically important molecules significantly control their biological functions.

Many molecular structures can be considered to be derived from a small number of simple geometric figures. In fact, when speaking of the shapes of molecules and ions, the names of these simple geometric forms are often used. The major features of three of the solid shapes are described below.

## A.1 THE TETRAHEDRON

A symmetrical tetrahedron, Figure A.1, is constructed of four equilateral triangles. There are four vertices. Molecular structures derived from the tetrahedron usually have one atom in the center that is bonded to four others at the vertices, as illustrated in the methane molecule, $CH_4$, in Figure A.2. Each H—C—H angle is 109.5°. The molecule is said to be *tetrahedral.*

## A.2 THE TRIGONAL BIPYRAMID

A trigonal bipyramid, Figure A.3, is composed of two trigonal pyramids that share a common base. The apex of one trigonal pyramid is above the base; the apex of the other is below the base (a trigonal pyramid is a pyramid that has a triangular base—a tetrahedron is a special case of a trigonal pyramid). Notice that the trigonal bipyramid has five vertices. Molecular shapes derived from this figure usually possess one atom in the center that is bonded to five others at the vertices. The molecule $PCl_5$ is illustrated in Figure A.4.

Not all bond angles are equivalent in the trigonal bipyramid. Notice that the Cl—P—Cl angles in the horizontal *equatorial* triangular plane are 120°, while those formed between an *axial* Cl atom (one that lies directly above or below the triangular plane) and an equatorial Cl atom are only 90°.

## A.3 THE OCTAHEDRON

An octahedron, Figure A.5, is an eight-sided geometric figure that might also be called a square bipyramid (two square pyramids that share a common square base). The octahedron possesses six vertices. Molecules that have this geometry generally have an atom in the center that is bonded to six others. An example is $SF_6$ (Figure A.6). Note that all of the F—S—F bond angles are the same, 90°.

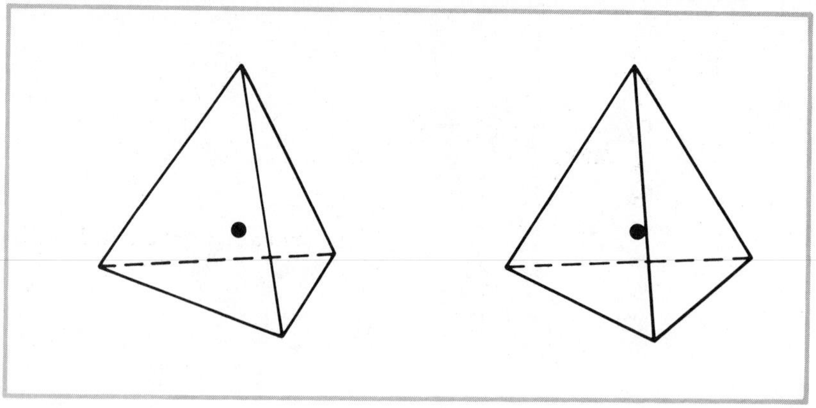

Figure A.1
*The tetrahedron.*

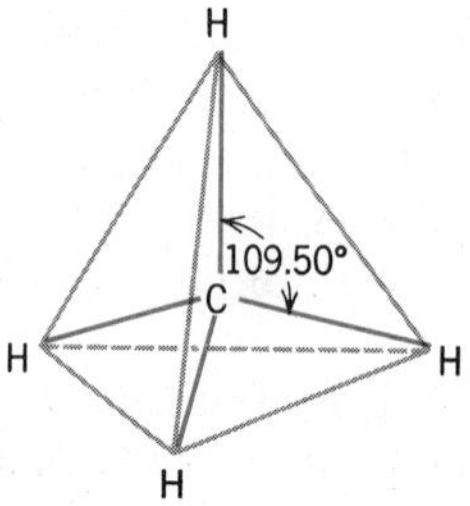


Figure A.2
*The tetrahedral methane molecule. The H—C—H angle is 109.5°.*

Figure A.3
*The trigonal bipyramid.*

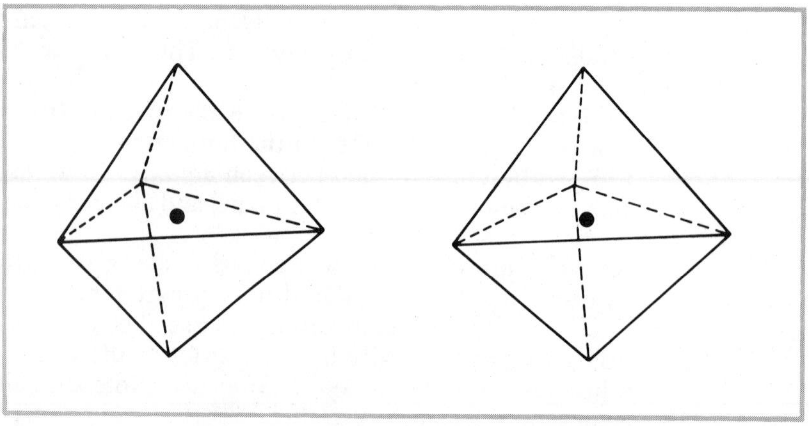

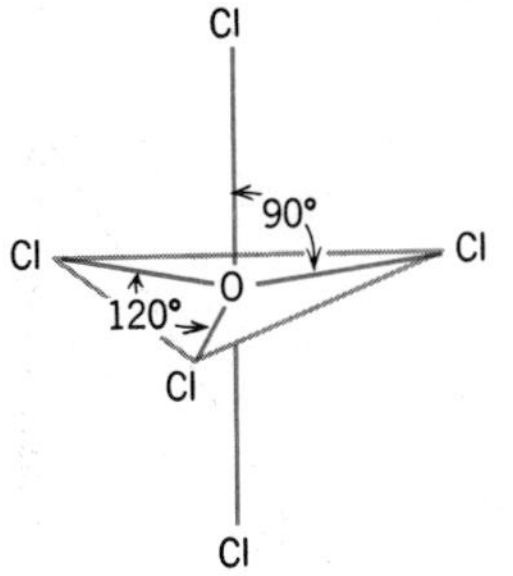


Figure A.4
*The trigonal bipyramidal* $PCl_5$ *molecule.*

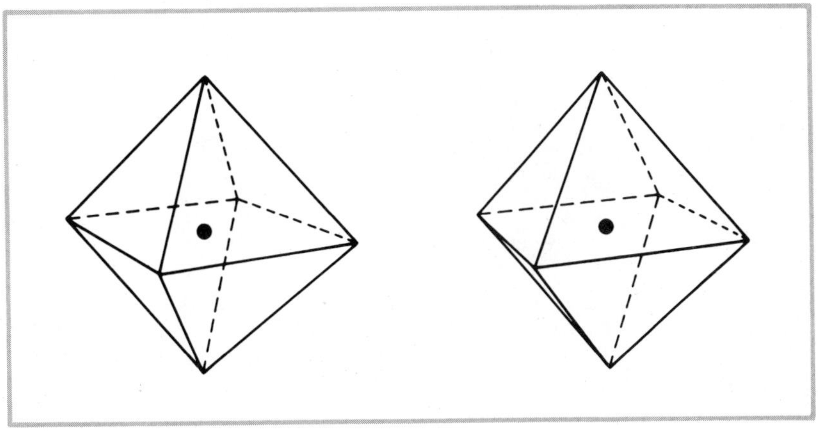

Figure A.5
*The octahedron.*

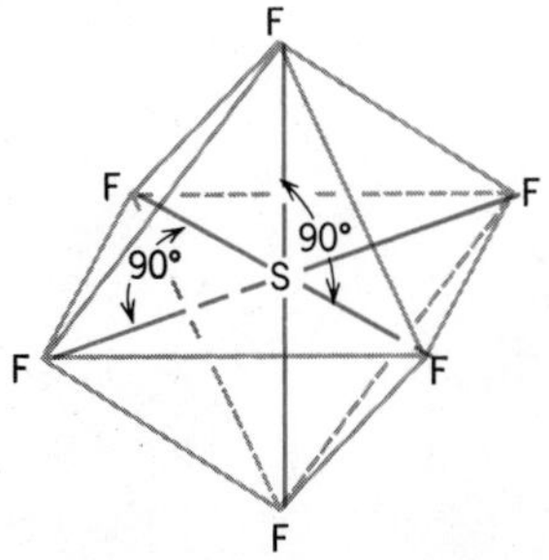


Figure A.6
*The* $SF_6$ *molecule.*

# APPENDIX B NAMING INORGANIC COMPOUNDS

This appendix gives the rules of nomenclature that apply to the naming of most simple inorganic compounds. The naming of complex inorganic substances and organic compounds will be found in Chapters 20 and 21, respectively.

## B.1 BINARY COMPOUNDS

A **binary compound** is composed of atoms of only two different elements. In naming these substances the less electronegative (more metallic) element is specified first by giving its ordinary English name. The name of the second element (almost always a nonmetal) is obtained by adding the suffix *ide* to its stem, as shown in Table B.1. Some typical examples are

| | |
|---|---|
| $NaCl$ | sodium chloride |
| $SrO$ | strontium oxide |
| $Al_2S_3$ | aluminum sulfide |
| $Mg_3P_2$ | magnesium phosphide |
| $HBr$ | hydrogen bromide |

Many elements are commonly found to exist in more than one positive oxidation state. When the element is a metal, there are two methods that may be used to indicate its oxidation state. In the older of these methods the suffixes *-ic* and *-ous* are used to differentiate between high and low oxidation states. Thus the +3 and +2 oxidation states of chromium would be specified as

| | | | |
|---|---|---|---|
| $Cr^{3+}$ | chromic; | $CrCl_3$ | chromic chloride |
| $Cr^{2+}$ | chromous; | $CrCl_2$ | chromous chloride |

When the metal has a symbol derived from the Latin name for the element, its Latin stem is generally used. For example, with iron there are two common oxidation states, $Fe^{3+}$ (ferric) and $Fe^{2+}$ (ferrous). Other common examples are found in Table B.2. Notice that this system *only* differentiates between high and low oxidation states; it does not specify what the oxidation state of the metal is.

The second and preferred method of indicating the oxidation state of the metal is called the *Stock system.* It involves placing a Roman numeral, equal to the oxidation number of the metal, in parentheses, following the regular English name of the element. Thus $Fe^{2+}$ and $Fe^{3+}$ would be iron(II) and iron(III), respectively. The alternative names for the compounds $FeCl_2$

Table B.1
*Names of anions derived from nonmetals*

| Group IVA | Group VA | Group VIA | Group VIIA |
|---|---|---|---|
| $C^{4-}$; *carb*ide[a] | $N^{3-}$ *nitr*ide | $O^{2-}$ *ox*ide | $F^-$ *fluor*ide |
| $Si^{4-}$; *silic*ide | $P^{3-}$ *phosph*ide | $S^{2-}$ *sulf*ide | $Cl^-$ *chlor*ide |
| | $As^{3-}$ *arsen*ide | $Se^{2-}$ *selen*ide | $Br^-$ *brom*ide |
| | | $Te^{2-}$ *tellur*ide | $I^-$ *iod*ide |

[a] Carbon also forms a number of complex carbides, for example, $C_2^{2-}$ in $CaC_2$.

and $FeCl_3$ are therefore

| | | | |
|---|---|---|---|
| $FeCl_2$ | ferrous chloride | or | iron(II) chloride |
| $FeCl_3$ | ferric chloride | or | iron(III) chloride |

Even though the Stock system is preferred today, it is necessary to know the older system as well. For example, if an experiment calls for $FeCl_3$, iron(III) chloride, it is likely that you will only find it in a reagent bottle labeled "ferric chloride."

When naming binary covalent compounds formed between two nonmetals, a third system of nomenclature is preferred in which the numbers of each atom in a molecule is specified by a Greek prefix: *di-* (2), *tri-* (3), *tetra-* (4), *penta-* (5), *hexa-* (6), *hepta-* (7), *octa-* (8), *nona-* (9), *deca-* (10), and so on. The Stock system is usually *not* used for these compounds because it does not distinguish between molecular formulas such as the first two examples below.

| | | |
|---|---|---|
| $NO_2$ | nitrogen(IV) oxide | nitrogen dioxide |
| $N_2O_4$ | nitrogen(IV) oxide | dinitrogen tetroxide |
| $N_2O_5$ | nitrogen(V) oxide | dinitrogen pentoxide |
| $PCl_3$ | phosphorus(III) chloride | phosphorus trichloride |
| $PCl_5$ | phosphorus(V) chloride | phosphorus pentachloride |

In some instances the prefix *mono-* (1) is also used to avoid ambiguity.

| | |
|---|---|
| $CO_2$ | carbon dioxide |
| CO | carbon monoxide |

## B.2 COMPOUNDS CONTAINING POLYATOMIC IONS

Many ions are found to contain more than one atom and are therefore referred to in general as polyatomic ions. These species enter into ionic com-

Table B.2
*Metals commonly found in two oxidation states*

| Chromium | Manganese | Iron | Cobalt | Lead |
|---|---|---|---|---|
| $Cr^{2+}$ chromous | $Mn^{2+}$ manganous | $Fe^{2+}$ ferrous | $Co^{2+}$ cobaltous | $Pb^{2+}$ plumbous |
| $Cr^{3+}$ chromic | $Mn^{3+}$ manganic | $Fe^{3+}$ ferric | $Co^{3+}$ cobaltic | $Pb^{4+}$ plumbic |

| Copper | Tin | Mercury |
|---|---|---|
| $Cu^+$ cuprous | $Sn^{2+}$ stannous | $Hg_2^{2+}$ mercurous (note that there are two Hg atoms) |
| $Cu^{2+}$ cupric | $Sn^{4+}$ stannic | $Hg^{2+}$ mercuric |

pounds as discrete units and generally stay intact in most chemical reactions. A list of these is given in Table 4.3 (p. 102). As with binary compounds, salts that contain these ions are always named with the positive ion first. Some examples are:

| | | | |
|---|---|---|---|
| $Na_2CO_3$ | sodium carbonate | $Ba(OH)_2$ | barium hydroxide |
| $Ca(C_2H_3O_2)_2$ | calcium acetate | $(NH_4)_2SO_4$ | ammonium sulfate |

The Stock system is also preferred when the metal can exist in more than one oxidation state.

| | Stock System | Old System |
|---|---|---|
| $MnSO_4$ | manganese(II) sulfate | manganous sulfate |
| $Fe_2(C_2O_4)_3$ | iron(III) oxalate | ferric oxalate |

## B.3 BINARY ACIDS

Acids can be described as substances that release $H^+$ when dissolved in water. A binary acid (also sometimes called a *hydro* acid) is a compound containing hydrogen and a nonmetal, having a general formula $H_nX$ (for example, HCl, $H_2S$). They are named as *hydro . . . ic acid,* where the stem of the name of the nonmetal is inserted in place of the dotted line. Examples are

| | |
|---|---|
| HF | hydro*fluor*ic acid |
| HCl | hydro*chlor*ic acid |
| HBr | hydro*brom*ic acid |
| HI | hydr*iod*ic acid |
| $H_2S$ | hydro*sulfur*ic acid |

When these acids are reacted with hydroxide ion (a reaction called **neutralization**), an anion is formed, for example,

$$\underset{hydro\mathbf{chlor}ic\ acid}{HCl} + OH^- \longrightarrow H_2O + \underset{\mathbf{chlor}ide}{Cl^-}$$

Note that *hydro . . . ic acids* give *. . . ide* salts, for example, sodium chloride.

## B.4 OXOACIDS

Oxoacids are acids that contain hydrogen, oxygen, and at least one other element. Sulfuric acid, $H_2SO_4$, is an example. Where the third element (S in the case of $H_2SO_4$) can exist in more than one oxidation state, more than one oxoacid is possible. For example, two common oxoacids of sulfur are $H_2SO_4$ and $H_2SO_3$, containing sulfur in the +6 and +4 oxidation states, respectively. The acid containing the element in the higher oxidation state is given the suffix *-ic,* whereas the acid having the element in the lower oxidation state is given the ending *-ous.* Thus we have

| | |
|---|---|
| $H_2SO_4$ | sulfuric acid |
| $H_2SO_3$ | sulfurous acid |

Salts produced by neutralization of these acids contain polyatomic anions (Table 4.3). The anion derived from the "ic" acid ends in *-ate,* whereas the anion from the "ous" acid ends in *-ite:*

| | | | |
|---|---|---|---|
| $H_2SO_4$ | sulfuric acid | $SO_4^{2-}$ | sulfate |
| $H_2SO_3$ | sulfurous acid | $SO_3^{2-}$ | sulfite |
| $HNO_3$ | nitric acid | $NO_3^-$ | nitrate |
| $HNO_2$ | nitrous acid | $NO_2^-$ | nitrite |
| $HClO_3$ | chloric acid | $ClO_3^-$ | chlorate |
| $HClO_2$ | chlorous acid | $ClO_2^-$ | chlorite |

Some elements form oxoacids in more than two oxidation states. In this case the prefixes *hypo-* and *per-* are used to designate a lower or higher oxidation state, respectively. A good example occurs among the oxoacids of the halogens.

| | | | |
|---|---|---|---|
| hypochlorous | $HClO$ | $ClO^-$ | hypochlorite |
| chlorous | $HClO_2$ | $ClO_2^-$ | chlorite |
| chloric | $HClO_3$ | $ClO_3^-$ | chlorate |
| perchloric | $HClO_4$ | $ClO_4^-$ | perchlorate |

## B.5 ACID SALTS

Partial neutralization of an acid that is capable of furnishing more than one $H^+$ per acid molecule gives salts that are called acid salts. Examples are

| Parent Acid | Typical Acid Salts |
|---|---|
| $H_2SO_4$ | $NaHSO_4$ |
| $H_2CO_3$ | $NaHCO_3$ |
| $H_3PO_4$ | $NaH_2PO_4$ |
| | $Na_2HPO_4$ |

When only one acid salt is formed (as with $H_2SO_4$ or $H_2CO_3$), the salt can be named by adding the prefix *bi-* to the name of the anion of the acid.

| | |
|---|---|
| $NaHSO_4$ | sodium **bi**sulfate |
| $NaHCO_3$ | sodium **bi**carbonate |

The salt can also be named by specifying the presence of the H by writing "hydrogen."

| | |
|---|---|
| $NaHSO_4$ | sodium hydrogen sulfate |
| $NaH_2PO_4$ | sodium dihydrogen phosphate |
| $Na_2HPO_4$ | sodium hydrogen phosphate (disodium hydrogen phosphate) |

Note the use of the prefix *di-* to indicate the number of hydrogen atoms (as well as to remove ambiguity as to the number of Na in the last formula).

# APPENDIX C MATHEMATICS FOR GENERAL CHEMISTRY

For many students, solving numerical problems is often the most difficult part of any chemistry course. In this appendix we review some of the mathematical concepts that you will find useful in your study of chemistry.

## C.1 THE FACTOR-LABEL METHOD OF PROBLEM SOLVING

Even after learning the principles of chemistry, students sometimes have difficulty in setting up the arithmetic correctly to give the proper numerical answer to a problem. The "factor-label" method uses the units associated with numbers as a guide in working out the arithmetic. The method is based on the idea that *units cancel from numerator and denominator in a fraction, just as numbers do.* For example, if the units *feet* appear in both numerator and denominator, they may be cancelled

$$\frac{3\,\cancel{\text{ft}}}{2\,\cancel{\text{ft}}} = \frac{3}{2}$$

Numerical problem solving uses this idea by employing valid relationships between units to create **conversion factors.** For instance, to convert 4 *yards* into *feet* the relationship,

$$3 \text{ ft} = 1 \text{ yard}$$

is used to create a conversion factor (a fraction) with which the 4 yards is multiplied. We do this by dividing both sides of this equation by "1 yard."

$$\frac{3 \text{ ft}}{1 \text{ yard}} = \frac{1\,\cancel{\text{yard}}}{1\,\cancel{\text{yard}}} = 1$$

Notice that this fraction is equal numerically to 1. If we multiply a quantity by 1, we do not change its magnitude. Thus we can multiply 4 yards by this fraction without altering the length. The only effect is to change units.

$$4\,\cancel{\text{yards}} \left(\frac{3 \text{ ft}}{1\,\cancel{\text{yard}}}\right) = 12 \text{ ft}$$

Note also that the factor was constructed deliberately so that *yards* cancelled and only the desired units, *feet,* remained. Had we inverted this conversion factor, we would obtain the wrong numerical answer *and* the wrong units,

$$4 \text{ yard} \left(\frac{1 \text{ yard}}{3 \text{ ft}}\right) = \frac{4}{3} \frac{\text{yard}^2}{\text{ft}}$$

Creation of a conversion factor can be accomplished from any *valid* relationship between a set of units. This can be an equality, as in the relationship between feet and yards (that is, 3 ft equal 1 yard). It can also be an equivalency. For instance, for a student who earns 5 dollars per hour, there is an equivalence between dollars and time.

5 dollars are equivalent to 1 hour

We shall use the symbol ~ to stand for "are equivalent to." Thus, in the example just cited,

$$5 \text{ dollars} \sim 1 \text{ hr}$$

If this student works 12 hr, we can use the dollar-hour relationship to construct a conversion factor that allows us to calculate his pay (before taxes!)

$$12 \cancel{\text{hrs}} \left(\frac{5 \text{ dollars}}{1 \cancel{\text{hr}}}\right) = 60 \text{ dollars}$$

Note that hours cancel. You will see many examples in the text in which numerical problems are solved using this "factor-label" technique.

## C.2 EXPONENTIAL NOTATION (SCIENTIFIC NOTATION)

Quite often in science it is necessary to deal with numbers that are very large, such as Avogadro's number,

602,300,000,000,000,000,000,000

or numbers that are very small, such as the mass of a single molecule of water,

0.000 000 000 000 000 000 000 03 g

These numbers are very cumbersome and difficult to work with without making mistakes in arithmetic computations. To aid us in handling these large and small numbers, a system called either **exponential notation** or **scientific notation** is employed. In this system, a number is expressed as a decimal part multiplied by 10 raised to an appropriate power. Thus

$$\begin{aligned} 200 &= 2 \times 10 \times 10 = 2 \times 10^2 \\ 205{,}000 &= 2.05 \times 100{,}000 = 2.05 \times (10 \times 10 \times 10 \times 10 \times 10) \\ &= 2.05 \times 10^5 \end{aligned}$$

To determine the exponent on the 10, we can also count the number of places the decimal must be moved to produce the number that precedes the 10 when the number is expressed in the scientific notation

205 000 ⊙ $= 2.05 \times 10^5$

5 places

Note that the exponent on the 10 is positive when the decimal is moved to the left. When it is moved to the right, the exponent is negative.

0 ⊙ 000000315 $= 3.15 \times 10^{-7}$

7 places

In carrying out arithmetic operations with numbers expressed in scientific notation, the following rules apply.

**MULTIPLICATION.** In multiplication, the decimal portions of the number are multiplied and the exponents on the 10 are *added* algebraically.

$$(2.0 \times 10^4) \times (3.0 \times 10^3) = (2.0 \times 3.0) \times 10^{(4+3)} = 6.0 \times 10^7$$
$$(4.0 \times 10^8) \times (-2.0 \times 10^{-5}) = (4.0 \times (-2.0)) \times 10^{(8+(-5))} = -8.0 \times 10^3$$

**DIVISION.** The decimal portions are divided, and the exponent on 10 in the denominator is *subtracted algebraically* from the exponent on 10 in the numerator.

$$\frac{8.0 \times 10^7}{4.0 \times 10^3} = \left(\frac{8.0}{4.0}\right) \times 10^{(7-3)} = 2.0 \times 10^4$$
$$\frac{6.0 \times 10^5}{2.0 \times 10^{-3}} = \left(\frac{6.0}{2.0}\right) \times 10^{(5-(-3))} = 3.0 \times 10^8$$
$$\frac{9.0 \times 10^{-4}}{3.0 \times 10^{-6}} = \left(\frac{9.0}{3.0}\right) \times 10^{(-4-(-6))} = 3.0 \times 10^2$$

You have probably noticed that the usual practice is to express a number with the decimal point located between the first and second digit. There are, of course, other ways that these numbers can be written that are all equivalent, and you will undoubtedly find occasions where it is convenient to use a number in other than its standard form. An example of a few equivalent expressions of the same number are

$$3.15 \times 10^{-7} = 315 \times 10^{-9} = 0.0315 \times 10^{-5}$$

Notice that in converting from one to another, one part of the number is increased while the other is decreased. For instance, to change $8.25 \times 10^6$ to $825 \times 10^4$, multiply *and* divide by 100 (or $10^2$)

$$8.25 \times 10^6 \left(\frac{100}{100}\right) = (8.25 \times 100) \times \left(\frac{10^6}{10^2}\right) = 825 \times 10^4$$

**ADDITION AND SUBTRACTION.** When carrying out addition and subtraction, each quantity must first be written with the same power of 10. Then addition or subtraction is performed on the decimal parts; the power of 10 remains the same. For example,

$$(2.17 \times 10^5) + (3.0 \times 10^4) = ?$$

If we express both numbers with the same power of 10, we have

$$\begin{array}{r} 2.17 \times 10^5 \\ +0.30 \times 10^5 \\ \hline 2.47 \times 10^5 \end{array} \quad \text{or} \quad \begin{array}{r} 21.7 \times 10^4 \\ +3.0 \times 10^4 \\ \hline 24.7 \times 10^4 \end{array}$$

**TAKING A ROOT.** To extract a root (e.g., the square root), the exponent on the 10 is made to be divisible by the desired root. For instance, to take the square root of $3.7 \times 10^7$, we first change the number so that the power of 10 is divisible by 2. Then we take the square root of the decimal part and divide the exponent by 2.

$$\sqrt{3.7 \times 10^7} = \sqrt{37 \times 10^6} = \sqrt{37} \times 10^3 = 6.1 \times 10^3$$

## C.3 LOGARITHMS

A logarithm is an exponent! **Common logarithms** are exponents to which 10 must be raised to give a specified number. For instance, the log (100) = 2 because $10^2 = 100$. Similarly, log (1000) = log ($10^3$) = 3.

Since logarithms are exponents, when we perform mathematical operations the same rules that apply to exponents also apply to logarithms. Thus we have

$$\text{Multiplication} \quad \begin{cases} \text{add exponents} \\ \text{add logarithms} \end{cases}$$

$$\text{Division} \quad \begin{cases} \text{subtract exponents} \\ \text{subtract logarithms} \end{cases}$$

For example,

$$\boxed{10^3 \times 10^4} = 10^{3+4} = \boxed{10^7}$$

$$\log(10^3 \times 10^4) = \log(10^3) + \log(10^4) = 3 + 4 = 7 = \log(10^7)$$

Similarly, for division

$$\boxed{\frac{10^8}{10^6}} = 10^{8-6} = \boxed{10^2}$$

$$\log\left(\frac{10^8}{10^6}\right) = \log(10^8) - \log(10^6) = 8 - 6 = 2 = \log(10^2)$$

For decimal numbers between 1 and 10 their logarithms lie between 0 and 1, since

$$\log(1) = 0 \qquad (1 = 10^0)$$
$$\log(10) = 1 \qquad (10 = 10^1)$$

For example, $\log 2 = 0.3010$ or

$$10^{0.3010} = 2$$

The logarithm of 2 and other numbers between 1 and 10 can be obtained from the table of logarithms in Appendix D.

To use this table to find the logarithm of a number, we use the extreme left column to locate the first two digits of the number, and the top horizontal row to locate the third digit. The value in the table corresponding to these is the logarithm of our number. For example, if we want log (4.61), we would locate 46 in the left column and proceed to the right until we were in the column headed by 1.

The answer is

$$\log(4.61) = 0.6637$$

This table is extremely easy to use as long as our numbers are expressed in this fashion, that is, as a decimal number between 1 and 10. If the number whose logarithm we seek does not appear this way, we can first express the number in exponential notation and then take its logarithm. For example, what is log(728)?

$$\log(728) = \log(7.28 \times 10^2)$$
$$\log(7.28 \times 10^2) = \log(7.28) + \log(10^2)$$

$$\log(7.28) = 0.8621 \qquad \text{(from table)}$$
$$+ \log(10^2) = \underline{2.0000}$$

therefore, $$\log(728) = 2.8621$$

What would be the value of log(0.00583)? Once again we first express the number in exponential notation:

$$\log(0.00583) = \log(5.83 \times 10^{-3})$$
$$\log(5.83 \times 10^{-3}) = \log(5.83) + \log(10^{-3})$$

$$\log 5.83 = +0.7657$$
$$+ \log 10^{-3} = \underline{-3.0000}$$

Adding these algebraically, we get

$$\log(0.00583) = -2.2343$$

Sometimes it is necessary to obtain the number whose logarithm is known. This is called taking the **antilogarithm.** The procedure is simply the reverse of that given above. For example, suppose that we wish to find the number whose logarithm is 3.253.

$$\log x = 3.253$$

First, we divide the number into two parts, a positive integer and a decimal.

$$3.253 = 3 + 0.253 = 0.253 + 3$$
$$\log x = (0.253 + 3)$$

We locate 0.253 in the body of the log table and find that it is the log of 1.79; we also know that 3 is the log of $10^3$. Therefore,

$$\log x = \log(1.79) + \log(10^3) = \log(1.79 \times 10^3)$$
$$x = 1.79 \times 10^3$$

NATURAL LOGARITHMS. A system of logarithms encountered frequently in the sciences, known as natural logarithms, has as its base $e = 2.71828. . . .$ In other words, natural logarithms are exponents to which $e$ must be raised to give a number. The relationship between common logs and natural logs is seen below

$$\log_{10}(10) = 1 \quad \text{or} \quad 10^1 = 10$$
$$\log_e(10) = 2.303 \quad \text{or} \quad e^{2.303} = 10$$

With common logarithms we usually omit the base and write simply, $\log 10 = 1$. With natural logarithms the base $e$ is omitted, and they are written

$$\ln 10 = 2.303$$

The conversion from base $e$ to base 10 logarithm is accomplished by the equation

$$\ln x = 2.303 \log x$$

## C.4 THE QUADRATIC EQUATION

When an equation can be written in the form

$$ax^2 + bx + c = 0$$

in which the coefficients $a$, $b$, and $c$ are known, two values (called roots) of the variable $x$ can be obtained by substituting the values of $a$, $b$, and $c$ into the expression

$$x = \frac{-b \pm \sqrt{b^2 - 4ac}}{2a}$$

For example, given the equation

$$x^2 - 5x + 4 = 0$$

what is the value of x? In this equation $a = 1$, $b = -5$, and $c = 4$. Thus

$$x = \frac{-(-5) \pm \sqrt{(-5)^2 - 4\,(1)(4)}}{2\,(1)} = \frac{5 \pm \sqrt{25 - 16}}{2}$$

$$= \frac{5 \pm \sqrt{9}}{2} = \frac{5 \pm 3}{2}$$

Therefore,

$$x = \frac{2}{2} = 1 \quad \text{and} \quad x = \frac{8}{2} = 4$$

Both values of x are mathematically correct. Usually when a quadratic equation is encountered in a chemical problem, only one of the roots has any real significance. Generally, the other root will be clearly meaningless; for instance, a negative concentration, which is impossible (you can't have a smaller amount of matter than no matter at all!).

## C.5 ELECTRONIC CALCULATORS

Today much of the tiresome work of arithmetic is relieved by the use of small, hand-held electronic calculators. Scientific calculators possessing remarkable computational power are available for less than $50 (the price of a good slide rule not too many years ago), but to get the most out of them you should be aware of some simple mathematical relationships. Those that are most useful to you in chemistry are mentioned below. Since operational procedures differ on various calculators, you will have to refer to the direction booklet that accompanies your calculator for specific instructions about how to apply these relationships.

**LOGARITHMS AND ANTILOGARITHMS.** We've seen that a logarithm is an exponent and that there are two systems of logarithms generally encountered, base *e* and base 10. If your calculator possesses logarithm capabilities, you will probably find a key labeled LN for base *e* (natural) logarithms and a key labeled LOG for base 10 (common) logarithms. Generally, if a number is entered and the LN key depressed, the display will show the natural log of that number. The common log would have appeared had you depressed the LOG key.

Useful relationships among logarithms are

$$10^{\log X} = X$$
$$e^{\ln X} = X$$

For example, log 2 = 0.3010, ln 2 = 0.6931.

$$10^{\log 2} = 10^{0.3010} = 2$$
$$e^{\ln 2} = e^{0.6931} = 2$$

These provide a means to obtain the antilogarithms. If you have the natural log of a number, enter it and depress the "$e^x$" key. If you have the common log, enter it and depress the "$10^x$" key. In each case you will obtain the antilogarithm.

If your calculator does not have a "$10^x$" key, but does have an "$x^y$" (or "$y^x$") key, enter *10* and raise it to an exponent that corresponds to the common log. The result will be the antilogarithm.

**EXPONENTS AND ROOTS.** Most calculators have $X^2$ and $\sqrt{X}$ keys, and these operations are simple. For higher powers and roots you can use either of two methods.

1. **Using the $x^y$ key.** To compute $X = a^b$, enter $a$ and raise it to the power $b$. To compute $X = \sqrt[b]{a}$, enter $a$ and raise it to the power, $1/b$. For example,

$$X = 2^3 = 8$$
$$X = \sqrt[3]{2} = 2^{1/3} = 2^{0.3333\ldots3} = 1.25992$$

2. **Using logarithms.** To compute $X = a^b$ with natural logarithms, we use the relationship that

$$\ln X = \ln a^b = b \ln a$$

Therefore,

$$X = e^{\ln X}$$
$$X = e^{b \ln a}$$

Let's suppose that we wished to compute $3^5$. On a typical calculator we would perform this computation in the following sequence:

| | |
|---|---|
| 1. Take ln 3 | $\ln 3 = 1.098612$ |
| 2. Multiply ln 3 by 5 | $5 \ln 3 = 5.493061$ |
| 3. Raise $e$ to this exponent | $e^{5 \ln 3} = 243$ |

$$3^5 = 243$$

These calculations, of course, can also be done using common logarithms, in which case

$$X = 10^{b \log a}$$

To compute a root, $X = \sqrt[b]{a}$, we find that

$$X = e^{(\ln a)/b}$$

For example, suppose that we wished to calculate $\sqrt[5]{12}$. The sequence of operations is

| | |
|---|---|
| 1. Take ln 12 | $\ln 12 = 2.484907$ |
| 2. Divide ln 12 by 5 | $\frac{(\ln 12)}{5} = 0.496981$ |
| 3. Raise $e$ to this exponent | $e^{(\ln 12)/5} = 1.643752$ |

$$\sqrt[5]{12} = 1.643752$$

# APPENDIX D LOGARITHMS

| | 0 | 1 | 2 | 3 | 4 | 5 | 6 | 7 | 8 | 9 |
|---|---|---|---|---|---|---|---|---|---|---|
| 10 | 0000 | 0043 | 0086 | 0128 | 0170 | 0212 | 0253 | 0294 | 0334 | 0374 |
| 11 | 0414 | 0453 | 0492 | 0531 | 0569 | 0607 | 0645 | 0682 | 0719 | 0755 |
| 12 | 0792 | 0828 | 0864 | 0899 | 0934 | 0969 | 1004 | 1038 | 1072 | 1106 |
| 13 | 1139 | 1173 | 1206 | 1239 | 1271 | 1303 | 1335 | 1367 | 1399 | 1430 |
| 14 | 1461 | 1492 | 1523 | 1553 | 1584 | 1614 | 1644 | 1673 | 1703 | 1732 |
| 15 | 1761 | 1790 | 1818 | 1847 | 1875 | 1903 | 1931 | 1959 | 1987 | 2014 |
| 16 | 2041 | 2068 | 2095 | 2122 | 2148 | 2175 | 2201 | 2227 | 2253 | 2279 |
| 17 | 2304 | 2330 | 2355 | 2380 | 2405 | 2430 | 2455 | 2480 | 2504 | 2529 |
| 18 | 2553 | 2577 | 2601 | 2625 | 2648 | 2672 | 2695 | 2718 | 2742 | 2765 |
| 19 | 2788 | 2810 | 2833 | 2856 | 2878 | 2900 | 2923 | 2945 | 2967 | 2989 |
| 20 | 3010 | 3032 | 3054 | 3075 | 3096 | 3118 | 3139 | 3160 | 3181 | 3201 |
| 21 | 3222 | 3243 | 3263 | 3284 | 3304 | 3324 | 3345 | 3365 | 3385 | 3404 |
| 22 | 3424 | 3444 | 3464 | 3483 | 3502 | 3522 | 3541 | 3560 | 3579 | 3598 |
| 23 | 3617 | 3636 | 3655 | 3674 | 3692 | 3711 | 3729 | 3747 | 3766 | 3784 |
| 24 | 3802 | 3820 | 3838 | 3856 | 3874 | 3892 | 3909 | 3927 | 3945 | 3962 |
| 25 | 3979 | 3997 | 4014 | 4031 | 4048 | 4065 | 4082 | 4099 | 4116 | 4133 |
| 26 | 4150 | 4166 | 4183 | 4200 | 4216 | 4232 | 4249 | 4265 | 4281 | 4298 |
| 27 | 4314 | 4330 | 4346 | 4362 | 4378 | 4393 | 4409 | 4425 | 4440 | 4456 |
| 28 | 4472 | 4487 | 4502 | 4518 | 4533 | 4548 | 4564 | 4579 | 4594 | 4609 |
| 29 | 4624 | 4639 | 4654 | 4669 | 4683 | 4698 | 4713 | 4728 | 4742 | 4757 |
| 30 | 4771 | 4786 | 4800 | 4814 | 4829 | 4843 | 4857 | 4871 | 4886 | 4900 |
| 31 | 4914 | 4928 | 4942 | 4955 | 4969 | 4983 | 4997 | 5011 | 5024 | 5038 |
| 32 | 5051 | 5065 | 5079 | 5092 | 5105 | 5119 | 5132 | 5145 | 5159 | 5172 |
| 33 | 5185 | 5198 | 5211 | 5224 | 5237 | 5250 | 5263 | 5276 | 5289 | 5302 |
| 34 | 5315 | 5328 | 5340 | 5353 | 5366 | 5378 | 5391 | 5403 | 5416 | 5428 |
| 35 | 5441 | 5453 | 5465 | 5478 | 5490 | 5502 | 5514 | 5527 | 5539 | 5551 |
| 36 | 5563 | 5575 | 5587 | 5599 | 5611 | 5623 | 5635 | 5647 | 5658 | 5670 |
| 37 | 5682 | 5694 | 5705 | 5717 | 5729 | 5740 | 5752 | 5763 | 5775 | 5786 |
| 38 | 5798 | 5809 | 5821 | 5832 | 5843 | 5855 | 5866 | 5877 | 5888 | 5899 |
| 39 | 5911 | 5922 | 5933 | 5944 | 5955 | 5966 | 5977 | 5988 | 5999 | 6010 |
| 40 | 6021 | 6031 | 6042 | 6053 | 6064 | 6075 | 6085 | 6096 | 6107 | 6117 |
| 41 | 6128 | 6138 | 6149 | 6160 | 6170 | 6180 | 6191 | 6201 | 6212 | 6222 |
| 42 | 6232 | 6243 | 6253 | 6263 | 6274 | 6284 | 6294 | 6304 | 6314 | 6325 |
| 43 | 6335 | 6345 | 6355 | 6365 | 6375 | 6385 | 6395 | 6405 | 6415 | 6425 |
| 44 | 6435 | 6444 | 6454 | 6464 | 6474 | 6484 | 6493 | 6503 | 6513 | 6522 |
| 45 | 6532 | 6542 | 6551 | 6561 | 6571 | 6580 | 6590 | 6599 | 6609 | 6618 |
| 46 | 6628 | 6637 | 6646 | 6656 | 6665 | 6675 | 6684 | 6693 | 6702 | 6712 |
| 47 | 6721 | 6730 | 6739 | 6749 | 6758 | 6767 | 6776 | 6785 | 6794 | 6803 |
| 48 | 6812 | 6821 | 6830 | 6839 | 6848 | 6857 | 6866 | 6875 | 6884 | 6893 |
| 49 | 6902 | 6911 | 6920 | 6928 | 6937 | 6946 | 6955 | 6964 | 6972 | 6981 |
| 50 | 6990 | 6998 | 7007 | 7016 | 7024 | 7033 | 7042 | 7050 | 7059 | 7067 |
| 51 | 7076 | 7084 | 7093 | 7101 | 7110 | 7118 | 7126 | 7135 | 7143 | 7152 |
| 52 | 7160 | 7168 | 7177 | 7185 | 7193 | 7202 | 7210 | 7218 | 7226 | 7235 |
| 53 | 7243 | 7251 | 7259 | 7267 | 7275 | 7284 | 7292 | 7300 | 7308 | 7316 |
| 54 | 7324 | 7332 | 7340 | 7348 | 7356 | 7364 | 7372 | 7380 | 7388 | 7396 |

| | 0 | 1 | 2 | 3 | 4 | 5 | 6 | 7 | 8 | 9 |
|---|---|---|---|---|---|---|---|---|---|---|
| 55 | 7404 | 7412 | 7419 | 7427 | 7435 | 7443 | 7451 | 7459 | 7466 | 7474 |
| 56 | 7482 | 7490 | 7497 | 7505 | 7513 | 7520 | 7528 | 7536 | 7543 | 7551 |
| 57 | 7559 | 7566 | 7574 | 7582 | 7589 | 7597 | 7604 | 7612 | 7619 | 7627 |
| 58 | 7634 | 7642 | 7649 | 7657 | 7664 | 7672 | 7679 | 7686 | 7694 | 7701 |
| 59 | 7709 | 7716 | 7723 | 7731 | 7738 | 7745 | 7752 | 7760 | 7767 | 7774 |
| 60 | 7782 | 7789 | 7796 | 7803 | 7810 | 7818 | 7825 | 7832 | 7839 | 7846 |
| 61 | 7853 | 7860 | 7868 | 7875 | 7882 | 7889 | 7896 | 7903 | 7910 | 7917 |
| 62 | 7924 | 7931 | 7938 | 7945 | 7952 | 7959 | 7966 | 7973 | 7980 | 7987 |
| 63 | 7993 | 8000 | 8007 | 8014 | 8021 | 8028 | 8035 | 8041 | 8048 | 8055 |
| 64 | 8062 | 8069 | 8075 | 8082 | 8089 | 8096 | 8102 | 8109 | 8116 | 8122 |
| 65 | 8129 | 8136 | 8142 | 8149 | 8156 | 8162 | 8169 | 8176 | 8182 | 8189 |
| 66 | 8195 | 8202 | 8209 | 8215 | 8222 | 8228 | 8235 | 8241 | 8248 | 8254 |
| 67 | 8261 | 8267 | 8274 | 8280 | 8287 | 8293 | 8299 | 8306 | 8312 | 8319 |
| 68 | 8325 | 8331 | 8338 | 8344 | 8351 | 8357 | 8363 | 8370 | 8376 | 8382 |
| 69 | 8388 | 8395 | 8401 | 8407 | 8414 | 8420 | 8426 | 8432 | 8439 | 8445 |
| 70 | 8451 | 8457 | 8463 | 8470 | 8476 | 8482 | 8488 | 8494 | 8500 | 8506 |
| 71 | 8513 | 8519 | 8525 | 8531 | 8537 | 8543 | 8549 | 8555 | 8561 | 8567 |
| 72 | 8573 | 8579 | 8585 | 8591 | 8597 | 8603 | 8609 | 8615 | 8621 | 8627 |
| 73 | 8633 | 8639 | 8645 | 8651 | 8657 | 8663 | 8669 | 8675 | 8681 | 8686 |
| 74 | 8692 | 8698 | 8704 | 8710 | 8716 | 8722 | 8727 | 8733 | 8739 | 8745 |
| 75 | 8751 | 8756 | 8762 | 8768 | 8774 | 8779 | 8785 | 8791 | 8797 | 8802 |
| 76 | 8808 | 8814 | 8820 | 8825 | 8831 | 8837 | 8842 | 8848 | 8854 | 8859 |
| 77 | 8865 | 8871 | 8876 | 8882 | 8887 | 8893 | 8899 | 8904 | 8910 | 8915 |
| 78 | 8921 | 8927 | 8932 | 8938 | 8943 | 8949 | 8954 | 8960 | 8965 | 8971 |
| 79 | 8976 | 8982 | 8987 | 8993 | 8998 | 9004 | 9009 | 9015 | 9020 | 9025 |
| 80 | 9031 | 9036 | 9042 | 9047 | 9053 | 9058 | 9063 | 9069 | 9074 | 9079 |
| 81 | 9085 | 9090 | 9096 | 9101 | 9106 | 9112 | 9117 | 9122 | 9128 | 9133 |
| 82 | 9138 | 9143 | 9149 | 9154 | 9159 | 9165 | 9170 | 9175 | 9180 | 9186 |
| 83 | 9191 | 9196 | 9201 | 9206 | 9212 | 9217 | 9222 | 9227 | 9232 | 9238 |
| 84 | 9243 | 9248 | 9253 | 9258 | 9263 | 9269 | 9274 | 9279 | 9284 | 9289 |
| 85 | 9294 | 9299 | 9304 | 9309 | 9315 | 9320 | 9325 | 9330 | 9335 | 9340 |
| 86 | 9345 | 9350 | 9355 | 9360 | 9365 | 9370 | 9375 | 9380 | 9385 | 9390 |
| 87 | 9395 | 9400 | 9405 | 9410 | 9415 | 9420 | 9425 | 9430 | 9435 | 9440 |
| 88 | 9445 | 9450 | 9455 | 9460 | 9465 | 9469 | 9474 | 9479 | 9484 | 9489 |
| 89 | 9494 | 9499 | 9504 | 9509 | 9513 | 9518 | 9523 | 9528 | 9533 | 9538 |
| 90 | 9542 | 9547 | 9552 | 9557 | 9562 | 9566 | 9571 | 9576 | 9581 | 9586 |
| 91 | 9590 | 9595 | 9600 | 9605 | 9609 | 9614 | 9619 | 9624 | 9628 | 9633 |
| 92 | 9638 | 9643 | 9647 | 9652 | 9657 | 9661 | 9666 | 9671 | 9675 | 9680 |
| 93 | 9685 | 9689 | 9694 | 9699 | 9703 | 9708 | 9713 | 9717 | 9722 | 9727 |
| 94 | 9731 | 9736 | 9741 | 9745 | 9750 | 9754 | 9759 | 9763 | 9768 | 9773 |
| 95 | 9777 | 9782 | 9786 | 9791 | 9795 | 9800 | 9805 | 9809 | 9814 | 9818 |
| 96 | 9823 | 9827 | 9832 | 9836 | 9841 | 9845 | 9850 | 9854 | 9859 | 9863 |
| 97 | 9868 | 9872 | 9877 | 9881 | 9886 | 9890 | 9894 | 9899 | 9903 | 9908 |
| 98 | 9912 | 9917 | 9921 | 9926 | 9930 | 9934 | 9939 | 9943 | 9948 | 9952 |
| 99 | 9956 | 9961 | 9965 | 9969 | 9974 | 9978 | 9983 | 9987 | 9991 | 9996 |

# APPENDIX E ANSWERS TO EVEN-NUMBERED NUMERICAL PROBLEMS

## CHAPTER 1

1.18 5, 3, 3, 4, 5

1.20 (a) $1.25 \times 10^3$ (b) $1.3 \times 10^7$ (c) $6.023 \times 10^{22}$ (d) $2.1457 \times 10^5$ (e) $3.147 \times 10^1$

1.22 (a) 30,000,000,000 (b) 0.0000254 (c) 1.22 (d) 0.00000034 (e) 32,500

1.24 (a) $5.56 \times 10^3$ (b) $2.9 \times 10^4$ (c) $1.49 \times 10^{10}$ (d) $3.8 \times 10^{-6}$ (e) $9.0 \times 10^{-31}$

1.26 (a) 2140 (b) 12100 (c) 41 (d) 5.9 (e) 6.2

1.28 (a) $1.40 \times 10^2$ (b) 2.8 m (c) 0.185 liter (d) $1.8 \times 10^{-2}$ kg (e) 8 $m^2$ (f) $6.34 \times 10^6$ m (g) 14 mi/hr (h) $4 \times 10^9$ $m^3$ (i) $2 \times 10^3$ cm/sec (j) 25 $dm^3$

1.30 56 km/hr

1.32 264 lb/yr

1.34 (g O in I)/(g O in II) = $\frac{1}{2}$

1.36 $7.2 \times 10^3$ J, $1.7 \times 10^3$ cal

1.38 86°F, 3601°F

1.40 g X/g Y = 0.560 in each sample

1.42 (g P in I)/(g P in II) = 5/3

1.44 3400°C

1.46 98.6°F = 37.0°C, 39°C = 102°F

1.48 −0.23 degrees C

1.50 $6.5 \times 10^5$ cal

1.52 °N = (°C − 80)100/138, 0°C = −58°N, 100°C = 14°N

## CHAPTER 2

2.12 (a) 40.31 (b) 110.98 (c) 208.22 (d) 135.02 (e) 163.94

2.14 262 g

2.16 1910 g $PbSO_4$

2.18 2.88 moles $NaHCO_3$

2.20 0.870 mole $H_2SO_4$

2.22 4.50 moles O

2.24 3.75 moles $BaSO_4$

2.26 10.5 moles S

2.28 $5.684 \times 10^{-22}$ g

2.30 $3.99 \times 10^{-15}$ g C

2.32 (a) 34.43% Fe, 65.57% Cl (b) 42.07% Na, 18.89% P, 39.04% O (c) 28.71% K, 0.74% H, 23.55% S, 47.00% O (d) 21.21% N, 6.87% H, 23.45% P, 48.46% O (e) 84.98% Hg, 15.02% Cl

2.34 5.6 g N

2.36 $1 \times 10^4$ styrene units

2.38 $NaBH_4$

2.40 $C_2H_6O$
2.42 $CH_3NO$
2.44 $C_3H_6NOCl_2$
2.46 0.50 g
2.48 (a) 0.177 mole $HClO_3$ (b) 8.50 g $H_2O$ (c) 4.43 g $HClO_3$ ($ClO_2$ limiting)
2.50 256 g alcohol
2.52 (a) 0.860 mole HCl (b) 18.2 g HCl (c) 0.400 mole HCl ($COCl_2$ limiting)
2.54 (a) $C_2H_2$ (b) 84.1 g $C_2H_3Cl$ (c) 1.90 g HCl (xs)
2.56 (a) 101 g $CCl_2F_2$($SbF_3$ limiting) (b) 21 g $CCl_4$
2.58 1.02 g $Ag_2S$ ($H_2S$ limiting)
2.60 0.59 g product, 510 g starting material
2.62 15.8 ton $H_2SO_4$

**CHAPTER 3**

3.56 151.9
3.58 207
3.60 $6 \times 10^{23}$
3.62 (a) 486.272 nm (b) 1094.11 nm
3.64 $1.5 \times 10^{15}$ Hz
3.66 $2 \times 10^{-11}$ erg
3.68 0.095 mi
3.70 75.76% $^{75}Cl$, 24.24% $^{37}Cl$
3.72 $3 \times 10^{19}$ sec

**CHAPTER 4**

4.52 101.5 kcal energy released
4.54 calc. EN $\approx$ (I.E. + E.A.)/2
4.56 E.A. = −81.7 kcal/mole

**CHAPTER 5**

5.42 (a) $2.5 \times 10^{-2}$ mole KCl (b) 2.31 moles $HClO_4$ (c) $2.5 \times 10^{-4}$ mole $HC_2H_3O_2$
5.44 6.41 g $Ba(OH)_2$
5.46 1.24 g/ml, 2.27 *M*
5.48 0.179 g AgCl
5.50 150 ml soln
5.52 (a) 116 (b) $C_6H_{12}O_2$
5.54 (a) 2.80 g $BaSO_4$ (b) $SO_4^{2-}$, 0.129 *M*; $Cl^-$, 0.253 *M*; $Fe^{3+}$, 0.171 *M*
5.56 $TiCl_3$
5.58 (a) $3Ba(OH)_2 + Al_2(SO_4)_3 \rightarrow 3BaSO_4(s) + 2Al(OH)_3(s)$ (b) 3.08 g (c) $Al^{3+}$, 0.143 *M*; $SO_4^{2-}$, 0.215 *M*
5.60 (a) 151 g (b) 75.5 g (c) 37.8 g (d) 30.2 g
5.62 (a) 49.00 g (b) 100.5 g (c) 32.98 g (d) 39.58 g (e) 26.00 g
5.64 45.0 g
5.66 10 N
5.68 0.064 g lactic acid, 0.036 g caproic acid
5.70 56.9%
5.72 38 ml
5.74 8.45 ml
5.76 250 ml
5.78 45.8 ml $H_2O$

CHAPTER 6
6.20 882 torr
6.22 288 ml
6.24 802 torr
6.26 1.88 liters
6.28 2.49 liters
6.30 648 K
6.32 659 torr
6.34 1.54 g/liter
6.36 33 psi
6.38 850 torr
6.40 680 torr
6.42 81.5 ml
6.44 73.9 ml
6.46 0.701 g $SO_2$
6.48 43.9 g/mole
6.50 16.0 g/mole
6.52 45.0 g/mole
6.54 (a) $CH_3$ (b) $C_2H_6$
6.56 (a) 15.7 ml $N_2$ (b) 7.2 ml $N_2$
6.58 5740 ml
6.60 600 mph
6.62 1.002 atm (real), 1.000 atm (ideal)
6.64 14.7 lb/in.$^2$ (14.7 psi)
6.66 (a) 0.423 (b) $X_{CO_2} = 0.43$, $X_{O_2} = 0.44$, $X_{N_2} = 0.13$ (c) 1.5 g $N_2$
6.68 $3.20 \times 10^{-2}$ g
6.70 1 liter atm = 101.325 J, 8.314 J mole$^{-1}$K$^{-1}$, 1.987 cal mole$^{-1}$K$^{-1}$
6.72 (a) $2.1 \times 10^{-7}$ (b) $2.1 \times 10^{-7}$ (c) $2.1 \times 10^{-7}$ (d) $4.7 \times 10^{-6}$ liter (e) $2.4 \times 10^{-4}$ ppm
6.74 (a) $1.36 \times 10^{-3}$ mole (b) 0.111 g (c) 44.4%

CHAPTER 7
7.32 2.06 Å, 1.53 Å, 1.21 Å
7.34 1.249 Å
7.36 1.4420 Å
7.38 1.76 Å
7.40 (a) 7.50 g/cm$^3$ (b) 9.70 g/cm$^3$ (c) 10.6 g/cm$^3$; fcc
7.42 0.48 Å$^3$, 0.50 Å$^3$, 0.75 Å$^3$
7.44 $d$ = 15.95 g/cm$^3$ if fcc, $d$ = 7.98 g/cm$^3$ if bcc; $d$ = 3.99 g/cm$^3$ if simple cubic
7.46 1.04 Å
7.48 2.4498 Å

CHAPTER 8
8.42 11.0 kcal
8.44 59.3 J
8.46 600 cal
8.48 41°C
8.50 7.04 kcal/mole or 29.4 kJ/mole
8.52 328 K (55°C)

CHAPTER 9
9.26 $X = 8.09 \times 10^{-2}$, 31.0%, 4.89 $m$
9.28 (a) 11.0% (b) 0.653 $m$ (c) 0.0116 (d) 0.643 $M$
9.30 18.0 $m$, 9.11 $M$
9.32 0.526%, 18.5 $m$

9.34 $X = 0.101$, wt. fract. = 0.268
9.36 −5760 cal
9.38 86 g
9.40 572 torr
9.42 92.0 torr
9.44 549 torr
9.46 MW = 128; $C_8H_4N_2$
9.48 10.9 g; 100.206°C
9.50 −0.200°C
9.52 2000 g/mole
9.54 $i = 1.03$
9.56 13 g
9.58 69 torr
9.60 (a) 0.0285 (b) 0.0294 (c) 280 g/mole
9.62 7.5%

CHAPTER 10
10.24 $4.11 \times 10^4$ J
10.26 $\Delta H = -15.6$ kcal; $\Delta E = -15.6$ kcal because $P\Delta V = -3.02 \times 10^{-5}$ kcal
10.28 $\Delta H_f = 11$ kcal/mole
10.30 $\Delta H = 10.5$ kcal/mole, $P\Delta V = 0.592$ kcal/mole
10.32 (a) −41.8 kcal (−175 kJ) (b) −31.6 kcal (−132 kJ) (c) −40.8 kcal (−171 kJ) (d) 9.9 kcal (41 kJ) (e) −683.7 kcal (−2861 kJ)
10.34 −267 kJ/mole (−63.7 kcal/mole)
10.36 −94 kcal
10.38 (a) −113 kJ (b) 306 kJ (c) −107 kJ
10.40 −326.7 kcal
10.42 500 g
10.44 $-3.1 \times 10^4$ kcal; 910 g $H_2$; 64.9 liters (17.1 gallons)
10.46 624 g glucose
10.48 188.1 kcal
10.50 57.9 kcal/mole (calc.); actal $\Delta H_f$ smaller by 38.1 kcal/mole; benzene is more stable than expected.
10.52 $\Delta S_{vap} = 26.1$ cal/mole K; $S_{fus} = 5.27$ cal/mole K
10.54 (a) , $S = -22.5$ cal/mole K
10.56 (a) −14.4 cal/mole K (b) −165.4 cal/mole K (c) −44.3 cal/mole K (d) 11.7 cal/mole K (e) −28.2 cal/mole K
10.58 −495.7 kcal; real processes are not reversible

CHAPTER 11
11.32 $\Delta[A]/\Delta t = -1.0 \times 10^{-2}$ mole liter$^{-1}$min$^{-1}$, $\Delta[B]/\Delta t = 2.0 \times 10^{-2}$ mole liter$^{-1}$min$^{-1}$; $\Delta[A]/\Delta t = -7.2 \times 10^{-3}$ mole liter$^{-1}$min$^{-1}$, $\Delta[B]/\Delta t = 1.35 \times 10^{-2}$ mole liter$^{-1}$min$^{-1}$; $\Delta[C]/\Delta t = 5.0 \times 10^{-3}$ mole liter$^{-1}$min$^{-1}$ (25 min); $3.5 \times 10^{-3}$ mole liter$^{-1}$min$^{-1}$ (40 min)
11.34 (a) $1.0 \times 10^{-2}$ (b) $2.0 \times 10^{-2}$ (c) $4.1 \times 10^{-2}$
11.36 (a) Rate = $k[NOCl]^2$ (b) $k = 4.0 \times 10^{-8}$ liter mole$^{-1}$sec$^{-1}$ (c) 2.25
11.38 $E_a = 39$ kcal/mole, $A = 1.36 \times 10^{16}$ liter$^2$mole$^{-2}$sec$^{-1}$
11.40 $1.19 \times 10^{-4}$ liter mole$^{-1}$sec$^{-1}$
11.42 $E_a = 59$ kcal/mole
11.44 Straight line obtained for $1/[A]$ *vs.* time, second order; $k = 2.5 \times 10^{-2}$ liter mole$^{-1}$min$^{-1}$

CHAPTER 12
12.22 $K_c = 0.22$ mole/liter
12.24 $K_P = 10.13$
12.26 $-2.59$ kcal/mole
12.28 $2.84 \times 10^{61}$
12.30 Not as equilibrium, reaction proceeds from right to left.
12.32 (a) 18.5 kcal (b) $2.70 \times 10^{-14}$ (c) $5.48 \times 10^{-3}$ atm
12.34 $K_P = 0.30$ atm$^{-1}$; $K_c = 2.4 \times 10^{-2}(T)$, where $T$ = abs. temperature
12.36 (a) $[CO] = 1 \times 10^{-4}\ M$; $[O_2] = 5 \times 10^{-5}\ M$ (b) 0.1
12.38 $[COCl_2] = 0.02\ M$, $[CO] = [Cl_2] = 2.09 \times 10^{-6}\ M$
12.40 $[HI] = 0.113\ M$, $[H_2] = [I_2] = 0.016\ M$
12.42 $p_{NO} = 0.0080$ atm
12.44 $[N_2O_4] = 0.39\ M$, $[NO_2] = 0.23\ M$

CHAPTER 14
14.16

| $[H^+]$ | $[OH^-]$ | pH |
|---|---|---|
| (a) $1.0 \times 10^{-3}$ | $1.0 \times 10^{-11}$ | 3.00 |
| (b) 0.125 | $8.00 \times 10^{-14}$ | 0.903 |
| (c) $3.2 \times 10^{-12}$ | $3.1 \times 10^{-3}$ | 11.49 |
| (d) $4.2 \times 10^{-13}$ | $2.4 \times 10^{-2}$ | 12.38 |
| (e) $2.1 \times 10^{-4}$ | $4.8 \times 10^{-11}$ | 3.68 |
| (f) $1.3 \times 10^{-5}$ | $7.7 \times 10^{-10}$ | 4.89 |
| (g) $1.2 \times 10^{-12}$ | $8.4 \times 10^{-3}$ | 11.92 |
| (h) $2.1 \times 10^{-13}$ | $4.8 \times 10^{-2}$ | 12.68 |

14.18 (a) 12.7 (b) 8.27 (c) 10.00 (d) 6.2 (e) 3.06 (f) 1.39
14.20 $1.4 \times 10^{-4}$
14.22 (a) $1.6 \times 10^{-3}$ (b) $5.8 \times 10^{-4}$ (c) $1.7 \times 10^{-2}$ (d) $6.2 \times 10^{-5}$ (e) $4.2 \times 10^{-6}$
14.24 (a) 11.20 (b) 10.76 (c) 12.23 (d) 9.79 (e) 8.62
14.26 $1.8 \times 10^{-10}$
14.28 (a) 1.3% (b) 3.7% (c) 0.014% (d) 0.63% (e) 0.025% (f) assume 100%
14.30 (a) $3.0 \times 10^{-5}$ (b) $1.8 \times 10^{-4}$ (c) $3.4 \times 10^{-4}$ (d) $3.3 \times 10^{-10}$ (e) $9.8 \times 10^{-9}$
14.32 9.1 g HCl
14.34 0.0796% (fraction = $7.96 \times 10^{-4}$)
14.36 3.43
14.38 0.11 $M$
14.40 $[H_3PO_4] \approx 1.0\ M$, $[H^+] = [H_2PO_4^-] = 8.7 \times 10^{-2}\ M$, $[HPO_4^{2-}] = 6.2 \times 10^{-8}\ M$, $[PO_4^{3-}] = 1.6 \times 10^{-18}\ M$
14.42 $[HCO_3^-] = 4.3 \times 10^{-5}\ M$, $[CO_3^{2-}] = 2.4 \times 10^{-12}\ M$
14.44 $[S^{2-}] = 1.8 \times 10^{-13}\ M$
14.46 (a) 9.26 (b) 5.05 (c) 8.41 (d) 0.70
14.48 5.6
14.50 $-0.17$ pH units
14.52 (a) 7.87 (b) 5.08 (c) 11.63 (d) 11.15 (e) 3.37
14.54 $2.0 \times 10^{-6}$
14.56 8.85
14.58 8.41
14.60 8.23
14.62 (a) thymol blue or phenolphthalein (b) thymol blue
14.64 6.8
14.66 $[HCO_2H] = 8.7 \times 10^{-3}\ M$, $[H^+] = [CO_2H^-] = 1.3 \times 10^{-3}$

14.68 1.3 ml
14.70 6.3

## CHAPTER 15

15.12 $3.2 \times 10^{-14}$
15.14 $1.0 \times 10^{-4}$
15.16 $3.2 \times 10^{-8}$
15.18 $2.9 \times 10^{-70}$
15.20 (a) $8 \times 10^{-14}$ (b) $7.9 \times 10^{-6}$ (c) $3.9 \times 10^{-5}$ (d) $8 \times 10^{-7}$ (e) $3 \times 10^{-9}$ (f) $9 \times 10^{-3}$
15.22 1 g
15.24 $1.8 \times 10^{-8}$
15.26 $3.6 \times 10^{-3}$
15.28 $1.7 \times 10^{-6}$
15.30 (a) no ppt (b) ppt forms (c) ppt forms
15.32 $[H^+] = 0.050\ M$, $[NO_3^-] = 0.10\ M$, $[Ag^+] = 0.050\ M$, $[Cl^-] = 3.4 \times 10^{-9}\ M$
15.34 $PbCrO_4$ will ppt first; $[Pb^{2+}] = 7.5 \times 10^{-7}$
15.36 $5.5 \times 10^{-3} \leqslant H^+ < 0.96\ M$; $[Zn^{2+}] = 3.2 \times 10^{-6}$
15.38 16 $M$
15.40 $6.1 \times 10^{-11}$
15.42 ppt will form (ion prod. $= 4.2 \times 10^{-3} > K_{sp}$)
15.44 0.52 mole
15.46 $[Mg^{2+}] = 1.2 \times 10^{-3}\ M$, pH = 10.00
15.48 2.0 g $Ni(OH)_2$, pH = 8.00

## CHAPTER 16

16.32 $6.24 \times 10^{18}$
16.34 (a) 0.0927 $\mathcal{F}$ (b) $4.66 \times 10^{-4}\ \mathcal{F}$ (c) $9.14 \times 10^{-2}\ \mathcal{F}$
16.36 (a) 223 min (b) 239 min (c) 132 min
16.38 172 g Na, 265 g $Cl_2$
16.40 946 g Cu
16.42 $1.91 \times 10^3$ sec
16.44 50 min
16.46 1.34 amp
16.48 0.0833 mole Cr
16.50 12.13
16.52 (a) 1.46 volts (b) 0.36 volt (c) 0.93 volt (d) 0.08 volt (e) 1.03 volts
16.54 (a) $5.2 \times 10^3$ (b) $1.3 \times 10^9$ (c) 3
16.56 (a) $1.39 \times 10^{-13}$ (b) $3.9 \times 10^{-51}$ (c) $4.49 \times 10^{21}$ (d) $6.37 \times 10^{-82}$ (e) $1.60 \times 10^{-9}$
16.58 (a) $7.33 \times 10^4$ J (b) $2.88 \times 10^5$ J (c) $-1.24 \times 10^5$ J (d) $4.63 \times 10^5$ J (e) $5.02 \times 10^4$ J
16.60 (a) $\mathcal{E}° = 1.46$ volts, $\mathcal{E} = 1.49$ volts, $\Delta G = -8.63 \times 10^5$ J
(b) $\mathcal{E}° = 0.11$ volt, $\mathcal{E} = 0.17$ volt, $\Delta G = -3.28 \times 10^4$ J
(c) $\mathcal{E}° = 1.28$ volts, $\mathcal{E} = 1.26$ volts, $\Delta G = -2.43 \times 10^5$ J
16.62 0.06 volt
16.64 0.19 volt
16.66 $2.15 \times 10^{-14}$
16.68 $1.6 \times 10^{-19}$ coul
16.70 1330 amp
16.72 0.30 volt
16.74 360 ml

16.76 $1.04 \times 10^3$ kJ, right to left
16.78 $[Cu^{2+}] = 3.0 \times 10^{-15}$, $K_{sp} = 8.4 \times 10^{-36}$

CHAPTER 18
18.44 $\Delta H_{hyd} > 1228$ kcal/mole
18.46 1680 K (1400°C)
18.48 104 J/mole K

CHAPTER 20
20.54 $d = 13.2$ g/cm$^3$

CHAPTER 23
23.26 (a) 0.500 g (b) 0.125 g (c) 0.0313 g
23.28 164 days
23.30 $2.50 \times 10^{-2}$ day
23.32 $8.3 \times 10^8$ yr
23.34 $1.64 \times 10^{-13}$ J
23.36 $4.75 \times 10^{10}$ kJ (8.8 MeV/nucleon)
23.38 $6.41 \times 10^{-3}$ hr$^{-1}$, 110 hr
23.40 2
23.42 $1.79 \times 10^6$ gal

# INDEX

## Physical Constants and Conversion Factors

### Constants

| | | | |
|---|---|---|---|
| Avogadro's Number | N | = | $6.023 \times 10^{23}$ |
| Electronic Charge | e | = | $1.60 \times 10^{-19}$ coul |
| Faraday Constant | $\mathcal{F}$ | = | 96,494 coul/mole of $e^-$ |
| | | = | 23,060 cal/volt |
| | | = | 96,494 J/volt |
| Gas Constant | $R$ | = | 0.0821 liter atm $mole^{-1}$ $K^{-1}$ |
| | | = | $6.24 \times 10^4$ ml torr $mole^{-1}$ $K^{-1}$ |
| | | = | 8.314 J $mole^{-1}$ $K^{-1}$ |
| | | = | 1.987 cal $mole^{-1}$ $K^{-1}$ |
| Planck's Constant | $h$ | = | $6.625 \times 10^{-27}$ erg sec |
| $\pi$ | | = | 3.1416... |
| e (base of natural logarithms) | | = | 2.71828... |
| Speed of Light | $c$ | = | $3.00 \times 10^{10}$ cm/sec |

### Conversion Factors

| | |
|---|---|
| mass: | 1 lb = 454 g = 0.454 kg; 1 kg = 2.205 lb |
| length: | 1 in = 2.54 cm; 1 m = 39.37 in |
| | 1 mi = 1.609 km; 1 km = 0.6214 mi |
| volume: | 1 qt = 946 ml; 1 liter = 1.057 qt |
| | 1 oz = 29.57 ml |
| energy: | 1 joule = 0.2389 cal; 1 cal = 4.184 joule (exactly) |
| | 1 joule = $10^7$ erg = 1 kg $m^2$ $sec^{-2}$; 1 erg = 1 g $cm^2$ $sec^{-2}$ |
| | 1 eV/molecule = 23.1 kcal/mole = 96.7 kJ/mole |
| | 1 liter-atm = 24.217 cal = 101.32 J |
| pressure: | 1 atm = 760 torr = 29.92 in. Hg |

### Other Useful Relationships

$\ln x = \log_e x = 2.303 \log_{10} x$

$T(\mathrm{K}) = t(^\circ\mathrm{C}) + 273$